Dynamic Business Law
THE ESSENTIALS

Fifth Edition

Nancy Kubasek
Bowling Green State University

M. Neil Browne
Bowling Green State University

Daniel J. Herron
Western Carolina University

Lucien J. Dhooge
Georgia Institute of Technology

Linda L. Barkacs
University of San Diego

mheonline.com/advancedplacement

Copyright © 2021 McGraw Hill

All rights reserved. No part of this publication may be reproduced or distributed in any form or by any means, or stored in a database or retrieval system, without the prior written consent of McGraw Hill, including, but not limited to, network storage or transmission, or broadcast for distance learning.

Send all inquiries to:
McGraw Hill
8787 Orion Place
Columbus, OH 43240

ISBN: 978-1-26-464740-8
MHID: 1-26-464740-9

Printed in the United States of America.

2 3 4 5 6 7 8 9 LKV 30 29 28 27 26 25 24 23

About the Authors

Nancy K. Kubasek received her JD from the University of Toledo College of Law in 1981 and her BA from Bowling Green State University in 1978. She joined the BGSU faculty in 1982 and became an associate professor in 1988 and a full professor in 1993.

During her tenure at Bowling Green State University, she has primarily taught courses in business law, legal environment of business, environmental law, health care law, and moral principles. She has published over 75 articles, primarily in law reviews and business journals. Most of her substantive articles focus on environmental questions, and she writes a quarterly column about environmental issues for the *Real Estate Law Journal.* She has helped get students involved in legal research, and a number of her articles are co-authored with students. She has also published a number of pedagogical articles in teaching journals, focusing primarily on teaching critical thinking and ethics.

She wrote the first environmental law text for undergraduate students, *Environmental Law,* and co-authored *The Legal Environment of Business: A Critical Thinking Approach.* She has written supplemental materials, such as study guides, test banks, and instructors' manuals.

Active in many professional organizations, she has served as president of the Academy of Legal Studies in Business, the national organization for professors of legal studies in colleges of business. She has also served as president of the Tri-State Academy of Legal Studies in Business, her regional professional association.

In her leisure time, she and her husband Neil Browne fish for halibut and salmon in Alaska and large-mouth bass in Florida. In addition, they are regular participants in polka, waltz, zydeco, and Cajun dance festivals in Europe and the United States. For almost 30 years, they have been successful tournament blackjack players as well. Both are avid exercisers—lifting weights, doing yoga, and running almost every day.

M. Neil Browne is senior lecturer and research associate and a distinguished teaching professor emeritus at Bowling Green State University. He received his BA in history and economics at the University of Houston, his PhD in economics at the University of Texas, and his JD from the University of Toledo. He has been a professor at Bowling Green for five decades.

Professor Browne teaches courses in jurisprudence, ethical reasoning, critical thinking, and economics at both the undergraduate and graduate levels. He has received recognition as the Silver Medalist National Professor of the Year, the Ohio Professor of the Year, and Distinguished Teacher and Master Teacher at Bowling Green State University, as well as numerous research awards from his university and from professional organizations. His consulting activities with corporate, government, and educational institutions focus on improving the quality of critical thinking in those organizations. In addition, he serves as a Rule 26 expert with respect to the quality of the reasoning used by expert witnesses called by the party opponent in legal actions.

Professor Browne has published 20 books and more than 130 professional journal articles in law journals and economics, sociology, and higher education journals. His current research interests focus on the relationship between orthodox economic thinking and legal policy. In addition, he is in the midst of writing books about the power of questionable assumptions in economics, the usefulness of asking questions as a learning strategy, and the importance of critical thinking in environmental arguments.

Professor Browne tries to find time for a broad array of outside activities. He and his wife Nancy Kubasek fish for halibut and salmon in Alaska and large-mouth bass in Florida as frequently as possible. In addition, they are regular participants in polka, waltz, zydeco, and Cajun dance festivals in Europe and the United States. For almost 30 years, they have been successful tournament blackjack players as well. Both are avid exercisers—lifting weights, doing yoga, and running almost every day.

Daniel J. Herron has taught business Legal Studies for more than thirty-five years. He is currently retired from Miami University in Oxford, Ohio, but is a visiting professor at Western Carolina University in Cullowhee, North Carolina. He is the executive secretary of the Academy of Legal Studies in Business. He has been married for over forty years to Deborah, and they have two children and seven grandchildren. His publication record included over twenty-two scholarly articles and over forty scholarly presentations. He currently lives in Cullowhee, North Carolina.

Lucien J. Dhooge is the Sue and John Staton Professor of Law at the Georgia Institute of Technology. He teaches business law, international business law, and ethics. After completing an undergraduate degree in history at the University of Colorado in 1980, Professor Dhooge attended the University of Denver College of Law, where he received his J.D. in 1983. He received his LL.M. in 1995 from Georgetown University Law Center where he specialized in international and comparative law. Before

coming to Georgia Tech, Professor Dhooge spent 11 years in private law practice and 12 years serving on the faculty of the University of the Pacific in California. Professor Dhooge has authored more than 60 scholarly articles, co-authored fourteen books, and presented research papers and courses throughout the United States as well as in Asia, Australia, Europe, and South America. Professor Dhooge is the recipient of numerous research awards given by the Academy of Legal Studies in Business, including seven Ralph C. Hoeber Awards for excellence in research. He was designated the outstanding junior business law faculty member in the United States by the Academy in 2002 and received the Kay Duffy Award for outstanding service in 2005. Professor Dhooge received the Academy's Distinguished Career Faculty Award in 2019. Professor Dhooge was the program chair for the Academy's 2009 international conference in Denver and served as the Academy's president from 2009 to 2010. He is a past editor-in-chief of the American Business Law Journal and the Journal of Legal Studies Education.

A native of Chicago but raised in Denver, Professor Dhooge enjoys spending time with his family and following the fortunes of the Chicago Cubs and Colorado Rockies professional baseball teams.

Linda L. Barkacs received her JD from the University of San Diego in 1993. She also has a BA in political science from San Diego State University and an AA in accounting from Irvine Valley College.

Upon graduating from law school and passing the California bar exam, Professor Barkacs became an associate at a downtown San Diego law firm. During her time with that firm, she was involved in a number of high-profile trials, including a sexual harassment case against the City of Oceanside that resulted in a $1.2 million verdict. In 1997, Professor Barkacs and her husband Craig (also a professor at USD) started their own law firm specializing in business and civil litigation (in both federal and state courts), employment law cases, and appeals. They were also involved in numerous mediations and arbitrations.

Professor Barkacs began teaching at USD in 1997 and was tenured in 2011. As an educator, she has designed and taught numerous courses on law, ethics, and negotiation. She teaches in USD's undergraduate and graduate programs, including the Master of Science in Executive Leadership (a Ken Blanchard program), the Master of Science in Global Leadership, and the Master of Science in Supply Chain Management. Professor Barkacs often teaches in USD's study-abroad classes and has traveled extensively throughout Europe, Asia, and South America.

Professor Barkacs has received numerous awards for her teaching at USD, including USD's 2010 Professor Impact (MSCM Program); the 2008 USD School of Business Outstanding Undergraduate Business Educator of the Year; and the 2007 and 2008 Professor of the Year, USD Senior Class (university-wide).

She and her husband are principals in The Barkacs Group (www.TheBarkacsGroup.com), a consulting firm that provides negotiation, ethics, and team training for the private sector. Professor Barkacs has published numerous journal articles in the areas of law, ethics, and negotiation. She and her husband are co-authoring a book on negotiation. She has been the president, vice president, conference chair, and treasurer for the Pacific Southwest Academy of Legal Studies in Business (www.pswalsb.net).

Professor Barkacs currently spends her time teaching, publishing, consulting for The Barkacs Group, and doing volunteer work for various civic causes. She enjoys walking, weight-lifting, and spending her free time with her two cats, Violet and Vanessa.

Preface

We wrote this book because our primary sense of who we are as professionals is that we are teachers. We play various roles in our careers, but we are especially dedicated to our students. We want them to listen, read, create, and evaluate more effectively as a result of their experience in a business law class. Above all, we want them to sense the importance and excitement of the law.

We want them to be aware that business resides in an atmosphere of legal rights and responsibilities. The more they have an understanding of relevant law, the more their business activities will flourish.

We tried to construct a book that contains the basics of business law but does not get bogged down in the kind of details that would be more appropriate in an upper-level law class.

This 5th edition continues and advances the pedagogical strengths of earlier editions. As difficult as it is to recall what it is like to know almost nothing about business law, we tried repeatedly to force ourselves to remember that the text is for students, not for us to reveal our previous training. Toward that end, we have simplified and clarified as often as we could.

Our pedagogical features that provide the distinctive value of this book. Each feature stands by itself as an aid to the kind of learning we hope to encourage. Yet the features are also a cohesive unit, contributing both to the liberal education of the students who read it and to their skills as decision makers in a market economy.

Specifically, we provide what competing texts deliver, an examination of the basic questions, concepts, and legal rules of business law. But this edition advances our treatment of these basics. We are very sensitive to the feedback we receive that emphasizes that this text is used to train future business leaders, not lawyers. Consequently, for each chapter we asked ourselves: what are the brief take-aways that capture the essence of the chapter for our primary audience—people who need guidance for achieving success in business. Then we answered that question by placing a section at the end of each chapter entitled "Practical Tips for Business Managers."

Decisions to trade and produce require trust—trust that consumers, firms, workers, financial institutions, and asset owners will do as they promise and that violations of such promises will be unacceptable in the marketplace. Without guarantees that promises will be kept, market exchanges would grind to a halt. Business law provides these guarantees and the boundaries within which certain promises can be made and enforced.

Market decisions are made in a context—a persistently changing context. The law, in turn, is dynamic in response. New technologies and business practices bring new disputes over rights and responsibilities in a business setting.

We knew that the 5th edition needed to be different from the 4th because law is evolving. Certain areas of business law experience recurring re-examination and revision. For example, we created new material for this edition emphasizing immigration law.

In addition, each author pledged to refresh our chapters with recent developments in business law.

Future business leaders need knowledge of existing business law as well as a set of skills permitting them to adjust efficiently and effectively to new legal issues as they arise over the course of their careers.

We are excited about the contents of our features and want to explain the function of each of them in preparing students for leadership positions in business.

NEW Connect's Business Law Application-Based Activities

New for the 5th edition are Application-Based Activities for Business Law. This digital feature provides students valuable practice using problem-solving skills to apply their knowledge to realistic scenarios. Students progress from understanding basic concepts to using their knowledge to analyze complex scenarios and solve problems. Application-Based Activities have been developed for the most often taught topics (as ranked by instructors) in the business law course. These unique activities are assignable and auto-gradable in Connect.

PRACTICAL TIPS for Business Managers

1. Remember that the meaning and implications of laws are a matter of interpretation. Therefore, it is wise, if possible, to consult with more than one lawyer when you are curious about the meaning of a law that is immensely important to your operations.

2. Try to stay current with your knowledge of law relevant to your industry and that of your suppliers. Law evolves, and yesterday's law may not be today's law.

3. When you engage in any international commercial activity, be sure to investigate the law of countries in which you are doing business.

NEW Practical Tips for Business Managers

This feature emphasizes that this book is a BUSINESS Law book, full of helpful ideas for modern business leaders. These are the practical takeaways highlighting the usefulness of what was learned in the chapter.

Case Openers and NEW Case Opener Wrap-ups

The Case Opener feature provides a vivid, practical introduction to the kinds of legal issues that arise in the new chapter. This approach is much more enticing than were we to simply launch immediately into the relevant legal principles.

The NEW Case Opener Wrap-ups conclude the chapters. They explain how courts resolve fact patterns such as those in the Case Openers.

BUT WHAT IF . . .

Business law is a set of rules and regulations that modern managers must obey, but those rules and regulations apply in specific situations to particular facts. Those facts create a special setting, one not exactly like any other business decision. We added a But What If feature to highlight for readers the need to think about the implications of specific facts for the application of business laws.

BUT WHAT IF . . .

WHAT IF THE FACTS OF THE CASE OPENER WERE DIFFERENT?

In the opening case, the EPA believed it did not have enough power to make rules regarding global climate change. Let's say that a member of the EPA argued that only Congress has the power to make this type of regulation, which would directly affect the way many businesses operate. Would the EPA member be correct?

In addition, this feature provides a reminder that applying the law is a complex activity, requiring a thorough knowledge of the law. An alert business manager needs to consider how each of the facts in the dilemma she or he faces may have legal implications that are not always clear when simply reading the law for the first time. The facts of a case guide the application.

Those who have formal legal training know that a common form of teaching law is to pose hypotheticals. The But What If feature mimics that approach.

In other words, at first glance the law seems to point in a certain direction, thereby giving business decision makers confidence that they know what is legally permissible or required, but changing a single fact can sometimes dramatically alter the eventual application of the law. An awareness of the significance of facts in the law can greatly enhance legal compliance and save firms huge expenses.

Global Boxes

This feature highlights the emerging, interconnected market. Where relevant, the chapters contain Global Context boxes. Because so many market decisions are made in an international context, learners need to familiarize themselves with the likelihood that a particular legal principle essential to doing business in one country may not be appropriate in other countries. The Global Context boxes illustrate how unique the law in a certain country often is. After reading these stories of difference, readers will certainly understand better the need to discover relevant law in all jurisdictions where their market decisions have legal implications.

©Digital Archive Japan/Alamy

GLOBAL Context

The Court Structure in England

Even though we trace the roots of the U.S. legal system back to England, changes have occurred in both countries' court structures; as a result, the court structures in the United States and England share similarities but also have distinct features. The lowest criminal courts in England are the magistrates courts, which hear minor offenses. More serious cases are tried before a judge and jury in the crown court, which also hears cases appealed from the magistrates courts on factual points. The high court (in the Queen's Branch Division) hears appeals on points of law, and the court of appeal in the Criminal Division hears appeals on sentences and convictions.

Civil cases are first heard in the county courts (for minor claims) or the high court, which is divided into three divisions: Queen's Bench, Family, and Chancery. Cases may be appealed to the court of appeal (Civil Division). Cases may also be appealed from the county court to the high court.

The House of Lords is the supreme court of appeal. Its cases are heard by up to 13 senior judges known as *law lords*. In addition to the courts, there are specialized tribunals, which hear appeals on decisions made by various public bodies and government departments in areas such as employment, immigration, social security, tax, and land.

We believe that students learn innumerable valuable lessons about American business law by contrasting the concepts of our business law system with those of our primary trading partners. We typically use Canada, Japan, China, Russia, Mexico, and the European Union for our comparisons because modern business managers will more likely be interacting with the law in those particular jurisdictions.

Critical Thinking

After each case in the book, we have provided critical-thinking questions to highlight the need to think critically about the reasoning the court used. In addition, we include in every chapter a Point/Counterpoint problem that encourages the reader to evaluate the conflicting reasoning surrounding a key issue in the chapter.

Critical Thinking and Business Law Appendix 1A

CASE 16-1

ALASKA PACIFIC TRADING CO. v. EAGON FOREST PRODUCTS INC.
Washington Appellate Division 1
933 P.2d 417 (1997)

FACTS: Alaska Pacific Trading Company (ALPAC) and Eagon Forest Products Inc. contracted to buy and sell raw logs. ALPAC and Eagon engaged in months of communications about a shipment of 15,000 cubic meters of logs from Argentina to Korea between the end of July and the end of August 1993. The delivery date passed without ALPAC shipping the logs. Eagon canceled the contract, alleging that ALPAC had breached the agreement by failing to deliver. ALPAC alleged that its failure to deliver was not a material breach and that the parties had modified the delivery date. Alternatively, ALPAC argued that Eagon breached the contract by failing to provide adequate assurances or repudiating the contract. The miscommunication between the parties occurred after the market for logs began to soften, making the contract less attractive to Eagon. ALPAC was reluctant to ship the goods because it was concerned that Eagon might not accept the shipment. However, Eagon never stated that it would not accept the cargo.

ISSUE: The question for the court is whether to apply the common law doctrine of material breach or the UCC doctrine of perfect tender.

REASONING: In the court's own words:

> As a contract for the sale of goods, this contract is governed by the Uniform Commercial Code, Article II (UCC II) which replaced the common law of material breach, on which ALPAC relies, with the "perfect tender" rule. Under this rule, "If the goods or the tender of delivery fail in any respect to conform to the contract, the buyer may . . . reject the whole." . . . Both the plain language of the rule and the official comments clearly state that, if the tender of the goods differs from the terms of the contract in any way, the seller breaches the contract and the buyer is released from its duty to accept the goods.

DECISION AND REMEDY: The court applied the literal language of the UCC, holding that the perfect tender rule applied and that ALPAC's failure to deliver was a breach of contract.

SIGNIFICANCE OF THE CASE: Though difficult to measure at times, the perfect tender rule is a departure from the common law rule of substantial performance. Any deviation from the contract may indeed constitute a breach of the perfect tender rule, materiality notwithstanding.

CRITICAL THINKING

Here, Eagon got lucky. The company got out of a contract that was unfavorable to it, given the softening market for logs. In what way did the judge simplify the case? Is it fair to say the judge oversimplified the case?

ETHICAL DECISION MAKING

What ethical norm or value underlies the judge's decision? Explain.

We do much more than just ask a lot of critical-thinking questions at particular locations throughout the chapters. We encourage the use of a step-by-step critical thinking approach that has been developed and used in classrooms in many countries. We do not just repeatedly urge students to think critically. Instead, we describe for them what is meant by that phrase in the context of business law.

We include this step-by-step approach in Appendix 1A at the end of Chapter 1. Instructors who wish to emphasize critical thinking can use that appendix as a structured approach for learning how to evaluate legal reasoning.

Ethical Reasoning

Our book emphasizes consideration of all stakeholder interests in every market decision. Business ethics should never be an afterthought or something firms consider only because they think they must. In response to requests from our users, we created a separate chapter, Chapter 2, devoted entirely to business ethics. We expanded the discussion of business ethics while simplifying the framework for applying ethical reasoning.

BUSINESS Ethics Flashpoint 2.5

The Dofasco Steel Company's Approach to Workers

Dofasco is the second-largest steel producer in Canada and sells a wide range of steel products. In the 1980s, Dofasco experienced shrinking profits and decided to change its approach to business.

One of the ways that Dofasco changed was in its approach to its workers. Specifically, Dofasco attempted to implement a philosophy taking into account the whole person at work. In other words, this approach sought to include wellness as an important characteristic of the employee experience.

In line with this new philosophy, the Dofasco management claimed it would abide by the following policy statement issued in its 1996 guides of the company's health, safety, and lifestyle activities: "At Dofasco, there is nothing more important than the health and safety of our people."

Consistent with this policy statement, Dofasco provided training for employees in problem solving, manufacturing processes, and customer service. Dofasco also built on its tradition of providing its workers with ergonomic and fitness resources as well as preventive medical services.

Yet another way that Dofasco sought to focus its energies on employee needs was by encouraging healthy lifestyle practices at work. A healthy lifestyle at work would be achieved by preventing workplace accidents and maintaining a safe environment. In addition, Dofasco assessed employee needs in response to worker suggestions.

Dofasco believes that its focus on a safe and healthy workplace has improved its business. Specifically, the company claims that its policies have reduced lost time associated with work injuries and minimized Workers Safety Insurance Board payments.

In addition to focusing on worker safety and well-being, Dofasco also has a history of being environmentally responsible. For example, Dofasco was the first corporation in Canada voluntarily to sign an Environmental Management Agreement with Environment Canada and Ontario's Ministry of the Environment. In this agreement, Dofasco made commitments to abide by specific air and water quality standards and sustainable energy use and waste management. Through this agreement, Dofasco expressed its commitment to the community.

In addition, Dofasco created a Sustainable Development Strategy Team and is a founding sponsor of the Sustainable Enterprise Academy at York University's Schulich School of Business. The mission of this academy is to provide executives with training in how to manage businesses with an eye for sustainability.

Dofasco's organizational practices resulted in being named one of the world's most sustainable companies in the Dow Jones Sustainability Group Index.

Sources: Dan Corbett, "Excellence in Canada: Healthy Organizations–Achieve Results by Acting Responsibly," *Journal of Business Ethics* 55, no. 2 (December 2004), pp. 124, 130; Joan Enric Ricart, et. al. "The Sustainable Enterprise: Learning from DJSI Leaders", *Fundacion BBVA* (2005).

Instead, business ethics is what provides the social legitimacy for markets, what distinguishes markets from the life of the jungle. Although market decisions are calculating and purposeful, they must at the same time reflect awareness that the good and the right provide social borders that elevate those decisions above simple greed and egoism.

Ethical discussion focuses on the basic observation that we are socially and globally interdependent as entrepreneurs, asset owners, workers, businesspeople, and consumers. Our inescapable contact with one another requires our aspirations to be defined, at least in part, by their impact on others.

Our text has several ethical reasoning possibilities in each chapter. But for the reader to use this emphasis requires a practical step-by-step approach. In other words, students need more than just a discussion about values or ethics.

They need to have some sense that the discussion is headed somewhere. They want to know, "How will my behavior be any more ethical after I have read the chapter and participated in the class discussions?" Our text answers their question.

Chapter 2 provides a clear explanation of our approach—an approach that students can use on a regular basis. The language and organization of our model of ethical reasoning leans implicitly on standard ethical theories, but it meets the challenge of a fast-paced business world. It pushes stakeholders to the forefront of market decisions, where they belong, and does so in a manner that is both powerful and doable without becoming tedious.

Business ethics serve as the guidelines we use to shape the world we wish to create. As such, they provide guidance for the kind of business behavior we want to reinforce. After each case discussion, we pause to think about the ethics of business law by asking a question derived from the practical approach to business ethics developed in Chapter 2.

E-Commerce Boxes

A central feature of modern business decisions is new technology, specifically the rapid spread of electronic commerce. This development has created new challenges and opportunities that were unforeseeable until very recently. By including boxes that call attention to these challenges, we think we can convince students most effectively of the pervasive influence of this new, complicating aspect of business decisions.

E-COMMERCE and the Law

E-Government: Changing the Relationship between the Government and Citizens

According to the United Nations, *e-government*, which is defined as the delivery of government services through digital information technologies, including the Internet, is important because it reduces the cost of government while significantly improving the quality of services and citizen access. According to the latest UN report about this topic, e-government makes public administration more proactive, efficient, transparent, and service-oriented. Annual surveys by the UN indicate that nations throughout the world have made significant progress in e-government development, with the United States and Europe being among the leaders in this development.

The UN has characterized the development of e-government into the following stages:

Stage I—Emerging: A government's online presence mainly consists of a web page and/or an official website; links to ministries or departments of education, health, social welfare, labor, and finance may or may not exist. Much of the information is static, and there is little interaction with citizens.

Stage II—Enhanced: Governments provide more information on public policy and governance. They have created links that enable citizens to access archived information such as documents, forms, reports, laws and regulations, and newsletters easily.

Stage III—Interactive: Governments deliver online services such as downloadable forms for tax payments and applications for license renewals. In addition, the beginnings of an interactive portal or website with services to enhance the convenience of citizens are evident.

Stage IV—Transactional: Governments begin to transform themselves by introducing two-way interactions between citizen and government (C to G). This includes options for paying taxes and applying for ID cards, birth certificates, passports, and license renewals as well as similar C-to-G interactions, and it allows the citizen to access these services online 24/7.

Stage V—Connected: Governments transform themselves into a connected entity that responds to the needs of its citizens by developing an integrated back office infrastructure. This is the most sophisticated level of online e-government initiatives, and it is characterized by:

1. Horizontal connections (among government agencies).
2. Vertical connections (central and local government agencies).
3. Infrastructure connections (interoperability issues).
4. Connections between governments and citizens.
5. Connections among stakeholders (government, private sector, academic institutions, NGOs, and civil society).

In addition, governments support and encourage e-participation and citizen engagement in the decision-making process.[9]

The most recent ranking of nations in terms of their progress toward development of effective e-government was as follows: (1) Sweden, (2) Denmark, (3) Norway, (4) United States, (5) Netherlands, (6) Republic of Korea, (7) Canada, (8) Australia, (9) France, and (10) United Kingdom.[10] Sadly, the United States was at the top of the list in 2001, but because the Bush administration failed to make a serious commitment to transforming the federal government through information technology and failed to develop an overall e-government plan detailing where the government wants to go and how it wants to get there, other nations have made much more rapid progress in this area than has the United States.[11]

Online Assignments and Resources

We have designed our features around the things we do in our classes to encourage excitement about business law. We believe they provide Business Law Plus. We also have a variety of supplementary material available for instructors, to aid in course development, and for students, to provide additional study.

NEW Connect's Business Law Application-Based Activities (ABAs)
Application-Based Activities for Business Law provide students valuable practice using problem-solving skills to apply their knowledge to realistic scenarios. Students progress from understanding basic concepts to using their knowledge to analyze complex scenarios and solve problems. Application-Based Activities have been developed for the most often taught topics (as ranked by instructors) in the business law course. These unique activities are assignable and auto-gradable in Connect.

SmartBook 2.0 Available within Connect, SmartBook 2.0 is an adaptive learning solution that provides personalized learning to individual student needs, continually adapting to pinpoint knowledge gaps and focus learning on concepts requiring additional study. SmartBook 2.0 fosters more productive learning, taking the guesswork out of what to study, and helps students better prepare for class. With the ReadAnywhere mobile app, students can now read and complete SmartBook 2.0 assignments both online and off-line. For instructors, SmartBook 2.0 provides more granular control over assignments with content selection now available at the concept level. SmartBook 2.0 also includes advanced reporting features that enable instructors to track student progress with actionable insights that guide teaching strategies and advanced instruction, for a more dynamic class experience.

ROGER | CPA Review (Powered by UWorld) McGraw-Hill has partnered with Roger CPA Review (Powered by UWorld), a global leader in CPA Exam preparation, to provide students a smooth transition from the accounting classroom to successful completion of the CPA Exam. While many aspiring accountants wait until they have completed their academic studies to begin preparing for the CPA Exam, research shows that those who become familiar with exam content earlier in the process have a stronger chance of successfully passing the CPA Exam. Accordingly, students using these McGraw-Hill materials will have access to Roger CPA Review multiple choice questions supported by explanations written by CPAs focused on exam preparation. McGraw-Hill and Roger CPA Review are dedicated to supporting every accounting student along their journey, ultimately helping them achieve career success in the accounting profession. For more information about the full Roger CPA Review program, exam requirements, and exam content, visit www.rogercpareview.com.

Interactive Applications McGraw-Hill Connect's Interactive Applications offer a variety of automatically graded exercises that require students to apply key concepts. Whether the assignment includes a *click and drag*, *video case*, or *decision generator*, these applications provide instant feedback and progress tracking for students and detailed results for the instructor.

Instructor's Manual Written by our co-author Neil Browne, this resource includes lecture notes, case briefs, answers to all questions in each chapter, assignment ideas, teaching assistance (emphasizing practical tips that new or part-time teachers can try right away), and suggested handouts.

Test Bank and Quizzes Our test bank and quizzes contain a variety of true/false, multiple-choice, and essay questions as well as scenario-based questions, which are application-based and use a situation described in a narrative, with three to five multiple-choice test questions based on the situation.

PowerPoint Presentation Slides This edition's revised PowerPoints contain an easy-to-follow lecture outline summarizing key points for every chapter.

You Be the Judge Online This interactive product features case videos that showcase courtroom arguments of business law cases. These case videos give students the opportunity to watch profile interviews of the plaintiff and defendant; read background information; hear each case; review the evidence; make their decisions; and then access an actual, unscripted judge's decision and reasoning.

Videos Links to brief videos for classroom use are provided.

Business Law Newsletter McGraw-Hill Education's monthly Business Law newsletter, *Proceedings,* is designed specifically with the Business Law educator in mind. *Proceedings* incorporates "hot topics" in business law, video suggestions, an ethical dilemma, teaching tips, and a "chapter key" cross-referencing newsletter topics with the various McGraw-Hill Education business law textbooks. *Proceedings* is delivered via email to business law instructors each month and can be found here: http://successinhighered.com/businesslaw/newsletter.

Tegrity: Lectures 24/7

Tegrity in Connect is a tool that makes class time available 24/7 by automatically capturing every lecture. With a simple one-click start-and-stop process, you capture all computer screens and corresponding audio in a format that is easy to search, frame by frame. Students can replay any part of any class with easy-to-use, browser-based viewing on a PC, Mac, iPod, or other mobile device.

Educators know that the more students can see, hear, and experience class resources, the better they learn. In fact, studies prove it. Tegrity's unique search feature helps students efficiently find what they need, when they need it, across an entire semester of class recordings. Help turn your students' study time into learning moments immediately supported by your lecture. With Tegrity, you also increase intent listening and class participation by easing students' concerns about note-taking. Using Tegrity in Connect will make it more likely you will see students' faces, not the tops of their heads.

AACSB Statement

McGraw-Hill Education is a proud corporate member of AACSB International. The authors of *Dynamic Business Law: The Essentials,* Fifth Edition understand the importance and value of AACSB accreditation and recognize the curricula guidelines detailed in the AACSB standards for business accreditation by connecting selected questions in the test bank to the general knowledge and skill guidelines in the AACSB standards.

The statements contained in *Dynamic Business Law: The Essentials,* Fifth Edition, are provided only as a guide for the users of this textbook. The AACSB leaves content coverage and assessment within the purview of individual schools, the mission of the school, and the faculty. Although *Dynamic Business Law: The Essentials,* Fifth Edition, and the teaching package make no claim of any specific AACSB qualification or evaluation, we have within *Dynamic Business Law: The Essentials,* Fifth Edition, labeled selected questions according to the general knowledge and skill areas.

Preface xi

Important Changes in This Edition

Following are the updates that were made to the 5th edition. Note that we have multiple chapters and cases where we integrated two new themes: privacy law and small business.

Overall Updates Incorporated into Almost Every Chapter in the 5th Edition

- **NEW Practical Tips for Business Managers** is a feature which emphasizes that this book is a BUSINESS Law book, full of helpful ideas for modern business leaders. These are the practical takeaways highlighting the usefulness of what was learned in the chapter.
- **A new major case** is an inclusion signifying our recognition that students want fresh cases.
- **A new Case Opener Wrap-up** feature which explains how courts handle the issues introduced in the Case Opener.
- **Three new case problems** are included in the Questions & Problems section in recognition of our pledge to include fresh materials throughout.
- **NEW Application-Based Activities** for Business Law provide students valuable practice using problem-solving skills to apply their knowledge to realistic scenarios. Students progress from understanding basic concepts to using their knowledge to analyze complex scenarios and solve problems. Application-Based Activities have been developed for the most often taught topics (as ranked by instructors) in the business law course. These unique activities are assignable and auto-gradable in Connect.

CHAPTER 1: An Introduction to the Fundamentals of Dynamic Business Law

- Clarified the meaning of "commercial" and "supreme law of the land" because reviewers said their students were not familiar with the terms.
- Broadened the treatment of the boundaries of "criminal law" in response to reviewer's concern that the previous discussion of criminal law had focused too much on insider trading.
- Revised the discussion of the importance of comparative law by replacing the outdated material based on Google's conflicts with China by discussing the restrictions on American firms resulting from the new 2017 cybersecurity law. This addition contributes to our expanded treatment of cyberlaw.
- Explained in the description of critical thinking the significance of the "But What If" feature in DBL5 for being alert to the impact of single facts on success or failure in a legal dispute.

CHAPTER 2: Business Ethics and Social Responsibility

- Added a discussion of Core Values because so many businesses tie their mission statements and discussions about new directions to their core values. Then the core values of CVS and Adidas are contrasted with those of American Express.
- Included a chart of well-known instances of ethical and unethical business behavior featuring CVS, Starbucks, Wells Fargo, Bernie Madoff, and Microsoft in response to reviewers' desire for more visuals in the text.
- Made explicit the open-ended nature of ethical decisions. In other words, we should never expect universal agreement about how to answer ethical questions. But by learning the various ways that thoughtful people have learned to talk about ethical dilemmas, businesspeople can play a leadership role in sustaining an ethical market system.
- Emphasized the practicality of the WH approach to Business Ethics in response to reviewers' requests.

CHAPTER 3: The U.S. Legal System and Alternative Dispute Resolution

- Removed discussion of quasi in rem jurisdiction at the suggestion of a reviewer who felt that this concept was complicated and beyond the needs of the typical business student.
- Deleted the older case 3-1 about in personam jurisdiction based on the internet and replaced it with a more current Case Nugget, Minimum Contacts Via the Internet, that illustrates the same point. Before the Nugget some additional commentary was added about minimum contacts related to a case involving a university, which should be of interest to students.
- Added a sentence to the E-Commerce Box making it clear to students why they should be concerned about the subject matter of the Box, the sliding scale for internet transactions.
- At the suggestion of a reviewer, added two new exhibits: a sample forum selection clause and choice of law clause. While these clauses may seem self-explanatory, it will be helpful for the students to see what they actually look like.
- Updated Exhibit 3-6 to include the newest Supreme Court Justice's name and photo.
- Added a sentence clarifying that the map of the Circuit Courts of Appeal did not include the DC Circuit in response to a reviewer's concern about a potential conflict between the text, which correctly specifies that there are 12 Circuit Courts of Appeal, and the map, that showed only 11 circuits.
- Added the potential defendant's option of filing a motion to dismiss in lieu of an answer.
- Replaced the 1996 J.E.B. case with a 2016 case to provide a more current illustration of the concept of the Bateson challenge

CHAPTER 4: Administrative Law

- Added *NASA v. Nelson,* focusing on agency background checks of employees and informational privacy, including the nondisclosure requirement of the Privacy Act of 1974.
- Included new Case Nugget, *Open Communities Alliance v. Carson.*
- Retitled section previously called "How Are Agencies Run?" to "How Do Agencies Operate?" The new title, suggested by a reviewer, better reflects the content of that section.
- Changed Point/Counterpoint to "Is Administrative Regulation of Business a Good Thing?"

CHAPTER 5: Constitutional Law

- Inserted a new Case Nugget, "Can States Require On-line Sellers Without a Physical Presence in a State to Collect State Sales Tax?", based on the 2018 Supreme Court case *South Dakota v. Wayfair.* This case is extremely important because it changes the conditions under which states can tax internet sales transaction.
- Replaced old case 5-3, *Riley v. California,* with a newer case that shows the continuing evolutions of Fourth Amendment protections related to cell phone usage. In *Riley,* which we now just discuss briefly in the chapter, the high court ruled that warrantless searches from a cell phone taken from an arrestee at the time of his arrest constitute a violation of the Fourth Amendment. In the new case 5-3, *Carpenter v. United States,* the high court now holds that authorities must now seek a warrant for cell tower location information.

CHAPTER 6: Criminal Law and Business

- Created examples of the different states of mind a defendant could have when committing a crime. These illustrations will help the students more clearly see the differences among the states of mind.
- Included a new case, *Salmon v. US,* that clarifies when the U.S. Supreme Court will find an insider liable of being a tippee. A new example of embezzlement of money from a Girl Scout troop, demonstrating how even nonprofit corporations can be victims of embezzlement.

- Created Case Nugget, "Can Corporate Executives be Charged with Murder or Manslaughter?", highlighting the case recently filed against those responsible for the decapitation of a ten-year-old boy at a waterpark. Not only is this a current case that the students will find interesting, but it also raises an interesting and controversial issue that may lead to good classroom discussion.
- Changed the Point/Counterpoint to ask the question of whether insider trading should be legal. Insider trading is in the news a lot, and so students may be able to relate the discussion to situations they may heave read about but which aren't in their text.

CHAPTER 7: Tort Law

- Added new Case Nugget, *Jacobus v. Trump,* in response to a reviewer suggestion that we expand discussion of the distinction between fact and opinion. The case involves actions taken on Twitter, and thus is in line with our increased emphasis on cyber law.
- Provided a new case 7-1, *Mazda Motor Corporation v. Hurst,* providing a more updated scenario that shows the application of the *Gore* guideposts. a discussion about the innovative attempt by five California cities to use the tort of public nuisance to sue the Big 5 oil companies for damages caused by climate change, arguing that these firms created a public nuisance by misleading the public as to the effects of using fossil fuels.

CHAPTER 8: Real, Personal, and Intellectual Property

- Added new Exhibit 8-1 to illustrate types of property.
- Replaced the appellate decision with the new U.S. Supreme Court decision to have the latest holding in the case.
- Added an update after case 8-2 explaining how that Supreme Court case had affected a pending appeal by the Washington Redskins as to the legitimacy of their trademark, showing how U.S. Supreme Court cases not only affect the parties before the high court but may also affect other pending cases.
- Created new Exhibit 8-5, illustrating a few common trademarks.
- Added new Exhibit 8-6, which gives students some common ways to identify trademark and copyright protected materials.
- Designed new Global Box discussing liability in Japan of online providers for copyright violation. This addition is in keeping with our attempts to integrate more references to online activity.

CHAPTER 9: Introduction to Contracts and Agreement

- Replaced case 9-2 with a more current case, *Tripani v. Elliott,* demonstrating the elements of an implied contract. Offers and Acceptances in Saudi Arabia.
- Added a new Global Context box, "Offers and Acceptances in Saudi Arabia," which illustrates a slightly different way of forming an agreement.
- Changed the Point/Counterpoint to, Should the Business World Embrace Smart Contracts? This issue introduces students to the smart contracts, which is a new development in contract law.

CHAPTER 10: Consideration

- Included new Case Nugget in "Promissory Estoppel" section: *Ruzicka v. Conde Nast Publishing.*
- Added Case in "Preexisting Duty" section: Case 15-4: *Margeson v. Artis.*

CHAPTER 11: Capacity and Legality

- Replaced Case 11-3 with *Riley v. Iron Gate,* a more recent case that illustrates how the courts determine whether an exculpatory clause violates public policy and therefore is illegal.

- Added a cartoon to the section about Legality, focusing on a severed contract. This image may help students remember the meaning of "severed contract."
- Added a new Point/Counterpoint, "Should Minors Have Full Capacity to Enter into Binding Legal Contracts at the Age of 15?"

CHAPTER 12: Reality of Assent

- Added a cartoon to the Mistake section to get students thinking about what a mistake means.
- Expanded the discussion of undue influence and included four clarifying questions that can be used to determine whether undue influence exists.
- Replaced old case 12-3 with a new case, *Garage Solutions, LLC v. Monty J. Pearson.*

CHAPTER 13: Contracts in Writing and Third-Party Contracts

- Revised the Discussion of the Equal Dignities Rule in the Contracts Falling Within the Statute of Frauds Section. A reviewer thought the original version was somewhat confusing, and the revision is much clearer and contains an example to make the rule much more easily understandable.
- Added new Case Nugget, *Aurigemma v. New Castle Care, LLC,* to illustrate the difficulty of knowing when a writing is needed.
- Deleted section "integrated contracts" and moved the information from that section into the parole evidence rule section where it fit better.
- Added a new case, *Barker v. Price,* to illustrate how making a slight error in terms of what you include in your contract can make a difference in whether the parole evidence rule applies to your written contract.

CHAPTER 14: Discharge and Remedies

- Inserted a new case 14-3, *Hemlock Semiconductor Operations, LLC v. Solarworld Industries Sachsen Gmbh,* which provides a more recent illustration of the discharge of a contract due to commercial impracticability.
- Added Exhibit 14-1, a new graphic, to highlight differences among types of conditions.
- Included new Global Context Box, "Damages for Breach of Contract: Not the Preferred Remedy in Latvia," at the start of the remedies section. This box is an important addition because the preference in Latvia for equitable relief is the exact opposite of the preference in the United States for legal (non-equitable) relief.

CHAPTER 15: Formation and Performance of Sales and Lease Contracts

- Wrote new Chapter opener based on *Baker v. Zicron.*
- Created Case Nugget "Risk of Loss—Seller Breach" based on *Chicago Prime Packers, Inc. v. Northam Food Trading Co.*

CHAPTER 16: Sales and Lease Contracts: Performance, Warranties, and Remedies

- Created new Chapter opener based on *Accettura v. Vacationland, Inc.*
- Designed new Case Nugget about customary and usual practices.
- Included new case about direct v consequential damages.

CHAPTER 17: Negotiable Instruments: Negotiability and Transferability

- Added case 17-1, *Amplify Federal Credit Union v. Garcia.* This case examines the elements determining whether a contract is a negotiable instrument.
- Created a Global Box explaining the unique understanding of negotiable instruments in India. A concurrent traditional form of negotiable instruments governs particular kinds of instruments, side by side with a system of negotiable instruments similar to those in the United States.

CHAPTER 18: Holder in Due Course, Liability, and Defenses

- Included case 18-1, *Walker v. Probandt and Raynor.* This case shows how defenses are affected by the transfer of a note, and how defenses come into play when deciding liability. It also discusses how holder in due course status affects defenses.
- Added a new graphic to visualize the requirements to acquire holder-in-due-course status to supplement the prose description.
- Included a new graphic describing the requirements that must be met for there to be notice of such a defect. To say that a negotiable instrument was accepted without notice of defects is to say that a party was not knowledgeable of any defects in the instrument at the time of acceptance.

CHAPTER 19: Secured Transactions and Bankruptcy

- Replaced Case Nugget with a case addressing the definition of the terms "intangibles" and "proceeds" in a financing statement and their effect upon a debtor's subsequent award of damages (*Bayer Cropscience, LLC v. Stearns Bank Nat'l Ass'n.*
- Replaced case 19-2 with a case addressing the status of student loans in bankruptcy and maintenance of a "minimal standard of living" (*In re Murray*).
- Updated bankruptcy exemptions in Exhibit 19-4 and Chapters 13 and 12 filing requirements for individual repayment plans, family-farmer petitions, and family-fisherman petitions.

CHAPTER 20: Agency and Liability to Third Parties

- Added case 20-2, *Doe v. Uber Technologies, Inc.* This case provides a good illustration of how a court might determine whether a worker is an employee or independent contractor as well as whether an employer may be liable under *respondeat superior.*

CHAPTER 21: Forms of Business Organization

- Added new Point/Counterpoint: Should Aspiring Business Owners Form an LLC Instead of a Corporation?
- Included new case 21-2, *Patterson v. Domino's Pizza.*

CHAPTER 22: Corporations: Formation and Organization

- Clarified that *ultra vires* is uncommon, but important, in response to a reviewer's suggestion.
- Clarified the extent of the requirements of the Uniform Partnership Act in response to a reviewer's suggestion.
- Added clarification as to why LLCs are now such a popular form of business in response to a reviewer's suggestion.
- Inserted an additional description regarding franchises in response to a reviewer's suggestion.
- Added new case 22-2, *KDW Restructuring & Liquidation v. Greenfield,* which represents a clear statement of the contemporary application of the business judgment rule.

CHAPTER 23: Securities Regulation

- Added new case 23-2, *Levista, Inc. v. Ranbaxy Pharmaceuticals, Inc.*
- Added new Point/Counterpoint: Should Previous Dealings of the Parties Control the Contractual Relationship?

CHAPTER 24: Employment and Discrimination Law

- Added two new learning objectives and sections.
- Included the question in 24-7: "May employers use social media in employment decisions?" The issue of the use of social media by employers in hiring and firing decisions is a hot topic and one that is fraught with potential pitfalls. Savvy employers must be informed as to

when and how using social media for such purposes is wise and when it should be avoided (privacy issue).

- Added the question in 24-8: What are the rights and responsibilities of employers when hiring foreign workers? The recent executive orders related to immigration have made this a hot topic. United States' employers want to be able to access the best talent worldwide but also avoid legal entanglements.
- Added two major developments in the law regarding sexual orientation discrimination – an agency case and a federal circuit court case extending the definition of "sex" discrimination under Title VII to include discrimination based on sexual orientation.
- Added new case 24-2, *Donald Baldwin, Complainant v. Anthony Foxx, Secretary, Dept. Of Transportation (FAA), Agency,* First time the EEOC (as an agency) interprets "sex" discrimination under Title VII as including discrimination based on sexual orientation.
- Included a new section re: U.S. Supreme Court's 2015 decision granting marriage equality.
- Provided a new case 24-3, *Nickel v. Staples Contract and Commercial,* an age discrimination case in which "substantial evidence" of discrimination prevailed over allegations of employee theft, resulting in $16M verdict.
- Added new Case Nugget: *Roe v. Teletech Customer Care Mgmt. (Colo.) LLC* on whether an employee may be fired for using medical marijuana.
- Added new Point/Counterpoint: Should Employers Be Permitted to Use Social Media in Hiring and/or Firing Decisions?

CHAPTER 25: Consumer Law
- Clarified the rulemaking authority of the FTC in response to a reviewer's comment that the concept had not been introduced at the time it first appeared in the chapter.
- Added a section about data mining and privacy.
- Designed a discussion of the reduced regulations placed on data mining by legislation during President Trump's first hundred days.
- Included a new Point/Counterpoint: Should Prescription Drugs be Advertised Directly to Consumers?
- Added new case 25-2, *Crawford v. LVNV Funding.*

Acknowledgments

This final component of the Preface contains a palpable tone of gratitude and humility. Any project the scope of *Dynamic Business Law: The Essentials,* Fifth Edition, is a collective activity; the authors are but the visible component of a remarkably large joint effort. We want to thank several contributors by name, but there are doubtlessly many other students, colleagues, and friends who made essential contributions to these pages.

Our largest gratitude goes to the dozens of business law colleagues who saved us from many embarrassing errors while tolerating our stubborn reluctance to adhere to certain of their suggestions. In addition, the book could not have been written without the competent and dedicated research assistance we received from Arata-Enrique Kaku, Alex Jacobs, and Caitlyn Reeder. Their desire to improve our book is a compliment to their character.

We'd like to thank the following people for helping us with their comments and feedback on various stages of our manuscript:

Joan Alexander, *Nassau Community College of the SUNY*
James Blair, *Diablo Valley College*
William H. Bradley, *Azusa Pacific University*
Carol Brady, *Milwaukee Area Technical College*
Peter D. Churchill, *Diablo Valley Community College*
Nanette Clinch, *San Jose State University*
Richard Alan Cooper, *Lindenwood University*
Patrick Creehan, *Flagler College-Tallahassee Branch*

Peter Dawson, *Collin College*
Cheryl Davisson Gracie, *Washtenaw Community College*
Laura Drake, *Cincinnati State*
Pamela Evers, *University of North Carolina Wilmington*
Ronald Feinberg, *Suffolk County Community College*
Armand Grunberger, *American University-Washington, D.C.*
Shelly Grunsted, *University of Oklahoma*
Marc Hall, *Auburn University-Montgomery*
Mark E. Handley, *Delaware Technical & Community College*
Mo Hassan, *Cabrillo College*
Cheryl L. Head, *Central Piedmont Community College*
Lynda Hodge, *Guilford Technical Community College*
Sarah Holcomb, *Northern Arizona University*
Russell E. Holmes, *Des Moines Area Community College*
Catherine G. Jones-Rikkers, *Grand Valley State University*
Steven C. Kaber, *Baldwin-Wallace College*
Raymond Keener, *Marshall University*
Edward M. Kissling, *Ocean County College*
Diana Lamphiere, *Davenport University*
Christine W. Lewis, *Auburn University-Montgomery*
Darryll Lewis, *University of Nebraska-Omaha*
Jessica A. Magaldi, *Pace University*
Jason Bryan Malone, *Virginia Tech*
Lisa Mayer, *Bergen Community College*

Linda B. Meltzer, *Queensborough Community College*
Christine Mooney, *Queensborough Community College*
Sandra Mullings, *Baruch College, CUNY*
Tonia Murphy, *University of Notre Dame*
Michael J. O'Hara, *University of Nebraska-Omaha*
Gary Patterson, *California State University, San Bernardino*
James Alan Pilant, *Northwest Arkansas Community College*
Tara Radin, *George Washington University*
Michael Rogers, *Albany State University*
Sean D. Scott, *St. Petersburg College*
Stanley Seat, *University of Texas at Arlington*
Bonnie S. Shane, *Sinclair Community College*
Chris Simcox, *Bucks County Community College*
Tisha Simmons, *Davenport University*
Harvey Slentz, *Florida State College-Jacksonville*
Michael Spagnoli, *American International College*
Laura Sullivan, *Sam Houston State University*
Jennifer Thor, *Oakland University*
Mary Tripp, *Wisconsin Indianhead Technical College*
Timothy Welch, *Southeastern University*
Kim Wong, *Central New Mexico Community College*
Dr. Robert J. Woods, *Texas Woman's University*
Stephen Yoder, *University of Alabama at Birmingham*

Finally, this content was but a raw, unsold manuscript until the skilled publishing team refined it. Our manuscript benefited immeasurably from the guidance of the multiple levels of skill provided to us by McGraw Hill. We respect and honor our Managing Director, Tim Vertovec; our Product Developers, Jaroslaw Szymanski and Michael McCormick; Executive Portfolio Manager, Kathleen Klehr; Marketing Director, Natalie King; Marketing Manager, Claire McLemore; and Content Project Manager, Pat Frederickson and Angela Norris.

FOR INSTRUCTORS

You're in the driver's seat.

Want to build your own course? No problem. Prefer to use our turnkey, prebuilt course? Easy. Want to make changes throughout the semester? Sure. And you'll save time with Connect's auto-grading too.

65%
Less Time Grading

Laptop: McGraw-Hill; Woman/dog: George Doyle/Getty Images

They'll thank you for it.

Adaptive study resources like SmartBook® 2.0 help your students be better prepared in less time. You can transform your class time from dull definitions to dynamic debates. Find out more about the powerful personalized learning experience available in SmartBook 2.0 at **www.mheducation.com/highered/connect/smartbook**

Make it simple, make it affordable.

Connect makes it easy with seamless integration using any of the major Learning Management Systems—Blackboard®, Canvas, and D2L, among others—to let you organize your course in one convenient location. Give your students access to digital materials at a discount with our inclusive access program. Ask your McGraw-Hill representative for more information.

Solutions for your challenges.

A product isn't a solution. Real solutions are affordable, reliable, and come with training and ongoing support when you need it and how you want it. Our Customer Experience Group can also help you troubleshoot tech problems—although Connect's 99% uptime means you might not need to call them. See for yourself at **status.mheducation.com**

FOR STUDENTS

Effective, efficient studying.

Connect helps you be more productive with your study time and get better grades using tools like SmartBook 2.0, which highlights key concepts and creates a personalized study plan. Connect sets you up for success, so you walk into class with confidence and walk out with better grades.

Study anytime, anywhere.

Download the free ReadAnywhere app and access your online eBook or SmartBook 2.0 assignments when it's convenient, even if you're offline. And since the app automatically syncs with your eBook and SmartBook 2.0 assignments in Connect, all of your work is available every time you open it. Find out more at **www.mheducation.com/readanywhere**

"I really liked this app—it made it easy to study when you don't have your textbook in front of you."

- Jordan Cunningham, Eastern Washington University

Calendar: owattaphotos/Getty Images

No surprises.

The Connect Calendar and Reports tools keep you on track with the work you need to get done and your assignment scores. Life gets busy; Connect tools help you keep learning through it all.

Learning for everyone.

McGraw-Hill works directly with Accessibility Services Departments and faculty to meet the learning needs of all students. Please contact your Accessibility Services office and ask them to email accessibility@mheducation.com, or visit **www.mheducation.com/about/accessibility** for more information.

Top: Jenner Images/Getty Images, Left: Hero Images/Getty Images, Right: Hero Images/Getty Images

Brief Contents

PART 1
The Legal Environment of Business

chapter 1 An Introduction to the Fundamentals of Dynamic Business Law 1
chapter 2 Business Ethics and Social Responsibility 14
chapter 3 The U.S. Legal System and Alternative Dispute Resolution 31
chapter 4 Administrative Law 64
chapter 5 Constitutional Law 85
chapter 6 Criminal Law and Business 112
chapter 7 Tort Law 141
chapter 8 Real, Personal, and Intellectual Property 171

PART 2
Contract Law

chapter 9 Introduction to Contracts and Agreement 196
chapter 10 Consideration 216
chapter 11 Capacity and Legality 233
chapter 12 Reality of Assent 254
chapter 13 Contracts in Writing and Third-Party Contracts 272
chapter 14 Discharge and Remedies 302

PART 3
Domestic and International Sales Law

chapter 15 Formation and Performance of Sales and Lease Contracts 325
chapter 16 Sales and Lease Contracts: Performance, Warranties, and Remedies 351

PART 4
Negotiable Instruments and Banking

chapter 17 Negotiable Instruments: Negotiability and Transferability 374
chapter 18 Holder in Due Course, Liability, and Defenses 402

PART 5
Creditors' Rights and Bankruptcy

chapter 19 Secured Transactions and Bankruptcy 427

PART 6
Agency

chapter 20 Agency and Liability to Third Parties 455

PART 7
Business Organizations

chapter 21 Forms of Business Organization 486
chapter 22 Corporations: Formation and Organization 509
chapter 23 Securities Regulation 542

PART 8
Government Regulation

chapter 24 Employment and Discrimination Law 565
chapter 25 Consumer Law 599

APPENDIXES

appendix A The Constitution of the United States of America A-1
appendix B Sarbanes-Oxley Act of 2002 B

Glossary G
Index I-1

Contents

Preface v

PART 1
The Legal Environment of Business

1 An Introduction to the Fundamentals of Dynamic Business Law 1

Law and Its Purposes 3
Classification of the Law 3
Sources of Business Law 4
 Constitutions 4
 Statutes 5
 Cases 5
 Administrative Law 7
 Treaties 7
 Executive Orders 8
 Schools of Legal Interpretation 8
Global and Comparative Law 9
Questions & Problems 10

2 Business Ethics and Social Responsibility 14

Business Law and Business Ethics 16
 Ethical Failures Can be Bad for Business 19
What Are Values? 19
How do Values Provide a Starting Point for Thinking about Business Ethics? 21
The WH Framework for Business Ethics 23
 Who *are the Relevant Stakeholders? 23*
 How *do We make Ethical Decisions? 24*
Point/Counterpoint 28
Questions & Problems 28

3 The U.S. Legal System and Alternative Dispute Resolution 31

Case Opener: Questionable Jurisdiction over Caterpillar 31
Jurisdiction 32
 Original Versus Appellate Jurisdiction 32
 Jurisdiction Over Persons and Property 32
 Subject-Matter Jurisdiction 35

Case 3-1: Hertz Corporation v. Friend United States Supreme Court 36
Venue 37
Forum Selection and Choice of Law 38
Case 3-2: Atlantic Marine Construction Company, Inc. v. United States District Court for the Western District of Texas Et Al. 38
The Structure of the Court System 40
 The Federal Court System 40
 State Court Systems 41
Threshold Requirements 43
 Standing 43
 Case or Controversy 44
 Ripeness 44
Steps in Civil Litigation 44
 The Pretrial Stage 44
 The Trial 48
Case 3-3: Timothy Tyrone Foster, Petitioner v. Bruce Chatman, Warden United States Supreme Court 49
 Posttrial Motions 51
 Appellate Procedure 51
Alternative Dispute Resolution 53
Primary Forms of ADR 54
 Negotiation 54
 Mediation 54
 Arbitration 55
Other ADR Methods 57
 Med-arb 57
 Summary Jury Trial 58
 Minitrial 58
 Early Neutral Case Evaluation 58
 Private Trials 58
Summary 60
Point/Counterpoint 60
Questions & Problems 61

4 Administrative Law 64

Case Opener: Does the EPA Have an Obligation to Regulate Automobile Emissions? 64
Introduction to Administrative Law 65
 What is Administrative Law 65
 Why and How are Agencies Created? 65
Case 4-1: Doe v. United States Sec U.S. District Court for the Northern District of California 66

Types of Administrative Agencies 68
 Federal and State Administrative Agencies 68
How Do Agencies Operate? 69
 Informal Rule Making 69
 Formal Rule Making 70
 Hybrid Rule Making 71
 Exempted Rule Making 72
 Regulated Negotiation 72
 Problems Associated with Rule Making 72
 Other Administrative Activities 73
Case 4-2: Yan Ju Wang v. George Valverde Court of Appeal of California, Second Appellate District 73
Limitations on Agency Powers 74
 Political Limitations 75
 Statutory Limitations 75
 Judicial Limitations 75
Case 4-3: Northwestern University and College Athletes Players Association (CAPA) 76
 Informational Limitations 78
Case 4-4: Nasa v. Nelson 78
Summary 81
Point/Counterpoint 83
Questions & Problems 83

5 Constitutional Law 85

Case Opener: The Constitutionality of the Affordable Care Act 85
The U.S. Constitution 86
 Judicial Review 86
The Supremacy Clause and Federal Preemption 87
The Commerce Clause 88
 The Commerce Clause as a Source of Authority for the Federal Government 88
Case 5-1: Christy Brzonkala v. Antonio J. Morrison Et Al. United States Supreme Court 89
 The Commerce Clause as a Restriction on State Authority 90
Case 5-2: Black Star Farms, LLC, Et Al. v. Jerry Oliver, Arizona Department of Liquor License and Control Et Al. U.S. Court of Appeals for the Ninth Circuit 91
Taxing and Spending Powers of the Federal Government 93
Other Constitutional Restrictions on Government 93
 The Privileges and Immunities Clause 93
 The Full Faith and Credit Clause 94
 The Contract Clause 94
The Amendments to the Constitution 94
 The First Amendment 94
 The Fourth Amendment 99
Case 5-3: David Leon Riley v. California and United States v. Wurie Supreme Court of the United States 100
 The Fifth Amendment 102
 The Ninth Amendment 104
 The Fourteenth Amendment 105
Summary 106
Point/Counterpoint 108
Questions & Problems 108

6 Criminal Law and Business 112

Case Opener: A Questionable Stock Investment 112
The Nature and Elements of a Crime 113
 Elements of a Crime 113
 Classification of Crimes 114
White-Collar Crime 114
 Specific White-Collar Crimes 115
Case 6-1: Salman v. United States United States Supreme Court 118
Case 6-2: United States of America v. Thomas Tanke United States Circuit Court of Appeals for the Fourth Circuit 119
Corporate Criminal Liability 125
Case 6-3: United States v. Park United States Supreme Court 126
Criminal Procedure 128
 Pretrial Procedure 128
 Trial Procedure 130
 Posttrial Procedure 130
Tools for Fighting Business Crime 131
 RICO 132
 The False Claims Act 132
 The Sarbanes-Oxley Act 133
 State Whistle-Blower Protection Laws 134
Summary 136
Point/Counterpoint 137
Questions & Problems 138

7 Tort Law 141

Case Opener: Plastic Surgeon Defamation 141
Introduction to Tort Law 142
Classification of Torts 143
Intentional Torts 143
 Intentional Torts Against Persons 143
Case 7-1: Hancock v. Variyam Supreme Court of Texas 145
Case 7-2: Aaron Olson, Appellant, v. Centurylink Respondent Court of Appeals of Minnesota 151
 Intentional Torts Against Property 152
 Intentional Torts Against Economic Interests 153
Negligence 154
 Duty 155
 Breach of Duty 156
 Causation 156
 Damages 157

Plaintiff's Doctrines 158
 Res Ipsa Loquitur 158
Case 7-3: Swoboda v. Fontanetta Supreme Court of the State of New York 158
 Negligence Per Se 159
Defenses to Negligence 160
 Contributory Negligence 160
 Comparative Negligence 161
 Assumption Of The Risk 161
 Special Defenses To Negligence 162
Strict Liability 162
Damages Available in Tort Cases 162
 Compensatory Damages 163
 Nominal Damages 163
 Punitive Damages 163
Case 7-4: Mazda Motor Corporation v. Hurst Alabama State Supreme Court 165
Summary 167
Point/Counterpoint 167
Questions & Problems 168

8 Real, Personal, and Intellectual Property 171

Case Opener: Smartphone Wars: Focus on Technology Companies 171
The Nature of Real, Personal, and Intellectual Property 172
Real Property 173
 Extent of Ownership 173
 Interests in Real Property 174
 Property Transfer 176
Case 8-1: Kelo v. City of New London United States Supreme Court 177
Personal Property 179
 Voluntary Transfer of Property 179
 Involuntary Transfer of Personal Property 180
 Other Means of Acquiring Ownership of Property 181
Intellectual Property 181
 Trademarks 181
Case 8-2: Matal, Interim Director, United States Patent and Trademark Office v. Tam United States Supreme Court 183
Case 8-3: Toys "R" Us, Inc. v. Canarsie Kiddie Shop, Inc. District Court of the Eastern District of New York 185
 Copyrights 186
 Patents 188
 The America Invents Act 190
 Trade secrets 190
Summary 192
Point/Counterpoint 192
Questions & Problems 193

PART 2
Contract Law

9 Introduction to Contracts and Agreement 196

Case Opener: The Problematic Promotion 196
The Definition of a Contract 197
 Elements of a Contract 197
 Defenses to the Enforcement of a Contract 197
Sources of Contract Law 198
 Common Law 198
 Uniform Commercial Code 198
Classification of Contracts 199
 Bilateral Versus Unilateral Contracts 199
Case 9-1: Zappos.com Inc., Customer Data Security Breach Litigation United States District Court of Nevada 199
 Express Versus Implied Contracts 201
Case 9-2: Trapani Const. v. Elliott Group Inc. Appellate Court of Illinois, First Division 201
 Quasi-Contracts 202
 Valid, Void, Voidable, and Unenforceable Contracts 202
 Executed Versus Executory Contracts 203
 Formal Versus Informal Contracts 203
The Agreement 203
Elements of the Offer 205
 Intent 205
Case 9-3: Lucy v. Zehmer Supreme Court of Virgina 205
 Definite and Certain Terms 207
 Communication to the Offeree 207
Termination of the Offer 208
 Revocation by the Offeror 208
 Rejection by the Offeree 208
 Death or Incapacity of the Offeror 209
 Destruction or Subsequent Illegality of the Subject Matter 209
 Lapse of Time or Failure of Another Condition Specified in the Offer 209
Elements of the Acceptance 209
 Manifestation of Intent to Be Bound to the Contract 209
 Acceptance of Definite and Certain Terms: the Mirror-Image Rule 210
 Communication to the Offeror 211
Summary 212
Point/Counterpoint 213
Questions & Problems 214

10 Consideration 216

Case Opener: Upper Deck—Contract Liability or Gift? 216
What Is Consideration? 217
Case 10-1: Hamer v. Sidway New York Court of Appeals 217

Rules of Consideration 218
 Lack of Consideration 218
 Adequacy of Consideration 220
Case 10-2: Thelma Agnes Smith v. David Phillip Riley Court of Appeals of Tennessee, Eastern Section, at Knoxville 220
 Illusory Promise 222
 Past Consideration 222
Case 10-3: Simmons v. Simmons United States District Court of Appeals for Arkansas 223
 Preexisting Duty 224
Case10-4: Margeson v. Artis Supreme Court of Iowa (2009) 225
Partial Payment of a Debt 226
Summary 229
Point/Counterpoint 229
Questions & Problems 230

11 Capacity and Legality 233

Case Opener: Apple's Questionable Contract 233
Capacity 234
 Minors 234
Case 11-1: Adrian Lopez v. Kmart Corporation United States District Court for the Northern District of California 235
 Mentally Incapacitated Persons 238
 Intoxicated Persons 239
Legality 240
 Contracts That Violate State or Federal Statutes 240
Case 11-2: King v. Riedl Alabama Civil Court of Appeals 241
 Agreements That Violate Public Policy 243
Case 11-3: Riley And v. Stone Gate Court of Appeals of Washington, Division 2 246
 Effect of Illegal Agreements 248
Summary 250
Point/Counterpoint 251
Questions & Problems 252

12 Reality of Assent 254

Case Opener: Disagreement Over an Agreement 254
The Importance of Genuine Assent 255
Mistake 255
 Unilateral Mistake 256
 Mutual Mistake 257
Case 12-1: Simkin v. Blank Court of Appeals of New York 258
Misrepresentation 259
 Innocent Misrepresentation 259
 Negligent Misrepresentation 260
 Fraudulent Misrepresentation 260
Case 12-2: Fantastic Sams Salons Corp. v. Pstevo, LLC and Jeremy Baker United States District Court for the Northern District of Illinois, Eastern Division 261
Undue Influence 263
Duress 265
Case 12-3: Garage Solutions v. Monty J. Pearson Court of Appeals of Louisiana, Third Circuit 266
Summary 267
Point/Counterpoint 268
Questions & Problems 269

13 Contracts in Writing and Third-Party Contracts 272

Case Opener: Fallout from a Forgettable Fight 272
Statute of Frauds 273
Contracts Falling within the Statute of Frauds 273
 Contracts whose Terms Prevent Possible Performance within One Year 274
 Promises Made in Consideration of Marriage 275
 Contracts for One Party to Pay the Debt of Another if the Initial Party Fails to Pay 276
 Contracts Related to an Interest in Land 276
 Contracts for the Sale of Goods Totaling more than $500 278
 Further Requirements Specific to Certain States 278
Exceptions to the Statute of Frauds 279
 Admission 279
 Partial Performance 279
 Promissory Estoppel 279
 Exceptions Under the UCC 280
Sufficiency of the Writing 280
Case 13-1: Heritage Constructors, Inc. v. Chrietzberg Electric, Inc. and Richard Marc Chrietzberg: Court of Appeals of Texas, Sixth District 280
Parol Evidence Rule 282
Exceptions to the Parol Evidence Rule 283
 Contracts that have been Subsequently Modified 283
 Contracts Conditioned on Orally Agreed-On Terms 283
 Nonfinalized, Partially Written and Partially Oral, Contracts 284
 Contracts Containing Ambiguous Terms 284
 Incomplete Contracts 284
Case 13-2: Baker v. Price Indiana Court of Appeals 284
 Contracts with Obvious Typographical Errors 285
 Void or Voidable Contracts 285
 Evidence of Prior Dealings or Usage of Trade 285
Third-Party Rights to Contracts 286
Assignments and Delegations 286
 Assignment 286
 Delegation 289
Assignment of the Contract 291
Third-Party Beneficiary Contracts 291
 Intended Beneficiaries 292
Case 13-3: Lawrence v. Fox Court of Appeals of New York 293
 Incidental Beneficiaries 295

Summary 297
Point/Counterpoint 298
Questions & Problems 299

14 Discharge and Remedies 302

Case Opener: Impossible Wine Bottles 302
Methods of Discharging a Contract 303
 Conditions 303
Case 14-1: Mind & Motion Utah Investments, LLC, Appellee v. Celtic Bank Corporation Supreme Court of the State of Utah 2015 304
Discharge by Performance 305
 Discharge by Material Breach 307
Case 14-2: Hamilton v. State Farm Fire & Casualty Insurance Company United States Court of Appeals for the Fifth Circuit 307
 Discharge by Mutual Agreement 308
 Discharge by Operation of Law 309
Case 14-3: Thrifty Rent-A-Car System v. South Florida Transport U.S. District Court for the Northern District of Oklahoma 311
Remedies 313
 Legal Remedies (Monetary Damages) 314
Case 14-4: S. Brooke Purll, Inc., T/A Purll Construction v. Patrick Darrell Vailes District of Columbia Court of Appeals 317
 Equitable Remedies 318
Summary 321
Point/Counterpoint 321
Questions & Problems 322

PART 3
Domestic and International Sales Law

15 Formation and Performance of Sales and Lease Contracts 325

Case Opener: On Your Mark, Get Set, Go! 325
The Uniform Commercial Code 326
 The Scope of the UCC 326
 The Significance of the UCC 326
Articles 2 and 2(A) of the UCC 327
 Article 2 of the UCC 327
Article 2(A) of the UCC 330
Contracts for the International Sale of Goods 331
 The Scope and Significance of the CISG 331
How Sales and Lease Contracts Are Formed under the UCC 332
 Formation in General 332
 Offer and Acceptance 332
 Consideration 333
 Battle of the Forms 333
 Requirements under the Statute of Frauds 334
 Unconscionability 334
Case 15-1: Larson v. Burton Constr., Inc., 335
Title 336
 Three Kinds of Title 336
Acquiring Good Title 336
Case 15-2: Alfonso Candela, Appellant, v. Port Motors Inc., Respondent, Et Al., Defendant Supreme Court of New York, Appellate Division, Second Department 337
 Voidable Title 337
 Entrustment 338
Case 15-3: Frenk v Solomon 338
 Title Versus Other Interests in the Goods 340
Types of Sales Contracts 341
 Simple Delivery Contract 342
 Common-Carrier Delivery Contract 342
 Goods-in-Bailment Contract 343
 Conditional Sales Contract 344
Risk of Loss during a Breach of Contract 344
 When the Seller is in Breach 344
 When the Buyer is in Breach 345
Summary 346
Point/Counterpoint 348
Questions & Problems 348

16 Sales and Lease Contracts: Performance, Warranties, and Remedies 351

Case Opener: A Nightmare of a Vacation 351
The Basic Performance Obligation 352
 Good Faith 352
Specific Obligations of Sellers and Lessors 353
 The Perfect Tender Rule 353
Case 16-1: Alaska Pacific Trading Co. v. Eagon Forest Products Inc. Washington Appellate Division 1 355
 Exceptions to the Perfect Tender Rule 356
Specific Obligations of Buyers and Lessees 357
 The Basic Obligation: Inspection, Payment, and Acceptance 357
 Exceptions to the Basic Obligation 357
Warranties 358
 Warranties of Title 358
 Express Warranties 358
 Implied Warranties of Quality 359
Case 16-2: Priscilla D. Webster v. Blue Ship Tea Room, Inc. Supreme Judicial Court of Massachusetts 359
Warranty Rights of Third Parties 360
Warranty Disclaimers and Waivers 361
The Goal of Contract Remedies 362

Remedies Available to Sellers and Lessors under the UCC 362
 Cancel the Contract 362
 Withhold Delivery 362
 Resell or Dispose of the Goods 363
 Sue to Get the Benefit of the Bargain 363
 Stop Delivery 363
 Reclaim the Goods 363
Remedies Available to Buyers and Lessees under the UCC 363
 Cancel the Contract 363
 Obtain Cover 363
 Sue to Recover Damages 364
Case 16-3: Stern Oil Co. v. Brown, 2018 S.D. 15 364
 Recover the Goods 365
 Obtain Specific Performance 366
 Reject Nonconforming Goods 367
 Revoke Acceptance of Nonconforming Goods 367
 Accept the Nonconforming Goods and Seek Damages 367
Modifications or Limitations to the Remedies Otherwise Provided by the UCC 367
Summary 369
Point/Counterpoint 370
Questions & Problems 371

PART 4
Negotiable Instruments and Banking

17 Negotiable Instruments: Negotiability and Transferability 374

Case Opener: Oral Agreements and Negotiable Instruments 374
The Need for Negotiable Instruments 375
 Contracts as Commercial Paper 375
 Problems with Commercial Paper 375
Types of Negotiable Instruments 376
An Overview of the Law of Negotiable Instruments 376
Case 17-1: Amplify Federal Credit Union v. Garcia Court of Appeals of Texas: Court of Appeals, 3rd District (2017) 377
 Requirements for Negotiability 378
Negotiation 382
 Delivery 382
 Endorsement 382
 Noncriminal Endorsement Problems 383
Case 17-2: Jones v. Wells Fargo Bank 383
Banking 385

Checks 386
 Cashier's Checks 387
 Traveler's Checks 387
 Money Orders 387
 Certified Checks 388
 Why Use Cashier's or Certified CHECKS? 388
 Lost, Stolen, or Destroyed Cashier's or Certified Checks 389
Accepting Deposits 390
 The Check Collection Process 390
 Electronic Check Presentment 391
 Availability Schedule for Deposited Checks 391
 The Truth-in-Savings Act 392
When a Bank May Charge a Customer's Account 392
 Wrongful Dishonor 392
 Stop-Payment Order 393
 Postdated Checks 393
 Stale Checks 393
 Forgeries and Alterations 393
Electronic Fund Transfers 395
 Consumer Fund Transfers 395
 Commercial Fund Transfers 395
E-Money and Online Banking 395
 Online Banking Services 396
 Regulatory Compliance 396
 Privacy Protection 396
Summary 397
Point/Counterpoint 399
Questions & Problems 399

18 Holder in Due Course, Liability, and Defenses 402

Case Opener: Dishonored Check and Holder-in-Due-Course Status 402
Holder-in-Due-Course Doctrine 403
 Reason for Holder-in-Due-Course Status 403
Requirements for Holder-in-Due-Course Status 403
 Be a Holder of a Complete and Authentic Negotiable Instrument 404
 Take Instrument for Value 405
 Take Instrument in Good Faith 406
Case 18-1: Wawel Savings Bank v. Jersey Tractor Trailer Training, Inc. United States Court of Appeals for the Third District 406
 Take Instrument without Notice 407
Case 18-2: Creative Ventures, LLC v. Jim Ward & Associates 410
The Shelter Principle and HDC 410
Liability 411
Signature Liability 411
 Primary Liability of Makers and Acceptors 412
 Secondary Liability of Drawers and Endorsers 412

Accommodation Parties 414
Unauthorized Signatures and Endorsements 414
Warranty Liability 416
Transfer Warranty 416
Presentment Warranty 417
Avoiding Liability for Negotiable Instruments 417
Defenses to Liability 417
Discharge of Liability on Instruments 419
Summary 422
Point/Counterpoint 423
Questions & Problems 424

PART 5
Creditors' Rights and Bankruptcy

19 Secured Transactions and Bankruptcy 427

Case Opener: A Commercially Reasonable Sale? 427
Important Definitions Associated with Secured Transactions 428
Creation of Secured Interests 429
Written Agreement 429
Value 429
Debtor Rights in the Collateral 429
Purchase-Money Security Interest 429
Perfected Security Interest 430
Perfection by Filing 430
Perfection by Possession 430
Automatic Perfection 431
Perfection of Movable Collateral 431
Perfection of Security Interests in Automobiles and Boats 431
The Scope of a Security Interest 431
After-Acquired Property 432
Proceeds 432
Termination of a Security Interest 432
Priority Disputes 433
Secured Versus Unsecured Creditors 433
Secured Versus Secured Creditors 433
Secured Party Versus Buyer 434
Default 435
Taking Possession of the Collateral 435
Case 19-1: Price Auto Sales, Inc. v. Sanders Texas Court of Appeals 436
Proceeding to Judgment 437
Bankruptcy and Reorganization 438
The Bankruptcy Act and Its Goals 438
Title 11 of The United States Code 438
Bankruptcy Proceedings 438

Specific Types of Relief Available 439
Chapter 7: Liquidation Proceedings 439
Case 19-2: In Re Murray U.S. Bankruptcy Court, District of Kansas 444
Chapter 11: Reorganizations 445
Chapter 13: Individual Repayment Plans 446
Chapter 12: Family-Farmer and Family-Fisherman Plans 447
Summary 449
Point/Counterpoint 450
Questions & Problems 451

PART 6
Agency

20 Agency and Liability to Third Parties 455

Case Opener: FedEx and Independent Contractors 455
Introduction to Agency Law 456
Creation of the Agency Relationship 456
Expressed Agency (Agency By Agreement) 457
Agency By Implied Authority 458
Apparent Agency (Agency By Estoppel) 459
Case 20-1: Ackerman v. Sobol Family Partnership, LLP, Supreme Court of Connecticut 460
Agency By Ratification 461
Recent Developments in Agent/Employer Law 461
Agency Relationships 461
Principal-Agent Relationship 461
Employer-Employee Relationship 462
Employer-Independent Contractor Relationship 462
Case 20-2: Doe v. Uber Technologies, Inc., United States District Court for the Northern District of California 463
Duties of the Agent and the Principal 464
Principal's Duties to the Agent 465
Agent's Duties to the Principal 466
Rights and Remedies 468
Principal's Rights and Remedies Against The Agent 468
Agent's Rights and Remedies Against The Principal 469
Authority of the Agent: The Link to the Principal's Liability 469
Express Authority 470
Implied Authority 470
Apparent Authority and Estoppel 470
Contractual Liability of the Principal and Agent 471
Classification of the Principal 471
Authorized Acts 471
Unauthorized Acts 472

Tort Liability and the Agency Relationship 473
 Principal's Tortious Conduct 473
Case 20-3: Iglesia Cristiana La Casa Del Senor, Inc., ETC. v. L.M. Court of Appeal of Florida, Third District 474
 Agent Misrepresentation 475
 Principal's Liability and The Independent Contractor 475
Termination of the Agency Relationship 476
 Termination by Acts of Parties 477
 Termination by Operation of Law 478
Summary 480
Point/Counterpoint 482
Questions & Problems 483

PART 7
Business Organizations

21 Forms of Business Organization 486

Case Opener: The Dunkin' Donuts Franchise Agreement 486
Major Forms of Business Organization 487
 Sole Proprietorship 487
 Partnership 488
 Different Types of Partnerships 492
 Corporation 494
 Limited Liability Company 494
Specialized Forms of Business Organization 496
 Cooperative 496
 Joint Stock Company 497
 Business Trust 497
 Syndicate 498
 Joint Venture 498
Case 21-1: Meyer v. Christie U.S. Court of Appeals for the Tenth Circuit 498
 Franchise 499
Case 21-2: Live Cryo, LLC v. Cryousa Import and Sales, LLC United States Distrct Court, E.d. Michigan, 501
Case 21-3: Cousins Subs Systems, Inc. v. Michael R. McKinney U.S. District Court for the Eastern District of Wisconsin 503
Summary 504
Point/Counterpoint 505
Questions & Problems 505

22 Corporations: Formation and Organization 509

Case Opener: The Formation of the Facebook Corporation 509
Characteristics of Corporations 510
 Legal Entity 510
 Rights as a Person and a Citizen 510
Case 22-1: Protect My Check, Inc. v. Craig C. Dilger, John Steffen United States District Court, E.D. Kentucky, Central Division. Frankfort 510
 Creature of the State 511
 Limited Liability 511
 Free Transferability of Corporate Shares 511
 Perpetual Existence 511
 Centralized Management 512
 Corporate Taxation 512
 Liability for Officers and Employees 512
Corporate Powers 513
 Express and Implied Powers 513
 Ultra Vires Act 513
Classification of Corporations 513
 Public or Private 513
 Profit or Nonprofit 514
 Domestic, Foreign, and Alien Corporations 514
 Publicly Held or Closely Held 514
 Subchapter S Corporation 514
 Professional Corporation 515
Formation of the Corporation 515
 Organizing and Promoting the Corporation 515
 Selecting a State for Incorporation 516
Legal Process of Incorporation 516
 Selection of Corporate Name 516
 Incorporators 516
 Articles of Incorporation 516
 First Organizational Meeting 517
Potential Problems with Formation of the Corporation 517
 Responses to Defective Incorporation 517
Corporate Financing 519
 Debt Securities 519
 Equity Securities 520
Importance of Regulating Interactions among Directors, Officers, and Shareholders within a Corporation 520
Roles of Directors, Officers, and Shareholders 520
 Directors' Roles 521
 Officers's Roles 521
 Shareholders' Roles 521
Duties of Directors, Officers, and Shareholders 523
 Duties of Directors and Officers 523
 Duties of Shareholders 526
Case 22-2: McCann v. McCann Supreme Court of Idaho 526
Liabilities of Directors, Officers, and Shareholders 527
 Liability of Directors and Officers 527
Case 22-3: KDW Restructuring & Liquidation v. Greenfield United States District Court, S.D. New York 528
 Liability of Shareholders 529

Rights of Directors, Officers, and Shareholders 530
 Directors' Rights 530
 Officers's Rights 530
 Shareholders' Rights 530
Purchase of Stock 532
The Nature of Takeovers 533
 Types of Takeovers 533
 Response to Takeovers 533
Case 22-4: Campbell, Kesser, and Williams v. Pothas Corporation U.S. Court of Appeals for the Sixth Circuit 534
Response to Termination 535
 Dissolution 535
Summary 536
Point/Counterpoint 538
Questions & Problems 539

23 Securities Regulation 542

Case Opener: The Martha Stewart Case 542
What Is a Security? 543
Case 23-1: Securities and Exchange Commission, v. Arcturus Corporation United States District Court, N.d. Texas, Dallas Division 543
Securities Regulation 544
 The Securities and Exchange Commission 544
The Securities Act of 1933 546
 Registration Statement 546
 Prospectus 546
 Periods of the Filing Process 546
 Special Registration Provisions 547
 Exemptions under the 1933 Act 547
 Violations and Liability 550
The Securities Exchange Act of 1934 550
 Section 10(B) and Rule 10B-5 551
 Insider Trading 552
Case 23-2: Securities and Exchange Commission v. Texas Gulf Sulphur Co. U.S. Court of Appeals for the Second Circuit 552
 The Private Securities Litigation Reform Act of 1995 553
 Outsiders and Insider Trading 553
Case 23-3: United States of America v. Matthew Kluger United States Court of Appeals for the Third Circuit 554
 Section 16(B) 555
 Proxy Solicitations 556
 Violations Of The 1934 Act 556
Regulation of Investment Companies 557
State Securities Laws 558
Summary 559
Point/Counterpoint 561
Questions & Problems 562

PART 8
Government Regulation

24 Employment and Discrimination Law 565

Case Opener: Brad Gets Fired from So Clean! 565
When May an Employee Be Fired? 566
Federal Employment Discrimination Laws Governing Employers 566
Civil Rights Act—Title VII 566
 Disparate-Treatment Discrimination Under Title VII 567
 Disparate-Impact Discrimination under Title VII 568
 Sexual Harassment under Title VII 569
Case 24-1: Brennan v. Townsend & O'leary Enterprises, Inc. Court of Appeals of California, Fourth Appellate District 570
 Harassment of Other Protected Classes Under Title VII 571
 Pregnancy Discrimination Act of 1987— An Amendment to Title VII 571
 Defenses to Claims Under Title VII 572
 Remedies under Title VII 572
 Procedure for Filing a Claim under Title VII 573
Discrimination Based on Sexual Orientation—Actionable? 573
Case 24-2: Donald Baldwin, Complainant v. Anthony Foxx, Secretary of Dept. of Transportation (FAA), Agency 574
Discrimination Based on Marriage 575
Age Discrimination in Employment Act of 1967 576
Case 24-3: Bobby Dean Nickel v. Staples Contract & Commercial Inc. 576
Americans with Disabilities Act 578
May an Employer Discriminate against a Smoker? 578
Equal Pay Act of 1963 579
Additional Laws Governing the Employment Relationship 579
 Fair Labor Standards Act 580
 Family and Medical Leave Act 580
 Unemployment Compensation 580
 Workers's Compensation Laws 580
 Consolidated Omnibus Budget Reconciliation Act of 1985 581
 Employee Retirement Income Security Act of 1974 581
 Occupational Safety and Health Act of 1970 581
Employee Privacy in the Workplace 582
 Electronic Monitoring and Communication 582
Case 24-4: Ehling v. Monmouth-Ocean Hosp. Serv. Corp. 582
Use of Social Media in Employment Decisions 584
 Use of Social Media in Hiring Decisions 584
 Use of Social Media in Firing Decisions 585

Responsibilities of Employers in Hiring Foreign Workers 586
Labor Laws and Unions 586
 The Wagner Act of 1935 590
 The Taft-Hartley Act of 1947 591
 The Landrum-Griffin Act of 1959 592
 The National Labor Relations Board 592
 The Collective Bargaining Process 592
Summary 593
Point/Counterpoint 595
Questions & Problems 595

25 Consumer Law 599

Case Opener: Deceptive Advertising and the Ultimate Weight Loss Cure 599
The Federal Trade Commission 600
 How the FTC Brings an Action 600
 Trade Regulation Rules 601
Deceptive Advertising 601
Case 25-1: Gotlin v. Lederman United States Court of Appeals for the Second Circuit 603
 Bait-and-Switch Advertising 604
 FTC Actions against Deceptive Advertising 605
 Telemarketing and Electronic Advertising 605
 Data Mining and Privacy 606
 Tobacco Advertising 608
Labeling and Packaging Laws 609
Sales 610
 Door-to-Door Sales 610
 Telephone and Mail-Order Sales 610
 FTC Regulation of Sales in Specific Industries 611
Credit Protection 612
 The Truth in Lending Act 612
 The Fair Credit Reporting Act 614
 The Fair Debt Collection Practices Act 615
Case 25-2: Crawford v. LVNV Funding, LLC United States Court of Appeals, Eleventh Circuit 615
 The Credit Card Fraud Act 616
 The Fair Credit Billing Act 617
 The Fair and Accurate Credit Transactions Act 617
 The Credit Card Holders' Bill of Rights Act 617
Consumer Health and Safety 618
 The Federal Food, Drug, and Cosmetic Act 618
Case 25-3: United States v. Kaplan, United States Court of Appeals, Ninth Circuit 836 F. 618
 The Consumer Product Safety Act 619
Summary 621
Point/Counterpoint 622
Questions & Problems 623

Appendixes

APPENDIX A
The Constitution of the United States of America A-1
APPENDIX B
Sarbanes-Oxley Act of 2002 B

Glossary G
Index I-1

Part 1 The Legal Environment of Business

CHAPTER 1

An Introduction to the Fundamentals of Dynamic Business Law

LEARNING OBJECTIVES

- LO 1-1 Define business law.
- LO 1-2 Relate the functional areas of business to the relevant areas of business law.
- LO 1-3 Recall the purposes of law.
- LO 1-4 Distinguish among types of law.
- LO 1-5 Differentiate between sources of the law.
- LO 1-6 Identify the various schools of jurisprudence.

LO 1-1 Define business law.

This book is for future business managers, especially those who wish to be leaders. An essential factor in preparing for that career is awareness of the legal issues arising in business. Businesses need to finance capital growth, purchase inputs, and hire and develop employees. In addition, they must sell to consumers, please owners, and comply with government rules. All these activities are full of potential legal conflicts.

A theme of this book will be a surprising number of legal implications that result from what on the surface might look like a simple, everyday business decision. Think along with us about the desire of a young entrepreneur who has a new idea and wants to develop its market potential.

> In the seemingly ever-expanding market for smart phone apps, individuals are often rewarded for creating apps that make lives simpler or that entertain people. You have watched the business success of hundreds of smart phone applications. You have certainly downloaded your share of them. You have an idea for a phone app that you believe has the potential to become the next Vine or Instagram.
>
> Your proposed app will allow users to search for lawyers based on the location and specialization that the user desires. The app will also rate different lawyers and provide personal information about the lawyers who are being rated. Your goal is to make it easier for uninformed consumers to locate the lawyer best suited for their needs.
>
> You and a friend, who has a background in computer science, develop software to run the application. You also create a design for the app: what the app will look like and how it will function.
>
> After you complete the design of the app, you decide that you would like to upload it to the Apple App Store. You contact Apple, and a representative tells you that you will be able to upload the app if you pay a fee and sign a few forms.

In this textbook, you will find legal concepts and terms that will allow you to see this scenario with different eyes. What you will learn will warn you about the dangers of letting the excitement of the moment—your big moment—cause you to forget that business is, in part, a series of conflicts. And, in those struggles, you need to understand your interests and responsibilities. From the instant you first interact with another party in a business relationship, insights from your study of business law should kick in to protect you.

Let's walk through just a small sample of the legal issues you will face as you try to make your app the next big thing.

Because you need a platform on which to sell your app, you will need to establish a contract with an app store such as the Apple App Store. This book will teach you about the elements required to establish a legal contract. In other words, only certain kinds of legal agreements will provide legal protection if a later dispute arises. Whether these elements are met will dictate whether the party with whom you are contracting is obligated to perform a certain duty. For example, if you believe that Apple has a legal duty to put your app in its App Store or, perhaps, place your app in a particular position in its online store, you would have to prove that a contract creating such a duty exists.

The knowledge you will gain regarding contract law will also help you understand the complexity of the relationship between you and the user purchasing your app. For example, if the app you created malfunctions after a user has purchased it, how will you and the customer settle a dispute? Will you go to court, or would you prefer disputes to be settled through alternate dispute resolution methods such as arbitration? Will you create a binding arbitration clause to require consumer disputes to be resolved through arbitration? After reading this textbook, you will become more aware of the various conflict resolution mechanisms available to businesses and consumers.

In addition to the contractual issues underlying your app business, you will also need to be aware of possible tort issues related to your app. (A tort, at its simplest level, is legal language for a civil [non-criminal] harm committed against another in which the injured party can sue for damages.) For example, because your app will compare the desirability of certain lawyers to others, the lawyers whom you are rating might feel threatened by your app. Will these lawyers be able to file a suit in tort against you for defamation or privacy torts such as false light or public disclosure of private facts?

Alternatively, what can you do when a competing business writes a column in a newspaper mocking and generally trashing your new app? To answer that question, you need a good understanding of the torts of disparagement or slander of quality, two torts that you might be able to use to defend your

business against fraudulent claims. The sections in this textbook on tort law will enable you to ask questions about potential tort claims related to the eventual marketing of your app.

Later, when you market your app, you will also have to be aware of the restrictions on advertising that exist under consumer protection laws. Will you make any promises to consumers regarding your app in your advertisements? Will these advertisements constitute legally permitted puffery, or will they be deceptive advertisement under the Federal Trade Commission Act? As you read about consumer law in this textbook, you will learn about the restrictions the law places on advertising.

When you design your app, you will have to make sure that you do not infringe on the intellectual property—such as patents or trademarks—of other app businesses. What words and symbols will you use? Do those words and symbols constitute trademark infringement? This book will explain the intellectual property law that governs the invention of an app.

Finally, once you have studied this book, you will realize the importance of carefully selecting the form of business organization you and your friend choose for operating this app. For example, will you operate the app under a sole proprietorship or a partnership agreement? As you will learn, each of these forms of business organization provide you with different advantages and disadvantages.

Business law consists of the enforceable rules of conduct that govern commercial relationships ("commercial" meaning "for the purpose of making a profit"). For example, a firm is required by law to obey the antitrust laws when it considers merging with another firm. In other words, buyers and sellers interact in market exchanges within the rules that specify the boundaries of legal business behavior. Constitutions, legislatures, regulatory bodies, and courts spell out what market participants may and may not legally do. These rules and responsibilities provide the stability required in a thriving market economy.

business law
The enforceable rules of conduct that govern the actions of buyers and sellers in market exchanges.

LO 1-2 Relate the functional areas of business to the relevant areas of business law.

Understanding business law is necessary for future businesspeople because there simply is no market transaction that occurs outside legal guidelines. All contracts, employment decisions, and payments to a supplier are limited and protected by business law. Each of the six functional areas of business—management, production and transportation, marketing, research and development, accounting and finance, and human resource management—sits on a foundation of business law.

Law and Its Purposes

LO 1-3 Recall the purposes of law.

Many of us might like to impose rules on others, defining their rights and responsibilities. Few of us can do this as individuals, but a majority of citizens in a democracy can agree to establish rules for business behavior. They can permit certain authorities to make and enforce rules describing what behavior is permitted and encouraged in their community. These rules are what we refer to as the law, and they are enforceable in the courts of that community. Exhibit 1-1 lists a few of the numerous purposes fulfilled by the law.

Exhibit 1-1 is a reminder of why we are very proud when we say we are a society of laws. The respect we give to the law as a source of authority is in part a recognition of the fact that in the absence of law, we would rely solely on the goodwill and dependability of one another. Most of us greatly prefer the law to that option.

law
Rules of conduct in any organized society that are enforced by the governing authority of the community.

Classification of the Law

LO 1-4 Distinguish among types of law.

There are many ways of thinking about the law. For example, we can divide law into *national* versus *international* law, *federal* versus *state* law, and *public* versus *private* law. **Private law** involves disputes between private individuals or groups. As an illustration, if a businessperson owns a

private law
Law that involves suits between private individuals or groups.

Exhibit 1-1

Purposes of the Law

- Providing order such that one can depend on a promise or an expectation of obligations.
- Serving as an alternative to fighting.
- Facilitating a sense that change is possible, but only after a rational consideration of options.
- Encouraging social justice.
- Guaranteeing personal freedoms.
- Serving as a moral guide by indicating minimal expectations of citizens and organizations.

public law
Law that involves suits between private individuals or groups and their government.

civil law
The body of laws that govern the rights and responsibilities either between persons or between persons and their government.

criminal law
The body of laws that involve the rights and responsibilities an individual has with respect to the public as a whole.

computer equipment store and is delinquent in paying rent to the landlord, the dispute between them entails private law. **Public law** involves disputes between private individuals or groups and their government. For instance, if a computer store dumps waste behind its building in violation of local, state, or federal environmental regulations, the resulting dispute focuses on public law.

Another distinction among types of laws is civil law versus criminal law. **Civil law** involves the rights and responsibilities found in relationships between persons and between persons and their government. It also involves the remedies available when someone's rights are violated. For example, in 2009, Mississippi Valley Silica Co. was ordered to pay a plaintiff $9 million because the court ruled that it sold sand to the plaintiff's employer with the knowledge that using that sand on a regular basis would expose a worker to a form of cancer. Yet, Mississippi Valley did not alert those who bought the sand about the risk.

In contrast to civil law, **criminal law** applies to situations in which someone commits an act against the public as a whole. There are several business activities that can be considered criminal, such as money laundering, bribery, mail fraud, racketeering, price fixing, insurance fraud, embezzlement, and theft of trade secrets. These crimes are prosecuted not by individuals but by the state or federal government.

One example of a criminal law is the prohibition against insider trading on the stock exchange. Insider trading occurs when an individual uses insider, or secret, company information to increase her or his own finances or those of family or friends. For example, several years ago, an IBM secretary allegedly told her husband, who told several other people, that the company was going to take over operations of Lotus Development. This information was spread among a number of individuals before it was publicly announced.

On the basis of this leak, 25 people bought stock that increased greatly in value following IBM's public announcement. The Securities and Exchange Commission filed charges against the 25 stock purchasers because they were creating an unfair trading environment for the public.

Sources of Business Law

LO 1-5 Differentiate between sources of the law.

How is law created, and where do we look to find the laws?

Constitutions

The United States Constitution and the constitution of each state establish the fundamental principles and rules by which the United States and the individual states are governed. The term **constitutional law** refers to the general limits and powers of the federal and state governments as stated in their written constitutions. The U.S. Constitution is the supreme law of the land, meaning that it overrules all federal, state, and local laws that conflict with it, and is the foundation for all laws in the United States. It is the primary authority to study when trying to identify the relationship between business organizations and government.

constitutional law
The general limits and powers of a government as interpreted from its written constitution.

Statutes

Legislative actions, called *statutes,* are another important source of law. The assortment of rules and regulations put forth by legislatures is what we call statutory law. These legislative acts can be found in the U.S. Code when they are passed by Congress or in the various state codes when they are enacted by state legislatures. The codes are a collection of all the laws in one convenient location.

Because so much business activity occurs within the jurisdiction of state courts, business managers must be familiar with the local city and county ordinances that govern matters not covered by federal or state codes. These ordinances address important business considerations such as local taxes, environmental standards, zoning ordinances, and building codes.

For example, if you want to open a Krispy Kreme franchise in Santa Fe, New Mexico, you must follow local guidelines regarding the area where you may build your store, the materials that you may use to build, and the state minimum wage that you must pay to employees making donuts. The regulations will be different if you want to open your franchise in Toledo, Ohio, or in Seattle, Washington.

Although they are not a source of law in the same sense as constitutions and statutory law, model or uniform laws serve as a basis for some statutory law at the state level. Business activity is made more difficult when state laws vary. To prevent such problems, a group of legal scholars and lawyers formed the National Conference of Commissioners on Uniform State Laws (NCC). The NCC regularly urges states to enact model laws to provide greater uniformity of law. The response is entirely in the hands of the state legislatures. They can ignore a suggestion or adopt part or all of the proposed model law.

The proposals of the NCC, though not laws themselves, have been adopted on more than 200 occasions by state legislatures. The NCC is an especially important influence on business law. Paired with the publications of the American Law Institute, the NCC became the source of the *Uniform Commercial Code (UCC).* The UCC is a body of law so significant for business activities that it will be the focus of intensive study in several chapters of this text. The UCC laws include sales laws and other regulations affecting commerce, such as bank deposits and collections, title documents, and warranties. For example, these laws govern the different types of warranties that companies such as Microsoft, Sony, and Honda provide with their products.

statutory law
The assortment of rules and regulations put forth by legislatures.

model laws
Laws created to account for the variability of laws among states. These laws serve to standardize the otherwise different interstate laws. Also called *uniform laws.*

Cases

Constitutions, legislatures, and administrative agencies encourage certain behavior and prevent other actions. But the boundaries of these laws are seldom self-explanatory. Consequently, law must be interpreted.

Case law is the collection of legal interpretations made by judges. An alternative name for case law is **common law.** These interpretations are law unless they are revoked later by new statutory law.

Judicial opinions are especially significant for businesses because a modern business often operates in multiple legal jurisdictions. Because statutory laws are subject to interpretation, one court may have interpreted particular laws one way at one business location, and a second court may interpret a similarly worded statute differently at a second business location.

We will often refer to these judicial opinions as "cases." Be sure to distinguish judicial opinions in your mind from the business cases you may study in your other business classes, where the cases refer to business situations where you are being asked to apply the business principles you have learned.

Courts issue judicial decisions that often include interpretations of statutes and administrative regulations. These decisions contain the reasoning the courts use to arrive at their decisions. The reasoning depends heavily on precedent, the use of past decisions to guide future decisions. An earlier decision in a similar fact pattern is a precedent that guides later decisions, thereby providing greater stability and predictability to the law.

The 2014 case of *Air Wisconsin Airlines Corp. v. Hoeper* (No. 12-315, slip op. Sup. Ct. January 27, 2014) provides an illustration of how important knowledge of business law and precedent can be. Mr. Hoeper was a pilot for Wisconsin Air. He needed to be certified to fly a new kind of aircraft

case law
The collection of legal interpretations made by judges. They are considered to be law unless otherwise revoked by a statutory law. Also known as *common law.*

precedent
A tool used by judges to make rulings on cases on the basis of key similarities to previous cases.

his employer bought. At the conclusion of his fourth failed attempt to be certified, Mr. Hoeper raised his voice, tossed his headset, used profanity, and accused the instructor of railroading him.

The manager who booked a return flight for Hoeper discussed Hoeper's behavior with officials of Wisconsin Air. They knew that Hoeper was licensed to carry a firearm. Consequently, an airline executive warned the Transportation Security Administration (TSA). During the call, the executive said that the airline was concerned about Hoeper's mental stability and the whereabouts of his firearm.

The TSA removed the pilot from the airplane, searched him, and questioned him about the whereabouts of his firearm. Hoeper sued for defamation. The jury found for Hoeper, and the Colorado Supreme Court agreed with the jury on appeal.

When the U.S. Supreme Court heard the case, it reversed the decision, finding against Hoeper's defamation claim. In the course of that opinion, Justice Sotomayor, writing for the Court, relied heavily on the earlier *Masson v. New Yorker Magazine, Inc.,* 501 U.S. 496 (1991) case for precedent.

Hoeper maintained that the Wisconsin Air executive who warned the TSA should have qualified the concern about the location of Hoeper's weapon by saying that "it had no reason to think he actually was armed." But Justice Sotomayor pointed out that *Masson* provided that in a defamation action a materially false statement would have to be one that would have had a different effect on the mind of the listener from that which the truth would have produced and, in this case, a reasonable TSA officer would have searched the pilot for a gun after hearing that the pilot was upset.

It is crucial for business managers to pay attention to changes in the law and cases in which new precedents are set. These precedents must be taken into account when making future business decisions. For instance, courts sometimes make new rulings regarding the kinds of warnings companies should be required to provide to consumers about potential harm that could result from their products.

One example was a case filed against McDonald's. After a woman was severely burned by very hot coffee, the company was found negligent for failing to provide a warning label on its hot-beverage cups. Now many retailers of hot beverages provide warning labels on their beverage cups because of the precedent that was set by this case.

stare decisis
"Standing by the decision" a principle stating that rulings made in higher courts are binding precedent for lower courts.

When courts rely on precedent, they are obeying **stare decisis** ("standing by the decision"). Following stare decisis creates greater predictability for both businesses and individuals that look to the courts for the rules on which they should rely when they engage in market exchanges. In accordance with stare decisis, rulings that are made in higher courts become binding precedent for lower courts.

Even though this practice is meant to create a consistent and reliable justice system, different judges may view the facts of a case in different ways. In addition, courts are sometimes presented with a new issue and do not have a binding decision to follow. In such instances, they may look to decisions made in similar cases by nonbinding courts in other states or jurisdictions.

If a new issue comes before two state courts and there is no binding decision from the state supreme court, both state courts need to look for other rulings on similar cases. They are not bound by each other's decision, so they might have different decisions on the same subject. The decisions in these lower courts can be appealed to the state appeals court, however, and the appeals court's decision can be appealed to the state supreme court.

One example of a case that has been used in accordance with stare decisis as a binding precedent is *Brown v. Board of Education*,[1] which abolished discriminatory policies for individuals of different racial backgrounds. For instance, in *Regents of the University of California v. Bakke*,[2] the plaintiff, a white male, had applied to the University of California at Davis medical school two years in a row and been denied admittance. He alleged that the admissions process was discriminatory because out of 100 slots, 16 were reserved for members of minority races.

The U.S. Supreme Court found that the school's admissions policy was not legal, referencing *Brown* and stating that the basic principle behind it and similar cases was that individuals could not be excluded on the basis of race or ethnicity. The Court wrote, "Preferring members of any one group for no reason other than race or ethnic origin is discrimination for its own sake."

[1] 347 U.S. 483 (1954).
[2] 438 U.S. 265 (1978).

Just as state statutes have been strongly influenced by the suggestions of the NCC, common law evolves with the assistance of a mechanism called Restatements of the Law. These Restatements are summaries of the common law rules in a particular area of the law that have been enacted by most states.

In addition to the Restatements, many influences are at work in the minds of judges when they interpret constitutions, statutes, and regulations. For example, the values and social backgrounds of the judges function as lights and shadows, moving the judges toward particular legal decisions.

Restatements of the Law
Summaries of common law rules in a particular area of the law. Restatements do not carry the weight of law but can be used to guide interpretations of particular cases.

Administrative Law

Constitutions and statutes are never complete in the sense of covering all the detailed rules that affect government and business relations. The federal government, as well as state and local governments, has dozens of administrative agencies whose task is to perform a particular government function. For example, the Environmental Protection Agency (EPA) has broad responsibilities to enforce federal statutes in the area of environmental protection.

Administrative law is the collection of rules and decisions made by all these administrative agencies. Just glance at Exhibit 1-2 to get a sense of the scope of a few of the major federal administrative agencies.

Businesses function within the rules established by agencies like these. For example, the Occupational Safety and Health Administration (OSHA) oversees health and workplace safety and makes sure that employees are working in conditions that are not hazardous.

In 2015, illustrating the implications of OSHA for business managers, Walmart was fined $7,000 when a store in Valley Stream, New York, did not "furnish a place of employment which was free from recognized hazards that were causing or likely to cause death or serious physical harm to employees in that employees were exposed to the hazards of asphyxiation or being struck due to crowd crush, crowd surge or crowd trampling." Walmart's managers and employees failed to provide effective crowd management training and did not use appropriate crowd management techniques to safely manage a large crowd of approximately 2,000 customers.

administrative law
The collection of rules and decisions made by administrative agencies to fill in particular details missing from constitutions and statutes.

Treaties

A *treaty* is a binding agreement between two states or international organizations. Treaties may be called several things: international agreements, covenants, exchanges of letters, conventions, or protocols. In the United States, a treaty is generally negotiated by the executive branch. To be binding, it must then be approved by two-thirds of the Senate.

A treaty is similar to a contract in two important ways. Both treaties and contracts are attempts by parties to determine rights and obligations among themselves. In addition, when a party fails to obey a treaty or an international contract, international law imposes liability on that party.

INDEPENDENT AGENCIES	EXECUTIVE AGENCIES
Commodity Futures Trading Commission (CFTC)	Federal Deposit Insurance Corporation (FDIC)
Consumer Product Safety Commission (CPSC)	Occupational Safety and Health Administration (OSHA)
Equal Employment Opportunity Commission (EEOC)	General Services Administration (GSA)
Federal Trade Commission (FTC)	National Aeronautics and Space Administration (NASA)
Federal Communications Commission (FCC)	Small Business Administration (SBA)
Interstate Commerce Commission (ICC)	International Development Cooperative Agency (IDCA)
National Labor Relations Board (NLRB)	
National Transportation Safety Board (NTSB)	National Science Foundation (NSF)
Nuclear Regulatory Commission (NRC)	Veterans Administration (VA)
Securities and Exchange Commission (SEC)	Office of Personnel Management (OPM)

Exhibit 1-2
Major Federal Administrative Agencies

Executive Orders

The president and state governors can issue directives requiring officials in the executive branch to perform their functions in a particular manner. The Code of Federal Regulations (CFR) contains all the executive orders created by the president. (The CFR is online at https://www.govinfo.gov/help/cfr.) Presidents claim the power to issue such orders on the basis of their Article II, Section 1, constitutional power to "take care that the laws be faithfully executed."

An illustration of an especially controversial executive order is Order 9066, issued by President Franklin Roosevelt during World War II. On the basis of this order, Japanese-Americans on the West Coast, as well as thousands of Italian-American and German-American families, were sent to internment camps for the duration of the war.

Schools of Legal Interpretation

LO 1-6 Identify the various schools of jurisprudence.

When legislators or courts make law, they do so guided by certain habits of mind and specific beliefs about human nature. These views guide them toward particular legal solutions and away from others. This section briefly describes several of the more common guides to legal interpretation.

The point of learning about these alternative methods for interpreting the implications of particular legal facts is to encourage you to see the law as a human creation shaped by many perspectives and approaches to what a court decision or statute means.

identification with the vulnerable
A school of jurisprudence that holds that society should be fair. Particular attention is therefore paid to the poor, the ill, and the elderly.

Identification with the Vulnerable Some thoughtful legal scholars claim that our very nature as humans compels us to advancing the law on the basis of **identification with the vulnerable.** Some members of our society can take care of themselves in terms of most life situations. Others, especially the ill, children, the aged, the disabled, and the poor, require assistance to meet their fundamental needs of life, health, and education.

This guide to legal change is tied closely to the pursuit of fairness in our society. The metaphor of a level playing field is linked with some higher law or body of moral principles that connects all of us in the human community. We might look at a particular employment contract, for example, and react by observing that "it is just not fair." Our caring impulse as a human feels outrage at that legal arrangement. That outrage can be a stimulus for legal change. One example of identification with the vulnerable is minimum-wage laws. They reflect the beliefs that workers should receive a minimum hourly wage and that employers should not be allowed to pay them less.

historical school
A school of jurisprudence that uses traditions as the model for future laws and behavior. Also called *tradition* or *custom*.

Historical School: Tradition One of the most often used guidelines for shaping the law is **tradition,** or *custom,* which is also called the **historical school.** Stare decisis is rooted in this perspective. When we follow tradition, we attempt to link our future behavior to the behavior of those who faced similar problems in earlier historical periods. The logic of the approach is that we need not reinvent the wheel each time a legal problem arises. Past practice is assumed to have been the product of careful thought.

legal realism
A school of jurisprudence that holds that context must be considered as well as law. Context includes factors such as economic conditions and social conditions.

Legal Realism Legal realism is based on the idea that, when ruling on a case, judges need to consider more than just the law. This school of thought dictates that they also take factors such as social and economic conditions into consideration when making a judgment. Followers of legal realism argue that the law must not be the sole factor in deciding a case since legal guidelines were designed by humans and exist in an ever-changing society.

Judges who follow this school of thought are more likely to depart from past court decisions to account for the fact that our society is constantly shifting and evolving. Those who subscribe to legal realism also believe that the law can never be enforced with complete consistency. They argue that because judges are human, they will bring different methods of reasoning to very similar cases.

One example of a law that has been enacted to reflect changes in our society is the Family and Medical Leave Act (FMLA). This act mandates that businesses employing more than 50 people must provide their workers with up to 12 weeks' unpaid leave every year to take care of

family-related affairs such as caring for ill parents, adopting a child, or having a child. The FMLA protects pregnant women who take time off work because their employers must provide them with the same pay and the same or an equivalent job when they return to work. This act reflects the fact that more mothers are working outside the home and more women are returning to work soon after they have a child. The FMLA also protects new parents against some types of employer discrimination that might occur after they return to work.

Cost-Benefit Analysis From this perspective, we need to examine all the costs and benefits for alternative laws or decisions and place monetary values on those costs and benefits. If we possessed those calculations, we could use cost-benefit analysis as a guide to legal change, choosing the legal alternatives that maximize the ratio of benefits to costs. For instance, in a contract dispute, someone using this approach would attempt to attach responsibility for the problem in such a way that total benefit is maximized in relation to costs.

> **cost-benefit analysis**
> An economic school of jurisprudence in which all costs and benefits of a law are given monetary values. Laws with the highest ratios of benefits to costs are then preferable to those with lower ratios.

This approach to legal change is tied closely to the pursuit of efficiency. If a law yields more benefits than costs, then we have saved resources. Those resources can, in turn, be used to provide us with more goods and services. Our economy is thus more efficient in the sense of producing more for less.

For example, regulations enacted by the EPA can affect costs and benefits in the national economy. Polluted land is an economic loss because it cannot be used for farming or recreation. Polluted water can be toxic for fish and cannot be used for drinking. Polluted air can cause health problems and result in higher health care costs. Although EPA regulations may end up costing companies more initially to control pollutants, the cost of environmental cleanup, and the loss of productivity in the economy as a whole, may be greater than the resources companies expend to control pollution.

Global and Comparative Law

Comparative and international law is extremely important for future business managers. Because of advances in technology and transportation, trade with other countries is far easier today than it was in past years. Now it is possible to make different components for the same product in various countries all over the world and then assemble them in another country. It is possible to operate an antique store in Poughkeepsie but sell to customers in Moscow or Taipei through a website.

As a result of this ease in trade, business managers must be familiar with global trade laws that regulate business practices among nations. For instance, the United States and other countries have entered into agreements such as the North American Free Trade Agreement (NAFTA) and the General Agreement on Tariffs and Trade (GATT). These agreements help establish the conditions of trade between countries.

In 1995, the World Trade Organization (WTO) was created to replace GATT. The WTO stimulates negotiations among member nations to reduce barriers to international trade. The WTO attempts to enforce trade agreements governing trade among its 160 member nations. In this role, it serves as a world court adjudicating trade disagreements that threaten to disrupt international trade.

It is also important for future business managers to understand comparative law, comparing and studying the laws in different countries. Business managers need to be aware of various trade laws and restrictions in different countries so that they can act accordingly when they set out to do business in those countries. The European Union, for example, has laws that differ from those of the United States in terms of regulating the taxes on Internet sales and the amount of pollution that can be released into the environment. Companies doing business in the EU must take these standards into account.

For example, contrast the extent that the Chinese and American governments explicitly gather data on their citizens. The Chinese government does not want its citizens to have access to certain information and websites. A new Chinese cybersecurity law that went into effect in June 2017 requires firms that do business in China to store in China all data that firms have on Chinese citizens and businesses. The law includes further strict guidelines, including ensuring that the data

contains no "state secrets prior to transferring the data outside the country" (though the Chinese government decides what information is considered a "state secret"). The law also requires any "critical network equipment" and "specialized cybersecurity products" to be approved by the Chinese government before they can be sold in China (likely included to ensure that the products cannot bypass China's internet restrictions). This law will require swift and sweeping changes to the operating policies of foreign businesses in China, and many businesses may find compliance too expensive or too contrary to their core values to continue doing business in China.

As an additional illustration of the importance of being aware of the diversity of law in different countries, just imagine the problems that could arise if you were to need space for your business in Vietnam, but the only property law you knew was U.S. property law.

In the United States, we take for granted the right to purchase a piece of property if we have the money to do so. Property is not so freely available and transferable in all nations. Vietnam's new constitution, written in 1992, provides guidelines for the allocation, transfer, and sale of private property. However, it still asserts that the people, or the state, own all the land.

Thus, if individuals or private enterprises want to use land, they must pay tax on it as a form of rent and are granted a "use of right" that entitles them to extended use and the freedom to transfer the property. Technically, they are transferring not the *property* but, rather, the right to use it. Transference of property rights can occur only with the approval of a state official. The official ensures that the new owner intends to use the land for the original, state-approved purpose.

Moreover, the new owner can never be given a longer term of right or more extensive rights over the land than the original owner had. Finally, the state official determines the price that the property will be transferred for. The government has specified certain prices depending on how the land is used.

Modern business managers must have an ongoing fascination with the law to function effectively. Business law tells business managers the basic rules of the business game. Play any game without having first studied the rules, and you will probably fail.

But, unlike an ordinary game, the business game has a rule book that is changing dynamically. In addition, increased globalization requires business leaders to be alert to legal differences among national jurisdictions.

PRACTICAL TIPS for Business Managers

1. Remember that the meaning and implications of laws are a matter of interpretation. Therefore, it is wise, if possible, to consult with more than one lawyer when you are curious about the meaning of a law that is immensely important to your operations.

2. Try to stay current with your knowledge of law relevant to your industry and that of your suppliers. Law evolves, and yesterday's law may not be today's law.

3. When you engage in any international commercial activity, be sure to investigate the law of countries in which you are doing business.

Questions & Problems

1. What is the difference between private law and public law?
2. What are the main distinctions between the different schools of legal interpretation?
3. If there are different ways to interpret statutes and case decisions, would it make sense to say that all interpretations of the law are equally appropriate?
4. Presidential elections are partially about who will be able to select Supreme Court judges. As you look at the various alternative ways to interpret and apply the law, which of those methods of interpretation would you hope is held by your favorite candidate for president?

5. Most business managers know they need to consult a lawyer when they make important business decisions. If that assumption is correct, why would a manager or entrepreneur need to have any knowledge of business law? Couldn't she or he just ask the lawyer?

6. Appendix 1A provides guidance for becoming a better critical thinker. A recent *Forbes* poll of business leaders found that the most important skill business leaders are seeking in their employees is critical thinking. As you study the appendix, why do you think this skill of critical thinking is so valuable to business managers?

7. This chapter distinguishes between private and public law. Why are the interests of the public identified with the behavior of the government? Why would public law not be a term referring to the interests of corporations, for example?

8. What is the need for a special area of law for administrative agencies? Why would the law as it applies to businesses apply not just as well to administrative agencies?

9. Choose an area of law that fascinates you. Spend some time on the Internet searching for differences between how the United States and other countries respond to conflicts among people within that area of law, just as we did in the text in the illustration of real property law in Vietnam.

10. Jarold Daniel Friedman worked as a temporary computer contractor for a pharmaceutical warehouse. The warehouse offered him a permanent position, but the warehouse required him to get a mumps vaccine, grown in chicken embryos, as a condition of his permanent employment. Friedman, a vegan, believed that the vaccination would violate his religious beliefs and declined to be vaccinated. As a result, the warehouse withdrew its offer of employment. Friedman claimed that the warehouse discriminated against him on the basis of religion. Do you agree with Friedman? Do employers have a duty to respect the beliefs of their employees? If so, what happens when that duty conflicts with employers' duty to provide a safe and healthy work environment?

Source: *Friedman v. Southern California Permanente Medical Group,* 102 Cal. App. 4th 39 (2002).

APPENDIX 1A

Critical Thinking and Business Law

Success in the modern business firm requires the development of **critical-thinking skills**—the ability to understand what someone is saying and then to apply a set of evaluative criteria to determine the worth of what was said. In other words, businesspeople need to be able to sort sense from nonsense by evaluating arguments about relevant business law.

A critical thinker is always asking, "Now does that make sense?" And there is no better context in which to develop critical-thinking skills than in the study of the laws that affect business. Critical-thinking skills learned in the study of business law will be easily transferred to the decisions of business stakeholders.

Legal reasoning is similar to other reasoning in some ways and different in others. When people, including lawyers and judges, reason, they do so for a purpose. Some problem or dilemma bothers them. The stimulus that gets them thinking is a *legal issue.* It is stated as a question because it is a call for action. It requires us to *do* something, to think about answers.

For instance, we may be interested in such issues as the following:

- When are union organizers permitted under the National Labor Relations Act to trespass on an employer's property?
- Do tobacco manufacturers have liability for the deaths of smokers?
- Must a business fulfill a contract when the contract is made with an unlicensed contractor in a state requiring all contractors to be licensed?

Each of these questions has several potential answers. Which answer best accomplishes a particular business objective? Which answers are consistent with the law? Here is where critical thinking is essential to business success. Some answers can get the decision maker into trouble; others will advance the intended purpose. Each answer is called a *conclusion.*

critical-thinking skills
The ability to understand the structure and worth of an argument by evaluating the facts, issue, reasons, and conclusion of that argument.

Business firms often come into contact with such conclusions in the form of laws or court decisions. Business managers are therefore both consumers of and contributors to these legal conclusions. As businesspeople learn about and react to decisions or conclusions made by courts, they have two primary methods of response:

1. Memorizing and understanding the conclusions or rules of law as a guide for future business decisions.
2. Making judgments about the quality of the conclusions.

This book encourages you to do both. There are many forms of critical thinking, but they all share one characteristic: they focus on the quality of someone's reasoning. Critical thinking is active; it challenges each of us to form judgments about the quality of the link between a set of reasons and the conclusion supposedly derived from them. In particular, we will be focusing on the link between a court's reasons and its conclusions.

Our reactions to legal arguments shape our efforts either to support the status quo in the legal environment of business or offer support for particular changes.

The following structure for critical thinking is a thoroughly tested method that successful market decision makers use. Every time you read a case, try to follow this pattern of careful thinking.

1. Find the facts.

 Facts: Here, we are looking for the most basic building blocks in a legal decision or argument. These building blocks, or facts, provide the environment or context in which the legal issue is to be resolved. Certain events occurred, certain actions were or were not taken, and particular persons behaved or failed to behave in specific ways. All of these and many more possibilities together make up the intricate setting for the playing out of the issue in question. We always wonder, "What happened in this case?" A major feature of this book is called "But What If." This feature highlights the importance of facts in legal reasoning by focusing attention on the huge effect of changes in single facts on success or failure in a legal dispute.

2. Look for the issue.

 Issue: In almost any legal conflict, finding and expressing the issue is an important step in forming our reaction. The issue is the question that caused the lawyers and their clients to enter the legal system. Usually, there are several reasonable perspectives concerning the correct way to word the issue in dispute. Don't let the possibility of *multiple* useful ways to word the issue cause you any confusion. The legal issue is certainly not just anything that we say it is.

3. Identify the judge's reasons and conclusion.

 Reasons and conclusion: Judges do not form legal conclusions on the basis of whim. They have support for their decisions. That support consists of their reasons. When we ask someone *why* a judge formed a particular conclusion, we are showing our respect for reasons as the proper basis for any assertion. We want a world rich with opinions so that we can have a broad field of choice. But we should agree with only those legal opinions that have convincing reasons supporting the conclusion. Asking "Why?" is our way of saying, "I want to believe you, but you have an obligation to help me by sharing the reasons for your conclusion."

4. Locate in the decision the rules of law that govern the judge's reasoning.

 Rules of law: Judges cannot offer just any reasoning that they please. They must always look back over their shoulders at the laws and previous court decisions that together provide an anchor for current and future decisions. What makes legal reasoning so complex is that statutes and legal findings are never *crystal* clear. They may seem very clear, but judges and businesspeople have room for interpretive flexibility in their reasoning.

5. Apply critical thinking to the reasoning.

 Evaluation of the reasoning: A judge's reasoning, once it has been laid before us by following the steps discussed here, is a message that we may either accept or reject. One of the most exciting things about our legal system is its potential for change. **Critical thinking in the legal context** consists of examining the legal opinion in search of potential problems in the reasoning.

critical thinking in the legal context
Examining a legal opinion to search for potential problems in the reasoning.

Here is a small sample of some especially useful critical-thinking guidelines for business managers to apply when thinking about business law:

- *Look for potential ambiguity in the reasoning.* The term *ambiguity* refers to a lack of clarity in a word or phrase in the reasoning. Many words have multiple meanings; until the intended meaning is discovered, we cannot tell whether we want to agree or disagree with the reasoning.
- *Ask whether the analogies used in the decision are strong.* When judges follow particular precedents, they are saying that the key similarities between the facts in the precedent and those in the case at hand are so similar that it makes sense to apply the same rule of law in both. Are there key differences in the factual situations that raise questions about the quality of that analogy?
- *Check the quality of the judge's use of evidence in his or her reasoning.* Is the evidence used by the judge to support the opinion both abundant enough and reliable enough that we should agree with the reasoning?
- *Think about the extent to which important missing information prevents you from being totally confident about the judge's reasoning.* Is there important missing information that you would need to have before making up your mind?
- *Consider the possibility of rival causes.* When the judge claims that one action caused another, think about whether some alternative cause may have been responsible.

Part 1 The Legal Environment of Business

CHAPTER 2

Business Ethics and Social Responsibility

LEARNING OBJECTIVES

- **LO 2-1** Distinguish between business ethics and the social responsibility of business.
- **LO 2-2** Recognize the relationship between business law and business ethics.
- **LO 2-3** Define values.
- **LO 2-4** Explain how values provide a starting point for thinking about business ethics.
- **LO 2-5** Analyze business decisions using the WH framework.

LO 2-1 Distinguish between business ethics and the social responsibility of business.

What business managers should do when their decisions have impacts on a wide assortment of stakeholders is not altogether clear. Ethical conversation is less about finding the one and only right thing to do than it is about finding the better thing to do. Whatever you choose to do, some stakeholders will be hurt and others will benefit. Business ethics requires a thought process that weighs the benefits of a decision compared to its harm, consideration of the duties we owe other human beings, the characteristics possessed by a virtuous person, and the care we have for the well-being of those affected by business decisions.

To think about the ethics of a decision requires us to have a language that accelerates consideration of what is good and right. A central objective of this chapter is to make you aware of the ideas and words we use when we tackle an ethical dilemma.

This chapter provides some assistance for thinking systematically about issues of right and wrong in business conduct. Initially, we need to sort through the meaning of key terms like *business ethics* and *social responsibility*. Then, because it is helpful to have a useful approach to ethical decision making, we provide a practical method by which future business managers can think more carefully about the ethical dilemmas they will face during their careers.

A business has a huge assortment of responsibilities. Each responsibility involves a relationship with other people in the community. Some of the responsibilities are more obvious than others. Firms have a responsibility to devote their energies to fulfilling the interests of their owners, certainly. The pursuit of profits is both a stimulus for incredible innovation and efficiency and a temptation to use whatever power the firm has to conceal information from consumers and the public.

In addition, business firms owe safe working conditions to the employees who make it possible for the firm to thrive. Increasing the safety of the workplace can diminish profits, at least in the short run. Business ethics is full of dilemmas such as the extent to which short-term profits can be achieved at the cost of reduced safety for employees.

Other responsibilities are just as real but perhaps less obvious. A business firm must obey the laws in the countries where it offers its products and services. This course spells out many of those responsibilities, and the list is quite extensive. In addition, firms have a much more abstract ethical responsibility to obey the standards of conduct consistent with prevailing ethics and expectations in the communities where they practice their trade.

Ethics is the study and practice of decisions about what is good or right. Ethics guides us when we are wondering what we should be doing in a particular situation. **Business ethics** is the application of ethics to the special problems and opportunities businesspeople experience. An **ethical dilemma** is a problem about what a firm should do, for which no clear, right decision is available. Reasonable people can expect to disagree about optimal solutions to ethical dilemmas.

Closely related to ethical dilemmas arising in business situations is the concept of the social responsibility of a business. The **social responsibility of business** consists of the expectations the community imposes on firms doing business within its borders. These expectations must be honored to a certain extent, even when a firm wishes to ignore them, because firms are always subject to the implicit threat that legislation will impose social obligations on them. So, if the community expects businesses to obey certain standards of fairness even when the standards interfere with profit maximization, firms that choose to ignore this expectation do so at their peril.

The analogy here is to the role of a good citizen. The community guarantees you certain protections and rights. In return for that support network, we owe responsibilities of civic engagement and obedience to the law. A business firm gets the protection of laws designed to provide stability and productivity in their activities. Its social responsibilities extend to the expectations the community has for how a business should behave.

The definition of business ethics refers to *standards* of business conduct. *It does not result in a set of correct decisions.* Business ethics can improve business decisions by serving as a reminder not to simply choose the first business option that comes to mind or the one that enriches the firm in the short run. But business ethics can never produce a list of correct business decisions that all ethical businesses will make.

ethics
The study and practice of decisions about what is good or right.

business ethics
The use of ethics and ethical principles to solve business dilemmas.

ethical dilemma
A question about how one should behave that requires one to reflect on the advantages and disadvantages of the optional choices for various stakeholders.

social responsibility of business
The expectations that a community places on the actions of firms inside that community's borders.

Exhibit 2-1 Business Ethics Issues Take Many Forms

Business Ethics Issues Take Many Forms

Small Business
Hayner Hoyt Corporation Case:
The Hayner Hoyt Corporation establishes a token president to defraud the government of special contracts.

Consumer Protection
Company Doe v. Inez Tenebaum et al.
Anonymous corporation sues to suppress a product complaint posted by a customer to a government watchdog site. Perhaps the first case where a company was allowed to use a pseudonym.

Professional Ethics
The Goodyear Tire & Rubber Co. v. Hager et al.
After Goodyear settled a case with an injured party, it came to light that two Goodyear lawyers withheld vital testing information that had been requested during discovery, and that the lawyers repeatedly and deliberately attempted to frustrate the injured part's case.

Securities Exchange
USA v. Shkreli et al.
Martin Shkreli is accused of looting his former company Retrophin to pay off investors he defrauded in a hedge fund he ran. Allegedly, Shkreli concealed his ownership of free Retrophin shares.

Agency Law
Paul Bishop et al. v. Wells Fargo & Co. et al.
Wachovia Bank (later acquired by Wells Fargo) was accused by a whistleblower of using an off-record "Black Box" to hide certain transactions and loans from regulators. Perhaps more significantly, this case introduced several new protections for whistleblowing employees.

Real Property
Jordan v. Nationstar Mortgage, LLC
Nationstar mortgage found to be engaging in unethical and illegal behavior by securing properties and changing the locks on doors before foreclosure was completion, essentially locking families out of their homes far before they were supposed to vacate.

Exhibit 2-1 provides a picture of the broad scope of business ethics. Many decisions made by managers and entrepreneurs affect a huge array of people inside and outside the walls of the business. Those effects raise issues of what is right, fair, and acceptable. Consequently, business ethics is an essential part of business planning and strategy.

Business Law and Business Ethics

LO 2-2 Recognize the relationship between business law and business ethics.

Before business managers consider the social responsibilities of firms in their communities, they need to gather all the relevant facts. Experienced managers know that assembling the facts is just the beginning of a thoughtful business decision. Next, it makes sense to ask, *Is it legal to go forward with this decision?*

The legality of the decision is the minimal standard that must be met. But the existence of that minimum standard is essential for the development of business ethics. Think of the law as the most recent consensus about the lower level of business ethics. We can expect that minimum to evolve alongside new technologies, emerging demographic patterns, and continuing growth in our empathy for marginalized voices in our country.

To make this point, let's take a look at the growing practice of bribery in the absence of legal standards. In some countries, businesses must pay bribes to receive legitimate supplies. Although the businessperson may be morally opposed to paying the bribes, the supplies are necessary to stay in business, and there may be no other means of obtaining them.

Thus, foreign companies face an ethical dilemma: they must decide whether to pay bribes or find alternative sources of supplies. For instance, when McDonald's opened its doors in Moscow, it made arrangements to receive its supplies from foreign providers. These arrangements ensured that the franchise did not have to engage in questionable business practices.

At the same time that business ethics guides decisions within firms, ethics helps guide the law. Law and business ethics serve as an interactive system—informing and assessing each other. For example, our ethical inclination to encourage trust, dependability, and efficiency in market

BUSINESS Ethics Flashpoint 2.1

Chevron in Ecuador

In 1964, Texaco Petroleum (now known as Chevron), along with Gulf Oil Corporation (Gulf), was invited by the Ecuadoran government to conduct exploratory activities in the eastern region of Ecuador. This region was known as the Oriente.

In 1967, Texaco and Gulf discovered large quantities of oil in the Oriente. After building a pipeline that would allow for the transfer of oil to the Pacific coast, Texaco began exporting the oil.

Texaco acted as the operator in a partnership, known as a consortium, between Texaco and Gulf. In 1973, production reached 200,000 barrels of oil per day. During this period, Ecuador's gross national product (GNP) had doubled.

During the early 1970s, the military seized control of the Ecuadoran government. This takeover resulted in a series of governmental reforms that limited the size of concession areas granted to foreign oil companies such as Texaco. Texaco, therefore, relinquished some of its concessions to Ecuador's state-run oil company, Petroecuador. In addition, Petroecuador joined the consortium between Texaco and Gulf.

Soon after Petroecuador joined the consortium, Gulf left. The consortium, now consisting of only Texaco and Petroecuador, lasted from 1977 until 1990. In 1992, Ecuador did not renew Texaco's concessions. As a consequence, Texaco's interest in the oil industry came to an end.

During the period in which it operated in Ecuador, Texaco's oil-related operations exacted a great toll on the residents of the Oriente.

For example, Texaco dumped billions of gallons of toxic wastewater into surrounding rivers and streams of the Amazon. Texaco is also reported to have spilled about 17 million gallons of crude oil into surrounding lands in the Oriente. Another source reports that about 2.5 million acres of land were affected by the leaking of oil into soil and groundwater, as well as by oil discharges into wetlands, streams, and rivers. In addition, Texaco had allegedly stored toxic chemicals in hundreds of unlined pits. The waters that indigenous groups relied on for drinking, cooking, bathing, and fishing became receptacles for Texaco's garbage.

Another source of environmental damage in the Oriente region was gas flaring and the burning of crude oil. These acts resulted in toxic fumes that polluted the air.

The contaminated water and polluted air in the region of the Oriente had significant effects on the health of the residents living in the Oriente. A little more than 8000 residents suffered one or more diseases attributable to the contaminants with which Texaco polluted the area. The people of the Oriente reported higher rates of cancer than the general population of Ecuador and in other Amazonian provinces. Birth defects were also attributed to the environmental pollution. However, despite these adverse effects on the surrounding population, Texaco did not endeavor to address these problems during its operations.

In addition to suffering from diseases related to toxic contaminants, the indigenous population in the Oriente suffered economic losses. For example, 75 percent of Oriente residents reported that they had suffered from loss of their crops as well as loss of animals on which they relied because of Texaco's toxic discharges into the environment. Texaco's clearing substantial parts of natural habitat to make room for oil operations also contributed to the elimination of the region's forests. Thus, the contamination of the natural habitat in the Oriente led not only to disease but also to economic and cultural losses.

In 1993, forty-six residents of Ecuador who belonged to numerous indigenous tribes in the Oriente filed a suit against Texaco (now known as Chevron due to a merger) seeking redress for Texaco's actions. Texaco denied allegations of a causal connection between its oil operations and the diseases reported.

Additionally, Chevron endeavored to exclude Ecuador from trade negotiations with the United States until Ecuador shut down the lawsuit. To this action, then-Senators Barack Obama and Patrick Leahy responded: "While we are not prejudging the outcome of the case, we do believe the 30,000 indigenous residents of Ecuador deserve their day in court..."

After numerous proceedings that began in New York and then moved to Ecuador, the plaintiffs were awarded $19 billion in damages—later reduced to $9.5 billion—in February 2011, for purposes of restoring the environment, providing health care treatment, and obtaining clean water. However, Chevron refused to pay this fine.

On March 4, 2014, a U.S. federal judge blocked U.S. courts from being used to collect the $9.5 billion judgment against Chevron. The U.S. court concluded that the Ecuadoran courts had been influenced by corrupt relations between judges and attorneys representing the indigenous groups. As a consequence, Chevron's fine was essentially revoked.

Sources: Trudie Styler, "Seeking Justice in Ecuador," March 5, 2014, HuffingtonPost.com; Lucien J. Dhooge, Aguinda v. Chevrontexaco: Mandatory Grounds for the Non-Recognition of Foreign Judgments for Environmental Injury in the United States, 19 J. of Transnational Law & Policy 1 (2009).

Exhibit 2-2 Examples of Unethical and Ethical Business Executives and Businesses

Examples of Unethical Business Executives

Bernard Madoff
In 2009 Bernard L. Madoff admitted to organizing the largest private ponzi scheme in history. Using his reputation as a highly successful stock broker and financial advisor, Madoff was able to defraud his nearly 5000 wealthy clients out of a combined $64.8 billion by promising them big returns on low-risk investments that, in reality, had been completely fabricated by Madoff. Madoff was sentenced to 150 years for his crimes of securities fraud, investment-advisor trust fraud, mail fraud, money laundering, wire fraud, issuing false statements to the police, perjury, making false filings to the SEC, and ordered to pay $170 billion in restitution.

Harvey Weinstein
Co-founder of the weinstein company and former American film producer, Harvey Weinstein, allegedly sexually abused over 80 women, many of them actors of film-set staff, throughout the course of his career as a film producer. After being expelled from the Academy of Motion Picture and Arts Sciences and publicly disgraced because of these allegations of sexual misconduct, Weinstein was arrested and charged with rape and a number of other offenses on May 25th, 2018.

Examples of Ethical Businesses and Business Executives

Microsoft
Given that Microsoft's founder, Bill Gates, is one of the most generous philanthropists on the planet, it should come as no surprise that Microsoft and its employees donate over $1 billion a year to various charities and non-profit organizations. Additionally, through Microsoft's TEAL program, Microsoft employees are encouraged to volunteer their time at local schools instructing students in computer sciences.

Patagonia
Patagonia, a winter and outdoor clothing company, began the "1% for the Planet Pledge" in 1985 and has encouraged other companies to also donate 1% of their sales to help protect the environment. Additionally, Patagonia is so dedicated to the preservation of our environment that they actually frequently ask their customers to refrain from purchasing Patagonia products if the customers don't really need them.

Starbucks
Starbucks should be praised for the actions the company takes to behave environmentally friendly. In addition to ensuring they use as little water as possible in the production of their coffee and switching to cups made from recycled materials, Starbucks also offers customers who bring their own reusable cup to the store a discount on their purchase. Additionally, Starbucks also takes good care of their baristas, offering them competitive wages, full health insurance benefits, and awards in the form of stock in the company.

exchanges shapes many of our business laws. The principles of contract law, for instance, facilitate market exchanges and trade because the parties to an exchange can count on the enforceability of agreements. Legal rules that govern the exchange have been shaped in large part by our sense of commercial ethics.

At this point it might be helpful to get a stronger sense of the range of ethical and unethical business behavior by studying the chart in Exhibit 2-2.

Of course, different ethical understandings prevail in different countries. Thus, ethical conceptions shape business law and business relationships uniquely in each country. Increasingly, business leaders require sensitivity to the differences in legal guidelines in the various countries in which they operate. These differences are based on somewhat different understandings of ethical behavior among business people in diverse countries.

As we mentioned earlier, business ethics does not yield one correct decision. So how are business managers to chart their way through the ethical decision-making process? One source of assistance consists of the general theories and schools of thought about ethics. Each ethical system provides a method for resolving ethical dilemmas by examining duties, consequences, or virtues.

Ethical Failures Can be Bad for Business

Business behavior that might look attractive for profits can be a nightmare for a firm and the business leaders involved in making the decisions that led to the questionable behavior. For example, in 2016 a report documented that Wells Fargo employees had opened more than 2 million accounts for customers without their permission, and often applied fees and other charges to these extra accounts without the customers knowing. An investigation revealed that Wells Fargo employees were under extreme pressure from the company to meet certain sales quotas, and often the only way to meet these quotas was to fraudulently open extra accounts.

Wells Fargo was required to return $2.6 million to consumers and, due to the severity of Wells Fargo's wrongdoing, the Consumer Financial Protection Bureau (CFPB) ordered Wells Fargo to pay the largest fine the CFPB has ever imposed, $185 million. In addition, Wells Fargo CEO John Stumpf stepped down from in position as a result of the scandal.

Perhaps what is most interesting about this case, however, is the scope of the reaction that it inspired. Not only did the CFPB order an enormous fine, but several U.S. cities, including Seattle, Chicago, and Sacramento, largely or entirely divested their funds from Wells Fargo accounts in protest over the scandal. This largely unprecedented move may represent a growing demand for business accountability and transparency, among public and private consumers alike.

Another well-known 2016 business ethics case involved the drug company Mylan and its EpiPen product. An EpiPen is a pocket-sized device used for the self-injection of epinephrine, an antidote for anaphylactic shock brought on by allergy reactions. Having access to an EpiPen can truly be a matter of life or death for someone prone to extreme allergic reactions. Thus, when Mylan suddenly increased the price of an EpiPen by 500 percent, making the price of each EpiPen pack $600, there was a strong outcry from consumers, especially considering that generic alternatives to the device were practically non-existent. Several state attorneys general launched fraud and antitrust investigations, and the House Oversight Committee got involved as well, in the form of a hearing with Mylan's CEO. The bad publicity sank Mylan's stock, and Mylan eventually reduced the price of the EpiPen (although the price remains relatively high compared to the original price).

In their drive to fulfill some of their responsibilities, business firms can trample on other ethical responsibilities they have. In the interest of providing future business managers with a practical approach to business ethics that they can use to avoid these ethical errors, we suggest a three-step process: the **WH approach.** This approach provides future business managers with some **ethical guidelines,** or practical steps, that serve as a dependable stimulus to ethical reasoning in a business context.

What is your reaction to Chevron's behavior in Ecuador? How do we begin to think about what is right and wrong in a situation like this one?

The starting point in wrestling with dilemmas about what is right and good is the idea of values.

WH approach (to ethical decision making)
A set of ethical guidelines that urges us to consider whom an action affects, the purpose of the action, and how we view its morality.

ethical guideline
A simple tool that helps determine whether an action is moral.

What Are Values?

LO 2-3 Define values.

When we think about the ultimate reason or purpose for which we make decisions in a business firm, we turn to the basic unit of business ethics—values. **Values** are positive abstractions that capture our sense of what is good or desirable. They are *ideas* that underlie conversations about business ethics. We derive our ethics from the interplay of values. Values represent our understanding of the purposes we will fulfill by making particular decisions.

For example, we value honesty. We want to live in communities where the trust we associate with honesty prevails in our negotiations with one another. Business depends on the maintenance of a high degree of trust. No contract can protect us completely against every possible contingency. So we need some element of trust in one another when we buy and sell.

values
Positive abstractions that capture our sense of what is good and desirable.

Exhibit 2-3

American Express' Core Values

AMERICAN EXPRESS' CORE VALUES

- Customer Commitment
- Quality
- Integrity
- Teamwork
- Respect for People
- Good Citizenship
- A Will to Win
- Personal Accountability

Exhibit 2-4

Adidas' Core Values

ADIDAS' CORE VALUES

- We are committed to continuously strengthening our brands and products to improve our competitive position.
- We are consumer-focused and therefore we continuously improve the quality, look, feel and image of our products and our organizational structures to match and exceed consumer expectations and to provide them with the highest value.
- We are innovation and design leaders who seek to help athletes of all skill levels achieve peak performance with every product we bring to market.
- We are dedicated to consistently delivering outstanding financial results.
- We are a global organization that is socially and environmentally responsible, that embraces creativity and diversity and is financially rewarding for our employees and shareholders.

Take a look at American Express' core values listed in Exhibit 2-3. Think about how each abstraction would play out through the business' behavior.

If we think about the definition of values for a moment, we realize two things. First, a huge number of values pull and push our decisions. Second, to state that a value is important in a particular situation is to start a conversation about what is meant by that particular value. To return to the Chevron case, when someone says we wanted to maximize efficiency, environmental sustainability, or justice, we are only slightly closer to any understanding of the complexity of the dilemma.

Sometimes a firm's core values seem less ambiguous, but they still leave a lot of ambiguity and flexibility for business decision making. See Adidas' core values in Exhibit 2-4 as an example.

Sometimes, staying true to a company's values can mean making decisions that do not necessarily seem good for business. For example, the pharmacy chain CVS considers itself in the business of health care and prides itself on "helping people on their path to better health." However, CVS also sold cigarettes—a product long known to cause serious health problems. So, in 2014, CVS's CEO Larry Merlo made the tough decision to stop selling cigarettes at every single CVS location across the nation. Merlo knew that this decision could have substantial costs for CVS as smokers took their business elsewhere, but he stated that "[e]nding the sale of cigarettes and

VALUE	ALTERNATIVE MEANINGS
Freedom	1. To act without restriction from rules imposed by others 2. To possess the capacity or resources to act as one wishes 3. To escape the cares and demands of this world entirely
Security	1. To possess a large-enough supply of goods and services to meet basic needs 2. To be safe from those wishing to interfere with your property rights 3. To achieve the psychological condition of self-confidence to such an extent that risks are welcome
Justice	1. To receive the products of your labor 2. To treat all humans identically, regardless of race, class, gender, age, and sexual preference 3. To provide resources in proportion to need 4. To possess anything that someone else is willing to grant you
Efficiency	1. To maximize the amount of wealth in society 2. To get the most from a particular output 3. To minimize costs

Exhibit 2-5

Primary Values and Business Ethics

tobacco products at CVS Pharmacy is simply the right thing to do for the good of our customers and our company." By making this move, CVS also firmly established its dedication to health in the minds of its customers. Balancing values and profits is never easy, but it can be integral to running a successful business.

How do Values Provide a Starting Point for Thinking about Business Ethics?

LO 2-4 Explain how values provide a starting point for thinking about business ethics.

To help make WH useful to you as a manager, Exhibit 2-5 makes clear the central role of clarifying what version of a value one feels loyalty toward. Without that clarification, we lose sight of why something is said to be good or right. The exhibit identifies four of the most important values influencing business ethics and presents alternative meanings for each. This table should not only help clarify the importance of values in your own mind but also enable you to question others who claim to be acting ethically. For instance, a manager might be deciding whether to fire an employee whose performance is less than impressive. In making the decision, the manager explores alternative visions of key values such as justice and efficiency and then makes choices about which action to take. Values and their alternative meanings are the foundation for different ethical decisions.

Where do our value priorities come from? That question is another way of asking: At the foundational level, what causes us to place a higher weight on decisions that reflect some values rather than others?

Value priorities can stem from many sources. Tradition, family, culture, reasoning about the consequences of alternative choices, and a sense of duty or obligation can all fuel the value priorities we bring to an ethical dilemma.

BUSINESS Ethics Flashpoint 2.2

Dole Food Company and Its Workers

Dole Food Company (Dole) dominates the market for bananas, holding a 25 percent share in the world market. The majority of bananas from Dole come from Latin America. Specifically, Dole operates banana plantations in several Latin American countries, including Ecuador, Nicaragua, Costa Rica, Guatemala, Honduras, and Panama.

An important step in the process of growing bananas is the application of pesticides to prevent disease and pathogens. Fumazone was among the most toxic pesticides used in banana cultivation during the 1960s through the 1980s.

The main ingredient in Fumazone is dibromochloropropane (DBCP). DBCP is known to have harmful effects on male reproductive organs and sperm counts. However, despite knowledge of this danger, the U.S. government approved its use on farms in 1964.

During a brief period after 1964, about 60 workers in California that manufactured Fumazone were found to be sterile. As a consequence, in 1979, the U.S. Environmental Protection Agency banned the use of DBCP in the United States. However, Dole continued using DBCP in its Latin American plantations well into the mid-1980s.

Dole admitted to knowing that DBCP could harm the workers but chose not to inform its workers in Latin America. As a consequence of Dole's use of DBCP on its plantations, the health of Latin American plantation workers was threatened.

DBCP posed a health hazard to plantation workers in that it was known to cause male sterility as well as genetic disorders in the children of men and women exposed to DBCP. In fact, the National Institutes of Health claims that DBCP is toxic, carcinogenic, and mutagenic. Further, the State of California's Safe Drinking Water and Toxic Enforcement Act of 1986 concluded that DBCP caused reproductive and developmental harm.

Although the toxicity of DBCP was well-known, banana plantation workers were reportedly not given equipment or training to minimize exposure to the harmful chemical. Unfortunately, the task of applying DBCP on banana plants fell to the most marginalized of workers because few workers desired to risk exposure to DBCP.

Another problem with DBCP is that it negatively affected the environment. For example, the application of Fumazone to banana roots caused DBCP to seep into the groundwater. This contamination of the environment killed organisms living in the surrounding habitat, including fish, mammals, and plants.

In 2004, a small breakthrough was made. Twelve banana plantation workers filed a suit against Dole in a California court. Six of these workers were found to have been made sterile due to the poor working conditions of Dole's banana plantations. During these proceedings, evidence was presented showing that Dole knew that DBCP could harm the workers and nevertheless insisted on using DBCP. In addition, Dole had ignored safety recommendations by the manufacturer of the pesticide Fumazone.

On November 5, 2007, the suit filed by the twelve workers resulted in Dole being ordered to pay damages to six of the twelve workers. Specifically, the Los Angeles Superior Court awarded the plantation workers $3.3 million in damages.

On November 15, 2007, a jury ordered Dole also to pay $2.5 million in punitive damages to five of the workers—that is, damages paid in excess of the plaintiffs' proved injuries. Later, however, a judge vacated the punitive damages and lowered the actual damages of $3.3 million to $1.58 million.

As a result of the use of DBCP on banana plantations, thousands of banana pickers have filed suits against produce companies, such as Dole Food Co., Chiquita Brands International Inc., and Fresh Del Monte Produce Inc., and chemical companies, such as Dow Chemical Company and Shell Chemical Company.

However, thousands of workers continue to suffer from their work in the banana industry in Latin America. In addition, many of the banana workers are children; roughly 70 percent of these children work on Dole plantations, according to a Human Rights Watch report.

Sources: Armin Rosencranz, Stephen Roblin, and Nicole Balloffet, "Doling Out Environmental Justice to Nicaraguan Banana Workers: The Jose Adolfo Tellez v. Dole Food Company Litigation in the U.S. Courts," 3 Golden Gate U. Envtl. L.J.161 (2009); Human Rights Watch, "Ecuador: Widespread Labor Abuse on Banana Plantations," April 25, 2002, HRW.org.

For many people, their preeminent values derive from religious teachings. For example, Catholic social doctrine teaches that the moral quality of a society depends on how it treats its most vulnerable members. Similarly, the United Methodist Church in its 2009–2012 Social Principles document pledges Methodists "to support the poor and challenge the rich."

Try to express the struggle between Dole and its workers as a struggle between two sets of values rather than simply a contest between people in white hats and villains in black hats. If

GLOBAL Context

©Digital Archive Japan/Alamy

Business Ethics and Swedish Advertising Law

Recall from the WH approach to ethical decision making that when businesses make decisions, they should consider *all* the relevant stakeholders. One relevant stakeholder is the customer, or consumer. In Sweden, consumers are assumed to be vulnerable in the marketplace; therefore, protecting the consumer is considered the job of the government. Unlike in the United States, in Sweden public policy acknowledges that "there exist large groups of consumers who, owing to low incomes, deficient education and knowledge of the market, etc., are less well equipped than the average citizen for their role as consumers." One group of consumers less equipped than the average citizen in Sweden is children.

The government considers children extra vulnerable and impressionable and thus in need of even more protection than the average consumer. Swedish laws ban all television advertising targeting children below the age of 12. The ban includes not only toy advertisements but also commercials for foods high in fat and sugar, such as sweets and fast food. Advertisements for such products can be aired if the targets of the advertisements are not children. For example, commercials for toys, sweets, and foods high in fat may be aired if they are designed to attract and inform adults as opposed to children. What may surprise many Americans is that almost 90 percent of Swedish advertising professionals are in favor of the ban on advertising to children because they recognize that such advertising is misleading since children do not understand the commercial nature and purpose of advertising.

you can, you will be more fair in your assessment of Dole's behavior. To be fair is not to condone but, rather, to be open to understanding the logic and purposes of all the relevant players in the ethical flashpoint.

The WH Framework for Business Ethics

LO 2-5 Analyze business decisions using the WH framework.

A useful set of ethical guidelines requires the recognition that managerial decisions must meet the following primary criteria (Exhibit 2-6):

- The decisions affect particular groups of stakeholders in the operations of the firm. The pertinent question is, thus, w*hom* would this decision affect?
- The decisions must meet the standards of action-oriented business behavior. Managers need a doable set of guidelines for *how* to make ethical decisions.

Who are the Relevant Stakeholders?

The **stakeholders** of a firm are the many groups of people affected by the firm's decisions. Any given managerial decision affects, in varying degrees, the following stakeholders:

1. Owners or shareholders.
2. Employees.
3. Customers.
4. Management.
5. The general community where the firm operates.
6. Future generations.

Managers should make sure they consider all relevant stakeholders when they engage in ethical reasoning.

stakeholders
The groups of people affected by a firm's decisions.

Exhibit 2-6
The WH Process of Ethical Decision Making

1. **W—WHO (Stakeholders):**
 Customers
 Owners or investors
 Management
 Employees
 Community
 Future generations

2. **H—HOW (Guidelines):**
 Public disclosure
 Universalization
 Golden rule

BUSINESS Ethics Flashpoint 2.3

The WorldCom Accounting Scandal

WorldCom Inc. was a telecommunications company that began operating in the state of Mississippi in the 1980s. In 1989, WorldCom became a publicly traded corporation; that is, it began selling shares of its company to investors.

In the 1990s, WorldCom became the second largest U.S. long-distance telephone service carrier. In addition, it was a leading operator of Internet services and placed forty-second among the *Fortune* 500 companies in 2001.

As the 1990s came to an end, telecommunications was expanding and becoming more competitive. Numerous telecommunications companies entered the long-distance communications market, including the local telephone companies and the Regional Bell Operating Companies, and many companies entered the market for Internet services.

Eventually, the supply of telecommunications services outgrew the demand for those services, and the industry experienced a recession—that is, a drop in economic activity. WorldCom's revenue began to fall.

During this time, WorldCom used fraudulent accounting methods that allowed it to overstate its financial status to its shareholders. Specifically, CEO Bernard Ebbers presented a false picture of WorldCom's net worth, informing the Board of Directors of WorldCom that the company would continue growing vigorously. However, Ebbers knew that his projections were inconsistent with WorldCom's declining position.

One reason for boosting the reported revenues of WorldCom was to increase shareholder confidence. In other words, the logic behind Ebbers' fraudulent accounting methods was that if he could deceive shareholders into believing that WorldCom was growing, shareholders would continue to inject money into the company.

The fraud at WorldCom was propagated not just by Ebbers but also by several other officials at WorldCom, including CFO Scott Sullivan and WorldCom's comptroller, David Myers. Both of these individuals aided in secretly making entries of hundreds of millions of dollars into WorldCom's accounting report without any basis for such entries.

The fraud committed by the executive officials at WorldCom resulted in overstating the company's income by $11 billion.

In 2005, Ebbers was found guilty in federal court of fraud and received a sentence of twenty-five years in prison. Scott Sullivan received five years in prison, pleading guilty to fraud, conspiracy, and making false financial statements. David Myers received a sentence of one year and one day in prison.

As a result of the accounting fraud that occurred in the early 2000s, WorldCom was forced to declare bankruptcy. When this occurred, thousands of workers at WorldCom lost their jobs, and shareholders lost billions of dollars as the worth of WorldCom plummeted.

WorldCom not only affected its workers and shareholders but the entire telecommunications industry. Specifically, competing telecommunications companies that had once trusted WorldCom's false accounting reports now realized that they had been duped into relying on false data when making business decisions.

Sources: Dennis R. Beresford, Nicholas deB. Katzenbach, and C. B. Rogers, "Report of Investigation by the Special Investigative Committee of the Board of Directors of Worldcom, Inc." (March 31, 2003), FindLaw.com; Jennifer Bayot and Roben Farzad, "Ex-WorldCom Officer Sentenced to 5 Years in Accounting," August 12, 2005, NYTimes.com.

Study Business Ethics Flashpoint 2.3. First, identify the stakeholders directly affected by the behavior of WorldCom and then tackle the harder task of identifying the many stakeholders who are in the background, perhaps, but were nevertheless affected immensely by WorldCom's actions.

Here is another practice exercise to get you into the habit of looking for the values involved in a dilemma and the stakeholders affected by the dilemma.

Some of the more perplexing ethical dilemmas have emerged as a result of our increasing dependence on e-commerce.

How do We make Ethical Decisions?

Making ethical decisions has always been one of our most confusing *and* important human challenges. In the process of meeting this challenge, we have discovered a few general ethical guidelines to assist us. An *ethical guideline* provides one path to ethical conduct. Notice that all three of the following ethical guidelines reflect a central principle of business ethics: consideration of stakeholders.

E-COMMERCE and the Law

Technological Records and Ethics

Journalists use different technological media to contact and store information about confidential sources, including government whistleblowers. One of the most important parts of journalism is to protect confidential informants so that news sources can still enlighten the public with important and sensitive information. Ethical issues arise from this challenge. Documents containing information about informants' identities may not be collected by the government without permission from the Deputy Attorney General. In other words, the First Amendment protects the privacy and individual ownership of these email and phone records.

For example, in 2004, the Federal Bureau of Investigation (FBI) illegally obtained phone records related to the confidential informants of the *New York Times'* and *Washington Post*'s bureaus in Indonesia. The executive editors of both newspapers were called by the FBI, and both received apologies. The Justice Department looked into the illegal investigation by the FBI after the FBI went ahead with what it called "emergency" gathering of private information related to an alleged terrorist investigation. The editors were never told why their records were needed.

Now, news corporations are putting numerous security barriers and encryptions around their records so that nobody could have access to them even if they tried. In fact, Wikileaks, the website that publicizes secret government documents, has probably the strongest security procedures protecting its sources. Instant messages are expertly encrypted, as are all files passed between people. A tool that enables users to communicate completely anonymously, the Tor Project, completely hides all servers. Other news companies are attempting to mimic such standards and adjust privacy policies to reserve the right not to release any information about a source "to law enforcement authorities or to a requesting third party without notice."

CASE Nugget

Tyson Foods' Bribery Charges

In 2004, memos containing information regarding illegal bribes were sent from a Tyson Foods plant manager in Mexico to corporate officials in the United States. Basically, company officials in Mexico were sending roughly $2,700 a month to the wives of two veterinarians involved in safety and quality of the food produced at the plant. At the time, Tyson Foods was attempting to increase its national export. However, countries involved in the importation of these products require experts to certify that these food products meet certain safety and sanitary standards. Essentially, the two veterinarians were being sent extra payments so that they would sign off on the quality of the products produced at the plant regardless of whether the products were actually up to par with the standards imposed. This act is especially dangerous considering the reputation poultry products have of passing on disease, including salmonella, to consumers.

The company officials in Arkansas, including the president of Tyson International and vice president of operations, realized that bribing foreign officials, the biggest issue at hand, was a felony according to the Foreign Corrupt Policies Act. Faced with the option of investigating the allegations and bringing the illegal activities to light, the company officials instead found a way to keep sending the bribes but make the payments look legal. Two years later, Tyson officials finally hired a law firm to do an internal investigation of the bribery. In 2011, Tyson was charged with conspiracy and violating the Foreign Corrupt Practices Act. To avoid trial, Tyson was forced to pay $4 million as a criminal penalty and $1.2 million to the Securities Exchange Commission for the charge of maintaining illegal records.

BUSINESS Ethics Flashpoint 2.4

The Health Focus of Revolution Foods

Revolution Foods started as a small business venture in Oakland, California. The start-up's objective is to provide healthy prepackaged lunch meals for children.

Currently, the market for prepackaged lunch meals is dominated by Kraft Foods. Kraft Foods sales of Lunchables account for approximately 76 percent of sales in the prepackaged lunch market niche.

Kids love Lunchables because these meals allow them to create their own food. Simply, these meals are fun. Parents also benefit from Lunchables in the sense that these meals require no preparation.

However, Lunchables are not necessarily healthy. According to food critics, including the founders of Revolution Foods, Lunchables contain an excess of fat, sugar, salt, and preservatives that are unhealthy.

For example, Kraft previously had a Maxed Out line of Lunchables that was considered one of the Five Worst Packaged Lunchbox Meals by the Cancer Project. Kraft eventually stopped producing this line of Lunchables—with a whopping 660 calories and 22 grams of fat—and endeavored to create more healthy meals.

In response to the apparent need for more nutritious lunch kits, Revolution Foods claims to offer a healthy alternative.

Kristin Richmond, one of the founders of Revolution Foods, believes that the objective of Revolution Foods is to provide higher-quality ingredients to cultivate healthier lifestyles. Making foods with real meat, real cheese, and other recognizable ingredients appears to be the driving force behind the company.

In other words, rather than focusing solely on profits, Revolution Foods attempts to focus on what will also help the consumer lead a healthier lifestyle.

One example of the types of prepackaged lunch kits that Revolution Foods offer is the Ham and Cheddar Meal Kit, consisting of ham from animals raised with no antibiotics. This lunch kit also has no preservatives, no high-fructose corn syrup (a commonly used artificial sweetener), and no artificial coloring.

On average, Revolution Foods' Meal Kits are lower in fat, salt, and sugar than Kraft's Lunchables.

The director of nutrition policy at the Center for Science in the Public Interest has claimed that Revolution Foods, although not necessarily healthy, is nutritionally improved. The fact that Revolution Foods provides foods that are locally grown, natural, and organic without preservatives appeals to those consumers who desire to cultivate healthy dietary habits.

Revolution Foods' Meal Kits can now be found in more than 1,000 stores, including Kroger's, Target, and Whole Foods.

Sources: Stephanie Strom, "Lunchables, The Lunchbox King, Faces a Rival Vowing Higher-Quality Fare", *N.Y. Times,* August 21, 2013; Marc Gunther, "Revolution Foods: a healthier competitor to Lunchables", March 11, 2014, The Guardian.com.

The Golden Rule The idea that we should interact with other people in a manner consistent with the way we would like them to interact with us has deep historical roots. Both Confucius and Aristotle suggested versions of that guideline. One scholar has identified six ways the Golden Rule can be interpreted:

1. Do to others as you want them to gratify you.
2. Be considerate of others' feelings as you want them to be considerate of yours.
3. Treat others as persons of rational dignity like you.
4. Extend brotherly or sisterly love to others, as you would want them to do to you.
5. Treat others according to moral insight, as you would have others treat you.
6. Do to others as God wants you to do to them.

Regardless of the version of the Golden Rule we use, this guideline urges us to be aware that other people—their rights and needs—matter. The focus on others that is the foundation of the Golden Rule is also clearly reflected in a second ethical guideline: the public disclosure test.

Public Disclosure Test One way to think of the public disclosure test is to view it as a ray of sunlight that makes our actions visible rather than obscured. The public disclosure test is sometimes called the television test, for it requires us to imagine that our actions are being broadcast on national television. The premise behind this general guideline is that ethics is hard work, labor

BUSINESS Ethics Flashpoint 2.5

The Dofasco Steel Company's Approach to Workers

Dofasco is the second-largest steel producer in Canada and sells a wide range of steel products. In the 1980s, Dofasco experienced shrinking profits and decided to change its approach to business.

One of the ways that Dofasco changed was in its approach to its workers. Specifically, Dofasco attempted to implement a philosophy taking into account the whole person at work. In other words, this approach sought to include wellness as an important characteristic of the employee experience.

In line with this new philosophy, the Dofasco management claimed it would abide by the following policy statement issued in its 1996 guides of the company's health, safety, and lifestyle activities: "At Dofasco, there is nothing more important than the health and safety of our people."

Consistent with this policy statement, Dofasco provided training for employees in problem solving, manufacturing processes, and customer service. Dofasco also built on its tradition of providing its workers with ergonomic and fitness resources as well as preventive medical services.

Yet another way that Dofasco sought to focus its energies on employee needs was by encouraging healthy lifestyle practices at work. A healthy lifestyle at work would be achieved by preventing workplace accidents and maintaining a safe environment. In addition, Dofasco assessed employee needs in response to worker suggestions.

Dofasco believes that its focus on a safe and healthy workplace has improved its business. Specifically, the company claims that its policies have reduced lost time associated with work injuries and minimized Workers Safety Insurance Board payments.

In addition to focusing on worker safety and well-being, Dofasco also has a history of being environmentally responsible. For example, Dofasco was the first corporation in Canada voluntarily to sign an Environmental Management Agreement with Environment Canada and Ontario's Ministry of the Environment. In this agreement, Dofasco made commitments to abide by specific air and water quality standards and sustainable energy use and waste management. Through this agreement, Dofasco expressed its commitment to the community.

In addition, Dofasco created a Sustainable Development Strategy Team and is a founding sponsor of the Sustainable Enterprise Academy at York University's Schulich School of Business. The mission of this academy is to provide executives with training in how to manage businesses with an eye for sustainability.

Dofasco's organizational practices resulted in being named one of the world's most sustainable companies in the Dow Jones Sustainability Group Index.

Sources: Dan Corbett, "Excellence in Canada: Healthy Organizations—Achieve Results by Acting Responsibly," *Journal of Business Ethics* 55, no. 2 (December 2004), pp. 124, 130; Joan Enric Ricart, et. al. "The Sustainable Enterprise: Learning from DJSI Leaders", *Fundacion BBVA* (2005).

that we might resist if we did not have frequent reminders that we live in a community. Given our role as a member of a community, our self-concept is tied, at least in part, to how that community perceives us.

Universalization Test A third general guideline shares with the other two a focus on the "other"—the stakeholders whom our actions affect. Before we act, the **universalization test** asks us to consider what the world would be like if our decision was copied by everyone else. Applying the universalization test causes us to wonder aloud: "Is what I am about to do the kind of action that, *if others followed my example,* makes the world a better place for me and those I love?"

This chapter concludes with two Business Ethics Flashpoint examples: Flashpoints 2.4 and 2.5. For the first one, try to word what the firm did in terms of the ethical guidelines. Did its behavior show potential consistency with all three of the guidelines?

As you study this next Business Ethics Flashpoint, start by asking yourself what values are in conflict in this business scenario. After you have established which values are in tension, work through both steps in the WH process for making ethical business decisions.

In summary, business managers can apply the WH approach to most ethical dilemmas. The WH framework provides a practical process suited to the frequently complex ethical dilemmas that business managers must address *quickly* in today's society.

> **universalization test**
> The ethical guideline that urges us to consider, before we act, what the world would be like if everyone acted in this way.

Point/Counterpoint

The Social Responsibility of Corporations

Do a Firm's Ethical Responsibilities Extend Beyond Maximization of Profits?	
NO	**YES**
The basic purpose of a business firm is to provide a service or product to consumers that makes profits for its owners. Capitalist theory depends on the incentive of the profit motive as a guarantee that firms will meet the desires of consumers. Please those consumers, and the market is an outstanding institution. Displease them, and you go out of business. This sanitizing process relies on the assumption that firms will seek to maximize profits. When firms decide for themselves that they will have goals that conflict with profit maximization, they undercut the meaning and the benefits of capitalism. Democracies have a process for deciding which of the countless potential good deeds a community will pledge to do. Tax money is never bountiful enough to satisfy our human desire to assist one another. We must pick and choose which needs to meet and which to postpone for another day. But the point is that we make these decisions AS A COMMUNITY. When a corporation takes it upon itself to use money it earned by charging consumers a price for a good or service, that firm is seizing the role of government BECAUSE IT CAN. No community has ceded the right to make social decisions to the corporation. Instead, the firm uses its powers to spend the community's money AS THE CORPORATION SEES FIT. This behavior is an affront to democracy and basically says to the community, "We know best what the community's priorities should be." If firms have so many excess profits that they can become a "good deeds" agency as well as a business firm, they should honor their true role and lower prices so more consumers can buy their products. By being what they claim is socially responsible, they are implicitly taxing their consumers who pay for a good or service AND then an extra fee for the firm to do what it designates as our top social priorities.	Communities have dozens of needs that, when met, assist citizens in flourishing. What sense does it make to take an important sector like corporate America and deny them the opportunity to assist in meeting these needs? When a computer firm decides to enhance the educational mission of the town where they produce their product by giving computers to the school, they do so in an environment where schools are often pinched for funds. Those students need computers, and they would take them from anyone who is kind enough to provide them. Business firms are major citizens. They require utilities, roads, airports, police, sewage systems, power plants, and fire protection. These services impose heavy costs on the community. Firms, like other citizens, have a responsibility to assist in paying for these public services. Their tax payments certainly serve this purpose. But from the community's perspective, should we not encourage citizens of all kinds to give back to their communities as often and as generously as they are willing to provide? Applauding the social responsibility of firms when they give back to their communities strengthens the ties of the firm to that community. A sense of mutual respect blooms between the corporation and the community where so many of its managers and workers have agreed to live.

Questions & Problems

1. What ethical guidelines could help an American business manager working in another country decide whether she should engage in behavior that is ethical where she works but unethical in the United States?

2. What is the relationship between business behavior that is legal and business decisions that are ethical?

3. How do business ethics and business law interact with each other?

4. How does the WH approach to ethics respond to an ethical problem?

5. In 2000, Donald Stern incorporated a startup company named Grail in order to market a new and revolutionary chip design that Stern had invented. While witnesses expressed the view that Stern's innovation would be the "Holy Grail of memory technology," the company was unable to find investors, and by March,

2001, Grail had run out of money. On April 19th, 2001, the following month, Stern met with Ryuichi Matsuo and Kazutoshi Hirayama, two representatives from Mitsubishi. Matsuo signed a nondisclosure agreement which required Mitsubishi "to keep in strict confidence and trust and not use, disclose or make available to others, including any of its affiliates or third parties any 'Proprietary Information' and 'Company Documents and Materials' (together 'INFORMATION') without the prior written consent of [Grail]." In the meeting, Stern showed Hirayama 16 items of proprietary information, including the core concept behind his miracle memory chip, of which Hirayama took extensive notes. Nothing came from the meeting between Grail and Mitsubishi.

In 2003, Mitsubishi and Hitachi formed Renasas Technology in a joint venture. Hirayama was one of the employees transferred from Mitsubishi. Stern suspected that Renasas was using his innovations and that Mitsubishi had broken the NDA because of an article he read promoting Renasas' new technology. In the ensuing 2012 trial, expert witnesses found that "of the 16 items of confidential information disclosed to Mitsubishi, 11 had been implemented" in Renasas' new technology. On May 16, 2012, the jury found that Mitsubishi had violated the NDA and determined damages to be $123,898,889.00. Mitsubishi has filed an appeal for a new trial to address the damages, which is ongoing now.

How would Mitsubishi's violation of the NDA affect the decisions of other innovative tech companies to approach potential investors?

Which aspects of the WH process did Mitsubishi neglect in their decision to break the NDA?

Source: *Grail Semiconductor, Inc. v. Mitsubishi Electric & Electronics USA, Inc.*, 225 Cal. App. 4th 786 (2014).

6. Tyco International, a manufacturing company, sued its former CEO, Dennis Kozlowski, in regard to his receipt of unauthorized bonuses from the company profits. Among his other wrongdoings, Kozlowski awarded unauthorized special bonuses to himself and more than 40 other Tyco employees. In sum, he misappropriated from company funds over $100 million that he was not authorized to receive. Included in this misappropriation was the use of company funds for personal expenditures, such as $18 million to buy and redecorate his New York City apartment and $1 million for his wife's 40th birthday party, held on the Italian island of Sardinia. The extravagant party featured an ice sculpture of the statue of David urinating Stolichnaya vodka. Based on these and other corporate misdeeds in related criminal proceedings, Kozlowski is currently serving 8.33 to 25 years at the Mid-State Correctional Facility in New York. How do you think Kozlowski would have acted had he considered the public disclosure test before engaging in these corporate misdeeds? What if he applied the universalization test?

Source: *Tyco Intl, Ltd. v. Kozlowski,* 2005 DNH 68; 2005 U.S. Dist. LEXIS 6867 (April 21, 2005).

7. In Arkansas, methamphetamine (meth) was manufactured primarily in small toxic labs (STLs) located in homes, tents, barns, or hotel rooms. According to the plaintiffs (a number of counties in Arkansas), Arkansas had one of the highest numbers of STLs in the nation. The counties alleged that they had spent significant amounts of taxpayer dollars combating the manufacture of methamphetamine, including law enforcement costs for locating, eliminating, and cleaning up STLs; prison and jail costs for housing illegal users, dealers, and manufacturers; addiction treatment costs for users and addicts; payments to family service agencies for housing and treating children whose parents were arrested on meth-related charges; and costs for treating the physical side effects of meth use and exposure to its production. To recoup these costs, the plaintiffs sought compensation from Pfizer, Inc., and other manufacturers of cold and allergy medicines. They brought the suit against these manufacturers because meth cannot be produced without ingredients from their cold and allergy medicines. What ethical duties, if any, do you think the drug manufacturers owe to the state or people in this case? Apply the WH framework to help reach your conclusions.

Source: *Ashley County et al. v. Pfizer, Inc.,* 552 F.3d 659; 2009 U.S. App. LEXIS 19.

8. Todd Russell Dolezal is a bail bondsman who was slapped with a misdemeanor, a $1000 fine, and a probationary period of 1 year for soliciting bail in violation of California Code of Regulations, title 10, 170*170 section 2079.1. Section 2079.1 bans bail bondsmen from directly soliciting bail in jail facilities or court facilities. Furthermore, "any solicitation [regarding bail] of an arrestee himself... shall be only after a bona fide request for bail services has been received from the arrestee..." In 2013, Dolezal tried to appeal on grounds that the regulation violated his First Amendment right, but the appeals court affirmed the previous result.

What purpose does section 2079.1 serve within the context of the WH process? Who does this law protect? Why would a law like this be necessary?

Source: *People v. Dolezal,* 221 Cal. App. 4Th 167 (2013).

9. Dragon, a speech recognition technology company considered to be "a leader in speech technology products" and valued at $600 million, was in perilous straits. Despite being valued so highly, the

company lost money in every year of its existence except for one year. Dragon was sorely in need of capital. Dragon considered merging with another company to sustain itself until it could develop its technologies further and make them profitable. To go about this business, Dragon entered into a contract with Goldman Sachs on December 8, 1999, to "provide [Dragon] with financial advice and assistance in connection with this potential transaction, which may include performing valuation analysis, searching for a purchaser acceptable to [Dragon], coordinating visits of potential purchasers and assisting [Dragon] in negotiating the financial aspects of the transaction."

Goldman formed a four-member team to conduct financial due diligence on Lernout & Hauspie Speech Products N.V. ("L & H"), a company that had offered to buy Dragon for $580 million paid in stock. Goldman's assessment of L & H was "positive in the overall" but "did not give an opinion pro or con on the transaction." In June 2000, the Dragon board approved the merger with L & H and ceased to become an independent entity.

Unfortunately for the ex-board members of Dragon, the Wall Street Journal reported that L & H had vastly overstated its revenues in an August 2000 article. The SEC promptly began investigating L & H's financial statements. L & H had improperly recorded $373 million in revenue and was forced to restate its financials for the past two and a half years. In November 2000, L & H filed for bankruptcy and the share of L & H stock used to pay for the merger became worthless.

Naturally, the ex-board members of Dragon took Goldman Sachs to trial for negligent performance of services, gross negligence, and intentional and negligent misrepresentation, among other claims. However, the court ruled in favor of Goldman on January 23, 2013, and an appeal for a new trial by the plaintiffs on the grounds of evidentiary errors and erroneous jury instructions was rejected by the First Circuit Appeals Court on November 12, 2014.

L & H's fraudulent actions obviously caused harm to the shareholders of Dragon. Who are some less obvious stakeholders, as defined by the WH Process, that were negatively affected by L & H?

During the trial, it was noted that Goldman's team had not disclosed their concern about the status of due diligence of L & H to Dragon because "the client did not ask." How ethical was this course of action?

Source: *James and Janet Baker; Paul G. Bamberg and Robert Roth v. Goldman, Sachs & Co.,* 771 F.3d 37 (1st Cir. 2014).

10. Titan Distribution Inc. had subcontracted with a company that provided employees, but Titan later decided to hire its own employees and end the subcontracting contract. Robert Chalfant worked for the subcontracted company, and when Titan made this change, Chalfant sought the same position with Titan that he had held while employed by the subcontractor. Although Titan initially indicated that Chalfant passed his physical examination and would be hired, the company subsequently decided that he failed the physical and did not hire him. The evidence indicated that Titan mistakenly believed that Chalfant's physical ailments substantially limited his ability to work, that he previously performed and remained able to perform the functions of the job, and that the refusal to hire him on the basis of the allegedly failed physical was linked to his disability. Further, Titan's inconsistent and unexplained decision-making process, with knowledge of the federal discrimination laws, was sufficient to support an award of punitive damages. Later you will learn about discrimination, but right now, consider the ethical issues of this case. What values are at odds in the dilemma outlined in this case?

Source: *Chalfant v. Titan Distribution, Inc.,* 475 F.3d 982 (2007).

Part 1 *The Legal Environment of Business*

CHAPTER 3

The U.S. Legal System and Alternative Dispute Resolution

CASE OPENER

Questionable Jurisdiction over Caterpillar

James Lewis, a resident of Kentucky, sustained an injury while operating a Caterpillar bulldozer. He filed suit against Caterpillar, a company incorporated in Delaware but with its principal place of business in Illinois. Lewis also filed suit against the supplier of the bulldozer, Whayne Supply Company, whose principal place of business was Kentucky. Lewis filed his case in a Kentucky state court, alleging defective manufacture, negligence, failure to warn, and breach of warranty. Lewis and Whayne Supply Company agreed to settle out of court. Caterpillar then filed a motion to exercise its right of removal (its right to move the case from the state to the federal court system), arguing that the federal court had jurisdiction over the case because Caterpillar and Lewis were from different states. Lewis disagreed with Caterpillar's contention, claiming that because he had not completed his settlement with Whayne, the case still included a defendant (Whayne) from Lewis's state, Kentucky. Thus, Lewis argued, federal courts did not have jurisdiction over the case.

 The court agreed with Caterpillar's argument and moved the case to a federal district court. Shortly thereafter, Lewis and Whayne finalized their settlement agreement, and the district court dismissed Whayne from the lawsuit. The federal district court granted Caterpillar a favorable judgment. Lewis, however, appealed the district court's decision, renewing his argument that the district court did not have jurisdiction over the case. The court of appeals agreed with Lewis, holding that because Whayne was a defendant in the case at the time that Caterpillar moved the case from state to federal court, the diversity of citizenship necessary to give the federal court jurisdiction over the case was absent. Thus, a state court should have resolved the dispute. Consequently, the appellate court vacated the district court's decision. Caterpillar then appealed to the U.S. Supreme Court.

1. What factors determine whether the state or federal court system hears a case?
2. If you were a businessperson with Caterpillar, why might you prefer a federal court to hear the dispute with Lewis, rather than a state court?

LEARNING OBJECTIVES

LO 3-1 Explain the different types of jurisdiction a court must have before it can render a binding decision in a case.

LO 3-2 Recall the rules related to choice of venue.

LO 3-3 Identify the threshold requirements that must be met before a court will hear a case.

LO 3-4 Recognize the steps in civil litigation.

LO 3-5 Compare the various forms of alternative dispute resolution used by businesses today.

As the Case Opener illustrates, when a dispute arises, parties in this country do not simply "go to court." They often must choose between federal and state court systems. This chapter examines these systems as well as the trial procedures that apply in civil cases.

Jurisdiction

A useful way to understand jurisdiction is to think of it as courts' power to hear cases and render decisions that bind the parties before them. A court must have several types of jurisdiction to decide any particular case.

Original Versus Appellate Jurisdiction

Trial courts, or **courts of original jurisdiction,** have the power to hear and decide cases when they first enter the legal system. In these courts, the parties present evidence and call witnesses to testify. In most states, these courts are called *courts of common pleas* or *county courts*. In the federal system, these trial courts are called *federal district courts*.

Courts of appellate jurisdiction, or **appellate courts,** have the power to review previous judicial decisions to determine whether trial courts erred in their decisions. Appellate courts do not hold trials. Instead, appellate judges review transcripts of trial court proceedings and occasionally consider additional oral and written arguments from each party.

Appellate courts handle only questions of law, not questions of fact. A **question of law** is an issue concerning the interpretation or application of a law. In contrast, a **question of fact** is a question about an event or characteristic in the case. For example, whether a white student yelled racial slurs on a college campus is a question of fact. On the other hand, whether the First Amendment protects the student's right to utter racial slurs is a question of law.

Only judges can decide questions of law. Trial courts determine questions of fact. In a bench trial (a trial with no jury), the judge decides questions of fact; in a jury trial, the jury decides questions of fact. Appellate courts can, however, overrule trial courts' decisions on questions of fact, but only when the trial court's finding was clearly erroneous or when no trial evidence supports the trial court's finding.

Jurisdiction Over Persons and Property

LO 3-1 Explain the different types of jurisdiction a court must have before it can render a binding decision in a case.

In personam jurisdiction (literally, "jurisdiction over the person") is a court's power to render a decision affecting the rights of the specific persons before the court. Generally, a court's power to

trial court
A court in which most civil or criminal cases start when they first enter the legal system. The parties present evidence and call witnesses to testify. Trial courts are referred to as *courts of common pleas* or *county courts* in state court systems and as *district courts* in the federal system. Also called *court of original jurisdiction* and *court of first instance*.

appellate court
A higher court, usually consisting of more than one judge, that reviews the decision and results of a lower court (either a trial court or a lower-level appellate court) when a losing party files for an appeal. Appellate courts do not hold trials but may request additional oral and written arguments from each party; they issue written decisions, which collectively constitute case law or the common law. Also called *court of appellate jurisdiction*.

question of law
An issue concerning the interpretation or application of a law.

question of fact
A question about an event or characteristic in a case.

exercise *in personam* jurisdiction extends only over a specific geographic region. In the state court system, a court's *in personam* jurisdiction usually extends to the state's borders. In the federal system, each court's jurisdiction extends across its geographic district.

A court acquires *in personam* jurisdiction over a person (the **plaintiff**) when she files a lawsuit with the court. The court acquires jurisdiction over the person the plaintiff is suing (the **defendant**) when it gives him a copy of the complaint and a summons. The **complaint** specifies the factual and legal basis for the lawsuit and the relief the plaintiff seeks. The **summons** is a court order that notifies the defendant of the lawsuit and explains how and when to respond to the complaint.

Service of process is the procedure by which courts present these documents to defendants. Traditionally, courts used personal service: An officer of the court handed the summons and complaint to the defendant. Over time, however, courts began to employ other methods of service, including *residential service,* in which a court representative leaves the summons and complaint with a responsible adult at the defendant's home, and *service by certified* or *ordinary mail.*

Increasingly, courts are authorizing service by email, especially in situations where it is difficult to find the defendant's physical address. The reasons for recognizing this form of service are that (1) it is reasonably calculated to provide notice of the pending action, and (2) email is not substantially less likely to provide notice than the customary methods of providing notice. Recently, a few jurisdictions have even recognized service through Facebook accounts when traditional methods of service did not work.[1]

If the defendant is a corporation, courts generally serve either the president of the corporation or an agent whom the corporation has appointed to receive service. Most states require corporations to appoint an agent for service when they incorporate. Corporations are subject to *in personam* jurisdiction in three locations: the state of their incorporation, the location of their main offices, and the geographic areas in which they conduct business.

Courts have *in personam* jurisdiction over persons only within a specific geographic region. In the past, a state court could not acquire *in personam* jurisdiction over out-of-state defendants unless it served the defendants within the court's home state. Thus, defendants who injured plaintiffs could evade legal action by leaving the state and remaining outside its borders.

To alleviate this problem, most states have enacted **long-arm statutes** that enable the court to serve defendants outside the state as long as the defendant has sufficient minimum contacts within the state. Each state has its own minimum-contact requirements, but most state statutes hold that acts such as committing a tort or doing business in the state are sufficient to allow the state to serve a defendant. In the Caterpillar case, the company sold products in Kentucky, and its products caused an injury in that state. These two facts were sufficient minimum contacts to allow the Kentucky court to serve Caterpillar, even though it was an out-of-state company.

BUT WHAT IF...

WHAT IF THE FACTS OF THE CASE OPENER WERE DIFFERENT?

Recall that in the Case Opener, the company sold products in Kentucky, and its products caused an injury in that state. Let's say in the Case Opener that Caterpillar did not sell products in the state of Kentucky, but its products did cause an injury in that state. Could a Kentucky court still exercise *in personam* jurisdiction over Caterpillar and serve the company? Why or why not?

In contrast to the situation in the opening scenario, the Florida appellate court did not find minimum contacts with that state to enable a foreign corporation to sue Columbia University, located in New York City, for a tort that allegedly occurred in New York. The court found that the fact that Columbia had alumni associations in Florida, owned some interactive classrooms in that state, and offered some online classes to residents did not constitute sufficient minimum contacts for a lawsuit in which none of the tortious acts were alleged to have occurred in the state.[2]

[1] How extensive recognition of service via social media will become remains to be seen.
[2] Trustees of Columbia University v. Ocean World, SA 2009 WL 1212229, Ct. App. Fla.

in personam jurisdiction
The power of a court to require a party (usually the defendant) or a witness to come before the court. The court must have personal jurisdiction to enforce its judgments or orders against a party. *In personam* jurisdiction extends only to the state's borders in the state court system and across the court's geographic district in the federal system.

plaintiff
The person or party who initiates a lawsuit (also known as an *action*) before a court by filing a complaint with the clerk of the court against the defendant(s). Also known as *claimant* or *complainant*.

defendant
The person or party against whom a civil or criminal lawsuit is filed in a court of law.

complaint
A formal written document that begins a civil lawsuit; contains the plaintiff's list of allegations against the defendant along with the damages the plaintiff seeks.

summons
A legal document issued by a court and addressed to a defendant that notifies him or her of the lawsuit and how and when to respond to the complaint. A summons may be used in both civil and criminal proceedings.

long-arm statute
A statute that enables a court to obtain jurisdiction against an out-of-state defendant as long as the defendant has sufficient minimum contacts within the state, such as committing a tort or doing business in the state.

E-COMMERCE and the Law

The Sliding-Scale Standard for Internet Transactions

Does a business that has Internet contact with a plaintiff in a different state satisfy the minimum-contacts standard? Anyone who engages in transactions over the internet should be concerned about this question so that they do not unexpectedly find themselves forced to answer to a complaint in a very inconvenient location.

A federal district court established the following "sliding-scale" standard the a 1997 case *Zippo Mfg. Co. v. Zippo Dot Com, Inc.*:[3]

> [T]he likelihood that personal jurisdiction can be constitutionally exercised is directly proportional to the nature and quality of commercial activity that an entity conducts over the Internet. This sliding scale is consistent with well developed personal jurisdiction principles.

At one end of the spectrum are situations where a defendant clearly does business over the Internet. If the defendant enters into contracts with residents of a foreign jurisdiction that involve the knowing and repeated transmission of computer files over the Internet, personal jurisdiction is proper.

At the opposite end are situations where a defendant has simply posted information on an Internet website that is accessible to users in foreign jurisdictions. A passive website that does little more than make information available to those who are interested in it is not grounds for the exercise of personal jurisdiction.

The middle ground is occupied by interactive websites through which a user can exchange information with the host computer. In these cases, the exercise of jurisdiction is determined by examining the level of interactivity and the commercial nature of the exchange of information that occurs on the website.

As the E-Commerce and the Law box that follows illustrates, the Internet has created complications for the courts in determining when an Internet presence creates in personam jurisdiction over a website operator. There is a need to protect companies from being dragged to an infinite number of courts even though their customer bases may be nationwide; at the same time, there is a need to not impinge on the sovereignty of states by curtailing the conditions under which defendants may be commanded to appear in a state's court.

CASE Nugget

Minimum Contacts via the Internet

Burdick v. Superior Court

233 Cal. App. (Ct. App 2015)

When a defendant acts on the internet to an online audience, it is difficult to know whether the court will find sufficient connections with an out of state defendant to allow the court to have in personam jurisdiction. Burdick, a consultant working for Nerium International, a skincare company, posted defamatory statements about two California bloggers, Sanderson and Taylor, who had questioned the safety and efficacy of one of Nerium's products on their blog. Sanderson and Taylor attempted to bring suit in California against Burdick, a resident of Illinois. He moved to quash the California summons on grounds that he did not have minimum contacts with the state of California. The court ruled that Burdick's post was clearly defamatory, but the post did not specifically target the forum state (California) and therefore Burdick did not have the minimal connection to the state necessary to try him there. The court noted that although Sanderson and Taylor were affected by the defamatory statements in California, the pair could not show that the post targeted or might be read by a significant number of other California residents.

If a defendant has property in a state, a plaintiff may file suit against the defendant's property instead of against the owner. For example, suppose a Utah resident had not paid property taxes on

[3] *Zippo Mfg. Co. v. Zippo Dot Com, Inc.*, 952 F. Supp. 1119 at 1124 (1997).

a piece of land she owned in Idaho. Idaho courts have *in rem* jurisdiction (Latin for "jurisdiction over the thing") over the property. Thus, an Idaho state court has the power to seize the property and sell it to pay the property taxes in an *in rem* proceeding.

Subject-Matter Jurisdiction

Subject-matter jurisdiction is a court's power to hear certain kinds of cases. Most industrialized countries have a single court system, with courts that have the power to hear both national law cases and local law cases. In contrast, the United States has both a state and a federal court system. Subject-matter jurisdiction determines which court system may hear a particular case. Cases may fall under state jurisdiction, exclusive federal jurisdiction, or concurrent jurisdiction. Exhibit 3-1 illustrates the subject-matter-jurisdiction divisions.

Exclusive Federal Jurisdiction The federal court system has exclusive jurisdiction over very few cases: admiralty cases; bankruptcy cases; federal criminal prosecutions; lawsuits in which one state sues another state; claims against the United States; and patent, trademark, and copyright cases. Additionally, federal courts have exclusive jurisdiction over claims arising under federal statutes that specify exclusive federal jurisdiction.

State Jurisdiction The state court system has a broad range of jurisdiction; state courts have the power to hear all cases not within the exclusive jurisdiction of the federal court system. State courts also have exclusive jurisdiction over certain cases, such as cases concerning adoption and divorce. Most cases, therefore, fall under state court jurisdiction. The Caterpillar case fell under state court jurisdiction because its subject matter—product liability and negligence—did not place the case under the exclusive jurisdiction of the federal court system.

Concurrent Federal Jurisdiction Concurrent federal jurisdiction means that both state and federal courts have jurisdiction over a case. Concurrent jurisdiction covers two types of cases: federal-question and diversity-of-citizenship cases. *Federal-question* cases require an interpretation of the U.S. Constitution, a federal statute, or a federal treaty. For example, suppose a plaintiff alleges that a Florida campaign-financing law violates his First Amendment free speech rights. Because this case raises a federal question, it falls under concurrent jurisdiction, and both state and federal courts have the power to hear it.

A *diversity-of-citizenship* case must satisfy two conditions: (1) The plaintiff(s) does (do) not reside in the same state as the defendant(s) and (2) the controversy concerns an amount in excess

in rem jurisdiction
The power of a court over the property or status of an out-of-state defendant when that property or status is within the court's jurisdiction area.

subject-matter jurisdiction
The power of a court over the type of case presented to it.

Exhibit 3-1

Subject-Matter Jurisdiction Divisions

Exclusive Federal Jurisdiction
- Admiralty cases
- Bankruptcy cases
- Federal criminal prosecutions
- Cases in which one state sues another state
- Claims against the United States
- Other claims involving federal statutes that specify exclusive federal jurisdiction

Concurrent Jurisdiction
- Federal-question cases
- Diversity-of-citizenship cases

State Jurisdiction
- All cases not falling within exclusive federal jurisdiction

of $75,000. Courts use the location of a party's residence to determine whether diversity of citizenship exists. Most federal court cases are based on diversity of citizenship.

A business may reside in two states: the state of its incorporation and the state of its principal place of business. Thus, in the Case Opener, Caterpillar was a resident of Delaware, the state where it incorporated, and of Illinois, the state of its primary place of business.

Although the rule for determining what state a defendant is a citizen of may sound simple, determining where a corporation's principal place of business is caused a lot of controversy initially. Was it where most of the states' business activities occurred? Or was it where its headquarters was located? In Case 3-1, the Supreme Court finally clarified how courts should determine a corporation's principal place of business.

CASE 3-1

HERTZ CORPORATION v. FRIEND
United States Supreme Court
130 S. Ct. 1181 (2010)

FACTS: *California employees sued their employer, Hertz Corporation, in state court, alleging violations of the state's wage and hour laws. Defendant Hertz, claiming diversity of citizenship, removed the case to the federal district court. Hertz argued that it was a citizen of New Jersey because New Jersey is its principal place of business. It is its principal place of business because Hertz has facilities in 44 states, but its leadership is at its corporate headquarters in New Jersey, and its core executive and administrative functions are primarily carried out there. Hertz also stated that California accounted for only a small portion of its business activities.*

The District Court concluded that it lacked diversity jurisdiction because Hertz was a California citizen under Ninth Circuit precedent, which asks whether the amount of the corporation's business activity is "significantly larger" or "substantially predominates" in one state. Finding that California was Hertz's "principal place of business" under that test because a plurality of the relevant business activity occurred there, the District Court remanded the case to state court. Hertz appealed, and the Ninth Circuit Court of Appeals affirmed. Hertz then appealed to the U.S. Supreme Court.

ISSUE: For purposes of establishing citizenship of a corporation to determine whether there is diversity of citizenship to give the federal court jurisdiction, what test should the courts use to determine a corporation's principal place of business, a nerve center test or a predominance of business activity test?

REASONING: "Principal place of business" is best read as referring to where a corporation's officers direct, control, and coordinate the corporation's activities. In practice, it should normally be where the corporation maintains its headquarters—provided that the headquarters is the actual center of direction, control, and coordination, that is, the nerve center, and not simply an office where the corporation holds its board meetings.

Three reasons support use of the nerve center approach. The first is that the approach is consistent with the language of the jurisdictional statute. The statute's word "place" is singular, not plural. Its word "principal" requires the main, prominent, or most important place to be chosen. And the fact that the word "place" follows the words "State where" means that the "place" is a place within a State, not the State itself. A corporation's nerve center, usually its main headquarters, is a single place. The public often considers it the corporation's main place of business. And it is a place within a State. By contrast, the application of a more general business activities test has led some courts to look not at a particular place within a State but incorrectly at the State itself, measuring the total amount of business activity the corporation conducts there and determining whether it is significantly larger than in the next-ranking State.

Second, administrative simplicity is desired in a jurisdictional statute. A nerve center approach, which ordinarily equates that center with a corporation's headquarters, is simple to apply, comparatively speaking.

Third, the statute's legislative history suggests that the words "principal place of business" should be interpreted to be no more complex than an earlier, numerical test that was criticized as too complex and impractical to apply. A nerve center test offers such a possibility. A general business activities test does not.

This test is relatively easier to apply and does not require courts to weigh corporate functions, assets, or revenues different in kind, one from the other. This test may produce results that seem to cut against the basic rationale of diversity jurisdiction; accepting occasionally counterintuitive results is the price the legal system must pay to avoid overly complex jurisdictional administration while producing the benefits that accompany a more uniform legal system.

DECISION AND REMEDY: For purposes of federal court diversity jurisdiction, a corporation's principal

(continued)

place of business refers to the place, often called nerve center, from which the corporation's high-level officers directed, controlled, and coordinated corporation's activities. Although it appeared from the evidence that Hertz corporate offices and its nerve center are located in New Jersey, the respondents should have a fair opportunity on remand to litigate this issue in light of this holding as to the applicable rule.

SIGNIFICANCE OF THE CASE: The case settles a long-standing conflict among the circuits about what test to apply to determine where a corporation's primary place of business is.

CRITICAL THINKING

Which critical-thinking skill is the focal point of this case? Which words are not clear?

ETHICAL DECISION MAKING

What primary value can you hear in the court's reasoning?

It is important to note that diversity must be complete for a case to fall under concurrent jurisdiction. In the Caterpillar case, Lewis argued that diversity was not complete because both he and the supply company, the second defendant he originally sued, were residents of Kentucky. The appellate court agreed with his argument and overturned the district court's decision because the district court lacked subject-matter jurisdiction.

When a plaintiff files a case involving concurrent jurisdiction in state court, the defendant has a *right of removal*. This right entitles the defendant to transfer the case to federal court. Thus, either party to a case involving concurrent jurisdiction can put the case in the federal court system: the plaintiff can file the case in federal court initially, or, if the case was initially filed in state court, the defendant can transfer the case to federal court by exercising her right of removal. In the Case Opener, Caterpillar exercised its right of removal, and the state trial court moved the case to a federal district court.

BUT WHAT IF . . .

Recall that in the Case Opener, Catepillar filed a motion to exercise its right of removal to move the case to federal court. If it had not filed that motion, would the case have been held in state court, even though it fell under concurrent jurisdiction?

Venue

LO 3-2 Recall the rules related to choice of venue.

Once a case is in the proper court system, **venue** determines which trial court in the system will hear the case. Venue is a matter of geographic location that each state's statutes determine. Usually, the trial court for the county where the defendant resides is the appropriate venue. If a case involves property, the trial court where the property is located is also an appropriate venue. Finally, if the focus of the case is a particular incident, the trial court where the dispute occurred is an appropriate venue. The plaintiff initially chooses from among the appropriate venues when she files the case.

If the location of the court where the plaintiff filed the case is an inconvenience to the defendant or if the defendant believes it will be difficult to select an unbiased jury in that venue, he may

venue
(1) The place where a hearing takes place. Its geographic location is determined by each state's statutes and based on where the parties live or where the event occurred or the alleged wrong was committed. (2) A legal doctrine relating to the selection of a court with subject-matter and personal jurisdiction that is the most appropriate geographic location for the resolution of the dispute.

request that the judge move the case by filing a motion for a change of venue. The judge has the discretion to grant or deny the motion.

Forum Selection and Choice of Law

Businesses that are involved in frequent litigation realize that sometimes time and money can be wasted in arguing over which court should hear the case and which law should apply. To minimize extra costs associated with future litigation, many businesses will include in their contracts a forum-selection clause, a choice-of-law clause, or both.

In a forum-selection clause, the parties specify which court would hear a dispute over an alleged breach of the contract. While initially there was some uncertainty as to the enforceability of these clauses, the following U.S. Supreme Court case provides some guidance as to their enforceability.

CASE 3-2

ATLANTIC MARINE CONSTRUCTION COMPANY, INC. v. UNITED STATES DISTRICT COURT FOR THE WESTERN DISTRICT OF TEXAS ET AL.
United States Supreme Court
134 S. Ct. 568 (2013)

FACTS: *Atlantic Construction Company, located in Virginia, subcontracted with Texas-based J-Crew Management. Despite the contract containing a clause stating that any dispute would be litigated in Virginia, J-Crew brought suit against Atlantic in the District Court for the Western District of Texas for failure to pay J-Crew for work it performed. Atlantic moved to dismiss the suit, arguing that the forum selection clause required the suit to be brought in Virginia. The District Court denied Atlantic's motions, claiming that Atlantic did not show why it was in the interest of justice to hear the trial in Virginia. Atlantic appealed, and the court of appeals denied Atlantic's motions on the ground that a Texas court was proper venue given that the parties entered into and performed the agreement in that district. Atlantic appealed to the Supreme Court.*

ISSUE: Did the forum-selection clause give Atlantic the ability to dismiss the case?

REASONING: The ruling in this case revolved around USC §1404(a), which states:

"[f]or the convenience of parties and witnesses, in the interest of justice, a district court may transfer any civil action to any other district or division where it might have been brought or to any district or division to which all parties have consented."

According to the Court, the statute holds that a forum selection clause should be enforced with a motion to transfer rather than the motion to dismiss that Atlantic sought. The Supreme Court argued that the district and appeals courts erred in outright dismissing the case. Under the Court's own precedent set by *Stewart Organization, Inc. v Ricoh Corp.*, a case may only be dismissed when a venue is "wrong" or "improper." And even if a forum-selection clause renders a venue to be "wrong" or "improper," the clause in question in this case is to be enforced under the doctrine of *forum non conveniens*, meaning the case is to be transferred to another federal court should the current forum prove inappropriate.

To put another dent in the lower courts' decisions, the Court held that *Stewart* places upon the plaintiff the burden of establishing that transfer to the contractual forum is unwarranted. Thus, the district court erred in placing the burden of proving that the preselected forum was appropriate upon Atlantic instead of requiring J-Crew to prove that a transfer was disfavored due to public-interest factors.

DECISION: The Supreme Court reversed and remanded the court of appeals decision; the lower court was to decide whether Atlantic had grounds to *transfer* the case instead of dismissing it.

SIGNIFICANCE: This case serves as an example of a challenge to the appropriateness of the venue, and demonstrates who has the burden of proof in such a dispute.

(continued)

CRITICAL THINKING

What is the single fact that has the greatest impact on this decision? What is the logic of relying on this fact to such a large extent?

ETHICAL DECISION MAKING

Building on your answer to the critical thinking question above, identify the value that the court is leaning on to prop up its reasoning about why the courts should have acted to enforce the original agreement between the two contending parties.

BUT WHAT IF...

Facebook has its principal place of business in Palo Alto, California. During the process of signing up for a Facebook account, you fill out a lot of personal information and click "Sign Up." A line then appears that says, "By clicking Sign Up, you are indicating that you have read and agree to the Terms of Service." A hyperlink is available, and if you click on it, it takes you to the Terms, which include a provision that all disputes between Facebook and the user will be resolved in a federal or state court in Palo Alto, California. You live in New York City. If you file a claim against Facebook in the Federal District Court located in New York City and Facebook files a motion to dismiss your case or transfer it to the Federal District Court in Palo Alto, what do you think the outcome will be, and why?

SAMPLE FORUM SELECTION CLAUSE

Any lawsuit, claim or other action to enforce the terms of this Agreement shall be brought exclusively in the state courts of Nye County, Nevada. The parties to this Agreement hereby submit to the jurisdiction of the foregoing courts and waive any objection either may have to said courts asserting jurisdiction over them or their assets and property.

A choice-of-law clause is what its name implies. It is a clause inserted into the contract that specifies which law will be applied to resolve any disputes that arise under the contract. This kind of clause can be especially useful when you have parties from different states or different countries entering into a contract. While a court does not have to uphold a choice-of-law clause, normally they will uphold the clause as long as there is some connection between the choice of law and the location of the transaction or one of the parties to the dispute. For example, a Michigan farmer and a Delaware corporation that does business in several states may include a choice-of-law clause that specifies that the state law of Michigan will apply to any disputes arising under the contract. The parties might also include a choice-of-forum clause stating that any disputes will be resolved in the Federal District Court located in Wayne County, Michigan.

Exhibit 3-2
Sample Forum Selection Clause

SAMPLE CHOICE OF LAW CLAUSE

This Agreement shall be governed by and construed in accordance with the laws of the State of Ohio, as such laws are applied to contracts entered into and performed in such State.

Exhibit 3-3
Sample Choice of Law Clause

The Structure of the Court System

The U.S. legal system has two parallel court structures: a federal system and a state system. Once a plaintiff files a case in one of the systems, the case remains in that system throughout the appeals process. The only exception to this rule occurs when a party to a lawsuit appeals the decision of a state supreme court to the U.S. Supreme Court.

The Federal Court System

The federal court system derives its power from Article III, Section 2, of the U.S. Constitution and consists of three main levels: trial courts, intermediate appellate courts, and the court of last resort. Exhibit 3-4 illustrates this system.

Federal Trial Courts In the federal court system, the trial courts, or courts of original jurisdiction, are U.S. district courts. The United States has 94 districts; each district has at least one trial court of general jurisdiction. Courts of general jurisdiction have the power to hear a wide range of cases and can grant almost any type of remedy. Almost every case in the federal system begins in one of these courts.

A small number of cases, however, do not begin in trial courts of general jurisdiction. For cases concerning certain subject matter, Congress has established special trial courts of limited jurisdiction. The types of cases heard by these special trial courts include bankruptcy cases; claims against the U.S. government; and copyright, patent, and trademark cases.

In an extremely limited number of cases, the U.S. Supreme Court functions as a trial court of limited jurisdiction. These cases include controversies between states and lawsuits against foreign ambassadors.

Intermediate Courts of Appeal The U.S. circuit courts of appeal make up the second level of courts in the federal system. The United States has 11 numbered circuit courts, plus a federal circuit court federal circuit court of appeals that hears appeals from government administrative agencies. Each numbered circuit court hears appeals from district courts in its geographic area. Exhibit 3-5 illustrates the geographic circuit divisions.

Exhibit 3-4 The Federal Court System

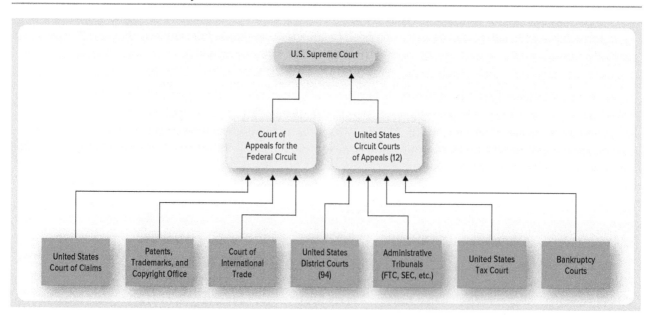

Exhibit 3-5 The Circuits of the Federal Court System

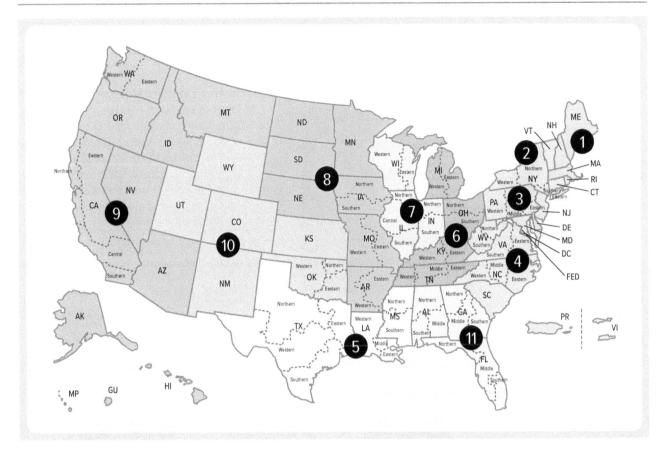

The Court of Last Resort The U.S. Supreme Court is the final appellate court in the federal system. Nine justices, who have lifetime appointments, make up the high court. Exhibit 3-6 identifies the nine justices on the U.S. Supreme Court.

The U.S. Supreme Court hears appeals of cases from the court of last resort in a state system. The Court will not, however, hear cases considering questions of pure state law. The Court also functions as a trial court on rare occasions.

State Court Systems

No uniform state court structure exists because each state has devised its own court system, but most states have a structure similar to the federal court system's structure.

State Trial Courts In state court systems, most cases begin in a trial court of general jurisdiction. As in the federal system, state trial courts of general jurisdiction have the power to hear all cases over which the state court system has jurisdiction except those cases for which the state has established special trial courts of limited jurisdiction. Most states have a trial court of general jurisdiction in each county. The names of these courts vary by state, but most states refer to them as *courts of common pleas* or *county courts.* In some states, these courts have specialized divisions such as domestic relations and probate.

Most states also have trial courts of limited jurisdiction. Usually, these courts can grant only certain remedies. For example, small claims courts, a common type of court of limited jurisdiction in most states, may not grant damage awards larger than a specified amount. Other courts of limited jurisdiction have the power to hear only certain types of cases. For example, probate courts hear only cases about asset and obligation transfers after an individual's death.

GLOBAL Context

©Digital Archive Japan/Alamy

The Court Structure in England

Even though we trace the roots of the U.S. legal system back to England, changes have occurred in both countries' court structures; as a result, the court structures in the United States and England share similarities but also have distinct features. The lowest criminal courts in England are the magistrates courts, which hear minor offenses. More serious cases are tried before a judge and jury in the crown court, which also hears cases appealed from the magistrates courts on factual points. The high court (in the Queen's Branch Division) hears appeals on points of law, and the court of appeal in the Criminal Division hears appeals on sentences and convictions.

Civil cases are first heard in the county courts (for minor claims) or the high court, which is divided into three divisions: Queen's Bench, Family, and Chancery. Cases may be appealed to the court of appeal (Civil Division). Cases may also be appealed from the county court to the high court.

The House of Lords is the supreme court of appeal. Its cases are heard by up to 13 senior judges known as *law lords*. In addition to the courts, there are specialized tribunals, which hear appeals on decisions made by various public bodies and government departments in areas such as employment, immigration, social security, tax, and land.

Exhibit 3-6

U.S. Supreme Court Justices

Source: Steven Petteway, Collection of the Supreme Court of the United States.

Source: Franz Jantzen, Collection of the Supreme Court of the United States.

Source: Fred Schilling, Collection of the Supreme Court of the United States.

Source: Steven Petteway, Collection of the Supreme Court of the United States.

Source: Steven Petteway, Collection of the Supreme Court of the United States.

Source: Steven Petteway, Collection of the Supreme Court of the United States.

Source: Steven Petteway, Collection of the Supreme Court of the United States.

Source: Steven Petteway, Collection of the Supreme Court of the United States.

Source: Steven Petteway, Collection of the Supreme Court of the United States.

Associate Justice

Elena Kagan
Appointed in 2010 by President Obama

Associate Justice

Neil Gorsuch
Appointed in 2017 by President Trump

Associate Justice

Brett Kavanaugh
Appointed in 2018 by President Trump

Associate Justice

Sonia Sotomayor
Appointed in 2009 by President Obama

Chief Justice

John G. Roberts
Appointed in 2005 by President G.W. Bush

Associate Justice

Clarence Thomas
Appointed in 1991 by President G.H.W. Bush

Associate Justice

Ruth Bader Ginsburg
Appointed in 1993 by President Clinton

Associate Justice

Stephen G. Breyer
Appointed in 1994 by President Clinton

Associate Justice

Samuel Alito
Appointed in 2006 by President G.W. Bush

Intermediate Courts of Appeal Intermediate courts of appeal, analogous to federal circuit courts of appeal, exist in approximately half the states. These courts usually have broad jurisdiction, hearing appeals from courts of general and limited jurisdictions as well as from state administrative agencies. The names of these courts vary by state, but most states call them *courts of appeal* or *superior courts.*

Courts of Last Resort Appeals from the state intermediate courts of appeal lead cases to the state court of last resort. Most states call this court the *supreme court,* although some states refer to it as the *court of appeals.* Because approximately half of the states lack intermediate courts of appeal, appeals from trial courts in these states go directly to the state court of last resort.

Threshold Requirements

LO 3-3 Identify the threshold requirements that must be met before a court will hear a case.

Before a case makes it to court, it must meet three *threshold requirements.* These requirements ensure that courts hear only cases that genuinely require adjudication. The three requirements are standing, case or controversy, and ripeness (see Exhibit 3-7).

Standing

A person who has the legal right to bring an action in court has **standing** (or *standing to sue*). For a person to have standing, the outcome of a case must personally affect him or her. Thus, if you hire a landscaper to mow your lawn every week and he fails to show up every other week, you have standing to sue your landscaper. But if your friend hired the landscaper, you lack the standing to sue on your friend's behalf because you do not have a personal stake in the outcome of the case.

The reason the American legal system requires this personal stake in the outcome is the belief that the plaintiff's personal stake stimulates her to present the best possible case. The requirements for standing, according to the 2000 U.S. Supreme Court case *Friends of the Earth v. Laidlaw Environmental Services,*[4] are (1) the plaintiff must have an injury in fact that is concrete and actual or imminent; (2) the injury must be fairly traceable to the challenged action of the defendant; and

standing
The legal right of a party or an individual to bring a lawsuit by demonstrating to the court sufficient connection to and harm from the law or action challenged (i.e., the plaintiff has to demonstrate that he or she is harmed or will be harmed). Otherwise, the court will dismiss the case, ruling that the plaintiff "lacks standing" to bring the suit.

Exhibit 3-7
Threshold Requirements

[4] 120 S. Ct. 923 (2000).

(3) it must be likely that the injury will be redressed by a favorable decision.[5] In applying those criteria to the Laidlaw case, the Supreme Court found that FOE members' testimony that they were afraid to fish and swim in a river they previously enjoyed satisfied the first two criteria. The Court held that although the FOE members would not directly receive money from a penalty against Laidlaw, they would benefit because the penalties would deter Laidlaw and other companies from polluting the river in the future.[6] The Court ruled in FOE's favor and assessed Laidlaw a $405,800 penalty payable to the U.S. Treasury.

Case or Controversy

case or controversy
A term used in the U.S. Constitution to describe the structure and requirements of conflicting claims of individuals that can be brought before a federal court for resolution. A case or controversy requires an actual dispute between parties over their legal rights that remains in conflict at the time the case is presented and that is a proper matter for judicial determination. Also referred to as *justifiable controversy*.

The **case-or-controversy** (or *justifiable controversy*) requirement ensures that courts do not render advisory opinions. Three criteria are necessary for a case or controversy to exist. First, the relationship between the plaintiff and the defendant must be adverse. Second, actual or threatened actions of at least one of the parties must give rise to an actual legal dispute. Third, courts must have the ability to render a decision that will resolve the dispute. In other words, courts can give final judgments that solve existing problems; they cannot provide rulings about hypothetical situations.

Ripeness

ripeness
The readiness of a case for a decision to be made. The goal is to prevent premature litigation for a dispute that is insufficiently developed. A claim is not ripe for litigation if it rests on contingent future events that may not occur as anticipated or may not occur at all.

The case-or-controversy requirement is closely linked to the **ripeness** requirement. A case is *ripe* if a judge's decision is capable of affecting the parties immediately. Usually the issue of ripeness arises when one party claims that the case is moot—in other words, there is no point in the court's hearing the case because no judgment can affect the situation between the parties.

In the *Laidlaw* case, Laidlaw argued that the case was moot because by the time the case went to trial, the company had complied with the requirements of its discharge permits. The Supreme Court disagreed, ruling that the fact that a defendant voluntarily ceases a practice once litigation has commenced does not deprive a federal court of its power to determine the legality of the practice, because such a ruling would leave the defendant free to return to his old unlawful practices. Thus, the Court found the case was not moot because imposing a penalty on the defendant would have an important deterrent effect.[7]

Steps in Civil Litigation

LO 3-4 Recognize the steps in civil litigation.

The U.S. litigation system is an adversary system: A neutral fact finder—a judge or jury—hears evidence and arguments that opposing sides present and then decides the case on the basis of the facts and law. Strict rules govern the types of evidence fact finders may consider. Theoretically, fact finders make informed and impartial rulings because each party has an incentive to find all relevant evidence and make the strongest possible arguments on behalf of her or his position.

Critics of the adversary system, however, point out several drawbacks: the time and expense each lawsuit requires, the damage a suit may cause to the litigating parties' relationship, and the unfair advantage to those with wealth and experience using the court system.

The Pretrial Stage

The rules of civil procedure govern civil case proceedings. The Federal Rules of Civil Procedure apply in all federal courts. Each state has its own set of rules, but most states' rules are very similar to the Federal Rules of Civil Procedure. In addition, each court usually has its own set of local court rules.

Informal Negotiations The initial attempt to resolve a business dispute is usually informal: a discussion or negotiation among the parties to try to find a solution. If the parties are unable to

[5] Ibid.
[6] Ibid.
[7] Ibid.

resolve their dispute, one party often seeks an attorney's advice. Together, the attorney and client may be able to resolve the dispute informally with the other party.

Pleadings The first formal stage of a lawsuit is the *pleading stage*. The plaintiff's attorney initiates a lawsuit by filing a *complaint* in the appropriate court. The complaint states the names of the parties to the action, the basis for the court's subject-matter jurisdiction, the facts on which the plaintiff bases her claim, and the relief the plaintiff seeks. The pleadings prevent surprises at trial; they allow attorneys to prepare arguments to counter the other side's claims. Exhibit 3-8 shows a typical complaint.

Service of Process To obtain *in personam* jurisdiction over a defendant and to satisfy due process, a court must notify the defendant of the pending lawsuit. Service of process occurs when a representative of the court serves (delivers) a copy of the complaint and a summons to the defendant.

The complaint explains the basis of the lawsuit to the defendant. The summons tells the defendant that if he or she does not respond to the lawsuit within a certain period of time, the plaintiff will receive a default judgment. A default judgment is a judgment in favor of the plaintiff that occurs when the defendant fails to answer the complaint and the plaintiff's complaint alleges facts that would support such a judgment.

Defendant's Response The defendant typically responds to the complaint with an answer. In this document, the defendant denies, affirms, or claims no knowledge of the accuracy of the plaintiff's allegations. However, in some cases, when the defendant believes that the even if the facts the plaintiffs alleges are true, the plaintiff would not have proven a case against the defendant, the defendant would file a motion to dismiss. For example, if a plaintiff filed a breach of contract case against the defendant but failed to allege that their was a contract between the two of them, the defendant could file a motion to dismiss.

A defendant uses an *affirmative defense* when his or her answer admits that the facts contained in the complaint are accurate but also includes additional facts that justify the defendant's actions and provide a legally sound reason to deny relief to the plaintiff. For example, if a woman sued a man for battery because he punched her in the face, he might claim that he hit her only because she aimed a gun at him and threatened to shoot. His claim that he was acting in self-defense is an affirmative defense.

If the defendant plans to raise an affirmative defense, he must raise it in his answer to give the plaintiff adequate notice. If he fails to raise an affirmative defense in the answer, the judge will likely not allow him to raise it during the trial.

Upon receiving the complaint, if the defendant believes that even though all the plaintiff's factual allegations are true, the law does not entitle the plaintiff to a favorable judgment, the defendant may file a motion to dismiss, or *demurrer*. (A motion is a request by a party for the court to do something; in this instance, the request is to dismiss the case.) In deciding whether to grant a motion to dismiss, a judge accepts the facts as stated by the plaintiff and rules on the legal issues in the case. Judges generally grant motions to dismiss only when it appears beyond a doubt that the plaintiff cannot prove any set of facts to justify granting the judgment she seeks.

If the defendant believes he has a claim against the plaintiff, he includes this counterclaim with the answer. As Exhibit 3-9 shows, the form of a counterclaim is identical to the form of a complaint. The defendant states the facts supporting his claim and asks for relief.

If the defendant files a counterclaim, the plaintiff generally files a reply. A reply is an answer to a counterclaim. In the reply, the plaintiff admits, denies, or claims a lack of knowledge of the accuracy of the facts of the defendant's counterclaim. If the plaintiff plans to use an affirmative defense, she must raise it in the reply.

Pretrial Motions The early pleadings establish the legal and factual issues of the case. After the pleadings, the plaintiff or defendant may file a motion to conclude the case early, eliminate some claims, or gain some advantage. A party may move, or request, the court to do almost anything pertaining to the case. For example, if the plaintiff files a suit about the right to a piece of

service of process
The procedure by which a court delivers a copy of the statement of claim or other legal documents, such as a summons, complaint, or subpoena, to the defendant.

default judgment
Judgment for the plaintiff that occurs when the defendant fails to respond to the complaint.

answer
The response of the defendant to the plaintiff's complaint.

motion to dismiss
A request by the defendant that asks a judge or a court in a civil case to dismiss the case because even if all the allegations are true, the plaintiff is not entitled to any legal relief. Also called *demurrer*.

motion
An application by a party to a judge or a court in a civil case requesting an order in favor of the applicant.

counterclaim
A claim made by the defendant against the plaintiff that is filed along with the defendant's answer.

reply
A response by the plaintiff to the defendant's counterclaim.

Exhibit 3-8

A Typical Complaint

THE COURT OF COMMON PLEAS OF CLARK COUNTY, NEVADA

Bob Lyons and Sue Lyons, Plaintiffs

v.

Christine Collins, Defendant

COMPLAINT FOR NEGLIGENCE

Case No. _____

Now come the plaintiffs, Bob Lyons and Sue Lyons, and, for their complaint, allege as follows:

1. Plaintiffs, Bob Lyons and Sue Lyons, both of 825 Havercamp Street, are citizens of Clark County, in the state of Nevada, and defendant, Christine Collins, 947 Rainbow Ave., is a citizen of Clark County in the state of Nevada.
2. On May 1, 2001, the defendant built a wooden hanging bridge across a stream that runs through the plaintiffs' property at 825 Havercamp Street.
3. Defendant negligently used ropes in the construction of the bridge that were not thick enough to sustain human traffic on the bridge.
4. At approximately 4:00 p.m., on May 20, 2001, the plaintiffs were attempting to carry a box of landscaping stones across the bridge when the ropes broke, and the bridge collapsed, causing plaintiffs to fall seven feet into the stream.
5. As a result of the fall, plaintiff, Bob Lyons, suffered a broken arm, a broken leg, and a skull fracture, incurring $160,000 in medical expenses.
6. As a result of the fall, plaintiff, Sue Lyons, suffered two broken cervical vertebrae, and a skull fracture, incurring $300,000 in medical expenses.
7. As a result of the fall, the landscaping stones, which had cost $1,200, were destroyed.
8. As a result of the foregoing injuries, plaintiff, Bob Lyons, was required to miss eight weeks of work, resulting in a loss of $2,400 in wages.
9. As a result of the foregoing injuries, plaintiff, Sue Lyons, was required to miss 12 weeks of work, resulting in a loss of $3,600 in wages.

WHEREFORE, plaintiffs demand judgment in the amount of $467,200, plus costs of this action.

Harlon Elliot

Attorney for plaintiffs

824 Sahara Ave.

Las Vegas, Nevada, 89117

JURY DEMAND

Plaintiffs demand a trial by jury in this matter.

property, she may move that the court prohibit the current possessor of the land from selling it. Courts may grant or deny such motions at their discretion.

When a party files a motion with the court, the court sends a copy to the opposing attorney, who may respond to the motion, usually by requesting that the judge deny the motion. In many cases, the judge rules on the motion immediately. In other cases, the judge holds a hearing at which the attorneys for both sides argue how the judge should decide the motion.

Two primary pretrial motions are a motion for judgment on the pleadings and a motion for summary judgment. Once the parties file the pleadings, either party can file a **motion for judgment on the pleadings**. The motion is a request for the court to consider that all the facts in the pleadings are true and to apply the law to those facts. The court grants the motion if, after this process, it finds that the only reasonable decision is in favor of the moving party.

motion for judgment on the pleadings
In a civil case, a request made by either party, after pleadings have been entered, that asks a judge or a court to issue a judgment.

> **Exhibit 3-9**
>
> Defendant's Answer and Counterclaim

THE COURT OF COMMON PLEAS OF CLARK COUNTY, NEVADA

Bob Lyons and Sue Lyons, Plaintiffs v. Christine Collins, Defendant
ANSWER AND COUNTERCLAIM FOR BREACH OF CONTRACT
Case No. _____

Now comes the defendant, Christine Collins, and answers the complaint of plaintiffs herein as follows:

FIRST DEFENSE

1. Admits the allegations in paragraphs 1 and 2.
2. Denies the allegation in paragraph 3.
3. Is without knowledge as to the truth or falsity of the allegations contained in paragraphs 4, 5, 6, 7, 8, and 9.

SECOND DEFENSE

4. If the court believes the allegations contained in paragraph 3, which the defendant expressly denies, plaintiffs should still be denied recovery because they were informed prior to the construction of the bridge that there should be no more than one person on the bridge at one time and that no individual weighing more than 200 pounds should be allowed to walk on the bridge.

COUNTERCLAIM

5. On April 15, the parties agreed that the defendant would build a wooden hanging bridge across a stream that runs through the defendant's property at 825 Havercamp Street, in exchange for which plaintiffs would pay defendant $2,000 upon completion of construction.
6. On May 1, 2001, the defendant built the agreed upon ornament, wooden, hanging bridge across a stream that runs through the defendant's property at 825 Havercamp Street, but plaintiffs failed to pay the agreed upon price for the bridge.
7. By their failure to pay, plaintiffs breached their contract and are liable to defendant for the contract price of $2,000.

WHEREFORE, defendant prays for a judgment dismissing the plaintiffs' complaint and granting the defendant a judgment against plaintiffs in the amount of $2,000 plus costs of this action.

———————

Melissa Davenport
Attorney for Defendant

777 Decatur Ave.
Las Vegas, Nevada 89117

discovery
The pretrial phase in a lawsuit during which each party requests relevant documents and other evidence from the other side in an attempt to "discover" pertinent facts and avoid any surprises in the courtroom during the trial. Discovery tools include requests for admissions, interrogatories, depositions, requests for inspection, and document production requests.

interrogatories
A formal set of written questions that one party to a lawsuit asks the opposing party as part of the pretrial discovery process in order to clarify matters of evidence and help determine in advance what facts will be presented at any trial in the case. The questions must be answered in writing under oath or under penalty of perjury within a specified time. Also called *requests for further information*.

request to produce documents
In a lawsuit, the right of a party to examine and copy papers of the opposing party that are relevant to the case. A legal request may be made, and the categories of the documents must be stated to allow the other party to know what documents he or she must produce.

Discovery After filing the initial pleadings and motions, the parties gather information from each other through **discovery**. The discovery process enables the parties to learn about facts surrounding the case so that they are not surprised in the courtroom. Three common discovery tools are interrogatories, requests to produce documents, and depositions.

Interrogatories are written questions that one party sends to the other to answer under oath. Frequently, a *request to admit certain facts* accompanies interrogatories. Attorneys work with their clients to answer interrogatories and requested admissions of facts. Sometimes interrogatories are called *requests for further information*.

A **request to produce documents** (or other items) forces the opposing party to produce (turn over) certain information unless it is privileged or irrelevant to the case. Parties may request documents such as photographs, contracts, written estimates, medical records, tax forms, and other government documents. In tort cases, the defendant frequently asks the plaintiff to submit a mental- or physical-examination report.

deposition
A pretrial sworn and recorded testimony of a witness that is acquired out of court with no judge present.

motion for summary judgment
In a civil case, a request made by either party that asks a judge or a court to promptly and expeditiously dispose of a case without a trial. Any evidence or information that would be admissible at trial under the rules of evidence, such as affidavits, interrogatories, depositions, and admissions, may be considered on a motion for summary judgment. A court may hold oral arguments or decide the motion on the basis of the parties' briefs and supporting documentation alone.

pretrial conference
An informal meeting of the judge with the attorneys representing the parties before the actual trial begins.

Finally, the parties may obtain testimony from a witness before trial through a **deposition**. At a deposition, attorneys examine a witness under oath. A court reporter (stenographer) records every word the witness and attorneys speak. Both parties receive a copy of the testimony in document form. Depositions provide information and may also set up inconsistencies between a witness' testimony at the deposition and his testimony at trial. If a party discovers an inconsistency in the testimony of one of the other party's witnesses, she can bring the inconsistency to the fact finder's attention to diminish the witness' credibility. The parties may also use depositions when a witness is elderly, moving, or ill and thus may be unavailable at the time of the trial.

If a party does not comply with requests for discovery, the court may admit the facts the other party sought to discover. Attorneys who feel that certain material is outside the scope of the case often argue that the material is irrelevant to the case. If the court disagrees, however, the party must supply the requested information. Although these discovery tools are important in the United States, many countries do not have a discovery process.

After the discovery process, either party can file a **motion for summary judgment**. The motion asserts that no factual disputes exist and that if the judge applied the law to the undisputed facts, her only reasonable decision would be in favor of the moving party. The difference between this motion and a motion for judgment on the pleadings is that in a motion for summary judgment, the moving party may use affidavits (sworn statements from the parties or witnesses), relevant documents, and depositions or interrogatories (a party's sworn answers to written questions) to support his motion. The judge grants the motion if, after examining the evidence, she finds no factual disputes. If, however, she finds any factual issues about which the parties disagree, she denies the motion and sends the case to trial.

Pretrial Conference A pretrial conference precedes the trial. A **pretrial conference** is an informal meeting of the judge with the attorneys representing the parties. During this conference, the parties try to narrow the legal and factual issues and possibly work out a settlement. If the parties cannot reach a settlement, the attorneys and the judge discuss the administrative details of the trial: its length, witnesses, and any pretrial stipulations of fact or law to which the parties agree.

The Trial

If a plaintiff seeks at least $20 in monetary damages, the Seventh Amendment to the U.S. Constitution entitles the parties to a jury trial. The plaintiff must, however, demand a jury trial in her or his complaint. Some types of trials require 12-person juries, but 6-person juries hear most civil cases. If the plaintiff seeks an equitable remedy (an injunction or other court order) or if the parties have waived their right to a jury, a judge serves as the fact finder in the case.

Trials have six stages: jury selection, opening statements, examination of witnesses, closing arguments, conference on jury instructions, and posttrial motions. The following sections describe these stages.

Jury Selection The jury selection process begins when the clerk of the courts randomly selects a number of potential jurors from the citizens within the court's jurisdiction. Once the potential jurors have reported for jury duty, the **voir dire**, or jury selection, process begins. The voir dire process selects the jurors who will decide the case, as well as two or three alternate jurors who will watch the trial and be available to replace any juror who, for some legitimate reason, must leave jury duty before the trial ends.

voir dire
The process of questioning potential jurors to ensure that the jury will be made up of unbiased individuals.

peremptory challenge
In a jury trial, the right of the plaintiff and the defendant in jury selection to reject, without stating a reason, a certain number of potential jurors who appear to have an unfavorable bias.

During voir dire, the judge and/or attorneys question potential jurors to determine whether they are able to render an unbiased opinion in the case. If a potential juror's response to a question indicates that she or he may be biased, either attorney may challenge, or ask the court to remove, that potential juror "for cause." For example, a lawyer could challenge for cause a potential juror who was a college roommate of the defendant. In most states, each party has a certain number of **peremptory challenges**. These peremptory challenges allow a party to challenge a certain number of potential jurors without giving a reason.

Peremptory challenges, however, may lead to abuse. For example, in the past, attorneys have used peremptory challenges to eliminate a certain class, ethnic group, or gender from the jury. In the 1986 case, *Batson v. Kentucky*,[8] the U.S. Supreme Court ruled that race-based peremptory challenges

[8] 476 U.S. 79 (1986).

in criminal cases violate the equal protection clause of the Fourteenth Amendment to the U.S. Constitution. (Chapter 5 discusses the amendments to the Constitution in more detail.) The Supreme Court later extended the ban on race-based challenges to civil cases.[9] Case 3-3 illustrates the successful use of a Batson challenge to a discriminatory use of peremptory challenges. Remember that while this illustrative case is a criminal case, the principle applies the same way to civil cases.

[9] J.E.B V. ALABAMA, EX. REL. T.B. UNITED STATES SUPREME COURT.

CASE 3-3

TIMOTHY TYRONE FOSTER, PETITIONER v. BRUCE CHATMAN, WARDEN
United States Supreme Court
P136 S. Ct. 1737 (2016)

FACTS: *Timothy Foster was convicted of capital murder and sentenced to death in Georgia. During jury selection, the prosecutors used their peremptory challenges to prevent all four prospective black jurors from serving on the jury. Foster raised a Batson challenge, arguing that prosecutors' use of those strikes were racially motivated. The state habeas court denied relief and the Georgia Supreme Court denied Foster the Certificate of Probable Cause required to file an appeal. Foster's only recourse was to file petition for a writ of certiorari seeking review of the denial. The United States Supreme Court granted certiorari.*

ISSUE: Was there prohibited, race-based discrimination in the jury selection process of Foster's trial?

REASONING: To understand the reasoning of this case, one must first be aware of *Batson v. Kentucky* 106 S. Ct. 1712 (2016). In that case, the Supreme Court ruled that jurors cannot be preemptively dismissed (i.e., not asked to serve jury duty) based solely on their race. In *Foster v. Chatman*, the argument before the Court was whether *Batson* was violated in Foster's trial.

Batson set up three criteria for the defense to show that a violation has occurred. First, the defense must show that the prosecution used preemptory challenges discriminatorily (often by showing that less-qualified jurors were admitted before more-qualified jurors of a particular race). Second, the prosecutor must provide plausible, race-neutral reasons for why they made the challenged dismissals. Third, the defense must show that the prosecutor intentionally discriminated in the jury-selection process.

In *Foster,* the Supreme Court found that the third step of *Batson* was met. Although the prosecutor provided a long list of reasons as to why the black jurors were struck, evidence shows that the prosecution had put the black jurors on a "definite NOs" list early on in the jury selection process. On top of this fact, several white jurors with similar traits, and even some who were less qualified to serve, were admitted, and the prosecution's reasons for striking the black jurors changed over time. Thus, it was clear that the prosecution had violated *Batson* in its juror selection process.

DECISION: The Supreme Court reversed and remanded for further proceedings.

SIGNIFICANCE: This case demonstrates how *Batson* is applied in the juror selection process as well as how vital the juror selection process is to the outcome of a case.

CRITICAL THINKING

The Supreme Court in *Batson v. Kentucky* provided a three-step process for determining when a strike is discriminatory: "First, a defendant must make a prima facie showing that a peremptory challenge has been exercised on the basis of race; second, if that showing has been made, the prosecution must offer a race-neutral basis for striking the juror in question; and third, in light of the parties' submissions, the trial court must determine whether the defendant has shown purposeful discrimination." What possible problems might arise that prevent a court from rightfully determining a strike is discriminatory as happened in this case? What changes would you make to the three-step process to prevent further transgressions?

ETHICAL DECISION MAKING

Which values does this decision tend to emphasize?

Despite *Batson,* however, it is still not easy to challenge a dismissal as being based on race or gender. And in a 2006 case, the U.S. Supreme Court said that to find a dismissal unconstitutional, there must be no alternative to a gender- or race-based reason for dismissing a juror.[10]

The voir dire process has become more sophisticated over time. In cases involving significant amounts of money, rather than relying on their instinct or experience during jury selection, attorneys use professional jury selection services to identify demographic data to help select ideal jurors. Jury selection firms also provide additional services, including **mock trials,** in which mock jurors who are demographically similar to the real jurors hear the arguments and give feedback prior to the case, and **shadow juries,** in which shadow jurors sit in during the actual trial and deliberate at the end of each day to evaluate for the attorneys how each side is doing. Critics argue that jury selection services give an unfair advantage to one side when only one party can afford these services.

Opening Statements Once the attorneys have impaneled, or selected, a jury, the case begins with opening statements. Each party's attorney explains to the judge and jury what facts he or she intends to prove, the legal conclusions to which these facts lead, and how the fact finder should decide the case based on those facts.

The Examination of Witnesses and Presentation of Evidence Following opening statements, the plaintiff and defendant, in turn, present their cases-in-chief by examining witnesses and presenting evidence. The plaintiff has the burden of proving the case, meaning that if neither side presents a convincing case, the fact finder must rule in favor of the defendant. Thus, the plaintiff presents her case first.

The procedure for each witness is the same. First, the plaintiff's attorney questions the witness in *direct examination.* The plaintiff's attorney asks the witness questions to elicit facts that support the plaintiff's case-in-chief. Questions must relate to matters about which the witness has direct knowledge. Attorneys cannot elicit hearsay from the witnesses. *Hearsay* is testimony about what a witness heard another person say. Hearsay is impermissible because the opposing attorney cannot question the person who made the original statement to determine the statement's veracity.

The federal rules of evidence also prohibit attorneys from asking leading questions. Leading questions are questions that imply a specific answer. For example, an attorney cannot ask a witness, "Did the defendant come to your office and ask you to purchase stock from him?" Instead, attorneys must ask questions such as, "When did you first encounter the defendant?"

After direct examination, opposing counsel may *cross-examine* the witness. Opposing counsel, however, may ask only questions related to the witness' direct examination. On cross-examination, attorneys can ask leading questions. Attorneys try to show inconsistencies in the witness' testimony, cast doubt on the claims of the plaintiff's case, and elicit information to support the defendant's case.

After cross-examination, the plaintiff's attorney may conduct a *redirect examination,* a series of questions aimed at repairing damage done by the cross-examination. At the judge's discretion, opposing counsel has an opportunity to *re-cross* the witness to question her testimony on redirect examination. The parties follow this procedure for each of the plaintiff's witnesses.

Immediately following the plaintiff's presentation of her case, the defendant may move for a **directed verdict.** This motion is a request for the court to direct a verdict for the defendant because even if the jury accepted all the evidence and testimony presented by the plaintiff as true, the jury would still have no legal basis for a decision in favor of the plaintiff. The federal court system refers to a motion for a directed verdict as a *motion for a judgment as a matter of law.* Courts rarely grant motions for a directed verdict because plaintiffs almost always present at least *some* evidence to support each element of the cause of action.

If the court denies the defendant's motion for a directed verdict, the defendant then presents his case. The parties question the defendant's witnesses in the same manner as they questioned the plaintiff's witnesses, except that the defendant's attorney conducts direct and redirect examination and the plaintiff's attorney conducts cross-examination and re-cross-examination.

[10] *Rice, Warden et al. v. Collins,* 126 S. Ct. 969 (2006).

GLOBAL Context

©Digital Archive Japan/Alamy

Trials in Japan

Civil procedure in Japan differs significantly from American civil procedure. The Japanese legal system has no juries and no distinct pretrial stage. Instead, a trial is a series of discrete meetings between the parties and the judge. At the first meeting, the parties identify the most critical and contested issues. They choose one and recess to gather evidence and marshal arguments on the issue.

At the next meeting, the judge rules on the chosen issue. If the judge decides against the plaintiff, the case is over. If the plaintiff wins, the process continues with the next issue. The process continues until the plaintiff loses an issue or until the judge decides all issues in the plaintiff's favor, resulting in a verdict for the plaintiff.

In addition, the discovery process in the Japanese court system is not as simple as it is in the United States. To obtain evidence, parties must convince the judge to order others to testify or produce documents. The judge can fine or jail parties who refuse to comply with such orders. Additionally, if a party does not comply with the judge's requests for discovery, the judge may admit the facts the other party sought to discover.

Closing Arguments After the defendant's case, the attorneys present closing arguments. In closing arguments, each attorney summarizes evidence from the trial in a manner consistent with his or her client's case. The plaintiff's attorney presents her closing argument first, followed by the defendant's attorney, and the plaintiff has the option to present a rebuttal of the defendant's closing argument.

Jury Instructions In a jury trial, the judge "charges the jury" by instructing the jurors on how the law applies to the facts of the case. Both sides' attorneys submit statements to the judge explaining how they believe she should charge the jury. The judge's instructions are usually a combination of both sides' suggestions.

After the judge charges the jury, the jurors retire to the jury room to deliberate. Once they reach a decision, they return to the courtroom, where the judge reads their verdict and discharges them from their duty. Trial procedures in the United States are quite different from trial procedures in other countries, as the Global Context box illustrates.

Posttrial Motions

Once the trial ends, the party who received the favorable verdict files a *motion for a judgment in accordance with the verdict.* Until the judge enters the judgment, the court has not issued a legally binding decision for the case.

The party who loses at trial has a number of available options. One option is to file a *motion for a judgment notwithstanding the verdict,* or *judgment non obstante verdicto,* asking the judge to issue a judgment contrary to the jury's verdict. To grant the motion, the judge must find that, when viewing the evidence in the light most favorable to the nonmoving party, a reasonable jury could not have found in favor of that party. In other words, as a matter of law, the judge must determine that the trial did not produce sufficient evidence to support the jury's verdict. This motion is similar to a motion for a directed verdict, except the parties cannot make this motion until *after* the jury issues a verdict. The federal court system refers to this motion as a *motion for judgment as a matter of law.*

The losing party can also file a *motion for a new trial.* Judges grant motions for a new trial only if they believe the jury's decision was clearly erroneous but they are not sure that the other side should necessarily have won the case. A judge often grants a motion for a new trial when the parties discover new evidence, when the judge made an erroneous ruling, or when misconduct during the trial may have prevented the jury from reaching a fair decision.

Appellate Procedure

Either party may appeal the judge's decision on posttrial motions, or on her or his final judgment. Sometimes, both parties appeal the same decision. For example, if a jury awarded the

plaintiff $10,000 in damages, the plaintiff and the defendant may both appeal the amount of the judgment. Appellate courts, however, reverse only about 1 out of every 10 trial court decisions on appeal.

To be eligible for appeal, the losing party must argue that a *prejudicial error of law* occurred during the trial. A **prejudicial error** is a mistake so significant that it likely affected the outcome of the case. For example, a prejudicial error could occur if the judge improperly allowed hearsay evidence that enabled the plaintiff to prove an element of her case.

To appeal a case, the attorney for the appealing party (the appellant) files a notice of appeal with the clerk of the trial court within a prescribed time. The clerk then forwards the record of appeal to the appeals court. The record of appeal typically contains a number of items: the pleadings, a trial transcript, copies of the trial exhibits, copies of the judge's rulings on the parties' motions, the attorneys' arguments, jury instructions, the jury's verdict, posttrial motions, and the judgment order.

The appellant then files a **brief**, or written argument, with the court. Appellants file briefs to explain why the judgment in the lower court was erroneous and why the appeals court should reverse it. The attorney for the party who won in the lower court (the appellee) files an answering brief. The appellant may then file a reply brief in response to the appellee's brief. Generally, however, appellants do not file reply briefs.

The appeals court then usually allows the attorneys to present oral arguments before the court. The court considers these arguments, reviews the record of the case, and renders a decision.

There are four possible decisions an appellate court can render. The court can accept the lower court's judgment by **affirming** its decision. Alternatively, the appellate court may conclude that the lower court's decision was correct but the remedy was inappropriate, in which case it **modifies** the remedy. If the appellate court decides that the lower court was incorrect, it **reverses** the lower court's decision. Finally, if the appeals court thinks the lower court committed an error but does not know how that error affected the outcome of the case, it **remands** the case to the lower court for a new trial.

An appellate court usually has a bench with at least three judges. Appellate courts do not have juries; rather, the judges decide the case by majority vote. One of the judges who votes with the majority records the court's decision and its reasons in the *majority opinion*. These decisions have precedential value—that is, judges use these prior appellate court decisions to make decisions in future cases. Also, these decisions establish new guidelines in the law that all citizens must follow. If a judge agrees with the majority's decision but for different reasons, she may write a *concurring opinion,* stating the reasons she used to reach the majority's conclusion. Finally, judges disagreeing with the majority may write a *dissenting opinion,* giving their reasons for reaching a contrary conclusion. Attorneys arguing that a court should change the law frequently cite dissenting opinions from previous cases in their briefs. Likewise, appellate judges who change the law often cite dissenting opinions from past cases.

For most cases, only one appeal is available. In states with both an intermediate and a final court of appeals, a losing party may appeal from the intermediate appellate court to the state supreme court. In a limited number of cases, the losing party can appeal the decision of a state supreme court or a federal circuit court of appeals to the U.S. Supreme Court.

Appeal to the U.S. Supreme Court Every year thousands of individuals file appeals with the U.S. Supreme Court, but the Court hears, on average, only 80 to 90 cases each year. To file an appeal to the U.S. Supreme Court, a party files a petition asking the Court to issue a **writ of certiorari,** an order to the lower court to send to the Supreme Court the record of the case. The Court issues very few writs.

The justices review petitions and issue a writ only when at least four justices vote to hear the case (the *rule of four).* The court is most likely to issue a writ in four instances: (1) The case presents a substantial federal question that the Supreme Court has not yet addressed; (2) multiple circuit courts of appeal have decided the issue of the case in different ways; (3) a state court of last resort has ruled that a federal law is invalid or has upheld a state law that may violate federal law; or (4) a federal court has ruled that an act of Congress is unconstitutional. If the Supreme Court does not issue a writ of certiorari, the lower court's decision stands.

For a quick overview of the steps of a civil lawsuit, see Exhibit 3-10.

prejudicial error
An error of law that is so significant that it affects the outcome of the case.

brief
A written legal argument, which a party presents to a court, that explains why that party to the case should prevail.

affirm
An appellate court decision that accepts a lower court's judgment in a case that has been appealed.

modify
An appellate court decision that grants an alternative remedy in a case; rendered when the court finds that the decision of the lower court was correct but the remedy was not.

reverse
An appellate court decision that overturns the judgment of a lower court, concluding that the lower court was incorrect and its verdict cannot be allowed to stand.

remand
An appellate court decision that returns a case to the trial court for a new trial or for limited hearing on a specified subject matter; rendered when the court decides that an error was committed that may have affected the outcome of a case.

writ of certiorari
A Supreme Court order, issued after the Court decides to hear an appeal, mandating that the lower court send to the Supreme Court the record of the appealed case.

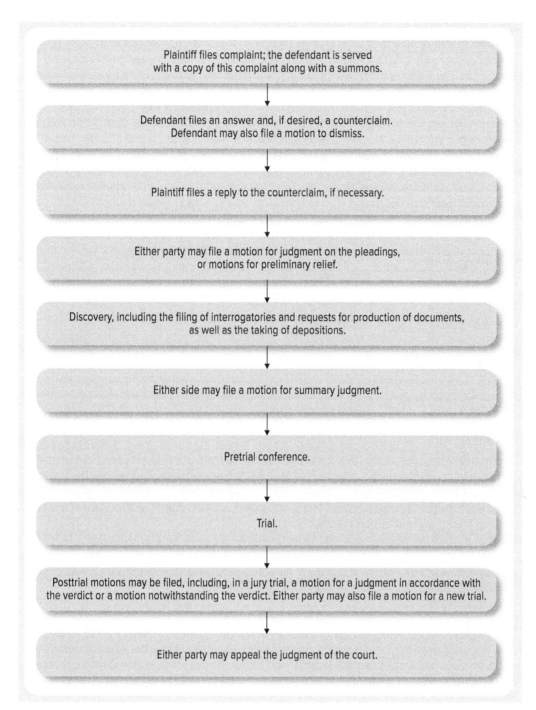

Exhibit 3-10

Steps in a Civil Lawsuit

Alternative Dispute Resolution

Many firms find that using **alternative dispute resolution (ADR)** methods to resolve their legal problems offers many benefits. The term *ADR* refers to the resolution of legal disputes through methods other than litigation, such as negotiation, mediation, arbitration, summary jury trials, minitrials, neutral case evaluations, and private trials.

Why might a business prefer ADR to litigation? First, ADR methods are generally faster and cheaper. According to the National Arbitration Forum, the average time from filing a complaint to judgment through litigation is 25 months.[11] Because ADR is faster, it is usually cheaper.

alternative dispute resolution (ADR)
The resolution of legal problems through methods other than litigation.

[11] National Arbitration Forum, *Business-to-Business Mediation/Arbitration vs. Litigation: What Courts, Statistics, & Public Perceptions Show about How Commercial Mediation and Commercial Arbitration Compare to the Litigation System*, January 2005, p. 3.

According to the American Intellectual Property Law Association, for cases valued in the $1 million to $25 million range, the average total cost of patent litigation for each party through the close of discovery is $1.9 million.[12] Through the end of trial, the average cost to each party is $3.5 million. Thus, if a party can resolve a dispute through alternative dispute resolution, this can save a significant amount of money.

Second, a business may want to avoid the uncertainty associated with a jury decision; many forms of ADR give the participants more control over the resolution of the dispute. Specifically, the parties can select a neutral third party, frequently a person with expertise in the area of the dispute, to help facilitate resolution of the case. Third, a business may want to avoid setting a precedent through a court decision. Fourth, a business may prefer ADR because it is confidential. Fifth, because many forms of ADR are less adversarial than litigation, ADR allows the parties to preserve a business relationship.

Courts also generally support the use of ADR, which alleviates some of the pressure on the overwhelming court dockets. Congress has recognized the benefits of ADR methods through its enactment of the Alternative Dispute Resolution Act of 1998. This act requires federal district courts to have an ADR program along with a set of rules regarding the program. Additional evidence of congressional support for ADR comes from the passage of the Administrative Dispute Resolution Act, which mandates that federal agencies must create internal ADR programs.

Primary Forms of ADR

LO 3-5 Compare the various forms of alternative dispute resolution used by businesses today.

Negotiation

negotiation
A bargaining process in which disputing parties interact informally to attempt to resolve their dispute.

Many business managers make frequent use of **negotiation,** a bargaining process in which disputing parties interact informally, either with or without lawyers, to attempt to resolve their dispute. No neutral third party is involved. Thus, negotiation differs from other methods of dispute resolution because the parties maintain high levels of autonomy. Some courts require parties to negotiate before they bring their dispute to trial.

Before negotiation begins, each side must determine its goals for the negotiation. Moreover, each side must identify the information it is willing to give the other party. Because negotiation generally occurs in every case before a more formal dispute resolution method is chosen, negotiation is not necessarily considered an alternative to litigation.

Mediation

mediation
A type of intensive negotiation in which disputing parties select a neutral party to help facilitate communication and suggest ways for the parties to solve their dispute.

An extension of negotiation is mediation. In **mediation,** the disputing parties select a neutral party to help facilitate communication and suggest ways for the parties to solve their dispute. Therefore, the distinguishing feature of mediation is that the parties voluntarily select a neutral third party to help them work together to resolve the dispute. The neutral third party frequently has expertise in the area of the dispute.

Mediation begins when parties select a mediator. Each party then typically writes a mediation brief to explain why it should win. An important feature of mediation is that it allows multiple parties to participate in a dispute. The parties take turns explaining the dispute. One of the mediator's main goals is to help each party listen carefully to the opposing party's concerns. The mediator asks the parties to identify any additional concerns. The parties begin generating alternatives or solutions for the disputed points. The mediator helps the parties evaluate the alternatives by comparing the alternatives with the disputed points and interests identified earlier. Finally, the mediator assists the parties in agreeing on a solution.

The mediation concludes when the agreement between the parties is reached. The agreement is then usually put into the form of a contract and signed by the parties. The mediator may

[12] AIPLA, Report of the Economic Survey, pp. 1-109–1-110.

participate in the drafting of the contract. If one of the parties does not follow the agreement, that party can be sued for breach of contract. However, parties typically abide by the agreement because they helped to create it.

If mediation is not successful, the parties can turn to litigation or arbitration to resolve their dispute. However, nothing said during the mediation can be used in another dispute resolution method; the mediation process is confidential.

Advantages and Disadvantages of Mediation The primary advantage of mediation is that it helps the disputing parties preserve their relationships; this is especially attractive for businesses with a working relationship they would like to continue. A second advantage is the possibility of finding creative solutions. The goal of mediation is to find a compromise between the needs of various parties rather than to find one party right and all the others wrong. A third advantage is the high level of autonomy mediation gives the participants. Instead of a neutral third party pronouncing a solution, the interested parties work together to create a solution, and this can make them more committed to following the agreement afterward.

These benefits can obviously be very worthwhile. However, we need to pay attention to the critics of the mediation process. One criticism of mediation is that it creates an image of equal parties working toward an equitable solution and thereby hides power imbalances that can lead to the party with greater power getting an agreement of greater benefit. A second criticism is that some people who enter mediation have no intention of finding a solution but, instead, use mediation as a tactic to draw out the dispute.

Uses of Mediation Mediation is most commonly used in collective bargaining disputes because it allows the workers to maintain a relationship with their employer while still having their needs addressed. Under the National Labor Relations Act (NLRA), a union must contact the Federal Mediation and Conciliation Services to attempt to mediate its demands before beginning a strike to achieve higher wages or better working hours.

Similarly, the Equal Employment Opportunity Commission (EEOC) encourages the mediation of employment discrimination claims. The EEOC has a mediation program that uses mediators employed by the EEOC as well as external mediators trained in mediation and discrimination law. Mediation is also used extensively in environmental law because environmental disputes are often best served by finding a compromise between the frequently multiple parties.

Arbitration

One of the most frequently used methods of dispute resolution is **arbitration**, the resolution of a dispute by a neutral third party outside the judicial setting. Arbitration is frequently used in disagreements between employees and employers, and it is increasingly being used between consumers and businesses. Arbitration is often a voluntary process in that parties have a contractual agreement to arbitrate any disputes. This agreement may stipulate how the arbitrator will be selected and how the hearing will be administered.

Lawyers, professors, and other professionals typically serve as arbitrators. The general qualifications for being an arbitrator are honesty, impartiality, and subject-matter competence. Additionally, arbitrators are expected to follow the Arbitrator's Code of Ethics. Typically, parties choose arbitrators from the Federal Mediation and Conciliation Services (FMCS), a government agency, or the American Arbitration Association (AAA), a private, nonprofit organization.

arbitration
A type of alternative dispute resolution in which disputes are submitted for resolution to private nonofficial persons selected in a manner provided by law or the agreement of the parties.

The Arbitration Hearing The arbitration hearing is similar to a trial. Both parties present their cases to a neutral third party; parties may represent themselves or use legal counsel. During this presentation, the parties may introduce witnesses and documentation, cross-examine the witnesses, and offer closing statements. The fact finder offers a legally binding decision. However, arbitration is also different from a trial in several ways. First, the arbitrator often takes a much more active role in an arbitration hearing than a judge takes in a trial, and the arbitrator can question witnesses. Second, no official written record of the hearing is kept in most arbitrations. Third, the rules of evidence applicable in a trial are typically relaxed in arbitration.

The Arbitrator's Award The arbitrator typically provides a decision within 30 days of the arbitration hearing. The arbitrator's decision is called an *award,* even if no monetary compensation is awarded. The arbitrator's decision differs from a judge's decision in several ways. The arbitrator does not have to state any findings of fact, conclusions of law, or reasons to support the award, and he or she is not as bound by precedent as a judge is. Also, because the arbitrator was hired to resolve a dispute between two parties, the arbitrator is more likely to make a compromise ruling instead of a win-lose ruling.

Unlike the case in most other forms of ADR, the arbitrator's decision is legally binding. In certain cases, a decision may be appealed to the district court. However, few of these cases are appealed. The courts give extreme deference to arbitrators' decisions.

The Federal Arbitration Act (FAA), the federal law enacted to encourage the use of arbitration, explicitly lists four grounds on which an arbitrator's award may be set aside: (1) The award was the result of corruption, fraud, or other undue means; (2) the arbitrator displayed bias or corruption; (3) the arbitrator refused to postpone the hearing despite sufficient cause, refused to hear relevant evidence, or otherwise misbehaved to prejudice the rights of one of the parties; (4) the arbitrator exceeded his or her authority or failed to use that authority to make a mutual, final, and definite award. In a 2008 decision, the U.S. Supreme Court held that these grounds are the only grounds for appeal and that the parties in an arbitration proceeding do not have the ability to give the courts additional grounds to set aside a decision.[13]

Consequently, in the United States, arbitration decisions are generally upheld. In fact, the Fifth Circuit recently held that "manifest disregard of the law and contrary to public policy are the only nonstatutory bases recognized by this circuit for the vacatur of an arbitration award."[14]

Advantages and Disadvantages of Arbitration Arbitration may be preferable to litigation for several reasons. First, arbitration is more efficient and less expensive. Second, parties have more control over the process of dispute resolution through arbitration. They choose the arbitrator and determine how formal the process will be. Third, the parties can choose someone to serve as the arbitrator who has expertise in the specific subject matter. Fourth, the arbitrator has greater flexibility in decision making than a judge has. Unlike judges, who are bound by precedent, arbitrators generally do not have to offer reasons for their decisions.

However, arbitration is not without its critics. First, arbitration panels are being used more frequently, resulting in a loss of some of the prior advantages of arbitration, such as efficiency and lower cost. Second, because appealing an arbitration award is so difficult, some scholars argue that injustice is more likely to occur. Third, some individuals are concerned that by agreeing to give up one's right to litigate, one may be losing important civil rights or giving up important potential remedies without really understanding which rights are being given up. Fourth, some scholars are afraid that if more and more employers and institutions turn to mandatory arbitration, it will become more like litigation. Fifth, some scholars are concerned about the privacy associated with arbitration. Companies and employers are able to hide their disputes through arbitration.

Binding Arbitration Clause Given the benefits associated with arbitration, parties may voluntarily submit their cases to arbitration. The primary method of securing arbitration is through a **binding arbitration clause,** a provision in a contract that mandates that all disputes arising under the contract must be settled by arbitration. The clause also typically states how the arbitrator will be selected. If parties have a binding arbitration agreement, the parties *must* resolve the dispute through arbitration. Both federal and state courts must uphold agreements to arbitrate. The Case Nugget illustrates the courts' willingness to enforce binding arbitration agreements.

A constraint on binding arbitration clauses is that they must be drafted in a way that ensures that the courts do not see them as unconscionable. An *unconscionable contract provision* has been defined as one in which the terms are "manifestly unfair or oppressive and are dictated by a dominant party."[15] The doctrine has been used most often to strike down binding arbitration clauses in consumer and employment contracts. Exhibit 3-11 offers tips on creating a binding arbitration clause.

binding arbitration clause
A contract provision mandating that all disputes arising under the contract must be settled by arbitration.

[13] *Hall Street Associates v. Mattell,* 2008 U.S. LEXIS 2911.
[14] *Kergosien et al. v. Ocean Energy, Inc.,* 390.F.3d 346 (5th Cir.2004).
[15] *Farris v. County of Camden,* 61 F. Supp. 2d 307, 341 (D. N.J. 1999).

CASE Nugget

The Validity of Binding Arbitration Agreements

Preston v. Ferrer United States Supreme Court

128 S. Ct. 978 (2008)

Under the FAA, is an administrative proceeding the same as a trial?

It has long been held that under the FAA, even if state statutes refer a state law controversy initially to a judicial forum, if the parties have agreed to submit that dispute to arbitration, the dispute must be arbitrated. Until 2008, however, the question existed of whether the same rule would apply when the state law directed that the first mode of resolution must be an administrative proceeding rather than a court hearing. In this case, the U.S. Supreme Court ruled that for purposes of the FAA, an administrative proceeding was the equivalent of a judicial proceeding. In *Preston v. Ferrer,* the two had included a binding arbitration agreement in a contract for Preston to represent Ferrer. When Preston attempted to collect his fees, a question arose about whether the matter had to be resolved by the California Labor Commission, which had jurisdiction over issues arising under the California Talent Agencies Act. The high court agreed with Ferrer that the dispute must be resolved through arbitration because the FAA preempted the state law. Even if a state law provides that an administrative forum has jurisdiction over a specific type of dispute, a binding arbitration agreement will supersede the forum declared by the state law.

Overall, make sure the clause treats both parties fairly.

1. Be clear and unmistakable about what you wish to arbitrate.
2. The arbitration clause must be bilateral.
3. State explicitly which party will pay the arbitrator's fees, and make sure that it will not cost the non-drafting party more to arbitrate than it would have cost to litigate. Specify how the arbitrator will be selected.
4. Spell out the costs associated with the arbitration.
5. Avoid limitations on the remedies available to the parties.
6. Consider potential parties when determining where to hold the arbitration.

Exhibit 3-11

Tips for Creating a Binding Arbitration Clause

Other ADR Methods

Med-arb

Med-arb is a dispute resolution process in which the parties agree to start out in mediation and, if the mediation is unsuccessful on one or more points, to move on to arbitration. In some cases, the same neutral third party may participate in both the mediation and the arbitration. Some critics argue that if parties know that the mediator may become the ultimate decision maker, they will be less likely to disclose information during the mediation stage. Others argue that having the same neutral mediator-arbitrator offers faster resolution because the third party is familiar with the facts of the case.[16]

med-arb
A type of dispute resolution process in which both parties agree to start out in mediation and, if unsuccessful, to move on to arbitration.

[16] See Gerald F. Phillips, "Same Neutral Med-Arb: What Does the Future Hold?" *Dispute Resolution Journal* 60 (May–July 2005), p. 24.

Summary Jury Trial

summary jury trial
An abbreviated trial that leads to a nonbinding jury verdict.

A **summary jury trial** is an abbreviated trial that leads to a nonbinding jury verdict. Two advantages are inherent in this method of dispute resolution. First, it is quick; a summary jury trial lasts only a day. Second, because a jury offers a verdict, both parties get a chance to see how their case would fare before a jury of their peers. Each judge can set his or her own rules. The judge advises the jury on the law, and each party's lawyer presents an opening statement and a limited amount of evidence before the jury. The lawyers have limited time for their presentations, and witnesses do not usually testify. The jury reaches an advisory verdict, although the jury does not know that its verdict is nonbinding. The parties then enter into a settlement conference, where they decide to accept the jury verdict, reject the verdict, or settle on some compromise. Approximately 95 percent of cases are settled at this time. However, if the case is not settled, it will go to a regular trial. At that trial, nothing from the summary jury trial is admissible as evidence.

Minitrial

minitrial
A type of conflict resolution in which lawyers for each side present their arguments to a neutral adviser, who then offers an opinion as to what the verdict would be if the case went to trial. This decision is not binding.

A **minitrial** is similar to arbitration and mediation because it involves a neutral third party. However, despite the presence of a neutral third party, business representatives of the disputing corporations participate and have settlement authority. Lawyers for each side present their arguments before these executives and the neutral adviser, who then offers an opinion as to what the verdict would be if the case went to trial. The neutral adviser's opinion, like the jury's verdict in a summary jury trial, is not binding. Next, the corporate executives discuss settlement options. If they reach an agreement, they enter into a contract that reflects the terms of the settlement.

Early Neutral Case Evaluation

With *early neutral case evaluation,* the parties select a neutral third party and explain their respective positions to this neutral, who then evaluates the strengths and weaknesses of the case. The parties use this evaluation to reach a settlement. Eighteen federal district courts currently use early neutral case evaluation.[17]

Private Trials

private trial
An ADR method in which a referee is selected and paid by the disputing parties to offer a legally binding judgment in a dispute.

Several states allow **private trials,** an ADR method in which a referee is selected and paid by the disputing parties to offer a legally binding judgment in a dispute. The referees do not have to have

Exhibit 3-12

FORMS OF ADR

NEGOTIATION	Informal discussions between disputing parties in an attempt to resolve the conflict.
MEDIATION	A neutral third party facilitates discussion between disputants in an attempt to help them reach a resolution of their dispute.
ARBITRATION	Resolution of the dispute by a neutral third party outside the judicial setting.
MED-ARB	Parties agree to attempt to solve the dispute through mediation, but will shift to arbitration if mediation is not successful.
SUMMARY JURY TRIAL	An abbreviated trial that results in a nonbinding jury verdict.
MINITRIAL	A form of dispute resolution available only to corporations that uses corporate executives, along with a neutral third party, to resolve the dispute.
EARLY NEUTRAL CASE EVALUATION	Use of a neutral third party to evaluate the strengths and weakness of each side to aid them in negotiating a settlement.
PRIVATE TRIALS	A dispute process similar to litigation, except a retired judge is used to solve the dispute and the process is private.

[17] Michael H. Diamant et al., *Strategies for Mediation, Arbitration, and Other Forms of Alternative Dispute Resolution,* SK074 ALI-ABA 205 (2005) [citing the CPR Institute for Dispute Resolution, www.cpradr.org].

any specific training; however, because retired judges often serve as referees, this method is often referred to as "rent-a-judge." The cases are often heard privately to ensure confidentiality. The referee writes a report and files it with the trial judge, but a dissatisfied party reserves the right to request a new trial before a trial court judge. Private jury trials with experienced jury members are becoming more popular today as well. Private trials have been criticized because they seem to provide faster and cheaper justice for those who can afford the initial fee and because they hide the trial from the public eye.

CASE OPENER WRAP-UP

Questionable Jurisdiction Over Caterpillar

The timing of events was crucial to the outcome of the Caterpillar case. At the time Lewis filed the case in the state court system, one of the defendants and the plaintiff were from the same state, so the state court system had jurisdiction. Once the supply company reached an agreement with Lewis, the other defendant, Caterpillar, filed a motion to exercise its right of removal because diversity of citizenship existed in the absence of the Kentucky defendant. But the agreement was not final at the time of the motion because the agreement was subject to the insurer's approval. Thus, the appellate court ruled that because the supply company was still a party to the agreement, the federal court system could not exercise jurisdiction over the case.

The U.S. Supreme Court, however, overruled the appellate court. The Supreme Court ruled that the state court should not have granted Caterpillar's initial motion to remove the case because at the time of removal, the insurer had not accepted the settlement agreement, the supply company remained a party in the case, and, therefore, the diversity of citizenship was not complete. The Supreme Court held further, however, that the district court's error in hearing the case was not fatal because the settlement agreement was approved and the case satisfied the jurisdictional requirements by the time the federal court issued its decision. The Court ruled that to require the district court to send the case back to the state system would be an undue waste of judicial resources.

Why might Caterpillar have wanted to move the case to the federal court system? First, the case involved product liability claims. Data suggest that average damage awards in product liability cases tend to be higher in state courts than in federal courts. Second, Caterpillar may have feared local prejudice. While all judges must strive for neutrality, out-of-state defendants may fear that state judges are slightly biased in favor of in-state parties.

PRACTICAL TIPS for Business Managers

1. As a manager, take some time to familiarize yourself with basic legal terms and concepts (reading this book is a great way to do so!). Your primary job is to help run a business, sure, but even when things seem to be going well, legal disputes involving you, your employees, or your business can arise without warning. Should a situation occur such that you need to be involved in litigation, you do not want to be going in blind. You also want to be able to understand what your lawyer means when he speaks to you using legal terminology.

2. For better or for worse, the U.S. legal system values justice over efficiency. Thus, trials and litigation are time-consuming and expensive processes that are tedious for just about everyone involved. Many trials, especially those involving contract or employment disputes, can be largely avoided if proper care is taken to ensure that contracts and agreements are properly and thoroughly drafted. Insist that all contracts and agreements are properly and thoroughly drafted. Insist that all contracts and agreements your business makes are in writing and consider hiring a lawyer to make sure they cover most contingencies. You may be saving yourself a lot of time and money down the road.

3. Remember that although ADR is generally less resource-intensive than a trial, it still has certain disadvantages when compared to trial litigation. The spirit of ADR is one of compromise, whereas trials have a definite winner and loser. If you think your business has a strong case and stands to gain much more from a trial than ADR despite the costs of a trial, litigation may actually be the better option, so do not be hesitant to go to trial when circumstances dictate that litigation is the best option for your firm.

SUMMARY

Jurisdiction	*In personam jurisdiction* is the power of a court to render a decision affecting a person's legal rights.
	Subject-matter jurisdiction is the power of a court to render a decision in a particular type of case. The three forms of subject matter jurisdiction are state, exclusive federal, and concurrent.
Venue	*Venue* is the geographic location of the trial.
The Structure of the Court System	The United States has two parallel court structures: the state and federal systems. The federal structure has *district courts* (trial courts), *circuit courts of appeal,* and the *U.S. Supreme Court.* The state court structure varies by state but generally includes courts of common pleas (trial courts), state courts of appeal, and a state supreme court.
Threshold Requirements	*Standing:* For a party to have the legal right to file a case, the outcome of the case must personally affect that party.
	Case or controversy: There must be an issue before the court that a judicial decision is capable of resolving. Parties cannot ask the judge for an advisory opinion.
	Ripeness: The case cannot be moot; it must be ready for a decision to be made.
Steps in Civil Litigation	The stages of a civil trial include the pretrial, trial, posttrial, and appellate stages.
	Pretrial includes consultation with attorneys, pleadings, the discovery process, and the pretrial conference.
	The *trial* begins with jury selection, followed by opening statements, the plaintiff's case, the defendant's case, closing arguments, jury instructions, jury deliberations, the jury's verdict, and the judgment.
	After the trial, parties may file *posttrial motions*.
	The parties may then file *appeals* to the appropriate appellate court and, in some cases, to the U.S. Supreme Court.
Alternative Dispute Resolution	*Alternative dispute resolution* is a way to settle problems without having to go to a costly and time-consuming trial.
Primary Forms of ADR	The primary forms of ADR are *negotiation, mediation,* and *arbitration.*
Other ADR Methods	Other methods of ADR include *med-arb, summary jury trials, minitrials, early neutral case evaluation,* and *private trials.*

Point/Counterpoint

Should Courts Increase Their Use of Service by Social Media?	
YES	NO
The traditional methods of service of process are antiquated and expensive, especially if we are talking about serving someone who is out of the country. Think about how much it costs to track down someone's physical address and then serve him in person or by registered or regular mail.	Service of process is designed to ensure that a person's due process right of notice of a potential deprivation of his property is about to occur. Given the importance of this right, courts need to think very carefully before rushing to allow service by social media.

It is a lot easier to find someone's electronic address than it is to find their physical address. Just type their name into Google and you will immediately get their Facebook or Linked In page, their Twitter or Instagram account, or their blog.

And everyone regularly is on Facebook or checks their Twitter feed. People have their smartphones with them all the time, so the service would be almost instant. Who even checks their regular mailbox anymore?

And when we are talking about serving businesses, an increasing number of business don't even have a posted physical address. They are primarily a web presence.

We do everything online these days . . . shopping, banking, getting a mortgage. Shouldn't courts move into the twenty-first century and start using social media as a preferred method of service.

The traditional methods of service of process have been shown over time to ensure that defendants have adequate notice of lawsuits filed against them. It is too easy for someone to set up a fraudulent Facebook page, and then the notice would go to someone to whom it was not intended. Besides, a number of people may have the same name, so there is no guarantee that the "Jim Smith" you serve via his Facebook page is the "Jim Smith" you intended to serve.

Another problem is that many people just get overwhelmed by social media. They may regularly check their Facebook posts and then may decide they need to quit wasting so much time, so they quit reading their Facebook posts for a while. What if they receive service during that time period?

And, of course, those who advocate for the use of service by social media are focused on those who have access. We cannot forget the digital divide. Even today, not everyone owns a smartphone or a computer. But most people have a home and a mailbox.

The traditional methods of service have been working just fine for decades. If something isn't broken, there is no need to try to fix it.

Questions & Problems

1. Explain the two types of jurisdiction that a court must have to hear a case and render a binding decision over the parties.
2. Explain the differences between trial courts and appellate courts.
3. Identify and define the alternative tools of discovery.
4. Explain the three threshold requirements a plaintiff must meet before he or she can file a lawsuit.
5. The citizens of each state elect state court judges. The president, on the other hand, appoints federal court judges for lifetime positions. Why might this difference lead to different rulings on similar cases? Which method—election or appointment—do you like better? Why? What are the advantages and disadvantages of each method?
6. Missouri was International Shoe Corporation's principal place of business, but the company employed between 11 and 13 salespersons in the state of Washington, who exhibited samples and solicited orders for shoes from prospective buyers in Washington. The state of Washington assessed the company for contributions to a state unemployment fund. The state served the assessment on one of International Shoe Corporation's sales representatives in Washington and sent a copy by registered mail to the company's Missouri headquarters. International Shoe's representative challenged the assessment on numerous grounds, arguing that the state had not properly served the corporation. Is the corporation's defense valid? Why or why not?

Source: *International Shoe Co. v. Washington,* 326 U.S. 310 (1945).

7. The Robinsons, residents of New York, bought a new Audi car from Seaway Volkswagen Corp., a retailer that was incorporated in New York and had its principal place of business there. World-Wide Volkswagen, a company incorporated in New York and doing business in New York, New Jersey, and Connecticut, distributed the car to Seaway. Neither Seaway nor World-Wide did business in Oklahoma, and neither company shipped cars there. The Robinsons were driving through Oklahoma when another vehicle struck their Audi in the rear. The gas tank of the Audi exploded, injuring several members of the family. The Robinsons brought a product liability suit against the manufacturer, distributor, and retailer of the car in an Oklahoma state court. Seaway and World-Wide argued that the Oklahoma state court did not have *in personam* jurisdiction over them. After the state's trial court and supreme court held that the state did have

in personam jurisdiction over Seaway and World-Wide, they appealed to the U.S. Supreme Court. How do you think the Court decided in this case? Why?

Source: *World-Wide Volkswagen Corp. v. Woodson,* 444 U.S. 286 (1980).

8. Plaintiff Wolfe was a New Jersey resident who owned a collection of classic automobiles, one of which was a rare Alfa Romeo 6C 3000 Competizione that was to be auctioned. In the federal District Court in New Jersey, Wolfe filed suit for defamation and trade libel against Gooding and Company, a California corporation that auctions automobiles, and generates business, in part, by sending catalogs to New Jersey residents, but whose principle place of business is Santa Monica, California. The plaintiff claimed that the defendant had made defamatory comments about him in three places: the company's website, the company's catalog of cars listed for auctions, and at the company's auction in California. The defamatory statements allegedly had the effect of lowering the value of the car that the plaintiff was trying to sell. The defendant filed a motion to dismiss on grounds that the company did not have sufficient minimum contacts with the state of New Jersey. Do you think the trial court granted the defendant's motion to dismiss? Why or why not?

Source: *Wolfe v. Gooding Co.* 2015 U.S. Dist. LEXIS 26165.

9. Increasingly, we are seeing email used in cases involving defendants located in foreign countries. Plaintiffs filed suit against four Defendants: Qingdao Sunflare New Energy Co. Skone Lighting Co., Sailing Motor Co., and Guangzhou LF Sky Energy Technology Co., serving all defendants via email. Plaintiff subsequently filed a motion with the court requesting an order deeming the service effectuated via email. The court granted the motion with respect to defendants Guangzhou and Qingdao, but denied the motion with respect to the other two defendants without prejudice.

With permission of the court, plaintiffs then submitted the following supplemental evidence as to why the motion should also be granted for the other two defendants: Lighting held out the email address at issue as its preferred contact information. The same address was provided to a website domain registrar in its website registration materials. Additionally, email tracking measures utilized by plaintiffs indicate that emails to the Skone Lighting email address were sent successfully, and no "bounceback" was received suggesting that the email was not delivered.

The situation involving Sailing Motor was similar. The email address appears on Sailing Motor's price sheet. A man who identified himself as the owner of the company provided the email address as the company's contact information to plaintiffs' investigators, and plaintiffs' investigators communicated via that email address to facilitate the purchase of allegedly infringing products.

Upon reconsideration, do you think the court granted or denied the plaintiff's motion? Why or why not?

Source: *D.Light Design, Inc., et al., v. Boxin Solar Co., Ltd., et al.,* 2015 U.S. Dist. LEXIS 14683.

10. In 2012, Janet Napolitano, U.S. Secretary of Homeland Security, released a memo outlining a policy which became known as Deferred Action for Childhood Arrivals (DACA). The policy directed relevant agencies to defer the removal of illegal aliens who were eligible for the DACA. In 2014, Napolitano's successor, Jeh Johnson, revised the DACA program to extend the duration of deferred removal and outlined a new policy known as DAPA that enabled eligible parents of American citizens to be deferred from removal.

In November 2014, President Obama announced the changes to DACA and the new DAPA. That very day, Joseph Arpaio, the elected sheriff of Maricopa County, Arizona, sued Obama and other federal officials to enjoin the deferred action policies. Arpaio asserted that DACA and DAPA were unconstitutional, arbitrary and capricious, and invalid under the Administrative Procedures Act. For the court to review the claim, Arpaio must have Article III standing.

Sheriff Arpaio argues that he has standing because the deferred action policy will act as a magnet and draw more undocumented aliens into Maricopa County, where they will commit crimes, and deferred action policies will decrease the total deportations of six million eligible undocumented aliens so that more individuals will be present in Maricopa County and commit crimes. Arpaio's argument is based on the premise that more people causing more crimes will harm him because he will be forced to spend more money policing the county and running its jails.

The judge ruled against Arpaio, reasoning that he had failed to establish the three elements of standing.

What are the three elements of standing? Based on the three elements, why did the judges rule against Sheriff Arpaio? Give a reason for each element.

Source: *Joseph M. Arpaio v. Barack Obama,* 2015 U.S. App., LEXIS 14270.

11. Martina Beverly sued her former employer, Abbott Laboratories (Abbott), for employment discrimination and retaliation. During a private mediation, the parties signed a handwritten agreement stating that Beverly demanded $210,000 and mediation costs in exchange for dismissing the lawsuit. Abbott later accepted Beverly's demand and circulated a more formal settlement proposal. After Beverly refused to execute this draft proposal, Abbott moved to enforce the original handwritten agreement.

 The district court found that the parties entered into a binding settlement agreement and granted Abbott's motion to enforce. Beverly appealed this decision, arguing that the handwritten agreement was merely a preliminary document that captured the parties' intention to execute a binding settlement agreement in the future, and that Abbott intended to be bound only by the terms of the typewritten final agreement. How do you think the appellate court should have decided this case? Why?

 Source: *Beverly v. Abbott Laboratories,* 817 F.3d 328 (2016).

12. Asbestos, a fibrous type of material once used in insulation products, is now widely known to be correlated with health problems. Walter Brown was exposed to asbestos while he was an Air Force mechanic stationed in Europe and the United States. Additionally, Walter Brown was diagnosed with and died from malignant mesothelioma, a cancer thought to be caused by asbestos exposure. Prior to passing away, Walter Brown brought suit against Lockheed Martin and thirteen other companies in the Connecticut Superior Court, seeking to recover in tort for injuries sustained while working as an Air Force mechanic. Cindy S. Brown, Walter Brown's daughter, took up the mantle and replaced Walter Brown as the plaintiff after his passing.

 Lockheed Martin is an American global aerospace, defense, security, and advanced technologies company with worldwide interests. Having physical presence in Connecticut since 1982, Lockheed finally obtained a formal certificate to do business in the state in 1995. The company does not own property in the state, but has leased the same space in New London, Connecticut, since 1997 in addition to at least three other properties.

 Lockheed Martin, citing its status as a federal contractor, removed the case to the Connecticut District Court and subsequently moved to dismiss the case for lack of personal jurisdiction. The Connecticut District Court dismissed the case, and Cindy Brown filed an appeal with the Second Circuit Court. Do you think the Connecticut District Court has jurisdiction over Lockheed Martin? Why or why not?

 Source: *Brown v. Lockheed Martin Corp.,* 814 F. 3d 619 (2016).

Part 1 *The Legal Environment of Business*

CHAPTER 4
Administrative Law

CASE OPENER

Does the EPA Have an Obligation to Regulate Automobile Emissions?

On October 20, 1999, a group of 19 private organizations filed a rule-making petition asking the Environmental Protection Agency (EPA) to regulate "greenhouse gas emissions from new motor vehicles" under the Clean Air Act.[1] The petitioners cited the fact that 1998 was the "warmest year on record," that greenhouse gas emissions have significantly accelerated climate change, and that carbon dioxide is the most important man-made contribution to climate change. Fifteen months after the petition was filed, the EPA requested public comment on the issues. Then, on September 8, 2003, the EPA entered an order denying the rule-making petition, citing two reasons: (1) The Clean Air Act does not authorize the EPA to issue mandatory regulations to address global climate change; and (2) even if the agency had authority, it would be unwise to do so at this time. The case, *Massachusetts v. EPA*, eventually worked its way up to the U.S. Supreme Court, which had to decide whether the EPA was improperly failing to regulate carbon dioxide gas in automobile exhaust as a climate-changing pollutant.

1. What government entity is the final determiner of whether a particular type of rule will be issued by an agency?
2. Did the EPA err in not regulating greenhouse gases under the Clean Air Act?

LEARNING OBJECTIVES

- **LO 4-1** Define administrative law.
- **LO 4-2** Define administrative agency.
- **LO 4-3** Describe the enabling legislation powers that are granted to agencies.
- **LO 4-4** Identify the types of powers administrative agencies have.
- **LO 4-5** Differentiate between executive and independent agencies.
- **LO 4-6** Define Administrative Procedures Act.
- **LO 4-7** Distinguish between formal and informal rule making.
- **LO 4-8** List the four limits on agency power.

[1] *Massachusetts v. EPA*, 549 U.S. 497 (2007).

Introduction to Administrative Law

What is Administrative Law

LO 4-1 Define administrative law.

As a business owner or manager, you will need to be aware of regulations that affect your business. In addition to learning about laws passed by Congress, you will also need to know about rules passed by administrative agencies. **Administrative law** consists of the substantive and procedural rules created by **administrative agencies** (bodies of the city, county, state, or federal government), involving applications, licenses, permits, available information, hearings, appeals, and decision making.

administrative law
The collection of rules and decisions made by administrative agencies to fill in particular details missing from constitutions and statutes.

LO 4-2 Define administrative agency.

An administrative agency is generally defined as any body created by the legislative branch (e.g., Congress, a state legislature, or a city council) to carry out specific duties. Agencies have three types of power: *legislative, judicial, and executive.* They may make rules for an entire industry, adjudicate individual cases, and investigate corporate misconduct. Because legislative, judicial, and executive powers have traditionally been placed in separate branches of government by the Constitution, the role of administrative agencies has led some to refer to agencies as the unofficial fourth branch of government. Although there is a semblance of truth to that characterization, administrative agencies are not in fact another branch, primarily because all their authority is simply delegated to them, and they remain under the control of the three traditional branches of government.

administrative agency
Any government body created by the legislative branch (e.g., Congress, a state legislature, or a city council) to carry out specific duties.

The first federal administrative agency, the **Interstate Commerce Commission (ICC)**, was created by Congress near the end of the nineteenth century. Congress felt that the anticompetitive conduct of railroads could best be controlled by a regulatory body. The ICC no longer exists as a separate agency,[2] but for more than 100 years, the ICC regulated passenger and freight transportation. Following the crash of the stock market and the Great Depression of the 1930s, Congress saw a need for additional agencies to regulate business in the public interest. Since then, numerous agencies have been created whenever Congress believed an area required more intense regulation than Congress could provide. In fact, after the Enron scandal, there was talk that Congress might create a new agency to regulate the accounting industry. To date, no such agency has materialized.

Interstate Commerce Commission (ICC)
The first federal administrative agency; created to regulate the anticompetitive conduct of railroads.

Why and How are Agencies Created?

LO 4-3 Describe the enabling legislation powers that are granted to agencies.

When Congress sees a problem that it believes needs regulation, it may create an administrative agency to deal with that problem. The idea is that the agency can be staffed with people who have special expertise in the area the agency is regulating and therefore know what types of regulations are necessary to protect the citizens in that area. Agencies typically act more swiftly than Congress in creating and enacting new laws. Today, administrative agencies actually create more rules than Congress and the courts combined.

Congress creates administrative agencies through passage of **enabling legislation,** which is a statute that specifies the name, functions, and specific powers of the administrative agency. Enabling statutes grant agencies broad powers for the purpose of serving the "public interest, convenience, and necessity." These broad powers include rule making, investigation, and adjudication.

enabling legislation
A statute that specifies the name, functions, and specific powers of an administrative agency and grants the agency broad powers for the purpose of serving the public interest, convenience, and necessity.

[2] The functions of the ICC were transferred to the Transportation Department by Congress as part of a cost-saving measure.

BUT WHAT IF . . .

WHAT IF THE FACTS OF THE CASE OPENER WERE DIFFERENT?

In the opening case, the EPA believed it did not have enough power to make rules regarding global climate change. Let's say that a member of the EPA argued that only Congress has the power to make this type of regulation, which would directly affect the way many businesses operate. Would the EPA member be correct?

LO 4-4 Identify the types of powers administrative agencies have.

Rule Making Enabling statutes permit administrative agencies to issue rules that control individual and business behavior. These rules have the same effect as laws. If an individual or business fails to comply with agency rules, there are often civil, as well as criminal, penalties. Agencies may enact three types of rules: procedural, interpretive, and legislative.

Procedural rules are rules regarding the internal operations of an agency. *Interpretive rules* are rules that explain how the agency views the meaning of the statutes for which the agency has administrative responsibility. Finally, *legislative rules* are policy expressions that have the effect of law. The various rule-making processes are discussed later in this chapter.

subpoena
An order to appear at a particular time and place and provide testimony.

subpoena duces tecum
An order to appear and bring specified documents.

Investigation Enabling statutes grant executive power to agencies to investigate potential violations of rules or statutes. Many times, companies cooperate with agencies and voluntarily furnish information. Other times, however, agencies must use their investigative powers, defined in their enabling legislation, to gather information. Such powers typically include the power to issue a subpoena (i.e., an order to appear at a particular time and place and provide testimony) and a subpoena *duces tecum* (i.e., an order to appear and bring specified documents). For more on the investigatory powers given to administrative agencies, see Case 4-1.

CASE 4-1

DOE v. UNITED STATES SEC
U.S. District Court for the Northern District of California
U.S. Dist. Lexis 114382 (2011)

FACTS: *The administrative subpoena at issue seeks to discover the identity of the owner of the email address "aurorapartners@gmail.com," which the SEC has reason to believe may have been used to tout stock for Jammin Java Corp. as part of a "pump and dump" scheme. "Pump and dump" schemes generally involve the touting of a company's stock (typically microcap companies) through false and misleading statements to the marketplace. After pumping the stock, fraudsters make huge profits by selling their cheap stock into the market. The SEC is investigating a rapid increase in Jammin Java's price, which coincided with the wide dissemination of online newsletters touting its stock. The SEC is investigating the online newsletters to determine whether they contained materially false information about Jammin Java, and/or failed to disclose the disseminator's financial interest in the company or compensation the company paid them. The SEC represents that it "has obtained information indicating that an individual using the email address 'aurorapartners@gmail.com'* may be involved in the touting activity at issue in the investigation." *Pursuant to its investigative powers and the Electronic Communications Privacy Act (ECPA), the SEC served an administrative subpoena on Google seeking the identity and contact information of the owner of the email address "aurorapartners@gmail.com." Google gave the owner of the email address notice of the subpoena, the owner hired counsel, and counsel moved to quash the subpoena.*

ISSUE: May an agency investigating possible violation of agency rules obtain the identity and contact information of the owner of an email address?

REASONING: The ECPA requires "a provider of electronic communication service or remote computing device" to "disclose to a governmental entity the (A) name; (B) address; (C) local and long distance telephone connection records, or records of session times and durations; (D) length of service . . . and types of service utilized; (E)

(continued)

telephone or instrument number or other subscriber number or identity . . .; and (F) means and source of payment for such service, of a subscriber when the governmental entity uses an administrative subpoena authorized by Federal or State statute. . . ." There is no provision in the ECPA allowing a customer whose non-content records are being sought to prevent disclosure, although where the governmental entity is seeking the *contents* of customer communications, customers may challenge disclosure.

Movant argues the subpoena should be quashed because disclosure of his identity would violate his right to anonymous speech, which is protected by the First Amendment to the United States Constitution. Movant has used the email address "to speak anonymously on the Internet. Specifically, I use the email address to post my opinions on various political blogs, including about the policies of specific members of the U.S. Congress. I have made these political statements under the belief that my anonymity will be protected."

The SEC directs the Court to *Brock v. Local 375, Plumbers Int'l Union of America*. . . for the uncontroversial proposition that administrative agencies may investigate on the mere suspicion the law is being violated. Brock sets out the three-part standard of review courts in the Ninth Circuit apply in an agency subpoena enforcement proceeding: (1) Has Congress granted the agency the authority to investigate; (2) have procedural requirements been followed; and (3) is the information sought relevant and material to the investigation? *Brock* recognizes that First Amendment protections may apply to information sought by an administrative subpoena and provides that where a party opposing disclosure makes a *prima facie* showing of arguable First Amendment infringement, the burden will shift to the government to show the information sought is rationally related to a compelling governmental interest and the subpoena is the "least restrictive means of obtaining the desired information." Movant contends the *Brock* test does not apply here because a "more rigorous standard" applies to anonymous speech cases than to cases involving the right to free association at issue in *Brock*.

Movant's argument relies exclusively on cases analyzing standards for disclosure in civil actions involving private parties embroiled in discovery disputes—not cases balancing the needs of a legitimate government investigation with the interest of a party wishing to remain anonymous. Congress recognizes the need for government agencies to investigate wrongdoing and accordingly has granted these agencies investigatory powers that are not available to civil parties. The ECPA, for example, allows government agencies like the SEC to issue administrative subpoenas requesting information from ISPs or to obtain court orders compelling disclosure of subscriber information.

DECISION AND REMEDY: Movant has not established that his addressing information is protected under the First Amendment, and the SEC sufficiently has established that the ownership of the email address "aurorapartners@gmail.com" is relevant to the Jammin Java investigation. Accordingly, the motion to quash the subpoena directed at Google requesting this information is denied.

Affirmed, and respondent recovered the costs of her appeal.

SIGNIFICANCE OF THE CASE: Administrative agencies have broad investigative powers that are not available to parties in civil cases.

CRITICAL THINKING

As we see again and again, the meaning of key words will often push a decision in a particular direction. How could the distinction between subpoenas requesting identity and those requesting content be made more clear in this case?

ETHICAL DECISION MAKING

Do you believe it is ethical to engage in anonymous online postings that are critical of others and/or their organizations? If so, is the golden rule obsolete? In other words, if you are okay with anonymous postings about others, would you be okay if someone anonymously criticized you online?

Adjudication Enabling statutes delegate judicial power to agencies to settle or adjudicate individual disputes that an agency may have with businesses or individuals. After investigation, the agency holds an administrative hearing before an **administrative law judge (ALJ).** The ALJ will try to convince the parties to reach a settlement via a **consent order,** but the judge also has the authority to render a binding decision (an **order**) after a hearing (administrative law matters are heard only by an ALJ because there is no right to a jury trial in administrative agencies). An

administrative law judge (ALJ)
A judge who presides over an administrative hearing; may attempt to get the parties to settle but has the power to issue a binding decision.

consent order
A statement in which a company agrees to stop disputed behavior but does not admit that it broke the law.

order
A binding decision rendered by a judge.

appeal to the full commission or the head of an agency may then be filed. That decision may be appealed to the circuit court of appeals. If there are no appeals, the ALJ's initial order becomes the final order. Decisions of ALJs are typically upheld.

Example of an Administrative Problem. The EPA administrator, using the congressional mandate under the Clean Air Act, sets forth rules governing the amount of certain hazardous air pollutants that may be emitted into the atmosphere. Using these standards, another branch of the EPA sends investigators to inspect a plant suspected of violating the act. If the inspector finds a violation and the EPA imposes a penalty, the plant operator will most likely contest the imposition of the fine, and a hearing will be held before an ALJ employed in another division of the EPA. If the matter is not settled at the hearing, the ALJ will preside over another hearing and render a binding order. That order may be appealed within the agency and finally to the federal court. The courts, however, typically defer to the expertise of the agency and the associated ALJ. In other words, most orders by an ALJ are upheld.

Types of Administrative Agencies

executive agency
An agency that is typically located within the executive branch, under one of the cabinet-level departments. The agency head is appointed by the president with the advice and consent of the Senate.

independent agency
An agency that is typically not located within a government department. It is governed by a board of commissioners, who are appointed by the president with the advice and consent of the Senate.

Agencies are classified as either executive or independent. The administrative head of an **executive agency** is appointed by the president with the advice and consent of the U.S. Senate. Executive-agency heads may be discharged by the president at any time, for any reason. When a new president is elected, he will typically place his appointees in charge of executive agencies. These agencies are generally located within the executive branch, under one of the cabinet-level departments. Hence, executive agencies are referred to as *cabinet-level agencies*. Examples of traditional executive agencies are the Federal Aviation Agency (FAA), located within the Department of Transportation, and the Food and Drug Administration (FDA), located within the Department of Health and Human Services.

Independent agencies are governed by a board of commissioners, one of whom is the chair. The president appoints the commissioners of independent agencies with the advice and consent of the Senate, but these commissioners serve fixed terms and cannot be removed except for cause. No more than a simple majority of an independent agency can be members of any single political party (e.g., if the board consists of seven members, no more than four may be from the same political party). Serving fixed terms is said to make the commissioners less accountable to the will of the executive (thus the term *independent* agency). These agencies are generally not located within any department. Examples of independent agencies are the Federal Trade Commission (FTC), the Securities and Exchange Commission (SEC), and the Federal Communications Commission (FCC).

LO 4-5 Differentiate between executive and independent agencies.

Another difference between these two types of agencies is the scope of their regulatory authority. Executive agencies tend to have responsibility for making rules covering a broad spectrum of industries and activities. Independent agencies, often called *commissions,* tend to have more narrow authority over many facets of a particular industry, focusing on such activities as rate making and licensing. Executive agencies have a tendency to focus more on *social* regulation, whereas independent agencies are more often focused on what we refer to as *economic* regulation.

Federal and State Administrative Agencies

Currently, more than 100 federal agencies are in operation, as well as countless state agencies. Often, when there is a federal agency, there are also comparable state agencies to which the federal agency delegates much of its work. For example, the most important federal agency affecting environmental matters is the Environmental Protection Agency (EPA). Every state has a state environmental protection agency to which the federal EPA delegates primary authority for enforcing environmental protection laws. However, if at any time the state agency fails to enforce these laws, the federal EPA will step in to enforce them. Exhibit 4-1 lists the major administrative agencies.

INDEPENDENT AGENCIES	EXECUTIVE AGENCIES
Commodity Futures Trading Commission (CFTC)	Federal Deposit Insurance Corporation (FDIC)
Consumer Product Safety Commission (CPSC)	General Services Administration (GSA)
Equal Employment Opportunity Commission (EEOC)	International Development Corporation Agency (IDCA)
Federal Communications Commission (FCC)	National Aeronautics and Space Administration (NASA)
Federal Trade Commission (FTC)	National Science Foundation (NSF)
National Labor Relations Board (NLRB)	Occupational Safety and Health Administration (OSHA)
Nuclear Regulatory Commission (NRC)	Office of Personnel Management (OPM)
Securities and Exchange Commission (SEC)	Small Business Administration (SBA)
	Veterans Administration (VA)

Exhibit 4-1

Major Administrative Agencies

LO 4-6 Define Administrative Procedures Act.

Some agencies do not fall clearly into one classification or the other. These agencies are typically referred to as **hybrid agencies.** Created as one type of agency, the body may share characteristics of the other. The EPA, for example, was created as an independent agency, not located within any department of the executive branch. Yet it is headed by a single administrator who serves at the whim of the president. During the early 1990s, in fact, there were discussions of the need to transform the EPA into a cabinet-level executive agency. (These initiatives did not get beyond the discussion stage.) Another example is the "independent" Federal Energy Regulation Commission (FERC), which has the typical structure of an independent agency yet is located within the Department of Energy.

hybrid agency
An agency that has characteristics of both executive and independent agencies.

How Do Agencies Operate?

In 1946, Congress passed the **Administrative Procedures Act (APA)** as a major limitation on how agencies are run. Prior to passage of the APA, agencies could decide on their own how to make rules, conduct investigations, and hold hearings and trials. Under the APA, there are very specific guidelines on rule making by agencies. The two most common types of rule making are *informal* and *formal*; a third type is known as *hybrid*. Each is discussed below, along with a few exemptions.

Informal Rule Making

LO 4-7 Distinguish between formal and informal rule making.

The primary type of rule making administrative agencies use is **informal rule making,** or **notice-and-comment rule making.** Informal rule making applies in all situations in which the agency's enabling legislation or other congressional directives do not require another form. An agency initiates informal rule making by publishing the proposed rule in the *Federal Register*, along with an explanation of the legal authority for issuing the rule and a description of how the public can participate in the rule-making process. The *Federal Register* is the official daily publication for rules, proposed rules, and notices of federal agencies and organizations, as well as executive orders and other presidential documents.[3]

After publication, opportunity is provided for all interested parties to submit written comments. The comments may contain data, arguments, or other information a person believes might influence the agency in its decision making. Although the agency is not required to hold hearings,

Administrative Procedures Act (APA)
Federal legislation that places limitations on how agencies are run and contains very specific guidelines on rule making by agencies.

informal rule making
A type of rule making in which an agency publishes a proposed rule in the *Federal Register*, considers public comments, and then publishes the final rule. Also called *notice-and-comment rule making*.

Federal Register
The government publication in which an agency publishes each proposed rule, along with an explanation of the legal authority for issuing the rule and a description of how the public can participate in the rule-making process, and later publishes the final rule.

[3] Federal Register, www.federalregister.gov.

it has the discretion to receive oral testimony if it wishes to do so. Although the agency is not required to respond to all comments it receives, it is required to respond to comments that significantly concern the proposed rule. After considering the comments, the agency may alter the rule. It publishes the final rule, with a statement of its basis and purpose, in the *Federal Register*. This publication also includes the date on which the rule becomes effective, which must be at least 30 days after publication. Informal rule making is most often used because it is more efficient for the agency in terms of time and cost. No formal public hearing is required, and no formal record need be established. Some people believe that informal rule making is unfair because parties who are interested in the proposed rule have no idea what types of evidence the agency has received from other sources with respect to that rule. Thus, if the agency is relying on what one party might perceive as flawed or biased data, that party has no way to challenge that data.

BUT WHAT IF . . .

WHAT IF THE FACTS OF THE CASE OPENER WERE DIFFERENT?

Let's say, in the Case Opener, that the EPA decides it would like to make a rule governing the carbon dioxide emissions of new vehicles. The EPA publishes the rule in the *Federal Register* along with an explanation of the legal authority of the rule. Is this enough information for the *Federal Register?* What is the missing third prong?

Formal Rule Making

LO 4-7 How do formal, informal, and hybrid rule making differ from one another?

formal rule making
A type of rule making that is used when legislation requires a formal hearing process with a complete transcript; consists of publication of the proposed rule in the *Federal Register*, a public hearing, publication of formal findings, and publication of the final rule if adopted.

The APA requires **formal rule making** when an enabling statute or other legislation mandates all regulations or rules to be enacted by an agency as part of a formal hearing process that includes a complete transcript. The first step in formal rule making is publication in the *Federal Register* of a notice of proposed rule making by the agency.

The second step is a public hearing at which witnesses give testimony on the pros and cons of the proposed rule and are subject to cross-examination. An official transcript of the hearing is kept. On the basis of information received at the hearing, the agency makes and publishes formal findings. On the basis of these findings, an agency may or may not promulgate a regulation. If a regulation is adopted, the final rule is published in the *Federal Register*. Because of the expense and time involved in obtaining a formal transcript and record, most enabling statutes do not require a formal rule-making procedure when promulgating regulations. If a statute is drafted in a manner that is at all ambiguous with respect to the type of rule making required, the court will *not* interpret the law as requiring formal rule making.

In the following case, the court had to determine whether an agency effectively may suspend a duly promulgated regulation without observing the procedures or identifying relevant factual criteria that the law requires to effect such a change.

Notice and Comment Rulemaking

Open Communities Alliance v. Carson

2017 U.S. Dist. LEXIS 21131 (2017)

Section 8 of the Fair Housing Act of 1968 serves two statutory purposes: (1) aiding low-income families in obtaining a decent place to live and (2) promoting economically mixed housing. This case, however, is not about what is good housing policy. This case is about

(continued)

the rule of law—whether an agency effectively may suspend a duly promulgated regulation without observing the procedures or identifying relevant factual criteria that the law requires to effect such a change. The U.S. Department of Housing and Urban Development (HUD), without notice and comment or any evidence, delayed almost entirely by two years implementation of a rule requiring local Public Housing Authorities in 24 metropolitan areas, to calculate housing vouchers' values based on local, rather than metropolitan-wide, prevailing market rents. HUD argued that the Secretary had broad authority to delay the regulation's implementation wholesale whenever the Secretary determines he has reason to do so. The plaintiffs are two African-American women living in metropolitan areas the Rule would affect, as well as Open Communities Alliance ("OCA"), a nonprofit organization devoted to providing social and economic opportunities for low-income people. Plaintiff Crystal Carter is a voucher holder who would like to move from Hartford, Connecticut, to nearby Simsbury, Connecticut, to be near her children's schools and to provide her family a safer home and better opportunities. Plaintiff Tiara Moore would like to move from Chicago, Illinois, to the high-opportunity area of DuPage County, Illinois, to provide a safer neighborhood and better opportunities for her daughter. Not only does DuPage County have better schools than Chicago, Ms. Moore's mother lives in DuPage County and could provide childcare while Ms. Moore works. Neither plaintiff can afford to move without implementation of the rule. The plaintiffs sought a preliminary injunction to require HUD to implement the rule on its effective date of January 1, 2018. The court granted the preliminary injunction, holding that HUD's suspension of the rule was done without proper notice and comment procedure, did not meet any of the three exceptions to the requirement of notice and comment, and that the action was arbitrary and capricious.

Hybrid Rule Making

LO 4-7 How do formal, informal, and hybrid rule making differ from one another?

After agencies began regularly making rules in accordance with the appropriate procedures, the flaws of each type of rule making became increasingly apparent. In response to these problems, a form of hybrid rule making became acceptable to the courts and legislature. **Hybrid rule making** is an attempt to combine the best features of both formal and informal rule making. The starting point, publication in the *Federal Register*, is the same. Publication is followed by the opportunity for submission of written comments, and then there is an informal public hearing with a more restricted opportunity for cross-examination than that in formal rule making. The publication of the final rule is done in the same manner as for other forms of rule making.

Exhibit 4-2 compares formal, informal, and hybrid rule-making procedures.

hybrid rule making
A type of rule making that combines features of formal and informal rule making; consists of publication in the *Federal Register*, a written-comment period, and an informal public hearing with restricted cross-examination.

Exhibit 4-2

Rule-Making Procedures

PROCEDURE	Type of Rule Making		
	FORMAL	INFORMAL	HYBRID
Public hearing	Yes	No	Yes
Formal record	Yes	No	No
Publication of proposed rule in *Federal Register*	Yes	Yes	Yes
Written comments from the public and interested parties	Yes	Yes	Yes
Oral testimony and cross-examination	Yes	No (agency discretion)	Yes (limited)
Publication of final rule in *Federal Register*	Yes	Yes	Yes

EXEMPTED RULE MAKING

The APA contains an exemption from rule making that allows an agency to decide whether public participation will be allowed. Exemptions include rule-making proceedings with regard to "military or foreign affairs" and "agency management or personnel." Exemptions are also granted for rule-making proceedings relating to "public property, loans, grants, benefits, or contracts" of an agency. Military and foreign affairs often need speed and secrecy, which are incompatible with public notice and hearings. Other exemptions are becoming more difficult to justify in the eyes of the courts unless they meet one of the exemptions of the Freedom of Information Act (discussed later in this chapter).

Also exempted from the rule-making procedures are interpretive rules and general policy statements. An **interpretive rule** is a rule that does not create any new rights or duties but is merely a detailed statement of the agency's interpretation of an existing law. Interpretive rules are generally very detailed, step-by-step statements of what actions a party is to take to be considered in compliance with an existing law.

Policy statements are general statements about the directions in which an agency intends to proceed with respect to its rule-making or enforcement activities. Again, these statements have no binding impact on anyone; they do not directly affect anyone's legal rights or responsibilities.

A final exemption applies when public notice and comment procedures are "impracticable, unnecessary, or contrary to the public interest." This exemption is used most commonly either when the issue is so trivial that there would probably be very little, if any, public input, or when the nature of the rule necessitates immediate action. Whenever an agency chooses to use this exemption, it must make a good-cause finding and include in its publication of the final rule a statement explaining why there was no public participation in the process.

The following case demonstrates the broad powers given to agencies to investigate willful violations of agency rules.

interpretive rule
A rule that does not create any new rights or duties but is merely a detailed statement of an agency's interpretation of an existing law, including the actions a party is to take to be in compliance with the law.

policy statement
A general statement about the directions in which an agency intends to proceed with respect to its rule-making or enforcement activities; has no binding impact on anyone.

reg-neg
A type of rule making in which representatives of concerned interest groups and of the involved government agency participate in mediated bargaining sessions to reach an agreement, which is forwarded to the agency.

Regulated Negotiation

The exceedingly high number of challenges to regulations, as well as a growing belief that structured bargaining among competing interest groups might be the most efficient way to develop rules, has stimulated interest among a number of agencies in a relatively new form of rule making, often referred to as **reg-neg**. Each concerned interest group and the agency itself sends a representative to bargaining sessions led by a mediator. After the parties achieve a consensus, that agreement is forwarded to the agency.

The agency is then expected to publish the compromise as a proposed rule in the *Federal Register* and follow through with the appropriate rule-making procedures. The agency, however, is not bound to do so. If it does not agree with the proposal the group negotiated, the agency is free to try to promulgate a completely different rule or a modification of the one obtained through the negotiation. The reasoning behind reg-neg is similar to that supporting the increased use of mediation. If the parties can sit down and try to work out a compromise solution together, that solution is much more likely to be accepted than one handed down by some authority. The parties who hammered out the agreement now have a stake in making it work because they helped to create it.

Admittedly, reg-neg is not possible in all situations. If there is an unmanageably large group of interests that would have to be represented, if any possible compromise would have to result from one group's backing away from a fundamental principle, or if two groups feel so antagonistic toward each other that they would be unable to sit down and talk rationally, reg-neg would probably not even be worth trying.

Problems Associated with Rule Making

Agency employees are not subject to the same political pressures as legislators, but they are also not unbiased. Often the people with the necessary expertise to regulate specific areas come from the industry they will now be regulating. There is some concern that it will be difficult for regulators to ignore their past ties to industry and pass regulations that are in the public interest, especially when the regulations would increase costs to the industry or are opposed by the industry for other reasons. When people are discussing an agency in which they perceive this problem as existing, they often refer to the agency as being a captured agency. The counterargument is that

those who have been deeply involved in an industry know it best. To prevent such actions during his administration, President Obama issued the *Executive Order on Ethics Commitments by Executive Branch Personnel*. The order prohibits executive-branch employees from accepting gifts from lobbyists; closes the revolving door that allowed government officials to move to and from private-sector jobs in ways that gave that sector undue influence over government; and requires government hiring to be based on qualifications, competence, and experience, not political connections.[4] Regulators must be sensitive to the role the economy plays in the public's willingness to accept or support certain regulations. When the economy is flourishing and unemployment is low, there is much greater acceptance of regulations. When unemployment is high, people are much more reluctant to accept regulations that they believe might cause some workers to lose their jobs and/or cause the prices of products to rise.

Other Administrative Activities

Agencies perform a variety of less well known but equally important tasks. These include advising, conducting research, issuing permits, and managing property. One of the most common situations in which individuals come into contact with agencies occurs when an agency advises businesses and individuals on whether the agency considers an activity legal or illegal. Agencies also conduct studies of industries and markets. For example, the FTC, OSHA, and FDA conduct studies to determine, respectively, the level of economic concentration, safety in the workplace, and safety of drugs. Also, agencies provide information to the general public on various matters through hotlines, publications, and seminars. Agencies also devote much of their time to issuing licenses or permits. The EPA, for example, helps protect the environment by requiring certain environmentally sound activities before granting permits. Local agencies are responsible for issuing liquor licenses and cabaret (dancing) permits to local bars and restaurants. Finally, agencies often are responsible for managing government property.

Case 4-2 illustrates limitations on an agency's use of its powers to issue and, in this case, revoke a driver's license.

[4] For a full text of the executive order, see www.whitehouse.gov.

CASE 4-2

YAN JU WANG V. GEORGE VALVERDE
Court of Appeal of California, Second Appellate District
162 Cal. App. 4th 616 (2008)

FACTS: *The trial court granted plaintiff Wang's petition for a writ of administrative mandate compelling the defendant, the Department of Motor Vehicles (DMV), to set aside its revocation of Wang's class C driver's license. On appeal, the DMV argued that, as a matter of law, it had the authority to revoke Wang's class C license because it allegedly caught her cheating on an examination for a class B license. Wang, the holder of a valid class C (noncommercial) driver's license, applied for a class B (commercial) driver's license. She was given a written examination to determine whether she was qualified for a class B license, but she was not permitted to complete that examination because she was allegedly cheating while taking the examination by using crib notes. Wang was never criminally prosecuted for using crib notes, under Vehicle Code section 14610.5, because the DMV determined that there was insufficient evidence to support criminal action. The only administrative action taken against Wang by the DMV was to order the revocation of her class C (noncommercial) driver's license.*

ISSUE: Does the Vehicle Code empower the state to revoke a driver's license if the driver cheated on the test for a different driver's license?

REASONING: The DMV appealed the Appeals Court's concurring opinion that "[n]o such action is authorized by the Vehicle Code." The DMV argues that, as a matter of law, the DMV does have authority to revoke Wang's class C license because it allegedly caught her cheating on the examination for a class B license. The DMV purports to derive that authority from the Vehicle Code as follows: (1) Section 13359 provides that the DMV "may suspend or revoke the privilege of any person to operate a motor vehicle upon any of the grounds which authorize the refusal to issue a license;" (2) section 12809, subdivision (d), provides that the DMV may refuse to issue a license to any person who has "committed any fraud in any application;" (3) an examination is part of an application; (4) the use of a crib sheet in taking an examination

(continued)

is a fraudulent act; so (5) by using the crib sheet, Wang committed a fraud in an application, which therefore authorized refusal to issue a class B license, which therefore authorized revocation of her class C license, because any ground for refusal to issue a license is also a ground for revocation of a license.

The DMV's argument thus depends upon the DMV's contention that under section 13359, *any* ground for refusal to issue *one* license is also sufficient to justify revocation of a *different* license. The DMV cites no authority for its construction of section 13359. We have found no case on point, but we conclude that the DMV's interpretation cannot be correct because it would lead to untenable results. In general, the DMV's interpretation of the statute would authorize revocation of a class C license held by *any* unsuccessful applicant for a class B license. The statutory requirements for a class B license, however, are more demanding than the statutory requirements for a class C license, so statutory ineligibility for a class B license has no tendency, in itself, to show lack of statutory entitlement to a class C license.... Because the DMV's interpretation of section 13359 turns every unsuccessful application for a class B license into an authorization to revoke the applicant's class C license, it effectively negates the lower statutory threshold for entitlement to a class C license.

DECISION AND REMEDY: Affirmed, and respondent recovered the costs of her appeal.

SIGNIFICANCE OF THE CASE: The administrative agency is limited to the powers provided in the statute governing its operations.

CRITICAL THINKING

What facts, had they existed, would have altered this decision?

ETHICAL DECISION MAKING

Who, besides the parties directly involved, have interests in the results of this decision? In other words, who are the stakeholders that an ethical manager should consider?

Limitations on Agency Powers

LO 4-8 List the four limits on agency power.

There are four basic limits on agency power: political, statutory, judicial, and informational (see Exhibit 4-3). The hope is that these limitations will keep the agencies and their thousands of employees from abusing their discretion.

Exhibit 4-3

Limits on Agency Power

TYPE OF LIMIT	SPECIFIC LIMITATIONS
Political	The Senate must approve agency heads; Congress has power over agency budgets.
Statutory	Congress may create or eliminate agencies and amend enabling legislation (i.e., powers of agencies); Congress reviews and may override agency rules.
Judicial	Interested parties may challenge administrative rules in the courts, which may review the agency's finding of facts, its interpretation of the rule, and the scope of the agency's power in making the rule.
Informational	The Freedom of Information Act, Government in Sunshine Act, and Privacy Act of 1974 specify agencies' responsibilities regarding public access to information.

Political Limitations

As discussed earlier in this chapter, the president appoints the administrative heads of all executive agencies with the advice and consent of the Senate. If the president is unhappy with the head of such an agency, that person may be fired at any time for any (or no) reason. As such, executive agencies are particularly accountable to the executive branch. Moreover, the political slant (liberal or conservative) of the agency is significantly influenced by the president.

Congress also has significant control over agencies because the Senate must approve presidential nominees for them to become the administrative heads of agencies. Moreover, Congress has control over the budgets of all agencies. If Congress decides that a particular agency is not performing as it wishes, that agency's budget may be cut or even defunded. Just before the Enron crisis, Congress, after being heavily lobbied by the accounting industry, threatened to defund the Securities and Exchange Commission (SEC). In effect, Congress did not like the SEC's proposal that stock options should be charged as earnings. Arthur Levitt, then head of the SEC, heard the warning loud and clear and decided to walk away from his firmly held position. In later interviews, Levitt said it was his biggest regret as head of the SEC.[5]

Statutory Limitations

Congress has the power to create or dissolve an agency. If Congress is unhappy with an agency, it may amend the agency's enabling legislation and limit the agency's power. Moreover, in 1996, legislation was signed into law that gives Congress 60 days to review proposed agency rules. Congress may override those rules before they become effective. In addition, the APA (passed by Congress in 1946) sets forth guidelines that all agencies must follow when engaged in rule making.

Judicial Limitations

An individual or business that believes itself harmed by an administrative rule may challenge that rule in federal court after all administrative procedures have been exhausted.[6] This is probably the biggest constraint on agency power. If a rule is subjected to judicial review, the court will consider the following:

1. *The facts of the case:* Courts typically defer to an agency's fact finding. The facts must be supported by **substantial evidence**.
2. *The agency's interpretation of the rule:* Once again, the courts typically defer to the expertise of the agency and uphold the agency's interpretation of the rule.
3. *The scope of the agency's authority:* Has the agency exceeded the authority granted it by its enabling legislation?

substantial evidence
Refers to evidence that a reasonable mind could accept as adequate to support a conclusion. It is defined as "more than a scintilla but less than preponderance" and consists of "such relevant evidence as a reasonable person would accept as adequate to support a conclusion."

Although we typically think of judicial limitations as reining in overly zealous administrative regulators, the Case Opener illustrates that sometimes the judiciary is called on to restrict inaction by an agency. In *Massachusetts v. EPA*, the agency had resisted regulating greenhouse gases, arguing that carbon dioxide and similar gases are not pollutants under the Clean Air Act and, therefore, that the agency had no power to regulate them. When the high court reviewed the legislative mandate, however, it held that U.S. motor-vehicle emissions make a "meaningful contribution to greenhouse gas concentrations" and hence to global warming—and therefore the agency did have the authority to regulate such gases as air pollutants under the Clean Air Act.

In June 2009, the EPA granted a Clean Air Act waiver of preemption to California, allowing California to implement its own, more stringent greenhouse gas emission standards for motor vehicles beginning with model year 2009. By mid-2010, with the agreement of the auto industry, the California standard became the federal standard.[7]

[5] PBS video, *Bigger than Enron—How Greed and Politics Undercut America's Watchdogs*, 1999.
[6] In a few situations, a court may not review an agency action. These include situations involving politically sensitive issues and those in which the agency's enabling legislation prohibits judicial review.
[7] Tony McAdams, Nancy Neslund, and Kiren Zucker, *Law, Business, & Society*, 10th ed. (New York: McGraw-Hill, 2012), p. 745.

The National Labor Relations Board (NLRB) is an independent administrative agency (see Exhibit 4-1). As you read the following case, Northwestern University and College Athletes Players Association, are you convinced there is *substantial evidence* that student athletes should be treated as employees under the law? Moreover, when this administrative decision is appealed to the courts, should the courts defer to the NLRB's interpretation of what it means to be an "employee"?

CASE 4-3

NORTHWESTERN UNIVERSITY AND COLLEGE ATHLETES PLAYERS ASSOCIATION (CAPA)

Case 13-RC-121359, National Labor Relations Boards (NLRB)

(March 26, 2014)

FACTS: *Upon a petition duly filed under Section 9(c) of the National Labor Relations Act, as amended ("the Act"), a hearing was held before a hearing officer of the National Labor Relations Board ("the Board"). Pursuant to the provisions of Section 3(b) of the Act, the Board has delegated to the undersigned its authority in this proceeding.*

The Employer [Northwestern University] is a private, nonprofit, nonsectarian, coeducational teaching university chartered by the State of Illinois, with three campuses, including one located in Evanston, Illinois. It currently has an undergraduate enrollment of about 8,400 students. . . . The Employer maintains an intercollegiate athletic program and is a member of the National Collegiate Athletic Association (NCAA). The NCAA is responsible for formulating and enforcing rules governing intercollegiate sports for participating colleges. The Employer is also a member of the Big Ten Conference, and its students compete against the other 11 member schools (as well as nonconference opponents) in various sports. There are currently 19 varsity sports, which the Employer's students can participate in at the Division I level, including 8 varsity sports for men and 11 varsity sports for women. In total, there are about 500 students who compete in one of these sports each year for the Employer. . . . As part of its athletic program, the Employer has a varsity football team that competes in games against other universities. The team is considered a Football Bowl Subdivision (FBS) Division I program. Since 2006, the head football coach has been Patrick Fitzgerald, Jr., and he has been successful in taking his team to five bowl games. His football staff comprises a Director of Football Operations, Director of Player Personnel, Director of Player Development, nine full-time assistant coaches, and four graduate assistant coaches who assist him with his various duties. There are also five full-time strength coaches, two full-time video staff employees, two administrative assistants, and various interns who report to him. In turn, Head Coach Fitzgerald reports to Athletic Director James J. Phillips and President Dr. Morton Shapiro. The Employer's football team is composed of about 112 players of which there are 85 players who receive football grant-in-aid scholarships that pay for their tuition, fees, room, board, and books. The players on a scholarship typically receive grant-in-aid totaling $61,000 each academic year. . . .

ISSUE: Under the Act, should student athletes who play football be deemed employees and given the legal right to bargain collectively?

REASONING: A party seeking to exclude an otherwise eligible employee from the coverage of the Act bears the burden of establishing a justification for the exclusion. Accordingly, it was the Employer's burden to justify denying its scholarship football players employee status. I find that the Employer failed to carry its burden. Section 2(3) of the Act provides in relevant part that the "term 'employee' shall include any employee. . . ." The U.S. Supreme Court has held that in applying this broad definition of "employee," it is necessary to consider the common law definition of "employee." *NLRB v. Town & Country Electric*, 516 U.S. 85, 94 (1995). Under the common law definition, an employee is a person who performs services for another under a contract of hire, subject to the other's control or right of control, and in return for payment. As a result, the Board has subsequently applied the common law test to determine whether individuals are indeed statutory employees. As the record demonstrates, players receiving scholarships to perform football-related services for the Employer under a contract for hire in return for compensation are subject to the Employer's control and are therefore employees within the meaning of the Act.

Clearly, the Employer's players perform valuable services for their Employer. Monetarily, the Employer's football program generated revenues of approximately $235 million, during the nine-year period from 2003 to

(continued)

2012 through its participation in the NCAA Division I and Big Ten Conference, that were generated through ticket sales, television contracts, merchandise sales, and licensing agreements. The Employer was able to use this economic benefit provided by the services of its football team in any manner it chose. Less quantifiable but also of great benefit to the Employer is the immeasurable positive impact to Northwestern's reputation a winning football team may have on alumni giving and increase in number of applicants for enrollment at the University.... Equally important, the type of compensation that is provided to the players is set forth in a tender that they are required to sign before the beginning of each period of the scholarship. This tender serves as an employment contract and gives the players detailed information concerning the duration and conditions under which the compensation will be provided to them. Because NCAA rules do not permit the players to receive any additional compensation or otherwise profit from their athletic ability and/or reputation, the scholarship players are truly dependent on their scholarships to pay for basic necessities, including food and shelter....

In the instant case, the record establishes that the players who receive scholarships are under strict and exacting control by their Employer throughout the entire year.... The coaches have control over nearly every aspect of the players' private lives because they must follow many rules under threat of discipline and/or the loss of a scholarship. The players have restrictions placed on them and/or have to obtain permission from the coaches before they can (1) make their living arrangements, (2) apply for outside employment, (3) drive personal vehicles, (4) travel off campus, (5) post items on the Internet, (6) speak to the media, (7) use alcohol and drugs, and (8) engage in gambling. The fact that some of these rules are put in place to protect the players and the Employer from running afoul of NCAA rules does not detract from the amount of control the coaches exert over the players' daily lives.... In sum, based on the entire record in this case, I find that the Employer's football players who receive scholarships fall squarely within the Act's broad definition of "employee" when one considers the common law definition of "employee."

DECISION AND REMEDY: For the reasons discussed in detail below, I find that players receiving scholarships from the Employer are "employees" under Section 2(3) of the Act. Accordingly, IT IS HEREBY ORDERED that an election be conducted under the direction of the Regional Director for Region 13 in the following appropriate bargaining unit: Eligible to vote are all football players receiving football grant-in-aid scholarship and not having exhausted their playing eligibility employed by the Employer located at 1501 Central Street, Evanston, Illinois, but excluding office clerical employees and guards, professional employees and supervisors as defined in the Act.

Peter Sung Ohr, Regional Director
National Labor Relations Board, Region 13

SIGNIFICANCE OF THE CASE: The case was appealed and on August 17, 2015 the NLRB unanimously decided as follows: "We conclude that asserting jurisdiction in this case would not serve to promote stability in labor relations. Our decision today is limited to the grant-in aid scholarship football players covered by the petition in this particular case." In a clear victory for the college sports establishment, the five-member board preserved, for now, one of the N.C.A.A.'s core principles: that college athletes are primarily students.

CRITICAL THINKING

There is extensive listing in this decision of the various aspects of the student athlete's life that are controlled by the coaching staff. What is the purpose of that listing as justification for the eventual decision? In other words, how does this listing shape the reasoning?

ETHICAL DECISION MAKING

One of the groups of stakeholders who are often ignored in an ethical decision are future generations. Those of us alive now do not know the names of future generations, have by definition never met them, and consequently have a difficult time considering them relevant when we make a decision *even when* these people are going to be affected in major ways by the decision. What groups of people in future generations will be affected by this decision? How will they be affected?

Informational Limitations

Agency power is limited by the Freedom of Information Act (FOIA), the Government in Sunshine Act, and the Privacy Act of 1974.

Freedom of Information Act (FOIA)
Federal legislation that mandates and facilitates public access to government information and records, including records about oneself. Sensitive information (e.g., on national security) is excluded.

Freedom of Information Act The Freedom of Information Act (FOIA), passed in 1966, requires federal agencies to publish in the *Federal Register* places where the public can get information from the agencies. The act requires similar publication of proposed rules and policy statements. It requires agencies to make such items as staff manuals and interpretations of policies available for copying to individuals, on request. Finally, all federal government agencies must publish records electronically.

Any individual or business may make a FOIA request to a federal government agency. Information may be obtained regarding how the agency gets and spends its money. Statistics and/or information collected by the agency on a given topic are also available. Perhaps most important, citizens are entitled to any records that the government has about them. For example, you could contact the Internal Revenue Service (IRS) and request all information that it has collected about you. FOIA does not apply to Congress, the federal courts, the executive staff of the White House, state or local governments, and private businesses. Exemptions to FOIA include, but are not limited to, national security, internal agency matters (e.g., personnel issues), criminal investigations, financial institutions, and an individual's private life.

Immediately after taking office, President Obama issued a Memorandum for the Heads of Executive Departments and Agencies, indicating that FOIA should be administered with a clear presumption: In the face of doubt, openness prevails.[8] The memorandum directs that information should not be withheld simply because it is legal to do so and that if an agency cannot make full disclosure of information, it should consider making a partial disclosure.

Government in Sunshine Act The Government in Sunshine Act requires agency business meetings to be open to the public if the agency is headed by a collegiate body. A *collegiate body* consists of two or more persons, the majority of whom are appointed by the president with the advice and consent of the Senate. This open-meeting requirement applies only when a quorum is present. The law also requires agencies to keep records of closed meetings.

Privacy Act Under the Privacy Act of 1974, a federal agency may not disclose information about an individual to other agencies or organizations without that individual's written consent. This law guarantees three primary rights:

1. The right to see records about oneself, subject to the Privacy Act's exemptions.
2. The right to amend a nonexempt record if it is inaccurate, irrelevant, untimely, or incomplete.
3. The right to sue the government for violations of the statute, such as permitting unauthorized individuals to read your records.

Case 4-4 demonstrates the privacy protections given to individuals with regard to their personal information.

[8] U.S. Department of Justice, www.usdoj.gov/ag/foia-memo-march2009.pdf.

NASA v. NELSON
562 U.S. 134 (2011)

FACTS: *The National Aeronautics and Space Administration (NASA) has a work force of both federal civil servants and Government contract employees. Respondents are contract employees at NASA's Jet Propulsion Laboratory (JPL), which is operated by the California Institute of Technology (Cal Tech). Respondents were not subject to Government background*

(continued)

checks at the time they were hired, but that changed when the President ordered the adoption of uniform identification standards for both federal civil servants and contractor employees. The Department of Commerce mandated that contract employees with long-term access to federal facilities complete a standard background check by October 2007. NASA modified its contract with Cal Tech to reflect the new requirement, and JPL announced that employees who did not complete the process in time would be denied access to JPL and face termination by Cal Tech. Standard Form-85 asks whether an employee has "used, possessed, supplied, or manufactured illegal drugs" in the last year. If so, the employee must provide details, including information about "treatment or counseling received." The employee must also sign a release authorizing the Government to obtain personal information from schools, employers, and others during its investigation. With the deadline for completing the process drawing near, respondents brought suit claiming violation of privacy.

ISSUE: Whether the background-check process violates a constitutional right to informational privacy and whether the information collected is adequately protected with regard to confidentiality under the Privacy Act of 1974.

REASONING: In two cases decided over 30 years ago, this Court referred broadly to a constitutional privacy "interest in avoiding disclosure of personal matters." The Court upheld a New York law permitting the collection of names and addresses of persons prescribed dangerous drugs, finding that the statute's "security provisions," which protected against "public disclosure" of patient information, were sufficient to protect a privacy interest "arguably . . . root[ed] in the Constitution." In a separate case, the Court upheld a law requiring former President Nixon to turn over his presidential papers and tape recordings for archival review and screening, concluding that the federal Act at issue, like the statute in the prior case, had protections against "undue dissemination of private materials." We assume, without deciding, that the Constitution protects a privacy right of the sort mentioned in those two prior cases. We hold, however, that the challenged portions of the Government's background check do not violate this right in the present case. The Government's interests as employer and proprietor in managing its internal operations, combined with the protections against public dissemination provided by the Privacy Act of 1974, satisfy any interest in avoiding disclosure that may arguably have its roots in the Constitution.

DECISION AND REMEDY: The judgment of the Court of Appeals is reversed, and the case is remanded for further proceedings consistent with this opinion.

SIGNIFICANCE OF THE CASE: Statutory exceptions to the Privacy Act's disclosure bar may, in some cases, leave its protections too porous to supply a meaningful check against "unwarranted disclosures." Conversely, the mere fact that the Privacy Act's nondisclosure requirement is subject to exceptions does not show that the statute provides insufficient protection against public disclosure.

CRITICAL THINKING

Do you agree with the court's decision to require long term contract employees to undergo government background checks in order to keep their jobs? Why or why not?

ETHICAL DECISION MAKING

Ethics requires an initial consideration of the values that are in play in this particular decision. What values are in play with the court's decision to uphold the President's order to treat contract employees the same as government employees with regards to background checks?

E-COMMERCE and the Law

E-Government: Changing the Relationship between the Government and Citizens

According to the United Nations, *e-government*, which is defined as the delivery of government services through digital information technologies, including the Internet, is important because it reduces the cost of government while significantly improving the quality of services and citizen access. According to the latest UN report about this topic, e-government makes public administration more proactive, efficient, transparent, and service-oriented. Annual surveys by the UN indicate that nations throughout the world have made significant progress in e-government development, with the United States and Europe being among the leaders in this development.

The UN has characterized the development of e-government into the following stages:

Stage I–Emerging: A government's online presence mainly consists of a web page and/or an official website; links to ministries or departments of education, health, social welfare, labor, and finance may or may not exist. Much of the information is static, and there is little interaction with citizens.

Stage II–Enhanced: Governments provide more information on public policy and governance. They have created links that enable citizens to access archived information such as documents, forms, reports, laws and regulations, and newsletters easily.

Stage III–Interactive: Governments deliver online services such as downloadable forms for tax payments and applications for license renewals. In addition, the beginnings of an interactive portal or website with services to enhance the convenience of citizens are evident.

Stage IV–Transactional: Governments begin to transform themselves by introducing two-way interactions between citizen and government (C to G). This includes options for paying taxes and applying for ID cards, birth certificates, passports, and license renewals as well as similar C-to-G interactions, and it allows the citizen to access these services online 24/7.

Stage V–Connected: Governments transform themselves into a connected entity that responds to the needs of its citizens by developing an integrated back office infrastructure. This is the most sophisticated level of online e-government initiatives, and it is characterized by:

1. Horizontal connections (among government agencies).
2. Vertical connections (central and local government agencies).
3. Infrastructure connections (interoperability issues).
4. Connections between governments and citizens.
5. Connections among stakeholders (government, private sector, academic institutions, NGOs, and civil society).

In addition, governments support and encourage e-participation and citizen engagement in the decision-making process.[9]

The most recent ranking of nations in terms of their progress toward development of effective e-government was as follows: (1) Sweden, (2) Denmark, (3) Norway, (4) United States, (5) Netherlands, (6) Republic of Korea, (7) Canada, (8) Australia, (9) France, and (10) United Kingdom.[10] Sadly, the United States was at the top of the list in 2001, but because the Bush administration failed to make a serious commitment to transforming the federal government through information technology and failed to develop an overall e-government plan detailing where the government wants to go and how it wants to get there, other nations have made much more rapid progress in this area than has the United States.[11]

The Privacy Act applies only to records about individuals maintained by agencies in the executive branch of the federal government. There are 10 exemptions to the Privacy Act under which an agency can withhold certain kinds of information from you. Examples of exempt records are those containing classified information on national security and those concerning criminal investigations.[12]

[9] United Nations, "E-Government Survey 2008," http://unpan1.un.org/intradoc/groups/public/documents/UN/UNPAN028607.pdf.
[10] Ibid., p. 20.
[11] Robert D. Atkinson (Progressive Policy Institute), "Unsatisfactory Progress: The Bush Administration's Performance on E-Government Initiatives," October 14, 2004.
[12] Ibid.

CASE OPENER WRAP-UP

Does the EPA Have an Obligation to Regulate Automobile Emissions?

Nineteen private organizations had petitioned the EPA to require that it regulate carbon dioxide from automobile emissions. After holding hearings and requesting comments from the public, the EPA declined the request to regulate, arguing that they did not have statutory authority to do so and that executive policy addressing global warming warranted their refusal to do so. The matter was appealed and eventually worked its way to the US Supreme Court. The Bush White House filed an *amicus curie* ("friend of the court") brief arguing that the EPA was attempting to get the automobile industry to voluntarily reduce emissions. The Alliance of Automobile Manufacturers also came to the EPA's defense, arguing that the agency, as well as the states, had no authority to regulate automobile emissions.

In a 5–4 decision, the US Supreme Court disagreed, holding that the Clean Air Act authorizes the EPA to regulate greenhouse gas emissions from new motor vehicles in the event that it forms a "judgment" that such admissions contribute to climate change. Moreover, the Court held that the Clean Air Act's definition of *air pollutant* includes carbon dioxide.

In June 2009, the EPA granted a Clean Air Act waiver of preemption to California, allowing California to implement its own, more stringent greenhouse gas emission standards for motor vehicles beginning with model-year 2009. By mid-2010, with the agreement of the auto industry, the California standard became the federal standard.

SUMMARY

Introduction to Administrative Law	*Administrative law* consists of the substantive and procedural rules created by administrative agencies (government bodies of the city, county, state, or federal government) involving applications, licenses, permits, available information, hearings, appeals, and decision making.
	An *administrative agency* is generally defined as any body created by the legislative branch (e.g., Congress, a state legislature, or a city council) to carry out specific duties.
	Administrative agencies are created by Congress through passage of *enabling legislation*, which is a statute that specifies the name, functions, and specific powers of the administrative agency. Enabling statutes grant agencies broad powers for the purpose of serving the "public interest, convenience, and necessity."
	An agency holds an administrative hearing before an *administrative law judge (ALJ)*. The ALJ may attempt to get the parties to settle but has the power to issue a binding decision.
Types of Administrative Agencies	*Executive agency:* The administrative head of an executive agency is appointed by the president with the advice and consent of the U.S. Senate. Executive-agency heads may be discharged by the president at any time, for any reason. Executive agencies are generally located within the executive branch, under one of the cabinet-level departments (and thus are referred to as *cabinet-level agencies*). Examples are the Federal Aviation Agency (FAA) and the Food and Drug Administration (FDA).
	Independent agency: Independent agencies are governed by a board of commissioners. The president appoints the commissioners with the advice and consent of the Senate. Commissioners serve fixed terms and cannot be removed except for cause. No more than a simple majority of an independent agency can be members of any single political party.

(continued)

Hybrid agency: Hybrid agencies do not fall clearly into classification as an independent or an executive agency. Created as one type of agency, the body may share characteristics of the other type. The EPA, for example, was created as an independent agency, not located within any department of the executive branch, yet the head of the EPA serves at the whim of the president.

How Do Agencies Operate?

In 1946, Congress passed the *Administrative Procedures Act (APA)* as a major limitation on how agencies are run. The APA provides very specific guidelines on rule making by agencies.

- *Informal rule making:* Under the APA, the proposed rule is published in the *Federal Register,* and there is opportunity for public comment.
- *Formal rule making:* All rules must be enacted by an agency as part of a formal hearing process that includes a complete transcript. The process begins with publication in the *Federal Register* of a notice of proposed rule making by the agency. Next, there is a public hearing at which witnesses give testimony on the pros and cons of the proposed rule and are subject to cross-examination. The agency makes and publishes formal findings. If a regulation is adopted, the final rule is published in the *Federal Register.*
- *Hybrid rule making:* Hybrid rule making is an attempt to combine the best features of formal and informal rule making. The starting point is publication in the *Federal Register,* followed by the opportunity for submission of written comments and then an informal public hearing with a more restricted opportunity for cross-examination than that in formal rule making.
- *Exempted rule making:* The APA contains an exemption from rule making that allows an agency to decide whether public participation will be allowed. Exemptions include rule-making proceedings with regard to military or foreign affairs, agency management or personnel, and public property, loans, grants, benefits, or contracts of an agency.

Interpretive rules are rules that do not create any new rights or duties but merely state the agency's interpretation of an existing law. They are generally very detailed, step-by-step statements of what actions a party is to take to be considered in compliance with an existing law.

In *reg-neg,* each concerned interest group and the agency itself sends a representative to bargaining sessions led by a mediator. If the parties achieve a consensus, that agreement is forwarded to the agency. The agency is then expected to publish the compromise as a proposed rule in the *Federal Register* and follow through with the appropriate rule-making procedures. If the agency does not agree with the proposal the group negotiated, the agency is free to try to modify or replace it.

Limitations on Agency Powers

There are four basic limits on agency power: political, statutory, judicial, and informational.

Freedom of Information Act: Passed in 1966, FOIA requires federal agencies to publish in the *Federal Register* places where the public can get information from the agencies. Any individual or business may make a FOIA request to a federal agency for information on how the agency gets and spends its money, statistics and/or information collected by the agency on a given topic, and any records the government has about the individual seeking the information. Exemptions to FOIA include, but are not limited to, national security, internal agency matters (e.g., personnel issues), criminal investigations, financial institutions, and an individual's private life.

Government in Sunshine Act: This act requires agency business meetings to be open to the public if the agency is headed by a collegiate body (i.e., two or more persons, the majority of whom are appointed by the president with the advice and consent of the Senate). The law also requires agencies to keep records of closed meetings.

Privacy Act: Under the Privacy Act, a federal agency may not disclose information about an individual to other agencies or organizations without that individual's written consent.

Federal and State Administrative Agencies

Administrative agencies exist at the federal and state levels.

Point/Counterpoint

Is Administrative Regulation of Business a Good Thing?	
YES	NO
It looks out for the safety of consumers. It protects the safety and health of the general public as well as the environment. It looks after the stability of the economy. It keeps in mind the well-being of the people. It handles serious social problems and crises that are beyond the expertise of legislators and judges.	It creates a government bureaucracy that stifles growth. It can create monopolies that cause consumers to pay more. It squashes innovation by over-regulating. It can be extremely costly to tax payers. It can be difficult to follow one government regulation without breaking another.

Questions & Problems

1. What is enabling legislation?
2. What are the limits on agency power?
3. What are the three main powers given to agencies?
4. List the three categories of agencies.
5. Describe the various types of rule making.
6. In 1999 and 2001, the Department of Labor's (DOL) Wage and Hour Division issued letters opining that mortgage-loan officers do not qualify for the administrative exemption to overtime pay requirements under the Fair Labor Standards Act (FLSA). In 2004, the DOL issued new regulations regarding the exemption. Mortgage Bankers Association (MBA) requested a new interpretation of the revised regulations as they applied to mortgage-loan officers. In 2006, an opinion letter found that mortgage-loan officers fell within the administrative exemption. In 2010, the DOL again altered its interpretation of the administrative exemption. Without notice or an opportunity for comment, the DOL withdrew the 2006 opinion letter and issued an Administrator's Interpretation concluding that mortgage-loan officers do not qualify for the administrative exemption. MBA filed a lawsuit contending that the Administrator's Interpretation was procedurally invalid based on case law which holds that an agency must use the APA's notice-and-comment procedures when it wishes to issue a new interpretation of a regulation that deviates significantly from a previously adopted interpretation. Is a repeal of a rule subject to the notice-and-comment requirements of the APA?

 Source: *Perez v. Mortgage Bankers Association,* 135 S. Ct. 1199 (2015).

7. A trade association asked the Federal Communications Commission (FCC) to clarify language in 47 U.S.C.S. § 332(c)(7)(B)(ii) which required state and local governments to act on any request to place, construct, or modify personal wireless service facilities "within a reasonable period of time." The FCC issued a ruling that a "reasonable period of time" under the statute was presumptively (but rebuttably) 90 days to process a collocation application and 150 days to process all other applications. Several cities challenged the FCC's ruling. The Fifth Circuit found that the FCC possessed statutory authority to adopt the 90- and 150-day time frames and that the FCC's ruling was not arbitrary and capricious. The Supreme Court granted certiorari to review the issue. Is an agency's interpretation of a statutory ambiguity that concerns the scope of its regulatory authority (that is, its jurisdiction) is entitled to deference? Why or why not?

 Source: *City of Arlington v. FCC,* 569 U.S. 290 (2013).

8. Morales, a native and citizen of Mexico, was arrested in 1994 for entering the United States without inspection. He was released and served with a mail-out order to show cause why he should not be sent back to Mexico. Eventually, a removal hearing was scheduled, and Morales was notified by certified mail of the time and place of the hearing. When Morales failed to attend the hearing, he was ordered removed in absentia. The Immigration and Naturalization Service (INS) apprehended and removed Morales from the United States in 1998. He attempted to reenter illegally in January 2001—this time using a false border-crossing card. He was apprehended at the port of entry and was expeditiously removed. Undaunted, Morales reentered the United States undetected the following day. Sometime between his 1998 and 2001 removals, Morales married a U.S. citizen. In March 2001, Morales's wife filed an I-130 alien relative petition based on his marriage to a U.S. citizen. When Morales and his wife met with the INS in January 2003, an immigration officer served them with a denial of the I-130 petition and a notice of intent to reinstate Morales's removal order. The case came before a three-judge panel, which held that the regulation authorizing

immigration officers to issue reinstatement orders is invalid, and Morales's removal order could be reinstated only by an immigration judge. Until 1997, removal orders could be reinstated only by immigration judges (i.e., not immigration officers). In 1997, the attorney general changed the applicable regulation to delegate this authority, in most cases, to immigration officers. Does the attorney general have the authority to change an INS regulation? Why or why not?

Source: *Raul Morales-Izuierdo v. Alberto R. Gonzales, Attorney General,* 2007 U.S. App. LEXIS 10865 (9th Cir. 2007).

9. Six conservation organizations filed a petition in Thurston County Superior Court, asking that court to hold, pursuant to a provision in the Administrative Procedures Act (APA), that the Washington Forest Practices Board, the Washington Department of Ecology, and the Washington Department of Natural Resources (agencies) "failed to promulgate forest practice rules that advanced the environmental protection purposes and policies of the Forest Practices Act of 1974." The agencies countered by arguing that a formal petition for rule making must precede any petition for judicial review. Must an interest group exhaust administrative remedies before suing in a state or federal court? Why or why not?

Source: *Northwest Ecosystem Alliance v. Washington Forest Practices Board,* 66 P.3d 614 (2003).

10. Congress established an independent agency, the Consumer Financial Protection Bureau (CFPB), to curb fraud and promote transparency in consumer loans, home mortgages, personal credit cards, and retail banking. Under the Dodd-Frank Wall Street Reform and Consumer Protection Act, Congress gave the CFPB a single Director protected against removal by the President without cause. The Director of this independent agency may only be fired only for "inefficiency, neglect of duty, or malfeasance in office." That design choice is challenged here as an unconstitutional impediment to the President's power. Does removal of an independent agency's director (i.e., the Director of the CFPB) only for cause violate the President's powers under Article II?

Source: *PHH Corp. v. Consumer Fin. Prot. Bureau,* 881 F.3d 75 (2018).

11. The county board of commissioners (the Board) entered into a memorandum of understanding (MOU) with a professional baseball team. The MOU obligated the team, among other things, to relocate to the city for spring training. The MOU called for the renovation of a stadium complex. In addition, the MOU called for renovations at the team's minor league spring training facilities. Several informational briefings for individual members of the Board were conducted privately, and emails were circulated among Board members, regarding the negotiations and agreement with the team. Sarasota Citizens for Responsible Government sued, alleging that the privately conducted meetings and emails by Board members were a violation of the Government in Sunshine Law requiring that:

> All meetings of any board or commission of any state agency or authority or of any agency or authority of any county, municipal corporation, or political subdivision, except as otherwise provided in the Constitution, at which official acts are to be taken are declared to be public meetings open to the public at all times, and no resolution, rule, or formal action shall be considered binding except as taken or made at such meeting. The board or commission must provide reasonable notice of all such meetings.

How should the court rule? Why?

Source: *Sarasota Citizens for Responsible Gov't v. City of Sarasota,* 48 So. 3d 755 (2010).

12. The respondent, the Public Company Accounting Oversight Board, was created as part of a series of accounting reforms in the Sarbanes-Oxley Act of 2002. The board is composed of five members appointed by the Securities and Exchange Commission (SEC). It was modeled on private, self-regulatory organizations in the securities industry—such as the New York Stock Exchange—that investigate and discipline their own members subject to SEC oversight. Unlike these organizations, however, the board is a government-created entity with expansive powers to govern an entire industry. Every accounting firm that audits public companies under the securities laws must register with the board, pay it an annual fee, and comply with its rules and oversight. The board may inspect registered firms, initiate formal investigations, and issue severe sanctions in its disciplinary proceedings. The parties agree that the board is "part of the Government" for constitutional purposes. Although the SEC has oversight of the board, it cannot remove board members at will but can do so only "for good cause shown," "in accordance with" specified procedures. The parties also agree that the commissioners, in turn, cannot themselves be removed by the president except for "inefficiency, neglect of duty, or malfeasance in office." The board inspected the petitioner accounting firm, released a report critical of its auditing procedures, and began a formal investigation. The firm and the petitioner, Free Enterprise Fund, a nonprofit organization of which the firm is a member, sued the board and its members, seeking, *inter alia,* a declaratory judgment that the board is unconstitutional and an injunction preventing the board from exercising its powers. The petitioners argued that the Sarbanes-Oxley Act contravened the separation of powers by conferring executive power on board members without subjecting them to presidential control. Is the Public Company Accounting Oversight Board constitutional? Why or why not?

Source: *Free Enter. Fund v. Pub. Co. Accounting Oversight Bd.,* 130 S. Ct. 3138 (2010).

Part 1 The Legal Environment of Business

CHAPTER 5

Constitutional Law

CASE OPENER

The Constitutionality of the Affordable Care Act

In 2010, in an attempt to increase the number of Americans covered by health insurance and reduce the cost of health care, Congress passed the Patient Protection and Affordable Care Act. A key provision of the act was the "individual mandate," a provision requiring most Americans to obtain "minimum essential" health insurance. By 2014, those not exempt or covered by insurance provided by their employer or a government program would have to buy insurance from a private insurer, or they would be assessed a shared responsibility payment by the Internal Revenue Service. This penalty would be assessed and collected in the same manner as any other tax penalty.

Twenty-six states, several individuals, and the National Federation of Independent Business brought an action challenging the constitutionality of the individual mandate, among other provisions of the act. The 11th Circuit Court of Appeals upheld the other challenged provisions of the act but found that Congress lacked the authority to pass the individual mandate. The appellate court went on to hold that the individual mandate was severable from the rest of the act, despite the argument that the framework of the act depended on the individual mandate, and so upheld the rest of the act.

1. Did Congress have the authority to pass the act under the commerce clause?
2. Did Congress have the authority to pass the act under its power to lay and collect taxes?

LEARNING OBJECTIVES

LO 5-1 Define federalism.
LO 5-2 Explain the function of each branch of government and how each branch serves to check the others.
LO 5-3 Describe the relationship between the supremacy clause and preemption.
LO 5-4 Describe how the commerce clause affects the power of both the state and federal governments to regulate.
LO 5-5 Differentiate between commercial and political speech.
LO 5-6 Recognize how the rights guaranteed by the Bill of Rights affect business managers.

The U.S. Constitution sets forth the framework of our nation's government; it establishes a system of government that divides power between the federal government and the states. This system of government provides the focus for this chapter.

In this chapter, we examine the constitutional provisions that affect business. Then we turn our focus to the primary source of the federal government's authority to regulate business: the commerce clause. We next examine the federal government's authority to tax and spend. Finally, in the last part of the chapter, we focus on how several amendments to the Constitution affect business.

The U.S. Constitution

LO 5-1 Define federalism.

federalism
A system of government in which power is divided between a central authority and constituent political units.

The U.S. Constitution establishes a system of government based on the principle of **federalism**, according to which the authority to govern is divided between federal and state governments. According to the Tenth Amendment to the Constitution, all powers that the Constitution neither gives exclusively to the federal government nor takes from the states are reserved for the states. Because the federal government has only those powers granted to it by the Constitution, federal legislation that affects business must be based on an expressed constitutional grant of authority.

In addition to allocating authority between state and federal governments, the Constitution allocates the power of the federal government among the three branches of government. The first three articles of the Constitution establish three independent branches of the federal government: the legislative, executive, and judicial branches. The Constitution ensures that each branch maintains a separate sphere of power to prevent any one branch from obtaining undue power and monopolizing control of government.

The Constitution also establishes a system of *checks and balances*. Each branch's powers keep the other branches from dominating the government. For example, Congress, the legislative or lawmaking branch, has the power to enact legislation, but the president can veto a law that Congress passes. The legislature, however, can overturn a presidential veto with a two-thirds vote of the members of Congress. And if Congress passes a bill and the president signs it, the judiciary can strike it down as unconstitutional. Exhibit 5-1 illustrates the system of checks and balances in more detail.

Judicial Review

LO 5-2 Explain the function of each branch of government and how each branch serves to check the others.

judicial review
The power of a court to review legislative and executive actions, such as a law or an official act of a government employee or agent, to determine whether they are constitutional.

Although the Constitution does not explicitly allow courts to review legislative and executive actions to determine whether they are constitutional, early common law established this process, known as **judicial review**. In the landmark 1803 U.S. Supreme Court case *Marbury v. Madison*, Chief Justice John Marshall wrote for the majority:

> [I]f a law be in opposition to the constitution; if both the law and the constitution apply to a particular case, so that the court must either decide that case [conforms] to the law, disregarding the constitution; or [conforms] to the constitution, disregarding the law; the court must determine which of these conflicting rules governs the case. This is of the very essence of judicial duty.

Judicial review also allows courts to review the constitutionality of lower courts's decisions.

Exhibit 5-1 The System of Checks and Balances

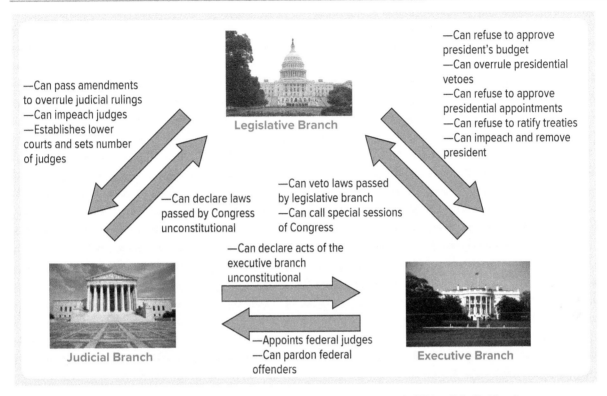

(top) ©Image Source, all rights reserved; (bottom right) ©McGraw-Hill Education/Jill Braaten; (bottom left) ©Hisham F. Ibrahim/Getty Images.

The Supremacy Clause and Federal Preemption

LO 5-3 Describe the relationship between the supremacy clause and preemption.

The **supremacy clause**, located in Article VI of the Constitution, provides that the Constitution, laws, and treaties of the United States constitute the supreme law of the land, "any Thing in the Constitution or Laws of any State to the Contrary notwithstanding." Any state or local law that directly conflicts with the Constitution, federal laws, or treaties is void. Federal laws include rules passed by federal administrative agencies. For example, the Federal Aviation Administration has banned the use of cell phones on airplanes. If a Missouri law permitted cell phone usage on airplanes, it would be unconstitutional according to the supremacy clause.

In some areas, the state and federal governments have concurrent authority; that is, both governments have the power to regulate the same subject matter. In such cases, the states may regulate in the area as long as a person's compliance with the state regulation would not cause her or him to be in violation of a federal regulation. For example, Congress has established a number of environmental standards, but some states have passed even more protective standards.

Sometimes, however, when the state and federal governments have concurrent authority, the federal government can decide to regulate that area exclusively. In such a situation, according to the doctrine of **federal preemption,** the state law is unconstitutional. To determine whether Congress intended to provide exclusive regulation, courts look to the language of the statute and transcripts of congressional hearings.

supremacy clause
Article VI, Paragraph 2, of the U.S. Constitution, which states that the Constitution and all laws and treaties of the United States constitute the supreme law of the land. Thus, any state or local law that directly conflicts with the U.S. Constitution or federal laws or treaties is void.

federal preemption
A principle asserting the supremacy of federal legislation over state legislation when both pertain to the same matter.

The Commerce Clause

commerce clause
Clause 3 of Article I, Section 8, of the U.S. Constitution, which authorizes Congress to "regulate Commerce with foreign Nations, and among the several States, and with the Indian Tribes."

The primary source of authority for federal regulation of business is the **commerce clause**, located in Article I, Section 8, of the Constitution. This clause states that the U.S. Congress has the power to "regulate Commerce with foreign Nations, and among the several States, and with the Indian Tribes." This allocation of authority simultaneously empowers the federal government and restricts the power of state governments.

The Commerce Clause as a Source of Authority for the Federal Government

LO 5-4 Describe how the commerce clause affects the power of both the state and federal governments to regulate.

Today, most federal regulations are exercises of congressional authority according to the commerce clause. As long as a law affects commerce among the states, or interstate commerce, in some way, the regulation is generally constitutional. Congress, however, has not always enjoyed such extensive power.

In many cases, legislatures change the law by replacing statutes with different ones. In many other cases, however, the law changes even when statutes do not change because courts often interpret statutes differently over time. For example, although the text of the commerce clause has not changed since 1787, courts's interpretations of the meaning of the commerce clause have morphed over time. In particular, the phrase "among the several States" has been subject to changing interpretations throughout U.S. history. Prior to the 1930s, courts interpreted the clause very strictly, requiring the regulated activity actually to involve trade between states. This interpretation limited federal regulation of business.

In the 1930s, however, the Supreme Court began to interpret the commerce clause more broadly. The 1937 case *NLRB v. Jones & Laughlin Steel Corp.* was a turning point in the Supreme Court's interpretation of the commerce clause. In that case, the Court ruled that Congress could regulate labor relations at a manufacturing plant because a work stoppage at the plant would seriously affect interstate commerce. The Court stated that "[a]lthough activities may be intrastate in character when separately considered, if they have such a close and substantial relationship to interstate commerce that their control is essential or appropriate to protect that commerce from burdens or obstructions, Congress cannot be denied the power to exercise that control." Since that case, Congress has regulated a broad range of business activities according to the commerce clause, through legislation such as the Federal Mine Safety and Health Act, which sets standards for safety in coal mines; the Americans with Disabilities Act, which prohibits firms from discriminating against employees and potential employees who have disabilities; and the Consumer Protection Act, which criminalizes certain loan-sharking activities. Although businesses have challenged these statutes as being beyond the scope of congressional power, courts have upheld the statutes as valid exercises of congressional authority according to the commerce clause.

When the law changes, however, its movement is not always unidirectional. In fact, sometimes courts revert to standards and methods of interpretation that they rejected in the past. For example, the 1995 case *United States v. Lopez* marked a significant change in the nearly 60-year trend in the Supreme Court's interpretation of the commerce clause. In *Lopez,* the Court ruled that Congress had exceeded its commerce clause authority when it passed the Gun-Free School Zone Act, a law banning the possession of guns within 1,000 feet of any school. In its ruling, the Court said that Congress could not regulate in an area that had "nothing to do with commerce, or any sort of enterprise."

Case 5-1 illustrates how the Supreme Court applies the commerce clause to determine the constitutionality of statutes Congress passes.

CASE 5-1

CHRISTY BRZONKALA v. ANTONIO J. MORRISON ET AL.
United States Supreme Court
120 S. Ct. 1740 (2000)

FACTS: *Petitioner Christy Brzonkala and respondents Antonio Morrison and James Crawford were students at Virginia Polytechnic Institute (Virginia Tech). Brzonkala met the respondents at a campus party, where they allegedly assaulted and raped her; according to Brzonkala, Morrison subsequently made boasting, debasing, and vulgar remarks on campus about what he would do to women. Brzonkala claimed to be emotionally disturbed and depressed as a result. She was treated by a university psychiatrist, and soon after the rape she withdrew from the university.*

Brzonkala filed a complaint against the respondents under Virginia Tech's Sexual Assault Policy. Morrison was initially found guilty and suspended for two semesters, but his punishment was ultimately set aside.

Brzonkala then sued Morrison, Crawford, and Virginia Tech in federal court, alleging, among other claims, that Morrison's and Crawford's attack violated the Violence against Women Act. Morrison and Crawford moved to dismiss this complaint on the grounds that it failed to state a claim and that the act's (§ 13981) civil remedy was unconstitutional.

The district court dismissed the complaint on grounds that Congress lacked authority to enact the section. The U.S. court of appeals affirmed the district court's conclusion.

ISSUE: Did Congress have the power under the commerce clause to enact the Violence against Women Act?

REASONING: Congress can regulate three broad categories of activities under its commerce powers. It can regulate the use of the channels of interstate commerce. It can regulate and protect the instrumentalities of interstate commerce, or persons or things in interstate commerce, even though the threat may come only from intrastate activities. It can regulate those activities that substantially affect interstate commerce. The act is properly analyzed as a potential regulation of an activity that substantially affects commerce.

The most relevant precedent is *Lopez*, in which the Supreme Court held that the Gun-Free School Zones Act of 1990, which made it a federal crime to knowingly possess a firearm in a school zone, exceeded Congress's authority under the commerce clause. That statute was struck down because, first, by its terms it was unrelated to commerce, and whenever the courts have sustained federal regulation of intrastate activity based on the activity's substantial effects on interstate commerce, the activity in question has been some sort of economic endeavor.

Second, the statute had no express jurisdictional element to limit its reach to a discrete set of firearm possessions that additionally have an explicit connection with or effect on interstate commerce. Third, Congress had no findings about the effect of gun possession at school on interstate commerce. Fourth, the link between gun possession and a substantial effect on interstate commerce was attenuated. To uphold the law would have meant that there were no limits on what Congress could regulate.

Likewise, gender-motivated crimes of violence are not economic activity, and there was no jurisdictional element showing that the federal cause of action is in pursuance of Congress's power to regulate interstate commerce. There were some statistics showing the impact of gender-related crime on victims and their families, but the Court had to decide whether these statistics demonstrate a substantial impact on interstate commerce, and the Court did not believe they do. The Court found that the congressional finding relied on a method of reasoning the Court had already rejected as unworkable if any limit on congressional powers is to be maintained. It is not sufficient to find that gender-motivated violence affects interstate commerce "by deterring potential victims from traveling interstate, from engaging in employment in interstate business, and from transacting with business, and in places involved in interstate commerce; . . . by diminishing national productivity, increasing medical and other costs, and decreasing the supply of and the demand for interstate products." Allowing such reasoning would give Congress the power to regulate almost any crime.

DECISION AND REMEDY: The Court affirmed the lower courts's decisions, holding that Congress may not regulate noneconomic, violent criminal conduct solely on the basis of that conduct's aggregate effect on interstate commerce.

SIGNIFICANCE OF THE CASE: The case clearly demonstrates an increasing willingness of the Court to limit congressional authority under the commerce clause. Note that the Court is not passing judgment on the quality of the law but is just saying that such a law would need to be passed by the state, not the federal, legislature.

CRITICAL THINKING

This case was a 5-4 decision. What reasoning do you think supported the dissent's opinion?

(continued)

> **ETHICAL DECISION MAKING**
>
> Recall the WH framework. What values led the majority to rule that Congress lacks authority to enact § 13981 of the Violence against Women Act? Do you think the values you listed are appropriate in this situation in spite of the individuals who are harmed by the decision?

The issue of whether the commerce clause has authorized Congress to act still arises today. One such recent illustration is in the individual mandate provision of the Affordable Care Act (ACA), described in the opening scenario. Despite cases such as *Lopez* and *Brzonkala*, many commentators had thought that the ACA individual mandate would be seen as a regulation affecting interstate commerce. The implementation of the act would certainly lead to an increase in purchases of health care insurance, thereby significantly affecting the market for it and for health care; more people would be able to obtain additional health care once they had insurance. However, Justice Roberts disagreed, making a new distinction between regulating existing activity and creating new activity, with the former being seen as affecting interstate commerce and the latter not doing so. Thus, the individual mandate provision contained in the ACA could not be justified as authorized by the commerce clause.

> **BUT WHAT IF? . . .**
>
> WHAT IF THE FACTS OF THE CASE OPENER WERE DIFFERENT?
>
> Recall that in the Case Opener, Obama's mandate required people to purchase health insurance who did not currently have. What if Obama's act had imposed certain rules on only those who currently had health insurance, simply mandating that their insurers could not drop them from coverage and requiring that individuals who currently had coverage could not drop that coverage?

The Commerce Clause as a Restriction on State Authority

police power
Power retained by each state to pass laws that protect the health, safety, and welfare of its citizens.

dormant commerce clause
A restriction on states's authority that is implied in the commerce clause of the U.S. Constitution: The power given to Congress to enact legislation that affects interstate commerce in effect prohibits a state from passing legislation that improperly burdens interstate commerce.

The federal government's authority to regulate interstate commerce sometimes conflicts with the states's authority to regulate intrastate commerce. Courts have attempted to resolve this conflict by distinguishing between regulations of commerce and regulations under states's police power. **Police power** consists of the residual powers retained by each state to safeguard the health and welfare of its citizenry. Typical exercises of a state's police power include state criminal laws, building codes, zoning laws, sanitation standards for restaurants, and regulations for the practice of medicine.

Sometimes a state's use of its police power affects interstate commerce. If the purpose of a state law is to regulate interstate commerce or to discriminate against interstate commerce, the law is usually unconstitutional. Likewise, if a law substantially interferes with interstate commerce, it is generally unconstitutional. This restriction on states's authority to pass laws that substantially affect interstate commerce is called the **dormant commerce clause**.

Most cases are not so simple, however, and courts must balance the states's interest in protecting their citizens against the impact on interstate commerce. In balancing these competing interests, a court generally asks whether the state regulation is rationally related to a legitimate state end. If it is, the court then asks whether the regulatory burden imposed on interstate commerce is

outweighed by the state's interest in enforcing the legislation. The court may also inquire whether there is a less drastic alternative available to attain the legitimate state purpose.

Although the supremacy clause establishes the sovereignty of federal law, courts generally presume that laws passed in accordance with states's police power are valid. For example, the city of Chicago passed an ordinance banning spray paint in the city as a means of reducing graffiti. Paint manufacturers challenged the legislation as a violation of the dormant commerce clause, but the U.S. court of appeals upheld the legislation.

Case 5-2 illustrates an attempt to challenge a state regulation on the grounds that it places an undue burden on interstate commerce.

CASE 5-2

BLACK STAR FARMS, LLC, ET AL. v. JERRY OLIVER, ARIZONA DEPARTMENT OF LIQUOR LICENSE AND CONTROL ET AL.
U.S. Court of Appeals for the Ninth Circuit
600 F.3d 1225 (9th Cir. 2010)

FACTS: *Arizona generally requires all alcoholic beverages sold to consumers in the state to pass through a three-tier distribution system comprising producers, wholesalers, and retailers. However, Arizona has carved out two exceptions to its system that allow wineries under specified circumstances to bypass the three-tier distribution system. First, all wineries that produce fewer than 20,000 gallons of wine per year—whether located in state or out of state—are allowed to ship an unlimited amount of wine directly to consumers, regardless of how the order is placed, and to sell directly to retailers. Second, all wineries—whether located in state or out of state—are allowed to ship two cases of wine per year directly to consumers who purchase wine while they are physically present at the winery.*

Black Star Farms, a Michigan winery, produced approximately 35,000 gallons of wine in 2006. Thus, the small-winery exception was not available to Black Star Farms, and wholesalers would not handle its wine in a cost-effective manner because it did not sell enough in the state. Nevertheless, the in-person exception did allow Black Star Farms to ship up to two cases per year to a consumer who purchased the wine while he or she was physically present at the winery in Michigan. Black Star Farms contended that these challenged exceptions to the three-tier system violated the dormant commerce clause and sued the State Liquor Control Board because the winery could not ship directly to consumers who do not physically go to Michigan to purchase the wine.

They were joined in their action by a couple who wished to purchase Black Star Farms wine in Arizona without going to the Michigan winery. These consumer plaintiffs contended that it was economically and practically impossible for them to visit wineries located far from Arizona, including those in Michigan, Napa, and Oregon, to purchase wine in person and have it shipped home.

Black Star Farms and defendants each filed motions for summary judgment, and the district court denied Black Farms's motion but granted defendants's cross-motions for summary judgment, concluding that Black Star Farms failed to carry its burden of showing that the challenged statutory scheme discriminates against out-of-state wineries in practical effect. Black Farms appealed.

ISSUE: Do the exceptions to the state of Arizona's three-tier distribution system violate the dormant commerce clause?

REASONING: For the purposes of the dormant commerce clause, discrimination means differential treatment of in-state and out-of-state economic interests that benefits the former and burdens the latter. If such discrimination is found, the court will apply strict scrutiny. The party challenging the statute must prove discrimination.

This act at issue is intended to provide for a separate method of regulating only the sale and delivery of wine produced by small wineries and allowing small wineries to bypass the wholesalers. Before the changes to the law, the small-wineries exception applied only to small wineries at which 75 percent of the grapes used were grown within the state of Arizona, making that exception facially unconstitutional because it treated in-state and out-of-state wineries differently. The exception now applies to all small wineries regardless of where their grapes are grown, so the law is facially neutral. The in-person exception likewise applies to in- and out-of-state wineries and is also facially neutral; it thus appears facially constitutional.

Although Black Star Farms Winery contends that the law is discriminatory in its effect, (1) there was no indication that the exception creates a system under which local goods constitute a larger share, and goods with an

(continued)

out-of-state source constitute a smaller share, of the total sales in the market, and (2) more than half of all U.S. wineries can take advantage of the exception, and as of the time the case was heard, more out-of-state than in-state wineries had taken advantage of it. Further, there is no indication whatsoever that the in-person exception has any effect on the flow of interstate commerce.

Thus, the law does not discriminate on its face or practically against interstate commerce.

DECISION AND REMEDY: The court of appeals concluded that Arizona's statutory exceptions to its three-tier distribution system, which treat similarly situated in-state and out-of-state wineries the same and impose no new impermissible burdens on out-of-state wineries, do not have the practical effect of favoring in-state economic interests over out-of-state interests. Therefore, the appellate court affirmed the district court's decision.

SIGNIFICANCE OF THE CASE: The case illustrates how the court responds to a challenge to a state statute based on grounds that the law violates the dormant commerce clause.

CRITICAL THINKING

What different facts might have led the court to a different conclusion?

ETHICAL DECISION MAKING

Identify the primary stakeholders helped and hurt by the outcome of this case.

CASE Nugget

Can States Require On-line Sellers Without a Physical Presence in a State to Collect State Sales Tax?

South Dakota v. Wayfair United States Supreme Court

_____ S. Ct. _____ (2018)

Ever since the 1992 decision in the case of *Quill Corporation v. North Dakota*, online retailers could be required to collect state and local sales taxes only in jurisdictions where they operated a warehouse or office or had some other physical presence. Small on-line retailers obviously appreciated this rule, but many states felt they were losing out on significant revenue, and stores with physical locations in many states felt the on-line retailers received an unfair competitive advantage due to this rule.

In 2018, in a case that the Government Accounting Office estimates could generate $13 billion in state and local tax revenue, the United State Supreme Court struck down this physical presence rule. The case, *South Dakota v. Wayfair*, arose after the state passed a law that required out-of-state sellers that deliver goods or services into South Dakota to collect state sales tax on those transactions. Sellers who sold less than $100,000 worth of goods in the state or engaged in 200 or fewer separate transactions for the delivery of goods or services into the state were exempt from the law. The majority felt that the emergence of the internet as a mainstream medium for interstate commerce caused the physical presence rule to become further removed from economic reality and resulted in significant revenue losses to the States. While this ruling only directly affects those doing business in South Dakota, most commentators predict that other states will now follow suit and require on-line retailers without a physical presence in their states to now collect sales tax.

Taxing and Spending Powers of the Federal Government

No government can function without a source of revenue. Article I, Section 8, of the Constitution gives the federal government the "Power to lay and collect Taxes, Duties, Imports, and Excises." The taxes laid by Congress, however, must be uniform across the states. In other words, the government cannot impose higher taxes on residents of one state than it imposes on those of another.

Although tax collection allows the government to provide essential services, the government can also use taxes for other purposes. For example, to encourage the development of certain industries and discourage the development of others, the government can provide tax credits for firms entering favored industries. As long as the "motive of Congress and the effect of its legislative action are to secure revenue for the benefit of the general government," the credit is constitutional. The fact that it also has a regulatory impact does not affect the constitutionality of the credit.

Although we may all think we know what a tax is, we may be surprised by what the courts call a tax. For example, many were surprised by the Supreme Court's determination that the penalty imposed under the Affordable Care Act on those who do not obtain health care insurance is a tax. In explaining that finding, Chief Justice Roberts explained that although the payment was intended to encourage the purchase of insurance, nothing in the mandate made the failure to purchase the insurance unlawful; the law could be read as simply imposing a tax, collected by the IRS like other taxes, on those choosing not to purchase health insurance. The chief justice found no problem with the fact that the statute referred to the payment as a penalty and not as a tax.

BUT WHAT IF? . . .

What if the facts had been different and the language of the ACA had clearly stated that although the payment was to be made through the IRS, the payment for failure to purchase health care insurance was a penalty and not a tax? Would that language have necessarily affected the Supreme Court's interpretation of the constitutionality of the statute?

Article I, Section 8, also grants Congress spending power by authorizing it to "pay the Debts and provide for the common Defense and general Welfare of the United States." As with its power to tax, Congress can use its spending power to achieve social welfare objectives. For example, in *South Dakota v. Dole,* the Supreme Court upheld a federal statute that grants federal funds for state highways to only those states in which 21 is the legal drinking age.

Other Constitutional Restrictions on Government

The Privileges and Immunities Clause

Article IV, Section 2, of the Constitution states that "Citizens of each State shall be entitled to all Privileges and Immunities of Citizens in the several States." This provision, called the **privileges and immunities clause,** prohibits states from discriminating against citizens of other states when those nonresidents engage in ordinary and essential activities. These activities include buying and selling property, seeking employment, and using the court system. States may treat residents and nonresidents differently only when they have substantial reason for doing so.

For example, according to the privileges and immunities clause, a state cannot prohibit nonresidents from opening restaurants in the state. States can, however, allow state universities to

privileges and immunities clause
Clause in the U.S. Constitution requiring a state to grant citizens of other states the same legal benefits that it grants its own citizens.

charge higher tuition to out-of-state students because residents pay taxes that fund state universities, whereas out-of-state students do not.

The Full Faith and Credit Clause

full faith and credit clause
The clause of the U.S. Constitution (Article IV, Section 1) mandating that each state must recognize, respect, and enforce the public records, legislative acts, and judicial decisions of the other states.

Article IV, Section 1, of the Constitution contains the **full faith and credit clause.** This clause states, "Full Faith and Credit shall be given in each State to the public Acts, Records, and judicial Proceedings of every other State." This provision requires courts in all states to uphold contracts and public acts established in other states. For example, this clause protects wills, marriage and divorce decrees, and judgments in civil courts. Courts have held, however, that states do not have to give full faith and credit to laws that violate their public policy. Thus, for example, although Massachusetts permits same-sex marriage, the full faith and credit clause does not require other states to recognize such marriages.

The Contract Clause

contract clause
The clause in the U.S. Constitution that prohibits the government from unreasonably interfering with an existing contract.

Article I, Section 10, contains the **contract clause,** which states that government may not pass any "Law impairing the Obligation of Contracts." In application, courts interpret this clause to mean that no law can be passed that will unreasonably interfere with existing contracts. For example, in the 1934 U.S. Supreme Court case *Home Building & Loan Association v. Blaisdell,* the Home Building & Loan Association challenged Minnesota's Mortgage Moratorium Act as a violation of the contract clause. The act, implemented temporarily during the Great Depression, authorized courts to extend the redemption periods of mortgages to delay foreclosures of mortgages on real estate. The Court ruled that the act's provisions were within the state's police power to protect its citizens and did not violate the contract clause. Although the act impaired contractual obligations between lenders and borrowers, the Court held that courts must balance even substantial contractual impairments against states' interest in protecting the welfare of their citizens.

The Amendments to the Constitution

The first ten amendments to the U.S. Constitution, known as the Bill of Rights, substantially affect government regulation of business. These amendments prohibit the federal government from infringing on individual freedoms. Moreover, the Fourteenth Amendment extends most of the provisions in the Bill of Rights to the states, prohibiting state interference in citizens' exercise of their rights. Thus, the federal and state governments cannot deprive individuals of the freedoms protected by the Bill of Rights. Remember, however, that the amendments are a limit only on actions of the government—they do *not* limit the behavior of private employers. This distinction is commonly misunderstood by many citizens.

Many other countries do not have constitutional provisions to protect citizens from the government. The Australian constitution, for example, only creates the framework for government in Australia. In many of the countries that do have individual protections in their constitutions, these protections have emerged recently. Some other countries' constitutions provide rights that American citizens do not have, as the Global Context box on Belarus's constitution illustrates.

Courts apply many amendments to corporations because corporations are treated, in most cases, as artificial persons. The remainder of this chapter describes the amendments that most significantly affect the regulatory environment of business. Exhibit 5-2 summarizes the first 10 amendments.

The First Amendment

Freedom of Speech and Assembly The First Amendment guarantees freedom of speech, including gestures and other forms of expression, and of the press. It prohibits abridgment of the right to assemble peacefully and to petition the government for redress of grievances. Finally, it prohibits the government from aiding the establishment of religion and from interfering with the free exercise of religion.

GLOBAL Context

Constitution of Belarus

©Digital Archive Japan/Alamy

We think of the Constitution as establishing our system of government and guaranteeing certain rights to us, but some nations' constitutions actually impose duties on their citizens. For example, the constitution of Belarus requires everyone to protect the environment, preserve the nation's historical and cultural heritage and other cultural treasures, and respect the dignity, rights, liberties, and legitimate interests of others.

In addition, although the U.S. Constitution's Bill of Rights primarily restricts the government from interfering with our freedoms, some constitutions go much further in terms of giving people rights. For example, under the Belarus constitution, everyone has the right to an education, housing, and health care. Working people are entitled to work no more than 40 hours per week.

Exhibit 5-2

Summary of the Bill of Rights

AMENDMENT	PROVISIONS
First	• Protects freedom of religion, press, speech, and peaceable assembly. • Ensures that citizens have the right to ask the government to redress grievances.
Second	• Finds that a well-regulated militia is necessary for security. • States that government cannot infringe on citizens' right to bear arms.
Third	• States that government cannot house soldiers in private residences during peacetime or during war except for provisions in the law.
Fourth	• Protects citizens from unreasonable search and seizure. • Ensures that government issues warrants only with probable cause.
Fifth	• Ensures that government does not put citizens on trial except by the indictment of a grand jury. • Gives citizens the right not to testify against themselves. • Prevents government from trying citizens twice for the same crime. • Creates due process rights. • States that government cannot take private property for public use without just compensation.
Sixth	• Provides the right to a speedy public trial with an impartial jury, the right to know what criminal accusations a citizen faces, the right to have witnesses both against and for the accused, and the right to have an attorney.
Seventh	• States that in common lawsuits in which the monetary value exceeds $20, citizens have the right to a trial by jury.
Eighth	• States that government will not set bail at excessive levels. • Prohibits government imposition of excessive fines. • Prohibits cruel and unusual punishment.
Ninth	• States that although the Bill of Rights names certain rights, this does not remove other rights retained by citizens.
Tenth	• States that powers that the Constitution does not give to the federal government are reserved for the states.

Like other rights, First Amendment rights are not absolute. For example, a person does not have the right to yell "Fire!" in a crowded theater. Nor does the First Amendment protect false statements about another that are injurious to that person's reputation. Due to the difficulty of determining the boundaries of individual rights, courts hear a large number of First Amendment cases.

LO 5-5 Differentiate between commercial and political speech.

Political Speech. The First Amendment protections also apply to corporations. Courts do not, however, treat all corporate speech the same. Sometimes corporations engage in **political speech;** that is, they support political candidates or referenda. At one time, states restricted firms' political advertising because they feared that corporations, with their large assets, would drown out other voices. In the 1978 case *First National Bank of Boston v. Bellotti,* however, the U.S. Supreme Court considered the constitutionality of a Massachusetts law that prohibited certain corporations from spending money to influence voters on issues that did not materially affect the corporation. The Court held that the law was unconstitutional, stating, "The concept that the government may restrict speech of some elements of our society in order to enhance the relative voice of others is wholly foreign to the First Amendment." The Court ruled that the First Amendment protects corporate political speech to the same extent as ordinary citizens's political speech.

LO 5-6 Recognize how the rights guaranteed by the Bill of Rights affect business managers.

Commercial Speech. Not all corporate speech is political speech. **Commercial speech** is speech that conveys information related to the sale of goods and services. Courts analyze government restrictions on commercial speech according to a four-part test established in *Central Hudson Gas & Electric Corp. v. Public Service Commission of New York.* The *Central Hudson* test is illustrated in Exhibit 5-3.

political speech
Speech that is used to support political candidates or referenda. It is given a high level of protection by the First Amendment as compared to other types of speech.

commercial speech
Speech made by businesses about commercial matters, such as the sale of goods and services. It is protected by the First Amendment.

Exhibit 5-3

The Central Hudson Test for Commercial Speech

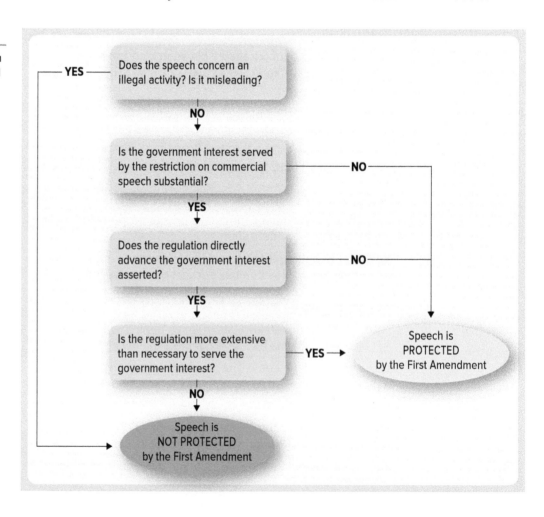

Commercial speech is broader than just advertisements. For example, in the case of *Bad Frog Brewery v. New York Liquor Authority,* the speech at issue was a label prominently featuring a drawing of a frog holding up its middle "finger" and a slogan such as "The beer so good. . . it's bad." When the New York State Liquor Authority rejected the Brewery's application for brand label approval and registration on grounds that the label encourages combative behavior, fosters a defiance to the health warning on the label, entices underage drinkers, and invites the public not to heed conventional wisdom and to disobey standards of decorum, the company sued to enjoin the state from prohibiting Bad Frog from selling its beer under the controversial label.

The court of Appeals applied the Central Hudson test to uphold the right of Bad Frog to use its label. In terms of meeting the first standard, that the speech be lawful and not deceptive, the courts said that the consumption of beer is legal in New York, and the labels cannot be said to be deceptive, even if they are offensive. The court found the state's two stated interests in banning the labels, (i) "protecting children from vulgar and profane advertising," and (ii) "acting consistently to promote temperance, i.e., the moderate and responsible use of alcohol among those above the legal drinking age and abstention among those below the legal drinking age," to be "substantial" interests.

However, the court found that the state failed to meet the third criterion of advancing its interests to a material degree. Merely prohibiting a single vulgar label on a product children could not even purchase could not be said to be part of any comprehensive program to protect children from vulgarity, and without any empirical evidence that the label could encourage underage drinking and disregard of the health warnings against alcohol use, the state did not demonstrate that their interest in promoting temperance could be materially advanced. Since the third prong of the test was not met, the court did not need to address the final question of whether the restriction was narrowly tailored, although the Brewery had advanced several less restrictive alternatives.

Unprotected Speech. The First Amendment right to free speech is not absolute. In the 1942 case *Chaplinsky v. New Hampshire,* the U.S. Supreme Court held:

> There are certain well-defined and narrowly limited classes of speech, the prevention and punishment of which have never been thought to raise any Constitutional problem. These include the lewd and obscene, the profane, the libelous, and the insulting or fighting words—those which by their very utterance inflict injury or tend to incite an immediate breach of the peace.

Thus, for example, the First Amendment does not protect defamation, or speech that harms the reputation of another. Moreover, as you will learn in Chapter 6, courts may require an individual who uses such speech to compensate the person whose reputation was harmed by the speech.

The First Amendment does not protect obscenity, either, although in many cases courts find it difficult to determine whether particular speech constitutes obscenity. In the 1973 case *Miller v. California,* the U.S. Supreme Court established a three-part standard to determine whether speech is obscene:

1. Would the average person, applying contemporary community standards, find that the speech, taken as a whole, appeals to the prurient (marked by or arousing an immoderate or unwholesome interest or desire) interest?
2. Does the speech depict or describe, in a patently offensive way, sexual conduct specifically defined by law?
3. Does the speech, taken as a whole, lack serious literary, artistic, political, or scientific value?

If the answer to all three questions is yes, the First Amendment does not protect the speech in question. Can you identify any significant ambiguities in the *Miller* standard?

Fighting words are a third class of unprotected speech. In *Chaplinsky v. New Hampshire,* a man protesting the government called the city marshal a "damned racketeer" and a "damned Fascist." The city arrested him for violating a statute prohibiting the use of offensive, derisive, or annoying words toward another in a public place. The U.S. Supreme Court held: "The English language has a number of words and expressions which by general consent are 'fighting words' when said without a disarming smile. . . . Such words, as ordinary men know, are likely to cause a fight." The Court further held that "'damned racketeer' and 'damned Fascist' are epithets likely to provoke the average person to retaliation, and thereby cause a breach of the peace." Thus, the Court determined that the First Amendment did not protect the man's speech.

Many universities believe that hate speech, or derogatory speech directed at members of another group, such as another race, satisfies the definition of fighting words. Thus, 60 percent of universities have banned verbal abuse and verbal harassment, and 28 percent have banned advocacy of an offensive viewpoint. State and federal appellate courts have struck down many of these hate-speech codes, but no cases have reached the Supreme Court yet. The international community, however, is less protective of hate speech. For example, a United Nations declaration and a number of foreign laws state that hate speech is not a protected form of expression.

Freedom of Religion The First Amendment contains two provisions that protect citizens's freedom of religion. The **establishment clause** maintains that government "shall make no law respecting an establishment of religion." In the 1971 case *Lemon v. Kurtzman,* the U.S. Supreme Court codified the following three tests to determine whether a particular government statute violates the establishment clause:

1. Does the statute have a secular legislative purpose?
2. Does the statute's principal or primary effect either advance or inhibit religion?
3. Does the statute foster an excessive government entanglement with religion?

To determine whether the statute fosters an excessive government entanglement with religion, courts examine the character and purposes of the institutions benefited, the nature of the aid that the government provides, and the resulting relationship between the government and the religious authority.

The **free-exercise clause** states that government cannot make a law "prohibiting the free exercise" of religion. Although government must remain neutral in matters of religion, determining whether a government action advances religion or merely allows free exercise of religion is often difficult. Likewise, determining whether a government action establishes religion or simply avoids interference with free exercise of religion is also difficult.

Issues concerning the establishment clause and the free-exercise clause often arise in workplace settings. In government workplaces, this conflict sometimes raises difficult issues. For example, in 1966, Tucker, an employee of the California Department of Education, insisted on signing office memos with his name and the letters "SOTLJC," an abbreviation for "Servant of the Lord Jesus Christ." In an attempt to avoid workplace disruptions and the appearance of government support for religion, Tucker's supervisor prohibited all displays of religious symbols in the workplace. The supervisor suspended Tucker for refusing to comply with the restrictions. When Tucker challenged the suspension on grounds that the rules interfered with his free exercise of religion, the appellate court agreed.

In 2014, free exercise of religion was in the headlines, as the U.S. Supreme Court heard another challenge to the Affordable Care Act (ACA). This challenge involved the requirement under the ACA that corporations provide health insurance coverage for certain types of contraception to their employees. The owners of Hobby Lobby, a closely held corporation, argued that complying with this regulation would place a burden on their sincerely held religious beliefs related to contraception, and thus violated the Religious Freedom Restoration Act of 1993 (RFRA), which provides that "government shall not substantially burden a person's exercise of religion even if the burden results from a rule of general applicability."[1]

In resolving conflicting circuit court decisions on this issue, the U.S. Supreme Court, in *Burwell v. Hobby Lobby,*[2] ruled that the regulations violated the RFRA, which prohibited the federal government from taking any action that substantially burdened the exercise of religion unless it constituted the least restrictive means of serving a compelling government interest. The court concluded that the challenged regulations substantially burdened the exercise of religion because compliance was contrary to the owners' religious objections to abortion and there was a heavy financial penalty for noncompliance. Assuming that the regulations served a compelling government interest, the court found that they were not the least restrictive means of serving that interest because there were other ways to ensure that every woman had cost-free access to certain

establishment clause
One of two provisions in the First Amendment of the U.S. Constitution that protect citizens's freedom of religion. It prohibits (i) the establishment of a national religion by Congress and (ii) the preference of one religion over another or of religion over nonreligious philosophies in general.

free-exercise clause
A clause in the First Amendment of the U.S. Constitution that states that government (state and federal) cannot make a law "prohibiting the free exercise" of religion. This clause is interpreted to include absolute freedom to believe and freedom to act that may face state restriction.

[1] 42 U.S.C., 2000bb et seq.
[2] 42 U.S.C., 2000bb et seq.

GLOBAL Context

©Digital Archive Japan/Alamy

Free Speech in China

The Chinese constitution does not guarantee freedom of speech or assembly, nor does it recognize any form of natural rights or human rights. Instead, it recognizes citizens' rights, which are specifically enumerated in the Chinese constitution or laws. Not only does the Chinese constitution not provide protections for expressive activity, but numerous Chinese laws prohibit citizens from engaging in political acts directed against the regime—acts that are protected in the United States. For example, Article 25 of China's Publishing Control Act prohibits the publication of any material that opposes the basic rules of the constitution. Articles 7 and 12 of the Law on Assemblies, Processions, and Demonstrations prohibit assemblies, processions, and demonstrations that oppose these basic rules.

contraceptives. Two alternatives Justice Alito put forth were that the Government could assume the cost of providing the four contraceptives to women unable to obtain coverage due to their employers' religious objections, or it could extend the accommodation already established for religious nonprofit organizations to nonprofit employers with religious objections to the contraceptive mandate. Ultimately, a divided Supreme Court held that corporations have the right to free exercise of religion, and therefore the Court held that closely held, for-profit corporations are not required to provide certain forms of birth control mandated by the Affordable Care Act.

In private workplaces, issues related to free exercise of religion most often arise under Title VII, the federal law prohibiting employment discrimination. We discuss this important legislation in detail in Chapter 24.

The Fourth Amendment

Freedom from Unreasonable Searches and Seizures The Fourth Amendment guarantees citizens the right to be "secure in their persons, their homes, and their personal property." Thus, it prohibits government from conducting unreasonable searches of individuals and seizing their property to use as evidence against them.

A search is unreasonable if the government official conducting the search does not first obtain a search warrant from a court. A **search warrant** is a court order that authorizes law enforcement agents to search for or seize items specifically described in the warrant. Government officials can obtain search warrants only if they can show *probable cause* to believe that the search will uncover specific evidence of criminal activity. In other words, the government officials must have a sufficient reason based on known facts to obtain a warrant.

The Supreme Court has ruled, however, that in certain circumstances, government officials do not need a search warrant. For example, when law enforcement officials believe it is likely that the items sought will be removed before they can obtain a warrant, they may conduct a search without a warrant. Law enforcement officials frequently conduct automobile searches without a warrant according to this rule.

In most other cases, though, law enforcement officials must obtain a warrant before conducting a search. For example, an Ohio law required buyers of five or more beer kegs to provide the beer distributor with the address of the party where the kegs would be consumed. Additionally, the law required the buyers to sign a form allowing police and liquor agents to enter their property without a warrant to search the premises to enforce state liquor laws. In 2001, a college professor challenged the law on grounds that it infringed on citizens's Fourth Amendment rights. Before the court decided the case, Ohio repealed the law.

Technological changes have raised interesting issues in the application of the Fourth Amendment. For example, because growing marijuana indoors requires heat lamps that use large amounts of electricity, when police found that a suspected marijuana grower had unusually high electric bills, they used an instrument that detects heat emissions to provide them with the evidence necessary to obtain a warrant to search his house physically. In determining whether using

search warrant
A court order that authorizes law enforcement agents to search for or seize items specifically described in the warrant.

a thermal-imaging instrument on private property constituted an unlawful search, the Court asked whether using a thermal-imaging instrument is more like going through someone's garbage, which had been previously held not to violate the Fourth Amendment, or more like using a high-powered telescope to look through someone's window, which had been previously found to be a violation.

In finding the use of this new technology more analogous to the latter type of search, the U.S. Supreme Court ruled that police use of thermal-imaging devices to detect heat patterns emanating from private homes constitutes a search that requires a warrant and further held that the warrant requirement applies also to any "more sophisticated systems" that give the police knowledge that in the past would have required physical entry of the home.

Somewhat more recently, drug-sniffing dogs have had a surprisingly prominent role in the Supreme Court's holdings related to the Fourth Amendment. In the 2013 case of *Florida v. Jardines*,[3] the high court held that bringing a drug-sniffing dog onto someone's front porch without a warrant violates the Fourth Amendment's protection against warrantless searches. Justice Scalia wrote that a person's minimal expectation of privacy extends not only to his house but to its immediate surroundings such as the porch.

The outcome is somewhat different, however, if the drug-sniffing dog is accompanying an officer to a person's vehicle for a routine traffic stop. In *Florida v. Harris*,[4] an officer made a routine traffic stop of a truck, and when he walked up to the vehicle with the drug-sniffing dog, the dog, through its behavior, indicated the presence of illegal drugs in the door of the truck. The officer treated the dog's behavior as giving him probable cause to search the interior of the door of the truck, where he subsequently found 200 pseudoephedrine pills and other methamphetamine precursors. In this case, the high court upheld the search.

In 2014, the Fourth Amendment was again interpreted by the high court in the following case, as it addressed the issue of what limitations there should be on cell phone searches pursuant to an arrest.

[3] 133 S. Ct. 1409 (2013).
[4] 133 S. Ct. 1050 (2013).

DAVID LEON RILEY v. CALIFORNIA AND UNITED STATES v. WURIE

Supreme Court of the United States

134 S. Ct. 2473 (2014)

FACTS: *The case heard by the Supreme Court as* Riley v. California *was actually a combination of two cases,* Riley v. California *and* United States v. Wurie. *Both cases involved the seizure of a cell phone.*

Riley v. California *began in 2009, when David Leon Riley was pulled over for expired registration tags on his car. It was then discovered that Riley's license had been suspended, which authorized impounding of Riley's car. Upon an immediate search of the vehicle, illegal weapons were found, and Riley was placed under arrest. The police then seized and searched Riley's phone without a warrant, and pictures and videos on the phone indicated that Riley was involved in a recent gang-related shooting. Police charged Riley in connection with the shooting. Riley moved to suppress the evidence that was found on his phone.*

In United States v. Wurie, *Brima Wurie was arrested after police observed him participate in a drug sale. At the police station, two of Wurie's cell phones were confiscated. Information on the phones led police to search Wurie's home, where illegal drugs and weapons were found. Wurie was subsequently charged with drug and weapons offenses. Wurie moved to suppress the evidence that was found through the seizure of his phones.*

ISSUE: Was the evidence admitted at trial from Riley's and Wuries's cell phones discovered through a search that

(continued)

violated their Fourth Amendment right to be free from unreasonable searches?

REASONING: While deciding how to rule on the case, the Court considered the precedent set by *Chimel v. California,* a 1969 Supreme Court case. In *Chimel v. California,* the Court ruled that if the police arrest someone, the officers may search the body of the person and "the area into which he might reach" without a warrant in order to protect material evidence or the officers' safety. In *People v. Diaz,* the California Supreme Court interpreted this precedent to mean that police could conduct a full search of a cell phone whenever the phone is found near the suspect at the time of arrest.

Ultimately, however, the Court ruled in favor of Riley and Wurie. The Court argued that while protecting evidence and police officers is certainly important, there's no way the data stored on a phone can itself be used as a weapon to harm an officer or aid in an arrestee's escape. In his opinion, Chief Justice Roberts contended that while evidence stored on a phone may be destroyed through remote wiping or data encryption, these are basic security features of today's cell phones and a suspect could destroy any evidence long before the phone is actually searched. A warrantless search is unlikely to make much of a difference.

Chief Justice Roberts continued by arguing that today's cell phones hold huge amounts of personal information on "the privacies of life." Cell phones differ in both a quantitative and qualitative sense from most other objects that would be on someone's person at the time of arrest and therefore should not be subject to the search and seizure procedures allowed by *Chimel v. California.*

DECISION: The warrantless seizure of a cell phone is a violation of the Fourth Amendment. Riley's and Wurie's requests to suppress evidence gathered from their cell phones were granted.

SIGNIFICANCE: This case is an excellent example of the law adapting to changes in technology and society in order to preserve the rights of citizens in a new era.

CRITICAL THINKING

The court analogized between the search of a cell phone and the search of a house. How strong is this analogy?

ETHICAL REASONING

What ethical norm is furthered by this decision? What ethical norm is being sacrificed?

In addition to protecting individuals and their homes, the Fourth Amendment also protects corporations and places of business. This protection generally applies in criminal cases, but Fourth Amendment issues also arise when government regulations authorize administrative agencies to conduct warrantless searches.

Although administrative searches usually require search warrants, courts have established an exception to this rule: If an industry has a long history of pervasive regulation, a warrantless search is not unreasonable. In such industries, administrative agencies can use warrantless searches to ensure that firms uphold regulations.

This *pervasive-regulation exception,* however, is not always easy to interpret. Courts have ruled that warrantless searches authorized by the Federal Mine Safety and Health Act are legal because the federal regulatory presence is comprehensive and well defined. Thus, reasonable commercial-property owners ought to know that their property is subject to periodic inspections. A warrantless search based on the Occupational Safety and Health Act, however, may violate the Fourth Amendment because no significant legislation of working conditions existed before Congress passed that act in 1970. Hence, businesspeople covered by the law cannot reasonably anticipate warrantless searches.

CASE Nugget

When Is the Consent of a Co-owner Sufficient Consent?

Walter Fernandez v. California United States Supreme Court

134 S. Ct. 1126 (2014)

Police officers saw Fernandez, a suspect in a violent robbery, run into an apartment building and heard screams coming from one of the apartments. When they knocked on the apartment door, Roxanne Rojas, who appeared to be battered and bleeding, answered. As soon as the officers asked her to step out of the apartment so that they could conduct a protective sweep, Fernandez came to the door and objected. Suspecting that Fernandez had assaulted Rojas, the officers removed him from the apartment and placed him under arrest. He was then identified as the perpetrator in the earlier robbery and taken to the police station.

An officer later returned to the apartment and, after obtaining Rojas's oral and written consent, searched the premises, where he found several items linking Fernandez to the robbery. The trial court denied Fernandez's motion to suppress that evidence, and he was convicted.

In determining that the Fourth Amendment rights of Fernandez had been not been violated by the search, the U.S. Supreme Court reviewed the firmly established law that police officers may search jointly occupied premises if one of the occupants consents and reaffirmed the narrow exception to this rule, that the consent of one occupant is insufficient when another occupant is present and objects to the search. However, it held that this exception does not extend to a case like that of Fernandez, in which consent was provided by an abused woman well after her male partner had been removed from the apartment they shared. A past objection to a search by an absent party does not invalidate consent by a present party with whom authority is shared.

The Fifth Amendment

The Fifth Amendment protects individuals in several important ways. First, it protects against *self-incrimination,* meaning that in a criminal case, the defendant does not have to testify in court as a witness against himself or herself. The Fifth Amendment also protects against *double jeopardy.* Thus, government cannot try a person more than once for the same crime.

Due Process For businesspeople and corporations, the Fifth Amendment's **due process clause** provides extensive protection. This clause states that government cannot deprive a person of life, liberty, or property without *due process* of law.

The due process clause guarantees two types of due process: procedural and substantive. **Procedural due process** requires the government to use fair procedures when taking the life, liberty, or property of an individual or corporation. At a minimum, procedural due process entitles a person to notice of any legal action against her and to a hearing before an impartial tribunal. Originally, courts interpreted the due process clause as protecting an individual's right of procedural due process only in federal criminal proceedings. The subsequent passage of the Fourteenth Amendment extended the requirement of due process to criminal proceedings by state governments. Today, courts apply the due process clause to diverse situations, including the termination of welfare benefits, food stamps, or Social Security benefits; the suspension of a

due process clause
A clause in the Fifth Amendment of the U.S. Constitution providing that the government cannot deprive an individual of life, liberty, or property without a fair and just hearing.

procedural due process
The requirement that a government must use fair procedures before depriving a person of his or her life, liberty, or property.

driver's license; the discharge of a public employee from his job; and the suspension of a student from school.

The procedures that government must follow when taking an individual's life, liberty, or property vary according to the nature of the taking. Generally, more procedures are necessary as the magnitude of potential deprivation increases.

Can States Deny Same Sex Couples the Right to Marry?

Obergefell et al. v. Hodges United States Supreme Court

135 U.S. 2584(2015)

One of the most awaited Supreme Court decisions in 2013 was *United States v. Windsor*, in which the high court struck down the Defense of Marriage Act (DOMA), an act providing that for purposes of all federal laws, a marriage must be between opposite sex partners. While this case brought significant relief to those whose states recognized their same-sex marriages, it did nothing to improve conditions for those whose states did not recognize same-sex marriage. Same-sex couples in the rest of the states would have to wait until the 2015 case of *Obergefell v. Hodges* to have their relationships recognized.

The petitioners in this case were 14 same-sex couples and two men whose same-sex partners were deceased. All had filed suits in Federal District Courts in their home states, claiming that respondent state officials had violated the Fourteenth Amendment by denying them the right to marry or to have marriages lawfully performed in another state given full recognition. Each District Court ruled in petitioners's favor, but the Sixth Circuit consolidated the cases and reversed. The U.S. Supreme Court then reversed the Sixth Circuit decision.

In that decision, the high court ruled that under the due process and equal protection clauses of the Fourteenth Amendment, same-sex couples have a fundamental right to marry. And because same-sex couples can exercise the fundamental right to marry in all states, the court could find no lawful basis for a state to refuse to recognize a lawful same-sex marriage performed in another state on the ground of its same-sex character. Laws of Michigan, Kentucky, Ohio, and Tennessee were held invalid to the extent they excluded same-sex couples from civil marriage on the same terms and conditions as opposite-sex couples.

While many Americans celebrated the ruling, not everyone embraced the decision. For example, Kim Davis, a county clerk of Rowan County, Kentucky, made headlines in September, 2015, when she refused to issue marriage licenses to same-sex couples, claiming that issuing marriage licenses to same-sex couples violated her beliefs.

This ruling will have a significant impact on how businesses must now treat a significant number of employees, bringing them under the umbrella of several federal statutes.

Substantive due process refers to the basic fairness of laws that may deprive an individual of her life, liberty, or property. To satisfy the substantive due process requirement, government must have a proper purpose for enacting laws that restrict individuals's liberty or the use of their property. The standard for determining whether a law violates substantive due process depends on the nature of the potential deprivation. Laws affecting fundamental rights must bear a substantial relationship to a compelling government purpose. These fundamental rights generally include the rights protected in the Constitution: the right to vote, the right to travel freely from state to state, the right to privacy, and so on. Compelling state interests include, for example, public safety and national security.

substantive due process
The requirement for laws depriving an individual of life, liberty, or property to be fair and not arbitrary.

Not all laws, however, affect fundamental rights. To show that a law that does not affect fundamental rights satisfies the substantive due process requirement, government must prove only that the law bears a rational relationship to a legitimate state interest. Courts uphold most government regulations according to this *rational-basis test.* For example, courts have upheld minimum-wage laws, rent control laws, banking regulations, environmental laws, and regulations prohibiting unfair trade practices according to the rational-basis test.

The Prohibition against Uncompensated Takings The Fifth Amendment provides that when government takes private property for public use, it must pay the owner *just compensation,* or fair market value, for her property. This provision is called the **takings clause**, and it applies to corporations. Several significant issues have arisen with respect to the takings clause. For example, what constitutes a public use for which government can take private property? This issue is explored in depth in Chapter 6.

The takings clause has prompted other issues as well. What happens, for instance, when a government regulation interferes so substantially with an individual's use of her property that it effectively takes her property? For example, environmental regulations often affect the way landowners can use their property. Under the case of *Lucas v. South Carolina Coastal Commission,* the Supreme Court held that if a regulation is so extensive that it leaves the property owner with no economically viable use of his land, the regulation constitutes a taking, and the owner is entitled to compensation for the value of his property, even though the state did not seek to take title to his land. These so-called regulatory takings do not occur very frequently because most regulations merely limit the use one can make of their property, and limitations on use do not constitute takings.

The Privilege against Self-Incrimination Although most provisions of the Fifth Amendment apply to corporations, corporations do not enjoy the Fifth Amendment's protection against self-incrimination. Sole proprietors, however, are entitled to this protection. Thus, different businesses have different constitutional rights depending on their form of business organization.

For example, in the 1988 U.S. Supreme Court case *Braswell v. United States,* the Court distinguished between the rights of corporate-record custodians and sole proprietors. Braswell was both the operator and the sole shareholder of his business. When a grand jury issued a subpoena requiring him to produce corporate books and records, Braswell argued that the subpoena violated his Fifth Amendment privilege against self-incrimination. The Court denied Braswell's claim, writing that "subpoenaed business records are not privileged, and as a custodian for the records, the act of producing the records is in a representative capacity, not a personal one, so the records must be produced." The Court held that the subpoena would have violated Braswell's privilege against self-incrimination if his business had been a sole proprietorship.

The Ninth Amendment

The Right to Privacy The Ninth Amendment states, "The enumeration in the Constitution, of certain rights, shall not be construed to deny or disparage others retained by the people." Although this amendment does not expressly guarantee the right to privacy, courts have interpreted the Ninth Amendment, together with the First, Third, Fourth, and Fifth Amendments, as providing individuals with a right to privacy.

In the 1965 case *Griswold v. Connecticut,* the Supreme Court ruled that a Connecticut law prohibiting the use of contraceptives was unconstitutional because it violated individuals' right to privacy. Justice Douglas wrote:

> Specific guarantees in the Bill of Rights have penumbras [fringes]. . . . Various guarantees create zones of privacy. The right of association contained in the penumbra of the First Amendment is one. . . . The Third Amendment in its prohibition against the quartering of soldiers "in any house" in time of peace without the consent of the owner is another facet of that privacy. The Fourth Amendment explicitly affirms the "right of the people to be secure in their persons, houses,

takings clause
A clause in the Fifth Amendment of the U.S. Constitution requiring that when government uses its power to take private property for public use, it must pay the owner just compensation, or fair market value, for the property.

papers, and effects, against unreasonable searches and seizures." The Fifth Amendment in its Self-Incrimination Clause enables the citizen to create a zone of privacy, which government may not force him to surrender to his detriment.

The Fourteenth Amendment

Equal Protection The Fourteenth Amendment contains the **equal protection clause,** which prevents states from denying "the equal protection of the laws" to any citizen. This clause combats discrimination because it applies whenever government treats certain individuals differently than other similarly situated individuals, usually through a classification scheme.

As with the due process clause, to determine whether a law violates the equal protection clause, courts use different standards based on the nature of the rights the classification affects. Three standards of scrutiny apply: strict scrutiny, intermediate scrutiny, and the rational-basis test.

If a law prevents individuals from exercising a fundamental right, or if the law's classification scheme involves suspect classifications, the action will be subject to **strict scrutiny.** *Suspect classifications* include classifications based on race, national origin, and citizenship. Courts uphold suspect classifications only if they are necessary to promote a compelling state interest. In cases involving suspect classifications, courts do not begin their analysis with a presumption that the classification is constitutional, so few laws pass the strict-scrutiny standard. For example, in the 1954 case *Brown v. Board of Education,* the U.S. Supreme Court ruled that the classification scheme used to segregate public schools racially violated the equal protection clause.

Several courts have held, however, that in some cases, remedying past discrimination against a group is a compelling state interest. Chapter 24 discusses this issue in detail.

If the law's classification scheme is based on gender or on the legitimacy of children, courts use **intermediate scrutiny.** According to this standard, the law is constitutional only if it is substantially related to an important government objective.

When a classification scheme involves other matters, courts apply a **rational-basis test.** According to this test, courts ask whether there is any justifiable reason to believe that the classification scheme advances a legitimate government interest. Because courts begin their analysis with a strong presumption that the government action is constitutional, almost all laws pass this test.

equal protection clause
A clause in the Fourteenth Amendment of the U.S. Constitution that prevents states from denying "the equal protection of the laws" to any citizen. This clause implies that all citizens are created equal.

strict scrutiny
The most exacting standard of review used by the courts in determining the constitutionality of a statute; requires a compelling government interest and the least restrictive means of attaining that objective.

intermediate scrutiny
A standard of review under which a law must be necessary to achieve a substantial, or important, government interest and it must be narrowly tailored to that interest.

rational-basis test
The lowest standard of review; requires a law to be designed to protect a legitimate state interest and be rationally related to that interest.

CASE OPENER WRAP-UP

The Constitutionality of the Affordable Care Act

In a decision that surprised many, the United States Supreme Court upheld the individual mandate of the Affordable Care Act (ACA). However, it did so on the ground that it was a valid exercise of congressional power under the taxing clause.

The high court did not agree with the federal government's argument that the mandate was a lawful exercise of commerce clause authority. In rejecting this argument, Chief Justice Roberts said that the power to regulate presupposes the existence of the commercial activity to be regulated. The act was not regulating existing activity but, rather, was trying to require people to become active. Roberts feared that extending the reach of regulation under the commerce clause to individuals because they were doing nothing would open a new and potentially vast domain to congressional authority. He noted that the Constitution gave Congress the authority to regulate commerce, not compel it.

In her opinion, Justice Ginsburg disagreed with the chief justice's contention that the ACA was not a regulation of commerce. She pointed out that the uninsured are in fact very active in the market for health care services and that it is the legislature that should determine the borders of the market that it is attempting to

(continued)

regulate. She also pointed out that the high court has upheld the authority of Congress to direct the conduct of individuals today because of their prophesied conduct tomorrow, noting that as far back in the history of commerce clause interpretation as *Wickard v. Filburn,* the court upheld prohibiting a farmer's growing wheat for home consumption because in the future he might grow wheat for sale in the market. And in *Raisch,* home-grown marijuana could be regulated because at some point in the future, the price of marijuana in the market may lead the individual growing for home consumption to enter the market as a seller, just as the uninsured in the future might need medical care. And, in fact, the need for medical care at some future date is much more certain than entry in the market in either *Filburn* or *Raisch.*

Because the high court upheld the constitutionality of the individual mandate, it did not have to address the issue of whether it was severable from the rest of the act.

Just before this book went to press, Congress eliminated the individual mandate. Despite this change in the law, the analysis is still useful to our understanding of how the Commerce Clause is interpreted.

Also, remember that one of the issues in the case was whether the individual mandate was integral to the functioning of the law. The court did not have to address that issue, but now, as the mandate has been removed, we will see by what happens to the law whether in fact this provision was integral to the functioning of the legislation.

PRACTICAL TIPS for Business Managers

1. Remember that if an inspector from an administrative agency shows up to do a search of your business, you have the right to ask to see a warrant as long as your business in not in an industry that has a long history of pervasive regulation.
2. Remember that while corporations enjoy most of the protections of the Bill of Rights, corporations do not have the Fifth Amendment right against self-incrimination, which means that corporate managers may be compelled to testify against the corporation.
3. Because the courts have found that corporations have free speech rights, when a regulation limits your company's advertising, you always want to ask whether the regulation violates your firm's First Amendment right to free speech.

SUMMARY

The U.S. Constitution

Federalism: The authority to govern is divided between two sovereigns, or supreme lawmakers: the federal government and the states.

Separation of powers: The Constitution divides power among the legislative, executive, and judicial branches of government. The system of checks and balances allocates specific powers to each branch to keep the other branches from dominating government.

The Supremacy Clause and Federal Preemption

Federal supremacy: Any state or local law that directly conflicts with the U.S. Constitution, federal laws, or treaties is void.

Concurrent authority: Both state and federal governments have the power to regulate certain matters; generally, the federal government defers to the state.

Federal preemption: The federal government uses this doctrine to strike down laws that do not directly conflict with a federal law but attempt to regulate an area within federal legislative jurisdiction.

The Commerce Clause	The *commerce clause* grants the federal government the authority to pass regulations that significantly affect interstate commerce. Today, this clause provides the basis for most federal government regulations. *Police powers* are the residual powers retained by states to pass laws to safeguard the health and welfare of their citizens. The *dormant commerce clause* prohibits states from passing laws that significantly interfere with interstate commerce.
Taxing and Spending Powers of the Federal Government	Congressional taxes must be uniform across all states. Congress can use taxes and spending for purposes other than generating revenue; for example, it can use them to indirectly promote social goals.
Other Constitutional Restrictions on Government	The *privileges and immunities clause* prohibits states from discriminating against citizens of other states. The *full faith and credit clause* states that in civil matters, courts in all states must uphold rights established by legal documents. The *contract clause* states that Congress cannot pass laws that unreasonably interfere with existing contracts.
The Amendments to the Constitution	*First Amendment:* • Protects corporate speech in certain circumstances. It protects corporate political speech to the same extent that it protects individuals's political speech. The *Central Hudson* test determines whether the First Amendment protects particular corporate commercial speech. • Contains the *establishment clause,* which states that Congress may not make laws respecting an establishment of religion, and the *free-exercise clause,* which states that Congress may not make laws prohibiting the free exercise of religion. *Fourth Amendment:* • Protects both corporations and individuals from unreasonable government searches and seizures. Although administrative searches generally require a warrant, administrative agencies may inspect some industries without a warrant to ensure compliance with industry regulations. *Fifth Amendment:* • States that government cannot take an individual's life, liberty, or property without due process of law. There are two types of due process: *procedural due process,* which focuses on rules for enforcing laws and entitles individuals to notice of legal action against them, and *substantive due process,* which requires government to have a proper purpose for enacting laws that restrict individuals's liberty or the use of their property. • States that if government takes private property for public use, it must compensate the owner. The extent to which some government regulations constitute takings, however, generates much litigation. • Includes a privilege against self-incrimination, although the provision does not apply to corporations. Only individual citizens and sole proprietorships may exercise this right. • Guarantees individuals equal protection under the law. Courts use three standards of scrutiny in equal protection cases: (i) *strict scrutiny,* to analyze government actions that abridge fundamental rights or that include suspect classifications; (ii) *intermediate scrutiny,* to analyze classifications based on gender or on legitimacy of children; and (iii) the *rational-basis test,* to analyze classifications involving other matters. *Ninth Amendment:* • Along with the First, Third, Fourth, and Fifth Amendments, this amendment provides individuals with a right to privacy. *Fourteenth Amendment:* • Applies the due-process clause to state governments and contains the equal protection clause, which prohibits states from denying the equal protection of the laws to any citizens.

Point/Counterpoint

Should Warrantless Wiretapping Be Allowed under the Fourth Amendment?	
YES	**NO**
Warrantless wiretapping differs from standard wiretapping procedures only in the lack of a court-ordered warrant. In matters of national security and cases in which secrecy is of paramount importance, the acquisition of intelligence information must come before the procedural formalities of the legal system. Valuable surveillance necessary for the protection of U.S. citizens is conducted by warrantless wiretapping and should not be prevented because of a lack of a warrant. The Internet has spawned an expanded realm of communications that creates new risks to national security. Technologies such as email, instant messaging, and websites allow individuals or nations to communicate in ways that are difficult to monitor in real time while still trying to obtain a warrant. Internet technology has accelerated the pace of communication, and the methods for surveillance must accelerate as well to match. Some terrorist attacks have been planned not through traditional channels, for which it is reasonable to obtain warrants in advance, but through the Internet. New situations require new methods of surveillance. The events of recent years demand a reexamination of the traditional intent of the Fourth Amendment. The rise in terrorist attacks and the new dynamics of conflict have shifted the nature of the conflict from that of conflict between nations to that of conflict among factions within one nation or several. Traditional conceptions of how wiretapping was applied in a different time must be replaced to preserve the overall intent of the Fourth Amendment: to keep people "secure in their persons."	The Fourth Amendment was designed to protect the right to privacy of all U.S. citizens. Breaching that privacy without a court-ordered warrant sets a dangerous precedent for the powers given to the federal government. The act of wiretapping without a warrant from the court violates the separation of powers doctrine by placing full authority for search and seizure in the hands of the executive branch of the federal government. Moreover, the Internet has brought forth a wide array of additional methods of communication that can potentially be "wiretapped." Email, instant messaging, websites, and other forms of electronic communication compound the problem by rendering private citizens even more vulnerable to surveillance by the federal government. Because electronic wiretaps are easier to obtain, monitor, and record than traditional wiretaps, the volume of information that the federal government can gather and use is significantly larger than it was before the advent of the Internet. Questions have been raised as to the efficacy of the warrantless wiretapping program. Since the program remains shrouded in secrecy, it is hard to demonstrate a reasonable benefit from the wiretapping. The program was initially designed for monitoring communications into and out of other countries, not within the borders of the United States. Nevertheless, the program has recently been applied to U.S. citizens communicating within U.S. borders. This marks a vast expansion of the power of the federal government regarding surveillance of the civilian population.

Questions & Problems

1. Explain how each branch of the government checks the power of the other branches.
2. How can both Sue and Sam be correct when Sue claims the commerce clause increases government's power and Sam claims the commerce clause reduces government's power?
3. What is the purpose of the contract clause?
4. How does the contract clause affect state regulation?
5. How does the First Amendment protection of corporate political speech differ from the protection of corporate commercial speech?
6. Taxpayers in the state of Kentucky alleged that a tax on out-of-state municipal bonds that excluded interest from in-state bonds was a violation of the commerce clause. Representatives of the state argued that the tax reflected a traditional government function that could persist without any differential treatment of local interests against the similar interests of out-of-state entities. Additionally, because Kentucky itself participates in the bond market, its potential discrimination should be found allowable. How would you have ruled in this case? Why?

 Source: *Department of Revenue of Kentucky v. George W. Davis*, 553 U.S. 328 (2008).

7. The town of Gilbert, Arizona, passed an ordinance that restricted the maximum size of a sign depending upon the content of the sign. Signs advertising political candidates were allowed to be as large as 32 square

feet and could stay in place indefinitely, and signs that advertised general ideas were limited to 20 square feet and were also allowed to be placed indefinitely. However, signs that advertised church services and similar events were limited to six square feet, could only be displayed just before and after an event, and only four were allowed on a property at a time. Good News Community Church and its pastor, Clyde Reed, whose Sunday church services are held at various temporary locations in and near Gilbert, posted signs early each Saturday bearing the church's name and the time and location of the next service and did not remove the signs until around midday Sunday. The church was cited for exceeding the time limits for displaying temporary directional signs and for failing to include an event date on the signs. Unable to reach an accommodation with the town, petitioners filed suit, claiming that the town code abridged their freedom of speech. The district court denied their motion for a preliminary injunction, and the Ninth Circuit affirmed. The church appealed to the U.S. Supreme Court. How do you think the high court ruled in this case and why?

Source: *Reed v. Gilbert, Arizona,* 135 S. Ct. 2218 (2015).

8. Delaware enacted an apprentice regulatory scheme to "develop and conduct employee training and registered apprenticeship programs" and to provide "for the establishment and furtherance of standards of apprenticeship and training to safeguard the welfare of apprentices and trainees." Under the state's prevailing wage regulations, contractors who participated in the program could pay their apprentices significantly lower wages on certain public works projects. One of the requirements for having a registered apprenticeship training program that would qualify a contractor to pay these lower wages was to be a Delaware Resident Contractor or hold and maintain a Delaware Resident Business License. "The Registrant must hold and maintain a permanent place of business, not to include site trailers or other facilities serving only one contract or related set of contracts." Thus, out-of-state contractors were basically excluded from participating in the program and qualifying for paying the lower wages. Tri-M, an out-of-state contractor who could not participate in the apprenticeship program and thus take advantage of paying lower wages to apprentices unless he established a permanent facility in the state of Delaware, sought a declaratory judgment and injunctive relief against enforcement of the Delaware prevailing wage law and the Delaware apprenticeship and training law. On what constitutional grounds do you think Tri-M sought this injunction? Why do you believe Tri-M should have won or lost its case?

Source: *Tri-M Group v. Sharp,* 638 F. 3d 406 (2011).

9. David Milam, a member of the Delta Tau Delta fraternity at the University of Kentucky, had leased a room at the fraternity house. University of Kentucky Police Detective John McBride received a tip that Milam was selling marijuana at the fraternity house, so with two other university police detectives, he went to the house to perform a "knock and talk" investigation. The detectives did not have a warrant to search the house. Upon arriving, the officers went to the back door, believing it was the front door. No one responded to their knocking and ringing the bell several times. Deciding that the fraternity house was more similar to an apartment complex than a private residence, they opened the unlocked and slightly ajar door and entered into the foyer, where they announced their presence. Shortly thereafter, another member of the fraternity, Neagli, entered the foyer area from an adjoining room. Without asking him who he was or whether he was affiliated with the fraternity, the detectives identified themselves as police officers and said that they were looking for Milam.

After a brief conversation, the detectives followed Neagli up the stairs to the second floor where the individual rooms were located. Once in the stairwell, the detectives could smell burnt marijuana. At the top of the stairwell, Neagli opened the door to the second floor. Neagli identified Milam's room by pointing, and the detectives then knocked on the door. Milam opened the door, revealing the strong smell of marijuana. A full jar of marijuana was sitting on the coffee table inside the room, in plain view of the detectives. Milam then consented to the detectives' searching his bedroom. During the search, they discovered marijuana, $1,700, Adderall pills, drug paraphernalia, and a fake driver's license.

Milam was charged with one count of trafficking in a controlled substance within 1,000 yards of a school, third-degree possession of a controlled substance, possession of drug paraphernalia, and second-degree criminal possession of a forged instrument. In the trial court, Milam argued that the detectives unlawfully entered and searched the house in violation of the Fourth Amendment. Several suppression hearings were held. Ultimately, the trial court denied Milam's motion to suppress the evidence discovered in his bedroom.

Thereafter, Milam entered a conditional guilty plea conditioned on the appeal of the denial of his suppression motion. Pursuant to that agreement, Appellant pled guilty to the trafficking charge and was sentenced to one year imprisonment and three years' probation. The Court of Appeals affirmed the trial court's denial of Milam's motion to suppress. Milam appealed again. How do you believe the Kentucky State Supreme Court ruled and why? Be sure to consider the Fourth Amendment in your reasoning.

Source: *Milam v. Commonwealth of Kentucky,* 2015 Ky. LEXIS 1617, May 14, 2015.

10. Members of the Old Order Groffdale Conference Mennonite Church have traditionally used horses and buggies for transportation. They began using tractors about 40 years ago, especially for hauling their agricultural products to market. To ensure that the tractors are not used to displace horses and buggies, the Order requires steel cleats that slow the speed of the tractors to be attached to the wheels of all tractors driven by church members. To minimize damage to the road from these cleats, the Order made them larger and mounted them on rubber belts, but in 2009, after determining that the cleats were damaging newly resurfaced roads, Mitchell County passed an ordinance providing that, "No tire on a vehicle moved on a highway is allowed to have any block, stud, flange, cleat or spike or any other protuberances of any material other than rubber." Zimmerman, a Mennonite, continued to operate his tractor on the road with the steel cleats attached to the tires and was cited for violating the ordinance. Zimmerman filed a motion to dismiss the charges on grounds that the law violated his First Amendment right to free exercise of religion, and the trial court denied his motion. How do you think the state's supreme court should have decided the appeal, and why?

Source: *Mitchell County v. Zimmerman,* 810 NW 2d 1 (2012).

11. On September 8, 2011, John Barth purchased a piece of property at 4 Lynn Street in Peabody, Massachusetts, for $1,000. Barth subsequently requested the City of Peabody for a variance to exempt his property from the local zoning ordinance. The City of Peabody's Zoning Board of Appeals held a public hearing and later voted to deny the variance.

 Following the refusal, Barth filed a *pro se* complaint alleging 23 "violations of law." Of the many claims Barth made, he specifically argued that the denial of his zoning variance request constituted a violation of the Fifth Amendment's Taking Clause. Barth contended that the denial amounted to a taking because it deprived him of the use and enjoyment of the land. Do you think the denial of the variance could legally be considered a taking?

Source: *Barth v. City of Peabody,* U.S. Dist. LEXIS 4080 (2017).

12. In 2010, a North Carolina state court dismissed a traffic ticket against Lester Gerard Packingham. Elated by the favorable outcome, Packingham posted a note on his Facebook profile, writing "Man God is Good! How about I got so much favor they dismissed the ticket before court even started? No fine, no court cost, no nothing spent....Praise be to GOD, WOW! Thanks JESUS!" Shortly after this post, Packingham was arrested.

 He was arrested for violating section 14-202.5, a statute that made it a felony for a registered sex offender to post to a social networking website known by the sex offender to permit children to maintain personal profiles. Packingham had been on the sex offender registry since 2001. At the time Packingham wrote his Facebook post, a member of the Durham Police Department was investigating possible violations of section 14-202.5. The investigating officer obtained a search warrant and discovered evidence that Packingham had indeed violated section 14-202.5.

 Packingham was indicted by a grand jury and eventually convicted by a trial court and given a suspended prison sentence. The state never alleged that Packingham had engaged in illicit contact with a minor during the trial or sentencing. Packingham appealed to the North Carolina court of appeals, arguing that the North Carolina Law was unconstitutional because it infringed his First Amendment rights. The appeals court accepted Packingham's argument and the decision was reversed. However, the North Carolina Supreme Court reversed the appellate court. Packingham appealed to the United States Supreme Court.

 Do you think limiting access to social networking websites such as Facebook infringes a person's First Amendment rights? How do you think the Supreme Court have decided this case? Why?

Source: *Packingham v. North Carolina,* 137 S. Ct. 1730 (2017).

13. Defendant was stopped by police inside the parking garage underneath his apartment building and subsequently arrested for operating while intoxicated. Defendant filed a motion to suppress, arguing that his Fourth Amendment rights were violated because the officers's conduct occurred during a warrantless entry into a constitutionally protected area—the curtilage of his home. The trial court denied the motion to suppress, and the court of appeals affirmed. How do you think the Wisconsin Supreme Court ruled and why?

Source: *State v. Dumstrey,* 2016 WI (Wis. Sup. Ct. 2016).

14. In April 1986, the FBI arrested Larry Dean Dusenbery in his home. After the FBI removed Dusenbery from the premises and placed him in custody, it obtained and executed a search warrant for his property. During the search, FBI agents seized drugs, drug paraphernalia, weapons, an automobile, and $21,939 in currency. Dusenbery pleaded guilty to a charge of possession with intent to distribute. The

trial court sentenced him to 12 years in prison and a 6-year special parole. Two years later, the FBI began administratively to forfeit the seized possessions. This process required the FBI to send written notice of the forfeiture procedures to Dusenbery. The FBI, in accordance with this policy, sent letters to Dusenbery by certified mail to the federal corrections institution where he was incarcerated. The FBI sent identical letters to his home address and his mother's address. The FBI received no response but nevertheless proceeded with the forfeiture of his assets. Five years later, Dusenbery filed a motion requesting the FBI to return his property to him. He received a response from the U.S. government informing him that the FBI had returned all property not used in the drug business and had forfeited the rest of it. Dusenbery filed a civil suit claiming that the FBI denied his right of due process when it forfeited his property because he had never received the letter from the FBI. The district court ruled that the FBI had not violated his due process rights; the FBI satisfied his rights when it sent the notice. How do you think the appellate court ruled in this case? Why?

Source: *Dusenbery v. United States,* 534 U.S. 161 (2002).

Part 1 *The Legal Environment of Business*

CHAPTER 6

Criminal Law and Business

CASE OPENER

A Questionable Stock Investment

For approximately 20 months, Robert E. McDonald perpetrated a scheme to solicit millions of dollars, purportedly for a $100 million purchase by the RAI Entities and certain other related corporate entities, of approximately 14 Marriott and Hilton branded hotels. He made false representations that the RAI Entities and/or related entities had already raised substantial capital assets toward the purchase of the Midwest Hotel Portfolio, that the RAI Entities and/or Related Entities had tens of millions of dollars on deposit at various financial institutions, including J.P. Morgan and Citibank, and that investor funds would be put toward the purchase of the Midwest Hotel Portfolio. The defendant knew that these representations were false. The defendant created phony bank statements that he provided to the hotel investor and failed to inform the hotel investor that he had been previously convicted for fraud offenses and was currently under federal supervised release.

After having obtained approximately $1,486,323 from a hotel investor and sending the investor stock certificates by Federal Express from New York, New York, the defendant caused the hotel investor's funds to be diverted to the defendant's own use, including through an interstate wire transfer of approximately $145,000 on or about January 15, 2010, to a Citibank account the defendant maintained in New York, New York.

1. Which crimes do you believe that McDonald committed?
2. Does it matter that the stock certificates were sent via Federal Express rather than the U.S. Mail?

LEARNING OBJECTIVES

LO 6-1 Explain the basic elements of a crime.
LO 6-2 Explain how are crimes classified.
LO 6-3 Recognize who can be held responsible when a corporate crime is committed.
LO 6-4 Outline the basic steps of a criminal proceeding.
LO 6-5 Describe the laws that attempt to prevent white-collar crime.

Crime comes in many forms, and sometimes the number of people affected can quickly grow. Although some crimes might at first seem victimless, there is always a victim, and, especially with white collar crimes, there are often many victims. Imagine, for example, if McDonald had not just offered his "investment opportunity" to one person, but to several. In many crimes involving securities, the perpetrator lures not just one victim into his fraudulent scheme, but many.

This chapter helps the future business manager recognize when conduct has become criminal. The discussion begins by introducing the elements of a crime and explaining some common crimes that occur in the business context. After discussing how liability can be imposed on corporations and corporate executives, the chapter sets forth the steps of criminal procedure. The chapter concludes by discussing some of the laws used to fight business crime.

The Nature and Elements of a Crime

LO 6-1 Explain the basic elements of a crime.

The purpose of criminal law is to punish an offender for causing harm to public health, safety, or morals. In a criminal proceeding, the *government* files the charges against the defendant. Thus, the government is always a party to a criminal action. Government involvement distinguishes criminal trials from civil proceedings. As you may recall from Chapter 3, an individual person or corporation can file a civil lawsuit. This difference exists because in a criminal trial, society is seen as the victim, whereas a civil trial involves an individual victim or victims.

In many situations, a particular course of conduct may lead a defendant to be prosecuted for committing a crime and to be sued in civil court for compensation by the victims of the crime. The outcome of the two cases may not be the same, however, for a number of reasons. As you will learn, the precise elements needed to prove the resulting criminal and civil cases will be different, and whereas many of the steps of civil and criminal procedure are the same, the defendant has additional protections in the criminal case. For example, the burden of proof in the criminal case is beyond a reasonable doubt, which is a much higher standard than that of a civil case, which, as you learned in Chapter 3, is a preponderance of the evidence. You will learn about many of the additional protections afforded criminal defendants later in this chapter.

Elements of a Crime

Criminal laws usually define criminal behavior and set guidelines for punishment. To punish an individual for criminal behavior, the government must demonstrate the two elements of a crime:

1. Wrongful behavior, that is, *actus reus* (guilty act).
2. Wrongful state of mind, also known as *mens rea* (guilty mind).

The government thus must show that a defendant committed a prohibited act with a wrongful intent.

To prove the first element, *actus reus* (guilty act), the government must establish the nonmental elements of the crime. That is, the government must demonstrate that a prohibited act or a prohibited consequence resulted because of the behavior of the defendant.

To prove the second element, *mens rea* (guilty state of mind), the government must prove that the defendant acted with the state of mind required by the law defining the relevant offense. The state of mind required by an offense may be purposeful, knowing, reckless, or negligent. The defendant's type of wrongful state of mind helps determine the seriousness of the punishment for the crime.

A defendant can *purposefully* commit a crime by engaging in a specific wrongful behavior to bring about a specific wrongful result. For example, a person who robs a bank at gunpoint to unlawfully gain money is acting purposely. A defendant can *knowingly* commit a wrongful act if the person knows the act is wrongful or believes that the act is wrongful yet does nothing to confirm or disconfirm this belief. The distinction between purposely and knowingly committing a crime can be subtle, but still distinct. For example, say a person wants to murder an enemy by blowing up the office building in which his enemy works. The perpetrator is aware that this act

actus reus
Latin for "guilty act;" a wrongful behavior that is associated with the physical act of a declared crime.

mens rea
Latin for "guilty mind;" the mental state accompanying a wrongful behavior.

will likely kill other people in the building but does not care. The perpetrator would be *purposely* murdering his enemy but *knowingly* murdering the other people in the building.

A defendant is *reckless* if a criminal act occurs when the individual consciously ignores substantial risk. Drinking and driving is a common example of a reckless crime. Driving while under the influence of alcohol carries significant and apparent risks for the driver and others; choosing to do despite those risks is acting recklessly. Finally, a defendant is *negligent* if he or she does not meet a standard of care that the reasonable person would use in the context that led to the criminal act. For example, a parent leaving his children in a hot car while the parent is having a good time at a bar would acting criminally negligent.

In certain cases, liability may be assessed even without establishing the guilty-mind criterion just mentioned. Only certain crimes, typically violations of regulatory statutes, allow punishment to be assessed without having to prove a guilty mind. When liability is assessed without the guilty-mind criterion, it is known as **liability without fault** or **strict liability**. Liability without fault involves actions that, regardless of the care taken, are deemed illegal to engage in; such acts are specifically prohibited regardless of state of mind. Liability without fault would apply, for example, if a business were to sell cigarettes or alcohol to a minor. Although the business might not have intended to sell cigarettes or alcohol to a minor (thus lacking a guilty mind), the statutory violation still allows liability to be imposed.

liability without fault
A legal term that imposes responsibility for damages regardless of the existence of negligence. Also called *strict liability*.

strict liability
A legal term that imposes responsibility for damages regardless of the existence of negligence. Also called *strict liability*.

Classification of Crimes

LO 6-2 Explain how crimes are classified.

Crimes are divided into categories based on the seriousness of the offense. Today, the categories used are felonies, misdemeanors, and petty crimes. **Felonies** include serious crimes, such as murder, that are punishable by imprisonment for more than one year in a penitentiary or by death. **Misdemeanors** are less serious crimes punishable by fines and/or imprisonment for less than one year. **Petty offenses,** such as violating a building code, are minor misdemeanors and are usually punishable by a jail sentence of less than six months or a small fine. The statute defining the crime usually establishes whether the crime is a felony, misdemeanor, or petty offense. Crimes may also be federal or state, depending on whether they are created by the state or federal legislature.

felony
A serious crime, such as murder, rape, or robbery, that is punishable by imprisonment for more than one year or by death.

misdemeanor
A crime that is less serious than a felony and is punishable by a fine and/or imprisonment for less than one year.

petty offense
A minor crime that is punishable by a small fine and/or imprisonment for less than six months in a jail.

white-collar crime
A variety of nonviolent illegal acts against society that occur most frequently in the business context.

White-Collar Crime

LO 6-3 Recognize who can be held responsible when a corporate crime is committed.

Originally, white-collar crimes were distinguished from other crimes because of the social status of the offender. The more modern approach is to define **white-collar crime** as a variety of nonviolent illegal acts against society that occur most frequently in the business context. Clearly, the crimes described in the Case Opener fall under this broad definition of white-collar crime.

Mail fraud, bribery, embezzlement, and computer crimes are examples of offenses typically classified as white-collar crimes. These crimes occur more frequently than you might think. According to some estimates, one in three American households is the victim of white-collar crime every year.

The consequences of white-collar crimes are far-reaching. First, the cost of white-collar crimes can be tremendous. According to some estimates, fraud in the health care industry alone costs society over $100 billion each year. Second, when company employees commit the crime, many companies fail to report the crime to avoid publicity. Third, white-collar crimes can be costly to the environment. For example, improper disposal of chemicals, such as dumping chemicals in a stream, has long-term repercussions for marine life, the surrounding ecosystem, and any humans who may come in contact with the contaminated water source.

Of course, the amount of money collected in fines from corporate executives engaged in criminal activity has not been insignificant either. For example, in 2010, the federal government collected $2.8 billion in restitution, criminal fines, and felony assessments in criminal

cases.¹ Fines, however, are only one way that those found guilty of fraud and other white-collar crimes might be punished.

There are at least four other ways to punish white-collar criminals. First, incarceration has been mandated for certain types of white-collar crimes. As the prevalence of white-collar crime has increased, so too has the number of white-collar offenders who are in jail. Second, white-collar criminals may be ordered to do community service, such as giving speeches about business crime. For instance, junk-bond king Michael Milliken was compelled to give speeches about corporate crime after he was convicted of insider trading. Third, judges may disqualify the individual from employment. For instance, the judge may prohibit the offender from engaging in an occupation in which the same or a similar criminal act could occur again. Fourth, the offender can be placed under house arrest. Individuals must wear a sensor so that the government can monitor whether they have left their houses.

With changes in technology, and governments that are increasingly pressed financially, governments are increasingly looking for sentences other than imprisonment for nonviolent offenders. People in the criminal justice field are also starting to recognize that there is something wrong with a system that results in a higher incarceration rate in the United States than in other industrialized nations. According to the most recent statistics, the United States was incarcerating 770 persons per 100,000, whereas the rate in Europe was around 110 per 100,000.

Specific White-Collar Crimes

Bribery One of the better-known white-collar crimes is bribery. Bribery is the offering, giving, soliciting, or receiving of money or any object of value for the purpose of influencing the judgment or conduct of a person in a position of trust. *Bribery of a public official* is a statutory offense under federal law. The purpose of this law against bribery is to maintain the integrity of the government. This statute also covers bribes offered to witnesses in exchange for testimony. Perhaps one of the most publicized examples of bribery was the 2002 Winter Olympics bribery scandal in Salt Lake City, Utah. The Salt Lake City Olympics Bid Committee allegedly gave International Olympic Committee members between $4 million and $7 million in cash and other benefits. These benefits included college tuition assistance payments, shopping trips for bathroom fixtures and doorknobs, and trips to the Super Bowl.

More common, however, is the simple cash bribe to obtain a government contract. For example, in 2008, the CEO of a multimedia communications company was convicted of paying over $200,000 in bribes to the technology director of Atlanta Public Schools to secure contracts to help connect schools in poor and rural areas to the Internet. He was sentenced to a five-year prison term and payment of restitution. The government official who had accepted the bribe, and ultimately testified against the man who bribed him, was sentenced to three years in prison.²

To demonstrate bribery under this statute, the government must show three elements: (1) something of value was offered, given, or promised to; (2) a federal public official with; (3) intent to influence that person's judgment or conduct. The "thing of value" element has been construed very liberally; actual commercial value is not necessary. For instance, had the Salt Lake City Olympics Bid Committee given a member of the International Olympic Committee shares of stock in a new Silicon Valley start-up called BellyUp.com but the shares turned out to be worthless because the start-up was never established, this act may constitute bribery even though the stock is commercially worthless. Ultimately, the Department of Justice was unable to prove any of the 15 charges of bribery that it had filed against two members of the Salt Lake City Olympics Bid Committee. Several members of the committee resigned, however. As a result of investigations into the matter, 10 members of the International Olympics Committee were fired and another 10 were sanctioned.

Additionally, the definition of "public official" is also expansive. The statute defines *public officials* as members of Congress, government officers and employees, and anyone "acting for or on behalf of" the federal government "in any official function, under or by authority of" a federal

bribery
A corrupt and illegal activity in which a person offers, gives, solicits, or receives money, services, or anything of value to gain an illicit advantage.

¹ Department of Justice, "U.S. Attorneys Contributed to $6.68 Billion in Civil & Criminal Collections in Fiscal Year 2010," December 16, 2010, https://www.justice.gov/archive/usao/fls/PressReleases/2010/101216-01.html.

² "Georgia Man Sentenced for E-Rate Bribery," AsianFanaticsForum, December 28, 2008, asianfanatics.net/forum/topic/612815-georgia-man-sentenced-for-e-rate-bribery/.

government department or agency. For example, individuals who work for private corporations that have some degree of responsibility for carrying out federal programs or policies are considered public officials.

Another type of bribery is *commercial bribery,* which includes a bribe in exchange for new information or payoffs. For example, suppose Comdac Computers is looking for a company that manufactures a certain type of computer part. Jane Devlon owns such a company and realizes that if Comdac gets the computer parts from her, she will potentially make a lot of money. When Comdac sends its contractor to the Devlon factory, Devlon offers the contractor $500,000 in exchange for the contractor's promise that Comdac will buy all such parts from only her factory for the next year. Alternatively, she might offer the contractor $5,000 to disclose the dollar amounts of the competing bids so that she can offer a better bid to earn the contract.

One final type of bribery is the *bribery of foreign officials.* The purpose of the Foreign Corrupt Practices Act (FCPA) is to combat such bribery. The FCPA prohibits payments to foreign officials to influence an official act or decision corruptly or to influence a foreign government. You might think of the FCPA as an extension of the law to prevent bribery of a public official. The FCPA allows the government to prosecute those who bribe foreign officials.

But Congress does not pass laws affecting foreign affairs in a vacuum. Laws governing white-collar crime, like all laws, are dynamic, and one source of change in these laws is international law. Multinational organizations in which the United States is a member often pressure the United States to alter American law so that it conforms to certain international standards. Thus, for example, Congress amended the FCPA in 1998 to conform to the antibribery convention adopted by the Organization for Economic Cooperation and Development (OECD). The amendments broaden the scope of actions the FCPA covers and the definition of a public official. As of this writing, 30 OECD countries—such as Japan, South Korea, and Brazil—are expected to adopt similar antibribery statutes.

Extortion Often confused with bribery, but completely different, is the crime of extortion. **Extortion,** otherwise known as *blackmail,* is the making of threats for the purpose of obtaining money or property. Whereas bribery is offering someone something to obtain a desired result, extortion involves the threat of doing something if the target of the extortion does not relinquish money or a specific piece of property.

Fraud Criminal **fraud** encompasses a variety of means by which an individual intentionally uses some sort of misrepresentation to gain an advantage over another person. Fraud generally requires the following three elements: (1) a material false representation made with intent to deceive *(scienter),* (2) a victim's reasonable reliance on the false representation, and (3) damages. Both the federal and the state governments have passed a number of statutes codifying specific types of fraudulent schemes. These schemes to defraud include a multitude of frauds, such as credit card fraud, insurance fraud, and securities fraud. Exhibit 6-1 summarizes the fraudulent acts that might occur in the corporate setting, many of which are discussed in this chapter.

One of the FBI's largest *securities fraud* crackdowns occurred in June 2000, when it arrested a group of mafia leaders for a series of scams that cost investors an estimated $25 million. Government officials claimed that the mob bought large stakes in small companies and then bribed and coerced brokers to promote the stocks to other investors at inflated prices.

The Enron scandal also involved securities fraud, with Enron's top executives allegedly overstating its earnings to maintain high stock prices. Complex accounting methods were used to hide the corporation's debt. Enron's top executives and big investors sold their stocks while encouraging employees to continue buying company stocks. With the drastic decline in the company's stock value, many Enron employees lost much of their 401(k) investments.

Another specific type of securities fraud in the Enron case is insider trading. **Insider trading** is generally the buying or selling of a security, in breach of a fiduciary duty or other relationship of trust and confidence, while in possession of material, nonpublic information about the security. Enron's big investors and top executives knew the stock prices were inflated due to overstated earnings. They used this information, which was not public knowledge, to sell off their stock before the price fell.

extortion
A criminal offense in which a person obtains money, property, and/or services from another by wrongfully threatening or inflicting harm on the other's person, property, or reputation. Also called *blackmail.*

fraud
An intentional deception that causes harm to another.

insider trading
The illegal buying or selling of a corporation's stock or other securities by corporate insiders, such as officers and directors, in breach of a fiduciary duty or some other relationship of trust and confidence, while in possession of material, nonpublic information about the security.

	Exhibit 6-1
	Selected Fraudulent Crimes

1. *Forgery:* The fraudulent making or altering of any writing in a way that changes the legal rights and liabilities of another.
2. *Defalcation:* The misappropriation of trust funds or money held in a fiduciary capacity.
3. *False entries:* The making of an entry into the books of a bank or corporation that is designed to represent the existence of funds that do not exist.
4. *False token:* A false document or sign of existence used to perpetrate a fraud, such as making counterfeit money.
5. *False pretense:* A designed misrepresentation of existing facts or conditions by which a person obtains another's money or goods, such as the writing of a worthless check.
6. *Fraudulent concealment:* The suppression of a material fact that a person is legally bound to disclose.
7. *Mail fraud:* The use of mails to defraud the public.
8. *Health care fraud:* Any fraudulent act committed in the provision of health care products or services.
9. *Telemarketing fraud:* Any scheme, including cramming and slamming, using the telephone to commit a fraudulent act.
10. *Ponzi scheme:* An investment swindle in which high profits are promised from fictitious sources and early investors are paid off with funds raised from later ones.
11. *Check kiting:* Drawing checks on an account in one bank and depositing them in an account in a second bank when neither account has sufficient funds to cover the amounts drawn. Just before the checks are returned for payment to the first bank, the kiter covers them by depositing checks drawn on the account in the second bank. Due to the delay created by the collection of funds by one bank from the other, known as the float time, an artificial balance is created.
12. *Pretexting:* Using fraudulent means to obtain information about someone's phone use.

In March 2011, another apparent insider-trading scheme came to light when it was alleged that a corporate lawyer and a trader operated an insider-trading scheme that lasted decades and earned the two $32 million. The lawyer worked for a mergers and acquisitions firm, and he gave nonpublic information about deals the firm was working on to his partner, the trader, so that he could make appropriate stock trades using this information. One of their alleged profitable inside trades occurred when information that the lawyer acquired in April 2009 about Oracle's proposed bid for Sun Microsystems led the trader to purchase more than 4 million shares of the target company, which he was then able to make a handsome profit on when the deal closed in 2010.[3]

Insider-trading violations may also include "tipping" such information and trading on it by the person "tipped."[4] A tippee who receives such information with the knowledge that its disclosure breached the tipper's duty acquires that duty and may be liable for securities fraud for any undisclosed trading on the information. Thus, if a person has inside information and does not trade on it himself but passes it on to another who uses it to trade, the person who passed on the information is guilty of insider trading.

Tippee liability requires that the tipper's disclosure breaches a fiduciary duty, which occurs when the tipper discloses the information *for a personal benefit*. A personal benefit may be inferred where the tipper receives something of value in exchange for the tip or makes a gift of confidential information to a trading relative or friend. The following case illustrates how the U.S. Supreme Court recently clarified how to establish a tippee's liability.

[3] Ben Protess, "Two Charged in Insider Trading Scheme Linked to Law Firms," *The New York Times,* April 6, 2011.
[4] U.S. Securities and Exchange Commission, https://www.sec.gov/fast-answers/answersinsiderhtm.html.

CASE 6-1

SALMAN v. UNITED STATES
United States Supreme Court
137 S. Ct. 420 (2016)

FACTS: *Petitioner Salman was indicted for federal securities-fraud crimes for trading on inside information he received from a friend and relative-by-marriage, Michael Kara, who, in turn, had received the information from his brother, Maher Kara, a former investment banker at Citigroup. Maher testified at Salman's trial that he had shared inside information with his brother Michael to benefit him and expected him to trade on it. Michael testified to sharing that information with Salman, who knew that it was from Maher. Salman was convicted.*

ISSUE: If an insider just gives inside information to an outsider without receiving a financial payment for that information, can the insider still be held liable for insider trading?

REASONING: The Securities Exchange Act of 1934 as well as SEC Rule 10b-5 prohibit undisclosed trading on inside corporate information. It is also prohibited for someone with insider information to give, or tip, that information to others for trading. The receiver of insider information, or tippee, acquires the tipper's duty to disclose the inside information or abstain from trading on it. In *Dirks v. SEC,* the Supreme Court ruled that the tippee's responsibility derives from the fact that the tipper breached a fiduciary duty when disclosing the information for personal benefit. This personal benefit can even include when the tipper is giving the confidential information as a gift to a relative or friend.

In this case, Salman argued that giving a gift of information to a relative does not constitute securities fraud. Salman contended that a tipper does not personally benefit unless he or she gains something of tangible value in exchange for the information. However, the Supreme Court ruled that *Dirks* is quite clear that when a tipper gives inside information to a trading relative or friend, a jury can infer that it is equivalent to a cash gift (the assumption being that the trades made on the information will make money in the future). In these cases, the tipper benefits personally in the same way that giving someone a tangible money gift benefits personally. When Maher tipped the information to Salman, Maher breached a duty of trust to Citigroup and its clients—a duty which Salman inherited and breached himself by trading on the information.

DECISION AND REMEDY: Yes, he can. And therefore Salman's conviction for insider trading was affirmed.

SIGNIFICANCE OF THE CASE: This case illustrates that tipping and/or trading based on insider information—regardless of the type of information, who tipped it, and the reasons for tipping—is invariably illegal; there really is not any way around insider trading restrictions.

CRITICAL THINKING

How would you evaluate the court's claim that giving a gift of trading information is the same thing as trading by the tipper followed by a gift of the proceeds?

ETHICAL DECISION MAKING

Which stakeholders benefit the most from the decision in this case?

Stock-option backdating is yet another type of securities fraud that the SEC has been increasingly cracking down on. Backdating occurs when an employee falsifies documents to make it appear as though the company had granted options on certain dates, but the dates are selected after the fact by looking backward for dates on which the stock price was low, thereby falsely inflating the net profits of the company. In 2007, Myron F. Olesnyckyj, the former general counsel of Monster Worldwide, Inc., pleaded guilty to securities fraud for participating in a multiyear scheme with other Monster executives to secretly backdate stock options granted to thousands of Monster officers, directors, and employees, including himself. By backdating and improperly accounting for options, Monster granted undisclosed compensation to its employees, failed to recognize compensation expenses, and overstated its net income by $340 million from 1997 through 2005. Olesnyckyj personally made $381,000 from the scheme, which he agreed to forfeit.

An additional type of fraud is false pretense. **False pretense** is the illegal obtaining of property belonging to another through materially false representation of an existing fact, with knowledge of the falsity of the representation and with the intent to defraud. For example, Jim goes door-to-door selling vacuum cleaners at an amazing discount. To obtain the discount, the customers must pay immediately, and the vacuum cleaner will be delivered in three to five business days. However, Jim possesses no vacuum cleaners and does not plan to deliver any. Jim has committed the crime of false pretense because he illegally obtained another's property (the money he received as payment) by making a false representation (telling the customers they would be receiving a vacuum) with the knowledge of the falsity (he knew there were no vacuum cleaners) and the intent to defraud (he did not plan to deliver any vacuum cleaners).

Forgery, the fraudulent making or altering of any writing in a way that changes the legal rights and liabilities of another, is another type of fraud. If you sign your colleague's name to the back of a check made out to your colleague, you have committed forgery.

One of the most frequently prosecuted frauds is *mail fraud*, which is the use of mails to defraud the public. The Mail Fraud Act of 1990 makes mail fraud a federal crime. To prove mail fraud, the government must demonstrate two elements: (1) an intent to defraud and (2) the use of or causing the use of mails to further the fraudulent scheme. Case 6-1 demonstrates how the courts address the second element.

false pretense
The illegal obtaining of another's property by making a materially false representation of an existing fact, with knowledge of the falsity of the representation and with the intent to defraud.

forgery
A criminal falsification by imitating, impersonating, or altering a real document with the intent to deceive or defraud.

CASE 6-2

UNITED STATES OF AMERICA v. THOMAS TANKE
United States Circuit Court of Appeals for the Fourth Circuit
743 F.3d 1296 (2014)

FACTS: Rafael Martin and his family ran two construction companies, Azteca and CERS. Thomas Tanke worked as the vice president of Azteca for about four years, during which time he had the authority to issue checks from Azteca. Tanke also ran a consulting business called Cedar Creek Associates. Over a 20-month period, Tanke embezzled more than $192,000 from Azteca and CERS, issuing 21 Azteca checks to cover personal expenses and diverting several checks from CERS into an account held by Cedar Creek. Tanke falsified records, creating at least six false invoices to make the Azteca checks look legitimate, and forging signatures and endorsements for the checks issued by CERS.

Martin discovered the embezzlement and sent Tanke an email one month after Tanke left Azteca, inquiring about a missing check. Tanke responded by stating that he had diverted the CERS check into the Cedar Creek account to ensure payment of "long overdue invoices" for consulting work that Cedar Creek had allegedly done for CERS and Azteca. Additionally, Tanke sent a letter from Cedar Creek to Martin by U.S. airmail on September 16, 2004. The letter stated that Azteca and CERS had previously paid invoices due to Cedar Creek and that these payments had been reversed due to false information that Azteca and CERS had provided to Cedar Creek's bank. The letter demanded that reversed payments be returned.

In 2009, the grand jury indicted Tanke on five counts of bank fraud and two counts of mail fraud. A jury found Tanke guilty on all counts, and the district court sentenced him to 70 months' imprisonment, 60 months' supervised release, and $243,403.96 in restitution. Tanke filed an appeal, contending, among other things, that there was insufficient evidence to convict him on the second count of mail fraud related to the letter sent on September 16, 2004.

ISSUE: When do mailings designed to avoid detection or responsibility for a fraudulent scheme fall within the mail fraud statute?

REASONING: The second prong of the test for mail fraud requires that the state prove that in executing the scheme, the defendant made use of or caused the use of the mails. In deciding when mailings designed to avoid detection or responsibility for a fraudulent scheme fall within the mail fraud statute, the mailing must be sent prior to the scheme's completion. In the more difficult cases, where it is not an ongoing scheme, the court must determine when the scheme ended. Applying a totality of the circumstances approach, the appellate court found that a reasonable jury could have inferred that the Cedar Creek cover story was part of Tanke's embezzlement scheme from the outset or as it evolved over time—just another misrepresentation to facilitate diversion of Azteca's and CERS's funds into Tanke's bank accounts.

DECISION AND REMEDY: The defendant's conviction was affirmed.

(continued)

SIGNIFICANCE OF THE CASE: This case makes the point that to support a mail fraud conviction, the mailing must have occurred before the fraudulent scheme was completed. It also illustrates how difficult it can sometimes be to determine exactly when a fraudulent scheme is completed. In addition, this case highlights the importance of relying on precedent that is consistent with a statute's plain language.

CRITICAL THINKING

What additional facts, if they existed, might have caused the court of appeals to overturn the conviction?

ETHICAL DECISION MAKING

What ethical norms are furthered by the appellate court's decision in this case?

Congress amended the mail fraud statute in 1994 to include commercial carriers and courier services as substitutes for the U.S. mail. Thus, using FedEx or UPS to further a fraudulent scheme can be prosecuted as mail fraud. Similarly, the wire fraud statute attempts to prevent the use of wire, radio, or television transmissions to defraud the public.

Other types of fraud include health care fraud, telemarketing fraud, and bankruptcy fraud. The Department of Justice recently named health care fraud its top priority after violent crime. An example of health care fraud is the submission of false claims to insurance plans such as Medicare or Medicaid. Additionally, some doctors prescribe unneeded equipment to patients and then receive kickbacks from the manufacturers of the equipment.

Telemarketing fraud has also gained much attention. The National Fraud Information Center reported that telephone cramming was the top telemarketing fraud of 1998. Cramming is a scheme in which companies bill consumers for optional services that the consumers did not order. Another frequent telemarketing scheme is slamming, in which consumers are tricked into changing their phone service to another carrier without their consent. In 2014, the FTC settled a long running case against a group of defendants who engaged in text message, robocall, and cramming schemes involving offers for "free" merchandise, recovering approximately $10 million.

The elderly, who have been more susceptible to telemarketing fraud, are often the specific targets of telemarketing schemes. Thus, the Department of Justice and the FBI have been collaborating on various undercover operations, such as Operation Senior Sentinel, to investigate and prosecute fraudulent telemarketers.

Another type of fraud involves bankruptcies. In 2014, 1,107,699 bankruptcy claims were filed. An individual can file for bankruptcy to be relieved of oppressive debt. Yet claims need to be reviewed carefully to prevent bankruptcy fraud. For example, an individual might hide some assets so that these will not be considered during the proceedings. Or, as one of the stars of the *Real Housewives of New Jersey* show did, a person might hide part of her income. In the *Real Housewives* case, the housewife claimed to be unemployed, even though she had allegedly just signed a contract for a second season of *Real Housewives,* and she did not reveal her income from website sales and personal and magazine appearances. Conversely, a creditor (a party to whom another owes money) might file a false claim against a debtor (the party who owes money to a creditor). Thus, bankruptcy fraud can occur on the part of the debtor or the creditor, although 70 percent of bankruptcy fraud claims involve concealment of assets by the debtor.

The U.S. Department of Justice estimates that 10 percent of all bankruptcy petitions contain some elements of fraud. Yet, according to a report by the U.S. Department of Justice, in fiscal year 2012, the United States Trustee Program (USTP) filed 2,120 criminal referrals, a 7.7 percent increase over the 1,968 criminal referrals made during fiscal year 2011. One way the Justice Department is attempting to fight bankruptcy fraud is through its Internet hotline, where the public can report suspected bankruptcy fraud to the USTP.

Bankruptcy fraud is punishable by a fine of up to $250,000 and/or up to five years in prison, but often the person who engages in bankruptcy fraud is also engaging in another crime, such as tax evasion, which can result in greater penalties. For example, in 2013, Larry Lake was sentenced to 168 months in prison and ordered to pay $550,000 in fines and approximately $25 million in taxes, interest, and penalties for bankruptcy fraud and tax evasion. The day before he filed for bankruptcy, he had knowingly and fraudulently transferred and concealed more than $3 million held in an online trading account and a bank account. He had transferred the funds through a series of bank deposits, wire transfers, and cashier's checks and used a shell company to help hide assets.

Sometimes fraud cases can be very complex and involve multiple types of fraud. For example, in one recent case, a Beverly Hills lawyer was convicted of three counts of mail fraud, seven counts of wire fraud, and five counts of lying in a court hearing. In an elaborate fraudulent scheme, the lawyer had purchased a yacht for $1.9 million, sold it to two partners to drive up its insurance value, and then had a company he owned repurchase the yacht and insure it for $3.5 million. Finally, he and his partners sank the yacht and tried to collect the insurance.

BUT WHAT IF . . .

Former Mets outfielder Lenny Dykstra went to court for fraud in 2012. While Dykstra owned a mansion formerly owned by Wayne Gretzky, he claimed that he was over $31 million in debt and his total assets amounted to only $50,000. Thus, he was seeking to file for bankruptcy. If Dykstra secretly hid and sold $400,000 worth of possessions yet on the stand said he had only a few items put into his storage unit, would he be accountable for only bankruptcy fraud or other types of fraud as well? Alternatively, if Dykstra had reviewed the sold, stored, and destroyed items with a bankruptcy trustee, would he be liable for fraud at all?

A fraudulent crime that has recently come into the public eye is pretexting. *Pretexting* can be defined as using or causing others to use false pretenses, fraudulent statements, fraudulent or stolen documents, or other misrepresentations, including posing as an account holder or employee of a telecommunications carrier, to obtain the telephone records of another. Pretexting entered the national lexicon in 2006 when news broke that the investigators hired by Hewlett-Packard's board of directors to locate the source of several leaks to the media were engaging in this practice. Under the federal Telephone Records and Privacy Protection Act of 2006, anyone convicted of employing fraudulent tactics to persuade phone companies to hand over confidential data about a customer's calling habits can be sent to prison for up to 10 years. Some states also have statutes outlawing pretexting.

Both individuals and businesses can be victims of fraud. To minimize the chances that your firm will be a victim of fraud, review the fraud prevention tips in Exhibit 6-2 very carefully.

Ponzi schemes are another form of fraud that Americans are increasingly seeing discussed in the news. A Ponzi scheme is a fraudulent operation in which the operator, an individual or organization, pays returns to its investors from new capital paid to the operators by new investors rather than from profit earned by the operator. In early 2009, Bernie Madoff, the perpetrator of one of the largest Ponzi schemes to date, was charged with 11 felony charges, including securities fraud, investment adviser fraud, mail fraud, wire fraud, three counts of money laundering, false statements, perjury, false filings with the United States Securities and Exchange Commission, and theft from an employee benefit plan.

Madoff ultimately pled guilty to all 11 counts and was subsequently sentenced to a term of imprisonment of 150 years for this crime, which was estimated to have cost investors losses of $18 billion. As of April 2014, the government had seized $2.35 billion in corporate and personal assets from Madoff that were to be distributed to victims of the fraud.

Even though Madoff was prosecuted in 2009, it wasn't until 2014 that five of his former aides were convicted on grounds that they had helped him conceal his massive fraud for years. In a trial that lasted five months, back-office director Daniel Bonventre, portfolio managers Annette Bongiorno and Joann Crupi, and computer programmers Jerome O'Hara and George Perez pled guilty on all counts, including securities fraud and conspiracy to defraud clients. Prior to this trial, nine other people pled guilty in connection with Madoff's fraud.

Exhibit 6-2

Fraud Prevention Tips

RED FLAGS SUGGESTING POSSIBLE FRAUD

When you are looking at a brochure offering an investment opportunity that seems almost too good to be true, it may indeed not be as good as it seems, especially if it contains these red flags of fraud:

- *No independent proof of profitability:* The brochure provides the investment strategy but includes no details that can be verified, such as names and addresses of specific companies in which it invests.
- *Control by a single person.*
- *No audited financial statements and no evidence of internal controls.*
- *Unusually high rates of return:* The rates of return are significantly higher than those for other funds with a similar investment strategy.
- *New investors's reliance primarily on existing investors in deciding to invest.*

False Assurances That an Investment Opportunity Is Not Fraudulent

- *Long-time existence of investment opportunity:* A Ponzi scheme can operate for a long time. As long as the company keeps getting new investors, it can continue to use the new investors to pay off the few who insist on collecting their returns. Many investors may be content to accumulate paper profits, especially if there is a financial incentive for reinvestment of profits. For example, the Financial Advisory Fund had been in business for almost 20 years before its fraudulent foundation was uncovered.
- *A list of names and addresses of satisfied investors you can contact:* All such schemes will have some satisfied investors who have received payments from the new investment money.
- *Membership in the Better Business Bureau:* Any company can pay to join. If most of the investors have not attempted to withdraw their funds and the company is still getting new investments, there will be no disgruntled investors to complain to the BBB.
- *A report of profitability from Dun & Bradstreet:* Dun & Bradstreet does not do an independent financial audit of a company's profits. As stated on its website, Dun & Bradstreet simply provides the information regarding profits that it receives from companies.
- *Acceptance of money rolled over from IRAs:* There is no government check on the soundness of firms that roll over IRAs.
- *Bank references:* If a firm simply has a large amount of money in a checking account in a bank and has never sought a loan from the bank, the bank would have no reason to do any due diligence on the firm. Also, if the owner of the firm is personable and the account has never been overdrawn, the bank personnel may well give the firm a good reference.
- *Glossy brochures and television ads:* All you need for both is some money and the knowledge that many people may be overly impressed by a fancy brochure or ad campaign.

These tips come from Barry Minkow, who went from starting his own carpet cleaning business at the age of 16, to taking this multimillion-dollar company public at the age of 20, to facing a 23-year prison sentence and $26 million victim restitution payment at the age of 21 for massive fraud and theft perpetuated on behalf of his company, to now being a pastor and the cofounder of the Fraud Discovery Institute.

The fascinating story of how Minkow, through fraudulent Ponzi schemes, check kiting, and theft, managed to build the multimillion-dollar ZZZZ Best Carpet Cleaning Company by the age of 20 is contained in his book, *Cleaning Up*. The book also provides insights into why some people may commit white-collar crimes, describes many of the long-running frauds Minkow helped to uncover since deciding to use his understanding of the perpetuation of fraud to help uncover and prevent it, and how he became a preacher. Unfortunately, Minkow's conversion was not permanent, and in 2014 he was sentenced to five years in prison for embezzling $3 million from his San Diego church congregation.

GLOBAL Context

©Digital Archive Japan/ Alamy

Embezzlement and Bribery in China

On March 8, 1982, in response to a concern about a rise in white-collar crimes, Chinese officials added a resolution entitled "Severely Punishing Criminals Who Do Great Damage to the Economy" to the Chinese Criminal Code. The provision targeted top-ranking officials by providing that any state functionaries who extort, accept bribes, or exploit their office will no longer receive the fixed-term punishment (usually 10 years) allocated for bribery and extortion. Instead, those officials found guilty of extortion and accepting bribes would be sentenced to life imprisonment or possibly put to death.

Bank fraud is another white-collar crime that may be punishable by death. In 2004, China executed four people, including employees of two of its Big Four state banks, for fraud totaling $15 million. Three of the cases involved China Construction Bank.

A former accounting officer at this bank worked with others to steal 20 million yuan ($2.4 million) from the bank by using fake papers. He and an accomplice were executed, along with another Construction Bank employee who was found to have taken 20 million yuan from the bank in an unrelated case. Additionally, an official employed by the Bank of China was executed for cheating.

The precise number of people executed in China is not made available to the public. Estimates range from 5,000 to 10,000 a year, for crimes including murder, corruption, and, on occasion, even bottom-pinching. Most are executed by a single shot in the back of the head.

Such punishments may seem extreme, but the reaction of the Chinese to the increase in white-collar crime reflects the culture's unusually great concern for social harmony.

A $300 million Ponzi scheme involving real estate investments led to Fred "Dave" Clark Jr., the founder and CEO of Cay Clubs, being convicted on three counts of bank fraud and making a false statement to a financial institution and one count of obstructing the SEC by lying to the agency. In 2016, he was sentenced to 40 years in prison, and fined $303.8 million for the bank fraud and $3.3 million for obstruction charge. He also was ordered to forfeit $2.6 million in overseas assets, and to to pay the U.S. Internal Revenue Service over $1.5 million in restitution for taxes owed on unreported income.

The way this crime unfolded was that Cay Clubs sold condos as condo-hotel vacation units in the Keys, Clearwater, and Las Vegas and promised renovations and big payouts to the investors, but, instead, used new buyers' money to pay lease-back fees owed to earlier buyers while extracting $22 million from the company. Clark used the money to buy waterfront homes, cars and planes. The scheme lasted almost four years before collapsing on itself.

Embezzlement Joe is Kathleen's attorney. Kathleen gave Joe $5,000 to put in escrow, an account to which Joe will have access, although the money is not his to use as he pleases. Suppose Joe takes some of that money out of the account and uses it to gamble, and he fixes the records to cover up the fact that he has used some of the money for his personal use. He has committed the crime of embezzlement, the wrongful conversion of another's property by one who is lawfully in possession of that property. Embezzlement is distinguished from larceny because the individual did not take the property from another; he was already in possession of the property.

Embezzlement usually occurs when an employee steals money. Thus, it is often employees in banks who commit the crime. However, as the recent embezzlement of $6.9 million from the American Cancer Society demonstrates, even nonprofit organizations are vulnerable to the crime of embezzlement. Even the Girls Scouts have been victims of embezzlement. A Girl Scout Troop leader pocketed the proceeds of her troop's annual Girl Scout Cookie Sale, as well as a portion of a donation made to the troop by a family member of a deceased troop leader. The embezzlement came to light only when the troop was unable to afford to participate in their ordinary scouting activities.[5]

White-collar crimes such as embezzlement are a problem in many countries. For example, the Chinese government, which has been focusing on economic crimes, caught employees in two major Chinese banks altering deposit slips and bank orders to direct the money to personal

embezzlement
A wrongful conversion of another's funds or property by one who is lawfully in possession of those funds or that property.

[5] http://legaldictionary.net/embezzlement/.

accounts throughout China. Those employees received death sentences for embezzlement, as discussed in the Global Context box.

computer crime
Crime committed using a computer.

Computer Crimes The term computer crime refers broadly to any wrongful act that (1) is directed against computers, (2) uses computers to commit a crime, or (3) involves computers. Computer crime is not necessarily a new kind of crime but is, instead, a new way of committing traditional crimes. Fraud, for example, is a crime that has existed for centuries. Today, online auctions, such as eBay, are a common source of fraud in the United States. But electronic fraud is not the only costly computer related crime. According to the FBI, dealing with viruses, spyware, PC theft, and other computer crimes cost U.S. businesses 67.2 Billion.

Computer crimes are often difficult to prosecute, in large part because they are difficult to detect. A computer crime could be committed by an insider, such as an employee, or by an outsider, such as a hacker—a person who illegally accesses or enters another person's or a company's computer system to obtain information or to steal money. In addition, computer systems are quite open to attack. The American Society for Industrial Security has reported that the losses from computer crimes in a single year were more than $250 billion for U.S. companies. Furthermore, the attacks are frequent. The IRS alone detects 800 to 1,200 cases of computer system misuse annually.

hacker
A person who illegally accesses or enters another person's or a company's computer system to obtain information or to steal money.

cyber terrorist
A hacker whose intention is the exploitation of a target computer or network to create a serious impact, such as the crippling of a communications network or the sabotage of a business or organization, which may have an impact on millions of citizens if the terrorist's attack is successful.

A cyber terrorist is a hacker whose intention is the exploitation of a target computer or network to create a serious impact, such as the crippling of a communications network or the sabotage of a business or organization. The activity of a cyber terrorist can have an impact on millions of citizens if the terrorist's attack is successful, particularly if it involves the computer systems of a major stock exchange, any bank, or any federal agency or government.

In an attempt to aid prosecutions of computer crimes, Congress passed the Counterfeit Access Device and Computer Fraud and Abuse Act of 1984 (also know as the Computer Fraud and Abuse Act, or CFAA). The act prohibits six broad categories of computer crimes:

1. Unauthorized use of or access to a computer to obtain classified military or foreign policy information with the intent to harm the United States or to benefit a foreign country.
2. Unauthorized use of a computer to collect financial or credit information protected under federal privacy law.
3. Unauthorized access to a federal computer and the use, modification, destruction, or disclosure of data it contains or the prevention of an authorized person's use of such data.
4. Alteration or modification of data in financial computers, causing a loss of $1,000 or more.
5. Modification of data that impedes medical treatment to individuals.
6. Fraudulent transfer of computer passwords or similar data that could aid unauthorized access that either (a) affects interstate commerce or (b) permits access to a government computer.

Computer crimes falling into the first category are felonies, whereas the others are misdemeanors.

The National Information Infrastructure Protection Act of 1996 amended the Computer Fraud and Abuse Act. One of the major changes to the CFAA is the substitution of the term *protected computers* for *federal interest computers* so that the statute now protects any computer attached to the Internet.

In a further attempt to aid prosecution of computer crimes, the U.S. Department of Justice formed the Computer Crime and Intellectual Property Section (CCIPS) in its Criminal Division. Attorneys in this division prosecute only federal computer crimes. They also coordinate their activities with numerous government entities, the private sector, scholars, and foreign representatives in an attempt to develop a global response to computer crime.

virus
A computer program that rearranges, damages, destroys, or replaces computer data.

Destruction of data is one of the most serious problems facing companies today. A virus is a computer program that rearranges, damages, destroys, or replaces computer data. Thus, if an employee or a hacker creates and releases a virus, a company can easily lose vital information.

Until recently, the most economically destructive computer crime to date was the creation of the Love Bug virus. In May 2000, the virus spread rapidly throughout the world by email. Once the email was opened, it destroyed files on the user's computer, and then the virus sent itself to every address in that computer's email address book. In the end, the virus caused more than $10 billion in damages and halted computers in major companies and government agencies worldwide.

E-COMMERCE and the Law

Computer Use and Ethics

The use of computers to store and transfer important business information has resulted in a new set of ethical concerns. How private is the computer screen? Are companies allowed to collect information about their customers? Who owns the information customers give to e-businesses? E-commerce law is gradually adjusting to such questions. But, as the beginning of Chapter 1 pointed out, knowing the law is just the first step in discovering the ethical business decision.

One case that considered questions related to e-commerce and privacy involved Toysmart.com and its privacy policy. In 2000, the Federal Trade Commission (FTC) filed a complaint against Toysmart.com, charging the online retailer with selling customer lists despite its earlier privacy statements that claimed its customers's personal data would never be shared with a third party. The customer lists were included as assets to be sold as part of the company's bankruptcy proceedings. The FTC issued a settlement with Toysmart.com allowing the lists to be sold as long as (1) the sale occurred before July 2001, (2) the lists would be sold to a family-oriented company, and (3) the buyer would agree to abide by the original Toysmart.com privacy policy. After the FTC imposed these restrictions, the result was that customers's personal information was not sold.

The Toysmart.com case makes it clear that, for both ethical and legal reasons, companies that adopt a privacy policy need to think that policy through before announcing it to customers.

However, over the weekend of May 12, 2017, unknown cyberhackers launched "Wanna Cry," believed to be, as of when this book went to press, the biggest online extortion attack ever recorded. The global "ransomware" attack struck governments, universities and private firms in over 150 countries, causing a wide range of problems. For example, in London, a fifth of the National Health Services that manage the hospitals were hit, leading to thousands of canceled appointments and operations.

Companies can try to prevent the destruction of data by installing virus detection programs on their computers. However, the detection programs recognize only previously existing harmful files. As a result, if someone creates a new virus, the detection program is useless.

When an employee uses his or her computer in a manner not authorized by the employer, the employee has committed a crime. Employers, then, must clearly communicate with employees about authorized versus unauthorized behavior. For example, if an employee uses her work computer to run her own personal business on the side, her employer may argue that she is engaging in theft because she is not using her computer in an authorized manner. Thus, when you become a business manager, you will want to make sure that you explicitly list acceptable computer uses, along with the penalties associated with unauthorized use.

Corporate Criminal Liability

LO 6-3 Who can be held responsible when a corporate crime is committed?

An individual or a corporation can be charged with and convicted of a crime. However, there has been much debate over whether corporations should be held criminally responsible. Under the common law, a corporation could not be considered a criminal because it was not an actual person and thus did not have a mind. Consequently, it could not meet the *mens rea* (guilty-mind) requirement for a crime. Slowly, however, courts began to impose liability on corporations for **strict-liability offenses,** those offenses that do not require state of mind. Next, courts imposed liability on corporations by imputing the state of mind of the employee to the corporation.

strict-liability offense
Offense for which no *mens rea* is required.

Currently, corporations can be held criminally accountable for almost any crime. However, they cannot be held liable for crimes that are punishable only by a prison sentence. For a corporation to be held criminally liable for the acts of an employee, the prosecutor must show that (1) the employee was acting within the scope of her or his employment; (2) the employee was acting with the purpose of benefiting the corporation; and (3) the act was imputed to the corporation.

In addition to corporations having criminal liability, corporate executives may also be personally liable for a business crime. The beneficiary of the crime is irrelevant because corporate

executives and officers can be found to be personally liable regardless of whether the crimes were committed for their personal benefit or for the benefit of the corporation.

Also, under the "responsible corporate officer" doctrine, a court may assess criminal liability on a corporate executive or officer even if he or she did not engage in, direct, or know about a specific criminal violation. As Case 6-3 demonstrates, corporate executives sometimes have the responsibility and power to ensure the company's compliance with the law. If the executive fails to meet this responsibility, the executive can be held criminally liable.

CASE 6-3

UNITED STATES v. PARK
United States Supreme Court
421 U.S. 658 (1975)

FACTS: *Defendant Park, the president of a national food-chain corporation, was charged, along with the corporation, with violating the Federal Food, Drug, and Cosmetic Act by allowing food in the warehouse to be exposed to rodent contamination. Park had conceded that his responsibility for the "entire operation" included warehouse sanitation, but he claimed that he had delegated the responsibility for sanitation to dependable subordinates. He admitted at trial that he had received a warning letter from the Food and Drug Administration regarding the unsanitary conditions at one of the company's warehouses. The trial court found the defendant guilty. The court of appeals reversed. The case was appealed to the U.S. Supreme Court.*

ISSUE: Can corporate executives be criminally liable for failing to ensure a company's compliance with the law?

REASONING: The Supreme Court has noted that the only way corporations can act is through their individual employees. The act in question, the Federal Food, Drug, and Cosmetic Act, requires stringent, affirmative actions on the part of corporate executives, but these standards are justified by the public interest in food safety. We expect executives such as Park to exercise appropriate authority and demonstrate supervisory responsibility. There is considerable evidence that Park had a duty to ensure sanitary conditions and that he violated that duty by failing to ensure that those to whom he delegated responsibility did their jobs. The Court said: "[T]he requirements of foresight and vigilance imposed on responsible corporate agents are beyond question, demanding, and perhaps onerous, but they are not more stringent than the public has a right to expect of those who voluntarily assume positions of authority in business enterprises whose services and products affect the health and well-being of the public that supports them."

DECISION AND REMEDY: The U.S. Supreme Court reversed the appellate court's reversal of the trial court's conviction, finding that corporate executives can be criminally liable for failing to ensure a company's compliance with the law.

SIGNIFICANCE OF THE CASE: This case signaled to corporate officials that they could no longer use ignorance as a defense, because a court will hold high-level corporate actors responsible for their company's failure to comply with the law, even if they were unaware of what was going on within their company.

CRITICAL THINKING

Given the rule laid down in this case, could the defendant have presented any evidence that would have led to a different decision?

ETHICAL DECISION MAKING

Suppose you were in Park's position in this case. You allegedly allowed food in your warehouse to be exposed to rodent contamination. If you were guided by the public disclosure test, what would you decide to do?

Since *United States v. Park,* the general rule that corporate executives may be held accountable for the crimes arising from their failure to meet their responsibility has remained intact. The rule has, in fact, broadened so that now executives can be held criminally liable for offenses that we generally do

not think of as crimes. For example, in *United States v. Iverson*,⁶ a corporate officer of a waste treatment facility was convicted and sentenced for violating the Clean Water Act after he ordered employees to dispose of wastewater illegally. This case demonstrates that corporate executives can be sentenced not only for committing the common law crimes that we have traditionally understood to be criminal offenses but also for violating provisions of regulatory statutes that permit criminal sanctions. In such cases, vicarious liability, or liability imposed on one person for the acts of another, is assessed to the corporate executive. In fact, courts have held that employers are vicariously liable for the wrongful acts of their employees if the employer directed, partook in, or authorized the wrongful act.

vicarious liability
The liability or responsibility imposed on a person, a party, or an organization for damages caused by another; most commonly used in relation to employment, with the employer held vicariously liable for damages caused by its employees.

CASE Nugget

Can a Corporate Executive be Held Liable for Murder or Manslaughter?

While we do not typically think of murder or manslaughter as being white-collar crimes, if corporate employees's decisions lead to the deaths of customers, the corporate decision makers may well find themselves charged with murder or manslaughter. A good illustration of this situation is the indictment in 2018 of five corporate employees and a corporation because of the death by decapitation of a 10 year old child was riding a 17-story waterslide.

The slide in question was required two or three riders to be strapped onto a raft, and, according to the park's website, the raft would "slide down a jaw-dropping 168-foot-7-inch structure, only to be blasted back up a second massive hill and then sent down yet another gut-wrenching 50 foot drop for the ultimate in water slide thrills." The child was decapitated when the raft few into the air and collided with overhead hoops and netting attached to the slide.

The indicted parties included the two designers of the slide, one of whom was a co-owner of the park, the former operations manager of the park, and the companies that constructed the slide and operated the park. Allegations in the indictment included a claim that one of the defendants who was responsible for many of the design decisions had no technical and engineering expertise and that there were multiple flaws in the design, some of which caused the design to violate longstanding safety standards in the industry. The indictment also alleged that the constructions was rushed, leading to its opening in an unfinished and unsafe condition, and that employees knew of similar accidents with the slide causing injuries to at least ten other riders, and that defendants lied to investigators.⁷

Cases charging corporate executives or corporations with murder or manslaughter are certainly uncommon. The first such charges were filed in a case arising out of the death of an employee from cyanide poisoning. The now defunct Film Recovery Systems was charged with involuntary manslaughter, and two corporate officials and two management-level employees were charged with murder. The indictments alleged that the company employed illegal workers who did not speak English, could not read warning labels, were not given the necessary safety equipment and were forced to work in conditions that were likely to lead to death or serious injury. While the prosecution was unsuccessful, and there have not been many attempt to pursue such charges since that time, perhaps the case against the Schlitterbahn Water Park is the beginning of a new turn in regulating corporate behavior.⁸

Source: Marilia Brocchetto and Melissa Gray., CNN, Kansas water park executives charged with Murder in boy's death, https://www.cnn.com/2018/03/27/us/schlitterbahn-waterpark-death-arrests/index.html (March 27, 2018); Schlitterbahn faces criminal case after Kansas boy decapitated on 17-story waterslide, Dallas News, https://www.dallasnews.com/news/us-news/2018/03/23/schlitterbahn-faces-criminal-case-after-kansas-boy-decapitated-17-story-waterslide.

⁶ 162 F.3d 1015 (1998).

⁷ Marilia Brocchetto and Melissa Gray., CNN, Kansas water park executives charged with Murder in boy's death, https://www.cnn.com/2018/03/27/us/schlitterbahn-waterpark-death-arrests/index.html (March 27, 2018); Schlitterbahn faces criminal case after Kansas boy decapitated on 17-story waterslide, Dallas News, https://www.dallasnews.com/news/us-news/2018/03/23/schlitterbahn-faces-criminal-case-after-kansas-boy-decapitated-17-story-waterslide.

⁸ E.R. Schipp, Can a Corporation Commit Murder? New York Times, May 19, 1989, https://www.nytimes.com/1985/05/19/weekinreview/can-a-corporation-commit-murder.html.

Criminal Procedure

LO 6-4 Outline the basic steps of a criminal proceeding.

Criminal procedure differs from civil procedure in several key ways. First, the government, referred to as the *prosecutor,* always brings the criminal case, whereas in a civil case, the party filing the case, the plaintiff, can be an individual, business, or government entity. Second, the outcome of each is different. In a criminal case, the objective is punishment, so the defendant may be fined or imprisoned; in a civil case, the objective is to remedy a wrong done to the plaintiff, so the defendant either will have to provide compensation to the plaintiff or may be subject to an equitable remedy such as an injunction or order for specific performance. Third, because the potential impact of losing a criminal case is so much more serious than the impact of losing a civil case, the Constitution provides a number of safeguards for the criminal defendant. These safeguards are set forth in Exhibit 6-3.

Pretrial Procedure

Prior to an arrest, grand juries may conduct criminal investigatory proceedings or issue a grand jury subpoena for company records. Criminal proceedings generally begin when an individual is **arrested** for a crime. A law enforcement officer must perform the arrest. This officer is often a police officer but may also be an agent of another government agency, such as the Bureau of Alcohol, Tobacco and Firearms; the Federal Bureau of Investigation; or the Immigration and Naturalization Service.

A law enforcement officer should obtain an arrest warrant before an individual is taken into custody. Ordinarily, to obtain an arrest warrant, the law enforcement agent must demonstrate that there is **probable cause,** or a likelihood, that a suspect committed or is planning to commit a crime. A *magistrate,* the lowest-ranking judicial official, issues the arrest warrant. In certain circumstances, however, courts have recognized that law enforcement agents can arrest a suspect without a warrant if the officer believes there is probable cause but not enough time to obtain the warrant.

When law enforcement agents arrest individuals, the officers must inform the individuals of their **Miranda rights.** If they fail to do so, any information a defendant offers at the time of the

arrest
The action in which a police officer or a person acting under the law takes a person into custody.

probable cause
Any essential element and/or standard by which a lawful officer may make a valid arrest, conduct a personal or property search, or obtain a warrant.

Miranda rights
The rights that are read to an arrested individual by a law enforcement agent before the individual is questioned about the commission of a crime.

Exhibit 6-3

Constitutional Provisions That Safeguard Criminal Defendants's Rights

AMENDMENT	PROVISIONS
Fourth	• Protection from unreasonable search and seizure. • Restrictions on warrants.
Fifth	• Prohibition of double jeopardy. • Right not to incriminate oneself. • Right to due process.
Sixth	• Right to a speedy and public trial. • Right to a trial by an impartial jury of one's peers. • Right to be informed of the accusations against oneself. • Right to confront witnesses. • Right to have witnesses on one's side. • Right to counsel at various stages of the proceedings.
Eighth	• Freedom from excessive bail. • Freedom from excessive fines. • Freedom from cruel and unusual punishment.
Fourteenth	• Extension of the right to due process in all state matters. • Extension of most constitutional rights to defendants at the state level.

arrest is not admissible at trial. To comply with the Supreme Court's requirements for protecting a citizen's rights, a law enforcement officer must inform the defendant of the following facts before questioning:

1. "You have the right to remain silent and refuse to answer any questions."
2. "Anything you say may be used against you in a court of law."
3. "You have the right to consult an attorney before speaking to the police and have an attorney present during any questioning now or in the future."
4. "If you cannot afford an attorney, one will be appointed for you before the questioning begins."
5. "If you do not have an attorney available, you have the right to remain silent until you have had an opportunity to consult with one."
6. "Now that I have advised you of your rights, are you willing to answer any questions without an attorney present?"

The Miranda rights are not absolute. In fact, the Supreme Court has several times created exceptions to the Miranda rights. For example, in *New York v. Quarles*,[9] the Court created a public-safety exception to the Miranda rights. This exception allows statements to be used at trial, even if a person was not informed of his or her Miranda rights, if the need to protect the public is served by the admissibility of the statements in question.

After being arrested and read their Miranda rights, defendants are taken to the police station for **booking,** a procedure during which the name of the defendant and the alleged crime are recorded in the investigating agency's or police department's records. After the prosecutor files the complaint, the defendant makes the **first appearance,** which is the appearance before a magistrate who determines whether there was probable cause for the arrest. If the magistrate ascertains that probable cause did not exist, the individual is freed.

If the defendant committed a minor offense and pleads guilty, the magistrate sentences the individual. However, if the defendant claims innocence, the magistrate ensures that the defendant has a lawyer, or appoints one, if necessary, for an indigent defendant, and sets bail. **Bail** is the amount of money defendants pay to the court on release from custody as security that they will return for trial.

Next, the prosecutor has a choice: Should the case be prosecuted? The *Principles of Federal Prosecution,* established by the U.S. Department of Justice, suggest that at the federal level the decision to prosecute depends on two primary factors: (1) whether the evidence is sufficient to obtain a conviction and (2) whether prosecuting the case serves a federal interest. If the prosecutor decides not to go forward with the case, the defendant may still be liable for his or her actions in civil court.

If the prosecutor chooses to proceed with a criminal action, the prosecutor must demonstrate the likelihood that the defendant's actions and intent meet the elements of a crime by charging the defendant with a crime through an *information* or an *indictment.* For a misdemeanor, the prosecutor must present to the magistrate evidence sufficient to justify prosecution of the defendant. This is done through a written document called an **information,** a formal accusation stating the facts and specifying the violation of criminal law.

However, for a felony, the prosecutor must present a grand jury with evidence adequate to justify bringing the defendant to trial. If the grand jury agrees that the evidence is adequate, it issues an **indictment,** a written accusation of the crime allegedly committed by the defendant.

Note that the grand jury does not determine guilt. It is simply a group of citizens who consider evidence of criminal conduct presented by the prosecutor and then determine whether there is enough evidence to try the defendant for the crime.

In federal cases, a defendant accused of a felony has a constitutional right to a grand jury indictment. However, a felony prosecution may proceed by information if the defendant waives that right. For instance, in a high-profile case in which the defense attorney is trying to work out a deal with the prosecution, the defendant may ask for the case to proceed by information. Federal misdemeanor cases may proceed by indictment or information.

booking
After an individual is arrested, the procedure during which the name of the defendant and the alleged crime are recorded in the investigating agency's or police department's records.

first appearance
The initial appearance of an arrested individual before a judge who determines whether there was probable cause for the arrest. If the judge ascertains that probable cause did not exist, the individual is freed.

bail
A thing of value, such as a money bail bond or any other form of property, that is given to the court to allow a person's temporary release from jail and to ensure his or her appearance in court.

information
A document, prepared by the prosecutor and presented to the magistrate, which demonstrates that there is enough evidence to charge the defendant and bring him or her to trial.

indictment
Finding by the grand jury that there is adequate evidence to charge the defendant and bring him or her to trial.

[9] 467 U.S. 649 (1984).

If the criminal trial takes place in state court, the defendant may or may not have access to a grand jury. The U.S. Supreme Court has held that indictment by a grand jury is not a fundamental right; thus, states are not required to offer grand juries. About half the states still require felony prosecutions to be initiated by grand jury indictments; most of the rest use information as the method for commencing prosecution.

If the grand jury issues an indictment, the defendant appears in court to answer the indictment in a proceeding called the **arraignment**. At this time, the defendant enters a plea of guilty or not guilty. A defendant may also enter a plea of **nolo contendere**, whereby the defendant does not admit guilt but agrees not to contest the charges. The advantage of a nolo contendere plea over a guilty plea is that the former cannot be used against the defendant in a civil suit. If the defendant pleads not guilty, his case will be heard before a **petit jury**, which is a fact-finding jury.

At any time, the prosecutor and defendant can make a **plea bargain**, an agreement in which the prosecutor agrees to reduce charges, drop charges, or recommend a certain sentence if the defendant pleads guilty. Plea bargaining benefits both parties. The defendant gets a lighter sentence, and the prosecution saves time and resources by not trying the case. Businesspeople who commit crimes that affect business often engage in plea bargaining to avoid the publicity associated with a trial and the risk of a severe sentence.

Trial Procedure

If the case goes to trial and the crime is a felony or a misdemeanor punishable by six months or more in prison, the defendant has a constitutional right to a jury trial. In most states, if the defendant waives the right to a jury trial, the judge will hear the case. When a judge is the fact finder in a case, the trial is called a **bench trial**.

In a criminal trial, the prosecutor has the burden of proof and the defendant does not have to prove anything. The **burden of proof** has two elements: the burden of production of evidence and the burden of persuasion. To meet the *burden of production of evidence,* the prosecution must produce any tangible evidence and testimony that prove the elements of the crime that the defendant allegedly committed. Along with producing evidence, the prosecution also has the *burden of persuasion:* The prosecutor must persuade the jury *beyond reasonable doubt* that the defendant committed the crime. The burden of proof, then, is higher in a criminal case than in a civil case because, in a civil case, the burden of persuasion requires only that the claim is supported by a preponderance of the evidence.

The prosecution presents the government's case first, introducing evidence and examining witnesses to prove beyond a reasonable doubt that the defendant committed the crime. Once the prosecution rests its case, it is the defendant's turn. The defendant either (1) can show, through examination of witnesses and introduction of evidence, that the prosecution did not meet its burden of proof or (2) can present an affirmative defense. An *affirmative defense* is a reason the defendant, even though he or she committed the physical acts alleged by the prosecution, should be excused from responsibility for those actions. Exhibit 6-4 lists some common affirmative defenses.

After the jury hears the case, it deliberates and tries to reach a verdict. A jury that is unable to reach a verdict is known as a hung jury. If the jury finds the defendant not guilty, the accused is acquitted and released. If the jury returns a guilty verdict, the judge sets a date for sentencing.

Posttrial Procedure

If the petit jury returns a verdict of not guilty, the government cannot appeal the acquittal. However, if the verdict is guilty, the defendant may appeal the verdict by claiming that a prejudicial error of law occurred at the original trial. If there is no appeal, the defendant will be sentenced after the judge has received additional information relevant to sentencing. Before 1991, sentencing was *indeterminate,* with judges allowed but not required to engage in free-form fact finding before selecting the appropriate punishment within broad statutory ranges. In 1991, however, there was a shift to *determinate sentencing,* under which federal sentences are determined largely by sentencing guidelines prescribing a specific range of possible penalties for each crime.

Judges are given certain factors to consider in sentencing, such as the defendant's criminal record. Guidelines have also been established for white-collar crimes, once again allowing judges

arraignment
The first appearance in court by the defendant, at which the defendant is advised of the pending charges, the right to counsel, and the right to trial by jury and he or she enters a plea to the charge.

nolo contendere
A plea in which the defendant does not admit guilt but agrees not to contest the charges.

petit jury
A group of 6 or 12 citizens who are summoned to and sworn by the court to hear evidence presented by both sides and render a verdict in a trial.

plea bargain
An agreement in which the prosecutor agrees to reduce charges, drop charges, or recommend a certain sentence if the defendant pleads guilty.

bench trial
A trial before a judge, with the judgment decided by the judge rather than a jury; occurs when the defendant has waived his or her right to a jury trial.

burden of proof
The duty of the plaintiff or prosecution to establish a claim or allegation by admissible evidence and to prove that the defendant committed all the essential elements of the crime to the jury's or court's satisfaction beyond any reasonable doubt, in order to convict the defendant.

Infancy	A person who is not yet a legal adult may use this defense to defuse the guilty-mind requirement of a crime.	**Exhibit 6-4** Affirmative Defenses to Crimes
Mistake of fact	The defendant says his or her honest and reasonable mistake about a fact relevant to the commission of the crime negates the idea of a guilty mind.	
Intoxication	The defendant says he or she took an intoxicant without awareness of its likely effect, mistook its identity, or took the intoxicant under force.	
Insanity	The defendant had a severe mental illness that substantially impaired the defendant's capacity to understand and appreciate the moral wrongfulness of the act.	
Duress	The defendant says he or she was forced to perform an act against his or her will.	
Entrapment	The defendant would not have committed the crime or broken the law if not induced or tricked into doing so by law enforcement officers.	
Necessity	The defendant was acting to prevent imminent harm, and there was no legal alternative to the action he or she took.	
Justifiable use of force	Use of force was necessary to prevent imminent death or great bodily harm to the defendant or to another or to prevent the imminent commission of a forcible felony.	

to consider individual factors, such as the company's history of past violations and cooperation with federal investigators.

On April 9, 2015, the United States Sentencing Commission approved new sentencing guidelines for business fraud crimes in response to "longstanding concerns [of judges and lawyers] that the guidelines do not appropriately account for harm to victims, individual culpability, and the offender's intent." One of the most significant changes involves how "intended loss," or the financial harm that was intended to result from the fraud, is understood. It is now interpreted as the loss that the defrauder intended to cause and not necessarily only the loss that was actually caused.

In another major change, the financial situation of the victims of the fraud is now being considered. This means that if a person defrauds another person who experienced "financial hardship" as a result of the fraud, such as bankruptcy, harm to the ability to gain credit, substantial losses in investments such as retirement or education, or substantial changes in living arrangements (such as having to move into a less-expensive home), harsher sentences may be invoked.

Many states have adopted sentencing guidelines similar to the federal guidelines.

Exhibit 6-5 summarizes the steps of a criminal procedure.

Tools for Fighting Business Crime

LO 6-5 Describe the laws that attempt to prevent white-collar crime.

Although clearly a significant number of business-related crimes can be committed, certain federal laws help fight such crime. Three federal laws that have been somewhat successful in this battle are RICO, the False Claims Act, and the Sarbanes-Oxley Act of 2002. Although these laws are currently extremely powerful white-collar-crime laws, RICO and the False Claims Act were created with different purposes in mind. Sarbanes-Oxley, however, was created specifically to combat white-collar crime.

Exhibit 6-5

Steps in a Criminal Procedure

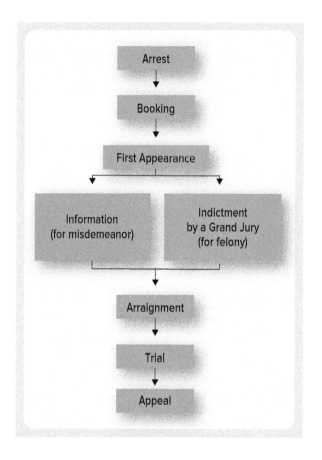

RICO

Racketeer Influenced and Corrupt Organizations (RICO) Act

A U.S. law that provides extended penalties for criminal acts performed as part of an ongoing criminal organization.

One of the most important tools for fighting white-collar crime is Title IX of the Organized Crime Control Act of 1970: the **Racketeer Influenced and Corrupt Organizations (RICO) Act**. Although the statute was originally enacted to combat organized crime, in effect it prevents legitimate businesses from serving as covers for racketeering. This statute prohibits persons employed by or associated with an enterprise from engaging in a pattern of racketeering activity. Anyone whose business or property has been damaged by this pattern of racketeering activity can sue to recover treble damages and attorney fees in a civil action.

Demonstrating a claim under RICO requires proof of a pattern of racketeering. Courts have defined a pattern as more than one action. Thus, a one-time violator could not be prosecuted under RICO because there could be no pattern as yet. Some courts additionally have found that a pattern requires continued criminal activity over a "substantial" period of time. Although *pattern* is restricted to more than one act, racketeering has been defined broadly to include almost all criminal actions, such as acts of violence, fraud, bribery, securities fraud, and the provision of illegal goods and services. Therefore, RICO is an extremely effective tool in combating white-collar crimes.

In addition to being held civilly liable under RICO, a violator may also be subject to RICO's criminal penalties. A person found to have violated the act may be subject to a fine of up to $25,000 per violation, imprisonment for up to 20 years, or both.

The False Claims Act

False Claims Act

An act that allows employees to sue employers on behalf of the federal government for fraud against the government. The employee retains a share of the recovery as a reward for his or her efforts.

Since 1986, private citizens have been using the **False Claims Act** to sue employers on behalf of the government for fraud against the government. For example, an employee in a health care facility might realize that his employer is submitting fraudulent claims to Medicare. The employee can bring a suit against the employer on behalf of the government for fraud against the government.

If an employee realizes that her employer is committing fraud against the government, she must first notify the government of her intent to file the case on behalf of the government. If the government chooses to intervene and prosecute the case itself, with the help of the employee, the citizen would receive 25 percent of the amount recovered. If the government does not opt to get involved, and the citizen continues with the case on behalf of the government, she receives 30 percent.

Certainly, an employee who brings a suit against his employer might be worried about retaliation, such as being fired or demoted. Thus, the act provides protection for employees who use the law. If an employer is found guilty of retaliation, the employer may be forced to pay the employee twice the amount of back pay plus special damages. In a 2018 case that was seen as a victor for employers,[10] however, the 10th Circuit Court of Appeals ruled that ". . .the False Claims Act's anti-retaliation provision unambiguously excludes relief for retaliatory acts occurring after the employee has left employment."

The biggest reward thus far under the act was the approximately $51 million share that four whistle-blowers and their attorneys received for blowing the whistle on a Swiss drug company, Serono, for participating in a marketing scheme that resulted in physicians illegally prescribing over $11.5 million worth of a drug that was paid for by Medicaid. The government settled the case against the company for $704 million. The reward in this case, however, was far above the norm. In 2010, the Justice Department recovered a total of $53.7 billion under the False Claims Act, up from $53.5 billion collected in 2009.[11]

Although some of these cases might not have been pursued had it not been for the False Claims Act, many people are opposed to the act. Some argue that the whistle-blowers are receiving money that should belong to the taxpayers; others say that the act prompts people to refrain from reporting fraud right away but wait until the value of the case grows. Still others complain that the act results in frivolous lawsuits as employees try to find an easy way to make money.

Defenders of the False Claims Act point out that the act has led to revelation of some significant cases of fraud that otherwise might have cost the government millions of dollars. They also point out that when an employee does report a fraudulent employer, it becomes very difficult for that employee to get a job in the same field in the future; thus, a significant incentive needs to be offered to convince employees to report such fraud.

A number of False Claims Act cases have been filed against universities, including an interesting case filed in 2006 against Chapman University. In this case, three faculty members alleged that the institution for years encouraged early dismissals, resulting in many students not getting the minimum classroom training required in several subjects. If the college had admitted that it had engaged in this practice, it would never have been accredited and thus never have received millions of dollars in federal grants and student aid, which the lawsuit claimed it took under false pretenses.[12] The case was ultimately dismissed by a Summary Judgment in 2007, and in 2011, the plaintiff's appeal to the U.S. Supreme Court was denied. Other False Claims Act cases against universities have been more successful. For example, in 2009, the University of Phoenix settled a False Claims Act suit for $67.5 million for violating federal law that prohibits colleges from paying admissions counselors incentives based on the number of students recruited.

The Sarbanes-Oxley Act

Unlike RICO and the False Claims Act, the Sarbanes-Oxley (SOX) Act of 2002 was intended to curb white-collar crime. Congress passed Sarbanes-Oxley largely in response to the business scandals of the early 2000s, such as Enron, WorldCom, and Global Crossing, as well as the accounting firm Arthur Andersen. Although much of the act consists of new rules and regulations for accounting firms, part of the act specifically addresses the issue of white-collar crime.

Sarbanes-Oxley Act
An act that criminalizes specific nonaudit services when provided by a registered accounting firm to an audit client; also increases the punishment for a number of white-collar offenses. Also known as the Public Company Accounting Reform and Investor Protection Act of 2002.

[10] Potts v. Center for Excellence in Higher Education, Inc., No. 17-1143 (10th Cir. Nov. 6, 2018).

[11] Fiscal Year 2012 Budget Request, Civil Division, http://www.justice.gov/jmd/2012justification/pdf/fy12-civ-justification.pdf (accessed July 1, 2010).

[12] Martin Van Der Werf, "Lawsuit U.: The Growing Reach of the False Claims Act Has Lawyers Fearing Trouble Everywhere," *Chronicle of Higher Education,* August 4, 2006, http://chronicle.com/weekly/v52/i48/48a02301.htm.

GLOBAL Context

©Digital Archive Japan/Alamy

Protecting Whistle-Blowers in the United Kingdom

One way in which the United Kingdom fights white-collar crime is through the use of a national whistle-blower statute, the Public Interest Disclosure Act of 1998 (PIDA), which protects the whistle-blower from firing or retaliation if he or she makes a good-faith disclosure based on genuine concerns about crime, civil offenses, miscarriage of justice, danger to health and safety or the environment, and the cover-up of any of these.

The emphasis of PIDA is on fixing the problem within the company. The most protected, and preferred, disclosure is one that is made to a manager within the company, so the first step is often to disclose the possible violation to a manager in the company. Sometimes such disclosures are not always advisable or possible, so disclosures to the appropriate ministry are protected, when applicable, as well as disclosures to the appropriate regulatory agency, such as the Health and Safety Executive. Rather than just having a suspicion of wrongdoing, whistle-blowers must believe that their allegations are substantially true before disclosing them to a regulator.

If the whistle-blower reasonably believed that he or she would be victimized for disclosure to the company or regulator, reasonably believed that a cover-up was likely and there was no prescribed regulator, or had already raised the issue internally or with a regulator and no action had been taken, he or she would be protected for a wider disclosure.

The whistle-blowers do not get monetary rewards, as do whistle-blowers in the United States under the False Claims Act. Instead, they may be protected and compensated if action is taken against them.

Sarbanes-Oxley makes it illegal for registered public accounting firms to provide nonaudit services to an audit client. Section 201 of the act specifically names as illegal:

1. Bookkeeping or other services related to the accounting records or financial statements of the audit client.
2. Financial information systems design and implementation.
3. Appraisal or valuation services, fairness opinions, or contribution-in-kind reports.
4. Actuarial services.
5. Internal audit outsourcing services.
6. Management functions or human resources.
7. Broker or dealer, investment adviser, or investment banking services.
8. Legal services and expert services unrelated to the audit.
9. Any other service that the board determines, by regulation, is impermissible.

In addition to criminalizing these activities, Sarbanes-Oxley has amended the penalties associated with other white-collar crimes. Under Sarbanes-Oxley, it is now a felony to fail willfully to maintain proper records of audits and work papers for at least five years, and the punishment for not maintaining these records is up to 10 years's imprisonment. Also, the destruction of documents received updated punishments. The destruction of documents involved in a federal bankruptcy investigation is now a felony, with possible sentences of up to 20 years's imprisonment. Similarly, the punishment for securities fraud has been increased to 25 years's imprisonment.

In addition, Sarbanes-Oxley has extended the statute of limitations regarding the discovery of fraud. The statute of limitations now extends to two years from the date of discovery of the fraud and five years from the criminal act. The old statute was one year from discovery and three from the act. Also, much like the False Claims Act, Sarbanes-Oxley has taken affirmative steps toward further protecting whistle-blowers.

State Whistle-Blower Protection Laws

Forty-seven states currently have whistle-blower protection laws for either public or private sector employees or both. However, these laws are very different, so it is important for both employers and employees to know the law in their state. The biggest difference among the states is to whom the violation may or must be reported. In many states, employees in the private sector receive

protection from retaliation only if they report the wrongdoing to an external governmental body. Other states take the opposite approach and require the whistle-blower to report his suspicions first internally to his supervisors. Almost no states protect employees who choose to go to the media with their reports of employer wrongdoing.

CASE Nugget

Does Whistle-blower Protection under Sarbanes-Oxley Extend to Employees of Private Contractors Serving Public Companies?

Lawson v. FMR LLC

134 S. Ct. 1158 (2012)

Plaintiffs were former employees of FMR, a group of private companies that contracted to advise or manage mutual funds. As is common in the industry, the mutual funds FMR served are public companies with no employees. Both plaintiffs allege that they blew the whistle on putative fraud relating to the mutual funds and, as a consequence, suffered retaliation by FMR. Each filed suit in federal court. Moving to dismiss the suits, FMR argued that the plaintiffs could state no claim under Sarbanes-Oxley's whistler-blower protection provision because that provision protects only employees of public companies, not employees of private companies that contract with public companies. On appeal from the District Court's denial of FMR's motion to dismiss, the First Circuit reversed, concluding that the term "an employee" under SOX refers only to employees of public companies.

The U.S. Supreme Court, however, reversed that decision, giving a broader interpretation of the statute and finding that the act's whistle-blower protection includes employees of a public company's private contractors and subcontractors. As the court pointed out, to hold otherwise would be to allow the entire mutual fund industry to insulate itself because virtually all mutual funds are structured so that they have no employees of their own; they are managed, instead, by independent investment advisors.

The high court also pointed out that this construction was analogous to the whistle-blower protection given to employees in the aviation industry, in which the term *employees* includes protection of employees of contractors and subcontractors of air carriers.

CASE OPENER WRAP-UP

A Questionable Stock Investment

1. The defendant was convicted of one count of securities fraud for his misrepresentations related to stock he sold the hotel investor. He was also convicted one count of mail fraud for sending the stock certificates via Federal Express, and one count of wire fraud for causing the approximately $1,486,323 from the hotel investor to be diverted to his own use through an interstate wire transfer of said funds to a Citibank account the defendant maintained in New York, New York. On appeal, the conviction was upheld in 2014.

2. Had this been an older case, the fact that the certificates were sent via Federal Express would have been important, but the mail fraud statute has now been amended to specifically include the use of commercial postal services including Federal Express and UPS.

PRACTICAL TIPS for Business Managers

1. White-collar crime can cripple your business or, at the very least, greatly mire its reputation. Thus, as a business manager, you will want to root out white-collar crime before it becomes a problem. Maintaining an internal audit service with effective and trained auditors can help you keep white-collar crime to a minimum.

2. Provide your employees with proper training so that they understand what is and is not considered ethical and legal activity within your business.

3. Screen potential employees properly and thoroughly to ensure that they have a clean record before appointing them to senior positions.

SUMMARY

The Nature and Elements of a Crime

The elements of a crime are:

- *Actus reus:* Wrongful behavior (guilty act).
- *Mens rea:* Wrongful state of mind or intent (guilty mind).

Crimes are classified into three categories:

- *Felonies:* Serious crimes punishable by imprisonment for more than one year or by death.
- *Misdemeanors:* Less serious crimes punishable by fines or imprisonment for less than one year.
- *Petty offenses:* Minor misdemeanors punishable by small fines or short jail sentences.

White-Collar Crime

White-collar crimes include:
1. Bribery,
2. Extortion,
3. Fraud,
4. Embezzlement, and
5. Computer crimes (destruction of computer data, unlawful appropriation of data or services).

Corporate Criminal Liability

Both corporations, as legal entities, and the corporate officers and managers can be held liable for crimes committed on behalf of the corporation.

Criminal Procedure

Pretrial procedure: The arrest, booking, first appearance, indictment, and arraignment.

Trial procedure: Jury selection, trial with burden of proof on prosecution, jury deliberations, jury verdict, and (if guilty) sentencing hearing.

Posttrial procedure: Appeal.

Tools for Fighting Business Crime

RICO: Prohibits persons employed by or associated with an enterprise from engaging in a pattern of racketeering activity. Anyone whose business or property has been damaged by this pattern of activity can sue under RICO to recover treble damages and attorney fees in a civil action.

False Claims Act: Allows employees to sue employers on behalf of the federal government for fraud against the government. The employee retains a share of the recovery as a reward for his or her efforts.

Sarbanes-Oxley Act: Criminalizes specific nonaudit services when provided by a registered accounting firm to an audit client; also increases the punishment for a number of white-collar offenses.

Point/Counterpoint

As you are considering this issue, you might want to think about statistics that would help you resolve it.

Should Insider Trading be Made Legal?	
YES	NO
The ban on insider trading is intended to level the playing field for all investors and make the securities market more fair. However, not only does the reality of banning insider trading fail to live up to its intentions, there are several ways that legalized insider trading could actually help the securities market as a whole. For example, as the law stands now, insiders who know that a stock is going to tank in the future can only sit by and do nothing as the price of the stock remains the same and unwitting investors continue to buy in. Take Enron, for example. Enron had been cooking its books for some time before its collapse, and because this practice had remained a secret, the price of Enron's stock remained largely unaffected. The insiders who knew about Enron's shady actions were effectively barred from selling their soon-to-be-worthless stock. If they had been able to sell their stock well before Enron's collapse, surely the market would take notice of the mass selling, and the stock would be much less attractive to investors. Instead of more investors unwittingly buying worthless stock prior to a crash, the price could have gradually decreased to its accurate value due to insider sell-offs, and future investors could have been warned, which would have prevented the whole Enron crisis in the first place. Moreover, white-collar crime tends to affect higher-income individuals, whereas street crime tends to affect lower-income individuals who lack the resources to protect themselves. Street crime disproportionately affects lower-income individuals who cannot afford the private security measures that higher-income individuals use to protect their persons and their property. The law ought to protect those who lack the power to protect themselves. A second reason is that other markets use insider trading and function just fine. The real-estate market is a good example of a market where insider trading is common. For example, say someone is selling a house, and you have insider information that something about the house or the property makes it more valuable than the price that the seller is offering. You pay the seller's price and obtain the house. Both parties walk away happy: the seller sold the house for the price they wanted, and you got the house for a lower price than you thought you would have to pay. A final reason is that it is incredibly hard to make insider trading charges stick. Being that we cannot read the mind of a defendant in an insider trading case, it is almost impossible to know what someone knew	Insider trading can do great harm to investors as well as to the securities market as a whole, and the practice has been made illegal for good reasons. First, insider trading almost necessarily leads to investors with insider information cheating other investors without insider information. If an insider knows that a stock's value is about to skyrocket, he can buy multiple shares for much less than what they are worth and give another investor a bad deal. If an insider knows that a stock's value is about to plummet, he can sell off all of his shares to unwitting investors who are now about to lose a lot of money on their investment. In either case, the insider unfairly profits. The above explanation leads directly into another reason insider trading is illegal: if investors were constantly worried about getting cheated by insider trading, the integrity of the securities market would be damaged, and many investors would be discouraged from participating in the market. Fewer investors means fewer buyers and sellers and, therefore, less money being invested into businesses. A final reason, which is closely tied to the previous reason, is that legalizing insider trading would endanger market transparency and corporate transparency. If an insider comes across nonpublic information, he is likely to hide this information not only from the public but from the corporate directors as well in order to maximize his profits. If the corporate directors are unaware of this information, they cannot provide accurate financial information to the public and the public cannot make well-informed investments. The market simply cannot properly operate if it lacks transparency and is full of trades made on secret information. In insider trading cases, the only people who truly win are the insiders.

and when they knew it. Investigating and prosecuting insider trading is inefficient and highly resource-consuming. During the 2009 investigation of the Galleon Group, the federal government pored over 14,000 wiretapped phone conversations looking for evidence of insider trading. Often, even the most high-profile insider trading cases end up with next to nothing: after all the noise made in Martha Stewart's insider trading case, she was convicted of obstruction of justice and lying to investigators, not insider trading. The government would be much better off using their resources to detect and prevent large-scale crises such as the Enron crash than going after something as victimless as insider trading.

Questions & Problems

1. Name and explain the two elements of a crime.
2. Identify the distinguishing features of white-collar crime.
3. List and define the primary affirmative defenses used in criminal cases.
4. List and describe the main types of fraud committed in the business context.
5. What is meant by reading a defendant his Miranda rights, and why are these rights important?
6. Explain the steps of a criminal prosecution.
7. Explain the federal laws that are currently being used to fight white-collar crime.
8. After Kuczora lost his finance job in 2007, he styled himself as the managing director of KCS Financial, a phony finance firm he ran from his Elgin, Illinois basement. Kuczora falsely represented to unwary investors that he could help them secure millions of dollars in financing; they paid him large sums of money to cover fees, which Kuczora then pocketed for personal use before disappearing. Based on these facts, to what crime do you think Kuczora pled guilty?

 Source: United States v. Kuczora, No. 17-2725 (7th Cir. 2018).

9. This case involved a blue-collar worker committing a white-collar crime. Dennis Dwayne Tucker was a long-distance trucker employed by MBM Moving Systems, LLC (MBM), headquartered in Greensboro, North Carolina. In December 2010, Tucker picked up a load of household goods belonging to a customer of MBM, Leah Plotkin. Upon delivery of the goods, Plotkin paid the outstanding balance for the move, $2,086.19 in cash.

 According to company policy, an MBM driver is required to convert cash payments into money orders and then use the company's FedEx account to send the payment to headquarters with the associated paperwork. Employees are also not allowed to spend any of the payment money.

 Subsequent to the Plotkin move, Tucker's commercial driver license expired. Unable to legally drive the truck, Tucker bought a plane ticket using part of the Plotkin payment and flew to North Carolina, where MBM headquarters was. Although Tucker eventually submitted the paperwork for the Plotkin move, he did not deposit the payment to MBM.

 In February 2011, Tucker stopped working for MBM. He still had not remitted the Plotkin payment at this point and did not return any attempts of communication by the vice president of MBM. At his trial, Tucker testified in his own defense, stating that he believed "in this case, the supervisor had given him permission to use the money he had received from Plotkin."

 Of what crime do you believe Tucker was tried and convicted? Tucker appealed his conviction. Do you think his conviction was overturned on appeal?

 Source: *State Of North Carolina V. Dennis Dwayne Tucker* 743 S.E.2d 55 (2013 N.C. App.).

10. After the financial crisis of 2008, Anchor BanCorp Wisconsin (ABCW), and its subsidiaries, AnchorBank and Investment Directions, Inc. (IDI) were in dire straits. Anchorbank and ABCW were trying to negotiate an extension for a $116 million loan from U.S. Bank with a large payment due by the end of March 2009.

 On December 29, 2008, David Weimert, the vice president of AnchorBank, and the president of IDI, was instructed to sell commercial properties held by IDI to acquire the cash to pay U.S. Bank in March. Weimart successfully closed a deal that resulted in $2 million more than he was told to sell for while

also reducing IDI's mortgage liability by $15 million, therefore freeing more money that could be sent to AnchorBank or ABCW. In closing the deal, however, Weimert saw an opportunity to insert himself into the transaction.

Weimert convinced the two prospective buyers that he would be a useful partner. Both buyers included a provision having Weimert buy a small percentage of the property in their offer letter. Testimony given at the subsequent trial revealed that Weimert had misled the board of directors that the buyers would not purchase the property unless he was included in the deal. Indeed, Weimert was convicted of five out of six counts of wire fraud on the theory that his actions were part of a scheme to obtain money or property by fraud.

Weimert appealed the ruling on grounds that his actions did not rise to the level of fraud.

Source: *United States v. Weimert,* 819 F. 3d 351 (2016).

11. From 2004 to 2010, Edward Weaver was the CEO of Vendstar, a corporation that sold candy-vending machines to approximately 7,000 customers who spent a total of about $62 million on the investments.

 Although Vendstar's promotional materials and salespeople represented that the investment opportunities would be lucrative, for the most part, the vending machines would earn Vendstar's customers almost no money. Customer complaints became so frequent that Weaver assigned a Vendstar employee to respond to those complaints full time. In the end, many customers lost virtually all of their investments in Vendstar. The size of the investments varied, and the most common transaction involved the victim paying $10,000 for a package that included 30 candy-vending machines and an initial supply of candy, although some customers spent significantly more than that.

 Under the Vendstar scheme, potential customers would respond to Internet or newspaper ads approved by Weaver. Vendstar would mail them a brochure, order form, and certain disclosure documents. The brochure, which Weaver also approved, included claims that the salespeople were "Vendstar vendors themselves" and promised that the investments had "little risk." Vendstar salespeople would then follow up with customers over the phone. During those phone conversations, salespeople routinely lied to customers, promising them utterly unrealistic earnings and claiming that the investments were sound. Salespeople also falsely stated that they owned their own Vendstar machines, a misrepresentation that encouraged customers to trust them and buy the machines. At trial, victims of the scam testified that those lies influenced their decision to purchase the machines.

 If customers agreed to purchase the machines, they signed a contract that included the following disclaimers in capital letters:

 Purchaser understands that seller has no affiliation or financial relationship with professional locating companies and that seller has no involvement whatsoever in securing retail locations. . . .

 Purchaser and seller agree that this purchase order contains the entire understanding of the agreement between the parties and there is no reliance upon any verbal representation whatsoever. Seller has not guaranteed any minimum or maximum earnings. . . .

 It is further acknowledged that no statements, promises[,] or agreements influenced this purchase or are expected other than anything contained in this purchase order. . . .

 What crime do you think Weaver was convicted of? Weaver appealed this conviction. What do you think his argument on appeal was? Do you think his conviction was upheld?

 Source: *United States v. Weaver,* 860 F.3d 90; 2017 U.S. App. LEXIS 10948.

12. On 17 occasions, Andrews asked friends and colleagues to loan him a total of roughly $2 million to purchase property in Indianapolis or to improve property that he owned in the area. He never owned, bought, or improved property in Indianapolis, and primarily used the money to fund a day-trading count with TD Ameritrade. He did pay some of the money back, but most of it disappeared, and his victims lost over $1.4 million. What crime do you think Andrews would have been convicted of? Why?

 Source: *United States v. Andrews,* (6th Cir. 2015), available http://caselaw.findlaw.com/us-6th-circuit/1715876.html.

13. Toby Scammel's girlfriend was briefly involved in Disney's acquisition of Marvel in 2009. Scammel bought 659 Marvel options for $5,465 in the days prior to the announcement of the acquisition. Within the nine days following the announcement, Scammel sold all his Marvel shares, amounting to a total of $192,496.61. Although Scammel's girlfriend would mention general facts about a big project she was working on, the couple claimed that she never gave him details about the acquisition. However, he had access to her electronic documents related to the acquisition and had begun piecing together the companies involved and the timing of the acquisition. Scammel's girlfriend had confidential information about the deal, including that Disney would pay $50 per share for Marvel and that the deal would be announced by Labor Day. Scammel claimed that his girlfriend had told him nothing and that he had done nothing wrong;

he had frequently been involved in the stock market and was simply trading shares normally as he had done before. The SEC, in its investigation, also learned just before he purchased many of the Marvel securities, Scammel searched the internet for the terms "insider trading," "material, non-public information," and "Rule 10b-5." What crime do you believe Scammel was charged with? Do you think there was enough evidence to warrant his conviction for that crime?

Source: *SEC v. Scammel,* 2011 WL 3506153 (C.A.C.D. Aug. 11, 2011).

14. On November 11, 2011, the window of Jamie Williams's house was allegedly broken by someone who then entered through the broken window and removed several items, including a laptop computer. One day after the incident, Williams filed a report, naming Sammie Smith as the perpetrator.

At the trial, Williams testified that he saw Smith running away from his house on the night of the burglary. Williams' neighbor, Dacorious Clark, also testified, stating that on November 11, Smith showed up at his house with a laptop computer tucked into his pants. Smith allegedly tried to sell the computer to Clark. Clark declined the offer but was asked to keep the device anyway. After opening the computer and looking at the login screen, Clark realized that the laptop belonged to Williams, and returned it to him one week after the night of the incident.

Smith was the only witness for the defense and testified that he was at a football game on the night of the burglary. According to Smith, he met his cousin and exchanged a bag of marijuana for the computer and then traded the computer with Clark for $60 worth of cocaine.

Based on the testimony presented at the trial, what crime do you believe Smith was convicted of? On appeal, do you believe the appellate court found that there was sufficient evidence to convict Smith?

Source: *Sammie Smith v. State of Mississippi,* 150 So. 3d 122 (2014).

Part 1 The Legal Environment of Business

CHAPTER 7

Tort Law

CASE OPENER

Plastic Surgeon Defamation

Dr. Walter Sullivan was one of several plastic surgeons in Las Vegas whom Julie Jones visited. Jones, an exotic dancer, sought plastic surgery to improve her ability to make money in her profession. After visiting Sullivan for a consultation, she then visited Dr. Joseph Bongiovi, Jr. During her consultation with Bongiovi, Jones mentioned her earlier visit with Sullivan. Bongiovi then told her that a patient of Sullivan's died the previous week during the same procedure Jones sought. Bongiovi told her the death was the direct result of Sullivan's negligence.

Despite Bongiovi's allegations, Jones saw Sullivan again and scheduled the surgery with him. Jones did, however, attend a prescheduled appointment with Bongiovi. During the appointment, at Jones's prompting, Bongiovi confirmed what he had said before—that Sullivan had recently been responsible for a patient's death during the same procedure Jones sought.

On the basis of the confirmation from Bongiovi, Jones called to cancel her surgical appointment with Sullivan. When Sullivan's office manager asked why she was canceling the appointment, Jones said she had been told that Sullivan was under investigation for a patient's death. When Sullivan learned of the cancellation, he called Jones to find out who had made the statements; he was unsuccessful in obtaining a name. After speaking with Sullivan, Jones again called Bongiovi's office to receive confirmation about the allegation. Bongiovi's assistant confirmed that the statements were true.

When Sullivan eventually learned the identity of Bongiovi, he filed suit for defamation. According to Sullivan, Bongiovi's statements were slanderous per se. At the conclusion of a trial, a jury found in favor of Sullivan and awarded him $250,000 in compensatory damages and $250,000 in punitive damages. Bongiovi appealed, arguing that the jury should have been instructed that actual malice was the standard because Sullivan was a public figure. Furthermore, Bongiovi argued that the compensatory and punitive damages awarded were exorbitant.

1. What defenses, if any, could Bongiovi have presented to prevent the damage awards?
2. Under which ethical system, if any, should Bongiovi be required to pay damages to Sullivan? Why?

LEARNING OBJECTIVES

- **LO 7-1** Classify types of torts.
- **LO 7-2** Summarize the intentional torts against persons, property, and economic interests.
- **LO 7-3** Identify the elements that must be proved to win a negligence case.
- **LO 7-4** Recognize the doctrines that help a plaintiff establish a case of negligence.
- **LO 7-5** Recall the defenses to a claim of negligence.
- **LO 7-6** Explain how a party proves a strict-liability case.
- **LO 7-7** Recall the types of damages available in tort cases.

tort

A violation of another person's rights or a civil wrongdoing that does not arise out of a contract or statute; primary types are intentional, negligent, and strict-liability torts.

As a future business manager, you will likely be involved in a situation in which one party believes he or she has been injured by the actions of another party, in the same way Dr. Sullivan believed he had been injured by Dr. Bongiovi's allegations about the death of one of Dr. Sullivan's patients due to the doctor's negligence. Wrongs such as these are often referred to as *torts*. A **tort** is commonly defined as a wrong or injury to another, other than a breach of contract. In fact, *tort* is a French word meaning "wrong." This chapter first examines the goals of tort law and the three primary classifications of torts. Next, the chapter examines each of the categories of torts. It concludes by discussing damages that may be available in tort cases.

Introduction to Tort Law

Chapter 6 introduced criminal law and explained the punishment individuals may receive for committing crimes. Most of the actions that give rise to criminal prosecutions also provide the basis for a tort claim by the victim of the crime.

Although the primary objectives of criminal law are to punish wrongdoers and preserve order in society, the primary objective of tort law is to provide compensation for injured parties. Tort law also contributes to maintaining order in society because it discourages private retaliation by injured persons and their friends. After all, we do not want to live in a community where vigilantes with tempers are roaming about, righting some harm they believe has occurred to them.

A third objective of tort law is to give citizens a sense that they live in a just society. Our collective sense of right and wrong suggests that someone who creates harm should make things right by compensating those who were harmed. The recognition that one will have to pay for the personal injuries she or he causes may also serve to deter the commission of torts.

Although this chapter discusses torts as if they were the same everywhere, tort law is primarily state law, so states may have slightly different definitions of each tort. In describing torts, this chapter uses the definitions common in most states, noting when there seems to be a significant difference in the way certain states define the tort.

Despite the public impression of a tort litigation explosion often conveyed by the media, because tort law is state law, no entity regularly compiles and publishes an annual report of the total number of tort cases filed nationwide. Writers for a legal publication, *Lawyer's Weekly USA*, however, do keep track of the top 10 jury verdicts every year. What they have found is that after years of decline since 2000, the top 10 jury verdicts rose dramatically in 2008 and increased by a smaller number in 2009, with the average of the top 10 jury verdicts increasing from $112 million to nearly $145 million. But the very top award in 2009 was slightly lower, at $370 million, versus $388 million in 2008. In 2010, the average increased less than the prior year, however, rising from nearly $145 million to just under $157 million, but the top award was significantly higher in 2010. A Las Vegas attorney won a judgment of $505 million, with $500 million being punitive damages,

against a manufacturer and distributor whose oversized vials of a drug led to an outbreak of hepatitis C at outpatient surgical centers.[1] Given what we do know about tort litigation and damages, we know that potential tort liability should concern a competent business manager.

Classification of Torts

LO 7-1 Classify types of torts.

Torts are most commonly classified as intentional, negligent, or strict-liability torts. Each category differs in terms of the elements needed to prove the tort, the available damages, the available defenses, and the degree of willfulness of the actor. **Intentional torts** occur when the defendant takes an action intending that certain consequences will result or knowing certain consequences are likely to result. **Negligent torts** occur when the defendant acts in a way that subjects other people to an unreasonable risk of harm. In other words, the defendant is careless to someone else's detriment. Finally, **strict-liability torts** occur when the defendant takes an action that is inherently dangerous and cannot ever be undertaken safely, no matter what precautions the defendant takes. It is important to remember that when we discuss these classifications, we are referring to their use in the United States. The Chinese legal system, for example, narrowly defines the activities actionable under tort law.

Intentional Torts

Intentional torts are the most willful of torts. Intentional torts are predicated on the common element of intent. The intent at issue is not intent to harm but, rather, to engage in a specific act, which ultimately results in an injury, physical or economic, to another. In fact, one does not need to establish a motive when proving liability in an intentional tort case. Moreover, in tort law, it is assumed that people intend what could be considered the normal consequences of their actions. For example, if Rob throws a rock toward a group of people, it would be assumed under the law that he intended to hit someone with the rock and that the person hit would be hurt, regardless of Rob's intention merely to scare the group of people.

As a general rule, each specific intentional tort has a set of elements that must be proved to establish the tort, along with specific defenses that can be raised against each tort. However, not all harms intentionally committed will fall neatly into an existing category of torts. Therefore, a general theory of intentional tort liability has been created to aid judges in their decision making. In Section 870 of the Restatement (Second) of Torts, the general theory is explained this way:

> One who intentionally causes injury to another is subject to liability to the other for that injury, if his conduct is generally culpable and not justifiable under the circumstances. This liability may be imposed although the actor's conduct does not come within a traditional category of tort liability.

Intentional torts are divided into the following three categories: (1) torts against persons, (2) torts against property, and (3) torts against economic interests. The following sections discuss a number of specific torts that fall into each category, along with the defenses for each.

Intentional Torts Against Persons

LO 7-2 Summarize the intentional torts against persons, property, and economic interests.

Torts against persons are intentional acts that harm an individual's physical or mental integrity. There are a significant number of these torts. Assault and battery are two of the most common. Imagine that after searching for a parking space for 20 minutes, you finally pull into a spot.

intentional tort
A civil wrong resulting from an intentional act committed on the person, property, or economic interest of another. Intentional torts include assault, battery, conversion, false imprisonment, intentional infliction of emotional distress, trespass to land, and trespass to chattels.

negligent tort
A civil wrong that occurs when the defendant acts in a way that subjects other people to an unreasonable risk of harm (i.e., the defendant is careless, to someone else's detriment). Negligence claims are usually used to achieve compensation for accidents and injuries.

strict-liability tort
A civil wrong that occurs when the defendant takes an action that is inherently dangerous and cannot ever be undertaken safely, no matter what precautions the defendant takes. In such situations, a defendant is liable for the plaintiff's damages without any requirement for the plaintiff to prove that the defendant was negligent.

[1] Lawyers USA Online, "Top Ten Jury Verdicts of 2010," http://lawyersusaonline.com/free-white-paper-top-ten-jury-verdicts-of-2010.

However, as soon as you turn off your car, a man who looks like Mike Tyson starts pounding on your car window. He angrily yells, "You took my spot! If you don't move your car now, I'm going to hit you so hard you won't remember what your car looks like!" The man has just assaulted you.

An **assault** occurs when one person places another in fear or apprehension of an immediate, offensive bodily contact. Therefore, in the above example, if you think the man is just joking and you start laughing, no assault has taken place because there is no fear or apprehension. Not only must you feel fear or apprehension, but that feeling must be reasonable under the circumstances.

An assault is often, but not always, followed by a **battery,** an intentional, unwanted, offensive bodily contact. Almost any unwanted, intentional contact constitutes a battery. Even contacts that are harmless, if unwanted, are batteries. And even if a touch is intended as a joke, if the reasonable person would be offended, the contact is deemed offensive.[2] To return to the example of the parking-space incident, if the man actually hit you, his action would constitute a battery. In contrast, if you both happened to be getting out of your respective cars at the same time and consequently bumped into each other, no battery would have occurred because there was no *intentional* bodily contact.

assault

A civil wrong that occurs when one person intentionally and voluntarily places another in fear or apprehension of an immediate, offensive physical harm. Assault does not require actual contact.

battery

A civil wrong that occurs when one person intentionally and voluntarily brings about a non-consented harmful or offensive contact with a person or something closely associated with him or her. Battery requires an actual contact.

BUT WHAT IF...

WHAT IF THE FACTS OF THE CASE OPENER WERE DIFFERENT?

Recall in the Case Opener that Dr. Sullivan sued Dr. Bongiovi because of the false statement he made. What if, instead, Dr. Sullivan had called Dr. Bongiovi and said, "If you ever tell a potential client a lie like that about me again, I will slit your throat!" Would Dr. Sullivan have committed an assault or battery? Why or why not?

A limited number of defenses are available to an action for a battery. A party charged with a battery may argue that the offended party consented to the contact. *Consent,* as a defense, mitigates the element of unwanted. A person cannot commit a battery if the contact was consented to and therefore wanted.

The most common defense to battery is *self-defense,* responding to the force of another with comparable force to defend oneself. In our parking-space example, if the man took a swing at you and you shoved him to try to keep him from hitting you, causing him to fall backward and hit his head on the street, he would not succeed in a case against you for battery because you could raise the defense of self-defense. To use this defense, you cannot use a greater level of force than is being used against you.

A third defense, *defense of others,* allows you to defend another by using the same degree of force that you could use to defend yourself. A final defense, *defense of property,* allows you to use reasonable force to defend your property from an intruder. The use of deadly force in defense of property is rarely, if ever, considered justified.

defamation

A false statement or an action that harms the reputation or character of an individual, business, product, group, government, or nation.

The tort alleged in this chapter's opening case was **defamation,** which is the intentional publication (communication to a third party) of a false statement harmful to an individual's reputation. In addition to the person who publishes a false statement, anyone who republishes, or in any manner repeats, a defamatory statement is also liable for defamation, even if he or she cites the original source of the defamation.

If the defamation is published in a permanent form, such as printed in a magazine or newspaper, it is known as *libel.* Defamation through television and radio broadcasts is also considered libel because the broadcasts are permanently recorded. In the case of libel, general damages are presumed. Thus, the victim would be entitled to compensation for the damages that are presumed to flow from defamation yet are hard to prove, such as the humiliation the victim would feel.

If the defamation is made orally, it is *slander.* To recover damages in a case of slander, the plaintiff must prove special damages; that is, the plaintiff must show specific monetary loss that

[2] Restatement (Second) of Torts, sec. 19.

resulted from the defamatory statements. Although libel is contained in a permanent form, slander, by virtue of being spoken, is not. It is the lack of permanence that gives way to the special damages involved with slander. If the people who heard the slander do not act in a way to cause harm to the slandered person, there is no cause for compensation, which is one of the main goals of tort law.

One exception to the requirement of special damages occurs if the false statements constitute *slander per se.* Slander per se statements are considered so inherently harmful that general damages are presumed. The kinds of statements considered slander per se are claims that an individual (1) has a loathsome, communicable disease (traditionally, venereal disease or leprosy); (2) has committed a crime for which imprisonment is a possibility; (3) is professionally incompetent; or (4) if a woman, has engaged in sexual misconduct.

Case 7-1 illustrates how a plaintiff attempted to use defamation per se under the category of professional incompetence. As you read the decision, think of how it might be applicable to the case described in the opening scenario.

CASE 7-1

HANCOCK v. VARIYAM
Supreme Court of Texas
400 S.W.3D 59 2013

FACTS: Dr. Variyam was the Chief of the Gastroenterology Division of the Texas Tech University Health Sciences Center (the Division). Dr. Hancock was an associate professor under Variyam. A dispute arose between the two over the transfer of patients from Hancock's care to Variyam's. Variyam sent a letter to Hancock complaining about Hancock's disregard for patient care and detailing the alleged manner in which Hancock's transfer of patients violated the Division's policy and sent a copy to the Chair of the Department of Internal Medicine (the Department). The letter gave Hancock an opportunity to respond before Variyam lodged a formal complaint.

Hancock responded by sending a letter the same day to the Chair of the Department, the Dean of the School of Medicine, a Division colleague, and the entity reviewing the Division's application for accreditation for its gastroenterology fellowship. In the letter, Hancock resigned his faculty position under Variyam, stated that Variyam had a "reputation for lack of veracity" and "deals in half-truths, which legally is the same as a lie." The Division's fellowship was not accredited, and, in February 2006, the Chair of the Department removed Variyam as Chief of the Division. Variyam sued Hancock for defamation and sought damages for his removal as Chair, loss of reputation, and mental anguish.

Hancock moved for partial summary judgment on Variyam's claim for damages for removal as Chair, which the trial court granted. The trial court subsequently granted a directed verdict that Hancock's letter was defamatory per se. The court of appeals affirmed, reasoning that accusations that someone is a liar are "so obviously hurtful to the person aggrieved that no proof of [their] injurious character is required to make [them] actionable." Hancock appealed.

ISSUE: Can the statements that a physician is untruthful in the performance of his job duties constitute defamation per se?

REASONING: Hancock's primary argument was that his statements about Variyam's untruthfulness were not defamation per se because they did not injure Variyam's profession. Hancock contended that his statements concerned Variyam as a person and not his competency or reputation as a physician. Variyam countered by arguing that his profession requires a truthful reputation in both interacting with patients and colleagues as well as when it comes to teaching and publishing.

The court pointed to precedent that defines defamation per se to include statements that are so obviously hurtful to a person's reputation that general damages result due to loss of reputation. The court went on to say that a statement that injures a person in his or her office, profession, or occupation is typically considered to be defamatory per se.

Ultimately, the court agreed with Hancock that his statements were not defamatory per se. First, Hancock's did not claim that Variyam is an incompetent physician, nor did Hancock attack Variyam's fitness as a professional. Instead, Hancock claimed that Variyam "lacked veracity", which the court argued had nothing to do with Variyam's profession. Second, even if the statements *were* about Variyam's competency as a physician, Variyam failed to

(continued)

prove that any actual damages resulted from Hancock's statements.

DECISION AND REMEDY: Reversed in favor of Hancock.

SIGNIFICANCE OF THE CASE: This case provides a good example of the difference between defamation and defamation per se and how the latter distinction only applies to very specific situations.

CRITICAL THINKING

How would you argue that a physician's veracity is relevant to his reputation as a professional?

ETHICAL DECISION MAKING

What values are furthered by this decision?

If you say your boss is a tyrant or your roommate is a slob, are you in danger of being sued for defamation? You probably are not because such statements are not really statements of fact; rather, they are opinions, and subjective opinions that cannot be proven are generally not actionable.

That distinction between fact and opinion may become more important, as well as perhaps more unclear, as the use of social media continues to expand. Consider, for example, the court's determination that statements made via Twitter were mere opinion in the following Case Nugget.

The Increasingly Important Role of Context in Distinguishing Fact From Opinion

Cheryl Jacobus v. Donald J. Trump, Corey Lewandowski, and Donald J. Trump for President, Inc. Manhattan State Supreme Court

2017 NY Slip Op 08625

According to allegations of the plaintiff's complaint taken as fact for purposes of Defendant's motion to dismiss, Plaintiff is a "political strategist and public relations consultant" and frequent commentator on television news channels and other media outlets, offering "political opinion and analysis from the Republican perspective." In May of 2015, she received a message from a Trump campaign employee, asking if she would be interested in becoming the campaign's communications director. She then met with campaign employees Dornan and Lewandowski, who expressed interest in working with her, with Lewandowski asking for her salary requirements. Later that day, Dornan sent a message to Plaintiff, stating that Lewandowski wanted to meet with her again. By email to Lewandowski, Plaintiff provided her salary requirements and indicated her interest in a position with the campaign.

(continued)

On June 9, Plaintiff met with Dornan and Lewandowski again. At this meeting, during a discussion about communications issues, Lewandowski became agitated, loud, and rude, exclaiming that the FOX television network would do whatever the campaign wanted, and telling Plaintiff that she had no idea how FOX works. The meeting ended soon after, and Plaintiff texted Dornan that she could not work with Lewandowski. They had no further meetings.

After Trump's formal announcement of his presidential campaign on June 16, 2015, Plaintiff frequently appeared on television as a commentator and posted comments on social media sites, including Twitter, both defending and criticizing Trump.

In late January 2017, Plaintiff appeared on a CNN cable news program commenting on the Trump campaign and criticized Trump's debating skills and knowledge. The next day, on an MSNBC program, Lewandowski referenced Plaintiff's comments about Trump, stating that "[t]his is the same person who came to the office on multiple occasions trying to get a job from the Trump campaign, and when she wasn't hired clearly she went off and was upset by that."

In February, Plaintiff appeared once again on a news show criticizing Trump. Later that night, Trump posted the following on Twitter: "Great job on @donlemon tonight @ kayleighmcenany @cherijacobus begged us for a job. We said no and she went hostile. A real dummy! @CNN." A day later, on February 3, 2016, Plaintiff's then lawyer sent Trump a cease and desist letter. Two days after that, on February 5, 2016, Trump posted the following tweet about plaintiff: "Really dumb @CheriJacobus. Begged my people for a job. Turned her down twice and she went hostile. Major loser, zero credibility!" Many of the President's followers responded with tweets attacking the plaintiff.

The plaintiff sued Trump, Lewandowski, and the campaign for defamation. In a decision that illustrates the increasing difficulty of distinguishing fact from opinion in statements made via social media, the court dismissed the plaintiff complaint, writing that "Loose, figurative or hyperbolic statements, even if deprecating the plaintiff, are not actionable."

The court emphasized the importance of context, stating that, "Words that are challenged as defamatory 'must be construed in the context of the entire statement or publication as a whole, tested against the understanding of the average reader.'" The court further stated that "defamatory statements advanced during the course of a heated public debate, during which an audience would reasonably anticipate the use of 'epithets, fiery rhetoric or hyperbole,' are not actionable."

In its analysis, the court said, "Trump's characterization of plaintiff as having 'begged' for a job is reasonably viewed as a loose, figurative, and hyperbolic reference to plaintiff's a state of mind and is therefore, not susceptible of objective verification. . . . To the extent that the word 'begged' can be proven to be a false representation of plaintiff's interest in the position, the defensive tone of the tweet, having followed plaintiff's negative commentary about Trump, signals to readers that plaintiff and Trump were engaged in a petty quarrel. Lewandowski's comments, overall, are speculative and vague, and defendants' implication that plaintiff was retaliating against them for turning her down, notwithstanding the unmistakable reference to her professional integrity, is clearly a matter of speculation and opinion.

Moreover, the immediate context of defendant's statements is the familiar back and forth between a political commentator and the subject of her criticism, and the larger context is the Republican presidential primary and Trump's regular use of Twitter to circulate his positions and skewer his opponents and others who criticize him, including journalists and media organizations whose coverage he finds objectionable. . . . Thus, although the intemperate tweets are clearly intended to belittle and demean plaintiff, any reasonable reading of them makes it 'impossible to conclude that [what defendants said or implied] could subject [plaintiff] to contempt or aversion, induce any unsavory opinion of [her] or reflect adversely upon [her] work,' or otherwise damage her reputation as a partisan political consultant and commentator."

One of the important elements of defamation is that the defamatory statement must be damaging to someone's reputation. For example, when a convicted criminal serving three life sentences attempted to sue a journalist for alleged errors in an article detailing his criminal history, the journalist filed a motion to dismiss on grounds that the defendant was libel-proof as a matter of law regarding claims about his criminal background. Upholding the dismissal, the court said the plaintiff's public reputation at the time the articles were published was so diminished with respect to a specific subject (his kidnapping and murder convictions) that he could not be further injured by allegedly false statements on that subject.[3]

BUT WHAT IF...

WHAT IF THE FACTS OF THE CASE OPENER WERE DIFFERENT?

Recall in the Case Opener that Dr. Bongiovi knew Jones was contemplating going to Dr. Sullivan for treatment, and he wanted her business, so he told her negative stories about Dr. Sullivan. What if Dr. Bongiovi had no idea Jones had ever heard of Dr. Sullivan and was telling her the story just as a reminder that no surgery is completely risk-free?

The increase in communication over the Internet has presented new questions for the law of defamation to answer. First, does a false statement made over this information network constitute defamation? Second, who can be held liable if defamation does exist?

The court first attempted to answer these issues in the case of *Cubby v. CompuServ*.[4] In that case, CompuServ was sued because of defamatory statements published on one of the forums available through its online information service. In holding that CompuServ could not be held liable, the court made an analogy between an online information service provider such as CompuServ and bookstores, saying, "CompuServ's CIS product is in essence an electronic, for profit library." The court went on to say that once CompuServ decides to carry a given publication such as a news forum, it has little or no editorial control over that forum. It would therefore be no more feasible for CompuServ to examine every publication it carries for defamatory material than it would be for libraries or booksellers to do so.

Since *Cubby* was decided, the Communications Decency Act of 1996 was passed. One section of this act gives immunity to providers of interactive computer services for liability they might otherwise incur because of material they disseminate but others create.

A person who is accused of defamation can raise two defenses: truth and privilege. *Truth* is frequently considered an absolute defense. That is, one cannot be held liable for defamation, regardless of whether damages result, if the statement made was the truth. If I say Bill is a convicted felon, and he is, I have not committed defamation. Under ordinary circumstances, the fact that you *thought* a statement was true is not a defense. So if I honestly believe Bill is a convicted felon and I tell others he is, but he is not, then I have committed slander.

Privilege is an affirmative defense in a defamation action. An affirmative defense occurs when the defendant admits to the accusation but argues that there is a reason he should not be held liable. A privilege is either absolute or conditional. When an **absolute privilege** exists, one cannot be sued for defamation for any false statements made, regardless of intent or knowledge of the falsity of the claim. Absolute privilege arises in only a limited number of circumstances. The speech and debate clause of the U.S. Constitution gives an absolute privilege to individuals speaking on the House and Senate floors during congressional debate. This privilege exists because Congress wants to get to the truth of matters before it, and if people testifying before Congress had to fear they might be sued, they might be afraid to testify. Absolute privilege also arises during a judicial proceeding. Again, we do not want people to be afraid to testify in court, so we prohibit them from being sued for whatever occurs within the courtroom. As the following Case Nugget illustrates, the concept of a judicial proceeding can extend beyond a traditional courtroom.

absolute privilege
A special right, immunity, permission, or benefit given to certain individuals that allows them to make any statements about someone without being held liable for defamation for any false statement made, regardless of intent or knowledge of the falsity of the claim.

[3] *Thomas p. Lamb v. Tony Rizzo*, 391 F.3d 1133 (2004).

[4] 77b F. Supp. 135 (1991).

Conditional privilege is the second type. Under a conditional privilege, a party will not be held liable for defamation unless the false statement was made with actual malice. A statement is made with actual malice if it is made with *either* knowledge of its falsity or reckless disregard for its truth. Businesspersons should be most concerned about the conditional privilege that arises with respect to job recommendations. To encourage employers to give honest assessments of their former employees, this privilege protects an employer who makes a false statement about a former worker: The employer will not be held liable as long as the statement was made in good faith and was made only to those who had a legitimate interest in the information being communicated.

Another conditional privilege is the public figure privilege. Public figures are individuals who are in the public eye, typically politicians and entertainers. Because these individuals have a significant impact on our lives, we want to encourage free discussion about them, so we do not hold people liable for making false statements about them as long as the statements were not made with malice. This privilege does not seem to place an unfair burden on the public figure because such an individual can easily respond publicly to any false claims, given that he or she is already in the public eye and has appropriate outlets available for giving his or her opinion. In the opening scenario, Bongiovi argued on appeal that Sullivan was a public figure by nature of his profession. If Bongiovi could establish that Sullivan was a public figure, Sullivan would have to show that Bongiovi acted with actual malice while making the statements to the patient.

There are those who believe a conditional privilege should apply when the defamatory statement is posted somewhere on the Internet. The reasoning behind application of the privilege in this context is that the person who has been defamed over the Internet can respond to the defamatory remarks in the same forum with minimal effort. Thus, there is less need for the stronger legal protection we ordinarily give to private parties who are defamed. In addition, we want to encourage free expression and the exchange of ideas on the Internet. One way to do so is to allow people to respond openly to criticisms on the Internet. When people are overly concerned about making errors, free exchange is hindered. Relaxing defamation standards in regard to the Internet can encourage people to speak their minds freely. Thus far, however, no such privilege has been established. And when a person is found to have committed defamation on an online bulletin board or website, the damages can be significant. For example, a jury awarded $3 million to a university professor who sued a former student who had accused him of being a pedophile on a website she maintained.

conditional privilege
A special right, immunity, permission, or benefit given to certain individuals that allows them to make any statements about someone without being held liable for defamation for any false statement made without actual malice.

actual malice
In defamation, either a person's knowledge that his or her statement or published material is false or the person's reckless disregard for whether it is false.

public figure privilege
A special right, immunity, or permission given to people that allows them to make any statements about public figures, typically politicians and entertainers, without being held liable for defamation for any false statement made without malice.

CASE Nugget

The Applicability of an Absolute Privilege

Hartman and Swinehart v. Keri Supreme Court of Indiana

2008 Ind. LEXIS 265, 883 N.E.2d 774

On May 12, 2003, graduate students Virginia Hartman and Suzanne Swinehart filed formal complaints alleging sexual harassment by Dr. Gabe Keri, an assistant professor at Indiana University–Purdue University Fort Wayne (IPFW) from whom they had taken courses. IPFW charged Dr. Elaine Blakemore with the task of investigating the allegations; Blakemore submitted a report on June 30, 2003, that concluded Hartman and Swinehart's allegations were credible and commended their willingness to report the behavior. Furthermore, Blakemore suggested Keri be immediately removed from teaching positions. These findings and recommendations were further investigated by a three-person panel and Chancellor Wartell. Blakemore's report was supported and Keri was given a 100 percent research position with no use of his university office.

Keri appealed to the president of Purdue, who upheld the chancellor's decision. In response, Keri filed suit against Hartman and Swinehart in the Allen Superior Court, claiming libel, slander, and malicious interference. Hartman and Swinehart moved for summary judgment, claiming their statements were protected by absolute privilege. This motion was denied. Hartman and Swinehart appealed.

(continued)

In finding that the students' statements were entitled to absolute privilege, the state supreme court judges explained that administrative proceedings that are quasi-judicial in nature are entitled to absolute protection. They applied the factors courts use to determine whether a proceeding is quasi-judicial in nature. These factors include whether the body has the power to (1) exercise judgment and discretion; (2) hear and determine to ascertain facts and make decisions; (3) make binding orders and judgments; (4) affect the personal or property rights of private persons; (5) examine witnesses and hear the litigation of issues on a hearing; or (6) enforce decisions or impose penalties.

Privacy Torts. The fact that truth is an absolute defense to a defamation action does not mean people are free to reveal everything they know. Four distinct torts, collectively called *invasion of privacy torts,* protect the individual's right to keep certain things out of public view even if they are true. The four privacy torts are (1) false light, (2) public disclosure of private facts, (3) appropriation for commercial gain, and (4) intrusion on an individual's affairs or seclusion.

False light is closely related to defamation and occurs when publicity about a person creates an impression about that individual that is not valid. It could involve attributing characteristics or beliefs to a person that she does not possess or creating the impression that an individual has taken certain actions he has not taken.

Public disclosure of private facts occurs when someone publicizes a private fact about another that a reasonable person would find highly offensive. The individual must have not waived his or her right to privacy. Publication of information about someone's sex life or failure to pay debts would fall under this tort.

Appropriation for commercial gain occurs when someone uses another person's name, likeness, voice, or other identifying characteristic for commercial gain without that person's permission. Many businesses may choose to use celebrities in their advertisements as a way to appeal to consumers. However, when companies use a celebrity's likeness or image without his or her consent, they open themselves up to potential litigation and liability. For example, Woody Allen believed that American Apparel appropriated his image when they used an image of him from the movie *Annie Hall* in their billboards. Allen sued for $20 million but ended up settling, the day before the case was to go to trial, for $5 million.

The final privacy tort is **intrusion on an individual's affairs or seclusion,** which occurs when someone invades a person's solitude, seclusion, or personal affairs when the person has the right to expect privacy. Examples include wiretapping, using people's passwords to gain access to their email messages, and installing two-way mirrors in a women's dressing room at a gym.

You may recall the invasion-of-privacy lawsuit that made headlines in 2016, when a jury awarded television broadcaster Erin Andrews $55 million against a peeping Tom who made a video of her in the nude and posted it on the Internet and the hotel where the invasion of privacy took place. Michael Barnett had asked a hotel employee which room Andrews was staying in and then asked to be placed in the room next door. He removed peepholes in the door, filmed her in the nude with his cell phone, and then posted the video online. The jury allocated 51 percent of the responsibility to Barnett and 49 percent to the hotel, whose employees never told Andrews of Barnett's request to have the room next to hers.

One final tort against persons is the tort of intentional infliction of emotional distress. Sometimes called the tort of outrage, intentional infliction of emotional distress occurs when someone engages in outrageous, intentional conduct likely to cause extreme emotional distress to another party. For example, if a person calls his former employer and falsely says her son was just arrested for a double homicide after a botched robbery attempt, most courts would find that behavior outrageous enough to satisfy the first element of the tort.

Before damages are awarded in some jurisdictions, the plaintiff must demonstrate injury through physical symptoms directly related to the emotional distress. For instance, in the preceding example, if the employer fainted upon hearing the news, hitting her head on the table and cutting it as she passed out, she would have physical symptoms sufficient to justify a recovery. Other physical symptoms from emotional distress include headaches, a sudden onset of high blood pressure, hives, chills, inability to sleep, and inability to get out of bed.

Businesses frequently find themselves sued for this type of tort when they terminate someone's employment or fail to provide a service that a consumer expected. Many of these actions, like the one in Case 7-2, are unsuccessful, primarily because the bar for what is considered outrageous is high, as are the standards for emotional distress.

CASE 7-2

AARON OLSON, APPELLANT, v. CENTURYLINK
Respondent Court of Appeals of Minnesota
No. A12–0884 (UNPUBLISHED OPINION, February 4, 2013)

FACTS: On December 14, 2011, appellant Aaron Olson contracted to receive telephone service from respondent, CenturyLink, and applied for reduced-rate services that CenturyLink provides through Minnesota's Telephone Assistance Plan (TAP), a statewide program established to reduce telephone rates for low-income households.

The respondent did not apply the reduced rate to Olson's first bill but attached a blank copy of the TAP application. Olson completed the application and mailed it to the respondent, but his next bill also did not reflect a reduced rate. Olson called the respondent and learned that the company had not received his application. Olson then faxed another application to the respondent. A few days later, he contacted the respondent company but could not confirm whether his application had been received.

On January 31, 2012, the respondent company disconnected Olson's telephone service. When Olson called the respondent to resolve the dispute, the company representatives repeatedly hung up on him.

In April 2012, Olson filed suit, claiming that the respondent company violated the Minnesota Human Rights Act (MHRA) and committed intentional infliction of emotional distress. Because he did not have the money for filing fees, Olson also filed a petition to proceed in forma pauperis (IFP). The district court denied the petition, determining that the action was frivolous because the claims had no basis in law. Olson appealed.

ISSUE: When should a defendant be held liable in tort for intentional infliction of emotional distress?

REASONING: To prevail on this tort, a plaintiff must establish that the defendant's conduct (1) was extreme and outrageous, (2) was intentional or reckless, (3) caused emotional distress, and (4) the distress was severe. Extreme and outrageous conduct is behavior that is "so atrocious that it passes the boundaries of decency and is utterly intolerable to the civilized community." Furthermore, to hold a party liable in tort for intentional infliction of emotional distress, the emotional distress must be so severe "that no reasonable man could be expected to endure it."

In this case, the respondent's acts of failing to process Olson's application, disconnecting his telephone service, and hanging up on him during telephone conversations are not so atrocious that they pass the boundaries of decency. For example, previous case law *Langeslag v. KYMN Inc.* states that "insults, indignities, annoyances, petty oppressions, and other trivialities do not constitute extreme and outrageous conduct." The evidence provided by the appellant regarding the acts against him by the respondent mirror the acts cited by *Langeslag* that were previously determined not to constitute extreme and outrageous conduct.

DECISION AND REMEDY: The court affirmed the denial of the petition by the district court in favor of the defendant.

SIGNIFICANCE OF THE CASE: This case outlines the standards that must be met for a defendant to be held liable in tort for intentional infliction of emotional distress.

CRITICAL THINKING

Notice that case law can, as in this case, serve as a restriction on our emotional response to a case. Many of us have had trouble with our phone carrier. The alleged behavior of CenturyLink is at minimum annoying, but case law lays out standards that determine the availability of remedies in cases like these. Reading the case law, describe behavior that CenturyLink would have had to engaged in for the plaintiff to have prevailed.

ETHICAL DECISION MAKING

What value is the court upholding in finding against the plaintiff?

Intentional Torts Against Property

Because businesses own property, anyone who manages a business should be concerned about torts against property. The most common torts a businessperson might be involved with are trespass to realty, nuisance, trespass to personalty, and conversion.

Trespass to Realty The tort of **trespass to realty**, also called *trespass to real property*, occurs when a person intentionally (1) enters the land of another without permission; (2) causes an object to be placed on the land of another without the landowner's permission; (3) stays on the land of another when the owner tells him to depart; or (4) refuses to remove something he placed on the property that the landowner asked him to remove.

It is no defense for a person to argue that she thought she had a legal right to be on the property or she thought the land belonged to someone else. The intent refers to intentionally being on that particular piece of land. In a recent, unusual case heard in a small claims court in Westchester County, a plaintiff sued the defendant for trespass to realty when the defendant entered the plaintiff's property to serve the plaintiff with a reply affidavit for another legal action. The defendant had previously been barred from entering the plaintiff's property. The plaintiff argued that the defendant could not dictate how legal papers are served. The plaintiff sued for $3,000 in compensatory, nominal, and punitive damages. Although the court ruled that the defendant committed trespass to realty, it awarded only nominal damages in the sum of $1.

Because guests are welcomed onto one's property, they are not considered trespassers. However, if a guest is asked to leave and he or she refuses, the person immediately becomes a trespasser and no longer maintains a right to be on the property in question. If charges were brought against such a person, that person could not raise the defense that he or she was a guest.

Also, when a trespasser is on someone's property, the trespasser is liable for damages she or he might cause to the property. Furthermore, under common law, a trespasser cannot hold an owner of property liable for damages the trespasser sustains while on the property. However, as of late, courts have been shifting away from this common law rule. Now, courts typically maintain that owners owe a reasonable duty to anyone who may end up on their property. The specifics of the duty vary by jurisdiction as well as by the status of the parties. In some jurisdictions, it is possible, although rare, for owners to be liable for trespassers who were injured while trying to steal property from the owners.

Private Nuisance A **private nuisance** occurs when a person uses her property in an unreasonable manner that harms a neighbor's use or enjoyment of his property. Using one's property in a manner that caused the neighbor to be subjected to flooding, vibrations, excessive noise, or smoke could lead to a nuisance claim.

A public nuisance occurs when a party acts in a way that endangers the property, health, safety, or welfare of the community. Because a public nuisance is one that affects the public as a whole, actions for public nuisance must be brought by a public official. In late 2017, a number of municipalities in California tried an innovative use of the tort of public nuisance. In almost identical lawsuits, they each sued the big five oil companies, accusing them of creating a public nuisance by ignoring warnings that fossil fuels would cause catastrophic global warming and mounting campaigns to discredit climate change science. They are seeking compensation for the costs of sea walls and other infrastructure needed to protect against climate change.[5] Other cities, including Boulder, Colorado, and New York City have filed similar actions. Such litigation, however, is unlikely to be successful. In 2008, a similar suit was filed by an Alaskan village of Kivalina in federal court. A judge dismissed on grounds that the issue was governed by the Clean Air Act. And in a more severe blow to these cases, in 2018, a district court judge dealt a severe blow to this line of cases, dismissing a lawsuit filed by San Francisco and Oakland, stating that, "The problem deserves a solution on a more vast scale than can be supplied by a district judge or jury in a public nuisance case."

trespass to realty
A tort that occurs when someone goes on another's property without permission or places something on another's property without permission.

private nuisance
A nuisance that affects only a single person or a very limited number of individuals.

[5] California Cities Sue Big Oil for Climate Change, https://www.breitbart.com/politics/2017/09/21/california-cities-sue-oil-companies-over-climate-change/; San Francisco Sues Oil Companies over Climate Change,https://www.usnews.com/news/best-states/california/articles/2017-09-20/san-francisco-sues-oil-companies-over-climate-change.

Trespass to Personal Property A person commits **trespass to personal property**, also called *trespass to personalty*, by temporarily exerting control over another's personal property or interfering with the true owner's right to use the property. Under trespass to personal property, the trespasser is responsible for damages caused to the property as well as for damages caused to the owner related to the trespasser's wrongful possession of the property. For example, if I take someone's bike from his garage and use it for a week, I have committed trespass to personalty. If I return the bike after it has a flat tire, I will have to compensate the owner for the cost of repairing the tire and any other expenses that resulted from my having the bike for a week. Also, if the bike was the only means the person had for getting to work, I would be responsible for the person's lost wages resulting from the missed work, because I took the person's means of transportation.

trespass to personal property
Temporary interference with another's use or enjoyment of his or her personal property.

Conversion **Conversion** occurs when a person permanently removes personal property from the owner's possession and control. When conversion occurs, the true owner can no longer regain the property. The owner usually recovers damages for the full value of the converted item plus any additional damages that resulted from the loss.

It is not a defense to conversion for a person to argue that she believed she had a legal claim to the goods. For example, if Brittany accidentally takes Melvin's suitcase, believing it to be hers, and then loses it, she is still liable for conversion. Moreover, the possession of stolen goods also makes a person liable for conversion. Therefore, buying goods in good faith without knowledge of any impropriety also is not a defense to conversion. Even if a person bought the goods believing the purchase was legal, the person with the stolen goods is liable to the legal owner of the goods.

An illustration of conversion comes from a recent case heard in the Westchester County Supreme Court. An amateur race car driver left her race car at a service station. While the car was in the possession of the service station, an employee of the station with a known drinking problem apparently drove the car, wrecked it, and totally destroyed it. The car could never be returned in the condition it was in when brought to the station. The owner sued for conversion and recovered the value of the car in damages.

conversion
Permanent interference with another's use and enjoyment of his or her personal property.

Intentional Torts Against Economic Interests

All businesspersons should be familiar with the torts against economic interest. The five most common torts against economic interests, frequently referred to as "business torts," are disparagement, intentional interference with contract, unfair competition, misappropriation, and fraudulent misrepresentation. The first tort, **disparagement**, is similar to defamation, but it is a business' or product's reputation that has been tarnished.

The plaintiff in a disparagement case must prove that the defendant published a false statement of a material fact, about the plaintiff's product or service, that resulted in a loss of sales. When such statements are criticisms of the quality, honesty, or reputation of the business or product, the tort is sometimes called **slander of quality** (if spoken) or **trade libel** (if in printed form). If the statements relate to the ownership of the business property, the term **slander of title** is used.

Damages for disparaging are ordinarily based on a decrease in profits that can be linked to the publication of the false statement. An alternative, although less common, way to prove damages is to demonstrate that the plaintiff had been negotiating a contract with a third party but the third party lost interest shortly after the publication of the false statement. The profits the plaintiff would have made on the contract would be the damages.

Some interesting variations of the tort of disparagement have developed. For example, in 2007, California became the thirteenth state to recognize the tort of **food disparagement**, which critics call veggie libel. Such laws provide ranchers and farmers with a cause of action when someone knowingly makes false, damaging statements about a food product. The California law was drafted in response to an incident during 2006 in which Taco Bell executives wrongly identified green onions grown at Boskovich Farms in Oxnard as the source of an *E. coli* outbreak that sickened 70 of the fast-food chain's customers.

A second tort against economic interests is the tort of **intentional interference with contract**. To bring a claim of intentional interference with contract successfully, the plaintiff must prove that (1) a valid and enforceable contract between the two parties existed; (2) the defendant party

disparagement
A business tort that occurs when a statement is intentionally used to defame a business product or service.

slander of quality
A business tort that occurs when false spoken statements criticize a business product or service and result in a loss of sales.

trade libel
A business tort that occurs when false printed statements criticize a business product or service and result in a loss of sales.

slander of title
A business tort that occurs when false published statements are related to the ownership of the business property.

food disparagement
A tort that provides ranchers and farmers with a cause of action when someone spreads false information about the safety of a food product.

intentional interference with contract
The tort that occurs when someone intentionally takes an action that will cause a person to breach a contract that she or he has with another.

knew of the existence of the contract and its terms; (3) the defendant intentionally undertook steps to cause one of the parties to breach the contract; and (4) the plaintiff was injured as a result of the breach.[6]

When a contract exists, clear liability is placed on third parties for inducing a party to the contract to breach the contract. However, when a prospective contract exists, a third party might be liable for inducing a party to pull out of the contract before it is formed. Because the essence of business involves competition, simply offering a better deal is not enough to create liability when a prospective contract exists. However, if a party uses illegal means to cause another party not to enter into a contract, the party who acted illegally is liable for interfering with contractual relations.[7]

The most common situation involving intentional interference with a contract in the business context occurs when one employer tries to lure an employee away from another employer. Liability in such a situation, however, is limited to when the employee has a contract for a set period of time and the prospective employer actually knows of the contract.

When a third party interferes with a contract, injured parties may recover damages for what was directly lost through the breached contract. They may also recover any losses suffered related to the breached contract, in addition to damages for emotional distress and harm to reputation.[8]

A third tort against economic interests, **unfair competition,** exists because of American views about the purpose of a business. Americans believe a business is intended to make a profit, and the law protects businesses acting on this profit motive. Therefore, when someone enters a business with the sole intent of driving another firm out of business, the law punishes this act as unfair competition. For example, if there is only one jewelry store in town, Mark cannot come in and set up a store that makes no profits, just to drive the other store out of business so that an acquaintance of his can then move in and open up a legitimate jewelry store once the competition has been eliminated.

A fourth tort against economic interests is **fraudulent misrepresentation,** which occurs when one uses intentional deceit to facilitate personal gain. To establish that a fraudulent misrepresentation exists, a party must demonstrate all the following:

1. A party knowingly, or with reckless disregard for the truth, misrepresented material facts and conditions.
2. The party intended to have other parties rely on the misrepresentations.
3. The injured party reasonably relied on the misrepresentations.
4. The injured party suffered damages because of relying on the misrepresentations.
5. A direct link exists between the injuries suffered and a reliance on the misrepresentations.[9]

As in the criminal act of fraud, in the civil act of fraudulent misrepresentation, a party materially misrepresents something and thereby causes another party to suffer damages. Typically, fraudulent misrepresentation applies only to the misrepresentation of material facts. However, when a party with expert knowledge regarding a specific matter states an opinion, any party reasonably relying on the statement, although it is an opinion and not fact, may recover damages under the tort of fraudulent misrepresentation.

unfair competition
The act of competing with another not to make a profit but for the sole purpose of driving that other out of business.

fraudulent misrepresentation
The tort that occurs when a misrepresentation is made with the intent to facilitate personal gain and with the knowledge that it is false.

Negligence

LO 7-3 Identify the elements that must be proved to win a negligence case.

negligence
Behavior that creates an unreasonable risk of harm to others.

Negligence is behavior that creates an unreasonable risk of harm to others. Unlike intentional torts, which result from a person willfully taking actions that are likely to cause injury, negligent torts involve the failure to exercise reasonable care to protect another's person or property.

[6] Restatement (Second) of Torts, sec. 766.
[7] Restatement (Second) of Torts, sec. 766.
[8] Restatement (Second) of Torts, sec. 774A.
[9] Restatement (Second) of Torts, sec. 525.

Sometimes, however, harm occurs because an individual suffers an unfortunate accident, an incident that simply could not be avoided, even with reasonable care. For example, suppose Jonathan is driving on the highway when he suffers a stroke. Because of the stroke, he crashes into two other vehicles. He is not liable for damages caused by the accident because he was not negligent. Yet if Jonathan had some type of warning that the stroke was going to occur, he might be liable for the accident.

To win a negligence case, the plaintiff must prove four elements: (1) duty, (2) breach of duty, (3) causation, and (4) damages (see Exhibit 7-1). A plaintiff who cannot establish all four of these elements will be denied recovery.

Duty

The plaintiff must first establish that the defendant owes a *duty* to the plaintiff. In some particular situations, the law specifies the duty of care one individual owes to another. In most cases, however, the courts use the reasonable person standard to determine the defendant's duty of care. The **reasonable person standard** is a measurement of the way members of society expect an individual to act in a given situation. To determine the defendant's duty of care, the judge or jury must determine the degree of care and skill that a reasonable person would exercise under similar circumstances. The judge or jury then uses this standard to evaluate the actions of the individual in the case.

When courts attempt to determine whether a reasonable person would have owed a duty to others, they consider four questions:

1. How likely was it that the harm would occur?
2. How serious was the potential harm?
3. How socially beneficial was the defendant's conduct that posed the risk of harm?
4. What costs would have been necessary to reduce the risk of harm?

> **reasonable person standard**
> A measurement of the way members of society expect an individual to act in a given situation.

In many situations, it is far from clear what a reasonable person would do. For example, if a reasonable person saw an infant drowning in a shallow swimming pool, what would she do? In most situations like this one, the law holds that individuals have no duty to rescue strangers from perilous situations.

In some cases, however, the courts hold that individuals have a duty to aid strangers in certain types of peril. For example, if Sam negligently hits Janice with his car and, as a result, Janice is lying in the street, Sam has a duty to remove her from that dangerous position. Similarly, employers have a special duty to protect their employees from dangerous situations.

The courts generally hold that landowners have a duty of care to protect individuals on their property. Similarly, businesses have a duty of care to customers who enter business property. It is important, therefore, for future business managers to be knowledgeable about this duty. Businesses should warn customers about risks they may encounter on business property. Some risks, however, are obvious, and businesses need not warn customers about them. For example, a business need not inform customers that they could get a paper cut from the pages of a book.

The courts generally hold that businesses have a duty of care to protect their customers against foreseeable risks about which the owner knew or reasonably should have known. For example, in *Haywood v. Baseline Construction Company*, a woman who tripped over lumber on the front porch of the House of Blues restaurant in Los Angeles sued for negligence. The business' attempt to warn customers by marking the lumber with yellow construction tape was insufficient to avoid the determination of negligence; the woman was awarded $91,366 in damages.

To prove negligence, a plaintiff must demonstrate:

1. Duty
2. Breach of duty
3. Causation—actual and proximate
4. Damages

Exhibit 7-1

Elements of Negligence

E-COMMERCE and the Law

Negligence on the Internet

A commonly offered explanation for the increasing occurrence of violence is the increased violence portrayed in the media. Some plaintiffs try to hold owners of certain websites liable under negligence theories for violent acts committed by teenagers. For example, in *James v. Meow Media,* a 14-year-old boy took six guns to school and shot three of his classmates to death. The parents of the deceased classmates brought suit against several Internet websites and the creators and distributors of various video games. The parents argued that these defendants had a duty of ordinary care to the slain girls.

The courts have been consistent, however, in finding that it was not foreseeable that a boy who played certain video games and viewed certain websites would murder three of his classmates. In similar cases, courts have ruled that defendants (such as website owners, creators and distributors of video games, and directors and producers of movies) do not have a duty to protect a person from the criminal acts of a third party unless there is a special relationship that requires the defendant to act with that duty.

Although it appears that website owners, manufacturers, and producers will not be held liable, plaintiffs continue to bring suits against these groups of people. Can you think of an argument for why these groups of people might owe a duty of care to these plaintiffs?

The liability of businesses grew out of the traditional common law, under which the duty of owners and occupiers of real property depends primarily on the status of the injured party. The injured party may be an invitee, a person who enters another's premises as a result of an express or implied invitation of the owner for their mutual benefit, with a business invitee being one who is invited onto the property for the purpose of doing business; a licensee, a person who enters another's premises with the occupier's consent for his or her benefit alone; or a trespasser, one who is on the property without the owner's or occupier's permission, as described earlier in this chapter. Under this approach, the owner owes the greatest duty to the invitee, a duty to inspect the premises to ensure that they are reasonably safe and to warn of any hidden dangers. The owner has a duty to a licensee to warn of any hidden dangers of which he knows but has no affirmative duty to inspect for hidden problems. The owner owes no duty to trespassers except to do them no intentional harm.

In 1968, however, the California Supreme Court rejected this rigid division,[10] substituting a general duty of reasonable care in its place, under which a duty of reasonable care is owed any entrant onto the land, regardless of status. More than 25 states have now moved to this standard, with a few using this standard for both the invitees and the licensees, while maintaining the traditional approach to trespassers. The trend is toward the use of this unitary standard, but given the lack of uniformity among the states, it is important to know which approach is being used by the state in which you live and do business.

Professionals have more training than ordinary people. Thus, when professionals are serving in their professional capacity, courts generally hold that they have a higher duty of care to clients than does the ordinary person. A professional cannot defend against a negligence suit by claiming ignorance of generally accepted principles in her or his field of expertise. Clients who feel that they have suffered damages as a result of a professional's breach of her duty of care can bring a negligence case against her. These actions are frequently referred to as *malpractice cases.*

Breach of Duty

Once the plaintiff has established that the defendant owes her a duty of care, she must prove that the defendant's conduct violated that duty. This violation is called a *breach of duty.* For example, the driver of an automobile owes the other passengers in his car a duty of care to obey traffic signs. If he fails to stop at a stop sign, he has violated his duty to follow traffic signs and has therefore breached his duty of care.

Causation

Causation is the third element of a successful negligence claim, and it has two elements: actual cause and proximate cause. The plaintiff must prove both elements of causation to be able to recover damages.

[10] Rowland, 443 P.2d 561 (Cal. 1968).

GLOBAL Context

©Digital Archive Japan/Alamy

Negligence in Germany

German law is concerned with the defendant's ability to foresee, understand, and avoid danger. Both mental and physical capabilities are taken into account. For example, the duty-of-care standard stipulates that "physical and mental disabilities or defects, panic, or confusion" exempt the defendant from being found negligent. Also, although the distinction is not recognized by a statute, the courts distinguish between conscious and unconscious negligence. *Conscious negligence* requires knowledge that the offense is about to occur and that it is an actual offense. *Unconscious negligence* occurs when the defendant is either unaware that the act constitutes an offense or unaware that the act is occurring at all. In such cases, the defendant is found not liable by reason of unconscious negligence.

The first element, **actual cause** (also known as *cause in fact*), is the determination that the defendant's breach of duty resulted directly in the plaintiff's injury. The courts commonly determine whether a breach of duty actually caused the plaintiff's injury by asking whether the plaintiff would have been injured if the defendant had fulfilled his or her duty. If the answer is no, then the actual cause of the plaintiff's injury was the defendant's breach. Actual cause is sometimes referred to as but-for causation because the plaintiff argues that the damages she suffered would not have occurred *but for* (except because of) the actions of the defendant. **Proximate cause,** sometimes referred to as *legal cause,* refers to the extent to which, as a matter of policy, a defendant may be held liable for the consequences of his actions. In most states, proximate cause is determined by foreseeability. Proximate cause is said to exist only when both the plaintiff and the plaintiff's damages were reasonably foreseeable at the time the defendant breached his duty to the plaintiff. Thus, if the defendant could not reasonably foresee the damages that the plaintiff suffered as a result of his action, the plaintiff's negligence claim will not be sustained because it lacks the element of proximate causation.

You must be able to foresee that either the plaintiff himself or herself or a person in the plaintiff's position would be hurt. If you are running a red light, it is foreseeable that a person will be crossing with the light. If you hit a person crossing with the light, we say the person hit was foreseeable. If it makes more sense, we could say "both a person in the plaintiff's position" and the "plaintiff's damages," but most people refer simply to "the plaintiff and the plaintiff's damages."

For example, if a defective tire on a vehicle blows out, it is foreseeable that the driver may lose control and hit a pedestrian. It is not foreseeable, however, that the pedestrian may be a scientist carrying a briefcase full of chemicals that may explode on impact, causing a third-floor window to shatter, injuring an accountant at his desk. In most states, the accountant would not succeed if he sued the tire manufacturer for negligence. The tire failure is not considered a proximate cause of the accountant's injury because the contents of the pedestrian's briefcase were highly unusual. The pedestrian, however, would be eligible to recover damages from the tire manufacturer because hitting a pedestrian is a foreseeable consequence of tire failure. Thus, the defect in the tire is a proximate cause of the pedestrian's injury.

Palsgraf v. Long Island Railroad Company is a well-known case in which the majority opinion set forth the rule of foreseeability in determining proximate cause. The dissent, however, set forth an alternative definition of proximate cause that is accepted today in a small minority of states. Courts following the minority definition do not distinguish actual cause from proximate cause. In these states, if the defendant's action constitutes an actual cause, it is also considered the proximate cause. Therefore, in these few states, both the pedestrian-scientist and the third-floor accountant would be able to recover damages from the tire manufacturer in the previous example.

Damages

Damages are the final required element of a negligence action. The plaintiff must have sustained compensable injury as a result of the defendant's actions. Because the purpose of tort law is to compensate individuals who suffer injuries as a result of another's action or inaction, a person

actual cause
The determination that the defendant's breach of duty resulted directly in the plaintiff's injury.

proximate cause
The extent to which, as a matter of policy, a defendant may be held liable for the consequences of his or her actions. In the majority of states, proximate cause requires the plaintiff and the plaintiff's damages to have been foreseeable at the time of the accident. In the minority of states, proximate cause exists if the defendant's actions led to the plaintiff's harm.

compensatory damages
Money awarded to a plaintiff as reimbursement for her or his losses; based on the amount of actual damage or harm to property, lost wages or profits, pain and suffering, medical expenses, disability, and so on.

punitive damages
Compensation awarded to a plaintiff that goes beyond reimbursement for actual losses and is imposed to punish the defendant and deter such conduct in the future.

gross negligence
An action committed with extreme reckless disregard for the property or life of another person.

res ipsa loquitur
A doctrine that allows the judge or jury to infer that, more likely than not, the defendant's negligence was the cause of the plaintiff's harm, even though there is no direct evidence of the defendant's lack of due care.

cannot bring an action in negligence seeking only nominal damages. Rather, a person must seek **compensatory damages** or damages intended to reimburse a plaintiff for her or his losses.

In typical negligence cases, courts rarely award **punitive damages** or *exemplary damages,* which are imposed to punish the offender and deter others from committing similar offenses. Instead, courts usually award punitive damages in cases in which the offender has committed **gross negligence,** an action committed with extreme reckless disregard for the property or life of another person.

Plaintiff's Doctrines

LO 7-4 Recognize the doctrines that help a plaintiff establish a case of negligence.

The plaintiff has the burden of proving all four elements of a negligence case. Direct evidence of negligence by the defendant, however, is not always available. For example, there may have been no witnesses to the negligent conduct, and other evidence may have been destroyed. Therefore, two doctrines have been adopted by courts to aid plaintiffs in establishing negligence claims: *res ipsa loquitur* and negligence per se.

Res Ipsa Loquitur

Res ipsa loquitur literally means "the thing speaks for itself." The plaintiff uses this doctrine to allow the judge or jury to infer that, more likely than not, the defendant's negligence was the cause of the plaintiff's harm, even though there is no direct evidence of the defendant's lack of due care. To establish *res ipsa loquitur* in most states, the plaintiff must demonstrate that:

1. The event was a kind that ordinarily does not occur in the absence of negligence.
2. Other responsible causes, including the conduct of third parties and the plaintiff, have been sufficiently eliminated.
3. The indicated negligence is within the scope of the defendant's duty to the plaintiff.

Proof of these three elements does not require a finding of negligence, however; it merely permits it. Once the plaintiff has demonstrated these three elements, the burden of proof shifts to the defendant, who must prove that he was not negligent to avoid liability.

One of the earliest uses of *res ipsa loquitur* was the case of *Escola v. Coca Cola.*[11] In that case, the plaintiff, a waitress, was injured when a bottle of Coca-Cola that she was removing from a case exploded in her hand. From the facts that (1) bottled soft drinks ordinarily do not spontaneously explode and (2) the bottles had been sitting in a case, undisturbed, in the restaurant for approximately 36 hours before the plaintiff simply removed the bottle from the case, the jury reasonably inferred that the defendant's negligence in the filling of the bottle resulted in its explosion. The plaintiff therefore recovered damages without direct proof of the defendant's negligence. Plaintiffs in numerous accident cases have subsequently used the doctrine when there has been no direct evidence of negligence. The defendant's best response to this doctrine is to demonstrate other possible causes of the accident. Case 7-3 illustrates the use of this doctrine.

[11] 24 Cal. 2d 453, 150 P.2d 436 (1944).

CASE 7-3

SWOBODA v. FONTANETTA
Supreme Court of the State of New York
131 A.D.3d 1042 (2015)

FACTS: On November 9, 2007, the defendant physician Adrian Philip Fontanetta, who was employed by the defendant Western Nassau Orthopedic Associates, P.C. (Western Nassau), performed an arthroscopic subacromial decompression of the right shoulder of the plaintiff Caroline Swoboda. The surgery was performed at the defendant Day-Op Center

(continued)

while the injured plaintiff was under general anesthesia administered by a physician employed by the defendant AMBI Medical Associates, P.C. (AMBI).

Upon leaving Day-Op, the plaintiff felt pain in her clavicle area, but did not immediately report this pain to Fontanetta because she thought that it was normal to feel pain after her surgery. On November 13, she went to Fontanetta's office because she was in great pain and had swelling in her clavicle area. Without examining her, Fontanetta told her that she was feeling pain because of a muscle and sent her home. On November 27, the injured plaintiff returned to Fontanetta's office, complaining of continuing pain in the mid-clavicle area. An examination revealed supraclavicular swelling, and an X-ray revealed a fracture of the mid-shaft of the injured plaintiff's clavicle. The injured plaintiff sued to recover damages for medical malpractice, contending that the fracture of the injured plaintiff's clavicle occurred during the surgery, and relied upon the doctrine of res ipsa loquitur. The lower court granted the motion to dismiss filed by all the defendants. Plaintiff appealed.

ISSUE: Should a plaintiff be allowed to use *res ipsa loquitur* in a simple medical malpractice case that requires no expert to enable the jury to reasonably conclude that the accident would not happen without negligence?

REASONING: A plaintiff should be able to use *res ipsa loquitur* when (1) the event is of a kind that ordinarily does not occur in the absence of someone's negligence; (2) the event is caused by an agent or instrumentality within the exclusive control of the defendant; and (3) the event was not caused by any voluntary action or contribution on the part of the plaintiff. A plaintiff can generally use the doctrine to establish a prima facie case when an unexplained injury in an area that is remote from the treatment site occurs while the patient is anesthetized. In a multiple defendant action, a plaintiff is not required to identify the negligent actor, especially in a situation in which the plaintiff has been anesthetized. To rely on *res ipsa loquitur*, a plaintiff need not conclusively eliminate the possibility of all other causes of the injury. It is enough that the evidence supporting the three conditions affords a rational basis for concluding that it is more likely than not that the injury was caused by the defendants' negligence. All that is required is that the likelihood of other possible causes of the injury be so reduced that the greater probability lies with the defendant.

DECISION AND REMEDY: Because the evidence submitted by the defendants revealed the existence of a triable issue of fact regarding the liability of the defendants under the doctrine of res ipsa loquitur, the defendants were not entitled to a judgment as a matter of law dismissing the complaint. The court reversed the trial court's granting of motion to dismiss and ordered a new trial at which *res ipsa loquitur* may be used.

SIGNIFICANCE OF THE CASE: This case provides an illustration of the type of conditions that can lead a jury to infer negligence.

CRITICAL THINKING

What evidence do you believe the defendant could have presented that might have prevented the plaintiffs from successfully using the doctrine of *res ipsa loquitur*? Why would this evidence have helped them?

ETHICAL DECISION MAKING

Which stakeholders's interests were furthered by the use of *res ipsa loquitur*?

Negligence Per Se

Negligence per se (literally, "negligence in or of itself") is another doctrine that helps plaintiffs succeed in negligence cases. Negligence per se applies to cases in which the defendant has violated a statute enacted to prevent a certain type of harm from befalling a specific group to which the plaintiff belongs. If the defendant's violation causes the plaintiff to suffer from the type of harm that the statute intends to prevent, the violation is deemed negligence per se. The plaintiff does not have to show that a reasonable person would exercise a certain duty of care toward the plaintiff. Instead, the plaintiff can offer evidence of the defendant's violation of the statute to establish proof of the negligence.

negligence per se
A doctrine that allows a judge or jury to infer duty and breach of duty from the fact that a defendant violated a criminal statute that was designed to prevent the type of harm that the plaintiff incurred.

For example, if Ohio passes a statute prohibiting the sale of alcohol to minors, and a minor runs a red light and kills two pedestrians while driving under the influence of alcohol sold to him illegally, the liquor store's violation of the statute prohibiting the sale of alcohol to minors establishes negligence per se on the part of the store. The families of the pedestrians do not need to establish that a reasonable person would have a duty not to sell alcohol to a minor. The Case Nugget provides another illustration of negligence per se.

Defenses to Negligence

LO 7-5 Recall the defenses to a claim of negligence.

The courts's doctrines of *res ipsa loquitur* and negligence per se help the plaintiff in a negligence case, but the courts permit certain defenses that relieve the defendant from liability even when the plaintiff has proved all four elements of negligence. Defendants can successfully rebut negligence claims with contributory negligence, comparative negligence, assumption of the risk, and other special negligence defenses.

A Clear Illustration of Negligence per se

O'Guin v. Bingham County Superior Court of Idaho

122 P.3d 308 (Sup. Ct. Id. 2005)

Shaun and Alex O'Guin cut across a field on their way home from school and entered the back of a landfill that was not fenced off, despite a law that required it to be fenced. A portion of a wall of the landfill collapsed, killing the boys. The boys' parents sued the county that operated the landfill, alleging that failure to have the landfill properly fenced constituted negligence per se.

The court agreed, explaining that the following four elements of negligence per se had been met: (1) The statute or regulation clearly defines the required standard of conduct, (2) the statute or regulation must have been intended to prevent the type of harm the defendant's act or omission caused, (3) the plaintiff must be a member of the class of persons the statute or regulation was designed to protect, and (4) the violation must have been the proximate cause of death.

Contributory Negligence

contributory negligence
A defense to negligence whereby the defendant can escape all liability by proving that the plaintiff failed to act in a way that would protect him or her from an unreasonable risk of harm and that the plaintiff's negligent behavior contributed in some way to the plaintiff's accident.

Contributory negligence, a defense once available in all states but replaced today in some states by the defense of comparative negligence (discussed in the next section), applies in cases in which both the defendant and the plaintiff were negligent. The defendant must prove that (1) the plaintiff's conduct fell below the standard of care needed to prevent unreasonable risk of harm and (2) the plaintiff's failure was a contributing cause of the plaintiff's injury. How can defendants use contributory negligence in a case? Some defense lawyers argue that if a plaintiff involved in a car accident failed to wear her seat belt, that failure constitutes contributory negligence because her action contributed to her injuries.

If the defendant successfully proves contributory negligence, no matter how slight the plaintiff's negligence, the plaintiff will be denied any recovery of damages. Because this defense seems unfair, many states have adopted the **last-clear-chance doctrine.** This doctrine allows the plaintiff

to recover damages despite proof of contributory negligence as long as the defendant had a final clear opportunity to avoid the action that injured the plaintiff.

For example, suppose that Samantha and Nicole, in their cars, are facing each other while stopped at a red light. The light turns green, and Nicole starts to turn left at the intersection. Samantha sees Nicole start to turn, but she still continues to travel straight through the intersection and crashes into Nicole's car. Although Samantha had the right-of-way at the intersection, she could have avoided hitting Nicole's car by braking or swerving. Thus, according to the last-clear-chance doctrine, Nicole could recover damages.

Comparative Negligence

Most states have replaced contributory negligence with comparative negligence because the adoption of the last-clear-chance doctrine still left many situations in which an extremely careless defendant can cause a great deal of harm to a plaintiff who is barred from recovery due to minimal contributory negligence. There are two forms of comparative negligence: pure and modified. According to a **pure comparative negligence** defense, the court determines the percentage of fault of the defendant. The defendant is then liable for that percentage of the plaintiff's damages.

Courts calculate damages according to **modified comparative negligence** in the same manner, except that the defendant must be more than 50 percent at fault before the plaintiff can recover.

Twenty-eight states have adopted modified comparative negligence, 13 have adopted pure comparative negligence, and 9 have adopted contributory negligence. Every state has adopted one of these three defenses. Thus, the parties to a negligence suit cannot choose among them.

Assumption Of The Risk

Another defense available to defendants facing negligence claims is called **assumption of the risk**. To use this defense successfully, a defendant must prove that the plaintiff voluntarily and unreasonably encountered the risk of the actual harm the defendant caused. In other words, the plaintiff willingly assumed as a risk the harm she suffered. There are two types of this defense. *Express assumption of the risk* occurs when the plaintiff expressly agrees (usually in a written contract) to assume the risk posed by the defendant's behavior. In contrast, *implied assumption of the risk* means that the plaintiff implicitly assumed a known risk.

The most difficult part of establishing this defense is showing that the plaintiff assumed the risk of the *actual* harm she suffered. A 1998 case against the Family Fitness Center illustrates an unsuccessful attempt to use assumption of the risk as a defense against a negligence claim.[12] In that case, the plaintiff was injured when a sauna bench on which he was lying collapsed beneath him at the defendant's facility. The trial court granted summary judgment in favor of the defendant on the basis of assumption of the risk. The plaintiff had signed a contract that included the following provision: "Buyer is aware that participation in a sport or physical exercise may result in accidents or injury, and Buyer assumes the risk connected with the participation in a sport or exercise and represents that Member is in good health and suffers from no physical impairment which would limit their use of FFC's facilities." The appellate court overturned the trial court's decision because the type of injury the plaintiff suffered was not the type of risk he had assumed. The court held that anyone signing a membership agreement could be deemed to have waived any hazard known to relate to the use of the health club facilities, such as the risk of a sprained ankle due to improper exercise or overexertion, a broken toe from a dropped weight, injuries due to malfunctioning exercise or sports equipment, or injuries from slipping in the locker room shower. No patron, however, could be charged with realistically appreciating the risk of injury from simply reclining on a sauna bench. Because the collapse of a sauna bench, when properly used, is not a known risk, the court concluded that the plaintiff did not assume the risk of this incident as a matter of law.

[12] *Leon v. Family Fitness Center, Inc.*, 61 Cal. App. 4th 1227 (1998).

last-clear-chance doctrine
A doctrine used by the plaintiff when the defendant establishes contributory negligence. If the plaintiff can establish that the defendant had the last opportunity to avoid the accident, the plaintiff may still recover, despite being contributorily negligent.

pure comparative negligence
A defense accepted in some states whereby the defendant is not liable for the percentage of harm that he or she can prove is due to the plaintiff's own negligence.

modified comparative negligence
A defense accepted in some states whereby the defendant is not liable for the percentage of harm that he or she can prove is due to the plaintiff's own negligence if the plaintiff's negligence is responsible for less than 50 percent of the harm; if the defendant establishes that the plaintiff's negligence caused more than 50 percent of the harm, the defendant has no liability.

assumption of the risk
A defense whereby the defendant must prove that the plaintiff voluntarily assumed the risk the defendant caused.

Special Defenses To Negligence

Many states have additional ways to defend against a claim of negligence. For example, laws in some states hold that people in peril who receive voluntary aid from others cannot hold those offering aid liable for negligence. These laws, commonly called **Good Samaritan statutes**, attempt to encourage selfless and courageous behavior by removing the threat of liability.

The defendant in a negligence suit can also avoid liability by establishing a superseding cause. A *superseding cause* is an unforeseeable event that interrupts the causal chain between the defendant's breach of duty and the damages the plaintiff suffered. For example, suppose Jennifer is improperly storing ammonia in her garage when a meteor strikes her garage, spilling the ammonia into a stream nearby. Will, living downstream, drinks water from the stream and becomes dangerously ill. Because the meteor was unforeseeable, Jennifer is not liable for Will's injuries, even though she breached her duty of care to Will. Superseding causes allow the defendant to avoid liability because they are evidence that the defendant's breach of duty was not the proximate cause of the plaintiff's injuries. In other words, superseding causes disprove the causation element necessary to sustain a negligence claim.

Good Samaritan statute
A statute that exempts from liability a person, such as a physician passerby, who voluntarily renders aid to an injured person but negligently, but not unreasonably negligently, causes injury while rendering the aid.

Strict Liability

LO 7-6 Explain how a party proves a strict-liability case.

Strict liability is liability without fault. The law holds an individual liable without fault when the activity in which she engages satisfies three conditions: (1) it involves a risk of serious harm to people or property; (2) it is so inherently dangerous that it cannot ever be safely undertaken; and (3) it is not usually performed in the immediate community. Instead of banning such activities, the law allows people to engage in these activities but holds them liable for all resulting harm.

Inherently dangerous activities include dynamite blasting in a populated area and keeping animals that have not been domesticated. If an animal has shown a vicious propensity, strict liability applies and the owner of the animal is responsible for any injuries suffered in an attack by the animal. If an individual keeps an animal that has shown vicious propensity, he has a duty to warn and protect individuals who come into contact with the animal. Today, the theory of strict liability is used most commonly when the plaintiff is claiming to have been injured by a product that was unreasonably dangerous.

strict liability
Liability in which responsibility for damages is imposed regardless of the existence of negligence. Also called *liability without fault*.

Damages Available in Tort Cases

LO 7-7 Recall the types of damages available in tort cases.

There are three types of damages available in tort cases: compensatory, nominal, and punitive (see Exhibit 7-2). You will see this system of classifying damages again when we talk about damages in other contexts, such as in cases involving the breach of a contract.

Exhibit 7-2
Types of Tort Damages

TYPE	PURPOSE	AMOUNT
Compensatory	To make the plaintiff whole again.	An amount equivalent to all losses caused by the tort, including compensation for pain and suffering, but not attorney fees.
Nominal	To recognize that the defendant committed a tort against the plaintiff.	A trivial amount, typically $1 to $5.
Punitive	To punish the defendant and deter future wrongdoers.	An amount based on two factors: the severity of the wrongful conduct and the wealth of the defendant.

Compensatory Damages

Because the primary objective of tort law is to compensate victims, the primary type of damages are *compensatory damages,* damages designed to compensate the victim for all the harm caused by the person who committed the tort, often referred to as the **tortfeasor**. Although we seem to hear a lot about runaway jury awards, there does not seem to be evidence of a trend toward increasingly huge jury awards. It is more accurate to say that there has been a gradual, but not consistent, increase in personal injury awards since 2008 and 2009, where the median award each year was $40,000. In 2010, the median award fell to $39,216, but it rose the next two years to $60,924 and $75,000, respectively. It fell again in 2013 to $70,000. However, in 2014, (the latest year for which data is available) the median award in personal injury cases was back up to $75,000.[13]

tortfeasor
A person who commits an intentional or through-negligence tort that causes a harm or loss for which a civil remedy may be sought.

Compensatory damages are typically awarded for pain and suffering, costs of repairing damaged property, medical expenses, and lost wages. In the case opener, Dr. Bongiovi argued on appeal that the compensatory damages were excessive, but the appellate court upheld them, noting that when one is defamed in their line of business, they are entitled to general damages for harm to their reputation. In light of the fact that after the allegations the number of exotic dancers seeking Dr. Sullivan's services declined, the amount did not seem unreasonable to the court. Surprisingly, attorney fees are *not* recoverable as compensatory damages, despite the fact that most plaintiffs could not bring an action against the tortfeasor without hiring an attorney. Because the plaintiffs in personal injury cases must usually pay their attorneys anywhere from one-third to one-half of their recovery, some argue that compensatory damages fail to meet the intended goal of properly compensating victims. Others point out that one of the ways plaintiffs can, in essence, recover their attorney fees is by increasing their pain and suffering damages enough to cover these expenses.

Nominal Damages

Nominal damages are a small amount of money given to recognize that a defendant did indeed commit a tort in a case in which no compensable damages were suffered by the plaintiff. A plaintiff may receive nominal damages by simply failing to prove actual damages.

nominal damages
Monetary damages awarded to a plaintiff in a very small amount, typically $1 to $5, to signify that the plaintiff has been wronged by the defendant even though the plaintiff suffered no compensable harm.

Punitive Damages

Punitive damages are awarded to punish the defendant. They are given only when the defendant's conduct is extremely outrageous. The purposes of punitive damages are both to punish the defendant and to deter him and others who are similarly situated from engaging in that kind of activity again. In awarding punitive damages, juries usually consider the egregiousness or willfulness of the tort and the wealth of the defendant. Obviously, the more wrongful the nature of the defendant's act, the greater the desire to send a message that such behavior will not be tolerated; and the greater the wealth of the defendant, the higher the damages must be to actually punish the defendant.

Although the threat of large punitive damages is seen by many groups, including consumer advocates, as a good method for encouraging manufacturers to produce the safest possible products, others disagree. They believe that no threat beyond compensatory damages is required; and they argue that the main effect of punitive damages is to discourage innovation because manufacturers will be afraid of the risk of producing a defective product, which could cost them millions in punitive damages.

Since the late 1970s, insurance companies and tort reform groups have been trying to limit the amount of punitive damages that can be assessed, using both the legislative process and the courts. The 1994 case of *Honda Motor Company v. Oberg*[14] provided tort reformers with their first judicial victory. In this case, the U.S. Supreme Court finally struck down a punitive-damage award as being a violation of due process. *Oberg,* however, was a limited victory because of two unique aspects of the case. First, the punitive damages were over 500 times the amount of the compensatory damages, which is extraordinarily rare. Second, the state law under which the damages were awarded was the only state law in the country that had no provision for judicial review of the amount of punitive-damage awards, and it was the denial of this safeguard that violated the due

[13] http://www.iii.org/fact-statistic/litigiousness
[14] 114 S. Ct. 2331 (1994).

GLOBAL Context

©Digital Archive Japan/Alamy

Punitive Damages in Canada and Japan

Those who believe the U.S. tort system is in need of reform with respect to its treatment of punitive damages may look to their Canadian neighbors with envy. Punitive-damage awards in Canada are both rare and small.

A study in 1990 reported on punitive-damage awards in Ontario. The researchers found that the highest award was $50,000. The majority of punitive-damage awards were less than $25,000; the median award was approximately 20 percent of the compensatory-damage award in the particular case.

Following English common law, Canadian courts have traditionally restricted punitive damages to two situations: cases involving oppressive, arbitrary, or unconstitutional actions by government servants and cases in which the defendant's conduct was calculated to have made a profit in excess of compensatory damages. In 1989, the Canadian Supreme Court recognized that punitive damages could also be awarded for conduct deserving punishment because of its "harsh, vindictive, reprehensible, and malicious manner."

Two reasons seem to explain the differences in Canadian and American treatment of punitive damages. First, Canadians see something undignified about the flamboyant punitive-damage awards in the United States. Second, civil juries are much less common in Canada, and, in general, judges tend to be much more conservative than juries in making punitive-damage awards.

Another country that would be even more heroic to those opposed to punitive damages is Japan. Several times, the Japanese Supreme Court ruled that punitive damages violated Japan's public policy. Recently, following the lead of the courts, the Japanese legislature forbade the acceptance of punitive damages, even those awarded by foreign courts.

process clause. Because of its unusual facts, *Oberg* did not provide much guidance about when punitive damages were so excessive that they violated due process.

A few years later, however, in *BMW v. Gore*,[15] the Supreme Court set forth a test that a number of commentators thought would substantially curb punitive-damage awards. The Court said three factors should be considered in determining whether an award was grossly excessive: "the degree of reprehensibility of the nondisclosure; the disparity between the harm or potential harm suffered by [the plaintiff] and his punitive damages award; and the difference between this remedy and the civil penalties authorized or imposed in comparable cases." However, a 1999 study found that the year after *BMW v. Gore*, punitive-damage awards across the country were not reduced any more frequently than they had been the year before the decision.[16]

Since *BMW v. Gore*, the Supreme Court has continued to encourage the courts to scrutinize punitive-damage awards carefully. In the 2001 case of *Cooper Industries, Inc. v. Leatherman Tool Group, Inc.*,[17] the high court ruled that appellate courts must review the trial court's decision on the constitutionality of an award *de novo*, meaning they should no longer give deference to the trial court's determination that the jury award was not unconstitutionally excessive and uphold it unless there was a clear abuse of discretion on the part of the trial court judge.

In the 2003 case *State Farm v. Campbell*,[18] the Supreme Court once again addressed the issue of how to determine punitive damages properly. The Campbells had won a jury verdict for $1 million in compensatory damages and $145 million in punitive damages after State Farm failed to settle what was clearly a valid claim. The high court ruled that punitive-damage awards should bear some relationship to the actual harm caused and should not focus on the wealth of the defendants. Although refusing to draw a firm line about what ratio of compensatories to punitives was acceptable, the Court did say that "in practice, few awards exceeding a single-digit ratio between punitive and compensatory damages, to a significant degree, will satisfy due process."[19] The Court also stated that damages should not focus on deterrence based on wealth or on actions unrelated to the case at hand. Case 7-4 demonstrates how the courts are applying these guidelines today.

[15] *BMW of North America v. Ira Gore, Jr.*, 116 S. Ct. 1589 (1995).
[16] James Dam, "Large Punitives Mostly Upheld, but $5B Award Overturned," *Lawyer's Weekly*, November 12, 2001, p. A1.
[17] 532 U.S. 424 (2001).
[18] 123 S. Ct. 1513 (2003).
[19] Ibid., p. 1524.

CASE 7-4

MAZDA MOTOR CORPORATION v. HURST
Alabama State Supreme Court
2017 LEXIS 66

FACTS: *In November 2010 16-year-old Sydney McLemore was driving with her 15-year-old friend Natalie in the passenger seat. Sydney was speeding and lost control of the vehicle, resulting in it spinning into a light pole. The vehicle burst into flames shortly after the collision. Sydney suffered third degree burns but escaped the vehicle. Unfortunately, Natalie did not survive the fire. In 2012, Natalie's parents filed a wrongful death suit against Mazda. The Hursts alleged that the 2008 Mazda3 that Sydney drove was not crashworthy because its fuel system was defectively designed. They argued that the fuel tank was positioned a half-inch from the steel muffler with sharp, protruding edges. The Hursts contended that this design caused the muffler to pierce the fuel tank upon collision and start the fire.*

At trial, the Hursts brought forth several expert witnesses. One expert, Jerry Wallingford, presented a strong case against Mazda. Based upon his decades of experience in the automotive testing industry, he testified that Mazda's fuel tank design goes against industry design safety standards that have been in place since the 1970's. The jury awarded the Hursts $6.9 million in wrongful-death and punitive damages. Mazda appealed. On appeal, Mazda argued that the fine was excessive because Mazda could not have reasonably foreseen the design causing a deadly fire.

ISSUE: Could Mazda have reasonably foreseen the fire risks of its design? Should the plaintiff's $3 million punitive-damage award be struck down as constitutionally excessive when the amount of the compensatory damages is $471,258.26?

REASONING: Mazda's arguments as to why the award should be reduced based on the "guideposts" discussed in *BMW of North America, Inc. v. Gore* are inadequate. Those guideposts are: '(1) the degree of reprehensibility of the defendant's conduct; (2) the ratio of the compensatory damages award to the punitive damages award; and (3) the difference between the punitive damages award and comparable awards in similar cases.'

With respect to second guidepost, the degree of reprehensibility of the defendant's conduct, Mazda's argument that because there is insufficient evidence of wanton misconduct in this case, there likewise is insufficient evidence of "reprehensibility" fails because this guidepost is also treated differently in the wrongful-death context. In these cases, the jury is authorized to award punitive damages on a negligence claim when other factors exist. In Alabama, the legislature has authorized the jury to ascertain an amount of damages appropriate to the goal sought to be achieved — preservation of life because of the enormity of the wrong, and the jury's consideration of the 'enormity of the wrong' includes assessing the finality of death, the propriety of punishing the wrongdoer or wrongdoers, whether the death could have been prevented, and, if so, the lack of difficulty that would have been involved in preventing the death, as well as the public's interest in deterring others from committing the same or similar wrongful conduct.as in this case where the plaintiffs demonstrated that it would not have been difficult to avoid the defect found by the jury to have caused the fire in question, the second guidepost is satisfied.

With respect to the second guidepost, the degree of reprehensibility of the defendnt's conduct, Mazda contends that "the punitive-to-compensatory 'ratio' counsels remittitur." The ratio facto does not apply in this case because in Cherokee Electric Cooperative, this Court observed that "Alabama law allows no compensatory damages in a wrongful death case."

Finally, with respect to the third guidepost, the difference between the punitive damages award and comparable awards in similar cases, Mazda's argument that the jury's punitive-damages award bears no relation to any civil penalty, sanction, or jury verdict in a comparable case because there haven't been any cases against Mazda fails because there have been comparable wrongful death cases where such awards have been upheld even when the defendant has no history of the defendant's being the subject of other actions. Becuase the punitive damages award is not out of line with awards in other wrongful death cases, this guidepost has been satisfied.

DECISION AND REMEDY: Because none of the factors Mazda raises weighs in favor of a remittitur, the court concludes that the trial court did not err in upholding the Hursts's award of $3.9 million for the death of their daughter.

SIGNIFICANCE OF THE CASE: This case provides an illustration of the application of the Gore guideposts to determine when punitive damages may be considered excessive.

CRITICAL THINKING

What evidence, had it existed, do you think might have convinced the justices to overturn the judgment of the lower court?

ETHICAL DECISION MAKING

What values are guiding the judge's decision that the punitive-damage award was not excessive? Do you think that the values promoted by the decision are appropriate for the situation? Why or why not?

CASE OPENER WRAP-UP

Plastic Surgeon Defamation

On appeal, Dr. Bongiovi argued that Dr. Sullivan was a public figure and, as a result, the jury should have been instructed that actual malice had to be shown. The supreme court of Nevada held that Sullivan did not qualify as a public figure because he "did not voluntarily interject himself into a public [medical] controversy." Additionally, the court found that the awarded compensatory and punitive damages were proper. Sullivan had lost income as the result of Bongiovi's statement, and the compensatory damages were equivalent to the income lost. According to the court, the punitive-damage award was "reasonable and proportionate to the amount of harm to Sullivan" and would adequately serve as a deterrent. The district court ruling was affirmed.

PRACTICAL TIPS for Business Managers

1. As tempting as it may be to badmouth your competitors and try to tarnish their reputation, you must be careful. Lying about your competitors in order to make them look bad can land you in court for libel or slander and cost you enormous amounts of money.

2. When advertising your business or publishing business-related materials, be sure that the materials are either entirely original or that you have the proper permissions to use them. It can be very easy to appropriate someone's image without necessarily knowing you did so illegally.

3. As a business manager, you need to not only worry about your customers's safety but your employees's safety as well, especially if your business involves potentially hazardous workplaces. To avoid negligence suits from injured employees, establish strict safety protocols and do daily checks to ensure that these protocols are being followed.

4. If your business involves high-stakes endeavors (e.g., an accounting firm or a medical clinic), it may be wise to invest in professional liability insurance. That way, if something happens to go wrong, you are not at risk of losing everything from one bad lawsuit.

SUMMARY

Introduction to Tort Law	*Tort:* A civil wrong giving the injured party the right to bring a lawsuit against the wrongdoer to recover compensation for the injuries.
Classification of Torts	Torts are classified as intentional, negligent, and strict liability.
Intentional Torts	*Torts against persons* include assault, battery, and defamation. *Torts against property* include trespass to realty, private nuisance, trespass to personal property, and conversion. *Torts against economic interest* include disparagement, intentional interference with contract, unfair competition, misappropriation, and fraudulent misrepresentation.
Negligence	The plaintiff must prove four elements: duty, breach of duty, causation (actual and proximate), and damages.
Plaintiff's Doctrines	Doctrines used to establish negligence: • *Res ipsa loquitur* • Negligence per se
Defenses to Negligence	• Contributory negligence • Modified comparative negligence • Pure comparative negligence • Assumption of the risk
Strict Liability	Persons who engage in activities that are so inherently dangerous that no amount of due care can make them safe are strictly liable, regardless of the degree of care they used when undertaking the activity.
Damages Available in Tort Cases	*Compensatory damages* to make the plaintiff whole again. *Punitive damages* to punish the defendant and deter future wrongdoers. *Nominal damages* to recognize that the defendant committed a tort against the plaintiff.

Point/Counterpoint

Should the Creators of Violent Media Owe a Duty of Care to Victims of Crimes Based on Violent Media?

YES	NO
Some people like to think that we are totally in control of who we are, that we are the only ones who decide what goes on in our own minds and the sole decision-makers in our lives. However, that isn't the full story. We are heavily influenced by the world outside our minds, from the books we read, to the people we converse with, to the movies we watch. It's simply not reasonable to argue that interacting with violent media has zero influence on our decisions. Can violent movies or video games pick up a weapon and kill someone? Of course not. But can violent media influence growing minds toward violent actions? This question isn't so open-and-shut. When the glorified heroes in our films and television shows	When a person or persons create media that contains scenes of violence, they rarely, if ever, have the intent of, "I sure hope someone who interacts with my work commits violent acts against others after the interaction." Not only is said intent not present, but how can media creators foresee violent actions that may result from their media? When someone creates a film containing violence for the purpose of entertaining others and then a viewer murders someone after seeing the film, it's unreasonable to put blame on the creator. The creator cannot account for the intents, mental states, personalities, etc., of every potential viewer of the media, and it's unreasonable to expect a creator to do so.

regularly use violence, or when our increasingly realistic video games reward players for how efficiently they can kill people, it's not too far-fetched to say that these media are teaching children to be violent or, if nothing else, to become desensitized to real-life violence. Is it really reasonable to argue that real-life violence is not foreseeable as a possible consequence of creating and distributing violent media?

In addition, creators of this media should bear some kind of responsibility for crimes with violent media influence because, unfortunately, most people will not take any responsibility for moderating the media that young minds consume. For example, a 2012 survey given by the UK-based gaming website playr2.com revealed that 64 percent of parents do not check age recommendations on video games before purchasing them for their children. Most parents do not monitor what their children watch or play, and, as a result, these children can be exposed to horribly violent media before they fully understand the implications of what they are interacting with. We cannot force parents to stop allowing their children to view such media, so unfortunately some of the burden of responsibility for crimes derived from violent media has to be on the creators of said media.

Finally, media creators, realizing the effect their media could potentially have on those who experience the media, have the choice to not create media that includes violence. There are countless examples of successful films, books, television shows, video games, etc., that contain absolutely no violence. Violence is not needed to sell a product. So, by creating media containing violence, a creator is accepting the risk that their product could inspire someone to commit a violent act, and that the creator can therefore be held responsible in some way.

Essentially, putting the burden of duty of care on creators of violent media will serve no other purpose than to unreasonably censor said creators. If media creators are held responsible for violent acts derived from violent media, then the only choice is to start creating media that includes little-to-no violence so that it cannot be claimed that a violent scene caused real-life violence. While this forced self-censorship would not directly violate First Amendment speech rights, it's still a massive impedance on the free speech of media creators and unreasonably unfair.

Finally, putting the burden of responsibility on the creators of media ignores the responsibilities of other involved parties. Should the parents of a child who commits a violent crime after viewing violent content not be held responsible for allowing their child to view such content? The parents should be the ones who control what media their children interact with. It's not the responsibility of a media creator to actively prevent certain people from viewing their media.

Questions & Problems

1. Distinguish the three types of damages available in tort cases.
2. Explain why some people see punitive damages as a necessary aspect of our tort system, whereas others want to restrict their availability.
3. List five intentional torts and explain the elements needed to prove each.
4. List and define the elements that are necessary to prove a case of negligence.
5. Explain the differences between contributory and comparative negligence.
6. Explain the relationship between negligence per se and *res ipsa loquitur*.
7. After a nonparty to the lawsuit allegedly told defendant that plaintiff was gay or bisexual, defendant relayed that information to a third-party, a close family friend of plaintiff's longtime girlfriend, with the hope that the girlfriend would be told. Plaintiff, maintaining that defendant's actions caused the deterioration and ultimate termination of his relationship with his girlfriend, filed an action alleging slander per se. The defendant filed a motion to dismiss, arguing that the case was legally insufficient due to plaintiff's failure to allege special damages, which involve "the loss of something having *economic* or *pecuniary* value." Do you think that the court ultimately found that the accusations of homosexuality or bisexuality amounted to defamation per se? Why or why not?

 Source: *Yomanty v. Mincola,* 97 A.D.3d 141, 2012 N.Y. App. Div. LEXIS 5397 (2012).

8. Former Major League Baseball player and radio game announcer Bob Uecker sought an injunction against Ann Ladd, alleging a six- or seven-year pattern of

harassment. Ladd, who claimed to be a devoted fan, was charged with felony stalking. An injunction was issued against her bothering Uecker, but the stalking charge was dropped when she agreed to cooperate. Later, Ladd sued Uecker for defamation, claiming that Uecker defamed her in the material he filed in support of his application for the injunction. The trial court dismissed her suit, and she appealed. What do you think the outcome of her appeal was and why?

Source: *Ladd vb. Uecker,* 780 N.W.2d 216 (Ct. App., Wisc., 2010).

9. The NFL hired the law firm of Paul, Weiss, Rifkind, Wharton & Garrison LLP and one of its partners, Theodore Wells to investigate allegations of bullying within the Dolphins organization. After receiving the firm's report, the Dolphins fired then offensive line coach James Turner. The investigation had centered on the bullying of a football player, Jonathan Martin, who had abruptly left the Dolphins team midway through the 2013 season. After several months of investigation, Paul, Weiss published a 144-page report that concluded that bullying by other Dolphins players contributed to Martin's decision to leave the team. The Report also included several references to Coach Turner and opined that Coach Turner's unprofessional conduct, including discouraging players from "snitching" on one another, played a role in Martin's struggles. After his termination, Turner sued Wells and the law firm that conducted the investigation for defamation. Defendants filed a motion to dismiss on grounds that (1) the report consisted of opinions and therefor was not actionable in a defamations case; (2) the complaint failed to identify any false statements of fact in the report; and (3) plaintiff was a public figure and failed to plead actual malice in the complaint. The trial court granted the Defendants' motion to dismiss and plaintiff appealed. How do you think this case handled by the appellate court and why? Why do you believe Turner was or was not a public figure?

Source: Turner v. Theodore V. Wells, Jr., Paul, Weiss, Rifkind, Wharton & Garrison, LLP, Case No. 16-15692 (11th Cir. Court of Appeals, 2018), available at https://cases.justia.com/federal/appellate-courts/ca11/16-15692/16-15692-2018-01-18.pdf?ts=1516303835.

10. Floyd Mayweather, Jr., and Shantel Jackson were a highly publicized couple for a number of years before their relationship frayed and eventually ended in discord. In early April 2013, Jackson decided to end her relationship with Mayweather. However, Mayweather convinced Jackson to return to his home in Las Vegas. After a few days, the couple resumed fighting and Jackson attempted to leave. Jackson alleged that Mayweather declared he would not let her leave. According to her, Mayweather would only let her leave the Las Vegas abode under the supervision of one of his employees.

Jackson eventually moved back to Los Angeles in June 2013. In November 2013, she became pregnant by Mayweather. A December 2013 sonogram revealed twins. A month later, she terminated the pregnancy. In April 2014, Mayweather demanded Jackson take down photographs she had posted on social media that featured her and a rapper at a basketball game. After she refused, Mayweather made posts on his Facebook and Instagram accounts, "the real reason me and Shantel Christine Jackson @ MissJackson broke up was because she got an abortion, and I'm totally against killing babies. She killed our twin babies. #ShantelJackson #FloydMayweather #TheMoneyTeam #TMT." In addition to the post, Mayweather posted pictures of the sonogram and a medical report regarding the pregnancy.

In September 2014, Jackson filed suit against Mayweather for, among other claims, invasion of privacy through the public disclosure of private facts and false light. Do you believe she has a basis for either of these claims?

Source: Jackson v. Mayweather, 10 Cal. App. 5th 1240 (2017).

11. In May 2009, Bryan Lyles was admitted to the LSU Health Sciences Center and underwent an anterior corpectomy and discectomy surgery. During the procedure, Lyles's C5 vertebrae was replaced with a Verte-Stack implant. Additionally, an Atlantis Plate was inserted to hold the Verte-Stack in place and promote its union with the C4 and C6 vertebrae. Lyles was discharged the day after the operation after having several x-ray images taken.

About a week later, Lyles returned to the hospital, complaining that the condition he had undergone the surgery for had not abated. Additional images of Lyles's spine were taken, and the doctor who conducted the surgery found slight displacement of the Atlantis Plate but no breakage. Lyles symptoms showed no improvement, and he eventually underwent another surgery.

Regardless of the surgery's results, Lyles sued MSD, the company that designed and created the Atlantis Plate, for defective design and defect claims under the assertion that the Atlantis Plate had broken. To support his claim, medical experts hired by Lyles alleged that the images taken shortly after the first surgery and after Lyles returned the first time showed that the Atlantis Plate had come apart. The surgical doctor maintained that at no time did the Atlantis Plate break.

Lyles made his liability claim under the theory of *res ipsa loquitur*. What kind of evidence would Lyles have to present to satisfy this theory? Do you think Lyles's expert's opinions are sufficient for him to win the case? How did the court rule?

Source: Lyles v. Medtronic Sofamor Danek, USA, Inc., 871 F. 3d 305 (2017).

12. At 2:08 a.m., on Friday, December 29, 2006, the defendant received by email the Lake in the Hills police department Daily Bulletin, and it reported that Carolene Eubanks had been charged with theft and obstruction of justice. As it normally does, the newspaper placed the information from the report in an article and then placed the article in line for publication in the upcoming issue of the newspaper. The article was eventually printed before 6 a.m. on January 2, 2007, and appeared in the defendant's newspaper on the same date.

The Lake in the Hills police department had also sent a second email on December 29, 2006, at 10:25 p.m., that said to remove the name of Carolene Eubanks and replace it with the name of Barbara Bradshaw. Because the employee who posted the article had already gone home and it was a long holiday weekend, no one was in when the second email arrived, so it wasn't read until January 2, 2007, at 10:17 a.m.

Consequently, the January 2, 2007, edition of the defendant's newspaper, printed before 6 a.m. on that date, published the article indicating that the plaintiff had been arrested and charged with theft and attempted obstruction of justice. The defendant published a retraction of the article in its January 3, 2007, newspaper, stating that the plaintiff was not the one charged with those crimes.

On June 15, 2007, the plaintiff filed her complaint for defamation based on the January 2, 2007, publication. The trial court granted the defendant's motion for summary judgment based on its exercise of a privilege. Plaintiff appealed. What privilege do you think the court relied on? What do you think the outcome of the appeal was and why?

Source: *Eubanks v. Northwest Herald Newspapers,* 397 Ill.App.3d 746, 922 N.E.2d 1196 (2010).

13. Defendant McPherson owned property adjacent to the first Presbyterian Church and wanted to build a gas station on the property. The church sued, arguing that the operation of the station would constitute a nuisance. The lower court found that the unusual noises emanating from the station would interfere with the church's operation to such an extent that it would be a nuisance and therefore granted the permanent injunction.

On appeal, do you believe the injunction was upheld? Why or why not?

Source: *McPherson v. First Presbyterian Church of Woodward et al.,* 120 Okla. 40; 248 P. 561 (2015).

14. Plaintiff withdrew $4,000 in cash from his T.D. Bank account on August 28, 2014, and proceeded to dinner at Prime restaurant. He left 36 one hundred-dollar bills in the center console of his car and gave exclusive custody and control of his vehicle to employees of the defendant restaurant valet service, Advanced Parking Services. The plaintiff retrieved his vehicle one hour and 40 minutes later, only to discover that $3,600 in cash was missing. The plaintiff filed a small claims action to recover said sum, asserting a number of causes of action, including negligence.

Regarding the negligence claim, defendants produced no witnesses or evidence, but claimed to have no knowledge concerning the matter. Plaintiffs argue that, using the doctrine of *res ipsa loquitur,* the happening of the loss permits, but does not compel, the court to draw an inference that the defendants were negligent. They claimed: the instrumentality that caused the accident was in the exclusive control of the defendant; the event does not ordinarily occur in the absence of someone's negligence; and the event was not due to any voluntary action or contribution on the part of the plaintiff. Plaintiff's sworn testimony and exhibits excluded the possibility that the plaintiff himself caused or contributed to the loss, as he was in the restaurant at the time the money disappeared. How do you believe the course resolved this dispute? Why?

Source: *Zinner v. Advanced Parking Servs.* (2015). https://law.justia.com/cases/new-york/other-courts/2015/2015-ny-slip-op-25246.html.

Part 1 The Legal Environment of Business

CHAPTER 8

Real, Personal, and Intellectual Property

CASE OPENER

Smartphone Wars: Focus on Technology Companies

In the past five years, there have been significant innovations in technology such as smartphones and tablets. Technology companies rely on intellectual property law (IP), such as patents, trademarks, copyright, and trade secrets, to protect their innovations. For example, Apple and Samsung, the two largest smartphone companies, have filed various patent infringement suits against each other in courts around the world. Apple and Samsung are strong competitors, so if one company were able to obtain an injunction preventing the other company from selling certain phones and tablets, this injunction could be devastating to the losing company. In one of these cases that proceeded to a jury trial, Apple argued that Samsung copied Apple's design of the iPhone and iPad—specifically, that Samsung copied the rounded rectangle shape of the iPhone, in violation of Apple's design patent and the iPhone's trade dress. Steve Jobs, the former CEO of Apple, emphasized the importance of neat, clean design for Apple products, and Apple has thus sought and received hundreds of design patents, along with trademarks, protecting its products. Similarly, Samsung has asserted a number of its own patents against Apple, including a patent covering the ability to send photos from a camera phone.

Critics argue that these patents and trademarks are too broad (i.e., that something as simple as a rounded rectangle or sending a photo from a phone should not be protected by patents or trademarks). Critics further argue that such broad IP stifles innovation and prevents smaller players from entering the market because they cannot withstand the costs of litigation associated with competing in the smartphone and tablet markets. In response, the patent owners argue that they have expended significant resources in creating their products and deserve to avail themselves of the protection afforded by IP. Critics further argue that technology companies spend millions of dollars on litigation rather than on developing and improving products. Technology companies respond that they have no choice but to seek enforcement of their intellectual property through the legal system.

Furthermore, in January 2013, Apple Computer was granted a trademark for its retail store design and layout.[1] Specifically, the trademark covers the design, the shelves in the store, and the location of the tables displaying products. One of the main reasons this trademark was important to Apple was that fake Apple stores in China confuse customers.

1. Do you think technology companies should be able to receive protection for allegedly basic ideas (e.g., store design or shape or color of a product) and functionalities?
2. How does intellectual property litigation between competitors (such as Apple and Samsung) affect you?

[1] Erin Geiger Smith, "Apple Trademarks Design of Its Retail Stores," Reuters, January 29, 2013.

LEARNING OBJECTIVES

LO 8-1 Explain how the topics of real, personal, and intellectual property are related.
LO 8-2 Describe how real property law balances private and public rights.
LO 8-3 Outline the interests in real property that someone can hold.
LO 8-4 Identify the types of personal property.
LO 8-5 Describe how personal property is transferred.
LO 8-6 Describe how intellectual property law balances private and public rights.
LO 8-7 Differentiate among the methods for protecting various types of intellectual property.

Most Americans cherish the right to own property. For most Americans, this means the right to own real property, such as a house. Increasingly, Americans also cherish the right to own intellectual property, the fruits of the mind, such as a manuscript for a novel. In this chapter, we examine the nature of both real and intellectual property, as well as personal property. The chapter starts by exploring how the three areas of law are related. Then, for each of these types of property, the chapter considers what interests individuals can hold and how the laws governing these types of property balance private and public rights. The Case Opener considers an intellectual property issue. By the end of the chapter, you will see that disputes such as the Samsung–Apple dispute can arise not only when the property in question is intellectual but also when it is personal property.

The Nature of Real, Personal, and Intellectual Property

LO 8-1 Explain how the topics of real, personal, and intellectual property are related.

real property
Land and everything permanently attached to it.

Personal property
Tangible, movable objects.

intellectual property
Intangible property that is the product of one's mind and not one's hands.

The type of ownership interest a person has in a piece of property determines his or her rights to the property. **Real property,** commonly referred to as *realty,* is land and everything permanently attached to it. **Personal property** consists of tangible, movable objects. **Intellectual property** consists of the fruits of one's mind. The laws of intellectual property protect property that is primarily the result of mental creativity rather than physical effort. When we are thinking about property rights of any kind, it is important to remember that society, through laws, decides what rights individuals and communities have with regard to property.

One metaphor often used with regard to property is that of a bundle of sticks. Each stick represents a distinct right an individual can have with regard to property. The rights to possess property, use it, give it away, or take it away are all sticks. With regard to all types of property, one of the most important rights is the right to exclude others. For example, with regard to real property, an individual can put a fence around the land and ask others to stay off the land. With personal property, you may simply keep it in a place where no one else has access to it. For intellectual property, an individual can accomplish the same thing with a trademark or patent—the holder can put a virtual fence around the idea trademarked or patented and ask others to refrain from using the property.

With regard to real property, some other important sticks are the rights to sell, lease, or mortgage land and the right to allow someone limited access. The public also has important sticks,

The Nature of Real, Personal, and Intellectual Property

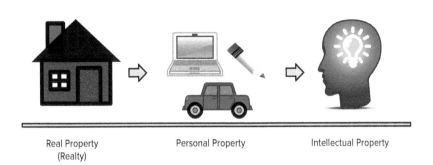

Exhibit 8-1

The Nature of Real, Personal, and Intellectual Property

Real Property (Realty) Personal Property Intellectual Property

such as the right to take land, regulate it, and/or tax it. In the next section, the various types of interests a person may have in real property are described in detail. These interests may be conveyed or transferred under legal guidelines for property rights. Most transfers of property are voluntary. However, for the benefit of the public, and to protect public health, safety, and welfare, the government may require involuntary transfers of property.

Real Property

LO 8-2 Describe how real property law balances private and public rights.

There are a number of types of interests in land, ranging from temporary to permanent to future. The duration of one's ownership interest and the power one has over using the land depend on the type of *estate* one is said to hold. The following section on interests in real property describes the various estates in detail. First, though, this chapter explores the landowner's rights that pertain to areas beyond the surface of the land itself.

Extent of Ownership

When a person owns land, she or he owns more than just the surface. The airspace above the land, extending to the atmosphere, is also part of the legal concept of real property. Rights to airspace generally do not generate much controversy, but occasionally disputes arise that involve aircraft flying too low over individuals's property. Alternatively, a tree's branches may hang over into the airspace of the property next door, and the owner of that airspace may want to cut the overhanging branches.

In dense, commercial urban areas, airspace may actually be an asset. Owners of two commercial buildings might want to build an overhead walkway joining their buildings across a parking lot that you own. They will have to pay you handsomely for the right to build in your airspace.

In addition to having airspace rights, the owner of real property also has *water rights,* the legal ability to use water flowing across or underneath the property. However, these rights are somewhat restricted; one cannot deprive the landowners downstream from the use of the water by diverting the water elsewhere.

Finally, ownership of real property extends to *mineral rights;* these rights involve the land below the surface. The landowner has the legal ability to dig or mine the materials from the earth, and he may sell or give these rights to another person. Ownership of these *subsurface rights* includes the right to enter onto the property to remove the underground materials.

All these additional rights are illustrated in Exhibit 8-2.

Exhibit 8-2

Extent of Rights

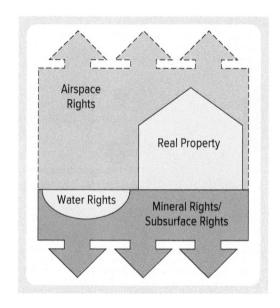

Interests in Real Property

LO 8-3 Outline the interests in real property that someone can hold.

Various interests in land are discussed in the following sections and summarized in Exhibit 8-3.

fee simple absolute
An ownership interest in which the holder has exclusive rights to ownership and possession of the land; the most comprehensive type of estate.

Fee Simple Absolute With regard to estates in land, a **fee simple absolute** is the most complete estate a person may have. Exclusive rights to ownership and possession of the land belong to the person who has a fee simple absolute. It is this type of estate to which most people refer when they speak of buying a house or piece of land. A fee-simple-absolute interest is passed to the owner's heirs when the owner of such an estate dies.

conditional estate
An ownership interest in which the holder has the same interest as that in a fee simple absolute except that this interest is subject to a condition.

Conditional Estate The owner of a **conditional estate** possesses the same interest as the owner of a fee simple absolute except that this interest is subject to a condition: If a certain required event fails to occur or a specific prohibited event occurs, the interest will be terminated. For example, Todd may be given property rights to a Victorian house on the condition that he is to preserve it in its original form. If he violates this condition by turning the house into a piano showroom or a beer hall, the house will either revert to the original owner or be transferred in accordance with the terms of the deed. (The *deed* is the instrument used to convey real property.)

life estate
An ownership interest in which the holder has the right to possess the property until her or his death.

Life Estate A **life estate** is granted for the lifetime of an individual; the right to possess the property terminates at the individual's death. On the death of the holder of the life estate, the property will go to another party as designated by the original grantor. Thus, this future owner has an interest in seeing that the life tenant does not *waste* the property; in other words, the life

Exhibit 8-3

Possessory Estates in Land

INTEREST	EXTENT OF RIGHTS
Leasehold estate	Right to possess property for an agreed-on period of time
Life estate	Right to possess property until death, subject to restriction against waste
Conditional estate	Right to possess property until death, subject to restriction against waste
Fee simple absolute	Right to possess property for life and pass it to heirs on death; the most encompassing interest

GLOBAL Context

Property Interests in Vietnam

©Digital Archive Japan/ Alamy

In the United States, we take for granted the right to go out and purchase a piece of property if we have the money to do so. Property is not so freely available and transferable in all nations. Vietnam's new constitution, written in 1992, provides guidelines for the allocation, transfer, and sale of private property. However, the constitution still asserts that the people, or the state, own all the land. Thus, if individuals or private enterprises want to use land, they must pay tax on it as a form of rent. Those who agree to pay the tax on the land are granted a use of right, which entitles them to extended use. The use of right also gives the renter the freedom to transfer the property. Technically, it is not the property that is being transferred but, rather, the right to use the property.

Transference of property can occur only with the approval of a state official. The official ensures that the new owner intends to use the land for the original, state-approved purpose. Moreover, the new owner can never be given a longer term of right or more extensive rights over the land than the original owner had. Finally, the state official determines the price at which the property will be transferred. The government has specified certain prices depending on how the land is used. Considering the involvement of the state in such property transactions, the extent to which the Vietnamese enjoy private property rights is limited.

holder is not allowed to neglect or abuse the property in a way that diminishes the property value. If the life tenant fails to make necessary repairs or uses the property in ways damaging to its future value, the future holder could bring legal action against the holder of the life estate to recover damages for waste.

Future Interest A future interest is a person's present right to property ownership and possession in the future. Such an interest usually exists in conjunction with a life estate or a conditional estate. For example, Joe owns a life estate in Oak Hills Apartments, and on his death the property will pass to Sarah with fee-simple-absolute rights. Sarah, in this case, holds a future interest in Oak Hills Apartments, and if Joe allows the buildings to deteriorate from neglect, Sarah may sue Joe to enjoin him from engaging in waste of the property.

> **future interest**
> A person's present right to property ownership and possession in the future.

Leasehold Estate The interest of a leasehold differs from the interests previously described in that the holder of such an estate has a possessory interest but not an ownership interest. This interest is transferred by a contract known as a *lease*. Both the owner of the property (the lessor, or landlord) and the tenant (the lessee) sign the lease. The contract generally specifies the property to be leased, the amount of the rent payments and when they are due, the duration of the leasehold, and any special rights or duties of either party. A leasehold gives the lessee, or tenant, exclusive rights to the use and possession of the land, including the right to exclude the property owner under most circumstances, for the term specified by the lease.

> **leasehold**
> A possessory interest, but not an ownership interest, transferred by contract.

The tenant's and landlord's rights and obligations may vary according to the lease, but some states have statutes requiring landlords to keep the property in good condition for the tenant's use and giving the tenant the right to withhold rent payments if the landlord fails to maintain the premises. However, if the tenant fails to make the agreed-on payments without such grounds, the landlord may evict the tenant. The landlord is not allowed to enter the property except when there is an emergency or when the tenant has given permission to enter to make repairs. Near the end of the leasehold, the landlord may enter with notice to the tenant for the purpose of showing the property to a potential tenant.

Subleasing of the property by the tenant to another party is permissible unless specifically prohibited by the lease. However, the initial tenant is liable throughout the entire term of the lease for payment of the rent to the landlord.

Nonpossessory Estates Although most people think of interests in land as being possessory in nature, some estates do not include the right to possess the property. Such interests include easements, profits, and licenses. Easements and profits are similar in that they are neither

easement

An irrevocable right to use some part of another's land for a specific purpose without taking anything from it.

profit

The right to go onto someone's land and take part of the land or a product of it away from the land.

ownership nor possessory interests, and they are subject to similar rules, but they are not exactly the same. An *easement* is an irrevocable right to use some part of another's land for a specific purpose without taking anything from it. A *profit* is the right to go onto someone's land and take part of the land or a product of it away from the land. For example, if Bill has the right to drive his car across Jenny's property to get to his property, he has an easement; but if he has the right to go onto her property and remove the topsoil he needs from it for his landscaping business, he has a profit. Easements and profits can be transferred in multiple ways: by express agreement, inheritance, necessity, implication, or prescription. If the land that benefits from the easement or profit is sold, the nonpossessory interest goes with the property. Thus, if Bill sold his land to Sonny, Sonny would also receive the easement across Jenny's property. Similarly, if Jenny sold her property, the new owner would have the burden of Bill's easement as long as it had been properly recorded. Of course, just as easements and profits can be created, they can also be terminated, with the most common method of termination being by agreement. The easement holder may simply deed the easement back to the property owner. If the easement arose by necessity and the necessity no longer exists, the easement would terminate.

license

A revocable right to use another's property temporarily.

A *license* is a right to use another's property that is both temporary and revocable. For example, when a person purchases a theater ticket, he or she has the right to a specific use of the property for a limited time, subject to good behavior. No property interest goes to the license holder.

Property Transfer

The owner's ability to transfer real property is part of the value of property. Any or all of the owner's property may generally be transferred to anyone for any price or no price, as the owner desires.

However, to effectuate such a transfer, the owner must follow certain legal procedures. These procedures are *execution, delivery, acceptance,* and *recording* (see Exhibit 8-4), the last of which is required to protect the recipient of the property. Such a transfer is presumed to be the conveyance of a fee simple absolute unless the contrary is stated.

The transfer of the owner's interest in real property is not always voluntary. Transfer of property may also take place without the owner's knowledge and even, in some cases, against his or her will by either adverse possession or condemnation. In *adverse possession,* a person takes ownership of property by treating a piece of real property as his or her own, without protest or permission from the owner. Most states have established statutes determining a length of time after which such a possessor receives the ownership interest in the property. The exact lengths of

adverse possession

An involuntary property transfer in which a person acquires ownership of property by treating a piece of real property as his or her own, without protest or permission from the owner.

Exhibit 8-4

Steps in a Voluntary Transfer

time vary from state to state, but adverse possession must be *actual* (the person lives on or uses the land as an owner would), *open* (not secretive), and *notorious* (without the owner's permission). In some states, the adverse possessor must have performed certain acts such as the payment of real estate taxes. In other states, such possession must have taken place under color of title; in other words, the adverse possessor was operating under the assumption that he or she actually held the title to the land.

Condemnation is the legal process by which a transfer of property is made against the protest of the property owner. As you will remember from Chapter 5's discussion of the takings clause of the Fifth Amendment, the government has a constitutional right to take private property for the use of the public provided it pays the owner fair compensation. This right may be exercised at any government level, and sometimes it is exercised on behalf of a private company operating to benefit the public.

condemnation
The legal process by which a transfer of property is made against the protest of the property owner.

When the government decides to exercise its power of eminent domain, it will first make an offer to purchase the property for what it believes is the fair market value. If the owner either does not want to sell or feels the price is too low, the government will initiate a condemnation proceeding, and the court will determine whether the government body bringing the action has a valid claim that the taking is for a legitimate public purpose. If the court determines that, in fact, the taking of the property will benefit the public interest, the court will then determine the fair market value of the property. This price will be paid to the owner, and the property will be transferred to the government.

It is sometimes difficult to determine whether the exercise of eminent domain is actually for the public's benefit when the property will be conveyed by the government to another private individual. Since the 1980s, businesses increasingly have been appealing to cities to use their power of eminent domain to assist the businesses in acquiring land that the owners have not wanted to sell. Because this transference of property from one private individual to another is for the alleged public use of economic development, an increasing number of state supreme courts began to address the issue of whether economic development was a public use under the constraints of the Fifth Amendment. Some state supreme courts found that job creation and expansion of the tax base did constitute public use, whereas other courts found that they did not constitute public use. The U.S. Supreme Court finally settled the issue—at least for the present time—in the 5-4 decision summarized in Case 8-1.

KELO v. CITY OF NEW LONDON
United States Supreme Court
126 S. Ct. 326 (2005)

FACTS: *New London, a city in Connecticut, used its eminent domain authority to seize private property to sell to private developers. The city said developing the land would create jobs and increase tax revenues. Susette Kelo and others whose property was seized sued New London in state court. The property owners argued that the city violated the Fifth Amendment's takings clause, which guarantees that the government will not take private property for public use without just compensation. Specifically, the property owners argued that taking private property to sell to private developers was not public use. The Connecticut Supreme Court ruled for New London.*

ISSUE: Does a city violate the Fifth Amendment's takings clause if the city takes private property and sells it for private development, with the hopes the development will help the city's bad economy?

REASONING:

- The city's taking of private property to sell for private development qualified as a public use within the meaning of the takings clause.
- The city was not taking the land simply to benefit a certain group of private individuals but was following an economic development plan. Such justifications should be given deference.
- The takings here qualified as public use even though the public would not use the land.

(continued)

- The Fifth Amendment did not require literal public use, the majority said, but the "broader and more natural interpretation of public use as 'public purpose.'"

O'Connor's dissent:

- An unelected (therefore voter-unaccountable) private nonprofit corporation was the primary beneficiary of the government taking.
- "Any property may now be taken for the benefit of another private party, but the fallout from this decision will not be random. The beneficiaries are likely to be those citizens with disproportionate influence and power in the political process, including large corporations and development firms."
- The decision eliminates "any distinction between private and public use of property—and thereby effectively delete[s] the words 'for public use' from the Takings Clause of the Fifth Amendment."

DECISION AND REMEDY: The city has not violated the Fifth Amendment. Affirmed, for New London.

SIGNIFICANCE OF THE CASE: The U.S. Supreme Court expanded significantly what constitutes public use. Many people believe the Supreme Court went too far—they agree with Justice O'Connor's opinion.

CRITICAL THINKING

When you examine the reasoning in both the majority and dissenting opinions, you can see that a significant conflict between the opinions exists because of the ambiguity of a key term. Identify this term and explain how its interpretation affects the reasoning. How do you think the courts should define the term?

ETHICAL DECISION MAKING

Who are the primary stakeholders involved in this case? How are they affected by the ruling? What are the implications of this ruling in terms of the distinction between public and private property? Further, this case highlights an important value conflict. Can you explain how certain values would lead one to support the majority opinion, whereas different values would lead one to support the dissenting opinion? How might holding a different value determine how one predicts the implications of this ruling?

Whereas the *Kelo* case involves government attempts to take property from individuals to sell to companies, sometimes governments seek to take the property of one company for use by another company. For instance, in June 2000, Costco, a large retail chain, wanted to expand its warehouse in Lancaster, California. Afraid of losing the large store, which threatened to move if it could not expand, the city considered using its eminent domain powers to acquire an adjacent 99 Cents Only store so that Costco could obtain the property. This example demonstrates that even businesses may be at risk of losing their property to larger, more powerful businesses.

The response to the *Kelo* decision was widespread. President George W. Bush issued an executive order limiting the taking of private property by the federal government to "situations in which the taking is for public use, with just compensation, and for the purpose of benefiting the general public, and not merely for the purpose of advancing the economic interest of private parties to be given ownership or use of the property taken." During the year and a half after *Kelo,* 30 state legislatures enacted statutes limiting the use of eminent domain. Additionally, in 2006, nine states passed ballot measures restricting the definition of public use for purposes of eminent domain. Thus, it appears that the ultimate impact of the *Kelo* decision will be the opposite of what that decision held; by statute, a much more restrictive definition of public use will govern government takings. The case is also important because it shows how constitutional issues directly affect areas of the law, such as property law.

GLOBAL Context

©Digital Archive Japan/Alamy

Eminent Domain in Germany

Germany has both federal and state *(Lander)* regulations governing eminent domain. The federal system is similar to that in the United States, but one difference is that most of its eminent domain laws are incorporated through statutes, not through cases as in the United States.

The German federal government may expropriate land for the public good as long as it provides compensation. Germany's basic law provides that "Such compensation shall be determined by establishing an equitable balance between the public interest and the interests of those affected." Compensation in the German system is not based on fair market value alone but must also consider the interests of those affected. Germans may dispute the amount of compensation in the ordinary courts.

A taking requires the government to have the legal authority for the taking. The taking must also be appropriate for accomplishing the government's purpose and must be done in the least intrusive way possible, meaning that the government cannot take property before it needs it, and it can take only the property it needs. The taking must also pass a balancing test, showing that the public's interest is greater than the owner's interest.

One other difference in Germany is that if the German government no longer needs the property, it must revert to its former owner. In the United States, once the government has provided just compensation for the property, it belongs to the government.

Personal Property

LO 8-4 Identify the types of personal property.

All property that is not land or not permanently affixed to land is personal property. Personal property may be either tangible or intangible. *Tangible property* is property that can be identified by the senses. It is property that you can see or touch. Tangible property includes items such as furniture, cars, and other goods.

Books are typically thought of as personal property. The owner of a book may write in it and generally do what he or she wants with it, and the law protects the owner against having the book taken by someone else. However, the growing popularity of e-books has raised questions about what rights e-book users have to their e-books. In July 2009, Amazon deleted copies of George Orwell's *1984* and *Animal Farm* directly off people's Kindles (Amazon's e-book reader) and refunded their money. This action has sparked massive controversy over whether Amazon has the right to remove content that customers had bought from their Kindles. At least one lawsuit has already been filed as a result of this incident. Much of the outrage has been because people do not see e-books and physical books as different types of property. Businesses that offer digital content should be clear about what rights customers have to the content, and they should make that information well known and visible to avoid problems like the one Amazon is faced.

Intangible property includes such items as bank accounts, stocks, and insurance policies. Because most intangibles (with the exception of some that are classified as intellectual property and discussed later in this chapter) are evidenced by writings, most of the following discussion applies to both tangible and intangible property. The primary issues that arise in conjunction with personal property involve the means of acquiring ownership of the property, which is discussed in the next section.

Voluntary Transfer of Property

LO 8-5 Describe how personal property is transferred.

Voluntary transfer, as a result of either a purchase or a gift, is the most common means by which property is acquired. Ownership of property is referred to as *title,* and title to property passes when the parties so intend. When transfer of the property is by purchase, the acquiring party gives some consideration to the seller in exchange for title to the property.

Such a transfer of ownership usually requires no formalities, but in a few cases, changes of ownership must be registered with a government agency. Sales of motor vehicles, watercraft, and airplanes are among the primary transfers requiring registration. To transfer such property, a certificate of title must be signed by the seller, taken to the appropriate government agency, and then reissued in the name of the new owner.

Gifts are another voluntary means of transferring ownership. They differ from purchases in that a promise to make a gift is unenforceable. Once properly made, however, a gift is irrevocable.

Three elements are necessary for a valid gift. First, there must be a *delivery* of the gift. Delivery may be actual, which is the physical presentation of the gift itself, or constructive, which entails the delivery of an item that gives access to the gift or represents it, such as handing over the keys to a car. Second, the delivery must be made with *donative intent* to make an immediate gift. The donor makes the delivery with the purpose of turning over ownership at the time of delivery. Third, there must be *acceptance,* a willingness of the donee to take the gift from the donor. Usually, acceptance is not a problem, although a donee may not want to accept a gift because of a desire not to feel obligated to the donor or because of a concern that ownership of the gift may impose some unwanted legal liability.

Sometimes, however, it can be difficult to determine whether something is a gift or a loan, especially when proper documentation is not filed at the time of the transaction. This was the scenario faced by Don Whittington when he disputed ownership of a 1979 Kremer Porsche displayed at the Indianapolis Motor Speedway Hall of Fame Museum. Whittington believed that he was the owner of the Porsche and that he had merely loaned it to the museum in 1980. The Indianapolis Motor Speedway Foundation believed that Whittington had given the car as a gift to the museum, and the foundation had even applied for and been granted a title to the car in 2001. In 2004, when Whittington requested the return of the car, he had to prove that he had a possessory interest in the car. Although the museum did not have a record of receiving the car as a gift or a loan, and Whittington had never applied for a tax reduction for giving the car as a gift and was able to produce one document in which he listed the car as an asset, the trial court found in favor of the foundation. The court determined that Whittington's behavior after the museum took possession of the car was more consistent with giving the car as a gift than as a loan.

The gifts we have been discussing so far have been what are called *inter vivos* gifts, gifts that are made by a person during his or her lifetime. Another type of gift that can be made is a gift *causa mortis,* a gift that is made in contemplation of one's immediate death. It can be revoked any time before the death of the donor, and it is automatically revoked if the donor recovers.

Litigation over gifts *causa mortis* often arises because the three elements of delivery, donative intent, and acceptance still have to occur before the gift is complete, and that means before the death of the donor. It can sometimes be difficult to determine whether in fact all the elements of a gift *causa mortis* have been met.

Involuntary Transfer of Personal Property

Involuntary transfers of ownership occur when property has been abandoned, lost, or mislaid. The finder of such property *may* acquire ownership rights to such property through possession.

Lost property is property that the true owner has unknowingly or accidentally dropped or left somewhere. He or she has no way of knowing how to retrieve it. In most states, the finder of lost property has title to the lost good against all except the true owner.

Mislaid property differs from lost property in that the owner has intentionally placed the property somewhere but has forgotten its location. The person who owns the realty on which the mislaid property was placed has the right to hold the mislaid property. The reason is that the true owner likely will return to the realty looking for the mislaid property.

In some states, the law requires that before becoming the owner of lost or mislaid property, a finder must place an ad in the paper that will give the true owner notice that the property has been found and/or must leave the property with the police for a statutorily established reasonable period of time.

Property that the original owner has discarded is *abandoned* property. Anyone finding such property becomes its owner by possessing it. For example, Karen is on vacation and decides that she doesn't really need her umbrella and doesn't want to carry it the rest of the trip, so she leaves

the umbrella on a seat in the airport. If Bill then picks it up and takes it with him, he is now the lawful owner of the umbrella.

Other Means of Acquiring Ownership of Property

There are additional means by which people acquire title to property. One such means is by creation: If a person creates a piece of property, then he or she owns that property. One exception to this rule occurs when a person is paid to create property for someone else, in which case the property is owned by the person who paid for its creation. Another means of acquiring ownership is by court order. In a number of types of cases, the court will determine who is entitled to ownership of property. For example, in a divorce case, the court may award ownership of certain property to different parties, or in a bankruptcy case, the court may award ownership of certain property to a creditor.

A far less common means of acquiring ownership is confusion, which involves only fungible goods. *Fungible goods* are goods for which one unit of the good is essentially the same as every other unit, such as grains of wheat or gallons of oil. If two people accidentally commingle their fungible goods, or if the goods are commingled because of the actions of a third party, each party is entitled to the percentage of the fungible goods that he contributed.

However, if one of the parties was responsible for the commingling and that person cannot prove what percentage of the commingled goods she contributed, then the innocent party acquires title to all the goods. For example, if a farmer had stored grain in a rented storage elevator and another farmer wrongfully added his grain to the elevator, the innocent farmer would be entitled to the entire amount in the silo.

Intellectual Property

LO 8-6 Describe how intellectual property law balances private and public rights.

Intellectual property, property that comes from creativity, also carries a bundle of sticks, both for the creator of the property and for the community. The creators of intellectual property often want to keep the property to themselves. For example, creators of copyrighted work (creative work in a fixed form) do not want others to copy that work without permission.

At times, intellectual property is more complicated than real property. For example, copyright often has an impact on freedom of speech. If a songwriter creates a song, others may think they have the right to express themselves through this song, but a copyright may prevent this expression.

Thus, copyright law and freedom of speech sometimes clash. Legislators and courts play an important role in responding to potential conflicts. They also balance the rights of innovators who create intellectual property with the interests of the community, especially interests in competition and freedom of expression. Courts and legislators decide what sticks to set aside for creators of intellectual property. They also decide what is left for the public, including artists and other businesses. The following sections describe specific types of intellectual property.

Trademarks

LO 8-7 Differentiate among the methods for protecting various types of intellectual property.

A **trademark** is a distinctive mark, word, design, picture, or arrangement that a producer uses in conjunction with a product and tends to cause consumers to identify the product with the producer. Even the shape of a product or package may be a trademark if it is nonfunctional.

Even though the description of a trademark is very broad, there has still been substantial litigation over precisely what features can and cannot serve as a trademark. In one case, discussed in the Case Nugget, the U.S. Supreme Court grappled with the issue of whether a color can be a trademark.

trademark
A distinctive mark, word, design, picture, or arrangement that is used by a producer in conjunction with a product and tends to cause consumers to identify the product with the producer.

Exhibit 8-5

Logos like those of McDonald's, Nike, and Coca Cola are all examples of well-known trademarks.

©Shutterstock/Rose Carson

CASE Nugget

Color as a Trademark?
Qualitex Co. v. Jacobson Products Co. United States Supreme Court

514 U.S. 159 (1995)

For years, the plaintiff, Qualitex Co., had colored the dry-cleaning press pads it manufactured a special shade of green-gold. When Jacobson Products, a competitor, started coloring its pads the same shade of green-gold, Qualitex sued Jacobson Products for trademark infringement. The defendant challenged the legitimacy of the trademark, arguing that color alone should not qualify for registration as a trademark.

The district court found in favor of the plaintiff, but the Ninth Circuit reversed, holding that color alone could not be registered as a trademark. The plaintiff appealed to the U.S. Supreme Court.

In writing for the majority that a color could constitute a trademark, Justice Breyer said that both the language of the Lanham Act and the basic underlying principles of trademark law would seem to include color within the universe of things that could qualify as a trademark. He noted that the broad language used to describe trademarks included "any word, name, symbol, or device, or any combination thereof." Human beings might use as a symbol or device almost anything at all that is capable of carrying meaning. Justice Breyer noted that a particular shape (of a Coca-Cola bottle), a particular sound (of NBC's three chimes), and even a particular scent (of plumeria blossoms on sewing thread) could be trademarked, and he reasoned that if a shape, a sound, and a fragrance can act as symbols, it made sense that so could a color.

He also noted that a color, if unusual enough in the context, could in fact come to identify goods with their source, just as descriptive words on a product could. The Court concluded that the green-gold color acts as a symbol that has developed secondary meaning (i.e., customers identified the green-gold color as Qualitex's) and identifies the press pads's source. The high court therefore reversed the decision in favor of the plaintiff.

Seventeen years later, courts considered whether a color can serve as a trademark in the fashion industry. Christian Louboutin designs and sells high-fashion shoes with red lacquered soles, and the red lacquered sole was registered as a trademark in 2008. In 2011, Yves Saint Laurent America, Inc. (YSL), planned to offer a line of monochromatic shoes whose upper part matched

the color of the sole, and one of these shoes had a red upper portion along with a red sole. When Louboutin learned of the red YSL shoe, Louboutin sued YSL for trademark infringement.

Relying on *Qualitex,* the district court held that a single color could never serve as a trademark in the fashion industry because single color marks are inherently functional. On appeal, the Second Circuit determined that there was nothing about the fashion industry that should warrant treatment different from other industries. However, the Second Circuit determined that Louboutin had put the color red "in a place that seemed unusual" *(Qualitex,* 514 U.S. at 162). Based on the evidence in the record, Louboutin's trademark should be limited to a red lacquer sole that *contrasts* with the upper portion of the shoe. Thus, YSL was permitted to sell its monochromatic shoe.[2]

A trademark used *intrastate* is protected under state common law. To be protected in *interstate* use, the trademark must be registered with the U.S. Patent Office under the Lanham Act of 1947. If a mark is registered, the holder of the mark may recover damages from an infringer who uses it to pass off goods as being those of the mark owner. The owner may also obtain an injunction prohibiting the infringer from using the mark. (Only the latter remedy is available for an unregistered mark.) Once the mark is registered, the registration must be renewed between the fifth and sixth years. After that initial renewal, the mark holder must renew the registration every 10 years. (If the mark was initially registered before 1990, however, renewal is necessary only every 20 years.)

To register a mark with the Patent Office, one must submit a drawing of the mark and indicate when it was first used in interstate commerce and how it is used. The Patent Office conducts an investigation to verify those facts and will register a trademark as long as it is not generic, descriptive, immoral, deceptive, the name of a person whose permission has not been obtained, or substantially similar to another's trademark. Until recently, the criterion that a mark not be deceptive was instead that the mark not be deceptive or disparaging, but the constitutionality of requiring that the mark not be disparaging was challenged in Case 8-2.

[2] *Christian Louboutin S.A. et al. v. Yves Saint Laurent America, Inc. (YSL) et al.,* 696 F.3d 206 (2d Cir. 2012).

MATAL, INTERIM DIRECTOR, UNITED STATES PATENT AND TRADEMARK OFFICE v. TAM
United States Supreme Court
582 U.S.1744(2017)

FACTS: *Simon Shiao Tam is the frontman for the Asian-American rock band The Slants. Tam claims that he named his band The Slants to "reclaim" and "take ownership" of Asian stereotypes. In 2011, Tam filed an application in order to register the mark "The Slants" for "Entertainment in the nature of live performances by a musical band." The examiner refused to register Tam's mark under Section 2(a) of the Lanham Act, which forbids marks that "may disparage" an identifiable group of Americans. The examiner claimed that, even though Tam had chosen the mark to "reappropriate the disparaging term," a substantial number of persons of Asian descent would likely find the mark offensive. The Trademark Trial and Appeal Board affirmed this decision, and Tam appealed to the U.S. Court of Appeals. On appeal, the Fourth Circuit found the disparagement provision of Section 2(a) to be unconstitutional. The case was then appealed to the United States Supreme Court by the Director of the Patent and Trademark Office.*

QUESTION: Should Tam be allowed to register The Slants as a trademark despite its potentially disparaging nature?

REASONING: This case turns on the legitimacy of Section 2(a) of the Lanham Act, which, in part, states that no trademark shall be refused registration unless it "Consists of or comprises immoral, deceptive, or scandalous matter; or matter which may disparage persons, living or dead, institutions, beliefs, or national symbols, or bring them into contempt, or disrepute . . ." The Court of Appeals did, indeed, decide that The Slants is a disparaging mark under Section 2(a). However, the question was not whether the mark was disparaging, but whether the ban on registration of disparaging marks by Section 2(a) violated the First Amendment.

The federal government put forward several arguments as to why Section 2(a) is constitutional. First, the federal government argued that Section 2(a) was content-and

(continued)

viewpoint-neutral. The Court rejected this argument, pointing out cases where identical marks where either rejected or accepted based upon whether they were used in a negative or positive way, respectively.

Next, the federal government argued that Section 2(a) doesn't implicate the First Amendment at all because it doesn't actually prohibit speech. Tam was free to name his band whatever he wished and use the name in commerce without registering a trademark. The Court rejected this argument as well, citing cases such as *Speiser v. Randall* which demonstrate that a denial of benefits is in effect a penalty or censorship of speech. The court noted that "federal trademark registration brings with it valuable substantive and procedural rights . . . denial of which has a 'chilling effect on speech.'"

The final strike against Section 2(a) came when the Court reasoned that the rule was unconstitutional because it failed the Central Hudson test. To pass the test, the commercial speech in question must concern lawful activity and not be misleading, and there must be a substantial government interest justifying the regulation. The Court ruled that the first prong was met, as there was no claim that disparaging trademarks are unlawful or mislead the public. However, the second prong was not met. Although the government argued that it had a legitimate interest in distancing itself from disparaging speech, the Court found that "disapproval is not a legitimate government interest where there is no plausible basis . . . [for the speech to be] reasonably attributed to the government by the public."

DECISION: The U.S. Supreme Court unanimously affirmed the decision of the appellate court that Section 2(a) was unconstitutional.

SIGNIFICANCE: This case illustrates a change in trademark law that has opened up the possibility of more controversial marks being granted protection.

CRITICAL THINKING

How would you criticize the Supreme Court's reasoning in this case?

ETHICAL DECISION MAKING

What values are upheld by this decision? What values would have been upheld had the decision instead went against Tam?

The foregoing case was especially important to the owners of the, Washington Redskins. In June of 2014, the U.S. Patent and Trademark Office had ruled that the Washington Redskins' trademarked name was offensive to Native Americans and therefore should be cancelled. The owners had appealed the decision, but in light of the Tam decision, the challenge to the Redskins' mark was dismissed.

Recently, a number of celebrities have sought trademark protection for what some see as marks that might not really serve the purpose of identifying a product with its producer. For example, in 2015, Donald Trump received trademark protection for "Make America Great Again." Taylor Swift sought trademark protection for the phrases, "nice to meet you" and "where you been" on guitar picks, jewelry cases, and other consumer goods. Athletes are joining the trademark game, with football player Tim Tebow seeking to own the phrase "Tebowing," and basketball player Jeremy Lin seeking to trademark "Linsanity."

It is sometimes difficult to determine whether a trademark will be protected. Also, once a trademark is registered, it is not always easy to predict whether a similar mark will be found to infringe on the registered trademark. The primary issue is whether consumers are likely to be confused by the two marks. This decision is based on factors such as the similarity of the marks; the similarity of the products or services in issue; the sophistication of consumers; and even the intent of the defendant to palm off its product as that of another.

Case 8-3 demonstrates a typical analysis used in a trademark infringement suit.

CASE 8-3

TOYS "R" US, INC. v. CANARSIE KIDDIE SHOP, INC.
District Court of the Eastern District of New York
559 F. Supp. 1189 (1983)

FACTS: *In 1961, plaintiff Toys "R" Us, Inc., obtained a trademark for "Toys 'R' Us" and aggressively advertised its products, children's toys and clothing, in stores across the country. In the late 1970s, defendant Canarsie Kiddie Shop opened two clothing shops called "Kids 'r' Us" within 2 miles of a Toys "R" Us shop. The defendant never tried to register its name. Toys "R" Us sued for trademark infringement in federal district court.*

ISSUE: Does "Kids 'r' Us" infringe on the "Toys 'R' Us" mark?

REASONING: To determine the likelihood of confusion associated with the defendant's mark, the court considered these factors:

- *Strength of the senior user's mark:* A mark can fall into one of four general categories, which, in order of ascending strength, are (1) generic (is entitled to no protection); (2) descriptive (has a secondary meaning and informs public about uses of product); (3) suggestive (suggests qualities of product but requires imagination); and (4) arbitrary or fanciful (does not describe the qualities of the product). The plaintiff's mark has a strong secondary meaning as a source of children's products, so the court determines the mark has medium strength. But it falls short of the arbitrary or fanciful protection.

- *Degree of similarity between the two marks:* The key question is whether consumers have the impression that purchasing from the junior user is purchasing from the senior user. Both parties use the phrase "R Us." However, the plaintiff uses an inverted capitalized "R," whereas the defendant uses a noninverted lowercase "r." The plaintiff uses "Toys" while the defendant uses "Kids." The plaintiff's letter is usually multicolored.

- *Proximity of the products:* Both the plaintiff and the defendant sell children's clothing and are direct competitors.

- *The likelihood that plaintiff will bridge the gap:* First, the senior user will expand sales efforts to compete with the junior user. Second, consumers will assume that both parties are related companies.

- *Evidence of actual confusion:* When there is proof that consumers are actually confused by the use of the name, the senior user has a stronger case.

- *Junior user's good faith:* The owner of Kids 'r' Us asserted that he did not recall whether he was aware of the plaintiff's mark. The court does not believe him and concludes that the defendant adopted the "Kids 'r' Us" mark with knowledge of the plaintiff's mark.

- *Quality of the junior user's product:* The court is much more concerned about the possible violation if the quality of the clothing is poor because such clothing would then reflect poorly on the senior user.

- *Sophistication of the purchasers:* Because children's clothing is not a major expenditure, customers will not exercise the same degree of care in evaluating the clothes.

- *Junior user's goodwill:* The junior user did not expend a lot of money advertising and promoting its own name. Thus, the defendant doesn't have a strong equitable interest in retaining its name.

DECISION AND REMEDY: Judgment for the plaintiff. The defendant's use of its mark creates the likelihood of confusion and constitutes trademark infringement.

SIGNIFICANCE OF THE CASE: This case clarifies the factors to consider in a trademark infringement case.

CRITICAL THINKING

How does this case depend to a large extent on the definitions of a few particular terms? How good are the definitions used? What standard are you using to determine whether the definitions are good?

ETHICAL DECISION MAKING

If you were the owner of Canarsie Kiddie Shop, Inc., how would your decision to refer to your store as "Kids 'r' Us" change if you were to act in accordance with the Golden Rule? What values are in conflict when considering the decisions of the owner of Canarsie Kiddie Shop, Inc., and of one who follows the Golden Rule?

E-COMMERCE and the Law

Domain Names and Trade Names

If you have a very strong trademark, what better domain name to have than that trademark? Unfortunately, the same trademark may be owned by two companies selling non-competing goods, yet there can be only one user of any single domain name. For example, "apple" is a trademark owned by both a computer company and the company that produces the Beatles' records. Both cannot establish a site identified as apple.com.

Domain names are important because they are the addresses by which people and businesses are located on the web. A domain name is made up of a series of domains separated by periods. Most websites have two domains. The first-level domain, the one that the address ends with, generally identifies the type of site. For example, if it is a government site, the domain name will end in *gov;* an educational site, in *edu;* a network site, in *net;* an organization site, in *org;* and a business site, in *com.* These top-level domain names are the same worldwide.

The second-level domain is usually the name of whoever maintains the site. For a college, for example, it would be an abbreviation of the college name, such as *bgsu.* Businesses generally want to use their firm name or some other trademark associated with their product because that name will obviously make it easiest for their customers to find them.

So how does a firm go about securing a domain name that reflects its trademark? Network Solutions, Inc. (NSI), which is funded by the National Science Foundation, is responsible for registering domain names. This entity acts on behalf of the Internet Network Information Center, which handles the daily administration of the domain name system in the United States.

Initially, a person applied to the NSI to use a specific domain name, and as long as no one else held that name, registration was granted. Since 1995, however, the NSI has been a little more careful in handing out names. Anyone seeking to register a domain name must now state in the application that the name will not infringe on anyone else's intellectual property rights and that the registrant intends to use the name on a regular basis on the Internet.

A party may lose registration of a name by not using it for more than 90 days. Or the domain name may be canceled if the registrant lied on the registration application. If you have a registered trademark and you find that a party is using your trademark as a domain name and that party does not also have ownership of that mark, you may give written notice to the NSI; under its domain dispute policy, it will most likely put the name on hold, meaning that no one can use the name. Of course, if that party had registered its domain name before you obtained the trademark, there is probably nothing you can do. The party will be entitled to retain the domain name.

Some firms have tried to get the domain name they desire by going to another country. That alternative is certainly a possibility. However, many countries require your firm to be incorporated within their borders before you can gain the right to the domain name there. An additional problem is that trademark law relating to domain names is even more unclear abroad.

For the new entrepreneur, the best advice is to apply for federal trademark protection and register the domain name simultaneously. And for those not yet on the web, the sooner you get your domain name, the more likely you are to get the name you want. If you feel that your mark is being violated by another's domain name, you may want to sue the party for infringement because the unauthorized use of another's trademark in a domain name has been found to be illegal. But you may be in for quite a fight because this is a new area of the law.

trade dress
The overall appearance and image of a product.

The term **trade dress** refers to the overall appearance and image of a product. Recall that, as described in the Chapter Opener, the shape and design of Apple's iPhone is an example of trade dress. Trade dress is entitled to the same protection as a trademark. The main focus of a case of trade-dress infringement is usually on whether there is likely to be consumer confusion. For example, in a recent case,[3] Tour 18, Ltd., a golf course, copied golf holes from famous golf courses without permission of the course owners. In copying a hole from one of the most famous courses in the country, Harbour Town Hole 19, Tour 18 even copied the Harbour Town lighthouse, which is the distinctive feature of that hole. In its advertising, Tour 18 prominently featured pictures of this hole, including the lighthouse. The operator of the Harbour Town course sued Tour 18 for trade-dress infringement. The court found that there was infringement and made Tour 18 remove the lighthouse and disclaim in its advertising any affiliation with the owner of the Harbour Town course.

copyright
The protection of the expression of a creative work; that is, protection of the fixed form that expresses the ideas.

Copyrights

Copyrights protect the *expression* of creative ideas. That is, they do not protect the ideas themselves but protect only the fixed form that expresses the ideas. Copyrights protect a diverse range of creative

[3] *Pebble Beach Co. v. Tour 18, Ltd.*, 942 F. Supp. 1513.

works, such as books, periodicals, musical compositions, plays, motion pictures, sound recordings, lectures, works of art, and computer programs. Titles and short phrases may not be copyrighted.

There are three criteria for a work to be copyrightable. First, it must be *fixed,* which means set out in a tangible medium of expression. Second, it must be *original.* Third, it must be *creative.* A copyright automatically arises under common law when the idea is expressed in tangible form. However, if the work is freely distributed without notice of copyright, the work falls into the public domain. A copyrighted work that is reproduced with the appropriate notice affixed is protected for the life of its creator plus 70 years.

Under the common law of copyright, an infringer may be enjoined only from reproducing a copyrighted work. For the creator to be able not only to seek an injunction but also recover damages arising from the infringement, the copyrighted work must be registered. One may register a work by filing a form with the Register of Copyrights and providing two copies of the copyrighted material to the Library of Congress. Whenever the work is reproduced, the appropriate notice of copyright *should* accompany it, although such notice is no longer required by law. Printed works, for example, should be published with the word *copyright* and the symbol © or the abbreviation *copr.,* followed by the first date of publication and the name of the copyright owner. Once the work is registered, as long as it is always accompanied by the notice of copyright when reproduced, the holder of the copyright has the additional right to sue an infringer for damages caused by the infringer's use of the copyrighted material and to recover any profits made by the infringer on the copyrighted material.

It is not always easy to determine whether a copyright has been infringed, even when two works are similar. After all, if we are talking about creative works, such as photographs of a famous scene, two people might independently take a very similar picture at completely different times and without even knowing of each other's work. The following Case Nugget illustrates the court's reasoning in a successful copyright infringement case.

A Question of Whether There Was Copying

Ty, Inc. v. GMA Accessories, Inc. U.S. Court of Appeals, Seventh Circuit

45 U.S.P.Q. 1519 (1998)

Ty began selling Beanie Babies, and then GMA started selling bean-bag stuffed animals three years later. Ty sought an injunction under the Copyright Act to stop the sale of GMA's Preston the Pig and Louie the Cow. Ty argued that these animals violated its copyrights for Squealer (a pig) and Daisy (a cow). Ty was successful in obtaining an injunction for copyright after the court asked the following questions: Did the alleged copier have access to the work he or she potentially copied? Did he or she actually copy the work? The court found that access (and copying) may be inferred when the two works are so similar and unlike anything in the public domain that it is likely the second user copied the first. This inference can be rebutted by disproving access or showing independent creation. Looking at pictures of the two cows compels an inference of copying. If GMA must have had access to Ty's cow, it must have had access to the pig, too. GMA's pig is strikingly similar to Ty's pig but not to anything in the public domain (e.g., a real pig). The only evidence GMA presented was an affidavit by the designer, who swears she never looked at Squealer. However, the actual bean-bag animal looks more like Ty's design than the designer's drawing.

The **fair-use doctrine** provides that a portion of a copyrighted work may be reproduced for purposes of "criticism, comment, news reporting, teaching (including multiple copies for classroom use), scholarships, and research." In determining whether the fair-use doctrine provides a valid defense to a claim of copyright infringement, Section 107 of the Copyright Act requires

fair-use doctrine
The lawful use of a limited portion of another's work for purposes of criticism, comment, news reporting, teaching, scholarship, and research that does not reduce the commercial value of the protected property.

E-COMMERCE and the Law

Changes to Copyright Law as Technology Evolves

Copyright law has also undergone multiple changes as new ways to reproduce and distribute copyrighted content multiply alongside frequent new developments in technology. For example, in 2014's *Broadcasting Cos. v. Aereo, Inc.*,[4] the Supreme Court had to decide whether or not a new twist on an old technology was violating copyright laws. To understand the case, one must first understand the technology Aereo was using. Aereo had set up banks of tiny digital antennas that would receive over-the-air television signals that were being broadcast by television companies. A computer attached to each antenna would then make a copy of the signal's data and then transmit this data to an Aereo user. This allowed Aereo subscribers to watch television shows on non-television devices using the Internet. While Aereo users paid for the service, Aereo did not pay the company that originally broadcast the copyrighted material. Aereo argued that because it created a unique, separate copy of a transmission for each user, and each copy was being sent to one user only, Aereo was not creating a "public performance" and therefore was not infringing on the broadcasters' copyrights. In his majority opinion, Justice Breyer disagreed, saying that Aereo "performed" by transmitting and that the "performance" was indeed public because, "when Aereo streams the same television program to multiple subscribers, it transmits . . . a performance to all of them." This type of case has become increasingly common in an age where communication technology is constantly evolving.

the court to weigh the following four factors: the purpose and character of the use, the nature of the copyrighted work, the amount and substantiality of the portion used in relation to the copyrighted work as a whole, and the effect of the use on the potential market for or value of the copyrighted work.

In the 1970s, Sony developed a format for videotaping called Betamax. Film studios brought suit against Sony in 1976, arguing that taping a television show was copyright infringement. The district court concluded that recording a show for personal, noncommercial use was fair use, but the appellate court disagreed and found Sony liable. The Supreme Court ultimately ruled on this issue in 1984, agreeing that recording a television program for the noncommercial purpose of time shifting is fair use.[5]

In 1994, the Supreme Court decided that commercial parody may be fair use. In that case, 2 Live Crew, a rap group known for its explicit lyrics, recorded and released a song called "Pretty Woman," which was a parody of Roy Orbison's "Oh, Pretty Woman." The Supreme Court considered the four factors listed previously. With respect to the purpose and character of the use, the more transformative the new work is, the less courts will need to focus on the other three factors.

As technology has continued to develop, copyright enforcement has remained an important concern. For example, Napster was a peer-to-peer file sharing network by which one user could download a file from another user for free. Files that were typically transferred were digital music files. The record industry brought suit against Napster and received a preliminary injunction to stop the exchange of digital music files. Napster appealed to the Ninth Circuit Court of Appeals, which affirmed the preliminary injunction.[6] Napster made several arguments regarding fair use. Like the time shifting purpose in the Betamax case, Napster argued that users were engaging in space shifting (i.e., transferring music files from CDs they had purchased to their computers). However, the court found that although the user may have been transferring the format (from CD to computer file) for personal use, the user was necessarily sharing the files with other users, which was not a fair use.

Patents

patent
Protection that grants the holder the exclusive right to produce, sell, and use the object of the patent for 20 years; can be obtained for a product, process, invention, machine, or a plant produced by asexual reproduction.

A **patent** protects a product, process, invention, machine, or a plant produced by asexual reproduction. For this protection to be granted, three criteria must be satisfied. First, the object of the patent must be *novel*, or new. No one else must have previously made or published the plans for this object. Second, the object must be *useful* unless it is a design. It must provide some utility to

[4] *Broadcasting Cos. v. Aereo, Inc.*, 134 S.Ct. 2498 (2014).
[5] *Sony Corp. of America v. Universal City Studios, Inc.*, 464 U.S. 417 (1984).
[6] *A&M Records, Inc. v. Napster, Inc.*, 239 F.3d 1004 (2001).

GLOBAL Context

Liability of Online Service Providers in Japan

©Digital Archive Japan/Alamy

Japan's key case that set its precedent on the liability online service providers (OSPs) is, interestingly, an offline case involving a karaoke bar. In that case, a karaoke bar was held liable for copyright infringement for not paying royalties on the public performances of the karaoke songs. However, the company that leased the karaoke machine to the bar was also held secondarily liable because it should have reasonably known that most of the songs would be subject to copyright but did not perform due diligence to ensure that the bar had a royalty agreement in place.

The decision in the karaoke bar case has been used to shut down various file-sharing and TV-streaming services in Japan. Using the same logic as the original case, the courts often rule that the provider, while not necessarily downloading or directly hosting the content, should know that the content passing through its service is subject to copyright and that users are violating copyright law.

HOW DO I KNOW IF AN IMAGE HAS BEEN PROTECTED BY A TRADEMARK OR COPYRIGHT?

Images that have been legally deemed a trademark or copyright protected often include small indications that the image in question is not available for public use. The absence of such symbols, however, should be be relied upon to definitively indicate that the image is not protected.

These indicators include:
- The copyright symbol: ©
- The trademark symbol: ™
- The registered trademark symbol: ®

Exhibit 8-6

Trademark and copyright symbols

society. Third, the object must be *nonobvious*. The invention must not be one that a person of ordinary skill in the trade could have easily discovered. When a patent is issued for an object, it gives its holder the exclusive right to produce, sell, and use the object of the patent for 20 years from the date of application. The holder of the patent may *license,* or allow others to manufacture and sell, the patented object. In most cases, patents are licensed in exchange for the payment of *royalties,* a sum of money paid for each use of the patented process.

BUT WHAT IF . . .

WHAT IF THE FACTS OF THE CASE OPENER WERE DIFFERENT?

What if Samsung had presented evidence at the trial that half of the phones on the market at the time of the trial that were made by companies other than Apple or Samsung also shared the iPhone's design of being rectangular with rounded corners?

tying arrangement
An illegal agreement in which the sale of one product is tied to the sale of another.

cross-licensing
An illegal contractual arrangement in which two or more parties license each other to use their specified intellectual property *only* on the condition that neither licenses anyone else to use the property without the other's consent.

The only restriction on the patent holder is that he or she may not use the patent for an illegal purpose. The two most common illegal purposes are tying arrangements and cross-licensing. A **tying arrangement** occurs when the holder issues a license to use the patented object *only* if the licensee agrees to buy some nonpatented product from the holder. **Cross-licensing** occurs when two patent holders license each other to use their patents *only* on the condition that neither licenses anyone else to use his or her patent without the other's consent. Both of these activities are unlawful because they tend to reduce competition.

To obtain a patent, one generally contacts an attorney licensed to practice before the U.S. Patent Office. The attorney does a *patent search* to make sure that no similar patent exists. If none

exist, the attorney fills out a patent application and files it with the Patent Office. As part of the application process, the inventor has a duty to disclose any *prior art* he or she is aware of related to the invention, including other patents. The Patent Office evaluates the application, and if the object meets the criteria already described, a patent is issued within approximately two years. Although two years may seem like a long time, it is short compared to the six years it typically takes to secure a patent in Japan.

Once the patent is issued, the holder may bring a patent infringement suit in a federal court against anyone who uses, sells, or manufactures the patented invention without the permission of the patent holder. A successful action may result in an injunction prohibiting further use of the patented item by the infringer and an award of damages. Sometimes, however, the result of the case is that the holder loses the patent. This loss would occur if the infringer can prove that the Patent Office should not have issued the patent in the first place.

Another way in which a potential patent infringer may attempt to challenge a patent is to ask the United States Patent and Trademark Office (USPTO) to reexamine the patent. Through the reexamination process, a party presents what it argues is invalidating prior art (i.e., patents, articles, or products that disclosed all features of the invention *before* the patent-at-issue was granted). The USPTO then considers whether the patent should have, in fact, been issued. Reexamination requests are frequently granted, and it may take several years for the USPTO to consider whether a patent should have been issued. Thus, reexaminations are tools defendants frequently use to slow down or even pause litigation because a court may prefer the USPTO to complete the reexamination before litigation may proceed.

BUT WHAT IF . . .

WHAT IF THE FACTS OF THE CASE OPENER WERE DIFFERENT?

What if Samsung had been able to bring in evidence that Samsung, in fact, sold a phone with a design similar to the iPhone's before the iPhone had come out? Would that fact have affected the outcome of the case with respect to Apple's claim of infringement on its design patent?

The America Invents Act

In 2011, Congress passed the America Invents Act, which changed the U.S. patent system in several important ways. First, the first person to file a patent application in the United States for an invention will be deemed to have rights to the invention. This is a change from the previous policy, by which the first person to *invent* had the rights to the invention. This change is important because in the past, employees could keep diligent records and take their time applying for the patent. The European and Japanese patent systems also have a first-to-file rule. Second, the Act expands the types of prior art that may be considered by the USPTO when granting a patent, which would arguably make it more difficult to obtain a patent. Third, the Act makes changes to the reexamination process, making the process more like litigation by allowing discovery requests during the process. Fourth, the Act prohibits a plaintiff from filing one suit against many unrelated defendants, which has typically increased the number of patent infringement cases filed. Finally, the Act makes it easier and more affordable for small businesses and sole entrepreneurs to apply for patents.

Trade secrets

trade secret
A process, product, method of operation, or compilation of information that gives a businessperson an advantage over his or her competitors.

A **trade secret** is a process, product, method of operation, or compilation of information that gives a businessperson an advantage over his or her competitors. Inventions and designs may also be considered trade secrets. Furthermore, a company's client list (or an employee's contact list) could be considered a trade secret. A trade secret is protected by the common law from unlawful appropriation by competitors as long as it is kept secret and consists of elements not generally known in the trade.

Competitors may discover a secret by any lawful means, such as by doing reverse engineering or by going on public tours of plants and observing the use of trade secrets. Lawful discovery of a secret means there is no longer a trade secret to be protected.

When an employee leaves a company to work for a competing company, the original company frequently investigates whether the employee has taken any trade secrets to the competitor. According to one study, 59 percent of departing employees take confidential information from the employer when leaving.[7] To ensure that confidential information is protected, companies should consider creating a policy stipulating that any trade secret information must be labeled Confidential or Trade Secret by employees (which would help the employer maintain a trade secret claim) and limit access to this confidential information within the company.

If a competitor acquires a trade secret by unlawful means, the originator of the secret can take legal action. To enjoin such a competitor from continuing the use of a trade secret and/or to recover damages caused by the use of the secret, a plaintiff must prove that a trade secret actually existed, that the defendant acquired it through unlawful means, and that the defendant used the trade secret without the plaintiff's permission.

A common dilemma facing an inventor is whether to protect an invention through patent or trade-secret law. If the inventor successfully patents the invention and defends the patent, the inventor has a guaranteed exclusive monopoly on the use of the invention for 20 years, a substantial period of time. The problem is that once this period is over, the patented good goes into the public domain and everyone has access to it. There is also the risk that the patent may be successfully challenged and the protection lost prematurely.

Trade-secret law, on the other hand, could protect the invention in perpetuity. The problem is that once someone discovers the secret lawfully, the protection is lost.

[7] "Protecting Trade Secrets When Employees Depart," *LAW 360*, September 18, 2009.

CASE OPENER WRAP-UP

Smartphone Wars: Focus on Intellectual Property Rights

In August 2012, Apple and Samsung participated in a four-week jury trial regarding patent infringement and trademark claims. Both Apple and Samsung asserted patents in this trial; however, the jury found that (1) Samsung infringed Apple's patents, (2) Apple did not infringe Samsung's patents, and (3) the trade dress of Apple's iPhone was diluted by some of Samsung's products. The jury awarded Apple over $1 billion in damages, which was one of the highest patent infringement awards in history. Prior to the trial, the judge issued a preliminary injunction that prevented Samsung from selling one of its tablets. However, following the trial, the injunction was lifted.

PRACTICAL TIPS for Business Managers

1. Before starting a business in a certain area of your city, be sure to learn about the local zoning ordinances to ensure that it is even legal to operate a business in that part of town in the first place. This tip is especially important if you decide to run a business out of your home; many zoning laws do not allow businesses in residential areas except under very particular circumstances.

2. When deciding to start a new business, you need to decide whether to lease or purchase the real property on which you pan to do business. The answer to this question relies on several factors. Do you think your business will outgrow the space? Do you want to deal with maintaining the property? How long do you intend to operate your business on that property? As a rule of thumb, it is generally cheaper in the long run to buy if you plan to operate a business on a piece of property for longer than seven years.

3. If you or an employee invent a new piece of intellectual property, it is very important that you patent it, register it as a trademark, or receive copyright protections as soon as possible. Failure to do so could lead to a competitor learning about the idea, beating you to the Patent and Trademark Office, and profiting from your idea.

4. Be sure to keep intellectual property currently in development a secret. Hire employees you can trust, and enforce strict internal security policies to ensure that your competitors do not learn about your ideas and secure a patent before you can.

5. Remember that, once you secure a trademark, it is your duty to uphold its uniqueness by preventing dilution and actively fighting against parties that use the trademark without permission. Failure to do so can result in the abandonment of the trademark, meaning your firm no longer holds the rights to the mark.

SUMMARY

The Nature of Real and Intellectual Property

Real property is land and anything permanently affixed to the land.
Intellectual property is property that is primarily the result of mental creativity rather than physical effort.

Real Property

Fee simple absolute: Right to possess for life and devise to heirs on death; the most encompassing interest.

Conditional estate: Right that is comparable to a fee simple absolute except that the interest will terminate on the happening or nonhappening of a specified condition.

Leasehold estate: Right to possess property for an agreed-on period of time.

Easement: Irrevocable right to use a portion of someone else's land for a specified purpose.
License: Right to use another's property temporarily.

Personal property

Personal property is all property that is not land or permanently affixed to land.
Voluntary transfer of personal property: Property can be transferred by purchase or by gift.

Involuntary transfer of property: If property is lost, mislaid, or abandoned, a new owner acquires title to the property through possession.

Intellectual Property

A *trademark* is a distinctive mark, word, design, picture, or arrangement that a producer uses in conjunction with a product and tends to cause the consumer to identify the product with the producer. Trademarks used in interstate commerce can be protected under the Lanham Act.

A *copyright* protects the fixed form of the expression of an original, creative idea. The most common defense to an allegation of copyright infringement is the fair-use doctrine, which provides that a portion of a copyrighted work may be reproduced for purposes of "criticism, comment, news reporting, teaching (including multiple copies for classroom use), scholarships, and research."

A *patent* protects a product, process, invention, machine, or plant produced by asexual reproduction that meets the criteria of being novel, useful, and nonobvious. Obtaining a patent under the Lanham Act allows the holder to license the use of his or her patented idea for royalties as long as the holder does not enter into a tying arrangement or engage in cross-licensing.

Point/Counterpoint

The process of applying for a patent and keeping a patent can be difficult and costly. As you know, an alternative to patenting an invention is to protect it through trade-secret law.

Should Corporations be Allowed to Patent Genes?	
YES	**NO**
While they are often not the first thing that comes to mind when considering intellectual property, most synthetic genes are allowed to be patented. The patenting of genes brings with it many benefits for both businesses and consumers. First of all, allowing a business to patent a gene gives the firm exclusive rights to look at and experiment with the gene for the patent's duration. This exclusivity means that the firm with the patented gene would have ample time to examine the gene and develop its usefulness to the fullest without worrying about a competitor beating them to market with a product based on the gene. The time allowed by the patent may also help small businesses that discover a particular synthetic gene compete against larger businesses that would otherwise have the resources to always beat out the smaller company.	While the Supreme Court has ruled that naturally occurring human genes cannot be patented, it has allowed for synthetic genes not found in nature to be patented. However there are several reasons why even the patenting of synthetic genes goes too far. First, exclusive rights to a synthetic gene and the products derived from the gene means no competition. A lack of competition can, in turn, lead to arbitrarily high prices that consumers have to pay if they wish to treat an illness that can only be cured by a product that a single firm holds the rights to. People who could not afford treatment could very well die before the patent expires and another company can provide a cheaper alternative. Second, allowing only one company to interact with a particular gene simply means fewer individuals

Another benefit of gene patenting is that the promise of exclusive rights to a synthetic gene sequence could spur innovation. As with most other industries, gene patents would protect the rights of innovators and guarantee that the time and resources spent developing a test or drug from a gene won't be undercut by a competitor. The potential increase in innovation could lead to a greater quantity and quality of discoveries brought upon by researching the genes.

Finally, gene patents make what is normally an unprofitable area of research potentially profitable. Without the patents, there is nothing stopping a competitor from releasing a generic version of a drug or treatment developed by a firm after studying a gene. Not only would the patents help secure profits for innovating firms and therefore provide more incentive to pursue gene research, but the higher likelihood of profits could also mean more people would be willing to invest in gene research. Increased investment could lead to a cycle of better gene-derived products, which means higher profits, which again could mean more investment.

working with the gene and could therefore slow down innovation as a whole in regards to developing products with said gene. Gene patenting also effectively puts one firm in control of a gene, and whether or not cures or life-improving products are developed for consumers who need them is decided entirely by that single business for the duration of the patent.

Third, because naturally occurring genes are not allowed to be patented while synthetic genes are, there could be a massive shift away from studying natural human genes toward the more profitable synthetic genes. While there are many breakthroughs to be had in the realm of synthetic genes, there are still many important discoveries to be made in the natural genes area as well.

Questions & Problems

1. Explain the bundle-of-sticks idea as it relates to real property.
2. Explain the five possessory interests in land.
3. Explain the bundle-of-sticks idea as it relates to intellectual property.
4. How are trademarks and patents different?
5. Explain copyright infringement.
6. On September 8, 2011, John Barth purchased a piece of property at 4 Lynn Street in Peabody, Massachusetts for $1,000. Barth subsequently requested the City of Peabody for a variance to exempt his property from the local zoning ordinance. The City of Peabody's Zoning Board of Appeals held a public hearing and later voted to deny the variance.

 Following the refusal, Barth filed a *pro se* complaint alleging 23 "violations of law." Of the many claims Barth made, he specifically argued that the denial of his zoning variance request constituted a violation of the Fifth Amendment's Taking Clause. Barth contended that the denial amounted to a taking because it deprived him of the use and enjoyment of the land. Do you think the denial of the variance could legally be considered a taking?

 Source: *Barth v. City of Peabody,* U.S. Dist. LEXIS 4080 (2017).

7. Adrian Samalia fled a stolen vehicle during a lawful traffic stop. While Samalia successfully escaped, he left his cell phone in the stolen vehicle. Police officers searched the cell phone without a warrant and used the information contained in it to confirm Samalia's identification. The data from the phone was used as evidence in a trial against Samalia for possession of a stolen vehicle.

 Samalia moved to suppress the evidence, arguing that his constitutional rights were violated when they searched the phone without a warrant. The State contended that the warrantless search was allowed under the abandonment doctrine. The trial court denied the motion to suppress and eventually found Samalia guilty. After a series of appeals, the case worked its way up to the Supreme Court of the State of Washington.

 Under the abandonment doctrine, an individual loses his or her privacy interest in a property if he or she voluntarily abandons it. While Samalia argued that the abandonment doctrine should not be applicable to cell phones because they hold "private affairs," the Washington Supreme Court ruled that there would be no exception to the abandonment doctrine for cell phones. At this point for the State to prevail, it had to show that Samalia voluntarily abandoned his cell phone.

 Do you think the cell phone left by Samalia as he ran from the police was voluntarily abandoned? Explain why or why not. How did the Washington Supreme

Court rule on this issue and what was its reasoning? Do you agree? Why or why not?

Source: *State v. Samalia,* 375 P. 3d 1082 (2017).

8. Thomas Garrison transferred title on two cars to one of his sons, Jay Garrison, while on his deathbed. After Thomas passed, his widow Pamela Garrison filed a petition to recover the car as the representative of the estate. Pamela challenged Thomas's competency to make *inter vivos* gifts of the cars. The trial court ruled that the cars were actually gifts *causa mortis* and concluded that the evidence of Thomas's competency to make such a gift was evenly split. Because the evidence was equal, the trial court ruled the cars had to be returned to the estate. Jay appealed the decision, contending that the trial court should have concluded the cars were a gift *inter vivos* and not a gift *causa mortis*.

 The type of gift matters because different standards apply for each one. According to the appellate court, the burden of showing whether a gift qualifies as an *inter vivos* gift is a preponderance of the evidence. On the other hand, evidence must be clear and convincing to establish a gift as a gift *causa mortis*. Why do you think the standard for the gift *causa mortis* is more stringent than for the *inter vivos* gift?

 Source: *Garrison v. Garrison,* 71 NE 3d 423 (2017).

9. Defendant Tubbs met her fiancé, Church, over the Internet. After several years of correspondence and visits, they became engaged in February of 2000. Church and Tubbs planned to be married in Las Vegas in July 2000. Two months before the engagement, Church paid off $4,100 of Tubbs's credit card debt. He also gave Tubbs an engagement ring that he purchased for $7,274.42. On March 15, 2000, he deposited $194,852.56 in Tubbs's bank account to fund the purchase of land and a residential home in Michigan. Tubbs purchased the home in both of their names as joint tenants, and in April, on Church's instructions, she moved in. Church moved some personal property to the residence, including a family heirloom diamond ring.

 On June 5, 2000, Tubbs emailed Church stating that their relationship was over because she was horrified after seeing his "bizarre and abnormal behavior" on the Internet and because she had discovered that he led a "risqué lifestyle as a cross-dresser and bi-sexual." Tubbs rejected Church's demands to repay the $4,100 and to return the engagement ring, his personal property, and her interest in the Michigan home. On July 24, 2000, Church died in England. Church's estate subsequently filed suit to recover the property. The court entered a final judgment entitling the estate to the rings or a money judgment for their values; the real property, partitioned as a matter of law to account for its appreciation; complete right, title, interest, and possession of the land and residential home, free and clear of any claim, right, title, or interest of Tubbs; and a money judgment in the amount of $75,000 (the amount of Tubbs's home equity mortgage), less credits to Tubbs of $13,000 for property taxes she paid from 2000 through 2005, or a modified money judgment in the amount of $62,000. Tubbs appealed. What arguments might she have made on appeal? How do you think the appellate court ruled in this case, and why?

 Source: *Salens v. Tubbs,* 2008 WL 4072342 (C.A. 6 Mich.).

10. When William S. Thrasher died, he left a life estate in the family home to his wife, Gail Thrasher, and a remainder interest to his children, James Hall Thrasher and Sheila Reed. When Gail attempted to sell her life estate, the children sued, arguing that by selling the life estate, she was forfeiting the life estate because the buyer might waste the property, thereby devaluing the interests of the children. How do you think the courts resolved this issue? Should the holder of the life estate be able to transfer the life estate?

 Source: *Thrasher v. Thrasher,* 169 So. 3d 1043; 2014 Ala. Civ. App. LEXIS (2014).

11. Fuel Clothing Company manufactures and markets apparel for the action sports industry. Fuel was founded in 1992 and has had the "Fuel" trademark on the Principal Register for clothing goods since 1999. Mr. Gould, the owner of Fuel Clothing Company stated in his deposition testimony that his company has "a thousand different uses" for the "Fuel" mark. The company has also taken steps to protect its trademark through the use of litigation proceedings and settlement agreements with other companies allegedly infringing on the mark.

 On February 27, 2012, Fuel Clothing Company filed a trademark infringement action against Nike following the release of the FuelBand at Nike stores on February 22, 2012. The action was based on claims of federal trademark infringement, federal trademark dilution, federal unfair competition and false designation of origin, common law trademark infringement and unfair competition, and violations of the South Carolina Unfair Trade Practices Act (SCUTPA). On November 4, 2013, after a series of pretrial motions, the court held a hearing on Nike's motion for summary judgment.

 Nike moved for summary judgment on two separate grounds. The first argument that Nike proffered was that all of Fuel's claims failed because Fuel had abandoned its mark through naked licensing. Nike also contended that Fuel's claims based on federal trademark infringement, federal unfair competition and

false designation of origin, common law trademark infringement and unfair competition, and SCUTPA failed because there is no likelihood of confusion among relevant consumers. What are the elements of abandoning a mark through naked licensing? How did the court explain its ruling in regards to the first ground of summary judgment that Nike moved for? What are the nine likelihood of confusion factors that the court bases its argument on in section B.1? What conclusion does the court come to?

Source: *Fuel Clothing Company, Inc. vs. Nike, Inc.* (7 F. Supp. 3d 594; 2014 U.S. Dist.).

12. The plaintiff sold his photography online. He sued defendants for copyright infringement. He believed that the defendants had obtained his photos illegally and placed one of these photographs, without the plaintiff's permission, on a commercial Internet website they operated. After the defendants ignored a letter by the plaintiff demanding that the online photos be taken down, the plaintiff filed a copyright infringement case seeking $30,000 in damages. The owner also alleged that the defendants' conduct constituted a willful violation of his copyrights. The court considered the following factors in determining whether a damages remedy existed: the infringer's state of mind; the expenses saved by the infringer; the revenue lost by the copyright holder; the deterrent effect on the infringer and third parties; the infringer's cooperation in the case; and the attitude and conduct of the parties. Based on the allegations stated above, do you believe that these factors merit a damages reward of $30,000? In whose favor do you believe the court ruled? Explain your reasoning.

Source: *Oppenheimer v. Holt,* No. 1:14-CV-000208-MR, 2015 WL 2062189 (W.D.N.C. May 4, 2015).

13. Securities brokerage firm Charles Schwab filed suit against a former employee who left the company and started his own wealth management firm. Charles Schwab claimed that the names of its clients, along with information regarding their accounts and assets, were trade secrets. The employee therefore misappropriated these trade secrets when he contacted the clients to ask them to transfer their accounts to his new firm. Charles Schwab argued that the employee could not have learned of these individuals but for his work at the Charles Schwab brokerage. Do you think this customer information will be treated as a trade secret? What factors do you think the court will consider when determining whether the customer information is a trade secret?

Source: *Charles Schwab & Co. Inc. v. Douglas Castro,* Case No. 650053/2013 (E.D.N.Y. January 7, 2013).

14. Nike and Michael Jordan created the Air Jordan line in the 1980s. Jordan filed suit against a Chinese sportswear company, Qiaodan Sports, in China. Jordan is known in China as Qiaodan, and Qiaodan Sports uses a logo of a silhouette of a tall man holding a basketball, along with the number 23. The company registered "Qiaodan" as a trademark in China. Qiaodan also filed applications for trademarks on Jordan's sons' names. According to the complaint, Qiaodan had sales of approximately $450 million per year. Do you think Jordan will be successful in his suit against Qiaodan? Why?

Source: "A Chinese Sportswear Company Has Trademarked Michael Jordan's Sons' Names," http://qz.com/79234/michael-jordan-versus-qiaodan-sports/.

Part 2 *Contract Law*

CHAPTER 9

Introduction to Contracts and Agreement

CASE OPENER

The Problematic Promotion

Pepsi had a promotion whereby consumers were encouraged to collect Pepsi points by consuming Pepsi products. They could then redeem the points for merchandise. If they did not have quite enough points for the prize they wanted, they could buy the needed additional points for 10 cents each; however, at least 15 original Pepsi points had to accompany each order.

At the climax of an early commercial for the promotion, three young boys are sitting in front of a high school building, one reading his Pepsi Stuff catalog while the others drink Pepsi, all gazing in awe at an object rushing overhead as the military march in the background builds to a crescendo. A Harrier jet swings into view and lands by the side of the school building, next to a bicycle rack. Several students run for cover, and the velocity of the wind strips one hapless faculty member down to his underwear. While the faculty member is being deprived of his dignity, the voice-over announces: "Now the more Pepsi you drink, the more great stuff you're gonna get."

A teenager opens the cockpit of the fighter and can be seen, without a helmet, holding a Pepsi. He exclaims, "Sure beats the bus," and chortles. The military drumroll sounds a final time, as the following words appear: "Harrier Fighter 7,000,000 Pepsi Points." A few seconds later, the following appears in more stylized script: "Drink Pepsi—Get Stuff."

John Leonard decided to accept Pepsi's offer of the Harrier fighter jet for 7 million Pepsi points. He quickly realized that it would be easier to raise the money to buy points than to collect the 7 million points. In early March 1996, he filled out an order form requesting the jet and submitted it to Pepsi, along with 15 Pepsi points and a check for $700,000.

In response, Pepsi sent him a letter saying, "The item that you have requested is not part of the Pepsi Stuff collection. It is not included in the catalog or on the order form, and only catalog merchandise can be redeemed under this program." John then sued for breach of contract.

1. Did Pepsi offer to sell the Harrier jet for 7 million points?
2. Did Leonard's submission of the order form constitute an acceptance?
3. Was the alleged contract in this case a bilateral or unilateral contract?

LEARNING OBJECTIVES

LO 9-1 Describe what a contract is.
LO 9-2 Integrate common law and UCC.
LO 9-3 Recognize means of classifying contracts.
LO 9-4 Determine the rules that guide the interpretation of contracts.
LO 9-5 Identify elements of a valid offer and a valid acceptance.
LO 9-6 Explain how an offer may terminate.

The Definition of a Contract

LO 9-1 Describe what a contract is.

This part of the text focuses on contracts. A **contract**, according to the Restatement (Second) of Contracts, is "a promise or set of promises for the breach of which the law gives a remedy or the performance of which the law in some way recognizes a duty."[1] Another, perhaps simpler, way to think of a contract is that it is a set of legally enforceable promises.

Elements of a Contract

A contract has four elements: the agreement, the consideration, contractual capacity, and a legal object. The **agreement** consists of an **offer** by one party, called the *offeror,* to enter into a contract, and an **acceptance** of the terms of the offer by the other party, called the *offeree.* The agreement is discussed in more detail in the latter part of this chapter.

The second element of a contract is the **consideration,** which is defined as the bargained-for exchange. Another way to think of the consideration is that it is what each party gets in exchange for his or her promise under the contract. Consideration is further discussed in Chapter 10.

The third element is **contractual capacity,** the legal ability to enter into a binding agreement. Most adults over the age of majority have the legal ability to enter into binding contracts, but Chapter 10 explains when people have either limited or no capacity to enter into these agreements. Persons who do not have the capacity to enter into legally binding contracts include those who are under the age of majority, intoxicated, or suffering from mental illness. Chapter 10 also discusses the fourth element of a binding, legal contract: **legal object.** The contract cannot be either illegal or against public policy.

The elements of a contract, as well as defenses to its enforcement, are listed in Exhibit 9-1.

Defenses to the Enforcement of a Contract

Sometimes the parties may enter into what appears to be a legally binding and enforceable contract because all four elements of a contract are present, but one of the parties may have a defense

contract
A promise or set of promises for the breach of which the law gives a remedy or the performance of which the law in some way recognizes a duty.

agreement
One of the four elements necessary for a contract; consists of an offer made by one party, the offeror, and the acceptance of the offer by another party, the offeree.

offer
A key factor in the agreement element of a contract; consists of the terms and conditions set by one party, the offeror, and presented to another party, the offeree.

acceptance
A key factor in the agreement element of a contract; consists of the agreement of one party, the offeree, to the terms of the offer in the contract made by the other party, the offeror.

ELEMENTS OF A CONTRACT	DEFENSES TO A CONTRACT
Agreement	Lack of genuine assent
Consideration	Lack of proper form
Legal purpose	
Capacity	

Exhibit 9-1

A Valid Contract Has Four Elements, and No Defenses Can Be Raised against It

[1] Restatement (Second) of Contracts, sec. 1.

consideration
The bargained-for exchange; what each party gets in exchange for his or her promise under a contract.

contractual capacity
The legal ability to enter into a binding agreement.

legal object
A purpose that does not violate a statute or public policy.

lack of genuine assent
A defense to the agreement of a contract in which the offeree claims that the offeror secured the agreement through improper means, such as duress, fraud, undue influence, or misrepresentation.

to the contract's enforcement. Such defenses fall into two categories. The first, discussed in Chapter 12, is a **lack of genuine assent.** A contract is supposed to be entered into freely by both parties, but sometimes the offeror secures the acceptance of the agreement through improper means, such as fraud, duress, undue influence, or misrepresentation. In these situations, there really is no genuine assent to the contract, and the offeree may be able to raise the lack of genuine assent as a defense to enforcement of the agreement.

The second defense, discussed in Chapter 13, is that the contract lacks the *proper form,* which typically means it lacks a writing. Certain contracts require a writing, and if there is none, the agreement will not be enforced. Chapter 14 discusses the performance and discharge of contracts as well as remedies arising when a contract is breached.

Sources of Contract Law

LO 9-2 Integrate common law and UCC.

There are two primary sources of contract law: case law and the Uniform Commercial Code (UCC). A third source of law, which has become more important with increasing globalization, is the Convention on Contracts for International Sales of Goods (CISG). Chapters 9 through 14 of this book focus primarily on the law of contracts as established by the common law, and Chapters 15 and 16 focus on the law as established by the UCC and CSIG.

Common Law

Today's law of contracts actually originated in judicial decisions in England, which were followed by the early courts in the United States. Since then, contract law has been modified by our legislatures and by rulings of our courts. The law of contracts is primarily common law, so to find out what the law is, one could go to the Reporters and read the decisions, but an easier way to know the law would be to go to the Restatement (Second) of the Law of Contracts. Prominent legal scholars, recruited by the American Law Institute, organized the principles of the common law of contracts into the original *Restatement of the Law, Contracts.* The compilation has subsequently been revised and published as *Restatement of the Law Second, Contracts.*

The Restatement (Second) is not actually the law itself, although judges frequently cite it in cases because it is an authoritative statement of what the law is. As the common law of contracts evolved in the various states, not all states interpreted all aspects of the law in the same way, so although we can generalize about the law of contracts, one would of course always want to know exactly what the law at issue is in one's own state. In the Restatement (Second), the drafters often explain what the law about a particular matter is in the majority of states and then provide alternative approaches that other states have adopted.

Uniform Commercial Code

Uniform Commercial Code (UCC)
A statutory source of contract law in the United States applicable to transactions involving the sale of goods. The UCC was created in 1952 and adopted by all 50 states, the District of Columbia, and the Virgin Islands; it may be modified by each state to reflect the wishes of the state legislature.

Having different laws governing contracts in different states did not make interstate commerce flow smoothly. To remedy some of the difficulties created by a patchwork of different laws governing commercial transactions, the National Conference of Commissioners on Uniform State Laws and the American Law Institute drafted a set of commercial laws that could be applicable to all states. This effort was called the **Uniform Commercial Code (UCC).** The UCC became law in each state that adopted it completely or partially as part of its state code. Thus, for example, if a firm enters into a contract governed by the Uniform Commercial Code in Ohio, it would be operating under the Ohio Uniform Commercial Code.

The part of the Uniform Commercial Code that is relevant to contracts is Article 2, which governs contracts for the sale (exchange for a price) of goods (tangible, movable objects). This part of the book will sometimes point out important differences between the UCC and the common law, but the UCC is discussed primarily in Chapters 15 and 16.

GLOBAL Context

©Digital Archive Japan/ Alamy

Contract Law in the European Union

If you are doing business with a member state of the European Union, there is one source to turn to in order to understand how the various member states regulate contracts, and that is the *Principles of European Contract Law*. The principles were drafted by an independent body of experts from each member state of the European Union under a project supported by the European Commission. The principles are stated in the form of articles with a detailed commentary explaining the purpose and operation of each article. Comments to each article provide illustrations, short cases that show how the rules are to operate in practice. Each article also has comparative notes surveying the national laws and other international provisions on the topic.

Classification of Contracts

LO 9-3 Recognize means of classifying contracts.

Contracts are classified in a number of ways. Different classifications are useful for different purposes, as the following discussion demonstrates.

Bilateral Versus Unilateral Contracts

All contracts can be classified as either bilateral or unilateral. Knowing whether a contract is bilateral or unilateral is important because that classification determines when the offeree is legally bound to perform. Whether a contract is bilateral or unilateral depends on what response the offeror (the party proposing the contract) expects from the offeree (the person agreeing to or accepting the contract).

If the offeror wants a promise from the offeree to form a binding contract, the contract is a **bilateral contract**. A bilateral contract is commonly defined as a promise in exchange for a promise. As soon as the promises are exchanged, a contract is formed and the parties's legal obligations arise. For example, when Shannon promises to pay Gary $1,000 in exchange for his promise to paint her car on July 1, they have a bilateral contract. If either party fails to perform, the other may sue for breach.

bilateral contract
A promise exchanged for a promise.

ZAPPOS.COM INC., CUSTOMER DATA SECURITY BREACH LITIGATION
United States District Court of Nevada
WL 4466660 (2013)

FACTS: *The defendant, Zappos.com, is a popular website known mainly for its discounted shoe sales. In 2012, a hacker hacked into the defendant website in an effort to obtain the personal account information of Zappos shoppers. After releasing news of the breach, the defendant faced numerous lawsuits from unhappy customers. Subsequently, the defendant moved to compel arbitration as mandated in its Terms of Use listed on its website. The defendant argued that it and its customers were in a bilateral agreement stating that arbitration must be used in the event of a dispute between the two parties, as supported by its customer terms of use. However, also in the terms of use, the defendant stipulated that it could change its terms of use and all of its agreements anytime at its own discretion.*

Hence, plaintiff customers argued that the agreement was not bilateral and was, in fact, unfairly unilateral. Specifically, the plaintiff customers argued that the defendant was not actually agreeing to anything and made no promise to its customers regarding dispute resolution. Therefore, the plaintiff customers argued that they should not have to use arbitration and instead should be able to file their class action lawsuit against the defendant.

ISSUE: Is an arbitration clause illusory, and therefore unenforceable, when a defendant company can avoid its promise to arbitrate by amending a provision, whereas customers are simultaneously bound to arbitration?

(continued)

REASONING: When the Terms of Use binds consumers to arbitration while leaving the defendant company free to litigate or arbitrate whenever it sees fit, no mutuality of obligation exists. When there is no mutuality of agreement, said arbitration agreement is illusory and therefore unenforceable. In this case, the Terms of Use gave the defendant the right to change the Terms of Use, including the Arbitration Clause, at any time without notice to the plaintiff customers. On one side, the Terms of Use purportedly bound any user of the defendant website to mandatory arbitration. However, if a customer sought to invoke arbitration pursuant to the Terms of Use, nothing would prevent the defendant from unilaterally changing the Terms and making those changes applicable to that pending dispute if it determined that arbitration was no longer in its interest. In effect, the agreement allowed the defendant to hold its customers and users to the promise to arbitrate while reserving its own escape hatch. By the provisions of the Terms of Use, the defendant company was free at any time to require a consumer to arbitrate and/or litigate anywhere it sees fit, while consumers are required to submit to arbitration in Las Vegas, Nevada.

DECISION AND REMEDY: The court denied the defendant's Motion to Compel Arbitration.

SIGNIFICANCE OF THIS CASE: This case provides explanation of when an arbitration clause is illusory and therefore unenforceable.

CRITICAL THINKING

Is there enough ambiguity with the word *agreement* that Zappos could argue that it had an agreement with its customers? If Zappos could change any rule or promise it made to a customer at any time, how could Zappos argue that it was an agreement to anything?

ETHICAL DECISION MAKING

When a court makes a decision in a contract case, what values are being elevated by the court? In other words, the court is anchoring its reasoning on a preference for a particular value or set of values. What is that value or set of values in a contract case?

unilateral contract
A promise exchanged for an act.

In a **unilateral contract**, the offeror wants a performance to form the contract. The offeror wants the offeree to do something, not promise to do something. The most common unilateral offer is a reward. If Jim loses his dog, he may post a sign stating "$50 reward for the safe return of my poodle, Frenchie." If Rita calls Jim and says, "Don't worry, I'll find your dog," she is not making a contract because the unilateral offer calls for an action, not a promise. Once she finds Frenchie and takes the dog to Jim, a contract is formed and Jim must pay.

Just as the offeree is under no obligation actually to do the act called for by the offeror, the offeror may revoke the offer at any time before performance. However, to prevent injustice, once an offeree begins performance, the offeror must hold the offer open for a reasonable time to allow the offeree to complete the performance.

In the Case Opener, the plaintiff would argue that Pepsi appeared to be making an offer for a unilateral contract. In its television commercial, Pepsi promised to provide a prize for a performance, the presentation of a set number of points. By submitting the requisite number of points, the plaintiff believed he was accepting a unilateral offer.

BUT WHAT IF . . .

WHAT IF THE FACTS OF THE CASE OPENER WERE DIFFERENT?

Recall that in the Case Opener, PepsiCo required contestants first to mail in Pepsi points before the company would mail the contestant a prize. But what if the Pepsi points contract required only a promise from contestants to send in Pepsi points on a certain date and PepsiCo would send the prize up front? Would the second scenario be a unilateral or a bilateral contract?

Express Versus Implied Contracts

Contracts are classified as express or implied, depending on how they are created. The terms in **express contracts** are clearly set forth in either written or spoken words. The contract in the Case Opener was an express contract; the terms were set forth in the ad and the catalog. **Implied contracts,** in contrast, arise not from words but from the conduct of the parties. For instance, when you have a dental emergency and your dentist pulls your severely infected tooth without prior negotiation about payment, or even any mention of payment, you have an implied contract for payment for his services.

As a general rule, three conditions must be met for the courts to find an implied, or as it is sometimes called, an *implied-in-fact,* contract. First, the plaintiff provided some property or service to the defendant. Second, the plaintiff expected to be paid for such property or service and a reasonable person in the position of the defendant would have expected to pay for such property or services. Third, the defendant had an opportunity to reject the property or services but did not. In Case 9-2, the Court had to decide whether the facts gave rise to an implied-in-fact contract.

express contract
A contract in which all the terms are clearly set forth in either written or spoken words.

implied contract
A contract that arises not from words of agreement but from the conduct of the parties.

TRAPANI CONST. v. ELLIOTT GROUP INC.
Appellate Court of Illinois, First Division
64 N.E, 3d 132 (1016)

FACTS: *In February 2007, The Elliot Group, Inc., an Illinois real estate developer, made a deal with the Village of Arlington Heights to develop property in that village. Arlington Market, LLC, was subsequently formed by The Elliot Group to develop, manage, and sell the property, a common practice in the residential development industry. Since Arlington Market had no employees, members of the Elliot Group instead managed the LLC.*

In March 2007, the Trapani Construction Company sent a proposal to The Elliot Group to provide construction services for the Arlington property. Trapani had previously performed work for The Elliot Group. Between March and July of 2007, Trapani sent a total of five draft contracts to the Elliot Group in which the Elliot Group were listed as the contracting party. Although none of the draft contracts were ever signed, Trapani began construction at the Arlington site in early July.

The case in question arose after Trapani finished its work at the Arlington site in accordance with the terms of the draft contract and sought payment. Despite having paid Trapani for previous payment requests on the Arlington project, The Elliot Group did not pay the final invoice. Trapani sued for breach of implied-in-fact contract, and The Elliot Group asserted that there was no contract between them and the plaintiff. The trial court ruled in favor of Trapani and The Elliot Group appealed.

QUESTION: Was there an enforceable implied contract between Trapai and the Elliott Group?

REASONING: The Elliot Group argued that, first and foremost, there was no contract between it and Trapani because Elliot never actually accepted Trapani's offer. The court pointed out, however, that an implied contract based on the parties' actions can exist even without clear acceptance. Implied contracts are true, binding contracts that contain the typical elements of offer, acceptance, and consideration. Upon examination of the facts, the court found that all these elements existed.

First, an offer existed on the part of Trapani. Trapani often used unsigned draft contracts when working with The Elliot Group. In this case, Trapani negotiated several drafts and proposals with The Elliot Group before and during the project. Second, consideration was provided by both parties. Not only did Trapani provide The Elliot Group with several different construction services during this project, but The Elliot Group even paid for most of the project before suddenly stopping and claiming that no contract existed. Third, acceptance was apparent regardless of The Elliot Group's claims. Trapani accepted The Elliot Group's terms when they began work on the project. The Elliot Group, despite its arguments, accepted the contract when it refused to reject Trapani's work while being fully aware it was taking place (and even paying for the majority of it). Being that the three elements of a contract existed, and especially because of The Elliot Group's unwillingness to cease Trapani's work, the court ruled that an implied contract existed between the parties and that The Elliot Group is liable for the costs of the project.

DECISION: The Court of Appeals agreed with the lower court that there was in fact an implied contract between the parties, and affirmed the decision in favor of plaintiff Tripani.

SIGNIFICANCE: This case clearly outlines the requirements for an implied contract and shows that contracts can still exist in the absence of a signed, written contract.

(continued)

> **CRITICAL THINKING**
>
> Although in this case, it may seem obvious that facts and circumstances point to the mutual intent to contract, what kind of situations would this be less clear? What kind of evidence would be sufficient, and what kind of evidence would merely be necessary to show the mutual intent to contract needed to show an implied-in-fact contract?
>
> **ETHICAL DECISION MAKING**
>
> Which stakeholders' interests are helped by this decision? Which stakeholders' interests are harmed?

Quasi-Contracts

quasi-contract
A court-imposed contractual obligation to prevent unjust enrichment.

Quasi-contracts are sometimes called *implied-in-law contracts,* but they are not actually contracts. Rather, to prevent one party from being unjustly enriched at the expense of another, the courts impose contractual obligations on one of the parties as if that party had entered into a contract.

For example, assume Jones hears a noise out in his driveway. He looks out and sees a group of workers apparently getting ready to resurface his driveway. The doorbell rings, and he does not answer it. He goes down into his basement office and stays there until the workers have gone and he has a resurfaced driveway. When he receives a bill from the paving company, he refuses to pay on grounds that he did not ask to have the driveway done. In such a case, where Jones knew that the company was getting ready to bestow on him a benefit to which he was not entitled, the court will probably impose a quasi-contract, requiring Jones to pay the paving company the fair market value of the resurfacing. Imposing such a duty prevents Jones from being unjustly enriched at the expense of the paving company.

There are limits to the doctrine, however; specifically, the enrichment must be unjust. Sometimes a benefit may simply be conferred on you because of a mistake by the other party, and the courts will not make people pay for others' mistakes. In our example, if Jones had been out of town when his driveway was repaved, he would have just gotten lucky. The courts would not make him pay for the pavers's mistake when he could have done nothing to prevent the benefit from being bestowed on him.

Valid, Void, Voidable, and Unenforceable Contracts

valid
A term applied to a contract that includes all four elements of a contract: agreement (offer and acceptance), consideration, contractual capacity, and legal object.

What everyone hopes to enter into, of course, is a valid contract. A **valid** contract is one that contains all the legal elements of a contract as set forth in Exhibit 9-1. As a general rule, a valid contract is one that will be enforced. However, sometimes a contract may be valid yet unenforceable.

unenforceable
A term applied to a contract that, because of a law, cannot be enforced by the courts.

A valid contract may be **unenforceable** when some law prohibits the courts from enforcing it. For example, the statute of frauds, discussed in Chapter 13, requires certain contracts to be evidenced by a writing before they can be enforced. Similarly, the statute of limitations mandates that an action for breach of contract must be brought within a set period of time, thereby limiting the enforceability of the contract.

void
A term applied to a contract that is not valid because its object is illegal or it has some defect that is so serious that it is not a contract.

A **void** contract is, in effect, not a contract at all. Either its object is illegal or it has some defect that is so serious that it is not a contract. If you entered into a contract with an assassin to kill your business law professor, that would be a void contract because it is obviously illegal to carry out the terms of the agreement.

voidable
A term applied to a contract that one or both parties have the ability to either withdraw from or enforce.

A contract is **voidable** if one or both of the parties can either withdraw from the contract or enforce it. If the parties discover that the contract is voidable after one or both have partially performed, and one party chooses to terminate the contract, both parties must return anything they had already exchanged under the agreement so that they will be returned to the condition they were in at the time they entered into the agreement.

Certain types of errors in the formation of a contract lead to it being voidable. Typically, the person who can void the contract is the person whom the court is attempting to protect or the

party the court believes might be taken advantage of by the other party. For example, contracts by minors are usually voidable by the minor, as discussed in Chapter 11. Contracts entered into as a result of fraud, duress, or undue influence—a problem with the validity of the acceptance, as described in Chapter 12—may be voided by the innocent party.

Executed Versus Executory Contracts

Once all the terms of the contract have been fully performed, the contract is said to have been **executed**. As long as some of the duties under the contract have not yet been performed, the contract is considered **executory**. For example, if Randolph hires Carmine to paint his garage on Saturday for $800, with $200 paid as a down payment and the balance due on completion of the job, the contract becomes executory as soon as the agreement is reached. When the down payment has been made and the painting is halfway completed, the contract is still executory. Once the painting has been finished and the final payment made, the contract is an executed contract.

executed
A term applied to a contract in which all of the terms have been fully performed.

executory
A term applied to a contract in which not all of the terms have been fully performed.

Formal Versus Informal Contracts

LO 9-4 Determine the rules that guide the interpretation of contracts.

Contracts may also be classified as formal or informal. **Formal contracts** are those that have a special form or must be created in a specific manner. The Restatement (Second) of Contracts identifies the following four types of formal contracts: (1) contracts under seal, (2) recognizances, (3) letters of credit, and (4) negotiable instruments.

When people hear the term *formal contract,* what often comes to mind is a **contract under seal**. The term *under seal* comes from the days when a contract was literally sealed by a piece of soft wax into which an impression was made. Today, sealed contracts may still be literally sealed with wax or some other soft substance, but they are more likely to be simply identified with the word *seal* or the letters *L.S.* (an abbreviation for *locus sigilli,* which means "the place for the seal") at the end of the document. Preprinted contract forms with a seal printed on them can also be purchased today, and parties using such documents are presumed, without evidence to the contrary, to be adopting the seal for the contract.

States today do not require contracts to be under seal. However, 10 states still allow a contract without consideration to be enforced if it is under seal.

A **recognizance** arises when a party acknowledges in court that he or she will perform some specified act and/or pay a price on failure to do so. An example of a recognizance is a bond used as bail in a criminal case. The person agrees to return to court for trial or forfeit the bond.

A **letter of credit** is an agreement by the person or institution that issues the letter to pay a sum of money on receipt of an invoice and other documents (typically documents indicating that the goods were insured against damage in transit). The Uniform Commercial Code governs letters of credit.

A **negotiable instrument** is a written document signed by a person who makes an unconditional promise to pay a specific sum of money on demand or at a certain time to the holder of the instrument. The most common forms of negotiable instruments are checks, notes, drafts, and certificates of deposit. They are governed primarily by the UCC. (Negotiable instruments are discussed in greater detail in Chapters 17 and 18.)

Any contract that is not a formal contract is an **informal contract**, also called a *simple contract.* Informal contracts may in fact be quite complex, but they are called simple because no formalities are required in making them.

formal contract
A contract that must have a special form or must be created in a specific manner.

contract under seal
A contract that has a seal certifying its legality. Such contracts require no consideration for them to be legal.

recognizance
An obligation in which a party acknowledges in court that he or she will perform some specified act and/or pay a price on failure to do so.

letter of credit
A binding document that a buyer can request from a bank to guarantee that the payment for goods will be made to the seller.

negotiable instrument
A written document signed by a person who makes an unconditional promise to pay a specific sum of money on demand or at a certain time to the holder of the instrument; an acceptable medium for exchanging value from one person to another.

informal contract
A contract that requires no formalities.

The Agreement

LO 9-5 Identify elements of a valid offer and a valid acceptance.

The first element of a contract is the agreement. As noted earlier, this chapter focuses on contracts governed by common law, that is, contracts other than those for the sale of goods. Under the common law, the agreement begins when one party, the offeror, makes an offer to another party, the offeree.

GLOBAL Context

Offers and Acceptances in Saudi America

Saudi Arabian Flag

In most common law countries, including the U.S., advertisements, displays of goods, etc. are considered invitations to buyers to make an offer. Under Saudi Arabia's version of Sharia law, however, advertising a good or a price is considered a binding offer (*ljab*) by the seller that is accepted by a buyer. Thus, sellers are required to state their offers in the past tense (i.e., "I am selling this car for $5,000" becomes "I sold this car for $5,000") to demonstrate that, in the mind of the seller, the price is already set and the product is a good as sold. Once a buyer accepts the offer, the seller has no right to revoke the offer. Saudi law also requires agreement and acceptance to all parts of a contract to be completed all at once. If any part of a contract is not accepted during the first negotiation, the entire contract is invalid. This stipulation comes from rule of Sharia law that forbids *gharar*, or uncertainty, in contracts.

Exhibit 9-2

The Formation of a Contract

This exhibit illustrates the first element of the contract, the agreement. For a contract to exist, the parties also must have legal capacity to enter into a contract, exchange valid consideration, and be entering into a contract with a legal purpose. The contract is formed as soon as the second party makes his or her promise.

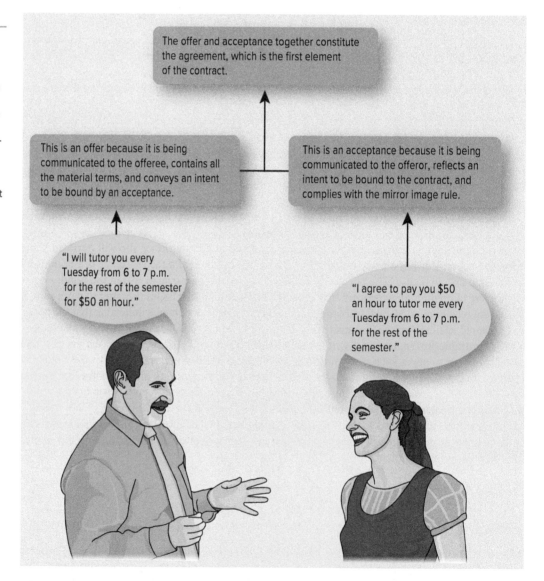

Elements of the Offer

An offer has three elements. Under the common law, these elements are intent, definite and certain terms, and communication to the offeree.

Intent

The first element of the offer is **intent**. The offeror must manifest an intent to be bound by the offeree's acceptance. Contracts are interpreted using an objective standard, meaning that the courts are concerned with only the party's outward manifestations of his or her intent, not what is going on in that party's mind. The courts interpret the words and actions of the parties the way a reasonable person would interpret them.

Thus, if an individual is clearly joking or speaking out in anger, the reasonable person would not think that the individual seriously intended to make an offer, and the courts would consequently not treat the words as an offer. If someone attempts to accept such an offer, the courts will not find that a contract has been made.

Sometimes an offeror may try to avoid a contract by later claiming that he was only joking when he made the offer, but the courts are not interested in one's hidden intent. As Case 9-3 demonstrates, if you joke too well, you may find yourself in an unwanted contract.

intent
The intended purpose or goal of an action, especially in a contract.

LUCY v. ZEHMER
Supreme Court of Virginia
196 Va. 493, 84 S.E.2d 516 (1954)

FACTS: *Plaintiffs W. O. and J. C. Lucy had wanted to purchase Ferguson Farm from the Zehmers for at least eight years. One night, Lucy stopped by the establishment the Zehmers operated and said that he bet Zehmer wouldn't accept $50,000 for the place. Zehmer replied that he would, but he bet that Lucy would not pay $50,000 for it. Over the course of the evening, the parties drank whiskey and engaged in casual conversation, with the talk repeatedly returning to the sale of Ferguson Farm. Eventually Lucy got Zehmer to draw up a contract for the sale of the farm for $50,000.*

When Lucy later attempted to enforce the agreement, Zehmer refused to complete the sale, arguing that he had been drunk and that the agreement to sell the property had been made in jest. Lucy sued to enforce the agreement.

ISSUE: Had Lucy and Zehmer entered into a legally binding agreement?

REASONING: The mental assent of the parties is not a requirement for the formation of a contract. If the words or other acts of one of the parties have but one reasonable meaning, that party's undisclosed intention is immaterial unless the other party somehow knows that the party does not mean what his words and actions would clearly indicate to any reasonable observer.

In this case, not only did Lucy believe that Zehmer was serious when he drew up the contract, but there was no evidence to indicate that any action taken by Zehmer would have indicated that he was not serious. When Zehmer asked his wife to sign the writing that he drew up containing the terms of the sale, he whispered to her that he was joking, but whispered so quietly that Lucy could not and did not hear it. What happened appeared to be a good-faith offer and a good-faith acceptance, followed by the execution and apparent delivery of a written contract. Therefore, the parties would be held to their outward manifestations of entering into a binding agreement.

DECISION AND REMEDY: The parties had entered into a binding agreement, and the court enforced the agreement.

SIGNIFICANCE OF THE CASE: This case demonstrates that the courts look at the circumstances surrounding the alleged making of a contract and focus on the outward manifestations of a person's intent. It clearly sends the message that people need to be careful when joking about entering into a contract because a person who jokes too well might find herself or himself bound to an agreement that she or he did not really want to make.

CRITICAL THINKING

Why must a joke be visibly a joke to a reasonable observer for there to be no acceptance?

ETHICAL DECISION MAKING

What stakeholders are being protected by the reasoning that supports the rule the court applied in this case?

BUT WHAT IF . . .

WHAT IF THE FACTS OF THE CASE OPENER WERE DIFFERENT?

Recall that in the case opener, the Pepsi commercial shows a fighter jet jokingly dropping a boy off at school. But what if the Pepsi commercial featured a man fully capable of owning and operating a fighter jet who turned in an appropriate and realistic number of number of Pepsi points, and Pepsi Company officials subsequently handed him the keys and deeds to such a prize? What if the commercial announcer also turned to the audience and seriously announced that such a prize was included in their prize catalog and could be theirs? Do both scenarios constitute a realistic offer, neither scenario, or only the second scenario?

Preliminary Negotiations An invitation to negotiate or an expression of possible interest in an exchange is not an offer. For example, if Rachael asked Bill whether he would sell his car for $5,000, she is not making an offer; she is just inquiring about his potential willingness to sell. Likewise, when a firm or government entity requests bids for a construction project, the request is just an invitation for contractors to make offers. Thus the bids would be the offers.

Advertisements Another illustration of the invitation to make an offer is the advertisement. If a custom furniture maker places an advertisement in the paper that reads, "Old-fashioned, hand-crafted cedar rocking chairs only $250 the first week in May," the store is merely inviting potential customers to come to the store and offer $250 for a rocker. Because no reasonable person would expect the store to be able to sell a rocking chair to every person who might see the ad, the court would interpret the intent of the store as being to invite readers to make an offer.

Under limited circumstances, however, an ad can be treated as an offer. If it appears from the wording of the ad that the store did in fact intend to make an offer, the courts will treat it as an offer. In general, when the ad specifies a limited quantity and provides a specific means by which the offer can be accepted, the courts will treat the ad as an offer, as demonstrated by the Case Nugget that follows.

The plaintiff in the Case Opener tried to rely on the *Lefkowitz* case to argue that the Pepsi commercial was an offer because it was "clear, definite, explicit and left nothing to negotiation." After all, the commercial clearly stated that 7 million points earned a Harrier jet, and the catalog provided an additional means of buying the points for cash.

The court in that case, however, found that the commercial could not be regarded as sufficiently definite because it specifically reserved the details of the offer to a separate writing, the catalog. Also, the commercial itself made no mention of the steps a potential offeree would be required to take to accept the alleged offer of a Harrier jet.

The court further found that the only offer in this scenario was the plaintiff's letter of March 27, 1996, along with the order form and the appropriate number of Pepsi points. Because Pepsi rejected this offer with its letter, there was no contract.

To prevent possible bait-and-switch advertising, some states have consumer protection laws requiring advertisers to state in their ads either that quantities of the item are limited to the first X number of people or that rain checks will be available if the item sells out.

CASE Nugget

When Is an Ad an Offer?

Lefkowitz v. Great Minneapolis Surplus Store, Inc. Supreme Court of Minnesota

251 Minn. 188, 86 N.W.2d 689 (1957)

In this case, the defendant had published a newspaper announcement stating: "Saturday 9 AM Sharp, 3 Brand New Fur Coats, Worth up to $1,000.00, First Come First Served $1 Each." Morris Lefkowitz arrived at the store, dollar in hand, but was informed that under the defendant's house rules the offer was open to ladies but not gentlemen. The court ruled that because the plaintiff had fulfilled all the terms of the advertisement and the advertisement was specific and left nothing open for negotiation, a contract had been formed.

From this case came the often-quoted exception to the rule that advertisements do not create any power of acceptance in potential offerees: an advertisement that is "clear, definite, and explicit, and leaves nothing open for negotiation." In that circumstance, "it constitutes an offer, acceptance of which will complete the contract."

Auctions Another situation in which what seems to be an offer may not be is the auction. When a person places a good with an auctioneer for sale by auction, is the seller making an offer or is the bidder? The answer depends on what kind of auction is taking place.

If nothing is stated to the contrary in the terms of the auction, an auction is presumed to be *with reserve*. In an auction with reserve, the seller is merely expressing an intent to receive offers. The auctioneer may withdraw the item from auction at any time before the auctioneer's hammer falls, signaling the acceptance of the bid. Similarly, at any time before the hammer falls, the bidder may also revoke the bid.

In an auction *without reserve,* the seller is treated as making an offer to accept the highest bid. Thus, the seller must accept the highest bid. Not surprisingly, very few auctions are without reserve.

Definite and Certain Terms

Under the common law, the terms of the offer must be definite and certain. In other words, all the material terms must be included. The *material terms* are those terms that allow a court to determine what the damages are in the event that one of the parties breaches the contract.

Sometimes an offer contains not the material term itself but a method for determining the term. For example, Guy's Sailboats is building Sara a sailboat, and the parties want to make it possible for her to pay one-third of the price of the boat in advance, one-third on delivery, and one-third in 12 monthly payments, with interest, beginning a month after delivery. Rather than stipulating an interest rate to be charged on the monthly payments, the contract might specify an external standard according to which the interest rate would be set through the course of the 12-month payment period.

The question of whether the terms of an alleged offer were adequate for the formation of a valid contract often arises in cases when one party believes that a contract had been formed and the other believes that there had been no contract because the terms of the alleged offer were not definite enough. It is therefore really important when drafting a contract to ensure that all the terms are included and clearly stated.

Communication to the Offeree

The offer must be communicated to the offeree or to the offeree's agent. Only the offeree to whom the offer was directed can accept the offer. If Bill overhears Sam offer to sell his car to Helen for $5,000, he cannot walk over and form a contract by accepting the offer. If he says to Sam, "I'll give you $5,000 for your car," he is not accepting the offer but, rather, is making a new offer.

Termination of the Offer

LO 9-6 Explain how an offer may terminate.

At some point in time, offers terminate; when an offer terminates, it can no longer be accepted to form a binding contract. There are five ways to terminate an offer: revocation by the offeror, rejection by the offeree, death or incapacity of the offeror, destruction or subsequent illegality of the subject matter of the offer, and lapse of time or failure of other conditions stated in the offer.

Revocation by the Offeror

The offeror is said to be the master of his offer and, as such, can revoke it at any time, even if he says he will hold the offer open for a stated period of time. If Jim sends Carol a letter offering to mow her yard every week during the summer for the price of $20 a week as long as she responds to his offer within the next month, he can still change his mind and tell her at any time before she responds that he is no longer interested in working for her, thereby revoking his offer.

If a person wishes to ensure that an offer will in fact be held open for a set period of time, the person may do so by entering into an option contract with the offeror. An **option contract** is an agreement whereby the offeree gives the offeror a piece of consideration in exchange for the offeror's agreement to hold the offer open for the specified period of time.

There are a number of situations when you might want to enter into an option contract. For example, if the person who made the offer is ill, and you are afraid he or she might die, you would want to enter into an option contract because you would then still have the time of the option to enter into the contract with the estate of the offeror; otherwise, the offer dies with the offeror. You might also want to enter into an option contract if you really want the object of the offer but need time to make sure you can finance it, and you do not want the offeror to sell the item to another party. For example, if the offeror offers to sell you a valuable antique car, but you are not sure you can afford the financing, you might want to get an option contract to give you time to secure financing.

There is no requirement as to the value of the consideration. If the consideration is money, the parties may agree that if the offer is eventually accepted and a contract is formed, the consideration for the option contract will become part of the payment for the contract. This situation frequently arises in contracts involving real estate. Jones may be considering opening a restaurant and would like to have the option of purchasing a lot owned by Smith, so he gives Smith $1,000 for a 30-day option to purchase the lot, with the provision that the $1,000 will be deducted from the purchase price if Jones purchases the property. In the event that Jones does not buy the property, Smith will keep the $1,000.

The UCC imposes an additional limitation on the offeror's ability to revoke the offer. A promise to hold an offer for the sale of a good open for a specified period cannot be revoked if it is made in writing and signed by a merchant.

As a general rule, the revocation of an offer is effective when it is received by the offeree. Because of this rule, if it is really important to the offeror to know that his offer has been revoked, he should personally deliver the revocation to the offeree.

Rejection by the Offeree

The second means by which an offer can be terminated is rejection by the offeree. Regardless of how long the offer was stated to be open, once the offeree rejects it, it is terminated. In our earlier example, if Carol calls Jim and says that she is not interested in his working for her this summer or any summer because of the poor quality of the work he has done for her in the past but then calls him back an hour later to say that she has changed her mind and would like to hire him in accordance with his proposed terms, it is too late. There is no offer for her to accept because her rejection terminated the offer.

option contract
An agreement whereby the offeree gives the offeror a piece of consideration in exchange for the offeror's agreement to hold the offer open for the specified period of time.

Death or Incapacity of the Offeror

The offer terminates immediately if the offeror dies or loses the legal capacity to enter into the contract. This termination occurs even if the offeree does not know of the terminating event. The exception to this rule occurs when the parties had already entered into an option contract to hold the offer open for a set period of time. If an option contract exists, the administrator of the offeror's estate or the guardian of the offeror must hold the offer open until it expires in accordance with the option contract.

Destruction or Subsequent Illegality of the Subject Matter

If the subject matter of the offer is destroyed or becomes illegal, the offer immediately terminates. For example, if Jamie offers Bill a job managing the riverboat casino he plans to open on January 1, but, before Bill accepts the offer, the state decides no longer to allow riverboat casinos to operate in the state, the offer of employment terminates.

Lapse of Time or Failure of Another Condition Specified in the Offer

As noted earlier, the offeror has the power to revoke the offer at any time, even if the offer states that it will be held open a set amount of time. But if the offer states that it will be held open for only a certain time, the offer will terminate when that time expires. In the absence of a time condition in the offer, the offer will expire after the lapse of a reasonable amount of time. What constitutes a reasonable amount of time varies, depending on the subject matter of the offer. For example, an offer by a retailer to purchase seasonal goods from a wholesaler would lapse before an offer to purchase goods that could be easily sold all year long.

Elements of the Acceptance

Once an offer has been made, the offeree has the power to accept that offer and form a contract. In examining the acceptance under the common law, the basic requirements for a valid acceptance parallel those for a valid offer. There should be a manifestation of intent to be bound by the acceptance to the contract, agreement to the definite and certain terms of the offer, and communication to the offeror.

Manifestation of Intent to Be Bound to the Contract

In general, there are two ways an offeree can manifest her or his intent to enter into the contract: by performance or by a return promise. The offeree must either do something or say something to form the contract.

Recall the distinction between a bilateral contract and a unilateral contract. If the offer is for a unilateral contract, the offeree can accept only by providing the requested performance. If Bill offered to pay $500 to anyone who returned his lost dog to him, Mary could accept the offer only by returning the dog. Bill did not want her promise, and if she called and promised to return the dog to him, that promise would have no legal effect because the only way to accept a unilateral offer is by performance.

With a bilateral contract, what the offeror wants to form the contract is not performance but, rather, a return promise. Sometimes, however, it is not clear whether the offeror wants performance or a return promise. In such a case, the offeree has the option of either performing or making a return promise.

Silence as a Form of Acceptance Silence, as a general rule, cannot be used to form a contract. For example, Lisa and Marie both work at a local diner where the manager is very flexible and lets the waitresses trade shifts. Marie calls Lisa and leaves her a voice-mail message saying,

©Digital Archive Japan/Alamy

GLOBAL Context

Contracts in Japan

The Japanese tend to view contracts as ongoing relationships in which the parties work with each other to smooth out any problems that arise in performance of the contract. The Japanese tend to be suspicious of long, detailed contracts. They have a distinct preference for short, flexible contracts that leave a number of terms to be decided later.

When negotiating a contract with a Japanese businessperson, an American businessperson must recognize how the Japanese will interpret his attempts to include more details in the agreement. The American should think about which details are essential for inclusion.

"I can't work my three night shifts this week. If you can cover them for me, I'll pay you an extra $40 on top of the money you'll receive from the boss for working my shifts. If I don't hear from you by 7 p.m. tomorrow, I'll assume we have a deal and you will work for me. Thanks so much!" If Lisa does not call back, no contract has been formed because silence under these circumstances would not constitute an acceptance.

The Restatement of Contracts, however, offers three circumstances under which silence can be an acceptance. First, the offeree receives the benefits of the offered services with reasonable opportunity to reject them and knowledge that some form of compensation is expected, yet the offeree remains silent. Second, the offeror tells the offeree that silence or inaction will constitute an acceptance, and the offeree, by remaining silent, *intends* to accept. Third, the parties, by their previous course of dealing with each other, have established a pattern of behavior whereby it would be reasonable to assume that silence was intended to communicate acceptance. For example, a wholesaler and retailer have a long-standing relationship in which the wholesaler routinely ships a certain type of merchandise to the retailer, who rejects a shipment if it does not meet his needs. After such a pattern of behavior has been established, it is reasonable for the wholesaler to assume that when a shipment is not sent back, the retailer means to accept it.

Acceptance of Definite and Certain Terms: the Mirror-Image Rule

mirror-image rule
The principle that holds that the terms of the acceptance must mirror the terms of the offer; if the terms of the acceptance do not mirror the terms of the offer, no contract is formed, and the attempted acceptance is a counteroffer.

When a bilateral contract is being formed under the common law, the mirror-image rule applies to the acceptance. The **mirror-image rule** says that the terms of the acceptance must mirror the terms of the offer. If the terms of the acceptance do not mirror the terms of the offer, no contract is formed. Instead, the attempted acceptance is a counteroffer.

The mirror-image rule has caused a significant amount of trouble for businesses because often a buyer would make an offer on one form that contained not only the essential terms of the contract but also a number of additional, somewhat minor terms. Generally, the offer was on a standard form that had a place where the essential terms were written in. The seller would often return his standard form with its preprinted terms, along with the essential terms written in to match the terms of the offer. The problem was that because these preprinted forms did not match, there was legally no contract. If both parties performed their contracts, no problems arose; but the fact that there was no contract meant that on occasion one party could decide simply to not perform, leaving the other party with no remedy. Or, if the parties did perform, the buyer might perform according to the terms on the buyer's form and be sued for breach by the seller, who believed that the terms on the seller's form were the terms of the agreement.

To eliminate this problem of the battle of the forms, UCC Section 2-207 significantly modified the mirror-image rule by providing that, as a general rule, an offeree may include in the acceptance terms that are additional to or different from the terms of the offer as long as (1) the offeror did not explicitly state in the offer that all terms of the offer must be accepted exactly as proposed; (2) the offeror does not promptly reject the new terms on receipt of the acceptance; and (3) the new terms do not materially change the terms of the original offer.

E-COMMERCE and the Law

Disclosure of Definite and Certain Terms

Many people would like to know what sort of terms they will be held to when buying a product. Unfortunately, many people who aren't aware of the importance of receiving prior disclosures also don't realize that online businesses commonly refrain from giving full disclosure of terms about a product until the payment information has already been given. If e-businesses do happen to provide these certain terms before a customer agrees to buy a product, the link to the terms themselves may be hard to find. Another aspect of the terms (if given) is that the language in which the contract is written is dense legal language, which may be hard for the ordinary person to understand fully. These e-business practices in regard to disclosure of terms can be very deceptive. A customer may inadvertently agree to a contract when he or she is given only some information about it. This situation would be as if a student were given a page of text that appeared to be whole and complete, and the student had completed an entire report on the text before being given an additional 50 pages that were also part of the text. Those customers who shop online should be aware of the potential for deception with online sellers and should use caution when making purchases.

In the case of additional terms—terms related to issues not addressed in the original offer—if none of the foregoing conditions occur, a contract is formed and the terms of the agreement are those of the acceptance, as long as the parties are merchants. If the parties are not merchants, the additional terms are merely proposals for additions to the contract and must be accepted by the offeror.

If the terms of the acceptance are different from those in the offer, various states treat the terms differently. In most states, the terms are said to cancel each other out, and the courts then look to the UCC for neutral terms to insert into the contract.

Communication to the Offeror

An offeror has the power to control the means by which the acceptance is communicated, and if the offeror specifies that only a certain means of communication will be accepted, then only that method of communication forms a valid offer. Suppose, for example, that Jennifer offers to paint Bill's car for $500 but says that he must accept the offer by telephone before midnight on Thursday. If he sends her an email Thursday morning accepting her offer, there is no valid contract. Even though email might be a valid means of accepting a contract offer if no means is specified, when the offer is limited to a specific means of communicating the acceptance, only that means results in a valid contract. Thus Bill's attempted acceptance was simply a new offer.

If no means of communicating the acceptance is specified, any reasonable means is generally acceptable. Telephone, telegraph, mail, fax, and email are all valid means of accepting an offer. If a person drafting an offer wants acceptance to be only by a particular means, the offer must make it clear that only a certain means is allowed.

The Mailbox Rule
Because not all acceptances are made in person, a rule needed to be developed to determine the point at which an acceptance made through the mail became effective. The courts settled on the **mailbox rule,** which provides that an acceptance is valid when it is placed in the mailbox, whereas a revocation is effective only when received by the offeree. Today, the mailbox rule has been expanded to apply to faxes in some jurisdictions, with courts holding that once the fax has been transmitted, the acceptance is effective. It is likewise applied to email transmissions in that once the acceptance is sent, it is considered accepted.

The Effect of an Acceptance after a Rejection
As previously stated, if an acceptance is received after a rejection is received, the acceptance is not valid because the rejection terminated the offer. However, sometimes a rejection is dispatched, but before it is received, the acceptance is communicated to the offeror. In that case, a valid contract has been formed because the rejection is not effective until it is received. Suppose Brenda emails an offer to Harry, and he puts a rejection in the mail. Then, before the rejection is received, Harry calls Brenda and tells her he accepts. A valid contract has been formed, and the rejection will have no effect when it is received. However, if the telephone call had been made after Brenda had received the rejection, there would have been no contract.

> **mailbox rule**
> The principle that holds that an acceptance is valid when it is placed in the mailbox, whereas a revocation is effective only when received by the offeree. In some jurisdictions, the mailbox rule has been expanded to faxes.

CASE OPENER WRAP-UP

The Problematic Promotion

Much to the plaintiff's dismay, the court in the Pepsi case found that the commercial could not be regarded as sufficiently definite to be an offer, because it specifically reserved the details of the offer to a separate writing, the catalog. Also, the commercial itself made no mention of the steps a potential offeree would be required to take to accept the alleged offer of a Harrier jet. As in most cases where a consumer attempts to place an order for an advertised item, the court regarded the plaintiff's purported acceptance as an offer. And it was an offer that Pepsi obviously rejected. And, while the court did not specifically mention this factor, common sense should have indicated to the plaintiff and his family that Pepsi did not really intend to give a Harrier jet as one of the promotional prizes.

PRACTICAL TIPS for Business Managers

1. When negotiating a contract with another business, be sure to take your time to discuss the nature and expectations of the proposed relationship and to prepare for any problematic contingencies. You could be saving yourself a lot of time and money by making sure that the contract is complete and clear now instead of fighting about the contract in court later.
2. Remember that contracts do not necessarily have to be in writing to be binding agreements. Whenever discussing an agreement with another party, keep in mind the criteria of an implied contract so you know whether you are agreeing to something that is binding.
3. *Lucy v. Zehmer* is a classic example of how an offer in jest can still be a reasonable and binding offer. While the authors are not suggesting that you never crack another joke, it is always wise to be careful and deliberate with your words, especially when those words may sound like an offer to a reasonable listener.

SUMMARY

The Definition of a Contract	Contracts at their simplest level are legally enforceable agreements. A *valid contract* is generally viewed as one that has the following elements: agreement, consideration, legal object, and parties with legal capacity.
Sources of Contract Law	There are two sources of contract law: the Uniform Commercial Code and state common law. The Uniform Commercial Code, in Article 2, governs contracts for the sale of goods. All other contracts are governed by the common law.
Classification of Contracts	Contracts may be classified in a number of ways. Every contract is either unilateral or bilateral; express or implied; valid, voidable, or void; executed or executory; and formal or informal.
The Agreement	The *agreement* is the first element of a contract. It begins when one party, the offeror, makes an offer to another party, the offeree.
Elements of the Offer	A valid offer contains the manifestation of the offeror's intent to be bound to the definite and certain terms that are communicated to the offeree.
Termination of the Offer	An offer can be terminated by revocation by the offeror, rejection by the offeree, death or incapacity of the offeror, destruction or subsequent illegality of the subject matter of the offer, or lapse of time or failure of other conditions stated in the offer.
Elements of the Acceptance	An acceptance is valid when a manifestation of intent to be bound to the terms of the offer is communicated to the offeror by the offeree.

Point/Counterpoint

Innovations in blockchain technology, the system by which cryptocurrencies such as Bitcoin operate, have made possible a new type of contract: the smart contract. Smart contracts are contracts written in computer code that automatically executes when a set of conditions is met. For example, say one party wants to rent an apartment from another. They can set up a smart contract that is coded so that when the landlord sends the renter the digital entry key to the apartment, sufficient funds will automatically be transferred from the renter's bank account to the landlord. If the key does not arrive, no funds are sent. The contract is encrypted and stored digitally on computers, and the contract's code cannot be altered by either party without alerting the other.

Should the Business World Embrace Smart Contracts?	
YES	NO
Smart contracts utilizing blockchain technology have many benefits over traditional contracts and will no doubt be the new standard in only a few years. Perhaps the biggest benefit of smart contracts is that they entirely cut out the middleman. No longer does someone need to draft a physical contract, track down a lawyer or notary, pay to have the contract notarized, and wait for all the paperwork to go through. With a smart contract, the parties simply need to decide on their terms and create a smart contract. The smart contract can contain all the rules, penalties, and terms of a conventional contract, but a smart contract will automatically enforce the obligations. Businesses can save time and money by switching over to smart contracts. Another benefit of smart contracts is that they are secure, both from interference from the contractual parties and from malignant third-parties. This means that there's no possibility for either party to change the contract in bad faith without alerting the other party, effectively nullifying the changes. Arguably, due to the smart contract's data being kept on a decentralized cloud ledger, smart contracts are more secure from hackers than traditional online transactions where the data is stored on one set of servers owned by a business. A final advantage of smart contracts is that they require accountability of both parties. The smart contract blockchain acts as a ledger that keeps track of all the actions taken by the parties concerning the contract. For example, say a business wants to hire a trucker to ship their products to a buyer. The trucker delivers the product, the delivery is recorded in the blockchain, and the smart contract then transfers the trucker's fee to his account. Later, the trucker tries to scam the business into paying him twice by claiming that he has not been paid. However, one needs only to look at the blockchain ledger to see that the payment did indeed go through. No trickery or confusion can occur in a smart contract, meaning the chances of breach are far lower and everyone is more likely to get what they want.	While great strides in blockchain technology have been made in the last few years, smart contracts simply are not ready to be adopted by the business world quite yet. First of all, while it is true that traditional contracts can contain many complicated clauses and terms, at least they are written in English. Smart contracts are written in code, meaning a business needs to hire a coder or programmer to write smart contracts. But coders are people too, and they can make mistakes. Whereas a traditional contract can be reviewed by the parties for errors, it is much harder for individuals untrained in a programming language to proofread a contract. This can even be used to gain an unfair advantage if one party can read the code and the other cannot. Another problem with smart contracts is that once they are written and implemented into the blockchain, they are immortal and unchangeable. A smart contract cannot be edited or revoked without writing an entirely new smart contract to override the previous one (and even then, the original smart contract still technically exists). This fact exacerbates any errors made in the "drafting" process of a smart contract, as any submitted errors are unfixable. His fact leads to several issues. First, if a government entity one day decides to regulate smart contracts, existing contracts may have provisions that are made illegal by new legislation (or, perhaps, the entire concept of a smart contract may be made illegal). Second, there is no set of governmental laws in place to make smart contracts legally binding. While the advocates of smart contracts tout that smart contracts are automatically enforceable and therefore either the deal goes through 100 percent or not at all, it seems shortsighted to think that smart contracts will inevitably work out in every circumstance and/or that no one will be wronged by a smart contract. Thus, it is probably unwise for a business to begin using smart contracts without a protective legal framework.

Questions & Problems

1. Explain why the first question a person should ask when getting ready to analyze a contract problem is, "Is this alleged contract a contract for the sale of a good?"

2. What is the difference between an offer for a unilateral contract and an offer for a bilateral contract? Why might that difference be important to understand?

3. What must a party prove to recover under the theory of quasi-contract?

4. What is the mirror-image rule?

5. What is the mailbox rule?

6. In 2014, Julie Gerstenecker, who was married to Adam Gerstenecker, had approximately $78,000 of student loan debt. Hearing Julie lament about the interest on her loans, Adam's mother, Janice Gerstenecker offered Julie an interest-free loan to pay off her student loan debt in return for monthly payments. In September 2014, Julie sent Janice an email with the details of Julie's student loan debt. The next day, Janice mailed checks covering the full amount of Julie's debt to the Department of Education Federal Loan Servicing and Sallie Mae. The agreement was not put into writing.

 In early 2015, Julie and Adam were going through divorce proceedings. An email exchange between Julie and Adam discussing divorce terms indicated that Julie's monthly payments to Janice would be reduced from $700 per month to $500 per month.

 In December 2015, after a year of non-payments, Janice sued Julie for breach of contract. In the trial, Janice testified that she had received four checks from Julie before the payments had ceased. Julie testified that she had no recollection or explanation as to why she had sent those checks to Janice. Ultimately, Julie argued that there was no contract between them because "[t]here is no tangible evidence supporting the assertion that Julie intended to agree to pay [Janice] back."

 What are the two ways an offeree can manifest intent to enter into the contract? Although there was no explicitly signed contract, what actions by Julie indicate that there was an agreement of the offer? How do you think the court ultimately resolved this case?

 Source: Gerstenecker v. Gerestenecker, Ala: Supreme Court (2017).

7. Timothy Reddick served in a two-year term appointment with the FDIC, with the term lasting from September 2010 to September 2012. In April 2012, prior to the end of the contract, the FDIC offered an extension to Reddick subject to the condition of Reddick's continued successful performance. Reddick accepted the offer several days after it was made.

 Unfortunately for Reddick, the FDIC revoked the extension offer in August 2012 due to his failing to meet the condition of continued successful performance by engaging in "highly questionable behavior." His employment with the agency terminated in September 2012. Reddick subsequently sued the FDIC. Do you think the FDIC was allowed to revoke the offer after Reddick's acceptance? Explain your reasoning.

 Source: Reddick v. FDIC, 809 F. 3d 1253 (2016).

8. NDWC Investment Properties owned six properties in DeKalb County, Georgia, which were secured by fully amortized 30-year loans. At some point before the trial, NDWC entered into negotiations with Branch Banking and Trust (BB&T) to refinance their loans. In the course of the negotiations, BB&T offered NDWC a lower interest rate on all six of the DeKalb County properties but only for a five-year term. NDWC expressed concerns about moving from 30-year fixed loans to a series of five-year loans. According to NDWC, BB&T assuaged NDWC's fears by assuring them that BB&T would continue to refinance the remaining balance on the loans in four successive five-year terms at the current market interest rate. NDWC made an agreement based on the oral representations by BB&T.

 In December 2009, NDWC transferred the six DeKalb properties along with the loans to a newly created company named Corso Properties. BB&T approved the loan transfer but explained to Corso that the term would be for one year instead of five because Corso was a new company. Corso asserted that it had made an oral agreement with BB&T that BB&T would refinance the remaining balance in successive five-year terms if Corso satisfied certain conditions. In 2011, BB&T refused to renew the loans despite Corso's claims that it had met the required conditions.

 Corso subsequently filed an action against BB&T, asserting claims of breach of Georgia's implied covenant of good faith and fair dealing when making contracts, fraud in the inducement, and fraud in the general. BB&T filed a motion to dismiss based on the plaintiff's failure to allege an independent breach of contract claim, as well as not pleading their fraud claims with enough details. The plaintiffs submitted an amended complaint which no longer contained the fraud in general claim but now included a breach of oral contract claim.

BB&T responded with a second motion to dismiss, asserting that the amended complaint failed to state a claim upon which relief could be granted. The plaintiffs did not respond to the defendant's second motion to dismiss and the defendant was deemed unopposed.

What arguments could the plaintiffs have made if they had responded to the defendant's second motion? Although the defendant won by being deemed unopposed, the court still reviewed the defendant's grounds for dismissal. How did the court rule on each of the three claims: Breach of Oral Contract, Breach of Duty of Good Faith and Fair Dealing, and Fraud in the Inducement? How do state laws affect this decision? Recall the WPH standards for Business Ethics. Although the court would have ruled in favor of BB&T by a matter of law, was the way BB&T acted ethical?

Source: *Corso Properties, LLC and WC Investment Property, Inc. v. Branch Banking And Trust Company,* (2013 U.S. Dist. Lexis 137955).

9. In April 2011, Elias Groisman entered into a contract with Shabsi Pfeifer to sell three real properties to Pfeifer on the condition that Pfeifer satisfied two liens on the properties against Groisman's name and upon payment of $55,000. The email which contained the offer did not have an end date by which Pfeifer had to complete performance.

 More than six months later, Groisman entered a contract of sale for the three properties with Maya Development, LLC. In December 2012, Pfeifer filed a complaint against Groisman and Maya Development, seeking specific performance of the contract for the sale of real properties.

 The trial court dismissed the case, and Pfeifer appealed to the New York Appellate Division Supreme Court. Did Groisman and Pfeifer enter into a valid contract on April 2011? What kind of agreement did both parties have instead? Based on your answers to the previous answers, who won the court case? Why?

 Source: *Shabsi Pfeifer v. Elias A. Groisman, et al.,* 997 N.Y.S.2d 706 (N.Y. App. Div. 2014).

10. Dr. Griffith allowed his life insurance to lapse after May 15, 2007. According to US Life's life insurance policy, he was granted a 31-day grace period, after which he would be able to reinstate his insurance by paying the balance of the unpaid bill and receiving written approval from *US Life* of the required evidence of insurability. According to a reminder notice, Griffith had 60 days to make a full payment. On or around June 15, 2007, the insurance provider sent Griffith a lapse notice that included a reinstatement form and a self-addressed envelope, and required payment and the reinstatement form to be received by the policy provider within 30 days from the lapsed coverage. On July 23, 2007, Griffith electronically directed payment to American Medical Association Insurance Agency (AMAIA), which acted as a third-party administrator for *US Life*. AMAIA acted on *US Life's* behalf to bill and collect premiums. AMAIA received the check from Griffith on July 30, 2007. On July 28, 2007, Griffith was kneeling beside his bicycle at Bethany Beach, Delaware, when he was struck by a car that had drifted off the road when the driver fell asleep at the wheel. Do you think Griffith was insured at the time of his death? Why or why not?

 Source: *US Life Insurance Company v. Wilson,* 2011 Md. App. LEXIS 52.

11. Adwoa Gyabaah was hit by a bus owned by Rivlab Transportation Company. She retained attorney Aronsky to represent her in negotiations with the bus company, their insurance company, and their attorneys for a contingency fee of one-third of her recovery. On October 1, 2010, defendant tendered Aronsky its $1 million policy limits for purposes of settlement. After Aronsky explained the settlement to her, Gyabaah said she accepted it and signed a general release on October 5, 2010. Aronsky said he would hold onto the release for her until she decided whether she wanted the settlement in one lump sum or paid over a period of time.

 On December 9, 2013, Gyabaah retained a new lawyer, Kenneth Wilhelm, and he advised Aronsky that Gyabaah did not wish to settle the case nor have the release sent over to the defendant. Aronsky filed a motion with the court seeking to enforce the $1 million settlement and setting his contingency fee at one-third of the recovery in accordance with the contingency agreement. The relevant issue in this case was whether there was a valid settlement agreement between the bus company and Gyabaah. Do you think the court found a valid agreement? Why or why not?

 Source: *Gyabaah v. Rivlab Transp. Corp.,* 22 N.Y.3d 1018, 4 N.E.3d 359, 2013 N.Y. LEXIS 3446 (2013).

Part 2 Contract Law

CHAPTER 10

Consideration

CASE OPENER

Upper Deck—Contract Liability or Gift?

In 1988, the Upper Deck Company was a baseball card company with an idea for a better baseball card: one that had a hologram on it. By the 1990s, Upper Deck was a major corporation whose value was at least a quarter of a billion dollars. In 1988, however, the outlook wasn't bright for Upper Deck, which lacked the funds for a $100,000 deposit it needed to buy some special paper by August 1. Without that deposit, its contract with the Major League Baseball Players Association would have been jeopardized.

Upper Deck's corporate attorney, Anthony Passante, Jr., loaned the company the money. That evening, the directors of the company accepted the loan and, in gratitude, agreed to give Passante 3 percent of the firm's stock. Passante never sought to collect the gift of stock, and later the company reneged on its promise. Passante sued for breach of oral contract.[1]

1. If you were on the jury, how would you decide the case? (The law of consideration will determine the outcome.) Was the offer of 3 percent of the firm's stock legal consideration for the loan? Or was it a mere gift?
2. Does Upper Deck have a moral obligation to give Passante the stock? If so, is this legally enforceable?

LEARNING OBJECTIVES

LO 10-1 Define consideration.
LO 10-2 Recognize when consideration is required.
LO 10-3 Describe what promissory estoppel is and when it can be used.
LO 10-4 Describe adequacy of consideration.
LO 10-5 Define illusory promise.
LO 10-6 Explain past consideration.
LO 10-7 Define preexisting duty.
LO 10-8 Differentiate between liquidated debt and unliquidated debt.
LO 10-9 Describe the concept of accord and satisfaction.

[1] Passante v. McWilliam, 53 Cal. App. 4th 1240 (1997).

TYPE OF CONSIDERATION	EXAMPLE
A benefit to the promisor	A promise to stay in a job until a particular project is complete (This is a benefit to the employer.)
A detriment to the promisee	A promise to your football coach to refrain from riding your motorcycle during football season even though you love riding it
A promise to do something	A promise to cook dinner for your roommate for the next six months
A promise to refrain from doing something	A promise to stop drinking alcohol during exam week

Exhibit 10-1

Examples of Consideration

What Is Consideration?

LO 10-1 Define consideration.

Rule: *Consideration is what a person will receive in return for performing a contract obligation.* Suppose, for example, that Dan agrees to purchase Mary's car for $1,000. Dan's payment of $1,000 is the consideration Mary will receive for her car. Title to and possession of the car is the consideration Dan will receive in exchange. Consideration can be anything, as long as it is the product of a bargained-for exchange. In a business context, it is often (but not always) money. Other examples of consideration are presented in Exhibit 10-1.

Is a promise to refrain from something you are legally entitled to do appropriate consideration for a contract? In Case 10-1, the court considers whether a nephew's promise to refrain from certain vices is adequate consideration for his uncle's promise to pay him $5,000.

HAMER v. SIDWAY

New York Court of Appeals

124 N.Y. 538 (1891)

FACTS: William E. Story and his nephew, William E. Story, 2nd, agreed that the uncle would pay his nephew $5,000 if the nephew would refrain from drinking, using tobacco, swearing, and playing cards and billiards for money until he turned 21. When the nephew turned 21, his uncle sent him a letter that indicated that the nephew had earned the $5,000 and that he would hold the money with interest until the nephew became capable of taking care of it responsibly. The nephew accepted the terms. The uncle died 12 years later without having transferred the funds to his nephew.

The nephew assigned the funds to Louisa Hamer, who brought suit against the executor of the uncle's estate, Franklin Sidway, to collect the money. Sidway, however, argued that the contract was without consideration to support it and, therefore, invalid. He asserted that the nephew, by refraining from the use of liquor and tobacco, was not harmed but benefited. He further asserted that what the nephew did was best for him to do independently of his uncle's promise and insisted that it followed that unless the uncle was benefited, the contract was without consideration.

At trial, judgment was entered in Hamer's favor but was reversed on appeal in Sidway's favor. Hamer appealed.

ISSUE: The issue before the court is whether a promise of forbearance (e.g., to refrain from particular activities) can be deemed valid and lawful consideration for a contract.

REASONING: Defendant contended that the contract was invalid because it lacked consideration and that there is no consideration unless the promisor is benefited. Such a rule could not be tolerated and is without foundation in the law. Consideration means not so much that one party is profiting as that the other abandons some legal right in the present or limits his legal freedom of action in the future. Now, applying this rule to the facts before us, the promisee

(continued)

used tobacco, occasionally drank liquor, and he had a legal right to do so. He abandoned that right for a period of years based upon the promise of his uncle that for such forbearance he would give him $5,000. We need not speculate on the effort which may have been required to give up the use of those stimulants. It is sufficient that he restricted his lawful freedom of action within certain prescribed limits upon the faith of his uncle's agreement. Now, having fully performed the conditions imposed, it makes no difference whether such performance was actually a benefit to the promisor, and the court will not inquire into it.

DECISION AND REMEDY: It is deemed established for the purposes of this appeal, that on January 31, 1875, defendant's testator was indebted to William E. Story, 2nd, in the sum of $5,000. All concur.

SIGNIFICANCE OF THE CASE: Forbearance is sufficient consideration for a valid and enforceable contract.

CRITICAL THINKING

What difference would it have made in this case had the nephew not had the legal right to drink or smoke? Why is this question crucial to the decision?

ETHICAL DECISION MAKING

What values is the idea of consideration upholding?

Rules of Consideration

LO 10-2 Recognize when consideration is required.

The key to understanding consideration is understanding the rules that govern it and their exceptions. We explore these principles in the following sections.

LACK OF CONSIDERATION

Rule: *For a promise to be enforced by the courts, there must be consideration.* A court will enforce one party's promise only if the other party promised something (an act, money, etc.) in exchange. For example, in a bilateral contract (a promise for a promise), the consideration for each promise is a return promise. Consider this example: Sue promises to pay Mike $2,000 for his car. Mike promises to sell Sue his car for $2,000. The exchange will be done tomorrow.

There is an oral contract. Sue's promise to pay Mike $2,000 for the car is her consideration to Mike. Mike's promise to sell Sue the car for $2,000 is his consideration to Sue. There has been a mutual exchange of something of value—a promise to pay $2,000 and a promise to sell the car for $2,000.

Another example of a bilateral contract, or a promise for a promise, occurred when the U.S. government seized control of insurance giant American International Group (AIG). The U.S. agreed to lend AIG up to $85 billion in exchange for nearly 80 percent of AIG's stock. The consideration AIG received was the promise of up to $85 billion in U.S. government loans. The consideration the U.S. received was the promise of almost 80 percent of AIG's stock.[2]

In a unilateral contract (a promise for an act), one party's consideration is the promise, and the other party's consideration is the act. Suppose your professor made the following statement in class: "If any student shows up at my house on Saturday and does the gardening, I will pay that student

[2] "U.S. Seizes Control of AIG with $85 Billion Emergency Loan," Washington Post, September 17, 2008, http://www.washingtonpost.com/wp-dyn/content/article/2008/09/16/AR2008091602174.html.

Exhibit 10-2
Type of Consideration Based on Type of Contract

TYPE OF CONTRACT	PROMISOR	PROMISEE
Bilateral	A promise	A promise
Unilateral	A promise	An act

$100." You show up and do the gardening. The professor's consideration to you is the promise of the payment of $100 on completion of the gardening, and your consideration to the professor is the act of completing the gardening. Once again, there has been a mutual exchange of something of value.

Exhibit 10-2 summarizes the consideration in bilateral and unilateral contracts.

Rule: *Promissory estoppel is the legal enforcement of an otherwise unenforceable contract due to a party's detrimental reliance on the contract.*

LO 10-3 Describe what promissory estoppel is and when it can be used.

One exception to the rule requiring consideration is **promissory estoppel**. Promissory estoppel occurs when three conditions are met:

- One party makes a promise knowing the other party will rely on it.
- The other party does rely on the promise.
- The only way to avoid injustice is to enforce the promise.

promissory estoppel
The legal enforcement of an otherwise unenforceable contract due to a party's detrimental reliance on the contract.

How does this work? Suppose that on graduation from college, Amanda receives a job offer across the country. She gives up her apartment, cancels all her other job interviews, and moves all her possessions. After arriving, she rents a new apartment and shows up for work. The only problem is that Amanda is told there is no job! May Amanda sue the employer? The answer in most states is yes, under the theory of promissory estoppel. She spent a lot of money in reliance on the job offer. She also moved across the country. Amanda may be able to recover her reliance damages (i.e., money she spent in reliance on the job offer). Promissory estoppel is not awarded regularly, but in the right case, it can provide a remedy when no other remedy exists.

In a recent case, the Ninth Circuit Court of Appeals held that Yahoo's promise to remove a nude photo from its website was subject to a claim of promissory estoppel. In that case, the plaintiff learned that her ex-boyfriend, pretending to be her, had posted nude photos of her on Yahoo. He also included all her contact information and an invitation for men to contact her for sexual purposes.[3] The plaintiff contacted Yahoo (in accordance with its established policies) and requested removal of the photo. Yahoo agreed but, despite repeated requests, did not remove the photo for six months. The court held that Yahoo's promise to de-post the profile meant that Yahoo had a duty to the plaintiff. As such, the plaintiff's claim of promissory estoppel could be maintained. If the plaintiff is able to prove that she reasonably relied on Yahoo's promise to her detriment, she may well prevail on her claim.

[3] "Do Interactive Websites Have a Legal Duty to Remove Malicious Content?" http://writ.news.findlaw.com/scripts/printer_friendly.pl?page=/ramasastry/20090519.html (retrieved May 25, 2009). This article discusses Barnes v. Yahoo, Inc., 2009 U.S. App. LEXIS 10940 (9th Cir. 2009).

CASE Nugget

Ruzicka v. Conde Nast Publishing

999 F.2d 1319 (1993)

Claudia Dreifus wrote an article for *Glamour Magazine* relating to therapist-patient sexual abuse. She promised she would not reveal the identity of the subject, Jill Ruzicka, whom she interviewed for the article. Ruzicka argued that Dreifus promised that she would not be "identified or identifiable" in the article. Dreifus refuted this, claiming that she only vaguely promised to do some masking. The article was published with Ruzicka's surname changed to "Lundquist," but included other information that made her "identifiable."

(continued)

In order to prevail on a claim for promissory estoppel, the plaintiff must prove; 1) that the promise was clear and definite; 2) that the promisor intended to induce reliance on the part of the promisee and that such reliance occurred to the promisee's detriment; and 3) that the promise must be enforced to prevent injustice. The district court granted motion for summary judgment against plaintiff, finding that the term "identifiable" was vague and therefore holding that the promise was not "clear and definite" enough to support recovery under promissory estoppel.

On appeal, the Eighth Circuit held that to be enforceable under promissory estoppel, a promise need not be as definite with respect to all details as promises forming conventional contracts. The appellate court also pointed out that Ruzicka approved an early draft of the article, but what was ultimately published revealed several identifying facts that Ruzicka had not provided to Dreifus. The Eighth Circuit Court of Appeals ruled in favor of Ruzicka. The court further opined that, based on U.S. Supreme Court precedent, enforcing the promise would not be unfair to the media defendant: "Where the press feels disclosure of [the] identity of a confidential source is valuable to the story and thus disregards its promise, the payment of compensatory damages is . . . simply a cost of acquiring newsworthy material to be published at a profit."

Adequacy of Consideration

LO 10-4 Describe adequacy of consideration.

Rule: *The court seldom considers adequacy of consideration.* This rule means that the court does not weigh whether you made a good bargain. Example: Donna purchases a stereo from Celia, a friend in her business law class. Donna pays $500 for the stereo but later realizes that the stereo is not worth even $100. May Donna sue Celia? Typically, the answer is no. It is Donna's responsibility to do her research and determine what price she should pay. The court will not set aside the sale because she made a bad deal.

There is an exception to this rule. Suppose that you are on the verge of bankruptcy, and your creditors are threatening legal action. You know that if you default on your mortgage, car payment, bank loans, credit card payments, and other bills, the court may order your assets to be sold to pay the creditors. Instead of risking this result, you decide to sell your assets. You sell your BMW to your best friend for $50 and your vacation house to your sister for $500. Now, asset-free, you declare bankruptcy, knowing that your creditors will be able to collect nothing. Is this permissible? The answer, of course, is no. In such a case, the court will look at adequacy of consideration. If it appears that the sale of the assets was done to avoid payment to creditors, the court may set aside each sale and sell the assets to pay your creditors. In Case 10-2, the court had to consider whether $1 plus love and affection was adequate consideration for the transfer of property.

THELMA AGNES SMITH v. DAVID PHILLIP RILEY
Court of Appeals of Tennessee, Eastern Section, at Knoxville
2002 Tenn. App. LEXIS 65

FACTS: Smith and Riley lived together out of wedlock for several years. They opened a joint checking account and, over time, both Smith and Riley deposited money into the account. In addition, Riley entered into a lease with Jerry Strickland and Wanda Strickland with respect to a residence owned by them. The lease was accompanied by an option to purchase. Almost four months later, Smith and Riley asked their attorney to prepare a bill of sale and an assignment. In the bill of sale, Riley transferred to Smith a one-half undivided interest in seven items of personal property. Riley also assigned to Smith a one-half undivided interest in the lease and option to purchase with the Stricklands; this interest included a right of survivorship in the one-half interest retained by Riley. The property Riley sold and assigned to Smith in the two agreements was stated in each to be "for and in consideration of the sum of One Dollar ($1.00) and other and good and

(continued)

valuable consideration, the sufficiency of which is hereby acknowledged." When the relationship ended, Smith filed suit against Riley in the trial court, seeking the dissolution of their domestic partnership and seeking to enforce the two written agreements with Riley regarding the sale and assignment of property to her. The trial court enforced the agreements and divided the parties's property. The defendant appealed, arguing that the agreements lacked consideration and were void as they were against public policy.

ISSUE: Is $1 plus love and affection adequate consideration for the two agreements between Smith and Riley?

REASONING: In the court's own words (abridged and slightly modified):

> It is a well-settled principle of contract law that in order for a contract to be binding, it must, among other things, be supported by sufficient consideration. In expounding on the adequacy of consideration, the court has stated that it is not necessary that the benefit conferred or the detriment suffered by the promisee shall be equal to the responsibility assumed. Any consideration, however small, will support a promise. In the absence of fraud, the courts will not undertake to regulate the amount of the consideration. The parties are left to contract for themselves, taking for granted that the consideration is one valuable in the eyes of the law.

Quoting the United States Supreme Court, the court went on to state that "[a] stipulation in consideration of $1 is just as effectual and valuable a consideration as a larger sum stipulated for or paid." Indeed, the consideration of love and affection has been deemed sufficient to support a conveyance. Both the bill of sale and the assignment recite that they are undertaken "for and in consideration of the sum of One Dollar ($1.00) and other and good and valuable consideration, the sufficiency of which is hereby acknowledged." Facially, the documents are therefore supported by sufficient consideration, as clearly recognized by the Supreme Court. Moreover, Smith's "society and consortium"—a concept comparable to the love and affection—is further evidence of sufficient consideration to support these conveyances.

DECISION AND REMEDY: The judgment of the trial court was affirmed. The case was remanded for enforcement of the judgment and for collection of costs. Costs on appeal were taxed to the appellant, Riley.

SIGNIFICANCE OF THE CASE: This case demonstrates that courts will generally not consider the adequacy of consideration.

CRITICAL THINKING

What is the reasoning of the appellant in terms of why the consideration was not adequate to cause the contracts to be enforceable? What key rule of law did this reasoning overlook?

ETHICAL DECISION MAKING

What values are being advanced by the logic of the relevant rule of law in this case? In other words, what values prevent the rule of law from being that consideration must be in an amount similar in value to the item or services being transferred for the contract to be enforceable?

CASE Nugget

Lack of Consideration

McLellan v. Charly Court of Appeals of Wisconsin

313 Wis. 2d 623, 758 N.W.2d 94 (2008)

The property in dispute is owned by Charly and located between the University of Wisconsin's Harlow Primate Psychology Laboratory and the Wisconsin National Primate Research Center. In July 2004, Rick Bogle, an animal rights activist, met with Charly to ask about purchasing the property. Bogle informed Charly he wanted to use the property for a National Primate Research Center Exhibition Hall, and Charly said he would sell the property to Bogle.

(continued)

Charly and McLellan executed a document titled "Option to Purchase" that granted McLellan the option of purchasing the property for $675,000 within 180 days. The agreement contained detailed provisions on the terms of the sale, including leaseback and repurchase provisions.

In June 2005, Bogle organized a rally at the property and posted a sign reading "Future Home of the National Primate Research Exhibition Hall," which contained a picture of a monkey with a device screwed into its skull. Charly had agreed the sign could be put up. Soon after the rally, a university representative called Charly to express its opposition to having the exhibition hall on the property and said the university was still interested in the property. Charly responded that he still wanted $1,000,000 for the property.

Shortly thereafter, Charly took the agreement he and McLellan had signed to another attorney. In a letter to McLellan on Charly's behalf, the attorney stated that the option was "voidable and void due to lack of consideration" and offered to enter into an option with different terms. McLellan declined to do that.

On August 1, 2005, Charly received a proposed option to purchase from the university under which the university had the option of purchasing the property for $1,000,000. On August 12, 2005, McLellan attempted to exercise his option to purchase the property, but the envelope was returned to him unopened. Through counsel, Charly informed McLellan there was no enforceable option and he declined to sell the property to him.

On appeal, Charly contends the circuit court erred in determining the leaseback and repurchase provisions were adequate consideration for the option. Among other reasons, he asserts the consideration required to make an option a binding contract must be separate from the consideration for the sale and therefore cannot be satisfied by the terms negotiated for the sale.

With respect to the claims for breach of the option, we conclude that, in order to make the option a binding contract, consideration is required that is separate from the consideration for the sale. We further conclude that the leaseback and repurchase provisions are not consideration separate from the consideration for the sale. Charly was therefore free to revoke the option.

Illusory Promise

LO 10-5 Define illusory promise.

Rule: *An illusory promise is not consideration.* What is an illusory promise? Suppose Shawn offers to sell Molly his skis for $300. Molly responds, "I'll look at them in the morning, and if I like them, I'll pay you." At this point, Molly has not committed to doing anything. The law considers this an **illusory promise**—that is, it is not a promise at all. Moreover, an illusory promise is not consideration.

Past Consideration

LO 10-6 Explain past consideration.

Rule: *Past consideration is no consideration at all.* Consideration is an essential element of a contract. For a court to enforce a promise, both sides must offer consideration. Imagine that you have graduated from college and gotten a great job. After you have worked at the company for five years, your boss says to you, "Because you have done such a great job the last five years, I am going to give you 5 percent of the company stock." Six months later, you still have not received the stock. May you sue your boss to enforce the promise? The answer is no. For a promise to be enforceable, there must be bargaining and an exchange. In this example, there was no exchange. Work done in the past, by definition, has already been performed. As such, you have given nothing in exchange, and the court will not enforce the promise. You are at the mercy of your boss' goodwill to keep her promise. A promise cannot be based on consideration that was provided before the promise was made.

In Case 10-3, the court must consider whether twenty-five years of marriage was legal consideration for a postnuptial agreement. As you read the case, consider how you would decide if you were the judge.

illusory promise
A situation in which a party appears to commit to something but really has not committed to anything. It is not a promise and thus not consideration.

CASE 10-3

SIMMONS v. SIMMONS
United States District Court of Appeals for Arkansas
249 S.W.3d 843 (2007)

FACTS: Vincent and Dorothy Simmons were married on October 9, 1976. On April 11, 1995, Vincent's mother, Louise Simmons, executed The Louise B. Simmons Trust in order to convey certain land in Florida to her children, Vincent and his sister, upon her death. Louise Simmons died on April 1, 1999, but the land remained in trust for several years after her death. After Louise died, Dorothy became concerned that she would not receive an interest in the Florida land if Vincent died before the trust was distributed, so she hired an attorney in Monticello, David Chambers, to prepare a document to protect her interest. After speaking with Dorothy, Chambers drafted an affidavit to be executed by Vincent, which stated in pertinent part as follows: "I have been married to my wife, Dorothy Simmons, for 25 years. It is my intention, through this affidavit, to convey to my said wife marital interest in said real property. If I should die prior to the above-stated Trust being dissolved, then my said wife shall receive my share of said real property as her own property. Otherwise, if said Trust is dissolved prior to my death, then my wife shall be entitled to her legal marital interest in said real property." The trust property was distributed to Vincent and his sister on November 1, 2002. On February 11, 2003, Dorothy filed a complaint for divorce. The parties reached agreement regarding the division of all property except for the Florida land. The trial court held a hearing regarding the character of this land as marital or non-marital on October 31, 2005.

On December 7, 2005, the trial court sent a letter to the parties's attorneys, stating that the letter was submitted as the court's findings on the issue of ownership of the land in Florida. While the trial court recognized that a person in a divorce action is not entitled to any interest in property inherited by his or her spouse, the court noted that this law can be negated by the inheriting spouse—that is, the inheriting spouse can convey the property to the non-inheriting spouse. The trial court found that, by signing the affidavit, Vincent conveyed one-half of his interest in the Florida land to Dorothy. The trial court determined that "[t]his was done in consideration of twenty-five years of marriage, and is found to be binding."

On appeal, Vincent argues that the trial court erred in finding that his affidavit constituted a contract to convey an interest in the Florida land to Dorothy.

ISSUE: May the postnuptial agreement between Vincent and Dorothy be supported by their past twenty-five years of marriage?

REASONING: A postnuptial agreement is an agreement entered into during marriage to define each spouse's property rights in the event of death or divorce. The term commonly refers to an agreement made at a time when separation or divorce is not imminent. The Arkansas Supreme Court recently addressed the validity of postnuptial agreements and held that they should be analyzed under the basic principles of contract law.

The parties disagree about whether Vincent's expressed intention to convey an interest in the Florida land to Dorothy was supported by legal consideration. While the issue of whether the marriage itself is adequate consideration to support a postnuptial agreement has not been specifically addressed in Arkansas, it has long been the law in Arkansas that past consideration will not support a current promise and is not adequate legal consideration. Other jurisdictions that have addressed this issue have held that an existing marriage is past consideration and will not support a postnuptial agreement. In Bratton, relied upon by the Arkansas Supreme Court in Stewart, the Tennessee Supreme Court stated that "the marriage itself cannot act as sufficient consideration because past consideration cannot support a current promise. Following the guidance of the Tennessee Supreme Court in Bratton and our own case law holding that past consideration will not support a current promise, we hold that the parties' marriage is not adequate legal consideration to support this agreement."

DECISION AND REMEDY: We reverse the trial court's order finding that the Florida land is marital property and awarding a one-half interest in the land to Dorothy. We remand to the trial court to enter an order consistent with this opinion.

SIGNIFICANCE OF THE CASE: Past consideration is not legal consideration to support a contract.

CRITICAL THINKING

Does the appellate court's reasoning make sense to you? What are the dangers, if any, of using past consideration in support of a contract?

ETHICAL DECISION MAKING

Using the ethical guidelines in the WH process, do you believe the appellate court's decision is the morally correct one? Should it matter whether Vincent intended to convey the interest in the land at the time he signed the contract?

GLOBAL Context

©Doug Armand/Getty Images

The Importance of Contracts in Other Countries

Contracts vary in importance, depending upon the countries using them. In the United States, businesspeople consider contracts to be very important. Contracts are highly detailed. If there is a dispute, the parties refer to the contract terms. If the parties cannot agree, a lawsuit may be filed.

The same is not true in many other countries. For example, in China, contracts exist for clarification purposes, not to enforce the underlying agreement. The Chinese believe that businesspeople should work together to solve their differences.

In Russia, businesspeople pay little attention to contracts, documenting only the essentials of the agreement. Parties resolve disputes on their own or attempt to enforce the contract through political and economic pressures. The table following gives an overview of the importance of contracts in these four countries: [4]

COUNTRY	IMPORTANCE OF CONTRACTS	DETAIL IN CONTRACT	LIKELIHOOD OF LEGAL ACTION	DEPENDABILITY OF LEGAL SYSTEM FOR CONTRACT ENFORCEMENT
U.S.	High	High	High	High
China	Low to Medium	Medium to High	Low	High
Russia	Low	Low to Medium	Low	Low to Medium

Preexisting Duty

LO 10-7 Define preexisting duty.

Rule: *A promise to do something that you are already obligated to do is not valid consideration.* There are actually two parts to this rule. First, *performance of a duty you are obligated to do under the law is not good consideration.* For example, a police officer is sworn to uphold the law. Part of that public duty involves catching suspected criminals. If someone offers a reward for the capture of a suspected criminal, the police officer may not collect the reward because he or she was already obligated to apprehend the suspect. Second, *performance of an existing contractual duty is not good consideration.* Gene decides to have a pool built in his backyard. Under the existing contract, the pool is to be completed by June 1, just in time for summer. The pool contractor comes to Gene and explains that due to a shortage of workers, the completion date cannot be met; however, if Gene were to pay an extra $5,000, additional workers could be hired and the pool would be completed on time. Gene tells the contractor that he will pay the $5,000. On June 1, the pool is completed, and the contractor asks for the additional $5,000 payment. Is Gene legally obligated to pay? The answer is no. The pool contractor had a preexisting contractual duty to complete the pool by June 1. Gene is under no obligation to pay the additional money.

In Case 10-4, the court must consider whether an addendum to a contract agreeing to pay more than the originally agreed to contract price requires additional consideration to be enforceable.

[4] Lothar Katz, *Global Contract Practices,* http://www.leadershipcrossroads.com/mat/Global%20Contract%20Practices.pdf.

CASE 10-4 — MARGESON v. ARTIS
Supreme Court of Iowa (2009)
776 N.W.2d 652

FACTS: *John and Jennifer Margeson entered into a contract to sell a weight-loss franchise business called "Inches-A-Weigh" to Theresa Artis. The parties memorialized their agreement in an "Asset Purchase Agreement" executed on October 1, 2004. The purchase price was $125,000, payable at the time of closing.*

The parties subsequently executed a second document entitled "Sales Agreement Addendum" (addendum). This addendum was signed on October 7, 2004. It set the price of the sale of the business at $155,000, with $135,000 payable at the time of the closing. Of the amount to be paid at closing, $125,000 was identified as the proceeds of a loan secured by Artis from First Bank, and $10,000 was to be paid in cash. The remaining portion of the purchase price was to be paid to the Margesons in monthly installments in amounts based on sales.

The closing was set for October 18, 2004. On that date, Artis tendered the $125,000 proceeds of the loan from First Bank, together with an additional $10,000 from two personal checks drawn on her bank.

The parties ran into some disputes following the closing. Artis stopped payment on one of the personal checks delivered at the time of closing and stopped making the monthly payments in March 2005. The Margesons responded by filing a lawsuit for breach of the addendum.

The Margesons eventually filed a motion for summary judgment. They claimed there was no genuine issue of material fact as to any of the elements of their claim for breach of contract. Artis asserted the addendum was unenforceable because it was not supported by consideration The district court found the addendum was supported by consideration. It granted summary judgment to the Margesons. Artis appealed, and we transferred the case to the court of appeals. The court of appeals affirmed the ruling of the district court. We granted further review.

ISSUE: Whether an addendum to a contract modifying the purchase price just prior to closing is supported by consideration?

REASONING: Justice Cady: . . . It is fundamental that a valid contract must consist of an offer, acceptance, and consideration. Generally, the element of consideration ensures the promise sought to be enforced was bargained for and given in exchange for a reciprocal promise or an act.

The Margesons seek to recover under the terms of the addendum. In doing so, they seek to enforce the promise by Artis to purchase the business for $155,000. Artis argues the terms of the addendum are not a legally binding part of the contract. More precisely, Artis argues the addendum, which was a modification of the original agreement, requires independent consideration to be binding. Artis argues there was no consideration in this case because the Margesons had a preexisting duty under the first agreement to sell the business to her for $125,000. The Margesons argue additional consideration was present to support the modification.

Generally, a promise to perform a preexisting duty does not constitute consideration. No consideration exists when the promisee has a preexisting duty to perform because a promisor is already entitled to receive the promise made by the promisee and the promisee has only made what amounts to a gratuitous promise. We have specifically applied this rule to preexisting contractual obligations when parties to an original contract agree to modify that contract. Artis argues the addendum was unenforceable because it arose only from a desire by the Margesons to obtain more money for their business than agreed in the original purchase agreement. Although Artis agrees she promised in the addendum to pay $30,000 more for the business, she claims the Margesons did not provide a return promise or performance in exchange for her promise to pay more money. The Margesons argue the addendum is supported by three instances of additional consideration: (1) a financing plan, (2) flexibility in payments, and (3) the ability to renegotiate the payment terms. The district court found these provisions to amount to independent consideration, as did the court of appeals.

New financing terms can constitute sufficient consideration to support the modification of a preexisting obligation. Yet, the new financing terms would need to apply to the preexisting obligations under the original agreement to satisfy the consideration requirement needed for the modification that the promisee promises to do something the promisee has no prior legal obligation to do or refrain from doing something that the promisee has a legal right to do. In this case, all three new financing terms pertained only to the promise made by Artis to pay the additional $30,000. Thus, the new financing terms under the modification were merely part of the new promise by Artis to pay an additional sum of money for the business. The terms did not establish that the Margesons promised to do something they were not otherwise already obligated to do (sell business for $125,000 due at closing) or promised to refrain from doing something they had a legal right to do. Of course, the Margesons would have furnished new consideration for the modification if the financing terms under the modification applied to part or all of the $125,000 due at the time of closing of the original agreement. Under such a case, the Margesons would have furnished consideration because they would have promised to refrain from doing something (giving up the right to receive the original purchase price in a lump sum at the time of closing) that they had a legal right to do.

Consequently, the modification was nothing more than a unilateral price hike. Here, Artis promised additional

(continued)

compensation for the same performance by the Margesons. Thus, the Margesons cannot enforce the promise by Artis to pay more money for the business because they failed to produce any evidence to show they promised to do something more than they had promised to do under the first agreement. The addendum does not reflect independent consideration. The Margesons were not entitled to summary judgment on their breach of contract claim.

DECISION AND REMEDY: Court of appeals vacated; district court judgment reversed; case remanded.

SIGNIFICANCE OF THE CASE: A change of the terms of consideration to an already negotiated contract (in this case, an addendum just prior to closing) can be precarious. If previously agreed to terms of consideration are changed for one, then the parties must be certain that there is new and/or additional consideration from all parties to the contract. Absent new and/or additional consideration by all, there is a risk that the court will see the contract as it did this one – unsupported by consideration and therefore unenforceable.

CRITICAL THINKING

Do you believe the court made the correct decision? How do you think most business owners would feel about the outcome?

ETHICAL DECISION MAKING

Is it ethical to use a legal argument to escape from a contract modification that you agreed to? Why or why not? How would you feel about the outcome if you were the sellers of this small business?

Exceptions to the Preexisting-Duty Rule There are exceptions to the preexisting-duty rule:

- Unforeseen circumstances
- Additional work
- UCC—sale of goods

unforeseen circumstances
Events that a reasonable person would not be expected to anticipate.

If **unforeseen circumstances** cause a party to make a promise regarding an unfinished project, that promise is valid consideration. Let's once again consider the pool example. Suppose that Gene's pool contractor has been building pools in Gene's neighborhood for the past 20 years. During all that time, the contractor has never had any problem with rocks—until now. While bulldozing the hole for the pool in Gene's backyard, the pool contractor hits solid rock. He estimates that it will cost an additional $5,000 to clear the rock by using jackhammers and possibly even dynamite. The contractor explains that he is sorry, but unless Gene agrees to pay the additional money, he will not be able to finish the pool. Gene agrees to pay. When the pool is completed, the contractor asks for the additional $5,000. Will a court enforce this promise? The answer is yes. Even though the contractor is completing only what he was obligated to do under the contract, neither party knew of the solid rock. The contractor has given additional consideration (removal of the rock), and Gene will be held to his promise to pay the additional money.

If a party to a contract agrees to do additional work (i.e., more than she is obligated to do under the contract), the promise is valid consideration. In the pool example, if the contractor asks Gene for an additional $10,000 but agrees to add a waterfall and a deck to the pool, the promise to do the additional work is consideration. If Gene agrees to pay the $10,000, that is his consideration. Both parties are now bound.

Partial Payment of a Debt

liquidated debt
Debt for which there is no dispute between the parties about the fact that money is owed and the amount of money owed.

LO 10-8 Differentiate between liquidated debt and unliquidated debt.

Partial payment of a debt may or may not be valid consideration, depending on whether the debt is liquidated or unliquidated. In a **liquidated debt**, there is no dispute about the fact that money is

owed and the amount of money owed. Example: Natalie owes $3,000 on her credit card. She calls the credit card company and explains that she is a poor student and cannot afford to pay the entire $3,000. The credit card company agrees to accept $2,000 as payment in full. The following month, Natalie receives her new credit card statement showing that she owes the remaining $1,000. May the credit card company collect the additional $1,000? Yes! A creditor's promise to accept less than owed, when the debtor is already obligated to pay the full amount, is not binding.

LO 10-9 Describe the concept of accord and satisfaction.

The exception to the rule regarding liquidated debt occurs when the debtor offers different performance. Suppose Natalie offered the credit card company her car in full settlement of the $3,000 debt. If the credit card company accepts, regardless of the value of the car, the debt is paid in full and the credit card company may not sue Natalie for any additional money.

In an **unliquidated debt,** the parties either dispute the fact that any money is owed or agree that some money is owed but dispute the amount. A dispute over an unliquidated debt may be settled for less than the full amount if the parties enter into an **accord and satisfaction.** For an accord and satisfaction to be enforceable, three requirements must be met:

1. The debt is unliquidated (i.e., the amount or existence of the debt is in dispute).
2. The creditor agrees to accept as full payment less than the creditor claims is owed.
3. The debtor pays the amount they have agree on.

unliquidated debt
Debt for which the parties either dispute the fact that any money is owed or agree that some money is owed but dispute the amount.

accord and satisfaction
An arrangement between contracting parties whereby one of the parties substitutes a different performance for his or her original duty under the contract.

CASE Nugget

Accord and Satisfaction

Haynes v. DaimlerChrysler Corp. Supreme Court of Appeals of West Virginia

228 W. Va. 441 (2011)

On January 22, 2006, Elgene Phillips was driving his 1998 Dodge Ram 1500 when the vehicle hydroplaned, ran off the left side of the road through a guardrail, and rolled over. Mr. Phillips died as a result of the accident. Gloria Phillips was a passenger in the truck. She testified that Mr. Phillips's seatbelt was latched prior to the accident but became unlatched during, and as a result of, the accident. An expert witness testified that Mr. Phillips would not have sustained fatal injuries if his seatbelt had remained latched during the accident. Ms. Phillips, whose seatbelt did remain latched during the accident, suffered only minor injuries.

As a result of mediation, the parties reached a settlement and entered into a handwritten agreement. Defendant was to pay plaintiff $150,000 for a full and final settlement of all claims, and the agreement was to be confidential. The plaintiff thereafter received two checks, one for $65,000 from Autoliv (the seatbelt manufacturer) and another for $85,000 from Chrysler. The second check, for $85,000, was returned for insufficient funds because Chrysler had just filed for bankruptcy. The handwritten settlement agreement provided that "Defendants" would pay the plaintiff $150,000. No apportionment between defendants was listed in the settlement agreement.

Plaintiff asked the court to force Autoliv to pay the remaining $85,000. Autoliv argued that plaintiff's cashing of the $65,000 check was an accord and satisfaction of Autoliv's portion of the debt; therefore, Autoliv was not responsible for the remaining $85,000. The plaintiff argued that when the $65,000 check was cashed, plaintiff was not aware that Autoliv meant the check to be the full and final payment of what was owed to plaintiff. The Supreme Court of Appeals of West Virginia agreed with the plaintiff and ordered Autoliv to pay the remaining $85,000.

Rule: *An accord and satisfaction is an arrangement between contracting parties whereby one of the parties substitutes a different performance for his or her original duty under the contract.* The promise to perform the new duty is the *accord,* and the actual performance of that new duty is the *satisfaction.*

Under such circumstances, the debt is fully discharged. The *accord* is the new agreement to pay less than the creditor claims. The *satisfaction* is the payment, by the debtor, of the reduced amount. If the debtor fails to pay the new amount, the creditor may then sue for the full amount of the original debt. It pays to keep your word.

One way that people sometimes attempt to create an accord and satisfaction is by sending a check to the creditor and writing "paid in full" on the check. Under common law, in many states, this did create an accord and satisfaction, and if the creditor cashed the check, he or she was bound to accept the smaller amount as payment in full. The UCC, however, has changed the scope of this rule. Under UCC Section 3-311, effective in 30 states, the rule remains in place but has two major exceptions.

First, in cases of business organizations, thousands of checks may be received each day. To protect themselves, businesses may notify their debtors that any offer to settle a claim for less than the amount owed must be sent to a particular address and/or person. If you check the terms printed on your credit card statement, you will likely find language directing you to send such payments to a different address and person than those to which regular payments are sent. This safeguard protects businesses from inadvertently creating accord-and-satisfaction agreements. Here is an example of the wording you might find on a credit card statement:

> *Conditional Payments:* Any payment check or other form of payment that you send us for less than the full balance that is marked "paid in full" or contains a similar notation, or that you otherwise tender in full satisfaction of a disputed amount, must be sent to [address omitted]. We reserve all rights regarding these payments (e.g., if it is determined that there is no valid dispute or if any such check is received at any other address, we may accept the check and you will still owe any remaining balance).[5]

Second, if a business does inadvertently cash a paid-in-full check, the business has 90 days from the date it cashed that check to offer repayment in the same amount to the debtor. For example, if John owed $3,000 to his credit card company and he sent the company a $2,000 check marked "paid in full" to the correct address and person, the credit card company has 90 days to offer to repay John the $2,000. Once the offer has been made by the business, no accord and satisfaction exists.

[5] From Chase Via Statement.

CASE OPENER WRAP-UP

Upper Deck—Contract Liability or Gift?

As you know from the Case Opener, Passante sued Upper Deck for breach of oral contract. At trial, the jury awarded him close to $33 million—the value of 3 percent of Upper Deck's stock at the time of the trial in 1993. Upper Deck appealed.

As a matter of law, any claim by Passante for breach of contract is necessarily based on the rule that consideration must result from a bargained-for exchange. In this case, the appellate court held that if the stock promised was truly bargained for, then Passante had an obligation to give Upper Deck the opportunity to have separate counsel represent it in the course of that bargaining. The legal profession has certain rules regarding business transactions with clients. Bargaining between the parties might have resulted in Passante's settling for just a reasonable finder's fee.

All Passante's services in arranging the $100,000 loan for Upper Deck had already been rendered (even though the board had not formally accepted the loan) before the idea of giving him stock came up. There was no evidence he had any expectation of receiving stock in return. If there is no expectation of payment by either party when services are rendered, the promise is a mere promise to make a gift and not enforceable. The promise of 3 percent of the stock represented a moral obligation but was legally unenforceable.

PRACTICAL TIPS for Business Managers

1. In many states, an option contract for the purchase of real property requires consideration that is separate from the consideration required for the sale of the land itself. When in doubt, consult a local real estate attorney.
2. Negotiate well. Absent fraud (or other extenuating circumstances) the courts will not review adequacy of consideration. If you make a bad deal, you are stuck with it.
3. Never cash a check marked "paid in full" (or with any similar language) unless you are happy to accept that amount and forego collection of any balance owed to you.

SUMMARY

What Is Consideration?	*Consideration* is something of value given in exchange for something else of value; it must be the product of a mutually bargained-for exchange.
Rules of Consideration	For a promise to be enforced by the courts, there must be *consideration*. A court will enforce one party's promise only if the other party promised something (an act, money, etc.) in exchange. *Promissory estoppel* occurs when one party makes a promise knowing the other party will rely on it, the other party does rely on it, and the only way to avoid injustice is to enforce the promise. The court seldom considers *adequacy of consideration*. This means that the court does not weigh whether you made a good bargain. An *illusory promise* is not consideration (i.e., it is not a promise at all). *Past consideration* is generally no consideration at all. A promise cannot be based on consideration that was provided before the promise was made. The *preexisting-duty rule* provides that when a party does what it is already legally obligated to do, there is no consideration because there has been no detriment.
Partial Payment of a Debt	In a *liquidated debt,* there is no dispute about the fact that money is owed and the amount of money owed. In an *unliquidated debt,* the parties either dispute the fact that any money is owed or agree that some money is owed but dispute the amount. For an *accord and satisfaction* to be enforceable, three requirements must be met: (1) The debt is unliquidated (i.e., the amount or existence of the debt is in dispute); (2) the creditor agrees to accept as full payment less than the creditor claims is owed; and (3) the debtor pays the amount they have agreed on. The accord is the agreement, and the satisfaction is the payment of the reduced amount.

Point/Counterpoint

Should Past Consideration Be Acceptable Consideration for a Contract?	
YES	NO
If a party to a contract wants to accept "past consideration" as consideration for a contract, then the law should permit that party to do so. It could be (and often is) easily labeled (or described) as such and acknowledged in the contract. Allowing past consideration as consideration for a contract would alleviate harsh outcomes in which one party clearly has the intent to reward another for past deeds but simply misunderstands the legal procedure to accomplish that task.	Consideration is and has always been a basic requirement for a valid and enforceable contract. Precedent should be followed. Without consideration, one party is simply making a gift to the other party. There are other legal procedures adequate to deal with gifts.

Questions & Problems

1. List the four types of consideration described in the text.
2. What is required to prove promissory estoppel when consideration is missing?
3. Can $1 be adequate consideration? Why or why not?
4. List and describe the three exceptions to the preexisting-duty rule.
5. List the three elements of accord and satisfaction.
6. Appellant is a temporary worker staffing company servicing client companies throughout the El Paso region. Jamrowski was a temporary worker for one of Appellant's client companies. Although the temporary workers work on-site for Appellant's client companies, the workers are ultimately employees of dmDickason and are subject to Appellant's corporate policies and procedures. The dmDickason Employee Handbook ("Handbook") contains an acknowledgement page at the outset of the document stating that the employee acknowledges the Handbook does not cover all corporate policies in detail. The acknowledgment page also contains the following language: "I also understand that the provisions in the Handbook may be changed at any time by dmDickason and that in certain circumstances; dmDickason may choose not to follow the provisions in the Handbook." Two copies of the Dispute Resolution and Arbitration Policy Agreement ("Agreement") are contained in the Handbook and listed in the Handbook's table of contents as attachments. One copy must be returned to dmDickason and one is for the employee's records. The Agreement requires employees to submit any disputes between them, dmDickason, or dmDickason's client companies "relating to or arising out of an employee's recruitment, hiring, employment, or . . . termination of employment[]" to binding arbitration. The Agreement contains the following provision: "dmDickason may change or modify this arbitration policy from time to time without advance notice or the consent of employees. However, with respect to any claim for which a demand for arbitration has been filed with the AAA, dmDickason will not modify or change the agreement between you and dmDickason to use final and binding arbitration to resolve employment-related disputes without notifying you and obtaining your agreement to such changes." Jamrowski filed suit in district court and brought employment claims against Appellant and Appellant's client company. Appellant moved to compel arbitration, which the trial court denied. An appeal followed. Is the arbitration clause an illusory promise? Why or why not?

 Source: *Temp. Alternatives, Inc. v. Jamrowski*, 511 S.W.3d 64 (2014).

7. The parties orally contracted for Pendergrass Construction to perform grading and other work on Ingram's property for the price of $2,500.00. After Pendergrass Construction began the work, Ingram requested additional work. The parties did not discuss any additional payment for this work. After the work was complete, Pendergrass Construction sent Ingram an invoice for $9,073.00. Ingram told Pendergrass Construction he would not pay that amount and sent them a cashier's check marked "pd in full " for $1,500.00, which was the balance due on the original $2,500.00 price. Pendergrass Construction marked through the "pd in full" notation on the check, cashed the check, and then notified Ingram that they considered the check to be a credit against the total amount owed. Pendergrass Construction later brought suit against Ingram to refute an accord and satisfaction and argued that the parties entered into two separate contracts, one for $2,500.00 and another for additional work above and beyond the original agreement. The trial court ruled in favor of Pendergrass Construction. This appeal followed. How should the court decide? Do you believe the parties entered into an accord and satisfaction? Why or why not?

 Source: *Pendergrass v. Ingram,* 2016 Tenn. App. LEXIS 440 (2016).

8. Vincent Simmons appeals from the trial court's order awarding to his wife, Dorothy Simmons, a one-half interest in land that he inherited from his parents. Vincent contends that the land is non-marital property and, consequently, should have remained his separate property. Vincent and Dorothy Simmons were married on October 9, 1976. On April 11, 1995, Vincent's mother, Louise Simmons, executed The Louise B. Simmons Trust in order to convey certain land in Florida to her children, Vincent and his sister, upon her death. Louise Simmons died on April 1, 1999, but the land remained in trust for several years after her death. After Louise died, Dorothy became concerned that she would not receive an interest in the Florida land if Vincent died before the trust was distributed, so she hired an attorney in Monticello, David Chambers, to prepare a document to protect her interest. After speaking with Dorothy, Mr. Chambers drafted an affidavit to be executed by Vincent, which stated, in pertinent part: "I have been married to my wife, Dorothy Simmons, for 25 years. It is my intention, through this affidavit, to convey to my said wife marital interest in said real property. If I should die prior to the above-stated Trust being dissolved, then my said wife shall receive my share of said real property as her own property. Otherwise, if said Trust is dissolved prior to my death, then my

wife shall be entitled to her legal marital interest in said real property." The trust property was distributed to Vincent and his sister on November 1, 2002. On February 11, 2003, Dorothy filed a complaint for divorce. The parties reached agreement regarding the division of all property except for the Florida land. On appeal, Vincent argues that the trial court erred in finding that his affidavit constituted a contract to convey an interest in the Florida land to Dorothy. He argues that there is a total absence of consideration to support a contract in this case. Dorothy argues that her ongoing marriage to Vincent constituted adequate consideration to support the contract. Should the appeals court uphold the trial court's ruling and grant one-half interest in the property to Dorothy? How would you rule? What rule of consideration applies in this case?

Source: *Simmons v. Simmons,* 98 Ark. App. 12, 249 S.W.3d 843, 2007 Ark. App. LEXIS 107 (Ark. Ct. App. 2007.).

9. Byers acquired and owned substantial tracts of ranchland, including the family ranch where Byers lived. For some time, Byers's son (and, thus, Bouton's brother) managed most of the ranching operations. In 2003, Byers became concerned about his son's conduct and how he was handling some of the business matters entrusted to him. Byers asked Bouton, his daughter, to help in assessing the status of the operations. At the time, Bouton held a tenure-track teaching position on the Washburn University School of Law faculty and earned about $100,000 a year. Bouton began helping Byers with the ranching enterprise. She went to the ranch several times a week and wrapped work on her father's business around her teaching duties. In late 2003, Byers showed Bouton a revised will and trust he had prepared leaving the ranch operations to her. According to Bouton, Byers said he wanted her to carry on the ranching business after his death. After two years, Byers convinced Bouton to quit her law professor job, move her family to the ranch, and run the business full time. He assured her that she needn't worry about money because he was bequeathing her land worth more than a million dollars. In December 2008, the two argued over issues related to the ranch and Byers told Bouton her services were no longer needed. The family left the ranch and Bouton returned to the Washburn Law School faculty in a part-time teaching position without any possibility of tenure. About 14 months later, Byers informed Bouton that he had contracted to sell the family ranch—the last of his land holdings except for 10 acres—for $1.2 million. So far as Bouton understood, Byers no longer owned any land that she might inherit. In November 2011, Byers signed a new trust that upon his death would distribute all of his assets to charitable foundations to provide college scholarships. Byers effectively disinherited Bouton. Bouton sued her father, Byer, to enforce his promise that she would inherit the family farm, worth over $1 million. On what legal theory did Bouton sue? How do you think the court resolved the matter?

Source: *Bouton v. Byers,* 50 Kan. App. 2d 34, 321 P.3d 780 (2014).

10. Joana Perez began working for Datamark in January 2005. She received two booklets at orientation, "Non-Staff Employee Handbook" and "Summary Plan Description." She did not read either one of them. According to the human resources director, Perez also received a "Problem Resolution Program" booklet (the PRP) that described company dispute resolution policies and procedure. Perez denied receiving it, but she did sign the "Receipt and Arbitration Acknowledgment" form, which was maintained in her personnel file. Her signature acknowledged that she had received and read (or had the opportunity to read) both "Summary Plan Description" and the PRP. She also acknowledged that an arbitration policy required the submission of all employee-related disputes to an arbitrator in accordance with the procedures described in the PRP. Datamark reserved the right to revoke or modify the PRP in writing at any time as long as the writing was signed by an officer of the company and articulated an intent to revoke or modify a policy. Perez learned she was pregnant in August 2005. While employed full-time, she began to miss work due to pregnancy difficulties. She was discharged on October 21, 2005, and filed suit alleging unlawful discrimination because of her gender and/or pregnancy. Perez also alleged that Datamark intentionally or recklessly engaged in extreme and outrageous behavior that caused her severe emotional distress. Datamark filed a motion to compel arbitration. In her response to the motion, Perez alleges that the arbitration agreement is illusory because Datamark could unilaterally change or terminate the agreement without prior notice to the employees. Do you believe the agreement to arbitrate is illusory? Why or why not?

Source: In *re Datamark, Inc., Relator,* Court of Appeals of Texas, Eighth District, El Paso, 296 S.W.3d 614; 2009 Tex. App. LEXIS 794 (2009).

11. Peter and Laureen were married on February 16, 1975; they are the parents of two adult children, Amanda and Matthew. In 1996, Peter and Laureen were on the verge of seeking a divorce. Peter had not been living in the marital home for almost two years. According to Laureen's testimony in the instant case, Peter had been verbally abusive to her; moreover, Peter admitted to having engaged in extramarital

affairs. When Peter requested permission to return to the marital home, several conversations took place between Peter and Laureen. In May of 1996, there was a family meeting regarding Peter's request. That meeting was attended by Peter, Laureen, and the two children. It was Laureen's testimony that, during this meeting, she agreed to allow Peter back into the marital home if he fulfilled three conditions: (1) continued alcohol treatment counseling; (2) contribution to family expenses; and (3) waiver of any claim to Laureen's pension. The agreement was made orally. Laureen and the two children testified that Peter assented to these three conditions. Peter, for his part, acknowledged that the family meeting had taken place and that he had agreed to the first two conditions, but he denied that they had discussed any waiver of his interest in Laureen's pension. Shortly after the family meeting, Peter returned to the marital home. In November of 2002, Laureen and Peter separated for the final time. Laureen obtained a restraining order against Peter, which called for his removal from the family home and prohibited him from having any contact with Laureen. Ultimately, the trial court awarded Laureen 80 percent of the family assets and denied Peter any portion of Laureen's pension. Peter claims that he never agreed to condition #3 and that he is entitled to a portion of Laureen's pension. Is the oral contract stated above enforceable? And are Peter's three alleged promises sufficient consideration to permit the court to uphold the conditions of the contract and deny Peter's request for a portion of Laureen's pension? How should the court rule?

Source: *Deangelis v. Deangelis,* 923 A.2d 1274 (R.I. 2007).

12. Charles Houser began working for appellee, 84 Lumber Company L.P. (84 Lumber), in 1985. In 1998, Houser became an outside salesman with 84 Lumber, and his compensation changed from a set salary to commission based on his sales. At that time, Houser signed a noncompete agreement, which prohibited him from engaging in sales activities with a competitor of 84 Lumber within a 25-mile radius of 84 Lumber's Macedonia store for a two-year period following the conclusion of his employment with 84 Lumber. In June 2008, Houser signed a contract providing a set weekly draw and yet another noncompete agreement. In March 2009, Houser left 84 Lumber and, almost immediately thereafter, began working for Carter Lumber, a competitor of 84 Lumber. 84 Lumber filed a lawsuit alleging that Houser had violated the noncompete agreement. The essential question is whether the 2008 noncompete agreement was supported by adequate consideration. "[A] restrictive covenant is enforceable if supported by new consideration, either in the form of an initial employment contract or a change in the conditions of employment." 84 Lumber Company argued that Houser's continued employment was adequate consideration for the new noncompete agreement. Do you agree? Why or why not?

Source: *84 Lumber Co., L.P. v. Houser,* 2011 Ohio 6852 (2011).

13. In December 2011, Jimenez began working for Cintas as a Fire Service Technician. She continued in this capacity until June 2012, when Cintas terminated her employment. On July 18, 2013, Jimenez filed a petition against Defendants in the Circuit Court of St. Louis County, challenging her termination and alleging that during her employment she suffered discrimination and harassment by Baker and Fitzsimmons, who acted in their capacity as managers for Cintas. On August 26, 2013, Defendants filed a motion to compel Jimenez to arbitrate her claims against Cintas, attaching a document entitled "Missouri Employment Agreement for Sales, Service and Marketing Personnel" ("Agreement"). The first page of the Agreement is dated December 12, 2011, and contains Jimenez's purported signature. Section 8 of the Agreement states that disputes "between Employee and Employer concerning whether either party at any time violated any duty, right, law, regulation, public policy, or provision of this Agreement" that cannot be resolved in good faith, must be "resolved through impartial and confidential arbitration." The Agreement requires the arbitration of all claims except those expressly excluded by Cintas. Defendants argue that the parties formed a valid arbitration agreement providing two separate types of consideration: (1) Cintas's offer of "new" or "future" at-will employment; and (2) Cintas's and Jimenez's mutual promises to arbitrate. Jimenez responds that the arbitration agreement is effectively devoid of consideration and therefore invalid because her employment relationship with Cintas was "at-will," and Cintas's promise to arbitrate was not mutually binding. How should the court rule? Do you think that new or future "at will" employment should be adequate consideration for a contract? Why or why not?

Source: *Jimenez v. Cintas Corp.,* 475 S.W.3d 679 (Mo. Ct. App. 2015).

Part 2 *Contract Law*

CHAPTER 11

Capacity and Legality

CASE OPENER

Apple's Questionable Contract

Parents of minors took Apple to court in 2012 for supplying game applications on iPhones that were free but through which users could purchase in-game currencies. Apparently, parents would log on to the games, but within a subsequent 15-minute time frame, the minors using the game would rack up bills ranging from $99.99 to $338.72 at a time.

Apple stated that although minors were downloading these applications and in-game currency, the contract in question was actually the Terms of Service between the parents and Apple. According to the Terms of Service, any unauthorized log-on on one's account or unauthorized purchases by anyone on the account was the responsibility of the account holder. On the other hand, the parents argued that all application and in-game purchases made by minors were separate contracts that may be disaffirmed by a parent or guardian. Apple also argued that a contract as the parents were describing it could not legally exist in that, as the parents described the scenario, the contractual offer was made to parents in the Terms of Service, yet accepted by the children, and consideration was provided by the parents (the original offerees).[1]

1. Do the individual purchases made by the minors indeed qualify as separate contracts between Apple and the minors?
2. Even if the purchases were contracts between the minors and Apple, could the parents void these contracts?

LEARNING OBJECTIVES

LO 11-1 Recognize circumstances wherein a minor would have a limited capacity to enter into a contract.

LO 11-2 Explain how a contract created when one is a minor may be ratified after one is no longer a minor.

LO 11-3 List the conditions under which a contract is unenforceable by law.

LO 11-4 Describe instances in which unfairness affects a contract's enforceability.

LO 11-5 Recall the legal effect of entering into a contract for illegal purpose.

LO 11-6 Describe the characteristics of a severable contract.

[1] In re *Apple In-App Purchase Litigation,* 5:11-CV-1758 (N.D. Cal.; March 31, 2012).

Capacity

capacity
The legal ability to enter into a binding contract.

This chapter examines the third and fourth elements of a contract. **Capacity** is the third element of a legally binding contract. A person who has legal capacity to contract is one who has the mental ability to understand his or her rights and obligations under a contract and therefore will presumably be able to understand how to comply with the terms of the agreement. *Incapacity,* or *incompetence,* as it is sometimes called, is some sort of mental or physical defect that prevents a natural person from being able to enter into a legally binding contract. Depending on the nature and extent of the defect, a person may have either no capacity, and therefore any attempted contract is void, or limited capacity, resulting in the ability to form only voidable contracts.

Historically, people with limited or no capacity included married women, minors, and insane persons. Through statutes, other categories were added for classes of people for whom guardians had been appointed, such as habitual drunkards, narcotic addicts, spendthrifts, the elderly, and convicts. Today, married women have been removed from the category of those lacking contractual capacity, although in a few states their capacity to enter into certain kinds of contracts is still limited. In this section of the chapter, we explain the current law limiting the capacity of some categories of persons to enter into legally binding agreements.

Minors

LO 11-1 Recognize circumstances wherein a minor would have a limited capacity to enter into a contract.

One of the oldest limitations on capacity is the ability of minors to enter into only voidable contracts. Today, in all but three states, a *minor* is someone under the age of 18.[2] In most states, however, a person is given full legal capacity to enter into contracts when he or she becomes emancipated before reaching the age of majority. *Emancipation* occurs when a minor's parents or legal guardians give up their right to exercise legal control over the minor, typically when the minor moves out of the parents' house and begins supporting himself or herself. Often the minor will petition the court for a declaration of emancipation. In most cases, when a minor marries, she or he is considered emancipated.

BUT WHAT IF . . .

WHAT IF THE FACTS OF THE CASE OPENER WERE DIFFERENT?

Let's say that in the Case Opener, the court decided that all of the purchases made by the minors were indeed contracts between the minors and Apple. Let's say that the parents attempted to void the contracts themselves because they are technically the minors' legal guardians. Would such a move be legal? Why or why not?

Disaffirmance of the Contract Because their contracts are voidable, minors have the right, until a reasonable time after reaching the age of majority, to disaffirm, or avoid, their contracts. No formalities are required to disaffirm the contract; the minor need only manifest an intention to rescind the contract, either by words or by actions. However, the minor must avoid the entire contract; he or she cannot choose to disaffirm only a portion of it. Case 11-1 provides an illustration of how the court sometimes grapples with the issue of whether to allow a minor to disaffirm a contract.

[2] In Alabama, Nebraska, and Wyoming, full capacity to contract does not arise until the person reaches the age of 19, which is the age of majority in those states. In Mississippi, the age of majority is still 21.

CASE 11-1

ADRIAN LOPEZ v. KMART CORPORATION
United States District Court for the Northern District of California
2015 U.S. Dist. Lexis 58328 (2015)

FACTS: In April of 2013, Kmart hired Adrian Lopez as a cashier in its Concord, California store. At the time, Lopez was 16 years old and a sophomore in high school. Before Lopez began working, he was required to participate in Kmart's online training program. During the program, Lopez was presented with an electronic arbitration agreement. The agreement stipulated any disputes Lopez might have with the company were to be handled out of court in arbitration. Lopez electronically signed the agreement and acknowledged his receipt of the agreement. In January of 2015, one month after his 18th birthday, Lopez filed a class action complaint against Kmart alleging violations of California wage and hour laws. Kmart filed a motion to compel arbitration. Lopez contended that, because he was a minor when he acknowledged the agreement, he is now entitled to disaffirm the agreement and render it void, thus allowing Lopez to pursue his claim in court.

QUESTION: Did Lopez have the right to disaffirm the agreement due to him being a minor when he signed it?

REASONING: To begin the case, the district court found that the arbitration agreement was indeed valid, but acknowledged that this fact hardly mattered if Lopez had the right to disaffirm the agreement. The court had to find whether Lopez was entitled to disaffirm the arbitration agreement or if he even had the capacity to enter into the agreement at all as a minor.

The court found that the California Family Code clearly provides that a minor has the capacity to contract, and that "[e]xcept as otherwise provided by statute, a contract of a minor may be disaffirmed by the minor before majority [age 18] or within a reasonable time afterwards[.]" Lopez claimed disaffirmance within a month of reaching the age of majority, which the court found to be a reasonable period of time. The court also cited *Fife v. Facebook*, a case that found that disaffirmance of a contract by a minor rescinds the entire contract, rendering it void. The court also noted that the California Family Code does specifically exclude certain types of contracts from disaffirmance, such as contracts for necessaries for the minor or the minor's family, and ruled that Lopez's contract with Kmart is not one of those types of contracts.

DECISION: Lopez had the right to disaffirm the agreement and did so in a timely manner, rendering the agreement void. Kmart's movement to compel arbitration was denied and Lopez was free to pursue his class action claim in court.

SIGNIFICANCE: This case demonstrates the power a minor has to disaffirm any contracts he or she enters into and may serve as a caution to those entering into contracts with minors.

CRITICAL THINKING

Kmart's attorneys argued unsuccessfully that by allowing the minor to disaffirm the employment, the court was discouraging employers from hiring minors in the future. Do you think the court should have given more weight to this argument?

ETHICAL DECISION MAKING

What values were in conflict when the court made this decision?

The obligations of the minor on disaffirmance vary from state to state. Traditionally, most states simply required the minor to notify the competent party of the intent to disaffirm and return to the competent party any consideration received under the contract that was still under the minor's control, regardless of the condition of the consideration. If the consideration has been destroyed or damaged, the minor returns whatever is left of it, and the other party has no recourse against the minor under most circumstances. For instance, if William, a minor, purchased a stereo receiver from Sound Systems, Inc., under a six-month contract and dropped it a week after he took it home, he could return it to the store in its broken condition and tell the owner that he wished to rescind the contract. He would be entitled to the return of his down payment and would owe no further obligations to the store.

GLOBAL Context

Capacity of a Minor in China

©Digital Archive Japan/ Alamy

United States law allows minors to void certain contracts. The mentality behind this law is that individuals below a certain age cannot be entirely accountable for their decisions. Chinese law holds minors accountable for their contracts in more situations than contracts for necessities. Children between 10 and 18 are deemed competent for entering into certain contracts, appropriate to each child's mental state. The validity of the contracts made by these minors is normally conditioned on the retrospective approach of a parent or guardian unless the contract is merely beneficial to the child or is an appropriate contract for the child's age, intelligence, and mental state. Some of these contracts involve agreements for necessities such as food or clothing, whereas others might involve agreements to purchase a bicycle, depending on the minor's intelligence and mental state. The contract is void when the mental state of a minor is not sufficient to claim that the minor is a competent party to the contract.

Such a rule makes sense if you view minors as innocents in need of protection from competent adults who would otherwise take advantage of minors. However, the rule does not encourage competent parties to enter into contracts with minors, and some observers argue that it allows a knowledgeable and unethical minor to take advantage of a competent party. Consequently, a number of states have modified the duty of the minor on disaffirmance, holding that the minor has a duty of restitution that requires her or him to place the competent party back in the position that party was in at the time the contract was made.

Thus, continuing with the stereo receiver example, in some states, William would have a duty of restoration that would require him to compensate the store owner for the difference between the value of the receiver in the condition it was in when given to William and its value in the condition it was in when returned.

The disaffirmance must occur before or within a reasonable time of the minor reaching the age of majority. What constitutes a reasonable time is determined on a case-by-case basis.

Exceptions to the Minor's Right to Disaffirm the Contract The minor's right to disaffirm is designed to protect the minor from competent parties that might otherwise take advantage of the minor. But there are certain situations in which, primarily for public policy reasons, either the courts or state legislatures have determined that the minor should not have the right to disaffirm the contract. As a general rule, most states will not allow a minor to disaffirm contracts for life insurance, health insurance, psychological counseling, the performance of duties related to stock and bond transfers and bank accounts, education loan contracts, child support contracts, marriage contracts, and contracts to enlist in the armed services.

Most of these exceptions apply in most, but not all, states. Another contractual issue on which the states disagree is the minor's misrepresentation of his or her age. Although the majority rule is that misrepresentation does not affect a minor's right to disaffirm a contract, some states hold that when a minor who appears to be of the age of majority misrepresents his or her age and a competent party relies on that misrepresentation in good faith, the minor gives up the right to disaffirm the agreement. One justification for this rule is that any minor who intends to misrepresent his or her age does not need the protection that disaffirmance is designed to provide.

Other states have found a compromise between the two extreme positions either by requiring the minor to restore the competent party to that party's precontract position before allowing the disaffirmance or by allowing the minor to disaffirm but then giving the competent party the right to sue the minor in tort and recover damages for fraud.

Of course, as the previous paragraphs indicate, the laws created to protect minors from being victimized by competent adults do not necessarily protect competent adults from being taken advantage of by knowledgeable and unethical minors. Thus, individuals operating or working in businesses that are subject to laws requiring their customers to be the age of majority or older must familiarize themselves with the laws pertaining to minors because, often, the responsibility of making sure that a business is dealing with people who are of legal age falls on the employees

and the owner. However, because some minors use false identification or misrepresent themselves as adults, it is sometimes difficult for business owners and employees to recognize which customers are truly of age.

For example, as CEO of Girls Gone Wild (GGW), Joe Francis runs a business that requires him to be familiar with the laws surrounding minors. In fact, Francis has said that GGW has very specific procedures in place to prevent filming underage girls and even teaches its camera crew ways to ensure that the girls the crew is selecting to appear in GGW spring break videos are of age. During the selection procedure, the GGW camera crew is required to check the IDs of girls wanting to be filmed, obtain signed written release forms in which the girls give their consent to be filmed, and videotape the girls's IDs as well as the actual process of signing the release forms.

Despite his company's strict policies, Francis found himself in the middle of a heated legal battle in 2003 when seven girls claimed that they were underage when the GGW camera crew filmed them on vacation in Panama City, Florida. Francis fought back, saying that the girls misrepresented themselves, knowingly sought out the GGW crew, and wanted to exploit the company to obtain a monetary settlement. After four years of court proceedings and intense media coverage, Francis reached an undisclosed settlement with the women, who reportedly wanted a total of $70 million.

Liability of Minors for Necessaries Contracts for necessaries are sometimes considered an exception to the rule that minors can disaffirm their contracts. Technically, a minor can disaffirm contracts for necessaries; however, the minor will be held liable for the reasonable value of the necessary. The purpose of this limitation is to ensure that minors can obtain the basic necessities of life when their parents will not provide them.

A *contract for a necessary* is a contract that supplies the minor with the basic necessities of life, generally thought of as food, clothing, shelter, and basic medical services. It is sometimes difficult, however, to determine whether something is in fact a necessary. Some courts define a necessary as what is needed for the minor to maintain his or her standard of living and financial and social status, but this can lead to a problem when an item that may be considered a necessary for a child of upper-income parents might be considered a luxury to a child of lower-income parents.

Whether an item is considered a necessary is also related to whether the minor's parents are willing to provide the item in question for the minor. For example, in a recent case, a minor chose to live independently, but her parents kept her bedroom open for her and were willing to have her return home any time. When she chose to return home after having fulfilled only two months of a year-long lease, the court found that the lease for the apartment was not a necessary because her parents had always been willing to provide a home for her.[3]

LO 11-2 Explain how a contract created when one is a minor may be ratified after one is no longer a minor.

Ratification Once a person reaches the age of majority, he or she may ratify, or legally affirm, contracts made as a minor. Once ratified, the contract is no longer voidable. Ratification may be either express or implied.

An *express ratification* occurs when, after reaching the age of majority, the person states, either orally or in writing, that he or she intends to be bound by the contract entered into as a minor. For example, when she is 17, Marcy enters into an agreement to purchase an automobile from Sam for 10 monthly payments of $1,000. After making the fifth payment, Marcy turns 18 and decides to move out of state. She emails Sam and tells him not to worry because, even though she is moving, she still intends to make her monthly payments to purchase the car. She has expressly ratified the contract.

An *implied ratification* occurs when the former minor takes some action after reaching the age of majority that is consistent with an intent to ratify the contract. Going back to the previous example, if, the day after she turns 18, Marcy enters into an agreement with Joe to sell him the car in six months, that action is obviously consistent with an intent to finish purchasing the car, so she has impliedly ratified the contract. Most courts find that continuing to act in accordance with the contract, such as continuing to make regular payments under a contract after reaching the age of majority, constitutes ratification. For example, if Marcy continued using the car and making

[3] *Kim Young v. Phillip Weaver*, 883 So. 2d 234 (Ala. Civ. App. 2003).

payments on it for several months after reaching the age of majority, the courts would probably find that she had ratified the contract.

Parents's Liability for Their Children's Contracts, Necessaries, and Torts As a general rule, parents are not liable for the contracts entered into by their minor children. Because of this rule, merchants are often reluctant to enter into contracts with minors unless some competent person is willing to agree to cosign the contract. In that way, the competent person will be legally bound to perform the obligations undertaken by the minor if the minor no longer wishes to live up to the terms of the contract.

Parents do, however, have a legal duty to provide their children with the basic necessities of life, such as food, clothing, and shelter. Thus, they may be held liable in some states for the reasonable value of necessaries for which their children enter into contracts.

In most states, parents are not liable for the torts of their minor children; minors are liable for their own personal torts. In many states, however, parents may be liable when a child causes harm because the parent failed to supervise the child properly, thereby subjecting others to an unreasonable risk of harm from the child.

Mentally Incapacitated Persons

Persons suffering from a mental illness or deficiency may have full, limited, or no legal capacity to enter into a binding contract, depending on the nature and extent of their mental deficiency. If a person suffers from mental problems yet still understands the nature of the contract and the obligations imposed by it, the person may enter into a binding, legal agreement. Suppose, for example, that Gina suffers from the delusion that she is a rock star. When an encyclopedia salesperson comes to her door, she buys a set of encyclopedias from him because she believes that it is important for her to be knowledgeable to set a good example for her fans. As long as she understands that she is binding herself through a contract to make monthly payments for two years to pay for the encyclopedias, she is bound to the contract. If, after making a year of payments, she no longer suffers from her delusions and wishes to disaffirm the contract, she will not be able to do so because her delusions did not affect her understanding of what she was legally agreeing to do.

However, a person has only limited capacity to enter into a contract if he suffers from a mental illness or deficiency that prevents him from understanding the nature and obligations of the transaction he is entering into. If, in the preceding scenario, Gina's delusions prevented her from understanding that she was signing a contract, and she thought she was giving the salesperson her autograph when she signed the contract, the contract is voidable. She may disaffirm it at any time until a reasonable time after she no longer suffers from the mental deficiency. Once the mental deficiency has been removed, she may also choose to ratify the contract.

As with contracts of minors, a contract of a person suffering from a mental deficiency that is for necessaries can be enforced for the reasonable value of the necessary.

If a person's mental deficiencies have resulted in him being adjudicated insane and a guardian has been appointed for him, he has no capacity to enter into contracts, and any contract he attempts to enter into is void. Guardians may be appointed not only for those who are adjudicated insane but also for those who are adjudicated habitual drunkards and those whose judgment has been impaired because of a condition such as Alzheimer's. In any case, although the person for whom the guardian has been appointed no longer has the legal capacity to enter into contracts, the guardian has the legal capacity to enter into contracts on that person's behalf.

BUT WHAT IF . . .

WHAT IF THE FACTS OF THE CASE OPENER WERE DIFFERENT?

Let's say in the Case Opener that it was not a child making a purchase but a mentally disabled person named Chloe. Let's say that Chloe was convinced that she owned a dog, even though she didn't, and ordered $1,000 worth of dog food. Which kind of legal capacity would Chloe have? Could she then avoid her contract? What if Chloe had been adjudicated insane and committed to a mental institution but was staying with her sister on a weekend visitation pass when she made the purchase?

Intoxicated Persons

For purposes of determining capacity, intoxicated persons include those under the influence of alcohol or drugs. There is some variation among states in the treatment of the capacity of intoxicated persons to enter into contracts. As a general rule, most states follow the Restatement of Contracts, Section 16, which provides that contracts of an intoxicated person are voidable by the intoxicated person if the other party had reason to know that because of the intoxicated person's condition, that person was unable to understand the nature and consequences of the transaction or was unable to act in a reasonable manner in relation to the transaction. If the intoxication merely causes someone to exercise poor judgment, the person's capacity to enter into a legally binding contract is not affected.

Likewise, if one party has no way of knowing that the other was intoxicated at the time the agreement was made, and the agreement is a fair one, it will be upheld by most courts. For example, Lisa emails Rob and offers to buy his antique car from him for $8,000. Rob has just broken up with his girlfriend and has been drinking nonstop all day. He gets the email message and immediately responds in the affirmative. Lisa had no way of knowing Rob was intoxicated, so they would have a valid contract in most states.

Once sober, the intoxicant has the ability either to ratify or disaffirm the contract. Because public policy does not favor intoxication, the courts tend to be unsympathetic to intoxicants and will fairly liberally interpret behavior that seems like ratification as ratifying the contract. If Jim became intoxicated at a bar one evening, and Randi took advantage of the situation by getting him to sign a contract to sell her his 2004 SUV for $8,000, any act Jim takes consistent with ratification after becoming sober will result in a binding contract. If Randi appears at his house the next morning with the cash, shows him the contract drafted on a napkin that he signed, and asks for the keys and the title, then by giving her the keys and saying, "I knew I shouldn't have drunk that much," Jim has entered into a binding contract.

If the contract is disaffirmed on the basis of intoxication, each party to the contract must return the other to the condition she or he was in at the time the contract was entered into. Also, as with contracts of persons who have limited capacity, a contract of an intoxicated person for necessaries will be enforced for the reasonable value of the necessaries.

Exhibit 11-1

The Three I's of Incapacity: General Rules

TYPE OF INCAPACITY	CONSIDERATIONS	IF THE ANSWER IS YES, THE GENERAL RULE IS:
Infancy	Is the person under the age of majority (a minor)?	The contract is voidable.
Insanity	Is the person suffering from mental deficiencies that prevent him from understanding his legal obligations under the contract he is entering into?	The contract is voidable.
	Does the person's mental deficiency simply impair her judgment about the desirability of the contract but not prohibit her from understanding her obligations under it?	The contract is valid.
	Is the person adjudicated insane?	The contract is void.
Intoxication	Is the sober party aware that the intoxicated person is so impaired that he is unable to understand his legal obligations under the contract he is entering into?	The contract is voidable.
	Is the intoxication such that it impairs only the intoxicated person's judgment but not her understanding of her contractual obligations?	The contract is valid.
	Has the intoxicated person been adjudicated a habitual drunkard?	The contract is void.

Legality

LO 11-3 List the conditions under which a contract is unenforceable by law.

To be enforceable, a contract must have a legal exchange as its subject and must be able to be performed legally. In the event of a contract with an illegal subject, it is the agreement or the bargain that is illegal, not the contract. By definition, contracts are legal. That is why when a contract is overturned due to being contrary to law, the contract is frequently declared void, as if it never existed, as opposed to being declared illegal. It is important to recognize, however, that contracts can be determined illegal even if they are not against a specific statute. When contracts are against generally accepted public policy, these agreements are determined to be **unenforceable,** and not void. No remedy is available for unenforceable agreements.

> **unenforceable**
> A term applied to a contract that, because of a law, cannot be enforced by the courts.

Contracts That Violate State Or Federal Statutes

As previously stated, certain agreements are illegal because either the subject or the performance of the agreement violates a state or federal statute. Given the various state and federal statutes, there are a large number of possible ways in which contracts can violate a statute. Some of the more common ways are discussed below.

Agreements to Commit a Crime or Tort Contracts cannot be for an illegal subject or be carried out by illegal means. Therefore, any agreement to commit a crime or tort is illegal and unenforceable. However, if a legal contract is formed and then the subject of the contract becomes illegal under a new statute, the contract is considered discharged by law. For example, Jim agrees to work for Heather for one year as a poker dealer at her casino. After Jim and Heather enter into this employment agreement, the state government passes a law saying that poker may no longer be offered in casinos in the state. Because it would now be illegal for Jim to work as a dealer, the contract is discharged by law.

Licensing Statutes All 50 states have statutes requiring people who work in certain professions to obtain a license before practicing their craft. For example, doctors of all varieties, plumbers, cosmetologists, lawyers, electricians, teachers, and stockbrokers are all required to obtain a license before practicing. Although this list of professions is far from exhaustive, it demonstrates how widespread the licensing requirement can be in various states.

When licensing is required by law, people in certain fields must obtain a license before they can legally enter into an agreement to perform the restricted services. Frequently, such licenses can be obtained only after extensive schooling or training. The requirement of schooling and licensing demonstrates the concern society places on proper performance of duties in the designated professions.

In addition to indicating concern with proper performance, licensing statutes have two main purposes. The first is to give the government an avenue by which to regulate the specific industries. By requiring professionals to obtain a license, the government has some say over who can perform which jobs as well as how many people can perform these jobs.

The second main purpose of licensing statutes is more closely related to the public interest. This second purpose is to protect the public's health, safety, and welfare. By imposing legal standards on a profession, the government is attempting to prevent people from being harmed due to substandard work. For example, surgery is a delicate, complicated process. It is not in the public's best interest to allow an unqualified person to perform surgery. Therefore, to limit the number of people who might be harmed during surgery, the government steps in and requires surgeons, even after extensive schooling, to obtain a license.

Given the different reasons for licensing various professionals, there are different outcomes when someone enters into an agreement with a person who is not lawfully licensed. The state in which the unlicensed person is practicing is also relevant because many licensing statutes occur at the state level and thus vary from state to state. For example, in some states, the statutes specify "no license, no contract." These states will not enforce any agreement with an unlicensed professional.

However, in a state that does not explicitly bar the enforcement of all contracts with unlicensed professionals, the courts typically examine the profession in question to determine why professionals in that field are required to be licensed. If the purpose of licensing is to provide government control over the profession, most states allow enforcement of the contract. Although the unlicensed professional is acting in violation of the law, there are no grave reasons that the contract should not be carried out. Nonetheless, because the professional is in violation of the law, he or she is usually required to pay a fine for working without the necessary license.

Conversely, if the licensing statute is intended to protect the public's health, safety, or welfare, the agreement is typically deemed illegal and unenforceable. Licensing statutes do not protect the public if people without the necessary licenses are allowed to practice the restricted professions. Returning to the example of surgeons, the public is made safer when the government allows only licensed professionals to perform surgery. Therefore, people cannot enter into a contract for professional service with unlicensed professionals when the law requires a license and the statute is intended to protect the public. Case 11-2 illustrates how failure to obtain a license can preclude a party from suing to enforce a contract.

KING v. RIEDL
Alabama Civil Court of Appeals
58 So. 3d 190 (2010)

FACTS: *The defendants, Roseann and Bryan Riedl, entered into a contract with the plaintiff, Jim King of King Home Services, to make improvements to the Riedls' house. The improvements entailed work throughout the entire property, including work to the yard and demolition and installation work to the house. The defendants paid the plaintiff a total of $14,075 for some, but not all, of the work specified in the contract. The plaintiff was not at any time a licensee of the Alabama Home Builders Licensure Board.*

The defendants were unsatisfied with the work contracted for and performed by the plaintiff. Consequently, they filed a small-claims complaint in the small-claims division of the Madison District Court, alleging that the plaintiff had damaged their house. The defendants sought compensation for repair work performed by other parties. In response, the plaintiff filed an action in the Madison Circuit Court against the defendants, alleging a number of claims, including breach of contract. In the district court, the plaintiff requested the district-court action to be consolidated with the circuit-court action. After the defendants also requested consolidation, the district-court action was transferred to the circuit court, and the two actions were consolidated.

The defendants filed a motion for a summary judgment in the circuit court, asserting that the plaintiff lacked standing to institute the circuit-court action against them because he was an unlicensed home builder. In response, the plaintiff claimed that he did not need a license to enforce his contract with the defendants. After a hearing, the circuit court entered a summary judgment in favor of the defendants, dismissing all of the plaintiff's claims against them. The plaintiff filed an appeal of the summary judgment entered in the circuit-court action, which was transferred to the Court of Civil Appeals of Alabama.

ISSUE: In Alabama, can a plaintiff home builder bring a breach of contract claim if that home builder is not licensed by the Alabama Builders Licensure Board?

REASONING: In Alabama, a residential home builder is defined as a home builder when the cost of the undertaking exceeds $10,000. All residential home builders must be licensed by the Alabama Home Builders Licensure Board to be permitted to bring a breach of contract claim against a customer. In this case, the plaintiff argued that the cost of the undertaking in the present case was less than $10,000 because the work done to the defendants' porch, doggie doors, and fence should not have been included in calculating the cost of the undertaking. The plaintiff further argued that the defendants had full control over the subcontractors and, thus, that he should have been exempted from obtaining a license by § 34-14A-6(5).

However, the plaintiff's argument that work done to the porch, doggie doors, and fence should be considered separately from work performed on the house in calculating the total cost of the undertaking, as well as his argument that he did not have sufficient control over the subcontractors, falls within an exemption to the licensing scheme. As such, the cost of the undertaking of all the work performed by the plaintiff was considered greater than $10,000. Because there was evidence that the cost of the undertaking was greater than $10,000, there was reason to believe that the

(continued)

plaintiff violated Alabama law and, thus, had no standing to bring his breach-of-contract claim against the defendants.

DECISION AND REMEDY: The Court of Civil Appeals affirmed the decision in favor of the defendants.

SIGNIFICANCE OF THE CASE: This case explains the importance of a home builder being licensed by the state for that home builder to be able to file any necessary breach of contract claims against a customer.

CRITICAL THINKING

When a judge or judicial panel makes a decision, it must do so on the basis of the information available to it. What is an illustration of a fact that, were it true, would have caused this case to have provided King standing with respect to his breach of contract action? What fact was especially important in moving the decision in the direction of summary judgment?

ETHICAL DECISION MAKING

The statute spelling out the need for a residential home builder license has an exemption for small contracts. What stakeholders are being especially advantaged by the exemption?

usury
The lending of money at an exorbitant or unlawful rate of interest.

Usury Almost as widespread as licensing statutes, statutes prohibiting usury exist in nearly every state. **Usury** occurs when a party gives a loan at an interest rate exceeding the legal maximum. The legal maximum interest rate varies from state to state, but it is easy to determine the rate in any given state.

Although usury statutes act as a ceiling in terms of setting a maximum possible rate, there are a few legal exceptions whereby loans may exceed the predetermined maximum. One exception relates to corporate loans. Most states with usury statutes allow corporations to lend and borrow at rates exceeding the maximum. The rationale behind the corporation exception is that if a business needs money to expand and is willing to pay the higher interest rate, the corporation should be afforded the opportunity to borrow. The converse is that if a corporation is willing to borrow at a high interest rate, parties should be allowed to lend at that rate for corporations only. The intent is to facilitate business transactions to keep the economy in a healthy state.

Another exception to usury statutes applies to smaller loans. Many states allow parties to make small loans at rates above the maximum to parties that cannot obtain a needed loan at the statutory maximum. The belief is that if people need money and the statutory maximum is not inducing people to lend, certain parties will make the necessary loans at a higher rate as long as the loan is small, which is defined however the state chooses to define it. This exception is the principle that allows cash advancement institutions to operate.

If there is no exception that allows a usurious loan, the legal outcome varies by state. A minority of states declare all usurious loans void, which means the lending party is not entitled to recover either the interest or the principal. A larger number of states allows lenders to recover the principal amount they loaned, but these parties cannot collect any interest on the loan. In addition, a final group of states allows lenders to recover the principal as well as an amount of interest up to, but not exceeding, the statutory maximum. In essence, these states allow the lenders to recover all but the excess interest, as if the loan were given at the statutory maximum rate.

gambling
An agreement in which a party pays consideration (money placed during bets) for the chance, or opportunity, to obtain an amount of money or property.

Gambling As with licensing statutes, all states engage in at least some regulation of gambling. In this chapter, the term **gambling** refers to agreements in which parties pay consideration (money placed during bets) for the chance, or opportunity, to obtain an amount of money or property.

Although gambling is still illegal in the vast majority of states, many states allow casino gambling. The states most notable for casino gambling are Nevada, New Jersey, and Louisiana. Some states allow certain other types of gambling, either intentionally or through legal loopholes. For example, given California's definition of gambling, betting on draw poker is legal. Some states

E-COMMERCE and the Law

Observing Sabbath Days Online

Sabbath days stem from the religious traditions that were so widespread in America's early days. Among the religions that still observe Sabbath days today is Judaism, which has very strict laws in relation to refraining from conducting business on the Sabbath. This prohibition even extends to the online realm. Not only are those of the Jewish faith told to abstain from conducting online transactions on the day of the Sabbath, but they are also prohibited from making transactions online even if the individual in question has no awareness of it. For example, if an individual schedules to pay for an item when it ships, and the item happens to be shipped on the day of the Sabbath, that individual is considered to have broken Sabbath law. Laws against conducting certain types of business are still active even today, though they may not be rigidly enforced.

Source: Rabbi Harvey Belovski, "Must Your Online Shop Shut On Shabbat?" Belogski.blogspot.com, March 6, 2008.

make other exceptions, such as for horse tracks, casinos located on Native American reservations, or legal state-run lotteries.

Sabbath Laws A large number of states still have what are known as *Sabbath, Sunday,* or *blue* laws on the books. Sabbath laws limit the types of business activities in which parties can legally engage on Sundays.

Sabbath laws
Laws prohibiting the performance of certain activities on Sundays.

In most states with Sabbath law restrictions, these laws date back to colonial times. In fact, the term *blue law* is a reference to the blue paper on which New Haven, Connecticut, printed its Sabbath laws in 1781. New Haven's 1781 law prohibited shops from being open and prohibited all work on the "Lord's day" (Sunday).

The activities restricted today by Sabbath laws vary from state to state. Most Sabbath laws pertain to the sale of alcoholic beverages. These laws tend to prohibit the sale of all alcohol or at least specific types either all day or at restricted times on Sundays. Some Sabbath laws go as far as to make it illegal to enter into any contractual relationship on a Sunday. However, an executed, or fully performed, contract created on a Sunday cannot be rescinded.

There are exceptions to Sabbath laws. Most states with Sabbath laws allow the performance of charity work on Sundays. In addition, Sabbath laws do not typically apply to contracts involving procurement of necessities, typically defined as prescription medication, food, or anything else related to health or survival.

Regardless of how widespread Sabbath laws are, the vast majority of states do not enforce all, or at least some, of their Sabbath laws. In fact, some Sabbath laws have in the past been deemed unconstitutional according to the First Amendment. Nevertheless, if Sabbath laws are on the books, despite convention to the contrary, they can be applied, and some states do apply such laws. As a businessperson, you should always find out whether Sabbath laws exist and, if they do, whether authorities enforce these laws.

Agreements That Violate Public Policy

Unlike agreements that are precluded by statute, some types of agreements are not illegal per se, because they are not in violation of any statute or legal code, but are unenforceable because courts have deemed them to be against commonly held public policy. Public policy involves both the government's concern for its citizens and the beliefs people hold regarding the proper subject of business transactions. The focus is on what is in society's best interest.

Contracts in Restraint of Trade It is a widely held belief in economics, as well as in America in general, that competition drives down prices, which is good for consumers. Accordingly, agreements that restrain trade are viewed as being harmful to consumers and thus against public policy. These restraints on trade are also known as *anticompetitive agreements.* Not only are anticompetitive agreements against public policy, they also frequently violate antitrust laws.

There is an exception that allows specific types of restraints on trade. When a restraint on trade is reasonable, as determined by the courts, and the restraint is part of a subordinate, or ancillary, clause, the restraint is typically allowed. Such restraints are known as **covenants not to compete,** or *restrictive covenants.* Covenants not to compete tend to come in two varieties. The first involves

covenant not to compete
An agreement not to compete against a party for a set period of time within a designated geographic area.

noncompetition in the sale of an ongoing business, and the second involves noncompetition in employment, both of which are discussed below.

The first category of permissible restraints on trade is covenants not to compete that involve the sale of an ongoing business. *Ongoing* refers to the status of the business being sold: The business is still running and will still run on its own; it is not being merged and is not closing down.

The public policy argument in favor of supporting restrictions with the sale of a business involves the fairness of the sale. Let us examine a hypothetical situation to see why this restraint on trade is allowed. Suppose you purchase a jewelry store from Ann Smith. Ann is a well-respected member of the community, and her business has been around for many years. The people in the community know the store, and they trust Ann to provide fair exchanges. As a well-informed businessperson, you know about Ann's reputation. In fact, her good reputation made the purchase more appealing.

Now suppose Ann turns around and opens another jewelry store a block away one month later. Ann's loyal customers are likely to go to her new store because they still trust her. In the meantime, Ann's good name is no longer associated with your store, and your business suffers accordingly. You entered into the sales agreement thinking you would benefit from Ann's good name, but in the end you overpaid for a business that lacks the benefit of Ann's name because Ann took her name with her when she went into competition with your store.

In the interest of self-proclaimed fairness, courts are willing to impose restrictions preventing Ann, or others in her position, from going into immediate competition with you or others in your position. Public policy requests fairness in business transactions, and this fairness does not occur when people profit from a business transaction and then start a new business that destroys the business they just sold.

Consequently, the restraints to trade being discussed with the sale of an ongoing business involve the seller's ability to open up another business within a certain geographic location within a designated time period. As previously mentioned, to be enforceable, a covenant not to compete needs to be in a *subordinate* clause in the contract. If the covenant not to compete is part of the main agreement, and therefore not subordinate, the agreement is typically considered unenforceable and void because it goes against public policy by providing unreasonable restraints to trade. The unreasonableness arises because when the covenant is an integral part of the contract, it cannot be objected to. In contrast, when the covenant is subordinate, the specific noncompetition clause can be removed and the agreement can go forward as planned.

The second category of permissible restraints on trade is covenants not to compete that involve employment contracts. The covenants in employment contracts are similar to those in sales contracts. However, with employment, the employee is agreeing not to compete with her boss, if she leaves, for a designated period of time within a designated geographic area.

Covenants not to compete in employment contracts are legal in most states. However, to be legal, the covenant needs to protect a legitimate business interest. Furthermore, the covenant needs to be for a reasonable period of time and geographic area so that it does not unlawfully impinge on the employee's right to earn a living. The agreement must not be for a longer period of time or larger geographic region than is necessary to protect the employer's legitimate business interest.

Given the variations from state to state in other areas of the law, it should come as no surprise that the enforceability of covenants not to compete also varies from state to state. For example, California does not allow any covenants not to compete. On a different note, Texas requires the employee to gain a specific benefit from a contract, beyond employment, before its courts will enforce a covenant not to compete. The employee must be given, or gain, something even if the covenant not to compete is for a reasonable period of time and geographic region.

LO 11-4 Describe instances in which unfairness affects a contract's enforceability.

Unconscionable Contracts or Clauses When courts are asked to review contracts, fairness is not usually high on the courts' lists of things for which to look. Instead, courts typically assume that the contracting parties are intelligent, responsible adults. The belief is, barring proof of coercion, that parties enter into contracts because they want to do so. Nevertheless, there are times when agreements are so one-sided, or heavy-handed against one party, that the courts will not

make the harmed, innocent person fulfill his or her contractual duties. These heavily one-sided agreements are known as **unconscionable** agreements. The term *unconscionable* refers to the fact that the agreement in question is so unfair that it is void of conscience.

Rules against unconscionable contracts exist in both the Restatement (Second) of Contracts and the Uniform Commercial Code. However, before the Restatement and the UCC existed, the common law would not enforce contracts the courts deemed too unfair, or unconscionable. Now, UCC Section 2-302 states:

> (1) If the court as a matter of law finds the contract or any clause of the contract to have been unconscionable at the time it was made, the court may refuse to enforce the contract, or it may enforce the remainder of the contract without the clause, or it may so limit the application of any unconscionable clause as to avoid any unconscionable result; (2) When it is claimed or appears to the court that the contract or any clause thereof may be unconscionable, the parties shall be afforded a reasonable opportunity to present evidence as to its commercial setting, purpose, and effect to aid the court in making the determination.

Every state, with the exceptions of California and Louisiana, has incorporated the preceding UCC section into its state's UCC. Section 208 of the Restatement also incorporates this section from the UCC.

There are two main types of unconscionable agreements, procedural and substantive. **Procedural unconscionability** relates to conditions that would impair one party's understanding of a contract and to the integration of terms into a contract. Regarding impaired understanding of contractual terms, the factors can be anything from tiny, hard-to-read print on the back of an agreement to excessive use of legalese (unnecessarily technical legal language) or even the inability of a person to read over a contract fully and ask any questions before he or she was required to sign.

Most frequently, procedural unconscionability arises when one party presents the other with an adhesion contract. An **adhesion contract** is a contract created by a party to an agreement that is presented to the other party on a take-it-or-leave-it basis. That is, the contract is presented as complete and as the only chance the presented party (known as the *adhering party*) will have to enter into the agreement. Although adhesion contracts are themselves legal, the presence of one does raise some potential problem indicators for courts. Given that adhesion contracts usually do not allow for debate regarding the agreement, courts tend to examine adhesion contracts to see whether they can determine how voluntary the agreement really was.

Conversely, **substantive unconscionability** involves overly harsh or lopsided substance in an agreement. That is, if an agreement is terribly one-sided, it is probably invalid on the basis of substantive notions of fairness and fair dealing. Courts would find the following, for example, to be substantively unconscionable: high differences between cost and price in a sales agreement; agreements in which one party gains vastly more than the other; agreements in which one party is prevented from having any sort of equal benefit; agreements in which one party has little to no legal recourse, according to the agreement; and portions of an agreement that are completely unrelated to anything having to do with either party's business risk.

Exculpatory Clauses

An **exculpatory clause** is a statement releasing one of the parties to an agreement from all liability, regardless of who is at fault or what the injury suffered is. Because tort law attempts to return the wronged party to the state he or she was in before the wrong occurred, anything preventing this corrective mechanism is against public policy. It does not benefit society to allow some parties to get away with not having to pay for wrongs they commit simply because they state they will not be liable in various contracts. In fact, the patently unfair nature of an exculpatory clause is closely tied to the idea of unconscionable contracts.

Exculpatory clauses frequently show up in rental agreements for commercial or residential property. These clauses are often found to be against public policy. The reasoning is that sometimes people are injured while on property they are renting from another, and sometimes the injuries are due to the carelessness, negligence, or other wrongdoings of the owner. Once again, it does not serve the public's interest to allow landlords, especially when it comes to residential property, simply to disavow all liabilities in advance. If they were allowed to do so, landlords would not be required to fix problems in their rental units, including potentially lethal problems such as faulty electric wiring or lead-based paint.

unconscionable
A term applied to a contract in which one party has so much more bargaining power than the other party that the powerful party dictates the terms of the agreement and eliminates the other party's free will.

procedural unconscionability
Unconscionability that derives from the process of making a contract.

adhesion contract
A contract created by a party to an agreement that is presented to the other party on a take-it-or-leave-it basis. Such contracts are legal but are sometimes rescinded on the grounds of unconscionability and the absence of one party's free will to enter a contract.

substantive unconscionability
Unconscionability that derives from one-sided, unjust, or overly harsh substance in a contract.

exculpatory clause
A statement in a contract that frees one party (usually the drafter of the agreement) from all liability arising out of performance of the contract, generally based on factors such as consumer ignorance or a great deal of unexplained fine print that serve to deprive the less powerful party of a meaningful choice.

A basic test to determine whether an exculpatory clause is unenforceable is to see whether the enforcing party engages in a business directly related to the public interest. Examples of such businesses include banks, transportation providers, and public utilities. Courts believe it is against the public interest to release businesses engaging in work in the public's interest from accountability to the public they are serving.

Another concern with large businesses that are serving the public interest is the unfair bargaining power they can possess in negotiating a contract. Given the power these businesses have, they could simply demand that all customers must accept the exculpatory clause, thus escaping all possible liability. Although the idea of these businesses not being liable is bad, what is worse is that there would be no punitive financial motive for them to conduct their business operations carefully, or so the argument goes. As a result, the potential for increased accidents would be large. Obviously, it is not in the public's interest to have unsafe businesses that are not accountable to the public. Thus, these businesses cannot enforce exculpatory clauses.

It is often difficult to know when a court is going to find that an exculpatory clause exists. Case 11-3 details a situation in which the court found that an illegal exculpatory clause did not exist.

CASE 11-3

RILEY AND v. STONE GATE
Court of Appeals of Washington, Division 2
395 P.3d 1059 (2017)

FACTS: *Larry Riley formed a rental agreement for a self-storage unit with Iron Gate Self Storage in December 2003. The agreement stated that a maximum of $5,000 of personal property could be stored in the unit and that Iron Gate could only be held liable for up to $5,000. Over the course of the lease, Riley fell behind on payments.*

From May 2010 to July 2010, Iron Gate sent Riley three past-due notices. In July, Iron Gate sent Riley a notice of auction and the auction occurred seven days later. A bidder paid $2,000 for the contents of Riley's storage unit. Although Iron Gate believed it acted lawfully, Washington state law required auctions in such cases to occur no less than fourteen days before the notice.

Two days after the auction, Riley sent Iron Gate a letter asserting that the notice was invalid and the auction should not have occurred. Iron Gate recovered many of the sold auction items by repurchasing them and held them in a storage unit free of charge until Riley retrieved them. However, Riley filed a complaint claiming Iron Gate had violated the Storage Act and the CPA and alleging $1.5 million in damages.

ISSUE: Was clause limiting the liability of Iron Gate to $5,000 an illegal exculpatory clause?

REASONING: To determine whether Iron Gate's exculpatory clause violated public policy, the court considered six criteria: (1) whether the agreement concerns a situation generally suitable for public regulation; (2) whether Iron Gate performs a service that is of great importance or necessity to the public; (3) whether Iron Gate holds itself out as willing to perform its services for any member of the public who seeks it; (4) whether Iron Gate possesses a decisive advantage in bargaining power over a member of the public who seeks its services; (5) in exercising its superior bargaining power, Iron Gate effectively forces people seeking its services into agreeing to the exculpatory clause; and (6) whether the property of people seeking such services must be placed under control of Iron Gate.

The court compared the facts of the case to each criterion. First, Iron Gate is in the self-storage industry, which is in fact a highly-regulated industry. Iron Gate is subject to numerous requirements contained within the Storage Act. Second, Iron Gate's self-storage services are not an essential public service. Third, Iron Gate advertises itself to the general public as willing to rent units to any person who seeks it. Fourth, Iron Gate does not have a advantage in bargaining power because Riley had the freedom to take his business elsewhere if he disagreed with the agreement's provisions. Fifth, Iron Gate did not force Riley into the exculpatory clause. Iron Gate provided Riley with an opportunity to purchase insurance that would cover the entire value of his property, but he chose not to do so. Finally, Riley had exclusive control over his unit and the property within. Riley used his own lock on the unit and Iron Gate was only allowed to enter with written notice, in case of emergency, or if Riley defaulted.

In light of the facts, the court concluded that Iron Gate's exculpatory clause did not violate public policy. Per the signed agreement, Iron Gate reasonably believed that the contents of Riley's unit did not exceed $5,000. Thus, Iron

(continued)

Gate is not liable for the $1.5 million in value that the contents of the unit allegedly had.

DECISION AND REMEDY: The lower court's finding that the clause did not violate public policy and therefore was not illegal was affirmed in favor of the defendant, and therefore its liability was limited to $5,000.

SIGNIFICANCE OF THE CASE: In this case, the court clearly lays out the requirements for an exculpatory clause to violate public policy and explains how the clause in question does not meet these requirements. Thus, this case serves as an excellent illustration of when such a clause is and is not binding.

CRITICAL THINKING

Consider the sixth item in the test applied by Washington state courts. In this case, the court asserted Riley had exclusive control of his storage unit because he placed his own lock on the unit and "Iron Gate could only enter the unit with written notice, in the case of an emergency, or if Riley defaulted." Do you think the sixth item in the test could be problematic if applied to apartment renters?

ETHICAL DECISION MAKING

Examine the actions of each party leading up to this dispute. Who behaved in a blameworthy fashion, and who in a praiseworthy fashion? What facts from the case and what ethical theories or guidelines support your claim?

Now consider each party's stance in the legal dispute. Does either one appear more or less ethical, relative to that party's earlier actions? Why or why not?

Although businesses closely linked to the public interest cannot enforce exculpatory clauses, not all exculpatory clauses are unlawful. To be able to enforce an exculpatory clause, the party seeking enforcement must be a private business or an individual and must *not* be important to the public interest. Such private businesses or individuals provide nonessential services and thus do not have the same bargaining power as the previously discussed groups, such as banks, utilities, and airlines. Given their lack of huge bargaining power, it is assumed that private businesses and individuals will enter on relatively equal terms and will voluntarily agree on or decline the final contract.

Examples of private businesses that can enforce exculpatory clauses include skiing facilities such as resorts or rental places, private gyms or health clubs, any business offering skydiving or bungee jumping, and amusement parks, to name a few. Because the services provided by all these businesses, as well as others in this category, are not something related to the public interest and not something in which people must engage, these parties are allowed to deny liability if the party with which they are contracting agrees to the exculpatory clause. Just because these parties *might* be able to enforce an exculpatory clause, however, does not mean the clause is always automatically enforceable.

The following Case Nugget looks at an interesting situation involving an exculpatory clause.

CASE Nugget

Can Parents Enter into Contracts on a Minor's Behalf?

Trent Woodman v. Kera LLC Supreme Court of Michigan

486 Mich. 228 (2010)

Five-year-old Trent Woodman was injured on his birthday at Bounce Party, an indoor play area operated by Kera LLC. Before Trent began playing at the play area, his father signed a liability waiver including an exculpatory clause on Trent's behalf. Mr. Woodman

(continued)

signed the form as the parent, and Trent's name was printed on the form as "participant," meaning he was the one participating in the activity for which the exculpatory clause was drafted. When Trent broke his leg after jumping off an inflatable slide, Trent's mother, as next friend, filed suit against Kera LLC, on behalf of her son, alleging negligence, gross negligence, and violation of the Michigan Consumer Protection Act (MCPA). Kera LLC filed a motion for summary judgment, claiming that the claims were barred by the exculpatory clause.

Michigan common law prevents parents from entering their children into contracts that waive, release, or compromise a minor's claims. Because parents' authority of a minor is limited to the care and custody of the minor, any contract that a parent might enter into on his or her minor's behalf is unenforceable.

The Michigan Supreme Court upheld the common law in this case by finding that the waiver that Trent's father signed was unenforceable. The court refused to change common law because it believed that the court lacked the tools to measure the risks involved in such a policy change adequately. Therefore, the better vehicle for changing the common law is through legislature. Trent's case was remanded for further proceedings.

Effect of Illegal Agreements

LO 11-5 Recall the legal effect of entering into a contract for illegal purpose.

Generally speaking, when an agreement is deemed illegal, courts will label the contract void. The reason most illegal agreements are void is that, in most cases, both parties are equally responsible for the illegal agreement. In the law, when both parties are equally responsible for an illegal agreement, the parties are said to be *in pari delicto*.

in pari delicto
In equal fault.

When both parties are at fault, it does not make sense for the courts to attempt to salvage the agreement or reward either party. Therefore, most illegal agreements are void, meaning neither party can enforce the agreement and neither party is entitled to recovery.

However, it is not always the case that both parties are at fault. If one party is not responsible for an illegal agreement, or special circumstances forgive the illegality, it makes sense sometimes to allow one party to an illegal agreement to recover various damages.

The first exception to the general rule occurs when a member of a protected class is involved in an agreement that contradicts a statute intended to protect the specific class. In the event of such a contradiction, the member of the protected class is allowed to sue for performance. The reasoning is that if a statute is intended to protect a specific class, people outside the class should not be allowed to use the statute to harm those inside the class.

For example, a work agreement between a truck driver and his or her employer may specify that the driver gets paid for the number of hours worked. Yet certain statutes limit the number of hours truck drivers may drive in a given time period. If the truck driver accidentally drives more than the allowable hours, the driver has technically violated a statute. However, this violation does not allow the employer to refuse to pay the driver for the extra hours. Rather, the driver may sue the employer to enforce the work agreement.

The second exception to the voiding of illegal agreements occurs when there is a justifiable ignorance of facts. A *justifiable ignorance of facts* is one party's lack of knowledge regarding a provision of the agreement that would make it illegal. Although ignorance of the law does not excuse illegal behavior, not knowing that the other party to a contract intended to fulfill the agreement through illegal means does function as an excuse.

When one of the parties is relatively innocent in the whole deal, the court may give back any consideration that party gave or may require an exchange for partial performance such that both parties can be returned to the positions they were in before they entered into the agreement. Conversely, if one party is completely innocent of any illegality and has completed his or her portion of the contract, then—depending on the reason the contract is considered illegal and depending on which state's laws are in question—the court might enforce the entire agreement.

A third exception to the general rule occurs when one of the parties withdraws from an illegal agreement. The key to any recovery is that the party must have withdrawn before any illegality occurred. If this is the case and if part or full performance has occurred, the party may recover value for whatever portion has been completed. However, if the party is involved in any way in the illegal activity, the party cannot recover at all.

LO 11-6 Describe the characteristics of a severable contract.

Severable Contracts A **severable contract**, also known as a *divisible contract*, contains multiple parts that can each be performed separately. In addition, separate consideration is offered for each individual part. In essence, a severable contract is like numerous contracts in one. Conversely, an **indivisible contract** requires complete performance by both parties, even if it appears as though the contract contains multiple parts, similar to a severable contract.

With respect to illegality, severable contracts have a huge advantage over indivisible contracts: If a severable contract has both legal and illegal portions, the court has the option of declaring void only those sections of the agreement that are illegal. The court can then enforce the remaining, legal portions of the contract. Indivisible contracts must be enforced or rejected in their entirety. For a court to be able to enforce parts of a severable contract, the parts that are enforced must still represent the main purpose of the original agreement. If declaring parts of a contract void substantially alters the contract, the court is not likely to enforce the remaining portions of the agreement. Courts ultimately want to facilitate business transactions and enforce the legal wishes of parties, and severable contracts enable courts to do so. And, of course, since severable agreements are subject to a court interpreting each section for legality, it is probably unwise to enter into such an agreement when the legality of some sections is questionable.

severable contract
A contract that contains multiple parts that can each be performed separately.

indivisible contract
A contract that requires complete performance by both parties.

Used by permission of Stu Rees, Stu's Views.

CASE OPENER WRAP-UP

Apple's Questionable Contract

The US District Court for the Northern District of California denied Apple's motion to dismiss the lawsuit. The court stated that the complaint could not be dismissed because no case law was provided to prove that Apple's Terms of Service served as a contract for all subsequent transactions. Apple then constructed a settlement with all the plaintiffs in the class action lawsuit that had to be court-approved.

First, Apple would have to immediately send notices to 23 million customers with iTunes accounts, notifying the customers of parental controls that can block extra purchases. Second, customers in the lawsuit were to receive a $5 credit for use at the iTunes store. Customers whose children spent $30 or more would be entitled to a cash refund instead of credits. Finally, Apple would pay $1.3 million in attorney fees.

PRACTICAL TIPS for Business Managers

1. Remember that people don't always look their age, so always make sure to verify the age of anyone with whom you are entering into a contract. You do not want to find out you have entered into a contract with a minor by the minor telling you he or she wants to disaffirm the agreement.

2. There may be times when you need an exculpatory clause in one of your contracts. To ensure that the exculpatory agreement is enforceable, it is a good idea to make sure the clause is printed in a large font and initialed by both parties. If possible, provide options to the other party, so that the clause is clearly negotiated.

3. When drafting a covenant not to compete in conjunction with the sale of a business, make sure that it is clearly reasonable both in terms of the length of time it is valid for and the geographic area. One way to determine reasonableness of both is to rely on your customer list, both in terms of where your customers live and how frequently they use your services. Because of the value of this information if you should sell your business, it is a good idea to maintain good customer lists because you never know when you may decide to sell.

SUMMARY

Capacity

Natural persons over the age of majority are presumed to have the full legal capacity to enter into binding, legal contracts.

A person has only limited capacity to enter into a legally binding contract if the person is:

- A minor.
- Suffering from a mental deficiency that prevents the person from understanding the nature and obligations of contracts.
- Intoxicated.

A person has no capacity to enter into a contract if the person:

- Has been adjudicated insane.
- Has been adjudicated a habitual drunkard.
- Has had a legal guardian appointed to enter into contracts on his or her behalf.

Legality

Contracts that do not have a legal object are not valid.

Contracts may lack a legal object because they violate a statute or violate public policy.

If the illegal portion of a contract can be severed from the legal portion, the courts may simply refuse to enforce the illegal parts and enforce the remaining legal parts.

Point/Counterpoint

Should Minors Have Full Capacity to Enter into Binding Legal Contracts at the Age of 15?	
YES	NO
The infancy doctrine, which holds that minors are not developed enough to make fully rational decisions and therefore do not have the capacity to enter into contracts, is not only antiquated, but makes faulty assumptions about the minds of minors of certain ages. No one is arguing that a two-year-old toddler has the capacity to make rational business decisions that are in her interests. The capacity of a sixteen-year-old, however, is a different story. Many developmental psychologists, including the famed Jean Piaget, have found that minors of ages fifteen and up have essentially the same rational capacity as adults and can utilize theoretical, hypothetical, and counterfactual thinking, as well as abstract logic and reasoning. While contract law does not recognize minors as fully rational beings, there are other areas of law that do. For example, many state criminal codes allow minors to be considered adults when being tried for heinous crimes. Approximately 250,000 minors are tried, sentenced, or convicted as adults every year across the United States, and until the Supreme Court effectively outlawed the practice in 2005, states could give minors the death penalty. The law is inconsistent in that it can assume that minors are rational adults when committing a crime, but not when entering into a contract. Another problem with the infancy doctrine is that it assumes that minors will necessarily be the victims of unfair contracts and therefore allows them the ability to disaffirm at will. However, this means that contracts with minors are voidable by the minor, but not void. In other words, the contracts are one-sided in that the minor can enforce them against the other party, but the other party cannot enforce them against the minor. In today's digital marketplace, where minors are becoming increasingly involved and may even provide content or a service to an adult, the infancy doctrine gives minors an unfair advantage. In fact, it is not unheard of for the minors of this generation to use their tech savvy to prey on unsuspecting consumers. The infancy doctrine in the age of the digital marketplace creates another problem: online, it is almost impossible to truly verify with whom one is doing business. An online businessperson will never truly know whether their online client is a minor or not. This issue creates all sorts of risks for such a businessperson because they have no idea whether any digital-sales contract, for example, is voidable by the other party at any time. The infancy doctrine is founded on questionable assumptions and simply does not work in today's world.	The infancy doctrine is more important than ever in today's marketplace. Even though many children have a technological prowess that overshadows that of many adults, children are not immune from abusive contracts and are likely not developed enough to enter into contracts in the first place. Some arguments in favor of doing away with the infancy doctrine cite as a primary reason that some teenagers are cognitively developed enough to be able to rationalize on the same level as adults. Regardless of whether this is true, children and teenagers are more impulsive and lack the experience and foresight to make certain decisions in their favor. In other words, while minors can sometimes rationalize a situation as well as an adult, that does not necessarily mean that they will make as good a decision as an adult would. Thus, minors are more likely to brazenly enter into unfair contracts without having the foresight to think of the long-term consequences of doing so. Another common argument against the infancy doctrine is that minors can abuse their right to disaffirmance and enter into one-sided contracts in their favor. A classic example is one where a minor enters into a sales contract to purchase a car, immediately proceeds to wreck the car, disaffirms the contract, and demands a refund for the car while returning the totaled vehicle. However, many courts have instituted exceptions to the infancy doctrine for situations just like the one described. Often, if a minor reaps some kind of benefit from the contract, they cannot disaffirm the contract without returning that benefit to the other party in full. In the case of the wrecked car, the minor would not be able to disaffirm the contract without returning a car to the seller that was in the same condition in which it was purchased. Those who argue against the infancy doctrine fail to understand the balance of power between corporations and minors. Over the past few decades, minors have become a class of consumer all of their own, and the digital commerce age has brought the market to minors in ways that were inconceivable even at the time when the infancy doctrine first came about. Corporations specifically utilize the inexperience, naiveté, and rash decision-making of minors to finely target their advertisements and sell their products. Is it fair, for example, for a business to enter into a contract with a minor for a credit card with a 20 percent interest rate when that minor has not had the time or experience to develop financial literacy? Businesses target minors because they know minors do not have the same level of education or decision-making skills as adults, and thus minors require the protection that the infancy doctrine provides.

Questions & Problems

1. Why does it matter that a contract with a minor is for a necessary?

2. Explain the obligations of a minor who chooses to disaffirm a contract.

3. Go back to the discussion of contracts that cannot be disaffirmed by minors and explain the policy reasons that support each of the exceptions. Can you make an argument for any additional kinds of contracts that should not be subject to disaffirmance by minors?

4. If all you know about a man is that his neighbors think he is crazy, you do not know whether a contract he entered into was valid, voidable, or void. Why not?

5. What factors determine whether a covenant not to compete is legal or illegal?

6. What is the relationship between contracts in restraint of trade and unconscionable contracts?

7. Three salesmen worked for Sentient Jet, a small luxury airline charter service. They signed a noncompete agreement, promising not to go to work for a competing employer within a year after working for Sentient and not to take any confidential information with them when they left the firm. When there was a change in the CEO of their firm and talk of the company possibly being bought out, the employees left the firm and went to work for Apollo Jets, a competitor, and allegedly took proprietary information with them that allowed them to solicit former Sentient clients. The plaintiff sought damages and an injunction to ban the employees from working for a competitor for a year. Defendants argued that material changes in circumstances should have made the noncompete agreement unenforceable. How do you think the jury decided in this case?

 Source: *Sentient Jet v. MacKenzie,* January 2013.

8. Valve Corporation makes and distributes its own games as well as games produced by third-party companies on a platform called Steam. In addition to Steam, Valve hosts a community driven market called Steam Marketplace where members can buy, sell, and trade digital items such as games and cosmetic effects.

 Plaintiffs in this case, minor children, who use Steam Marketplace, and their parents, are suing Valve for allegedly supporting "'illegal gambling' by 'allowing millions of Americans, including Plaintiffs, to link their individual Steam accounts to third-party websites' and by 'allowing third-party sites to operate their gambling transactions within Steam marketplace.'" Valve responded by filing a motion to compel arbitration as conspicuously set forth in the Steam Service Agreement that one must accept to set up and use a Steam Account.

 Recall question 2, where you explained what the obligations of a minor who chooses to disaffirm a contract are. Do you think the minor plaintiffs in this case can preclude the arbitration clause by disaffirming the agreement? Why or why not? Do you think the arbitration clause in the Steam Service Agreement binds the minor's parents to arbitrate? How do you think the court should rule in this case?

 Source: GG v. VALVE CORPORATION, Dist. Court, (2017).

9. Washington State resident Patty Gandee entered into a debt adjustment contract with Freedom Enterprises. She subsequently sought to file a class action against them for violations of the state debt adjusting act and the Consumer Protection Act. The company sought to compel arbitration based on a binding arbitration clause she had signed. The clause provided that any disputes under the contract were to be submitted to arbitration that would take place in Orange County, California, under American Arbitration Association rules, and the prevailing party would be entitled to reasonable legal fees and costs, including attorney fees. Both the trial court and the Washington Supreme Court refused to enforce the binding arbitration clause. Explain why they would not enforce the clause.[4]

 Source: *Patty J. Gandee v. LDL Freedom Enterprises,* Washington State Supreme Court, Case No. 87674-6, February 7, 2013.

10. In November 2011, Rosemarie Roopchand made a $200,000 loan to Raheek Mohammed. The parties executed a promissory note stating that Mohammed would repay Roopchand the principle sum of $200,000 with interest at the rate of 100 percent for the term of the loan, i.e. 50 percent per year for two years. The note also included a provision where, if Mohammed was unable to make payments, his company, Medina Petroleum Corporation would make Roopchand whole.

 Roopchand invariably filed an action against Mohammed, his wife, and Medina Petroleum after Mohammed fail to complete payment on the promissory note. In response, the defendants filed a motion for summary judgment on the ground that the terms of the loan were usurious. The court dismissed the case in favor of the defendants and Roopchand appealed. In its ruling, the appeals court stated that

[4] http://lawyersusaonline.com/wp-files/pdfs-5/gandee-v-ldl-freedom-enterprises.pdf

"where a loan agreement is usurious on its face, usurious intent will be implied and usury will be found as a matter of law." What do you think of this criteria? How would you define the limits of "usurious on its face"? In this particular case, the maximum interest rate allowed by the state for a loan was 16 percent. If the interest rate in the promissory note was 17 percent would it be "usurious on its face"? What about at 18 percent? 20 percent? 25 percent? How would this case be different if the loan had been made directly to Medina Petroleum Corporation?

Source: Roopchand v. Mohammed, 154 AD 3d 986 (2017).

11. William Jones applied to work for a Waffle House in Florida but was rejected. Some ten months later, Jones sued the Waffle House and data-reporting companies used by the restaurant, alleging that the defendants violated the FCRA by failing to give him a copy of the background checks that were run on him in relation to his employment at Waffle House and by failing to give him a chance to dispute the background checks.

While the lawsuit was pending, Jones still sought employment at Waffle House and was eventually hired at a Waffle House in Missouri. Jones signed a contract with the Missouri Waffle House to gain employment. The contract contained an arbitration clause compelling both parties to resolve all claims and controversies through arbitration. At no point before he signed the contract did Jones inform his lawyer that he intended to work at the Missouri Waffle House. Jones also did not inform the manager of the Missouri Waffle House that he was engaged in a lawsuit with Waffle House's parent company. Additionally, the manager at the Missouri Waffle House simply did not know Jones was involved in a lawsuit with Waffle House.

About two months after Jones was hired, Waffle House filed a motion to compel arbitration in the pending lawsuit. The trial court denied the motion on the grounds that the arbitration agreement was unconscionable. Waffle House timely appealed and the appeals court reversed the decision. What are the two types of unconscionability that courts look at to determine whether a contract is unconscionable? Why do you think the appellate court ruled in favor of Waffle House? Do you think if the hiring manager knew Jones was in litigation with Waffle House it would have affected the appellate court's decision?

Source: Jones v. Waffle House, Inc., 866 F. 3d 1257 (2017).

Part 2 *Contract Law*

CHAPTER 12

Reality of Assent

CASE OPENER

Disagreement Over an Agreement

In spring 1989, Michael Jordan and the Chicago Bulls were in Indianapolis, Indiana, to play against the Indiana Pacers. At the same time, Karla Knafel was singing with a band at a hotel in Indianapolis. After Knafel's performance, a National Basketball Association referee approached her and introduced her to Jordan by telephone. Knafel and Jordan began a long-distance telephone relationship that continued for several months.

In December 1989, Knafel traveled to Chicago to meet with Jordan, where the couple had unprotected sex for the first time. In November 1990, the couple had unprotected sex again while in Phoenix, Arizona. Shortly after this second meeting, Knafel learned that she was pregnant. Knafel was convinced that she was "carrying Jordan's baby" despite having had sex with other male partners. Later, during spring 1991, Knafel informed Jordan she was pregnant with his child.

As a result of several conversations about the baby, Knafel alleged that the two had agreed that Jordan would pay her $5 million when he retired from professional basketball. In return, Knafel promised she would not file a paternity suit against him and would keep their relationship a secret.

In July 1991, the baby was born. Jordan paid some hospital bills and medical costs, and he paid Knafel $250,000 for "her mental pain and anguish arising from her relationship with him." Knafel continued to keep the relationship and paternity a secret.

After Jordan retired from professional basketball, a lawsuit arose between the parties in 2000. Jordan sought declaratory judgment and an injunction against Knafel, who had been approaching him for the $5 million. Knafel filed a counterclaim for Jordan's alleged breach of contract. The trial court dismissed all claims, but the appellate court remanded Knafel's claim for breach of contract. Although Jordan had originally denied the existence of the agreement, on remand he did not contest the existence of the alleged settlement agreement. Instead, Jordan argued that the alleged agreement was not enforceable because it either was fraudulently induced or was based on a mutual mistake of fact. In support of his argument, Jordan produced the affidavit of Dr. Storm, who, after DNA testing, concluded that Jordan was not the child's father.

In response to Jordan's argument, Knafel claimed that the paternity of the child was irrelevant to the enforceability of the alleged agreement. An obstetrician had told Knafel that the baby was conceived on November 19 or 20, 1990 (while she was in Phoenix with Jordan). As a result of this information, Knafel believed that the baby was Jordan's. Additionally, Knafel asserted that the paternity was irrelevant because Jordan entered into the agreement knowing that she had been having sex with other men.

The trial court ruled in favor of Jordan, finding that "as a result of Knafel's fraudulent misrepresentation to Jordan that he was the child's father or, alternatively, as a result of a mutual mistake of fact, the alleged settlement contract is voidable and is therefore unenforceable against Jordan." Knafel appealed.

1. Imagine you are the judge in this case. Do you think that both parties were able to assent to the agreement legally?
2. Under which ethical system, if any, should Knafel be able to recover the $5 million for breach of contract?

LEARNING OBJECTIVES

LO 12-1 Describe the importance of legal assent.
LO 12-2 Recall the elements of mistake.
LO 12-3 Recall the elements of misrepresentation.
LO 12-4 Recall the elements of undue influence.
LO 12-5 Recall the elements of duress.

The Importance of Genuine Assent

LO 12-1 Describe the importance of legal assent.

When two people talk to each other in the hope that an exchange will take place, all kinds of things can go wrong. Global business needs dependability. Just imagine what it would be like if "Yes" meant "Maybe!" Deals would be closed only to be reopened again and again. The costs of all purchases would soar. Businesses would be forced to charge extra to pay for all the extra time they had to spend to finally get to the point when "Yes" really meant "Yes."

To make business transactions smoother and more dependable, the law has developed rules about when "Yes" means "Yes." As Chapter 9 explained, you need an offer and an acceptance to form a contract. Sometimes, however, there is a problem with the acceptance. Certain problems with the acceptance, or the *yes,* may give rise to the claim that the acceptance was not really genuine and, therefore, that the resulting contract is not valid. In this chapter, we examine the problems that may arise with an acceptance that will make the resulting contract **voidable**. When a contract is voidable, it may be **rescinded,** or canceled.

The cancellation of a contract permits the person who canceled it to require the return of everything she gave the other party. At the same time, the person who rescinds the contract must herself return whatever she has received from the other party. A major theme of this chapter is that good business managers aim for genuine assent in their contracts. This chapter shows you how to achieve genuine assent. It explains the major obstacles to genuine assent: *mistake, misrepresentation, undue influence,* and *duress.* Knowing about these potential problems will help you avoid them.

voidable
A term applied to a contract that one or both parties have the ability to either withdraw from or enforce.

rescind
To cancel a contract.

Mistake

LO 12-2 Recall the elements of mistake.

When people agree to buy or sell, they do so with a particular understanding about the nature of the good or service they are about to exchange. However, one or both parties may think they consented to exchange a particular thing only to find out later that no meeting of the minds had

occurred. People may misunderstand either some fact about the deal or the value of what is being exchanged. We focus on misunderstandings about facts because they are the only issues that raise the potential of rescission in American courts.

European courts take a different approach to mistakes about *the value of performance* of a contract. In general, they agree with the reluctance of American courts to interfere with a contract just because the value of the item in question has changed since the agreement. The parties are assumed to have accepted the risk, when they made the contract, that the value might change later. However, European courts permit rescission of the contract for a mistake of value when the mistake involves more than 50 percent of the value at the time of the contract.

Legal assent is absent when a legal mistake occurs. The term *mistake* in contract law has a special meaning that deserves special attention because it is not quite the same as the ordinary use of the term. A **mistake** is an erroneous belief about the facts of a contract *at the time the contract is concluded*.

There is one more important aspect of the definition of mistake. Later in this chapter, when we discuss misrepresentation, our focus will be on incorrect beliefs about the facts of the contract caused by the other party's untrue statements. Mistakes in contract law do not result from the untrue statements of the other party to the contract.

Mistakes may be **unilateral**, the result of an error by *one* party about a material fact, that is, a fact that is important in the context of the particular contract. Alternatively, they may be **mutual**, shared by both parties to the agreement. This distinction is important in determining which contracts are voidable.

mistake
An erroneous belief about the facts of a contract at the time the contract is concluded. When a mistake occurs, legal assent is absent.

unilateral mistake
The result of an error by one party about a material fact, that is, one that is important in the context of a particular contract.

mutual mistake
The result of an error by both parties about a material fact, that is, one that is important in the context of a particular contract.

Used by permission of Stu Rees, Stu's Views.

Unilateral Mistake

In general, a unilateral mistake does not void a contract. Courts are hesitant to interfere with a contract when one of the parties has a correct understanding of the material facts of the agreement. For instance, a widow seeking to rescind her and her husband's election to have his retirement benefits paid out over *his life* was not permitted to receive survivor's pension benefits. The court held that representatives of the retirement system had provided sufficient information to the plaintiff and her husband before they elected that particular form of payout.[1]

Sometimes, however, rescission is permitted for even unilateral mistakes. Because our economic well-being depends so heavily on reliable contracts, we want to be fully aware of the circumstances

[1] *Ricks v. Missouri Local Government Employees Retirement System*, 1999 WL 663217 (Mo. App. WD).

under which unilateral mistakes permit rescission. Any of the following conditions would permit a court to invalidate a contract on grounds of unilateral mistake:

1. One party made a mistake about a material fact and the other party either knew or had reason to know about the mistake.
2. The mistake was caused by a clerical error that did not result from gross negligence.
3. The mistake was so serious that the contract is unconscionable, that is, so unreasonable that it is outrageous.

These situations are rare, but it is important to be aware of them because any rescission can be costly in terms of time and lost opportunities.

Mutual Mistake

When both parties to a contract are mistaken about either a current or a past material fact, either party can choose to rescind the contract. Rescinding such a contract is fair because any agreement between the parties was an illusion: An ambiguity in some key fact prevented the parties from being in actual agreement.

The famous story of the ship *Peerless*[2] has taught generations of students the importance of being very clear in defining material facts in any contract. The parties to the contract that led to the contract disagreement had agreed that the vessel *Peerless* would deliver the cotton they were exchanging. Unfortunately for them, there were two ships named *Peerless*. So when the deal was made, one party had one *Peerless* in mind while the other meant the second *Peerless*. The times the ships sailed were materially different, so the court rescinded the contract.

A Questionable Mistake

Mary W. Scott (Respondent-Appellant) v. Mid-Carolina Homes, Inc. (Appellant-Respondent) Court of Appeals of South Carolina

293 S.C. 191 (1987)

Mary Scott signed a contract to purchase a repossessed 1984 mobile home from Mid-Carolina Homes Inc. for $5,644, to be paid in full before delivery. She gave the salesperson a check for $2,913.71 and agreed to pay the balance before the end of the month. Within the next week, the salesman called and told her he could not sell her the home because it had a bent frame, and the South Carolina Manufactured Housing Board would not permit him to sell a home with a bent frame. She offered to buy it as is and sign a waiver, but the salesman said that would not be legal. A few weeks later, Mid-Carolina sold the mobile home to another couple for $9,220. Scott sued and was awarded $3,600 actual damages and $6,400 punitive damages for breach of contract accompanied by a fraudulent act and $3,000 actual damages for violation of a state consumer protection law. The appeals court upheld the award.

On appeal, Mid-Carolina argued that it was entitled to rescind the contract because the salesperson was acting under a mistake of fact when he gave Scott the sales price. In upholding the award, the state supreme court explained that a contract may be rescinded for unilateral mistake only when the mistake has been induced by fraud, deceit, misrepresentation, concealment, or imposition of the party opposed to the rescission, without negligence on the part of the party claiming rescission, or when the mistake is accompanied by very strong and extraordinary circumstances that would make it a great wrong to enforce the agreement. Mid-Carolina had not demonstrated the presence of any of the circumstances that would justify a rescission. The salesperson was in the superior bargaining position to know the price, and the buyer's reliance on a salesperson's representation of the price was reasonable.

[2] *Raffles v. Wichelhaus,* 159 Eng. Rep. 375 (1864).

Warning: Anticipate ambiguity in the material facts and clarify them in advance to save yourself headaches later.

For a mutual mistake to interfere with legal consent, it must involve all the following:

1. A basic assumption about the subject matter of the contract.
2. A material effect on the agreement.
3. An adverse effect on a party that did not agree to bear the risk of mistake at the time of the agreement.

Courts will not void contracts for reason of mutual mistake if even one of the preceding attributes is missing.

For instance, to rise to the level of a basic assumption, the mistake would need to be about the existence, quality, or quantity of the items to be exchanged. To meet the second condition, the mistake must involve the essence of the agreement. A fact is material when it provides a basis for a person's agreement to enter into the contract. In the case opener, Knafel was basing her appeal on the argument that the paternity of her baby was not material to Jordan's decision to enter into the contract. Do you think she has a good argument? The second element is not satisfied when the person attempting to avoid the contract simply claims that the item to be exchanged is not the one he had intended to exchange.

The third condition is necessary to protect those who bargain with someone who agreed, at the time of the agreement, to bear the risk of mistake but then later wants to avoid that risk when the contract does not work out as well as he or she had planned. This situation might arise, for instance, if the adversely affected party had agreed in the contract to accept the items as-is but later felt they were not worth the price paid. Case 12-1 illustrates the application of this doctrine of mutual mistake.

SIMKIN v. BLANK
Court of Appeals of New York
19 N.Y.3d 46 (2012)

FACTS: When the plaintiff and defendant divorced in 2006, they split their $13.5 million in assets. Most of the plaintiff's $5.4 million share of the settlement was invested in Bernie Madoff's Ponzi scheme, whereas the defendant received a cash settlement. Then, in 2008, the plaintiff thought the terms of the divorce contract should be renegotiated because he lost almost all of his divorce proceeds when it came to light that his investment in Madoff's business turned out to be fraudulent. The plaintiff further argued that both he and his ex-wife had shared in the mistake of investing funds into Madoff's project, yet only the plaintiff received the invested funds in the divorce settlement, and the defendant received cash. The plaintiff also argued that because his funds never existed as an investment because they had already vanished in the Ponzi scheme, he never really received an equal share of their existing assets. Thus, he asked the defendant whether the two could renegotiate the contract. When she refused, the plaintiff sued. A lower court granted the plaintiff the right to sue. The defendant appealed this decision.

ISSUE: Should the plaintiff be permitted to use a mutual mistake claim in the context of a marital settlement agreement?

REASONING: A plaintiff may be permitted to use a mutual mistake claim in the context of a marital settlement agreement if the mutual mistake existed at the time that the contract was entered into and if the mistake is "material." In this case, the core allegation underpinning husband's mutual mistake claim—that the Madoff account was nonexistent when the parties executed their settlement agreement in June 2006—does not amount to a material mistake of fact as required by our case law. The premise of the husband's argument is that the parties mistakenly believed that they had an investment account with Bernard Madoff when, in fact, no account ever existed. In the husband's view, this case is no different from one in which parties are under a misimpression that they own a piece of real or personal property but later discover that they never obtained rightful ownership, such that a distribution would not have been possible at the time of the agreement. But that analogy is not apt here. The husband does not dispute that, until the Ponzi scheme began to unravel in late 2008—more than two years after the property division was completed—it would have been possible for him to redeem all or part of the investment. In fact, the amended complaint contains an

(continued)

admission that the husband was able to withdraw funds (the amount is undisclosed) from the account in 2006 to partially pay his distributive payment to the wife. Given that the mutual mistake must have existed at the time the agreement was executed in 2006, the fact that the husband could no longer withdraw funds years later is not determinative.

Given the extensive and carefully negotiated nature of the settlement agreement, we do not believe that this presents one of those exceptional situations warranting reformation or rescission of a divorce settlement after all marital assets have been distributed.

DECISION AND REMEDY: The Court of Appeals found in favor of the defendant and reversed the decision of the lower court.

SIGNIFICANCE OF THE CASE: This case provides illustration of the criteria courts use to determine whether a plaintiff can use a mutual mistake claim.

CRITICAL THINKING

When Simkin claimed that the contract terms should be altered based on a mutual mistake when the two entered into the divorce contract, he referenced their mistake as the settlement terms allocated to him being nonexistent due to the Ponzi scheme. Should Simkin have defined his use of "nonexistent" according to a timeframe? That is, the funds did not exist two years after the divorce settlement but did exist in 2006, during the divorce.

ETHICAL DECISION MAKING

Could a decision in favor of Simkin altering the terms of the marriage contract create copycat suits? That is, courts typically leave divorce terms intact except in rare circumstances, but many couples could attempt to claim that an outside force affected their divorce settlement assets.

Misrepresentation

LO 12-3 Recall the elements of misrepresentation.

Misrepresentations are similar to mistakes. In both cases, at least one of the parties is in error about a fact material to the agreement. But a **misrepresentation** is an untruthful assertion by one of the parties about a material fact. One party said something that prevented the parties from having the mental agreement necessary for a legal contract. The parties only appeared to agree, so their contract lacked legal assent.

The courts insist that there must be a meeting of the minds for a valid contract. Thus, they might rescind a contract even though the person making the false assertion was entirely innocent of any intentional deception.

misrepresentation
An untruthful assertion by one of the parties about a material fact.

Innocent Misrepresentation

An **innocent misrepresentation** is a false statement about a fact material to an agreement that the person who made the statement believed to be true. The person who made the false statement had no knowledge of the falsity of the claim. We say the person lacked **scienter**. In the case opener, given that Knafel's doctor had estimated the time she became pregnant to be when she was in Phoenix with Jordan, it is likely that the court would find that she did not realize that she was misrepresenting who the father of her child was.

Innocent misrepresentations permit the party that was misled by the false statement to rescind the contract. However, because the party that made the false statement had no intent to mislead, the aggrieved party cannot sue for damages. The reasoning in such cases is similar to the arguments in a mutual-mistake case, as you might expect.

innocent misrepresentation
A false statement about a fact material to an agreement that the person who made the statement believed to be true.

scienter
Intent or knowledge.

> **BUT WHAT IF . . .**
>
> **WHAT IF THE FACTS OF THE CASE OPENER WERE DIFFERENT?**
>
> Let's say that in the Case Opener, Knafel was incorrectly told by her doctor that the baby was Jordan's. She then told Jordan that she indeed was carrying his baby. Thus, Jordan signed the contract and paid Knafel a large sum of money. Later, the two found out the baby was not Jordan's. What kind of misrepresentation occurred? Can Jordan sue for damages?

Negligent Misrepresentation

In some contract negotiations, one party makes a statement of material fact that he thinks is true, but he is negligent in making the assertion. Such a **negligent misrepresentation** results when the party making the statement would have known the truth about the fact had he used reasonable care to discover or reveal it.

Even though there was no actual intent to deceive, the party making the false statement is treated in contract law as if the intent were present. If this standard seems unfair to you, remember that the courts find negligent misrepresentation only when the party making the false statement should have known the truth had she used the skills and competence required of a person in her position or profession. The impact of negligent misrepresentation is identical to that of fraudulent misrepresentation, discussed next.

negligent misrepresentation
A false statement of material fact made by a party who thinks it is true but who would have known the truth about the fact had he or she used reasonable care to discover or reveal it.

Fraudulent Misrepresentation

Any fraud on the part of a party to a contract provides a basis for rescission. The parties cannot be said to have assented when one of the parties was tricked into the agreement by a fraudulent misrepresentation. Thus, the agreement was not voluntary and can be rescinded on the grounds that there was no meeting of the minds.

Even in countries that are trying to encourage joint ventures and global commercial activity, such as China, fraudulent claims can end the country's hospitality to agreements with outsiders. In China, accusations of outsiders' fraudulent misrepresentation have resulted in heavy fines and even refusals to allow the fraudulent party to enter into any more agreements with Chinese firms. The point is that in most, if not all, cultures there is little judicial sympathy for those who consciously mislead others in commercial activities.

A **fraudulent misrepresentation** is a false representation of a material fact that is consciously false and is intended to mislead the other party. Fraudulent misrepresentation is **intentional misrepresentation**. Here, scienter is clear. The party making the misrepresentation either knows or believes that the factual claim is false or knows that there is no basis for the assertion.

To understand the requirements for a finding of fraudulent misrepresentation, start with the two elements from the definition:

fraudulent misrepresentation
A false representation of a material fact that is consciously false and is intended to mislead the other party.

intentional misrepresentation
A false representation of a material fact made by a party who either knows or believes that the factual claim is false or knows that there is no basis for the assertion. Scienter is clear.

1. *A false statement about a past or existing fact that is material to the contract.*
2. *Intent to deceive.* Intent can be inferred from the particular circumstances.

Then add a third necessary element:

3. *Justifiable reliance on the false statement by the innocent party to the agreement.* Justifiable reliance is generally present unless the injured party knew, or should have known by the extravagance of the claim, that the false statement was indeed false. For example, a person could not justifiably rely on a claim by another that a hair on her head, although worth $1,000, was available for $10.

Finally, if damages are sought, the defrauded party must have been injured by the misrepresentation.

BUT WHAT IF...

WHAT IF THE FACTS OF THE CASE OPENER WERE DIFFERENT?

Let's say that in the Case Opener, Knafel had Jordan sign a contract stipulating that he must make monthly payments to her for the baby they both believed was genetically theirs together. Then, Knafel discovered the baby was not in fact Jordan's but did not relay the information to him. What kind of fraudulent misrepresentation occurred here?

Each of these elements can become a source of debate in any attempt to rescind a contract on grounds of fraudulent misrepresentation. Thus, it is your responsibility as a person who will be involved with dozens of contracts in your business activities to know these elements. A rescinded contract is a time-consuming and expensive business opportunity that has gone wrong.

Pay careful attention to the facts in Case 12-2 and decide whether you believe the defendant committed fraud.

FANTASTIC SAMS SALONS CORP. v. PSTEVO, LLC AND JEREMY BAKER

United States District Court for the Northern District of Illinois, Eastern Division

2016 U.S. Dist. LEXIS 5590 (2016)

FACTS: *Fantastic Sams Salons entered into a franchise agreement with PSTEVO, LLC and Jeremy Baker (collectively, "PSTEVO") in which Fantastic Sams allowed PSTEVO to operate a Fantastic Sams franchised salon. Later, Fantastic Sams initiated a lawsuit against PSTEVO seeking declaratory relief with regard to the franchise agreement. In other words, Fantastic Sams wanted a court to clearly spell out the rights that Fantastic Sams held under the agreement. PSTEVO then asserted several counterclaims, including a claim of fraudulent misrepresentation by Fantastic Sams. PSTEVO alleged that, prior to entering into the franchise agreement, Baker met with two agents of Fantastic Sams who presented Baker with several financial disclosure documents. The documents allegedly stated that PSTEVO only needed three months worth of working capital to open the salon, and that the salon would be profitable after this three-month period. PSTEVO alleged that the agents and Fantastic Sams knew that it would take longer than three months for the salon to be profitable, and that these misrepresentations were material because PSTEVO relied on them when deciding whether to enter into the agreement. Fantastic Sams moved to dismiss the fraudulent misrepresentation claims.*

ISSUE: Did Fantastic Sams have the right, under the franchise agreement, to dismiss the misrepresentation claims?

REASONING: Fantastic Sams moved to dismiss PSTEVO's fraudulent misrepresentation claim by way of Federal Rule of Civil Procedure 12(b)(6). This means that Fantastic Sams was essentially arguing that PSTEVO failed to allege one or more of the key elements of a fraudulent misrepresentation claim, and therefore PSTEVO's claims were not sufficient enough to bring to trial. The court relied on the criteria for FRCP 12(b)(6) set forth in *Ashcroft v. Iqbal*, which stated that for a claim to survive this rule, it must be facially plausible: "A claim has facial plausibility when the plaintiff pleads factual content that allows the court to draw the reasonable inference that the defendant is liable for the misconduct alleged."

When arguing why PSTEVO's claims were insufficient, Fantastic Sams pointed to a disclaimer in the franchise agreement that read, "No oral, written or visual claim or representation which stated or suggested any sales, income or profit levels was made to me, except:", after which Baker had written "none" and initialed his response. Fantastic Sams argued that Baker had signified that no claims regarding the franchise's profitability were made to Baker. Thus, PSTEVO was precluded from claiming that it relied on Fantastic Sams' or its agents' representations of the franchise's profitability because no such representations were made. Being that reliance is a necessary element of a fraudulent misrepresentation claim, Fantastic Sams argued that PSTEVO's claim necessarily fails because PSTEVO could not argue reliance.

The court considered that Baker had, indeed, affirmed that no representations of profitability were made to him,

(continued)

as evidenced by his initials. Baker could therefore not argue that he relied on claims of profitability. The court ruled that PSTEVO's claims were not facially plausible because, lacking this reliance, PSTEVO was precluded from making a convincing claim that Fantastic Sams was liable for fraudulent misrepresentation.

DECISION AND REMEDY: The court granted Fantastic Sams' motion to dismiss the fraudulent misrepresentation claims under FRCP 12(b)(6).

SIGNIFICANCE OF THE CASE: This case important because it demonstrates the importance of being able to prove all of the elements of misrepresentation.

CRITICAL THINKING

According to the judge's reasoning, it was PSTEVO's own actions that led to the case being dismissed. Should PSTEVO have done differently?

ETHICAL DECISION MAKING

What general values might the court be interested in protecting in this ruling? How are they similar to values upheld by other cases in this chapter? How are they different? What opposing values are less important in these rulings?

Let's now revisit each of the elements of fraudulent misrepresentation. Although the elements may have seemed relatively straightforward when they were presented, they become more complicated in the context of actual disagreements among parties to a contract.

False Assertion of Fact For fraudulent misrepresentation to be the basis for a contract rescission, the statement of fact need not be an actual assertion. Either concealment or nondisclosure can be treated as the equivalent of an actual assertion. **Concealment** involves the *active* hiding of the truth about a material fact, for example, removing 20,000 miles from the odometer on your car before selling it to me. **Nondisclosure** is different because it involves a failure to provide pertinent information about the projected contract. The courts have until recently been hesitant to use nondisclosure as a basis for rescinding a contract because it is a passive form of misleading conduct. Under ordinary situations associated with a legal bargain, it is not the obligation of one party to bring up any and all facts he or she might possess. Each individual is, to a large extent, treated as a responsible decision maker.

concealment
The active hiding of the truth about a material fact.

nondisclosure
A failure to provide pertinent information about a projected contract.

However, courts will now find nondisclosure as having the same legal effect as an actual false assertion under certain conditions:

1. *A relationship of trust exists between the parties to the contract.* In this situation, the relationship provides a reasonable basis for one person's expectation that the other would never act to defraud him or her.
2. *There is failure to correct assertions of fact that are no longer true in light of events that have occurred since the initial consent to the terms of the agreement.* An illustration is my failure to inform you of the recent outbreak of rust on my rust-free car that you have agreed to purchase next month.

Nondisclosure is especially likely to provide the basis for rescission when one party has information about a basic assumption of the deal that is unavailable to the other party. As a result of this logic, sellers have a special duty to disclose because they know more about the structural makeup of the item being purchased.

GLOBAL Context

©Digital Archive Japan/ Alamy

Consumer Contracts Law in Japan

In 1997, after studying the application of civil law in the country, the Japanese Social Policy Council, an advisory body to the prime minister, recognized that the environment surrounding consumers was growing more diversified, and there was a significant gap between consumers and businesses in their access to information and knowledge as well as in their negotiating power. Because it cannot honestly be said that consumers and businesses are equal, as contracting parties are presumed to be under the country's Civil Code, the council developed a special Consumer Contracts Law. This legislation is considered to place consumers and businesses on a more equal footing in transactions.

Under the Consumer Contracts Law, if a business fails to provide a consumer with information or makes misrepresentations concerning basic or other important contractual matters that are necessary for the consumer to make a judgment about the contract, and if the consumer would not have entered into the contract had the information been provided or the misrepresentation not been made, the consumer may cancel the contract. This provision is applicable whenever a business, in trying to induce a consumer to enter into a contract, (1) fails to provide information about the contents of the contract, (2) fails to provide important information necessary for the consumer to make the decision to enter into the contract, or (3) makes misrepresentations. In many such cases, the consumer would not have been entitled to relief under the Civil Code because of its strict requirements for the application of fraud.

Intent to Deceive Scienter is present when the party accused of making the fraudulent assertion believed that the assertion was false or made the claim without any regard for whether it was true or false. Alternatively, intent to deceive is present when the party making the false statement claims or implies that he or she has personal knowledge of the accuracy of the assertion. Any resulting assent is not legal because the injured party was not allowed to join the mind of the deceiving party. The party with scienter or intent to deceive wanted the contract to be fulfilled on the basis of a falsehood.

Justifiable Reliance on the False Assertion What responsibilities does the injured party have when a false assertion was made by the other party? As we said earlier, the injured party has no justifiable claim of fraud when he or she relied on assertions that should have been obviously false. Anyone who pays for a house in reliance on the claim that it was built before the founding of our country cannot later try to rescind the contract on grounds of fraudulent misrepresentation.

In addition, parties to contracts cannot successfully claim they justifiably relied on a false assertion of fact when the error in the statement would have been clear to anyone who had inspected the item being exchanged. However, even this duty to inspect is declining in modern contract law. Increasing responsibility is being placed on the person who made the erroneous assertion.

Undue Influence

LO 12-4 Recall the elements of undue influence.

When legal assent is present, both parties to the agreement are assumed to have made their own choices based on complete freedom to consent to or reject the terms of the bargain. However, many factors in our lives can work together to make our choices anything but free. **Undue influence** refers to those special relationships in which one person has taken advantage of his or her dominant position in the relationship to persuade the other person to enter into a contract. The persuasive efforts of the dominant person must have interfered with the ability of the weaker person to make his or her own decision. When people are bargaining with their attorney, doctor, guardian, relative, or anyone else in a relationship involving a high degree of trust, they can be persuaded by unusual pressures unique to that relationship, resulting in a contract that is voidable due to undue influence.

undue influence
The situation in which one person has taken advantage of his or her dominant position in a relationship to persuade the other person unduly to the extent that the dominant person's persuasive efforts have interfered with the other person's ability to make his or her own decision.

E-COMMERCE and the Law

Fraudulent Misrepresentation in Social Media

Online social media offer a prime opportunity for individuals to create false identities. In some cases, these false identities are then used to communicate with unsuspecting individuals, often causing the real individuals who communicate with the misrepresented identities to experience financial or emotional harm. A tragic example of this is the case of *United States v. Lori Drew,* in which Drew created the false identity of a boy on the MySpace networking site. Megan Meier, a young girl, was lured by Drew and Drew's daughter to strike up a relationship with the false identity.

When the so-called boy broke up with Meier, Meier unfortunately committed suicide. The prosecution of Drew was unsuccessful. A major issue with the prosecution was that there was no federal statute against cyberbullying, and the judge did not feel comfortable relying on a breach of contract premise, which would make this case of fraudulent misrepresentation equivalent to a case of computer hacking. As the likelihood of fraudulently misrepresenting an individual increases, misrepresentation must be addressed by statutes that cover all possible aspects in which fraudulent representation could occur.

Source: Kim Zetter, "Judge Acquits Lori Drew in Cyberbullying Case, Overrules Jury," Wired.com, July 2, 2009.

If someone believes undue influence has occurred, asking the following questions can help establish whether there was undue influence:

1. Was the dominant party rushing the other party to consent?
2. Did the dominant party gain undue enrichment from the agreement?
3. Was the non-dominant party isolated from other advisers at the time of the agreement?
4. Is the contract unreasonable because it overwhelmingly benefits the dominant party?

The more of these questions that are answered in the affirmative, the more likely a court is to undue influence. The Case Nugget below provides an illustration of undue influence.

CASE Nugget

A Case of Undue Influence

Evan Rothberg v. Walt Disney Pictures

1999 US App 1472

Robert Jahn was a senior executive at Walt Disney Pictures until he died of complications from AIDS. Days before his death, a Disney official visited him at the hospital and convinced him to sign a release that waived his rights to approximately $2 million in employee benefits, including life insurance, stock options, bonuses, and deferred compensation. After his death, his estate sued to recover the benefits waived in the release. Disney received a motion for summary judgment, and the plaintiff appealed. In reversing the motion for summary judgment, the court ruled that the question of whether the release had been procured by undue influence was a question for a jury. The court pointed out that undue influence requires (1) undue susceptibility on the part of the weaker party and (2) application of excessive pressure by the stronger party. In this case, the first fact was self-evident. The defendant was in the hospital and was fearful that Disney would expose information that would destroy his reputation. Regarding the second element, however, the court noted that in most undue-influence cases, and this case was no exception, direct evidence is rarely obtainable and thus the jury must decide the issue on the basis of inferences drawn from all the facts and circumstances. Thus, the court said that summary judgment was improper.

GLOBAL Context

©Digital Archive Japan/ Alamy

Duress in India

In Indian law, duress is referred to as *coercion,* or the committing of or threatening to commit any act forbidden by the Indian Penal Code to force a person into entering an agreement. To prove coercion, the plaintiff must establish that (1) there was a clear utterance of threat; (2) the threat referred to an act forbidden by the Indian Penal Code; (3) the act was committed with the intention to cause the plaintiff to enter into an agreement; (4) there was unlawful detaining of or threat to detain any property; (5) coercion was employed by a party or by his agent; and (6) the plaintiff is a party or any other person with some interest in the agreement.

Indian contract law distinguishes between physical coercion and moral coercion. Physical coercion may include threats to commit suicide. In contrast, moral coercion is synonymous with *undue influence.* The Indian Contract Act states that a contract is induced by undue influence when the relations subsisting between the parties are such that one of the parties is in a position to dominate the will of the other and the dominating party uses that position to obtain an unfair advantage over the other.

Quite unlike the concept of duress in the United States, the Indian concept of moral coercion, or undue influence, includes the unique consideration of *pardanashin ladies* in contract law. In India, a pardanashin woman is a woman who lives in seclusion without communication with any male except for specific male relatives. Indian law gives protection to pardanashin ladies on the grounds that "they are ignorant so far as the worldly knowledge goes." Any contract with a pardanashin woman is automatically presumed to have been induced by undue influence.

In India, a major difference between acts of coercion and acts of undue influence is the legal consequence of each action. In cases of coercion, the agreement obtained by coercion is automatically voidable. In cases of undue influence, the contract is either voidable or is set aside. Factors that enter into the finding of undue influence include the following:

1. Inadequacy of consideration.
2. Inequality between the parties in regard to age, intelligence, or social status.
3. Fiduciary relationship between the parties.
4. Unconscionable bargains.
5. Pardanashin women.

The more of these factors present, the more likely a court is to rescind the contract on grounds of undue influence.

Duress

LO 12-5 Recall the elements of duress.

Duress is a much more visible and active interference with free will than is undue influence. **Duress** is found when one party was forced into the agreement by the wrongful act of another.

The wrongful act may come in various forms. Any of the following would trigger a successful request for rescission on grounds of duress:

- One party threatens physical harm or extortion to gain consent to a contract.
- One party threatens to file a criminal lawsuit unless consent is given to the terms of the contract. (Threats to file a civil case against a party do not constitute duress unless the suit is frivolous.)
- One party threatens the other's economic interests (this is known as *economic duress*). For instance, Jim refuses to perform according to a contract unless Ellen either signs another contract with him or pays him a higher price than that specified in the original agreement.

The injured party makes the case for duress by demonstrating that the threat left him or her no reasonable alternatives. The point is that the free will necessary for legal consent had been removed by the specifics of the threat. The following Case 12-3 is an illustration of a successful use of economic duress.

duress
Any unlawful act or threat exercised on a person whereby the person is forced to enter into an agreement or to perform some other act against his or her will.

GLOBAL Context

Defects in Assent in the Philippines

©Digital Archive Japan/Alamy

The law in the Philippines is somewhat unusual in the way it limits the time for rescinding a contract. In the Philippines, if consent is given through mistake, violence, intimidation, undue influence, or fraud, contracts are valid until they are annulled. The period within which to annul such a contract is four years. In cases of intimidation, violence, or undue influence, the time period is measured from the time the defect in the assent ceases. In cases of mistake or fraud, it is measured from the time of the discovery of the defect.

CASE 12-3

GARAGE SOLUTIONS v. MONTY J. PEARSON
Court of Appeals of Louisiana, Third Circuit
201 So.3D 962 (2016)

FACTS: *Defendant Monty J. Pearson began working for Garage Solutions, LLC, in March 2015. Three months into his employment, Pearson was sent by the owner of Garage Solutions Mark Fontenot to Rexburg, Idaho, and Little Rock, Arkansas to complete corporate training.*

Two days before he was scheduled to leave for training, Pearson was presented with three contracts. One was a non-competition contract, and the remaining two were training reimbursement contracts. The reimbursement contracts stated that Pearson would refund the cost of the training if he left the company within one year of completing the training. While Pearson never signed the non-competition contract, he signed the two reimbursement contracts one month after the conclusion of his corporate training.

In August 2015, Pearson left Garage Solutions, and the company filed a suit seeking compensation from Pearson. While Pearson did not raise an affirmative claim of duress, the trial court determined that the contract was unenforceable because Pearson was indeed under duress when he signed the contracts. The plaintiffs appealed.

ISSUE: Was Pearson under econoicduress when he signed the contracts?

REASONING: The court noted that the contracts signed by Pearson were themselves textually sound. However, after reviewing the facts of the case, the court found the validity of the contracts to be in question. Fontenot had discussed with Pearson the possibility of Pearson signing a noncompetition agreement if he received training. However, Pearson wasn't presented the contracts until two days before he was scheduled to leave for training. Upon initially receiving the contracts, Pearson refused to sign them. Pearson later testified that he only signed the contracts upon returning because Fontenot refused to pay him his wages until he signed. The court then pointed to precedent holding that "[F]ear of economic deprivation has been recognized as a type of duress that will vitiate consent." Although Pearson did not initially raise the defense of duress, the court found there to be evidence enough to enlarge the scope of the pleadings, especially since Garage Solutions did not object. Being that Pearson was under duress, he did not lawfully consent to the agreements. Because there was no consent, there was no contract as well as harm to the company's goodwill and reputation.

DECISION AND REMEDY: Affirmed in favor of Pearson.

SIGNIFICANCE OF THE CASE: This case shows that even a properly-written contract can be rendered entirely invalid if the circumstances surrounding the agreement are unfair.

CRITICAL THINKING

Do you think the outcome of the case could have been affected if Fontenot had presented the contracts earlier? Is the timing of the contracts a nonissue?

ETHICAL DECISION MAKING

Do you think it is fair to the plaintiffs for the court to enlarge the defendant's pleadings to include the duress claims? What ethical values does this decision support?

CASE OPENER WRAP-UP

A Disagreement over an Agreement

The trial court agreed with Michael Jordan's argument regarding a mutual mistake or fraudulent misrepresentation in the contract. Knafel appealed, and the court affirmed the lower court's decision. The court held that Knafel's representation to Jordan that he was the father met the requirements of being (1) a material fact; (2) made for the purpose of inducing Jordan to act; (3) that either was known by Knafel to be false or was not actually believed by her on reasonable grounds to be true, but Jordan reasonably believed it to be true; and (4) that was relied on by Jordan to his own detriment. Thus, the appellate court found that Knafel's representation that Jordan *was* the father constituted fraud. The agreement can be rescinded because Jordan would not have entered into the agreement but for the fraudulent representation made by Knafel. Even if Knafel did not act fraudulently, at the time the agreement was created both parties believed that the child was Jordan's. After the paternity tests revealed that the baby was *not* Jordan's, the agreement could still be rescinded based on a mutual mistake of fact.

PRACTICAL TIPS for Business Managers

1. Costly mistakes can often arise because someone is making an assumption that the other party is not making. It is a good idea, when you are about to either draft or sign a contract, to ask yourself whether there are any questionable assumptions you are making that might not be shared by the other party to the potential contract.

2. You do not want to be sued for misrepresentation, so before you make any assertions of material fact related to a contract, make sure that you are certain of those facts.

3. If you are about to enter into a contract with a party with whom you have what might be perceived as a dominant-subservient relationship, and you could be perceived as the dominant party, advise the weaker party to get legal advice before entering into the contract, to ensure the contract is enforceable.

SUMMARY

The Importance of Genuine Assent	If assent is not genuine, or legal, a contract may be voidable. For purposes of planning, it is important for people to understand the circumstances under which failure of assent may render their contracts voidable.
Mistake	*Mistakes* are erroneous beliefs about the material facts of a contract at the time the agreement is made. They may be either unilateral or mutual. Only under certain rare conditions is a unilateral mistake a basis for rescinding a contract. However, if both parties to a contract are mistaken about a material fact, either can opt to rescind the contract. In cases of mutual mistake, the agreement was not based on a meeting of the minds, a basic criterion for legal assent.
Misrepresentation	*Misrepresentation* is an intentionally untruthful assertion by one of the parties about a material fact. An innocent misrepresentation occurs when the party making the false assertion believes it to be true. The misled party may rescind the contract. When a misrepresentation is fraudulent, any assent that is given is gained by deceit. The courts permit rescission for fraudulent misrepresentation. In addition to requiring false assertion and the intent to deceive, fraudulent misrepresentation also requires justifiable reliance on the assertion by the innocent party. "If the misrepresentation is negligent, that is, made when the defendant would have known the statement was false had he taken reasonable care before making the statement, the misrepresentation is treated the same as a fraudulent misrepresentation."

(continued)

(concluded)

Undue Influence *Undue influence* refers to the persuasive efforts of a dominant party who uses a special relationship with another party to interfere with that other's free choice of the terms of a contract. Any relationship in which one party has an unusual degree of trust in the other can trigger concern about undue influence in gaining the assent of the more dependent party.

Duress *Duress* occurs when one party threatens the other with a wrongful act unless assent is given. Such assent is not legal assent because coercion interferes with the threatened party's free will. For a court to rescind the agreement, the injured party must demonstrate that the duress left no reasonable alternatives to agreeing to the contract.

Point/Counterpoint

The principles of legal assent are based on the belief that, in the main, each person can look out for himself or herself. Thus, parties are usually held to the agreements they make on the grounds that the terms were reached by decisions guided by the free wills of both parties. But we live in a world of rapid technological change, where chemicals, fibers, and services come in a blinding array of complexity that could make decision-making more difficult for some people.

Should Contracts Made through Tweets be Legally Binding?	
YES	**NO**
Whether it's spoken or handwritten, transmitted to a television screen or broadcast over the airwaves, shouted from the top of a building or tweeted in 140 characters or less, speech is speech. No matter what form speech comes in, people are legally liable for the thoughts that they express. The First Amendment grants certain protections for speech, but it is not intended to be a shield for those who wish to spew forth whatever crosses their mind without fear of consequence. When one communicates an offer under sound mind, they are, at least to others, expressing to the rest of the world that they are willing to enter into a contract. Regardless of the mode that was used to express the offer, the offer was made and the speaker is liable for that speech. Tweets are no exception to this doctrine. Besides a possible difference in size of audience, tweets are practically no different than putting an offer in writing and posting it around town for the public to see. Moreover, by not regarding tweets as potentially contractually binding, the law will essentially be allowing people to feign offers and then rescind them even after the offer has been accepted. Any kind of agreement through Twitter is effectively one-sided in favor of the offerer because the legal system would not recognize the agreement as binding, whereas in nearly any other circumstance, written or not, the agreement *would* be binding. Thus, Twitter users can effectively get away with making agreements with impudence through tweets. Some would argue that, due to the ambiguous nature of text-based communication on the Internet, it's hard to tell whether or not someone is serious about what they "say" online. Therefore, we shouldn't take anything said on the Internet, especially Twitter,	With today's rapidly advancing technology, the ways in which we communicate with each other are changing almost every day. Indeed, with its 140-character limit and focus on tiny spurts of communication, it would have been nearly impossible to predict a form of conversation as unique as Twitter. In fact, it is this uniqueness that sets Twitter apart from essentially all other types of communication. Sure, you can speak to another person in short bursts, à la a tweet, but it's not quite the same thing as a mode of articulation that forces miniature messages that has spawned a culture and attitude all its own. In fact, it is this culture of accepted insignificance or nonseriousness of a single tweet that truly sets it apart from other forms of communication and therefore does not make it fit for legally binding contracts. The problem with treating tweets as serious legal speech fit for contracts is that tweets, as of yet, have not been been accepted as anything more than curt expressions of opinion or personality, or perhaps as a means to share small bits of information with a particular group of Twitter users. Sure, people often have arguments through tweets or attempt to make profound personal statements in their petite allotment of words, but these types of tweets are often mocked for taking Twitter too seriously. The point is that Twitter has yet to be recognized as a serious form of communication, even by its users, and thus is no place for serious legally binding agreements to be invoked. Perhaps Twitter could evolve to someday be a shining star of intelligent, serious conversation, but legal discourse is simply counterintuitive to the purpose of Twitter as it stands today. In fact, if Twitter *were* suddenly to be considered an apt forum for legal discourse and contractual

seriously. In reality, though, the law should take just the opposite stance: everything said online that does not include some kind of disclaimer should be taken seriously. It's simply too dangerous not to in some cases. For example, if a person threatens to kill another through Twitter, the authorities do not stop to consider whether the poster is joking, they take action. Sitting around and debating whether the threat is serious could lead to someone being killed. While the need for tweets to be taken seriously under other circumstances can be seen more clearly, the law should remain consistent and consider apparent contracts as legally serious as well if it cannot be clearly determined in the court's eyes whether or not the poster was serious.

obligation, its lackadaisical tone could lead to difficulties in the area of making offers and forming contracts. For example, sarcasm is a rather popular way of expressing one's self on the Internet. Unfortunately, due to the lack of speech-based communication, with its varying intonations and volumes that aid in expressing emotion, it can be very difficult to decipher when a tweet is intended to be taken seriously or in jest. If I were to jokingly tweet, "Boy, I'm so hungry, I would give $1,000 to anyone who brought me a doughnut," have I just made an offer that someone can accept and contractually bind me to? Now of course, written contracts also lack speech to determine the intent of what is written, but said contracts are practically universally written to say just what they intend. Tweets, on the other hand, do not always intend what they actually say. Imposing a sudden expectation of contractual obligation to tweets (as has been done) simply makes no sense given the nature of tweets themselves.

Questions & Problems

1. Explain the difference between a unilateral mistake and a mutual mistake.
2. Explain when a unilateral mistake can lead to rendering a contract voidable.
3. Distinguish innocent misrepresentation from fraudulent misrepresentation.
4. Explain how nondisclosure can be treated as misrepresentation.
5. Explain the primary differences between duress and undue influence.
6. Kyle Connaughton was actively pursuing potential buyers for a ramen restaurant concept that he had created. Eventually, Chipotle's CEO, Steven Ells displayed interest in Connaughton's plan. Connaughton then tried to cater the design to fit Chipotle's needs. Ells initially offered to purchase the ramen restaurant concept, but eventually formed an agreement with Connaughton whereby Connaughton would continue to develop and implement the concept as a Culinary Director employed by Chipotle.

 Connaughton was on track to launch a restaurant in New York towards the end of his second year of employment. However, he learned from Chipotle's chief marketing officer (CMO) that Ells had a nondisclosure agreement (NDA) with another chef who had previously worked on the ramen restaurant concept with Chipotle. The previous arrangement ended after the two parties could not find mutually acceptable terms and the CMO informed Connaughton that the other chef would sue if the restaurant was launched. Connaughton brought up the NDA to Ells, who told Connaughton to continue working on the restaurant. Connaughton refused and was fired shortly thereafter. Following his dismissal, Connaughton sued Chipotle and Ells for fraudulent inducement, asserting that if it were not for his reasonable reliance on Ell's omissions about the previously contracted work on the ramen restaurant concept, he would not have agreed to work with Chipotle. Furthermore, Connaughton claimed he was damaged for lost business opportunities in connection to the ramen restaurant concept and that working for Chipotle had opened him up for liability to the chef that had signed an NDA. The trial court sided with Chipotle and the plaintiff appealed. How do you think the appeals court ruled? Why?

 Source: Connaughton v. CHIPOTLE INC., 29 NY 3d 137 (2017).

7. The parties' marriage was dissolved on August 27, 2010. The parties had an agreement which the court incorporated into its judgment of dissolution. On February 11, 2015, Defendant filed a Motion to Open the Judgment, alleging that he was subjected to continuous threatening and harassing behavior by Plaintiff, which resulted in his concession to the terms of the Separation Agreement that was incorporated by reference in the Judgment. Defendant claimed that he had suffered with Attention Deficit Hyperactivity Disorder (ADHD) for years and therefore is more susceptible to manipulation, coercion, and duress than is the average reasonable person.

Defendant had been diagnosed with ADHD by a therapist who had initially seen him for about six months in 2012. Having not seen the therapist in 2013 and 2014, Defendant returned to the therapist in January 2015 with questions regarding the effect of the ADHD on his divorce. The therapist did not know Defendant in 2010, but he said that an adult such as Defendant with the Defendant's diagnosis and symptoms was more susceptible to manipulation and coercion than would be someone without ADHD. He based his opinion about the impact of the ADHD on Defendant's ability to withstand manipulation and coercion and on Defendant's relating to him that "he just wanted it off of his plate" because he could not deal with the stress.

Defendant provided examples of the stress he underwent around the time of the divorce. Approximately one year prior to the filing of the complaint, Plaintiff was diagnosed with a life-threatening illness. Three months later, the parties separated when Plaintiff told Defendant that she needed time for herself. An action for legal separation was commenced by Plaintiff in August 2009. Plaintiff had engaged an attorney, after which she presented Defendant with a proposal for settlement. Plaintiff suggested to Defendant that he did not need an attorney and that the parties could go to a mediator or use her attorney. On advice from a friend, Defendant engaged an attorney. His attorney strongly advised against agreeing to Plaintiff's proposal. Based on these facts, do you believe that the court reopened the agreement on grounds of duress? Why or why not?

Source: *Martha Rochelle v. Hartley Mead,* 2015 Conn. Super. LEXIS 2587 (2015).

8. In 2006, Jack Kelly, LLC, entered into a commercial lease agreement for a second floor space of a building located in Manhattan. The lease stated that "Tenant shall use and occupy the Demised Premises for general offices of an executive recruiting firm. . . [and] for no other purpose." Additionally, the lease also stated that tenants could not use the rented space in any manner that violated the certificate of occupancy for the building or violated any laws or regulations.

In 2007, Jack Kelly discovered that the certificate of occupancy for the building required that the rented premises be used only for residential purposes. Jack Kelly subsequently asked the leasing company to amend the certificate of occupancy, but the the leasing company did not comply. Following the leasing company's failure to amend the certificate of occupancy, Jack Kelly vacated the space in 2009 and filed a claim of action for rescission, seeking declaratory judgement that the lease was invalid and unenforceable, among other things.

Although the trial court granted summary judgement to the defendants, in the ensuing appeal, the appellate court reversed the decision. What elements must be present to prove mutual mistake? How does Jack Kelly's case fulfill all the elements.

Source: Jack Kelly LLC v. Zegelstein, 140 AD 3d 79 (2016).

9. Audrey Vokes was a 51-year-old widow who wanted to become an accomplished dancer. She was invited to attend a dance party at J. P. Davenports's School of Dancing, an Arthur Murray franchise. She subsequently signed up for dance classes, at which she received elaborate praise. Her instructor initially sold her eight half-hour dance lessons for $14.50 each, to be used one each month. Eventually, after being continually told that she had excellent potential and that she was developing into a beautiful dancer—when in fact she was not developing her dance ability and had no aptitude for dance—she ended up purchasing a total of 2,302 hours's worth of dance lessons for a total of $31,090.45. When it finally became clear to Vokes that she was not developing her dance skills, in part because she had trouble even hearing the musical beat, she sued Arthur Murray. What would be the basis of her argument? Her case was initially dismissed by the trial court. What do you think the result of her appeal was?

Source: Vokes v. Arthur Murray, 212 So. 2d 906 (1968).

10. Plaintiff Murray-Calloway County Public Hospital Corporation (MCCH) hired Cheryl Coors and Coors Executive Resources LTD to recruit an ENT (ear-nose-throat) doctor for an opening at the hospital. Coors successfully recruited Defendant Dr. Samuel E. Sprehe, and in January 2005, he signed a contract and agreed to open an ENT practice at MCCH. Sprehe's time at the hospital was short-lived, however, and he left the practice after approximately one year. MCCH then sued Sprehe to recover money advanced under the terms of their contract. In response to the lawsuit, Sprehe filed a counterclaim against MCCH and a third-party claim against Coors and MCCH's former CEO, Jon O'Shaugnessy. Sprehe's sole cause of action in the third-party complaint was fraudulent misrepresentation. He alleges that Coors made multiple fraudulent statements about the potential for an ENT practice at MCCH and that he relied on those statements to his financial detriment.

The fraudulent misrepresentations alleged by Sprehe fall into two broad categories. First, in the "market" category, Sprehe alleges that Coors misrepresented the potential level of patient volumes and the absence of competition for an ENT practice in Murray,

Kentucky. Specifically, Sprehe claimed that Coors told him that Murray and the surrounding communities were in "desperate" need of a local ENT practice, that there was a "bonanza" of potential patients, and that there were no other ENT doctors in the relevant market area. Additionally, Sprehe alleges that Coors misrepresented the research she had done to substantiate her claims about Murray's ENT market. Second, in the "finance" category, Sprehe claims that Coors told him that MCCH was going to "bankroll" his practice by providing an office, staff, and all of the necessary equipment to have a successful ENT practice.

Sprehe claimed that he relied on these fraudulent misrepresentations to his financial detriment and that Coors is liable for those losses. Plaintiff MCCH filed a motion for summary judgment. How do you believe the court ruled and why?

Source: *Murray-Calloway County Public Hospital Corporation, Plaintiff v. Samuel E. Sprehe, M.D., Defendant/Third-Party Plaintiff V. Jon O'Shaughnessy; Cheryl Coors, Individually; And Coors Executive Resources, Ltd, Third-Party Defendants,* 2012 U.S. Dist. LEXIS 42724 (2012).

11. The plaintiff wanted to open a Victoria's Secret franchise in Puerto Rico. Defendant Domingo, a business broker, told the plaintiff that the franchise was a "done deal" if the plaintiff hired the defendant's firm and also hired former United States Senator Birch Bayh's firm to assist. In reliance on the defendant's representations, the plaintiff entered into retainer agreements with the two firms, paying them $225,000 and $400,000, respectively. Several months later, someone from Bayh's firm emailed the plaintiff that he could not get the franchise because Victoria's Secret did not use a franchise system and instead owned and operated its own stores. The plaintiff sued Domingo, Bayh, and their firms. Bayh and his law firm settled for an undisclosed amount, and a jury awarded damages of $625,000 against Domingo and his firm. On appeal, the judgment was affirmed. What do you believe was the legal basis for the lawsuit? Why do you agree or disagree with the outcome of this case?

Source: *Portugues-Santana v. Rekomdiv International, Inc.,* 725 F.3d 17, 2013 U.S. App. Lexis 15331 (2013).

12. In 2016, a father of two minor children in Cuyahoga county, Ohio, filed a claim to determine custody of the children. At the pretrial, the father informed the magistrate he wanted to be named a legal custodian because he paid child support. The magistrate appointed a guardian ad litem, a third party that considers the best interests of the children, to work with the couple and come to a resolution.

On the day of the the trial, the mother and father informed the court they had settled the matter. The mother and father signed an agreement that was approved by the guardian ad litem. The agreement detailed that the children would live with the mother, subject to the father's visitation rights as set out in the agreement. The agreement was incorporated into the magistrate's decision, with the magistrate designating the mother as the legal custodian and residential parent.

The father appealed the decision, asserting that because of a unilateral error on his part, where he thought he was signing a shared parenting agreement, the judgement should be overturned. Do you think the father has grounds to assert a unilateral mistake? What kind of evidence would the father need to successfully make his case?

Source: IN RE JW, 2017 Ohio 8486 (2017).

Part 2 *Contract Law*

CHAPTER 13
Contracts in Writing and Third-Party Contracts

CASE OPENER

Fallout from a Forgettable Fight

On June 28, 1997, in Las Vegas, Nevada, heavyweight boxers Mike Tyson and Evander Holyfield met for what was to be a night to remember. During the third round of the fight, a desperate Tyson bit off a piece of Holyfield's ear. The fight continued until moments later when Tyson bit Holyfield's other ear.[1] Tyson was disqualified, and numerous fans were unhappy with the outcome. Some fans were so outraged they decided to sue Tyson, the fight promoters, and the telecasters, seeking a refund.[2] The fans sued on the basis of, among other theories, a claim to being third-party beneficiaries to various contracts into which the defendants had entered.

1. Are the fans entitled to refunds based on a theory of third-party beneficiary rights? What type of beneficiaries would the fans have to be to enforce contractual rights?
2. If you were one of the fight promoters, what sorts of contractual duties would you have to the viewers?

LEARNING OBJECTIVES

LO 13-1 Identify the types of contracts that fall within the statute of frauds.
LO 13-2 Recognize the elements required for a writing to satisfy the statute of frauds.
LO 13-3 Describe the purpose of the parol evidence rule.
LO 13-4 Differentiate between an assignment and a delegation.
LO 13-5 Recall the rights and duties of an assignee.
LO 13-6 Differentiate between an intended and an incidental beneficiary.
LO 13-7 Differentiate between a creditor beneficiary and a donee beneficiary.

[1] SI, "Year in Review 1997," https://www.si.com/more-sports/2012/06/29/tyson-holyfield-oral-history.
[2] *Castillo v. Tyson,* 268 A.D.2d 336 (2000).

Written contracts provide certain advantages that oral contracts lack. When disputes may arise regarding the specifics of the terms of an oral contract, such disputes are easier to settle when the contractual terms are solidified in writing. In addition, when parties to a contract write down their terms, the moment of writing allows both parties to reconsider their terms and ensure that the parties are advocating what they desire in the contract. In general, written contracts help business contracting flow more smoothly.

The idea of requiring certain contracts to be in writing comes from the English Act for the Prevention of Frauds and Perjuries, passed in 1677. To prevent disputes over the existence of a contract, the act required that certain types of contracts, to be legally enforced, must be in writing and signed by the party denying the existence of the contract. In the United States, the term *statute of frauds* is somewhat misleading because there is no federal statute of frauds; rather, such legislation exists at the state level, with the exception of Louisiana. Maryland and New Mexico have statutes of frauds created by judicial decision rather than by legislation. Interestingly, the English have repealed almost all their requirements for writing, but American states and courts are still expanding the requirements for what falls within the statute of frauds. The term *statute of frauds* is also misleading in another way: The statute of frauds does not relate to fraudulent contracts, nor does it address the issue of illegal contracts. Rather, the statute addresses the enforceability of contracts that fail to meet the requirements set forth in the statute. Furthermore, the statute serves to protect promisors from poorly considered oral contracts by requiring certain contracts to be in writing.

This chapter addresses some commonalities among the statutes of frauds of different states, while referring simply to the "statute of frauds" as if it were a unitary law. The chapter examines which contracts need to be in writing as well as exceptions to the rule. Then the chapter examines the parol evidence rule, which discusses what types of oral evidence are admissible, and when, as related to contracts within the scope of the statute of frauds.

Although a contract is typically an agreement between two parties, sometimes third parties have rights in the contract as well, either through assignments or as beneficiaries. The second half of this chapter focuses on how third parties gain rights to contracts.

Statute of Frauds

The **statute of frauds** has three main purposes. The first is to ease contractual negotiations by requiring sufficiently reliable evidence to prove the existence and specific terms of a contract. The second is to prevent unreliable oral evidence from interfering with a contractual relationship. Requiring a written contract prevents the admittance of oral evidence denying or substantially altering the contract. (This chapter discusses the admissibility, or lack thereof, of oral evidence later, in the section on the parol evidence rule.) The third is to prevent parties from entering into contracts with which they do not agree. The statute's requirements impose on parties the need to consider the terms carefully, agree to them, write them down, and sign the contract.

statute of frauds
State-level legislation that addresses the enforceability of contracts that fail to meet the requirements set forth in the statute; serves to protect promisors from poorly considered oral contracts by requiring certain contracts to be in writing.

Contracts Falling within the Statute of Frauds

LO 13-1 Identify the types of contracts that fall within the statute of frauds.

As previously stated, only specific types of contracts fall within the statute of frauds and thus are required to be evidenced in writing. These types of contracts are (1) contracts whose terms prevent possible performance within one year, (2) promises made in consideration of marriage, (3) contracts for one party to pay the debt of another if the initial party fails to pay, and (4) contracts related to an interest in land. Although required to be in writing under the Uniform Commercial Code (UCC) and not the statute of frauds, a related fifth category is contracts for the sale of goods totaling more than $500.[3]

[3] UCC § 2-201.

Contracts whose Terms Prevent Possible Performance within One Year

Contracts whose performance, based on the terms of the contract, could not possibly occur within one year fall within the statute of frauds and, therefore, must be in writing.[4] It is important to realize that the one-year period begins to run the day after the contract is created, *not* when the contract is scheduled to begin.

The test for compliance with the one-year rule is not related to the likelihood of the completion of the contract within one year. Rather, the test considers the *possibility* of completing the contract within one year. For example, if Robert contracts with Elise for lifetime employment, they do not have to write and sign the agreement because it is possible for the contract to be completed within one year: Robert could die after two days of work. Moreover, Robert and Elise's contract, if oral, would be enforceable because it is not within the statute of frauds. The possibility that a contract's terms could be performed within one year removes the contract from the statute's written requirements.

Similarly, contracts for complex construction projects need not be in writing because, theoretically, they can be completed within one year if the crew works around the clock every day, even if the scenario is highly unlikely.

Used by permission of Stu Rees, Stu's Views.

The following Case Nugget illustrates how important the facts are when ascertaining whether a contract cannot be performed within a year.

[4] Restatement (Second) of Contracts, sec. 130.

CASE Nugget

Aurigemma v. New Castle Care, LLC
2006 Del. Super. LEXIS 337, June 12, 2006

Dr. Ralph M. Aurigemma filed suit against New Castle Care, LLC, alleging breach of an oral contract. New Castle Care operated the Arbors Rehabilitation Center, the facility where Aurigemma worked. Aurigemma claimed that after the medical director for the

(continued)

Arbors unexpectedly died, he and many other doctors expressed an interest in filling the newly vacant position. Aurigemma stated that individuals from New Castle Care made him interim medical director and, on September 4, 2003, created an oral contract under which he agreed to serve as permanent medical director from October 1, 2003, until October 1, 2004.

New Castle Care claimed that it made no such oral contract with Aurigemma and stated that even if it had, Aurigemma's oral contract would not be enforceable because the terms of the contract, which was created on September 4, 2003, and intended to go until October 1, 2004, could not possibly be completed within a year. Therefore, New Castle Care claimed that Aurigemma's alleged oral contract fell within the statue of frauds and thus required a writing to be enforceable. Additionally, New Castle Care claimed it had been clear that the company had not intended to have Aurigemma act as permanent medical director, and it cited a written contract it had created with another doctor on September 15, 2003, as proof that no oral contract existed.

Aurigemma countered by saying that because he began to assume the duties of medical director, he had already partially performed the terms of the contract. Thus, because partial performance sometimes creates an exception to the statue of frauds, Aurigemma argued that the oral contract did not need to be in writing.

On June 15, 2006, New Castle Care filed a motion for summary judgment; shortly after, a Delaware superior court granted summary judgment in favor of the company on both counts. In conclusion, the court stated that New Castle Care was correct in its argument. Because the terms of the oral contract were for a time period of more than a year, the oral contract would have had to have been in writing in order to have been valid. The court also pointed out that in Delaware the partial-performance exception to the statute of frauds does not apply to oral contracts incapable of being performed within a year. Therefore, Aurigemma's alleged oral contract was not included in the partial-performance exception to the statute of frauds and was accordingly unenforceable.

Legal Principle: If a contract can possibly be performed within a year, even if such performance is highly unlikely, then the contract does not need a writing to be enforceable.

Promises Made in Consideration of Marriage

Agreements regarding marriage in which one party is gaining something other than a return on his or her promise to marry are within the statute of frauds and must be in writing.[5] In other words, when one party promises something to the other as part of an offer of marriage, the contract must be in writing to be enforceable.

It is important to realize that mutual promises to marry do not fall within the statute of frauds. For example, if Ed and Jeanie promise each other they will get married, this agreement does not need to be in writing because it does not fall within the statute.

Although a mutual promise to marry does not fall within the statute of frauds, prenuptial agreements do. A **prenuptial agreement** is an agreement two parties enter into before marriage that clearly states the ownership rights each party enjoys in the other party's property. To be enforceable, prenuptial agreements must be in writing. Furthermore, although consideration is not legally required, courts tend to privilege prenuptial agreements that involve consideration. The reason consideration is privileged is that a prenuptial agreement is not automatically enforceable just because it is in writing. Although writing is required, it is not sufficient to establish enforceability. Consideration at least offers evidence that both parties understand and agree to the terms of the contract and that the contract is not biased in favor of one of the parties.

prenuptial agreement
An agreement two parties enter into before marriage that clearly states the ownership rights each party enjoys in the other party's property. To be enforceable, the agreement must be in writing.

[5] Restatement (Second) of Contracts, sec. 124.

Contracts for One Party to Pay the Debt of Another if the Initial Party Fails to Pay

The contracts involving promises to pay a debt that are within the statute of frauds are of a very limited kind. These promises are referred to as *secondary promises, collateral promises,* or *suretyship promises.* All of these terms refer to agreements in which a third party agrees to pay the debt owed by another party if that other party fails to pay his or her debt. For example, Helen enters into a contract with Tom to sell Tom her car. Subsequently, Diana agrees to pay Tom's debt if he fails to pay Helen the money he owes her. To be enforceable, Diana's promise needs to be in writing because it is within the statute.

It is important to understand the distinction between primary and secondary obligations to know when the statute necessitates a written agreement. *Primary obligations* are debts entered into in an initial contract. In our car-sale example, the primary obligation is Tom's promise to pay Helen for the car. Primary obligations are not within the statute of frauds and, therefore, need not be in writing to be enforceable.

Secondary obligations, however, are within the statute and need to be in writing. A *secondary obligation,* as explained earlier, arises when a party outside a primary agreement promises to fulfill one of the original party's (the primary debtor's) obligations if the original party (primary debtor) fails to fulfill his or her obligation.

A specific instance of a secondary obligation involves the administrator or executor of an estate. Administrators and executors of estates are responsible for paying the debts of an estate and then dividing the remaining wealth among the heirs. Although an agreement to pay the estate's debts with the estate's funds need not be in writing, promises the administrator or executor makes to pay those debts personally are within the statute of frauds and must be in writing. Because the administrator or executor is paying with his or her own money, and not the estate's, the promises must be in writing to be enforceable—the administrator or executor has assumed a secondary obligation.

There is an exception concerning when a promise to pay someone else's debt needs to be in writing: *the main-purpose rule.* If the main purpose for incurring a secondary obligation is to obtain a personal benefit, the promise does not fall within the statute and need not be in writing.[6] The assumption is that a party attempting to achieve a personal benefit will not back out of the agreement; thus there is no need for a written record of the agreement. The court's job is to use the context surrounding the agreement to determine the third party's main purpose for entering the agreement.

Contracts Related to an Interest in Land

Under the statute of frauds, land is considered real property, such as the land and soil itself, as well as anything attached to the land, such as trees or buildings. The statute is intended to prevent oral claims to the existence of a contract for the sale of land. That is, because the statute requires a writing as evidence of the contract, a claim to an oral contract for the sale of land is not enough to prove such a contract existed.

In addition to contracts for the sale of land, contracts transferring other interests in land are also within the statute of frauds. Mortgages and leases are within the statute of frauds because they are considered transfers of interest in land.

Determining exactly what constitutes an interest in land under the statute of frauds is difficult. Although the sale of land and the transfer of mortgages and leases are within the statute, a number of things that seem as if they are interests in land do not fall within the statute. For example, promises to sell crops annually, agreements between parties for profit sharing from the sale of real property, and boundary disputes that have been settled through the use of land are not within the statute of frauds and, therefore, do not require evidence in writing. The Case Nugget presents a case related to land in which the parties disagreed about whether a writing was needed.

[6] Restatement (Second) of Contracts, sec. 116.

GLOBAL Context

England and the Statute of Frauds

©Digital Archive Japan/Alamy

As stated in the text, England gave birth to the statute of frauds. Although the United States, in addition to numerous other Western countries, has subsequently adopted versions of the 1677 English act, the English have gone in the opposite direction. Instead of expanding the powers and use of the 1677 act, the English have severely limited the number of cases falling within their statute of frauds.

Although formal complaints were levied against the English statute of frauds as early as 1937, no action was taken until 1953. In that year, the Law Reform Committee addressed numerous arguments that had been made in 1937 in favor of repeal of the statute of frauds. The 1953 committee recommended that Parliament repeal Section 4 of the 1677 act, the section that identified the specific types of contracts that must be in writing. The Law Reform Act of 1954 subsequently repealed Section 4, with one cautionary exception: The 1954 act still required promises to pay for the debt of others to be in writing. In other words, the only provision still active (requiring a writing) under the statute of frauds in the country that invented the statute is what we would call *suretyship* or *collateral promise*.

What Is an "Interest in Land?"

Shelby's, Inc. v. Sierra Bravo, Inc.

Court of Appeals of Missouri, Southern District
68 S.W.3d 604 (2002)

Shelby's and Sierra entered into a written agreement in which Shelby's granted Sierra permission to use Shelby's land as disposal for waste and debris Sierra removed as part of the construction of a new highway. Shelby's claimed that the parties also entered into an oral contract for Sierra to construct a waterway and building pad on Shelby's property. Sierra never completed the construction and denied that an oral contract existed. Shelby's sued, and the jury found in its favor.

Sierra appealed on the basis that the oral agreement was within the statute of frauds and therefore unenforceable. Sierra saw the alleged oral agreement as involving a sale of an interest in land, which is within the statute of frauds. Therefore, the agreement, to be enforceable, would have had to be in writing.

The court firmly disagreed with Sierra's argument. The court stated:

> We agree with the well-reasoned argument of Respondent [Shelby's]. The contract in this case was not a "sale," much less a sale of an interest in lands.... Here, there was no transfer of ownership or title. The written agreement gave Appellant [Sierra] permission to deposit debris and soil on Respondent's land, not the right to do so. The oral contract was for the construction of a waterway and building pad and passed no interest in the land....
> We decline to create a new category to which the statute of frauds applies, that of a contract for services for the deposit of dirt and soil on land. The trial court did not err in denying Appellant's motion for judgment notwithstanding the verdict. Appellant's point is denied and the judgment of the trial court is affirmed.

Contracts for the Sale of Goods Totaling more than $500

Agreements in which the total price for a sale is $500 or more are required by the UCC, Section 2-201, to be recorded in a written contract or a memorandum. To satisfy this UCC requirement, the contract or memorandum need only state the quantity to be sold. Buyer, seller, price, and method of payment do not need to be included in the writing according to the UCC. Contractual terms beyond quantity may be written, but they need not be. In fact, terms other than quantity can be inexact or left out of the writing as long as what is written does not contradict the parties' agreement. The contract will be enforceable for the stated quantity and not a unit more. Furthermore, for the contract to be enforceable against a party, the party against whom action is sought must have signed the written document.

In addition to sales contracts for $500 or more, other specific situations under the UCC for contracts need to be in writing. These situations are contracts involving the sale of securities and personal property if the sale is greater than $5,000 in price.[7,8]

Further Requirements Specific to Certain States

As discussed at the beginning of the chapter, the statute of frauds is actually state law. Consequently, certain states have various requirements not found in other states. Although these requirements are not universal, it is important to note some of the distinctive elements of the statutes of frauds in a minority of the states.

Within their statutes of frauds, a minority of states have special provisions related to promises to pay debt. One example pertains to a promise to pay a debt that has already been discharged because of bankruptcy. For such an agreement to be enforceable, it must be in writing to prevent the promisor from hiding behind the fact that the debt has been discharged. Another example involves a promise to pay a debt where collection of the debt is barred by a statute of limitations. As in the first example, if the agreement is not in writing, the promisor can easily prove that he or she does not need to pay. Therefore, the statutes of frauds in certain states require both of these types of promises to be in writing to be enforceable.

Equal Dignities Rule According to this rule, which is used in some states, if one party (the principal) hires another person (the agent) to negotiate a contract that must be in writing under the Statute of Frauds, the contract giving the agent the power to act for the principal must also be in writing. Thus, if Roberto contracts with Carmine to purchase Essex Farms for him, both the contract between Roberto and Carmine and the contract for the purchase of the farm must be in writing. The former must be in writing because of the Equal Dignities Rule and the latter because of the Statute of Frauds.

Exhibit 13-1 summarizes the contracts that fall within the statute of frauds and presents a mnemonic for remembering them.

Exhibit 13-1

A Mnemonic for Remembering Which Contracts Fall within the Statute of Frauds

CIRCUMSTANCES IN WHICH THE STATUTE OF FRAUDS APPLIES (MY LEGS)		
M	Marriage	Contracts made in consideration of marriage
Y	Year	Contracts whose terms prevent possible performance within one year
L	Land	Contracts related to an interest in land
E	Executor	Contracts in which the executor promises to pay the debt of an estate with the executor's own money
G	Goods	Contracts for the sale of goods totaling more than $500
S	Suretyship	Contracts involving secondary obligations (promise to pay the debt of another) or suretyships

[7] UCC § 8-319.
[8] UCC § 81-206.

Exceptions to the Statute of Frauds

As with most legal rules, there are certain exceptions to which the statute of frauds does not apply, even though under normal circumstances it would. These exceptions are (1) admission, (2) partial performance, (3) promissory estoppel, and (4) although not under the statute of frauds, exceptions under the UCC.

Admission

An **admission** is a statement made in court, under oath, or at some stage during a legal proceeding in which a party against whom charges have been brought admits that an oral contract existed, even though the contract was required to be in writing.[9] All states except Louisiana and California adhere to the admission exception. To the extent that the statute of frauds is intended to necessitate proper evidence of agreements, the admission exception is well reasoned. However, to the extent that the statute of frauds is intended to cause people to be cautious about clearly establishing the specifics of their agreements, the admission exception seems to unnecessarily punish honest parties while rewarding dishonest ones.

Similar to the statute of frauds, the UCC makes a comparable exception in cases when parties admit to the existence of an oral contract. However, unlike the statute of frauds, the UCC specifies that an oral contract required to be in writing but admitted to in court will be enforceable only for the quantity admitted.[10]

> **admission**
> A statement made in court, under oath, or at some stage during a legal proceeding in which a party against whom charges have been brought admits that an oral contract existed, even though the contract was required to be in writing.

Partial Performance

Although the statute of frauds requires a writing for sales of interests in land, there is an exception based on the theory of partial performance. Under **partial performance,** if the buyer in an alleged contract for the sale of land has paid any portion of the sales price and either has begun to improve the land permanently or has taken possession of the land, the courts will consider the contract partially performed, and this partial performance will amount to proof of the contract.

Accordingly, partial performance can override the statute's requirement for a written agreement. The logic is that the actions of both parties demonstrate the existence of their agreement, so the agreement no longer needs to be in writing to be enforceable even though a writing would otherwise be needed. Under similar sections of the UCC, a buyer or seller is responsible for an oral contract that should have been in writing to the extent that he or she accepts payment or delivery of the goods in question.[11]

> **partial performance**
> An exception to the statute of frauds in which the performance of portions of an unwritten agreement by one or both parties can constitute proof that an oral contract exists between the parties.

Promissory Estoppel

Under certain circumstances, when a party relies on an oral contract that is within the statute of frauds, and therefore should have been in writing, the reliance can create a situation in which the contract is enforceable. **Promissory estoppel** is the legal enforcement of an otherwise unenforceable contract due to a party's detrimental reliance on the contract.

For promissory estoppel to be in effect, the party's reliance must be to his or her own detriment. Furthermore, the reliance must have been reasonably foreseeable; that is, the party who did not rely on the contract should have known that the other party would rely on it.[12]

Suppose that, after accepting an offer on your current house, you enter into a contract to buy a house. However, the new house costs more than the sales price of your old house, and the person from whom you are buying knows about the sale of your old house and the difference in the prices. To pay for the price difference, you sell your collection of rare coins. Unfortunately, during this whole process, you forgot to create a written contract for the purchase of the new house. The other person now refuses to sell the house to you. You are now homeless and have sold off your

> **Promissory estoppel**
> The legal enforcement of an otherwise unenforceable contract due to a party's detrimental reliance on the contract.

[9] Restatement (Second) of Contracts, sec. 133.
[10] UCC § 2-201(3)(b).
[11] UCC § 2-201(3)(c).
[12] Restatement (Second) of Contracts, sec. 139.

only real assets. Because the other person reasonably should have known you were relying on the contract, and because you did so to your own detriment, under promissory estoppel you could win performance of the oral sales contract.

Exceptions Under the UCC

In addition to the exceptions under the statute of frauds, exceptions also exist under the UCC about when a contract ordinarily required to be in writing need not be in writing. For instance, under the UCC, oral contracts between merchants need not be in writing to be enforceable. Therefore, if one merchant agrees to sell goods to another, the contract is enforceable even if it is not in writing.

Likewise, oral contracts for customized goods are enforceable even if they would normally have to be in writing. The reasoning is that customized goods are not likely to be able to be sold to a general audience. Accordingly, if one party backs out of an oral arrangement for the sale of customized goods, the party that did not back out probably incurred unreasonable costs due to the negated contract. To prevent one party from bearing the cost of a canceled contract, the UCC allows oral agreements to be enforced when they are for customized goods.

Sufficiency of the Writing

LO 13-2 Recognize the elements required for a writing to satisfy the statute of frauds.

Although there are no specific requirements for the form of a written contract under the statute of frauds, certain elements need to be present for a writing to constitute proper evidence of a written contract under the statute. It is worth mentioning that, in the absence of a single standard form, multiple documents can together evidence the existence of a written agreement under the statute.

Required elements in the writing include the identification of the parties to the contract, the subject of the agreement, the consideration if any is given, and any pertinent terms of the contract. Furthermore, the contract must be signed, but the signature does not need to be located at the end, which is the traditional location for a signature. In fact, the signature need not be the full signature; a mark, such as an initial, is permissible as long as the mark is placed with the intent to function as a signature. Although it is standard for both parties to sign the agreement, because the writing is being offered as proof of an agreement, only the party against whom action is sought needs to have signed the writing. If only one party signed, the agreement is enforceable against the signing party but not against the nonsigning party. The required elements can be contained in a memorandum, a written document, or a compilation of several written documents.

While multiple documents can together evidence the existence of a written agreement under the statute, as long as the documents reference one another, it is not always easy to use multiple documents to satisfy the statute. Case 13-1 illustrates how difficult that task can be.

HERITAGE CONSTRUCTORS, INC. v. CHRIETZBERG ELECTRIC, INC. AND RICHARD MARC CHRIETZBERG

Court of Appeals of Texas, Sixth District

2015 Tex. App. LEXIS 1974 (2015)

FACTS: In 2011, Heritage Constructors Inc., a general contractor that specializes in building water treatment plants, was preparing a bid to become the general contractor for the TexAmericas Center-East wastewater treatment plant improvement project. Heritage estimated that the project would take about 19 months to complete and intended to complete eighty to

(continued)

eighty-five percent of the project itself, but the project's electrical work necessitated the use of a subcontractor. Before Heritage's bid was due, Heritage was contacted by two electrical companies attempting to bid on a subcontract for the electrical work. Electrique Corporation bid $886,400 to perform electrical work on the project, and Chrietzberg Electric bid $704,857. With Chrietzberg being the lower bidder, Heritage took that bid and included Chrietzberg on the bid to TexAmericas. TexAmericas eventually awarded the bid to Heritage, and Heritage and Chrietzberg congratulated each other on the successful bid. After receiving an executed contract from TexAmericas, Heritage sent Chrietzberg a proposed subcontract. Shortly thereafter, Chrietzberg notified Heritage that it was withdrawing its bid. Heritage asked Chrietzberg to reconsider, without success, and so Heritage accepted Electrique's more expensive contract after accepting certain conditions of Electrique's. Heritage sued Chrietzberg for breach of contract, promissory estoppel, and negligent misrepresentation. The trial court ruled in favor of Heritage, but awarded Heritage only twenty-eight percent of the fees it had asked for. Heritage appealed, and Chrietzberg cross-appealed, claiming that all of Heritage's causes of action were barred by the statute of frauds.

QUESTION: Were Heritage's claims of breach and promissory estoppel invalid due to the statute of frauds?

REASONING: This case primarily revolved around the requirement that contracts that cannot be completed in under a year must be in writing is The Court reviewed the original project proposal, which indicated that the agreement between Heritage and Chrietzberg began December 31, 2011, and also indicated a projected project completion date of July 31, 2013. The Court ruled that the agreement would have required more than a year to complete and thus fell under the statute of frauds.

Next, the Court had to determine whether the agreement was enforceable under the statute of frauds. The Court found a precedent for this case in a similar case, *Cohen v. McCutchin*. In that case, the United States Supreme Court had ruled that because there was no writing that identified the plaintiff as a party to the agreement in question, the agreement did not satisfy the statute of frauds. Similarly, in the bid documents, which is where the agreement between Heritage and Chrietzberg supposedly laid, Chrietzberg was not once mentioned as a party to the agreement. Heritage also failed to point out any actions it took that are unequivocally referable to its agreement with Chrietzberg. The agreement fell under the statute of frauds but failed to satisfy it, thus the Court ruled that Chrietzberg cannot be held liable for breach of the agreement.

The Court then determined that there was no evidence that distinguished any of the damages supposedly caused by breach from damage caused by promissory estoppel. Therefore, because Chrietzberg was not liable for breach damages, it was also not liable for promissory estoppel damages. Similarly, the negligent misrepresentation damages were indistinguishable from the other damages and so too were denied.

DECISION: The Court reversed the trial court's ruling on Heritage's breach of contract and promissory estoppel claims and affirmed their denial of of negligent misrepresentation claims.

SIGNIFICANCE: This case exhibits the statute of frauds in action and demonstrates the significance of keeping all important agreements in detailed writing.

CRITICAL THINKING

Notice that the court relied on an analogy in coming to its decision that the writing was insufficient. Do you think the analogy is strong? Why or why not? Can you think of any other argument that Heritage might have made to obtain compensation in this case?

ETHICAL DECISION MAKING

What values are enhanced by this decision? What values were given less weight by the judges in this decision?

The requirements under the UCC for sales contracts vary slightly from those under the statute of frauds. The UCC does not require the parties to be named in the agreement. Rather, the UCC requires the writing to state clearly the quantity to be sold. Similar to the statute of frauds, the UCC allows a variety of written documents to constitute a writing. Some documents that count include faxes, emails, invoices, bills of lading, sales slips, checks, or any combination of these documents.

Exhibit 13-2

Requirements of a Writing Sufficient to Satisfy the Statute of Frauds under the Common Law

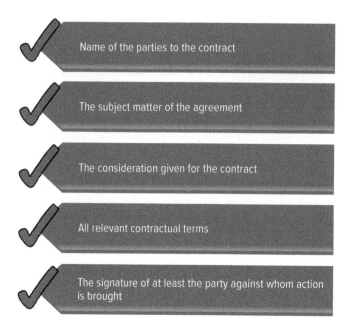

- Name of the parties to the contract
- The subject matter of the agreement
- The consideration given for the contract
- All relevant contractual terms
- The signature of at least the party against whom action is brought

Also noteworthy is the fact that oral testimony can be elicited that establishes the existence of a writing that would meet the requirements under the statute of frauds. For example, testimony regarding an invoice for products sold, when the actual invoice is not produced, is enough in certain states to meet the requirements under the statute.

Exhibit 13-2 provides a checklist to help you remember the elements needed for a writing that satisfies the statute of frauds.

Parol Evidence Rule

LO 13-3 Describe the purpose of the parol evidence rule.

A problem arises with a written contract when a party asserts that the writing is in some way deficient. To smooth transactions by limiting the types of evidence admissible to address the written agreement, the courts rely heavily on the parol evidence rule. The **parol evidence rule** is a common law rule that specifically addresses the admissibility of oral evidence as it relates to written contracts. The parol evidence rule states that oral evidence of an agreement made prior to or contemporaneously with a written agreement is inadmissible when the parties intend the written agreement to be the complete and final version of their agreement.[13] *Parol* in *parol evidence rule* means "speech" or "words," specifically words outside the original writing.

The purpose of the parol evidence rule is to restrict evidence from being admitted that substantially contradicts the agreement in its written form. Therefore, evidence of prior agreements and negotiations, as well as of contemporaneous agreements and negotiations, is typically excluded under the parol evidence rule. A written agreement is assumed to be complete, and evidence contradicting this final agreement usually impedes business transactions, which is why the parol evidence rule exists. In other words, the parole evidence rule applies when you have a *fully integrated agreement*.

However, when a court determines that the written agreement does *not* represent a complete and final version of the agreement, evidence may be admissible to further the court's understanding of the agreement. The admissible evidence is limited to additional elements, missing in the writing, that are consistent with the written agreement. These elements may be terms typically included in similar transactions that were left out of the specific writing, or they may be separate agreements in which consideration was offered.

parol evidence rule
A common law rule that states that oral evidence of an agreement made prior to or contemporaneously with a written agreement is inadmissible when the parties intend the written agreement to be the complete and final version of their agreement.

[13] Restatement (Second) of Contracts, sec. 213.

A cautionary note is in order: "Parol evidence rule," on its face, is a misnomer. First and foremost, parol evidence applies to both parol (spoken words) and written speech. The parol evidence rule covers evidence extrinsic to the original writing and is not limited to spoken words. Second, the parol evidence rule is *not* a rule of evidence but, rather, relates to substantive legal issues, namely, what constitutes a legally binding agreement and how we know what the agreement is. Third, the parol evidence rule is not a unitary concept or rule but an amalgamation of different rules and conditions.

Although the parol evidence rule applies to writings created at the same time as the written agreement, these writings tend to be treated differently than prior or contemporaneous oral agreements. That is, the writings are more readily admitted as part of the written agreement than is oral evidence regarding conditions or terms in the final agreement. The reasoning is that judges are allowed discretion in determining which documents constitute the final contract. As long as the contemporaneous written documents do not substantially contradict what is in the final writing, judges can deem these other writings part of the final written agreement. Consequently, the parol evidence rule does not usually exclude extrinsic written evidence.

Sometimes parties take the initiative and attempt to signal to judges that the written agreement is intended to be the final and complete statement of the agreement. A **merger clause** is a clause parties include in a written agreement within the statute of frauds that states that the written agreement accurately reflects the final, complete version of the agreement. Not all courts consider merger clauses to be conclusive proof of a contract. Nonetheless, where merger clauses are accepted, they greatly reduce the amount of guesswork courts must do in determining whether a written agreement was in fact intended to be the final statement of the agreement.

merger clause
A clause in a written agreement within the statute of frauds that states that the written agreement accurately reflects the final, complete version of the agreement.

Exceptions to the Parol Evidence Rule

As with the statute of frauds, there are exceptions to the general applicability of the parol evidence rule. That is, there are situations in which parol evidence, which is normally excluded, may be admissible in court. These exceptions pertain to (1) contracts that are subsequently modified, (2) contracts conditioned on orally agreed-on terms, (3) contracts that are not final because they are part-written and part-oral, (4) contracts with ambiguous terms, (5) incomplete contracts, (6) contracts with obvious typographical errors, (7) voidable or void contracts, and (8) evidence of prior dealings or usage of trade.

Contracts that have been Subsequently Modified

Although parol evidence contradictory to the final terms is inadmissible, evidence regarding a contract's subsequent modification is admissible. To be admissible, the modification must have been made after the writing, and the evidence must clearly indicate this later modification.

Despite the allowance of evidence to demonstrate modifications, not all evidence of modification is admissible. If the agreement is required to be in writing under the statute of frauds, oral modifications are unenforceable. However, oral evidence of a subsequent written agreement is admissible. In addition, if the contract's terms require modification to be in writing, oral modifications are inadmissible and unenforceable.[14]

Contracts Conditioned on Orally Agreed-On Terms

The parol evidence rule does not prevent parties from introducing evidence proving the written agreement was conditioned on terms agreed to orally. The reason the parol evidence rule does not apply is that the evidence being elicited does not substantially modify the written agreement. Rather, what is at issue with such evidence is the enforceability of the contract as written. No terms are altered, so the parol evidence rule does not apply.

[14] UCC § 2-209(2), (3).

condition precedent
In a contract, an event that must occur for a party's duty to arise.

When an entire contract is conditioned on something occurring first, that first thing is known as a **condition precedent**. Evidence of the existence of a condition precedent agreed to orally is admissible, as stated previously, because the contract is not modified by such evidence; rather, its enforceability is called into question. Since the statute of frauds is concerned primarily with the enforceability of agreements, it logically follows that the parol evidence rule does not apply to evidence of condition precedents.

Nonfinalized, Partially Written and Partially Oral, Contracts

When a contract consists of both written and oral elements, judges tend to treat the agreement as nonfinalized. Accordingly, the judge makes the assumption that the parties do not intend the written agreement to represent the entire agreement. Therefore, oral evidence related to the contract is admissible because the written document is not the complete and final representation of the agreement.

Contracts Containing Ambiguous Terms

When a contract contains what the court deems to be ambiguous terms, the court is faced with a dilemma in interpretation. To attempt to reach the most accurate interpretation of the original agreement, the court allows evidence, even if it is oral, for the sole purpose of clarifying ambiguous terms. As with evidence regarding orally agreed-on condition precedents, evidence used to clarify ambiguity is not believed to modify the contract but, rather, to clarify it. Accordingly, oral evidence related to ambiguity is admissible *only* to clarify, and not to change, any of the contractual terms.

Incomplete Contracts

When a contract is fundamentally flawed because it is missing critical information, courts can allow parol evidence to provide the missing information. Typically, the missing information is related to essential terms of the contract. Parol evidence will be admitted to clarify the contract by filling in the missing parts while not modifying, in any substantial way, the written agreement. Parol evidence is used to facilitate business transactions as opposed to forcing the parties to enter into a new, complete agreement. Case 13-2 is an illustration of a case where the completeness of the contract becomes an issue.

CASE 13-2

BAKER v. PRICE
Indiana Court of Appeals
2015 WL 9478049

FACTS: In 2014, Barker contacted Price about a van Price had advertised for sale. The advertisement described the van as a 1994 Ford E-350. Barker and Price agreed to meet, and, on April 9, Barker inspected the van. Following Barker's inspection, he orally agreed to purchase the van from Price for $15,000. The parties then agreed in writing that Barker would make an immediate $2,000 deposit and Price would provide Barker "title by 4/14/14 or deposit will be refunded in full". The deposit agreement described the van as a Ford E-350 but did not specify the model year.

Sometime prior to April 14, 2014, Price provided Barker with a certificate of title for the van. The certificate described the van as a 1993 model rather than a 1994 model. Barker refused to accept the certificate of title and demanded a refund of his $2,000. Price, in turn, refused to refund the deposit.

In his complaint, Barker alleged that Price had failed to provide a certificate of title in accordance with the deposit agreement, which entitled Barker to a refund of the deposit. After a hearing, the trial court entered summary judgment for Price. In relevant part, the trial court concluded that the year of the van was not a term material to the deposit agreement because that "agreement makes no reference to the year of the vehicle;" that Barker had accepted the van when he inspected it and paid the deposit; and that the certificate of title that Price tendered satisfied his obligation under the deposit agreement. This appeal ensued.

ISSUE: Did the deposit agreement contain the entire agreement between Barker and Price and, if not, should parol evidence be considered?

(continued)

REASONING: To begin, the appellate court examined the deposit agreement to determine whether it contained the entire agreement. The deposit agreement stated that, in exchange for a $2,000 deposit, Price "will have title by 4/14/14" for a "Ford E-350 Van". Notably, the deposit agreement does not contain the model year of the vehicle. Price argued that the model year was not a term that was material to the sale, and thus its inclusion was not necessary. The court disagreed with this reasoning. In effect, the court claimed, Price was asserting that he could have produced a certificate of title for a Ford E-350 van manufactured in any year without affecting his deal with Barker. Indeed, the court noted, model year plays a material role in most auto sales. In addition, the court found that the deposit agreement lacks at least one essential term, that being the sale price for the vehicle. Thus, the deposit agreement cannot be the entire agreement, and because model year is a material term, it can be considered as parol evidence in determining the true intent of the agreement.

DECISION AND REMEDY: Reversed in favor of Barker and remanded.

SIGNIFICANCE OF THE CASE: This case shows how seemingly minor or well-understood parts of a contract can actually be crucial to the proper enforcement of the contract. If the parties had been more thorough in drafting their agreement, this case probably would not have occurred.

CRITICAL THINKING

What types of evidence, if it exists. do you think Price can offer when the case is retried that might lead to his being successful in the new trial?

ETHICAL DECISION MAKING

What values are furthered by the court's ruling in this case?

Contracts with Obvious Typographical Errors

Whenever a written agreement under the statute of frauds contains a serious, and obvious, typographical error, parol evidence is admissible to demonstrate that the typo was a typo and to set forth the proper term. Allowing parol evidence to correct typos does not fundamentally alter the written agreement because the typo is not an accurate reflection of the parties' agreement. The agreement is not altered by correcting a typo; rather, the agreement is clarified.

Void or Voidable Contracts

Certain conditions can make an otherwise valid contract void or voidable. (Refer to Chapter 9 for an in-depth discussion of what makes a contract void or voidable.) Although the contract does not list these conditions, the courts allow parol evidence to demonstrate that the contract is void or voidable. As with the vast majority of the exceptions to the parol evidence rule, allowing evidence that proves a contract is void or voidable does not fundamentally alter the terms of the contract. Rather, the parol evidence would address the enforceability of the agreement, which is not the same as changing the terms of the agreement. Furthermore, evidence of a defense against a contract is admissible to prove a contract is void or voidable.

Evidence of Prior Dealings or Usage of Trade

This final exception, although directly related to those already mentioned, actually falls under the UCC and not the statute of frauds. Regardless, this exception also allows otherwise inadmissible parol evidence for a specific reason.

According to the UCC, parol evidence is admissible for the sake of clarification if the evidence addresses prior dealings between the parties or usages of trade in the same business field as that

which the parties are in.[15] Evidence related to past dealings can help clarify missing or ambiguous terms by examining how the parties had previously interacted. The assumption is that parties will continue to interact in a similar manner because repeated interactions tend to indicate pleasant interactions between the parties. Therefore, if a term is missing or ambiguous, the courts rely on evidence of what was done between the parties in the past to attempt to gauge what was intended in the contract in question.

Similarly, when a contract is ambiguous or incomplete, the courts examine the standard practices in the business. Here, the courts are assuming that businesses intend to engage in standard transactions. Furthermore, the courts believe that common business practices might sometimes be overlooked and, therefore, not be included when these common practices accurately reflect the intentions of the parties involved. Once again, an exception to the applicability of the parol evidence rule is made to allow evidence to clarify a contract as opposed to changing any material terms of the contract.

Third-Party Rights to Contracts

LO 13-4 Differentiate between an assignment and a delegation.

As you read in Chapter 9, contracts involve agreements between two parties in which each party agrees to give something to or do something for the other party. Contracts are typically private agreements, in the sense that the focus of the contract is the two parties involved and no one else. Accordingly, parties not in *privity of contract* (i.e., parties other than the contracting parties) usually do not have rights to a contract. However, as frequently is the case in the law, there are exceptions to the general rule.

There are two main situations in which a third party gains rights to a contract to which she or he is not a party. In the first, one of the contracting parties transfers rights or duties to a third party. In the second, the third party is a direct beneficiary of a contract involving two other parties. The remainder of this chapter examines both of these situations.

Assignments and Delegations

obligor
A contractual party who agreed to do something for the other party.

obligee
A contractual party who agreed to receive something from the other party.

Assignment
A contracting party's transfer of his or her rights to a contract to a third party.

assignor
In a contract, the party who transfers his or her rights to the contract to a third party (an assignee), giving the assignee the right to collect what was contractually agreed on in the original contract.

assignee
In a contract, the party who receives the rights of another party (an assignor) to collect what was contractually agreed on in the original contract.

Contracts typically involve an agreement between two parties in which each party agrees to do something for the other. That is, both parties are **obligors** (contractual parties who agreed to do something for each other) and **obligees** (contractual parties who agreed to receive something from each other). Consequently, contracts create a situation in which each party has a duty to perform the agreed-on action and a right to be the recipient of the other party's duty. These rights and duties can be transferred to *third parties*—parties not part of the original contract. The following sections discuss both the transfer of rights (assignment) and the transfer of duties (delegation) to third parties.

Assignment

Assignment occurs when a party to a contract (an **assignor**) transfers her rights to a contract to a third party (an **assignee**). In other words, the assignor gives to the assignee the right to collect what was contractually agreed on in the original contract (see Exhibit 13-3). For example, Lisa agrees to sell her car to Luke for $8,000. Lisa then assigns her right to receive Luke's payment to Kelly. Kelly, who was not part of the original contract between Luke and Lisa, is an assignee and now has the right to receive payment from Luke for Lisa's car.

LO 13-5 Recall the rights and duties of an assignee.

When an assignor transfers her rights to an assignee, the assignor legally gives up all rights she previously had to collect on the contract.[16] Now the assignee may legally demand performance from the other party to the original contract. In our example, once Lisa transfers her right to

[15] UCC §§ 1-205 and 2-202.
[16] Restatement (Second) of Contracts, sec. 317.

GLOBAL Context

Assignment of Rights in Russia

©Digital Archive Japan/Alamy

Almost all developed market economies permit the free assignability of contract rights. Assignments are important because they play a crucial role in business financing. Banks and businesses make loans and pay debts through assignments. In Russia, civil law is operated primarily by the Russian Civil Code. The Russian law does not address the assignment of a contract in the same manner as does U.S. law. Instead, the Russian Civil Code refers to the assignment of rights as an assignment of claim. The economic or monetary rights of the assignor may be transferred to another person; however, the Russian Civil Code states that "rights which are closely related with the personality of the creditor, such as alimony rights or rights for compensation of personal injury," are not permitted to be assigned to another person. Unlike the case in the United States, where assignments can be given orally or in writing, in Russia, the Civil Code requires assignments to be produced in writing. The debtor (or obligor) is not required to perform the obligation to the new assignee until he or she receives written proof that the assignment took place. The assignment of rights, or assignment of claims, in Russia is highly regulated by the Russian Civil Code.

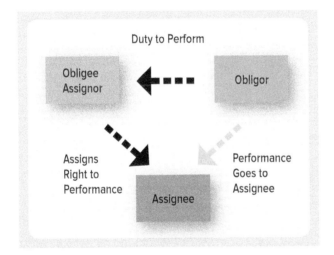

Exhibit 13-3

Assignment of Rights

Kelly, Lisa can no longer require Luke to pay her for her car; however, Kelly can request Luke to pay her for Lisa's car.

The assignee essentially fills in for the assignor as the legal recipient of the contractual duties. As such, the assignee acquires the same rights the assignor had. The assignee is offered no additional protections. Moreover, the obligor (the other party to the contract, who owes a duty to the assignee) may raise any of the same defenses for nonperformance to the assignee as he would have been able to raise against the assignor. Thus, if Lisa failed to deliver her car to Luke, he can legally refuse to pay Kelly on the basis of Lisa's breach of contract. It does not matter that Kelly had no duty in the original contract; she is subject to the same defense Luke had against Lisa; therefore, Kelly would not be paid in this situation.

Although there is no legally required wording to be used or forms to be filled out for assignments to be valid, certain restrictions exist. First, although assignments may be given orally or in writing, the UCC requires assignments to be in writing when the amount being assigned is greater than $5,000.[17,18] Furthermore, assignments covered by the statute of frauds also must be in writing. Because it is difficult to prove the existence of assignments given orally, it is usually suggested that assignments be in writing.

Second, an assignee must agree to accept the assigned rights. An assignee may decline an assignment if he has not legally agreed to the assignment and he declines in a timely fashion after

[17] Restatement (Second) of Contracts, sec. 324.
[18] UCC § 1-206.

Exhibit 13-4

Contractual Rights That Cannot Be Assigned

1. Rights that are personal in nature.
2. Rights that increase the obligor's risk or duties.
3. Rights whose assignment is prohibited by contract.
4. Rights whose assignment is prohibited by law or public policy.

learning about the assignment and its terms.[19] There is no protocol an assignee must follow to reject an assignment, but once the assignment is rejected, it is considered rejected from the time it was first offered. Third, in some situations contractual rights cannot be assigned.[20]

Rights That Cannot Be Assigned Exhibit 13-4 lists the four situations in which contractual rights cannot be assigned to a third party. We discuss each of them here.

First, the rights to a contract cannot be assigned when the contract is personal in nature. For a contract to be personal in nature, the obligor must have promised something specific to the person receiving it. Because the subject matter of the contract is personal, a third party cannot legally become the recipient. The one exception occurs when the only part of a contract left to be fulfilled is the payment, because rights to payment can always be assigned.[21]

Second, rights cannot be assigned when the assignment would increase the risk or duties the obligor would face in fulfilling the original contract. For example, Ben agrees to replace the siding on Erin's two-bedroom ranch house. Erin cannot assign her right to Ben's services to Chris, who lives in a three-story, five-bedroom house, because Ben's (the obligor's) duties would be greatly increased from the work required in the original contract with Erin to the work required in an agreement with Chris.

Third, rights cannot be assigned when the contract expressly forbids assignments. When parties include an *antiassignment clause* in their contract, the parties are attempting to limit their ability to assign their rights under the contract. However, the wording of the antiassignment clause is determinative regarding the effectiveness of the clause. That is, if worded improperly or ambiguously, the clause does not effectively limit assignments.

Most courts consider contractual prohibitions against assignments to be promises. Consequently, courts consider an assignment made despite an antiassignment clause to be effective, but the party who makes the assignment will still be liable for breaching the terms of the contract. Moreover, unless the clause is very specific in what it prohibits, the court will generally interpret the antiassignment clause as preventing delegation of duties, not assignment of rights.[22] A clause stating, "All assignments are void under this contract," would be considered effective. In contrast, when a contract includes a clause explicitly permitting assignments, the parties may assign rights, even when assignments would normally be considered improper because of an increased duty, risk, or burden to the obligor.[23]

Even in the presence of an antiassignment clause, there are exceptions in which assignments can still be made. For instance, antiassignment clauses do not affect assignments made by operation of law. That is, if a law necessitates an assignment, such as in bankruptcy cases, the assignment is effective regardless of any contractual agreement to the contrary. Likewise, as previously stated, the right to assign monetary payments cannot be denied. Therefore, even when a contract has an antiassignment clause, either party may still assign his or her right to receive payment.[24] Part of the reason the law does not bar the right to receive payment is that businesses often transfer rights to payments in the regular course of business. Also, one's duty to pay is not affected when the party to whom one must make a payment changes. That is, no added burden is placed on the obligor when the recipient of a payment changes, requiring that payment be sent to a different party.

[19] Restatement (Second) of Contracts, sec. 327.
[20] Restatement (Second) of Contracts, sec. 317(2).
[21] Restatement (Second) of Contracts, secs. 317 and 318.
[22] Restatement (Second) of Contracts, sec. 323(1), and UCC § 2-210(3).
[23] Restatement (Second) of Contracts, sec. 323(1).
[24] UCC § 9-318(4).

In addition, assignments for the right to receive damages for a breach of contract to sell goods or services are unaffected by antiassignment clauses.[25] Therefore, if one party breaches the contract, the other party can sue the breaching party and transfer the right to recovery to a third party.

Fourth, rights cannot be assigned when law or public policy forbids such assignments. Various state and federal statutes prohibit the assigning of specific rights. Even when no statute prohibits an assignment, if the assignment is determined to be against public policy, the assignment is deemed ineffective.

Except as outlined in this section, all other rights are presumed to be assignable. Once it has been established that an assignment is valid, notice should be given to the obligor regarding the assignment.

BUT WHAT IF...

WHAT IF THE FACTS OF THE CASE OPENER WERE DIFFERENT?

Let's say in the Case Opener that Tyson was performing in the match for a large sum of money. He wanted to assign his monetary rights to his mother. However, the contract contained an anti-assignment clause. Would Tyson still be able to assign his financial rights to his mother?

Notice of Assignment Although notice need not be given for a valid assignment, it is usually a good idea for the assignor or the assignee to notify the obligor of the assignment. Assignments are effective immediately, regardless of notice, but by providing notice, the assignor can help avoid two serious complications with assignments.

The first complication involves the obligor's fulfilling of the contractual duty. Without notice of assignment, the obligor can discharge his contractual duties simply by fulfilling his duties to the assignor. Because fulfilling the contract discharges the obligor's duties, the act also constitutes a discharge of the assignee's claim on the assignor's right. However, after the obligor has been given notice, the obligor can discharge his contractual obligations only by fulfilling the contract for the assignee.

The second complication regarding a lack of notice involves the granting of multiple assignments of the same right. When an assignor assigns two or more parties the same right, confusion arises about which party has the right to the contract. Most states use the **first-assignment-in-time rule,** which states that the first party granted the assignment is the party correctly entitled to the contractual right. By giving proper notice, a party can ensure that there is no confusion over when the assignment was made. Furthermore, a minority of states have adopted the **English rule,** which states that the first assignee to give notice of assignment to the obligor is the party with rights to the contract. Consequently, especially when living in a state using the English rule, parties are well advised to give notice of assignments to ensure that they maintain their assigned rights.

The Restatement (Second) of Contracts states a view between the first-assignment-in-time rule and the English rule.[26] The Restatement grants legal right to the first assignee in most situations. However, if the first assignment is legally voidable or revocable by the assignor, subsequent assignments are considered evidence of the voiding or revocation of the first assignment. In such situations, the later assignee has legal right to the contract. Also, the later assignee is considered the legal owner of the contractual right if she offers something to the assignor as consideration and then obtains (1) performance by the obligor on his duty, (2) judgment requiring performance by the obligor, (3) a new contract with the obligor, or (4) evidence frequently used to signify a contractual right (e.g., a writing indicating a contractual obligation).

Delegation

A **delegation** occurs when a party to a contract (a **delegator**) transfers her duty to perform to a third party who is not part of the original contract (a **delegatee**). Whereas assignments transfer rights to a contract, delegations transfer duties. (See Exhibit 13-5.) Instead of the third party

first-assignment-in-time rule
A rule that states that the first party granted the assignment is the party correctly entitled to the contractual right.

English rule
A rule that states that the first assignee to give notice of assignment to the obligor is the party with rights to the contract.

delegation
A contracting party's (a delegator's) transfer of his or her duty to perform to a third party who is not part of the original contract (a delegatee).

delegator
A party in a contract who transfers his or her duties to perform to a third party who is not part of the original contract (a delegatee).

delegatee
A third party who is not part of the original contract but to whom duties to perform are transferred by one of the contracting parties (a delegator).

[25] UCC § 2-210(2).
[26] Sec. 342.

Exhibit 13-5

Assignment of Duties

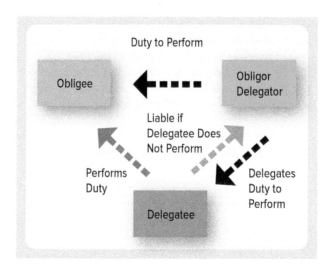

receiving something, as in an assignment, delegations require the delegatee to fulfill the delegator's contractual obligation to the obligee—the party to the contract to whom a duty is owed. For example, John contracts with Teresa to have her deliver machinery to his factory. Teresa then delegates her duty to Bill, who delivers the machinery to John.

One important distinction between assignments and delegations involves the rights of the transferring party—the assignor or the delegator. In assignments, after the assignment is made, the assignor has no right left to the original contract. Conversely, with delegations, the delegator is not relieved of his duty to perform by making a delegation. If the delegatee fails to fulfill the contract, the delegator is still liable to the obligee for fulfillment of the contract. In our example, if Bill fails to deliver the machinery to John, Teresa is liable to John for damages.

Duties That Cannot Be Delegated As with assignments, the starting assumption is that duties to a contract can be delegated. However, courts tend to examine delegations more closely than assignments. The reasoning is that an assignment usually does not affect the party to the contract who is not involved in the assignment (the obligor), whereas a delegation forces the party to the contract who is not involved in the delegation (the obligee) to receive performance of the contract from a party with whom the obligee did not directly contract.

Also, just as with assignments, in certain situations, duties cannot be delegated to a third party (see Exhibit 13-6).[27] The first type of duty that cannot be delegated, mirroring assignments, is any duty of a personal nature. A duty is personal in nature when it requires the specific talents, skills, or expertise of the obligor. An interesting situation arises when the initial contract bears an implicit assumption that work will be performed by others. In such situations, if supervision is important to the task, the supervision could be considered a personal duty that the obligor may not delegate to others. For example, suppose you are planning to have a new building built and you contract with a well-respected construction manager. He will not build the building; the contract is for his management skills. His contractual duties would be considered personal and therefore cannot be delegated.

Delegation for personal duties is admissible where otherwise not allowed when there is an explicit or implicit contractual agreement to allow delegations. Usually, for a delegation of personal duties to be effective, the contract must state that delegations are permitted.

Although personal duties typically cannot be delegated, any nonpersonal duties in a contract can be delegated. For example, the delivering of goods, mowing of one's lawn, paying of money, or painting of a house are all considered nonpersonal duties. These duties are nonpersonal because they do not require particular skill or expertise and therefore can be completed by most people. Accordingly, one can delegate these nonpersonal duties.

[27] Restatement (Second) of Contracts, sec. 318, and UCC § 2-210.

1. Duties that are personal in nature.
2. Duties for which the delegatee's performance will vary significantly from the delegator's.
3. Duties in contracts that forbid delegations.

Exhibit 13-6

Duties That Cannot Be Delegated

The second type of duty one cannot delegate is a duty whose performance by the delegatee would vary significantly from its performance by the delegator. To protect the obligee, who is part of the original contract, when performance would differ substantially from that to which the obligee contractually has the right, courts will rule that the delegation is ineffective.

The third type of duty one cannot delegate is any duty whose delegation is prohibited by the contract. When parties include agreements not to delegate, the courts typically treat these inclusions as indications of the parties's desire to consider the contractual obligations as personal. Consequently, the courts frequently find otherwise allowable delegations to be inappropriate when an agreement not to delegate is in the contract. However, even if a clause prohibiting delegations exists, the courts will probably allow delegations to occur if they are impersonal. For example, courts consider the payment of money impersonal, and therefore delegable, even if a contract prohibits delegations.

BUT WHAT IF . . .

WHAT IF THE FACTS OF THE CASE OPENER WERE DIFFERENT?

Assume that Mike Tyson had been hired for the match because of his fame and his enhanced skills. If Tyson tried to delegate his duties to another athlete who was neither famous nor highly skilled, would such a delegation of duties be lawful?

Assignment of the Contract

Frequently, contracts contain ambiguous language that makes it unclear what is being assigned or delegated. Examples of ambiguous language are "I assign the contract" or "I assign all of my rights under the contract." When a court cannot clearly tell what the parties intended, the court usually considers the assignment to be both of rights and of duties. In other words, when ambiguous language is used, the court considers that the assignor has assigned his rights and delegated any duties he had under the contract (see Exhibit 13-7). Doing so removes any right the assignor had to collect under the contract, but he is still liable to the obligee for any duties the delegatee, who is also the assignee, fails to perform.

Third-Party Beneficiary Contracts

We have thus far learned about one of the ways in which third parties may obtain rights to a contract: through assignments or delegations. We now move on to the other way, which is that the third party is an intended beneficiary of a third-party beneficiary contract. A **third-party beneficiary contract** is created when two parties enter into a contract with the intended purpose of benefiting a third party. To be an intended third-party beneficiary, the third party does not need to be explicitly named in the contract as long as the terms of the contract or events occurring after the creation of the contract make it clear who the intended recipient of the contractual benefits is.

third-party beneficiary contract

A contract created when two parties enter into a contract agreement with the intended purpose of benefiting a third party.

Exhibit 13-7

Assignment of the Contract

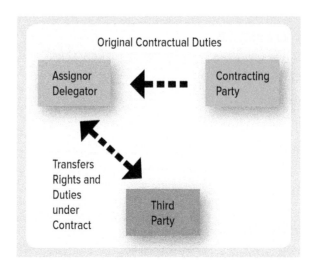

Intended Beneficiaries

LO 13-6 Differentiate between an intended and an incidental beneficiary.

Early in the common law, courts had difficulty when contracts were written to benefit third parties. The courts were not sure how to treat these contracts and usually deemed third parties to have no rights to contracts to which they were not in privity. Now, however, third parties who are intended beneficiaries have the right to enforce contracts. An **intended beneficiary** is a third party to a contract whom the contracting parties intended to benefit directly from their contract.

Intended beneficiaries may enforce their rights to a contract when both parties to the contract intended for the third party to benefit. In determining whether a third party is an intended beneficiary, courts ask whether the contracting parties intended the third party to be the "direct," "primary," or "express" beneficiary of the contract. A good example of an intended third party beneficiary contract would be in a contract between a woman and a car dealership in which she paid the dealership to deliver a car to her daughter for a birthday present.

intended beneficiary
A third party to a contract whom the contracting parties intended to benefit directly from their contact.

BUT WHAT IF...

WHAT IF THE FACTS OF THE CASE OPENER WERE DIFFERENT?

Let's say that a venue owner contacted Tyson to come perform in a match at his venue. Tyson will volunteer to compete in a match and all of the money from the ticket sales will go to charity. Is there a third-party beneficiary in this scenario?

Also, who is the promisor? What is the promise?

promisor
In a third-party beneficiary contract, the party to the contract who made the promise that benefits the third party.

promisee
In a third-party beneficiary contract, the party to the contract who owes something to the promisor in exchange for the promise made to the third party.

The **promisor** in a third-party beneficiary contract is the party to the contract who made the promise that benefits the third party. The **promisee** is the party to the contract who owes something to the promisor in exchange for the promise made to the third-party beneficiary. For example, Amanda contracts with Alex to clean his house. In exchange, Alex will pay Amanda's credit card debt. The credit card company is the third-party beneficiary because the contract is created to benefit the company. Alex is the promisor, for he made the promise to pay the third-party beneficiary. Amanda is the promisee because she owes a duty to the promisor, Alex.

In a third-party beneficiary contract, the intended beneficiary may sue the promisor to enforce the contract. Although the promisee typically owes something to the third-party beneficiary before the contract with the promisor exists, if the third party sues the promisee after the promisor does not fulfill his or her obligations, the promisee can then sue the promisor for breach of contract. Therefore, courts allow the third-party beneficiary to sue the promisor, thus eliminating the litigation that would ensue if the promisee sued the promisor. In our example, if Alex fails to pay the

credit card company for Amanda's debt, the credit card company has the right to sue Alex, even though it is Amanda's debt.

Let us return to the Case Opener. The fans of the Tyson fight argued that they were third-party beneficiaries and therefore had rights under several contracts that were violated when Tyson was disqualified early in the fight. Boxing matches are widely viewed events, and one can argue that the fights are organized for the enjoyment of the fans. Does the idea of boxing matches being organized for the fans make the fans the "direct," "primary," or "express" beneficiaries of contracts involved in the fight? In other words, are the fans of the Tyson fight intended beneficiaries to the contracts Tyson entered when agreeing to the fight? What else do we need to know before we can determine whether the fans have a legal right to a refund?

There are two types of intended beneficiaries: creditor beneficiaries and donee beneficiaries.

Creditor Beneficiaries A **creditor beneficiary** is a third party who benefits from a contract in which the promisor agrees to pay the promisee's debt. In our previous example, because Alex (the promisor) agreed to pay the debt of Amanda (the promisee), Amanda's credit card company is a creditor beneficiary.

Case 13-3 is a famous dispute in which the courts began to recognize the rights of third parties to sue promisors for performance.

creditor beneficiary
A third party who benefits from a contract in which the promisor agrees to pay the promisee's debt.

LAWRENCE v. FOX
Court of Appeals of New York
20 N.Y. 268 (1859)

FACTS: *In November 1857, Holly, at the request of Fox, loaned him $300. Before loaning the money, Holly informed Fox that Holly owed Lawrence $300, due the next day. In consideration of the loan, at the time of the loan, Fox agreed to pay Lawrence for Holly the next day. Fox did not pay, and Lawrence sued him. Fox sought to dismiss the charges because there was no proof to show that Holly was indebted to Lawrence, Fox's agreement with Holly to pay Lawrence was void for want of consideration, and there was no privity between Lawrence and Fox. Fox's motion to dismiss was denied. The jury ultimately found in favor of Lawrence for the sum of the loan plus interest. Fox appealed and the judgment was affirmed. Fox then appealed again.*

ISSUE: Does Lawrence have the right to sue Fox, the promisor, for payment?

REASONING: There is a principle of law, found in English law as well as in the law of many states, that if one person makes a promise to another for the purpose of benefiting a third party, that third person may "maintain an action upon it." Although many of the preceding cases in which this principle was applied were trust cases, nothing about the law requires the case to be a trust case. Fox's duty to pay Lawrence in return for the loan he received from Holly was a clear condition of the loan. The spoken promise is equal to a written promise in this case. Therefore, Lawrence should receive payment from Fox in the interest of justice, even if the law could be more narrowly interpreted.

DECISION AND REMEDY: The decision of the earlier court was affirmed. Fox was required to pay Lawrence as he had promised Holly.

SIGNIFICANCE OF THE CASE: This case was one of the earliest cases that upheld the rights of third parties to sue promisors to have agreements enforced.

CRITICAL THINKING

Form an opinion about the contemporary relevance of this ruling. Given the widespread and extremely significant changes in U.S. society and the world over the past 150 years, what is the justification for consideration of such an old case? How does this type of process aid our legal system? How might it detract from the reasonableness of modern deliberations?

ETHICAL DECISION MAKING

Consider the actions of Fox, both leading up to and through the course of the legal dispute. From an ethical standpoint, which stakeholders was Fox considering? What values was he attempting to uphold?

The fans in the Tyson case argued that they are intended beneficiaries. If the fans are correct, and we know Tyson did not have a debt to the fans, then they must be donee beneficiaries. Does an agreement to perform create a situation in which the audience members become the intended beneficiaries of the performance?

Donee Beneficiaries The other type of intended beneficiaries is donee beneficiaries. **Donee beneficiaries** are third parties who benefit from a contract in which a promisor agrees to give a gift to the third party. The most common form of donee beneficiary contracts is life insurance policies. Someone (the promisee) pays premiums on a life insurance plan to have the insurer (the promisor) agree to pay a third party (the donee beneficiary) on the promisee's death.

Donee beneficiary
A third party who benefits from a contract in which a promisor agrees to give a gift to the third party.

vest
To mature, as in the maturation of rights that allow a party to act legally on the rights.

Vesting of Rights Although an intended beneficiary can enforce her rights to a contract, she cannot do so until her rights to the contract **vest,** that is, until the rights mature so that a party can legally act on the rights. Before a third party's rights have vested, the original contracting parties can make changes to the original contract without the permission of the third party. For example, third-party rights in a life insurance policy do not vest until the promisee's death. Consequently, Jane (the promisee) can change the intended beneficiary of her life insurance policy from Mary to Peter. Jane does not need Mary's permission, and Mary cannot sue Jane because Mary's rights have not vested.

Generally, one of three things must occur for a third party's right to a contract to vest.[28] First, under certain circumstances, third-party rights vest immediately. When rights vest immediately, the third party can enforce the contract at any time. These rights take effect instantaneously, even if the beneficiary does not know about the contract.

Second, rights may vest when the beneficiary decides to accept the rights to the contract. Sometimes the beneficiary must positively accept the rights, such as by notifying the contracting parties of acceptance. However, in the absence of an overt act of rejecting the rights to a contract, acceptance is assumed when the beneficiary becomes aware of the contract.

Third, rights may vest when the beneficiary changes his position because of reliance on the contractual rights. In other words, the beneficiary must take some action he would not have taken otherwise because he is expecting to benefit from the contract. For example, Vince is a third-party beneficiary to a contract. When he finds out about the contract, he decides to lease a new car because he is expecting to benefit from the contract. Obtaining the lease would cause his rights to the contract to vest because doing so demonstrates a change in position based on reliance on the contract.

If a contract specifies that the original contracting parties maintain the right to alter or rescind the contract, the vesting of the third party's rights does not preclude the promisor or the promisee from changing the original contract.

Many states hold that donee beneficiary rights vest before creditor beneficiary rights. The rationale is that even if the contract is altered when there is a creditor beneficiary, the creditor beneficiary still maintains her rights against the debtor (the promisee). Donee beneficiaries do not have the same rights as creditor beneficiaries; thus many states allow their rights to vest more quickly than those of creditor beneficiaries.

LO 13-7 Differentiate between a creditor beneficiary and a donee beneficiary.

Creditor versus Donee Beneficiaries There are two main distinctions between creditor beneficiaries and donee beneficiaries (see Exhibit 13-8). The first distinction is based on the reason the third-party beneficiary contract was created. If the promise in the contract is intended to release a party from an obligation to a third party, such as the paying of one's debt, the contract involves a creditor beneficiary. Conversely, if the contract intends to grant a gift to a third party, the third party is a donee beneficiary.

The second distinction involves the conditions under which an intended beneficiary can enforce her rights under a contract. Creditor beneficiaries can enforce their rights under a contract whenever the contract is valid. Donee beneficiaries can enforce their rights to most contracts. However, some jurisdictions do not allow donee beneficiaries to enforce their contractual rights in all situations.

[28] Restatement (Second) of Contracts, sec. 311.

Exhibit 13-8

Creditor versus Donee Beneficiaries

CREDITOR BENEFICIARY	DONEE BENEFICIARY
Purpose of the Contract	
Contractual performance fulfills an obligation to a third party.	Contractual performance gives a gift to a third party.
Enforcement of Rights	
Beneficiary can enforce rights to a contract if the contract is valid and the rights have vested.	Beneficiary has limited ability to enforce contracts, depending on the jurisdiction.
Beneficiary can enforce rights against the promisor or the promisee.	Beneficiary can enforce rights against the promisor.

When a donee beneficiary can enforce his rights under a contract, he may do so only against the promisor. The reason the donee beneficiary cannot sue the promisee is that the promisee had no duty to the donee beneficiary. In contrast, creditor beneficiaries can sue the promisor or the promisee for performance because both parties owed a duty to the creditor beneficiary. However, if the creditor beneficiary wins a judgment against one party, she may not seek judgment against the other party. In addition, if a creditor beneficiary wins judgment against the promisee, the promisee may sue the promisor to recover under a theory of breach of contract.[29]

It is not always easy to determine when someone is a creditor or a donee beneficiary. Sometimes a contract is created for reasons that are intended both to be charitable and to pay a debt. Given the lack of clear distinction of when someone is a creditor beneficiary and when he is a donee beneficiary, the Restatement (Second) of Contracts takes a different approach.[30] The Restatement focuses on the distinction between intended and incidental beneficiaries.

Creditor and donee beneficiaries are both intended beneficiaries, and, according to the Restatement, intended beneficiaries have the right to enforce a contract. When it is clear that the contract was created for the benefit of a third party and that performance of the contractual duties will pay off the payee's debt or give a gift as the payee intended, the third party is an intended beneficiary.

Incidental Beneficiaries

Similar to intended beneficiaries, incidental beneficiaries are third-party beneficiaries to contracts. However, unlike the case with intended beneficiaries, the contracting parties do not *intend* to benefit incidental beneficiaries. An **incidental beneficiary** is one who unintentionally gains a benefit from a contract between other parties (see Exhibit 13-9). One significant difference between intended and incidental beneficiaries is that incidental beneficiaries cannot sue to enforce a contract. Incidental beneficiaries maintain no rights to enforce other people's contracts. Simply put, they are incidental beneficiaries.

In determining whether a party is an incidental beneficiary, the courts take a reasonable person approach: They ask whether a reasonable person in the position of the party in question

incidental beneficiary
One who unintentionally gains a benefit from a contract between other parties.

Exhibit 13-9

Intended versus Incidental Beneficiaries

INTENDED BENEFICIARIES	INCIDENTAL BENEFICIARIES
Contracting parties intended to benefit the third party with their contract.	Contracting parties did not intend to benefit the third party with their contract.
Beneficiary has the right to enforce the contract.	Beneficiary does not have the right to enforce the contract.
Beneficiary benefits from direct reception of contractual performance.	Beneficiary benefits from indirect circumstances created by contractual performance.

[29] Restatement (Second) of Contracts, sec. 310.
[30] Restatement (Second) of Contracts, sec. 302.

would believe that the contracting parties intended to benefit him or her. If the reasonable person would believe that the contracting parties consciously intended to benefit him or her, the courts will consider the party an intended beneficiary. If the reasonable person would not find such intent, the third party is an incidental beneficiary.

As a thought experiment, consider the reasonable person test in the context of the Tyson dispute in the Case Opener. For the fans to be able to receive refunds, the reasonable person in the position of the fans would have to believe that Tyson intended to benefit him or her with Tyson's performance. Do the fans meet the reasonable person test?

The courts consider other elements besides the reasonable person approach when determining whether a party is an intended or incidental beneficiary. One factor the courts consider is whether performance of the contract is done directly to the third party. If nothing is explicitly done for or given to the third party, she or he cannot be an intentional beneficiary. The courts also examine the third party's ability to control the specifics of performance of the contract. That is, if a third party has the ability to provide input regarding how the contractual duties are fulfilled, the third party is probably an intended beneficiary. For example, Dianne (the promisee) agrees to pay Charles (the promisor) to paint Henry's (the third party's) house. Henry tells Charles what color he wants the house and when Charles should be there to paint. Henry's ability to control how Charles paints the house demonstrates his status as an intended beneficiary. In addition, note that Charles renders performance directly to Henry, so the explicit-performance factor also indicates Henry's status as an intended beneficiary.

A third factor the courts examine in determining the type of third-party beneficiary is whether the contract directly states that the third party is the benefiting party. In our example, because Charles agreed in the contract to paint Henry's house, the contract lists Henry as the beneficiary. Consequently, Henry is an intended beneficiary. Although Henry meets all three additional tests (besides the reasonable person test), it is not necessary to meet all three. If a third party meets at least one of the last three tests, he is usually an intended beneficiary.

CASE OPENER WRAP-UP

Fallout from a Unforgettable Fight

The court hearing the Tyson case quickly dismissed the claims. It held that the fans were in no way third-party beneficiaries to any contract into which Tyson, the promoters, or the telecasters had entered. The fans cannot meet any of the tests for an intended beneficiary. Simply put, they are incidental beneficiaries.

PRACTICAL TIPS for Business Managers

1. Even though only certain contracts need to be in writing to be enforceable, it is a good idea as a matter of policy to ALWAYS put contracts in writing.
2. Whenever entering into a contract, to ensure that you do not have a parol evidence problem, include a simple statement at the end of the contract stating that the document intended to be the complete and final version of the contract.
3. If a party makes you an oral promise just before you sign a written agreement, and that promise is not in the written agreement, insist that the additional promise be handwritten on the contract and signed by both parties to prevent a potential parole evidence problem.
4. When you are the assignee of a contract, always notify the obligor immediately to ensure that the assignor doesn't try to assign the right to another party.
5. Even though a writing is not necessary for many assignments, it is always a good idea to put all assignments in writing in case a dispute arises over the assignment.

SUMMARY

Statute of Frauds — The term *statute of frauds* refers to various state laws modeled after the 1677 English Act for the Prevention of Frauds and Perjuries. These state laws are intended to (1) ease contractual negotiations by requiring sufficient reliable evidence to prove the existence and specific terms of a contract, (2) prevent unreliable oral evidence from interfering with a contractual relationship, and (3) prevent parties from entering into contracts with which they do not agree.

Contracts That Require a Writing
1. Contracts whose terms prevent possible performance within one year.
2. Promises made in consideration of marriage.
3. Contracts for one party to pay the debt of another if the initial party fails to pay.
4. Contracts related to an interest in land.
5. Under the Uniform Commercial Code, contracts for the sale of goods totaling more than $500.

Exceptions to the Statute of Frauds — The exceptions include (1) admission, (2) partial performance, (3) promissory estoppel, and (4) various exceptions under the UCC.

Sufficiency of the Writing — A sufficient writing under the statute of frauds must clearly indicate (1) the parties to the contract, (2) the subject of the agreement, (3) the consideration given for the contract, (4) all relevant contractual terms, and (5) the signature of at least the party against whom action is brought.

Under the UCC, a writing must clearly indicate (1) the quantity to be sold and (2) the signature of the party being sued.

Under neither the statute of frauds nor the UCC must the writing be contained within one document.

Parol Evidence Rule — The *parol evidence rule* is a common law rule stating that oral evidence of an agreement made prior to or contemporaneously with a written agreement is inadmissible when the parties intend to have the written agreement be the complete and final version of their agreement.

Exceptions to the Parol Evidence Rule
1. Contracts that are subsequently modified.
2. Contracts conditioned on orally agreed-on terms.
3. Contracts that are not final because they are part written and part oral.
4. Contracts with ambiguous terms.
5. Incomplete contracts.
6. Contracts with obvious typographical errors.
7. Voidable or void contracts.
8. Evidence of prior dealings or usage of trade.

Integrated Contracts — *Integrated contracts* are written contracts, within the statute of frauds, intended to be the complete and final representation of the parties' agreement, thus precluding the admissibility of parol evidence other than in the exceptions previously listed.

Third-Party Rights to Contracts — There are two main situations in which a third party gains rights to a contract to which she or he is not a party: (1) One of the contracting parties transfers rights or duties to the third party. (2) The third party is a direct beneficiary of a contract involving two other parties.

Assignments and Delegations — Contracts typically involve an agreement between two parties:

Obligor: Contractual party who owes a duty to the other party in privity of the contract.

Obligee: Contractual party who is owed a duty from the other party in privity of the contract.

An *assignment* is the transfer of rights under a contract to a third party.

Assignor: The party to a contract who transfers his or her rights to a third party.

Assignee: A party not in privity to a contract who is the recipient of a transfer of rights to a contract.

Contractual rights that cannot be assigned:
1. Rights that are personal in nature.
2. Rights that would increase the obligor's risks or duties.
3. Rights in a contract that expressly forbids assignment.

A *delegation* is the transfer of a duty under a contract to a third party.

Delegator: The party to a contract who transfers his or her duty to a third party.

Delegatee: A party not in privity to a contract who is the recipient of a transfer of a duty in a contract.

Contractual duties that cannot be assigned:
1. Duties that are personal in nature.
2. Duties whose performance by the delegatee would be substantially different from that which the obligee originally contracted.
3. Duties in a contract that expressly forbids delegation.

(continued)

(concluded)

Assignment of the Contract

When ambiguous language is used, courts interpret the transfer to consist of an assignment of rights and a delegation of duties.

Third-Party Beneficiary Contracts

An *intended beneficiary* is a third party to a contract whom the contracting parties intended to benefit directly from their contract.

 Promisor: A party to a contract who made the promise that benefits the third party.

 Promisee: A party to a contract who owes something to the promisor in exchange for the promise made to the third-party beneficiary.

 Creditor beneficiary: A third party who benefits from a contract in which the promisor agrees to pay the promisee's debt.

 Donee beneficiary: A third party who benefits from a contract in which a promisor agrees to give a gift to the third party.

 Vest: The maturation of rights such that a party can legally act upon the rights.

An incidental *beneficiary* is a third party who unintentionally gains a benefit from a contract between other parties. That is, it was never the conscious objective of the contracting parties to benefit the third party.

Point/Counterpoint

Should Obligors Have a Say in Who Becomes an Assignee?	
YES	**NO**
When two parties enter into a contract together, an agreement is being made between two consenting parties, often for the mutual benefit of each party. In other words, each party often knows what the party is getting into and has no objections to helping the other party in order to receive benefit. However, in the case of assignment, a party may end up owing a duty to another party whom it does not wish to support. For example, if Bob makes a contract with Business X, a business run by a good friend of Bob's, Business X could assign its contract to Business Y, a company that provides a product of the same quality but uses business practices that Bob does not agree with. In this case, Bob is obligated to pay or otherwise benefit a party that he strongly opposes, which is hardly fair to Bob. Bob may have entered into the contract with the intention of supporting his friend's business, and while he is still indirectly doing so, he must help a business that he detests in order to fulfill his side of the contract. Second, in a case where Business X is assigning the contract in order to avoid breach, Business X is essentially admitting that it cannot fulfill the contract and so Business X is the party that is potentially failing to uphold the contract. Shouldn't Bob, assuming he has upheld his side of the deal, get to pick the business to replace Business X in the contract instead of the party that will fail to honor the deal if Business X doesn't assign? An even more fundamental reason as to why Bob should have a say in his contract with Business X is that he is a party to the contract. If Business X has a say in who it assigns its contract to, shouldn't Bob, who is entitled to the same rights under the contract, have some kind of voice in the matter as well?	Allowing the obligor to decide whether the obligor agrees with the assignor's choice of assignee unnecessarily restricts the rights of the assignor. An assignor isn't going to go through the trouble of assigning its contract to another party unless it has a particular need to. Perhaps the assignor can no longer fulfill the contract and is assigning the contract in order to prevent breach, or perhaps the obligor owes a debt to the assignee and wishes to pay it back through assignment. In other words, if the obligor could approve or disapprove of an assignee choice, it could infringe upon the interests of the assignor. Allowing the obligor a say in a potential assignee is redundant, considering the obligor is protected by a guarantee that performance of the contract will not be affected by the assignment. Does it really matter that the obligor is receiving a product or service through a new party as long as it's the same product or service stipulated in the contract? The reasons an obligor would have to reject a potential assignee would likely be entirely arbitrary. Finally, being that the assignor is already going through the burden of having to assign the contract to another party, which effectively surrenders any of the benefits they would have received from the contract, it's unfair to add the additional burden of having the obligor approve or disapprove of the assignee, especially if the obligor will only approve assignments that do not benefit the assignor.

Questions & Problems

1. Explain how the requirements of the statute of frauds under the UCC are different from those under the common law.
2. List the kinds of contracts that require a writing under the statute of frauds.
3. Identify the exceptions to the parol evidence rule and explain why some people might argue that the parol rule is not very effective.
4. Integrate the concept of assignments with the concept of delegations.
5. Why is it that incidental beneficiaries cannot enforce rights under a contract? Should they be able to enforce such rights?
6. Explain the difference between an assignor's liability and a delegator's liability after rights have been transferred to a third party.
7. Explain the desirability to business of allowing a transfer of rights or duties to third parties.
8. Plaintiff filed a petition for preliminary injunction asking the court to enjoin the sale of a 1991 Ferrari F40 to Karam Automotive Inc., as well as a complaint seeking specific performance of the alleged contract to sell the Ferrari to plaintiff for $800,000, or in the alternative, damages for defendants's breach of the alleged contract.

 Michael Haber alleged the following: on December 28, 2013, he met with Mr. Moore, who held himself out as the Executor of Mr. Brehaut's Estate and expressed to Mr. Moore his interest in purchasing a Ferrari that was part of the estate. On January 1, 2014, Mr. Moore informed Mr. Haber that the Estate of Mr. Brehaut would accept $800,000 for the Ferrari and said to him, "It's your car." On January 6, 2014, following an inspection by Mr. Haber's mechanic, Mr. Moore shook Mr. Haber's hand and said, "We have a deal." After the oral agreement was reached, Mr. Moore advised Mr. Haber that there was an outstanding loan against the Ferrari, and that he would have to contact the bank to satisfy the loan. Later, when Mr. Haber arrived at his office, he received a phone call from a credit union employee stating that $75,554.30 was required to pay off the loan. On January 10, 2014, in a phone conversation, Mr. Moore told Mr. Haber, "You're going to be pissed at me;" someone had offered him a higher amount for the car, and so he did not agree to sell the car to Mr. Haber.

 In his answer, the defendant denied that an agreement had been reached and pleaded the statute of frauds. Do you think the court applied the statute of frauds in this case to bar the action by the plaintiff? Why or why not?

 Source: *Professional Sales, Inc. v. The Estate Of Joseph S. Brehaut, C. Christopher Moore, Jr., Executor And C. Christopher Moore, Jr. Individually And Joseph S. Brehaut And C. Christopher Moore, Jr. Partnership T/D/B/A Moll's Garage,* 2014 Pa. Dist. & Cnty. Dec. LEXIS 768 (2015).

9. The Laths were the owners of a farm that they wanted to sell. Mrs. Mitchell considered purchasing the land but found that an ice house located across the road was objectionable. Mitchell argued that the Laths orally agreed to remove the ice house in consideration of her promise to purchase the property, which she agreed to purchase for $8,400 in a written contract. After Mitchell moved into her new home and made several improvements to the land, the Laths never removed the ice house and expressly communicated their intention not to do so. Mitchell sued the Laths for breach of contract. What effect does the parol evidence rule have on the admissibility of the oral agreement to remove the ice house? Why this effect?

 Source: *Mitchell v. Lath,* 247 N.Y. 377 (1928).

10. St. John's Holdings (SJH), the buyer, and Two Electronics, the seller, were negotiating the terms of a land sale through email. Over the course of the exchange, SJH sent four unsigned drafts of a letter of intent to Two Electronics. After the final draft, Two Electronics's agent sent SJH's agent a text message stating that Two Electronics would sign the letter if SJH signed first and provided a check. SJH signed the letter of intent and gave a check to its agent to deliver to Two Electronics' agent. After the signing, the SJH agent sent a text to Two Electronics' agent stating, "I have the signed LOI and check it is 424 [PM] where can I meet you?" The two agents spoke over the phone later that day and agreed that SJH's agent would bring the signed letter and check to Two Electronics' agent's office. On the same day of the texts and phone calls, Two Electronics accepted a third party's offer to buy the land. Two Electronics never signed the letter of intent or told SJH that it had sold the land to a third party. SJH sued to enforce the letter of intent arguing that it was binding based on the emails and texts between the agents and that those communications satisfied the essential terms of the statute of frauds. If you were the judge in this case, how would you rule? Can text messages fulfill the writing requirements of the statute of frauds?

 Source: *St. John's Holdings, LLC v. Two Electronics, LLC,* MISC 16-000090, April 14, 2016.

11. Oncology Tech is a manufacturer of cancer therapy devices called compensators. CMS was a company that sold cancer treatment planning software for use

in purchasing cancer treatment devices. In 2006, CMS and Oncology Tech entered into an oral contract to make Oncology Tech's compensators available for purchase using CMS software. The contract had no specified duration; it was agreed that it would be indefinite. Plaintiff filed suit against the defendants for breach of that contract in 2011. Defendants argued that the oral contract was unenforceable against them by operation of the statute of frauds. The statute of frauds requires that any agreement that cannot be completed within one year must be in writing and signed by the person charged with performance. Neither party contends, nor has any evidence been produced indicating that the parties agreed upon a specific duration for the agreement. Do you think the court found that the plaintiff's action was barred by the statute of limitations?

Source: *Oncology Tech, LLC, Plaintiff, v. Elekta, Inc., A Wholly Owned Subsidiary Of Elekta AB; And Impac Medical Systems, Inc.,* 2013 U.S. Dist. LEXIS 189856 (2013).

12. Edward Hosch entered into four loan agreements with Citicapital Commercial Corporation to finance the purchase of heavy construction equipment. Four months later, Citicapital merged into Citicorp Leasing Corporation Inc., which was then renamed GE Capital Commercial Inc. A year later, GE Capital assigned most of the loans to Colonial Pacific Leasing Corporation via a written assignment. Hosch subsequently defaulted on the loans, and Colonial provided a notice of default and demanded payment. Hosch failed to repay the loans, so Colonial sued to collect them, submitting a copy of the written assignment as evidence, but not submitting any evidence that Hosch had been notified of the assignment. The trial court granted summary judgment to Colonial and entered final judgment against Hosch. On appeal, could Horsch argue that the assignment was invalid because he had not been notified of the assignment?

Source: *Hosch v. Colonial Pacific Leasing Corp.,* 113 Ga. App. 873 (2012).

13. Plaintiff entered into a contract with a general contractor to construct a facility. The general contractor subcontracted the roofing installation to Defendant. When the roof began to leak, Plaintiff filed a complaint against the general contractor and Defendant, alleging breach of contract, among other claims. The superior court granted Defendant's motion for summary judgment, holding that Plaintiff was only an incidental beneficiary, as opposed to an intended beneficiary, of the subcontract between Defendant and the general contractor. Plaintiff appealed. How do you think the state supreme court ultimately decided this case? Why?

Source: *Hexagon Holdings, Inc. v. Carlisle Syntec Inc.,* Rhode Island Supreme Court, Docket No. 17-175, January 17, 2019, available at https://law.justia.com/cases/rhode-island/supreme-court/2019/17-175.html?utm_source=summary-newsletters&utm_medium=email&utm_campaign=2019-01-18-contracts-510993353a&utm_content=text-case-title-10.

14. In 2006, Marshall Johnson, then a minor, was injured in a car accident. In 2008, the tortfeasor's insurer, State Farm, and Johnson's guardian settled a personal injury claim on Johnson's behalf through a structured settlement agreement (SSA). The SSA contained an antiassignment clause that prevented Johnson, but not State Farm, from assigning the right to receive payments from the SSA. State Farm went on to assign its rights to MetLife. In 2013, 23-year-old Johnson was short on cash. He contacted J. G. Wentworth, a factoring company, about selling his $29,000 annuity (which was to be doled out in smaller payments until 2020) for $17,250 cash. Wentworth filed a petition to obtain approval to have the SSA rights transferred to Wentworth. MetLife filed objections, citing the antiassignment clause. A hearing on the petition was held and the trial court allowed the transfer due to Johnson's need for the funds. MetLife appealed the trial court's decision. How do you think the appellate court ruled? Does MetLife have a say in the matter?

Source: *Johnson v. J.G. Wentworth Originations, LLC,* Case No. A156843, – P.3d –, 2017 WL 767659, 2017 Ore. App. Lexis 280 (Or. Ct. App. Mar. 1, 2017).

15. Potential expert witnesses made an oral agreement with a plaintiff's attorney to testify on behalf of the plaintiff for $60,000. However, after receiving the money, the witnesses backed out of the agreement and said they would not testify in court or pay back the money. The witnesses claimed that the plaintiff could not sue them because their oral contract was with the attorney and not the plaintiff. Yet, the plaintiff claimed she could sue because she was a third-party beneficiary to the contract. The court made a landmark decision regarding expert witnesses and third-party beneficiary case law. What do you believe the surprising decision was in this case? What rationale supports that outcome?

Source: *Isbell v. Friedman,* United States Court of Appeals, Fourth Circuit. No. 11-2113 (2012).

16. Richard Brumitt, owner of Southway Systems, Inc., agreed to sell his stock in the company to DTSG, Ltd. DTSG's owner, Don Oprea, contacted Tim Duffy, president of First Bank's Small Business Administration loans division, to secure financing for the deal. Oprea urgently expressed to Duffy that he needed to close on the loan by the year's end because Brumitt had pending offers from other interested

buyers. Duffy assured Oprea that the loan could be closed by then. However, things did not go as planned. For over fourteen months, First Bank scheduled and postponed numerous closings. During this time, Duffy continued to promise Oprea that the loan would go through, and Southway continued to sink. Oprea finally sought financing elsewhere, but he never obtained a loan nor acquired Southway. By the time of the trial, Southway had already failed.

Brumitt attempted to sue First Bank as a third-party beneficiary to the agreement between First Bank and DTSG. As evidence of his third-party beneficiary status, Brumitt submitted the three loan commitment letters executed by DTSG and First Bank. First Bank pointed out, however, that the language of the letters was unambiguous and did not clearly express the parties' intent to make Brumitt a third-party beneficiary. Indeed, the letters did not mention Brumitt at all. Does Brumitt have standing to sue as a third-party beneficiary? How do you think the court ruled on the issue?

Source: *First Bank v. Brumitt,* Supreme Court of Texas No. 15-0844 (2017).

Part 2 Contract Law

CHAPTER 14

Discharge and Remedies

CASE OPENER

Impossible Wine Bottles

The Anchor Glass Container Corporation and its parent company, Consumers Packaging, Inc. (CPI), entered into a series of agreements with Encore Glass, Inc., to supply glass containers of a specific type and quality for the wine industry. On June 24, 1999, Encore entered into an amended agreement with Anchor and CPI. In the amended agreement, the parties agreed that the products would be manufactured at CPI's Lavington plant. Additionally, the amended agreement gave Encore a generous rebate schedule ranging from 1 to 2.5 percent and a new discount schedule.

In May 2001, CPI filed for bankruptcy. As a result of the bankruptcy proceedings, the Lavington plant was sold in August 2001. The new owners of the Lavington plant did not assume CPI's obligations under the amended agreement. The Lavington plant could no longer be used to supply the glass containers to Encore. As a result of the sale, Anchor notified Encore on October 12, 2001, that it considered itself relieved of its obligations under the agreement due to its impossibility to perform. Encore took its business to another company, which did not offer the same rebates and discounts as had Anchor.

When Anchor filed for bankruptcy in 2002, Encore filed a claim to recover the $6,102,912.60 it lost when Anchor stopped providing it with rebates and discounts under the contract. The bankruptcy court ruled against Encore, finding that it was impossible for Anchor to perform after the Lavington plant was sold. Encore appealed.

1. Should Anchor be required to honor the contract despite the loss of the Lavington plant? Why or why not?
2. What ethical system, if any, would permit Encore to recover the lost rebates and discounts?

LEARNING OBJECTIVES

After reading this chapter, you will be able to answer the following questions:

LO 14-1 Identify the primary methods of discharging a contract.
LO 14-2 Recognize the primary legal remedies available for a breach of contract.
LO 14-3 Recall the primary equitable remedies available for a breach of contract.

Methods of Discharging a Contract

LO 14-1 Identify the primary methods of discharging a contract.

The previous five chapters focus primarily on how parties enter into a legally binding agreement. Once a party has entered into a binding agreement, how does the party terminate his or her obligation under the contract? That question is the focus of this chapter. When a party's obligations under a contract are terminated, they are said to be discharged. There are a number of ways by which a party's contractual obligations can be terminated, or discharged. The first, and the one most parties hope to secure from the other when they enter into an agreement, is performance. The others are the occurrence of a condition or its failure to occur, material breach by one or both parties, agreement of the parties, or operation of law. The chapter explains each of these methods.

Conditions

Under ordinary circumstances, a party's duty to perform the promise agreed to in a contract is absolute. Sometimes, however, a party's duty to perform may be affected by whether a certain condition occurs. Contracts containing conditions affecting the performance obligations of the parties are called **conditional contracts**. The conditions may be either implied by law or expressly inserted into the contract by the parties.

conditional contract
A contract that becomes enforceable only on the happening or termination of a specified condition.

Discharge by Conditions Precedent, Subsequent, and Concurrent There are three types of conditions: condition precedent, condition subsequent, and concurrent conditions, which are summarized in Exhibit 14-1. A **condition precedent** is a particular event that must occur for a party's duty to arise. If the event does not occur, the party's duty to perform does not arise. Frequently, real estate contracts are conditioned on an event such as the buyer's ability to sell his current home by a certain date. If the home does not sell, the condition does not arise, and the parties have no duty to perform and are thus discharged from the contract.

condition precedent
In a contract, an event that must occur for a party's duty to arise.

Another common example of a contract containing a condition precedent is an insurance contract. If Bill purchases a life insurance contract, he is obligated to pay the monthly premiums specified in the contract, but the insurance company's obligation to perform arises only when he dies. Bill's death is the condition that triggers the company's duty to pay his beneficiary. The key to this concept is that the condition must occur before a duty to perform arises.

A **condition subsequent** is a future event that terminates the obligations of the parties when it occurs. For example, John may enter into an agreement to lease an apartment for five years, conditioned on him not being called to active duty in the National Guard. If he is called to serve, his obligation to be bound by the lease is discharged.

condition subsequent
In a contract, a future event that terminates the obligations of the parties when it occurs.

Exhibit 14-1
Types of conditions

Condition Precedent
An event that must occur or an action that must be taken before a party's duty arises.
A life insurance company is only required to pay out the beneficiaries of a life insurance plan after the contract-holder has died

Condition Subsequent
A future event that terminates a party's duty when it arises.
Non-compete clauses (NCC) between employers and their current or former employees often have expiration dates.

Concurrent Conditions
When each party's duty is conditioned upon the performance of the other, meaning each party's duty must be performed simultaneously.
A buyer's duty to pay for a good is conditioned upon a seller's duty to deliver that good, and vice versa.

Concurrent conditions
In a contract, terms under which each party's performance is conditioned on the performance of the other; occur only when the parties are required to perform for each other simultaneously.

Express condition
A condition specifically and explicitly stated in a contract and usually preceded by words such as *conditioned on, if, provided that,* or *when.*

Implied condition
A condition that is not specifically and explicitly stated but is inferred from the nature and language of the contract.

Concurrent conditions occur when each party's performance is conditioned on the performance of the other. They occur only when the parties are required to perform for each other simultaneously. For example, when a buyer is supposed to pay for goods on delivery, the buyer's duty to pay is implied based on the seller's duty to deliver the goods, and the seller's duty to deliver the goods is implied based on the buyer's duty to pay for the goods. The legal effect of a concurrently conditioned contract is that each party must offer to perform before being able to sue the other for nonperformance.

Express and Implied Conditions Conditions in contracts are also described as being express or implied. **Express conditions** are explicitly stated in the contract and are usually preceded by words such as *conditioned on, if, provided that,* or *when.* For example, in a situation involving a potential sale of a house, the offer expressly required the buyer to make a deposit of $1,000 on acceptance. The buyer wrote "accepted" on the offer and returned it but did not include the deposit. No deposit of money was ever made. The seller then canceled the transaction. Several weeks later, the buyer attempted to tender payment to the seller. The court found that under the terms of the contract, payment of the $1,000 was an express condition of acceptance and, because the acceptance was incomplete, there was no contract.[1]

Implied conditions are those that are not explicitly stated but are inferred from the nature and language of the contract. For example, if a man enters into a contract with a builder to replace the windows in his house, there is an implied condition that the builder will be given access to the home so that she may fulfill her obligations under the contract.

BUT WHAT IF . . .

WHAT IF THE FACTS OF THE CASE OPENER WERE DIFFERENT?

Let's say in the Case Opener that the contract between Walgreen and Sara Creek contained an additional statement that the prohibition against leasing to another pharmacy was to remain in effect only as long as the population of the town did not increase by more than 10,000 residents. What kind of a condition would that be? If the census showed an increase of 9,500, do you think the courts would allow Sara Creek to breach the prohibition and lease the property for a store that would contain a pharmacy?

Sometimes conditions are confused with promises the parties are legally required to perform under the contract, sometimes referred to as covenants. Case 14-1 illustrates how the courts distinguish between conditions that must occur before a party's duties under the contract, and covenants, the breach of which discharges the nonbreaching parties and gives rise to the payment of damages to that party.

[1] *Smith v. Holmwood,* 231 Cal. App. 2d 549 (1965).

MIND & MOTION UTAH INVESTMENTS, LLC, APPELLEE v. CELTIC BANK CORPORATION
Supreme Court of the State of Utah 2015
UT 94 (December 2015)

FACTS: *Mind & Motion entered into a real estate purchase contract (REPC) with Celtic Bank to buy a large piece of the bank's foreclosed property. Although the county had approved plans to construct condominiums on the land, the former owner had not recorded the plats for the first phase of development, so the REPC required Celtic Bank to record the plats by a certain date. However, it also gave Mind & Motion sole discretion to extend the recording deadline*

(continued)

as necessary to allow Celtic Bank enough time to record. It further provided that any extension of the recording deadline automatically extended the deadline to complete the transaction. After extending the recording deadline once, Mind & Motion declined to extend it a second time and sued Celtic Bank for breach of contract.

The district court granted summary judgment in Mind & Motion's favor, concluding that the recording provision was unambiguously a covenant, not a condition. It then awarded Mind & Motion $100,000 in liquidated damages and more than $200,000 in attorney fees, as well as the return of Mind & Motion's $100,000 earnest money deposit. On appeal, Celtic Bank argued that summary judgment was improper because the recording provision is unambiguously a condition.

ISSUE: Was the term in the contract that the Celtic Bank would record the plots of land being sold a condition that would give rise to a duty on the buyer's part, or a covenant of the contract, the breach of which would discharge the buyer and give him the right to sue the bank for breach of contract?

REASONING: Conditions are typically phrased using explicitly conditional terms. In this contract, the parties employed explicitly mandatory language to characterize the recording provision, while using explicitly conditional language elsewhere in the agreement. Therefore, there is no plausible way to read the recording provision as anything other than a covenant.

CONCLUSION: The Court of Appeals upheld the district court's grant of summary judgment in Mind & Motion's favor, concluding that the recording provision was unambiguously a covenant, not a condition.

CRITICAL THINKING

How would one argue that the clause in question is a condition rather than a covenant? What additional facts might have led to a different outcome in this case?

ETHICAL DECISION MAKING

What ethical norms are furthered by this decision? How?

Discharge by Performance

In most situations, parties discharge their obligations by doing what they respectively agreed to do under the terms of the contract; this is called *discharge by performance*. Parties also discharge their duty by making an offer to perform and being ready, willing, and able to perform. This offer of performance is known as a **tender.** If a painter shows up at Sherry's house with his paint and ladders and is ready to start painting the garage, he has tendered performance. If Sherry refuses to let him start, the painter has now discharged his duties under the contract by his tender of performance and he may sue Sherry for material breach (discussed in detail later in this chapter).

Types of Performance There are two primary kinds of performance: complete performance and substantial performance. Performance may also be conditioned on the satisfaction of a party to the contract or of a third party.

Complete performance occurs when every aspect of the parties's duties under the contract is carried out perfectly. In many instances, complete performance is difficult, if not impossible, to attain, and courts today generally require only substantial performance.

Substantial performance occurs when the following conditions have been met: (1) completion of nearly all the terms of the agreement, (2) an honest effort to complete all the terms, and (3) no willful departure from the terms of the agreement. Substantial performance discharges the party's responsibilities under the contract, although the court may require the party to compensate the other party for any loss in value caused by the failure to meet all the standards set forth in the contract. For example, if a contract called for all bedrooms of a house to be painted blue but one was

tender
An offer by a party in a contract to perform, along with being ready, willing, and able to perform, a duty outlined in that contract.

Complete performance
Contract performance that occurs when all aspects of the parties' duties under the contract are carried out perfectly.

Substantial performance
Contract performance that occurs when nearly all the terms of the agreement have been met, there has been an honest effort to complete all the terms, and there has been no willful departure from the terms of the agreement.

inadvertently painted green, the court may require the contractor to compensate the buyer with the amount that it will cost the buyer to have that room repainted. The difficulty of determining whether there has in fact been substantial performance is illustrated in the following Case Nugget.

CASE Nugget

A Substantial or Insubstantial Performance?

Larry Pettit et al. v. Hampton and Beech, Inc., et al. Court of Appeals of Connecticut

101 Conn. App. 502, 922 A.2d 300 (2007)

The plaintiffs contracted for construction of a new house with the defendant. They wanted their home to be a reverse image of the house depicted in the defendant's advertisement; they also modified the plans by widening the kitchen and great room and extending the rear of the house. There were some delays in the construction. At the time of the issuance of the certificate of occupancy, which was one day after the closing date specified in the contract, the plaintiffs gave the defendants a punch card of items to be fixed, placed $20,000 of the purchase price in escrow, refused to set a new closing date, moved into a motel, and sued the defendant, claiming the defendant failed to substantially perform the contract.

The trial court found, pursuant to expert testimony, that reading and comprehending the blueprints provided by the plaintiffs were difficult for the framers and probably led to some of the shortcomings in the construction process. The court found that there were surface cracks in the concrete front porch because expansion joints were not installed and that the front steps, which were built to industry standards, had settled. The court further found that the porch could be repaired by resurfacing it and that the gap between the steps and the porch was a minor repair. The court awarded $3,238 to the plaintiffs to repair both problems.

There were cracks in the foundation, not uncommon after construction, which could be repaired for $700. Rather than following the blueprint requirement, the defendant used an alternate construction that provided better structural support and complied with the building code. The court awarded $105 to attach a joist to an existing one to add additional support, although this might not be necessary structurally. The court found that a beam needed to be installed under the kitchen and awarded $250 for the repair. Joint hangers were missing on the header and stringers of the stairway to the basement. Their omission was not a structural defect, and they easily could be installed at a cost of $75. Bridging and blocking were omitted or not nailed where ducts and pipes passed through joists, another nonstructural defect, and this could be remedied at a cost of $60. Settlement cracks in the walls, a common postconstruction phenomenon, could be repaired for $1,265. The court also awarded the plaintiffs $219.34 to replace a window over the garage.

The plaintiffs appealed on grounds that the proper standard for substantial performance was not applied and thus they were entitled to more damages. In affirming the lower court's judgment that the defendant had substantially performed, the appeals judge said that issuance of a certificate of occupancy may be evidence of substantial performance; it is not dispositive of the question. Other factors include the extent to which the injured party will be deprived of the benefit reasonably expected, the extent to which that party can be adequately compensated for the deficiency of performance, the extent to which the performing party will suffer forfeiture, the likelihood that the performing party will cure his failure in light of the circumstances and his reasonable assurances, and the extent of good faith and fair dealing on the part of the performing party. Applying the factors to the case, the ruling of the trial court judge was not clearly erroneous, and therefore his judgment was affirmed.

Performance Subject to Satisfaction of a Contracting Party Sometimes the performance of the contract is subject to the satisfaction of one of the contracting parties. In such a case, a party is not discharged from the contract until the other party is satisfied. Satisfaction is considered an express condition that must be met before the other party's obligation to pay for the performance arises.

Satisfaction may be judged according to either a subjective or an objective standard. When the judgment involved is a matter of personal taste, such as when a woman is having a dress custom-made for her, the courts apply a subjective satisfaction standard. As long as the person, in good faith, is not satisfied, the other party is deemed not to have met the condition.

If the performance is related to a mechanical or utility standard, the objective satisfaction standard applies. Also, if the contract does not clearly specify that the satisfaction is to be personal, the objective standard applies. When an objective standard is used, the courts ask whether the reasonable person would be satisfied with the performance.

Sometimes a contract is conditioned on the satisfaction of a third party. Usually, such provisions arise in construction contracts specifying that before a buyer accepts a building, an architect must provide a certificate stating that the building was constructed according to the plans and specifications.

Discharge by Material Breach

A *breach* occurs whenever a party fails to perform her obligations under the contract. If the breach is a minor one, it may entitle the nonbreaching party to damages, but it does not discharge the nonbreaching party from the contract.

A **material breach,** however, discharges the nonbreaching party from his obligations under the contract. A material breach occurs when a party unjustifiably fails to perform his obligations substantially under the contract. It is often difficult to know when the court will determine that a breach is material. Case 14-2 demonstrates the analysis a court may use to determine whether a defendant's behavior constitutes material breach.

material breach
A substantial breach of a significant term or terms of a contract that excuses the nonbreaching party from further performance under the contract and gives the nonbreaching party the right to recover damages.

HAMILTON v. STATE FARM FIRE & CASUALTY INSURANCE COMPANY
United States Court of Appeals for the Fifth Circuit
U.S. App LEXIS 8744 (2012)

FACTS: *When the plaintiffs' home was ruined by a hurricane, the couple moved out of the home to a new residence. After they moved out of the house, one of the insureds reported the theft of a storage truck. The plaintiffs provided the defendant insurance company with several documents and submitted an insurance claim. However, the plaintiffs refused to allow representatives from the defendant company to inspect their other homes or give the company other vital documents proving the damage of certain assets. Thus, the defendant company denied the plaintiffs' insurance claim, arguing that the couple materially breached the contract because they did not comply with the terms of the policy. The district court granted summary judgment for the defendant company. The plaintiffs appealed.*

ISSUE: Should a plaintiff be permitted to recover penalties under an insurance policy if they do not comply with the cooperation clause of the insurance policy?

REASONING: Failing to comply with the terms of a cooperation clause constitutes a material breach of said policy, and, under this circumstance, a plaintiff cannot recover penalties under that same insurance policy. Louisiana law provides that an insurance policy is a contract between the parties and should be construed by using the general rules of interpretation of contracts set forth in the Louisiana Civil Code. "If the policy wording at issue is clear and unambiguously expresses the parties' intent, the insurance contract must be enforced as written." Cooperation clauses in insurance contracts "fulfill the reasonable purpose of enabling the insurer to obtain relevant information concerning the loss while the information is fresh." Compliance with insurance policy provisions is a condition precedent to recovery under that policy, which must be fulfilled before an insured may proceed with a lawsuit.

(continued)

In this case, the Hamiltons reported the alleged theft to local law enforcement, submitted their claim to State Farm, and returned the PPIFs to State Farm as requested; the Hamiltons failed to provide most of the supporting documentation of their loss as requested by State Farm, with the exception of a few duplicate receipts. When asked for the additional documentation, the Hamiltons simply provided their sworn statements pertaining to the losses claimed, without providing the requested additional supporting documentation.

The Hamiltons' failure to comply with State Farm's request to examine the separate residences in which they lived, while not expressly required under the policy's cooperation clause, appears from the record to have been the event that prompted State Farm to request the examinations under oath. The Hamiltons, however, failed to respond to State Farm's multiple verbal and written requests for examinations under oath. Their failure to do so was in direct violation of the policy's cooperation clause provision, Section 1–Conditions, (2)(d)(3)(b), and is thus considered a material breach of the contract.

Because the Hamiltons materially breached the terms of the policy by failing to comply with the terms of the cooperation clause, they were precluded from recovering under the policy. Furthermore, considering that State Farm's denial of the claim was due to the Hamiltons' material breach of the policy, the Hamiltons are also precluded from recovering penalties and attorney fees.

DECISION AND REMEDY: The Court of Appeals ruled in favor of the defendant and affirmed the decision of the district court.

SIGNIFICANCE OF THE CASE: This case provides an example of when a plaintiff cannot recover penalties from an insurance company.

CRITICAL THINKING

What kind of evidence would have persuaded State Farm that the Hamiltons had a legitimate insurance claim? What is the reasoning of the court in supporting State Farm's expectations under the cooperation clause in the policy?

ETHICAL DECISION MAKING

How would the universalization principal have aided the Hamiltons in understanding the logic being used by Stae Farm?

BUT WHAT IF . . .

WHAT IF THE FACTS OF THE CASE OPENER WERE DIFFERENT?

Recall, in the Case Opener, that CPI manufactured Encore's products in its factory. Let's say that CPI simply stopped allowing the products in the factory. Would Encore be able to be discharged from the contract, or would Encore have to stay in the contract and find some sort of resolution?

Anticipatory Repudiation Sometimes, a contracting party may decide not to complete the contract before the actual time of performance. This situation often arises when market conditions change and one party realizes that it will not be profitable to carry out the terms of the contract. The breaching party may convey the anticipatory breach to the nonbreaching party either by making an express indication of her intent not to perform or by taking an action that would be inconsistent with her ability to carry out the contract when performance was due.

Once the contract has been anticipatorily repudiated, the nonbreaching party is discharged from his obligations under the contract. He is free to go ahead and sue for breach and to find similar contract elsewhere. However, if the nonbreaching party wishes, he may decide to give the party who repudiated the opportunity to change her mind and still perform.

Discharge by Mutual Agreement

Sometimes the parties to a contract agree to discharge each other from their obligations. They may do so through four primary means: discharge by mutual rescission, discharge by a substituted contract, discharge by accord and satisfaction, or discharge by novation.

Mutual Rescission Parties may agree that they simply wish to discharge each other from their mutual obligations and therefore may rescind or cancel the contract. For example, if James agreed to cater a graduation reception for Bill's son but it appears that the child is not going to graduate when planned, James could agree not to hold Bill responsible for paying him the agreed-on cost for the catering in exchange for Bill's agreement not to expect James to cater the reception.

Substituted Contract Sometimes, instead of canceling the contract and terminating their relationship, the parties want to substitute a new agreement in place of the original. The substituted contract immediately discharges the parties from their obligations under the old contract and replaces those obligations with the new obligations imposed by the substituted contract.

Accord and Satisfaction An accord and satisfaction is used when one of the parties wants to substitute a different performance for his original duty under the contract. The promise to perform the new duty is called the *accord,* and the actual performance of that new duty is called the *satisfaction.* The party's duty under the contract is not discharged until the new duty is actually performed. Thus, it is the satisfaction that discharges the party.

Novation Sometimes the parties to the agreement want to replace one of the parties with a third party. This substitution of a party is called a **novation**. The original duties remain the same under the contract, but one party is discharged and the third party now takes that original party's place. All three parties must agree to the novation for it to be valid.

novation
In a contract, the substitution of a third party for one of the original parties; the duties remain the same under the contract, but one original party is discharged and the third party takes that original party's place.

Discharge by Operation of Law

Sometimes a contract may be discharged not by anything the parties do but, rather, by operation of law. Alteration of the contract, bankruptcy, tolling of the statute of limitations, impossibility, commercial impracticability, and frustration of purpose are all situations in which a contract may be discharged by operation of law.

Alteration of the Contract The courts want to uphold the sanctity of contracts. Therefore, if one of the parties materially alters a written contract without the knowledge of the other party, the courts have held that such alteration allows the innocent party to be discharged from the contract. For example, if a seller, without knowledge of the buyer, changes the price of the contract, the buyer can treat the contract as terminated.

Bankruptcy When a party files bankruptcy, the court allocates the assets of the bankrupt among the bankrupt's creditors and then issues the party a discharge in bankruptcy. Once the assets have been distributed, all the bankrupt's debts are discharged. (Bankruptcy is discussed in detail in Chapter 18.)

Tolling of the Statute of Limitations The tolling of the statute of limitations, that is, the expiration of the time allowed by the statute, does not technically discharge a party's obligations under the contract. However, once the statute of limitations has tolled, neither party can any longer sue the other for breach, so for all practical purposes, the parties are no longer bound to perform.

Impossibility of Performance Sometimes an unforeseen event occurs that makes it physically or legally impossible for a party to carry out the terms of the contract. In such a situation, the parties will be discharged on grounds of impossibility of performance. Courts distinguish between *objective impossibility,* meaning it is in fact not possible to lawfully carry out one's contractual obligations, and *subjective impossibility,* meaning it would be very difficult to carry out the contract. Objective impossibility but not subjective impossibility discharges the parties's obligations under the contract.

For example, if farmer Gray has a contract with the Hunts Corporation to provide it with 100 bushels of tomatoes on August 30 and a flood wipes out Gray's crop, it is not physically impossible for him to comply with the agreement. He has to go out on the market and purchase 100

bushels of tomatoes to ship to the Hunts Corporation. It may be inconvenient, and perhaps subjectively impossible, but it is not objectively impossible.

In contrast, suppose farmer Jones owns a historic farmhouse built in 1827 and he agrees to sell it to Smith, but the night before the parties are to exchange money for the title, lightning strikes the farmhouse and it burns to the ground. It is now objectively impossible to comply with the terms of the contract, so the parties are discharged from their obligations. The historic farmhouse is not like tomatoes; the subject matter of the contract is forever destroyed and cannot be recreated.

There are three main situations in which the courts find objective impossibility. The first is *destruction of the subject matter*, as in the example of the historic farmhouse destroyed by fire. If we go back to the example of the tomatoes, note that we said the farmer still had to perform because it was still possible for him to obtain tomatoes elsewhere. To protect himself in the event that his crop was destroyed, farmer Gray could have drafted the contract to identify the subject matter as 100 bushels of tomatoes grown on the Smith family farm. In that case, if Smith's fields were flooded, it would be objectively impossible to comply with the contract because no tomatoes from the Smith farm exist.

The second situation of objective impossibility is the *death or incapacity of a party whose personal services are necessary* to fulfill the terms of the contract. For example, if a famous artist is commissioned to paint a portrait and the artist dies, the contract is discharged. The artist's style is unique, and there is no way for anyone to take over the artist's role.

The third situation is *subsequent illegality*. If the law changes after the contract was made, rendering the performance of the contract illegal, the contract is discharged. For example, Bill orders a case of a nutritional supplement from Osco Drugs and Supplements. Before his order can be filled, the nutritional supplement is banned because of recently discovered harmful side effects. The parties are now discharged from their duties because to sell the banned substance would be to violate the law.

The opening scenario provides another example of impossibility of performance. The parties in the opening scenario do not dispute that the contract provided that "[t]he parties contemplate that the products (as hereinafter defined) shall be manufactured at CPI's Lavington facility" (the Lavington plant). When the Lavington plant was no longer available for production, it became impossible for Anchor to fulfill the terms of the contract. Ultimately, Anchor informed Encore that it considered itself discharged from the contract.

Commercial Impracticability Commercial impracticability is seen by some as a response to a somewhat unfair harshness of the objective impossibility standard. Commercial impracticability is used when performance is still objectively possible but would be extraordinarily injurious or expensive to one party. Commercial impracticability arises when, because of an unforeseeable event, one party would incur unreasonable expense, injury, or loss if that party were forced to carry out the terms of the agreement.

According to the Restatement (Second) of Contracts, Section 261 (1981), discharge by reason of impracticability requires the party claiming discharge to prove the following three elements:

1. That an event occurred whose nonoccurrence was a basic assumption of the contract.
2. That continued performance is not commercially practicable.
3. That the party claiming discharge did not expressly or impliedly agree to performance in spite of impracticability that would otherwise justify his nonperformance.

It is sometimes difficult to know whether the potential harm to the party seeking to avoid the contract is sufficient to give rise to the use of commercial impracticability. The doctrine is most commonly used when raw materials needed for manufacturing goods under the contract become extraordinarily expensive or difficult to obtain because of an embargo, war, crop failure, or unexpected closure of a plant. Case 14-3 illustrates how the courts sometimes struggle to determine whether to apply the doctrine of commercial impracticability to discharge a contract.

CASE 14-3

THRIFTY RENT-A-CAR SYSTEM v. SOUTH FLORIDA TRANSPORT
U.S. District Court for the Northern District of Oklahoma
2005 U.S. Dist. LEXIS 38489

FACTS: *The plaintiffs, Thrifty Rent-A-Car System and its affiliates DTG and Rental Car Finance Corp., allowed South Florida Transport (SFT), the defendant, to establish a Thrifty franchise. In 2003, they entered into four agreements, providing SFT the right to use Thrifty's trademark and business methods in exchange for payment to Thrifty of licensing and administrative fees and requiring SFT to maintain a fleet of automobiles for rent.*

In July 2004, SFT provided DTG with a check as payment, but it was returned for insufficient funds. SFT continued to make delinquent payments, and by August, SFT owed Thrifty and DTG $1,134,819.40. Due to SFT's failure to make payments, Thrifty and DTG informed SFT that they would terminate the licensing agreements and repossess the vehicles, to which DTG had legal title. However, DTG agreed to postpone repossession due to predictions of severe weather and allowed SFT to continue renting vehicles until repossession was completed.

When DTG repossessed the vehicles, DTG noticed that numerous cars were missing. SFT notified DTG that it had sold 51 vehicles without authorization. By August 2005, SFT owed Thrifty and DTG $4,238,249.53. SFT claimed that several hurricanes rendered its business operations commercially impracticable. The plaintiffs filed a motion for summary judgment, seeking full reimbursement for the debts SFT owed.

ISSUE: Under what circumstances will commercial impracticability excuse nonperformance of a contract?

REASONING: In the words of the court:

> The doctrine of commercial impracticability is typically invoked in cases involving the sale of goods.
> Codified in section 2-615 of the Uniform Commercial Code (UCC), which has been adopted by the Oklahoma legislature, the doctrine of commercial impracticability provides a defense to a seller for a delay in delivery or nondelivery of promised goods if performance has been made impracticable by a contingency, the nonoccurrence of which is an assumption of the contract.
> Commercial impracticability may excuse a party from performance of his obligations under a contract where performance has become commercially impracticable because of unforeseen supervening circumstances not within the contemplation of the parties at the time of contracting. UCC commentary provides that a party pleading commercial impracticability must demonstrate the "basic assumption" prong of the test also found in the impracticability of performance context, that is, that the nonoccurrence of the supervening event was a basic assumption of the parties at the time of contracting. A rise or a collapse in the market standing alone does not constitute a justification for failure to perform. A contract is deemed commercially impracticable when, due to unforeseen events, performance may only be obtained at "an excessive and unreasonable cost . . . or when all means of performance are commercially senseless." In applying the doctrine of commercial impracticability, the crucial question is "whether the cost of performance has in fact become so excessive and unreasonable that failure to excuse performance would result in grave injustice."
> . . . The evidence strongly suggests that the non-occurrence of hurricanes was not a basic assumption of the parties' agreements. Moreover, defendant provides no evidence to support a suggestion that the event of the hurricanes made the cost of performance of the terms of the agreements unduly burdensome, or even remotely more expensive. Finally, the Court observes, again, that SFT was behind on its payments to Thrifty and DTG before the arrival of the hurricanes in August 2004. No genuine issue of material fact exists, and the Court holds that the defense of commercial impracticability is unavailable to defendant.

DECISION AND REMEDY: The plaintiffs' motion for summary judgment was granted.

SIGNIFICANCE OF THE CASE: This case illustrates how strictly courts look at the circumstances to determine whether to apply the doctrine of commercial impracticability.

CRITICAL THINKING

Especially given the increased rates of severe weather and natural disasters witnessed recently around the world, to what extent can parties entering into contracts reasonably be expected to plan for the effects of a rapidly changing global climate?

ETHICAL DECISION MAKING

What values do you think are furthered by the decision in this case? What values are being subordinated by this decision?

BUT WHAT IF...

WHAT IF THE FACTS OF THE CASE OPENER WERE DIFFERENT?

Let's say that CPI left its factory to Anchor so that Anchor could still manufacture products for Encore as stated in the contract. However, manufacturing the products as CPI did would be outrageously expensive for Anchor. Does Anchor still have to remain in the contract with Encore?

Frustration of Purpose Closely related to impracticability is frustration of purpose. Sometimes, when a contract is entered into, both parties recognize that the contract is intended to fulfill a particular purpose, and the occurrence of that purpose is said to be a basic assumption on which the contract is made. If, due to factors beyond the control of the parties, the event does not occur, and neither party had assumed the risk of the event's nonoccurrence, the contract may be discharged.

This doctrine arose from the so-called coronation cases in England. Numerous parties had contracted for rooms along the parade route for the king's coronation, but the king became ill and the coronation was canceled. The courts held that the parties's duties under the room contracts should be discharged and that any payments made in advance should be returned because the essential purposes of the contracts could no longer be fulfilled, through no fault of any of the parties.

Again, this doctrine is not frequently used. For example, if you contract for an organist to play at your daughter's wedding but the groom gets cold feet at the last moment and the wedding is canceled, you cannot use frustration of purpose to discharge the contract because the groom's reversal was a foreseeable event, even though it was unlikely. The following Case Nugget illustrates how the courts sometimes struggle with whether to apply the frustration of purpose document.

CASE Nugget

An Unforeseeable Event?

Liggett Restaurant Group, Inc. v. City of Pontiac Michigan Court of Appeals

260 Mich. App. 127, 676 N.W.2d 633 (Mich. App. 2003)

Elias Brothers Restaurants Inc. had a contract with the defendant, City of Pontiac Stadium Building Authority, to provide concessions at the Silverdome until 2000. The parties renegotiated the contract in 1990, and Elias Brothers agreed to pay additional consideration for the option to extend the contract until 2005 to coordinate with the end of the Detroit Lions's sublease. The additional consideration involved paying the city a higher percentage on profits from sales. This option was exercised on December 1, 1998, and the Detroit Lions prematurely discontinued playing in the Silverdome after the 2001 football season.

The plaintiff sought to use the frustration-of-purpose doctrine to discharge its obligations under the contract extension and therefore have returned to it the additional consideration it had paid under the extension. The plaintiff argued that the contract was made on the assumption that the Lions would play in the Silverdome until their lease ran out and, thus, that their early departure frustrated the purpose of the extension.

The court said the doctrine was inapplicable in this case. The court first set forth the conditions under which the doctrine applied: (1) The contract must be at least partially executory; (2) the frustrated party's purpose in making the contract must have been

(continued)

known to both parties when the contract was made; and (3) this purpose must have been basically frustrated by an event not reasonably foreseeable at the time the contract was made, the occurrence of which was not due to the fault of the frustrated party and the risk of which was not assumed by him.

Then the court noted that the situation clearly did not meet the third criterion. Far from being an unforeseen event, the Lions leaving prematurely was expressly addressed in the original contract by a paragraph specifying a reduction in the guaranteed minimum annual payment for each year in which the Lions did not play a minimum of eight games in the stadium.

Discharge by conditions	If precedent, concurrent, implied, and express conditions are not met or subsequent condition occurs.	**Exhibit 14-2** Methods of Discharging a Contract
Discharge by performance	If a party performs the terms of the contract or makes a tender (an offer to perform), or if the party performs to the satisfaction of the contracting party.	
Discharge by material breach	If a party fails to substantially perform his obligations, thereby justifying that the nonbreaching party be discharged from the contract.	
Discharge by mutual agreement	If the parties mutually agree to discharge one another, substitute a new contract, substitute a party, or substitute a different performance.	
Discharge by operation of the law	If one of the following occurs: alteration of the contract, bankruptcy, tolling of the statute of limitations, impossibility, commercial impracticability, or frustration of purpose.	

Remedies

LO 14-2 Recognize the primary legal remedies available for a breach of contract.

The fact that one party has breached a contract does not necessarily mean that the nonbreaching party will sue. A number of factors go into the decision of whether it makes sense to file suit. Some of those considerations are (1) the likelihood of success, (2) the desire or need to maintain an ongoing relationship with the potential defendant, (3) the possibility of getting a better or faster resolution through some form of alternative dispute resolution, and (4) the cost of litigation or some form of alternative dispute resolution (**ADR**) compared to the value of the likely remedy.

The remedies the potential plaintiff will be thinking about can generally be classified as either *legal remedies* (also known as *money damages*) or *equitable remedies,* some form of court-ordered action. The distinction between legal and equitable remedies can be traced back to a time in our legal system's English roots when, instead of one unitary legal system, there were two separate courts, a court of law and a court of equity. When parties were seeking money damages, they went to the court of law; but when parties needed any remedy other than money damages, they went to the High Court of Chancery, which was a court of equity. When the United States was establishing its legal system, it combined both these types of powers in a unitary system. The reasons for this joinder are not known, but it seems likely that the primary reason was that the early colonists simply did not have the resources to support separate systems. The courts did, however, still maintain the distinction between legal and equitable remedies. However, unlike judges in the old English courts, judges in the U.S. system have the power to award both legal and equitable remedies in the same case. This section discusses these various remedies.

©Sunflowerr/Shutterstock

GLOBAL Context

Damages for Breach of Contract: Not the Preferred Remedy in Latvia

In the United States, equitable remedies are rarely given; legal remedies are awarded whenever possible. Consequently, some parties will opt to voluntarily breach a contract and pay damages rather than fulfill a contract because doing so may prove to be more cost-effective. In Latvia, however, each party is expected and legally obligated to fulfill its side of the bargain. Damages for breach are an option only in the most extreme cases where fulfilling the contract would prove impossible. Effectively, parties are barred from backing out of a contract, as doing so is not only illegal but considered highly immoral. Thus, specific performance is often the only remedy available in Latvian contract disputes.

Legal Remedies (Monetary Damages)

Monetary damages are also referred to as *legal damages,* and they include compensatory, punitive, nominal, and liquidated damages. Whenever possible, courts award monetary damages rather than some form of equitable relief.

compensatory damages
Money awarded to a plaintiff as reimbursement for her or his losses; based on the amount of actual damage or harm to property, lost wages or profits, pain and suffering, medical expenses, disability, and so on.

Compensatory Damages The most frequently awarded damages are **compensatory damages,** damages designed to put the plaintiff in the position he would have been in had the contract been fully performed. These damages are said to compensate the plaintiff for his loss of the benefit of the bargain. He can recover, however, only for those provable losses that were foreseeable at the time the contract was entered into. Sometimes, the plaintiff actually may have no losses. Suppose, for example, that Dr. Wilcox hires Jeremy to work exclusively as his research assistant during the fall semester, for a salary of $2,000 per month. If Wilcox breaches the contract and terminates Jeremy for no reason with two months left on the contract, and the only job Jeremy can get as a substitute pays only $500 per month, Jeremy is entitled to compensatory damages of $3,000. However, if Jeremy gets a new job that pays $2,500 per month, he is actually better off now, so he has no compensatory damages. Sometimes these damages are referred to as *expectation damages* because they compensate a person for the benefit she or he expected to gain as a result of entering into the contract.

In addition to losing the benefit of the bargain, the plaintiff may suffer other losses caused directly by the breach. These losses may be compensated as *incidental damages.* For example, because Jeremy was unfairly terminated before his contractual term was over, he may have to spend money to find another job. His job search expenditures would be considered incidental damages.

Some kinds of contracts have special rules for determining their compensatory damages, namely, contracts for the sale of goods, contracts for the sale of land, and construction contracts. Each of these is discussed in a little more detail in the following sections.

Contracts for the sale of goods are governed today by the Uniform Commercial Code. If the seller breaches the contract, compensatory damages are generally calculated as the difference between the contract price and the market price on the day the goods were supposed to be delivered, plus any incidental damages resulting from the breach.[2] In other words, this measure of damages is the difference between what the buyer would have paid for the goods under the contract and what he or she now must pay to obtain the goods from another seller. Occasionally, however, the buyer may have no damages because the market price of the goods is lower than the parties had anticipated it would be at the date of delivery, so the buyer can now actually purchase the goods at a lower price than the contract price.

If the buyer breaches before accepting the goods, the seller would be able to resell the goods and recover as compensatory damages the difference between the price he sold the goods for and the contract price, plus any incidental expenses associated with the sale.[3] If the seller is unable to

[2] UCC §§ 2-708 and 2-713.
[3] UCC §§ 2-706 and 2-710.

sell the goods to another buyer, as might be the case, for example, with shirts embroidered with a company's monogram, then the seller may be entitled to the contract price as damages. If the buyer breaches before the goods are even manufactured, the seller's damages would typically be based on the profits that would have been made from the sale.

In construction contracts, contracts whereby an owner enters into an agreement to have a building constructed, damages are calculated differently depending on who the breaching party is and what stage the construction is in when the breach occurs. If the contract is breached by the owner before the construction is begun, damages are simply lost profits, which are calculated by subtracting the projected costs of construction from the contract price. For example, if Cameron Construction Company anticipates building a warehouse for the Johnson Corporation and the contract price is $500,000 and the cost of raw materials and labor is $420,000, Cameron could recover $80,000 in lost profits if Johnson were to breach the contract before performance had begun.

If, however, Cameron Construction had already expended $20,000 on materials and labor on the job when the breach occurred, the company would be able to recover $100,000 in damages because the amount of damages when construction is in progress is measured by the lost profits plus any money already invested in the project. And if the breach by Johnson had occurred after construction was completed, then Cameron would be entitled to recover the entire contract price, plus interest from the time payment for the project was due.

If the construction company, or contractor, breaches the contract before or during the construction, the owner's damages are generally measured by the cost of hiring another company to complete the project, plus any incidental costs associated with obtaining a new contractor and any costs arising from delays in the construction project. If the contractor completes the job but finishes after the date for completion, the owner is entitled to damages for the loss of the use of the building that she would have had if the contract had been completed in a timely manner.

Consequential Damages It should be apparent by now that contract law requires greater certainty in the proof of damages than does tort law. Damages are not recoverable for breach of contract unless they can be proved with a high degree of certainty. One type of damages in contract cases that is often especially difficult to prove is what are called **consequential** or **special damages**. Consequential damages are foreseeable damages that result from special facts and circumstances arising outside the contract itself. These damages must be within the contemplation of the parties at the time the breach occurs.

In the Case Nugget that follows, the court distinguishes consequential damages from the damages that arise naturally from a breach of contract.

Punitive Damages Just as in tort law, **punitive damages** in contract law are designed to punish the defendant and deter him and others from engaging in similar behavior in the future. Because the primary objective of contract law, however, is to ensure that parties' expectations are met, punitive, or exemplary, damages are rarely awarded. Most jurisdictions award them only when the defendant has engaged in reprehensible conduct such as fraud. The primary factor in determining the amount of punitive damages is how much is necessary to punish the defendant; thus, the amount depends on matters such as the wealth and income of the defendant.

Nominal Damages If no actual damages resulted from the breach of contract, the court may award the plaintiff **nominal damages**. The award is typically for $1 or $5, but it serves to signify that the plaintiff has been wronged by the defendant.

Liquidated Damages Typically, the court determines the amount of damages to which a nonbreaching party is entitled. Sometimes, however, the parties recognize that if there is a breach of contract, it will probably be somewhat difficult for the court to determine exactly what the damages are. To prevent a difficult court battle, the parties specify in advance what the **liquidated damages** will be if there is a particular kind of breach. The parties specify these damages in what is called a *liquidated-* or *stipulated-damage clause* in the contract. The damages may be specified as either a fixed amount or a formula for determining how much money is due. Such clauses are

consequential damages
In a contract, foreseeable damages that result from special facts and circumstances arising outside the contract itself. The damages must be within the contemplation of the parties at the time the breach occurs. Also called *special damages*.

punitive damages
Compensation awarded to a plaintiff that goes beyond reimbursement for actual losses and is imposed to punish the defendant and deter such conduct in the future.

nominal damages
Monetary damages awarded to a plaintiff in a very small amount, typically $1 to $5, to signify that the plaintiff has been wronged by the defendant even though the plaintiff suffered no compensable harm.

liquidated damages
Damages specified as a term of the contract, before a breach of contract occurs.

frequently used in construction contracts when the buyer needs to know that the property will be available by a specific date so that she can make her plans for moving in. In such a case, the parties may estimate in advance what it will cost the buyer for storage and temporary housing if the property is not ready by the specified date. The courts generally enforce such clauses as long as they appear to bear a reasonable relationship to what the actual costs will be. If the amount specified is so unreasonable as not to bear any logical relationship to foreseeable costs, the courts declare the clause a penalty clause and do not enforce it. In Case 14-4, the court had to decide whether the clause at issue was an unenforceable penalty clause or a valid liquidated-damage clause.

CASE Nugget

A Classic Case of Consequential Damages

Hadley v. Baxendale Court of Exchequer

156 Eng. Rep. 145 (1854)

The plaintiffs were millers whose mill was stopped when the crankshaft of the mill broke. They sent the shaft to Greenwich to be used as a model for a new crank. The plaintiffs's servant took the shaft to the defendant, a common carrier, and told the defendant's clerk that the mill was stopped and that the shaft must be sent immediately. At the time, however, most mills would have had extra crankshafts around somewhere, so without more being said, the clerk would not have known that the mill would not be soon back and running on a spare shaft. The clerk said the shaft would be delivered at Greenwich on the following day. The defendant's clerk was told that a special entry, if required, should be made to hasten the shaft's delivery. The delivery of the shaft at Greenwich was delayed by some neglect, and, consequently, the plaintiffs did not receive the new shaft for several days after they would otherwise have received it. During that time, the mill was shut down, and the plaintiffs thereby lost the profits they would have received had the shaft been delivered on time. They sought to recover damages for lost profits during that time. The defendant argued that the lost profits were "too remote." The court decided for the plaintiffs and allowed the jury to consider the lost profits in awarding damages. The defendant appealed.

On appeal, the appellate judge set forth what has come to be the standard for consequential damages:

> Where two parties have made a contract which one of them has broken, the damages which the other party ought to receive in respect of such breach of contract should be such as may fairly and reasonably be considered either arising naturally, that is, according to the usual course of things, from such breach of contract itself, or such as may reasonably be supposed to have been in the contemplation of both parties, at the time they made the contract, as the probable result of the breach of it. Now, if the special circumstances under which the contract was actually made were communicated by the plaintiffs to the defendants, and thus known to both parties, the damages resulting from the breach of such a contract, which they would reasonably contemplate, would be the amount of injury which would ordinarily follow from a breach of contract under these special circumstances so known and communicated. But, on the other hand, if these special circumstances were wholly unknown to the party breaking the contract, he, at the most, could only be supposed to have had in his contemplation the amount of injury which would arise generally, and in the great multitude of cases not affected by any special circumstances, from such a breach of contract. For, had the special circumstances been known, the parties might have specially provided for the breach of contract by special terms as to the damages in that case; and of this advantage it would be very unjust to deprive them.

In this case, the judge did not believe that the special circumstances had been communicated to the defendant. Thus it would have been unjust to hold the defendant liable for the damages, which he would not have contemplated as arising from the delayed delivery of a shaft.

CASE 14-4

S. BROOKE PURLL, INC., T/A PURLL CONSTRUCTION v. PATRICK DARRELL

Vailes District of Columbia Court of Appeals

850 A.2d 1135 (2004)

FACTS: *The plaintiff hired the defendant to renovate his house and paid him a $5,000 initial payment under the construction contract but failed to do initial demolition work required under the contract. The contract contained a clause stating:*

> The [owner] further agrees that if he shall cancel this Contract for any reason, whatsoever, then he shall pay to the contractor as fixed and liquidated damages, without proof of loss, the sum of money equal to Thirty Five Percent (35%) of the full contract price hereinabove stated. Furthermore, if it becomes necessary for the Contractor to file suit or take other legal action on this Contract because of a breach on the part of the party of the second part, the party of the second part agrees to pay reasonable attorney fees and court cost.

The plaintiff sued the defendant contractor for return of his deposit, and the defendant counterclaimed for $36,102 and attorney fees, the amount that would be required under the liquidated-damage clause. The trial court held that the plaintiff had breached the contract, but it found the liquidated-damage clause to be an unenforceable penalty clause and awarded the contractor $1,722. The defendant appealed.

ISSUE: Was the clause imposing a payment of 35 percent of the contract value for breach by the owner a penalty clause or a valid liquidated-damage clause?

REASONING: Today, the trend is toward furthering freedom of contract through the enforcement of stipulated-damage provisions as long as they do not clearly disregard the principle of compensation. Such clauses simplify the resolution of a breach-of-contract dispute and give the parties an opportunity to resolve the damage question without resorting to litigation, by setting the measure of damages at the outset, before a breach even occurs. Such provisions are especially appropriate when the parties enter into a contract in which the damages are uncertain in amount and cannot be easily ascertained. Thus, the courts tend to uphold these clauses unless they are clearly unreasonable. The liquidated sum must simply bear a reasonable relation to the damages foreseeable at the time the contract was made. The burden of proof is on the breaching party to demonstrate that the damage clause is disproportionate to the foreseeable damages.

In this case, the plaintiff provided no such evidence. The full contract price was $103,148.71. The contractor testified without contradiction that two-thirds of the contract price represented materials and labor and one-third represented his profit. The contractor explained that he consulted the Ames Construction Guide for the D.C. metropolitan area, which was customarily used in the industry for pricing construction jobs, in establishing his price. Thus, there was no evidence that the clause was unreasonable. The trial court had inappropriately struck the clause down because the contractor had failed to introduce evidence of his out-of-pocket expenses; such proof is not required to uphold a liquidated-damage clause.

DECISION AND REMEDY: The lower court's decision was reversed, in favor of the defendant contractor.

SIGNIFICANCE OF THE CASE: This case illustrates the preference for upholding liquidated-damage clauses.

CRITICAL THINKING

What evidence could the plaintiff have presented that might have changed the outcome of this case?

ETHICAL DECISION MAKING

The courts place the burden of proving the unreasonableness of a liquidated-damage clause on the breaching party. Which stakeholders benefit by this approach? Which stakeholders are disadvantaged by it?

Mitigation of Damages When a contract has been breached, the nonbreaching party is often angry at the breaching party and may want to make the breaching party pay through the nose. However, the courts do not allow a nonbreaching party to increase his damages intentionally. In fact, to recover damages in a breach-of-contract case, the plaintiff must demonstrate that he used reasonable efforts to minimize the damage resulting from the breach. This obligation is referred to as the *duty to mitigate one's damages.*

GLOBAL Context

©Digital Archive Japan/ Alamy

Liquidated Damages in Japan

Under Japanese contract law, there are basically two types of contract damages; those that would ordinarily arise from the breach and those that arise because of some special circumstances that should have been reasonably foreseeable to the parties at the time of contract. In regard to the latter type, Japanese law provides for liquidated damages, similar to those available in the United States.

Under Article 420 of the Japanese Code, parties to a contract may specify in advance the amount of compensation payable when contractual obligation is not performed at all or not performed by a specified date. However, unlike in the United States, a party can raise a defense that he did not intentionally or negligently breach the contract. Using a form of comparative negligence, the court may reduce the amount of the liquidated damages to reflect the degree of negligence of the complainant. Japanese courts have also, on rare occasions, eliminated or reduced liquidated damages when they believed the amount violated public policy or constituted a windfall profit.

Thus, if you are the manager of a hotel, and a person who had booked 10 rooms for the week calls to cancel all the reservations, you have a duty to attempt to rent the rooms to minimize the damages. The mitigation must be reasonable, however, and no one is expected to settle for less than what was contemplated under the contract in order to mitigate the damages.

One area in which interesting mitigation issues arise is when an employee is wrongfully discharged and must seek new employment to mitigate her damages. If the employee does not seek alternative employment, the amount of lost wages recovered as damages will be reduced by the amount the employee reasonably could have earned in another job. If the employee does not find another job, the court must decide whether the employee could have found comparable alternative employment with reasonable effort. For example, in 1970, the California Supreme Court considered a classic case that involved an actress who was contracted to star in a musical motion picture that was to be filmed in the United States but was canceled prior to production. The court held that the requirement to mitigate damages did not require the actress to accept as an alternative a starring role in a western drama being filmed in Australia.[4]

Equitable Remedies

LO 14-3 Recall the primary equitable remedies available for a breach of contract.

As noted earlier, equitable remedies grew out of the English court's authority to fashion remedies when the existing laws did not provide adequate ones. These remedies were typically unique solutions specifically crafted to the demands of the situation. Today, the most common equitable remedies include rescission and restitution, orders for specific performance, and injunctions.

As a carryover from the days of the English courts of law and equity, a party seeking equitable relief must meet five requirements. The party must prove that (1) no adequate legal remedy is available; (2) irreparable harm to the plaintiff may result if the equitable remedy is not granted; (3) the contract is legally valid (except when seeking relief in quasi-contract); (4) the contract terms are clear and unambiguous; and (5) the plaintiff has "clean hands," that is, has not been deceitful or done anything in breach of the contract.

rescission
The termination of a contract.

restitution
The return of any property given up under a contract.

Rescission and Restitution Sometimes the parties simply want to be returned to their pre-contract status; they want to have the contract terminated and to have any transferred property returned to its original owner. That is, they want rescission and restitution. **Rescission** is the termination of the contract, and **restitution** is the return of any property given up under the contract.

[4] *Parker v. Twentieth Century-Fox Film Corp.*, 3 Cal. 3d 176, 474 P.2d 689 (1970).

Restitution and rescission are most frequently awarded in situations in which there is a lack of genuine assent (discussed in Chapter 12). When a party enters into a contract because of fraud, duress, undue influence, or a bilateral mistake, the contract is voidable and the party who wants out may seek to avoid the contract or, in other words, seek rescission and restitution.

Specific Performance Specific performance is sometimes called *specific enforcement*. It is an order requiring the breaching party to fulfill the terms of the agreement. Courts are very reluctant to grant specific performance and will do so only when monetary damages simply are not adequate, typically because the subject matter of the contract is unique. If the subject matter of the contract is unique, then even if the nonbreaching party is given compensation, he cannot go elsewhere to buy the object from someone else, so this renders any kind of money damages inadequate. Primarily for historical reasons, every piece of real property is considered unique. Therefore, an order for specific performance would often be the appropriate remedy for the breach of a contract for the sale of a piece of real estate.

Under the UCC, orders for specific performance of a contract for the sale of goods may be awarded "where the goods are unique or in other proper circumstances."[5] Although the other circumstances have not been delineated by the UCC, the case law reveals that a commonly accepted circumstance is one in which the good is scarce or is no longer being produced and therefore would be difficult to obtain on the open market.

specific performance
An order of the court requiring a breaching party to fulfill the terms of the contract.

Injunction An **injunction** is an order either forcing a person to do something or prohibiting a person from doing something. Most commonly, injunctions are prohibitions against actions. Such an injunction might be used, for example, as a remedy in a contract case involving a personal service. Mandy is a lounge singer, and she has a contract to perform at JZ's Lounge every weekend night from January through June. Two months into her contract, she decides to go work for Bally's Lounge instead because Bally's will pay her twice as much money. There is no way to calculate the damages adequately that would arise from the singer going over to the other club to perform, so money damages would not really be an adequate remedy. Instead, the owner of JZ's may obtain an injunction prohibiting Mandy from performing in any lounge until the end of June, when her term of performance under the contract will have been completed.

injunction
A court order either forcing a party to do something or prohibiting a party from doing something.

In the Walgreen dispute in the Case Opener, the appellate court agreed with the trial court judge that it would be extremely difficult for the court to calculate adequately how to determine the damages that Walgreen would sustain if Sara Creek were allowed simply to violate the contract's exclusivity clause and lease space to another drugstore. The court would have had to determine how much business the new store would take from Walgreen over the course of the next 10 years, and it felt that such a calculation would be almost impossible to figure. Hence, the court agreed that money damages would not be adequate and that an injunction prohibiting the leasing of the space in violation of the exclusivity clause would be the only adequate remedy.

Sometimes, when a party is suing another for breach of contract, one of the parties is concerned that before the court has had a chance to decide the case, the other party will do something to make it impossible for the concerned party to get the relief he would be entitled to. In such a situation, the concerned party may ask for a preliminary injunction to prohibit the other party from taking any action during the course of the lawsuit that would cause irreparable harm to any of the parties to the contract. For example, Jim agrees to sell Bob a very rare antique car for $15,000 but then says he will not comply with the terms of the agreement. Bob sues Jim for breach of contract, but before the case goes to trial, Bob finds out that Sara has told Jim that she is willing to pay him $20,000 for the car. Bob may seek a preliminary injunction to prohibit Jim from selling the car to anyone else until the court decides whether Bob is entitled to an order for specific performance forcing Jim to sell the car to him. Thus, the preliminary injunction fulfills the purpose of maintaining the status quo until the case can be finally decided.

Reformation Sometimes a written contract does not reflect the parties' actual agreement, or there may be inconsistencies in the contract, such as the price being listed as "$200,000 (twenty thousand dollars)." In such a case, the written document may be rewritten to reflect what the parties had agreed on.

[5] UCC § 2-716.

GLOBAL Context

©Sunflowerr/Shutterstock

Remedies for Breach of Contract in Latvia

In the United States, breach of contract is a common occurrence. In fact, many parties will opt to voluntarily breach a contract and pay damages rather than fulfill a contract because doing so may prove to be more cost-effective. In Latvia, however, each party is expected and legally obligated to fulfill its side of the bargain. Damages for breach are an option only in the most extreme cases where fulfilling the contract would prove impossible. Effectively, parties are barred from backing out of a contract, as doing so is not only illegal but considered highly immoral. Thus, specific performance is often the only remedy available in Latvian contract disputes.

Recovery Based on Quasi-Contract When an enforceable contract does not in fact exist, the court may grant a recovery based on quasi-contract; that is, the court may impose a contract-like obligation on a party to prevent an injustice from occurring. Recovery in quasi-contract is often sought when a party thought a valid contract existed and gave up something of value in relying on the existence of a contract. To recover under a theory of quasi-contract, a plaintiff must prove that (1) the plaintiff conferred a benefit on the defendant; (2) the plaintiff had reasonably expected to be compensated for the benefit conferred on the defendant; and (3) the defendant would be unjustly enriched from receiving the benefit without compensating the plaintiff for it.

CASE OPENER WRAP-UP

Impossible Wine Bottles

Under the amended agreement between Anchor and Encore, the production of bottles was to take place at the Lavington plant. The Lavington plant was the only facility owned by CPI that was capable of producing the specific type and quality of glass container that is required by the wine industry. According to the court's ruling, when the Lavington plant was sold and the new owners did not take over CPI's obligations, the terms of the contract became impossible for Anchor to meet. As a result of the impossibility of performance, Anchor was discharged from the contract.

PRACTICAL TIPS for Business Managers

1. If your need for a particular contract is contingent on the occurrence or nonoccurence of a particular event, then make sure you make that contract a conditional one. Include a clause in the contract making it conditioned on that event either happening or no longer happening.
2. When you know that you will have damages if a contract is not performed on time, and the damages may be difficult to calculate or prove, be sure to include a liquidated damages clause in the contract.
3. When drafting a liquidated damages clause, to be safe, err on the side of making the amount of the award on the low side because you do not want the award to be thrown out as being an unreasonable penalty clause. The better you can substantiate the basis for your stipulated award, the more likely it is to be upheld.

SUMMARY

Methods of Discharging a Contract

Contracts may be discharged in a number of ways, including:

- The occurrence of a condition subsequent or the nonhappening of a condition precedent.
- Complete or substantial performance.
- Material breach.
- Mutual agreement.
- Operation of law.

Remedies

Courts may grant parties in a breach-of-contract action legal or equitable remedies. *Legal remedies,* or money damages, include:

- Compensatory damages.
- Nominal damages.
- Punitive damages.
- Liquidated damages.

Equitable remedies, which are granted only when legal remedies are inadequate, include:

- Restitution and rescission.
- Specific performance.
- Injunction.

Point/Counterpoint

Should Specific Performance Automatically be Granted in Every Breach of Contract for the Sale of Real Property?	
YES	NO
Clearly, monetary damages can be used to compensate a person for almost any claim that a piece of real estate is so unique that only that specific piece of real estate will satisfy the party's expectations for a contract. Monetary damages can be added to the value of almost any real estate transfer to equalize the value of the new piece of real estate with the value of the disputed land. Imagine, for example, a new subdivision with 500 lots on which homes will be built. It is standard practice to recognize that some of these lots are identical in value, while others require a monetary premium to make them more comparable in value. You may argue that you don't want just any plot in the subdivision, but one with some specific feature, such as a certain amount of trees. Almost any feature you want on your real estate can be created for the right price. So you figure out how much it will cost you to turn a piece of property you can purchase into the property of your dreams, and that amount would be your damages. There is no reason to assume that every piece of property is unique and, therefore, that all real property buyers are entitled to specific performance.	The uniqueness of each piece of real estate is clear. Any buyer of real estate has an extensive array of choices. They may, in some instances, take years to make their selection of a specific piece of real estate. Why does the process of purchasing real estate involve so much deliberation and comparison of alternatives? The buyer of real estate is not looking for just any piece of real estate. The buyer is looking for one distinct property. The buyer considers future value, neighboring properties, revenue possibilities, and maintenance costs, among multiple other factors, before settling on one special, unique property. No other property satisfies those criteria for that buyer. Therefore, every piece of real estate is unique, which entitles the buyer to specific performance for a seller's breach.

Questions & Problems

1. Explain the difference between legal and equitable remedies.

2. Explain how the existence of conditions subsequent and precedent affects the discharge of a contract.

3. Explain the relationship between commercial impracticability and frustration of purpose.

4. Developer Foxwood Estates (Developer) subdivided most of its property into 38 single-family homesites, called the Foxwood Estates subdivision. The land in dispute is the remaining property, which was a 30-acre undeveloped parcel immediately adjacent to the subdivision. In marketing and selling the lots, the Developer made specific representations as to the land in dispute, both orally and in the documents distributed. Developer provided prospective buyers with topographic and setback maps detailing the numbered lots and labeling the land in dispute as either "Outlot 1" or "Outlot 1 [and] Outlot 2." The setback map was attached to the Declarations of Restrictions and Covenants (the Declarations) which were given to prospective buyers and became part of an actual buyer's purchase contract. The Declarations expressly stated that the Homeowners' Association, not the Developer, would own the outlots.

 The Developer also provided prospective buyers with a brochure that referred to the secluded, natural feel of the subdivision, stating: "Foxwood Estates is an exceptional community designed and reserved for individuals who desire a lifestyle amidst the grandeur of nature, and who will appreciate the preservation of this natural habitat." The brochure included an aerial view showing the subdivision, including the land in dispute, bordered in red.

 When one of the homeowners read in the newspapers that the land in dispute had been approved for multifamily housing, he notified the Homeowners' Association, and the association filed suit against the developer, seeking specific performance of its promise to transfer title to the land in dispute to the Homeowners' Association.

 The trial court ordered defendant Foxwood Estates LLC (the Developer) to convey to Foxwood Estates Homeowners' Association the land in dispute to the Foxwood Estates subdivision. The developer appealed. How do you think the court of appeals resolved the appeal? Why?

 Source: *Foxwood Estates Homeowners' Association, Inc. v. Foxwood Estates, LLC,* 364 Wis. 2d 405; 866 N.W.2d 40 (2015).

5. Janelle Rederford suffered from systemic lupus erythematosus when she was terminated from US Airways. Rederford brought suit against US Airways under the Americans with Disabilities Act (ADA), alleging that she had been wrongfully terminated because of her condition. Concurrently, US Airways filed a petition for relief under Chapter 11 with the bankruptcy court. Through the bankruptcy proceedings, Rederford was required to request a hearing to establish her personal injury tort, which was covered by US Airways' insurance policy. Rederford did not file any such request with the bankruptcy court. The bankruptcy court accordingly disallowed any litigation pertaining to personal injuries US Airways caused prior to its bankruptcy proceedings. Rederford later filed suit in federal court against US Airways to seek compensation. US Airways filed a motion to dismiss, claiming that the case had been discharged in bankruptcy court. Rederford argued that because she sought an equitable remedy of reinstatement to her position that her case could not have been resolved in bankruptcy court. Do you find Rederford's argument convincing? Why or why not?

 Source: *Rederford v. US Airways, Inc.,* 2009 U.S. App. Lexis 27258.

6. On April 20, 2009, Severn and IFC entered into a Pesticide Application Agreement (PAA) requiring IFC to use phosphine, a pesticide, to fumigate a peanut storage dome owned by Severn and located in Severn, North Carolina. The PAA required IFC to apply the pesticide "in a manner consistent with instructions . . . and precautions set forth in [its] labeling."

 In return for IFC's services, Severn promised to pay IFC $8,604, plus applicable sales taxes. The contract specified that this $8,604 sum was not "sufficient to warrant IFC assuming any risk of incidental or consequential damages" to Severn's "property, product, equipment, downtime, or loss of business."

 On August 4, 2009, IFC improperly dumped approximately 49,000 tablets of Fumitoxin into Severn's peanut dome through a single access hatch, which ultimately led to an explosion, and the peanut dome sustained extensive structural damage. The company sued IFC for breach of contract, among other claims. Severn argues that the PAA's consequential damages exclusion does not bar its breach of contract claim for damage to its dome and peanuts and its associated remediation and lost business costs. Do you think the appellate court ultimately agreed or disagreed with Severn? Why or why not?

 Source: *Severn Peanut Co, Inc, et al v. Industrial Fumigant Co. & Rollins, Inc.* (2015).[7]

7. In 1991, EnerSys entered into an agreement with Exide Technologies to purchase substantially all of Exide's battery business for about $135 million. To formalize this sale, the parties entered into more than 23 agreements.

[7] http://caselaw.findlaw.com/us-4th-circuit/1719778.html.

Under the collection of agreements, Exide agreed to allow EnerSys to use the Exide trademark without royalties in the industrial battery business, and Exide retained the right to use the Exide trademark outside the industrial battery business. In 2000, Exide wished to reenter the North American industrial battery business. Both parties agreed to the termination of the 10-year noncompetition agreement, but Exide was unable to regain the Exide trademark from EnerSys. In 2002, when Exide filed for bankruptcy protection, Exide attempted to reject various agreements it had made with EnerSys in 1991, including the right to use Exide's trademark. The bankruptcy court granted an order finding that the agreement was an executory contract, a contract that was so underperformed that failure of either party to complete performance would result in a material breach excusing performance of the other. The Bankruptcy Code allows debtors to reject any executor contract. EnerSys appealed the bankruptcy court's decision, arguing that it had substantially performed its obligations under the 1991 agreement, including paying the full $135 million and operating the industrial battery business the 11 years prior to the 2002 bankruptcy ruling. The district court affirmed the bankruptcy court's holding that EnerSys' obligation for trademark use and quality standards provision amounted to elements of the agreement that would cause a material breach if they were not observed. EnerSys again appealed, arguing that any obligations EnerSys had under the 1991 agreement were not material. How do you think the court of appeals ruled? What standard would you use to determine whether an element of the agreement between the two parties was material?

Source: In *re Exide Techs.*, 2010 U.S. App. LEXIS 11029.

8. On November 7, 2005, Briarwood signed an agreement to sell Toll Brothers a planned 66-acre, 41-lot subdivision property in the village of Pomona, New York, for $13,325,000. The agreement expressly conditioned Toll's payment obligations on Briarwood's delivery, at its sole cost and expense, of final, unappealable subdivision approval of the property in accordance with the subdivision plan and the satisfaction by Briarwood of any conditions of the final approval, such that upon posting of customary security and payment of application and inspection fees by Toll Brothers, the company would be able to file the plat and commence infrastructure improvements and apply for and obtain building permits.

The agreement stated that the conditions set forth in the approval shall be subject only to "such conditions as Toll may approve at its sole discretion, which approval shall not be unreasonably withheld with respect to those modifications which do not have a material adverse effect on the proposed development." Closing was to take place 30 days following the date upon which all conditions to closing set forth in the approval had been satisfied and "[i]f on or before the date of closing all contingencies and conditions specified herein are not or cannot be satisfied, then Toll shall have the option of . . . cancelling this Agreement."

On January 12, 2006, Briarwood obtained preliminary subdivision approval of the property from the Pomona Planning Board. Final subdivision approval of the property, however, was subject to a number of conditions. By letter dated December 22, 2006, Toll notified Briarwood that five of the conditions would have a material adverse effect on the proposed development and that it would not accept them.

Toll Brothers claimed that Briarwood's December 28, 2006, response to Toll's December 22, 2006, letter was an anticipatory repudiation of the agreement.

In response to Toll's five objections, Briarwood (1) offered to post a bond to cover the potential cost of repaving Klinger Court; (2) pointed out that the steep slope condition restated the Village Code requirement to secure a site development plan permit for each lot within a subdivision; (3) agreed to pay any cost differential resulting from use of a 4 percent rather than a 10 percent grade for the cul-de-sac; (4) proposed two alternative solutions to the drainage system problem; and (5) pointed out that the landscaping plan condition, like the steep slope condition, restated a Village Code requirement.

Toll Brothers treated the letter from Briarwood as an anticipatory repudiation of the contract and refused to complete the transaction, so Briarwood sued for breach of contract. Both parties filed motions for summary judgment.

The district found that no reasonable reader could construe Briarwood's response as a positive and unequivocal repudiation of the agreement and that it did not signal that Briarwood was unwilling to comply with ¶ 16(a)(iii) or any other provision of the agreement. The court therefore granted summary judgment to Briarwood. Toll Brothers appealed. How do you believe the court of appeals ruled, and why?

Source: *Briarwood Farms, Inc. v. Toll Bros., Inc.,* 452 Fed. Appx. 59, 2011 WL 6415185 (C.A.2 (N.Y.) 2011).

9. The two companies had an agreement by which Sunrish was providing a food product to Nutrisoya. However, Sunrich allegedly did not deliver the products covered in the contract, and Nutrisoya took the company to court for material breach of contract. The seller wanted the district court to instruct the jury on the difference between a breach of a single installment and a breach of an entire installment contract, arguing that a single installment breach did not disregard the value of the whole contract. However, the district court did not give such an instruction and left the defendant to argue the point. How do you think the jury decided?

Source: *Nutrisoya v. Sunrich* 641 F.3d 282; 2011 U.S. App. LEXIS 11561.

10. Environs Inc. is an architectural firm that designs commercial and business buildings. SAMS Hotel Group LLC signed a contract with Environs, under which Environs would provide architectural services for the design of a six-story hotel in Fort Wayne, Indiana, that SAMS planned to build. The contract contained a liquidated-damages clause that limited Environs's liability for breach of contract to $70,000. When the building was almost complete, serious structural defects were discovered, the building department condemned the building, and it had to be destroyed.

 SAMS sued Environs, alleging they breached the contract by providing negligent architectural services, and seeking to recover $4.2 million in damages. The district court found that Environs had indeed been negligent and breached their contract, but because of the liquidated-damages clause, awarded the plaintiff only $70,000. SAMS appealed. How do you believe the appellate court ruled and why?

 Source: *SAMS Hotel Group LLC, Inc. v. Environs,* 716 F.3d 432, 2013 US App. LEXIS 11047 (2013).

11. Defendant Practice approached the Hospital about its desire to employ Dr. Rasheed and asked the Hospital to help facilitate his recruitment. The three entered into an agreement that Dr. Rasheed would relocate his practice and practice medicine in Jackson on a full-time basis for 36 months. Under the the agreement, "[a]ny material breach of this Agreement by Physician and/or Practice, or a failure by Physician and/or Practice to fulfill any material provisions of this Agreement shall entitle Hospital, at its option, to terminate this Agreement immediately." The Practice agreed to be jointly and severally liable for any amounts owed to the Hospital in the event of any breach by the Practice or Dr. Rasheed.

 The agreement guaranteed Dr. Rasheed a monthly cash collection of $23,350.00 for the first year. The Hospital advanced $180,187.43, to the Practice and Dr. Rasheed as cash collections guarantee payments during the 12-month guarantee period, an amount that the Practice and Dr. Rasheed were jointly and severally liable to repay at the end of the guarantee period. The agreement provided that the debt would be completely or partially forgiven if Dr. Rasheed continued to practice full-time in Jackson for two years at the end of the guarantee period, but if he failed to continue practicing for that time period, he or the practice would immediately reimburse the hospital the amount advanced.

 Less than one year into the agreement period, Dr. Rasheed stopped his full-time medical practice in Jackson, so the hospital requested reimbursements of the advanced payments pursuant to the terms of the agreement. The Practice did not respond to the Hospital's demand for payment.

 The hospital then sued the Practice for breach of contract, seeking to recover the amount owed under the agreement. The Practice answered that the doctor's actions were the cause of the breach, and therefore he alone was liable, not the Practice.

 Following discovery, the Hospital filed a motion for summary. In its response, the Practice stated that Dr. Rasheed left the area after being named as the subject of sexual misconduct allegations involving the staff and child patients, for which he was being pursued civilly and criminally. Therefore, it was impossible for Dr. Rasheed to practice and impossible for the Practice to allow him to continue practicing. Because there was an issue as to whether Dr. Rasheed had the ability to perform under the agreement, the Practice argued that summary judgment was inappropriate. The court granted the Practice's motion and the hospital appealed. How do think the appellate court decided the case and why?

 Source: *Jackson Hosp. Corp. v. United Clinics of Ky., LLC,* 545 S.W.3d 327 (2018).

12. Plaintiff, a tenant in common in a piece of real property, had agreed to sell his interest therein to defendant, the other tenant in common. The contract of sale provided that defendant was to pay plaintiff $25,000 upon the execution of a bargain and sale deed transferring title to the defendant as the sole owner of the real property. The contract of sale further provided that plaintiff could continue to store his equipment at the property "for a maximum of 60 days from the date" of the contract of sale, and that "[i]mmediately upon the permanent removal of all of the equipment from the premises, Evelyn A. Cuggino shall pay the sum of ten thousand ($10,000.00) dollars . . . to Peter J. Magnani, which shall occur no later than 60 days from the date of this agreement." At a nonjury trial, the evidence established that plaintiff had not removed his equipment within the 60-day period, yet defendant paid plaintiff the sum of $2,000 prior to plaintiff's removal of the equipment, but not the remaining $8,000. Plaintiff sued to recover $8,000 for breach of a contract for the sale of real property.

 The trial court dismissed plaintiff's action. The court found that plaintiff was not entitled to recover the remaining $8,000 due under the agreement, since removal of the equipment within the 60-day period was a **condition precedent** to plaintiff's entitlement to be paid the sum of $10,000 by defendant. Plaintiff appealed this dismissal, claiming the contract of sale did not provide for a forfeiture of plaintiff's right to recover the $10,000 if he failed to remove the equipment within the 60-day period, and therefore it was not a condition precedent. How do you think the case was resolved on appeal? Why?

 Source: *Magnani v. Cuggino,* 57 Misc.3d. 1; 2017 WL 2380647.

Part 3 Domestic and International Sales Law

CHAPTER 15

Formation and Performance of Sales and Lease Contracts

On Your Mark, Get Set, Go!

Kim Baker purchased four 1960's-era sporting race cars from Lilo Ben Zicron. The cars apparently did not conform to the terms of the original sales agreement. To that end, Baker went into mediation with Zicron. They agreed to a $950,000 settlement to be held in abeyance in order to give Zicron time to sell on consignment the four vehicles in order to apply the sale of the proceeds to the $950,000 due. The first vehicle, a 1969 Lola, was required to have been sold by October 20, 2016. Failure to do so would trigger immediately enforcement of the $950,000 agreement.

On October 19, defendant presented with Baker with a purchase agreement that stated the buyer, Gideon Group Holdings, LLC (Gideon), to pay the defendant in three installments: a $10,000 deposit upon execution of the purchase agreement, a $132,500 payment due January 15, 2017, and a final payment of $132,500 due March 15, 2017. The purchase agreement stated that the defendant would retain possession of the vehicle and its "ownership documents" until the final payment was made.

Plaintiff's counsel replied by sending the defendant a notice of breach, stating that the purchase agreement did not constitute a timely "sale" by October 20, 2016, as mandated by the agreement. The defendant denied breaching any part of the agreement. As a result, the plaintiff filed a motion for entry of judgment

The questions presented here are:

1. Is the consignment a "sale" under Article 2 of the Uniform Commercial Code?
2. Is a sale defined solely as a "transfer of title for a price" or is the existence of a purchase agreement sufficient to constitute a "sale" under the UCC.

LEARNING OBJECTIVES

After reading this chapter, you will be able to answer the following questions:

- **LO 15-1** Identify the kinds of contracts that fall under UCC interpretations
- **LO 15-2** Recall how sales contracts are formed under the UCC.
- **LO 15-3** Define what a merchant is, and why the designation matters.
- **LO 15-4** Describe the characteristics of a lease contract.
- **LO 15-5** Outline the focus and purpose of The United Nations Convention on Contracts for the International Sale of Goods (CISG).
- **LO 15-6** Describe the concept of title, how title passes.
- **LO 15-7** Explain risk of loss.

Businesses and organizations that purchase products need to be aware of which laws govern their purchases because the laws can differ. Three sources of laws that interpret sales contracts exist: state common law, the Uniform Commercial Code, and state statutory law. There is very little, if any, federal law that governs contracts for the buying and selling of items, and when such law does exist, it is highly specialized (e.g., the buying and selling of stock under laws created by the Securities and Exchange Commission).

This chapter introduces the Uniform Commercial Code. It explains the scope and significance of the UCC, discusses sections of the UCC that govern both sales and lease contracts, and reviews how these contracts are formed and executed.

The Uniform Commercial Code

The Scope of the UCC

In some areas of law, federal statutes ensure that the same law applies in all the states. Federal statutes do not govern the formation of sales and lease contracts. Instead, each state passes its own laws to outline rules in this area. Consequently, laws vary from state to state. Although all states (except Louisiana) follow the English common law, over 200 years of legal precedent that has developed in each state can create differences in contract interpretation and application. With the development of interstate commercial activities, these differences began to pose problems for parties from different states entering into a contract with each other. What was needed was some kind of uniform law for business transactions that all the states could adopt.

Uniform Commercial Code (UCC)
A statutory source of contract law in the United States applicable to transactions involving the sale of goods. The UCC was created in 1952 and adopted by all 50 states, the District of Columbia, and the Virgin Islands; it may be modified by each state to reflect the wishes of the state legislature.

The **Uniform Commercial Code (UCC)** was created in 1952 and has been adopted in part by all 50 states, the District of Columbia, and the Virgin Islands. Louisiana, not having the English common law heritage, has adopted only portions of the UCC that do not conflict with its historically French-based code law. When a state adopts the UCC, that code becomes part of the law of that particular state; it becomes the commercial code for that state. The UCC is divided into sections known as *articles*. These articles cover a wide range of topics, from sales contracts to secured transactions (see Exhibit 15-1).

The Significance of the UCC

The UCC is significant because it clarifies sales law and makes this area of law more predictable to businesses that engage in transactions in more than one state. Essentially, the UCC facilitates commercial transactions. This is an important point to remember. When in doubt, judges will try to interpret the UCC to keep an agreement in force. As the drafters of the UCC intended, the UCC is designed to facilitate commercial transactions, not construct ways to impede them.

Exhibit 15-1

An Outline of the UCC

ARTICLE	TOPIC
1	General provisions
2	Sales
2(A)	Leases
3	Negotiable instruments
4	Bank deposits and collections
4(A)	Wire transfers
5	Letters of credit
6	Bulk transfers
7	Documents of title
8	Investment securities
9	Secured transactions

However, it is important to emphasize that the UCC does not pertain to all business transactions. Under UCC Article 2, the subject matter of this chapter, the UCC explains the creation and interpretation of sales contracts.

Articles 2 and 2(A) of the UCC

LO 15-1 Identify the kinds of contracts that fall under UCC interpretations

Article 2 of the UCC governs sales contracts; Article 2(A) governs lease contracts.[1] Specifically, Article 2 focuses on contracts for the sale of goods; Article 2(A) focuses on contracts for the lease of goods. For the purpose of selling something, the UCC divides all the items that can be bought and sold into three categories: goods, realty, and services (including intangible goods such as securities). Article 2 pertains only to the sale of goods. Nonetheless, Article 2 is not a comprehensive guide to sales contract formation.

When Article 2 is silent on an issue of sales contract formation or interpretation, the common law rules apply. Of course, if a state has passed statutory law regarding contracts, that law always supersedes the common law. Additionally, it is important to note that under the UCC, the rules for transactions involving merchants differ from those for transactions involving regular buyers and sellers. Merchants will generally be held to a higher standard of care and behavior than nonmerchants. Every state except Louisiana has adopted Article 2 of the UCC.[2]

If you are not sure whether a contract falls under common law or under the UCC, the decision tree in Exhibit 15-2 can help you make a determination. To use this rubric, consider the following definitions of many of its terms.

Article 2 of the UCC

LO 15-2 Recall how sales contracts are formed under the UCC.

Sale Section 2-106(1) of the UCC states that a **sale** "consists of the passing of title from the seller to the buyer for a price."

sale
The passing of title from a seller to a buyer for a price.

Goods Section 2-105 of the UCC defines **goods** as "all [tangible] things which are movable at the time of identification to the contract for sale." Items are tangible if they exist physically. In the case opener, clearly "cars" would be tangible and moveable items thus qualifying as goods. Would

goods
All physically existing things that are movable at the time of identification to the contract for sale.

[1] See http://www.law.cornell.edu/ucc/2/ for a basic understanding of the Article. For recent changes consult the current version of the UCC.
[2] Louisiana's civil tradition is based on the Code Napoleon, or the French Civil Code.

Exhibit 15-2

Common Law or UCC?

1. Does the transaction involve a sale or lease (as opposed to, let's say, a gift)?
 If yes → go to question 2.
 If no → apply the appropriate area of common or statutory law.

2. Is the transaction for the sale or lease of a good?
 If yes → go to question 3.
 If no → apply common law.

3. If the contract is for the sale or lease of both goods and nongoods, which type of item is predominant in the contract?
 If goods → UCC Article 2 applies; go to question 4.
 If nongoods (real estate or services) → common law applies.

4. If this is a UCC Article 2 transaction, is either party a merchant?
 If yes → pay close attention to when the special rules on merchants apply.
 If no → then only "reasonable" care applies to the parties. From the UCC's perspective, everything that can be bought and sold falls into three categories: goods, services, and realty. Think of it this way:

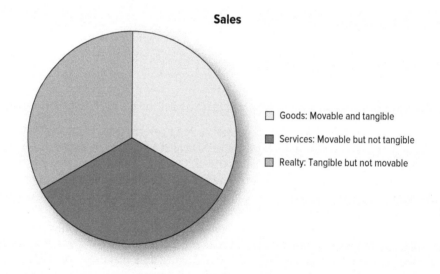

a contract for the purchase of automobiles before the cars are assembles qualify as a contract for the sale of goods. It is only when the goods are identified to the contract that the contract springs into existence. In the case opener, that would be when the specific cars themselves are identified specifically to the contract. Some items are not goods under Article 2. For example, corporate stocks and copyrights are not tangible, so they are not goods. Real estate cannot be moved, so it is not a good. Yet items attached to real estate that are used for business activities are known as *trade fixtures* and are treated as goods under the UCC.

Items taken from real estate may be treated as goods. Minerals, clay, and soil can all be treated as goods, with their sales contract governed by Article 2, if the owner takes these items out of the ground and then sells them to the buyer. If the owner sells the buyer the right to come and remove the items, the contract would be governed not by the UCC but by common law for the sale of an interest in realty, in this case, an interest known as a profit. Crops that are sold while still growing in the field are also considered goods, and their sales contracts are subject to UCC interpretation. Exhibit 15-3 gives a synopsis of these issues.

Exhibit 15-3
What is a Good?

ITEM	A "GOOD" FOR UCC/CISG PURPOSES	NOT A "GOOD"; COMMON LAW CONTRACT RULES APPLY
Grains in the field	X	
Items attached to real estate and meant to be permanent (swimming pool anchored in the ground)		X
Items attached to business realty (shelving, for example)	X	
Minerals, clay, shale, oil, etc., if removed by owner of land	X	
Mineral, clay, shale, oil, etc., if removed by buyer		X

BUT WHAT IF . . .

WHAT IF THE FACTS OF THE CASE OPENER WERE DIFFERENT?

What if the language regarding transfer of title had been omitted? Are there default provisions regarding transfer of title. Or, is transfer of ownership always tied to the actual physical or digital transfer of the title document? Remember, for automobiles, there is always a document of title, unless it is an abandoned vehicle, which these were not.

Consider the following Case Nugget. Could or should the Petersons have done anything as soon as they found the problem stalks? Is the loss caused by the storm or by the genetic composition? Who has this risk of loss? Are the seeds the goods, is the corn the goods, or is this even a UCC contract?

CASE Nugget

Peterson v. North American Plant Breeders
218 Neb. 258, 354 N.W.2d 625 (1984)

In spring 1981, Robert and William Peterson bought Migro Seed Company Inc. corn seeds from John Sandall. Both Petersons were local farmers, and John Sandall, also a farmer, was Migro's area dealer. Migro produces hybrid seeds for sale, and the corn seeds the Petersons bought were a "product of scientific genetic cross-breeding to produce a seed having desirable germination, growing, and production qualities."

Around July 23, 1981, the Petersons discovered that nearly 70 percent of their corn grown from the Migro seeds had broken off at the ear level after a typical Nebraskan thunderstorm the evening before. However, corn growing from non-Migro seeds had not suffered such damage.

It turns out that these plants had absorbed silica in the sandy soil in this particular area of Nebraska, causing them to develop a brittle stalk and thus break at about the first ear-of-corn level. No other variety of corn seed had this problem. When John Sandall, the neighboring farmer and Migro dealer, sold the Petersons the 102 bags of seed, the literature accompanying the seed indicated that one of the Migro seed's attributes was excellent stalk quality. When the crop was harvested, the Petersons harvested approximately 19½ bushels per acre of the Migro corn while averaging 113¾ bushels per acre of corn grown from seed provided by other seed companies. The Petersons brought a lawsuit for breach of contract.

E-COMMERCE and the Law

The UCC and the Internet

The Uniform Commercial Code was designed to address the sale of tangible goods such as scooters, electric power tools, and jeans. Because the UCC was written long before the Internet and e-commerce boom, this legislation was not designed for transactions related to less tangible things, such as software and information. Not surprisingly, the UCC does not respond well to certain legal issues that arise in today's marketplace. For example, the UCC does not tell us the answers to these questions: What rules should govern computer software, which is likely to be licensed rather than sold as a good? What about downloadable software files? What rules should govern information providers like Google or MSN which provide a continuing service? What rules should govern the exchange of information, such as stock quotes? What rules cover travel reservations a person makes online?

The National Conference of Commissioners on Uniform State Laws (NCCUSL) has adopted the Uniform Commercial Information Transactions Act (UCITA), which answers the preceding questions and many more. This law promises to do for electronic contracting what the UCC did for transactions in physical goods: protect consumers by providing predictability, uniformity, and clear rules. As with the UCC, states will choose whether to adopt UCITA. In 2001, Virginia became the first state to enact UCITA, probably because major Internet-related companies (i.e., AOL, UUNET Technologies) are headquartered in northern Virginia.

Sources: Robert Holleyman, "Updating Contract Law for the Digital Age," USA Today, March 1, 2000 (2000 WL 9014770); Scott W. Burt, "Controversial New Rules for Computer Contracts," Metropolitan Corporate Counsel 8 (June 2000), col. 1.

mixed sale
A contract that combines a good with a service or real estate.

Mixed Goods and Services Contracts Sometimes, it is not easy to tell whether something is a good because a tangible item is tied to or mixed with something intangible, such as a service. A contract that combines a good with a service is a **mixed sale**. Article 2 applies to mixed sales if the goods are the predominant part of the transaction.

LO 15-3 Define what a merchant is, and why the designation matters.

merchant
Under the UCC, a person who regularly sells goods as his or her business or trade or employs people to sell those goods; anyone who works for a person selling those goods; or anyone who holds himself or herself out as a "merchant" or a "dealer."

Merchants UCC Article 2 pertains to anyone buying and selling goods. However, the UCC distinguishes **merchants** from regular buyers and sellers; it contains provisions that either apply only to merchants or impose greater duties on merchants. The drafters of the UCC assumed that merchants have a greater ability to look out for themselves than do ordinary buyers and sellers. The definition in the margin can be used to determine whether a seller or a buyer is a merchant under the UCC.

Consider the facts in the preceding Case Nugget. Think of questions about merchants as they relate to the transaction between the Petersons, Migro Seed Company, and John Sandall. Specifically, can any part of UCC Section 2-104(1) be used to argue that John Sandall is a merchant? What about the Petersons? Also, consider the possibility that if the Petersons are merchants, a court will assume that they need less protection than they would if they were nonmerchant buyers of the seeds. There is no question that Migro Seed Company is a merchant. However, if the Petersons sue John Sandall as well, is he a merchant and thus subject to a higher standard of care? If the Petersons are merchants, then as merchants in the business of farming, should they have known that their sandy soil, rich in silica, could pose a problem?

lease
A transfer of the right to possession and use of goods for a term in return for consideration.

Article 2(A) of the UCC

Article 2(A) covers contracts for the lease of goods. This section of the UCC is increasingly important because consumers (both individuals and businesses) are more likely to lease goods today than ever before. Consumers lease cars, equipment, and machines. Article 2(A) does not cover leases related to real property.

lessor
A person who transfers the right to possession and use of goods under a lease.

LO 15-4 Describe the characteristics of a lease contract.

lessee
A person who acquires the right to possession and use of goods under a lease.

Leases UCC Section 2A-103(j) defines a **lease** as "a transfer of the right to possession and use of goods for a term in return for consideration." A **lessor** is "a person who transfers the right to possession and use of goods under a lease."[3] A **lessee** is "a person who acquires the right to possession and use of goods under a lease."[4] Thus, if you lease a car, the company that leases the car

[3] UCC § 2A-103(p).
[4] UCC § 2A-103(o).

to you (the lessor) transfers the right of possession and use of the car to you (the lessee) in return for consideration (money).

Contracts for the International Sale of Goods

LO 15-5 Outline the focus and purpose of The United Nations Convention on Contracts for the International Sale of Goods (CISG).

The Scope and Significance of the CISG

By the 1970s, it became clear that, increasingly, businesses planned to conduct commercial transactions not only across state lines but also across country borders. In 1980, the **United Nations Convention on Contracts for the International Sale of Goods (CISG)** was offered as a treaty that countries could sign, indicating their willingness to allow this treaty to govern international business-to-business sales contracts.

The CISG is important because if a problem arises with an international sale and a party to the transaction initiates litigation, the UCC does not provide guidance in the litigation; instead, the CISG preempts the UCC. The CISG covers the same general topics as the UCC. For instance, the CISG covers offers, acceptances, and other contract topics.

There are some fundamental and subtle differences between the UCC and the CISG. It seems that the CISG tends to be "friendlier" to the seller, but that determination will be fact-specific on the circumstances of the situation presented. Specific provisions of the CISG differ from the UCC as illustrated by Exhibit 15-4.

See the Point/Counterpoint at the end of this chapter for the significant question whether the United States should adopt the CISG.

United Nations Convention on Contracts for the International Sale of Goods (CISG) The legal structure for international sales, including business-to-business sales contracts.

Exhibit 15-4

UCC vs. CISG

UCC	CISG
Applies to consumer and commercial contracts for the sale of goods	Consumer contracts are not covered by the CISG
Statute of frauds writing requirement for sale of goods over $500	No writing requirement necessary in order to prove contract
Parties may agree on open terms in price and even specific quantity and still create an enforceable contract	Specificity is required
Parol evidence rule is strictly applied	Parol evidence is admissible to show party's intent even if contradictory to written documentation
Revocation of an offer is valid up to acceptance	Revocation of an offer is only possible if there has been no reliance, regardless of the status of the acceptance
Battle of the forms impliedly ratify changes presented in the contract	No battle of the forms is binding if material terms of the agreement are altered
Perfect tender is required or a breach occurs	Breaching party has the right to cure so long as the cure is done within a "reasonable" time
Specific warranty disclaimers are required to negate merchantability	No such disclaimer
Lost profit may or may not be included in consequential damages	Damages nearly always include lost profit
No such provision in the UCC	If goods suffer a decrease in value, buyer may reduce purchase price accordingly

How Sales and Lease Contracts Are Formed under the UCC

Earlier in this textbook, you studied the common law of contracts, from the rules for agreements to the remedies for breaches. Sales and lease contracts require the same components as general contracts, but some UCC provisions that govern contracts for the sale or lease of goods are not identical to the common law requirements you learned about in Part 2. This section highlights the most important provisions of the UCC with regard to the formation of sales and lease contracts. Differences between the UCC and the CISG will be noted.

Formation in General

Under both the UCC and the CISG, contracts for the sale or lease of goods may be made in any manner sufficient to show agreement.[5] A court will uphold a contract for the sale or lease of goods as long as the parties intended to make a contract and there is a reasonably certain basis for giving an appropriate remedy.[6]

Offer and Acceptance

Offer Under the UCC, if certain contract terms are left open, it is acceptable to fill them in. Exhibit 15-5 indicates what generally happens under the UCC when certain terms of a contract or lease are left open. Although the quantity element is usually required, since without it the courts will have no guideline to determine the intent of the parties, the UCC does permit "output" and "requirement" terminology to determine quantity. If a farmer agrees to sell all the grain his fields will produce but is unable to determine a specific amount, then "quantity" has been satisfied under the UCC as an *output* contract. Conversely, if a buyer agrees to buy all of her "requirements" for a certain good but is unable to identify an exact amount, then the UCC recognizes this as an enforceable *requirements* contract. See the following Case Nugget for an interesting example of this.

CASE Nugget

PMC Corporation v. Houston Wire & Cable Company
Supreme Court of New Hampshire
147 N.H. 685; 797 A.2d 125; 2002 N.H. LEXIS 68; 47 U.C.C. Rep. Serv. 2d (Callaghan) 1427 (2002)

PMC sells thermocouple wire. Houston entered into an agreement to buy a substantial or "major portion" of its wire from PMC. In the middle of the contract period, Houston informed PMC that it would still purchase wire from PMC but that the "bulk" of its purchase would be made through a competitor of PMC. PMC brought suit for breach of contract. The court refused the defendant's motion for summary judgment, and then the defendant appealed. The reviewing court found that there was sufficient evidence to find that the quantity element was sufficient for the contract to be enforceable under the statute of frauds. The writing expressed a quantity term when the distributor stated that it intended to purchase a major portion of the product from the supplier. This term was sufficient to afford a basis for believing that the offered evidence rested on a real transaction. There was sufficient evidence to determine the terms reasonably, including the quantity of the contract. The reviewing court found that the agreement contained language from which a jury could have determined that the distributor agreed to purchase exclusively from the supplier up to a certain quantity.

[5] UCC §§ 2-204(1) and 2A-204(1).
[6] Ibid.

Exhibit 15-5

The UCC and Open Terms

TERM LEFT OPEN	INTERPRETATION UNDER UCC
Price	A "reasonable price" is supplied at the time of delivery.
Payment	Payment is due at the time and place at which the buyer is to receive the goods.
Delivery	The place for delivery is the seller's place of business.
Time	The contract must be performed within a reasonable time.
Duration	The party that wants to terminate an ongoing contract must use good faith and give reasonable notification.
Quantity	Courts generally have no basis for determining a remedy.

The UCC also creates a new category of offers: the firm offer. Under UCC Section 2-205, offers made by merchants are considered **firm offers** if the offer (1) is made in writing and (2) gives assurances that it will be irrevocable for up to three months despite a lack of consideration for the irrevocability. If a firm offer is silent regarding time, the UCC assumes a three-month irrevocability period. This contrasts sharply with the common law, under which an offer is revocable at any time prior to acceptance unless a period of irrevocability (also known as an *option*) is supported by some kind of consideration.

firm offer
An offer made in writing and giving assurances that it will be irrevocable for not longer than three months despite a lack of consideration for the irrevocability.

Acceptance Under the UCC, an acceptance may be made by any reasonable means of communication, and it is effective when dispatched.[7] It is important to note that the **mirror-image rule** that applies under common law does not apply under the UCC. Recall that the mirror-image rule states that an offeree's acceptance must be on the exact terms of the offer. If the acceptance includes additional terms, the acceptance becomes a counteroffer instead of an acceptance.

Under the UCC, additional terms are permitted in contracts for the sale or lease of goods. Under UCC Section 2-207(1), additional terms do not negate acceptance unless acceptance is made expressly conditional on assent to the additional terms.

mirror-image rule
The principle that holds that the terms of the acceptance must mirror the terms of the offer; if the terms of the acceptance do not mirror the terms of the offer, no contract is formed, and the attempted acceptance is a counteroffer.

Consideration

Sales and lease contracts require consideration. Under the common law, when a contract is modified, it must be supported by new consideration. The UCC eliminates that requirement for the modification of sales and lease contracts.[8] The UCC requires only that modifications must be made in good faith.[9]

Battle of the Forms

When parties are negotiating and offers and counteroffers are being exchanged, the UCC recognizes something we call the battle of the forms. Essentially, it works like this: If the last set of negotiations is not specifically rejected, then the last terms standing, so to speak, are the ones that are applicable to the final transaction. The CISG tends to be less lenient with the battle of the forms and relies more on the old common law mirror-image rule; that is, if the parties did not expressly agree to certain terms, then those terms cannot be assumed. The UCC, focusing on its predisposition to facilitate commercial transactions, assumes that the terms found in the battle of the forms are part of the contract whether expressly agreed to or not.

[7] UCC §§ 2-206(1) and 2A-206(1).
[8] UCC §§ 2-209(1) and 2A-208(1).
[9] UCC § 1-203.

GLOBAL Context

E-Commerce Sales in the Middle East

©Digital Archive Japan/Alamy

The Middle East seems to provide a potentially fertile ground for e-business start-ups. Why? Just one country, the United Arab Emirates (UAE), is a huge source of buyers interested in purchasing online. Commentator Dan Stuart notes that:

> Eighty percent of respondents have a credit card; 70% of respondents use their credit card for online purchases; 40% of respondents buy online once per month or more; 56% of respondents are somewhat to very comfortable buying online from Europe, 59% from North America, but with the largest group of respondents from the GCC and Levant saying that buying from this region just hasn't even been an option; 49% of respondents cite the chief reason for hesitation or lack of online shopping from this region is the lack of options.

However, the fly in the ointment is the costs of the available payment providers. Consider that in North America, the requirements for setting up an e-commerce payment provider (such as PayPal) are $0 of collateral on account as a deposit with the merchant account provider, $199 setup fee, $420 yearly maintenance fee, and 2.6 percent and $0.25 on every transaction. In contrast, in the Middle East, these innocuous costs jump to $85,000 of collateral on account as a deposit with the merchant account provider (subtract from your operating capital), $1,000 setup fee, $5,000 yearly maintenance fee, and 3 percent on every transaction.

Source: Dan Stuart, "The State and Opportunity for E-commerce in the Middle East," danstuart.me, February 24, 2010.

Requirements under the Statute of Frauds

Under common law, the statute of frauds requires all material terms to a contract to be in writing. Under the UCC, contracts for the sale of goods must be in writing if the goods are priced at $500 or more; lease contracts that require payments of $1,000 or more must be in writing to be enforceable.[10,11] There is a significant difference between the statute of frauds and the UCC about what constitutes a **writing** that satisfies the statute. Common law requires some kind of writing created or signed by the party who is contesting the enforceability of the contract.

This rule holds true under the UCC unless the parties are merchants. If two merchants have an oral agreement, a written memo from either party to the other is deemed to satisfy the statute of frauds, even if it is not acknowledged by the receiving party. If the memo is not objected to within 10 days of receipt, the oral agreement, memorialized by the memo, is binding. The CISG has eliminated the statute of frauds. The CISG requires only credible and reasonable proof that a contract exists, regardless of whether the proof is documented or verbal.

writing
A type of documentation that shows contractual intent and satisfies the statute of frauds requirement.

Unconscionability

A contract or contract provision is *unconscionable* if it is so unfair that a court would be unreasonable if it enforced the contract. The UCC outlines actions a court can take if it discovers that a contract or lease provision or the contract or lease as a whole is unconscionable.[12] If a court finds that a contract or lease was unconscionable, in whole or in part, when it was made, the court can either refuse to enforce the contract or lease or enforce the parts that are fair.

In other aspects of contract formation, Common Law prevails when Article 2 and 2(a) are silent. Essentially, the common law interpretations of unconscionability apply. The same is true when undue influence, fraud, mistake, innocent misrepresentation, and coercion is present. Consider the following case where the court must decide if there is a mutual mistake in the contract formation, thus negating either the contract or the offending clause.or if there is a breach of contract.

[10] UCC § 2-201(1).
[11] UCC § 2A-201(1).
[12] UCC §§ 2-302 and 2A-108.

LARSON v. BURTON CONSTR., INC.
2018 WY 74 (2018)

FACTS: In December 2015, Ken Burton, on behalf of Burton Construction, Inc., and Mr. Larson entered into a written contract (the Contract) in which Mr. Larson agreed to purchase from Burton a 2015 Skyline mobile home for $43,000. The Contract was a "form" contract, consisting of boilerplate provisions and blank lines on which parties could enter information. Mr. Larson's agent completed the form by hand, obtained Mr. Larson's signature, and delivered the document to Burton with $500 in earnest money. Burton went through the Contract carefully, made a few minor changes, and countersigned it.

Although the Skyline mobile home was new and the parties intended to convey a new mobile home, the form was designed for the sale of a used mobile home. Thus, the Contract required Burton, as the "Owner" of the mobile home, to execute and deliver a Wyoming title at closing. The contract stated:

6. At closing, Owner shall execute and deliver a Wyoming title and Bill of Sale free and clear of all liens and encumbrances except: all will be free and clear.

7. Title shall be conveyed to the following named purchaser(s): Justin James Larson as Sole Owner. Title and the Bill of Sale shall be executed but escrowed if sold under contract.

8. Title shall be merchantable in the Seller, except as stated in this paragraph. Subject to payment or tender as above provided and compliance with other terms and conditions hereunder by Purchaser, the Seller shall execute and deliver a good and sufficient title in favor of those persons named in Paragraph 7 above, and a good and sufficient Bill of Sale, and deliver the same to said Purchaser at closing, which shall occur on or before January 15, 2016, conveying said personal properties free and clear of all liens and encumbrances, except:

(a) Personal property taxes for 2016, which shall be prorated to date of closing;

(b) Purchaser shall be responsible for sales tax, if any.

Receiving no title at closing, the buyer attempts to revoke the sale and sues for breach of contract. The Seller claims that there is no breach as the custom in Wyoming was to deliver the title at a later time.

The seller Burton claimed that despite the Contract's clear language, Burton testified that it was not his custom to deliver a Wyoming title to purchasers of new mobile homes at closing. Accordingly, Burton never intended to deliver a Wyoming title to Mr. Larson at closing because "[t]hat's not how you do it." Mr. Larson, as the second titleholder, would be deprived of the warranty.

ISSUE: Is this Mutual Mistake or Breach of Contract? Burton brought suit in circuit court for breach of contract, seeking either specific performance or damages, as well as attorney fees and costs. After a bench trial, the circuit court found that the parties made a mutual mistake, cancelled the Contract, and ordered Burton to return Mr. Larson's earnest money. On appeal, the district court reversed, ruling that the circuit court erroneously applied the doctrine of mutual mistake, that it was "factually unknown" whether Burton delivered a Wyoming title at closing, and that Mr. Larson breached the Contract when he refused to attend closing and to pay sales tax. Pursuant to W.R.A.P. 13, we granted Mr. Larson's petition for writ of review

COURT'S RATIONALE: There is no reason to venture beyond the four corners of the Contract to determine that Burton was required to deliver a Wyoming title at closing. The parties argued, and both lower courts found, that the Contract's language is unambiguous, and we agree. The Contract states that closing must occur by January 15, 2016, at which time Burton must deliver a Wyoming title: "At closing, Owner shall execute and deliver a Wyoming title and Bill of Sale free and clear of all liens and encumbrances."

We will construe contract language "in the context in which it was written, looking to the surrounding circumstances, the subject matter, and the purpose of the agreement to ascertain the intent of the parties at the time the agreement was made." However, we will not "rewrite contracts under the guise of interpretation, and so long as there is no ambiguity, we are bound to apply contracts as they have been scrivened." "[C]ourts apply an objective theory of intent whereby one party's intention is deemed to be that meaning a reasonable person in the position of the other contracting party would ascribe to the first party's manifestations of assent." A reasonable layperson, such as Mr. Larson, would not be privy to a dealer's process of transferring title. Further, although parties to a contract are presumed to know the law, we do not attribute to them a complete understanding of the statutes governing sales of mobile homes. "According to our established standards for interpretation of contracts, the words used in the contract are afforded the plain meaning that a reasonable person would give to them."

(continued)

HOLDING: We affirm the district court's reversal of the circuit court's finding of a mutual mistake of the parties. We reverse the district court's decision to hold Mr. Larson in breach of the Contract and conversely find that Burton breached the Contract. We remand for further proceedings consistent with this opinion.

SIGNIFICANCE: The courts give great defference to the literal language of the contract and will avoid any reformation of the parties' intentions.

CRITICAL THINKING

Think about the concept of breach of contract which you will be studying in the next chapter. But, as this case illustrates and which is expanded in the Ethical Decision Making box below, does "intention" of the parties impact a court's consideration of breach. We will look at this more in the next chapter, but it is wise to anticipate what is actually forthcoming. For a preview of this issue, see the case nugget on page 327.

ETHICAL DECISION MAKING

It is fairly clear here from the facts that neither party deliberately attempted to breach the contract. It seems to be honest mistakes about what the standard practice and procedures were in the physical transferring of title in Wyoming. Why do you think the court was relatively harsh in its ruling? Or was the court harsh at all? Consider the Practical Tips for Business Managers below when formulating your response.

Title

Once the sales contract has been created and the parties are obligated to perform, the next issue to consider is the concept of title. *Title* means ownership. The most important aspect of title is that a person must have title to pass title. In other words, you can't sell something that you don't own. Moreover, you can't give it away or use it as collateral for a loan. So what does it mean to have title?

Three Kinds of Title

good title
Title acquired from someone who already owns the goods free and clear.

void title
Not true title; for example, the title held by someone who knowingly or unknowingly purchased stolen goods.

voidable title
Title that occurs in situations in which the contract between the original parties would be void, but the goods have already been sold to a third party.

There are three kinds of title: good title, void title, and voidable title. First, **good title** is title that is acquired from someone who already owns the goods free and clear. Next, **void title** is not true title. Someone who purchases stolen goods, knowingly or unknowingly, has void title. Finally, **voidable title** occurs in certain situations in which the contract between the original parties would be void but the goods have already been sold to a third party. The next few sections of the chapter discuss these types of title in detail.

Acquiring Good Title

LO 15-6 Describe the concept of title, how title passes.

The most obvious way of attaining good title is acquiring it from someone who has good title, that is, the person who owns it free and clear, without any qualifications. In contrast, someone who has come into possession of stolen goods never has good title and can never pass a good title. A person in possession of stolen goods has void title.

However, problems may arise when someone thinks he or she has good title but actually has void title. Consider Case 15-2, which concerns the sale of a stolen vehicle and whether a car dealership's title was voidable or void.

ALFONSO CANDELA, APPELLANT, V. PORT MOTORS INC., RESPONDENT, ET AL.

Defendant Supreme Court of New York, Appellate Division, Second Department

208 A.D.2d 486; 617 N.Y.S.2d 49 (1994)

FACTS: *Candela purchased a car for approximately $19,000 from Port Motors Inc. The police later seized the car, claiming that it may have been stolen. Candela brought suit against Port Motors for breach of warranty of title. Port Motors's defense was that it indeed passed a good title to Candela. New York's Supreme Court, which is New York's trial court level, granted Port Motors' motion for summary judgment, agreeing with Port that the car dealer had voidable title. The case went up on appeal and considered the issue of whether good title can ever pass for stolen goods.*

ISSUE: Can a good-faith seller of stolen goods pass a voidable or even good title to a good-faith innocent buyer?

REASONING: In a word, no. Good faith and innocence have no bearing whatsoever on this matter. Someone who sells stolen goods, regardless of the degree of good faith, can never pass a voidable or good title to the buyer. The buyer has a void title, and therefore the rightful owner can reclaim the goods.

DECISION AND REMEDY: The court of appeals held that the only remedy is that, if the buyer did not receive a good title to the goods from the seller, the buyer could sue the seller for the money paid.

SIGNIFICANCE OF THE CASE: The significance of the decision is that if Port Motors purchased the vehicle from a thief, or from the successor to a thief, Port Motors' title would be void and Port Motors could not have conveyed good title to Candela.

CRITICAL THINKING

The appellate court reversed the trial court's decision, which had granted a summary judgment to Port Motors. Candela will have the opportunity to have a court rule on whether Port Motors passed good title to him. What key facts will determine whether the title was void or voidable?

ETHICAL DECISION MAKING

Why would Port Motors Inc. fight Candela in court? Shouldn't the car dealer want to make its customers happy? What ethical norm must be underlying Port Motors' behavior in contesting Candela's claim?

The key point to remember, which many people misunderstand, is that good faith is actually irrelevant when passing a void title. If a person has a void title (as in the best, and most frequent, example: a person has possession of stolen goods), then no matter how honorable the intentions of the seller are, that good-faith seller cannot pass anything to the buyer but another void title. The only exception to this is the entrustment situation, which is discussed later in the chapter.

Voidable Title

When a seller transfers goods to a buyer, normally the buyer gets good title. However, the buyer gets only voidable title if any of the following apply:

- The buyer deceived the seller regarding his or her true identity.
- The buyer wrote a bad check for the goods.
- The buyer committed criminal fraud in securing the goods.
- The buyer and seller agreed that title would not pass until some later time.
- The buyer is a minor.

The first four of these situations that can create a voidable title are articulated in Section 2-403 of Article 2 of the UCC; the final one, the buyer is a minor, comes from common law. A seller

who discovers any of these situations has the right to cancel the contract and reclaim the goods, even if they have already been delivered to the buyer. That is why the title is called *voidable*.

Entrustment

entrustment
The transfer of goods to a merchant who ordinarily deals in that type of goods. If the merchant subsequently sells them to a good-faith third-party purchaser, the buyer acquires good title to the goods.

If an owner *entrusts* the possession of goods to a merchant who deals in goods of that kind, the merchant can transfer all rights in the goods to a buyer in the ordinary course of business. One form of **entrustment** occurs when someone entrusts the possession of a good (e.g., a car) to a merchant and asks him to repair that good. Suppose the merchant fraudulently or accidentally sells the item as if it were part of his inventory.

If the purchaser is a good-faith purchaser (i.e., did not know that the item belonged to someone else and paid a fair market value, not some unreasonable discounted value), then that purchaser gets a good title. The original owner's only recourse is to bring suit against the merchant. Again, the rationale behind this concept is the facilitation of commercial activity regarding good-faith purchasers in the marketplace. In Case 15-3 below characterizes issues with title from one of the most tragic occurrences in human history.

CASE 15-3

FRENK v SOLOMON
2018 N.Y. Misc. LEXIS 3846 (2018)

FACTS: This matter sadly begins with the Nazi Holocaust, "the most tragic event of our time." This action involves the **art** collection of the late Paul Westheim, a Jewish **art** critic, who specialized in German expressionist **art**. Among over 3,000 paintings, sculptures, watercolors, drawings, and graphics, Mr. Westheim's collection allegedly included the five artworks at issue in this case: Paul Klee's *Opus 18;* Otto Mueller's *Drei Akte* (also known as *The Bathers*); Max Pechstein's *Portrait of Paul Westheim*(also known as *Portrait of a Standing Man*); Edgar Jene's *Plastische Imagination;* and Erich Heckel's *The Violinist* (collectively, Five Works).

In 1933, Mr. Westheim fled Nazi Germany for Paris and left his collection with Charlotte Weidler, an **art** dealer in Berlin. During World War II, Mr. Westheim fled from France to Spain, then Portugal. Ultimately, he settled in Mexico in 1941, where he met and was later married to Marianna Westheim-Frenk, plaintiff's mother. He also corresponded extensively, and in code, with Ms. Weidler during the war; however, those "vivid" communications terminated, according to Mr. Westheim, after the war when he inquired as to the whereabouts of his **art** collection. Indeed, Mr. Westheim's restitution claim, filed in West Germany in the 1950s, was thwarted by an affidavit of Ms. Weidler's sister, who affirmed that Mr. Westheim's entire collection had been destroyed during the war in a bombing; thus, Mr. Westheim was prevented from establishing Nazi persecution as the cause of the loss.

In this action, plaintiff asserts that Mr. Westheim's **art** collection was not destroyed; rather, she alleges that Ms. Weidler stole the collection and began to sell artworks from it soon after Mr. Westheim died. Westheim passed away in 1963, Ms. Weidler passed away in 1983, and Ms. Westheim-Frenk passed away in 2004.

<u>1973 Action and Release</u>

In 1973, Ms. Westheim-Frenk learned that Ms. Weidler had sold a painting from Mr. Westheim's collection, *Portrait of Dr. Robert Freund,* to a gallery in New York City. Ms. Westheim-Frenk commenced an action against Ms. Weidler and others (1973 Action) for damages and for "possession of all items of [Mr. Westheim's] **art** collection" in Ms. Weidler's custody. (Ms. Westheim-Frenk's 1973 Action unverified complaint, *Marianna Westheim v Serge Sabarsky Gallery, Inc. et al.,* Index No. 16877/1973 [Sup Ct, NY County], NYSCEF Doc. No. 138.) In the 1973 Action, Ms. Westheim-Frenk also sought an account of Ms. Weidler's actions regarding Mr. Westheim's **art** collection and "whatever sums of money are found to be due . . . on the basis of saidaccount." (*Id.*)

A pre-answer motion to dismiss was served in the 1973 Action, but—before the motion was decided or any discovery was exchanged—the 1973 Action was discontinued "with prejudice" by stipulation, dated March 21, 1974. (*See* NYSCEF Doc. Nos. 138, 141-142.) Ms. Westheim-Frenk, who was represented by New York counsel, contemporaneously executed a blanket release (Release) discharging Ms. Weidler and Ms. Weidler's

"heirs, executors, administrators, successors and assigns of and from all manner of actions, causes of action, suits, debts, dues, sums of money, accounts, reckoning, bonds, bills, specialties, covenants, contracts, controversies, agreements, promises, variances, trespasses, damages, judgments, extents, executions, claims and demands whatsoever, in law,

(continued)

in admiralty, or in equity which against . . . [Ms.] Weidler, [that Ms. Westheim-Frenk] ever had, now have, or which [Ms. Westheim-Frenk] or [her] heirs, executors, or administrators, hereafter can, shall or may have for, upon or by reason of any matter, cause or thing whatsoever from the beginning of the world to the day of the date of these presents."

The Release specified that it "maynot be changed orally" (*id.*), and no other documents set forth the terms of the settlement of the 1973 Action. In consideration for the Release, Ms. Westheim-Frenk received $7,500.

Ms. Weistheim-Frenk's daughter now brings an action against Ms. Weidler's daughter, Ms. Solomon, and her two sons for the return of the five paintings.

ISSUE: What is the validity of documents so old that there is no one living to verify the legitimacy of the papers?

REASONING: Here, all parties ask this court to consider documents dating back to the 1930s; for example, to establish that the Five Works belonged to Mr. Westheim, plaintiff relies on two documents Paul Westheim drafted in the 1960s. Defendants' conclusory attack on plaintiff's documents as fraudulent, inaccurate, and invalid is without merit and lacks an admissible factual basis. Notably, in violation of Commercial Division Rule 13, the opinion of defendants' expert (Expert Opinion) is unsigned and undated, except for a translator's certificate of accuracy, dated December 5, 2016; thus, for the purposes of this motion, the court deems the various documents submitted by both parties, and those referenced in the Expert Opinion, to be ancient and admissible.

"Under the 'ancient document' rule, a record or document which is found to be more than 30 years of age and which is proven to have come from proper custody and is itself free from any indication of fraud or invalidity 'proves itself.' This rule dispenses with the proof of the execution of a record or document on the proof of its antiquity. It presumes that the entrant of the record or document is dead after the passage of 30 years. If the genuineness of an ancient document is established, it may be received to prove the truth of the facts that it recites."

Thus, Mr. Westheim's ownership of the five paintings is established.

ISSUE #2 Isn't Ms. Weidler holding of the paintings an entrustment and may pass a good title on to a buyer in the ordinary course of business even when she is not the title-holder?

DECISION AND REMEDY: Defendants admitted that, between 1977 and 1983, Ms. Weidler sold, traded, or gave four of the Five Works to defendant Ms. Solomon. Further, Ms. Solomon admitted that, in 1998, she helped **art** collector Gour Shoshan sell the fifth artwork, *The Violinist*. Defendant Yris Rabenou Corporation is listed on the auction's schedule of property as the "OWNER" of *The Violinist*.

Citing the Uniform Commercial Code's Doctrine of Entrustment, defendants argue that Ms. Weidler held the power to transfer good title of the Five Works to defendants. Under this theory, because Mr. Westheim entrusted his **art** collection to Ms. Weidler, a merchant of **art,** Ms. Weidler had the power to transfer Mr. Westheim's full interest in the collection to buyers, like defendants, acting in the ordinary course of business.

"A defendant's status as a buyer in the ordinary course is an affirmative defense." However, defendants failed to plead this affirmative defense in their answer. In any event, even if defendants had timely raised that defense, Mr. Westheim left the collection with Ms. Weidler when he fled persecution in Nazi Germany; that does not constitute the ordinary, measured entrustment contemplated in the UCC [noting that a "signature at gunpoint cannot lead to a valid conveyance"].) Any conveyances by gift, sale, or otherwise to Ms. Weidler under those circumstances are suspect. The inadmissible opinion of defendants' expert does not establish that Mr. Westheim intended that Ms. Weidler would sell his **art** collection and send him the proceeds; moreover, the Expert Opinion is based on "decod[ing] the special language that Weidler and Westheim had been using" and necessarily conclusory. The court rejects defendants' entrustment argument.

ISSUE #3 Is the release enforceable?

DECISION AND REMEDY: Plaintiff argues that the Release does not bar this action because the Release was limited to only the painting, entitled *Portrait of Dr. Robert Freund,* the sale of which by the New York **art** gallery caused Ms. Westheim-Frenk to commence the 1973 Action. Plaintiff contends that the Five Works are not covered by the Release, and, even if they are so covered, she argues that a subsequent 1976 transaction constitutes waiver of the Release. Alternatively, plaintiff argues that Ms. Westheim-Frenkwas fraudulently induced by Ms. Weidler to execute the Release, and also asserts that defendants should be equitably estopped from using the Release as a shield because plaintiff alleges that Ms. Weidler is a thief.

None of plaintiff's allegations, read as a whole and afforded the light most favorable to plaintiff, are sufficient to raise a triable issue of fact as to fraud and fairness at the time of the Release. The alleged fraudulent misrepresentation was precisely the subject of the terminated 1973 Action: whether there were more artworks from Mr. Westheim's collection. "[A] party that releases a fraud claim may later challenge that release as fraudulently induced only if it can identify a separate fraud from the subject of the release." "Were this not the case, no party could ever settle a fraud claim with any finality."

Plaintiff also fails to establish a misrepresentation by Ms. Weidler at the time of the Release. Ms. Westheim-Frenk failed to condition the Release on the truth of the information, if any, that was provided to her–*i.e.,* that there were no other artworks from Mr. Westheim's collection. (*See e.g. id.* [noting that "plaintiffs were allegedly aware

(continued)

that they lacked a full picture of Conecel's internal finances and that the relationship between the parties had become adversarial, yet they failed to condition the release on the truth of the information supplied by defendants, obtain representations or warranties to that effect, or insist on viewing additional information"].) While Ms. Westheim-Frenk may have been unsophisticated in such matters, she was represented by New York counsel during the 1973 Action and the Release. Moreover, there was no apparent reason to rush to settlement.

Finally, although plaintiff also urges the court to deny the motion under the doctrines of equitable estoppel and public policy, "a court is not free to alter [an unambiguous] contract to reflect its personal notions of fairness and equity."

SIGNIFICANCE OF THE CASE: The court sympathizes with plaintiff's position; nevertheless, plaintiff fails to rebut defendants' prima facie case by raising a question of material fact requiring a trial. It is for this reason that summary judgment is granted and the case is dismissed.

CRITICAL THINKING

This case truly demonstrates the long-lasting effects of the horror of the holocaust. As late as 2018, the long shadow of Nazism still casts a pall on society. Do you agree with the court's rationale. Is the court too "technical" in its interpretation of the release contract. Should the court have considered the doctrine of unconscionability?

ETHICAL DECISION MAKING

Does this case create any emotional feelings in you? Is this just a case in commercial transaction? Is this yet another case where parties must be acutely aware of the language present in any contract?

Title Versus Other Interests in the Goods

The determination of who has good title does not always result in the expected and equitable solution to a problem. For example, suppose you purchased a couch at a furniture store. It is pretty safe to say that once you identify the couch and pay for it, you have title to it. Now suppose that when the store attempts to deliver the couch to your home, lightning hits the store's delivery truck and destroys the couch. Is the store legally obligated to replace the couch? Most of us would intuitively answer that it is. After all, you never took possession of the couch. However, as previously noted, *you* have title. Under pre-UCC law, this loss could have fallen on whoever had title at that time—and that person would be you. Under the UCC, if the store is a merchant, the risk of loss remains with the seller until the couch is actually delivered to you.

As a result of such dilemmas, the UCC breaks up the various issues traditionally correlated with title and treats them separately. Several issues normally are thought of under the concept of title:

- *Ownership:* When does title actually transfer from the seller to the buyer, since the right to transfer ownership of the goods, whether through a subsequent sale or gift, is tied to title?
- *Encumbrance:* The right to encumber goods as collateral for a debt depends on who is holding title. When title passes is important because having title means that one can then sell or encumber the goods. In other words, having title means that one can pass title.
- *Loss:* In regard to the right to indemnification if the goods are damaged, when the risk of loss attaches is important. This is because, regardless of title passing, we need to know the seller's and buyer's responsibility to each other in the event that the goods are damaged or destroyed prior to the buyer taking complete possession of the goods.
- *Insurable interest:* An *insurable interest* is the right to insure the goods against any risk exposure such as damage or destruction. When an insurable interest is created in the goods is important. Both the buyer and the seller can insure themselves for potential loss in the event the goods are damaged or destroyed at some point in the transaction. A key point is identifying the earliest time in the transaction that the buyer can claim an insurable interest. The following Case Nugget addresses that issue.

CASE Nugget

When Does Insurable Interest Arise?

National Compressor Corp. v. Carrow and McGee Federal Circuit Court of Appeals

417 F.2d 97 (1969)

National Compressor bought a large compressor from Davis, who had bought the compressor from Carrow and McGee. Title to the compressor was not to pass to National Compressor until the compressor was removed from Carrow and McGee's property. Prior to the compressor being moved (and title passing to National Compressor), a fire broke out at the site, destroying the compressor. National Compressor had already paid the $12,000 purchase price. The issue before the court was whether National Compressor had any kind of insurable interest since clearly title had not yet passed. The defendants claimed that National Compressor had no standing to bring the lawsuit because it had no interest in the property since title had yet to pass.

The court ruled otherwise, stating that National Compressor had an insurable and special interest in the good, thus giving it standing to sue. The court ruled:

> Where a third party so deals with goods *which have been identified to a contract* for sale as to cause actionable injury to a party to that contract (a) a right of action against the third party is in either party to the contract for sale who has title to or a security interest or a special property or an insurable interest in the goods; and if the goods have been destroyed or converted a right of action is also in the party who either bore the risk of loss under the contract for sale or has since the injury assumed that risk as against the other....

The buyer obtains a special property and an insurable interest in goods by identification of existing goods as goods to which the contract refers, even though the goods so identified are nonconforming and he has an option to return or reject them. Such identification can be made at any time and in any manner explicitly agreed to by the parties. In the absence of explicit agreement, identification occurs (a) when the contract is made if it is for the sale of goods already existing and identified.

One very important point to note is that the parties are always free to create a contract that lays out and defines such issues as when title passes and when risk of loss passes. The UCC's rules are essentially the default rules for contracts that do not clearly spell out such provisions.

Let's look at each one of these issues within the context of the four kinds of sales contracts that Article 2 creates.

Types of Sales Contracts

The UCC lays out essentially four broad factual scenarios for the sale of goods:

1. *Simple delivery contract:* A simple delivery contract occurs when the purchased goods are transferred to the buyer from the seller at either the time of the sale or some time later by the seller's delivery.
2. *Common-carrier delivery contract:* This type of contract occurs when the goods are delivered to the buyer by a common carrier, such as a trucking line.
3. *Goods-in-bailment contract:* This type of contract occurs when the purchased goods are in some kind of storage under the control of a third party, such as a warehouseman.
4. *Conditional sales contract:* A conditional sales contract occurs when the sale itself is contingent on approval, for example.

In the sections that follow, this chapter answers questions about ownership, encumbrance, loss, and insurable interest as they relate to these four types of sales contracts.

Simple Delivery Contract

simple delivery contract
Type of contract in which purchased goods are transferred to the buyer from the seller either at the time of the sale or at some time later by the seller's delivery.

With a **simple delivery contract**, the buyer and seller typically execute an agreement, and the buyer leaves with the goods. To most of us, it appears that title, risk of loss, and insurable interest all pass to the buyer at the moment the transaction is consummated and the buyer walks out with the goods. However, under the UCC, there are three distinct steps: (1) Title transfers to the buyer when the goods are identified to the contract, that is, when the contract is executed; (2) risk of loss transfers to the buyer when the buyer takes possession; and (3) insurable interest is created in the buyer when the goods are identified to the contract, in other words, at the same time that title passes.

But what happens if the buyer comes back later to pick up the goods or arranges to have the seller, or an agent of the seller, deliver the goods at a later time? For the purposes of title and insurable interest, nothing really changes. The dilemma occurs with risk of loss.

Let's suppose a buyer and seller execute a contract in which the seller will deliver a refrigerator later in the day to the buyer. Through no fault of the seller, the refrigerator is damaged in a fire at the seller's store. Who has the risk of loss if neither party is at fault? In this case, the issue rests on the seller's status. If the seller is a merchant, as in this instance, the risk of loss remains with the seller until the goods are actually delivered to the buyer. If the seller is not a merchant, the risk of loss remains with the buyer under the rule of **tender of delivery**.

tender of delivery
Requirement that the seller or lessor have and hold conforming goods at the disposal of the buyer or lessee and give the buyer or lessee reasonable notification to enable him or her to take delivery.

Simply put, tender of delivery is the moment the goods were available for the buyer to take. Consider this example: You purchase a dresser at a garage sale, but you want to go home and get your truck so that you can get it home easily. Unfortunately, a car hits the dresser and destroys it. You, the buyer, cannot get your money back because (1) the law does not consider the seller a merchant and (2) you could have taken the dresser with you when you bought it.

Note that the results are different if either party is at fault for the damage. In that case, the responsible person is liable under tort law for the damage caused. With a simple delivery contract whereby a seller transfers goods to a buyer without the middle-delivery common carrier, the various interests transfer as the following shows. (Note: Even if the seller has its agent deliver the goods to the buyer, this is still a simple delivery.)

Simple delivery: seller → buyer

1. Title transfers on identification of the goods to the contract.
2. If the seller is a merchant, risk of loss transfers on delivery of the goods to the buyer; if the seller is not a merchant, risk of loss transfers when the goods are made available for the buyer to possess (tender of delivery).
3. The parties may buy insurance on their goods if they hold title or have any risk of loss or other economic interest.

Common-Carrier Delivery Contract

LO 15-7 Explain risk of loss.

common-carrier delivery contract
A type of contract in which purchased goods are delivered to the buyer via an independent contractor, such as a trucking line.

If a buyer and seller execute a contract and the seller subsequently places the goods with a common carrier for delivery to the buyer, the parties have executed a **common-carrier delivery contract**. Note that a common carrier is an independent contractor and not an agent of the seller. What makes the common carrier an independent contractor, rather than an agent, is that the carrier controls the primary aspects of performance, such as how the goods are actually delivered.

The UCC names two kinds of delivery contracts in this category: origin, or shipment, contracts and destination contracts. *Shipment contracts* require the seller to ship the goods to the buyer by a common carrier. The seller is required to make proper shipping arrangements and deliver the goods into the common carrier's hands. Title passes to the buyer at the time and place of shipment. Thus, the buyer bears the risk of loss while the goods are in transit. *Destination contracts* require the seller to deliver the goods to the destination stipulated in the sales contract. This may be the buyer's place of business or some other location. The seller bears the risk of loss until that time. Exhibit 15-6 identifies shipping terms that create the conditions of transit and delivery.

TERM	EXPLANATION
FOB (free on board)	The selling price includes transportation costs, and the seller carries the risk of loss to either the place of shipment or the place of destination.
FAS (free alongside)	The seller, at seller's expense, delivers the goods alongside the ship before the risk passes to the buyer.
CIF or C&F (cost, insurance, and freight; cost and freight)	The seller puts the goods in possession of a carrier before the risk passes to the buyer. Contracts are usually shipment contracts rather than destination contracts.
Delivery ex-ship (delivery from the carrying vessel)	Risk of loss passes to the buyer when the goods leave the ship.

Exhibit 15-6

Shipping Terms Specifying Requirements for Delivery

With a shipment contract, whereby a seller transfers goods to a buyer with delivery of the goods effected by a common carrier, the various interests transfer as follows:

Shipment contract: seller → common carrier → buyer

1. If the shipment contract is an origin contract (if the contract is vague or ambiguous, an origin contract will be presumed), the title passes to the buyer when the goods are turned over by the seller to the common carrier.
2. If the shipment contract is a destination contract, the title transfers from the seller to the buyer when the common carrier delivers the goods to the buyer.
3. The risk of loss in either situation—origin or destination contract—transfers from the seller to the buyer simultaneously with the title.
4. An insurable interest is created when the buyer and/or seller holds title or retains a risk of loss.

Goods-in-Bailment Contract

A **goods-in-bailment contract** involves goods that are in some kind of storage (e.g., in a warehouse or on board a ship), so the seller cannot transfer physical possession of them. Instead, the seller has one of three documents indicating ownership of the goods: a negotiable document of title, a nonnegotiable document of title, or a contract or other instrument showing ownership that is not a negotiable or nonnegotiable document of title. If the seller has a negotiable document (i.e., a document containing the words "deliver to the order of [seller]"), then both title and risk of loss transfer from the seller to the buyer as soon as that negotiable instrument is endorsed over to the buyer. On the other hand, if the document is nonnegotiable (i.e., a document lacking the words "to the order of"), then the title passes with the instrument of title, but the risk of loss does not pass to the buyer until the bailee (the custodian of the goods) is notified of the transfer or a reasonable time has elapsed since the transaction. Finally, if there is neither a negotiable or a nonnegotiable document of title, then the title passes at the time the sales contract is executed, but the risk does not pass to the buyer until the bailee is notified of the transaction and acknowledges such notification.

goods-in-bailment contract
A type of contract in which the purchased goods are in some kind of storage under the control of a third party, such as a warehouseman.

With goods in bailment, the three interests—title, risk of loss, insurable interest—pass in the following way:

Seller → buyer, but the goods are elsewhere in some kind of storage and in a third party's possession and care.

1. Title passes from the seller to the buyer when the document of title (e.g., a warehouse receipt or a bill of lading) is actually endorsed or signed over to the buyer. If there is no document of title, then title passes when the goods are identified to the contract and the contract is executed.

2. Risk of loss passes to the buyer simultaneously with the document of title, provided that the document of title is a negotiable one. If it is nonnegotiable, the risk does not pass until the bailee (the possessor or custodian of the goods) is notified or a reasonable time elapses. If there is no document of title, risk passes to the buyer on notification and acknowledgment by the bailee.
3. Insurable interest is created when either party has title, risk of loss, or some other economic interest attached to the goods (e.g., a creditor who secures a loan by taking the goods as collateral).

Conditional Sales Contract

Conditional sales contracts are either sale-on-approval contracts or sale-or-return contracts. A contract is a **sale-on-approval contract** if the seller allows the buyer to take possession of the goods before deciding whether to complete the contract by making the purchase. Title and risk of loss remain with the seller until the buyer notifies the seller about the approval of the contract.

A **sale-or-return contract** occurs when the seller and buyer agree that the buyer may return the goods at a later time. Such contracts usually occur when the buyer is buying inventory to resell. For example, suppose the seller is a dress wholesaler and the buyer is a retailer who is purchasing the dresses to sell in her store. In their sale-or-return agreement, the insurable interest is created in the buyer once the goods are identified in the contract. Title and risk of loss depend on whether the goods are in bailment, delivered by common carrier, or delivered by the seller himself. Without an agreement to the contrary, if the buyer subsequently returns the dresses, she does so at her own expense and risk.

> **Conditional sales contract**
> A type of contract in which the sale itself is contingent on approval; can be either a sale-on-approval contract or a sale-or-return contract.
>
> **sale-on-approval contract**
> A contract in which the seller allows the buyer to take possession of the goods before deciding whether to complete the contract by making the purchase.
>
> **sale-or-return contract**
> A contract in which the buyer and seller agree that the buyer may return the goods at a later time.

Risk of Loss during a Breach of Contract

When the Seller is in Breach

The failure to deliver goods is the most common way a sales contract is breached. If the seller does not provide the goods that were described in the contract, the buyer may either (1) accept the nonconforming goods as-is or (2) reject the goods subject to the seller curing the deficiency in the goods. The buyer may reject the goods if no cure is possible or the seller fails to cure the deficiency within a reasonable time. In all these instances, the risk of loss remains with the seller until either the buyer accepts the goods or the seller cures the deficiency and provides the buyer with conforming goods.

If a cure is not possible or if the seller has failed to cure the deficiency within a reasonable time, the buyer has the option to revoke the contract. (The remedies are discussed in subsequent chapters.) However, the UCC creates a disincentive for buyers who do so: If the risk of loss would have transferred to the buyer had there not been a breach, the risk transfers to the buyer to the extent of any insurance the buyer has. The loss reverts to the breaching seller only to the extent that the buyer's insurance does not cover it. Consider the following case nugget regarding a seller who is accused of being in breach in breach.

CASE Nugget

Anyone for a Bar-B-Que?

Chicago Prime Packers, Inc. v. Northam Food Trading Co., United States Court of Appeals for the Seventh Circuit 408 F.3d 894 (2005)

Chicago Prime Packers Inc. is a meat wholesaler. Northam Food Trading Co. is a Canadian partnership that distributes meats to retailers and restaurants. Northam entered into a contract for the purchase of 40,500 pounds of pork back ribs for a purchase price of $178,200.

(continued)

Chicago Prime had title to the ribs, having purchased them from Brookfield Farms. Brookfield Farms held those ribs in B&B Cold Storage. So, we have a classic case of a contract for the sale of goods with shipment to take place upon the execution of the sales contract between Chicago Prime and Northam. When Chicago Prime and Northam inked the contract, Brown Brother's Trucking Company picked up the frozen ribs with its freezer trucks and delivered the ribs to Northam.

Northam discovered that the ribs were in an "off condition" and rejected the ribs as non-conforming. The USDA inspector found them to be spoiled and inedible. Northam refused to pay the purchase price, and Chicago Prime sued. Chicago Prime claimed that when the frozen ribs were put into Brown Brother's possession, the risk of loss transferred to Northam and that Northam owes the purchase price to Chicago Prime. Northam refuses to pay.

1. Do title passing and risk of loss passing occur at different times? Since the buyers could not show any negligence on the part of Chicago Prime, title and risk of loss passed to Northram upon transfer from Chicago Prime to Brown's Trucking. Northram has both title and risk now. The key is that the seller was actually not in breach, though Northram argued that the seller was but cold not prove it.
2. Could Brown Brother's Trucking be at fault here? Absolutely! Both Chicago Prime and Northram are non-breaching parties. The fault seems to lie with the shipper.
3. Northram is ordered to pay Chicago Prime the purchase price plus interest.

When the Buyer is in Breach

Most buyer breaches occur when a buyer refuses to accept conforming goods from the seller and then the goods are subsequently lost or damaged. With an origin or shipment contract, the risk would have already transferred to the breaching buyer. However, if the contract is a destination contract, the risk remains with the seller. To encourage sellers to create origin contracts, the UCC requires the risk of loss to remain with the seller to the extent of the seller's insurance. If the seller does not have insurance or the loss exceeds the seller's insurance, the remainder transfers to the breaching buyer.

CASE OPENER WRAP-UP

On Your Mark, Get Set, Go!

It is uncontested that the agreement is a valid and enforceable contract between the parties. Rather, the heart of the parties' dispute is the construction of the terms "sale" and "sell," which are undefined in the agreement. The defendant argues that the terms "sale" and "sell" should be construed to include installment sales such as the one reflected by the purchase agreement, thereby making that a timely sale under the agreement, even though the entire purchase price would not be paid until five months later. We are unpersuaded.

The parties acknowledge that the sale of the 1969 Lola was a consignment sale under the terms of the agreement. Such a sale is governed by Article 2 of the Uniform Commercial Code (UCC). According to the UCC, "a 'sale' consists in the passing of title from the seller to the buyer for a price."

Thus, construing the purchase agreement pursuant to Article 2 of the UCC, the sale of the 1969 Lola would not occur until title passed to Gideon in March of 2017, well outside the October 20, 2016, deadline. Therefore, we conclude that the mere execution of the purchase agreement between the defendant and Gideon did not constitute a timely "sale" under the terms of the agreement.

PRACTICAL TIPS for Business Managers

1. Save yourself potential headaches by making clear when negotiating any contract as to whether you are interpreting the transaction as being governed by the Uniform Commercial Code. Also, make sure that the other party understands this! Plus, put that expectation into the contract, i.e. "The parties agree that this agreement is governed by Article 2 or 2(a) of the Uniform Commercial Code."
2. Language is very, very important. Even when a mistake is made is complete good faith, language come back and bite you if you are not very careful. One of the reasons to have experienced attorneys review all legal documents, managers need to develop a habit to closely read what representations are made on any document.
3. Speaking of language, parties are free to draft the language in any contract. The UCC provides the "gap" meaning or assumptions when contracts are silent. For example, the parties are free to determine when title actually passes. If the contract is silent, then the UCC fills in the "gap."

SUMMARY

The Uniform Commercial Code

Scope of the UCC: The UCC is a uniform or model law that governs commercial transactions, from the sale of goods to secured transactions.

Significance of the UCC: The UCC adds clarity and predictability to sales law.

Articles 2 and 2(A) of the UCC

Article 2 (Sales) covers contracts for the sale of goods.

- *Sale:* The passing of title from the seller to the buyer for a price.
- *Goods:* Tangible things that can be moved.
- *Mixed goods and services contracts:* Contracts that include both goods and services. Courts apply Article 2 if the goods are the predominant part of the transaction.
- *Merchants:* Buyers or sellers who (1) deal in goods of the kind, (2) by occupation, hold themselves out as having knowledge and skill unique to the practices or goods involved in the transaction, or (3) employ a merchant as a broker, agent, or other intermediary. Various provisions of the UCC distinguish merchants from ordinary buyers and sellers.

Article 2(A)(Leases) covers contracts for the lease of goods.

- *Leases:* Transfers of the right to possession and use of goods for a term in return for consideration.
- *Special kinds of leases:* Consumer leases and finance leases. The UCC outlines special rules and protections for each kind of lease.

Contracts for the International Sale of Goods

Scope of the CISG: The CISG is a treaty countries can sign to govern international business-to-business sales contracts. Many major trading nations have signed the CISG.

Significance of the CISG: The CISG is important because it, rather than the UCC, governs international sales contracts. The CISG provides clarity, predictability, and uniformity for businesses that operate in the global economy.

How Sales and Lease Contracts Are Formed under the UCC

Formation in general: The UCC is more lenient than common law regarding contract formation. Courts look at the intent of the parties to a sales or lease contract.

Offer and acceptance: Offers are valid even if terms are left open. The common law mirror-image rule does not apply to contracts for the sale or lease of goods. Courts look on a case-by-case basis to determine whether to allow additional terms.

Consideration: When sales and lease contracts are modified, these modifications do not need to be supported by new consideration.

Battle of the forms: The last form in either the offer or acceptance that a party sends to the other is presumed to control the final terms of the contract unless specifically objected to by the receiving party.

Statute of frauds: Contracts for the sale of goods must be in writing if the goods are priced at $500 or more. Lease contracts that require payments of $1,000 or more must be in writing. Exceptions exist for:

- Specifically manufactured goods.
- Contracts that parties admit exist.
- Situations in which partial performance has occurred.

Unconscionability: Under the UCC, a court can refuse to enforce the parts of a contract or lease that are unfair or one-sided.

Title

There are three kinds of title: good title, void title, and voidable title.

- *Good title* is title that is acquired from someone who already owns the goods free and clear.
- *Void* title is not true title. Someone who purchases stolen goods has void title.
- *Voidable title* occurs in certain situations in which the contract between the original parties would be void but the goods have already been sold to a third party.

Acquiring Good Title

Article 2 of the Uniform Commercial Code covers issues related to acquiring good title.

- The most obvious way to attain *good title* is to acquire it from someone who has good title.
- Someone who has come into possession of stolen goods never has title and can pass only *void title*.
- A buyer gets *voidable title* if he or she has deceived the seller regarding his or her true identity, written a bad check for the goods, committed criminal fraud in securing the goods, or is a minor, or if the buyer and seller agreed that title would not pass until some later time.

Third-party purchasers generally get good title.

If an owner *entrusts* the possession of goods to a merchant who deals in goods of that kind, the merchant can transfer all rights in the goods to a buyer in the ordinary course of business.

The UCC provides recourse for situations in which good title may not be enough for an equitable result. The UCC responds to issues related to the following:

- *Ownership* refers to transfer of title.
- *Encumbrance* refers to when goods may be used as collateral for a debt.
- *Loss* refers to who has the risk of loss, which matters when someone is seeking indemnification for damaged goods.
- *Insurable interest* refers to the right to insure goods against any risk exposure.

Types of Sales Contracts

A *simple delivery contract* is formed when the buyer and seller execute an agreement and the buyer leaves with the goods. *Title* transfers to the buyer when the contract is executed. *Risk of loss* transfers to the buyer when the buyer takes possession. *Insurable interest* is created in the buyer at the same time title passes.

A *common-carrier delivery contract* exists when a buyer and seller execute a contract and the seller subsequently places the goods with a common carrier. There are two types of common-carrier delivery contracts: origin, or shipment, contracts and destination contracts.

- In *shipment contracts,* title transfers to the buyer at the time and place of shipment. The buyer bears the risk of loss while the goods are in transit.
- In *destination contracts,* the seller bears the risk of loss until the seller delivers the goods to the destination stipulated in the sales contract.

A *goods-in-bailment contract* is one that identifies goods that are in some kind of storage. Rules regarding passage of title, risk of loss, and insurable interest vary depending on whether the seller has a negotiable document of title, a nonnegotiable document of title, or a contract showing ownership that is neither a negotiable nor nonnegotiable document of title.

A *conditional sales contract* includes sale-on-approval or sale-or-return contracts.

- In a *sale-on-approval contract,* title and risk of loss remain with the seller until the buyer notifies the seller about the approval of the contract.
- In a *sale-or-return contract,* the insurable interest is created in the buyer once the goods are identified in the contract. Title and risk of loss depend on whether the goods are in bailment, delivered by common carrier, or delivered by the seller.

Risk of Loss during a Breach of Contract

When the seller is in breach by failing to deliver goods, the buyer may either accept the nonconforming goods as-is or reject the goods subject to the seller curing the deficiencies in the goods. Risk of loss remains with the seller until the buyer accepts the goods or the deficiencies are corrected.

When the buyer is in breach because he or she has refused to accept conforming goods and then the goods are subsequently lost or damaged, who bears the risk of loss depends on the type of contract that exists between the buyer and the seller.

Point/Counterpoint

Should the United States Adopt the CISG as a Substitute for the UCC?	
YES	**NO**
Most global companies that do business with American businesses are in countries that have adopted the CISG. Although similar to the UCC, the CISG does have some material differences and is not, unlike the UCC, based on common law contract. For example, the CISG does not recognize the parol evidence rule; another example is that the CISG only recognizes international trade usages, not national or local ones. However, with commercial activity becoming more and more globalized, the United States really does need to join the rest of the world to uniformize transactional activities. At the very least, the UCC should be modified to conform with the CISG.	The development of the common law rules of contract have a solid history and evolved for specific reasons and rationale. These rules, such as the parol evidence rule, evolved for specific reasons germane to the history and context of the common law. These rules are necessary in the context of American transactional activity. It seems that the two sets of rules are indeed justified: when the contractual parties are from the United States, the UCC applies and should apply because the UCC principles are relevant to domestic transactions; the CISG, with its differences, should apply to international contracts because the context is different, thus requiring a different set of rules.

Questions & Problems

1. Explain what determines whether a contract is covered by the UCC.

2. Under common law, most property issues can be settled by who has a good title. Explain why such traditional characteristics of property ownership, such as transfer, risk of loss, insurable interest, and right to encumber are "broken up" and subject to varying tests under the UCC.

3. Plaintiff and defendant enter into an agreement in which the plaintiff licenses the defendant to resell goods for the plaintiff. There is a fee of $419,000 in order for the defendant to obtain this license. A dispute arises over the fee among other issues and plaintiff sues the defendant in part for non-payment of this fee. The suit is brought more than four years after the cause of action arose but less than six years. The significance of these time periods is that the UCC mandates a statute of limitations of four years while common law in New York state is six years. Which applies: the UCC or Common Law?

 Source: *Calltroll Corp. v. DialConnection, LLC*, 2016 N.Y. Misc. Lexis 1794 (2016).

4. The defendant contracted with the plaintiff to design, maintain, and develop software and to host the defendant's web page. After a falling out, the defendant moved its web page to another site and refused to honor its contract with the plaintiff. The plaintiff sued. Did the contract fall under the UCC? Were the items—software, web hosting, website maintenance—goods under the UCC?

 Source: *Dennis Conwell and Frank Splittorff, D/B/A Piece of America, Appellant, v. Gray Loon Outdoor Marketing Group, Inc., Appellee*, No. 82S04-0806-CV-00309, Sup. Ct. Ind., 906 N.E.2d 805; 2009 Ind. LEXIS 465; 69 U.C.C. Rep. Serv. 2d (Callaghan) 71 (2009).

5. A foreclosure case saw creditors going after the business's real estate. The business was a diner, however, the diner itself was a prefabricated building. The

business owner claimed that the building itself was not realty and subject to the creditors' claims but was, instead, a trade fixture. Please discuss.

Source: *J.K.S.P Restaurant v. County of Nassau,* 513 N.Y.S.2nd 716 (N.Y. App. Div. 1987).

6. A company contracts with a marketing firm to construct software and create a business website. A quote is requested and accepted. Sometime later, the business asks for updates and revisions but fails to request a quote. The business subsequently refuses to pay for the changes, claiming no contract was formed under the requirements of the offer under the UCC. The marketing firm then took the website down for nonpayment. The marketing firm then sued for nonpayment of its invoice; the company countersues on the tort of conversion for the loss of the website. This intriguing case offers up several questions, the first being whether this dispute even falls under UCC subject matter? Does it? The second is whether the claim of conversion will lie. Will it? [*Dennis Conwell et al. v. Gray Look Outdoor Marketing Group, Inc.,* 906 N.E.2d 805; 2009 Ind. LEXIS 465; 69 U.C.C. Rep. Serv. 2d (Callaghan) 71 (Ind. Sup. Ct. 2009).] This appeal arises from a dispute between a law firm, respondent Freidberg & Parker, LLP (Freidberg), and a company that provides technical electronic services, appellant Digital Document Technologies, LLC (DDT). It is undisputed that Freidberg asked DDT to process electronic media for use in litigation, and that Freidberg had asked DDT to do other work before. This time, DDT's emails with Freidberg at least arguably suggested the work would result in a modest charge, but when the work was done, DDT charged over $50,000. After a court trial involving expert testimony as to the reasonable value of DDT's services, the trial court awarded DDT just over $8,000, but denied interest or attorney fees. On appeal, DDT faults the trial court for not finding that Freidberg was a "merchant" under the Uniform Commercial Code (**UCC**), and not understanding that DDT's pleading embraced a **UCC** contract claim, which, if sustained, would have entitled DDT to an award of attorney fees, costs, and interest. For this transaction, is the law firm a merchant?

Source: Digital Documents Tech v. Friedman & Parker. 2015 Cal App unpublished Lexis 8225 (2015).

7. A manufacturer based in Argentina and an importer based in New Jersey entered into an oral agreement under which the importer agreed to sell the manufacturer's products in the United States. The district court found that the manufacturer's claim that the importer had failed to pay the full amount due was barred by the United Nations Convention on Contracts for the International Sale of Goods (CISG) because the agreement was not in writing. Although the CISG eliminated formal writing requirements for contracts, Argentina had made a declaration opting out of that provision. However, the United States had not made such a declaration. The court of appeals held that it was necessary to consider New Jersey choice-of-law rules to determine whether New Jersey or Argentine form requirements governed the manufacturer's claim. Does the contract need documentation to be enforceable, or, if U.S. law applies with the adoption of the CISG, are the oral declarations satisfactory?

Source: *Forestal Guarani S.A., Appellant, v. Daros International, Inc.,* No. 08-4488, 3d Cir. Ct. of Appeals, 614 F.3d 395; 2010 U.S. App. LEXIS 14969.

8. Mark Moore (Moore) is the owner of the Moore gravel pit (Moore Pit) north of Red Lodge, Montana. Goran, LLC (Goran) is a Utah Limited Liability Company contracted by the Montana Department of Transportation (MDT) to build the Red Lodge Tied Projects. Moore and Goran entered into a contract (Contract) in which Moore agreed to provide Goran crushed aggregate material. Goran's form contract was used which stated that Moore would "[F]urnish F.O.B. Moore Pit an estimated quantity of materials. Goran disputes the tonnage and refuses to pay the invoice due. Goran claimed that the tonnage was reduced between the time of weighing at the Moore Pit and when Goran reweighed it at the Goran facility. How are the terms "F.O.B. Moore Pit" dispositive of this issue?

Source: Moore v. Goran, LLC, 2017 MT 208 (2017).

9. SMG was a frozen-poultry wholesaler, and Sanderson was one the suppliers. SMG contracted to sell 24 containers of frozen poultry to KVADRO, a Russian company. The shipping terms were CIF, cost insurance and freight. In connection with the business, SMG carried a one-year open cargo insurance policy from Lloyd's. SMG arranged for shipment to Russia through PO Nedlloyd. In April, the Russian government suspended all previously issued permits for the import of poultry into Russia from the United States. Due to this and the change of city it was to be shipped to, the shipment violated the 60-day rule and was subsequently seized. The investigation was ended and the shipment released, but before it was picked up, it was seized again, and SMG never received payment. SMG then filed a claim against the insurance policy with Lloyd's, which was rejected. Nedlloyd and SMG then sued either seeking freight charges and damages, respectively. Nedlloyd subsequently added Lloyd's to the suit, claiming to be a third-party beneficiary. Lloyd's moved for summary judgment because Nedlloyd was not an intended third party, SMG had no insurable interest at the time, and the

Lloyd's policy did not cover credit risk and seizure by customers and thus was excluded for the contract. A summary judgment was issued by the district court, claiming no insurable interest and Nedlloyd not as an intended beneficiary. SMG appealed the ruling. Does SMG have an insurable interest?

Source: *Nedlloyd v. Sanderson Farms, Inc.,* 2006 U.S. App Lexis 22227 (2006).

10. Marilyn Thomas purchased an installed pool heater from Sunkissed. The pool heater was delivered to Marilyn's residence, but the delivery slip was signed by Nancy Thomas. Marilyn did not know of anyone by that name. She called Sunkissed to advise the company to move the heater. The neighborhood was not safe, and she was worried that the heater would be taken. The heater remained in her driveway for approximately four days. When Marilyn noticed that the heater was no longer in her driveway, she again contacted Sunkissed, but she was told not to worry. Who was responsible for the loss of the heater? Did Sunkissed actually deliver the heater to Marilyn? How do you think the court decided?

Source: *In re Marilyn Thomas,* 182 B.R. 347 (1995).

11. New furniture is delivered to the buyer, and it is badly damaged. The purchaser notified the furniture store of the damage, and the store sent a representative to the buyer. The representative explained that the furniture could be repaired and restored. After hearing the details, the buyer refused and asked for her money back. The store refused, claiming that she had accepted the goods and was required to permit the seller to cure the defects. Who prevails?

Source: *Clark v. Zaid, Inc.,* 263 Md. 127 (1971).

12. The buyer contracted with the seller for a customized machine. The seller filed to have the machine ready by the agreed-upon deadline. Twice the buyer agreed to an extension, neither of which was met by the seller. The seller then terminated the contract. The buyer sued for breach claiming that it had not been given sufficient time to effect a cure. Who prevails?

Source: *Star Machine, Inc. v. Ford Motor Co.,* United States Court of Appeals for the Sixth Circuit, 1998 U.S. App. LEXIS 15392 (1998).

13. Treibacher, an Austrian vendor of hard-metal powders, agreed to two contracts with the defendant TDY to sell specified quantities of tantalum carbide (TaC), a hard-metal powder, to TDY Industries Inc., for delivery to consignment. TDY planned to use the TaC in manufacturing tungsten-graded carbide powders at its plant in Gurney, Alabama. After it had received some of the amount of TaC specified in the November 2000 contract, TDY refused to take delivery of the balance of the TaC specified in both contracts and, in a letter to Treibacher dated August 23, 2001, denied that it had a binding obligation to take delivery of or pay for any TaC that it did not want to use. Unbeknownst to Treibacher, TDY had purchased the TaC it needed from another vendor at lower prices than those specified in its contracts with Treibacher. Treibacher eventually sold the quantities of TaC that TDY had refused to take delivery of, but at lower prices than those specified in its contracts with TDY. Treibacher then filed suit against TDY, seeking to recover the balance of the amount Treibacher would have received if TDY had paid for all of the TaC specified in the November and December 2000 contracts. What is the appropriate remedy here for Treibacher if TDY is in breach? Does this case fall under the CISG or the UCC? Is there any significance to applying the CISG rather than the UCC?

Source: *Treibacher Industrie, A.G., Plaintiff-Appellee, v. Allegheny Technologies, Inc., a Pennsylvania Corporation et al., Defendants, TDY Industries, Inc., Defendant-Appellant,* 464 F.3d 1235 (11th Cir. 2006); 2006 U.S. App. LEXIS 23252; 19 Fla. L. Weekly Fed. C 1046 (2006).

14. The defendant, Nwabuoku, purchased $1,500 worth of furniture from the plaintiff, Y&N Furniture. Through an arrangement with the plaintiff, the defendant financed the purchase through a financing company named Beneficial. On receipt of the furniture, the defendant was to notify Beneficial that receipt was effected, and Beneficial would pay Y&N the $1,500 purchase price. Nwabuoku would then begin paying Beneficial the amount due plus interest, according to their financing agreement. Nwabuoku refused to acknowledge receipt to Beneficial and eventually rejected the goods, claiming that he did not want the furniture. Is this rejection enforceable?

Source: *Y & N Furniture Inc. v. Nwabuoku,* 734 N.Y.S.2d 392 (2001).

15. The plaintiff and the defendant entered into a three-year contract in which the defendant would be the sole supplier of steel parts that the plaintiff used in its products. A dispute arose after the defendant sought to surcharge the parts sold to reflect increased costs. The plaintiff filed a lawsuit for breach of contract, and the jury returned a verdict in favor of the defendant, finding on a special verdict that there had been a valid modification to the contract, based solely on emails between the parties. Is this evidence enough to support a reformation of the contract?

Source: *Copeland Corporation, Plaintiff-Appellant, v. Choice Fabricators Incorporated, Defendant-Appellee, No. 08-3194,* 6th Cir. Ct. of Appeals, 09a0622n.06; 345 Fed. Appx. 74; 2009 U.S. App. LEXIS 19986; 2009 FED App. 0622N (6th Cir.); 69 U.C.C. Rep. Serv. 2d (Callaghan) 802.

Part 3 Domestic and International Sales Law

CHAPTER 16

Sales and Lease Contracts: Performance, Warranties, and Remedies

CASE OPENER

A Nightmare of a Vacation

On April 19, 2014, plaintiffs bought the RV, a new 2014 Palomino trailer, from defendant for $26,000.25. On April 25, 2014, plaintiffs took possession of the RV. In June 2014, plaintiffs discovered water leaking into the RV from the emergency-exit window. Plaintiffs brought the RV to defendant for repair; defendant repaired the RV to plaintiffs's satisfaction, at no charge.

In July 2014, during a trip to Michigan, plaintiffs discovered a different leak in the RV. During a rainstorm, water leaked into the dinette area, damaging the walls and causing electrical failure. Plaintiffs brought the RV to defendant for repair on July 14, 2014. Defendant told plaintiffs that the RV needed to be sent to the manufacturer for repair. Defendant told plaintiffs that it could not estimate how long the manufacturer would take to repair the RV. On August 2, 2014, plaintiffs verbally revoked acceptance of the RV. The manufacturer had the RV in repair from approximately August 4 through September 23, 2014. On September 28, 2014, plaintiffs's attorney sent defendant a letter revoking acceptance of the RV.

On October 29, 2014, plaintiffs filed a four-count complaint against defendant, alleging the following. Since they purchased the RV, it had experienced numerous mechanical problems, including (a) water leakage through a defective emergency-exit window, (b) water leakage through a defective dinette window, (c) water leakage into a paneled wall, (d) an inoperative electrical system, (e) and "generally massive water leaks," which "have the potential of causing mold and serious health issues." Further, these "defects cannot be repaired. The [RV] was in repair for almost the entire summer of 2014, and still has not been repaired properly. Prior to filing this suit, Plaintiffs [sic] revoked their acceptance of the RV and canceled their contract. Defendant refused to return Plaintiffs' money."

The questions that this case raises are:

1. Are these incidents a breach of contract? Or are they not material?
2. Does the perfect tender rule apply here?
3. Do these incidents constitute a breach of any of the implied warranties?
4. Does the remedy of cure apply?
5. To what extent does "cure" apply?

LEARNING OBJECTIVES

After reading this chapter, you will be able to answer the following questions:

LO 16-1 Differentiate between conforming and nonconforming goods.
LO 16-2 Outline the perfect tender rule.
LO 16-3 Define the right to cure.
LO 16-4 Define commercial impracticability.
LO 16-5 Define implied warranties of title, merchantability, and particular purpose.
LO 16-6 Recognize the issues around third parties and warranties.
LO 16-7 Describe what a warranty disclaimer is, and why it is used.
LO 16-8 Describe the concept of cover and resale.
LO 16-9 Recall the money damages that are available for breach of contract.

When individuals enter into sales contracts or leases with others, they need to know their rights and obligations so that they can engage in efficient business transactions. This chapter explains the performance obligations of sellers and buyers. In addition, it explains the performance obligations of lessors and lessees, which are similar to those of buyers and sellers. The chapter explains the perfect tender rule and exceptions to this rule. It also explains the buyer's general obligation to inspect, pay for, and accept goods, and it discusses exceptions to this general obligation. Moreover, it explains performance obligations under both express and implied warranties, breach of contract, and remedies available in the event of breach.

The Basic Performance Obligation

LO 16-1 Differentiate between conforming and nonconforming goods.

The obligations of sellers/lessors and buyers/lessees are determined by (1) terms the parties outline in agreements, (2) custom, and (3) rules outlined by the Uniform Commercial Code (UCC). This chapter focuses on rules outlined by the UCC.

Under the UCC, sellers and lessors are *obligated to transfer and deliver conforming goods.* Buyers and lessees are *obligated to accept and pay for conforming goods in accordance with the contract.* Courts rely on UCC rules to clarify these obligations when the contract or lease the parties agreed to is unclear. In the Case Opener, was the motor home a "conforming" good? Should the buyers have returned the motor home to the sellers or is there an obligation to permit a cure? Is any kind of deviation of conforming goods a breach of contract? Must the deviation be material? This chapter addresses that topic and others after explaining how the idea of good faith places transactions in a particular context. What if there was an issue of mistake? Remember that at common law, a unilateral mistake may still constitute a binding contract, whereas a bilateral mistake will not. Issues of mistake often raise the corollary issue of unconscionability. In the Case Opener, is the seller of the motor home obligated to provide a perfect vehicle? If not, will the seller be liable? These questions will be addressed in the following discussions on good faith, performance requirements, and the available remedies and warranties that attach to a sale.

good faith
Honesty in fact.

commercial reasonableness
Reasonable commercial standards of fair dealing, required of merchants in addition to honesty in fact.

Good Faith

UCC Section 1-304 requires good faith in the performance and enforcement of every contract. *Good faith* means honesty in fact. When the parties are merchants, the UCC imposes a higher standard. Between merchants, the UCC imposes not only honesty in fact but also reasonable commercial standards of fair dealing. This second requirement is often called **commercial reasonableness**.

In the context of good faith, courts decide the specific obligations of sellers/lessors and buyers/lessees. The next few sections of the chapter discuss these specific obligations.

Reconsider the Case Opener. How can you argue that both good faith and commercial reasonableness may be an issue in the transaction? Could an argument be made for a "simple mistake" or immaterial non-conformity as noted in the following Case Nugget?

Do Mistake and Unconscionability Affect the Enforceability of a Contract?

Donovan v. RRL Corporation California Supreme Court

26 Cal. 4th 261 (2001)

The Donovans showed up at the defendant car dealership. After hearing the sales pitch, Mr. Donovan stated that he'd take the car at the price quoted in the newspaper, $25,995. The horrified salesman said that he could go as low as $37,995 but not $25,995. At that point, Donovan showed the salesman a copy of an ad that had been running in the local newspaper identifying the very same automobile for $25,995. The salesman responded that the advertisement had to be a mistake. Donovan said that he wanted the car at the $25,995 price. He subsequently brought an action in the municipal court in Orange County. The trial court found for RRL Corp. on mistake. However, the California court of appeals reversed, stating that all the material elements of a contract were clearly stated in the advertisement and that Donovan's acceptance of those terms constituted a good contract.

The California Supreme Court heard the case, reversing it on a combination of mistake and unconscionability. The supreme court found that a contract was indeed created but that the contract could be rescinded since the evidence showed that (1) the defendant's unilateral mistake was made in good faith; (2) the defendant did not bear the risk of the mistake; and (3) the enforcement of a contract with an erroneous price would be unconscionable as a matter of law.

Specific Obligations of Sellers and Lessors

LO 16-2 Outline the perfect tender rule.

The Perfect Tender Rule

The UCC requires sellers and lessors to tender conforming goods to buyers and lessees. UCC Sections 2-503(1) and 2A-508(1) state that **tender of delivery** requires the seller/lessor to have and hold conforming goods at the disposal of the buyer/lessee and give the buyer/lessee reasonable notification to enable the buyer/lessee to take delivery. **Conforming goods** are goods that conform to contract specifications. However, Article 49(1) of the CISG allows *avoidance* (the term used for the buyer's refusal to accept nonconforming goods) only if there is substantial and foreseeable nonconformity of the goods. In other words, the CISG follows more closely the common law concept of substantial performance than it does the perfect tender rule. In our case opener, are the various leaks in the motor home a violation or violations of the perfect tender rule? Does it matter how "bad" the leaks are? Is it subject to some kind of sliding scale of seriousness of "leaks"?

tender of delivery
Requirement that the seller or lessor have and hold conforming goods at the disposal of the buyer or lessee and give the buyer or lessee reasonable notification to enable him or her to take delivery.

conforming goods
Goods that conform to contract specifications.

GLOBAL Context

Choice of Law: "Religious Laws"?

©Doug Armand/Getty Images

It is a common practice for parties to any contract to indicate in the contract itself which laws will govern the transaction a dispute arises. For example, if a merchant in Germany purchases goods from a seller in the United States and the goods are in a warehouse in France, the contract may state that the laws of France will govern the contract. It could just as easily state the laws of Germany or the United States. In some situations, the contract could state some other jurisdiction, however remote or tenuous the jurisdiction's connection may be to the parties or goods. Which laws govern a contract is up to the will of the parties unless the jurisdiction has specific laws that decide the choice of law for the parties regardless of their intent. Hence, it is always important to know the local laws governing transactions.

An interesting issue that is starting to arise is whether parties may choose religious laws to govern disputes. For example, may the parties to a contract choose *Sharia* law, that is, the religious law based on the Koran in the Muslim religion? We've not yet seen the use of Sharia law in UCC or CISG contracts, but a Florida trial judge ruled in March 2011 that a dispute regarding the ouster of trustees of a mosque should be determined under Sharia law. The issue raises questions; if, for example, a mediation clause in a contract calls for mediation by canon law, should, or even could, a court adhere to that provision? In our ever-shrinking world, this issue does give one cause for thought.

Source: William R. Levesque, "Judge Orders Use of Islamic Law in Tampa Lawsuit Over Mosque Leadership," *Tampa Bay Times*, March 21, 2011.

perfect tender rule
Requirement that the seller deliver the goods in conformity with the contract, down to the last detail.

A common law rule known as the **perfect tender rule** required the seller to deliver goods in conformity with the terms of the contract, right down to the last detail. UCC Sections 2-601 and 2A-509 embrace the perfect tender rule. These sections indicate that if goods or tender of delivery fails *in any respect* to conform to the contract, the buyer/lessee has the right to accept the goods, reject the entire shipment, or accept part and reject part. Common law usually substitutes perfect tender with the doctrine of substantial performance. Substantial performance occurs when all the material elements of a contract are satisfied even if some nonmaterial requirements may not be satisfied. The perfect tender rule would not recognize the distinction between material and immaterial contractual requirements. The next Case Nugget considers how the perfect tender rule is affected when good faith is an issue.

CASE Nugget

How Does Good Faith Affect Perfect Tender?

Y&N Furniture Inc. v. Nwabuoku New York Court of Appeals

734 N.Y.S.2d 392 (2001)

The defendant, Nwabuoku, purchased $1,500 worth of furniture from the plaintiff, Y&N Furniture. Through an arrangement with the plaintiff, the defendant financed the purchase through a financing company named Beneficial. On receipt of the furniture, the defendant was to notify Beneficial that receipt was effected, and Beneficial would pay Y&N the $1,500 purchase price. Nwabuoku would then begin paying Beneficial the amount due plus interest according to their financing agreement.

Nwabuoku refused to acknowledge receipt to Beneficial and eventually rejected the goods, claiming that he did not want the furniture. The civil court of the City of New York found that Nwabuoku had rejected the goods, and the court cited the perfect tender rule as requiring "exact performance by the seller." Moreover, the court held that "rejection of goods must be within a reasonable time of their delivery or tender." The court indicated that Nwabuoku did indeed make a rightful rejection. However, the court noted that a rejection of goods must be based on some claim that the goods are in some way

(continued)

nonconforming. The defendant in this case had a disagreement over the terms of the financing (though the terms were clearly stated in the financing statements). The defendant made no claim of nonconforming goods. Thus, the court found that "in this case, too, we have a rejection that is not only 'wrongful,' but also in bad faith." A wrongful rejection is not necessarily in bad faith; the buyer may honestly believe that the goods fail to conform but may simply be mistaken. But a rejection that does not even purport to rest on an honest assessment of conformity of the goods has "the effect of destroying or injuring the right of the seller to receive the fruits of the contract."

Consider Case 16-1, which provides an illustration of a situation in which the UCC's version of the perfect tender rule was relevant when compared to the common law rule of material breach under the doctrine of substantial performance. In the Case Opener, does the tender of a motor home with leaks constitute substantial performance under common law and perfect tender under the UCC? What about under the CISG?

CASE 16-1

ALASKA PACIFIC TRADING CO. v. EAGON FOREST PRODUCTS INC.
Washington Appellate Division 1
933 P.2d 417 (1997)

FACTS: *Alaska Pacific Trading Company (ALPAC) and Eagon Forest Products Inc. contracted to buy and sell raw logs. ALPAC and Eagon engaged in months of communications about a shipment of 15,000 cubic meters of logs from Argentina to Korea between the end of July and the end of August 1993. The delivery date passed without ALPAC shipping the logs. Eagon canceled the contract, alleging that ALPAC had breached the agreement by failing to deliver. ALPAC alleged that its failure to deliver was not a material breach and that the parties had modified the delivery date. Alternatively, ALPAC argued that Eagon breached the contract by failing to provide adequate assurances or repudiating the contract. The miscommunication between the parties occurred after the market for logs began to soften, making the contract less attractive to Eagon. ALPAC was reluctant to ship the goods because it was concerned that Eagon might not accept the shipment. However, Eagon never stated that it would not accept the cargo.*

ISSUE: The question for the court is whether to apply the common law doctrine of material breach or the UCC doctrine of perfect tender.

REASONING: In the court's own words:

> As a contract for the sale of goods, this contract is governed by the Uniform Commercial Code, Article II (UCC II) which replaced the common law of material breach, on which ALPAC relies, with the "perfect tender" rule. Under this rule, "If the goods or the tender of delivery fail in any respect to conform to the contract, the buyer may . . . reject the whole." . . . Both the plain language of the rule and the official comments clearly state that, if the tender of the goods differs from the terms of the contract in any way, the seller breaches the contract and the buyer is released from its duty to accept the goods.

DECISION AND REMEDY: The court applied the literal language of the UCC, holding that the perfect tender rule applied and that ALPAC's failure to deliver was a breach of contract.

SIGNIFICANCE OF THE CASE: Though difficult to measure at times, the perfect tender rule is a departure from the common law rule of substantial performance. Any deviation from the contract may indeed constitute a breach of the perfect tender rule, materiality notwithstanding.

CRITICAL THINKING

Here, Eagon got lucky. The company got out of a contract that was unfavorable to it, given the softening market for logs. In what way did the judge simplify the case? Is it fair to say the judge oversimplified the case?

ETHICAL DECISION MAKING

What ethical norm or value underlies the judge's decision? Explain.

Exceptions to the Perfect Tender Rule

The perfect tender rule is not as inflexible as it appears. Although the rule itself demands perfection, both courts and UCC drafters have created exceptions that reduce the rule's rigidity. These exceptions limit the seller's obligation to deliver conforming goods and/or limit the buyer's power to reject goods that do not conform. The exceptions allow sellers/lessors and buyers/lessees to ask questions such as:

- What are the norms in the particular industry, and/or what past dealings have the parties had with one another?
- What does the parties's agreement say?
- Is it possible for the seller/lessor to cure or correct problems?
- What if the goods have been destroyed?
- What if nonconformity substantially impairs the value of goods?
- What if unforeseen circumstances make contract performance commercially impracticable?

Norms in the Industry and Past Dealings between the Parties When the buyer alleges that goods failed to conform to contract specifications, the buyer does not automatically have the right to reject the goods. The UCC requires courts to consider norms in a particular trade. Sometimes, the norms for a particular trade do not permit a buyer to reject goods with minor flaws. UCC Section 1-205(2) defines **usage of trade** as any practice that members of an industry expect to be part of their dealings.

In addition to its requirement on usage of trade, the UCC requires courts to consider the ideas of course of dealing and course of performance. UCC 1-205(1) defines **course of dealing** as previous commercial transactions between the same parties. Under UCC 208(1), **course of performance** refers to the history of dealings between the parties in the particular contract at issue.

Exceptions Outlined in the Parties' Agreement Sometimes, language in the parties's agreement limits the rigidity of the perfect tender rule. For instance, parties may agree that the seller must have the opportunity to repair or replace nonconforming goods within a particular period of time. Alternatively, parties may agree with a level of performance that is less than perfect. They could indicate, by agreement, the expectation regarding performance.

LO 16-3 Define the right to cure.

The Seller's and Lessor's Right to Cure Under UCC Sections 2-508 and 2A-513, sellers and lessors have the right to **cure** or fix problems with nonconforming goods. In particular, sellers and lessors can repair, adjust, or replace defective or nonconforming goods as long as they give prompt notice of the intent to cure and then cure within the contract time for performance.

Destroyed Goods Under UCC Sections 2-613 and 2A-221, if goods are identified at the time the parties entered into a contract and the goods are destroyed through no fault of the parties before risk passes to the buyer or lessee, the parties are excused from performance. If the goods are only partially destroyed, the buyer can inspect the goods and decide whether to (1) treat the contract as void or (2) ask the seller for a reduction of the contract price and then accept the damaged goods.

Substantial Impairment Two sections of the UCC use the concept of **substantial impairment** to modify the perfect tender rule. The first applies when a buyer *revokes acceptance of goods*. UCC Section 2-608 indicates that a buyer who has accepted goods may later revoke the acceptance only if the buyer can show that the defects substantially impair the value of the goods. The second applies when the buyer and seller have entered into an *installment contract*. UCC 2-612(2) and 2A-510(1) indicate that a buyer/lessee may reject an installment of a particular item only if the defects substantially impair the value of the goods and cannot be cured.

usage of trade
Any practice that members of an industry expect to be part of their dealings.

course of dealing
Previous commercial transactions between the same parties.

course of performance
The history of dealings between the parties in the particular contract at issue.

cure
A breaching party's right to provide conforming goods when nonconforming goods were initially delivered; subject to a reasonable time test.

substantial impairment
A concept, used to modify the perfect tender rule, whereby a buyer can revoke acceptance of goods and, with installment contracts, a buyer/lessee can reject an installment of a particular item only if the defects substantially impair the value of the goods.

LO 16-4 Define commercial impracticability.

Commercial Impracticability UCC Sections 2-615(a) and 2A-405(a) state that a delay in delivery or nondelivery, in whole or in part, is not a breach in circumstances in which performance has been made impracticable because a contingency has occurred that was not contemplated when the parties reached an agreement. For example, this rule would be relevant if a change in government regulation that neither party contemplated forbids the import or export of a particular item the parties had agreed would be shipped.

Specific Obligations of Buyers and Lessees

The Basic Obligation: Inspection, Payment, and Acceptance

Under UCC Sections 2-301 and 2A-516(1), buyers and lessees are obligated to accept and pay for conforming goods in accordance with the contract. Before paying for and accepting the goods, buyers/lessees ordinarily inspect the goods to make sure they conform to the specifications in the parties's agreement.

Exceptions to the Basic Obligation

Buyers/lessees do not always end up accepting and paying for goods. Sometimes, on inspection, the buyer or lessee decides to reject the goods and refrain from paying for them. In this section, we take a look at exceptions to the buyer's/lessee's basic obligation. These exceptions allow sellers/lessors and buyers/lessees to ask questions such as:

- What forms of payment are allowed under the UCC?
- In what circumstances can a buyer reject goods?
- Is the buyer allowed to accept part but not all of the goods?
- In what circumstances can a buyer revoke acceptance of goods?
- How and when can a buyer reject nonconforming goods?
- Are installment contracts treated differently than other kinds of contracts?

Problems on Inspection If all goes well in a transaction over the sale or lease of goods, the buyer or lessee inspects the goods and then pays by any means the parties have agreed on, including payment by cash, check, or credit card. Unless the parties have agreed otherwise, the buyer or lessee typically inspects the goods before paying. Under UCC Sections 2-513(1) and 2A-515(1), the seller or lessor must provide an opportunity for inspection before enforcing payment.

The concept of reasonableness governs the inspection process. For example, inspection must take place at a reasonable time and place, in a reasonable way. Once the buyer or lessee inspects the goods, he or she decides whether to accept them. Sometimes, on inspection, the buyer or lessee decides not to accept the goods.

Problems with Acceptance When all goes well, UCC Sections 2-606(1) and 2A-515(1)(a) indicate that the buyer or lessee, after inspecting, signifies agreement to the seller or lessor that the goods are either (1) conforming or (2) acceptable even though they are nonconforming. UCC 2-602(1), 2-606(1), and 2A-515(1)(b) allow the seller or lessor to presume acceptance if the buyer or lessee fails to reject the goods within a reasonable period of time. UCC 2-601(c) and 2A-509(1) allow the buyer or lessee to make a partial acceptance when the goods are nonconforming and the seller or lessor has failed to cure the defects.

When goods are nonconforming, the buyer or lessee is allowed to revoke or withdraw acceptance of the goods. The previous section on specific obligations of sellers/lessors discussed this concept under the topic of substantial impairment. From the buyer's/lessee's perspective, the buyer or lessee may revoke acceptance if the nonconformity substantially impairs the value of the goods, but only if he or she had a legitimate reason for the initial acceptance.

Warranties

LO 16-5 Define implied warranties of title, merchantability, and particular purpose.

At common law, the obligations and responsibilities of the contractual parties are determined by the express terms of the contract. UCC contracts are the same but with a major addition: implied warranties. *Implied* means that the warranty or guarantee is automatically part of the contract; the parties do not negotiate for it. At common law, the only implied warranty in a contract is the implied warranty of assignability. However, the UCC changes this part of contract law dramatically.

Let's look at the whole range of warranties in contract law. **Warranties** generally arise in conjunction with a sale or lease. They impose certain duties on the seller or lessor, and if the seller or lessor fails to live up to these duties, he or she may be sued for breach of warranty. There are three basic categories of warranties: warranties of title, express warranties, and implied warranties. Each is discussed in the following sections.

> **warranty**
> An assurance, either express or implied, by one party that the other party can rely on its representations of fact; in sales, a binding promise regarding a product in the event that the product does not meet the manufacturer's or seller's promises.

Warranties of Title

Although no warranties automatically arise under the common law, the UCC assumes that the seller:

1. Has good and valid title to the goods
2. Has the right to transfer title free and clear of any liens, judgments, or infringements of intellectual property rights of which the buyer does not have knowledge.

The UCC specifically permits buyers to recover from sellers who have breached these **warranties of title**. The only exceptions to title warranties occur if they are disclaimed or modified by specific language in the contract or if the seller is obviously unable to guarantee title, as would be the case, for instance, at a sheriff's sale of seized goods.

Clearly, if the buyer is aware of any problem with the transfer of goods, the buyer is indeed purchasing them at her own risk. Still, if the seller transfers goods to an unknowing buyer for which no good title passes or on which there are encumbrances or patent claims, the buyer may treat the contract as being in breach. Under such circumstances, the buyer may then avail herself of the remedies available under a breach situation.

> **warranty of title**
> An assurance, inferred in every UCC sales transaction, that the seller has good and valid title to the goods and has the right to transfer title free and clear of any liens, judgments, or infringements of intellectual property rights of which the buyer does not have knowledge.

Express Warranties

Although the common law does not use the term *express warranties*, the concept and its application do exist in the common law. It seems only fair and equitable that promises made by a seller to induce a buyer to execute a sales contract should be enforceable. An **express warranty** is any description of a good's physical nature or its use, in either general or specific circumstances, that becomes part of the contract. To use common law language, an express warranty is a material term of the sale or lease contract.

Express warranties may be found in advertisements or brochures (e.g., "This electric saw comes with a lifetime guarantee"). Such a warranty may also be part of a written sales or lease contract; or it may be a salesperson's oral promise concerning the good, made while he is attempting to close the deal. A sample or model may also provide an express warranty. Generally speaking, if the buyer relies on representations, those representations become part of the contract in the form of express warranties.

> **express warranty**
> Any description of a good's physical nature or its use, in either general or specific circumstances, that becomes part of a contract.

BUT WHAT IF . . .

WHAT IF THE FACTS OF THE CASE OPENER WERE DIFFERENT?

Let's say that in the Case Opener, the seller says that all the windows had been sealed with an all-purpose sealant? What kind of a statement is this: express or implied? Alternatively, what if the seller said that as far as he knew there had never been any leaking issue with this vehicle? What does this statement qualify as? Which statement is an express warranty?

Implied Warranties of Quality

Implied warranties arise by operation of law under certain circumstances. Earlier, you read about the implied warranties of title that arise under the UCC. This section focuses on the three warranties of quality that arise under the UCC.

Implied Warranty of Merchantability Consider the following scenario: You purchase a toaster from a local discount store. When you use the toaster, all you get is either burnt toast or bread that is only slightly warm. You take the toaster back to the store and are met with this answer: "Well, we don't guarantee how well the toaster will work. After all, it does toast, either very, very lightly or very, very burnt." This answer, of course, is nonsense. There is a reasonable expectation of how a toaster will perform. That reasonable expectation is codified in the UCC *implied warranty of merchantability*.

To invoke this implied warranty, the buyer or lessee must have purchased or leased the good from a merchant. Thus, a dirt bike purchased at a bicycle shop is covered by the warranty of merchantability, but a bike that is bought from a neighbor is not, unless the neighbor is a bicycle merchant.

Under the UCC, the goods must be *merchantable,* meaning that they must:

1. Be able to pass without objection in the trade or market for similar goods.
2. In the case of fungible goods, be of fair average quality within the description.
3. Be fit for the ordinary purposes for which such goods are used.
4. Be produced, within the variations permitted by the agreement, with even kind, quality, and quantity in each unit and among all units involved.
5. Be adequately contained, packaged, and labeled as the agreement may require.
6. Conform to the promises or affirmations made on the container or label, if any.

implied warranty of merchantability
An assurance, inferred in every sale unless clearly disclaimed, that merchantable goods will conform to a reasonable performance expectation. The purchaser must have purchased or leased the good from a merchant.

The quintessential case defining and illustrating the implied warranty of merchantability is that of the Blue Ship Tea Room and Ms. Webster (see Case 16-2). The case concerned the merchantability of food.

PRISCILLA D. WEBSTER v. BLUE SHIP TEA ROOM, INC.
Supreme Judicial Court of Massachusetts
347 Mass. 421; 198 N.E.2d 309 (1964)

FACTS: *Priscilla Webster, a restaurant patron who ordered seafood chowder and choked on a fishbone, brought this case. She maintained that she would not have reasonably expected to find a bone in the chowder. As a result of the bone in the chowder, Webster suffered through several hospital procedures to remove the bone, at considerable cost and pain. At the trial, a jury found for Webster. The Blue Ship Tea Room, the defendant, appealed the case on the basis of the legal interpretation of the implied warranty of merchantability. The appellate decision has become a classic in American jurisprudential reasoning.*

ISSUE: Does a fishbone found in fish chowder render the chowder inedible, thereby breaching the implied warranty of merchantability?

REASONING: The court, in great detail, discussed the rich culinary history of New England seafood chowders. In going through the details of exactly how good chowder is made, the court praised the richness of thick, plump, delicious pieces of fish and shellfish. Completely ridding the chowder of any possibility of bones would require the cook to mince the seafood finely, depriving the dish of one of its hallmark characteristics. Further, the court argued that anyone eating good New England seafood chowder would know or should know that fishbones are indeed a possibility.

DECISION AND REMEDY: As lamentable as Webster's injuries were, the court found that the Blue Ship Tea Room had not breached its implied warranty of merchantability.

SIGNIFICANCE OF THE CASE: Common sense needs to apply in defining "merchantability." Clearly, a piece of bone in a hamburger is unreasonable and would violate the implied warranty. However, in cases when the buyer should expect a problem, such as a possible fishbone in a seafood dish, the buyer must be prepared to accept the situation.

> **CRITICAL THINKING**
>
> As with most legal decisions, the critical-thinking activity that is most obvious is the need to examine possible analogies the court could use in justifying its conclusion. Consider this analogy: Is the fishbone in the chowder the same as a shard of beef bone in a beef vegetable soup? What about a piece of chicken bone in a chicken salad sandwich? How about a small caterpillar in a fresh green salad? Would the aptness of the analogy depend at all on the size of the bone in the fish chowder?
>
> **ETHICAL DECISION MAKING**
>
> The judge mainly used assumption of risk to rule against Webster, though she suffered an injury in fact. Should the restaurant have somehow compensated her? What would the WH framework indicate should be done?

implied warranty of fitness for a particular purpose

An assurance, inferred in any UCC sale, whereby when a seller or lessor knows or has reason to know (1) why the buyer or lessee is purchasing or leasing the goods in question and (2) that the buyer or lessee is relying on him or her to make the selection, the buyer or lessee has an enforceable warranty if such assurance is false.

Implied Warranty of Fitness for a Particular Purpose Another important UCC implied warranty is the **implied warranty of fitness for a particular purpose**. This warranty comes about when a seller or lessor knows or has reason to know (1) why the buyer or lessee is purchasing or leasing the goods in question and (2) that the buyer or lessee is relying on him or her to make the selection. Under this warranty, a seller or lessor does not have to be a merchant.

An implied warranty of fitness for a particular purpose should not be confused with an express warranty. If the buyer walks into a store and the salesclerk says, "This saw will cut through metal," the seller has created an express warranty. However, if the buyer comes into the store and asks the salesclerk for a saw to cut through some copper tubing and the salesclerk refers the customer to a wall of different saws, it is reasonable for the buyer to assume that all the saws on the wall will satisfy the *particular purpose* that the buyer has indicated. Thus, an implied warranty of fitness for a particular purpose has been created.

implied warranty of trade usage

An assurance, inferred in the context of certain UCC sales, depending on the circumstances, that can be created through a well-accepted course of dealing or trade usage.

Implied Warranty of Trade Usage The UCC, always diligent in its goal to facilitate the flow and ease of commercial activity, recognizes that a well-accepted course of dealing or trade usage may create implied warranties dependent on the circumstances. For example, if it is generally accepted in the trade that a certain product is always preassembled and shrink-wrapped, the failure of the seller to deliver the goods in that condition would be a breach of the **implied warranty of trade usage**.

Warranty Rights of Third Parties

LO 16-6 Recognize the issues around third parties and warranties

The idea of a seller being in breach of an implied warranty raises an entirely new issue: Is the seller liable to anyone other than the buyer? This question may initially sound peculiar. After all, the seller and the buyer are bound together by contract and if either breaches, then the breaching party is liable to the nonbreaching party.

However, consider this possible scenario: Jane buys a blender from a local store. Prior to using the blender, she lends it to her cousin Valerie to use at a party. While Valerie is blending drinks at the party, the blades fly off the blender and injure her. What obligation, if any, does the seller have to the injured Valerie? No contractual relationship exists between Valerie and the seller. However, it seems to be patently unfair to conclude that Valerie has no cause of action against the seller. The UCC recognizes this unfairness and clearly states that Valerie may indeed have a cause of action based on breach of warranty against the seller. The states are given the following three choices regarding *third-party beneficiaries of warranties:*

1. Seller's warranties extend to the buyer's household members and guests.
2. Seller's warranties extend to any reasonable and foreseeable user.
3. Seller's warranties extend to anyone injured by the good.

Most states have adopted the first option, which was the original UCC provision. Only a handful of states have adopted the second or third options.

Warranty Disclaimers and Waivers

LO 16-7 Describe what a warranty disclaimer is, and why it is used.

There really is no question of whether an implied warranty may be disclaimed. The real question is *how* it is to be disclaimed. Generally speaking, if an implied warranty is to be disclaimed, the seller must do so in clear, unambiguous, conspicuous language. To disclaim the implied warranty of fitness for a particular purpose, the seller must disclaim the warranty in writing. The seller may disclaim the warranty of merchantability either orally or in writing; some states require the term *merchantability* to be used in the disclaimer. Under the CISG, however, the term *merchantability* does not need to be specifically expressed to disclaim it. Any reasonable language disclaiming the warranty is acceptable.

Implied or Express Warranty?

Old Mexican Foods, Inc. v. Hanson Staple Company

285 Ga. 288; 676 S.E.2d 169 (2009)

Old Mexican Foods contracted with Hanson Staple Company to purchase made-to-order packaging material. Old Mexican Foods refused the shipment, claiming that the packaging material failed to meet quality expectations and breached the implied warranties of merchantability and particular purchase. Litigation followed and the parties reached a settlement agreement for Old Mexican Foods to purchase a minimum amount of the packaging material. Hanson moved for a court order instructing Old Mexican Foods to purchase the minimum amount of packaging agreed upon in the settlement agreement, subject to the UCC standards of the implied warranty of particular purpose. Old Mexican Foods complied and ordered the minimum amount under the settlement agreement. Once again, Old Mexican Foods claimed that the packaging material failed to meet their quality expectations under the implied warranties of merchantability and particular purpose. Old Mexican Foods refused Hanson's demand for payment on the invoice. Hanson sued on the invoice.

This seems fairly straightforward. However, the Georgia Supreme Court had other ideas. Agreeing with the court of appeals, the Court questioned the purpose of the settlement agreement: Was it created as a contract for the sale of goods under UCC article 2, or was it created predominantly for the purpose of settling the first lawsuit? The court opined that since the original lawsuit was for an amount over $260,000, the settlement agreement requiring Old Mexican Foods to purchase only $130,000 meant that the predominant nature of this mixed settlement agreement did NOT fall under the UCC but under common law. As such, no implied warranties are present. The court held "If [Old Mexican Foods] had desired a warranty as to the quality of the [packaging], it could have bargained for inclusion of one in the agreement, having chosen to forego an express warranty during the contract negotiations, it must now live with the consequences."

To avoid potential liability, a seller of corn for feed, for example, might decide to include in any sales contract for feed corn a clause that states, "No guarantees are made, either expressly or implied, that this feed corn has not been contaminated by mold or the toxin Fumonisin.

Buyer is advised to test the corn to determine the presence of these potential contaminants." Of course, such a disclaimer might certainly have a negative impact on the sales of one's product.

The buyer may also waive both implied and express warranties. A buyer may waive these rights by (1) failing to examine goods for which an express warranty was created by a sample or model or (2) failing to comply with the seller's request to inspect the goods. Suppose, for example, that a printer asks the buyer to come into the shop to proof letterhead and envelopes. The buyer refuses, claiming that he is too busy, and tells the printer to go ahead and run the stationery. On receipt of the stationery, the buyer discovers that the numbers in the phone number are transposed, making the stationery useless. Unfortunately, the buyer has indeed waived his rights due to his failure to inspect.

A buyer may also waive her warranty rights under the contract by failing to comply with the statute of limitations. Under the UCC, the buyer or seller must bring a lawsuit on a breached contract within four years of when the breach occurred or when the nonbreaching party became aware of it. The buyer and seller are free to negotiate contractually a shorter time period (as long as it is not less than one year), but they are not free to negotiate a longer time period than the four years.

Although the UCC remains the primary codification of both state and federal laws regarding sellers's warranties, there has also been specific legislation pertaining to this issue. The 1975 federal law known as the Magnuson-Moss Act requires that if a seller decides to issue a written warranty for a consumer good (the seller is not required to do so), the seller must indicate whether that warranty is a *full* warranty or a *limited* warranty. This applies to any consumer good sold for more than $10. If the written warranty is silent, it is presumed to be a full warranty, which means that if the good fails or is defective, the good or its defective part will be replaced. If replacement cannot be timely effected, the buyer has the right to a refund or a full replacement.

If the good is sold for more than $15, the written warranty must disclose a number of items of information—names and addresses of the warrantors, any limitations on the warranty, and the procedures required to activate the warranty remedies—all in readable and easily understood language, in other words, not in *legalese*.

The Goal of Contract Remedies

Of course, laying out these various contractual obligations gives rise to the question of what remedies are available if those obligations are breached. The obligations of sellers/lessors and buyers/lessees are determined by (1) terms the parties outline in agreements, (2) custom, and (3) rules outlined by the Uniform Commercial Code, such as the various implied warranties described previously.

Remedies Available to Sellers and Lessors under the UCC

Cancel the Contract

UCC Sections 2-703(f) and 2A-523(1)(a) allow a seller or lessor to cancel the contract if the buyer or lessee is in breach. The UCC requires sellers/lessors to notify buyers/lessees of the cancellation. Then the seller or lessor pursues remedies available under the UCC. Remember, these remedies give the seller/lessor the benefit of the bargain and no more.

Withhold Delivery

Sometimes a buyer breaches the contract or lease before the seller has delivered the goods. For instance, the buyer or lessee might fail to pay according to the terms of the agreement. UCC Sections 2-703(a) and 2A-523(1)(c) allow sellers or lessors to withhold delivery of goods when the buyer or lessee is in breach.

Resell or Dispose of the Goods

Sellers or lessors are also allowed to resell or dispose of the goods when the buyer is in breach and the goods have not yet been delivered. The seller or lessor then holds the buyer or lessee liable for any loss. UCC Section 2-706 allows the seller to recover the difference between the resale price and the contract price, plus incidental damages and minus expenses saved. Although buyers are liable for these damages, the seller gets to keep any profits it makes on the resale. UCC 2A-527(2) outlines a similar rule for lease agreements. The lessor is allowed to lease the goods to another party and recover unpaid lease payments and any deficiency between the lease payments due under the original lease contract and those due under the new contract. The lessor can also seek incidental damages.

Sue to Get the Benefit of the Bargain

In trying to give the seller or lessor the benefit of the bargain, and nothing more, courts often grant damages to recover the purchase price or lease payments due.

Stop Delivery

UCC Sections 2-705(1) and 2A-526(1) allow a seller or lessor to stop delivery of goods that are in transit. *In transit* means that the seller or lessor has delivered the goods to a carrier or bailee but the carrier or bailee has not yet turned them over to the buyer. Of course, the seller/lessor must give timely notice to the carrier/bailee so that the carrier/bailee is able to stop delivery. Also, the rules are different for insolvent and solvent buyers and lessees. If the buyer/lessee is insolvent, the carrier/bailee can stop delivery regardless of the quantity shipped. If the buyer/lessee is solvent, however, the carrier/bailee can stop delivery only if the quantity shipped is a large shipment (e.g., a carload or truckload).

Reclaim the Goods

Under UCC Sections 2-709(1) and 2A-529(1), if the buyer or lessee has possession of the goods and is in breach, the seller or lessor can sue for the purchase price of the goods or for the lease payments due, plus incidental damages. In some circumstances, the UCC allows the seller or lessee to reclaim the goods. UCC 2-702(2) allows a seller to reclaim goods when the seller discovers the buyer is insolvent. UCC 2A-525(2) allows a lessor to reclaim goods when the lessee fails to make payments according to the lease terms.

Remedies Available to Buyers and Lessees under the UCC

Cancel the Contract

Sometimes, sellers or lessors fail to deliver the goods or to deliver conforming goods and thus are in breach. UCC Sections 2-711(1) and 2A-508(1)(a) allow buyers and lessees to cancel the contract and then seek remedies that give them the benefit of the bargain.

Obtain Cover

LO 16-8 Describe the concept of cover and resale.

Under UCC Sections 2-712 and 2A-518, buyers and lessees are allowed to **cover**, or substitute, goods for those due under the sales or lease agreement.

The buyer must (1) demonstrate good faith in obtaining the substitute goods, (2) pay a reasonable amount for the substitute goods, (3) act without unreasonable delay in purchasing the substitute goods, and (4) purchase goods that are reasonable substitutes.

cover
A buyer's right to substitute goods for those due under a sales or lease agreement when the seller provides nonconforming goods.

consequential damages
In a contract, foreseeable damages that result from special facts and circumstances arising outside the contract itself. The damages must be within the contemplation of the parties at the time the breach occurs. Also called *special damages*.

Sue to Recover Damages

LO 16-9 Recall the money damages that are available for breach of contract

Buyers and lessees, are entitled to incidental and consequential damages. **Consequential damages** include damages for lost profits as long as these damages are not too speculative. These monetary damages give the injured buyer or lessee the benefit of the bargain. There is also the availability of liquidated damages and even, under some situations, punitive damages. How does one differentiate between direct and consequential damages. Consider the following case.

CASE 16-3

STERN OIL CO. v. BROWN
2018 S.D. 15
908 N.W.2d 144 (2018)

FACTS: *Stern Oil is a fuel and petroleum distributor based in Freeman operated by Scott and Staci Stern and Scott's father, Gillas. The business supplies fuel at locations across the Midwest. Brown is a businessman from Gettysburg. Brown operates two convenience stores in North Sioux City, South Dakota: Goode to Go and Freeway Mobil.*

In 2005, Brown and Stern Oil entered into two ten-year Motor Fuel Supply Agreements (MFSAs) for Stern Oil to supply ExxonMobil branded fuel to Brown to sell at his two convenience stores. The MFSAs required Stern Oil to sell and deliver up to a contractually determined "Maximum Annual Volume" of fuel to Brown. Brown was obligated to purchase at least 75 percent of that amount annually. Approximately a year and a half into the ten-year agreements, Brown stopped purchasing fuel from Stern Oil.

Stern Oil sued Brown for breach of contract. Brown counterclaimed and asserted that Stern Oil fraudulently induced him to enter into the MFSAs by verbally guaranteeing Brown a five-cent profit on each gallon of fuel sold at his convenience stores. Brown also asserted defenses to the validity of the MFSAs. The circuit court granted Stern Oil's motion for summary judgment on its claims for breach of contract and on Brown's fraud claims. The parties waived a jury on the issue of damages, and the case proceeded to a bench trial in October 2009 and January 2010. The circuit court awarded Stern Oil lost profits in the amount of $925,317.

ISSUE: Whether the circuit court erred in instructing the jury that Stern Oil's damages had to be foreseeable to Brown.

REASONING: Stern Oil argues that lost profits resulting from an immediate payment discount given by ExxonMobil were recoverable as direct damages and not as consequential damages, and as such, the damages were not subject to a foreseeability requirement. Brown contends that any profits arising from the discount received from ExxonMobil were consequential to the breach of the MFSAs because they were based upon a third-party contractual agreement between Stern Oil and ExxonMobil. Brown was not a party to that agreement and claimed he was not aware of its terms. He maintains the jury was properly instructed. In the alternative, Brown claims that if an error occurred, it was harmless.

Because Stern Oil established the factual basis to seek lost profits as the alternative measure of damages, the remaining question is whether the circuit court erred in instructing the jury on consequential damages and the foreseeability of Stern Oil's lost profits to Brown at the time of contracting. The UCC does not distinguish between direct and consequential damages for lost profits claimed by a seller. . . Where the UCC is silent, SDCL 57A-1-103 allows other "principles of law and equity" to supplement the UCC's provisions. Because the UCC does not address when lost profits are direct or consequential, we apply general contract law in determining this question.

This Court has stated that "the ultimate purpose behind allowance of damages for breach of contract is to place the injured party in the position he or she would have occupied if the contract had been performed, or to 'make the injured party whole.'" The amount of recovery may not exceed the amount the plaintiff would have gained if the contract had been fully performed. This Court has allowed a party to recover lost profits for breach of contract. In general, to prove damages for lost profit, a plaintiff must establish "a reasonable relationship between the method used to calculate damages and the amount claimed." Damages must also be "reasonably certain and not speculative." *Id.* Further, we have required that damages be a direct consequence of the breach of contract and reasonably within the contemplation of the parties at the time of making the contract. To this end, "[c]onsequential damages must be reasonably foreseeable by the breaching party at the time of contracting." We have not, however, had an occasion to address

(continued)

the distinction between lost profits as direct or consequential damages. A number of courts have wrestled with this question.

DECISION AND REMEDY: The lost profits claimed by Stern Oil are direct damages, not consequential damages. We require a non-breaching party to show that consequential damages were foreseeable to the breaching party at the time of the contracting. In contrast, direct damages for breach of contract do not have an element of foreseeability because they are, by their very nature, foreseeable by the parties at the time of contracting. Because the lost profit damages flowed directly from the MFSAs, Stern Oil did not have an obligation to show foreseeability.

SIGNIFICANCE OF THE CASE: This case illustrates the limits to which damages may be assessed against a breaching party. Clearly lost profits are not just consequential but are direct and foreseeable damages under any kind of breach and foreseeability need not be proven by the plaintiff.

CRITICAL THINKING

Can you envision a situation where lost profits are not foreseeable? What if the buyer is not an obvious merchant? A home baker orders a supply of cake ingredients for her home bakery business. The seller fails to deliver the correct supplies and the home baker loses a number of lucrative wedding cake customers. The supplier had no idea it was dealing with a commercial baker but thought the delivery was a home delivery. Should lost profits be recoverable?

ETHICAL DECISION MAKING

Following up on the critical thinking issue above, should it make a difference at all whether a buyer or seller is a merchant or not. Should a merchant be held to a higher standard? If the seller above was a neighbor from whom our home baker were buying supplies from and who subsequently breached, should the neighbor be liable for lost profits?

Liquidated damages are damages identified *before* the breach occurs. The parties are free to negotiate, as part of the contract, a liquidated-damage clause in which the parties agree in advance what the damages will be for each party if a breach occurs. Generally speaking, a court will enforce a liquidated-damage clause as long as it is not so far out of reasonable range as to be punitive in nature. Liquidated-damage clauses that are deemed punitive in nature are not enforceable. The topic of exactly how to compute damages is one that has eluded specific definition at times. See the following Case Nugget for an excellent step-by-step explanation of such a computation.

The code provides for liquidated damages if the parties have not expressly negotiated a liquidated-damage clause. UCC Section 2-718 pertains to liquidated damages and allows the non-breaching seller to claim against a breaching buyer 20 percent of the purchase price or $500, whichever is less, as liquidated damages.

Although the UCC does not mention the availability of punitive damages, other than in its voiding of liquidated damages that are punitive in nature, an issue that remains unsettled is the awarding of punitive damages against a breaching party who intentionally or egregiously breaches the contract. You will remember from tort law that when a tort is committed either intentionally or recklessly, the court may infer legal malice and instruct a jury that it may consider the awarding of punitive damages in addition to compensatory damages. Although this concept is well settled in tort law, it has never been widely applied in contract law. Yet some argue that it should be, especially to deter intentional breaches of contract. See the Point/Counterpoint at the end of this chapter for two perspectives on this issue.

liquidated damages
Damages specified as a term of the contract, before a breach of contract occurs.

Recover the Goods

UCC Sections 2-502 and 2A-522 allow buyers and lessees to recover the goods identified in the contract if the seller or lessor becomes insolvent within 10 days after receiving the first payment due under the agreement. Buyers or lessees are obligated to pay the remaining balance according to the terms of the agreement.

Obtain Specific Performance

UCC Sections 2-716(1) and 2A-521(1) allow buyers and lessees to seek the remedy of specific performance when either (1) the goods are unique or (2) a remedy at law is inadequate. **Specific performance** usually requires the seller or lessor to deliver the particular goods identified in the contract.

specific performance
An order of the court requiring a nonbreaching party to fulfill the terms of the contract.

How to Compute Hot Damages

Detroit Radiant Products Company v. BSH Home Appliances Corporation United States Court of Appeals for the Sixth Circuit

473 F.3d 623; 2007 U.S. App. LEXIS 300; 2007 FED App. 0005P (6th Cir.); 61 U.C.C. Rep. Serv. 2d (Callaghan) 701 (2007)

Detroit Radiant is a manufacturing company that makes gas-fired infrared heaters that are used for industrial and commercial purposes. BSH requested a price quote based on an estimated annual usage of about 30,000 units. Two purchase orders were sent from BSH to Detroit Radiant on August 10, 2001, and January 8, 2003, for 15,000 and 16,000 units, respectively. BSH provided release schedules to Detroit Radiant for exact timing of the order. On April 5, 2004, BSH had received 12,886 units, had not contested quality at any point, and had scheduled 6,000 more units to be delivered before February 2005. On April 12, a new schedule was released, slowing production schedules. On June 24, Detroit Radiant filed a suit for the profits relating to 18,114 unsold units and unsellable inventory. BSH contended both that the purchase orders were estimates and that the 2003 order superseded the 2001 order. The 2001 order was a blanket order, which BSH claims is an industry standard, meaning estimated order. An unsigned supplier argument stated that all scheduled orders were nonbinding forecasts. On March 1, 2006, a district court ruled in favor of Detroit Radiant, claiming that the purchase orders were of a certain quantity and did not contain any language to infer that it was estimated.

Michigan, like 49 of the 50 states (except Louisiana), has adopted UCC Article 2 to regulate the sale of goods. The default measures of damages under breach of contract are contract price less the market price of items in the time and place, plus any incidental damages. However, "lost profits" is available in two situations, one of which is a "lost volume seller," which Detroit Radiant is not and does not claim to be in the case. The other situation occurs when the default measure does not adequately make the seller whole. This could be a result of no reliable market price. In this case, the part is a special order for BSH, but that does not in it itself make the products unsellable. However, BSH would not have allowed the products to be sold because it considered the part proprietary. The BSH argument in court was that Detroit Radiant did not respect the proprietary nature of the items because it had listed them in a catalog. However, the court found that did not in it itself prove that Detroit Radiant did not consider the part proprietary property of BSH. Detroit Radiant may have just simply been attempting to show the types of parts that it could make. Further, BSH had no proof that it had actually sold any of these units. Thus, Detroit Radiant had shown that there was no reliable market value for the part outside of BSH. Therefore, Detroit Radiant is entitled to lost profits in addition to the value of unsellable inventory that it now has.

E-COMMERCE and the Law

Computer Contracts

Sometimes, computer purchasers are surprised to find out they are bound to agreements that were created when sellers included contracts in the box in which the computer was delivered. Some of these agreements limit the purchaser's remedies. For instance, in *Hill v. Gateway 2000*,* Hill purchased a computer from Gateway 2000 by placing a telephone order. The computer arrived through the mail. Gateway had placed a contract in the computer's shipping box that indicated that the terms sent in the box were binding on the buyer unless the buyer returned the computer within 30 days. One of the terms in this contract was a provision that stated that any disputes between the parties would be resolved through arbitration. The order taker had not read any of the terms of the contract over the telephone when Hill placed the order. When the computer arrived, Hill did not read the contract. In an effort to avoid the arbitration clause, Hill asked the court to determine that Gateway 2000 could not limit buyers's remedies or avenues through which they seek remedies (i.e., through arbitration) by bundling hardware and legal documents. The court ruled that the contract was binding on the parties. The court stated, "A contract need not be read to be effective; people who accept take the risk that the unread terms may in retrospect prove unwelcome."†

* 105 F.3d 1157 (1997).
† Ibid., p. 1158.

BUT WHAT IF . . .

WHAT IF THE FACTS OF THE CASE OPENER WERE DIFFERENT?

What if the motor home was never "fixable." Could the buyers sue for lost vacation time, emotional distress, replacement costs for a loaner motor home? What are reasonable consequential damages? What would be unreasonable?

Reject Nonconforming Goods

UCC Sections 2-601 and 2A-519 allow the buyer or lessee to reject nonconforming goods. The buyer or lessee may then obtain cover or cancel the contract.

Revoke Acceptance of Nonconforming Goods

UCC Sections 2-608 and 2A-517 sometimes allow the buyer or lessee to revoke acceptance of nonconforming goods.

Accept the Nonconforming Goods and Seek Damages

Under UCC Sections 2-607, 2-715, and 2A-519, buyers or lessees are allowed to accept nonconforming goods and then seek monetary damages to give them the benefit of the bargain. The buyer or lessee must give the seller or lessor reasonable notice of the defect.

Modifications or Limitations to the Remedies Otherwise Provided by the UCC

Parties to sales and lease contracts are allowed to modify or limit remedies. Under UCC Sections 2-719 and 2A-503, parties are allowed to create agreements that make it clear that the remedies outlined in the agreement are exclusive remedies. Courts uphold modifications or limitations to remedies unless the remedies fail in their essential purpose.

CASE OPENER WRAP-UP

A Nightmare of a Vacation

Plaintiffs sought damages under the following theories: revocation of acceptance, pursuant to of the Magnuson-Moss Warranty Act (Magnuson-Moss Act), breach of implied warranty of merchantability, pursuant to of the Magnuson-Moss Act; and revocation of acceptance and cancellation of contract. They also sought to recover the purchase price. Plaintiffs attached the following documents to their complaint: (1) the first page of the parties's contract for the sale of the RV, an alleged expert's report regarding water leakage and mold, (3) the letter to defendant purporting to confirm the revocation of acceptance, and (4) rental rates for a 23-foot trailer.

Issues

Clearly the purchase of a recreational vehicle is a purchase of a "good" under the Uniform Commercial Code's Article 2. However, the following issues are presented:

1. Are the issues with the RV "non-conforming" defects rising to a breach of contract?
2. Do the plaintiffs have an obligation to allow the defendants to cure the defects?
3. If so, is there a limit to the amount of time given to affect a cure?
4. May the plaintiff's revoke their acceptance and demand their purchase price back?
5. Have the contractual warranties been breached?

In short, the court states these answers to the following questions:
1. Yes, these are non-conforming defects
2. Yes
3. A reasonable amount---usually determined by a jury
4. Yes, but ONLY after a reasonable time has been given for a cure
5. Yes, same as question #1, but only after a reasonable time

The court found there were two instances of non-conforming issues. The first one, dealing with leaking into the dinette area, had been remedied, at least to some degree. When the leaking reappeared in greater force, the court treated it as a new breach. While the RV had been sent to the manufacturer for repair, the owners revoked acceptance and demanded their purchase price back. The plaintiff claims that the unreasonable time for the cure should be calculated from the first leak's appearance. The court agreed with the defendants that the first leak's issue was satisfied and a new clock began running with the appearance of the second, more serious leak. Since it had only been two weeks since the second leak was reported, a reasonable time for cure, as a matter of law, had not expired yet.

PRACTICAL TIPS for Business Managers

1. Don't try to take advantage of a good faith mistake by another party. You might win, unlike in this case, but the cost may be considerable. This kind of action can impact your reputation. Moreover, it is just dishonest. Put yourself in position of the party who made the mistake: how would you want to be treated?
2. Remember if a seller wishes to disclaim an implied warranty, it must be done with clear, conspicuous language and must use the language of the warranty itself. e.g. if a seller is disclaiming the implied warranty of merchantability include the language "implied warranty of merchantability" in the disclaimer.
3. The UCC was created to ease and facilitate commercial transactions between buyer and seller. When a breach occurs, don't "lawyer up" and don't run immediately to litigation. Work it out with the other contractual party if at all possible. Remember the old adage that the only winners when a case goes to trial are the attorneys.

SUMMARY

The Basic Performance Obligation	Under the UCC, sellers and lessors are obligated to *transfer and deliver conforming goods*. Buyers and lessees are obligated to *accept and pay for conforming goods in accordance with the contract*. The UCC requires *good faith* in the performance and enforcement of every contract.
Specific Obligations of Sellers and Lessors	The *perfect tender rule* indicates that if goods or tender of delivery fail in any respect to conform to the contract, the buyer/lessee has the right to accept the goods, reject the entire shipment, or accept part and reject part. Exceptions to the perfect tender rule allow sellers/lessors and buyers/lessees to consider:

- Norms in the industry and past dealings between the parties.
- Exceptions outlined in the parties's agreement.
- The seller's/lessor's right to cure.
- Excuse from performance when identified goods are destroyed through no fault of the parties.
- The concept of substantial impairment as it relates to revocation of acceptance and installment contracts.
- The concept of commercial impracticability.

Specific Obligations of Buyers and Lessees	If all goes well in a transaction over the sale or lease of goods, the buyer or lessee inspects the goods and then *pays* according to the agreement. The seller or lessor must provide an opportunity for *inspection*.

- The concept of reasonableness governs the inspection process.
- After inspection, the buyer or lessee decides whether to accept the goods.

After inspecting, the buyer/lessee signifies *acceptance* or partial acceptance.

- Sellers or lessors sometimes presume acceptance.
- Partial acceptance is allowed in some circumstances.
- Buyers or lessees are allowed to revoke or withdraw acceptance of nonconforming goods.
- Buyers or lessees must issue reasonable notice if they decide to reject goods.

Cases in which a buyer decides to assert his or her right to reject nonconforming goods are often categorized under the heading of "rescission or revocation of acceptance by buyer."

Warranties	A *warranty* is a promise on the part of the seller with respect to certain characteristics of the good. *Warranties of title:*

- Passage of good title.
- Implied promise of no liens or judgments against title.
- Implied promise that title is not subject to any copyright, patent, or trademark infringement.

 Express warranties:

- Description of the good's physical nature or its use.
- Either general or specific.
- Material term of the contract.
- Reliance of buyer on representations.

IMPLIED WARRANTIES:
Implied warranty of merchantability: A warranty based on a reasonable expectation of performance of the purchased good. The goods must:

- Pass without objection.
- Be of fair quality within the description.
- Be fit for ordinary uses.
- Have even quality.
- Be adequately packaged.
- Conform to promises made on the label.

	Implied warranty of fitness: A warranty that arises when the seller knows the purpose for which the buyer is purchasing goods and the buyer relies on the seller's judgment. *Implied warranty of trade usage:* A warranty that arises as a result of generally accepted trade practices.
Warranty Rights of Third Parties	The states have three choices regarding the warranty rights of third-party beneficiaries: • Seller's warranties extend to the buyer's household members and guests. • Seller's warranties extend to any reasonable and foreseeable user. • Seller's warranties extend to anyone injured by the good.
Warranty Disclaimers and Waivers	Methods of waiving include: • Seller does not make warranties in the first place (express warranty). • Seller disclaims in clear, unambiguous, conspicuous language (implied warranty). • Buyer fails or refuses to examine goods. • Buyer fails to file suit within the time of the statute of limitations. *Magnuson-Moss Act:* If a seller decides to issue a written warranty for a consumer good, the seller must indicate whether the warranty is full or limited.
The Goal of Contract Remedies **Remedies Available to Sellers and Lessors under the UCC**	The goal of contract remedies is to give the parties the *benefit of the bargain they struck* and nothing more. When the buyer/lessee is in breach, the seller/lessor can: • Cancel the contract. • Withhold delivery. • Sell or dispose of the goods. • Sue to recover the purchase price, lease payments due, or some other measure of damages that gives the seller or lessor the benefit of the bargain. • Stop delivery. • Reclaim the goods.
Remedies Available to Buyers and Lessees under the UCC	When the seller/lessor is in breach, the buyer/lessee can: • Cancel the contract. • Obtain cover. • Sue to recover damages. • Recover the goods. • Obtain specific performance. • Reject nonconforming goods. • Revoke acceptance of nonconforming goods. • Accept the nonconforming goods and seek damages.
Modifications or Limitations to the Remedies Otherwise Provided by the UCC	Parties to sales and lease contracts are allowed to modify or limit remedies. Courts uphold modifications or limitations to remedies unless the remedies fail in their essential purpose.

Point/Counterpoint

Do Implied Warranties Diminish the Fundamental Free Market and the Right to Contract?	
YES	**NO**
Western economies are based on the notion of freedom of contract. Parties are free to enter into the marketplace and negotiate agreements. The only restraint which should be place on that activity is that a contract for illegal or tortious activity is not enforceable.	This is a no-brainer. Contracts for the sale of goods should include the implied warranties of merchantability and particular purpose. It is a logical and common sense development in contract law for two significant reasons.

If the parties wish to include the implied warranties of merchantability and particular purpose then they can do so as express warranties and terms of the agreement. Since most UCC Article 2 contracts are between merchants, these are parties who are well-versed in the machinations of sales contract formation. The current implied warranties are just another example of legal intrusion in the market process and are paternalistic and unneeded.

First, not all UCC Article 2 contracts are between merchants. Many are between merchants and consumers. As such, the law needs to consider and protect consumers who are less versed and experienced in contract formation. Second, it is just common sense that these implied warranties exist. Who would ever consider buying an item if it does not work as it should? This is the outdated caveat emptor that we abandoned for good reasons over a century ago. Keep the warranties. They make good sense.

Questions & Problems

1. What determines the obligations of sellers/lessors and buyers/lessees?

2. Explain what the good-faith and commercial reasonableness obligations are.

3. Linda Budd went searching for a new friend . . . and she found one for $400. A brand new puppy. She purchased the puppy from Bernadette Vicidomine, a person who regularly sells puppies. Budd took her new friend home but realized that he was not in the best of health. Already attached to him, she did not return the puppy to Vicidomine but, instead, took him to the veterinarian. Nearly $2,400 later, the puppy was medically mended. Budd then sued for the $400 purchase price and nearly $2,400 in vet bills, alleging breach of the implied warranty of merchantability.

 Source: *Linda Budd, Appellant v. Maureen Quinlan et al., Respondents.*, 2008 NY Slip Op 28156; 19 Misc. 3d 66; 860 N.Y.S.2d 802; 2008 N.Y. Misc. LEXIS 2472; 66 U.C.C. Rep. Serv. 2d (Callaghan) 358 (2008).

4. An Art Gallery and Plaintiff entered into an Art Administration Agreement, dated March 18, 2004. Pursuant to the Agreement, the Gallery was to administer artworks owned by and acquired for Plaintiff, who would retain title to them. If the Gallery transferred an artwork belonging to Plaintiff, it was required to obtain from the buyer and transmit to Plaintiff a written acknowledgment that Plaintiff was the owner. Transfer of title was required to be from Plaintiff to the buyer. All sales proceeds were to be transmitted to Plaintiff. The Art Gallery sells paintings to the defendant who claims that he takes good title regardless of the restriction of the agreement between the Art Gallery and Plaintiff. Does he?

 Source: *24 Art Holdings, LLC v. Berry-Hill Galleries, Inc.*, 2014 N.Y. Misc. Lexis 3562 (2014).

5. Poor Sarah Jane not only had a short-lived marriage, but found out that her supposedly $45,000 ring was worth only half that amount. It seems that the jeweler had misrepresented it to her now ex-husband. Sarah Jane now brings an action against the jeweler for breach of an express warranty. Can she sue even though she was not the purchaser or, in fact, any purchaser?

 Source: *Schaurer v. Mandarin Gems of California, Inc.*, 125 Cal. Rptr. 4th 949 (2005).

6. Does possible misuse constitute a waiver of the implied warranties? Consider the case of a lessee of a car who, after 18 months, found the car was running roughly. When taking the car back to the dealership, the lessee was told that several valves in the engine were bent due to misuse by the lessee. The lessee refused to pay the bill of $500+, claiming that it should be covered under the warranties. The car dealership's position was that misuse constituted a waiver of the warranties.

 Source: *LaBella v. Charlies Thomas, Inc., and Mercedes-Benz of North America*, 942 S.W2d 127.

7. Plaintiff, a tree trimmer, purchased a pair of Brahma brand men's work boots from Wal-Mart on October 19, 2003. The boots' packaging described the boots as "iron tough," "rugged leather . . . men's work boots." Plaintiff wore the boots eight to twelve hours per day, six days a week, for about nine months. He claims that as the sole of "the boots wear down[,] the yellow rubber piece tends to unglue itself and roll up as you are walking, making it very dangerous when working." Plaintiff states that this unglued piece of the sole of the boots caused him to trip, fall over, and injure his back. On July 28, 2004, Plaintiff was wearing the boots while at work cutting down dead tree limbs and removing the logs. When he began to move a log weighing about 150 pounds, the unglued sole of his boot got caught on debris, causing him to fall backwards and drop the log on top of himself. He immediately felt a sharp pain in his back. The next morning, he was unable to get out of bed due to his back pain, and was driven to the emergency room. He had x-rays

and an MRI, which showed that he had two ruptured or bulging discs. Following five or six months of physical therapy, Plaintiff eventually underwent back surgery. Plaintiff filed his complaint against Defendants alleging breach of warranties on September 20, 2007, about three years and two months after the accident. Wal-Mart claims that since the lawsuit for consequential medical damages and not just for breach of the warranties of the boots, the three years tort statute of limitations apply rather than the four year statute under the UCC. Who prevails on this issue?

Source: Badilla v. Wal-Mart Stores East Inc., 2015 NMSC 029 (New Mexico 2015).

8. Mrs. Dempsey paid $541 for a poodle named Mr. Dunphy. Unfortunately for Mr. Dunphy, he had an undescended testicle. Mrs. Dempsey could have "inspected" Mr. Dunphy but didn't. After examinations by multiple veterinarians, Mr. Dunphy's condition was confirmed. Mrs. Dempsey returned Mr. Dunphy to the seller and demanded her money back. The court began its analysis with, "The first test of merchantable quality is whether the goods 'pass without objection in the trade under the contract description' (U.C.C. Sec. 2- 314(2)(a)). This standard effectively sums up all of the various 'definitions' of merchantable quality since the principal function of the warranty of merchantability has been to give legal effect to a buyer's reasonable expectations based on trade understanding of the quality of goods normally supplied under such a contract. (51 N.Y. Jur, Sales Sec. 176)." Is an undescended testicle a breach of merchantability?

Source: *Dempsey v. Rosenthal,* 121 Misc.2d 612 (N.Y. 1983).

9. Are there limits to the right to cure? Does a breaching party have the right to cure the breach, or can the non-breaching party proceed directly to a damage claim? See *Dunleavy v. Paris Ceramics,* 47 Conn. Supp. 565 (2002) for an interesting discussion of the right to cure.

10. Aubrey Reeves purchased a computer system for his business from Radio Shack Computer Center. Radio Shack is the local retailer for products sold by Tandy, its parent company. During negotiations, it became clear that Reeves needed software that Radio Shack could not provide. A Radio Shack salesperson referred Reeves to a software source book, let Reeves know he could choose compatible software from the source book, and informed him that Tandy did not support or service software from the source book. A disclaimer to this effect was in the source book. Reeves eventually purchased computers from Tandy, some software from Tandy, and more specialized software from a company called Lizcon. The Lizcon software did not meet Reeves's needs. Reeves subsequently sent a letter to Tandy, asking for rescission of the contract and damages. Could Reeves rescind?

Source: *Aubrey's R.V. Center, Inc. v. Tandy Corporation,* 731 P.2d 1124 (1987).

11. Landrum sold Gappelberg a big-screen television. The television had many defects, some of which Landrum and his service representatives fixed. Three weeks after the sale, the television stopped working altogether, and Gappelberg let Landrum know. Landrum promised to fix the television, but he did not. Gappelberg then requested return of consideration and asked the service representative to pick up the set but not to repair it because he would not accept it or a replacement. The question in the case was whether Gappelberg was allowed to prevent Landrum from curing the nonconforming television by his refusal to accept a replacement set. What did the court decide? Are you more sympathetic to Landrum or Gappelberg?

Source: *Leitchfield Development Corp. v. Clark,* 757 S.W.2d 207 (1988).

12. The defendant, Sterile Technologies Inc., purchased a sterilizer from the plaintiff, Troy Boiler Works, on an installment payment plan. The defendant was to make installment payments charged with 1.5 percent interest per month. The sterilizer was delivered on August 23, 1996. The last payment was received on April 21, 1998. At the time of the last payment, the defendant still owed the plaintiff $112,615 as the balance due on the sterilizer; as of the time of the filing of the lawsuit, the defendant owed an additional $134,214 in finance charges. The plaintiff filed its lawsuit to collect on the account on November 20, 2002. The defendant moved to have the suit dismissed because it was filed after the four-year statute of limitations had run out. (April 21, 1998, to November 20, 2002, is four years, seven months, and one day.) Has the statute of limitations run out?

Source: *Troy Boiler Works, Inc. v. SterileTechnologies, Inc.,* 777 N.Y.S.2d 574 (2003).

13. Grimberg sold a famous painting to Cohen for $785,000. Possession of the painting passed to Cohen, but payment was never made. Despite various and sundry attempts to resolve the issue, Grimberg never received payment or got the painting back. Cohen was subsequently convicted of fraud. Grimberg filed an insurance claim with his insurance company, which denied payment because, it claimed, Grimberg had transferred title to Cohen. Grimberg claimed that he transferred a voidable title and thus still retained an insurable interest in the painting. Who won?

Source: *Zurich American Insurance Co., Subscribing to Policy Number IM 8338602, Certificate Number D279FL, Plaintiff-Counter-Defendant-Appellee, v. Felipe Grimberg Fine Art, Fine Arts Services, Inc., Doing Business as Felipe Grimberg Fine Art, Felipe Grimberg, John Does, A-Z, Defendants-Counter-Claimants-Appellants*, 08-1300-cv, 2d Cir. Ct. of Appeals, 324 Fed. Appx. 117; 2009 U.S. App. LEXIS 9937; 68 U.C.C. Rep. Serv. 2d (Callaghan) 690, May 7, 2009.

14. A restaurant called The Inn Between entered into a contract to purchase a used restaurant computer system. The contract included installation and training from Remanco Metropolitan Inc. The contract also required Remanco to keep the computer system in good operating order. The system was delivered and installed on March 29, 1995. The following day, the computer malfunctioned and was down for three hours. Between March 30 and July 3, the restaurant contacted Remanco 48 times to report malfunctions. Although Remanco responded to each of the problems, the computer system continued to break down. Inn Between brought an action to revoke its acceptance of the computer system. It viewed the system as a nonconforming good. Remanco counterclaimed, seeking the unpaid price under the system maintenance agreement for nonreturn of the system. Which side got the remedy it sought?

Source: *The Inn Between, Inc. v. Remanco Metropolitan, Inc.*, 662 N.Y.S.2d 1011 (1997).

15. JRM Hauling & Recycling Services, Inc. ("JRM"), and defendant, The Newark Group, Inc. ("Newark"), agreed to purchase and JRM agreed to sell "all secondary fiber produced by [JRM] at" JRM's location in Malden, Massachusetts (the "Agreement"). JRM claims that Newark wrongfully terminated the Agreement in January 2015. By its terms, the Agreement was to run for ten years from its execution on November 1, 2006 to October 31, 2016. Newark argues that impossibility to perform the contract arose in that performance would result in a $1.5 million dollar loss. Is this claim, if proven, justifiable reason to anticipatorily repudiate the contract?

Source: *JRM Hauling & Recycling Servs. v. Newark Grp., Inc.*, 2017 Mass. Super. LEXIS 233 (2017))15; 2006 U.S. App. LEXIS 22227; 60 U.C.C. Rep. Serv. 2d (Callaghan) 1051 (2006).

Part 4: Negotiable Instruments and Banking

CHAPTER 17

Negotiable Instruments: Negotiability and Transferability

CASE OPENER

Oral Agreements and Negotiable Instruments

As a gambling facility, MGM Desert Inn Inc. regularly holds and executes negotiable instruments. During a period of two months, customer William E. Shack, Jr., entered MGM and delivered eight checks to the casino in exchange for markers. These checks, which totaled $93,400, were signed by Shack and dated at the time of transfer. When MGM sent Shack's checks to the bank for payment, they were dishonored because the funds in Shack's account were insufficient.

MGM filed an action in district court to obtain the $93,400. Shack contended that a casino host had told him that he had sufficient remaining casino credit to receive the markers. The district court judge ruled in favor of MGM, affirming its argument that the checks were negotiable instruments and stating that no evidence of an oral agreement between the casino and Shack was provided. The checks contained Shack's signature, on a specified date, and were to be payable on demand from Shack's bank to MGM. Shack was ordered to pay MGM $5,000 for attorney fees in addition to the $93,400 originally owed on the checks.[1]

1. If you were employed at MGM, what would you do to avoid future disputes with your patrons about the nature of payment agreements?
2. Assuming that Shack's claims about an oral agreement with MGM were true, would this affect your decision of whether the checks's payments were currently due?

LEARNING OBJECTIVES

After reading this chapter, you will be able to answer the following questions:

- **LO 17-1** Describe the purpose of negotiable instruments.
- **LO 17-2** Identify the types of negotiable instruments recognized by the UCC.
- **LO 17-3** Recall the requirements of negotiability.
- **LO 17-4** Recognize the words of negotiability.
- **LO 17-5** Identify the components of a check.

[1] *MGM Desert Inn, Inc., dba Desert Inn Hotel & Casino v. William E. Shack*, U.S. District Court, District of Nevada, 809 F. Supp. 783 (1993).

LO 17-6 Explain the differences among various types of checks.

LO 17-7 Recognize how deposits are accepted at various locations.

LO 17-8 Describe instances that allow a bank to charge a customer's account.

Once a sales contract has been created and executed and the parties are aware of their respective obligations under the contract, the next phase is the *payment* by the buyer to the seller for the goods purchased. Payment is usually made in one of three ways: in cash, through credit arrangements (which are discussed in the chapter on secured transactions), or with a *substitute for cash*. It is this substitute for cash that is the focus of this chapter and the next one.

A substitute for cash, or a **negotiable instrument,** is a written document containing the signature of the creator, who makes an unconditional promise or order to pay a sum certain in money either at a time certain or on demand. Negotiable instruments are executed on a daily basis in the form of checks, certificates of deposit, drafts, and promissory notes in exchange for goods, services, or business financing.

negotiable instrument A written document signed by a person who makes an unconditional promise to pay a specific sum of money on demand or at a certain time to the holder of the instrument; an acceptable medium for exchanging value from one person to another.

The Need for Negotiable Instruments

LO 17-1 Describe the purpose of negotiable instruments.

It is easy to figure out why using documents as payments greatly facilitates commercial transactions, especially in times when actual currency or cash is in short supply or when it is dangerous to transfer large amounts of currency or precious metals. These documents of payment are generically called *commercial paper,* and under Article 3 of the UCC, they are specifically labeled *negotiable instruments.*

Contracts as Commercial Paper

We have already discussed one common form of commercial paper: contracts. It does not matter whether the contract is a common law contract or a UCC Article 2 contract. Both kinds are examples of commercial paper. As such, through the process of assignment, these contracts may be circulated and transferred throughout the commercial world. Here is an illustration of assignment:

> Bob sells a bushel of apples to Betty; in exchange, Betty executes a contract to pay Bob $5 on demand. A few days later, Bob buys some oranges from Pat for $5. Instead of paying Pat the $5 in cash, Bob *assigns* Betty's obligation to pay Bob the $5 to Pat. When Pat demands the money from Betty, Betty will pay Pat the $5, and everyone will be square.

This illustration demonstrates the point that any contractual obligation, except personal ones and nonassignable ones, may be transferred and thus is classified as commercial paper.

Problems with Commercial Paper

Even though the illustration with Bob, Betty, and Pat is a simple one, it still contains a potential problem. Consider that fact pattern again, only this time assume that the apples Bob sold to Betty were all rotten. Betty did not know the apples's condition at the time because they looked and felt fine. It was only after cutting into them that the problem became apparent.

When Pat demands the $5 from Betty, Betty naturally is going to refuse, claiming that Bob had breached their original contract by delivering defective apples to her. Betty would be justified and on legally safe ground to refuse to pay. Pat could do nothing against Betty and would have to go back and sue Bob. The commercial paper that Bob transferred to Pat was not acceptable. In other words, the whole reason for having the transferability of commercial paper as a substitute for currency is defeated under these circumstances.

Types of Negotiable Instruments

LO 17-2 Identify the types of negotiable instruments recognized by the UCC.

Under Article 3, the UCC recognizes four specific types of negotiable instruments: notes, certificates of deposits (a highly specialized type of note), drafts, and checks (a highly specialized type of draft) [UCC Section 3-104]. A **note** is a promise by the *maker* of the note to pay a payee [UCC 3-103(a)(9)]. A **draft** is an order by a *drawer* to a drawee to pay a payee (in our previous example, Bob could have drawn a draft ordering Betty to pay Pat $5, since Betty owed Bob $5) [UCC 3-103(a) (6)]. A note is a two-party instrument, whereas, by definition, a draft is a three-party instrument.

Notes and drafts can either be demand instruments or time instruments. With a **demand instrument,** the payee (or subsequent holder) can demand actual payment at any time. The UCC defines an instrument that is payable on demand as one that "(i) states that it is payable on demand or at sight, or otherwise indicates that it is payable at the will of the holder, or (ii) does not state any time of payment" [3-108(a)]. With a **time instrument,** payment can be made only at a specific time designated in the future. The UCC requires an "instrument payable at a definite time" to have a time easily determined from the document itself [3-108(b)].

Certificates of deposit and checks are specific illustrations of these distinctions. A **certificate of deposit** is a promise made by a bank to pay a payee a certain amount of money at a future time. The UCC defines a certificate of deposit as "an instrument containing an acknowledgment by a bank that a sum of money has been received by the bank and a promise by the bank to repay the sum of money. A certificate of deposit is a note of the bank" [3-104(j)]. Usually, a payee buys one of these instruments from a bank and then collects the principle plus a determined amount of interest in the future. However, because these instruments have a present value, even though they are not payable until some future time, the payee may transfer them or even sell them before the future date of payment.

A **check** is a specific draft, drawn by the owner of a checking account, ordering the bank to pay the payee from that drawer's account [UCC 3-104(f)]. A check is always a demand instrument and can never be a time instrument (postdating really does not affect the ability of the holder to cash the check before the post date). Types of checks include:

- *Cashier's check:* "[A] draft with respect to which the drawer and drawee are the same bank or branches of the same bank" [UCC 3-104(f)].
- *Traveler's check:* "[A]n instrument that (i) is payable on demand, (ii) is drawn on or payable at or through a bank, (iii) is designated by the term 'traveler's check' or by a substantially similar term, and (iv) requires, as a condition to payment, a countersignature by a person whose signature appears on the instrument" [UCC 3-104(i)].
- *Certified check:* "[A] check accepted by the bank on which it is drawn" [UCC 3-409(d)].

An Overview of the Law of Negotiable Instruments

For an instrument to be negotiable, specific requirements of *negotiability* must be met. However—and this is a very important concept that many people miss—if an instrument fails to qualify as a negotiable instrument, that does *not* mean that the instrument fails to be a perfectly good, perfectly enforceable contract. All it means is that the special rules regarding negotiable instruments do not apply. Once an instrument is a negotiable instrument, it has the potential of conferring on its possessor some special qualities. Let's go back to our example with Bob, Betty, and Pat. Use Exhibit 17-1 to follow the logic of the exchanges.

Case 17-1 explores the criteria that make an instrument a negotiable instrument.

note
A promise by the maker of the note to pay the payee of the note.

draft
An instrument validating an order by a drawer to a drawee to pay a payee.

demand instrument
A type of draft that allows the payee to demand payment at any time from a holder.

time instrument
A type of draft that allows the payee to collect payment only at a specific time in the future.

certificate of deposit
A document whereby a bank promises to pay a payee a certain amount of money at a future time.

check
A special draft that orders a bank (the drawee) to pay a specified sum of money to the payee from the drawer's account.

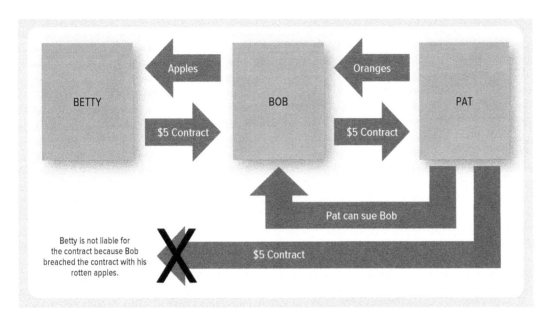

Exhibit 17-1

Potential Complexity of Negotiable Instruments

CASE 17-1

AMPLIFY FEDERAL CREDIT UNION v. GARCIA

Court of Appeals of Texas: Court of Appeals, 3rd District

(2017)

FACTS: In December 2008, Jason Garcia signed a motor vehicle sales contract with Mac Haik Dodge Chrysler Jeep, a dealer. In the contract, Garcia agreed to purchase a 2009 Dodge Ram 1500. The contract includes Garcia's agreement to pay a total of $62,370 spread out over 84 monthly payments of $742.50. Mac Haik Dodge Chrysler Jeep assigned the sales contract to Amplify Federal Credit Union (AFCU).

In 2010, after Garcia defaulted on payments, AFCU repossessed the vehicle, sold it, and applied the proceeds to the remaining balance. Garcia still owed $20,711 on the sales agreement and continued to accrue interest. In 2015, AFCU sued Garcia to recover over $35,000 and Garcia responded by asserting AFCU's claim was barred by the statute of limitations. Both parties moved for summary judgement. The trial court granted Garcia's motion for summary judgement and AFCU subsequently appealed.

ISSUE: Was the sales contract a negotiable instrument and therefore subject to a six-year statute of limitations?

DECISION: In its sole issue on appeal, AFCU argued that the sales contract for the vehicle was a negotiable instrument, which made it subject to a six-year statute of limitations as opposed to the four-year statute of limitations as ruled by the trial court. The court noted that Article 3 of the UCC is not intended to cover sales contracts. Thus, the instrument in question must be payable "to bearer" or "to order" to be considered negotiable.

The UCC provides that an instrument is payable to bearer if it states that it is payable to anyone in possession of the instrument or does not state the payee. However, the contract identifies Mac Haik Dodge Chrysler Jeep as the payee and was therefore not "payable to bearer." Nor was the contract "payable to order," as it provides only that Garcia promises to pay Mac Haik Dodge Chrysler Jeep the sales price for the vehicle. The court ruled that assigning the rights of the contract to AFCU had no bearing on whether the contract was a negotiable instrument. Because the contract was not a negotiable instrument, the four-year statute of limitations was appropriate. Affirmed in favor of Garcia.

SIGNIFICANCE: This case helps delineate the requirements for negotiability and indicate what is and is not a negotiable instrument.

(continued)

CRITICAL THINKING

What are the factual elements that caused the contract not to be payable to bearer?

ETHICAL DECISION MAKING

Which values are being given priority by the criteria making an instrument negotiable?

Betty buys apples from Bob for $5 and gives Bob a negotiable note promising to pay the $5 on demand. Bob then buys oranges from Pat for $5 and properly transfers (also known as *negotiates*) the $5 negotiable note to Pat. Pat presents the note to Betty for payment after Betty has found out that the apples she bought from Bob were all bad. Betty refuses to pay Pat the $5. Bob, that shady apple dealer, has disappeared. Pat sues Betty for the $5. Under the rules of negotiable instruments, Pat could very well prevail, and Betty, regardless of the awful apples she got from Bob, might just have to pay Pat.

That's an awful lot of information to jump in with right away, so we need to break down the steps. The following are the issues that have to be explored to understand this chain of events:

1. What constitutes a negotiable instrument?
2. How does one transfer a negotiable instrument?
3. What is the status of a holder of a negotiable instrument?
4. What happens when the person who created a negotiable instrument has a good defense for not honoring his or her payment on the instrument?

Clearly, when the contracts for such transactions are not in breach, everything works out fine. Potential problems, however, led to the evolution of the law surrounding negotiable commercial paper. The next section discusses the elements that make commercial paper a negotiable instrument. Remember that even when commercial paper does not meet the requirements for negotiability, it may still constitute an enforceable contract.

Requirements for Negotiability

LO 17-3 Recall the requirements of negotiability.

The UCC requires that for an instrument to be negotiable (do *not* equate *negotiable* with *enforceable*), the instrument must satisfy the six requirements in the following definition: A negotiable instrument is a written document that is signed by the maker or drawer with an unconditional promise or order to pay a sum certain in money on demand or at a time certain to the order of bearer [UCC 3-104(a)]. The six requirements are:

1. The instrument is a written document.
2. It is signed by the creator of the instrument.
3. The instrument has an unconditional promise or order to pay.
4. The amount to be paid is a sum certain in money.
5. Payment is to be made either on demand or at a fixed future time (a time *certain*).
6. The document must contain either the words of negotiability—"to the order of"—or words indicating that it is a bearer instrument.

As you read through the following explanations pertaining to the requirements of negotiability, use Exhibit 17-2 as an example of the requirements as they apply to personal checks.

E-COMMERCE and Negotiable Instruments

Can All Negotiable Instruments Be in Electronic Format?

Electronic negotiable instruments are often favorable compared to traditional negotiable instruments because electronic forms foster more efficient business practices by reducing the amount of time a document must wait to be processed. That being said, the very nature of electronically sent information means that certain types of negotiable instruments should not be in a virtual format. An excellent illustration of this concept is found within the 2008 Kenya Communications Act, which outlines exclusions for electronic negotiable instruments. For example, the Act excludes the creation or execution of a will from being upheld as a legal negotiable instrument in court. Exclusions such as the aforementioned case are stipulated because wills and certain other documents are only legally significant and binding if those documents have been signed in their physical form. Considering the ease with which a person could falsify an online signature, it is impractical to allow important documents such as wills to be completed in an electronic format. Specifying what is and is not legally valid in cyberspace is an important step in reducing legal ambiguities and possible conflicts.

Source: Roaa Al-Heeti, "Electronic Signatures - Case Notes Underwriters' Bulletin," Attorneys' Title Guaranty Fund, Inc. (ATGF.com).

Written Document The written-document requirement is not as simple as it may appear. Clearly, the law does not permit an *oral* negotiable instrument. Telling someone that you unconditionally promise to pay her $50 next Tuesday does not create a negotiable instrument. However, under the right factual circumstances and when these words are provable, such a statement may be a binding, enforceable contract.

What makes the writing requirement a little complex is the mandate that the written document must have two characteristics: **relative permanence** and **movability**. Writing a negotiable instrument in the mud, for example, clearly lacks permanence; it will likely disappear as the mud dries. Thus, it is not a negotiable instrument. Likewise, the mud is something that one cannot move about in a commercially reasonable or expected manner. After all, one of the reasons for negotiable instruments is the ease with which they can be moved about.

relative permanence
The quality of a negotiable instrument that ensures its longevity.

movability
The quality of a negotiable instrument that ensures it is mobile and available.

BUT WHAT IF . . .

WHAT IF THE FACTS OF THE CASE OPENER WERE DIFFERENT?

Recall that in the Case Opener, Shack contended that an oral agreement had been made between him and the plaintiff about rolling debt. What if Shack had presented a recording of this particular oral agreement actually being made? Would the judge still rely only on the written negotiable agreement?

Exhibit 17-2 Requirements of Negotiability in a Check

GLOBAL Context

©Digital Archive Japan/ Alamy

Special Negotiable Instruments in India

India's Negotiable Instrument Act of 1881 regulates the use and types of negotiable instruments in a manner similar to the rest of the Western world. However, running parallel to the Act is the hundi system, a much older system of negotiable instruments still in use today. While the Act technically regulates hundis, local traditions and customs typically hold more sway over their usage. Many different types of hundis exist; for example, firmanjog hundis are similar to checks and are payable to the order of the payee, whereas miadi hundis are payable only after a specified period of time. Shahjog hundis are payable only to a Shah (typically a member of the community with significant financial or political power and respect), and jokhmi hundis are drawn against shipments of goods and double as a bill of exchange and an insurance policy. Many other types of hundis exist, each with their own specific uses.

Signature of the Maker or Drawer The UCC and cases interpreting the signature requirement are fairly liberal in interpreting what constitutes a signature. Anyone's affirmative mark, from a full-blown signature to an *X,* will suffice, provided the party intended the mark to be placed on the instrument and uses that mark to identify himself [UCC 1-201(39)]. The UCC specifies that a signature "may be made (i) manually or by means of a device or machine, and (ii) by the use of any name, including a trade or assumed name, or by a word, mark, or symbol executed or adopted by a person with present intention to authenticate a writing" [3-401(b)].

Likewise, the signature of a duly authorized agent also satisfies this requirement. An agent's signature on behalf of her principal binds the principal and satisfies the signature requirement [UCC 3-401(b)].

A handwritten negotiable instrument satisfies the signature requirement even when no formal signature is present. The handwritten statement, "I, John Smith, promise to pay Roberta Alexander, or to bearer, the sum of $20 on Tuesday July 4, 2008," would satisfy the signature requirement because the handwriting affirms the maker's intent and, in that handwritten promise, the maker wrote his own name, John Smith.

Automated signatures, either by stamp or machine, satisfy the writing requirement as well. Even when the stamp or machine signature is fraudulently applied, such as by a person who had no authority to apply it, the signature may be valid if the maker or drawer intended his or her name to be affixed.

Unconditional Promise or Order to Pay The promise or order to pay must be specific and not implied. The language must be affirmative in nature [UCC 3-103(a)(9)]. For example, simply acknowledging a debt does not create language for payment; therefore, a common IOU is not a promise or an order to pay and thus cannot be a negotiable instrument.

The unconditional nature of the promise or order is often the controversial variable of this requirement of negotiability. Stated simply, the promise or order to pay cannot be contingent on something else happening. An instrument stating, "I promise to pay if the following occurs," is not a negotiable instrument. Remember: it may be a perfectly enforceable contract, but it fails to satisfy the terms of negotiability.

Sum Certain in Money Negotiable instruments must promise or order payment to be made in a national currency [UCC 3-104(a)]. For example, U.S. dollars, English pounds, euros, and Japanese yen all satisfy the currency requirement. Bushels of apples, gold, shares of stock, diamonds and rare gems, and the like are not currencies. Although promises to pay in apples or gold or stock may form a perfectly enforceable contract, the resulting instrument is not a negotiable instrument. An instrument promising payment in "German marks and rare French wine" would not be negotiable—even changing the *and* to an *or* would not salvage negotiability. Payment *must* be made in a currency.

E-COMMERCE and the Law

The End of the Float?

Consumers who rely on float (the time it takes for a check to go through the traditional check-clearing process) have a limited amount of time to enjoy the delayed payment afforded by the process. Businesses in many parts of the country are testing new technology that speeds up the check-clearing process. Soon, float might be an outdated tradition.

In Nevada, for instance, Nevada State Bank allows businesses that purchase a special service to scan checks, send them electronically, and get money for checks drawn on other banks much more quickly; the delay in check processing is cut by 40 percent. Other banks are testing similar products and services. In some states, businesses are substituting electronic images of checks for the checks themselves. Here's how the process works: (1) A customer gives a business a check in payment for a product or service, (2) the business scans the check and sends it to the bank providing the new check-clearing products and services, (3) the bank sends the image to the customer's bank, (4) the customer's bank prints a substitute check, and, quickly, (5) the customer's payment is deposited in the business' account at that business' bank.

It remains to be seen what kinds of litigation will emerge from this expedited process.

Payable at a Time Certain or on Demand A negotiable instrument must be payable on demand or at a specific time that can be computed from the instrument itself. Obviously, if the instrument states a specific date, the instrument has stated a time certain. If the instrument is dated and then states that "payment will be made 10 days after above date," the instrument is negotiable because the specific date can be calculated. A dated instrument stating that "payment is to be made at some future time after the above date" is clearly non-negotiable (although, again, it may be enforceable as a contract).

Likewise, an instrument that states that "payment will be made 10 days after delivery of the goods" but indicates nowhere in the instrument when delivery is to be made is not a negotiable instrument. (It might also be non-negotiable if such a reference is construed to be a condition of payment as well.)

Demand instruments, such as checks, are payable as soon as they are issued. If an instrument is silent on the time of payment, the UCC presumes that it is a demand instrument; thus, it retains its negotiable status [3-108(a)].

LO 17-4 Recognize the words of negotiability.

Words of Negotiability Finally, for an instrument to be negotiable, the instrument must indicate that it was created for the purpose of being transferred. How can an instrument, or, better yet, its maker or drawer, indicate this purpose? It can do so by having the phrase "to the order of" near the payee's name. When a specific payee is named, this is known as an *order* instrument [UCC 3-109(b)]. For the maker or drawer to indicate that she anticipates the instrument she is creating will be circulated, she writes the phrase *to the order of* and then the name of the payee, or some variation of that wording. Examples of order negotiable-instrument language are "Pay to the order of Jim Smith" and "Pay to Jim Smith or his order."

Negotiable instruments payable to whomever is bearing them are known as *bearer instruments* [UCC 3-109(a)]. Bearer instruments are treated like cash. Accordingly, one who comes into possession of a bearer instrument by any means, including theft, may claim the payment due on the instrument. Also, endorsing an order instrument, such as a check, converts the instrument into a bearer instrument that may be claimed by anyone in possession of it. Instruments payable to no one, to "X," or to "cash" are also considered bearer instruments.

The phrase *to the order of* is then necessary to create a negotiable instrument. Wordings such as "Pay to bearer," "Pay to Jim Smith or bearer," "Pay to cash," and "Pay to the order of cash [or bearer]" would all result in the paper being negotiable.

Once the six elements of negotiability are satisfied, the instrument created is indeed negotiable.

> **BUT WHAT IF...**
>
> **WHAT IF THE FACTS OF THE CASE OPENER WERE DIFFERENT?**
>
> Recall in the Case Opener that Shack contended that the terms of his old marker payment agreement were not enforceable because he could acquire a new marker to freshen the agreement. If Shack had signed a document that simply acknowledged that Shack had acquired a marker, and not that the payment for the marker was due to MGM, would this document qualify as a negotiable instrument?

Negotiation

One important characteristic of negotiable instruments is the ease associated with transferring the instrument to a third party. Once a negotiable instrument is created, it can be transferred to another party through negotiation. **Negotiation** is the transfer of possession to a third party, who becomes the holder of the negotiable instrument (UCC Section 3-201). A party who possesses a negotiable instrument payable to the party or to the bearer of the instrument is a **holder** of the instrument [UCC 1-201(b)(21)].

When the rights to a negotiable instrument are transferred from one party to another, such a transfer is called a *negotiation*. It is important to understand that the rules of negotiation differ slightly, depending on whether the instrument is an **order instrument** (i.e., an instrument payable to a specific, named payee) or a **bearer instrument** (an instrument payable to cash or to whoever is in possession of the instrument) (UCC 3-109). Bearer paper requires only a delivery of the instrument to the holder by the payee. In contrast, order paper requires delivery *and* an endorsement.

negotiation
The transfer of the rights to a negotiable instrument from one party to another.

holder
A party in possession of a negotiable instrument.

order instrument
An instrument payable to a specific, named payee.

bearer instrument
An instrument payable to cash or to whoever is in possession of the instrument.

Delivery

Delivery simply requires the physical handing of an instrument from the person who is entitled to it to the person who is intended to receive it. Thus, a bearer instrument that is blown off someone's desk by the wind, falls out a window, and lands in someone else's hand has not been properly negotiated. This lucky person cannot legally demand payment of that instrument because she is not a proper holder. However, she could pass it on to someone who could legally collect on it.

Drawers must be careful when delivering an instrument because if the drawer is negligent in how he or she makes the delivery, the drawer may be liable for notes paid with forged or unauthorized endorsements. For example, if the drawer of a check mails the check to the wrong business, the drawer may be liable for the amount of the check when the wrong business cashes the check.

Endorsement

Order paper must be *endorsed* as well as *delivered* to be negotiated. The person creating the endorsement is the **endorser,** and the person receiving the endorsement is the **endorsee.** Normally, there is a place on the negotiable instrument for endorsements (such as on the back of a check). If there is no room on the instrument or if all the room has been taken by previous endorsements, an **allonge** may be attached. An allonge is simply an additional piece of paper with the endorsements. It must be firmly attached (staples usually work; paper clips do not) [UCC 3-204(a)]. There are three kinds of endorsements that will affect the legal status of the negotiable instrument in different ways: unqualified, qualified, and restrictive endorsements.

endorser
One who issues an endorsement.

endorsee
One who receives an endorsement.

allonge
Accompanying a negotiable instrument, a piece of paper that provides room for an endorsement if no room is available on the instrument itself.

blank endorsement
A payee's or last endorsee's signature on a negotiable instrument.

special endorsement
An endorser's signature accompanied by the name of the endorsee.

Unqualified Endorsements: Blank and Special Endorsements There are two kinds of unqualified endorsements: blank and special. A **blank endorsement** is simply the payee's or last endorsee's signature, nothing else [UCC 3-205(b)]. The effect of this unqualified blank endorsement is that it turns an order paper into a bearer paper. An instrument with an unqualified blank endorsement may be negotiated by delivery only. A **special endorsement** consists of the signature of the endorser together with a named endorsee. An illustration would be "Pay to Jackie Jones" followed by the endorser's signature.

Note that in an endorsement, the words of negotiability, *to the order of,* are not needed. The instrument remains negotiable. However, adding those words keeps order paper as order paper and thus continues to require an endorsement and delivery for further negotiation.

Noncriminal Endorsement Problems

In situations involving fraud or forgery, it should come as no surprise that problems with endorsements exist. These criminal endorsement issues are examined later in this chapter. However, even in situations not involving criminal offenses, endorsement problems may arise. By way of introducing the type of problem that can arise, let's look at Case 17-2.

JONES v. WELLS FARGO BANK
666 F.3d 955 (5th Cir. 2012)

FACTS: *In September 2006, Adley Abdulwahab opened a business account with Wells Fargo Bank on behalf of W Financial Group LLC. Upon opening the account, Abdulwahab received a business account agreement that described the terms governing the account. In January 2007, Abdulwahab withdrew $1,701,250 from W Financial's account at Wells Fargo. Abdulwahab used the money to purchase a cashiers check payable to Lubna, Misba, Shahed, and Zahed Lateef. Later that day, Abdulwahab went to another Wells Fargo branch and deposited the cashier's check into the account of a separate entity, CA Houston Investment Center LLC. The Lateefs never received or endorsed the check. Rather, Wells Fargo stamped the back of the check and processed it. In 2008, the SEC brought an action against W Financial for allegedly selling false securities. Vernon T. Jones was appointed as a receiver for the purpose of recovering W Financial's assets for the benefit of investors. In the course of his duties, Jones filed an action against Wells Fargo, alleging that Wells Fargo was liable for conversion (that is, taking the personal property of a party without their permission) and responsible for the $1,701,250 in damages.*

ISSUE: Was Wells Fargo liable for the damages resulting from the processing of the bad check?

REASONING: In making their decision, the court relied on Texas's version of the UCC (which is almost identical to the UCC itself), particularly the provision that holds that a bank may be liable for conversion if "a bank makes or obtains payment with respect to the instrument for a person not entitled to enforce the instrument or receive payment." The court determined that CA Houston had no right to enforce the check and receive payment. The only way CA Houston could have been considered an enforcer was if the W Financial directly gave the check to CA Houston, and this was not what had occurred. Because Wells Fargo made payment on the check to CA Houston, an entity that had no right to enforce the check, Wells Fargo was liable for conversion.

The court also determined that Wells Fargo was liable for conversion because it deposited the check without the necessary endorsement. Wells Fargo accepted the check even though the check had not been endorsed by the Lateefs. Wells Fargo simply stamped "Credited to the account of within named payee Lack of Endorsement Guaranteed Wells Fargo Bank, N.A." and processed the check without an endorsement. Once again, the court considered the UCC in their decision. The court cited *Southwest Bank v. Information Support Concepts, Inc.,* a case where it was held that the UCC places the burden on the first bank in the collection chain to ensure an endorsement's authenticity.

DECISION: The Court of Appeals affirmed the district court's decision; Wells Fargo was liable for conversion and was held responsible for the $1,701,250.

SIGNIFICANCE: While there are exceptions, financial institutions are often liable for damages resulting from processing bad checks, and this case is just one example of this phenomenon.

CRITICAL THINKING

What reasoning supports the circuit judge's decision? Are there missing facts in the description of the case here that would better enable you to evaluate this reasoning if they were provided?

ETHICAL DECISION MAKING

What values does the court's decision promote? Does the golden rule assist us in deciding how we believe this case should have been decided?

GLOBAL Context

©Doug Armand/Getty Images

Negotiable Instruments in the Commonwealth

The law of negotiable instruments has its origins in English law. In the fourteenth century, merchants created the use of the draft, or bill of exchange, so that they could conduct business transactions without transporting large sums of money. Since 1882, negotiable-instrument transactions in England have been governed by the Bills of Exchange Act. It served as a model for the National Conference of Commissioners on Uniform State Laws (NCCUSL), which promulgated the Uniform Negotiable Instruments Law (NIL) in 1896. Subsequently, all of the U.S. states adopted the law by 1924, and the law of negotiable instruments and bank collections was revised and modernized through Articles 3 and 4 of the Uniform Commercial Code.

In the Commonwealth of Nations (formerly referred to as the British Commonwealth), almost all member states have codified their laws relating to negotiable instruments in Bills of Exchange acts, just as the United States used the original act when drafting its own negotiable-instrument laws. Notice the similarities between the UCC's regulations regarding negotiable instruments and the Bills of Exchange Act of England. The Bills of Exchange Act:

1. Defines a bill of exchange as "an unconditional order in writing, addressed by one person to another, signed by the person giving it, requiring the person to whom it is addressed to pay on demand, or at a fixed or determinable future time, a sum certain in money to or to the order of a specified person, or to bearer."

2. Defines a check as "a bill of exchange drawn on a banker payable on demand."

3. Defines a promissory note as "an unconditional promise in writing made by one person to another, signed by the maker, engaging to pay on demand, or at a fixed or determinable future time, a sum certain in money to or to the order of a specified person or to bearer."

Misspelled Name One problem with endorsements occurs when a negotiable instrument contains a misspelled name. In the event of a misspelled name, the holder may endorse the document with the misspelled name, the holder's actual name, or both. The typical practice is for the holder to endorse the instrument with the misspelled name followed by the holder's actual name, spelled properly.

Payable to a Legal Entity Another complication regarding endorsements involves instruments made payable to a legal entity. If the legal entity is an estate, organization, partnership, or the like, the instrument may be endorsed by any authorized representative of the entity. A variation of the same problem occurs when instruments are made payable to a public office. In such an event, the person holding the office may endorse the instrument. For example, when filing her local taxes, Sarah, not knowing whom to make her check payable to, filled in "pay to the order of County Tax Collector." Bill Deepockets, the county tax collector, may endorse the check because he is the person currently holding the office listed on the check.

Alternative or Joint Payees A third endorsement problem arises when a negotiable instrument has been made payable to more than one person. Two possibilities arise when the instrument is payable to more than one person. The first possibility is that there are *alternative payees*. When the instrument reads "Pay to the order of Jones *or* Smith," the instrument contains alternative payees. In the event of alternative payees, the endorsement of any one of the listed payees is sufficient.

The second possibility is that there are *joint payees*. Joint payees exist when an instrument lists more than one person as a payee and the instrument is to be payable to *all* the listed payees, typically indicated by the word *and*. In the event of joint payees, the instrument would read "Pay to the order of Smith *and* Jones." Instruments bearing joint payees require the endorsement of all listed payees before the instrument may be negotiated.

Courts have held that when an instrument is silent on the issue of whether the listed payees are joint or alternative, the instrument is to be interpreted as containing alternative payees. That is, when an instrument does not specify whether the payees are joint or alternative, the endorsement of only one listed payee is required to negotiate the instrument. The next Case Nugget illustrates a situation in which the check was silent on the issue of whether the payees were joint or alternative, and the court interpreted the ambiguity as containing alternative payees.

CASE Nugget

Check with Slash Mark between Payees's Names

New Wave Technologies, Inc. v. Legacy Bank of Texas U.S. Court of Appeals of Texas, Eighth District

WL 2553519, Ct. App. Tex. (2008)

Maxim Solutions Group Inc. and New Wave Technologies entered into a joint purchase agreement under which New Wave would sell computer equipment to Maxim and, afterward, Maxim would sell it to United Services Automobile Association (USAA). On July 7, 2004, USAA ordered scanners and computer equipment from Maxim/New Wave. The two purchase orders were supposed to be paid from USAA directly to New Wave via negotiable instrument. However, USAA mailed the checks to Maxim instead of New Wave.

The checks were made payable to "Maxim Solutions Group/New Wave Techn" for $134,656.16 and $52,558.73, respectively. The back of each check stated, "Each Payee Must Endorse Exactly As Drawn." The checks were received by the president of Maxim, after Maxim had gone out of business, and were subsequently deposited in Maxim's bank account. The checks were accepted for deposit by a teller at Legacy Bank of Texas. Each check had only Maxim's account number as an endorsement written on the back. The checks were not endorsed at all by the other payee, "New Wave Techn." Legacy Bank cashed the checks anyway and argued that it did not convert the checks since both Maxim and New Wave were alternative payees by law.

In the opinion written by Judge David Chew, the Texas Court of Appeals determined that the bank was not liable for conversion because the "payable to" line with multiple payees and a slash mark was ambiguous under Texas Business and Commerce Code Ann. § 3.110(d) (2002). The front of the check used a slash mark, which was interpreted to mean "or." Furthermore, the printed instructions on the back of the check only increased the ambiguity. Due to this ambiguity, the instruments were payable alternatively. Therefore, the bank did not take the instruments from a person not entitled to endorse them.

This case illustrates how it is particularly important when there are multiple payees for a negotiable instrument to be clear and explicit. By being less ambiguous, businesses attempt to avoid endorsement problems that arise when a negotiable instrument has been made payable to more than one person.

Banking

So far, this chapter has explained what a negotiable instrument is, how a negotiable instrument is transferred, and some of the problems associated with negotiable instruments. The following sections discuss the negotiable instruments with which you are probably most familiar: checks. You have most likely written, cashed, and deposited checks but, as you may have guessed, your relationship with the bank is more complicated than you originally thought.

When you open an account at a bank, you create a contractual relationship with the bank. This relationship is governed by Article 4 of the UCC. Within this relationship, both you, as the customer, and the bank have certain rights and duties. For example, you have the right to order a stop payment on a check for any or no reason. The corresponding duty of the bank is to follow this order.

As we explained earlier in this chapter, checks are considered negotiable instruments under Article 3 of the UCC. However, Article 4 of the UCC is also relevant; this section of the UCC governs the transfer of checks between banks. Thus, both Articles 3 and 4 are relevant to this chapter.

Article 3 of the UCC outlines the requirements that negotiable instruments, including checks, must meet. Article 3 also establishes the rights and responsibilities pertaining to parties to negotiable instruments. Article 4 creates a framework controlling deposit and checking agreements between banks and customers. In addition, Article 4 directs the relationships between banks as checks are processed among different banks. Moreover, according to UCC Section 4-102(a), when conflicts arise between rules in Articles 3 and 4, Article 4 is to take precedence.

In 2012, Americans wrote more than 23 million checks each day. Although this number is rapidly declining, clearly, checks are an enormous part of the bank–customer relationship.[2] In fact, of all the negotiable instruments regulated by the UCC, checks are the most common type used. Thus, we begin our discussion of banking by taking a closer look at different types of checks. Then we examine the process of check collection: If the bank accepts a check, how is the money from one account actually transferred to another account? Next, we consider when a bank may charge a customer's account in the context of potential problems with checks, such as stale, postdated, and forged checks. Finally, we turn to an increasingly important element of the banking process: the electronic transfer of funds.

drawer
The party that writes an order or the person who writes a check.

drawee
The party that must obey an order. In the context of banking, the drawee is the bank that must pay the funds ordered by a customer's check.

payee
The party that receives the benefit of an order (check, etc.).

Checks

LO 17-5 Identify the components of a check.

Although you have likely written a check, do you know what the actual characteristics of checks are? (The key terms and an illustration are provided in Exhibit 17-3 and Exhibit 17-4.) According to the UCC, a check is a special kind of draft. A *draft* is an instrument that is an order. Three parties are related to this order. First, a **drawer** is the party that gives the order. Second, a **drawee** is the party that must obey the order. For example, in Exhibit 17-4, the bank is the drawee. Finally, the **payee** is the party that receives the benefit of the order. Thus, when you write a check at the

Exhibit 17-3

Key Terms for Checks

Draft	An instrument whereby one party orders a second party to pay an amount of money to the party listed on the instrument. A check is a draft in which the bank is the drawee.
Drawer	The party giving the order to pay on a draft.
Drawee	The party ordered to pay on a draft.
Payee	The party receiving the money from the draft.

Exhibit 17-4

A Check

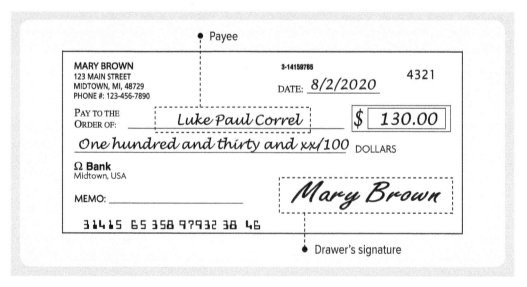

[2] Doreen Hemlock, "Still paying by check? There are fewer of us," *Sun Sentinel*, February 19, 2012, http://articles.sun-sentinel.com/2012-02-19/news/fl-checks-in-decline-20120217_1_paper-checks-electronic-payments-direct-deposit.

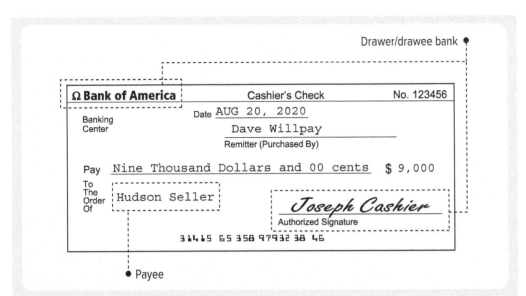

Exhibit 17-5

A Cashier's Check

grocery store, you are the drawer ordering the drawee (your bank) to make a payment to the payee (the grocery store).

A *check* is a special draft that orders the drawee, a bank, to pay a fixed amount of money on demand [UCC Section 3-104(f)]. The UCC defines a bank as "any business engaged in the business of banking" [4-105(1)]. Consequently, savings banks, savings and loans, credit unions, and trust companies are all considered banks. The drawer of a check writes the check and thus orders the bank to pay. The payee is the party to whom the check is written.

Cashier's Checks

LO 17-6 Explain the differences among various types of checks.

A **cashier's check** is a check for which both the drawer and the drawee are the same bank [UCC 3-104(g)]. (See Exhibit 17-5 for an example.) The payee of the check is a specific person. In other words, the bank is drawing on itself and thus assumes the responsibility for paying the check to that specific person. Customers often purchase cashier's checks to give to creditors who want to be sure the funds represented by the check are available. Cashier's checks are useful because they are considered by many in the business community to be the near equivalent of cash.

cashier's check
A check for which both drawer and drawee are the same bank.

Traveler's Checks

A **traveler's check** (see Exhibit 17-6) is an instrument that must (1) be payable on demand, (2) be drawn on or through a bank, (3) be designated by the phrase *traveler's check,* and (4) be countersigned by the person whose signature appears on the instrument [UCC 3-104(i)]. The drawer of a traveler's check is usually a large financial organization, such as American Express. The person who signs the traveler's check must sign it when she buys the checks; then, when she is ready to use the traveler's check to make some kind of payment, the same person must sign the traveler's check again, in the presence of the acceptor.

traveler's check
An order that is payable on demand, is drawn on or through a bank, is designated by the phrase *traveler's check,* and requires a countersignature by the person whose signature appears on the check.

Money Orders

Money orders (see Exhibit 17-7), particularly personal money orders, are usually in the same form as personal checks and are considered checks under UCC Section 3-104. Both banks and nonbanks sell money orders. The money order states that a certain amount of money is to be paid to a particular person. The amount of money to be paid is usually already imprinted on the money order. The person purchasing the money signs the money order as the drawer and fills in the name of the person who is to receive the money.

GLOBAL Context

Money Orders in India

Unlike in the United States, in India, money orders can be issued only by the post office. The money order can be verified by a signature, a thumb mark or fingerprint, or both. Money orders can be written in various languages, such as English, Hindi, or the language specific to the district of the post office. Although in the United States the individual purchasing the money order may be referred to as a purchaser, sender, or remitter, in India, the law refers to the individual who is writing the money order only as a remitter. The person who is receiving the money order would be referred to as the payee. After the money has been paid to the payee, the remitter always receives acknowledgment of payment to the payee. If the remitter wishes to stop payment on the money order, there are no stipulations to do so; the remitter simply returns to the post office and notifies the personnel of the payee's address and offers the receipt from the issued money order.

©Digital Archive Japan/Alamy

Exhibit 17-6

A Traveler's Check

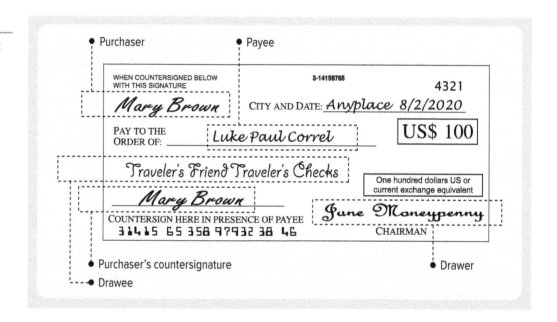

Certified Checks

certified check
A check that is accepted by the bank from which the funds are drawn.

A **certified check** is a check that is accepted at the bank at which it is drawn [UCC 3-409(d)]. For example, suppose Hope writes a check to Jeremiah from her account at Citizens National Bank (CNB). Hope asks CNB to certify her check. CNB then accepts the check, withdraws the money from Hope's account, and places that money in its certified-check account. CNB then signs or stamps the face of the check to indicate that it is certified. In other words, CNB is promising that funds are available to pay the check.

Banks are not required to certify checks [UCC 3-409(d)]. If a bank refuses to certify a check, the check is not considered dishonored; it merely lacks the extra protection of certification. However, once the bank does certify a check, the drawer of the check is no longer liable for the amount of the check [UCC 3-414(c)]. The bank has become primarily liable for the check.

Why Use Cashier's or Certified CHECKS?

There are a number of reasons to use a cashier's or certified check, as opposed to a regular check, when conducting business exchanges. Although all of these types of drafts are different, one thing they have in common is an increased guarantee of being paid. That is, a cashier's or certified check is less likely to be denied by a bank.

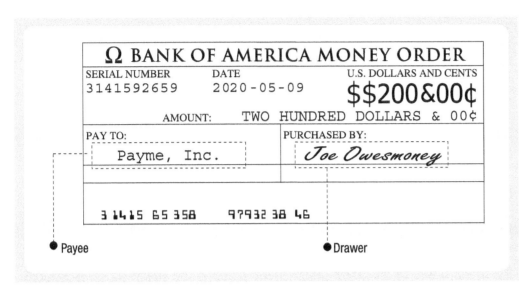

Exhibit 17-7

A Money Order

Cashier's checks are a valuable business tool because the bank, and not the individual, is the drawer as well as the drawee. When a cashier's check is presented for payment, the bank, and not the individual, must pay for the cashier's check. The added guarantee of knowing that the bank is paying for the cashier's check makes the cashier's check a veritable guarantee to pay. One downside of the cashier's check is that it must be paid for in advance, including a small fee. However, because the cashier's check is paid for first, it can be purchased at any bank, regardless of whether one has an account at the bank.

Certified checks are useful in business because when a bank certifies a check, it essentially says that it cannot refuse liability on the check. A certified check is one the bank sets aside money for and agrees to pay when the certified check is presented. Despite the added guarantee, there are two main drawbacks to a certified check. The first is that a person must have an account at a specific bank to obtain a certified check. If the person does not have an account at the bank, he or she cannot obtain a certified check from that bank.

The second drawback is that banks do not have to certify checks. Banks may refuse to certify any check for any reason. Refusing to certify is not the same as dishonoring a check, but it does not provide the same guarantee that a certified check provides. Although banks do not have to certify a check, they are not allowed to refuse to sell a cashier's check as long as the payment is valid.

Lost, Stolen, or Destroyed Cashier's or Certified Checks

In the event a cashier's or certified check is lost, stolen, or destroyed, the UCC allows for recovery. According to UCC Section 3-312, the remitter (the party who purchased the check) or the payee may request a refund because the check was lost, stolen, or destroyed. With proper identification, the party should be able to obtain a full refund for the amount on the check.

The claim is enforceable at the later of (1) the time the claim is made, or (2) the 90th day following the date on the check, if it is a cashier's check. For a certified check, it is enforceable at the later of (1) when the claim is made, or (2) the 90th day following the date of the acceptance of the check [UCC 3-312(b)(1)]. After the claim becomes enforceable, if no one presented the check for payment, a refund is issued and the bank is discharged of liability [UCC 3-312(b)(4)]. When a claim is made, the person making the claim warrants to the bank and any party who has an interest in the check that the check was really lost, stolen, or destroyed.

If the check was not lost, stolen, or destroyed, the holder barred from receiving payment on the check because of the claim may sue the person who made the claim for breach of warranty. A person filing a false claim is also subject to criminal penalties.

Accepting Deposits

LO 17-7 Recognize how deposits are accepted at various locations.

Charging a customer's account is only part of the bank-customer relationship; the bank must also credit a customer's account when the customer makes cash and check deposits to her account. This section considers the check collection process and examines several issues that focus on the availability of deposited money.

The Check Collection Process

Suppose Jack Blackstone gives Molly Whetfield a check to pay her for her consulting services. When Molly deposits that check into her account, how is the money from Jack's account actually transferred to Molly's account? This section considers the check collection process (see Exhibit 17-8) and examines several issues regarding deposits made to banks.

Check Collection within the Same Bank Sometimes the depositary bank is the same bank as the payor bank. When this is the case, the check is referred to as an on-us item. For example, suppose Molly's and Jack's accounts are at the same bank. When Molly deposits Jack's check into her account, the check does not have to be sent to another bank because she and Jack share the same bank. Instead, the bank gives a provisional credit to Molly's account. If this bank does not dishonor the check on the second day, the check is paid [UCC Section 4-215(e)(2)]. Finally, on the third day, the provisional credit becomes an actual payment.

Check Collection between Different Banks Suppose that Jack and Molly have accounts at different banks. Molly's account is in Los Angeles; Jack's account is in Miami. When Molly deposits Jack's check at her bank in Los Angeles, her bank is the depositary bank. When a depositary bank receives a check, it must present the check at the payor bank or send it through intermediary banks to reach the payor bank. Once a bank receives a check, it must pass the check on before midnight of the next day [UCC 4-202(b)]. When the check finally reaches the payor

Exhibit 17-8

Illustration of the Check Collection Process

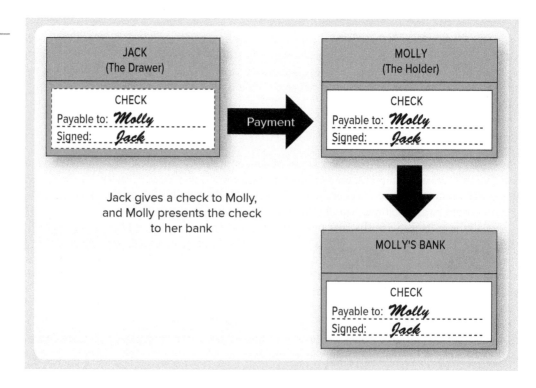

bank, the payor bank must respond to the check by dishonoring it or becoming liable for the face amount of the check [UCC 4-302].

The UCC allows banks to establish cutoff hours for making entries on their books. For example, a bank may determine that 2 p.m. or later is the cutoff hour for handling checks [UCC 4-108]. If the bank receives a check after this deadline, the bank will defer posting this check to its customer's account until the next day.

Electronic Check Presentment

In the past, checks were physically presented to each bank in the chain of collection. Now, checks are transmitted electronically from bank to bank [UCC 4-110]. Through electronic check presentment, a check can be processed on the day on which it is deposited. An item is encoded with information that is transferred from one bank's computer to another bank's computer. The person who enters the information into the computer (i.e., encodes the information) warrants that the information is correct [UCC 4-209]. Alternatively, the image of a check may be transmitted for payment to other banks.

Substitute Checks To help facilitate electronic presentment, in 2004, Congress passed the *Check Clearing for the 21st Century Act* (also known as *Check 21* or the *Check Truncation Act*). Check 21 allows banks to forgo sending original checks as part of the collection or return process and instead send a truncated version. In place of the original paper check, a bank may send (1) a substitute check or, (2) by agreement, an electronic image of the check along with data from the magnetic ink character recognition (MICR) line on the original check.

A substitute check is similar to the electronic image that may be sent in lieu of the original paper check. Check 21 defines a *substitute check* as a paper reproduction of the original check that conforms to the following requirements:

1. Contains a clear replication of the front and back of the original paper check.
2. Bears an MICR line with all the information on the original check's MICR line.
3. Conforms with generally applicable industry standards for paper stock, dimensions, and other general qualities.
4. Is suitable for automated processing in the same manner as the original paper check.

Check 21 provides the guidelines for the issuance and use of substitute checks. The act also allows the use of digital or paper substitutions for original paper checks.

Availability Schedule for Deposited Checks

Once a check is considered deposited in a customer's account, when are the funds available to the customer? In the past, banks placed extended holds (e.g., 10 days) on deposited funds to ensure that the payor bank would not dishonor the check. If the check was from out of state, the bank might place a two-week hold on the check.

Because customers were frustrated with such extended holds on checks, Congress enacted the Expedited Funds Availability Act of 1987. This act created explicit timelines that mandate when banks must make deposited funds available to customers. If the risk of dishonor is low, the funds must be made available very quickly. For example, the first $100 of any amount deposited must be available to the depositor on the business day following the day of deposit [12 C.F.R. Section 229.10(a)-(c)]. The availability of the rest of the funds depends on whether the check is local, what the amount of the check is, and how the depositor wants to withdraw the funds.

1. If the deposited check is drawn on a bank within the same Federal Reserve Bank area, the depositary bank must make the deposited funds available to the customer on the second business day following the deposit day [12 C.F.R. 229.10(b)].
2. If the deposited check is outside the same Federal Reserve Bank area, the depositary bank must make the funds available on the fifth business day following the day of deposit [12 C.F.R. 229.12(c)].

The Truth-in-Savings Act

In an effort to allow consumers to be better informed, Congress passed the Truth-in-Savings Act (TISA) in 1991. TISA requires depository institutions to disclose, in great detail, the terms and conditions of their accounts. Depository institutions include commercial banks, savings banks, credit unions, and savings and loan associations.

TISA requires every bank to inform its customers of a number of specific pieces of information relevant to the bank's accounts. This information is to be handed to the customer as a pamphlet, brochure, or handout before the customer opens a deposit account. The information that must be given to the customer includes:

1. The minimum balance required to open an account and to be paid interest.
2. The manner in which the balance of the account will be calculated.
3. The annual percentage yield of interest for the account.
4. The way interest on the account will be calculated.
5. Notification of any fees, charges, and penalties the account may be assessed and how they are calculated.
6. Notification of any limitations on withdrawals or deposits.

Furthermore, TISA and Regulation DD, which serves as the act's implementing regulation, require banks to pay interest based on the full balance of interest-bearing accounts daily. The requirement prevents banks from calculating interest on an account when the account is at its lowest balance.

When a Bank May Charge a Customer's Account

LO 17-8 Describe instances that allow a bank to charge a customer's account.

The following sections consider certain problems related to a bank accepting and paying a customer's check.

Wrongful Dishonor

When a customer opens a checking account, both the customer and the bank accept certain duties and rights. Generally, the customer assumes a duty to keep sufficient funds in her account to cover the checks written on her account. If the customer does not have enough funds to cover a check, the bank will dishonor the check and the customer becomes liable for the amount of the check.

Similarly, under the *properly payable rule,* the bank has a duty to pay checks from the customer's account as long as the checks are "properly payable" [UCC Section 4-401(a)]. In other words, a check must be authorized by the drawer and must not violate agreement between the bank and the customer. Generally, for a check to be considered properly payable, it must:

1. Have the drawer's authorized signature on the check.
2. Be paid to a person entitled to enforce the check.
3. Not have been altered.
4. Not have been completed by addition of unauthorized terms if the check was incomplete.
5. Be paid on or after the date of the check.
6. Not be subject to a stop payment from the drawer.

If the bank wrongfully fails to pay a check—wrongfully dishonoring the check—the bank may be liable to the customer for damages. The UCC clearly states that banks can be held liable, but

it does not cite a specific theory for recovery. Therefore, someone whose check was dishonored need prove only that the dishonoring was wrongful, and he or she will be entitled to recovery.

Stop-Payment Order

A customer can issue a **stop-payment order,** an order by a drawer that instructs the drawee bank not to pay a check that has been drawn on the customer's account (UCC 4-403). A customer issues a stop-payment order when he has issued a check that has not yet been accepted and he wants the check not to be accepted. If the bank pays a check in violation of a stop-payment order, the bank will incur liability for the damages suffered by the customer due to its violation of the stop payment [UCC 4-403(b)].

> **stop-payment order**
> An order by a drawer that instructs the drawee bank not to pay an issued check.

To be effective, the stop order must meet two requirements: (1) The customer must give the order in a reasonable time, and (2) the customer must describe the item with "reasonable certainty" [UCC 4-403(a)]. For example, the stop-payment order must be given promptly so that the bank has enough time to inform its tellers and other employees that they should not pay the check. Moreover, the UCC states that the stop-payment order must reach the bank by a certain cutoff time. Generally, the stop-payment order should list (1) the date of the check, (2) the name of the payee, (3) the amount of the check, (4) the number of the check, and (5) the checking account number.

Postdated Checks

Under previous versions of Articles 3 and 4 of the UCC, a check could not be charged to a customer's account until the date of the check. Thus, some customers attempted to hold off payment on a check by postdating the check and giving it to the payee. However, because banks generally use an automated system to process checks, checks are now frequently paid without regard to the date. If a bank pays a check before its date and thus depletes a customer's account, the bank could be liable to the customer for damages. Thus, some banks include certain clauses in their customer agreements that state they may pay checks regardless of the date.

The UCC presents a middle ground that protects a bank from liability while permitting customers to postdate checks. Section 4-401(c) states that customers can postdate checks but they must give the bank notice of the postdated check. Therefore, the bank can assume that it can pay all checks on presentment unless the bank has received notice. Most banks charge a processing fee for notice of a postdated check.

Stale Checks

If a check is not presented to a bank within six months of its date, the check is considered a **stale check.** Banks are not required to honor stale checks.

> **stale check**
> A check that is not presented to a bank within six months of its date.

Forgeries and Alterations

In 1999, attempted check fraud in the United States rose to approximately $2.2 billion; merchants and banks suffered actual losses approximating $679 million through check fraud. Thus, banks are clearly concerned about the acceptance of altered or forged checks.

Checks Bearing Forged Signatures Under the properly payable rule, the bank may pay a check only if it is authorized by the customer. Who is liable if a bank cashes a check signed by an unauthorized person? In other words, what happens when someone forges a drawer's signature on a check?

The UCC has established that a forged signature has no legal effect as a signature of the drawer [3-403(a)]. Consequently, in most cases, if a bank pays a check when the drawer's signature has been forged, the bank will be liable for the amount of the check.

However, there are some exceptions to this rule. First, if the customer's negligence substantially contributed to a forged signature and the bank pays the check in good faith, the bank is not required to repay the customer [UCC 3-406(a)]. For example, if the customer is an employer who keeps a rubber stamp of his signature in an unlocked drawer, and an employee uses the signature

E-COMMERCE and the Law

New Technology a Win-Win for ATM Users and Financial Institutions

VSoft Corporation provides technological solutions to financial institutions and has released software that captures images at ATM locations. These images are used for processing and image exchange. The significance of this software is that it makes ATM envelopes unnecessary. Financial institutions with the VSoft technology capture a check image and related data at the ATM and then integrate that information with validation from back-office operations. Banks can now engage in virtual sorting and deposit review. The software provides customers with a receipt that includes an image of the deposited check. This software promises service and convenience to customers. It also reduces the financial institution's operating costs and reduces opportunities for fraud.

Source: "VSoft Announces the New 5.20 Release of Its Centrum Gateway TM-ATM Software," *Business Wire*, September 17, 2008.

stamp to create a check payable to herself, the customer has substantially contributed to the forged signature and will likely not be able to recover the amount of the check.

Another exception to the forgery liability rule is related to the customer's duty to examine his or her bank statement. Banks make a customer's statement available to the customer approximately once a month. This statement lists or includes all the checks that have been charged against the customer's account over the past month. A customer must examine her bank statement reasonably promptly for any forgeries or unauthorized payments. If the customer discovers a forgery or unauthorized payment, she must notify the bank promptly [UCC 4-406(c)]. Under the UCC, if the customer does not notify the bank of an unauthorized signature within 30 days after the statement was made available, she cannot hold the bank liable for the payment [4-406(d)].

The duty to examine the bank statement is particularly important when there have been multiple forgeries by the same forger, or same wrongdoer. If a customer examines a statement and does not notify the bank of the first forgery within 30 days, the customer will be liable for future forgeries on the customer's account by the same wrongdoer.

Again, if the bank was negligent in paying the forged checks, the customer's liability for the checks may be diminished through comparative negligence [UCC 4-406(e)]. Regardless of the 30-day requirement for multiple forgeries or the care used in cashing a check, a customer must report a forgery within one year from the date the statement is available to the customer or he will lose the right to recover this money [UCC 4-406(f)].

Checks Bearing Forged Endorsements If a bank pays a check that has been fraudulently endorsed, who is responsible? In the same way that a drawer is not responsible for a check that bears the drawer's forged signature, the drawer is similarly not responsible for a check that has been fraudulently endorsed. Generally, when an endorsement has been forged, the first party to accept the forged instrument is ultimately liable for the loss because a forged endorsement does not legally transfer title [UCC 4-207(a)(2)]. However, the drawer again has a duty to examine her statement for fraudulent endorsements and then notify the bank. The drawer must report all forged endorsements within three years of receiving the returned forged items or her statement containing the forged items. If the drawer does not report the forgery within three years, the bank is no longer liable for that customer's loss (UCC 4-111).

Altered Checks Remember, under the properly payable rule, a bank is to pay only those checks that are authorized. When a party makes an unauthorized change or alteration to a check, the check becomes unauthorized. The UCC defines alteration as a change (without consent) that modifies the obligation of a party to the instrument. Generally, if a bank pays a check that has been altered, the bank will be liable for the alteration.

A customer's substantial contribution to the alteration will limit the customer's ability to require the bank to credit his or her account. In other words, if the customer leaves large blank spaces open on the check and thus another party may easily alter the check, the customer will likely be liable for the altered amount of the check. Similarly, under the UCC, any drawer who leaves the dollar amount blank may not later protest paying whatever amount has been written on the check [4-401(d)(2)].

alteration
An unauthorized change to an instrument that modifies the obligation of a party to the instrument.

As in the case of forged signatures, the customer's duty to examine his bank statement also applies to looking for altered checks. Thus, if the customer does not discover the altered check or does not report the altered check within a reasonable time, the bank's liability for the altered check is reduced. Similarly, if both the customer and the bank are negligent in contributing to and paying an altered check, both parties will be responsible for a portion of the check.

Electronic Fund Transfers

Consumer Fund Transfers

Consumer fund transactions are governed by the *Electronic Fund Transfer Act (EFTA) of 1978*. This act sets out the rights and liabilities of the parties involved in **electronic fund transfers (EFTs)**. Regulation E of the act allows the Federal Reserve Board to issue rules and regulations to enforce EFTA. The following transactions are considered consumer fund transactions: transactions in which a retail customer pays for an item with a debit card that allows the customer's bank account to be instantly charged, ATM transactions, and direct deposits of paychecks.

electronic fund transfer (EFT)
The transfer of funds by an electronic terminal, telephone, or computer.

Customer and Bank Rights and Responsibilities EFTA requires merchants to inform customers of their rights regarding EFTs. First, if a customer's ATM card is lost or stolen, the customer must notify the bank within two days. The customer is then liable for only the first $50 stolen. If the customer does not notify the bank, the customer will be held liable for up to $500 that is stolen. Second, the bank has a duty to provide a monthly statement that includes electronic fund transfers, and the customer has a duty to examine this bank statement for any unauthorized electronic fund transfers or errors. Third, the customer has a duty to notify the bank of any errors in the electronic transactions within 60 days of receiving the statement. Fourth, the bank is required to provide customers with receipts for electronic transactions. Fifth, the bank must notify the customer that preauthorized payments may be stopped; however, the customer must stop the payment by notifying the bank at any time up to three days before the preauthorized payment is scheduled. Although a customer may stop a preauthorized payment, a customer cannot order a stop payment on an EFT because such transfers occur instantaneously.

Unauthorized Transfers Under EFTA, an unauthorized electronic transfer is a federal felony punishable through criminal sanctions, such as a $10,000 fine or 10-year prison sentence. An electronic transfer is unauthorized if (1) it is initiated by a person who has no authority to transfer, (2) the customer receives no benefit from the transfer, and (3) the customer did not give his personal identification number to the unauthorized party.

When banks violate EFTA, consumers may recover actual and punitive damages, with the punitive damages totaling between $100 and $1,000.

Commercial Fund Transfers

Because EFTA did not cover all situations in which funds may be electronically transferred, Article 4(A) of the UCC was issued to address commercial fund transfers. An important type of commercial fund transfer is a wire transfer. Funds are wired between two commercial parties. Two major payment systems coordinate wire payments: the Federal Reserve wire transfer network (Fedwire) and the New York Clearing House Interbank Payments Systems (CHIPS). These two systems account for the transfer of more than $1 trillion daily. This sum is substantially more than is transferred by any other means.

E-Money and Online Banking

With the rapid advancements in technology, the banking system is finding itself changing rapidly. Electronic payments, or e-payments, are becoming increasingly prevalent in daily life. Increasingly, bank transactions are being conducted electronically, marking a shift away from physical currency. In fact, it is possible for electronic forms of money to replace physical currency, such as paper

Digital cash
Money stored electronically in place of physical currency.

e-money
Electronic money.

stored-value card
A plastic card that contains data regarding the value of the card, thereby allowing EFTs to be made.

and coin money, completely. **Digital cash,** money stored electronically on microchips, magnetic strips, or other computer media, would allow for the elimination of physical currency.

Helping to lead the digital banking revolution are the various forms of **e-money** (electronic money). The most common example of e-money is **stored-value cards.** Stored-value cards are typically plastic cards that contain magnetic strips. The magnetic strips, similar to those on credit cards or ATM cards, contain data regarding the value of the card.

Related to the expansion of technology in banking is the increase in use of online banking. With online banking, banks allow customers to have electronic access to their accounts so that they can check their accounts, transfer money, order payments, pay bills, check their investments, and, through some banks, even trade stock. Although services vary from bank to bank, many banks are offering at least some form of online banking to their customers.

Online Banking Services

Although online banking services vary among banks, there are three services most banks with online banking offer: (1) bill consolidation and payment, (2) transfer of funds from one account to another, and (3) loan applications (although an appearance at the bank to sign the loan is typically required to finalize the loan process). By offering these services, banks can cut down on their own costs as well as give customers greater control over their funds.

Despite the number of banking services offered online, not all banking services can be conducted over the Internet. For example, depositing and withdrawing funds are two services that cannot be conducted online.

Regulatory Compliance

In the main, banks are in favor of the increased use of online banking. Part of the reason banks like online banking is that it helps reduce a bank's operating cost and thus increases its profits. One way online banking reduces cost is through paperless billing. By not having to send paper statements to their customers, banks save on paper, ink, envelopes, and postage. The bank posts all the information to the user's account online, and this does not cost the bank much at all.

Another reason banks are in favor of online banking is that it decreases what is known as float time. *Float time* is the period between the time a check is written and the time it is presented for final payment, during which a customer can still use his or her funds. Because an electronic check does not have to transfer physically between banks, accounts can be credited or debited more quickly.

Privacy Protection

Are e-money institutions the same as traditional financial institutions? Do the same laws apply? These two questions hold the key to the question: How secure are e-payments and e-money? The answer is that we do not know.

E-Money Payment Information Which laws apply to e-money is still mostly untested legal ground. That is, there is little clarity regarding which laws do apply to e-money. The Federal Reserve has explicitly stated that Regulation E, which regulates traditional electronic fund transfers, does not apply to e-money transactions. Nonetheless, laws regarding computer files, not directly related to banking, might apply to e-money, such as laws against unauthorized access of electronic files or communication. There are several such laws, but it is not clear whether they apply to e-money.

E-Money Issuer's Financial Records In 1978, Congress passed the Right to Financial Privacy Act [12 U.S.C. Section 3401 et seq.]. Under this act, financial institutions, such as banks, may not give a federal agency information regarding a person's finances without either that person's explicit consent or a warrant. The Right to Financial Privacy Act may apply to digital cash providers if the provider is considered a legal bank or credit provider that supplies customers with a card considered similar to a credit or debit card. However, given the lack of a physical location

for digital cash providers, it is also possible that the Right to Financial Privacy Act does not apply, in which case digital cash providers may release a person's financial information freely to any federal agency.

Consumer Financial Data In an effort to protect people's financial privacy further, Congress passed the Financial Services Modernization Act, also known as the Gramm-Leach-Bliley Act [12 U.S.C. Sections 24a, 248b, 1820a, 1828b]. The act's purpose is to control how financial institutions handle customer information, thereby ultimately providing greater privacy protections to financial institution customers. Financial institutions are prohibited from disclosing personal information about their clients to third parties unless certain requirements set forth in the act are met. In addition, each financial institution is legally required to present customers with its privacy policies and practices.

CASE OPENER WRAP-UP

Oral Agreements and Negotiable Instruments

Referring to the dispute between MGM Desert Inn, Inc., and Shack, the district court held that the potential oral agreement was irrelevant to the negotiability of the checks. Instead, it focused on the criteria established within the UCC for negotiability, and it concluded that the checks were negotiable instruments. The checks were written documents, were signed by the maker, contained an unconditional promise to pay, specified the sum of money to be paid, were payable on demand, and contained words of negotiability, so they were negotiable instruments. Hence, MGM was considered the holder of the instruments.

PRACTICAL TIPS for Business Managers

1. Remember that all negotiable instruments are essentially contracts and that all the rules of contracts apply to negotiable instruments.
2. As has been stressed in this book several times before, it is always important to have written documentation for your business transactions. Negotiable instruments are always in writing per their nature, but, as the Case Opener shows, even the agreements surrounding a negotiable instrument should be in writing.
3. Whenever dealing with negotiable instruments, be sure to keep in mind the seven requirements for negotiability. A negotiable instrument can still be an enforceable contract without all seven factors, but it is not negotiable.

SUMMARY

The Need for Negotiable Instruments

Contracts as commercial paper: Any contractual obligation, except personal ones and nonassignable ones, may be transferred and thus is a form of commercial paper.
Problems with commercial paper: A breach of the contract invalidates the commercial paper.

Types of Negotiable Instruments

Note: A promise by the maker to pay a payee (e.g., a certificate of deposit).
Draft: An order by the drawer to a drawee to pay a payee (e.g., a check).
A note or draft can be either of the following:

- *Demand instrument:* Payee can demand actual payment at any time.
- *Time instrument:* Payment will be made only at a specifically designated time.

An Overview of the Law of Negotiable Instruments	*Requirements for negotiability:* 1. Written document: • Relative permanence • Movability 2. Signature of the maker or drawer: • Affirmative mark • Duly authorized agent • Handwritten, even without signature • Automated signature 3. Unconditional order or promise to pay: • Must be specific, not implied 4. Sum certain in money: • Currency only, any currency acceptable 5. Payable at a time certain or on demand: • Acceleration of payment • Extension of payment 6. Words of negotiability: • To the order of • Order instrument *The significance of negotiability:* More than a two-party transaction.
Negotiation	*Delivery:* The physical handing over of a negotiable instrument.
Banking	Article 3 of the UCC outlines the requirements that negotiable instruments, including checks, must meet, and it establishes the rights and responsibilities pertaining to parties to negotiable instruments. Article 4 creates a framework controlling deposit and checking agreements between banks and customers, and it directs the relationships between banks as checks are processed among different banks.
Checks	*Draft:* An instrument that is an order. *Drawer:* The party that gives the order. *Drawee:* The party that must obey the order. *Payee:* The party that receives the benefit of the order. *Check:* A special draft that orders the drawee, a bank, to pay a fixed amount of money on demand. *Cashier's check:* A check in which both the drawer and the drawee are the same bank. *Traveler's check:* An instrument that is payable on demand, is drawn on or through a bank, is designated by the phrase "traveler's check," and requires a countersignature by the person whose signature appears on the instrument. *Money order:* An instrument stating that a certain amount of money is to be paid to a particular person. *Certified check:* A check that is accepted at the bank at which it is drawn. *Accepting deposits:* The banking institution must take deposits for an existing account. Consequently, it must collect deposits and make the funds available for use.
When a Bank May Charge a Customer's Account	*Properly payable rule:* A bank may pay an instrument only when it is authorized by the drawer and does not violate an agreement between the bank and the customer. *Wrongful dishonor:* If a bank refuses to pay a properly payable check, the bank incurs liability. *Stop-payment order:* A drawer orders the drawee bank not to pay a check that has been drawn on the drawer's account. *Postdated check:* A customer can postdate a check but must give the bank notice of the postdated check. *Stale check:* A check is not presented to a bank within six months of its date. *Forgeries and alterations:* 1. *Check bearing a forged signature:* Generally, the drawer is not liable for a forged check unless the drawer substantially contributed to the forgery. 2. *Check bearing a forged endorsement:* Neither the drawer nor the drawer's bank is liable for a forged endorsement. 3. *Altered check:* If an unauthorized change modifies the obligation of a party to the instrument, the drawer is generally not liable for the altered amount unless he or she negligently contributed to the alteration.

	Electronic fund transfers: These transfers are transactions in which a retail customer pays for an item with a debit card that allows the customer's bank account to be instantly charged, ATM transactions, and direct deposits of paychecks.
E-Money and Online Banking	*Digital cash:* Money is stored electronically on microchips, magnetic strips, or other computer media.

Point/Counterpoint

<table>
<tr><th colspan="2">Should a Company Be Allowed to Require Employees to Receive Payment through Direct Deposit?</th></tr>
<tr><th>YES</th><th>NO</th></tr>
<tr>
<td>A company should be allowed to require employees to receive payment through direct deposit because direct-deposit payment is the most efficient form of payment for both employers and employees.

Direct deposit allows employers and employees security they would not have with mailed paychecks. Companies can keep track of employee payments more accurately when employees are paid through one payment method.

Through direct deposit, employers are assured that *all employees* are paid at the same time, on the same day. Neither employees nor employers need to be concerned with lost, stolen, delayed, or damaged paychecks.

Many people have already chosen the convenience of online banking for their other banking needs; adding direct deposit only simplifies their lives. In fact, studies show that the average worker spends between 8 1/2 and 24 hours each year cashing and depositing his or her payroll checks.

Direct deposit allows for a separate payroll account and more exact bookkeeping. The losses normally associated with stolen and doubled paychecks can be reinvested in the company, eventually allowing for potential salary/wage increases for all. Furthermore, a company saves a lot of money on paper costs alone by switching to direct deposit.

Many people willingly choose to deposit their paychecks directly. A company establishing a requirement of direct-deposit payment would only streamline and simplify the process for everyone involved.</td>
<td>A company should not be allowed to require its employees to receive payment through direct deposit. Although direct deposit is a good option for employers to provide to employees, the choice of payment form should be left to the employees.

First, very few guidelines have been established regarding direct-deposit procedures. Employees would not necessarily be provided an in-depth statement discussing the details of each pay period. With direct deposit, employees have more difficulty ensuring that pay statements are accurate.

Wage-based employees, for example, need to know exactly what they are paid per hour, for how many hours, so that the employees know whether they need to be paid overtime wages. Some direct-deposit statements list only the amount of money transferred to the employee's account.

Some employees simply prefer literally to hold and personally deposit a physical check. They also have a physical copy of their pay stub for paper records. The absence of a physical receipt creates gaps in an individual's paper records. These holes can create problems when an individual gathers documents in preparation for tax season.

Although companies can and should present to employees a list of the advantages of direct deposit, the ultimate decision should be left to the employees because the employees are most heavily affected if their paychecks are not deposited properly.</td>
</tr>
</table>

Questions & Problems

1. Explain the reason behind the need for negotiable instruments.
2. Are negotiable instruments more similar to money or contracts? Explain.
3. Identify and define each of the elements of negotiability.
4. Evaluate the following statement: "Order paper and bearer paper must be delivered to be negotiated."
5. Who are the three parties involved in the transfer of money through a check?
6. Explain the reason for the following policy: A customer has a duty to examine her or his bank statement.

7. Evaluate the following statement: "If a signature on a check is forged, the customer will never be responsible for the amount on the check."

8. Matthew Wiggins, Jr., filed a suit on January 14, 2015, seeking payment from Christopher Janousek and Madeleine Griffin for a promissory note that matured in April 2010. The defendants in this case argued that the statute of limitations barred Wiggins from collecting the payment. The crux of the matter is if the note is a negotiable instrument, which has a six-year statute of limitations, or if it is non-negotiable in which case it has a four-year statute of limitations.

 The relevant provisions of the note are as follow:

 "FOR VALUE RECEIVED, the undersigned hereby jointly and severally promise to pay to the order of Matthew D. Wiggins the sum of Fifty-Five Thousand Dollars ($55,000) together with interest thereon at the rate of ten percent (10%) per annum on the unpaid balance."

 This provision is followed by handwritten terms that state "Interest accrues only after cash advance; date; Interest only payable on the 1st day of month; No prepayment penalty; Total amount due on 4/1/10."

 Finally, the note states "All payments shall be first applied to interest and the balance to principal. All prepayments shall be applied in reverse order of maturity. This note may be prepaid, at any time, in whole or part, without penalty." Does this agreement fulfill the Sum Certain in Money element of negotiable instruments? How did the court rule and what was its reasoning?

 Source: *Wiggins v. Janousek,* Tex: Court of Appeals, 14th Dist. (2017).

9. On September 2006, Janice Logan executed a note for $304,000 in favor of IndyMac Bank, F.S.B. The note was secured by a mortgage on Logan's property. In March 2010, Deutsche Bank National Trust Co. began a foreclosure action against Logan for defaulting on the terms of payment. To win a summary judgement to foreclose a mortgage, the plaintiff must supply the note, the mortgage, and evidence of the default in payment.

 In her defense, Logan claimed because Deutsche Bank did not show how it came into possession of the note; it could not legally hold possession of it, thereby precluding it from foreclosing on the mortgage. Evaluate Logan's argument. How does the result change if the note has been endorsed in blank or if it were payable to the order to a named party?

 Source: *Deutsche Bank National Trust Company v. Logan,* 146 AD 3d 861 (2017).

10. T. Christian Cooper was a partner to Sanders and Richard Campbell d/b/a The Mullen Company. In 2001, Cooper helped bring about a management agreement between The Mullen Co. and Newnan Crossing Partnership. Newnan was a partnership of several families in Mexico and was represented by Roberto Segovia. Under the agreement, the Mullen Co. was responsible for developing and managing ten real estate locations.

 In 2006, Cooper learned that a third-party company was interested in making investments in the real estate properties related to the agreement between Newnan and the Mullen Co. Cooper took it upon himself in his individual capacity to negotiate a $600,000 loan from Segovia and become a partner in the third-party venture. Cooper agreed to apply all distributions from the partnership to the note. While he did receive distributions totaling over $1.4 million, he did not apply any of it to the note.

 In 2010, Segovia died and the new management at Newnan refused to pay the Mullen Co. $1.8 million in management fees that the Mullen Co. believed it was owed pursuant to the management agreement. The Mullen Co. sued Newnan and eventually reached a settlement in 2012. The settlement stipulated that the Mullen Co. would dismiss its lawsuit in exchange for $300,000 and an assignment of any causes of actions Newnan had against Cooper or the Mullen Co. As required in the agreement, Newnan assigned the $600,000 note to the Mullen Co.

 The Mullen Co. subsequently asserted a claim on the $600,000 promissory note against Cooper. Cooper responded by asserting the Mullen Co. had breached its fiduciary duty to Cooper as a partner and for breach of contract. The jury found in favor of both the Mullen Co. and Cooper on several counts and the final judgment awarded $1,431,000 in damages to the Mullen Co. and $519,300 to Cooper. Both parties appealed. In the appeal, Cooper argued, inter alia, that he could not be held personally liable according to the terms of the promissory note because the promissory note was also a non-recourse note.

 The nonrecourse provision is as follows: "Nothing in the Note or the Loan Documents to the contrary withstanding, Payee shall look solely to the Partnership Distributions (reduced by tax obligations of [Cooper], as provided above) and the Partnership Interest for payment on this Note. No deficiency judgment for amounts unsatisfied after application of such distributions and proceeds from sale of the Partnership Interest shall ever be instituted, sought, taken or obtained against any party for any amounts which become due and owing." Did the nonrecourse provision shield Cooper from having to pay beyond the agreed upon collateral? How do you think a

non-recourse provision coexists with the unconditional promise to pay required by negotiable instruments?

Source: *Cooper v. Campbell,* Tex: Court of Appeals, 5th Dist. (2016).

11. The plaintiff, U.S. Bank, filed a complaint to foreclose the defendant's mortgage because the defendant had failed to pay the monthly installments on the mortgage. However, the defendant purported to tender payment for the entire balance of the mortgage by an enclosed bonded bill of exchange. Attached to the bill were instructions on how to process it with the U.S. Treasury Department. The instructions stated that the defendant had established a personal treasury UCC contract trust account through the Treasury Department and that the bill was a negotiable instrument that should be mailed to the secretary of the treasury for redemption. However, the bill tendered by the defendant was not made "payable to bearer or to order." Does the bill discussed in this case meet the six requirements for negotiability? If not, which requirement is not met? How do you think the court ruled?

Source: *U.S. Bank N.A. v. Phillips,* 366 Ill. App. 3d 593 (2006).

12. On January 14, 2006, Sam and Odalis Groome entered into a sales contract with Alpacas of America LLC (AOA) to purchase an Alpaca named "Phasion Model" for $18,750. Almost exactly one year later, the Groomes purchased another Alpaca named "Black Thunder's Midnight" from AOA for $20,250. The Groomes had to finance the purchases and, together with the sales contract, the Groomes executed a promissory note which outlined a payment plan and contained a security agreement. In 2012, Alpacas of America LLC sued the Groomes for allegedly failing to make payments since October 2007. The Groomes filed a motion to dismiss AOA's claims, contending that the four-year statute of limitations for the sale of goods had expired. AOA fired back by arguing that they were basing their claim on promissory notes which were negotiable instruments; furthermore, the statute of limitations on negotiable instruments is six years long. The trial court granted the Groome's motion, reasoning that because the promissory notes enforced a sales contract, and because the promissory notes and the sales contract were part of the same transaction, the promissory notes were *not* negotiable instruments.

On February 4, 2014, the appeals court reversed the trial court decision. What did the results of the case hinge on? What was the appeals court's reasoning to reverse the decision in favor of Alpacas of America LLC?

Source: *Alpacas of America, LLC, v. Sam Groome et al.,* 179 Wn. App. 391 (2014).

13. Casey Anthony, a Florida resident, gained fame in the murder trial of her daughter, Kaylee. Among the charges brought against Anthony in court were check fraud and murder. The check fraud charge came about after an investigation revealed that Anthony had stolen her friend Amy Huizenga's checkbook and written checks amounting to $650. She wrote the checks to pay her bills and without Huizenga's knowledge. Anthony received no additional jail time and was ordered to serve one year probation. What area(s) of this chapter can we apply to this case?

Source: *Anthony v. State of Florida,* 11-2357, Florida District Court of Appeals, (2012).

14. In January 2008, $243,000 was transferred from the account of William Ryder's account at the Universal City Studios Credit Union to a bank in Hong Kong. The credit union said someone who identified himself as Ryder called to change a telephone number supposedly associated with the account at the credit union. The caller identified Ryder's date of birth, Social Security number, mother's maiden name, and other account activity. Five days later, a wire transfer of funds from Ryder's line of credit to the Hong Kong account was received. After verification of the signature on the fax, the credit union called the new telephone number to verify the request from a person who said he was Ryder. The transfer was made. The credit union submitted a claim to its insurance company for the amount of the transfer. The insurance company refused to pay because the security agreement with the credit union required wire transfers to be made through a secure telephone number that had been in place for at least 30 days prior to the transfer request. Would a court decision allowing the credit union to collect advance any values important to market exchange?

Source: *Universal City Studios Credit Union v. CUMIS Insurance Society, Inc.,* California Court of Appeal, Second Appellate District, B226868, (2012).

Part 4: Negotiable Instruments and Banking

CHAPTER 18

Holder in Due Course, Liability, and Defenses

CASE OPENER

Dishonored Check and Holder-in-Due-Course Status

In July 1993, Cigna Insurance Company issued James Mills a workers' compensation check for $484. Then Mills lied to Cigna and said he had not received the draft due to a change in his address. He requested that payment be stopped and a new draft issued. The insurer complied, stopped payment on the initial draft, and promptly issued a new check for Mills. However, Mills cashed the first check at Sun's Market before the stop-payment notation was placed on the draft. Sun's Market then presented the check for payment through its bank.

As a result of the stop payment on the initial draft, the bank dishonored the check, stamped it "stop payment," and returned the check to Sun's bank. After not receiving the cash for the bad check, Sun's Market tacked the check on its bulletin board. An individual, Robert Triffin, purchased the check from Sun's Market and obtained an assignment of Sun's interests in the instrument.

More than two years after the check was returned unpaid, Triffin filed a lawsuit against Cigna for payment on the check. Triffin argued that by purchasing the check from Sun's Market, he gained a special legal status, called *holder-in-due-course* status. This status entitled him to payment on the dishonored check. He argued that he received this special status under the *shelter principle* because the transfer by a holder in due course to a third party, even one with notice of the dishonor, transfers all rights of the holder in due course to the third party.

1. Who do you think should bear the dishonored check? In other words, should Cigna be required to pay Triffin for the dishonored workers' compensation check? Why or why not?
2. Suppose you are a business manager at Sun's Market. You learn that the court holds that Cigna Insurance Company does not have to pay for dishonored checks. Would you make any changes to your business policies? Would you be less likely to accept checks from insurance companies in the future?

CHAPTER 18 Holder in Due Course, Liability, and Defenses

> **LEARNING OBJECTIVES**
>
> **After reading this chapter, you will be able to answer the following questions:**
>
> LO 18-1 Define holder in due course.
> LO 18-2 Outline the requirements for holder-in-due-course status.
> LO 18-3 Define shelter principle.
> LO 18-4 Outline the necessary information to determine signature liability.
> LO 18-5 Define warranty liability.
> LO 18-6 Explain how to avoid liability for negotiable instruments.

Holder-in-Due-Course Doctrine

LO 18-1 Define holder in due course.

Suppose you want to purchase computers for your office building. You contract with a computer seller, Data Corp., to buy 50 computers. As partial payment, you give Data Corp. a note for $30,000. Data Corp. negotiates the note to Data Corp.'s landlord, Morgan, for payment of rent.

Suppose Morgan meets all the requirements (which you will soon learn) to be a holder in due course. You discover your computers are damaged, and you claim that Data Corp. breached its contract with you. If Data Corp. still held the $30,000 note, you could refuse to honor the note and claim as a defense that Data Corp. breached the contract. However, remember that Data Corp. negotiated the note to Morgan. Since Morgan is a holder in due course, you must pay Morgan because a holder in due course has higher rights to a negotiated instrument. If Morgan were simply a holder, you could use the defense of breach of contract against Morgan.

Reason for Holder-in-Due-Course Status

Do you think the elevated holder-in-due-course status is fair? The purpose of holder-in-due-course status is to protect a financial intermediary. In other words, someone who processes a payment should not be required to shoulder the transaction risks. Thus, the legal system offers greater legal protection to certain parties to encourage them to engage in financial transactions.

Requirements for Holder-in-Due-Course Status

LO 18-2 Outline the requirements for holder-in-due-course status.

To be considered a **holder in due course (HDC)**, a party must meet four requirements regarding the taking of the instrument. These requirements are established in UCC Section 3-302:

Meeting these requirements is very valuable, as the Case Nugget demonstrates.

The party must be a **holder** of a **complete** and **authentic** negotiable instrument, and the holder must have taken the instrument:

- **For value,**
- **In good faith,**
- **And without notice of any defects.**

holder in due course (HDC)
An individual who acquires a negotiable instrument in good faith.

Exhibit 18-1

Holder-in-Due-Course Requirements

CASE Nugget

HDC Status to the Rescue

Watson Coatings, Inc. v. American Express Travel Services, Inc. U.S. Court of Appeals for the Eighth Circuit

436 F.3d 1036 (2006)

Christine Mayfield used to work for Watson Coatings Inc., and part of her role there included that of company treasurer. During her time as treasurer, Mayfield wrote approximately 45-47 checks from Watson's account to American Express for personal debt totaling $745,969.39. With each check that was submitted, American Express credited Mayfield's personal account. After Mayfield was dismissed from Watson Coatings, the money she had taken was discovered and suit was filed against American Express for taking the money from Mayfield despite the checks having been clearly labeled as belonging to Watson Coatings Inc.

The district court granted American Express' motion for summary judgment, but Watson filed an appeal. In the opinion written by Judge Smith, the extent to which American Express can be considered a holder in due course of the checks written by Mayfield was considered. The judge explained that because a payee can be considered a holder in due course, American Express would qualify. However, the court also needed to decide whether holder-in-due-course status would offer any protection to American Express.

The appeals court found that even though Watson had brought forth several common law claims against American Express, the fact that American Express is a holder in due course can serve as a defense. The checks had been accepted by American Express in good faith, making it the holder in due course of the documents. Thus, the court affirmed the district court's grant of summary judgment to American Express.

BUT WHAT IF . . .

WHAT IF THE FACTS OF THE CASE OPENER WERE DIFFERENT?

Recall that, in the Case Opener, Mills was the original holder of the check from Cigna Insurance Company. Cigna sent Mills the check for $484, but Mills said that he never received the check, when in reality he cashed the check. So Cigna sent him a new check. Let's say, however, that Cigna realized prior to the cashing of the second check that the first check had been cashed, and due to breach of contract, Cigna now attempts to deny payment on the second check. Is Mills a holder of the check or a holder in due course? Depending on his status, will Cigna be able to refuse payment or not?

Be a Holder of a Complete and Authentic Negotiable Instrument

Party Must Be a Holder First, as we mentioned earlier in the chapter, to be a holder in due course, a party taking the instrument must be a **holder,** a party who is in possession of an instrument that is payable to the party or to the bearer of the instrument [UCC Section 1-201(20)]. For example, if Adam Brewer possesses a check that states, "Payable to Adam Brewer," Adam is a holder. However, suppose Adam asked his bank for a cashier's check to buy a boat from his friend, Corey Baum (see Exhibit 18-2). Even though Adam possesses the cashier's check and his name appears on the check, he is not a holder of the check because it is payable to his friend. When Adam gives the cashier's check to his friend, Corey, the friend becomes the first holder of the cashier's check.

holder
A party in possession of a negotiable instrument.

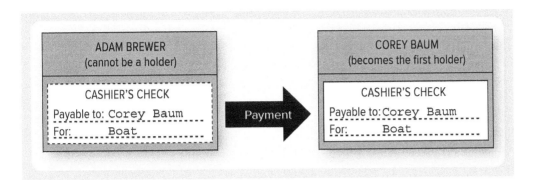

Exhibit 18-2

Illustration of the Relationship between Being a Payee and a Being a Holder

Similarly, if someone steals a check labeled "Payable to Adam Brewer" and forges Adam's signature on the back of the check, this thief is not a holder of the check (see Exhibit 18-3). Because the check is payable to Adam Brewer and not to the thief, the thief cannot be considered a holder, even if he forges Adam's name.

Instrument Must Be Negotiable Second, the instrument must be negotiable. A party cannot be a holder in due course of a non-negotiable instrument. If an instrument lacks any of the requirements for negotiability, the holder cannot be a holder in due course.

Instrument Must Be Complete and Authentic Third, the negotiable instrument must be complete and authentic [UCC 3-302(a)(1)]. When the instrument is issued or negotiated, the note must appear complete and authentic. What happens if the instrument is incomplete? For example, suppose that someone writing a check forgets to write the date on the check.

The UCC allows the holder to complete the check as long as the completion is consistent with the intent of the issuer (3-115). For example, suppose you write a check to a company to purchase some business supplies. Although you write the company's name, you forget to write the date. Under the UCC, the company may complete the instrument—in this instance, the check— by writing the date. However, if the person completes the instrument in a way that is inconsistent with the intent of the issuer, the instrument is considered materially altered.

If an instrument has clearly been materially altered or is so irregular or incomplete that its authenticity is called into question, the UCC bars a person taking this flawed instrument from becoming a holder in due course [3-302(a)(1)].

Take Instrument for Value

The holder in due course must take the negotiable instrument "for value." In other areas of the law, taking something for value usually means taking something with consideration, a bargained-for promise. However, the for-value requirement is more stringent than consideration: the party

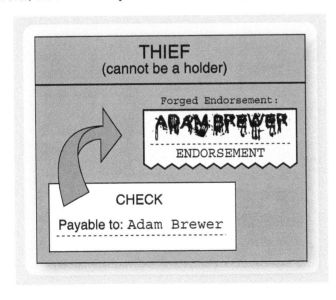

Exhibit 18-3

Illustration of a Failed Attempt to Become a Holder through Forgery

must take the instrument in exchange for a promise that has already been performed. The UCC explicitly excludes from "value" any promises that have not yet been performed. In other words, the party must suffer an out-of-pocket loss (3-303). Why? If a party has not yet performed the promise, he or she has not completely committed financially to the transaction and thus should not receive special legal protection. Therefore, when a party receives a negotiable instrument as a gift or acquires it through mistake, the party will be a holder instead of an HDC.

A holder can take an instrument for value if the holder:

1. Performs the promise for which the instrument was issued.
2. Acquires a security interest or some other lien in the instrument.
3. Takes the instrument for payment of a preceding claim.
4. Exchanges the instrument for another negotiable instrument.
5. Exchanges the instrument for an irrevocable obligation to a third party [UCC 3-303(a)].

BUT WHAT IF . . .

WHAT IF THE FACTS OF THE CASE OPENER WERE DIFFERENT?

Recall that in the Case Opener, Cigna sent Mills a workers' compensation check. What if the check was misprinted, and the monetary amount was never printed on the check? Although Mills knew the check was supposed to be for $484, he writes in the monetary amount of $1000. Would he be able to cash this check, considering Cigna made the mistake by not putting in a monetary amount?

Take Instrument in Good Faith

The HDC must take the negotiable instrument in good faith [UCC 3-302(a)(2)(ii)]. What exactly is good faith? Historically, there has been some debate about whether *good faith* had an objective or subjective definition. For example, when determining whether a holder took an instrument in good faith in the objective sense, some courts would decide whether the holder purchased the instrument with a proper degree of caution through a usual and ordinary manner of conducting business. Looking at good faith in an objective sense means considering what the reasonable holder would have done.

However, other courts looked at good faith in a subjective sense by asking whether the holder acted honestly when taking the instrument. Looking at good faith in a subjective sense means considering the holder's actual behavior.

The UCC's definition of good faith is somewhere in the middle of the subjective and objective standards. The UCC defines good faith as "honesty in fact and the observance of reasonable commercial standards of fair dealing" [3-103(a)(4)]. Therefore, to act in good faith, a holder must not deviate from the reasonable commercial standards of fair dealing. The UCC drafters were most concerned with the fairness of the dealing.

WAWEL SAVINGS BANK v. JERSEY TRACTOR TRAILER TRAINING, INC.

United States Court of Appeals for the Third District

500 F. 3D 147, U.S. APP. LEXIS 19597 (2009)

FACTS: *Wawel Savings Bank entered into a loan agreement with Jersey Tractor Trailer Training (JTTT) Inc. for the amount of $315,000. In the agreement, JTTT pledged all capital equipment and assets of the company as collateral, and Wawel filed Uniform Commercial Code financing statements with the New Jersey Department of the Treasury and the Bergen County Clerk's Office.*

(continued)

Sometime later, JTTT also entered into a factoring agreement with Yale Factors LLC, whereby JTTT agreed to sell the rights to its accounts receivable in return for a 61.5 percent up-front payment of the amount due on the particular accounts receivable.

On April 4, 2006, JTTT filed a voluntary petition for bankruptcy, and on June 29, 2006, Wawel brought action against JTTT, seeking declaratory relief that its lien on JTTT's accounts receivable had priority over that of Yale's. The bankruptcy court found in favor of Wawel, stating that Wawel did not authorize JTTT's factoring agreement with Yale and that Yale could not be considered a purchaser of instruments or a holder in due course because it did not establish that it acted in good faith by observing reasonable commercial standards of fair dealing. Yale appealed the decision.

REASONING: Judge Barry remanded the case based on the following reasoning:

> Yale asserts that it should be considered a holder in due course, or a purchaser of instruments, and therefore should have priority over Wawel's senior security interest. Yale argues, first, that because the security agreement accompanying Wawel's loan to JTTT did not expressly prohibit the sale of collateral, Wawel waived its security interest. That argument is without merit, especially given that in its agreement with Wawel, JTTT represented that it "w[ould] not settle any account for less than its full value without your written permission," and that it would "collect all accounts until [told] otherwise." JTTT's sale of its accounts receivable, therefore, ran afoul of the security agreement.

> Regardless of whether Wawel waived its security interest, Yale has priority over that interest if it is either a holder in due course or a purchaser of instruments. "A holder in due course is one who takes an instrument for value, in good faith, and without notice of dishonor or any defense against or claim to it on the part of any person." If those requirements are met, a holder in due course "take[s] priority over an earlier security interest, even if perfected. . . ." [UCC § 9-331(a)]. The same is true for a purchaser of instruments. To be considered a purchaser of instruments, Yale must have "give[n] value and take[n] possession of the instrument in good faith and without knowledge that the purchase violates the rights of the secured party." "Good faith" is defined in the UCC as "honesty in fact and the observance of reasonable commercial standards of fair dealing." The District Court is to remand this matter to the Bankruptcy Court to determine whether Yale qualifies as a holder in due course or a purchaser of instruments, and to resolve the good faith element of that analysis in accordance with this opinion.

DECISION AND REMEDY: On reviewing the appeal, Judge Barry remanded the case so that a determination could be made with regard to whether Yale qualifies as a holder in due course or a purchaser of instruments.

SIGNIFICANCE OF THE CASE: This case shows the way in which a court reasons with regard to the issue of determining who has priority over a security interest. Specifically, you should be able to see the privileges that a holder in due course is granted.

CRITICAL THINKING

The case entirely rested on the definition of the legal term, *good faith*. Suppose that one of your classmates believes the term *good faith* as outlined by the UCC is too ambiguous to make an informed ruling. Would you agree with your classmate about the ambiguity of the term? Why or why not?

ETHICAL DECISION MAKING

Some clues supporting Judge Barry's decision and the UCC's definition of good faith include honesty, transparency, and fairness. Can you think of any value preferences at the forefront of running a business that may conflict with the values outlined by the court and its definition of good faith?

Take Instrument without Notice

Finally, a holder must take an instrument without notice of various claims to or defects of the negotiable instrument. If the holder has notice or is aware of any of the following defects, the holder cannot be an HDC [UCC 3-302(a)]:

1. The instrument is overdue.
2. The instrument has been dishonored.
3. The instrument was issued as part of a series that is in default.

4. The instrument has been altered or contains an unauthorized signature.
5. There is a claim to the instrument. (Such claims are described in UCC Section 3-306.)
6. Another party has a defense or claim in recoupment to the instrument.

What Does It Mean to Have Notice?

BUT WHAT IF...

But suppose that Sun's Market, the check-cashing business from the Case Opener, received a letter from the insurance company that listed the numbers of checks that had stop-payment orders. As long as Sun's Market received this letter before it accidentally accepted the dishonored check, Sun's Market would have notice of a defect and thus could not be a holder in due course. The UCC states that "to be effective, notice must be received at a time and in a manner that gives a reasonable opportunity to act on it" [3-302(f)].

Suppose that a person receives an instrument that has clearly been altered; however, the person taking the instrument does not inspect the instrument and thus does not notice the obvious alteration. As stated earlier, one of the requirements of becoming a holder in due course is that the holder accepts a complete and authentic negotiable instrument [UCC 3-302(a)(1)]. Consequently, the question of a holder's awareness of such extreme irregularities does not matter; the very existence of such irregularities means that the holder cannot be a holder in due course.

If a person has notice of a defect before he or she gives value for the instrument but still gives value, the person will not be able to claim holder-in-due-course status.

Overdue Instruments Suppose you accept a check from a business associate. Unfortunately, the check falls behind your desk and is lost for the next four months. If you try to negotiate this instrument to another party, that party will not be permitted to claim holder-in-due-course status because the check is overdue. If a holder has notice that an instrument is overdue, the holder cannot claim holder-in-due-course status. How does a holder know whether an instrument is overdue? Overdue status depends on the type of instrument. The UCC explains how two types of instruments—demand and time instruments—may be overdue (3-304).

demand instrument
A type of draft that allows the payee to demand payment at any time from a holder.

Demand Instruments. A **demand instrument,** an instrument payable on demand, becomes overdue if it has been outstanding for an unreasonably long period of time after its date [UCC 3-304(a)(3)]. If the demand instrument is a check, the UCC states that the check is overdue 90 days after its date [3-304(a)(2)]. Thus, because your check has been outstanding for four months, it is overdue. The date on the check would give another party notice of its overdue status; thus, the party cannot claim HDC status.

time instrument
A type of draft that allows the payee to collect payment only at a specific time in the future.

Time Instruments. A **time instrument,** an instrument payable at a definite time, becomes overdue at any time after the expressed due date on the instrument. Most of the rules regarding whether a time instrument is overdue depend on the payment structure of the instrument. For example, a time instrument may or may not be payable in installments. If the instrument requires a party to make payment in a lump sum rather than in installments, the instrument is overdue if the party does not make the lump-sum payment by the due date [UCC 3-304(b)(2)].

Exhibit 18-4

According to the UCC, a person has notice of a fact if any of the following are true:

- The person has actual knowledge of the fact,
- The person receives notification of the fact,
- The facts and circumstances of the person in question indicate that the person should know the fact exists.

Essentially, to say that a negotiable instrument was accepted without notice of defects is to say that a party was not knowledgeable of any defects in the instrument at the time of acceptance.

However, how does nonpayment of installments affect whether the instrument is overdue? Overdue status depends on whether a late payment applies to the principal or the interest on the instrument. If a party misses payment of an installment on the principal of an instrument, the instrument is overdue until this installment is paid [UCC 3-304(b)(1)]. However, if a party misses a payment of interest on this instrument, the instrument is not overdue [UCC 3-304(c)].

Why might a party have reason to know that an installment payment of principal was not made, making the instrument overdue? This information is usually included on a credit report. For example, if a bank is considering purchasing a consumer note from a retailer, the bank can determine whether all payments have been made on the note by looking at the consumer's credit report. If the credit report indicates that an installment had not been paid, the bank would have notice that the instrument was overdue.

Dishonored Instruments If a holder is aware that an instrument he or she is taking has been dishonored, the holder cannot become an HDC. An instrument is **dishonored** when a party refuses to pay the instrument. For example, suppose you deposit a check from a customer into your company's account at a Wells Fargo bank. The customer has an account at Chase Manhattan. Wells Fargo credits your account and later presents the check for payment at Chase.

However, Chase refuses to pay because there are insufficient funds in your customer's account. Chase has dishonored the check and would likely stamp the phrase "insufficient funds" on your check. If you then tried to negotiate this check to another party, that party, seeing the words *insufficient funds*, would have notice that the check was dishonored.

However, if a party has no reason to know that a note has been dishonored, the party can become an HDC. Thus, if Chase did not stamp the phrase "insufficient funds" on the check, the new party would likely not have notice that the check was dishonored. Consequently, this new party could become an HDC.

Claims or Defenses If a party is aware of any claim or defense to an instrument, that party has notice and thus cannot become an HDC [UCC 3-302(a)(2)(v),(vi)]. However, if the party had no reason to know that various claims or defenses applied to an instrument, even if the claims and defenses existed, the party could become an HDC.

Exhibit 18-5 provides a summary of the defenses that do and do not apply to the holder in due course.

dishonored
Refused; specifically, a payment that has been refused despite a holder presenting an instrument in a timely and proper manner.

A holder in due course is generally free from personal defenses but is subject to real defenses. The holder in due course is *free* from the following personal defenses:

1. Lack or failure of consideration.
2. Breach of contract.
3. Fraud in the inducement in the underlying contract.
4. Incapacity.
5. Illegality.
6. Duress.
7. Unauthorized completion or material alteration of the instrument.
8. Unauthorized acquisition of the instrument.

The HDC is *subject to* the following real defenses:

1. Fraud in the essence.
2. Discharge of the party liable through bankruptcy.
3. Forgery.
4. Material alteration of a completed instrument.
5. Infancy—a party is below the legal age of consent.

Exhibit 18-5

Defenses and the Holder in Due Course

Case 18-2 demonstrates the importance of knowing whether you have met the standards for being a holder in due course.

CASE 18-2

CREATIVE VENTURES, LLC v. JIM WARD & ASSOCIATES
195 Cal. App. 4th 1430 Court of Appeals of California 2011

FACTS: *Creative Ventures (CV) borrowed roughly $3 million from Jim Ward & Associates (JW) to buy property in California. CV signed promissory notes, payable to JW. JW then sold the loan to 54 individual investors and kept possession of the promissory notes. CV made its payments to JW, and JW paid shares to the investors.*

California has a law making it illegal to charge an interest rate over 8 percent, except when the loan is made by a licensed real estate broker. JW claimed it was such a broker; it was not. CV learned of the lie and sued both JW and the investors for a refund of all interest it had already paid on the loan, as the California statute permitted. The investors claimed that, as holders in due course, they did not owe CV any interest because they had taken the note free of any claim against JW.

The trial court agreed with the investors; CV is appealing.

ISSUE: Does anyone claiming holder-in-due-course status need to physically possess the document for which the claim is being made?

REASONING: The Appeals Court pointed out that one of the criteria for being a holder in due course is possession. JW is the holder because it has the notes. The investors were benefitting from their relationship with JW, but they did not have the promissory notes. In short, the investors could not be a holder in due course because they are not holding the notes.

DECISION AND REMEDY: The only rights the investors have are those possessed by the person who offered them the original contract, JW. Therefore, CV is entitled to a full refund of the interest it has paid; the investors must return to CV all interest payments they have received.

SIGNIFICANCE OF THE CASE: Knowing all the elements necessary to be a holder in due course can have significant business implications.

CRITICAL THINKING

What missing information should the investors have asked JW at the point when they were purchasing the promissory notes?

ETHICAL DECISION MAKING

Make a list of the various stakeholders affected by this three-way transaction. In light of the Appeals Court decision, which stakeholders are they ranking as highly important and which are they ranking as not so important?

The Shelter Principle and HDC

LO 18-3 Define shelter principle.

shelter principle
The principle that when an item is transferred, the transferee acquires all the rights the transferor had to the item.

Generally, if an item is transferred from one person to another, the transferee acquires all the rights that the transferor had in the item. This idea is called the **shelter principle** [UCC Section 3-203(b)]. Therefore, following the shelter principle, even if a holder cannot attain holder-in-due-course status, the holder can acquire the rights and privileges of an HDC *if* the item is being transferred *from* an HDC. Note that the instrument does not need to be transferred directly from

GLOBAL Context

©Digital Archive Japan/Alamy

Negotiable Instruments in England and Similarities to the HDC Doctrine in the United States

In England, the doctrine of privity of contract declares that contracts are exclusive to those who created them. However, this doctrine often limits and complicates business transactions. The problems created by privity of contract in England can be countered by using a negotiable instrument. Unlike contracts, in England, bills of exchange, promissory notes, and checks can be negotiated between parties.

Negotiable instruments are used in business, in England as in the United States, to facilitate business transactions.

To ease the exchange and free flow of money in business transactions, businesses need to be able to transfer money or other forms of payment quickly. In the United States, companies can transfer their right to collect or their duty to perform under a contract. The English practice of privity prevents similar transfers. Accordingly, English businesses must be able to transfer money in a different way to conduct regular business. It is the ease of transferability of negotiable instruments and the rights that come with a transferred negotiable instrument, as outlined in this chapter, that make negotiable instruments an ideal way for English businesses to avoid the problems created by the privity requirement.

an HDC for the recipient to have the rights of an HDC. Under the shelter principle, as long as the holder of an instrument can demonstrate that someone through the line of transfers had obtained the rights of an HDC, then all subsequent holders have the rights of an HDC.

Therefore, in the Case Opener, plaintiff Triffin received the workers's compensation check from Sun's Market, a holder in due course. If Triffin did not qualify for HDC status on his own, he would have received the rights that Sun's Market, the transferor, had. In other words, Triffin is taking shelter in Sun's Market's status as an HDC.

One exception to the shelter principle involves cases of fraud or other illegal activities. If a person engages in fraud or other illegal interference with an instrument, he or she may not become an HDC, even if obtaining the instrument later from an HDC.

The shelter principle may at first seem contrary to the idea of the HDC principle; however, the purpose of the shelter principle is to encourage the marketability of instruments. The greater protection offered to an HDC is very appealing; thus, allowing parties to achieve the rights of an HDC through the shelter principle encourages financial interactions.

Liability

As you have seen, it is not always clear who should bear responsibility for the amount of a negotiable instrument. The following sections explain the various ways a party may be liable for a negotiable instrument. First, when a person signs a negotiable instrument, he or she is potentially liable for the instrument. This type of liability is called **signature liability**. Second, in contrast, a party may be liable if the transfer of the instrument breaches a warranty associated with the instrument. This type of liability is called **warranty liability**. After we consider both signature and warranty liability, we examine the defenses to these types of liability. Finally, we investigate how liability for a negotiable instrument may be discharged.

signature liability
Liability attributed because of a party's signature on an instrument.

warranty liability
Liability attributed when the transfer of an instrument breaches a warranty associated with an instrument.

Signature Liability

LO 18-4 Outline the necessary information to determine signature liability.

UCC Section 3-401(a) imposes liability if the party, or the party's agent, signs the instrument. If a party does not sign, the party cannot be held liable. Because the signature on an instrument leads to liability, it is important to know what counts as a signature. According to the UCC, a signature

411

E-COMMERCE and the Law

Electronic Signatures

Electronic signatures are now an accepted and rather standard method of authorizing payment. Any type of business with an online payment system puts its company at potential risk. Many banking companies have policies that deem the merchant rather than the buyer responsible for the charge if the payment made is fraudulent. For example, Visa tells those people with business accounts that make online transactions to know how to protect themselves because the business owners could be liable if payment taken from the customer is fraudulent. There are several options for businesses looking to ensure that their online transaction system can detect fraudulent payment. One of these options is to obtain chargeback insurance to prevent liability for fraud. The second option is to increase safety measures to prevent fraud, such as having online Completely Automated Public Turing Test to Tell Computers and Humans Apart (CAPTCHA) or requiring more extensive authentication procedures when a customer makes a payment. Because online business owners are at high risk for this type of fraud, owners should always properly authenticate the customer making a payment.

Source: Bahman Eslamboly, "Avoiding Problems When Using Electronic Signatures," LawGuru.com, January 25, 2013.

can be any name, word, mark, or symbol used by a party to authenticate a writing [3-401(b)]. Thus, if you wrote either your full name or an X with the intent to authenticate an instrument, either writing would constitute a signature.

Primary Liability of Makers and Acceptors

primarily liable
Liable for paying the amount designated on an instrument when it is presented for payment.

A party who is **primarily liable** for an instrument must pay the stated amount on the instrument when it is presented for payment. This liability for the stated amount begins as soon as the instrument is issued. Moreover, the primarily liable party must pay without resorting to any other party. For example, suppose you own a business and write a check drawing on your funds in your business account at First National Bank. First National has primary liability for the check; it must pay the stated amount when the check is presented for payment.

maker
A person who has promised to pay a set sum to the holder of a promissory note or certificate of deposit.

The UCC establishes that certain parties—makers and acceptors—are primarily liable. First, a **maker** is a party who has promised to pay. For example, the maker of a promissory note is primarily liable for the amount of the note because that party has promised to pay the amount of the instrument. Moreover, UCC Section 3-412 states that a party who signs as an issuer of an instrument is liable for the amount of the instrument as soon as it is issued. For example, a bank that issues a cashier's check is primarily liable for the amount of the check as soon as the cashier's check is created (UCC 4-412).

acceptor
A person (drawee) who accepts and signs a draft to agree to pay the draft when it is presented.

Second, an **acceptor**, a drawee of a draft who accepts and signs the draft to agree to pay the draft when it is presented, is primarily liable [UCC 3-413(a)]. A party who accepts a draft by signing on the face of the draft is primarily liable (UCC 3-413). For example, when a bank accepts a check, it is primarily liable for the amount of the check (UCC 3-409).

Secondary Liability of Drawers and Endorsers

secondarily liable
Liable for paying the amount designated on an instrument if the primarily liable party defaults.

A party who is **secondarily liable** for an instrument must pay the amount on the instrument if the primarily liable party defaults. Drawers and endorsers are secondarily liable parties. An **endorser** is a party who signs an instrument to restrict payment of it, negotiate it, or incur liability [UCC 3-204(b)]. A **drawer** is a person who signs as a party ordering payment [UCC 3-103(a)(5)].

endorser
One who issues an endorsement.

drawer
The party that writes an order or the person who writes a check.

For example, if you write a check from your bank account that is payable to the electric company, you are the drawer of the check, the bank is the drawee, and the electric company is the holder of the check. The holder of the check (the electric company) presents the check to the drawee (the bank) for payment. **Presentment** is defined in the UCC as making a demand for the drawee to pay [3-501(a)]. The UCC creates specific rules that govern the time and manner of presentment.

presentment
The act of making a demand for the drawee to pay.

Suppose the holder (the electric company) presented your check and the bank dishonored the check because of insufficient funds in your account. The UCC states that drawers of drafts are

liable for an instrument only after it has been dishonored (3-414, 3-415). Thus, you (the drawer) are now liable for the check.

Three conditions must be met for a drawer or endorser to become liable. First, the holder of the instrument must present the instrument in a proper and timely fashion. Second, the instrument must be dishonored. Third, notice of the dishonor must be given to the drawer.

Presentment An instrument must be presented in a proper and timely manner. First, the instrument must be presented to the proper party. If the instrument is a note, the holder must present the note to the maker of the note. In contrast, if the instrument is a draft, the holder must present the instrument to the drawee. Thus, continuing our electric company example, the electric company must present the check to the bank (the drawee).

Second, the instrument must be presented to the proper party in a proper way. UCC Section 3-501(b) states that an instrument can be presented (1) by any commercially reasonable means, (2) through a clearinghouse procedure, or (3) at the place designated in the instrument.

Third, the instrument must be presented to the proper party in a timely manner. Thus, if the instrument is a note, the holder must present the note to the maker on the note's due date. If the instrument is a draft, such as a check, the holder must present the instrument within a *reasonable* time. The failure to present an instrument on time is the most common reason that improper presentment occurs, ultimately discharging unqualified endorsers from secondary liability.

Dishonor When a holder presents an instrument within a timely and proper manner but acceptance or payment is refused, the instrument has been *dishonored*. The instrument must be explicitly dishonored; a refusal to pay does not necessarily mean that the instrument has been dishonored. For example, suppose you are a holder of a check that is payable to your business. You present this check for payment, but the bank refuses to pay the check because you cannot present identification [UCC 3-501(b)(2)]. Likewise, a bank may refuse to pay on an instrument because the endorsement of the instrument is not proper. Again, this refusal to pay does not dishonor the instrument. Remember, a secondarily liable party becomes liable *only* if a primarily liable party dishonors the instrument.

Notice of Dishonor The UCC provides a specific time line in which notice of dishonor of an instrument must be given to a secondarily liable party (3-503). (The process of determining secondary liability is summarized in Exhibit 18-6.) If the party that dishonors an instrument is a

	NOTE	DRAFT
To whom must the holder present the instrument?	Maker	Drawee
When should the instrument be presented?	On due date	Reasonable time; if a check, within 30 days of date of check or within 30 days of time of endorsement
If the instrument is presented and dishonored, who is usually now liable for the instrument?	Any endorser	Drawer or endorser
What are the requirements for an instrument to be officially dishonored so that a holder may then turn to secondarily liable parties?	1. Present to maker for payment 2. Maker dishonors 3. Holder gives notice of dishonor to secondarily liable parties (endorsers)	1. Present to drawee for payment 2. Drawee dishonors 3. Holder gives timely notice of dishonor to drawer or endorsers

Exhibit 18-6

Summary of Process of Determining Secondary Liability

collection bank, it must give notice before midnight of the next day [UCC 3-503(c)]. Other parties must give notice of the dishonor within 30 days of the day on which they receive notice of dishonor. This notice can be given in any commercially reasonable manner: oral, written, or electronic communication [UCC 3-503(b)]. The notice must identify the instrument in question and state that this instrument has been dishonored. If the word *dishonored* appears on the instrument, this writing is enough to constitute notice. As long as the holder gives notice to the secondarily liable parties about the dishonor of the instrument, the holder can sue the other parties.

If all three of these conditions are met, the holder can bring suit against a secondarily liable party. However, in most cases, although a secondarily liable party may have to pay a holder the amount of the instrument, the secondarily liable party can then seek recourse against the primarily liable party.

If an instrument contains more than one endorsement, each endorser is liable for the full amount to any subsequent endorser or to any holder. The secondary liability established through endorsement requires an endorser to pay anyone who endorses the instrument after him or her.

Accommodation Parties

Suppose that, after you graduate from college, you decide to start your own business. You need to borrow a significant amount of money from the bank, and you plan to create a promissory note. However, because you have never owned your own business and have little credit history, the bank is a little wary about whether you will be able to pay the note. Therefore, the bank decides to ask you to have a third party sign the note to ensure that the bank will be paid. Consequently, your business law professor cosigns your note. This third party is called an **accommodation party**, a party who signs an instrument to provide credit for another party who has also signed the instrument [UCC 3-419(a)].

accommodation party A party who signs an instrument to provide credit for another party who has also signed the instrument.

Accommodation parties may be primarily or secondarily liable for an instrument and can sign as makers, drawers, acceptors, or endorsers. However, accommodation parties more frequently sign as makers or endorsers. As a maker, an accommodation party has primary liability, but as an endorser, the party has secondary liability.

Suppose you, in the preceding example, cannot pay your note, and your business professor, as an accommodation party, pays the note instead. This professor has a right of reimbursement to recover the money from you, the accommodated party [UCC 3-419(e)]. If, however, you, the accommodated party, pay the note when it is due, you cannot force the professor to contribute to the amount due on the loan.

Unauthorized Signatures and Endorsements

As a general rule, if a signature to a negotiable instrument is unauthorized, the signature will not impose liability to the named party. This rule applies to two cases: forgery and unauthorized agents.

First, return to the preceding accommodation party example. Suppose you forged your business law professor's name to ensure that you would get your money to start your new business. If you could not pay on the note, your business law professor would not be forced to pay on the basis of the forged signature. Similarly, this rule applies to parties who forge the drawer's signature on a check. However, forgery in the employment context is treated differently, as we explain below in the section entitled "The Fictitious-Payee Rule."

Second, if an agent is not authorized to sign a negotiable instrument on behalf of a principal, the principal will generally not be liable for the instrument. Consequently, the agent would be personally liable for the instrument. However, if the principal decides to **ratify**, or approve of, the unauthorized agent's signature, the principal will then become liable for the instrument, and the agent will escape personal liability [UCC 3-403(a)].

ratify To approve an unauthorized agent's signature on an instrument.

How does a principal ratify an unauthorized signature? A principal could explicitly approve of the signature. For example, a Florida court held that a principal could not recover the amount of two checks he gave to his agent because the principal had ratified the signatures.[1] The agent, who

[1] *Fulka v. Florida Commercial Banks, Inc.*, 371 So. 2d 521 (Fla. Dist. Ct. App. 1979).

was supposed to deposit two checks into the principal's account, forged the principal's signature and deposited the checks into the agent's account. The agent ultimately told the principal about the forged checks and the location of the money. The principal did nothing until the agent later ran away with the money, and the court ruled that the principal's inaction was the same as his approval of his signature.

Alternatively, if the principal accepts the benefits associated with this unauthorized signature, the principal in effect ratifies the signature by his or her conduct. For example, in *Rakestraw v. Rodriguez,* a husband had forged his wife's signature to obtain a loan to start a grocery store.[2] A few days later, the wife discovered the forgery, but she did nothing to correct it and even participated in the business over the next few years. Part of running the business included sharing in the profits from the grocery store. When the business failed, the wife tried to avoid liability, claiming her husband forged her signature. The court ruled that her actions in sharing in profits and in helping run the business effectively ratified her signature.

This ratification of an unauthorized signature does not exclusively apply to the agent–principal relationship. Your business law professor could similarly choose to ratify your unauthorized signature of his name and thus would be liable for your promissory note.

However, these are some exceptions to the general rule that an unauthorized signature is not enforceable. Generally, the policy behind these exceptions is that courts want to place the burden on the parties who are in the best position to take a loss or take action to recover a loss. Moreover, particularly in regard to the last two rules that are discussed shortly (the imposter rule and the fictitious-payee rule), the court focuses on the intent of the party who is issuing the instrument.

Negligence In some cases, a party's negligence does not permit the party to escape liability for an unauthorized signature. If the party whose signature was forged behaved so negligently as to "substantially contribute to . . . the making of a forged signature," the party might not escape liability (UCC 3-406).

For example, in *Thompson Maple Products v. Citizens National Bank,* Thompson was a corporation that manufactured bowling pins from maple logs.[3] Thompson would accept loads of logs from timber owners. When a load arrived at the mill, a Thompson employee filled out a scaling slip that listed the name of the owner along with the quantity and grade of the logs. Thompson office employees then used the slips to prepare checks for the owners of the logs. A Thompson employee, Emery Albers, filled out blank scaling slips for fictitious loads of logs. The office employees, thinking the slips represented real loads of wood, then prepared checks for payment. Albers forged the name of the owner of the logs and cashed the checks or deposited them into his bank account at Citizens National Bank. The court ruled that Thompson "substantially contributed" to the forgeries because the blank scaling slips used to record loads of logs were easily accessible. In fact, office employees gave Albers two entire pads of these slips. Moreover, even though it was company policy for the slips to be initialed by authorized employees who were accepting the loads, Thompson office employees created checks even for slips that were not authorized. Thus, Thompson's negligence led to the conclusion that Thompson should be liable.

The Imposter Rule Suppose that Jamaar, a business manager, has been communicating through email with Carlie, a potential employee. Jamaar has scheduled a meeting with Carlie. However, Samantha, without Carlie's knowledge, decides to impersonate Carlie at the interview. Samantha (as Carlie) tells Jamaar she will strongly consider signing an employment agreement if Jamaar will issue her a $200 check as a presigning bonus. Jamaar agrees and issues the check to Carlie that day. Samantha forges Carlie's name and deposits the check into her own account. Will Jamaar be liable for the amount of the check?

Jamaar's signature has not been forged; he clearly signed the check with intent to transfer money to Carlie. But he did not know that Carlie was actually Samantha. Is Samantha's signature considered a forgery? No. Under the UCC's **imposter rule,** if a maker or drawer issues a negotiable instrument to an imposter, the imposter's endorsement will be effective [UCC 3-404(a)]. The court considers the intent of the drawer or maker when issuing the instrument. Because Jamaar intended for Samantha

imposter rule
A rule that holds that if one obtains a negotiable instrument by impersonating another and endorses it with the impersonated party's signature, the loss falls on the drawer of the instrument.

[2] 500 P.2d 1401 (Cal. 1972).
[3] 234 A.2d 32 (Pa. Super. Ct. 1967).

(as Carlie) to have the instrument, her endorsement of the instrument is considered valid. Moreover, it is easier for Jamaar as maker or drawer to identify the true identity of Carlie than it would be for a later holder of the check to do so. Perhaps you are surprised by the imposter rule. The UCC, as stated, places an immense responsibility on Jamaar to ensure that he is not being duped. What values are in conflict here? Should Jamaar be forced to shoulder this responsibility?

The Fictitious-Payee Rule Suppose now that Jamaar, who has been authorized to write checks from the company account, draws bonus checks from the company account for five more potential employees. Unfortunately, Jamaar never actually interviewed these employees; thus, these people are not entitled to the bonuses. Jamaar endorses these checks, which are made out to the fictitious potential employees, in their names and deposits the checks in his personal bank account. These potential employees have no interest (i.e., no right to payment) in the checks and are thus called **fictitious payees** (UCC 3-404, 3-405). As with the endorsement in the imposter case, Jamaar's endorsement of the fictitious payees is not considered forgery [UCC 3-404(b)(2)]. Jamaar's company will be liable for the checks.

Why is the company liable? Courts view the company as being in a better position to bear the loss of the checks. The loss has occurred because Jamaar, a company employee, has acted wrongly. Although the company is liable for the amounts of the checks, the company can recover the money from Jamaar.

> **fictitious payee**
> Someone having no right to payment. The UCC fictitious-payee rule governs the liability of companies and their endorsed checks: any check made out to a fictitious payee and endorsed must be honored and is not considered a forgery.

Warranty Liability

LO 18-5 Define warranty liability.

In the previous section, we explained how a party might be liable for an instrument on the basis of his or her signature on the instrument. In this section, we consider another type of liability: *warranty liability*. A party may be liable for an instrument because of a breach of warranty. There are two relevant types of warranties here: transfer warranties and presentment warranties.

Transfer Warranty

A negotiable instrument can be transferred from one party to another. A party who transfers a negotiable instrument to another party in good faith for consideration creates **transfer warranties** regarding the instrument and the transfer itself [UCC 3-416(a)]. Transfer warranties always apply to the party to whom the instrument is transferred (the transferee).

When a party transfers an instrument for consideration, he or she warrants that:

1. The transferor is entitled to enforce the negotiable instrument.
2. Signatures on the instrument are authentic and authorized.
3. The instrument has not been altered.
4. The instrument is not subject to a defense or claim in recoupment.
5. The transferor has no knowledge of insolvency proceedings against the maker, acceptor, or drawer of the instrument [UCC 3-416(a)].

> **transfer warranty**
> A warranty regarding a negotiable instrument and its transfer; created by a party who transfers the instrument.

If the transfer is through endorsement, these warranties apply to any future holders. However, if the transfer does not occur through endorsement, the warranties apply only to the transferee.

These rules on whether the warranties apply to future holders or only to the immediate transferee are important because liability can be imposed for breach of warranty. If the warranties apply and there is a breach of one of the warranties, the parties can bring suit against the transferor, the warrantor, for damages suffered as a result of the breach [UCC 3-416(b)].

As soon as a transferee discovers that a breach of warranty has occurred, he or she can bring suit against the transferor. However, the transferee must give notice of the breach-of-warranty claim to the transferor within 30 days of discovering the breach [UCC 3-416(c)]. If the transferee does not give notice within 30 days of discovering the breach, the warranty will be discharged to some extent. If the transferred instrument is a check, the warranties cannot be disclaimed [UCC 3-416(c)].

Although warranties on checks cannot be disclaimed, they can be disclaimed on other instruments. When parties agree to a disclaimer, an endorser can disclaim warranties by including in the endorsement the phrase *without warranties. Without warranties* disclaims warranty liability.

Presentment Warranty

In the "Signature Liability" section, we discussed the requirements for a negotiable instrument to be properly presented for payment. Certain warranties are associated with the presentment of an instrument. Remember, presentment occurs when a party properly presents an instrument for acceptance and the party to whom it was presented accepts the instrument or pays it in good faith.

Why are presentment warranties needed? Parties who accept or pay instruments may worry that they are not paying the proper party. Thus, whereas transfer warranties apply to the transferee, **presentment warranties** cover parties who accept instruments for payment. Both the party presenting the instrument and any previous transferor of the instrument make presentment warranties. Therefore, if there is a breach of presentment warranty, the acceptor can recover damages from the presenting party or previous transferors. As with the notice rule for transfer warranties, a party must give notice of a breach of presentment warranty within 30 days.

There are two types of presentment warranties. These types depend on what kind of instrument is being presented to a certain kind of party. When a party presents an unaccepted draft to a drawee, the holder guarantees:

1. The warrantor of the instrument is entitled to enforce the instrument.
2. The instrument has not been altered.
3. The warrantor has no knowledge that the drawer's signature or the draft is unauthorized [UCC 3-417(a)].

presentment warranty
A warranty covering the parties accepting an instrument for payment; created to ensure that the accepting or paying party is paying the proper party.

These warranties apply only to a drawee who pays or accepts a draft in good faith. If the instrument is not an unaccepted draft presented to a drawee, only one presentment warranty applies. The party presenting the instrument guarantees that the warrantor is or was entitled to payment or authorized to obtain payment [UCC 3-417(d)(1)]. In other words, only warranty 1 in the preceding list applies to presentments of instruments other than unaccepted drafts.

Avoiding Liability for Negotiable Instruments

LO 18-6 Explain how to avoid liability for negotiable instruments.

If a party tries to enforce a negotiable instrument, a defendant can try to avoid liability in two ways. First, the defendant can try to claim a defense to liability. Second, the defendant can try to claim that the liability has been discharged.

Defenses to Liability

Earlier in the chapter, we listed defenses to liability that do or do not apply to a holder in due course. Here, we return to these defenses to liability. There are two categories of defenses: real defenses and personal defenses. **Real defenses,** also called universal defenses, apply to all parties. **Personal defenses** do not apply to holders in due course.

real defense
A liability defense that applies universally to all parties.

personal defense
A liability defense that is not applicable to holders in due course.

Real Defenses A party's right to enforce a negotiable instrument is subject to the following real defenses:

1. Infancy (being below the legal age of consent), to the extent that it makes a contract void.
2. Duress, to the extent that it makes a contract void.
3. Lack of legal capacity, to the extent that it makes a contract void.
4. Illegality of the transaction, to the extent that it makes a contract void.
5. Fraud in the factum.

6. Discharge through insolvency proceedings (bankruptcy).
7. Forgery.
8. Material alteration.

The first six defenses are stated explicitly in UCC Section 3-305. As we discussed earlier in this chapter, the UCC establishes forgery as a defense to liability because a party must have signed the instrument to be held liable. Finally, a material alteration of an instrument discharges a party of a liability [UCC 3-407(a)].

Fraud in the Factum. When a party signs a negotiable instrument without knowing that it is, in fact, a negotiable instrument, the party can claim **fraud in the factum** (also called *fraud in the execution* and *fraud in the essence*) as a defense. For example, suppose Michael Jordan believes he is signing an autograph for a fan, but he is actually signing a promissory note. Because he did not intend to sign a negotiable instrument, he will not be held liable for the instrument.

Similarly, suppose you, a business manager, are negotiating with another company to purchase materials for your manufacturing business. After your negotiations, the company asks you to sign a document as a preorder for the materials. You hurriedly sign the document and leave. However, instead of signing a preorder, you have actually signed a note. Will you be held liable for this note? In this case, it depends on the extent to which you, as the signing party, had reason to know what you were signing.

For another illustration of these issues, see the Case Nugget.

fraud in the factum
A liability defense available to a party who signs a negotiable instrument without knowing that it is a negotiable instrument.

CASE Nugget

Defense of Ignorance?

Laborer's Pension Fund v. A & C Envtl., Inc. U.S. Court of Appeals for the Seventh Circuit

301 F.3d 762 (2002)

A & C Environmental Inc. is a corporation that transports and disposes of nonhazardous waste. In April 1999, A & C was asked to complete a job in Gary, Indiana, which prompted representatives from Laborer's Pension Fund to approach the company. Frattini of the fund asked Clark of A & C to sign a form that would guarantee the five individuals who would work in Gary, Indiana, the coverage of the local union. Clark was hesitant to sign the agreement because he feared that if someone within his company were covered under the union, the entire company would then be covered. It wasn't until after Frattini guaranteed Clark that the only employees of A & C who would be affected would be those working in Gary, Indiana, that the agreement was signed. When A & C did not pay dues for all of its employees, the fund brought suit for delinquent contributions.

The district court ruled against the fund as a result of the fraud-in-the-execution defense that was brought forth by A & C. The court decided that any reasonable juror would believe that Clark did not know that he was agreeing to pay the fund dues for each employee of A & C. The fund appealed to the U.S. Seventh Circuit Court of Appeals.

In the opinion written by Judge Ripple, the Seventh Circuit found that Clark may not have known what he was agreeing to. Unlike the district court, however, the appeals court found that Clark had a reasonable opportunity to review the document, which was written in English. Although Frattini had misrepresented the contents of the document to Clark, there was an opportunity to review the document that established dues for all employees of A & C. Thus, the court of appeals reversed the decision of the district court.

Although fraud in the factum is a real defense, courts have held that the signer's experience may determine whether the signer should have known what he or she was actually signing. Recall the situation with Jamaar in which he was solely responsible for ensuring the identity of the individual he was signing a check to. However, that level of responsibility is not required of Michael Jordan in the situation described earlier. Can you account for the difference in the two situations, pointing out the differing values and ethical norms?

As another example, consider *Schaeffer v. United Bank & Trust Co.*[4] United Bank sued Schaeffer to collect on a promissory note Schaeffer had signed as an accommodation maker. However, the Maryland court ultimately ruled that Schaeffer was not liable due to fraud in the factum. It turns out Schaeffer barely knew how to read, did not understand the document he was signing, and was lied to by the note's maker, who had told Schaeffer that Schaeffer's signature would serve as a character witness. The court ruled that United Bank was not a holder in due course and was subjected to Schaeffer's defense even if the bank were a holder in due course because the note was void due to fraud in the factum.

Material Alteration. The UCC defines a *material alteration* as "an unauthorized change in an instrument that purports to modify in any respect the obligation of a party, or an unauthorized addition of words or numbers or other change to an incomplete instrument relating to the obligation of a party" (3-407). Only unauthorized changes that affect the rights of the party are considered material alterations. Changes that typically fall under UCC Section 3-407 include changes to the parties to the instrument, the amount of the instrument, the date the instrument is due, and the applicable interest rate.

If the material alteration is fraudulent, the party whose rights have been affected by the change is completely discharged from the instrument [UCC 3-407(b)]. However, if the material alteration is not fraudulent, the instrument will be enforced only under the original terms.

Personal Defenses Personal defenses apply to holders, not holders in due course. Personal defenses can be divided into two categories. First, general defenses can be asserted against the defendant on general contract theory. These defenses include:

1. Breach of contract or warranty.
2. Lack or failure of consideration.
3. Fraud in the inducement. (See the following Case Nugget for an example of how the courts consider the defense of fraud in the inducement.)
4. Illegality.
5. Mental incapacity.

Second, the UCC lists specific personal defenses created by provisions of Article 3. These defenses, summarized in Exhibit 18-7, include the following:

1. Defense of nonissuance, conditional issuance, or issuance for a special purpose [3-105(b)].
2. Defense of modification or nullification of an obligation by a separate agreement (3-117).
3. Defense of nondelivery of the instrument [3-105(b)].
4. Defense of unauthorized, nonfraudulent completion of an instrument (3-115).

Discharge of Liability on Instruments

When a party's liability for a negotiable instrument is terminated, this party's liability has been **discharged.** In other words, the party is released from liability. Discharge can occur through a variety of ways. For example, as stated earlier, discharge of endorsers can occur if a party who has a right to enforce an instrument has materially altered the instrument. Keep in mind that discharge is not effective against a holder in due course (UCC 3-601).

discharged
Released from liability for a negotiable instrument.

[4] 360A.2d461 (Md. Ct. Spec. App. 1976).

Exhibit 18-7

Summary of Defenses to Liability for a Negotiable Instrument

Real (universal) defenses	Infancy (below the legal age of consent)*
	Duress*
	Lack of legal capacity*
	Illegality of the transaction*
	Fraud in the factum
	Discharge through Insolvency proceedings
	Forgery
	Material alteration
Personal defenses	*Contract defenses:*
	Breach of contract or warranty
	Lack or failure of consideration
	Fraud in the Inducement
	Illegality
	Mental Incapacity
	Article 3 defenses:
	Nonissuance of an instrument
	Modification of an obligation by a separate agreement
	Nondelivery of the instrument
	Unauthorized, nonfraudulent completion of the instrument

* To the extent that it makes a contract void.

CASE Nugget

Fraud in the Inducement

Hugh L. Caraway v. Land Design Studio Texas Court of Appeals

47 S.W.3d 696 (2001)

Internacional Realty hired Land Design to perform landscaping services on an apartment complex development known as the Deerfield Project, resulting in a debt owed to Land Design for work completed on the project. Land Design later told Caraway, the president of Realty, that it needed to remove Realty's account receivable from Land Design's books to continue receiving working capital from its lender. The parties created a promissory note that essentially replaced Realty's account receivable. In September 1998, the parties executed the note.

The note was to be paid in full no later than March 31, 1999; however, no payment was received. Land Design demanded payment and brought suit in July 1999 to collect the amount due. Land Design filed suit against Caraway. In response, Caraway denied liability, asserting several defenses, one being fraud in the inducement. Specifically, Caraway asserted that Land Design orally represented to him that the note would not be due until Realty closed on its Deerfield Project construction loan, thus fraudulently inducing him to sign the note. However, this type of parol evidence is generally inadmissible to alter the terms of a promissory note.

On the basis of the specifics of the case regarding parol evidence, the court determined that Caraway must show that some type of trickery, artifice, or device was employed by Land Design in addition to showing that Land Design represented to him that he would not be liable on the note. However, Caraway's summary judgment proof failed to present any evidence of trickery or artifice by Land Design. Caraway's only evidence was his repeated assertion in his affidavit (and two supplemental affidavits) that Land Design promised that the note would not be due until Realty closed on its construction loan. Even if correct, this was insufficient. A mere representation to a maker, without some accompanying trickery or artifice, will not defeat the application of the parol evidence rule.

(continued)

Caraway's claim of fraudulent inducement was further weakened by the fact that the promissory note itself makes a reference to closing on the loan. The note states: "Should Debtor close on the property prior to March 31, 1999, payment shall be made within three (3) days of Debtor's receipt of funds. . . ." The deadline for payment was March 31, 1999. Therefore, trickery is unlikely when the note references the event that Caraway claims is the event triggering his obligation to pay.

Because of Caraway's reliance on the assertions in his affidavit regarding Land Design's promise and the fact that such evidence is inadmissible under the parol evidence rule, Caraway offered no evidence to raise a genuine issue of material fact. Therefore, the court concluded that no factual question existed of fraudulent inducement.

Discharge through Payment and Tender of Payment Earlier in this chapter, we discussed how a party becomes liable for an instrument by signing the instrument. If a party (or another party on the first party's behalf) who has signed an instrument has an obligation to pay and then pays the full amount due, all parties who are liable will be discharged (UCC 3-602).

For example, Stuart creates a note in which he promises to pay Vanessa $1,000. If Stuart pays the $1,000 on the due date, he will be discharged from liability on the note. However, if Stuart makes the payment on the note to John, knowing that John stole the note from Vanessa and is wrongfully possessing it, Stuart's obligation will not be discharged. Paying John does not discharge Stuart's liability because John, who stole the note, is not a holder and is not entitled to the amount on the note [UCC 3-602(b)(2)].

Moreover, some parties's obligations on an instrument will be discharged if the obliged party tenders full payment on the due date but the holder of the instrument refuses to accept the money [UCC 3-603(b)]. If Stuart makes a proper tender of the full amount ($1,000) to Vanessa on the note's due date but she improperly refuses to accept the money, Stuart will still be liable for the $1,000 but will not have to pay interest on the amount. However, if any endorsers or accommodation parties are liable for Stuart's note, these parties's obligation will be discharged.

Discharge by Cancellation or Renunciation A party who is entitled to enforce an instrument may decide to cancel the instrument with or without consideration. Canceling the instrument discharges the obligation of a party who must pay the instrument (UCC 3-604). The party who decides to cancel the instrument may engage in an intentional voluntary act to cancel the instrument. For example, the party might write "Paid" on the instrument, intentionally destroy or mutilate the instrument, or give the instrument to the obliged party.

Alternatively, a party may renounce an instrument by promising not to sue to enforce the instrument. *Renunciation* occurs when a party agrees, in writing, not to sue the obliged party.

Discharge by Reacquisition *Reacquisition* occurs when a former holder of an instrument has the instrument transferred back to him or her by negotiation or other means. When reacquisition occurs, anyone who endorsed the instrument in between the initial acquisition and the reacquisition by the holder has his or her endorsement canceled. When an endorsement is canceled, discharge occurs. The holder who reacquired the instrument can further negotiate the instrument, but the intermediate endorsers will not be held liable (UCC 3-207).

Discharge by Impairment of Recourse A right to recourse is the ability of a party to seek reimbursement. Typically, when a holder presents an instrument to an endorser, the endorser can seek recourse from prior endorsers, the maker, the drawer, or accommodating parties. However, if the holder has in some way impaired the endorser's ability to seek recourse from any of these parties, the endorser is not liable on the instrument [UCC 3-605(i)].

Discharge by Impairment of Collateral If a party posts collateral to ensure his performance of the negotiable instrument and the holder of the collateral impairs the value of the collateral, the party to the instrument is discharged from the instrument to the extent of the damage to the collateral [UCC 3-605(d)].

CASE OPENER WRAP-UP

Dishonored Check and Holder-in-Due-Course Status

In this case, the court ruled in favor of Triffin and ordered Cigna to pay him $484 plus interest for the check. Even though Triffin knew the check was dishonored when he purchased it, he still was able to receive payment under the shelter principle and holder in due course. That is, the court ruled that the transfer by a holder in due course to a third party, even one with notice of the dishonor, transfers all rights of the holder in due course to the successor in interest.

As discussed earlier in the chapter, the purpose of the holder-in-due-course doctrine is to protect financial intermediaries and encourage them to continue to engage in financial transactions. If you were a manager for Sun's Market and discovered that the insurance company would not be responsible for the dishonored check, you would probably be less likely to accept checks in the future. The holder-in-due-course doctrine and shelter principle encourage market transactions and shield businesses like Sun's Market from unnecessary risks.

PRACTICAL TIPS for Business Managers

1. If you pay a party with a negotiable instrument and that party negotiates that note to a third party, HDC status means that you are obligated to pay that third party even if the first party wronged you. If you seek recourse, it must be against the party that wronged you, not the third party.
2. On the other hand, if someone is taking action against your business for an instrument you received from another party, remember the protections that HDC status provides you with and that the shelter principle can protect you even if you would normally be just a holder.
3. Some firms you try to do business with may try to abuse the protections of HDC status. If this attempt happens, remember that there are rules against certain abuses, such as taking instruments in bad faith. In other words, HDC status is not necessarily bulletproof, and you may have options of recourse.

SUMMARY

Holder-in-Due-Course Doctrine
The HDC doctrine provides incentive for financial intermediaries to engage in transactions because they receive greater legal protection.

Requirements for Holder-in-Due-Course Status
1. *Be a holder of a complete and authentic negotiable instrument.*
2. *Take the instrument for value:* The holder must suffer an out-of-pocket loss.
3. *Take the instrument in good faith:* The holder must take the instrument with "honesty in fact and the observance of reasonable commercial standards of fair dealing."
4. *Take the instrument without notice:* The holder must take the instrument without notice of the following defects: it is overdue, is dishonored, or is part of a series in default; it has been altered or has an unauthorized signature; or it is subject to claims or defenses.

The Shelter Principle and HDC
Shelter principle: If a holder cannot attain HDC status, the holder can acquire the rights and privileges of an HDC *if* the item is being transferred *from* an HDC.

Signature Liability
A party can be held liable for an instrument only if the party has signed the instrument.
Primary liability of makers and acceptors: They must pay the stated amount on the instrument when it is presented for payment.
Secondary liability of drawers and endorsers: They must pay the amount on the instrument if the primarily liable party dishonors the instrument and the following three conditions are met:
1. Presentment
2. Dishonor
3. Notice of dishonor

Accommodation parties: An accommodation party is one who signs an instrument to provide credit for another party who has also signed the instrument.

	Unauthorized signature: If a signature to a negotiable instrument is unauthorized, this unauthorized signature will not impose liability to the named party. 1. Negligence 2. Imposter rule 3. Fictitious-payee rule
Warranty Liability	*Transfer warranty:* When a party transfers an instrument to another party for consideration, the transferring party makes certain promises or warranties regarding the instrument and the transfer itself. *Presentment warranty:* When a party properly presents an instrument for acceptance, that party makes certain promises regarding the instrument and the party who is entitled to payment.
Avoiding Liability for Negotiable Instruments	1. *Defenses to liability:* The arguments for why a party should not be held liable for an instrument include: a. Real (universal) defenses. b. Personal defenses. 2. *Discharge of liability:* Release from liability can occur through: a. Discharge by payment or tender of payment. b. Discharge by cancellation or renunciation. c. Discharge by reacquisition. d. Discharge by impairment of recourse. e. Discharge by impairment of collateral.

Point/Counterpoint

Should Holder-in-Due-Course Laws Be Revised?	
YES	**NO**
While measures have been taken to reduce the potential abuse of the holder-in-due-course status, certain industries are still able to use it en masse to take unfair advantage of consumers, and the current justifications keeping the current statutes in place are outdated. When the holder-in-due-course rules were enacted, the use of negotiable instruments by consumers to make purchases was quite common (primarily in the form of checks). However, as modern consumers have largely abandoned checks in favor of bank lines of credit, the use of negotiable instruments has plummeted. Thus, the current laws are designed to protect a certain consumer behavior that has become obsolete. While the holder-in-due-course statutes have become less useful to consumers, they have become far more useful to certain industries, especially the subprime mortgage industry, for getting away with predatory loan practices. The use of assignees in the subprime loan market, which is a market where the buyer and seller are rarely on equal footing, has skyrocketed over the past two decades. The scenario in predatory subprime mortgage loans often involves targeting weak or desperate consumers (especially the elderly and those living in low-income areas), convincing them to agree to a loan with exorbitant interest and unfair terms, and assigning the promissory note to a third party that has far more leverage and means to mitigate the risk than the consumer. The current holder-in-due-course laws protect these practices by granting immunities to assignees and severely limiting the options of recourse for consumers.	The laws governing holder-in-due-course status work as intended and have many benefits that could be lost upon revision of these statutes. First, holder-in-due-course laws encourage commercial transactions by providing protections for assignees and instilling confidence in the negotiable instruments market. These factors make negotiable instruments more liquid and exchangeable as if they were cash. This increased liquidity and confidence can lead to benefits for society as well as the commercial realm. The holder-in-due-course status encourages third parties to buy notes, which can make obtaining credit cheaper and/or easier. For example, small banks have more incentive to make more home loans if they can resell their borrowers' promissory notes to investors or other banks. These third parties are more willing to buy the notes if the holder-in-due-course status means they can do so free from defenses. Thus, more people are able to obtain and use credit to better their lives with the current holder-in-due-course statutes.

Questions & Problems

1. Explain the rationale for the following statement: "The purpose of holder-in-due-course status is to encourage parties to engage in financial transactions."

2. What are the requirements of holder-in-due-course status?

3. What is the distinction between primary liability and secondary liability for a signed negotiable instrument?

4. Evaluate the following statement: "A party can never be held liable for a negotiable instrument if he or she did not sign the check."

5. What are the similarities and differences between transfer and presentment warranties?

6. How are real defenses different from personal defenses to liability for a negotiable instrument?

7. Brent Bentrim purchased a house by obtaining a loan from Bank of America. In 2002, Bentrim borrowed $182,700 from First Union National Bank, executing a mortgage and note in First Union's favor during the process. Bentrim used the loan to pay off the first loan to Bank of America in full.

 First Union subsequently merged with Wachovia Bank in 2006. Wachovia sold its interests in the note to The Money Store Service Corporation which was a subsidiary of The Money Store, LLC. This LLC was in turn a subsidiary of Wachovia Bank.

 In 2011, Bentrim filed a complaint against Wells Fargo, who allegedly held the note after merging with Wachovia in 2009. Wells Fargo in turn counterclaimed for foreclosure on the balance of the note. According to the court, Wachovia gained holder-in-due-course status after because of the sale of the note to a subsidiary of its subsidiary, The Money Store Corp. Do you think this is what lawmakers had in mind when they created this doctrine? Since Wachovia held HDC status on the promissory note from Bentrim, would Wells Fargo also be considered a holder in due course after gaining the note through its merger with Wachovia? How did the court decide?

 Source: *Bentrim v. Wells Fargo Bank, NA,* SC: Court of Appeals (2017).

8. In 2006, Wachovia Corporation merged with World Savings Bank, FSB's parent company and the successor company became known as Wachovia Mortgage, FSB. In 2008, Wells Fargo & Company acquired Wachovia and a year later, merged to become Wells Fargo Bank, N.A.

 Wells Fargo Bank is the appellee in this case. In 2004, Alice Vogel executed a promissory note with World Savings Bank. Vogel also executed a mortgage agreement with World Savings Bank. Vogel eventually defaulted on the loan and Wells Fargo filed a complaint seeking foreclosure. During the cross examination of Wells Fargo's vice president of loan documentation, it came to light that there were two endorsements on the last page of the original note. The first was a special endorsement from World Savings Bank to Fannie Mae, and the second was a blank endorsement. Neither endorsement was dated.

 Considering the series of mergers that led to Wells Fargo holding the note, what are the implications of the fact that there were two undated endorsements on the back of the note executed by Vogel?

 Source: *Vogel v. Wells Fargo Bank, NA,* 192 So. 3d 714 (2016).

9. In 2005, Lawrence Don, Victor Lissiak, and J. Randolph Light formed The Stretford at the Cascades Limited Partnership to develop a high rise condominium in Tyler, Texas. Between 2005 and 2007, Stillwater National Bank (SNB) loaned over $20 million to The Stretford. In 2008, SNB made a number of individual loans to the partners totaling $2.5 million. The loans were consolidated into one promissory note that year. SNB transferred the note to SW Loan OO, L.P. in 2011.

 After failing to reach an agreement to restructure their obligation, Light and Stretford filed actions against SNB and SW to gain declaratory judgement. SW subsequently filed a cross-claim against Lissiak to recover the balance on the $2.5 million promissory note. While Light, Stretford, and SNB managed to settle, SW and Lissiak were still embroiled in the lawsuit.

 The trial court eventually granted summary judgment in favor of SW and ordered that SW recover over $3 million. Lissiak appealed, arguing, among others, that there is a question of authenticity concerning his signature on the 2008 $2.5 million promissory note. As evidence, Lissiak submitted an affidavit where he stated he believed the note is not valid and and that he does not recall signing the promissory note. Do you think this is enough evidence to call into question the signature on the promissory note? What kind of evidence would be needed?

 Source: *Lissiak v. SW LOAN OO, LP,* 499 SW 3d 481 (2016).

10. Mary Lodge was one of eight African American first-time homebuyers who filed an action against United Homes LLC, lenders, appraisers, lawyers, and others, alleging that the defendants conspired to sell them overpriced, defective homes, financed with predatory loans.

 In 2002, Lodge purchased a home at 249 Halsey Street from United Homes for $419,000. Lodge took

out two mortgage loans with Olympia Mortgage Company after being told that she would be able to pay the for the mortgage by renting out rooms. After moving into the new house, Lodge found out that the house was in a state of disrepair; the floor of the house was rotten, the basement flooded during rain, and the house was not insulated.

Wachovia Bank, which had bought Lodge's first mortgage from Olympia, sought a summary judgement against Lodge's claims against the enforceability of the mortgage based on the defense that Wachovia was the holder in due course. However, the court found that Wachovia did not acquire the mortgage in good faith and denied Wachovia's motion for summary judgement.

The NYUCC defines "good faith" as "honesty in fact in the conduct or transaction concerned." Is this an objective or subjective definition of "good faith"? What evidence would the judge need to dismiss Wachovia's motion for summary judgement using the NYUCC definition of "good faith"?

Source: *Barkley v. Olympia Mortgage Company* [557 Fed. Appx. 22; (2014).

11. We meet Robert Triffin once again. On October 2012, the Child Support Enforcement Administration ("CSEA") issued a check in the amount of $544 to Rashad Jemal Christmas. The face of the check had a notation which stated "VOID AFTER 90 DAYS." Christmas cashed the check at Friendly Cash Checking Corporation ("Friendly") four days after being issued the check. Friendly later tendered the check to Bank of America for cashing. However, Friendly waited 131 days after the check was issued to cash the check. The check was subsequently dishonored by Bank of America and marked "REFER TO MAKER."

Triffin enters the picture here. On March 13, 2013, Triffin acquired the dishonored check from Friendly for value. Triffin then filed a complaint against CSEA and Christmas. Judgement was awarded against CSEA and Christmas for $756.41. CSEA then filed a motion to vacate the judgement on the grounds that Triffin was not a holder in due course because he did not acquire the check in good faith. Despite the fact that the court agreed with this claim, the judge maintained that Triffin was a holder in due course because of the shelter principle. When Christmas cashed the check at Friendly, he transferred his rights as holder in due course to Friendly, who then transferred the rights to Triffin. The court dismissed CSEA's motion to vacate judgement.

Do you agree with the court decision? Is Triffin abusing the shelter principle to make money off of dishonored checks? Are the benefits of encouraging the marketability of instrument of negotiation worth the potential issues created by the shelter principle?

Source: *Robert J. Triffin v. Maryland Child Support Enforcement Administration, and Rashad Jemal Christmas,* 436 N.J. Super. 621 (2014).

12. In this strange case, Albertson has already been convicted of passing checks that were ultimately dishonored because of insufficient funds in Albertson's checking account. Both Albertson and a witness claim that the person to whom the checks were issued had an agreement with Albertson that the drawee would hold on to the checks until Albertson had sufficient funds to cover the checks. The drawee was a used-car dealer who admitted in court that he had hold-check agreements with others but could not recall whether he had one with Albertson. Naturally, Albertson does not want to be liable for the checks that were later dishonored. Given the principles of law you learned in this chapter, what facts would need to be true for Albertson to have a reasonable argument on his behalf?

Source: *Albertson v. State,* 2013 WL 3233378 (Md. App. 2013).

13. A customer of Foster Bank named Choi deposited in her account a check for $133,026 that listed her as the payee. The check had been drawn on Wachovia Bank by a company called MediaEdge that had an account with that bank. Foster presented the check to Wachovia for payment. Wachovia paid Foster and debited MediaEdge's account. Now, as it happened, the actual payee of the check as originally issued was not Choi; it was a company called CMP Media. When CMP Media told MediaEdge that it had not received the check, an investigation ensued and revealed that Choi had somehow substituted her name for CMP Media on the check she deposited. By the time this was discovered, Choi had withdrawn the money from her account and vanished, and Wachovia had destroyed the paper check that Foster had presented to it for payment. It had done this in accordance with its normal practice, the lawfulness of which was not a question. It had retained a computer image of the check, but whether the image was of the original check drawn on Wachovia, with an alteration, or of a forged check could not be determined. The issuing bank's suit was based on the presentment warranty, under which, when a depositary bank presented a check for payment by the bank that issued the check, it warranted that the check had not been altered. On the basis of these facts and the information you know regarding warranties, who do you think should be liable for the amount of the check?

Source: *Wachovia Bank v. Foster Bancshares,* 457 F.3d 619 (2006).

14. Starting in November 15, 2010, Kenneth Flaska, a former partner to Dawda, Mann, Mulcahy & Sadler, P.L.C., defrauded the law firm by presenting and depositing checks drawn on the law firm's IOLTA account into Flaska's personal bank account at Bank of America. Flaska deposited nine checks over the course of three years, totaling $529,676.50. All nine of the checks were drawn from the law firm's IOLTA account and were payable on their face to "Bank of America." In order to recoup their losses, the law firm sued Bank of America, alleging that Bank of America did not make an inquiry into the checks before depositing them, therefore breaking the common law "duty of inquiry." The defendant contended that since the law firm is not one of its customers, it did not owe the law firm any duty. Bank of America also contended that it held the checks as a holder in due course. The judge rules in favor of the law firm, finding the holder in due course defense as inapplicable.

 Why did the judge decide that the holder in due course defense is inapplicable? Were there any irregularities in the checks which should have alerted Bank of America to possible fraud?

 Source: *Dawda, Mann, Mulcahy & Sadler, P.L.C. v. Bank of America, N.A.,* 85 U.C.C. Rep. Serv. 2d (Callaghan) 282 (2014).

15. A Chapter 7 bankruptcy trustee is filing an action to recover funds from the wife of a debtor. The wife signed a debtor's instrument in which her husband borrowed funds to construct utility barns and fencing necessary for his horse-raising business. The debtor and his wife used the property on which the horse-raising facilities were built as their residence. The purchase of the materials to construct the barns and fence was not made with any intention of increasing the value of their residential property. What additional information would you need to determine whether the wife was an accommodation party for purposes of determining liability regarding the debt instrument that she signed?

 Source: *In re Simpson,* 474 B.R. 656 (Bkrtcy. S.D. Ind. 2012).

Part 5 Creditors' Rights and Bankruptcy

CHAPTER 19

Secured Transactions and Bankruptcy

CASE OPENER

A Commercially Reasonable Sale?

On November 20, 2007, Shirley Bolden entered into a conditional sales contract which was assigned to M&T Bank for the purchase of a 2001 Mercedes-Benz CLK class automobile. Bolden took physical possession of the car, and M&T Bank retained a security interest in it. Bolden defaulted on the loan, and the car was repossessed in March 2011. Bolden received notice of M&T Bank's intent to sell the car at private sale. The car was sold by the Manheim Auto Auction, in Philadelphia, Pennsylvania, on April 12, 2011, for $3,900. The balance due on the loan at the time of the sale of the car was $15,668.86.

Manheim placed the car in a "dealers only" sale. In such a sale, only car dealers can attend the auction and purchase vehicles. There was no evidence as to how the sale was advertised or conducted, how many bidders were present, how many people actually bid on the car, or that the sale was done in accordance with the accepted industry practices. At the time that the car was sold by Manheim, the trade-in value of similar automobiles in good condition was $6,100. However, Bolden's car was in below average condition.[1]

M&T Bank sought recovery of the $11,768.86 deficiency balance left after reducing the total amount due on the loan by the sale proceeds.

1. If you were the judge deciding this case, would you grant M&T Bank the full $11,768.86 deficiency?
2. If you were the manager of M&T Bank, what steps would you undertake to secure the rights of your company to vehicles purchased on credit and at auction sales of vehicles subject to repossession?

LEARNING OBJECTIVES

LO 19-1 List the important definitions associated with secured transactions.
LO 19-2 Describe how secured interests are created.
LO 19-3 Describe how secured interests are perfected.
LO 19-4 Recall the scope of a security interest.

[1] *M&T Bank v. Bolden*, 2012 WL 6628947 (Del. Ct. Commo Pleas 2012).

LO 19-5 Explain how priority disputes are handled.
LO 19-6 Define default.
LO 19-7 Identify the goals of the Bankruptcy Act.
LO 19-8 Describe the specific types of relief available through bankruptcy.
LO 19-9 Explain how various creditors' claims are prioritized in bankruptcy.

The agreement made between M&T Bank and Bolden regarding the automobile is a common component of business transactions. Such agreements are called *secured transactions*. A **secured transaction** is a transaction in which the payment of a debt is guaranteed by personal property the debtor owns. In the Case Opener, Bolden guaranteed that she would pay the debt for the purchase of the automobile by giving M&T Bank rights to the vehicle, including the right to sell it upon repossession. In contrast, a creditor that does not have a secured interest must file a lawsuit, obtain a judgment, and execute on the judgment to recover money for an unpaid debt.

Article 9 of the Uniform Commercial Code (UCC) governs secured transactions in personal property (as opposed to real property). Thus, throughout the chapter, we refer to Article 9 of the UCC. As of 2014, with some variation, all states had enacted Article 9. Thus, although the law governing secured transactions is state law, the universal adoption of Article 9 permits us to discuss laws regarding secured transactions across state lines.

> **secured transaction**
> A transaction in which the payment of a debt is guaranteed by personal property the debtor owns.

Important Definitions Associated with Secured Transactions

LO 19-1 List the important definitions associated with secured transactions.

It is important to understand the definitions of the terms used in secured transactions to understand how such transactions are created. These definitions generally come from the UCC's definitions of the terms:

1. A **secured interest** is an "interest in personal property or fixtures which secures payment or performance of an obligation" [UCC 1-201(37)]. Suppose Best Buy sells you a laptop on credit. Best Buy retains a secured interest in the laptop, which means that the store can repossess the laptop if you fail to make payments.

2. A **secured party** is the person or party that holds the interest in the secured property. Thus, in the preceding example, Best Buy is the secured party. The secured party is also known as the *secured creditor*.

3. A **debtor** is the person or party that has an obligation to the secured party. You, the laptop owner, are the debtor because you have an obligation to make payments.

4. A **security agreement** is the agreement in which the debtor gives the secured interest to the secured party. Thus, when you made the agreement with Best Buy, you created a security agreement.

5. **Collateral** is the property that is subject to the security interest. In the Best Buy example, the laptop is the collateral. Collateral may include goods (consumer goods, farm products, inventory, and equipment), indispensable paper (documents of title, negotiable instruments, and chattel paper), intangibles (accounts, goodwill, and literary rights), and proceeds.

> **secured interest**
> An interest in personal property or fixtures that secures payment or performance to a creditor.
>
> **secured party**
> The party that holds the interest in a secured property.
>
> **debtor**
> A party that owes money to another party.
>
> **security agreement**
> An agreement in which the debtor gives the secured interest to the secured party.
>
> **collateral**
> The property that is subject to a secured interest.

Creation of Secured Interests

LO 19-2 Describe how secured interests are created.

How does a creditor become a secured party? To become a secured party, the creditor must gain a security interest in the collateral of the debtor. There are three criteria for creating the security interest:

1. The two parties create a security agreement and either (a) there is a record of the security agreement (usually a written agreement that describes the collateral and is signed by the debtor) or (b) the secured party is in possession of the collateral.
2. The secured party must give value to get the security agreement.
3. The debtor has a right in or to the collateral [UCC 9-203(b)].

Once these three criteria are met, the secured party's rights attach to the collateral. When *attachment* occurs, the creditor becomes a secured party with an interest in the collateral.

attachment
The point at which a creditor becomes the secured party that has a security interest in the collateral.

Written Agreement

The written agreement, often referred to as the *security agreement,* must be signed by the debtor. Moreover, the agreement must describe the collateral. The description must be accurate and detailed enough to identify the collateral. For example, in a description of a laptop that is serving as collateral, the serial number of the laptop might be listed in the written agreement. Let's revisit the M&T Bank case. The security agreement between Bolden and M&T Bank most likely included the following: (1) a statement that Bolden was buying the automobile on credit from M&T Bank; (2) a statement that M&T Bank is retaining a security interest in the automobile; (3) a description of the automobile and its vehicle identification number; (4) the price of the automobile and the amount of the monthly payments due to M&T Bank; and (5) a description of the process to be used in the event of Bolden's default and M&T Bank's subsequent repossession of the vehicle.

It is important for the collateral to be described clearly in the written agreement because the creditor could otherwise lose its rights to the collateral.

Value

The secured party must give value. What exactly does it mean to give value? According to the UCC, value is consideration. Thus, in the Case Opener, the use of the automobile was considered the value that M&T Bank gave to Bolden. Alternatively, suppose you receive a loan from a company. The money is the value given by the company, the secured party.

Debtor Rights in the Collateral

The final criterion necessary for the attachment of the security interest is the debtor's rights in the collateral. Bolden, the purchaser of the automobile, had a legal right to the vehicle (the collateral) after she signed the agreement.

Purchase-Money Security Interest

Now that we have discussed the criteria necessary for the creation of a security interest, we briefly consider one specific type of security interest, the **purchase-money security interest (PMSI)**, which is formed when a debtor uses borrowed money from the secured party to buy the collateral.

According to the UCC, a PMSI exists when a security interest is retained or taken by (1) the seller of the collateral to secure part or all of the purchase price *or* (2) a person who gives something of value to the debtor so that the debtor can gain rights to or use of the collateral [UCC 9-103]. In our laptop example, you bought the laptop on credit from Best Buy. Best Buy extended credit to you for the entire purchase price of the laptop. In other words, Best Buy is lending you the money to buy the laptop. Because the laptop is the collateral, Best Buy has a PMSI.

purchase-money security interest (PMSI)
A security interest formed when a debtor uses borrowed money from the secured party to buy the collateral.

Perfected Security Interest

LO 19-3 Describe how secured interests are perfected.

Suppose that you borrow money from a bank and the bank has a secured interest in your wedding ring. You are supposed to make monthly payments to the bank, but you lose your job and cannot make the payments. When you fail to make these payments, you **default** on the loan. Because the bank has a secured interest, it can repossess your ring.

But suppose you also borrowed money from another creditor, your boss, and you used your ring as collateral for that transaction as well. Your boss was unaware that you used the ring as collateral for your bank loan. Both your boss and your bank have an interest in your ring. Who gets the ring? The party that *perfects* its interest in the ring will have first claim.

When a party perfects its interest in collateral, it is legally protecting its claim to the collateral. **Perfection** is a series of legal steps a secured party takes to protect its right in the collateral from other creditors who want their debts satisfied through the same collateral. In the following sections, we consider the various methods of perfection.

> **default**
> Failure to make payments on a loan.

> **perfection**
> The series of legal steps a secured party takes to protect its right in the collateral from the other creditors who want their debt returned through the same collateral.

> **financing statement**
> A document that lists the names and addresses of all the parties involved in the transaction, a description of the collateral, and the signature of the debtor.

Perfection by Filing

The most common way to perfect an interest is to file a financing statement with a state agency. According to the UCC, a **financing statement** should list the names and addresses of all the parties involved, a description of the collateral, and the signature of the debtor [UCC 9-502]. First, the names and addresses of the statement are important elements because someone looking for the financing statement may want to contact the secured party or the debtor. Second, the financing statement must be filed under the name of the debtor whether the debtor is an individual, an unincorporated business association, or a corporation. If the statement is filed under an incorrect name, the perfection is likely not effective. Third, the financing statement must include a description of the collateral. The purpose of including a description of the collateral is to inform other potential parties who might want to lend money to the debtor. The description of the collateral may be replicated from the description of the collateral in the security agreement.

Once the financing statement is filed, the statement becomes public knowledge. Parties considering making a loan to another party are expected to search for notification of secured interests. Consequently, if you ask to borrow money from your boss, your boss should check with state agencies to ensure that you have not already used your ring as collateral. If your boss discovers that your bank has perfected its interest in your ring, your boss should not agree to loan you money with the ring as your collateral because the bank's perfection of the security interest means that the ring will go directly to the bank. Your boss would have a claim against the ring only as an unsecured creditor.

Place and Duration of Filing Suppose you need to file a financing statement. Where should you file? For how long is the filing effective? The place of filing depends on whether the debtor is an individual or a business. If the debtor is an individual, the secured party files the financing statement in the state in which the debtor resides. The actual place of filing differs from state to state. Sometimes the statements are filed with the secretary of state. Alternatively, the statements might be filed with the county clerk. Each state must establish a central filing office.

Once you have filed the financing statement with the correct agency, for how long is the statement valid? Under the UCC, the statement is valid for five years [UCC 9-515]. After five years, the statement expires, and the security interest is not protected. However, within six months of the expiration date of the financing statement, the secured party can file a continuation statement, which is valid for another five years.

Perfection by Possession

Sometimes a debtor gives a creditor the collateral to hold until the loan is paid off. For example, suppose that you borrow money from a bank and give the bank your diamond necklace to hold until you pay back the loan. The transfer of collateral to the secured party for the purpose of

perfection is called a **pledge**. When the bank takes possession of the necklace, it has perfected its interest without filing a financing statement. Once you pay off the loan, the necklace is returned to you.

Several advantages are associated with perfection by possession. First, a secured party does not have to file a financing statement. In fact, the parties do not even have to create a written security agreement. Second, there is little chance that another party will loan money to the debtor relying on the collateral that another secured party possesses. Third, if the debtor defaults on the loan, the secured party already has possession of the collateral, so no difficulties are associated with repossession.

Despite the many advantages of perfection by possession, it is often impractical because the debtor cannot benefit from the use of the collateral. For example, if you get a loan from a bank to purchase farm equipment, you likely need the farm equipment to produce crops that will enable you to make payments on the loan.

Certain types of collateral *must* be perfected through possession. These types of collateral include **instruments**—writings that serve as evidence of rights to payment of money, such as certificates of deposit—and stocks and bonds.

pledge
The transfer of collateral to the secured party.

instrument
A writing that serves as evidence of the right to payment of money.

Automatic Perfection

If a retailer had to file a financing statement every time it sold a laptop, a wide-screen television, or a washer and dryer on credit, the retailer would do nothing but file statements. Moreover, it does not make sense for the retailer to possess the collateral. Thus, when a creditor sells a consumer good to a debtor on a credit basis or a creditor extends a loan to a debtor for the purchase of a consumer good, the security interest in the good perfects automatically. Under the UCC, a **consumer good** is a good used or bought for use primarily for personal, family, or household purposes [UCC 9-102(23)]. Thus, if an item is purchased for business use, the security interest would not perfect automatically; the secured party would have to file a financing statement.

consumer good
A good used or bought for use primarily for personal, family, or household purposes.

Perfection of Movable Collateral

According to the UCC, a security interest in collateral that has been perfected in one state will generally transfer to another state for a period of four months from the date that the property is brought into the state. The secured party may reperfect the interest in the new state. However, if the interest is not reperfected, the secured party may lose its protection.

Perfection of Security Interests in Automobiles and Boats

We have now discussed various methods of perfection: perfection by filing, perfection by possession, automatic possession, and perfection of movable collateral. Exhibit 19-1 provides a summary of the methods of perfection. None of these methods apply to perfection of automobiles and boats. Each state has created special laws that pertain to perfection of motor vehicles.

Almost every state requires a certificate of title for any motor vehicle. Perfection of a security interest in a motor vehicle occurs when the secured party makes a notation of this interest on the certificate of title. The rationale for noting the interest on the title is that the title will follow the car owner everywhere. Thus, the title would be the best place to note a secured interest. If a creditor examines a title and discovers no notation of a secured interest in the vehicle, the creditor can assume that no other creditors have a secured interest in the vehicle.

The Scope of a Security Interest

LO 19-4 Recall the scope of a security interest.

Generally, once a secured party perfects its interest in collateral, the perfection is effective until the collateral is sold, exchanged, or transferred. Our examination of security interests to this point has been concerned with property the debtor currently possesses. However, a security interest can also apply to personal property that is not yet in the debtor's possession.

Exhibit 19-1
Summary of Methods of Perfection by Type of Collateral

Perfection by Filing	1. *Chattel paper:* Writing that indicates the debtor's monetary obligation as well as a secured interest
2. *Documents of title:* Papers that demonstrate the owner's possession of the goods (e.g., warehouse receipts)
3. *Accounts:* Rights to payments for goods sold or leased
4. *General intangibles:* Trademarks, copyrights, patents
5. *Equipment:* Goods purchased primarily for business use
6. *Farm products:* Products of livestock or crops
7. *Inventory:* Goods held for sale or lease
8. *Fixtures:* Goods that have become attached to real estate |
| **Automatic Perfection** | Purchase-money security interests in consumer goods |
| **Perfection by Possession** | 1. Chattel paper
2. Documents of title
3. Instruments (stocks, bonds, checks)
4. Pawnbroker holding jewelry or other valuables |
| **Perfection of Interests in Motor Vehicles** | Notation of secured interest on certificate of title |

After-Acquired Property

You borrow $60,000 from your bank to start an electronics store. However, you have recently graduated from college and have only your car, valued at $3,500, as collateral. The bank will take a security interest in your car, but it will also take a security interest in the materials that you will purchase to open your store. For example, you will need to create an inventory of televisions, computers, and stereos, which are **after-acquired property,** property acquired by the debtor after the security agreement is made. After-acquired property can be inventory, livestock, equipment, or almost any other kind of property. Under the UCC, a party may agree, through a clause in the security agreement, that the security interest will attach to after-acquired property. Whenever the debtor purchases new equipment or goods, the security interest is attached to the new good. Thus, whenever you buy a television to sell in your store, the bank's security interest in the television will attach.

after-acquired property Property acquired by a debtor after the security arrangement is made.

Proceeds

When a debtor sells collateral, he or she receives **proceeds,** something that is exchanged for the collateral. The secured party automatically has an interest in the proceeds. Why? If you use a good as collateral for a bank loan and then sell that good, the bank has nothing to continue to secure its loan. Consequently, the security interest in the good also applies to the proceeds from the good.

Under the UCC, the secured party's interest in the proceeds lasts only 10 days after the debtor receives the proceeds. At that time, the secured party will typically need to file a new financing statement. The parties may also agree in the security agreement that there will be extended coverage of interest in the proceeds.

proceeds Something that is exchanged for a debtor's sold collateral.

Termination of a Security Interest

Suppose that, rather than defaulting on her loan payment, Bolden paid M&T Bank the full amount for the automobile as required under the agreement. If a secured party has filed a financing statement and the debtor has repaid the secured party, the secured party must file a termination statement with the filing office. A **termination statement** is an amendment to a financing statement, which provides that the debtor has no further obligation to the secured party [UCC 9-513(a)].

termination statement An amendment to a financing statement, which states that the debtor has no obligation to the secured party.

Priority Disputes

LO 19-5 Explain how priority disputes are handled.

We have spent much time discussing the perfection of a security interest. Why? Remember, perfection of an interest is supposed to serve as protection of the secured party's interest in collateral from other creditors. In this section, we consider various conflicts between creditors who claim interest in the same collateral. A conflict may arise in many circumstances; however, priority disputes are most likely to arise when a debtor files for bankruptcy.

Secured Versus Unsecured Creditors

Generally, secured parties have priority over unsecured creditors. Thus, if two parties provide a loan based on the same collateral, the party with the secured interest will have priority in repossessing the collateral over the party with the unsecured interest. This priority to the collateral does not depend on the perfection of a secured interest. Rather, the priority depends on the *attachment* of the secured interest.

> **BUT WHAT IF?...**
>
> WHAT IF THE FACTS OF THE CASE OPENER WERE DIFFERENT?
>
> Let's say, in the Case Opener, that Bolden used her car as collateral not only with M&T Bank but also with a bank from which she received another loan. Let's say that M&T Bank is unsecured and the second bank is secured. Who would be able to claim the vehicle if Bolden defaults on both loans?

Secured Versus Secured Creditors

Who has priority to the collateral when both parties are secured creditors? If both parties are secured, the determination of priority for claim to collateral moves to considerations of time and perfection. The party that perfected its interest first will have priority in claim to the collateral [UCC 9-317]. Finally, if neither party has perfected its security interest, the party that attached its security interest first will have first claim to the collateral.

Exhibit 19-2 provides a summary of the priority of creditors' claims to collateral.

PMSI Conflicts There is an exception to these priority rules. Generally, if a PMSI is involved, the perfected PMSI will almost always have priority over other claims to collateral. The rules regarding PMSIs depend on whether the collateral is inventory or noninventory.

First, if the PMSI is in inventory, the perfected PMSI has priority over a previously perfected non-PMSI if the following two conditions are met: (1) The PMSI party perfects its interest before or at the same time that the debtor receives his inventory; and (2) the PMSI party checks for previous secured interests and gives written notice to the holders of any conflicting security interest [UCC 9-324(b)]. What happens if the collateral is *not* inventory? If the PMSI is in noninventory collateral, the PMSI has priority over any other secured perfected interests as long as the PMSI is perfected within 20 days of the debtor's possession of the collateral [UCC 9-324(a)].

DISPUTE	PREVAILING PARTY
Secured versus unsecured creditor	Secured creditor
Secured perfected creditor versus secured unperfected creditor	Secured perfected creditor
Secured perfected creditor versus secured perfected creditor	Party who perfected its interest first
Secured unperfected creditor versus secured unperfected creditor	Party who attached its interest first

Exhibit 19-2

Summary of Priority of Creditors' Claims to Collateral

Secured Party Versus Buyer

If a debtor sells collateral in which a secured party has an interest, the security interest generally remains in effect. Suppose you obtain a loan from a bank and use your boat as collateral. You need more money, so you sell your boat. The bank's secured interest in the boat remains with the boat. If you default on the loan, the bank can seize the boat from the buyer. However, the UCC provides some exceptions to this general rule.

buyer in the ordinary course of business
A person who routinely buys goods in good faith from a person who routinely sells such goods.

Buyers in the Ordinary Course of Business A buyer in the ordinary course of business is a person who routinely buys goods in good faith from a person who routinely sells such goods. Under the UCC, a buyer in the ordinary course of business can take the goods free of any security interest created by the seller of the goods *even if* the security interest is perfected [UCC 9-320(a)]. What is the rationale for this rule? Asking buyers to determine whether a security interest in inventory exists is burdensome. Thus, the purpose of this rule is to encourage commerce. For example, if your company buys products from an electronics store in the ordinary course of business, you can take possession of these products free of any security interest created by the party who originally sold the products to the electronics store.

Buyers of Consumer Goods Suppose you buy a digital camera for $800 on credit from Best Buy. Best Buy has a security agreement for a PMSI in the digital camera. (Remember, a PMSI in a consumer good perfects automatically.) However, you discover that you do not have enough money to pay your rent. Consequently, you sell your digital camera to your neighbor, who is unaware of Best Buy's secured interest in the camera. Can Best Buy repossess the camera? Under the UCC, as long as the buyer is not aware of the security interest, purchases the good for his or her personal use, and purchases the good before the secured party files a financial statement, the buyer obtains the good free of the security interest [UCC 9-320(b)]. Even though the secured party's interest perfects automatically, the secured party must have filed a financial statement to repossess a consumer good from another buyer.

chattel paper
A writing that indicates both a monetary obligation and a security interest in specific goods.

Buyers of Chattel Paper and Instruments If a buyer purchases chattel paper, a writing that indicates both a monetary obligation and a security interest in specific goods, or an *instrument,* a writing that demonstrates a right to payment of money, in the ordinary course of business, the buyer can obtain the good free of any security interest. The buyer must typically be unaware of the security interest in the good. Why is there an exception for buyers of chattel paper and instruments? Both chattel paper and instruments are easily transferable. Consequently, the UCC provides that these forms of collateral can be sold to a buyer free of the secured party's interest.

Sometimes, the issue with a secured creditor is not the simple repossession of collateral but the collection of monies relating to the collateral. The following Case Nugget examines the meaning of the term "proceeds of the collateral." Specifically, the issue was whether a settlement and insurance claim were within the meaning of the term "proceeds of the collateral" such that they belonged to the secured creditor.

Rights of a Secured Party

Bayer Cropscience, LLC v. Stearns Bank Nat'l Ass'n
Eighth Circuit Court of Appeals

837 F.3d 911 (8th Cir. 2016)

Bayer CropScience ("Bayer") brought an interpleader action to determine its obligations with regards to a settlement reached with Texana Rice Mill and Texana Rice, Inc. (collectively, "Texana"). The settlement came about as a result of lawsuits that arose when Bayer

(continued)

introduced genetically modified rice into the United States commercial long-grain rice supply. Texana alleged the contaminated rice damaged its property including its plant and equipment. The parties in this appeal, Stearns Bank National Association ("Stearns") and Amegy Bank National Association ("Amegy") were both bank creditors of debtor Texana. Texana settled its commercial tort claim against Bayer, and after disbursement of certain amounts, $933,697.90 remained. Stearns and Amegy claimed priority over those funds. The district court found for Amegy.

Stearns' loan was made in September 2002 and was secured, in part, by Texana's equipment and all proceeds from its sale, destruction, loss, or other disposition. Amegy's loan was made in February 2006 and subsequently secured by its claim against Bayer. Stearns and Amegy's security interests were perfected.

Stearns argued its security interest had priority because it filed a financing statement covering Texana's general intangibles before Amegy filed its statement covering the Bayer litigation. It also argued it was entitled to the settlement funds as proceeds from its original collateral, which included the equipment damaged by Bayer's negligence.

The court concluded Stearns had no interest in the proceeds of the Bayer lawsuit as a general intangible as Article 9 as adopted in Texas imposed heightened identification requirements to encumber commercial tort claims and required that the claim exist at the time of the security agreement. However, some portion of the proceeds were attributable to damage to Texana's equipment which was subject to Stearns' security interest. Article 9 as adopted in Texas allowed for this claim as it defined "proceeds," in part, as arising from claims arising out of the loss, nonconformity, or interference with the use of or damage to the collateral. Furthermore, a security interest in proceeds was perfected if the interest in the original collateral was perfected. Accordingly, Stearns held a security interest in the settlement proceeds as a right of recovery with respect to damage to its original collateral. The court remanded the case to the district court for determination of the amount of this interest.

Default

LO 19-6 Define default.

Generally, when a debtor fails to make payments on a loan or declares bankruptcy, the debtor has *defaulted* on the loan. However, the UCC does not define *default*. Consequently, each security agreement provides a specific definition of what is considered a default. Moreover, each agreement determines the procedures and consequences that occur in the event of default. Because the creditor is usually in a better bargaining position, the creditor usually determines the definition of default.

What are a secured party's remedies to recover its money when a debtor defaults on a loan? The secured party can (1) take possession of the collateral or (2) ignore its rights in the collateral and proceed to judgment. However, the party is not limited to just one of these remedies. If one method is unsuccessful, the party can attempt to pursue the other method. Note that these remedies are limited if the debtor has filed for bankruptcy.

Taking Possession of the Collateral

If a debtor defaults on a loan, the secured party can take possession of the collateral [UCC 9-609]. Although the secured party may act without any court order to retain possession of the property, the secured party may not breach the peace in repossessing the property. What exactly is *breaching the peace?* The UCC does not define the phrase. Generally, if the secured party can repossess the collateral without using force or committing trespass, the action is not a breach of the peace. If the secured party is unable to repossess the property without breaching the peace, the party

will file suit against the debtor to obtain a court order for the debtor to turn over the property. Consider Case 19-1, in which the debtor was injured during the creditor's attempt to repossess the collateral after the occurrence of a default.

Generally, the security agreement will state that if the debtor defaults on the loan, the debtor must turn over the collateral in a reasonable time and manner. If the debtor refuses to turn over the property, the secured party will typically hire a repossession company to obtain the collateral. Once the secured party has possession of the collateral, the party can choose to retain or dispose of the collateral.

CASE 19-1

PRICE AUTO SALES, INC. v. SANDERS
Texas Court of Appeals
2012 Tex. App. LEXIS 7274 (Tex. App. 2012)

FACTS: *Price Auto hired David Chad Kuykendall, doing business as Chad's Auto Recovery, to repossess a 1998 Cadillac DeVille owned by Carlous Sanders, who was in default on his motor vehicle retail installment sales contract. On May 11, 2009, Kuykendall went to the apartment complex where Sanders lived and located the Cadillac in a parking space facing Sanders' apartment. Using a key provided by Price Auto, Kuykendall entered the car and started the engine. Sanders testified that he was descending the stairs from his apartment when he first heard his car's ignition and then saw a person inside his car backing the vehicle out of its parking space. Believing his car to be in the process of being stolen, Sanders approached the vehicle. As Sanders approached the driver's side door, the car moved onto his foot, causing him to fall to the ground and suffer a head injury. The car then rolled over Sanders' leg as it exited the parking lot. Kuykendall did not stop to render aid and exited the apartment complex. Kuykendall proceeded to attach the vehicle to his tow truck and deliver it to Price Auto.*

ISSUE: Sanders sued Price Auto and Kuykendall for his injuries, alleging that they were liable for the breach of the peace that occurred during the repossession. Sanders was awarded $20,281.39 in damages at trial. Price Auto appealed on the basis that the repossession was complete before any alleged breach of the peace occurred.

REASONING: The court agreed with Sanders and upheld the trial court's award of damages. The court held that a secured party may take possession of collateral after a default without judicial process if it proceeds without breach of the peace. This rule recognizes that society's interest in preserving the peace is more important than a secured party's right to possession. When a secured party chooses to pursue a non-judicial repossession, it assumes the risk that a breach of the peace might occur. In this case, the evidence did not conclusively establish that the repossession was complete before Sanders suffered his head and leg injuries. Thus, Price Auto did not demonstrate reversible error in connection with the trial court's determination that a breach of the peace occurred in the process of the repossession. In reaching this conclusion, the court cited two cases. In the first case, the recovery agent towed the vehicle out of sight of the debtor's premises and subsequently discovered the debtor's two children inside the vehicle. He immediately returned to the debtor's home and returned the vehicle and children. There was no breach of the peace in this case as the recovery agent removed the vehicle from a public street without confrontation or trespassing on the debtor's premises and immediately terminated repossession efforts and returned the children upon learning of their presence. By contrast, in the second case, the vehicle's owner mistook the recovery agent for a car thief and, along with a coworker, attempted to stop her from leaving with the vehicle. The recovery agent struck the coworker and seriously injured him while driving away from the premises. In this case, the court held that the owner had a right to object to the attempted repossession as it was no longer peaceable. This case was very similar to the facts concerning Price Auto's repossession of Sanders' vehicle.

DECISION AND REMEDY: The court held that the evidence presented at trial was sufficient to support the trial court's decision and thus affirmed the judgment against Price Auto.

SIGNIFICANCE OF THE CASE: This case illustrates the difficulties creditors may encounter in attempting to repossess collateral upon the occurrence of a default. Although Sanders was in default on his obligations to Price Auto, it was nevertheless required to proceed in a manner to repossess the collateral without breaching the peace. As illustrated in this case, repossession can lead to confrontations which may turn physical and result in injuries to one or both parties. Adding to the complications is the use of recovery agents to perform the actual repossession. The actions of these individuals may be attributable to the secured party, as occurred in this case, and thus care in their selection, utilization, and supervision is important.

(continued)

CRITICAL THINKING

What is the argument that this judge is making? How good are his reasons? Is there any information that you feel he left out that would help you come to a conclusion on the issue presented in this case?

ETHICAL DECISION MAKING

Return to the materials on ethical decision making. Suppose you were the manager of Price Auto Sales in this situation. How might you have behaved differently if you were following the Golden Rule?

Disposition of the Collateral Under the UCC, the secured party can sell, lease, or transfer the collateral in any commercially reasonable method [UCC 9-610]. The secured party may sell the collateral in a private or public sale. Regardless of where the collateral is sold, the secured party must strive to receive the best price for the collateral. Striving to receive the best price is part of the requirement to conduct every aspect of the sale in a commercially reasonable manner. In addition to adhering to the commercial reasonableness requirement, the secured party must also notify the debtor of the sale. Additionally, the secured party must notify any other parties that have secured interests in the collateral.

Once the collateral is sold, the proceeds must be paid in the following order: (1) paying the reasonable expenses of retaking and disposing of the collateral (including attorney fees), (2) satisfying the debt of the secured party, and (3) satisfying remaining holders of junior security interests [UCC 9-615(a)].

Retention of the Collateral Instead of disposing of the collateral, the secured party may choose to keep the collateral in full or partial satisfaction of the debt. However, the secured party must notify the debtor of this intent by sending written notice. The debtor has 20 days to object to the secured party's retention of the collateral. If the debtor does not object to the retention, the secured party may retain the collateral [UCC 9-620-622]. By retaining the collateral for full satisfaction of debt, the secured party gives up any claim to the debt. In other words, the secured party cannot demand additional money from the debtor. Consequently, if the collateral is not valued at the full amount of the money due to the secured party, the secured party has no right to additional money from the debtor. If the debtor objects to the secured party's retention of the collateral, the secured party must sell or dispose of the collateral.

BUT WHAT IF? . . .

WHAT IF THE FACTS OF THE CASE OPENER WERE DIFFERENT?

Let's say, in the Case Opener, that M&T Bank took Bolden's car after she defaulted on her payments, and M&T Bank decided to keep her car. Is M&T Bank free to choose to keep the collateral in satisfaction of the debt?

Proceeding to Judgment

Another remedy for a defaulted loan is the secured party's rejection of the right in the collateral and the party's filing of a suit against the debtor. The secured party can sue the debtor for the entire amount of the debt. Rather than taking the time to organize the sale of the collateral, the secured party may choose to file suit. In contrast, an unsecured creditor has just one option: to file suit and seek a judgment in the amount of the debt.

Bankruptcy and Reorganization

When an entity is unable to pay its debts, bankruptcy law provides various options for the entity to resolve these debts. Bankruptcy remedies are available to individuals, partnerships, and corporations.

Generally, a debtor is an entity that owes money to another entity. However, the term *debtor* is defined differently under each type of bankruptcy remedy; thus, an individual or company that is eligible for one type of bankruptcy might not be eligible for another type of bankruptcy.

The remainder of this chapter explains the various bankruptcy remedies available to debtors. We begin with a discussion of the goals of bankruptcy law and an overview of the Bankruptcy Code. Then we consider specific types of bankruptcy relief under the code.

The Bankruptcy Act and Its Goals

LO 19-7 Identify the goals of the Bankruptcy Act.

There are two general goals of bankruptcy laws. First, bankruptcy laws provide protection to **creditors**—entities to whom a debtor owes money. Bankruptcy laws ensure that creditors competing for a debtor's assets are treated equally and receive a fair share of the debtor's assets. Second, bankruptcy laws provide opportunities for debtors to gain a fresh financial start. In summary, bankruptcy law provides an organized method by which **insolvent debtors**—debtors who cannot pay their debts in a timely fashion—respond to their debts.

Bankruptcy law is federal law. Article I, Section 8, of the Constitution states: "The Congress shall have the power . . . to establish . . . uniform laws on the subject of bankruptcies throughout the United States." Congress revised the Bankruptcy Code through the **Bankruptcy Abuse Prevention and Consumer Protection Act (BAPCPA) of 2005**. This revision included the most comprehensive changes to bankruptcy law in over 25 years.

Title 11 of The United States Code

Title 11 of the United States Code (U.S.C.) contains the Bankruptcy Code, which is divided into chapters. Chapters 1, 3, and 5 provide the general definitions and provisions concerning bankruptcy case administration and debtors. These chapters apply to all types of bankruptcy relief. Chapters 7, 9, 11, 12, 13, and 15, briefly described in Exhibit 19-3, create six specific types of bankruptcy proceedings.

Bankruptcy Proceedings

Most bankruptcy cases share a set of procedures or certain actions that must be taken in every bankruptcy case. These actions are as follows:

1. All bankruptcy cases begin with the filing of a petition for bankruptcy.
2. Once the petition is filed, the court grants an automatic stay for creditor actions against the debtor's estate. In other words, creditors' legal actions against the debtor must cease.

creditor
An entity to which a debtor owes money.

insolvent debtor
A debtor who cannot pay debts in a timely fashion.

Bankruptcy Abuse Prevention and Consumer Protection Act (BAPCPA) of 2005
A renovation in the system of bankruptcy enacted to address increased numbers of bankruptcy filings, significant losses associated with bankruptcy filings, loopholes and incentives that allowed for abuse, and the financial ability of debtors.

Exhibit 19-3

Types of Bankruptcy Relief by Chapter

Chapter 7	Sale of the debtor's assets by a trustee and the distribution of money to creditors
Chapter 9	Adjustment of a municipality's debts
Chapter 11	Reorganization of the debtor's financial affairs under supervision of the bankruptcy court
Chapter 12	Reorganization of a family farmer/fisherman's debts
Chapter 13	Reorganization of an individual's debts
Chapter 15	Recognition of insolvency proceedings pending in a foreign country and relief for foreign debtors

3. The court determines whether an order of relief should be granted.
4. The creditors meet with the debtor.
5. Some type of payment plan is created and approved, usually by the creditors and the court.
6. The payment plan is implemented by actions of the trustee and the debtor.
7. Debts remaining after the plan is implemented are usually discharged.

Specific Types of Relief Available

LO 19-8 Describe the specific types of relief available through bankruptcy.

Before a debtor files for one specific type of relief, the clerk of court must give the debtor written notice of the other types of relief available. This requirement helps to ensure that the debtor has full information about the bankruptcy process.

Under BAPCPA, an individual may not be considered a debtor under any chapter unless, within 180 days prior to filing, he or she receives credit counseling from a nonprofit budget and credit counseling agency. If the individual does not fulfill the credit-counseling requirement, the bankruptcy court will dismiss the bankruptcy petition—even if the individual faces immediate home foreclosure or wage garnishment.

Moreover, BAPCPA attempts to prevent repeat filers from filing one bankruptcy claim after another. If an individual was a debtor in a bankruptcy case that was dismissed within 180 days of the current case, the individual is generally not eligible to be a debtor under Chapters 7, 11, or 13. However, if the previous bankruptcy was completed rather than dismissed, the individual is generally permitted to file for bankruptcy again. If a party completes a Chapter 7 bankruptcy, the party is not permitted to seek a Chapter 7 bankruptcy again for eight years.

In the following sections we discuss the main types of relief available, including Chapter 7, 11, 12, and 13 bankruptcies.

Chapter 7: Liquidation Proceedings

The most familiar type of bankruptcy proceeding is liquidation, which is sometimes called *straight bankruptcy*. Liquidation occurs when a debtor turns over all assets to a trustee, an individual who takes over administration of the debtor's estate. Trustees are usually attorneys in private practice who specialize in bankruptcy law. Every Chapter 7 proceeding has a trustee who sells the nonexempt assets and distributes the proceeds of the sale among the creditors. Liquidation provides an organized method of selling the debtor's property to generate cash to pay creditors.

liquidation
The process in which a debtor turns over all assets to a trustee.

trustee
In bankruptcy proceedings, an individual who takes over administration of a debtor's estate.

Who is defined as a debtor for liquidation purposes? Under Chapter 7 liquidation proceedings, individuals, partnerships, and corporations are considered debtors. Railroads, insurance companies, banks, savings and loan associations, industrial banks, credit unions, and health maintenance organizations are not eligible for Chapter 7 relief.

Petition Filing As is the case with most bankruptcies, liquidation begins when the petition is filed. Under Chapter 7, liquidation may be voluntary or involuntary; thus, a voluntary or involuntary petition is filed. The person filing the petition is also responsible for paying filing fees.

Voluntary Liquidation Petition. When the debtor decides to file for bankruptcy, he or she files a voluntary petition. The debtor must state that he or she understands the other types of bankruptcy relief available and chooses liquidation. Once the petition is filed, all the debtor's prepetition assets form the bankruptcy estate. Assets that the debtor gains after filing the petition are generally not part of the bankruptcy estate unless they fall under an exemption.

bankruptcy estate
The assets that are collected from a debtor who files for bankruptcy.

Involuntary Liquidation Petition. If a debtor is not paying debts as they come due, creditors can attempt to force the debtor into bankruptcy by filing an involuntary petition under Chapter 7. By filing the petition, the creditors attempt to force the debtor to surrender his or her assets so that the proceeds from the sale of the assets may be distributed among the creditors. If the judge

believes that creditors are using involuntary liquidation proceedings frivolously, the court may force the creditors to pay the attorney costs, fees of the debtor, and even punitive damages.

Not everyone can be forced into bankruptcy through Chapter 7. Farmers, ranchers, and nonprofit organizations are examples of debtors that cannot be forced into liquidation. Also, debtors that are ineligible to file for Chapter 7 bankruptcy voluntarily (railroads, insurance companies, banks, savings and loan associations, credit unions, and health maintenance organizations) are excluded from forced bankruptcy.

Dismissal of Petition A bankruptcy judge may dismiss a voluntary or involuntary bankruptcy petition. Under BAPCPA, a bankruptcy judge may dismiss a petition for cause. One cause for dismissal is failure of the *means test*. If an individual's debt is primarily consumer debt and if the individual's income is above the median income in his or her state, the court may presume that the individual is abusing the bankruptcy provisions. However, the individual may continue with the petition under the presumption.

Automatic Stay Once a petition, voluntary or involuntary, is filed, the code provides for an **automatic stay**, or moratorium, for almost all creditor litigation against the debtor. During the stay, creditors cannot bring or continue legal action against the debtor or his property. For example, creditors cannot attempt to repossess property during bankruptcy proceedings. Moreover, if a creditor received a judgment against a debtor prior to the bankruptcy filing, the creditor may not enforce the judgment.

There are some exceptions to the stay. First, under BAPCPA, if the debtor was a debtor in a bankruptcy case that was dismissed within a year of the current bankruptcy case filing, the stay automatically terminates 30 days after the current filing. Second, legal actions to determine paternity or to collect child support or alimony payments are not subject to the stay. Third, the court may exclude secured creditors from the stay if they petition the court to show that they do not have adequate protection under the stay. If a creditor is aware of the stay and continues to engage in legal action against the debtor, the code provides that the debtor may recover damages, costs, attorney fees, and possibly even punitive damages.

Order of Relief After a bankruptcy petition has been filed, the next step is for the court to determine whether an order of relief is granted. An **order of relief** means that bankruptcy relief is ordered; that is, the bankruptcy proceedings can continue. If the filing of the voluntary petition is proper, the petition automatically becomes an order of relief. Similarly, if a debtor does not object to an involuntary bankruptcy, the order of relief is automatic.

If the debtor challenges the involuntary petition for bankruptcy, a hearing will be held. At the hearing, the judge generally will grant the order of relief for involuntary bankruptcy if one of two conditions occurs:

1. The debtor is not paying debts as they become due.
2. The custodian took possession of almost all the debtor's property within 120 days before filing the petition.

After the court enters the order of relief, a U.S. trustee, a government official appointed by the attorney general, selects an interim trustee. A *trustee* is the person responsible for collecting the debtor's available assets and liquidating the property into cash for the creditors. The interim trustee is responsible for organizing the creditors' meeting.

Creditors' Meeting Between 20 and 40 days after the order of relief has been granted, the interim trustee calls a **creditors' meeting**—a meeting of all the creditors listed in the Chapter 7 required schedule for liquidation. Although the debtor and the interim trustee also attend this meeting, the bankruptcy judge does not attend.

If the debtor fails to appear at the meeting, the court may refuse to grant the bankruptcy. The principal purpose of the creditors' meeting is to enable the creditors and trustee to examine the debtor under oath regarding her financial affairs. Creditors want to know more about the way the debtor is handling her assets and property. Moreover, they want to ensure that the debtor is not concealing property. Not only does the trustee ask the debtor questions regarding her financial status, but the

trustee also ensures that the debtor is aware of the other forms of bankruptcy relief as well as the consequences of filing for bankruptcy.

Another important purpose of the creditors' meeting is the election of a permanent trustee. The interim trustee might become the permanent trustee, or the creditors might elect a different one. The creditors elect the trustee because the trustee generally represents the creditors.

The Trustee The general duty of the trustee is to collect the debtor's available assets (i.e., the debtor's prefiling assets) and to liquidate the property to cash that will be distributed among the creditors. Thus, the trustee takes possession of the debtor's property and has it appraised. Moreover, the trustee examines the debtor's records and might even temporarily take over the debtor's business. If someone else holds the debtor's property, the trustee has the power to require the person to return that property. Next, the trustee separates the exempt property from the nonexempt property and sells the nonexempt property. In addition to selling the debtor's property, the trustee has a variety of powers that assist him in fulfilling his duties. First, the trustee has the power to sue and be sued. He can initiate collection actions but must also defend against creditor actions. He has the right to assume or reject executory contracts. Moreover, he has the right to obtain credit. Finally, the trustee has the power to void certain liens against the debtor's property. For example, a debtor's ability to exempt property is hindered by any liens against such possessions. Consequently, the trustee is permitted to void these liens.

Exempt Property A debtor is not required to give up her exempt property. The trustee sorts the exempt property from the nonexempt property. The exemptions are stated in the Bankruptcy Code and are adjusted every three years on the basis of the consumer price index.

There are federal exemptions as well as state exemptions. States, through legislation, can and have opted out of the federal exemptions to give debtors the option for state exemptions only. However, in some states, a debtor may choose whether to make state or federal exemptions; the debtor cannot make some state exemptions and some federal exemptions. Exhibit 19-4 lists the federal exemptions. In addition to having the exemptions in Exhibit 19-4, a debtor is entitled to 100 percent of Social Security benefits, veteran's benefits, and civil service retirement benefits.

Why has Congress created such exemptions for property? Suppose that a debtor was forced to sell all his property through liquidation. Remember, one purpose of bankruptcy is to provide a debtor with a fresh start. If a debtor were forced to sell all his property, he would likely fall right back into debt again. Thus, the exemptions are for items considered necessary to earning a living.

A debtor must file a list of the property that she or he claims is exempt. Under the Bankruptcy Code and the bankruptcy rules, creditors may file objections to the claimed exemptions. If a debtor improperly lists property as exempt and a creditor does not file a timely objection, the debtor will likely be permitted to claim the property as exempt.

Preferential Payments Because a major purpose of the Bankruptcy Code is to prevent debtors from making payments to one creditor and thus treating that creditor preferentially, the trustee

Exhibit 19-4

Federal Bankruptcy Exemptions

1. Up to $23,675 for residence
2. Interest in a motor vehicle (not necessarily an automobile) up to $3,775
3. Interests in personal and household goods and furnishings, clothing, appliances, books, animals, crops, and musical instruments
4. Interest in jewelry up to $1,600
5. $1,250 of any property the debtor chooses (functions as a wild-card exemption)
6. Tools of trade and professional books up to $2,375
7. Any unmatured life insurance contract owned by the debtor
8. Professionally prescribed health aids
9. The right to receive certain personal injury awards up to $23,675
10. Retirement funds in an IRA or SEP up to $1,283,025 per person

preferential payment
A payment made by an insolvent debtor that gives preferential treatment to one creditor over another.

has the power to recover **preferential payments,** or payments made by an insolvent debtor that give preferential treatment to one creditor over another. If the debtor made any payments within 90 days of the bankruptcy filing, the trustee can examine these payments as preferential payments. For a payment to be preferential, the trustee must show that the transfer gave the creditor more money than the creditor would have received through bankruptcy proceedings. Thus, other creditors are disadvantaged by the debtor's preferential payment.

If the preferred creditor is an insider, such as a relative or partner, the trustee has the power to recover payments made within one year before the debtor's filing for bankruptcy. However, in such cases, the trustee must demonstrate the debtor's insolvency rather than assume insolvency. In contrast, the trustee may not recover preferential payments made for alimony or child support.

fraudulent transfer
A transfer of property made with intent to defraud creditors or made for an amount significantly lower than the property's fair market value within two years of filing for bankruptcy.

Fraudulent Transfers A trustee can recover preferential payments because these payments are not fair to creditors. Similarly, a trustee can void **fraudulent transfers** of property. These transfers can be actually or constructively fraudulent. An *actual fraudulent transfer* would be made with intent to defraud creditors. Generally, there is no direct evidence of intent to defraud (e.g., an email written by a debtor who states that he is transferring assets to hide them). To show *constructive* fraud, a trustee would establish intent through circumstantial evidence.

Similarly, if a debtor transfers property for an amount significantly lower than its fair market value, he or she may have engaged in a fraudulent transfer. Suppose that you are unable to pay your debts and are preparing to file for bankruptcy. You decide to sell your $50,000 boat to your business partner for $50. The trustee of your case could recover the boat if the sale occurred within two years of your bankruptcy filing. Furthermore, you could be subject to criminal penalties for your attempt to hide your assets. By punishing debtors who attempt to hide their assets, the Bankruptcy Code provides protection for creditors.

Suppose that, instead of selling your boat to your business partner, you sold your boat to a creditor. Because you are making a payment to a creditor, the trustee would analyze this payment as a preferential payment (assessing how the other creditors would be harmed by this payment) rather than a fraudulent transfer (assessing the reasonableness of consideration for the transfer).

LO 19-9 Explain how various creditors' claims are prioritized in bankruptcy.

Creditors' Claims Within 90 days of the creditors' meeting, all creditors (except secured creditors) must file a proof of claim with the bankruptcy court clerk to receive a portion of the debtor's estate. This proof of claim lists the creditor's name and address and the amount of the debt owed to the creditor. If the creditor fails to file such a claim, the creditor may not receive a portion of the debt. However, the fact that the claim was filed does not mean that the creditor will automatically receive a portion of the proceeds from the sale of the debtor's assets. The trustee must permit the claim to be allowed, and there are numerous defenses to creditors' claims.

When the trustee is sorting and examining the property, the trustee must determine whether a creditor has a secured interest in the debtor's property. If a secured interest exists, the creditor has first claim to the property. The debtor may decide to surrender the property to the creditor to satisfy the debt. Alternatively, the creditor may foreclose on the property and use the proceeds of the sale to reduce the debt owed to the creditor. However, the secured party is secured only to the value of the collateral.

All creditors do not have equal claims to the proceeds of the sale. Secured parties have priority over all unsecured parties in receiving portions of the proceeds of the liquidation. Thus, secured parties are paid first. The code establishes classes of priority claims. One class must be completely paid before anyone in another class receives any payment. If there are insufficient funds to pay all the creditors in one class fully, one class is paid proportionately and the lower classes get nothing.

The classes of priority claims among unsecured creditors are shown in Exhibit 19-5.

discharge
A written federal court order signed by a bankruptcy judge stating that the debtor is immune from creditor actions to collect debt; that is, a release from liability.

Discharge If a debtor has honestly dealt with her creditors during bankruptcy proceedings, the debtor is likely eligible for a discharge of her remaining debts. A **discharge** is a written federal court order signed by a bankruptcy judge stating that the debtor is immune from creditor actions to collect debts. When a debt is discharged, the debtor is essentially no longer responsible for the debt.

Exhibit 19-5

Classes of Priority Claims for Distribution

Class 1	Any unpaid domestic support obligations (alimony or child support)
Class 2	Court costs, trustee fees, attorney fees, and other administrative claims among expenses associated with the bankruptcy
Class 3	Unsecured claims in involuntary bankruptcy that arise through the debtor's ordinary business expenses from the date of filing the petition to the date of the appointment of the trustee
Class 4	Unsecured claims for unpaid wages, salaries, and commissions earned within 180 days of filing the petition
Class 5	Unsecured claims for contributions to employee retirement plans
Class 6	Unsecured claims by farmers and fishermen against grain operators of grain storage facilities or fish storage or processing facilities
Class 7	Claims for deposits given to the debtor in connection with property or services never given
Class 8	Certain taxes and penalties due to government units
Class 9	Claims in bankruptcies related to federal depository institutions
Class 10	Unsecured claims for personal injuries and deaths caused by the debtor's operation of a motor vehicle while under the influence of alcohol or drugs

When is a discharge of debt appropriate? Discharge of debt is a privilege, not a right. A judge determines which debts are dischargeable. If the debtor has received a previous discharge of debt within eight years of the current filing for bankruptcy, the judge will likely refuse a new discharge of debt.

Exceptions to Discharge. Generally, most debts are discharged unless there are objections to the discharge or the debts are ineligible for discharge. If a major goal of bankruptcy law is to give debtors a fresh start, why are some debts nondischargeable? Some of the debts that are not dischargeable are debts to those who are in a weak bargaining position with the debtor. For example, child support and alimony payments are nondischargeable. Children and ex-spouses often depend on payments; consequently, the court will not permit the discharge of these debts. Exhibit 19-6 lists the debts that are nondischargeable as established by the Bankruptcy Code.

Exhibit 19-6

Nondischargeable Debts under the Bankruptcy Code

1. Claims for back taxes or government fines within three years of filing for bankruptcy
2. Claims for liabilities against the debtor for obtaining money or property under false pretenses, false representation, or fraud
3. Claims by creditors who weren't listed on the schedule and did not have notification of the bankruptcy proceedings
4. Claims based on fraud, embezzlement, and larceny by the debtor while she or he was acting in a fiduciary relationship
5. Alimony, child support, and some property settlements
6. Claims of willful or malicious conduct by the debtor that caused injury to another person or property
7. Specific student loans unless payment of the loans imposes undue hardship on the debtor
8. Judgments against a debtor for claims resulting from the debtor's drinking and driving
9. Debts not discharged in previous bankruptcies
10. Claims for money borrowed to pay a tax to the United States that would be nondischargeable
11. Cash advances on credit cards

Case 19-2 addresses one type of nondischargeable debt, specifically, loans made to a debtor for educational purposes.

Revocation of Discharge. Although a debt might be discharged, the discharge is not necessarily permanent. If the trustee or a creditor discovers that the debtor has acted fraudulently or dishonestly during the bankruptcy proceedings, the court may revoke the discharge within one year. The revocation of discharge allows the creditors to bring actions against the debtor.

CASE 19-2

IN RE MURRAY U.S. BANKRUPTCY COURT
DISTRICT OF KANSAS
563 B.R. 52 (Bankr. D. Kan. 2016)

FACTS: *Alan and Catherine Murray filed a voluntary petition under Chapter 13 of the Bankruptcy Code in September 2014. The case was later converted to Chapter 7. The Murrays had student loans totaling $77,524 in principal and more than $200,000 in accrued interest owing to Educational Credit Management Corporation (ECMC). The Murrays made regular payments and were never late, but the amount of their payments was insufficient to service the interest accruing at nine percent annually on the principal and accrued interest. Interest payments alone accrued at the rate of $65.89 per day.*

The Murrays claimed that they could not meet a minimal standard of living if they were forced to repay the principal amount of the student loans plus the accrued interest. The Murrays had a monthly net income of $6,100 and monthly expenses of $4,442 resulting in disposable monthly income of $1,658 at the time of their bankruptcy filing. This amount included no funds for emergencies, no savings for a down payment on a home, no savings for retirement, no vacations and $50 per month for entertainment other than cable television. ECMC opposed their request to discharge the interest payments.

ISSUE: Section 523(a)(8) of the Bankruptcy Code provides that an educational loan is not dischargeable in bankruptcy "unless excepting such debt from discharge would impose an undue hardship on the debtor and the debtor's dependents." The court was required to determine whether the Murrays would suffer undue hardship if required to repay the loans in full with interest and what factors constitute a minimal standard of living to which the Murrays were entitled.

REASONING: The bankruptcy court held that the Murrays were entitled to a discharge with respect to accrued interest payments but were required to repay the principal balance of their student loans. The court found that requiring the Murrays to repay the loans and accrued interest would provide them with "a minimal or spartan standard of living." The court defined "undue hardship" as consisting of three factors. The first factor is that the debtor cannot maintain, based on current income and expenses, a "minimal" standard of living if forced to repay the loans. The second factor is that the debtor's circumstances are likely to persist for a significant portion of the repayment period of the student loans. The final factor is that the debtor has made good faith efforts to repay the loans.

The court found that the Murrays met all three factors. According to the court, a "minimal standard of living" requires: (1) shelter including heating and air conditioning; (2) basic utilities such as electricity, natural gas, water, and telephone service; (3) food, personal hygiene products and decent clothing; (4) a motor vehicle, including expenses for gasoline, insurance and maintenance; (5) health insurance or the ability to pay for medical and dental expenses; and (6) recreation even if it is just watching television or keeping a pet. The Murrays would be deprived of some portion of these items if they were required to repay the student loans in their entirety with interest.

As to the second and third factors, the court found that the Murrays would not be able to pay the accrued interest on their loans in the foreseeable future as they were not recent graduates, the loans were more than twenty years old, the Murrays were in their late forties, raises and promotions at their jobs were unlikely, and medical and dental expenses would increase as they aged. Finally, the Murrays had made good faith efforts to repay the loans, were never late with a payment, and had reduced the principal by $54,000 over the course of twenty years.

However, the court also held that dischargeability of student loans is not an all-or-nothing proposition. When a debtor shows that the requirements of Section 523(a)(8) have been satisfied for a portion of the student loan debt, the bankruptcy court may exercise its equitable powers and discharge just that portion of the debt. In this case, the court found that the Murrays could afford to pay $900 per month on the student loan obligations while maintaining a minimal standard of living. Such payments would be sufficient to pay the original principal of the loans with interest at nine percent per year in twelve years when the Murrays would be sixty years old. Thus, the court discharged the accrued interest obligation in order to afford the Murrays a fresh start.

(continued)

DECISION AND REMEDY: The court held that the Murrays were entitled to a discharge with respect to the accrued interest payments but were still required to repay the principal balance of their student loans.

SIGNIFICANCE OF THE CASE: This case sets forth a detailed definition of "undue hardship" with respect to the Bankruptcy Code provisions relating to the potential discharge of educational loans. The Murrays met the court's three-part test for discharge with respect to the accrued interest obligation, specifically, that they could not maintain a minimal standard of living, that such circumstances were likely to persist for an extended period of time, and that they had made a good faith effort to repay the loans. The court's definition of what constitutes a "minimal standard of living" is particularly noteworthy. However, the court did not grant an outright discharge of all obligations relating to the Murrays' educational loans but rather forged a compromise requiring continued efforts by the debtors to repay the principal balance.

CRITICAL THINKING

Do you agree with the factors identified by the court as essential to a "minimal standard of living"? What factors would you include or exclude from the list of factors? Was the court's decision too harsh in requiring the Murrays to repay the principal balance? Why or why not?

ETHICAL DECISION MAKING

How would you evaluate the court's balancing of the creditor's right to full repayment versus the debtors' right to relief pursuant to the Bankruptcy Code? What values are reflected in the court's balance of these competing interests?

Reaffirmation of Debt A debtor may wish to repay a dischargeable debt. The debtor might owe money to a family member or a longtime business associate. To maintain a good relationship with these people, a debtor might choose to repay the debt instead of having the debt discharged. This repayment can occur through a **reaffirmation agreement,** an agreement in which the debtor agrees to pay the debt even though it could be discharged.

Unfortunately, creditors may attempt to pressure a debtor into reaffirming a debt. Congress was worried that debtors were unwisely surrendering their ability to discharge their debts by making reaffirmation agreements. Thus, Congress created the following extensive rules so that a decision for reaffirmation of debt cannot be an impulsive one:

reaffirmation agreement
An agreement in which the debtor agrees to pay a debt even though it could have been discharged in bankruptcy.

1. The reaffirmation agreement must be made before the debt is discharged.
2. The debtor must be able to cancel the agreement, and the agreement must contain explicit information regarding the time frame in which the debtor may cancel the agreement.
3. The agreement should contain a statement notifying the creditor that the law does not require the agreement.
4. The agreement must be filed with the bankruptcy court. Typically, the debtor's attorney files a form along with the agreement stating that the debtor is voluntarily entering the agreement and the agreement will not result in hardship to the debtor. Unless the attorney files this form, court approval for the agreement is needed.

Chapter 11: Reorganizations

The largest bankruptcy in U.S. history was the 2008 Lehman Brothers Holdings' bankruptcy filing. Lehman, an investment bank, had over $691 billion in assets at the time of filing. By contrast, WorldCom's 2002 bankruptcy was filed under Chapter 11, which allows the creditors and debtor to create a plan to reorganize the debtor's financial affairs under the supervision of the bankruptcy

court instead of liquidating the assets. Generally, the creditors and debtor agree that part of the debt will be discharged while the other part is or will be paid. Creditors generally agree to these plans because they believe that the value of the operating business is greater than the value of the business broken up and sold in pieces.

Who is eligible for Chapter 11 reorganization? Corporate debtors most frequently use Chapter 11 reorganization because they are permitted to remain in business. However, the debtor does not have to be a business entity to use reorganization; the debtor can be an individual unrelated to business. Stockbrokers, commodities brokers, banks, and savings and loan companies are not permitted to file under Chapter 11.

Most of the Chapter 11 reorganization procedures are similar to Chapter 7 liquidation procedures. Like liquidation, reorganization may be voluntary or involuntary. The reorganization process begins with the filing of the reorganization petition.

When the petition is filed, an automatic stay prohibits legal action against the debtor during the reorganization process. The debtor is required to file a list of creditors. Next, the order of relief is granted, and the court appoints a trustee, who appoints a creditors' committee of unsecured creditors. Because there may be hundreds of creditors, the creditors' committee is supposed to represent the interests of the range of creditors. The Bankruptcy Code contemplates seven creditors on the committee. The members of the creditors' committee have a fiduciary duty to the other creditors, and under BAPCPA, the creditors' committee must provide the other creditors with access to information.

Generally, the trustee is responsible for developing the reorganization plan to handle the creditors' claims. The *reorganization plan* is a contract between a debtor and his or her creditors. The goal of the plan is to rehabilitate the debtor while preserving the assets for the creditors. Reorganization plans usually state three things:

1. The classes of claims and interests in the debtor's property.
2. The treatment for each class of creditors.
3. A description of the means for execution of the agreement.

A debtor has an exclusive right to file a plan within the first 120 days of the case. If the debtor files a plan within this period, no one else may file a plan within the first 180 days after the debtor's filing. This period allows the debtor time to persuade the creditors to accept the plan. Once the plan has been developed, creditors must vote to accept the plan. For the plan to be accepted, two-thirds of the creditors of each class of creditors must vote to approve it. If the plan is approved, it will go before the court for confirmation. The court may decide to reject the plan if it is not in the best interests of the creditors. When the court confirms a plan, the debts not under the reorganization plan are discharged.

Collective Bargaining Agreements and Reorganization Through reorganization, a debtor can reject some contracts. In the early 1980s, legal scholars were concerned that debtors could use reorganization as a tool to avoid collective bargaining agreements. This issue became a serious concern when the Supreme Court held, in *National Labor Relations Board v. Bildisco & Bildisco,* that collective bargaining agreements are executory contracts and thus subject to rejection. In other words, a debtor was not required to engage in collective bargaining before rejecting part of the collective bargaining agreement through reorganization. However, in the opinion, the Court stated that rejection is not permitted unless the reorganization procedures would benefit through the rejection.

After the Supreme Court decision was handed down in 1984, Congress amended the code to prevent debtors from misusing Chapter 11 to reject collective bargaining agreements. The amendments set forth procedures under which collective bargaining agreements can be rejected or modified.

Chapter 13: Individual Repayment Plans

Chapter 13, "Adjustments of Debts for Individuals," permits individuals to pay their debts to creditors in installment plans under the supervision of the court. Repayment plans may seem similar to reorganization. Any debtor who files under Chapter 13 could also have filed under Chapter 11.

However, Chapter 13 repayment plans are usually simpler and less expensive. By statute, these plans last between 36 and 60 months.

Who is defined as a debtor for a Chapter 13 repayment plan? Only individuals are permitted to file under Chapter 13; partnerships and corporations are not eligible. Individuals must have a regular income and must owe less than $394,725 for fixed unsecured debts or $1,184,200 for fixed secured debts.

Filing the Petition As with the other forms of bankruptcy relief, the repayment process begins only when the debtor files the petition. However, repayment is distinct from other types of relief because repayment is voluntary only. A debtor cannot be forced into a repayment plan. In the petition, the debtor states that he is unable to pay his debts. The debtor might request an extension of time to pay the debts or ask the total amount of debt to be reduced.

Alternatively, the debtor might request a combination of those options. As in liquidation proceedings, the debtor commonly files various schedules listing the creditors as well as the debtor's assets. Furthermore, an automatic stay on litigation against the debtor is granted when the debtor files the petition. However, one of the benefits of Chapter 13 is that this stay applies to creditors' attempts to collect from co-debtors.

Creditors' Meeting and the Repayment Plan Proceedings After the debtor files the petition, the court calls a creditors' meeting. At this meeting, the debtor submits a plan of payment for her debts. This plan is not required to provide for the full payment of all claims. However, the plan must treat all same-class creditors equally, and the plan for repayment must not exceed a three-year period, unless the court approves otherwise.

Creditors do not vote to approve a Chapter 13 plan; if the court approves of the plan, the plan is accepted. The court holds a hearing to determine whether to confirm the plan. Unsecured creditors may object to the plan at the hearing. However, the court can overrule their objections if all of the debtor's projected disposable income is used to make payments. If the court believes that the plan was created in good faith, the plan is accepted.

However, if the plan modifies a creditor's secured claim while proposing that the debtor keep the property securing the claim, the creditor may oppose the proposed plan. Court approval of an opposed plan that changes a secured claim is called a "cram down." Since BAPCPA, it has been harder for debtors to make changes to secured claims over the objections of creditors.

After the court approves the plan, the court appoints a trustee to carry out the repayment plan. Every Chapter 13 case has a trustee. The debtor makes monthly payments to the trustee based on his or her disposable income. The trustee then disburses payments to the creditors. Moreover, as payment for services, the trustee receives a percentage of the funds distributed to creditors.

Thirty days after the debtor files the plan, the debtor must begin making payments. The trustee must ensure that the debtor is making the payments. If the debtor fails to make payments, the court may decide to transfer the case to a liquidation bankruptcy or to refuse to grant the repayment plan.

Discharge of Debts under Repayment Proceedings After a debtor makes all payments under a repayment plan, the remaining debts are discharged. However, even if all payments are not made by the expiration of the plan, the court might discharge some of the debts if the debtor has experienced a severe hardship. Like liquidation, some debts, such as alimony and child support debts, are not dischargeable.

Chapter 12: Family-Farmer and Family-Fisherman Plans

Chapter 12 provides for adjustment of the debts of family farmers or family fishermen. Chapter 12 was added to the code in the 1980s in the midst of severe debt among large numbers of farmers. Many farmers had borrowed large amounts of money to increase their production; however, during that time, there was a surplus of farm products. Consequently, many farmers faced serious financial problems. Congress addressed these problems with Chapter 12.

Not all farmers and fishermen are eligible for Chapter 12 relief. A family farmer or fisherman, under Chapter 12, must have regular annual income and be either (1) an individual or individual and spouse or (2) a corporation or partnership. A family farmer's or family fisherman's gross income must be at least 50 percent farm- or fishing-dependent. Moreover, 50 percent of the farmer's debt must be farm-related, and 80 percent of the fisherman's debt must be fishing-related. Finally, the total debt must be under $1.9 million for a family fisherman and $4.1 million for a family farmer.

What are the family-farm and family-fisherman relief proceedings? Congress modeled Chapter 12 after Chapter 13 relief. The procedure begins when the farmer or fisherman files a Chapter 12 petition. An automatic stay is granted to protect the farmer or fisherman from legal action. Within 90 days of the filing, the farmer or fisherman must file a plan. A trustee is appointed to oversee the financial affairs and is permitted to sell unnecessary assets. The court holds a hearing to rule on the proposed plan. Unsecured creditors are entitled to at least liquidation value of the debt owed to them. Once the farmer or fisherman fulfills his or her adjustment plan, the farmer or fisherman will likely receive a discharge from his or her debts and retain possession of the farm or fishery and related personal property.

CASE OPENER WRAP-UP

M&T Bank

The Delaware Court of Common Pleas held that, in order to receive a deficiency judgment on a secured loan, a secured creditor must establish that every aspect of the sale of the collateral was commercially reasonable. The burden of proof in this regard is entirely upon the secured creditor by a preponderance of the evidence. M&T Bank failed to meet its burden of proof. It was unfamiliar with the circumstances surrounding the auction of the car. The bank did not provide any evidence relating to the advertisement of the sale or how the sale was conducted. Furthermore, M&T Bank did not prove that the auction was an industry-approved method for disposing of repossessed vehicles. Therefore, it is not entitled to recover the deficiency balance on the loan.

PRACTICAL TIPS for Business Managers

1. Compliance with the technical requirements including filing rules and deadlines are imperative if a secured creditor wishes to preserve its position with respect to the collateral and in relation to other creditors. Failure to strictly follow the rules places a creditor in jeopardy of losing its secured position and ending up as an unsecured creditor.

2. It is imperative for secured creditors to conduct due diligence with respect to the debtor's business activities and status of the collateral. Sales and movement of the collateral may negatively affect an unsuspecting secured creditor's status.

3. Secured creditors must act in a reasonable manner when they seek to repossess collateral upon default. Creditors must keep in mind that the actions of those retained to conduct the actual repossession can and will be attributed to them and may result in unanticipated and unwelcome liability.

4. Laws relating to bankruptcy are extremely complex with many technical requirements including pre-petition credit counseling, document filing requirements, automatic stays, and notice provisions. Compliance is imperative from the perspective of debtors and creditors. Failure to strictly follow the rules places a debtor's ability to obtain a discharge or a creditor's ability to recover on its claim in jeopardy. Debtors and creditors should not navigate the intricacies of bankruptcy without at least the advice, if not representation, of legal counsel.

5. Truthfulness and full disclosure matter in bankruptcy. Bankruptcy is not the time for dishonesty, obfuscation or denial. Bankruptcy proceedings lay bare a debtor's financial position without exaggerations or misrepresentations. This may be discomforting or embarrassing for some debtors. However, it must be remembered that a discharge of debts is a privilege and not a right and may be denied to debtors who make less than complete and truthful financial disclosures.

6. Bankruptcy should be a last resort for debtors and creditors. Although the stigma of bankruptcy has been somewhat alleviated, it nevertheless has serious implications for a debtor's future financial dealings. Contrary to popular belief, debtors cannot declare bankruptcy, keep all of their assets, and discharge their debts. Creditors should try to avoid bankruptcy as well. Not only might creditors lose a portion of the value of their claims, but they are also subject to the rulings of a federal bankruptcy court with respect to repayment. Bankruptcy may be avoided in many situations by honest and open dialogue between creditors and debtors.

SUMMARY

Important Definitions Associated with Secured Transactions	*Secured interest:* An interest in personal property or fixtures that secures payment or performance of an obligation. *Secured party:* The person or party that holds the interest in the secured property. *Debtor:* The person or party that has an obligation to the secured party, the person who owns interest in the property. *Security agreement:* The agreement in which the debtor gives the secured interest to the secured party. *Collateral:* The property that is subject to the security interest.
Creation of Secured Interests	Attachment of a security interest requires these three elements: *Written agreement:* An agreement that describes the collateral and is signed by the debtor. *Value:* An item of value given from the creditor to the debtor. *Debtor rights in the collateral:* The rights of the debtor over the collateral.
Purchase-Money Security Interest	A *purchase-money security interest* is formed when a debtor uses borrowed money (e.g., buying on credit) from the secured party to buy the collateral.
Perfected Security Interest	A *perfected security interest* is a security interest in which the creditor has legally protected his or her claim to the collateral. Methods of perfection include: 1. *Perfection by filing:* Perfection of an interest by filing a financing statement with a state agency. • Place and duration of filing: Generally, the financial statements for consumer goods must be filed with the county clerk, and the statement is valid for five years. 2. *Perfection by possession:* Perfection of an interest by holding the collateral of the debtor until the loan is paid off. 3. *Automatic perfection:* Perfection that automatically occurs when a retailer sells a consumer good. 4. *Perfection of movable collateral:* Perfection of an interest in collateral moved to another state by reperfecting the interest after four months. 5. *Perfection of security interests in automobiles and boats:* Perfection of an interest in an automobile or boat by noting the interest on the certificate of title.
The Scope of a Security Interest	*After-acquired property:* A creditor has a security interest in property acquired by the debtor after the security agreement is made, if a clause to this effect is included in the agreement. *Proceeds:* A creditor automatically has rights to the proceeds from the sale of collateral for 10 days.
Termination of a Security Interest	*Termination statement:* An amendment to a financing statement, stating that the debtor has no obligation to the secured party.
Priority Disputes	A *priority dispute* occurs when two creditors claim rights to the same collateral. *Secured versus unsecured:* When an individual with a secured interest is disputing with an individual with an unsecured interest, the individual with the secured interest wins. *Secured versus secured:* When two individuals with secured interests are disputing, the individual who perfected his or her interest first wins. *PMSI conflicts:* If a party with a perfected purchase-money security interest disputes with another party, the PMSI party will almost always have a right to the collateral—regardless of when the agreement was perfected. *Secured party versus buyer:* If a debtor sells his collateral, the creditor may dispute with the buyer over the collateral. 1. *Buyers in the ordinary course of business:* If a person buys the collateral in the ordinary course of business without realizing that it is collateral, she has a right to the good. 2. *Buyers of consumer goods:* As long as the consumer does not know that the product is secured, the buyer's new product is free of any security interest. 3. *Buyers of chattel paper and instruments:* If the buyer purchases chattel paper and instruments, he is free from any security interest.

Default	*Default* occurs when a debtor fails to pay back her or his loan. Remedies include: 1. *Taking possession of the collateral:* If a debtor defaults on a loan, the secured party can take possession of the collateral. • *Disposition of the collateral:* The creditor may sell, lease, or transfer the collateral. • *Retention of the collateral:* The creditor may choose to keep the collateral as payment of the debt. 2. *Proceeding to judgment:* A secured party may sue the debtor for the entire amount of the debt instead of dealing with the collateral.
Bankruptcy and Reorganization	A *debtor* is an entity that owes money to another entity. Bankruptcy law provides various options for debtors to resolve their debts. Bankruptcy remedies are available to individuals, partnerships, and corporations.
The Bankruptcy Act and Its Goals	*Purpose:* 1. To provide debtors with an opportunity to achieve a fresh financial start. 2. To offer protection to creditors.
Bankruptcy Proceedings	1. Every case begins with the filing of a petition for bankruptcy. 2. The court grants an automatic stay for creditor actions against the debtor's estate. 3. The court determines whether an order of relief should be granted. 4. The creditors meet with the debtor. 5. A payment plan is created and approved, usually by the creditors and the court. 6. Debts remaining after the plan is carried out are usually discharged.
Specific Types of Relief Available	*Chapter 7:* Sale of nonexempt assets and the distribution of money to the creditors. *Chapter 11:* Reorganization of the debtor's financial affairs under supervision of the bankruptcy court. *Chapter 12:* Adjustment of family farmer's or fisherman's debts. *Chapter 13:* Adjustment of individual's debts.

Point/Counterpoint

Secured transactions give a creditor more assurance that it will be repaid even if the debtor is unable to pay. As you learned in this chapter, when a debtor has both secured and unsecured creditors, the secured creditors will be paid first. Some scholars have argued that secured credit unfairly transfers risk to unsecured creditors.

Should Secured Credit Be Limited?	
YES	**NO**
Suppose an unsecured creditor extended a loan to a debtor years before a secured creditor makes a loan. Later, a secured creditor extends credit to the debtor and gains an interest in the debtor's property. The debtor's financial situation has changed since the unsecured loan. Because of the secured creditor's interest, the likelihood that the unsecured creditor will be paid in full is now lower. The unsecured creditor's only response is to increase its interest rate; however, the unsecured creditor does not necessarily know about the secured creditor's interest in the collateral. Consequently, when the debtor is unable to pay, the unsecured creditor may get nothing. Moreover, the availability of secured credit may force a debtor to encumber her assets to obtain credit. By limiting the availability of secured credit, both unsecured creditors and debtors will be treated more fairly.	Debtors and creditors are rational actors who have to make choices. An unsecured creditor makes a choice in extending credit to a debtor. The unsecured creditor makes money on the basis of the interest it charges the debtor. Frequently, the interest rate on an unsecured loan is much higher than the interest rate on a secured loan. The unsecured creditor is compensated for the higher risk it may face. Moreover, a debtor has a choice of whether he wants to give the creditor an interest in his property. The debtor may choose a secured loan because of the lower interest rate. We should not restrict the debtor's choice. Finally, many people who receive secured credit are economically disadvantaged. Limiting the availability of secured credit would harm these parties' ability to access credit.

Questions & Problems

1. What are the required elements of a security interest?
2. Describe the various ways in which secured interests are perfected.
3. How does bankruptcy law benefit debtors and creditors?
4. Under what circumstances would Chapter 11 be used rather than Chapter 7?
5. Between July 2007 and July 2008, N&N Partners LLC executed three agreements to finance the purchase of commercial equipment. The collateral securing N&N's obligations under each agreement was the equipment being purchased. All three agreements were subsequently assigned to Colonial Pacific Leasing Corporation. N&N defaulted under the agreements by failing to make the payments as they came due in September 2009. The collateral was repossessed and sold in December 2009 and March 2010. Before each sale, N&N received notice and had an opportunity to redeem the collateral. After the sales, the proceeds were credited to N&N's account, but a debt remained. Colonial filed an action seeking a deficiency judgment in the amount of $101,094.14. N&N admitted its default but claimed that the sales of the collateral were not commercially reasonable due to technical flaws in the notices of sale, the fact that the sales occurred too early in the morning to allow for full public participation, and that Colonial failed to appraise the equipment before the sales, thereby preventing it from proving the fair market value of the equipment prior to the sales. Colonial alleged that the sales were commercially reasonable as N&N received timely notice; the sales occurred at a time, place, and in a manner typical for this type of collateral; and the proceeds equaled or even exceeded the collateral's fair market value. Should technical errors in the notices of sale render the sales commercially unreasonable? Should sales occur later in the day to maximize participation? Should the creditor have to provide an appraisal of all collateral sold in order to satisfy the commercially reasonable sale requirement?

 Source: *Colonial Pac. Leasing Corp. v. N&N Partners, LLC,* 981 F. Supp.2d 1345 (N.D. Ga. 2013).

6. In January 2005, Joseph Palmer executed a retail installment contract and security agreement with Suburban Tractor Company to finance the purchase of a compact utility tractor and front loader. Palmer stated in the loan documentation that the equipment would be used primarily for personal, family, or household purposes. Suburban subsequently assigned the contract to Deere & Company. Neither Deere nor Suburban filed a financing statement with the Ohio Secretary of State with respect to the lien claimed on the equipment. Palmer filed for bankruptcy and, in the creditors' meeting, testified that he purchased the equipment primarily for use in his greyhound training business, as it was needed to haul supplies and maintain the kennels on his property. Based upon this testimony, the bankruptcy trustee claimed that the security interest was not properly perfected and was subordinate to a hypothetical lien creditor as represented by the estate. Deere contended that it was entitled to rely on Palmer's representations in the loan documentation that the equipment would be used primarily for personal, family, or household purposes and, because the equipment constituted consumer goods, the filing of a financing statement was not required in order to perfect its purchase-money security interest. Was Deere entitled to rely on Palmer's representation in the loan documents that the equipment he financed was going to be used for personal rather than commercial purposes? Should Palmer's actual use of the equipment carry more evidentiary weight than his statements in the contract and security agreement? Why or why not?

 Source: *In re Palmer,* 365 B.R. 816 (S.D. Ohio 2007).

7. In 2007, Indiana Golf and Sports Turf, LLC ("Indiana Golf"), entered into two separate purchase agreements with Deere Credit, Inc. ("Deere") for five items of golf course maintenance equipment. One agreement was for $54,745 and the second agreement was for $16,975. Each agreement contained an individual guaranty by signed by Timothy D. Spitler ("Spitler"). In 2009, Indiana Golf failed to make payments pursuant to the agreements, Deere repossessed and sold the equipment. Deere sold the equipment on an on-line auction site only accessible to authorized Deere dealers and eighty used equipment brokers. Other types of bidders were excluded from this private sale. Deere did not employ an outside appraiser to value the collateral before resale but did provide Spitler with notice of the sale and had the collateral inspected by an independent third party. Deere subsequently sued Spitler for $31,946.05, the amount of the deficiency following the sale of the equipment. Spitler contended that the sale was commercially unreasonable. Did the on-line auction format and the limitation upon the number of participants render the sale commercially unreasonable? Why or why not?

 Source: *Deere Credit, Inc. v. Spitler,* 2014 WL 1343025 (Ohio Ct. App. 2014).

8. Joe and Maria Borges filed individual Chapter 11 voluntary petitions in bankruptcy in June 2010. The Borges did not complete the required credit counseling prior to filing their petitions but instead completed the counseling three days after their filings. The court issued a show cause order requiring the Borges to demonstrate why their petitions should not be dismissed. Joe suffered from numerous medical maladies including the aftereffects of a stroke, mild dementia, manic depression, and an episode of diabetic shock one month before his bankruptcy filing. Maria claimed that the need to care for Joe and run the family dairy business prevented her from completing the counseling in a timely manner. How should the court decide this case? Should it dismiss Joe and Maria's petitions or one but not the other?

 Source: *In re Borges,* 440 B.R. 551 (Bankr. D.N.M. 2010).

9. Fordyce Bank & Trust Company (Bank) issued several loans to Bean Timberland (Bean) so that Bean could purchase timber from various landowners. Bean gave the bank security interests in the purchased timber, and the proceeds from the sale of the timber were intended to repay the loans the Bank had made to Bean. The Bank, intending to perfect its security interests, filed financing statements with the Arkansas Secretary of State's Office. However, when Bean sold the timber to the various lumber mills with which it did business, Bean failed to remit the sales proceeds to the Bank. The Bank filed suit against these purchasers, alleging that the purchasers had "negligently entered into contracts" with Bean for the purchase of timber and had "failed to exercise good faith" in those transactions. The Bank further contended that the purchasers had been negligent in failing to request a lien search of the UCC records of the Arkansas Secretary of State's Office. If the purchasers had conducted a lien search, they would have discovered the Bank's financing statement and security agreement. The trial court granted the purchasers' motion for a directed verdict because they were purchasers in the ordinary course of business and were not required to perform a lien search on the timber purchased from Bean. The Bank filed an appeal. How would you decide this case on appeal? Do businesses have a duty to conduct lien searches with respect to those firms from whom they purchase goods? Should they have this duty? Why or why not?

 Source: *Fordyce Bank & Trust Co. v. Bean Timber-land, Inc.,* 251 S.W.3d 267 (Ark. 2007).

10. Simon and Clayton became involved in a fistfight outside of a diner in Lafayette, Louisiana, in October 2010. Simon claimed Clayton disrespected him by speaking with Simon's girlfriend who was seated in Simon's pickup truck while Simon was in the diner. Simon punched Clayton and then attempted to leave the parking lot. However, Clayton removed the keys from Simon's truck in order to keep him from leaving before the police arrived. Simon responded by striking Clayton, knocking him to ground and then repeatedly punching him. Clayton was hospitalized as a result of his injuries and incurred more than $14,000 in medical expenses. Clayton sued Simon in Louisiana state court and received a default judgment. Simon subsequently filed a Chapter 7 bankruptcy petition seeking to discharge Clayton's judgment. Clayton objected and claimed the damages award was non-dischargeable as his injuries resulted from Simon's willful and malicious acts. Simon claimed Clayton provoked the fight, he did not intend to harm Clayton but was only attempting to retrieve his keys and defend himself, and Clayton consented to the battery as a willing participant in the fight. Was Simon's conduct willful and malicious such as to prevent the discharge of the debt owed to Clayton? What effect should state tort law doctrines have on the outcome of bankruptcy proceedings?

 Source: *In re Simon, 2017 WL 4118284* (Bankr. W.D. La. Aug. 3, 2017).

11. Vogeler filed for Chapter 7 bankruptcy in March 2007. Vogeler's debts consisted of $11,000 in secured debt represented by his automobile and $35,925 in unsecured debt. Vogeler's assets on the petition date totaled approximately $2,785.00, all of which he claimed were exempt. Vogeler was unemployed at the time of filing and reported past earnings of $18,000 annually. Vogeler also reported monthly expenses of $1,617. In April 2007, Vogeler won $130,000.00 from the Kansas Lottery from which he received $90,692.89, after taxes. The bankruptcy trustee instructed Vogeler not to spend the lottery proceeds, but Vogeler subsequently spent the entire amount on new vehicles and non-emergency personal expenses. Vogeler continued his bankruptcy proceeding and requested a discharge of his pre-petition debts. The trustee sought to dismiss the bankruptcy petition on the basis that its continuation was in bad faith given the totality of the circumstances. Should the court grant the trustee's motion to dismiss Vogeler's bankruptcy petition? Does it make a difference that the lottery proceeds were post-petition and not part of the bankruptcy estate? Why or why not?

 Source: *In re Vogeler,* 393 B.R. 240 (Bankr. D. Kan. 2008).

12. Carlos and Suzanne Cortez filed their petition for Chapter 7 bankruptcy on April 8, 2004. At the time of filing, Suzanne Cortez was working, but Carlos was unemployed. Suzanne had reduced her work hours so

that her minimum net income per month was $750. Carlos accepted a job offer a few days after the filing of the bankruptcy petition that netted him $5,986 monthly. The Cortezes' new combined net income was $6,646 per month, which exceeded their expenses by $1,325 per month. The trustee appointed to the Cortezes' case moved to dismiss their petition because it appeared that the debtors had the means to repay a substantial portion of their debts through a Chapter 13 plan. The bankruptcy court denied this motion and ruled that it was limited to reviewing the Cortezes' financial circumstances as they existed when the petition was filed and that postpetition events could not be taken into account in determining the proper chapter under which the petition should have been filed. The trustee appealed the decision to the U.S. district court. How should this case be decided on appeal? Should postpetition events, especially those occurring within a few days of the filing, be relevant to the determination of whether a petition was properly filed? Why or why not?

Source: *In re Cortez,* 2005 U.S. Dist. LEXIS 39778 (N.D. Tex. March 9, 2005).

13. Nancy Polleys was a single mother living in Wyoming. She obtained federally guaranteed student loans totaling $51,000 in order to attend college. She ultimately obtained an accounting degree and worked for a short time in the field before losing her job. Polleys subsequently spent several years working a number of jobs that paid as much as $16,000 to as little as $3,000 annually. As a result, Polleys became unable to repay her student loans, repayment of which would have required $420 per month for over 20 years. Polleys sought to discharge her student loan indebtedness in bankruptcy on the basis of undue hardship. At the time of the filing of her petition, Polleys lived in rental property owned by her parents and paid no rent or utilities. Polleys owned a 1993 Subaru with significant body damage but little other property. Polleys qualified for food stamps, and her income was below federal poverty guidelines. Polleys had no health insurance. Although in good physical health, Polleys suffered from cyclothymic disorder, for which she was undergoing mental health treatment including the use of antidepressant medication. Her mental condition resulted in a suicide attempt and involuntarily commitment on one occasion. Educational Credit Management Corporation, which was charged by the U.S. Department of Education with collecting delinquent federally guaranteed student loans, objected to the discharge. The basis for the objection was Polleys' failure to prove that she could not maintain a minimal standard of living for herself and her daughter, that her current financial difficulties were most likely to continue for a significant amount of time, and the absence of a good faith effort to repay the loans. Should the court grant Polleys a discharge on the basis of undue hardship? What factors are relevant to making a hardship determination?

Source: *Educational Credit Management Corp. v. Polleys,* 356 F.3d 1302 (10th Cir. 2004).

14. Andres Soza and his spouse transferred $30,000 into an annuity fund on October 13, 2005. The Sozas filed a Chapter 7 petition for bankruptcy the following day. The Sozas claimed the annuity was exempt from inclusion in the bankruptcy estate. The money to purchase the annuity came from an inheritance. The Sozas purchased the annuity to safeguard the inheritance from creditors pending a determination of how the proceeds were to be distributed to Soza and his siblings. The Sozas further claimed that the annuity was exempt pursuant to provisions of the Texas Insurance Code. The trustee objected to the exemption claim and alleged that the conversion of nonexempt property into an exempt annuity on the eve of bankruptcy amounted to fraud detrimental to the creditors, expressly prohibited by the Code. The court was required to determine whether the Code prevented actual as well as constructive fraud by a debtor to the detriment of a creditor. The court focused its attention on the presence or absence of several badges of fraud. What is a badge of fraud? What type of conduct should be considered indicative of fraud in the context of bankruptcy proceedings?

Source: *In re Soza,* 542 F.3d 1060 (5th Cir. 2008).

15. Woodcock graduated from law school and finished his MBA in 1983. His student loans came due nine months later. Because he was a part-time student until 1990, he requested that payment be deferred, which the lender incorrectly approved. Because he was not in a degree program, payment should not have been deferred under the terms of the loan. Woodcock filed for bankruptcy in 1992, more than seven years after the loans first became due. Hence, that debt would be discharged unless there was an applicable suspension of the repayment period. Do you feel this mistaken extension is an applicable suspension? Should his student loans be discharged through filing for bankruptcy?

Source: *Woodcock v. Chemical Bank,* 144 F.3d 1340 (10th Cir. 1998).

16. Egebjerg filed a voluntary Chapter 7 bankruptcy petition in 2006. He had been employed for 27 years, earned a gross income of $6,115.56 per month, and had unsecured consumer debt of around $31,000.

About two years earlier, he took a loan from his 401(k), and he paid back this loan through automatic deductions to his paycheck in the amount of $733.90 per pay period. Egebjerg listed the 401(k) loan repayment in his bankruptcy petition as a necessary expense, leaving him with just $15.31 of disposal income per month. The U.S. trustee moved to dismiss his petition as presumptively abusive because the 401(k) loan repayment was not a necessary expense; thus, his filing failed the means test. Do you think the court agreed that the 401(k) loan repayment was a necessary expense and thus should be calculated in the debtor's monthly ability to pay? Why or why not?

Source: *In re Egebjerg*, 2009 WL 2357706 (9th Cir. 2009).

Part 6 *Agency*

CHAPTER 20

Agency and Liability to Third Parties

CASE OPENER

FedEx and Independent Contractors

In July 2006, the International Brotherhood of Teamsters, Local Union 25, filed two petitions and held an election for a collective bargaining representative at two FedEx locations. However, FedEx refused to bargain with the union. FedEx did not contest the vote count of the election; instead, FedEx refused to bargain collectively because it believed that its single-route drivers were not employees but, rather, independent contractors, and the rules of collective bargaining in this situation apply only to employees. However, the National Labor Relations Board concluded that FedEx committed an unfair labor practice by refusing to bargain with the union certified as the collective bargaining representative of the drivers. FedEx sought judicial review of the decision of the board, and the board cross-applied for enforcement of its order requiring the company to bargain.

1. Do you agree with the National Labor Relations Board's ruling that FedEx engaged in an unfair labor practice? Why?
2. What duties and rights would the FedEx drivers have if they were considered employees? How would their duties and rights differ if the drivers were considered independent contractors of the company?

LEARNING OBJECTIVES

After reading this chapter, you will be able to answer the following questions:

LO 20-1 Describe how an agency relationship is created.
LO 20-2 Recognize the different types of agency.
LO 20-3 Identify the different types of agency relationships.
LO 20-4 Describe the duties of the agent.
LO 20-5 Describe the duties of the principal.
LO 20-6 Identify the different types of authority of the agent.
LO 20-7 Explain the contractual liability between the principal and the agent.
LO 20-8 Outline the ways an agency relationship can be terminated.

In the Case Opener, it is important to determine whether single-route drivers are employees or independent contractors of FedEx because different legal obligations arise, depending on the nature of the relationship. One of the most important relationships in the business world is the *agency relationship,* in which the employee may be an agent of the employer, having the ability, among others, to bind that employer legally.

Agency relationships are crucial to running large businesses because the use of agents allows corporations to enter into contracts and therefore conduct business in multiple locations simultaneously. This chapter explores the nature and creation of the agency relationship as well as the legal obligations that the parties take on when they form such a relationship.

Introduction to Agency Law

agency relationship
The association between one party and an agent who acts on behalf of that party.

agent
A party who has the authority to act on behalf of and bind another party.

principal
The party that an agent's authority can bind or act on behalf of.

agency
The fiduciary relationship that arises when one person consents to have another act on his behalf and subject to his control and the other consents to do so.

fiduciary
A person who has a duty to act primarily for another person's benefit.

Agency is generally defined as a relationship between a principal and an agent. In an **agency relationship**, the **agent** is authorized to act for and on behalf of the **principal**, who hires the agent to represent him or her. The Restatement of Agency defines **agency** as "the fiduciary relation that results from the manifestation of consent by one person to another that the other shall act in his behalf and subject to his control, and consent by the other so to act."[1] (A **fiduciary** is a person who has a duty to act primarily for another person's benefit. A lawyer, for example, is a fiduciary for his or her client. We discuss fiduciaries in depth later in this chapter.)

Agency law is primarily state law. Thus, it can vary somewhat from state to state. As of July 2010, 40 states had enacted statutes governing the behavior of sports agents. The specific legislation that regulates the conduct of athlete agents in these 40 states is the Uniform Athlete Agents Act (UAAA). Non-UAAA laws also regulate athlete agents in California, Michigan, and Ohio. The UAAA provides for criminal, civil, and/or administrative penalties based on the particular state's statute. Twenty-three of those states impose civil penalties or damages on the agent for violations of the athlete agent statutes. Twenty-four states (including California and Florida) have established criminal penalties for sports agents who violate the state statute. In contrast, only five states have criminalized violations of the statute by the athletes themselves. In addition to state laws protecting athletes and agents, players associations's model contracts describe the nature of the services agents can perform on behalf of their principals and the duties the parties owe to one another. For example, the Major League Baseball Players Association's (MLBPA's) "Regulations Governing Player Agents" expressly states that agents act in a fiduciary capacity vis-à-vis their athlete clients.

Agency law is especially important for modern firms doing business in foreign countries. Although foreign countries offer fresh markets and eager consumers, American companies often run into legal difficulties due to language barriers or lack of knowledge about local laws. To avoid such problems, many companies hire agents familiar with local laws, customs, and customers to help them function smoothly in foreign markets.

Creation of the Agency Relationship

LO 20-1 Describe how an agency relationship is created.

Agency relationships are consensual relationships formed by informal oral agreements or formal written contracts. However, an agency relationship exists only when the principal takes action to ask another individual to act on behalf of the principal.

Under what circumstances can an agency relationship be created? First, an agency relationship can be created only for a lawful purpose in the same way that a contract cannot have an illegal purpose. Thus, a principal could not hire an agent to kill someone on behalf of the principal.[2] Second, almost anyone can act as an agent. A contract made by an agent is viewed as a contract

[1] Restatement (Second) of Agency, sec. 1(1). The Restatement is a valuable reference that is well respected in the legal profession. It is frequently cited by judges as well as attorneys and scholars in making legal arguments. The Restatement provides a summary of agency law.

[2] Restatement (Second) of Agency, sec. 19.

Exhibit 20-1
Creation of Agency Relationships

Expressed agency	Formed by making a written or oral agreement
Implied authority	Formed by implication through the conduct of the parties
Agency by estoppel	Formed when a principal leads a third party to believe that another individual serves as his or her agent but the principal had made no agreement with the so-called agent
Agency by ratification	Exists when an individual misrepresents himself or herself as an agent for another party and the principal accepts or ratifies the unauthorized act

between the principal and a third party. Therefore, a person can serve as an agent regardless of age or competency.[3] However, if an individual does not have contractual capacity, he or she could not hire an agent to make contracts on his or her behalf. For example, because a contract entered into by a minor is voidable, most states do not allow minors to hire agents.

LO 20-2 Recognize the different types of agency.

As long as these two criteria are met, agency relationships can be created on the basis of any of the following four forms of authority:

1. By expressed agency, or agency by agreement.
2. By implied authority.
3. By apparent agency, or agency by estoppel.
4. By ratification.

The following sections discuss the types of agency formation, and Exhibit 20-1 presents a summary.

Expressed Agency (Agency By Agreement)

When parties form an agency relationship by making a written or oral agreement, the agency is known as **expressed agency**, or agency by agreement. Expressed agency is the most common type of agency. As specified in the expressed agreement, the agent has the authority to contract on behalf of the principal. Although a contract is not necessary, if an agency agreement is formed through a contract, the contract must meet all the elements of a contract discussed in Chapter 9. If a principal agrees to hire no other agent for a period of time or until a particular job is done, the principal and agent have entered into an exclusive agency contract.

One legal document that establishes agency relationships is a **power of attorney**, a document that gives an agent authority to sign legal documents on behalf of the principal. Power of attorney can be classified in several ways. First, a power of attorney can be general or specific. For example, with a general power of attorney, a principal gives an agent broad authority to sign legal documents on behalf of the principal. In contrast, with a specific power of attorney, a principal gives authority to an agent for only the specific areas or purposes listed in the agreement.

Powers of attorney are often given for business and health care purposes. In the health care instance, a principal gives power of attorney to an agent so that the agent can make decisions about the principal's medical care in case the principal cannot make those decisions. Given that a principal must have the ability to enter into contracts to enter into an agency relationship, a principal may not enact a power of attorney after becoming incompetent. Therefore, before incapacitation, a principal may create a durable power of attorney.

A **durable power of attorney** is a written document, created by a principal, expressing his or her wishes for an agent's authority to be unaffected by the principal's subsequent incapacity. Alternatively, a durable power of attorney might become active only after a principal becomes incapacitated in any manner. Because the agreement is entered into before incapacitation, a power of attorney is legally binding after the principal becomes incapacitated.

expressed agency
An agency created in a written or oral agreement. Also called *agency by agreement*.

power of attorney
A specific form of express authority that grants an agent specific powers.

durable power of attorney
A document that specifies that an agent's authority is intended to continue beyond the principal's incapacitation.

[3] Restatement (Second) of Agency, sec. 21.

CASE Nugget

Durable Power of Attorney

Penny Garrison et al. v. The Superior Court of Los Angeles et al.

132 Cal. App. 4th 253 (2005)

On Ella Needham's request, her daughter, Penny Garrison, was designated Needham's agent through a durable power of attorney. Needham was later admitted to a residential care facility. As part of the admissions process, Garrison, acting under the durable power of attorney, executed two arbitration agreements. After Needham's death, Garrison and Needham's other daughters sought to sue the facility for a number of concerns the family had regarding the care their mother had received. The facility sought to enforce the two arbitration agreements. However, the family contended that the agreements were unenforceable because Garrison could not legally enter into them.

The durable power of attorney had given Garrison power to (1) make all health care decisions for Needham according to what she believed was in Needham's best interest and (2) make decisions relating to Needham's personal care, including but not limited to determining where she lived. Therefore, Garrison was legally in charge of picking the residential care facility, and she had the power to enter into agreements regarding Needham's care.

Nowhere in the enumerated legal powers did the durable power of attorney state that Garrison could not enter into arbitration clauses. Moreover, the arbitration clauses were optional to the original contract, and they allowed a 30-day period during which Garrison could cancel them. Because the durable power of attorney was legal and enforceable, Garrison could not cancel the agreements into which she entered.

It is important to note a few additional points. First, the Restatement states, "[A]n agency relation exists only if there has been a manifestation by the principal to the agent that the agent may act on his account, and consent by the agent so to act."[4] Therefore, in addition to the preceding criteria, there needs to be agreement by the principal to have the person act as an agent and by the agent to act for the principal. As noted shortly, this agreement can be reached in several ways.

Second, agency agreements usually do not need to be in writing. However, there are two important exceptions. First, agency agreements need to be in writing whenever an agreement is reached for an agent to enter into a contract that under the statute of frauds must be in writing. For example, Janet wants Phil to act as her agent. She grants him the power to enter into contracts. The statute of frauds, or, more specifically, the equal dignities rule, mandates that the type of contracts Phil is allowed to enter into must be in writing. Therefore, Phil's agreement with Janet must also be in writing. Also, whenever an agent is given power of attorney (discussed later), the agreement must be in writing.

Third, agents may perform their services gratuitously. A **gratuitous agent** is one who acts without consideration; that is, the agent is not paid for his or her services. Gratuitous agents function much like regular agents, with a few specific exceptions noted later in this chapter.

gratuitous agent An agent who acts without being paid for his or her services.

Agency By Implied Authority

In some cases, an agency relationship is not created by an expressed agreement but, rather, is implied by the conduct of the parties. The circumstances of a situation determine the extent of an agent's ability to conduct business on behalf of the principal. However, the implied authority cannot conflict with any express authority.

[4] Restatement (Second) of Agency, sec. 15.

GLOBAL Context

Formation of Power of Attorney in Luxembourg

In the United States and other common-law jurisdictions, a power of attorney authorizes the agent only to "conduct a series of transactions" under instruction from the principal. This limitation makes the power of attorney in common law distinctly different from that provided by civil law, which authorizes the agent to do "everything and anything which the principal himself could do."

Under the law of the Grand Duchy of Luxembourg, the person for whom the power of attorney is acting on behalf is called a donor instead of a principal. Under this Luxembourg law, there are two types of power of attorney from which a donor must choose. The first type is a power of attorney that is valid until death, and the second type is a power of attorney that has unlimited validity, even after death. The actions that are permitted to be performed by a power of attorney are completely determined by the law in Luxembourg.

The law in Luxembourg also requires the power of attorney to be authenticated by a public authority or a public notary by providing a certified identification card or passport as well as proof of residency.

Apparent Agency (Agency By Estoppel)

Suppose a principal leads a third party to believe that another individual serves as his or her agent but the principal has made no agreement with the so-called agent. Does an agency relationship exist? Yes, because by his or her conduct, the principal has created **apparent agency,** or agency by estoppel. According to the principal's conduct, the agent has apparent authority to act; thus, the principal is estopped, or prevented, from denying that the individual is an agent.[5] When a third party relies on the principal's conduct and makes an agreement with an apparent agent, the principal must uphold any agreements made by the agent. If the principal attempts to deny an agency relationship existed, the third party must demonstrate that he or she reasonably believed, on the basis of the principal's conduct, that an agency relationship existed. The court will consider the principal's conduct in determining whether an agency relationship existed.

apparent agency
An agency relationship created by operation of law when one party, by his actions, causes a third party to believe someone is his agent when that person actually has no authority. Also called *agency by estoppel.*

Used by permission of Stu Rees, Stu's Views.

[5] Restatement (Second) of Agency, sec. 8B.

Suppose an executive of Emporium Carpet is interviewing a potential salesperson named Jose. In the course of the interview, the employer introduces Jose to some clients who are in the building. During the conversation, he says, "I cannot wait to have Jose in our firm. The sooner he works for us, the happier I will be." For some reason, Jose is not hired. But Jose happens to meet these clients later, and in the course of their conversations, the clients place an order with Jose, believing him to be an employee of Emporium Carpet. Emporium now has an obligation to fulfill the order because the third party reasonably believed Jose was an agent of Emporium. In Case 20-1 below, pay close attention to how the court determines whether the party had apparent authority in this agency issue.

CASE 20-1

ACKERMAN v. SOBOL FAMILY PARTNERSHIP, LLP
Supreme Court of Connecticut
298 Conn. 495 (2010)

FACTS: During a two-day period in June 2008, Coe expressly assured the defendants' attorneys on separate occasions, in response to direct questioning on the issue, that the settlement offer proposed by him at that time was fully authorized by his clients as well as the additional plaintiff's attorney, and that if accepted by the defendants, it would resolve the litigation.

The final settlement demand by the plaintiffs' counsel to the Bank of America was the sum of $1.1 million. The Bank of America accepted the $1.1 million settlement proposal on July 1, 2008, prior to the 5 p.m. deadline. At this time, the global settlement offer had been accepted by all defendants. At no time prior to the acceptance of the settlement proposal were the defendants or their attorneys notified that the offer had been withdrawn, unauthorized, or otherwise ineffective.

After the proposal had been accepted by the defendants, the plaintiffs sued the defendants, claiming that the plaintiffs' lead attorney, Glenn Coe, lacked the authority to make several settlement proposals or bind them to a global settlement agreement with Bank of America and the other defendants. The defendants filed motions to enforce the settlement agreement, and the superior court granted the defendants' motions and rendered judgments for defendants. The plaintiffs appealed.

ISSUE: Did Coe have the apparent authority to settle with Bank of America on behalf of the plaintiffs?

REASONING: To determine whether Coe had apparent authority, the court used the two-part test developed in *Tomlinson v. Board of Education*. This test holds that apparent authority exists if (1) a principal has held out an agent as possessing the authority to undertake the act in question and therefore (2) the person dealing with the agent acting in good faith reasonably believed that the agent had the necessary authority.

The court found that Coe had accompanied the plaintiffs to mediation proceedings long before the settlement was reached and that settlement discussions took place during these mediations. It was clear that these settlement discussions took place with the obvious assent of the plaintiffs. Thus, Coe was certainly held out as having the authority to negotiate a settlement on behalf of the plaintiffs. Given these numerous negotiations involving the defendants, the plaintiffs, and Coe, it was reasonable for the defendants to believe that Coe had the authority to settle on behalf of the plaintiffs.

DECISION: Affirmed.

SIGNIFICANCE: When one party makes another party their agent, that agent has the authority to make binding agreements for the principal unless that power is clearly denied by the principal. This case serves as a cautionary example that an agent can deals a principal does not agree with even if the agent is acting in good faith.

CRITICAL THINKING

Is any important information missing from this decision that might further clarify the nature of the issue of agency between the concerned parties? Could it change the acceptability of the judge's reasoning?

ETHICAL DECISION MAKING

Does this ruling appear to follow a coherent ethical guideline? If so, what form does it take? Who are the stakeholders in this situation? Are they awarded proper consideration under the selected ethical guideline?

Agency By Ratification

Suppose Fred is driving home and sees a car with a For Sale sign in the window. He stops to look at the car because his friend wants to buy a used car. He's amazed at the price and quality of the car, so he tells the owner his friend wants to buy the car. The owner claims that another individual is coming who probably will buy the car in an hour. So, to ensure that his friend gets this car, Fred signs a contract to purchase the car, but he notes on the contract that he is an agent of his friend. Because he is not an agent for his friend, the friend is not required to uphold the contract.

However, if Fred's friend agrees to purchase the car, the friend has accepted him as the agent for the contract. Fred's friend is now bound by the contract, and Fred cannot be held liable for misrepresenting himself as the agent. This type of agency relationship is **agency by ratification**. As suggested in the preceding example, there are two requirements for agency by ratification:

1. An individual must misrepresent himself or herself as an agent for another party.
2. The principal accepts or ratifies the unauthorized act.

For ratification to be effective, two additional requirements must be met:

1. The principal must have complete knowledge of all material facts regarding the contract.
2. The principal must ratify the entirety of the agent's act. (The principal cannot accept certain parts and reject other parts of the agent's act.)

agency by ratification An agency relationship that arises when an individual misrepresents himself or herself as an agent for another party and that party accepts the unauthorized act.

Recent Developments in Agent/Employer Law

The laws that determine the status between an agent and an employer are constantly changing at both the state and federal level. In terms of federal law, one of the most significant potential developments comes in the form of proposed changes to the Fair Labor Standards Act. The new rules intend to increase the minimum salary required for overtime exemption for administrative and professional employees from $23,660 a year to $47,476, effectively increasing the number of employees eligible for overtime compensation nationwide. The proposed rule faced a large amount of pushback from the business community, which, in general, argued that being forced to disburse more overtime pay would lead to increased layoffs. The new rule was blocked by a federal judge in November 2016, but this injunction only served to suspend the rule; a new, modified version may very well appear sometime in 2017.

Another federal-level development involves instituting stricter measures on the classification of employees. With the rise of the "gig economy" (an economy where short-term contracts and freelance work prevail) and high-profile lawsuits against companies such as Uber and Amazon, the divisions between employees, independent contractors, temporary workers, and owner-operators have become increasingly blurry. Thus, the Department of Labor, National Labor Relations Board, and Equal Employment Opportunity Commission have each developed joint employer doctrines for companies that rely on third-party contractors that no longer safeguard these businesses from employment-related liability issues arising from third-party contractors.

Agency Relationships

LO 20-3 Identify the different types of agency relationships.

There are generally three types of business relationships to which agency laws are relevant: the principal–agent relationship, the employer–employee relationship, and the employer–independent contractor relationship. We discuss all three in the following sections and summarize them in Exhibit 20-2.

Principal-Agent Relationship

The principal–agent relationship typically exists when an employer hires an employee to enter into contracts on behalf of the employer. This relationship is the most basic type of agency relationship. For example, suppose a salesclerk at Abercrombie & Fitch sells Amanda a shirt. The clerk is acting on behalf of the owner; consequently, any sales the salesclerk makes are binding on the

Exhibit 20-2

Types of Business Relationships and Their Significance for Agency

RELATIONSHIP	HOW TO IDENTIFY	SIGNIFICANT FOR WHAT ISSUES?
Principal–agent	Parties have *agreed* that agent will have power to bind principal in contract.	Contract law
Employer–employee	Employer has right to *control* conduct of employees.	Tort law, tax law, wage law, discrimination law, copyright law
Employer–contractor	Employer has *no control* over details of conduct of independent contractor.	Tort law, tax law, wage law, discrimination law, copyright law

owner of Abercrombie & Fitch. Similarly, think of all the advertisements seen on television in which a professional athlete speaks on behalf of a product. The athlete usually hires an agent to find and make agreements on behalf of the athlete to promote products.

Employer-Employee Relationship

Whenever an employer hires an employee to perform some sort of physical service, the parties have created an employer–employee relationship. The employee is subject to the control of the employer.[6] Generally, all employees are considered agents of the employer. Even employees not legally authorized to enter into contracts binding their employer or to interact with third parties are considered agents. Although all employees are agents, the reverse is not true. That is, not all agents are employees.

Employer-Independent Contractor Relationship

independent contractor
A person who is hired to perform a task for another but who is not the other's employee.

Employers often hire **independent contractors,** persons who are not employees, to conduct certain tasks. The Restatement of Agency defines an independent contractor as "a person who contracts with another to do something for him but who is not controlled by the other nor subject to the other's right to control with respect to his physical conduct in the performance of the undertaking."[7] For example, building contractors, doctors, stockbrokers, and lawyers are independent contractors. Furthermore, building contractors, doctors, stockbrokers, and lawyers are agents but not employees. However, not all independent contractors are agents. Independent contractors cannot enter into contracts on behalf of the principal unless authorized to do so by the principal.

Employee or Independent Contractor? The question of whether a worker is an employee or an independent contractor has important implications in terms of workers's compensation, workplace safety, and unemployment statutes. The employer-employee relationship is subject to the workers's compensation, workplace safety, employment discrimination, and unemployment statutes, whereas the employer-independent contractor relationship is not. Furthermore, as discussed later in this chapter, employers are generally liable in tort for the actions of their employees, whereas they are generally not liable for the actions of independent contractors.

BUT WHAT IF . . .

WHAT IF THE FACTS OF THE CASE OPENER WERE DIFFERENT?

Let's say in the Case Opener that a single-route driver named Rick was an independent contractor for FedEx. He delivered a package to a woman named Sarah, but Sarah's package had been damaged at a FedEx center. Rick formed a contract for Sarah, telling her that FedEx would award her the cost of the item that was shipped. Is Rick legally correct in this scenario? Why or why not?

[6] Restatement (Second) of Agency, sec. 2.
[7] Restatement (Second) of Agency, sec. 2.

When courts are deciding whether a worker is an employee or an independent contractor, perhaps the most important issue they consider is how much control the employer exerts over the agent.[8] If the employer has substantial control over day-to-day operations of the worker, the worker is generally considered an employee.

Case 20-2 examines the criteria that determine whether a worker is an employee or an independent contractor.

[8] Restatement (Second) of Agency, sec. 2(3).

DOE v. UBER TECHNOLOGIES, INC.
United States District Court for the Northern District of California
184 F. Supp. 3D 774 (2016)

FACTS: This tort suit was brought by Jane Doe 1 and Jane Doe 2 against Uber Technologies, Inc. stemming from separate alleged sexual assault incidents.

In February 2015, in Boston, Massachusetts, Doe 1 and her friends used the Uber app to get a ride after they had gone to a party. An Uber driver named Abderrahim Dakiri picked up Doe 1 and her party. Dakiri dropped off Doe 1's friends first. With Doe 1 the sole remaining customer in the Uber, Dakiri subsequently began to sexually assault her, taking a circuitous route "in order to increase his opportunity to sexually assault her," and parking the car in a remote area.

In August 2015, in Charleston, South Carolina, Doe 2 and her friends arranged a ride using Uber's app. Uber driver Patrick Aiello drove Doe 2 and her group to a bar and someone in the group asked Aiello if he would agree to pick them up later. Aiello eventually drove Doe 2 and her friends to one of the group's apartment. While still in the Uber driver's car, Doe 2 mentioned she could not find her phone and wanted to search for it in the apartment. Doe 2 searched for the phone and then left, intending to walk back to her own apartment. When Doe 2 went outside, Aiello was waiting and offered to drive her home. Doe accepted under the assumption that Aiello was still acting as an Uber driver. Aiello subsequently raped Doe 2.

The plaintiffs brought suit against Uber under six claims for relief, four of which rely on a theory of respondeat superior. Uber moved to dismiss the complaint for failure to state a claim.

ISSUE: Are Uber's drivers employees, and, if so, did the sexual assault occur within the drivers's scopes of employment?

REASONING: The plaintiffs in this case brought their argument under the doctrine of *respondeat superior*, which holds that an employer can be liable for the illegal actions of its employees. However, this doctrine only applies to employer-employee relationships and not to independent contractors. The plaintiffs alleged that Uber controls many aspects of a driver's job, such as setting fare prices without driver input, retaining control over customer information, the fact that Uber can terminate drivers at will and that there are no specialized skills needed to drive for Uber. In addition, the court found that Uber controls the manner and means by which drivers offer rides through the Uber app and that drivers must accept all ride requests when logged into the app. Considering the amount of control Uber has over its drivers and recent precedent with similar conclusions, the court ruled that Uber's drivers were indeed employees.

Next came the question of whether the assaults were committed within the scope of employment, which is a necessary condition for *respondeat superior*. Uber pointed out that in Doe 2's case, the assault happened after the original ride had ended. The court pointed out, however, that Doe 2 would not have been in the position to take the second ride had she not hired the first one. In addition, sexual assault is, unfortunately, an action that Uber could reasonably foresee one of its employees taking, which is exactly why traditional taxi companies perform background checks on their drivers. Because of Uber's control over its employees and its unique position to help prevent these incidents from occurring, the court argued that a question of fact as to whether the drivers were acting within the scope of their employment existed.

DECISION: Uber's motion to dismiss was denied and the case was allowed to proceed.

SIGNIFICANCE: This case provides a good illustration of how a court might determine whether a worker is an employee or independent contractor as well as whether an employer may be liable under *respondeat superior*.

(continued)

CRITICAL THINKING

What could Uber have done to assure that their drivers are considered independent contractors?

ETHICAL DECISION MAKING

What values are being treated as relatively unimportant when Uber establishes its work environment for drivers?

Although the degree of control an employer maintains is the central factor courts look at in determining whether a person is an employee or an independent contractor, it is by no means the only factor examined. Exhibit 20-3 lists the additional criteria the court considers in deciding whether a worker is an employee or an independent contractor. These criteria were established in the Restatement of Agency.[9]

Duties of the Agent and the Principal

LO 20-4 Describe the duties of the agent.

An agency relationship is a fiduciary relationship of trust, confidence, and good faith. Thus, the formation of an agency relationship results in the creation of certain duties that the principal and agent owe each other. These duties are established in the agency agreement and are implied by the law. We discuss them in the following sections and present a summary in Exhibit 20-4.

Exhibit 20-3

Independent Contractor or Employee?

CRITERIA	EMPLOYEE	INDEPENDENT CONTRACTOR
Does the worker engage in a distinct occupation or an independently established business?	No	Yes
Is the work done under the employer's supervision, or does a specialist without supervision complete the work?	Employer supervision	Specialist without supervision
Does the employer supply the tools?	Yes	No
What skill is required for the occupation?	No specialized skill	Great degree of skill
What is the length of time for which the worker is employed?	Long time	Varies
Is the worker a regular part of the business of the employer?	Yes	No
How is the worker paid?	Regular payments according to time	When the job is completed

[9] Restatement (Second) of Agency, sec. 220.

GLOBAL Context

Special Types of Agency in the Middle East and Africa

The legal concept of agency has been around for centuries, but it was actually heavily influenced by an even older system known as *hawala* that is still highly utilized in the Muslim world. *Hawala* is essentially a way of transferring money from one party to the other through the use of *hawaladars,* or middlemen agents. For example, if a buyer wishes to purchase real estate from a seller in a different city, the buyer would first give the money to *hawaladar* X along with a special password. *Hawa* X would then call *hawa* Y in the seller's city and tells him the amount to be paid along with the password. The seller, who has been told the password, then approaches *hawaladar* Y, who pays the seller the stipulated amount, minus a small commission. *Hawaladar* X now owes *hawaladar* Y that amount of money, and the two *hawaladars* will settle their debt at a later date. The settlement of debt can take many forms (such as goods or services) and need not be paid out in cash.

Hawala is unique in that it involves no written contracts and relies entirely on an honor system between *hawaladars* to ensure debts are paid. Because there is no paper trail, the transaction goes untaxed by local governments and thus the practice has been made illegal in many of the countries where it is practiced. Despite the ban, it remains highly popular because many residents of these countries see it as a cheaper and more reliable alternative to exchanging money through banks. In fact, in developing parts of the world without a steady financial infrastructure, *hawala* is often the *only* option for secure fund transfers.

Principal's Duties to the Agent

LO 20-5 Describe the duties of the principal.

When a principal contractually agrees that another person can make agreements on his or her behalf, the principal owes certain duties to the agent. If these duties are not fulfilled, the principal has violated the agent's rights. Consequently, the agent can bring a suit against the principal. If the agent is successful in bringing a suit against the principal, the agent is entitled to contract or tort remedies. Furthermore, if the principal has failed to meet duties owed to the agent, the agent can refuse to act on behalf of the principal until the failure is remedied.

Duty of Compensation The principal has a **duty to compensate** the agent for services provided unless the parties have agreed that the agent will act gratuitously. The agency contract usually specifies the type and amount of compensation as well as the time at which the compensation will be given to the agent. Thus, if a person hires an attorney to represent her, she has a duty to pay that attorney, her agent. If there is no agreement on the amount the principal will compensate the agent, the courts suggest that compensation should be calculated according to the customary fee in the situation.[10]

duty to compensate
A principal's obligation to pay an agent for his or her services.

Duty of Reimbursement and Indemnification The principal has a **duty of reimbursement and indemnification** to the agent. If an agent makes authorized expenditures in the course of working on behalf of the principal, the principal has a duty to reimburse the agent for the amount of money spent.[11] Thus, if an agent takes a trip on behalf of the principal, the principal must have authorized this trip if the agent is to be reimbursed by the principal.

Similarly, the principal has the duty to indemnify or reimburse the agent for any losses the agent incurs while working within the scope of authority on the principal's behalf.[12] For example, suppose an agent makes an agreement with a third party on behalf of the principal and the principal fails to uphold the agreement. The third party could sue the agent for damages because of the failure to uphold the agreement. The principal has a duty to indemnify the agent for the losses the third party regained.

duty of reimbursement and indemnification
A principal's obligation to reimburse an agent for any expenses or losses incurred by the agent on the principal's behalf.

[10] Restatement (Second) of Agency, sec. 443.
[11] Restatement (Second) of Agency, sec. 438.
[12] Restatement (Second) of Agency secs. 438 and 439.

CASE Nugget

Duty to Compensate

Ralph T. Leonard et al. v. Jerry D. McMorris et al.

320 F.3d 1116 (2003)

NationsWay was one of the largest privately held trucking companies in the United States, with 3,200 employees operating in 43 states. In 1999, NationsWay filed for Chapter 11 bankruptcy and terminated most of its employees. Ralph Leonard and a number of the other employees who were terminated sued Jerry McMorris and other NationsWay executives, arguing they were personally liable for unpaid wages under their duty to compensate arising from the employer–employee relationship.

The defendants argued they could not be held personally liable for agreements made between the employees and the corporation. As the case began, NationsWay was continuing its bankruptcy filings, under which the former employees were to receive approximately $3 million in unpaid wages. However, the plaintiffs wanted additional amounts covering accrued vacation pay, sick-leave pay, holiday pay, and other nonwage compensation, as well as a 50 percent penalty and attorney fees.

In deciding the case, the court addressed "[w]hether officers of a corporation are individually liable for the wages of the corporation's former employees under the Colorado Wage Claim Act." The court concluded, "[U]nder Colorado's Wage Claim Act, the officers and agents of a corporation are *not* jointly and severally liable for payment of employee wages and other compensation the corporation owes to its employees under the employment contract and the Colorado Wage Claim Act." Although the corporation has a duty to compensate, the executives who were the defendants were not individually liable to the former employees for the unpaid wages.

duty of cooperation
A principal's obligation to assist an agent in the performance of the agent's duties and not to interfere with reasonable conduct of the agent.

duty to provide safe working conditions
A principal's obligation to provide proper equipment, suitable premises, and a safe working environment in accordance with federal and state statutes.

Duty of Cooperation The principal also owes a **duty of cooperation** to the agent. Therefore, the principal must assist the agent in the performance of his or her duties. Furthermore, the principal can do nothing to interfere with the reasonable conduct of an agent. For example, suppose Suzy hired someone to sell her car for her. Suzy must be willing to let the agent show her car to interested buyers.

Duty to Provide Safe Working Conditions The principal has a **duty to provide safe working conditions** for the agent. This includes equipment, premises, and other working conditions. If the principal is aware of unsafe working conditions, the principal has a duty to warn the agent of the potential danger and make the necessary repairs. Federal and state statutes, such as the Occupational Safety and Health Act (OSHA), set specific standards for the working environment. If an employer violates these standards, the employer may be subject to fines.

Agent's Duties to the Principal

When an agent agrees to act on behalf of a principal, the agent is a *fiduciary,* a person in a position of trust and confidence and, as such, owes certain duties to the principal. However, because the agent makes agreements on behalf of the principal, the agent is in a position to harm the principal. If, for example, an agent makes numerous contracts the principal could not possibly carry out all at once, it is likely the third parties would sue the principal for not carrying out the agreements. If the agent breaches his or her fiduciary duties, the principal can sue the agent, and the principal may be entitled to a variety of contract and tort remedies beyond those stated in the contract.

Duty of Loyalty Courts suggest that the **duty of loyalty** is perhaps the most important duty an agent owes to a principal. Because the agency relationship is a fiduciary relationship (i.e., a relationship of trust), the agent has a responsibility to act in the interest of the principal.[13] The agent carries out her duty of loyalty to the principal in a number of ways, including avoiding conflicts of interest and protecting the principal's confidentiality.

An agent can, in most situations, represent and act on behalf of only one principal in an agreement. An agent usually cannot represent both the principal and a third party (who would then become another principal if represented by the same agent) in an agreement because there could be a conflict of interest. In addition, the agent has a duty to notify the principal of any offers from third parties.

For example, suppose Tony has hired a real estate agent to make land purchases for him. A third party notifies the real estate agent that some of her property will soon be going up for sale, and the third party wants to know whether Tony, the principal, would be interested in buying the property. The real estate agent cannot decide to buy that property for himself until (1) the agent has communicated the offer to Tony and (2) Tony has considered and rejected the offer.

The duty of loyalty also requires the agent to keep confidential any information about the principal during the course of agency and after the agency relationship is terminated. The agent cannot disclose or misuse any information received during or after the agency relationship with the principal.

duty of loyalty
An agent's obligation to act in the interest of the principal.

Duty of Notification Not only does the agent have to communicate offers from third parties, but the agent must also communicate any information the agent thinks could be important to the principal. This duty is known as the **duty of notification**.[14] For example, if a third party has made an agreement through an agent with a principal and the third party has failed to meet the agreement, the agent must notify the principal of this information in a timely manner. It is critical for an agent to inform the principal of all relevant information because the law typically assumes the principal is aware of all information revealed to the agent, regardless of whether the agent shares all this information with the principal.

duty of notification
An agent's obligation to inform the principal of the agent's actions on the principal's behalf and of all relevant information.

Duty of Performance The agent owes a **duty of performance** to the principal. This duty is twofold. First, the agent must perform the duties as specified in the agency agreement. For example, suppose an insurance agent contacts Bethany about purchasing a car insurance policy. Bethany agrees to purchase the policy, but for some reason the agent never obtains the policy for her. Bethany discovers the insurance agent's mistake when she gets into a car accident. The insurance agent did not meet the duty of performance. Consequently, Bethany could bring a claim against the insurance agent.

Second, the agent must perform the specified duties with reasonable skill and care. The agent is expected to provide the same standards of skill, care, and professionalism that a reasonable person in the same situation would provide. For example, when an attorney advertises that he is a specialist in a certain type of law, this lawyer will be held to the reasonable standard of care in that specialty of law.[15] A gratuitous agent cannot be found liable for a breach of contract for failure to perform because no contract exists between the principal and the agent. However, if a gratuitous agent begins to act as a regular agent, and the principal affirms the relationship, a duty to perform arises insofar as the agent has begun a specific task for the principal.

duty of performance
An agent's obligation to perform the duties specified in the agency agreement and to do so with reasonable skill and care.

Duty of Obedience Under the **duty of obedience**, the agent must follow the lawful instruction and direction of the principal.[16] Thus, if the agent makes an unauthorized agreement, the agent has failed to meet the duty of obedience. However, if the principal gives unlawful or unethical instructions, the agent is not required to behave in accordance with those instructions. For example, let's say a principal tells an agent to sell a basketball autographed by Michael Jordan; however, the agent knows that the principal forged the signature on the basketball. The agent is not required to obey the instructions of the principal.

duty of obedience
An agent's obligation to follow the lawful instruction and direction of the principal.

[13] Restatement (Second) of Agency, sec. 401.
[14] Restatement (Second) of Agency, sec. 381.
[15] Restatement (Second) of Agency, sec. 379.
[16] Restatement (Second) of Agency, secs. 383 and 385.

Exhibit 20-4

Summary of Duties in an Agency Relationship

PRINCIPAL TO AGENT	AGENT TO PRINCIPAL
Duty of compensation	Duty of loyalty
Duty of reimbursement and indemnification	Duty of performance
Duty of cooperation	Duty of notification
Duty to provide safe working conditions	Duty of obedience
	Duty of accounting

duty of accounting
An agent's obligation to keep an accurate account of the transactions of money and property made on behalf of the principal.

Duty of Accounting Under the duty of accounting, the agent must keep an accurate account of the transactions of money and property made on behalf of the principal.[17] If the principal requests to see this accounting, the agent has a duty to provide the principal with the accounting. Part of the duty of accounting is the duty of the agent to keep separate accounts for the principal's funds and the agent's funds. The agent cannot allow the funds to mix.

Rights and Remedies

Principal's Rights and Remedies Against The Agent

Because the agency relationship generally *is* a contractual relationship, a principal has contract remedies available for breach of fiduciary duties. In addition, a principal may use tort remedies for an agent's misrepresentations, negligence, or other business failings that cause damage to the principal. Furthermore, when an agent breaches his or her fiduciary duties, the principal has the right to terminate the agency relationship. Although numerous possible remedies are available to the principal, the three main remedies are constructive trust, avoidance, and indemnification.

Constructive Trust Agency relationships exist primarily for the benefit of the principal. Therefore, principals are the legal owners of anything an agent may come to possess through the employment or agency relationship. Accordingly, when an agent through deceit or other means retains profits or goods obtained through the employment or agency relationship, which by law belong to the principal, the agent has breached his or her fiduciary duties. For example, Joy is an agent of Sarah's, selling real estate. Joy sells a piece of property for $2,000 more than Sarah anticipated. Joy then keeps the extra $2,000 and reports the sale at the price Sarah expected. By law, the profits belong to Sarah, and Joy has breached her fiduciary duties by keeping the money.

Also, an agent may not use the agency relationship to obtain goods or property for himself when the principal desired to obtain the same goods or property. In an agency relationship, the principal always has right of first refusal. That is, before an agent may use the relationship for a personal acquisition, the principal has the opportunity to buy whatever goods or properties are in question. Returning to Sarah and Joy, if Joy were to buy a piece of land she knew Sarah wanted to purchase, she again would have breached her fiduciary duties to Sarah.

When an agent illegally benefits from the agency relationship, the principal may enact a constructive trust on the profits, goods, or property in question. A constructive trust is an equitable trust imposed on a person who wrongfully obtains or holds legal right to property he or she should not possess. When a principal enacts a constructive trust, the court rules that the agent is merely holding the property or goods in trust for the principal, granting the principal legal right or possession.

constructive trust
An equitable trust imposed on a person who wrongfully obtains or holds legal right to property he or she should not possess.

Avoidance When an agency relationship exists through a contract and the agent breaches the agreement or his fiduciary duties, the principal may use her right of avoidance. *Avoidance* allows a principal to nullify any contract the agent negotiated. Which, if any, contracts are voided is at the discretion of the principal.

Indemnification Agents enter into contracts with third parties on behalf of the principal. If a third party believes that the agent is acting with actual or apparent authority, she may sue the principal for any breach of contract. However, when the breach was caused by the agent's negligence, the

[17] Restatement (Second) of Agency, sec. 382.

principal has a right to *indemnification.* That is, when sued by a third party, a principal may sue his agent to recover the amount assessed to the third party if the breach of contract is the agent's fault.

In addition to recovering for negligence, a principal can recover if an agent fails to follow the principal's instructions. Courts have had difficulty in determining when a principal gives limiting instructions and when he merely gives advice. Going against advice does not impose liability on an agent, but violating limiting instructions does. For a principal to avoid a potential lawsuit from a third party, he should notify the third party whenever a relationship with an agent ceases or limiting instructions are given.

Agent's Rights and Remedies Against The Principal

Although agency relationships are intended to benefit the principal, the agent is not without rights and remedies. Whenever a duty is imposed on the principal, a corresponding right exists for the agent. Agents have available tort and contract remedies in addition to the right to demand an accounting.

Tort and Contract Remedies Agents, much like principals, have certain tort and contract remedies available for situations in which a principal violates an agency agreement. These remedies are the same as standard tort and contract remedies.

Demand for an Accounting When an agent feels she is not being properly compensated, especially when working on commission, the agent may *demand an accounting.* When an agent demands an accounting, she may withhold further performance of her duties until the principal supplies appropriate accounting data. For example, Hal is a used-car salesman working for Not a Lemon Car Dealers. When Hal receives his pay, he believes he has been shorted the appropriate amount he made on commission. Hal can request Not a Lemon to obtain an auditor to perform an audit and determine whether he was in fact paid the proper amount for his sales.

BUT WHAT IF...

WHAT IF THE FACTS OF THE CASE OPENER WERE DIFFERENT?

Let's say in the Case Opener that Janelle, a truck driver, was an independent contractor for FedEx. She looked at her income over the course of the prior year and was quite certain that she was not being paid her promised compensation. She demanded that FedEx hire an auditor to go over the company's records and make sure she and other independent contractors in her area had been paid fairly. Does FedEx have a duty to Janelle to do this?

Specific Performance When a contract exists and a principal agrees to certain conditions but fails to perform, under contract remedies the agent may seek court assistance in forcing the principal to perform the contract as stipulated. However, when the agency relationship is not contractual or the contract is for personal services, an agent does not have the right to seek specific performance. In noncontractual relationships, an agent may recover for services rendered and/or future damages, but the agent may not force the principal to fulfill the specific contractual agreements or even to continue to employ the agent.

Authority of the Agent: The Link to the Principal's Liability

LO 20-6 Identify the different types of authority of the agent.

Earlier in the chapter, we discussed how an agency relationship and its resulting authority could be created. That section introduced the important concepts of (1) expressed agency, or agency by agreement; (2) implied agency; and (3) apparent agency, or agency by estoppel. Each of these avenues for creating agency includes a form of authority that attaches to that type of agency.

Express Authority

express authority
The authority of an agent that arises when the principal explicitly instructs the agent to do something. Also called *actual authority*.

In an expressed agency relationship, the agent has **express authority**, which is often referred to as *actual authority*. The principal has explicitly instructed the agent to do something. Therefore, if the principal requests her agent to sell her house for $100,000, the agent has explicit authority to sell the house for that amount. Express authority is binding in the sense that it gives the agent the power to make transactions and agreements with third parties. Because an agent with express authority has specific goals he or she is to achieve, the agent is also granted the use of other means, beyond the expressed agreement, reasonably necessary to achieve the desired goal.[18]

When express authority is granted for an agent to enter into a contract legally required to be in writing, most states require the grant of authority also to be in writing. This requirement is known as the **equal dignity rule**. When an agency agreement fails to adhere to the equal dignity rule, the subsequent contracts are typically considered voidable by the principal. This rule is based on the statute of frauds.

equal dignity rule
A rule requiring that contracts that would normally fall under the statute of frauds and need a writing if negotiated by the principal must be in writing even if negotiated by an agent.

There are two key exceptions to the equal dignity rule. The first pertains to executive officers. When an executive officer is conducting business in the usual course of her job, she need not obtain written approval from the corporation for every decision she makes. The second exception is rooted in common sense. The equal dignity rule does not apply to an agent when the agent is working in the direct presence of the principal.

Implied Authority

implied authority
The authority of an agent that arises by inference from the words and actions of the principal.

In an implied agency relationship, the agent has **implied authority**; that is, the relationship is inferred from the conduct of the parties. Consequently, the authority of the agent is implied on the basis of words and actions of the principal to the agent.[19] An agent's implied authority is derived from an agent's express authority and consists of what is reasonably necessary for carrying out the agent's grant of express authority.

One way an agent may have implied authority is through custom. Therefore, it is important for third parties to familiarize themselves with the trade. Furthermore, if an agent must make an agreement or transaction to carry out the transaction he or she has agreed to perform for the principal, we can generally infer the presence of the authority.

For example, suppose one agrees to serve as a manager for a store. The contract probably does not expressly give authority for every decision the manager might need to make. However, the manager has implied authority to make agreements and transactions that customarily go along with the manager's position.

Another example of implied authority revolves around anyone, such as a purchasing agent, who ordinarily is the person who everyone in the industry expects will make contracts for particular inputs. For someone to act as if the purchasing agent lacks authority to make such commitments on behalf of the firm would be to act contrary to the expected lines of authority.

Apparent Authority and Estoppel

As you learned earlier in this chapter, apparent agency exists when a third party reasonably believes, on the basis of a *principal's* actions, that an agency relationship exists between the principal and another individual. Thus, if the third party reasonably believes that the agent has authority to represent the principal, the principal must uphold any agreements made with the agent, who has apparent authority.[20] The principal is prevented (estopped) from acting as if the agent had no such authority.

[18] Restatement (Second) of Agency, sec. 35.
[19] Restatement (Second) of Agency, sec. 33.
[20] Restatement (Second) of Agency, secs. 27 and 159.

Contractual Liability of the Principal and Agent

LO 20-7 Explain the contractual liability between the principal and the agent.

In addition to identifying the type of authority an agent has, courts must identify a few other factors when making decisions about an agency relationship's liability to third parties. First, the court must determine the classification of the principal. Second, the court must decide whether the principal authorized the actions of the agent.

Classification of the Principal

Classifications are made from the perspective of the third party. In other words, the extent of the third party's knowledge about the principal determines the classification of the principal.

What are potential ethical issues arising from questions about the authority of a particular agent? The law of agency places special weight on the viewpoint of the agency relationship from the perspective of the third party (neither the agent nor the principal). But who are the other stakeholders in a managerial decision who might be affected by the assignment of a particular form of authority?

When a principal is classified as a **disclosed principal**, the third party is aware that the agent is making an agreement on behalf of a principal; the third party also knows the identity of the principal. In contrast, if a third party is aware that an agent is making an agreement on behalf of a principal but the third party is unaware of the identity of the principal, the principal is classified as a **partially disclosed principal** or an unidentified principal. Finally, if a third party does not know an agent is acting on behalf of a principal, the principal is classified as an **undisclosed principal**. Remember, the classification of the principal is important because it is a significant factor in determining the liability of the principal.[21]

Authorized Acts

When an agent acts within the scope of her authority on behalf of a disclosed or partially disclosed principal, the agent is not liable for the acts of the principal.[22] The principal is liable only if the agent has some kind of authority to act on the principal's behalf. With a disclosed principal, the agent is not liable because he or she is not a party to the transaction.

Yet, if the principal is partially disclosed, the agent can be held liable herself because the courts generally treat the agent as a party to the contract. As a party to the contract, the agent may be found liable to the third party for contractual nonperformance.[23] Whether disclosed or partially disclosed, apart from any liability the agent might have, the principal is liable for the agreements made with the third party.

When the agent acts within her authority on behalf of an undisclosed principal, the law will likely hold the agent liable for the agreement. Remember, with an undisclosed principal, the third party is not aware of the existence of the principal. Thus, in the eyes of the third party, the agent is the only possible person who could be liable. Yet, if the agent is liable to the third party, then the undisclosed principal is liable to the agent. However, there are certain situations in which the agent is the only party liable for the contract. These situations are:

1. The contract expressly excludes the principal from the contract. In that the principal was not a party to the contract, he or she has no liability to the agent.
2. The agent enters into a contract that is a negotiable instrument. The Uniform Commercial Code (UCC) governs negotiable instruments and states that another party, that is, a

disclosed principal
A principal whose identity is known to a third party. The third party is aware that the agent is making an agreement on behalf of the principal.

partially disclosed principal
A principal whose identity is not known by a third party, although the third party is aware that the agent is making an agreement on behalf of a principal. Also called *unidentified principal*.

undisclosed principal
A principal whose existence is not known by a third party. That is, the third party does not know an agent is acting on behalf of a principal.

[21] Restatement (Second) of Agency, sec. 4.
[22] Restatement (Second) of Agency, sec. 320.
[23] Restatement (Second) of Agency, sec. 321.

Exhibit 20-5

Contractual Liability to Third Parties for Authorized Agent Acts

CLASSIFICATION OF PRINCIPAL	AGENT LIABILITY?	PRINCIPAL LIABILITY?
Disclosed	No	Yes
Partially disclosed	Possibly	Yes
Undisclosed	Yes	Yes

principal, cannot be liable for a negotiable instrument if his name is not on the instrument or if the agent's signature does not indicate that the agreement was made in a representative capacity.[24]

3. The third party enters into a contract with the agent that requires the performance of the agent. In this case, the third party may reject the performance of the principal. For example, if the agent is a photographer and she enters into a contract for her principal without disclosing this fact, the third party may reject the principal's attempt to fulfill the contract by taking the third party's picture.

4. If the principal or agent knows a third party would not enter into a contract with the principal if the principal's identity were disclosed but the agent enters into a contract with the third party anyway, the agent will be the only party liable if the third party rescinds the contract.

In situations in which a principal was undisclosed, a judgment for a third party against an agent, when the third party comes to know of the undisclosed principal's identity, releases the principal from liability.[25] Also, a judgment against a previously undisclosed principal frees the agent from liability.[26]

Exhibit 20-5 summarizes contractual liability to third parties for authorized acts of the agent.

Unauthorized Acts

If an agent has no authority to act on behalf of a principal but the agent still enters into a contract with a third party, the principal, regardless of the classification, is not bound to the contract unless the principal ratifies the agreement.

When the agent exceeds his authority to act on behalf of the principal, the agent will likely be personally liable to the third party. Yet, when the third party is aware that the agent does not represent the principal, the law does not hold the agent liable for the agreement. In almost all other cases in which the agent claims to have authority to contract on behalf of the principal, the law holds the agent liable to the third party. If an agent enters into a contract knowingly misrepresenting his alleged authority, the agent is liable to the third party in a tort action.

Agents who go beyond their authority when the principal is disclosed or partially disclosed are liable for a breach of implied warranty, not for a breach of contract. The agent cannot be found liable for breach of contract because the agent is never an intended party to the contract, even when exceeding his or her authority. The agent can breach the implied warranty intentionally, through a knowing misrepresentation, or unintentionally, through a good-faith mistake, such as simply misjudging his or her authority. In either case, the agent is liable if the third party acted relying on the agent's alleged status.

Exhibit 20-6 summarizes the contractual liability of principals and agents to third parties for unauthorized acts of the agent.

Exhibit 20-6

Contractual Liability to Third Parties for Unauthorized Acts of the Agent

THIRD-PARTY BELIEF	AGENT LIABILITY?	PRINCIPAL LIABILITY?
Believes that the agent has authority	Yes	No
Believes that the agent is mistaken about her authority	No	No

[24] UCC § 3-402(b)(2).

[25] Restatement (Second) of Agency, sec. 210.

[26] Restatement (Second) of Agency, sec. 337.

Tort Liability and the Agency Relationship

If an agent commits a tort that injures a third party, the agent is personally liable for his or her actions, regardless of both the classification of the principal and the liability of the principal.[27] The principal may also be held liable for the agent's actions. This liability can arise from authorized or unauthorized acts. Furthermore, tortious liability of the principal can be established directly or indirectly. Finally, if an agent is an employee and the principal/employer controls the employee's behavior, the principal can be found liable. The next section explains these methods of establishing tortious liability.

Principal's Tortious Conduct

The law holds a principal directly responsible for the tortious conduct of his agent under two conditions. First, if the principal directs the agent to commit a tort, the principal is liable for any damages caused by that tort. The principal is authorizing the agent's unlawful behavior; thus, the principal is held responsible.[28] Similarly, the principal is liable for an agent's tortious act if the principal, although not condoning the agent's conduct, ratifies the agent's actions knowing the agent acted improperly.[29]

Second, if the principal fails to provide proper instruments or tools or gives inadequate instructions to the agent concerning the necessity to employ competent agents, the law then holds the principal liable to a third party for negligence. Under this provision of liability, a principal is liable for his or her negligent hiring of an agent. Thus, a principal must use proper care in selecting an agent for a job. This doctrine of negligent hiring has been used when an agent commits a tort against a customer, who often argues that the principal is liable because he or she should have taken more care in hiring the agent.

Respondeat Superior The doctrine of *respondeat superior* (a Latin phrase meaning "let the superior speak") is used in the context of the principal/employer–agent/employee relationship. The principal/employer holds **vicarious liability** (i.e., liability assigned without fault) for any harm caused by the agent/employee during the time the agent/employee is working for the principal. In other words, the principal/employer is liable not because he or she was personally at fault but because he or she negligently hired an agent.

The rationale behind this doctrine is that employers should be held liable for employees who commit torts because the employer is furthering his or her business through the work of the employee. If the employer is benefiting by the work of the employee, the employer should also be responsible for the harms caused by the employee.

Therefore, when a third party is injured through the negligence of an employee during the course of the employee's work, the third party can sue either the employee or the employer.[30] To establish employer liability, the third party must establish that the wrongful act occurred within the scope of employment. The courts consider the following elements in determining whether an act has occurred within the course and scope of employment:[31]

1. Did the employer authorize the employee's act?
2. Did the act occur within the time and space limits of employment?
3. Was the act performed, at least in part, on behalf of the employer?
4. To what extent were the employer's interests advanced by the act?
5. To what extent were the private interests of the employee involved?
6. Did the employer provide the means (i.e., tools) by which the act occurred?
7. Did the employee use force the employer did not expect?
8. Did the employer know that the act would involve the commission of a serious crime?

For example, if an employee is a delivery driver and negligently injures a third party while making deliveries on behalf of the employer, both the employee and the employer will be held liable.

Now suppose that the delivery driver is using the company vehicle when he stops at a drive-through at a fast-food restaurant to get some coffee. Could the employer be liable to a third party

respondeat superior
Latin for "let the superior speak," the principle by which liability for harm caused by an agent/employee is held by the principal/employer.

vicarious liability
The liability or responsibility imposed on a person, a party, or an organization for damages caused by another; most commonly used in relation to employment, with the employer held vicariously liable for damages caused by its employees.

[27] Restatement (Second) of Agency, sec. 343.
[28] Restatement (Second) of Agency, sec. 212.
[29] Restatement (Second) of Agency, sec. 218.
[30] Restatement (Second) of Agency, secs. 216 and 219.
[31] Restatement (Second) of Agency, sec. 229.

for an accident involving the delivery driver? If an agent makes a substantial departure from the course of the employer's business, the employer is not liable.

Courts often refer to a substantial departure as a "frolic of his own." However, if the deviation from the employer's business is *not* substantial, the employer can still be held liable. In Case 20-3 the court considers the scope of the employment relationship.

CASE 20-3

IGLESIA CRISTIANA LA CASA DEL SENOR, INC., ETC. v. L.M.
Court of Appeal of Florida, Third District
783 So. 2d 353 (2001)

FACTS: L.M. sued Ali Pacheco, the former pastor of Iglesia Cristiana la Casa del Senor Inc. (the Church), as well as the Church, alleging that Pacheco had sexually assaulted her in July 1991 when she was a minor. The allegation of sexual assault formed the basis of L.M.'s claims against the Church based on respondeat superior. When the criminal act occurred, L.M. was 16 years old.

Before the criminal act took place, Pacheco visited L.M.'s residence twice when L.M. had been left home alone. On another occasion, Pacheco visited L.M. at her school. L.M. told her mother about Pacheco's visit but did not advise anyone from the Church.

According to L.M., on July 8, 1991, Pacheco called her at work and invited her to lunch to discuss her parents' marital problems. L.M. accepted, and Pacheco picked her up from work. L.M. noticed a sandwich and soft drink in the car. Pacheco drove to a Marriott Hotel. L.M. testified that Pacheco led her to a room he had rented and told her not to worry because she would finally be cured. He then proceeded to sexually assault her. Pacheco testified that L.M. consented to having sex.

According to him, their meeting was prearranged. They had discussed the matter and had in fact been to the Marriot Hotel on the previous day, intending to have sexual relations but had decided against it. Pacheco testified that he knew what he was doing was wrong, but he explained that it was a great temptation in his life.

The jury returned a verdict in L.M.'s favor, finding the Church liable for Pacheco's criminal act on the grounds of respondeat superior. The Church appealed.

ISSUE: Was Pacheco acting as an employee of the Church when he assaulted L.M.?

REASONING: The court stated that for the doctrine of respondeat superior to apply, Pacheco's employer could not be liable for the tortious or criminal acts of Pacheco "unless the acts were committed during the course of the employment and to further a purpose or interest, however excessive or misguided, of the employer. An employee's conduct is within the scope of his employment, when (1) the conduct is of the kind he was employed to perform, (2) the conduct occurs substantially within the time and space limits authorized or required by the work to be performed, and (3) the conduct is activated at least in part by a purpose to serve the master."

In the case of L.M., the sexual assault did not occur on Church property, and what Pacheco did was not the kind of behavior the Church hired Pacheco to perform. In addition, his act was not an expression of Pacheco's desire to serve the Church. Little counseling occurred on the day of the crime. The Church would have tried very hard to prevent his actions had it known of his plans.

DECISION AND REMEDY: The court reversed the trial court's judgment and entered judgment for the Church.

SIGNIFICANCE: Respondeat superior does not mean that the employer is responsible for whatever an employee might do while at work. This case spells out the requirements for such liability to be imposed on the principal.

CRITICAL THINKING

Assume that L.M.'s account of the crime is true. Examine the exception to the scope-of-employment criteria mentioned by the judge. How could the plaintiff make an argument, using that exception, that Pacheco's conduct was within the scope of his employment?

ETHICAL DECISION MAKING

The judge in this case outlines a doctrine for determining the liability of an employer for the actions of employees. What value preference is highlighted by that doctrine?

If the third party is able to establish employee negligence such that the employer is liable, the employer has the right to recover from the employee any damages the employer paid to the third party as a result of the employee's negligence. The right to recover damages is referred to as the *right of indemnification.* However, if the employee is innocent of negligence, the employer is also free of liability.

Intentional Torts and *Respondeat Superior* The agent is liable for any torts he or she commits. In the same way that the principal is responsible for the negligent acts of the employee under the doctrine of *respondeat superior,* the principal may be liable for any intentional torts of the employee. Furthermore, an employer may be responsible for any tortious acts of the employee if the employer knew or should have known that the employee had a tendency to commit a tortious act. Hence, a principal may be liable for negligent hiring if he or she fails to do a background check to learn about the tendencies of potential employees.

For example, the principal of an employee with a criminal background may be held liable for tortious acts committed by his or her hired agent, even though the employee may not recognize the wrongfulness of his act. Therefore, employers will most likely purchase liability insurance in case particular employees engage in tortious activities.

Agent Misrepresentation

If an agent misrepresents himself or herself to a third party, the principal may be tortiously liable for the agent's misrepresentation. Unlike tort liability, which is based on whether the agent/employee was acting in the scope of employment, *misrepresentation liability* depends on whether the principal authorized the agent's act. If the principal authorizes the agent to engage in an act and the agent misrepresents herself intentionally or unintentionally, the principal is always tortiously liable to someone who relied on the agent's misrepresentation.

If an agent has misrepresented herself, the third party has two options:

1. The third party can cancel the contract with the principal and be compensated for any money lost.
2. The third party can affirm the contract and sue the principal to recover damages.

Principal's Liability and The Independent Contractor

As we discussed earlier in this chapter, an independent contractor is not an employee of the individual who hires him or her to do work. The individual doing the hiring does not control the details of the independent contractor's performance. Consequently, an individual who hires an independent contractor cannot be held liable for the independent contractor's tortious actions under the doctrine of *respondeat superior.*

Suppose that while working on the outside of the building he is renovating, an independent contractor accidentally injures an innocent bystander when he drops a pile of bricks on the bystander. The owner of the building is not liable for the innocent bystander's injuries; the independent contractor is liable.[32]

If, however, the independent contractor engages in extremely hazardous activities, such as blasting operations, for the principal, the principal will be responsible for any damages by the independent contractor. Certain activities are held strictly liable because of their inherently dangerous nature. Thus, an employer cannot escape the strict liability associated with those hazardous activities simply by hiring an independent contractor to complete the activities. In addition, the employer cannot escape liability for an independent contractor's tort if the employer directs the contractor to commit the tort.

The following Case Nugget demonstrates the role of tort principles in establishing the liability of those who employ independent contractors to engage in inherently dangerous activities.

[32] Restatement (Second) of Agency, sec. 250.

CASE Nugget

Liability When Hiring Independent Contractors

Larry S. Lawrence v. Bainbridge Apartments et al. Court of Appeals of Missouri, Western District

957 S.W.2d 400 (1997)

In 1989, Smart Way Janitorial offered a bid to Larry Lawrence to wash the windows of Bainbridge Apartments. In May 1989, when Larry went to look at the apartments, he was told that Bainbridge wanted the windows washed from outside the building so that the residents would not be disturbed. The Bainbridge property consisted of six buildings: two seven-story buildings and four four-story buildings. Even though Lawrence could not create a safety line for washing the outside of the windows of the four-story buildings, the building manager insisted that Lawrence wash the windows from the outside. When Lawrence started washing the windows from the outside, he fell from one of the shorter buildings and suffered injuries. He brought a suit against Bainbridge Apartments, arguing that Bainbridge was negligent based on the inherently-dangerous-activity exception to the doctrine that landowners are not vicariously liable for injuries caused by the negligence of an independent contractor or his employees.

The trial court ruled that because Lawrence had recovered workers's compensation benefits, the injury was not covered by the inherently-dangerous-activity exception. The trial court granted summary judgment to Bainbridge; however, when Lawrence appealed, the decision was reversed and remanded because the court of appeals ruled that Lawrence was not a covered employee entitled to workers's compensation benefits.

The court argued that in establishing the rule of liability in this case, it would look to which party can best avoid the harm and manage the risk of loss of the inherently dangerous activity in question. An independent contractor who knows he will not be compensated by the landowner for his injuries has a strong incentive to take additional care and avoid neglect in performing his duties. This encourages him to demand additional safety measures and compensation before he agrees to undertake an inherently dangerous activity. The independent contractor is free to bargain and negotiate without restraint with the landowner. An independent contractor holds himself out as an expert who is uniquely qualified and skilled to bid for and perform the acts in question. As an expert, he is in a better position to understand the risks and costs involved in a particular job, and he may demand sufficient remuneration and safety measures to cover what he believes the attendant risks to be. In return for his bargained-for price, he accepts the allocation of the risk. The court held that an injured independent contractor, although uninsured, cannot recover under the inherently-dangerous-activity exception.

Termination of the Agency Relationship

There are multiple ways to end an agency relationship. The parties may act to terminate the relationship. Alternatively, some agencies terminate automatically by the lapse of time, fulfillment of purpose, or operation of law. If the agency relationship has ended, the agent no longer has authority to make agreements on behalf of the principal. However, the agent's apparent authority continues until the principal notifies third parties that the agency relationship has ended. What terminates the agency relationship?

Notice of the termination of an agency relationship can be given in two general forms: actual or constructive. **Actual notice** of termination is given when third parties are directly informed, orally

actual notice
Notice of agency termination that is given by directly informing third parties, either orally or in writing.

or in writing, that an agency agreement has terminated.[33] Actual notice must be given to third parties who have had business interactions with the agent. Also, when the agent's authority was granted in writing, actual notice must be given in writing. Parties not directly related to an agency agreement may receive **constructive notice,** which is how notice of the termination of an agency agreement is generally announced.[34] Constructive notice is most frequently delivered through publication in a generally circulating newspaper for the area where the agency agreement existed.

constructive notice
Notice of agency termination that is usually given by publishing an announcement in a newspaper.

The issue of termination of the agency is especially important for global commerce. When parties are forming a contract of agency in a foreign jurisdiction, it is extremely important for the parties to include the conditions of termination within the contract.

Termination of an agency relationship involving persons from separate jurisdictions often results in complicated legal battles over compensation and liability. For example, principals must be aware of local labor laws because these laws can create agents' compensation and pension rights that are unknown to the principal. Local legal counsel should be especially knowledgeable about the provisions surrounding termination of agency in the particular jurisdiction and be able to help managers avoid unnecessary legal battles.

For example, if a manager has business transactions in one of the European Union countries and wants to terminate an agency relationship, he or she would want to have access to knowledge about the intricacies of Chapter IV of the Agency Relationship Law, which focuses on termination. After the termination of an agency contract in the EU countries, the agent receives help from the law in collecting damages. A released agent receives compensation if the agent has brought the principal new customers from whom the principal continues to profit, if the agent is unable otherwise to recover costs incurred through the performance of the contract, or if the agent dies.

Under certain circumstances, however, the law in European Union countries prohibits the agent from receiving compensation. For instance, if the principal has terminated the contract due to the incapacity of the agent, the principal is not obligated to dispense compensation. Also, compensation is not payable if the agent terminates the contract or assigns rights and duties under the agency contract to another person.

Termination by Acts of Parties

The agency relationship can be terminated following certain acts. Several possibilities are discussed in the following sections.

Lapse of Time If an agency agreement specifies that the agency relationship will exist for a certain amount of time, the relationship will end when the amount of time expires.[35] For example, an agency agreement might state that the agency relationship will begin on September 1 and end on September 30. Although the agent and principal can agree to continue the agency relationship through October, they will have to make a new agreement. The agent's express authority ends when the relationship ends; thus, it is the responsibility of the principal to notify third parties that the former agent no longer has the ability to act on the principal's behalf.

Fulfillment of Purpose Suppose a homeowner enters into an agreement with a real estate agent to sell his house. In other words, the very basis for the relationship is a function or purpose—the sale of a piece of real estate. Once the house is sold, the agent no longer has the authority to act on the seller's behalf. The agent has fulfilled the purpose of the agency relationship.[36]

Occurrence of a Specific Event An additional means of ending an agency relationship is the occurrence of a specific event. For example, John chooses Claire as an agent to move his bed to another location. Once the move has been made, the agency relationship would terminate.

Mutual Agreement by the Parties Remember, agency is a consensual agreement between two parties. Consequently, if the two parties mutually decide they do not wish to continue in the agency relationship, they can cancel the agreement and therefore terminate the agency relationship.

[33] Restatement (Second) of Agency, sec. 136(2).
[34] Restatement (Second) of Agency, sec. 136(3).
[35] Restatement (Second) of Agency, sec. 105.
[36] Restatement (Second) of Agency, sec. 106.

Revocation of Authority At any time, a principal can revoke an agent's authority.[37] Again, the basis of agency law is consensual agreement. If the principal does not want the agent to act on his behalf, the principal can revoke the agent's authority. However, the principal's revocation might constitute a breach of contract with the agent. Therefore, the principal could be liable for damages due to the breach of contract.[38] If the agent has somehow breached her fiduciary duty to the principal, the principal can revoke the agent's authority without being liable for damages to the agent.

Renunciation by the Agent In the same way that a principal terminates an agency relationship by revoking authority, the agent can terminate the agency relationship by renouncing the authority given to the agent. The agent can be liable for breach of contract if the agency agreement has stated a specific amount of time the agency relationship is to exist.

Agency Coupled with an Interest An agency coupled with an interest is a special kind of agency relationship. Unlike regular agency agreements, an agency coupled with an interest is created for the agent's benefit, not for the principal's. Given that the agent is the one who benefits, an agency coupled with an interest is also known as *power given as security*. Because the agency arrangement is created to benefit the agent, the principal may not terminate the relationship. Rather, the agency relationship is terminated when an event occurs that discharges the principal's obligation.

> **agency coupled with an interest**
> An agency relationship that is created for the benefit of the agent, not the principal.

Termination by Operation of Law

LO 20-8 Outline the ways an agency relationship can be terminated.

Certain events lead to automatic termination of the agency relationship. These events may involve a situation in which the agent is unable to fulfill his task or one in which the principal does not desire to continue with the performance. In addition, automatic termination may occur when further pursuit of the agency relationship's objectives would be illegal. These events are discussed in the following sections.

Death If either the principal or the agent dies, the agency relationship is automatically terminated. Even if one party is unaware of the other party's death, the agency relationship no longer exists. Suppose an agent has authority to buy antiques on behalf of a principal. However, without the agent knowing it, the principal dies, but the agent continues to purchase items on behalf of the principal. Those transactions are not binding on the principal's estate because as soon as the principal died, the agent's authority to act ended.

Insanity If a principal or agent becomes insane, the agency relationship is finished. Some states have modified this law so that unless the person has been adjudicated insane, the agency contract still exists.

Bankruptcy If the principal or agent files a bankruptcy petition, the agency relationship is generally no longer in existence. If the agent is filing for bankruptcy and the agent's credit is important to the agency relationship, the relationship will end. However, if the agent's financial history is not relevant to the agency relationship, the agency may continue. Insolvency, which is defined as the inability to pay debts or the condition of liabilities outweighing assets, does not necessarily result in the termination of the agency relationship.[39]

Changed Circumstances If an unusual change in circumstances leads the agent to believe that the principal's instructions do not apply, the agency relationship terminates.[40] For example, suppose Smith contracts someone to act as her agent to sell a painting she found in her great-aunt's attic. Smith authorizes the agent to sell the painting for $5,000. However, in the course of

[37] Restatement (Second) of Agency, sec. 119.
[38] Restatement (Second) of Agency, sec. 118.
[39] Restatement (Second) of Agency, sec. 113.
[40] Restatement (Second) of Agency, sec. 109.

E-COMMERCE and the Law

Agents and Electronic Contracts in Singapore

Many legislatures have found it necessary to add or create certain laws in response to technology. Singapore, for instance, addressed the growing use of electronic contracts with new legislation in 1997. Because of the possibility of email and electronic fraud, certain risks are associated with forming electronic contracts. The new legislation attempts to combat those risks and specifies the consequences of such fraud.

The fourth section of this legislation pertains to the formation of agency contracts. Agency contracts made electronically will be valid and enforceable if the principal or a principal's designated agent sent the contract. As for contracts made between third parties, certain circumstances must exist for the third party to make the assumption that the electronic record is that of the principal.

Either the third party follows a procedure of clarification previously agreed on by the principal or he or she sees the message originated from an agent who has been endorsed by the principal. Under either of these circumstances, the third party is legally allowed to assume that the electronic message originated with the principal.

If an agent in Singapore sends an electronic record not approved by the principal, the third party has the right to act as a result of that message. If such actions result in injuries or damages to the third party, the principal is responsible under law. The principal cannot claim that he or she was unaware of the agent's actions. Although there may have been no actual awareness, Singapore does not recognize the lack of awareness as a defense.

Singapore's legislation intends to protect third parties from the poor judgment of principals by creating this direct link between them. Making the principal answerable and liable to the third party increases the pressure to employ reliable agents.

showing the painting to several buyers, the agent learns the painting is a Van Gogh original. Thus, because the painting is worth much more than $5,000, the agent should infer that the principal does not want the original agency to continue.

Change in Law When a new law is passed subsequent to the formation of an agency agreement that makes the commission of the agency agreement illegal, the agency agreement is terminated. For example, Tyler hires Ryan to paint her house green. Then, after Ryan was hired, the city council passes a law making it illegal to paint one's house green. The new law automatically terminates the agency agreement.

Impossibility Suppose that while attempting to sell the painting in our earlier example, there is a fire in Smith's house and the painting is destroyed. Because it is impossible for the agent to sell the painting, it is impossible for the agency relationship to continue.[41]

In addition, if the agent loses qualifications needed to perform duties for the principal, the agency relationship ends because of impossibility. For example, Jackson hires a lawyer to serve as his agent. However, the lawyer has unfortunately engaged in a series of illegal actions and is disbarred. Because the lawyer can no longer fulfill the functions Jackson has authorized him to perform, the agency relationship is terminated.

Disloyalty of Agent An agency agreement is terminated whenever the agent, unknown to the principal, acquires an interest against the principal's interest. The agency agreement is also terminated if the agent breaches the duty of loyalty he or she has to the principal.[42] For example, Mary is an attorney representing Lola in her suit against a pharmaceutical company. If the pharmaceutical company offers Mary a job and she accepts, the agency agreement would terminate because Mary has acquired an interest opposed to Lola's interests.

War A principal has an agent in Iran authorized to conduct business dealings on the principal's behalf.[43] If the United States goes to war with Iran, this agency relationship will no longer exist because there is no way to enforce the rights of the parties.

[41] Restatement (Second) of Agency, sec. 124.
[42] Restatement (Second) of Agency, sec. 112.
[43] Restatement (Second) of Agency, sec. 115.

CASE OPENER WRAP-UP

FedEx and Independent Contractors

Ultimately, the court ruled in favor of FedEx and found that the board's decision was unenforceable because the drivers in question were independent contractors rather than employees. To determine whether the FedEx drivers should be classified as employees or independent contractors, the court applied traditional agency law principles. It discovered that FedEx "may not prescribe hours of work, whether or when the drivers take breaks, what routes they follow, or other details of performance;" drivers "are not subject to reprimands or other discipline;" and the owners of the FedEx stores (called *contractors*) are responsible for all the costs associated with operating and maintaining their vehicles. Therefore, FedEx does not exercise the degree of control necessary for the relationship to be considered employer-employee. Rather, in this situation, the route drivers are independent contractors who have "significant entrepreneurial opportunity for gain or loss" because they can operate multiple routes, hire additional drivers and helpers, sell routes without permission, and negotiate their price to deliver the packages. Therefore, the rights and duties of employees as agents discussed throughout the chapter do not apply to FedEx drivers.

This case illustrates the importance of understanding agency relations and whether a person is an employee or independent contractor. Although FedEx was successful in the case, this case suggests that it is essential for businesses to have knowledge of the kinds of agency relationships involved in their transactions. In the future, your knowledge about agency relationships could save you or your company large amounts of time and money that would have been spent on litigation.

PRACTICAL TIPS for Business Managers

1. It is very important to know the distinction between an employee and an independent contractor and whether your agent is an employee or an independent contractor under the law. Several businesses have gotten into major legal disputes for classifying their employees as independent contractors. Properly classifying your agents in the first place and fulfilling the obligations of each distinction will save you a lot of trouble.

2. Once you understand the distinction between an employee and an independent contractor, you must then decide which type of agent will best suit your business needs. Employees tend to provide more reliable work but come with numerous obligations. Independent contractors can be good for temporary jobs, but you must maintain a hands-off approach to how they work. See this chapter's Point/Counterpoint for a more in-depth discussion of the pros and cons of employees and independent contractors.

3. Remember that an agency relationship is a two-way street. You owe your agent certain duties as a principal, but the agent also owes certain duties to you. As with most agreements, it is wise to have a written agency contract that lays out these duties and provides options for settling any disputes.

SUMMARY

Introduction to Agency Law

Agency: The relationship between a principal and an agent.

Agent: One authorized to act for and on behalf of a principal.

Principal: One who hires an agent to represent him or her.

Fiduciary: One with a duty to act primarily for another person's benefit.

Creation of the Agency Relationship	*Expressed agency:* Agency formed by making a written or oral agreement. *Power of attorney:* Document giving an agent authority to sign legal documents on behalf of the principal. *Durable power of attorney:* Power of attorney intended to continue to be effective, or to take effect, after the principal has become incapacitated. *Agency by implied authority:* Agency formed by implication through the conduct of the parties. *Agency by estoppel:* Agency formed when a principal leads a third party to believe that another individual serves as his or her agent but the principal has made no agreement with the so-called agent. *Agency by ratification:* Agency that exists when an individual misrepresents himself or herself as an agent for another party, and the principal accepts or ratifies the unauthorized act.
Agency Relationships	An *agency relationship* is a fiduciary relationship (a relationship of trust) in which an agent acts on behalf of the principal. A *principal–agent relationship* exists when an employer hires an employee to enter into contracts on behalf of the employer. An *employer–employee relationship* exists when an employer hires an employee to perform some sort of physical service. An *employer–independent contractor relationship* exists when an employer hires a person other than an employee to conduct certain tasks.
Duties of the Principal and the Agent	*Principal:* - Duty of compensation - Duty of reimbursement and indemnification - Duty of cooperation - Duty of safe working conditions *Agent:* - Duty of loyalty - Duty of performance - Duty of notification - Duty of obedience - Duty of accounting
Rights and Remedies	*Principal:* - Constructive trust - Avoidance - Indemnification *Agent:* - Tort and contract remedies - Demand for an accounting - Specific performance
Authority of the Agent: The Link to the Principal's Liability	*Express authority:* The principal has explicitly instructed the agent to do something. *Implied authority:* The relationship is inferred from the conduct of the parties. *Apparent authority and estoppel:* Apparent agency exists when a third party reasonably believes on the basis of a principal's actions that an agency relationship exists between the principal and another individual.
Contractual Liability of the Principal and Agent	*Classification of the principal:* The principal must be classified as disclosed, partially disclosed, or undisclosed. *Authorized acts:* These are acts within the scope of the agent's authority. *Unauthorized acts:* These are acts that go beyond the scope of the agent's authority.

Tort Liability and the Agency Relationship	*Principal's tortious conduct:* The law holds a principal directly responsible for his or her own tortious conduct under two conditions: (1) the principal directs the agent to commit a tortious act or (2) the principal fails to provide proper instruments and tools or adequate instructions.
Agent misrepresentation: If an agent misrepresents himself or herself to a third party, the principal may be tortiously liable for the agent's misrepresentation.	
Respondeat superior: The principal/employer is liable not because he or she was personally at fault but because he or she negligently hired an agent.	
Principal's liability and the independent contractor: An individual who hires an independent contractor cannot be held liable for the independent contractor's tortious actions under the doctrine of *respondeat superior* unless the contractor engages in hazardous activities.	
Termination of the Agency Relationship	*Termination by acts of parties:* This may occur by lapse of time, fulfillment of purpose, occurrence of a specific event, mutual agreement by the parties, revocation of authority, or renunciation by the agent.
Termination by operation of law: The agency relationship may be terminated automatically due to death, insanity, bankruptcy, changed circumstances, change in law, impossibility, or disloyalty of the agent. |

Point/Counterpoint

Should a Business Owner Hire Independent Contractors Instead of Employees?

YES	NO
Although hiring full-time employees for a business is the more traditional route, hiring independent contractors to work for a business has several advantages that can outweigh the benefits of a full-time employee staff.	
First, hiring independent contractors often saves employers money in the long run. While contractors tend to be paid more per hour, there are several expenses that employers can forgo, such as employer-provided benefits and on-the-job expenses, when hiring contractors. Employers also have to contribute to an employee's Social Security and Medicare taxes, workers' compensation insurance, and unemployment insurance. These expenses add up quickly and can be entirely avoided by hiring independent contractors.
Second, hiring contractors gives an employer much more flexibility than hiring employees does. Hiring an employee is often intended to be a long-term commitment, and employees often require costly training. Independent contractors can be hired for a specific project and then will leave when the job is finished. This possibility is extremely beneficial to employers with fluctuating workloads as an employer doesn't have to deal with firings or layoffs when work is in lower demand. Also, contractors often bring specialized expertise with them, eliminating the need (and the cost) of training that comes with traditional employees. | Hiring independent contractors does seem to have advantages when compared to hiring full-time employees, but independent contractors also come with several disadvantages that make traditional employees the better choice.
First, employers have much less control over an independent contractor than they do an employee. Employers can closely monitor and supervise the work of an employee, but contractors enjoy an extended level of autonomy and can decide for themselves the best way to perform the task the employer has given them. In fact, if an employer interferes too much in the work of an independent contractor, the employer runs the risk of causing the contractor to be considered an employee, which comes with all of the expenses and liabilities that an employer must provide to employees.
Second, because of the temporary nature of independent contractors, workers will be constantly coming and going. Not only can this be inconvenient and disruptive, but an employer would not have a steady, dependable workforce. Independent contractors come in with their own skills and experience and can often do things their own way, which can lead to an uneven quality of work. Training a staff of employees can be more expensive, but these employees will do the same job with the same level of quality every day. To ensure a final product with consistent quality that is produced in a consistent amount of time, employers are better off hiring employees. |

Finally, traditional employees have a wide array of rights guaranteed to them under state and federal laws. These include the right to minimum wage, overtime compensation, unionization, sick leave, and maternity leave. Independent contractors are not entitled to these rights, which can make hiring, firing, and managing contractors a much easier process. Independent contractors also cannot sue their employers for wrongful termination like employees can. (There may be restrictions, however, on the firing of a contractor based upon the contract between the employer and contractor.)

Finally, there are several legal issues that come with the independent nature of contractors that can be avoided with full-time employees. For example, if an employee gets injured on the job, they are usually covered by workers' compensation. In this situation, an employee gives up their right to sue the company in exchange for receiving benefits. Independent contractors, on the other hand, are not covered by workers' compensation and therefore can sue an employer if they are injured on the job. Also, based upon the agreement an employer makes with an independent contractor, the employer may not have an unrestricted right to fire a contractor and may not have the rights to copyrighted material produced by the contractor. Employers have to take extra care when hiring independent contractors that often need not be taken when hiring employees.

Questions & Problems

1. Who can serve as an agent or a principal?
2. What are the similarities and differences between the types of agency relationships?
3. Why is it important to determine whether a worker is an independent contractor or an employee?
4. How is apparent agency, or agency by estoppel, different from expressed agency?
5. What are the different types of authority an agent might have, and why is it important to identify the type of an agent's authority?
6. When might a principal be liable for torts committed by an agent?
7. Explain how the doctrine of *respondeat superior* affects agency relationships.
8. Michael Savage was an employee of the City of Memphis Public Works Division who filed an equal protection claim against the City of Memphis after being terminated. The City's Civil Service commission upheld the termination. Savage subsequently filed a petition for writ of certiorari with the chancery court, which found substantial and material evidence to support the Commission's decision but remanded the case on the equal protection claims.

 Before the hearing took place, the assistant city attorney, Barbaralette Davis, acting with the approval of the city attorney, contacted Savage to discuss if he would be willing to resolve the dispute through a mediation. Savage agreed to mediate, and the two parties agreed to have former Judge Bernie Weinman serve as the mediator. Prior to the mediation, Judge Weinman informed the parties that each side had to bring someone with the authority to settle to the mediation. The mediation came to a successful close, and the parties reached an agreement to settle for $72,000. Davis had Savage sign the settlement and informed Savage that she would "get the ball rolling on getting the check request routed." The settlement agreement was eventually submitted to the city mayor, AC Wharton, to sign, despite the fact that the settlement agreement did not contain any language to indicate that enforceability of the agreement was subject to the mayor's approval. Wharton rejected the settlement agreement upon reviewing it, saying that he knew nothing about the case, and Savage subsequently filed a petition to enforce settlement. In the lawsuit, Savage asserted that the City of Memphis was bound by the settlement agreement because Davis had authority to settle the matter and that Mayor Wharton had delegated the authority to Davis. The city argued that it was not bound to the settlement because Davis had no actual or apparent authority to enter the settlement without the mayor's approval. The trial court granted summary judgement in favor of Savage, and the City of Memphis filed an appeal.

 How did the appeals court rule? What reasons did the judge give? Do you agree with this reasoning?

 Source: *Michael Savage v. City of Memphis,* 464 S.W.3d 326; 2015 Tenn. App.

9. Scheerer, a licensed real estate agent, and Fisher, a manager of commercial real estate, worked together on past real estate deals and knew each other from their business relationship. Scheerer gave Fisher a tip that some commercial real estate was on the market that Fisher might be interested in purchasing. Fisher made an offer to buy the properties and, with this offer, orally contracted for a two percent commission for Scheerer from the seller and another two

percent commission from himself. Unfortunately, that deal fell apart. Shortly thereafter, however, Fisher formed another company that purchased real estate and bought the very same property. Scheerer was not paid a commission on the deal and sued, contending that he was an agent for Fisher and that an implied contract for his services had been breached. Fisher argued that no formal contract in writing regarding the commission or Scheerer's agency relationship was ever created. Did the authority of the agent in this case need to be in writing to be legally binding? Does the fact that there was no separate, written agency contract between Fisher and Scheerer mean that an agency relationship did not exist?

Source: *Scheerer v. Fisher,* 688 S.E.2d 472 (2010).

10. In February 2011, Paula Stenlund traveled to Panama City with her husband. While in Panama City, the Stenlunds were guests at the Panama City Marriott Hotel. Before and during the trip to Panama City, the Stenlunds received Marriott Marketing material that promoted the Royal Casino as being connected to the hotel.

On February 13th, Paula Stenlund visited the Royal Casino. Stenlund followed an employee to locate a place where she could cash her winnings. Stenlund had to descend a set of stairs to reach the redemption area. However, she tripped over an electrical cable draped across the stairway and struck her face, head, and knees. The Royal Casino employee made effort to provide medical help. While the hotel manager sent a doctor to Paula's room, the doctor did not provide substantive care. The Stenlunds returned to the United States on February 15th and Paula commenced an action against Marriott International in May.

Paula alleged that Marriott International breached various duties it owed to her while staying at the Panama City Marriott Hotel and visiting the Royal Casino.

The Panama City Marriott Hotel is owned by Hotel Properties of Panama and the Marriott branding was attached under an agreement between Marriott International and Hotel Properties. In addition, all management at the hotel was done by Marriott International Services, Limited, a company based in Bermuda and separate from Marriott International, in an agreement solely between Marriott Limited and Hotel Properties.

In its opposition to Paula's motion for summary judgement, Marriott International claimed Paula had sued the wrong party. In response, Paula argued that because Marriott International "tightly control[led] virtually every aspect of the [H]otel and [C]asino where Plaintiff was injured," the level of control demonstrated that an agency relationship existed between Marriott International and Marriott Services. In essence, Paula argued that there was either actual agency or apparent agency between the two Marriott companies.

In this case, the court determined that it did not matter whether Panama law or Maryland law applied because the result would be the same. What did the court decide and why? Describe the court's reasoning for both the actual agency count and the apparent agency count.

Source: *Stenlund v. Marriott International, Inc.,* 172 F. Supp. 3d 874 (2016).

11. Kenneth Kinney and Kathleen Gallagher were in a personal relationship during which they agreed to purchase a house. While the couple did buy the house together, Kinney's name was not on the deed because he was participating in a tax repayment agreement with the IRS due to previously owed taxes. The couple separated in January 2005. In August 2005, Kinney filed an action against Gallagher in the New York State Court, alleging that he and Gallagher had orally agreed to put Kinney's name on the deed because he had allegedly contributed $10,000 for the down payment of the house.

In October 2007, Gallagher filed a voluntary Chapter 7 petition with the bankruptcy court. She listed Kinney as a holder of an unsecured claim and her attorney instructed Kinney's attorney to terminate the state court action because of the automatic stay established by the bankruptcy court. Although Kinney was listed on the schedule of debts, he did not file a claim for assets during the bankruptcy proceeding and therefore did not receive any distribution of funds from the estate.

In January 2008, Gallagher was granted a discharge of debts, and the bankruptcy case was closed in December 2011 after the trustee completed the administration of estate assets. Despite the discharge of debts, Kinney continued the state court action, and, on October 3, 2013, Gallagher filed a motion for contempt against Kinney, asserting that the continuation of the state action was a violation of the Chapter 7 discharge. Kinney argued that he was not a creditor because his action was based on a constructive trust for the return of property. The court explains the four elements of a constructive trust. What are they? What scenarios would a constructive trust based on these elements be relevant in a principal-agent relationship? How did the court rule and why?

Source: *Kenneth C. Kinney v. Kathleen M Gallagher,* 524 B.R. 455; 2015 U.S. Dist.

12. Jose Vargas and Luis Felipe Villalobos were a two-man team independently contracted to Eves Express, Inc., to drive tractor-trailers. FMI, Inc. is a

federally licensed motor carrier that conducts shipping business. FMI's shipping distribution center is located in San Pedro, California. In 2010, FMI hired Eves Express as a contractor to deliver cargo from California to New Jersey. Vargas and Villalobos were the drivers assigned for the job. On January 10, 2010, during the trip, Vargas was resting in the sleeper-berth while Villalobos was driving when Villalobos drifted to sleep at the wheel and rolled the tractor-trailer over, injuring Vargas. In April 2011, Vargas sued FMI, Eves Express, and Villalobos for negligence. Vargas contended that both FMI and Eves Express were under a nondelegable duty to members of the public, including himself, because they are were operating a vehicle under a public franchise; the defendants were therefore liable for the negligence of Villalobos in injuring Vargas. Furthermore, Vargas argued that since the FMI and Eves Express were engaged in interstate commerce, they were each vicariously liable for Villalobos's negligence. Finally, Vargas argued that Eves Express had failed to properly train, monitor, and supervise Villalobos which led to the accident. FMI and Eves Express filed a motion for summary judgment, contending that under California law, all workplace safety responsibilities and tort liabilities are implicitly delegated to an independent contractor; because Vargas was an independent contractor, they did not own Vargas a duty to provide a safe workplace. The court entered summary judgement for FMI and Eves Express and Vargas timely appealed.

Consider the California law which implicitly delegates workplace safety responsibilities and tort liabilities from the contracting party to the independent contractor in terms of the WPH framework. Who are the stakeholders of the law and how are they affected by the law? Think beyond the obvious stakeholders such as the independent contractor and think of individuals such as prospective employees. Do you agree with the trial court's decision? Why or why not? How did the appeals court rule? What was the appeals court's reasoning?

Source: *Jose Vargas, Plaintiff and Appellant, v. FMI, Inc., et al., Defendants and Respondents,* 233 Cal. App. 4th 638 (2015).

13. This case arises from a principal appealing the damages it was awarded by a court from its agent. The defendant William Pompa had breached his duty of loyalty to plaintiff Wall Systems, Inc., by working for both Wall Systems and a competitor at the same time. Pompa also accepted three kickbacks from subcontractors in connection with his work for Wall Systems. As part of the remedy, the trial court also placed a constructive trust on a joint bank account held by Pompa and his wife Jill Pompa.

During the appeal, Jill Pompa asserted that the trial court had erred in putting a constructive trust on the joint bank account. What is a constructive trust and under what conditions may a court impose one? Wall Systems acknowledged that while there was no evidence that William Pompa had deposited the profits he received from his disloyal activities, "that it is fair to infer that he did, because the defendant testified that he deposited his paychecks from MK Stucco there, and there was no evidence that he and Jill Pompa had other bank accounts." Is this argument enough to show that the trial court was correct?

Source: *Wall Systems, Inc. v. Pompa,* 154 A. 3d 989 (2017).

Part 7 *Business Organizations*

CHAPTER 21

Forms of Business Organization

CASE OPENER

The Dunkin' Donuts Franchise Agreement

Dunkin' Donuts Corporation operates numerous restaurants worldwide, organizing many of them as franchises. Dunkin' Donuts has the exclusive license to use and to license others to use its trademarks, service marks, and trade name. These marks and trade name have been used continuously since 1960 to identify Dunkin's doughnut shops as well as the doughnuts, pastries, coffee, and other products associated with those shops. Dipak N. Bhayani operated two Dunkin' Donuts franchises in Illinois for many years. Dunkin' Donuts later notified Bhayani that his two franchises had been violating parts of the franchise license agreement. After repeated incidents and failure to cure the violations over a substantial period of time, Dunkin' Donuts (the franchisor) demanded termination of both of Bhayani's franchises.

1. Did Dunkin' Donuts lawfully revoke Bhayani's franchises?
2. What are some potential problems that a franchisor and a franchisee might experience in their relationship?

LEARNING OBJECTIVES

After reading this chapter, you will be able to answer the following questions:

LO 21-1 Differentiate among the major forms of business organization.
LO 21-2 Describe the advantages of each alternative form of business organization.
LO 21-3 Explain the special significance of the operating agreement in a limited liability company.
LO 21-4 Recognize the differences among the different forms of business organization.
LO 21-5 List the specialized forms of business organization.

E-COMMERCE and the Law

Exploring Forms of Business Organization on the Internet

If you are considering starting a business, the Internet can provide much information to help you decide which form of business you should create. For example, at http://smallbusiness.findlaw.com/incorporation-and-legal-structures.html, you can read more about sole proprietorships, partnerships, and corporations. Furthermore, using the Internet, you can learn about the laws that affect the forms of business within your specific state. For example, at Texas Business Forms (http://www.sos.state.tx.us/corp/forms_option.shtml), you can read about and actually retrieve the forms required to create various business types in Texas.

Suppose that you get an idea to produce a novel product. You think that production of this new product could lead to enormous profits, but what is the best way to produce this product? Should you produce it yourself by creating your own business? Do you have enough money to create your own business? What are the legal ramifications for you if your business is not successful? What legal responsibilities do you have with respect to your business?

Maybe you share your idea for this new product with your best friend, who suggests that the two of you become partners in the production and sale of this product. What are the benefits associated with forming a partnership? What are the disadvantages? Are there other forms of businesses that you should consider?

Deciding what form of business to create is one of the most important decisions a business makes. The extent of liability, as well as the extent of the control the owner will have over the business, depends on the form of the business. However, the business world is not static, and businesses can, and do, change form over time. Accordingly, this chapter is relevant not only to new businesses but also to existing businesses. The first section in this chapter introduces the major types of business organization, describing how these forms are both created and ended. The second section considers several types of business organization that are less well known but important nevertheless.

Major Forms of Business Organization

LO 21-1 Differentiate among the major forms of business organization.

Sole Proprietorship

If you decide to go into business on your own, you are creating a **sole proprietorship**, a business organization in which you, as the **sole proprietor**, are in sole control of the management and the profits. Thus, if you wanted to open a lawn-mowing business or a sewing shop, you would likely be creating a sole proprietorship. It is a naturally occurring form of business organization in the sense that if you started doing business without thinking about the form, you would be a sole proprietor.

Why might an entrepreneur choose to create a sole proprietorship over other business organizations? First, a sole proprietorship requires very few legal formalities. Thus, one advantage of the sole proprietorship is the ease of creating such a business. Second, a sole proprietor has complete control of the management of the business. Consequently, the sole proprietor has great freedom to hire employees, determine business hours, and expand or change the nature of the business. Finally, the sole proprietor keeps all the profits from the business. These profits are taxed as the personal income of the sole proprietor.

However, sole proprietorships have disadvantages too. Suppose you are the sole proprietor of a restaurant in which someone is injured. This customer sues your business. You are personally liable for any losses or any obligations associated with the business. Consequently, if you accrue large debts because of your business, you might have to sell your home to cover those debts. Moreover, because the sole proprietorship is not considered a separate legal entity, you, as the

> **sole proprietorship**
> A business in which one person (sole proprietor) is in control of the management and profits.
>
> **sole proprietor**
> The single person at the head of a sole proprietorship.

Exhibit 21-1

Advantages and Disadvantages of the Sole Proprietorship

ADVANTAGES	DISADVANTAGES
1. Creation is easy.	1. Proprietor has personal liability for all losses.
2. Proprietor is in total control of management.	2. Funding is limited to personal funds and loans.
3. Proprietor keeps all profits.	

owner and sole proprietor, can be personally sued. Sole proprietorships are terminated, however, when the sole proprietor dies.

Not only are you personally liable for any debts of the business, but the funding for your business is limited to your personal funds as well as loans you might be able to obtain. Thus, sole proprietorships often struggle in the initial stages of business because of large start-up costs in relation to the profits they make.

Sole proprietorships are by far the most popular form of business organization in the United States. Exhibit 21-1 summarizes their advantages and disadvantages.

An alternative form of business organization that retains many of the advantages of the sole proprietorship but that addresses the funding drawback, in part, is the partnership.

Partnership

Suppose you and your best friend from college decide to create a business in which you plan to buy and sell used books and DVDs through the Internet. You both agree that you will share control of the Internet business and will split the profits equally. According to the Uniform Partnership Act (UPA), you have created a **partnership,** a voluntary association between two or more persons who co-own a business for profit. Except in a few cases, a partnership is not considered a separate legal entity and is dissolved when a partner dies. The UPA governs partnerships in most states in the absence of an express agreement and applies even to partnerships that do not formally file for a partnership or hold elections.

What are the advantages associated with creating a partnership? First, the creation of a partnership is easy. The partners, each of whom is considered an agent of the partnership, are generally not required to create an official or even written agreement to create the partnership. Second, because the partnership is in most cases not considered a separate legal entity, the income from the business is taxed as individual income for each partner. Because the income is taxed as personal income, the partners can deduct the business losses.

Although there are certain advantages to creating a partnership, there are also disadvantages associated with partnerships. Most important, the partners are personally liable for the debts of the partnership. For instance, suppose you are in a partnership with your best friend, who embezzles $50,000 through your partnership. Because of the partnership, you would likely be held personally liable for the debts of the partnership. In other words, you would likely be responsible for the $50,000. Exhibit 21-2 summarizes the advantages and disadvantages of partnerships.

Although it is not listed in Exhibit 21-2, one advantage of a partnership is the ability to choose the people with whom you want to be partners. As the Case Nugget demonstrates, the courts recognize the right to choose one's partners as a basis for partnership law.

partnership
A voluntary association between two or more people who co-own a business for profit.

Exhibit 21-2

Advantages and Disadvantages of Partnerships

ADVANTAGES	DISADVANTAGES
1. Creation is easy.	1. Each partner has personal liability for all losses, including those of another partner (in most cases).
2. Income of business is personal income.	
3. Business losses can be deducted from taxes.	

CASE Nugget

Do Partners Have a Duty to Remain Partners?

Colette Bohatch v. Butler & Binion Supreme Court of Texas

977 S.W.2d 543

Colette Bohatch became an associate in the Washington, D.C., office of Butler & Binion in 1986. John McDonald and Richard Powers, both partners, were the only other attorneys in the office. After Bohatch was made a partner in February 1990, she became concerned that McDonald was overbilling Pennzoil, the main client of the office. Bohatch met with the law firm's managing partner, Louis Paine, to report her concern regarding McDonald's overbilling. In July 1990, McDonald met with Bohatch to report that Pennzoil was dissatisfied with her work.

The next day, Bohatch relayed her concern to Paine as well as to two other members of the law firm's management committee. Paine led an investigation of Bohatch's complaint and discussed the billed hours with the in-house counsel at Pennzoil, who concluded that the bills were reasonable. In August 1990, Paine met with Bohatch, telling her that there was no basis for her claims against McDonald and that she should look for work elsewhere. The firm refused Bohatch a year-end partnership distribution for 1990. Finally, in August 1991, Bohatch was given until November to vacate her office. She filed suit in October 1991, and the firm voted to expel her from the partnership three days later.

At trial, the jury ruled that the firm breached the partnership agreement and its fiduciary duty and awarded Bohatch $57,000 for past lost wages, $250,000 for past mental anguish, $4,000,000 total in punitive damages (this amount was apportioned against several defendants), and attorney fees. Later, the trial court reduced the punitive damages to around $237,000. The court of appeals ruled that the firm's only duty to Bohatch was to avoid expelling her in bad faith. When the court found no evidence that the firm fired Bohatch for self-gain, it ruled that Bohatch could not recover for breach of fiduciary duty. The case was appealed to the state supreme court.

The supreme court ruled that the firm did not owe Bohatch a duty to remain partners. Although Bohatch presented a compelling argument that firms should not be permitted to expel partners who whistle-blow in good faith, the court could not ignore one basis of partnership law: a belief that partners should be allowed to choose their partners. The court found that forcing firms to keep whistle-blowing partners would cause partners to engage in business with people they could no longer trust, thereby undermining the viability of the partnership. Therefore, the Texas Supreme Court affirmed the ruling of the court of appeals.

Duties of Partners to One Another Most of the duties that partners hold to one another involve a duty to be loyal. They include the fiduciary duty to the other partners, the duty of obedience, and the duty of care.

Perhaps the most important duty of the partners is their fiduciary duty. They must, in good faith, work for the benefit of the partnership and refrain from taking any kind of action that will undermine the partnership. Consequently, the partners must not engage in any business that competes with the partnership.

Partners must disclose any material facts affecting the business. If a partner derives some kind of benefit from the partnership without the consent of the other partners, he or she must notify the partners of this benefit.

The second duty that the partners have to each other is the duty of obedience, to obey the partnership agreement. If they do not obey the agreement, they can be held liable for any losses that the partnership incurs.

The third duty the partners have is a duty of care to the other partners. Each partner must perform her management functions to the best of her abilities. If a partner makes an honest mistake in fulfilling her responsibilities to the partnership, she will not be held liable for the mistake.

Formation of the Partnership Although an explicit written agreement is not required to create a partnership, the partners are advised to create one. The written agreement ensures that the terms of the partnership will be upheld. For example, suppose that you and your partner orally agree that you will receive three-fourths of the profits because you are doing significantly more management tasks. However, when you distribute the funds, your partner sues you because you give him only one-fourth of the profits. Without a written partnership agreement, the courts will have a difficult time ruling in your favor. A written agreement that creates a partnership is called the **articles of partnership**.

What kind of information do the articles of partnership usually include? First, the partners's names, as well as the name of the partnership, should be listed on the document. Second, the agreement should address the duration of the partnership. The agreement could include the date or event on which the partnership agreement would expire. Alternatively, the agreement could specify that the term of the partnership is indefinite. Third, the agreement should state the division of profits and losses. Fourth, the articles of partnership should establish the division of management duties. Fifth, the agreement should state exactly what capital contributions will be made by each partner. Creating articles of partnership can prevent legal problems by explicitly establishing the terms of the partnership agreement.

In some cases, parties who are not named in partnership agreements can be considered partners. How? Suppose you create a partnership agreement with your best friend. When you interact with your first potential customer, you tell this customer that your parents are also partners in this business. On the basis of your parents's participation in the partnership, the customer decides to place an order to purchase certain goods from you. Your parents discover that you have reported that they are your partners, but they do not contact the customer to tell her that they are not your partners. When your business cannot afford to purchase these goods to sell them to the customer, the customer sues you as well as your parents. Because your parents were aware of the misrepresentation but did not correct it, they will be *estopped* from denying they are your partners. Although they will not be able to claim the rights associated with being a partner (e.g., sharing the profits), in many states they could be held liable for the costs of the damages to the customer.

Most states recognize two situations in which a *partnership by estoppel* exists. First, as in the preceding example, if a third party is aware of a misrepresentation of partnership and consents to the misrepresentation, a partnership by estoppel is present. Second, if a nonpartner has represented himself or herself as a partner, and a third party *reasonably relies* on this information to his or her detriment, the nonpartner can be held liable for the third party's damages.

Termination of the Partnership Before the termination of any partnership can be considered complete, a partnership must experience what are referred to as the *dissolution stage* and the *winding-up stage*. The first of these stages, dissolution, is considered complete when any partner stops fulfilling the role of a partner to the business (by choice or default). Section 29 of UPA defines **dissolution** as "the change in the relation of the partners caused by any partner's ceasing to be associated with the carrying on, as distinguished from the winding up"—the activity of completing unfinished partnership business, collecting and paying debts, collecting partnership assets, and taking inventory—"of the business."

Depending on the reason for the dissolution, the dissolution will be considered rightful or wrongful. Exhibit 21-3 lists reasons that may be given to dissolve a partnership rightfully.

Once a partnership is liquidated, the partners begin the process of **winding up**, the activity of completing unfinished partnership business, collecting and paying debts, collecting partnership assets, and taking inventory. During the winding up, the partners must still fulfill their fiduciary duty to one another, in the sense that they must disclose all information about the partnership assets. However, during the winding-up process, the partners can now engage in business that competes with the partnership business.

articles of partnership
The written agreement that creates the partnership.

dissolution
The first stage in the termination of a partnership.

winding up
The second stage in the termination of a partnership.

Exhibit 21-3

Reasons for Rightful Dissolution of a Partnership

1. The term established in the partnership agreement expires.
2. The partnership meets its established objective.
3. A partner withdraws from the partnership at will.
4. A partner withdraws in accordance with the partnership agreement.
5. A partner is expelled from the partnership in accordance with the partnership agreement.
6. A partner dies.
7. A partner is adjudicated bankrupt.
8. The business of the partnership becomes illegal.
9. A partner is adjudicated insane.
10. A partner becomes incapable of performing the duties as established by the partnership agreement.
11. The business of the partnership can be carried on only at a loss of profits.
12. A disagreement between the partners is such that it undermines the nature of the partnership.
13. Other circumstances of the partnership necessitate the dissolution.

Who can demand that the winding-up process begin? If a partnership has been rightfully dissolved, any partner can demand the winding-up stage to begin. However, if a partner wrongfully dissolves a partnership, that partner has no right to demand a winding up.

When Does a Partnership Dissolve?

In re Leah Beth Woskob, Debtor; Alex Woskob; Helen Woskob; the Estate of Victor Woskob v. Leah Beth Woskob, Appellant U.S. Court of Appeals for the Third Circuit

305 F.3d 177 (2002)

In 1996, Leah Beth Woskob and Victor Woskob formed a partnership, the Legends Partnership, to construct, own, and operate the Legends, an apartment building. They were married at the time that the partnership was formed; however, they separated and filed for divorce the following year. During the divorce proceedings, Victor prevented Leah from receiving any of the partnership proceeds. Leah was granted a petition for special relief and awarded the exclusive right to manage and derive income from the partnership. Shortly thereafter, Victor filed for bankruptcy. Leah continued to file tax returns on behalf of the partnership, each of which listed Victor as a general partner. When Victor died in a car accident in 1999, Leah gave his estate notice that she was exercising her right to buy out Victor's interest in the partnership. Victor's estate sued, claiming that the partnership had been dissolved previously and requesting that someone be appointed to oversee the winding up of the partnership and a full accounting of the company's assets. When Leah filed for bankruptcy, the suits were moved to the bankruptcy court. The bankruptcy court ruled in favor of Leah, finding that the partnership had dissolved on Victor's death. Victor's estate appealed to the district court, which found that the partnership had dissolved two years before Victor's death, making Leah's attempt to buy out Victor's interest untimely. Leah appealed.

(continued)

The task before the appeals court was to determine the timeliness of Leah's attempt to buy out Victor's interest in the partnership, which depended entirely on the date of the dissolution of the partnership. The court looked to the Uniform Partnership Act (UPA), which defined the dissolution of a partnership as "the change in the relation of the partners caused by any partner ceasing to be associated in the carrying on, as distinguished from the winding up, of the business." Victor's estate claimed that the dissolution of the partnership occurred at any one of three points, each at least 18 months prior to Victor's death. First, Victor excluded Leah from the partnership after they separated; second, Leah excluded Victor from the partnership after seeking special relief from the Court of Common Pleas; third, Victor filed for bankruptcy. The appeals court found that the exclusions of Leah and Victor from the partnership were not, in and of themselves, grounds for automatic dissolution of the partnership. Rather, the exclusions could have provided a basis for dissolution, had either Leah or Victor sought judicial decree of the dissolution after the exclusions. In addition, bankruptcy in and of itself is not grounds for automatic dissolution of the partnership. If the nondebtor partner does not consent to continue the partnership with the debtor, bankruptcy may be grounds for dissolution. However, Leah continued to list Victor as a general partner on the tax returns she filed for the partnership, even after he filed for bankruptcy. Thus, the appeals court found that the partnership had not dissolved prior to Victor's death in 1999 and that Leah's attempt to buy out Victor's interest in the partnership was therefore timely.

Different Types of Partnerships

general partnership
A partnership in which the partners divide profits and management responsibilities and share unlimited personal liability for the partnership's debts.

There are several types of partnerships: general partnerships, limited partnerships, and limited liability partnerships. A **general partnership** consists of an agreement that the partners will divide the profits (usually equally) and management responsibilities and share unlimited personal liability for the partnership's debts. Thus, in our Internet business example, you and your best friend would form a general partnership by agreeing to share the management responsibilities and profits as well as assuming unlimited personal liability.

Now imagine that your parents want to invest in your Internet business. Suppose that they want to share in the profits associated with the business, but they do not want to share in the management responsibilities or assume personal liability for the debts of the partnership. Your parents can join your business as limited partners, and your partnership would become a limited partnership. A **limited partnership (LP)** is an agreement between at least one general partner and at least one limited partner. The *general* partners, you and your best friend, assume unlimited personal liability for the debts of the partnership. However, your parents, the *limited* partners, assume no liability for the partnership beyond the capital they have invested in the business. Moreover, the limited partners do not have any part in the management of the company. However, limited partners pay taxes on their share of the business profit.

limited partnership (LP)
A partnership consisting of at least one general partner and at least one limited partner in which the general partners assume all liability for the partnership's debts, and the limited partners assume no liability beyond their originally invested capital.

If a limited partner dies, the limited partnership is usually unaffected. If a general partner dies, however, the limited partnership is usually dissolved.

The limited partnership must meet certain requirements that are not expected of general partnerships. First, the limited partnership must use the word *limited* in its title. Second, to create a limited partnership, the parties must file a certificate of partnership with a state office. If the certificate of partnership is incorrectly filed, or simply not filed at all, the courts will rule that a general partnership exists. Consequently, all parties will be held personally liable for all the debts of the partnership.

limited liability partnership (LLP)
A partnership in which all the partners assume liability for any partner's professional malpractice to the extent of the partnership's assets.

Now suppose that you are an attorney and a partner in a law firm with 30 other partners. Suppose further that one of your partners is sued because he was negligent in his duties as an attorney. This partner has unlimited liability because of professional malpractice. But will you and the other partners at the firm be held liable for this partner's malpractice?

If you and your fellow partners have created a **limited liability partnership (LLP)**, a different form of partnership than the LP, all the partners assume liability for any partner's professional

GLOBAL Context

©Digital Archive Japan/Alamy

Limited Liability Partnerships in Japan

The limited liability partnership was not introduced in Japan until 2005. Japan's LLP limits the liability of its members to their capital contribution. Regarding profits, the LLP requires the members to manage the business and negotiate among each other to determine how profits and losses should be distributed among individual members. When establishing the rules of the LLP, Japan requires either the agreement of all members or a majority of at least two-thirds of the members.

Japan also established additional limitations and regulations to ensure the preservation of assets of the LLPs. In Japan, the Limited Liability Partnership Law requires members to work to preserve the assets of limited liability partnerships by putting restrictions on which types of capital contribution are permitted and on the distribution of partnership assets.

malpractice to the extent of the partnership's assets. In other words, the limited liability partnership is different from other forms of partnership because the partners's liability for professional malpractice is limited to the partnership. If one partner in an LLP is guilty of malpractice, the other partners's personal assets cannot be taken. Therefore, professionals who do business together commonly use the LLP. It is the extra protection awarded partners in an LLP that makes the LLP a *separate* form of partnership from a limited partnership. Limited partnerships are not the same as limited liability partnerships.

BUT WHAT IF? . . .

WHAT IF THE FACTS OF THE CASE OPENER WERE DIFFERENT?

Let's say that in the Case Opener, two women, Sara and Melissa, owned Dunkin' Donuts as a partnership. Let's say that Melissa was embezzling funds from the business that made the company incur $50,000 in debt. Sara had all of the documentation to prove that Melissa had embezzled the funds from the business, and she wanted only Melissa to be liable for the debt. How would the court decide in this case?

Limited liability partnerships are fairly new; in 1991, Texas was the first state to enact a statute permitting the creation of LLPs. Almost all states now have similar statutes. Like the limited partnership, the LLP has several special requirements. First, the business name must include the phrase *Limited Liability Partnership* or the abbreviation *LLP*. Second, the parties must file a form with the secretary of the state to create the LLP.

The LLP is not considered a separate legal entity. Each partner pays taxes on his or her share of the income of the business.

Exhibit 21-4
What's the Difference: Types of Partnerships

GENERAL PARTNERSHIP	LIMITED PARTNERSHIP	LIMITED LIABILITY PARTNERSHIP
A general partnership consists of an agreement between the partners that profits and management responsibilities will be shared, as will all personal liability for the partnership's debts.	A limited partnership consists of at least one general partner, and at least one limited partner. In this arrangement, the general partners assumes all liability for the partnership's debts, while the liability of the limited partner is restricted to their original capital investment.	In a limited liability partnership, all partners assume liability for any partner's professional malpractice to the extent of the partner's assets.

Corporation

corporation
A legal entity formed by issuing stock to investors, who are the owners of the corporation.

shareholders
Investors in a corporation, who own the corporation.

When you hear the word *business,* you probably think about businesses such as Walmart, Kmart, McDonald's, and Nike. Perhaps the most dominant form of business organization is the **corporation,** a legal entity formed by issuing stock to investors, who are the owners of the corporation. The investor-owners are called **shareholders.** These shareholders elect a board of directors, which is responsible for managing the business. The board of directors, in turn, hires officers to run the day-to-day business.

The corporation is considered a separate legal entity; thus, the corporation can be sued. None of the other forms of business we have discussed are considered separate legal entities. How does a corporation become a separate legal entity? It must be created according to state law.

What are the consequences of a corporation's status as a separate legal entity? First, although the corporation can be held liable, shareholders cannot be held personally liable for the debts of the corporation. Their liability is usually limited to the amount they have invested. Second, the corporation is not dissolved when the shareholders die. Third, the corporation must pay taxes on its profits. In addition, the shareholders must pay taxes on the dividends they receive from the corporation.

S corporation
A corporation that enjoys the tax status of a partnership.

One way that a corporation can avoid this double taxation is by forming an **S corporation,** a business organization formed under federal tax law that is considered a corporation yet is taxed like a partnership as long as it follows certain regulations. For example, the S corporation cannot have more than 100 shareholders. Any income of the corporation is taxed when it is distributed to the shareholders, who must report the income on their personal income tax forms. S corporations are always formed under federal law. S corporations cannot be formed under state law, whereas other forms of corporations are created under state law.

Exhibit 21-5 summarizes the advantages and disadvantages of the corporate form of business.

Limited Liability Company

limited liability company (LLC)
An unincorporated business that is taxed like a partnership, with the members paying personal income taxes, and has the limited liability of a corporation.

One of the newest forms of business organization in the United States is the **limited liability company (LLC),** an unincorporated form of business organization that many people see as combining the most advantageous features of partnerships and corporations. It combines the tax advantages and management flexibility of a partnership with the limited liability of a corporation.

This form of business organization was first recognized in the United States in 1977 in Wyoming and is now recognized in every state, although the rules regarding this new form did not evolve uniformly. As with other areas of law, the National Conference of Commissioners on Uniform State Laws recognized the desirability of bringing some uniformity to this area of law and drafted the Uniform Limited Liability Company Act (ULLCA) in 1995. In 2006, the commissioners revised the ULLCA. This act provides a model for states to follow, but it has not been uniformly adopted, so it is always necessary when establishing an LLC to check the exact requirement of the law for LLCs in the state in which you wish to create your company.

LO 21-2 Describe the advantages of each alternative form of business organization.

Key Reasons for the Rapid Acceptance of LLCs As previously mentioned, the LLC form offers its members the same limited liability for business debts as that offered by the corporate form. However, unlike the corporate form, it does not require profits and losses to be allocated in proportion to ownership interests. Also unlike corporations, LLCs are not required to hold an

Exhibit 21-5

Advantages and Disadvantages of Corporations

ADVANTAGES	DISADVANTAGES
1. Limited liability for shareholders	1. Corporate income taxed twice
2. Ease of raising capital by issuing stock	2. Formalities required in establishing and maintaining corporate form
3. Profits taxed as income to the shareholders, not the partners	

annual meeting and draft meeting minutes, so record keeping is simpler and more flexible. Unlike the case in limited partnerships, to obtain limited liability, the owner (referred to as a *member*) does not have to give up his or her right to participate in management of the LLC. In fact, an additional advantage of the LLC form is the flexibility it offers members in terms of alternative ways to structure its management.

Because the LLC essentially provides the best parts of other forms of business organization, it has become a very popular method of doing business and has become the most common choice of organization for new small businesses

The most frequently cited advantage of the LLC is that the IRS generally treats it like a partnership or sole proprietorship, with members reporting their share of the profits and losses of the LLC on their personal tax returns, and no separate tax is assessed on the company itself, thereby allowing its members to avoid the double taxation that owners of a corporation pay. However, if an LLC's members prefer, they may elect to have the entity taxed like a corporation. In a situation in which most of the profits will be reinvested in the business, this option allows the profits to be taxed at the lower corporate rate. So, although we think of the opportunity to avoid double taxation as a key benefit of the LLC, more important, perhaps, is the fact that the members have the choice of how they want to be taxed.

In our global environment, an increasingly important advantage of LLCs is that members need not be citizens or permanent residents of the United States. Other organizational forms, such as the subchapter S corporation, are available only when all the owners are U.S. citizens. Finally, as with a corporation, ordinary business expenses, such as salaries paid to owners, can be deducted from the profits of an LLC before the LLC's income is allocated to its owners for tax purposes.

LO 21-3 Explain the special significance of the operating agreement in a limited liability company.

Formation and Management of LLCs A limited liability company is formed by filing articles of organization in the state in which members want to establish their LLC. Although precise requirements vary by state, typically the articles include the name of the business, which must include the words *Limited Liability Company* or the initials *LLC*, its principal business address, the name and address of a registered agent for service, the names of the owners, and information about how the company's management will be structured.

LLCs typically want to do business in states other than the state in which they are formed, and they usually need to register in every additional state in which they want to operate, a process typically referred to as *qualification*. Qualification simply entails filing a certificate of authority or a similar document and getting a business license in each additional state in which the business plans to operate. The LLC will usually be referred to as a *foreign company* in the additional states, and, under most state statutes, the LLC will be governed by the LLC rules of the state in which it was created, regardless of where it is transacting business.

For purposes of jurisdiction, however, an LLC is considered a citizen of every state in which its members reside. Remember that a party can be sued in federal court over a matter involving more than $75,000 when diversity of citizenship exists, that is, when no plaintiff and defendant are residents of the same state. For determining whether diversity exists, a corporation is considered a resident of both the state in which it is incorporated and the state in which it has its primary place of business. However, this rule does not apply to LLCs because their citizenship is determined by the residences of their members. Consequently, if parties want the flexibility to avail themselves of the federal courts, they may want to limit their membership to individuals of only one or a few states or use a different form of business organization.

When members form an LLC, they typically draft an operating agreement, which is the foundational contract among the entity's owners. It spells out such matters as how the company is to be managed, how the profits and losses will be allocated, how interests may be transferred, and how and when the LLC may be dissolved. Any matter not covered in the operating agreement will be resolved in accordance with the state LLC statute; if a matter is not covered by the relevant statute, the principles of partnership law are generally followed.

Although an LLC is not required to have a detailed, written operating agreement, to ensure the smooth functioning of the company, it is a good idea to have such an agreement. Failure to have a

detailed, written agreement may result in a court imposing standards on the company that may be very different from what the members had in mind when forming the company.

It is very important, however, to make sure that everyone understands the terms of the operating agreement, because once the parties enter into such an agreement, the courts will enforce it.

One of the key issues to be determined in the operating agreement is how the company is to be managed. There are two options: member-managed and manager-managed. Under the ULLCA, unless the operating agreement specifies otherwise, LCCs are member-managed. In a *member-managed LLC,* all members participate in management, with decisions in the ordinary course of business activities made by a majority vote. The consent of all members, however, is required to sell, lease, exchange, or otherwise dispose of all, or substantially all, of the company's property; approve a merger or conversion to a different form; undertake any other act outside the ordinary course of the company's activities; and amend the operating agreement. The members in such an LLC all have the apparent and actual authority to enter into contracts on behalf of the LLC.

In a *manager-managed LLC,* the members select a group of managers to manage the affairs of the company. The managers may be selected from the members or may be nonmembers. The managers have the apparent and actual authority to enter into contracts on behalf of the LLC, whereas the members who are not managers do not have such authority. The managers owe the LLC and its members the same fiduciary duties as the officers and directors of a corporation owe to the corporation and its shareholders.

BUT WHAT IF? . . .

WHAT IF THE FACTS OF THE CASE OPENER WERE DIFFERENT?

Let's say in the Case Opener that two doughnut shops, Dunkin' Donuts and Harry's Bagels, joined to produce one kind of doughnut that turned out to be a big hit. The two companies decided to form a partnership in producing this doughnut together. What kind of company would this be called?

LO 21-4 Recognize the differences among the different forms of business organization.

Exhibit 21-6 provides a comparison of the alternative forms of business organization.

Dissolution of LLCs Under the ULLCA, an LLC dissolves on the happening of any event that the operating agreement specifies will cause dissolution, the consent of all the members, the passage of 90 consecutive days during which the company has no members, or the issuance of a court order for dissolution. Under the ULLCA and most state LLC statutes, a member's voluntary withdrawal from the LLC, referred to as *dissociation,* does not terminate the LLC.

Specialized Forms of Business Organization

LO 21-5 List the specialized forms of business organization.

In addition to the more traditional forms of business organization mentioned, the following specialized forms of business organization have become increasingly important: cooperatives, joint stock companies, business trusts, syndicates, joint ventures, and franchises.

Cooperative

cooperative
An organization formed by individuals to market new products. Individuals in a cooperative pool their resources to gain an advantage in the market.

A **cooperative** is an organization formed by individuals to market products. The cooperative is a business organization in which the members usually pool their resources to gain some kind of advantage in the market. For instance, farmers might pool certain crops to ensure that they get a high market price for their crops. Usually, members of the cooperative receive dividends in proportion to how many times a year they engage in business with the cooperative.

Exhibit 21-6 Comparison of Alternative Forms of Business Organization

	SOLE PROPRIETORSHIP	GENERAL PARTNERSHIP	LIMITED PARTNERSHIP	CORPORATION	LIMITED LIABILITY COMPANY
Legal Position	Not a separate legal entity.	Not a separate legal entity in most states.	A separate legal entity.	A separate legal entity.	A separate legal entity for purposes of liability but not always for taxes.
Control Considerations	Sole proprietor has total control.	Each partner is entitled to equal control.	Each partner is entitled to equal control.	Separation of ownership and control.	Members control allocation of control.
Liability	Sole proprietor has unlimited personal liability.	Each partner has unlimited personal liability for partnership debts.	Each partner has liability limited up to his or her capital contribution.	Liability is limited to loss of capital contribution.	Liability is limited to loss of capital contribution.
Lifetime	Limited to life of proprietor.	Limited by life of partners.	Limited to life of general partners.	Can have unlimited life.	Operating agreement spells out the lifetime.
Taxation	Profits are taxed directly as income to the sole proprietor.	Profits are taxed as income for partners.	Profits are taxed as income for partners.	Profits are taxed as income to corporation and as income to partners in the form of dividends.	Operating agreement spells out whether profits will be taxed as a corporation or partnership.
Transferability of Ownership Interest	Nontransferable.	Nontransferable.	Nontransferable.	Generally unlimited transfer.	Operating agreement spells out the transferability.

Cooperatives may be incorporated or unincorporated. Unincorporated cooperatives are treated like partnerships, meaning that the members share joint liability for the cooperative's actions. Members of incorporated cooperatives, on the other hand, enjoy limited liability just as do the shareholders of a corporation.

Joint Stock Company

A **joint stock company** is a partnership agreement in which company members hold transferable shares whereas all the goods of the company are held in the names of the partners. Thus, the joint stock company is a mixture of a corporation and a partnership. As with the corporation, the members who hold shares of stock own the joint stock company. As with the partnership, the shareholders have personal liability, and in most cases, the company is not a separate legal entity. The joint stock company is formed by agreement rather than by statute.

Business Trust

A **business trust** is a business organization governed by a group of **trustees,** who operate the trust for the **beneficiaries.** A written trust agreement establishes the duties and powers of the trustees and the interests of the beneficiaries.

joint stock company
A partnership agreement in which company members hold transferable shares, whereas all the goods of the company are held in the names of the partners.

business trust
A business organization governed by a group of trustees, who operate the trust for beneficiaries.

trustee
A person who operates a trust for beneficiaries in a business trust.

beneficiary
A person who can expect to benefit from a relationship.

As with a corporation, the trustees and beneficiaries enjoy limited liability, and, in most states, business trusts are taxed like corporations.

Syndicate

syndicate
An investment group that comes together for the explicit purpose of financing a specific large project.

An investment group that comes together for the explicit purpose of financing a specific large project is a syndicate. Syndicates are often used in the purchase of professional sports teams. The syndicate is quite useful in the sense that it can raise large amounts of money in a small amount of time. Syndicates are usually considered a type of joint venture; thus, they are almost always governed by partnership law.

Joint Venture

joint venture
An association between two or more parties wherein the parties share profits and management responsibilities with respect to a specific project.

A joint venture is a relationship between two or more persons or corporations created for a specific business undertaking. This relationship may entail financing, producing, and selling goods, securities, and commodities. Participants in the joint venture usually share the profits and losses of the joint venture equally.

Joint ventures can be agreements between small businesses as well as agreements between very large businesses. For example, Nestlé and Häagen-Dazs entered into a joint venture to create Ice Cream Partners, USA. Generally, joint ventures are taxed like partnerships. In fact, from a legal standpoint, partnerships and joint ventures are virtually the same. Thus, courts frequently apply partnership law to joint ventures. Joint ventures differ from partnerships, however, because a joint venture is usually created for making and selling a single product, whereas a partnership creates an ongoing full business. Once all the single products are sold, the joint venture is usually terminated. The members, however, can determine when the joint venture will end.

Despite the similarities, there are several minor differences between partnerships and joint ventures in the eyes of the law. First, if one of the members of a joint venture dies, the joint venture is not automatically terminated. Second, the members of a joint venture have less authority than general partners because members of a joint venture are not agents of the other members.

The parties who comprise a joint venture usually share equal management of the task for which they have come together. They, however, can make an agreement to give one party greater management responsibilities. Furthermore, all the parties usually assume liability for the project. Each party can be held responsible for the liability of another party in the joint venture.

Like a partnership, a joint venture may be formed without drawing up a formal agreement. Case 21-1 provides a judicial discussion of the elements necessary for the establishment of a joint venture.

MEYER v. CHRISTIE
U.S. Court of Appeals for the Tenth Circuit
634 F. 3D 1152 (2011)

FACTS: *In March 2005, David Christie and Alexander Glen met with Alan Meyer and John Pratt and allegedly entered into an oral joint venture agreement to purchase and manage an area for residential housing. The four agreed that they would call the joint venture Junction City Partners, that Mr. Christie and Mr. Glen would be 50/50 partners with Mr. Meyer and Mr. Pratt, and that they would hire a contractor from Dovetail Builders. However, a few weeks later, Mr. Christie and Mr. Glen terminated their relationship with Mr. Meyer and Mr. Pratt and formed a corporation, The Bluffs LLC. Christie and Glen then gave 50 percent partnership interest in the The Bluffs LLC to two outside individuals.*

At this time, plaintiffs Meyer and Pratt, joined by Dovetail Builders, filed a claim of breach of the joint venture agreement and claims of breach of fiduciary duty and wrongful dissociation against the defendants, Christie and Glen. The jury found in favor of the plaintiffs on all of these claims and found more than $9 million in damages. The defendants appealed,

(continued)

asserting the lack of existence and enforceability of the alleged joint venture agreement.

ISSUE: Did the defendants breach the joint venture agreement by terminating their contract with the plaintiffs?

REASONING: The defendants argue that a joint venture does not exist because the fourth factor of the following five have not fully been met, and a party must prove these factors to prove that a joint venture exists according to *Terra Venture, Inc. v. JDN Real-Estate Overland Park, L.P.* The five factors for determining whether a joint venture exists are: (1) the joint ownership and control of property; (2) the sharing of expenses, profits, and losses and having and exercising some voice in determining the division of net earnings; (3) a community of control over and active participation in the management and direction of the business enterprise; (4) the intention of the parties, express or implied; and (5) the fixing of salaries by joint agreement.

The court was not persuaded by the defendants' argument. The court reasoned that the evidence presented in this case to prove that a joint venture existed was much stronger than the evidence available in *Terra Venture:*

> In *Terra Venture,* the only evidence suggesting the parties intended to form a joint venture was their agreement that they would refer to their project as a joint venture in press releases. In this case, by contrast, the evidence was that the parties agreed they would in fact form a joint venture, not just that they would refer to their project as such. The evidence also indicated that Plaintiffs and Defendants informed not just the public but also their employees and attorneys that they were jointly working on the project. Furthermore, although it is undisputed the parties did not actually share expenses, profits, and losses, the evidence introduced at trial was that they agreed they would share future profits and losses and would true-up expenses when the corporation paperwork had been completed.

DECISION AND REMEDY: Mr. Meyers and Mr. Pratt's claims against defendants were affirmed.

SIGNIFICANCE: This case discusses the elements that are required to prove that a joint venture exists. The case also provides an illustration of how a judge weighs the strengths and weaknesses of various pieces of evidence to determine that a joint venture exists.

CRITICAL THINKING

The process of critical thinking requires us to ask critical questions to evaluate whatever reasoning we encounter, even if, as in this case, the reasoning appears very convincing. Of the five criteria Judge McKay uses to determine whether a joint venture exists, which criterion do you think he provides the least justification for?

ETHICAL DECISION MAKING

Essentially, the court ruled in favor of the plaintiffs and determined that the defendants did in fact breach a joint venture agreement. What theory or theories of ethical decision making might lead a judge to rule in favor of the defendants instead? Who in the business community would be affected if this theory of ethical decision making was employed? Explain.

Franchise

When you go into McDonald's to eat lunch, what type of business are you patronizing? You are likely eating at a **franchise,** a business that exists because of an arrangement between the **franchisor,** an owner of a trade name or trademark, and the **franchisee,** a person who sells goods or services under the trade name or trademark.

What are the advantages and disadvantages for the franchisor and the franchisee? Exhibits 21-7 and 21-8 summarize these advantages and disadvantages. First, the franchisee enjoys the franchisor's help in opening the franchise. Second, think about how many times you have driven through a strange town and felt relieved when you saw a Burger King. Even though you may have never been to that town before and did not know who controlled that particular business, you know Burger King. Before you walk into the restaurant, you know almost exactly what will be on the menu. Thus, on the one hand, the franchisee benefits from the franchisor's strong trade name or trademark. Moreover, the franchisee benefits from the franchisor's worldwide advertising of the trade name or trademark. The franchisor, on the other hand, does not take a large risk in creating the franchise, yet it can greatly benefit from the income it receives from the franchisee.

franchise
A business arrangement between an owner of a trade name or trademark and a person who sells goods or services under the trade name or trademark.

franchisee
The seller of goods or services under a trade name or trademark in a franchise.

franchisor
The owner of the trade name or trademark in a franchise.

Exhibit 21-7

Advantages and Disadvantages to the Franchisee of Starting a Franchise

ADVANTAGES	DISADVANTAGES
1. Help from the franchisor in starting the franchise	1. Must meet the franchisor's standards or risk losing the franchise
2. Instant recognition due to the franchisor's strong trademark or trade name	2. Has little to no creative control over the business
3. Benefits from the franchisor's worldwide advertising	

Exhibit 21-8

Advantages and Disadvantages to the Franchisor of Starting a Franchise

ADVANTAGES	DISADVANTAGES
1. Low risk in starting a franchise	1. Has little control over the franchise
2. Increased income from franchises	2. Can become liable for the franchise if it exerts too much control

chain-style business operation
A type of franchise in which the franchise operates under the franchisor's business name and is required to follow the franchisor's standards and methods of business operation.

distributorship
A type of franchise in which the franchisor manufactures a product and licenses a dealer to sell the product in an exclusive territory.

manufacturing arrangement
A type of franchise in which the franchisor provides the franchisee with a formula or necessary ingredient to manufacture a product.

Generally, franchises fall into one of three categories. McDonald's and Burger King are examples of chain-style business operations. In a **chain-style business operation,** the franchise operates under the franchisor's business name and is required to follow the franchisor's standards and methods of business operation.

The second category of franchises is the distributorship. **Distributorships** are franchises in which the franchisor manufactures a product and licenses a dealer to sell the product in an exclusive territory. A car dealership is an example of a distributorship.

Finally, the third category of franchises is the manufacturing arrangement. In a **manufacturing arrangement,** the franchisor provides the franchisee with a formula or necessary ingredient to manufacture a product. The franchisee then manufactures the product and sells it according to the franchisor's standards.

It should be noted that a franchise is not a legal entity itself by default. For example, a franchisee can choose to incorporate their franchise or register their franchise as an LLC while operating under a franchise agreement.

Exhibit 21-9 describes the various types of franchises. Exhibit 21-10 provides a sense of just how important franchises are for the market economy.

Exhibit 21-9

Types of Franchises

FRANCHISE TYPE	DESCRIPTION
Chain-style business operation	Franchisor helps franchisee set up a business run under the franchisor's business name according to the franchisor's usual methods and standards.
Distributorship	Franchisor licenses franchisee to sell the franchisor's product in a specific area.
Manufacturing arrangement	Franchisor provides the franchisee with the technical knowledge needed to manufacture the franchisor's product.

Exhibit 21-10

The Top 10 Global Franchises, 2018

1. McDonalds
2. 7-Eleven Inc.
3. Dunkin' Doughnuts
4. The UPS Store
5. RE/MAX
6. Sonic Drive-In
7. Great Clips
8. Taco Bell
9. Hardee's
10. Sport Clips

Source: Ranked by *Entrepreneur Magazine* on the basis of financial strength and stability, growth rate, and size of the system; www.entrepreneur.com/franchise500/index.html.

Franchise Law Because franchisors are usually larger than franchisees and have more resources, they often have the upper hand in franchise relationships. Federal and state laws, however, have been established to protect the franchisee in the franchise relationship.

In addition, several laws that are more specific have been established to govern the franchise relationship. For example, the Federal Trade Commission has a franchise rule requiring franchisors to present prospective franchisees with the material facts necessary for the franchisee to make an informed decision about entering a franchise relationship. Moreover, the Automobile Dealers' Franchise Act of 1965 prohibits car dealership franchisors from terminating franchise relationships in bad faith. Thus, for example, DaimlerChrysler could not terminate its franchise relationship with a dealer because the dealer failed to meet impossible standards. Finally, the Petroleum Marketing Practices Act of 1979 outlines the reasons for which a franchisor may terminate a gas station franchise. Franchises must also be aware of federal antitrust laws. Antitrust laws prohibit specific forms of anticompetitive behavior and might be applicable to franchises.

Case 21-2 spells out the criteria courts use to determine whether a business is a franchise.

LIVE CRYO, LLC v. CRYOUSA IMPORT AND SALES, LLC
United States Distrct Court, E.d. Michigan, Southern Division (2017)

FACTS: *CryoUSA Import and Sales, CryoUSA Franchising, LLC, CryoUSA Holding, LLC, and CryoUSA Mobile, LLC are Texas corporations. Together, they obtain, supply, and provide training to operate cryotherapy chambers. Cryotherapy chambers use liquid nitrogen to maintain a temperature of 240 degrees below zero. Clients enter the cryotherapy chamber to allegedly gain health benefits such as more restful sleep, improved mood, and better athletic performance.*

Live Cryo, LLC, is a Michigan corporation that entered an agreement with the CryoUSA corporations on July 15, 2016. The agreement provided that Live Cryo could exclusively distribute cryotherapy chambers in Michigan for a 24-month period. The agreement also provided that Live Cryo had to purchase 15 chambers within the first six months to maintain the exclusive dealership status. In addition, CryoUSA was to control all marketing and training, with all cryotherapy chambers bearing its mark. Relevant to this case, the agreement also contained a choice-of-law provision that called for the application of Texas law in any legal dispute.

Live Cryo found that the cryotherapy chambers did not work as described. For example, an HVAC technician hired to examine the chambers concluded that it never reached 166 degrees below zero. Experiencing various difficulties, Live Cryo sued the CryoUSA corporations, alleging the defendants had engaged in fraud and were in violation of the Michigan Franchise Investment Law (MFIL). In its defense, CryoUSA argued that Texas law should apply because of the choice-of-law provision in the agreement. Live Cryo contended that MFIL made void the choice-of-law provision. In response, CryoUSA argued that MFIL did not apply because Live Cryo was not a franchisee.

ISSUE: Is Live Cryo a franchise of CryoUSA?

REASONING: Under Michigan law, there are three factors which determine whether a business is a franchise: (a) the franchisee sells or distributes goods under a marketing plan or system prescribed by the franchisor; (b) the franchisee is granted the right to sell or distribute goods or services associated with the franchisor's trademark, logo, advertising, or other commercial symbol; (c) the franchisee is required to pay, directly or indirectly, a franchise fee.

The court found that the first criterion was met. The agreement between the two parties stated that "[Live Cryo] shall be allowed to further the promotion, marketing, sale and other distribution of the Products in the Territory and to keep CryoUSA informed of marketing developments." The court also determined that the second criterion was met as all products sold by Live Cryo bore the CryoUSA name and logo. Live Cryo was also specifically allowed to use CryoUSA marks, including "Cryosense" and "CryoUSA ETablet." Finally, Live Cryo argued that it paid an indirect franchise fee because CryoUSA charged a "mark up" on the purchase of each chamber as well as a $10,000 fee to renew the agreement. Because Live Cryo had to pay this fee to do business with (and, arguably, on behalf of) CryoUSA, the court found the third criterion to be satisfied.

With these facts in mind, the court found that there was a question of fact as to whether the parties entered into a franchise agreement. As such, the court would not enforce the forum selection clause in the agreement and allowed Live Cryo to proceed with its claims.

DECISION AND REMEDY: In favor of Live Cryo.

SIGNIFICANCE OF THE CASE: This case delineates some of the common criteria for determining whether a business is a franchise or not. It also explores a legal implication of that distinction, i.e., whether a choice of law clause is valid based on the distinction.

(continued)

> **CRITICAL THINKING**
>
> What facts would have had to have been different for the court to have ruled that there was no franchise agreement in this case?
>
> **ETHICAL DECISION MAKING**
>
> What stakeholders would benefit from laws that made it more difficult for a business to be a franchise?

franchise agreement
A contract whereby a company (the franchisor) grants permission (a license) to another entity (the franchisee) to use the franchisor's name, trademark, or copyright in the operation of a business and associated sale of goods in return for payment.

Creation of the Franchise In the franchise relationship, the parties make a **franchise agreement** regarding various factors: the payment to the franchisor, the location of the franchise, the restrictions the franchisee must follow, and the method of termination of the franchise.

First, the franchise agreement usually states that the franchisee is to pay a large sum to the franchisor for use of the trade name or trademark. Moreover, the agreement determines what percentage of the income from the sales of the goods will go to the franchisor.

Second, the agreement usually determines where the franchise will be located. If the franchise requires a building, the agreement specifies who will pay for buying or renting the building. Similarly, if the building must be constructed, the agreement states who is responsible for the costs of the construction of the building.

Third, the franchisor usually includes in the agreement certain business practices that are forbidden and other business standards that must be met. For example, the franchisor might require the business to meet specific levels of cleanliness. Moreover, the franchisor can set certain sales quotas as well as requirements for record keeping by the franchisee. The franchisor can require the franchisee to purchase certain supplies from the franchisor at a set price, but the franchisor cannot attempt to establish the price at which the franchisee must sell the goods.

The disagreement in the opening scenario for this chapter arose because of the third factor in franchise agreements. Because many Dunkin' Donuts restaurants are owned by franchisees, Dunkin' Donuts established guidelines and policies that promote business practices that enhance the quality of food and services at each restaurant. Dunkin' Donuts also has quality, safety, and cleanliness standards for each of its franchises. The franchise agreement stipulated that Dunkin' Donuts could inspect Bhayani's restaurant at any reasonable time.

Although the franchisor has the legal authority to ensure that the franchisee maintains the quality of goods and services associated with the franchise, the franchisor must be cautious. If it exercises too much authority in the day-to-day affairs of the business, the franchisor could be held liable for the torts of the franchisee's employees.

Termination of the Franchise The franchise agreement establishes how the franchise will be terminated. The franchise is usually established for a trial period, such as a year. If the franchisee does not meet the requirements established in the franchise agreement, the franchisor can terminate the franchise agreement, but the franchisor must give the franchisee sufficient notice of the termination. Furthermore, the termination usually must have cause. Much of the litigation associated with franchises regards wrongful termination of a franchise. The typical agreement gives the franchisor broad authority to terminate a franchise. In recent years, however, many states have been giving the franchisee greater termination protection.

The courts usually rely heavily on the written franchise agreement when determining whether a franchise was wrongfully terminated. Look at Case 21-3, which illustrates the importance of the written franchise agreement.

CASE 21-3

COUSINS SUBS SYSTEMS, INC. v. MICHAEL R. MCKINNEY
U.S. District Court for the Eastern District of Wisconsin
59 F. Supp. 2d 816 (1999)

FACTS: *Cousins Subs Systems entered into an agreement with Michael McKinney, a Minnesota businessman who owned a company that operated a chain of gas stations, for McKinney to operate several Cousins submarine sandwich shops. The sub shops were to be placed in the gas stations. In April 1998, McKinney became disillusioned with the agreement and terminated the agreement. McKinney claimed that Cousins guaranteed him annual sales at each of his franchises ranging from $250,000 to $500,000. Furthermore, McKinney argued that Cousins promised to provide advertising for the franchises. McKinney also claimed that Cousins guaranteed it would provide assistance in recruiting other franchises. McKinney alleges that he terminated the agreement because Cousins failed to uphold these promises. In June 1998, Cousins filed suit against McKinney for wrongfully terminating the agreement with Cousins. Later in 1998, McKinney filed a counterclaim against Cousins. Cousins filed a motion to dismiss the counterclaim.*

ISSUE: Did Cousins Subs Systems make misleading statements of material facts?

REASONING: McKinney alleges that Cousins violated the Minnesota statute prohibiting businesses from using false statements of material facts (or omissions of material facts) to sell franchises. McKinney claims that Cousins made false verbal statements of material facts; however, when an alleged verbal statement contradicts a written contract, the written contract carries more weight. In this case, the written contracts, signed by McKinney and Cousins and later entered as exhibits, contradict the verbal statements McKinney claims Cousins made. The contracts state that no promises were made regarding the profit McKinney would earn, that advertising was McKinney's responsibility, and that the contracts contained all terms, conditions, and promises made by Cousins.

DECISION AND REMEDY: As demonstrated by the contracts McKinney signed, Cousins did not make any misleading statements of material facts. The counterclaim was dismissed.

SIGNIFICANCE: It is unclear whether Cousins made the verbal promises alleged by McKinney. Regardless, McKinney did sign contracts that contradicted the alleged verbal promises. This case demonstrates how important it is for businesspeople to read contracts thoroughly and to ensure that all terms, conditions, and promises are contained in those contracts.

CRITICAL THINKING

What are the primary facts of this case? How would you word the issue of the case in your own words?

The judge stated that the written terms of the contracts between Cousins and McKinney were inconsistent with any alleged oral agreements that they made. Do you agree that written contracts should overrule oral agreements in most instances? Why or why not?

ETHICAL DECISION MAKING

The court ruled in favor of Cousins. Who are the primary stakeholders affected by the court's ruling for Cousins?

CASE OPENER WRAP-UP

Bhayani's Dunkin' Donuts franchises violated the license agreement on multiple levels. Financially, under the franchise agreements, Bhayani agreed to pay a franchise fee of 4.9 percent of gross sales to Dunkin' Donuts and an advertising fee of 5 percent of gross sales to the Franchise Owners Advertising and Sales promotion fund. However, Bhayani fell behind on both financial payments on numerous occasions. By the time Dunkin' Donuts sent notice of termination, Bhayani owed $33,189.38 in delinquent fees and failed to cure the financial default.

(continued)

Furthermore, Bhayani's franchises were also in violation of health and safety standards from the license agreement. During its routine health inspections, Dunkin' Donuts found multiple health and sanitation violations such as pests and evidence of pests; improper storage, refrigeration, and cooking temperatures; improper food and chemical storage; unsanitized utensils; faulty faucets; unclean floors, walls, countertops, toilets, and sinks; insufficient employee hygiene; ill-kept trash areas; and various documentation deficiencies. While some of these were cured by the time of the next inspection, most were not. After each substandard inspection, Dunkin' Donuts sent a notice of default and notice to cure. On the basis of the perceived failure to cure these violations over a substantial period of time, Dunkin' Donuts sent Bhayani a supplemental notice of termination.

As a franchisor, Dunkin' Donuts Corporation is permitted to establish certain standards for franchisees. With regard to Bhayani's restaurants, Dunkin' Donuts established standards for cleanliness and also negotiated financial rates. In accordance with the provisions in the license agreement, Dunkin' Donuts terminated the franchises. Bhayani claimed that any breaches of the agreement by him—either financial breaches or health, sanitation, and safety violations—were directly caused by the bad-faith actions of Dunkin' Donuts. He argued that Dunkin' Donuts targeted him for his franchisee activism by classifying him as a "C" franchise and blocking his attempt to open another franchise. However, the court found that Bhayani did not show that his franchises were exceptional or that they were terminated on the basis of some sort of "pretext" of Dunkin' Donuts. Thus, Dunkin' Donuts was well within its legal rights to terminate the franchise agreement.

Disagreements regarding payments or health standards of franchises are examples of what could go wrong with a franchising agreement. Another example of potential problems between franchisors and franchisees is disagreement over the termination of the franchise. All of these problems existed in the Dunkin' Donuts case. Both parties probably would have benefited from a greater understanding of the responsibilities of the franchisor and franchisee.

PRACTICAL TIPS for Business Managers

1. As this chapter demonstrates, there are many, many ways to organize your business, each with its distinct advantages and disadvantages. Before starting a business, take ample time to think about what type of business model will best suit your needs. Consider questions such as, what type of product or service do I want to provide? Should I operate this business on my own, or do I need help running it? Will I need to raise funds or capital at some point in my business's life? Do I intend to manage this business for a long time? Thinking about these and other factors can help you decide how best to form your business.

2. Although there are statutes that cover all forms of doing business, each type of business organization has its own set of rules and regulations as well. For example, the laws surrounding partnerships are different from the ones concerning corporations. Be sure to know the law surrounding your particular type of business and do not assume that just because an LLC can do action X means that your franchise or joint venture can also do X.

3. The cases in this chapter should indicate to you that franchises, while beneficial in many ways, come with their own complex legal disputes concerning employees and agreements. If you decide to start a franchise, do your best to keep communication open between yourself, your franchisor, and your employees so that each party knows its duties and rights under the franchise agreement.

SUMMARY

Major Forms of Business Organization

Sole proprietorship: The owner has total control and unlimited personal liability, and profits are taxed directly as income to the sole proprietor.

General partnership: For most purposes, the partnership is not a legal entity, and each partner has equal control and unlimited liability, with profits that are taxed as income for partners.

Limited partnership: Limited partnerships are similar to general partnerships, except that limited partners's liability is limited to the extent of their capital contributions.

Corporation: A corporation is a separate legal entity wherein the owners's liability is limited to the amount of their contributions, and the profits are taxed as income to the corporation.

Limited liability company: A limited liability company combines the best tax and liability features of partnerships and corporations. In an LLC, all members may participate in management of the business entity while enjoying a shield of limited liability for their personal assets.

Specialized Forms of Business Organization

The specialized forms are cooperatives, joint stock companies, syndicates, business trusts, joint ventures, and franchises.

A *franchise* is a contractual relationship between the franchisor and the franchisee. Thus, contract law, and the Uniform Commercial Code in particular, applies to the franchise relationship. If the terms of the contract are not met, either side can sue for breach of contract.

Point/Counterpoint

Should Aspiring Business Owners Form an Llc Instead of a Corporation?

YES	NO
While traditional corporations certainly have their benefits, LLCs present a hybrid form of business that combines that best parts of corporations and partnerships. There are several reasons to form an LLC instead of a corporation. First, forming and running an LLC can be much easier than doing so for a corporation. LLCs require much less paperwork and registration to form, and LLCs can deduct business losses—such as those incurred while starting the business—on their personal tax returns. Corporations are also required to have a board of directors that oversee business decisions and are required to hold meetings and publish annual reports. LLCs have no such requirement and so are much more flexible in how they can be operated. In addition, LLCs can distribute ownership according to any agreement instead of based solely on capital put into the company. For example, an LLC can guarantee that all members receive equal share of future profits, even if the members have invested unequal amounts of money into the LLC. LLCs also enjoy several tax benefits that corporations do not. For example, LLCs get to avoid the double taxation that certain corporations can face. LLC members are only taxed on their individual incomes, whereas corporations have both their profits and dividends to shareholders taxed. In most situations, LLCs can save a lot of money by bypassing this double taxation. However, the flexibility inherent to LLCs allows them to be taxed as an S- or C-corporation under certain circumstances, which may prove to be beneficial in specific cases.	Although LLCs are a novel alternative to the traditional structure of a corporation, the benefits of a corporation can still outshine the benefits of an LLC in many cases. There are many reasons to start a corporation over an LLC, but perhaps the greatest advantage corporations have over LLCs is the ability to issue stock. Issuing stock is a reliable and time-tested method of increasing investment in a business and allowing it to grow. An LLC can get along without stock, but as the business grows, investors will likely begin to want tangible ownership in the business, and that's where stock comes in. A business owner is essentially capping the growth of his business if he decides to start an LLC instead of a corporation. Stock is not only an effective means of raising capital and growing a business, but it also serves as a way to acquire and maintain loyal employees. By offering stock options to a business's best employees, a company can provide an incentive for employees to not only work hard for the business but to remain working for the business. Other benefits corporations have over LLCs are the ability to leave money in the corporation (as opposed to all profits going directly to members and being taxed), the ability to provide benefits to shareholders who are also employees (and deduct the cost), and that the laws regarding corporations is much more established than the laws concerning LLCs and therefore it's easier to know how the law will react to a corporation than to an LLC.

Questions & Problems

1. What is the distinction between a general partnership and a limited partnership?
2. Explain why a cooperative could not claim to be a syndicate.
3. Suppose you were asked to review and assess a franchise agreement. What responsibilities would you expect to be included in that agreement?

4. Tracy Allen rented a home in which she had been residing since the summer of 1998. The landlord-owner of the house later failed to pay taxes on the property, and, on March 16, 2000, Hard Assets LLC acquired the property from the owner in lieu of foreclosure. When Hard Assets obtained title, it did not intend to lease the property, nor were its members aware that Allen was living at the property. Once Hard Assets became aware that Allen was residing at the property, Hard Assets advised her and her family that they were not supposed to be there and that they had 30 days to vacate the premises. After 30 days, Allen had not vacated the property. While residing at the property, the children living with Allen suffered elevated blood-lead levels and were allegedly injured from lead paint exposure. As a result, the children's mother filed a suit against one of the members of the LLC for violations of the city's housing code and negligence. Should the member of the LLC be held personally liable for the alleged injuries of the children who lived at Allen's home? Why or why not?

 Source: *Allen v. Dackman,* 964 A.2d 210 (2009).

5. P is a residential development company with a single member. The member's responsibilities "were to locate and purchase property, and then work with engineering companies and municipalities to have the property zoned and fully developed for residential living." P entered into a contract with D to excavate a property purchased by P. The member signed on behalf of P, and D signed on behalf of himself (Perrin) and Perrin Excavating.

 Shortly thereafter, Duray Development and Perrin entered into a new contract, which was supposed to supersede the former contract. This contract was between Duray Development and Outlaw—which is an excavation company that Perrin and Vining owned. Perrin and Vining signed the new contract on behalf of Outlaw. They both also held themselves as owners of Duray Development as well.

 D then began excavation, but did not perform as expected on time. Duray Development then sued D for breach of contract. D later learned that Outlaw did not obtain an adequate filing status as an LLC in time, thus Outlaw was not a valid LLC at the time of the second contract. D argues he isn't personally liable because he signed on behalf of Outlaw, which was treated by all parties as a properly formed LLC. In so doing he cites the de facto corporation doctrine.

 Limited Liability Act – "the existence of the limited liability company begins on the effective date of the articles of organization . . . the document is effective at the time it is endorsed." However, the Michigan Supreme Court stated that once the articles of operation are executed and acknowledged, they are a de facto corporation. The idea is to prevent a poorly formed corporation from not being recognized as a corporation because of the defect when the parties acknowledged and believed it to be one.

 What ambiguity in this Michigan Supreme Court decision is likely to lead to further litigation to sharpen the boundaries of this decision?

 Source: *Court of Appeals of Michigan,* 792 N.W.2d 749 (2010).

6. Mike Karimi has been in the hotel management business since the 1980s. The name of Karimi's company is MAK LLC, which manages and operates several hotels in the United States. In 2004, two brothers, the Khatris, who were friends of Karimi's, wanted him to take over the operation of one of their hotels. The problem with the hotel was that only 40 out of the 186 hotel rooms were suitable for guests to inhabit. This concerned Karimi, who nevertheless agreed to take over the hotel. Karimi received the rights to the hotel from the previous leasers. The hotel, Red Lion, wrote up a franchise agreement with Karimi. In the agreement, it states: "An 'Event of Default' will occur if you fail to satisfy or comply with any of the obligations, requirements, conditions, or terms set forth in this Agreement, the Manual, or any attachment to this Agreement. An 'Event of Default' will also occur if you make any misrepresentations to us, whether in entering into this Agreement, or in the performance of your obligations to us." After the renovations were underway, a representative of Red Lion Hotels decided to tour the hotel to check on the renovations. This representative decided that the hotel looked "old and tired." In early 2008, Red Lion sent Karimi a "notice of default and termination." Did Red Lion have the right to send this notice to Karimi?

 Source: *Red Lion Hotels Franchising Inc. v. MAK, LLC,* 663 F.3d 1080 (2011).

7. Brothers Aurelio and Hugo V. Garcia founded a trash removal business called Garcia's Inc. in the early 2000s. Hugo served as president and director. Garcia's decided to lease trucks and equipment to United Leasing. However, United required, as a part of the agreement, that the stock of Garcia's would be collateral and that United could vote Garcia's share in the company if Garcia's defaulted. Suddenly, United claimed that Garcia's had defaulted, and United liquidated the company's assets and took control. James C. Lehner, a former employee of United, formed the Lehner Family Business Trust in 2005, and shortly thereafter, Lehner offered Hugo $50,000 to assign all claims against United to the trust. Hugo executed the requested assignment, which assigned all of the legal claims, rights, and causes of action that Garcia's, "a Virginia corporation in dissolution," and Hugo,

"individually and as trustee in liquidation for Garcia's, Inc.," had under the equipment leases with United and otherwise. The Lehner Trust sued United Leasing for breach of contract, claiming that "[b]y assignment, the Trust acquired all of Garcia's claims, causes of action, choses in action, rights of action, rights and interests against [United Leasing] and all claims and rights, etc., relating to the Garcia's Leases." Also, the trust claimed that United had been secretly holding over $1 million from Garcia's, which rightfully deserved it. Was the Trust right in suing United Leasing?

Source: *United Leasing Corp. v. Lehner Family Business Trust,* 279 Va. 510 (2010).

8. 1-800-Got Junk? is a junk removal franchise business headquartered in Vancouver, British Columbia, Canada. In late 2003, Got Junk entered a franchise agreement with Millenium Asset Recovery Inc. The agreement stated that the franchisee, Millennium, would pay a percentage of its gross revenue to Got Junk on every junk removal job it performs. In 2007, Got Junk terminated Millennium's franchise on the grounds that Millennium deliberately had not reported certain jobs and the gross revenue derived from such jobs. Was Got Junk right in terminating the agreement? Why or why not?

Source: *1-800-Got Junk? LLC v. Superior Court (Millennium Asset Recovery, Inc.),* Cal.App.4th, Second Dist., Div. Three. (2010).

9. Yumilicious, a company that franchises frozen yogurt restaurants in Texas, entered a franchising agreement with Why Not, LLC. The agreement was also personally guaranteed by the principals of Why Not, Matthew Barrie, and Kelly and Brian Glynn. The agreement franchises two Yumilicious frozen yogurt stores in South Carolina.

Why Not faced difficulties turning a profit, partly because a supply agreement negotiated by Yumilicious fell through which meant Why Not could not economically purchase supplies in South Carolina. As a result of its difficulties, Why Not closed one of its stores without permission from Yumilicious and failed to make payment for royalties and products. Yumilicious subsequently sued.

Why Not countersued Yumilicious, alleging that it had violated the Texas Deceptive Trade Practices Act, specifically the provision that made it illegal for a franchisor to "represent[] that goods or services have ... characteristics [or] benefits ... which they do not have" or to "fail[] to disclose information concerning goods or services which was known at the time of the transaction if such failure to disclose such information was intended to induce the consumer into a transaction into which the consumer would not have entered had the information been disclosed."

As proof of the violations, Why Not alleged that Yumilicious had failed to provide an updated Franchise Disclosure Document. Furthermore, the Franchise Disclosure Document that was provided underestimated start up costs and failed to mention the out-of-state supply agreement that would supply Why Not could fall through.

Which category of franchise does the franchise in this case fall under? How did the court rule on Why Not's counterclaims and what was its reasoning?

Source: *Yumilicious Franchise, LLC v. Barrie,* 819 F. 3d 170 (2016).

10. In April 2008, The First Pentecostal Church of Beaumont settled an insurance claim for property damage from Hurricane Rita. The church received $1,094,611.02. During this time, the church was embroiled in a sexual harassment lawsuit. Concerned that the settlement money could be targeted in the lawsuit, Lonnie Treadway, the pastor of the church, decided to let its defending law firm, Lamb Law Firm, P.C. hold the settlement money. The money was deposited into an account in Lamb's name. However, over the course of a year, the funds were transferred by Kip Lamb, the owner of Lamb Law, to a separate account held by Lamb Law. The funds totaling $1.2 million were spent for Kip Lamb's personal expenses and Lamb Law's operating expenses. In 2011, Leigh Parker, an attorney contracted by Lamb Law to work on the church's case, informed the church's board of trustees. Parker had known that Lamb had spent the funds since the summer of 2010 but had not informed the church until the church had requested the transfer of the settlement funds. In 2012, the church sued Lamb and Parker. Lamb also faced criminal charges and was eventually sentenced to fifteen years in prison. In its case against Parker, the church asserted that Parker was jointly and severally liable to it because he was part of a joint venture with Lamb, among other things. Specifically, the church argued that Parker and Lamb were part of a joint venture to misappropriate the church's money and then hiding the fact. What are the four elements of a joint venture as listed by the Texas Supreme Court in this case? The court ruled against the church for this argument. What was its reasoning?

Source: *First United Pentecostal v. Parker,* 514 SW 3d 214 (2017).

11. Keith Hammond drafted and signed articles of organization for Jetmar Properties LLC in November 2002, but he did not file the articles with the secretary of state. Later that same year, Hammond, acting as president of Jetmar, offered to buy commercial property from Selwin Ortega. Hammond and Ortega signed a purchase agreement, but the sale did not

close because Hammond did not have the money. Nonetheless, based on Hammond's representation that he needed cash to develop the property into condominiums, Ortega gave Hammond a three-day, unsecured $200,000 loan, which Hammond failed to repay. In April 2003, Hammond met Dale Stone, a retiree, and convinced him to invest in several of Jetmar's "real estate ventures." During the course of their relationship, Stone gave Hammond more than $50,000 in cashier's checks for Jetmar's various projects. Sometime in May 2003, Hammond asked Stone to quitclaim a duplex to Jetmar to improve Jetmar's balance sheet, which would allow it to secure financing for a large condominium development. Stone was renting out the property to supplement his Social Security income. In exchange for the deed, Hammond promised Stone an interest in the development. Hammond also told Stone that he would deed the property back to Stone "free and clear" in sixty days and that Stone could continue to collect rent. On May 14, 2003, Stone quitclaimed the property to Jetmar. Hammond purported to accept the deed as Jetmar's president and recorded it the same day. On May 15, 2003, Hammond mortgaged the duplex to Ortega in exchange for an extension on the $200,000 loan. Ortega recorded the mortgage on May 20, 2003, after checking the title and determining that Jetmar had title to the property by virtue of a quitclaim deed from Stone. Hammond did not repay the loan or deed the property back to Stone. Sometime around December 2003, Ortega began foreclosure proceedings. Ortega sent the duplex tenants a notice of foreclosure, which they passed along to Stone. Stone confronted Hammond about the mortgage, but he was told that there would be time to redeem and regain title to the property. Based on Hammond's assurances, Stone did not alert Ortega of his claimed interest. On March 2, 2004, Ortega conducted a sale under the foreclosure-by-advertisement procedures. Because there were no higher bidders, Ortega purchased the property in exchange for the surrender of his $200,000 claim against Jetmar. Hammond filed Jetmar's articles of organization on March 11, 2004, and received a certificate of organization. Stone brought this action in October 2004, alleging that Hammond and Jetmar had defrauded him. Stone sought damages and a declaratory judgment that he was the owner of the duplex. What do you believe the court will rule?

Source: *Stone v. Jetmar Properties, LLC,* 733 N.W.2d 480 (Minn. App. 2007).

Part 7 *Business Organizations*

CHAPTER 22

Corporations: Formation and Organization

CASE OPENER

The Formation of the Facebook Corporation

On October 28, 2003, Mark Zuckerberg was a sophomore at Harvard University when he created a website called Facemash that was similar to an existing web service called Hot or Not. However, the following semester, in 2004, Zuckerberg began working on new code for a new website to be called Facebook. His friends Eduardo Saverin, Dustin Moskovitz, Andrew McCollum, and Chris Hughes joined Zuckerberg to promote the new social networking site. Membership on the site quickly grew from only students at Harvard College to students at most universities in the United States. In the summer of 2004, Facebook was incorporated. As a corporation, Facebook is an artificial person, a status with legal ramifications for both the corporate entity and its owners.

1. What are the legal implications of Facebook's status as a corporation?
2. How are corporations formed? What factors should a businessperson consider in forming a corporation?

LEARNING OBJECTIVES

After reading this chapter, you will be able to answer the following questions:

LO 22-1 Recall the powers granted to corporations by the states.
LO 22-2 Summarize the process of formation of the corporation.
LO 22-3 Explain the role of a director, an officer, and a shareholder in a corporation.
LO 22-4 List the duties of a director, an officer, and a shareholder in a corporation.
LO 22-5 Summarize the liabilities of a director, an officer, and a shareholder in a corporation.
LO 22-6 Explain the rights of a director, an officer, and a shareholder in a corporation.

This chapter explains the steps necessary to establish a corporate entity. Although state law generally governs corporations and each state has its own corporate regulatory statutes, the Revised Model Business Corporation Act (RMBCA) is the basis of most state statutes. More than 25 states have adopted at least part of RMBCA.

In addition, the chapter describes the legal relationships among the many groups of individuals with various priorities and agendas within a corporation. Not surprisingly, these priorities and agendas often come into conflict. To ensure that individuals, corporations, and the public achieve fair outcomes to conflicts within corporations, statutory laws have been designed to delegate particular roles, duties, and rights to each group of individuals within corporations.

Characteristics of Corporations

How are corporations different from other forms of business organization?

Legal Entity

In U.S. law, corporations are legal entities. In other words, corporations exist separately from their shareholders. Thus, corporations can sue or be sued by others.

Rights as a Person and a Citizen

Courts consider corporations to be legal persons. Thus, in most cases, corporations, like natural persons, have certain rights according to the Bill of Rights. Specifically, the Fifth and Fourteenth Amendments state that government cannot deprive any person of life, liberty, or property without due process. Courts have held that corporations are persons in this case and thus have a right to due process.

Furthermore, courts consider corporations to be persons with respect to the Fourth Amendment. Hence, the Fourth Amendment protects corporations from unreasonable searches and seizures. Finally, courts consider corporations to be persons that have free-speech rights protected by the First Amendment.

Case 22-1 discusses the conditions that limit the free speech rights of corporations.

PROTECT MY CHECK, INC. v. CRAIG C. DILGER, JOHN STEFFEN

United States District Court, E.D. Kentucky, Central Division, Frankfort

176 F. Supp. 3D 685 (2016)

FACTS: *Protect My Check, Inc. is a section 501(c)(4), an organization that must not be organized for profit and must be operated exclusively to promote social welfare. While it is based in Florida, PMC conducts business in Kentucky. PMC's purpose in Kentucky is to "expand" employee rights by promoting legislation for "right to work" protections. To carry out its mission, PMC makes both direct and indirect contributions to candidates for state and local offices, and to organizations that support such candidates.*

Section 150 of Kentucky's Constitution forbids corporations from contributing money or other assets to influence elections. PMC brought this action against the Kentucky Registry of Election Finance to obtain a preliminary injunction to seek relief from Kentucky's political corporate contribution laws. In its argument, PMC asserted that Section 150 and other related laws violated its right to free speech under the First Amendment of the federal Constitution.

ISSUE: Does forbidding PMC from making contributions to political campaigns violate its right to free speech?

REASONING: The First Amendment of the Constitution holds that Congress may pass no law that infringes upon freedom of speech. *Citizens United v. FEC* struck down a ban on independent corporate expenditures (i.e., corporations

(continued)

can indirectly make contributions to political campaigns through PACs). PMC argued that the ruling in *Citizens United* should also strike down Kentucky's ban on direct contributions because Kentucky law does not allow a PAC option. Kentucky argued that *Citizens United* only applies to independent expenditures and does not affect the law in question.

The court pointed out that there are four main reasons that a state has historically banned corporate contributions, the most relevant one here being in the interest of preventing corruption between the corporations and the candidates. Using *FEC v. Beaumont,* the court argued that the anti-corruption interest is a sufficient one to ban corporate contributions, and the court pointed out that this rationale was even directly affirmed by the *Citizens United* court. In addition, PMC was unable to show any evidence elsewhere that bans on direct contributions were found unconstitutional due to the lack of PAC option. Thus, Kentucky's ban was constitutional as it was put in place to serve the anti-corruption interest.

DECISION: Kentucky's motion granted in part and denied in part.

SIGNIFICANCE: This case demonstrates that speech can be constitutionally restricted in some cases to serve a significant public interest.

CRITICAL THINKING

How does this case heighten our awareness of the danger of interpreting a right as always and ever trumping all other rights?

ETHICAL DECISION MAKING

What values are responsible for limiting the speech rights of corporations?

Creature of the State

State incorporation statutes establish the requirements for corporate formation. Each individual corporation's charter creates a contract between that corporation and the state.

Limited Liability

Because corporations are legal entities distinct from their shareholders, corporations are liable for corporate actions. Shareholders' liability, therefore, is limited to their investment in the corporation. For example, in 1977, Big O Tire Dealers sued Goodyear Tire & Rubber Company. Big O Tire accused Goodyear of copying its BIGFOOT trademark on new tires. The court agreed and awarded Big O Tire several million dollars in damages. The Goodyear corporation, and not individual Goodyear shareholders, paid the damages. Although payment of damages to Big O Tire may have reduced the dividends shareholders received, the court did not hold the shareholders individually liable for any portion of the award.

Free Transferability of Corporate Shares

Generally, shareholders can freely transfer their corporate shares. For example, shareholders can sell their shares or give them to charity.

Perpetual Existence

The life of a corporation does not end when the lives of its owners end. If shareholders die, corporations do not dissolve. Similarly, if corporate directors or officers withdraw or die, the corporation continues to exist. In some cases, the *articles of incorporation,* a document a corporation files with the state explaining its organization, may include a restriction on the duration of the corporation. Otherwise, in most states, corporations continue to exist indefinitely.

A small number of states, however, set a maximum length of existence for corporations. After the maximum duration expires, corporations must formally renew their corporate existence.

Centralized Management

Unless the articles of incorporation specify otherwise, shareholders do not participate in corporate management. Instead, shareholders elect a board of directors. The board, in turn, selects officers to manage the day-to-day business of the corporation.

Corporate Taxation

Because corporations are separate legal entities, government taxes their income directly. (S corporations are an exception; this chapter discusses them, and their tax advantages, in detail later.) Corporations must pay federal and state taxes on their income.

Corporations, however, have control over their income. Corporations can distribute their income to shareholders in the form of **dividend,** although they do not receive tax deductions for distributing these dividends. When corporations distribute income to shareholders, the shareholders pay taxes on that income. Because the corporation pays income taxes and the shareholders are also paying taxes on their dividends, the dividends are subjected to double taxation, thus creating a disadvantage for corporations. Alternatively, corporations can keep profits, or **retained earnings,** to reinvest. Corporations' investment of retained earnings can lead to higher stock prices, thus benefiting shareholders when they sell their stock.

dividend
A distribution of corporate profits or income ordered by the directors and paid to the shareholders.

retained earnings
Profits that a corporation keeps.

> **BUT WHAT IF? . . .**
>
> **WHAT IF THE FACTS OF THE CASE OPENER WERE DIFFERENT?**
>
> Let's say in the Case Opener that after Zuckerberg had Facebook incorporated, he and the other corporate shareholders were on a private jet when it crashed, killing all the travelers. Without any of the shareholders, what would happen to Facebook? Does state or federal government put a limit on how long a corporation can exist?

Liability for Officers and Employees

LO 22-1 Recall the powers granted to corporations by the states.

Because the relationship between corporations and their directors, officers, and employees is an agency relationship, corporations are liable for torts and crimes committed by their agents during the scope of their employment. Courts refer to this liability as the doctrine of *respondeat superior* (Latin for "let the master answer"). Although in the past courts were reluctant to impose criminal liability on corporations, prosecutions today are much more common.

Exhibit 22-1 presents a summary of nine characteristics of corporations.

Exhibit 22-1

Characteristics of Corporations

1. Separate legal entity
2. Status as "legal person" and "citizen"
3. Creature of the state
4. Limited liability of shareholders
5. Free transferability of corporate shares
6. Perpetual existence
7. Centralized management separate from owners
8. Unique taxation method
9. Corporate liability for torts and crimes of agents

Corporate Powers

Because corporations are creatures of the state, they have only those powers states grant them. States give powers to corporations through state incorporation statutes and through each corporation's articles of incorporation.

Express and Implied Powers

State incorporation statutes typically grant the following express powers to corporations: the power to have perpetual existence, the power to sue and be sued in the corporation's name, the power to acquire property, the power to make contracts and borrow money, the power to lend money, the power to make charitable donations, and the power to establish rules for managing the corporation. In addition, corporations may take whatever actions are necessary to execute these express powers. Thus, corporations have implied powers. Generally, the statement of corporate purpose in each corporation's articles of incorporation gives each corporation its implied powers.

Ultra Vires Act

If corporations act beyond their express and implied powers, the act is called an *ultra vires* (Latin for "beyond powers") act. Corporations commit *ultra vires* acts most frequently when they create contracts outside the scope of their powers. While these acts are not common, they can be very important when they do occur.

Today, most state incorporation statutes permit businesses to incorporate for any lawful purpose; thus, most articles of incorporation do not limit corporate powers. Hence, the doctrine of *ultra vires* is much less significant today than it was in the past because very few corporate acts qualify as *ultra vires*.

Historically, courts have ruled that *ultra vires* acts are null and void. More recently, however, courts have permitted corporations to use the *ultra vires* defense if neither party to the contract has performed the terms of the contract. Yet courts uphold *ultra vires* contracts if one of the parties has executed her or his part of the contract.

For example, suppose a timber corporation's articles of incorporation specified that the corporation's purpose was to harvest and sell timber. Suppose further that the directors of the corporation entered into an *ultra vires* contract with a machinery company to purchase machines to mine coal on the corporation's cleared land. As long as the machinery company has not delivered the machinery to the timber corporation, courts will hold the contract null and void. If, however, the machinery company has delivered the machinery to the timber corporation, courts will uphold the *ultra vires* contract and require the timber corporation to fulfill the terms of the contract. Can you see which party the court is trying to protect by upholding the *ultra vires* contract in this situation?

If a corporation commits an *ultra vires* act, RMBCA provides several remedies:

1. Shareholders may sue to prohibit the corporation from fulfilling the *ultra vires* contract.
2. The corporation or shareholders may sue corporate directors or officers for the damages caused by the *ultra vires* act.
3. The state attorney general can have the corporation dissolved or prevent the corporation from fulfilling the *ultra vires* contract.

Classification of Corporations

Corporations can be classified as public or private; profit or nonprofit; domestic, foreign, or alien; publicly or closely held; S corporations; or professional corporations.

Public or Private

A **public corporation** is a corporation created by government to help administer law. Thus, public corporations often have specific government duties to fulfill. The Federal Deposit Insurance

public corporation
A corporation that is created by government to help administer law.

Corporation (FDIC) is an example of a public corporation. In contrast, private persons create **private corporations** for private purposes. Private corporations do not have government duties.

Profit or Nonprofit

Most corporations are **for-profit corporations**. Thus, their objective is to operate for profit. Shareholders seeking to make a profit purchase the stock these corporations issue. This profit can take two forms. First, shareholders receive dividends from the corporation. Second, the market price of the stock can increase. Shareholders can then sell their stock at a higher price than the purchase price of the stock.

Nonprofit corporations may earn profits, but they do not distribute these profits to shareholders. In fact, nonprofit corporations do not have shareholders. More important, nonprofit corporations' objective is not to earn profit. They do not issue stock. Instead, nonprofit corporations provide services to their members (not shareholders). These corporations reinvest most of their profits in the business. Churches and charitable organizations are examples of nonprofit corporations.

Domestic, Foreign, and Alien Corporations

Every corporation is incorporated in a particular state. A corporation is a **domestic corporation** in the state in which it is incorporated. Many corporations, however, do business in more than one state. A corporation that does business in states other than the state in which it is incorporated must obtain a certificate of authority in each state in which it does business. A corporation is a **foreign corporation** in states in which it conducts business but is not incorporated. For example, the McDonald's Corporation is incorporated in Delaware but does business in all 50 states. Thus, McDonald's is a domestic corporation in Delaware and a foreign corporation in the other 49 states.

An **alien corporation** is a business incorporated in another country. Thus, if a U.S. corporation wants to do business in Canada or Mexico, it is an alien corporation in those countries.

Publicly Held or Closely Held

The stock of **publicly held corporations** is available to the public. Thus, if you want to invest in a corporation, you could purchase stock in a publicly held corporation. Most publicly held corporations have many shareholders, and managers of these corporations usually do not own large percentages of the corporation's stock. Shareholders wishing to sell their shares do not face many transfer restrictions.

In contrast, **closely held corporations** (also called *close, family,* or *privately held corporations*) generally do not offer stock to the general public. Shareholders are usually family members and friends who are often active in the business. Controlling shareholders typically manage closely held corporations. Because closely held corporations are often family businesses, they often maintain restrictions on the transfer of shares to prevent outsiders from obtaining control of the business. Although they account for only a small fraction of corporate assets and revenues, most U.S. corporations are closely held corporations.

Subchapter S Corporation

S corporations are named after the subchapter of the Internal Revenue Code that provides for them. They are a particular type of closely held corporation. Government taxes S corporations differently from other corporations. Government taxes most corporations twice on their income: corporations must pay income tax, and shareholders must pay taxes on dividends they receive. S corporations, however, enjoy the tax status of partnerships. Thus, shareholders of S corporations report their income from the corporation as personal income.

In addition to the avoidance of double taxation, S corporations offer two more important tax advantages. First, shareholders in an S corporation are allowed to deduct corporate losses

from their personal income, ultimately lowering how much the shareholders have to pay in taxes. Second, depending on a shareholder's income, choosing to be an S corporation can result in the shareholder's paying a lower personal income tax rate than what the regular corporate tax rate would be.

To be classified as an S corporation, a corporation must meet certain requirements. First, it cannot have more than 100 shareholders. Second, only individuals, trusts, and (in certain circumstances) corporations can be shareholders (partnerships cannot be shareholders). Third, S corporations can issue only one class of shares, although not all shares must have identical voting rights. Fourth, all S corporations must be domestic corporations. Fifth, no shareholder of an S corporation can be a nonresident alien.

Professional Corporation

If a group of dentists, doctors, or other professionals wants to practice as a corporation, all 50 states permit them to incorporate. Because of the nature of professional work, however, courts sometimes alter the liability associated with such corporations. For example, courts often impose personal liability on doctors in professional corporations for medical malpractice performed under their oversight.

Formation of the Corporation

LO 22-2 Summarize the process of formation of the corporation.

The creation of a corporation involves two parts: general organizational activities and legal activities necessary for incorporation.

Organizing and Promoting the Corporation

Two groups of important players are responsible for the organization of the corporation: promoters and subscribers.

Promoters Promoters begin the corporate creation and organization process by arranging for necessary capital, financing, and licenses. Promoters raise capital for the infant corporation by making **subscription agreements** with **subscribers** (investors) in which the subscribers agree to purchase stock in the new corporation.

Promoters have other organizational responsibilities besides making subscription agreements. For example, promoters prepare the corporation's incorporation papers. Promoters can also enter into contracts as needed to establish the new corporation. For example, promoters can purchase or lease buildings for the corporation. Frank Seiberling was the promoter who founded the Goodyear Tire & Rubber Company. In 1898, Seiberling purchased Goodyear's first plant in Akron, Ohio, with $3,500 borrowed from his brother-in-law. Seiberling also established Goodyear workers's hourly wages between 13 and 25 cents.

When problems with preincorporation contracts arise, courts generally hold promoters liable and rule that these contracts do not bind infant corporations. Courts usually hold that promoters are not agents of the infant corporation because promoters cannot serve as agents for a principal (the new corporation) that does not yet exist.

Once incorporated, however, corporations can accept or reject preincorporation agreements. However, even if a corporation accepts a preincorporation agreement, courts usually still hold promoters liable for the contract.

Subscribers Subscribers make offers to purchase stock in a corporation in the incorporation process. A subscriber becomes a shareholder once the corporation incorporates or accepts her purchase offer, whichever occurs first.

promoter
A person who begins the corporate creation and organization process.

subscription agreement
An agreement between promoters (persons raising capital for a new corporation) and subscribers (investors), in which the subscribers agree to purchase stock in the new corporation.

subscriber
An investor who agrees to purchase stock in a new corporation.

Selecting a State for Incorporation

Next, an infant corporation must select a state in which to incorporate. Each state has different laws governing the incorporation process and different corporate tax rates. Other factors corporations consider when selecting a state for incorporation include the following:

- How much flexibility does the state grant to corporate management?
- What rights do state statutes give to shareholders?
- What restrictions does the state place on the distribution of dividends?
- Does the state offer any kind of protection against takeovers?

Although most corporations incorporate in the state in which they are located and do most of their business, more than half of all publicly held corporations, including more than half of the Fortune 500 companies, are incorporated in Delaware. Decades ago, Delaware law offered extremely low corporate tax rates and granted more extensive rights to management in the event of a takeover than did other states. Thus, in the 1940s and 1950s, many corporations changed their state of incorporation to Delaware. Since then, other states have made their corporate laws more attractive to corporations. Many corporations remain incorporated in Delaware, however, because Delaware courts are highly experienced in issues involving corporate law.

Closely held corporations and professional corporations, however, almost always incorporate in the state in which most of their stockholders live.

Although a corporation can incorporate in only one state, it can file a certificate of authority to do business in other states. Some states fine corporations that fail to obtain a certificate of authority before conducting business in the state. Other states fine directors and officers of these corporations directly and hold them personally liable for contracts made in the state.

Once a corporation chooses a state for incorporation, it can begin the formal legal process of incorporation.

Legal Process of Incorporation

Selection of Corporate Name

All state incorporation statutes require corporations to indicate in the name of the corporation that the business is incorporated. Every corporation must attach *Corporation, Company, Limited,* or *Incorporated,* or an abbreviation of one of these terms, to the end of its business name. In addition, every corporation must distinguish its name from the names of all other domestic or foreign corporations licensed to do business within the state. This requirement protects third parties from confusion about similar names. Once the corporation has chosen a name, this name is subject to the approval of the state.

Incorporators

incorporator
An individual who applies for incorporation on behalf of a corporation.

An **incorporator** is an individual who applies to the state for incorporation on behalf of a corporation. RMBCA requires only one incorporator to incorporate a business, although it permits more. Generally, the incorporators' only duty is to sign the articles of incorporation.

Articles of Incorporation

articles of incorporation
A document that contains basic information about a corporation and is filed with the state.

certificate of incorporation
A document certifying that a corporation is incorporated in the state and is authorized to conduct business.

One of the most important elements of the corporate formation process is the **articles of incorporation,** a document providing basic information about the corporation. According to RMBCA, a corporation's articles of incorporation must include (1) the name of the corporation, (2) the address of the registered office, (3) the name of the registered agent (i.e., the specific person who receives legal documents on behalf of the corporation), and (4) the names and addresses of the incorporators.

The incorporators must execute and sign the articles of incorporation and file the document with the secretary of state, including the required filing fee, to legally form the corporation. Once the incorporators file the document, it governs the corporation. Next, the secretary of state usually issues a **certificate of incorporation,** a document certifying that the corporation is incorporated in the state and is authorized to conduct business.

First Organizational Meeting

After the secretary of state issues the certificate of incorporation, the shareholders meet to elect the corporate board of directors, pass corporate bylaws, and carry out other corporate business. In some cases, however, shareholders name the board of directors before the first organizational meeting and list the board members in the articles of incorporation. In these situations, the directors usually run the meeting.

At the first organizational meeting, shareholders adopt a set of corporate bylaws, or rules and regulations that govern the corporation's internal management. The articles of incorporation determine who has the power to amend the corporate bylaws after the first organizational meeting: shareholders, directors, or both.

bylaws
Rules and regulations that govern a corporation's internal management.

Potential Problems with Formation of the Corporation

Most businesses incorporate to enjoy limited liability or the perpetual existence of the corporation. Shareholders enjoy these benefits, however, only if the promoters and incorporator formally and correctly incorporate the business. If the incorporator or promoters make an error or omission during the incorporation process, courts may rule that the organization is not a corporation. In this case, the organization is a defective corporation. Shareholders may be personally liable for a defective corporation's actions.

defective corporation
A corporation about which an error or omission was made during its incorporation process.

Responses to Defective Incorporation

Suppose an incorporator incorrectly indicates the address of the corporate office in the articles of incorporation. Does the corporation still exist? Depending on the seriousness of the error, courts may disregard an error in the articles of incorporation by recognizing the corporation as a *de jure* or *de facto* corporation.

De Jure Corporations A *de jure* corporation (literally, a "lawful corporation") has met the substantial elements of the incorporation process. A corporation that has received its certificate of incorporation has met the mandatory statutory provisions and is thus a *de jure* corporation. Courts usually hold that corporations that make minor errors in the incorporation process still enjoy *de jure* corporate status. Exhibit 22-2 illustrates the process for creating a *de jure* corporation.

Thus, even if the incorporator wrote the incorrect address of the corporate office in the articles of incorporation, courts would not revoke the corporation's limited liability. No party can question a *de jure* corporation's status as a corporate entity in court.

de jure corporation
Latin for "lawful corporation;" a corporation that has met the mandatory statutory provisions and thus received its certificate of incorporation.

De Facto Corporation Suppose, however, that the incorporator makes a more serious mistake or omission. For example, suppose the incorporator did not file the articles of incorporation with the secretary of state. In this case, courts may recognize the corporation as a *de facto* corporation (literally, a "corporation in fact"). A *de facto* corporation has not substantially met the requirements of the state incorporation statute, but courts recognize it as a corporation for most purposes to avoid

de facto corporation
Latin for "corporation in fact;" a corporation that has not substantially met the requirements of the state incorporation statutes.

Exhibit 22-2

De Jure Corporation Formation

INFANT CORPORATION + INCORPORATOR → Articles of Incorporation (incomplete) → DE JURE CORPORATION

GLOBAL Context

Corporate Structure in China

As you may expect, corporate governance varies among countries. Specifically, there are countries that have a one-tier board system (as in the United States), and there are others that have a two-tier board system. In a one-tier board, all directors (both executive directors and nonexecutive directors) form one board. A two-tier board consists of a management board (all executive directors) and a separate supervisory board (all nonexecutive directors). German corporation law, for example, requires all corporations to have two boards.

In recent years, China has followed the German model and adopted a two-tier board structure, with both a board of directors and a supervisory board. Corporations following this model are required to provide in their annual reports a supervisory board report (SBR). In China, the supervisory board has significant power to supervise directors, examine financial documents, and convene shareholder meetings. In these respects, supervisors exercise substantial control over corporate managers.

Although corporations in China are not required to have a supervisory board (unlike the case in Germany), there is usually a supervisory board that consists of shareholder representatives and an employee representative. However, the supervisory board does not always function at the level that many envisioned it would. In practice, supervisory boards in China have not functioned effectively because they do not have genuine power to select or discipline directors and managers. Furthermore, some critics argue that there is some overlap between the function of the supervisory board and that of the board of directors. What do you think are the advantages and disadvantages of the two-tier board system?

unfairness to third parties who believed it was properly incorporated. *De facto* corporations, regardless of whether the state has a general corporation statute, must meet the following requirements:

- The promoters, subscribers, and incorporator made a good-faith attempt to comply with the incorporation statute.
- The organization has already conducted business as a corporation.

The process for recognizing a corporation as a *de facto* corporation is depicted in Exhibit 22-3. Only the state can challenge a *de facto* corporation's existence as a corporate entity.

Corporation by Estoppel Defective corporations cannot escape corporate entity status due to mistakes or omissions in their incorporation procedures. For example, suppose a corporation's articles of incorporation do not include the name of its registered agent. Suppose further that the

Exhibit 22-3 *De Facto* Corporation Recognition Process

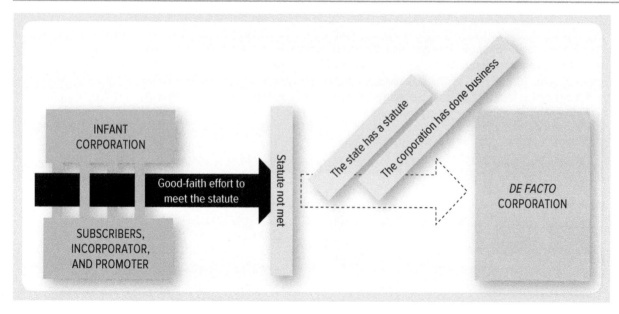

corporation's directors, managers, and shareholders are unaware of the mistake. If the corporation conducts business with a third party who later sues for breach of contract, the corporation cannot claim that it is not a corporate entity to escape liability.

In such cases, courts hold that the corporation is a **corporation by estoppel**; thus, courts *estop* (prevent) the corporation from denying its corporate status. Although a court ruling of corporation by estoppel prevents the business from denying corporate status, it does not remedy the error or grant the organization corporate status for conducting future business.

If a corporation makes a significant error in the incorporation process and is not a *de jure* or *de facto* corporation, and corporation by estoppel does not apply to the situation, courts usually deny the organization corporate entity status. Thus, the organization does not enjoy limited shareholder liability.

corporation by estoppel
A defective corporation that has conducted business with a third party and therefore cannot deny its status as a corporation to escape liability.

BUT WHAT IF? . . .

WHAT IF THE FACTS OF THE CASE OPENER WERE DIFFERENT?

Let's say in the Case Opener that when filling out documents for the incorporation of Facebook, Zuckerberg's attorney became confused and wrote the wrong address for company on the documents. Later, long after Facebook was incorporated, Zuckerberg saw the incorrect company address on the documents. Should Zuckerberg worry about the clerical error? What could be the effects of having the wrong company information on such documents?

Piercing the Corporate Veil In some cases, courts will deny limited liability to a corporation that would normally have *de jure* or *de facto* status because shareholders have used the corporation to engage in illegal or wrongful acts. Shareholders attempt to hide behind the corporate veil of limited liability to protect themselves from personal liability. In these cases, courts *pierce the corporate veil*, or impose personal liability on shareholders. Courts are likely to pierce the corporate veil when:

- A corporation lacked adequate capital when it initially formed.
- A corporation did not follow statutory mandates regarding corporate business.
- Shareholders' personal interests and corporate interests are commingled such that the corporation has no separate identity.
- Shareholders attempt to commit fraud through a corporation.

For example, corporations must keep corporate funds and records separate from shareholder funds and records. If a corporation does not carefully maintain these separate records, courts may pierce the corporate veil and impose personal liability on shareholders.

Corporate Financing

Corporations, like other businesses, need a source of funding. Corporations most commonly obtain financing by issuing and selling corporate securities: **debt securities** (bonds), which represent loans to a corporation, and **equity securities** (stocks), which represent ownership in a corporation.

debt securities
Securities that represent loans to a corporation.

equity securities
Securities that represent ownership in a corporation.

Debt Securities

Debt securities, or bonds, represent loans to a corporation from another party. Bonds are usually long-term loans on which the corporation promises to pay interest. Bonds frequently list a maturity date on which the corporation must repay the face amount of the loan. Before the maturity date, however, corporations usually pay bondholders fixed-dollar interest payments on a scheduled basis. Hence, bonds are sometimes called *fixed-income securities.*

Equity Securities

Stocks and bonds are not identical. Whereas bond owners have loaned money to a corporation, stock owners actually own part of a corporation. Thus, bond owners have no voice in corporate management; stockholders have a voice in control of the corporation. Not all corporations issue bonds, but all for-profit corporations issue stock. Common stock and preferred stock are the two major types of stock.

preferred stock
Stock that conveys preferences to its holders with respect to assets and dividends.

Preferred Stock Owners of **preferred stock,** or *preferred shares,* enjoy preferences with respect to assets and dividends. Corporations often have several classes of preferred shares. Preferred stock shares many characteristics with bonds. For example, preferred stock owners usually receive a percentage of dividends associated with the face value of their preferred stock. Furthermore, corporations pay dividends to owners of preferred stock before they pay owners of common stock. Some corporations limit preferred stock owners's voting rights.

common stock
Corporate stock that does not convey any preferences to its holders.

Common Stock Owners of **common stock,** or *common shares,* own a portion of a corporation but do not enjoy any preferences. A common stock owner is entitled to a portion of the corporation's dividends in proportion to the number of shares of common stock she owns. Common stock owners also have the right to vote in corporate elections. Each share is usually worth one vote in corporate elections. Thus, if you own 20,000 common shares of a corporation, you have 20,000 votes. In some cases, however, most notably the election of the board of directors, corporations use a method called *cumulative voting* to increase the influence of shareholders who own a small number of shares.

Common stock owners have the lowest priority when a corporation distributes dividends. Creditors and preferred stock owners receive dividends first. Once a corporation pays these groups, however, common stock owners have a claim to the remainder of the corporate earnings.

Importance of Regulating Interactions among Directors, Officers, and Shareholders within a Corporation

The three major groups of individuals within a corporation are *directors, officers,* and *shareholders.* All three groups have different interests, and, in many situations, the interests of one group conflict with the interests of another. Statutory law provides rules to ensure that the directors, officers, and shareholders within a corporation work together to the benefit of all involved.

Although directors and officers play different roles within the corporation, they share the same goal. Both directors and officers attempt to ensure that their institution survives and that they keep their jobs. Shareholders, on the other hand, have a different agenda. Their goal is to raise the value of the company's stock.

These different agendas can lead to conflict within corporations. If a corporation has an opportunity to make a decision that will raise the value of its stock quickly, shareholders will push the directors and officers to make this decision. However, if the directors and officers believe that the decision might jeopardize their jobs, they will resist the pressure of the shareholders. To resolve these conflicts, the law gives legal duties and rights to different groups within the corporation.

Roles of Directors, Officers, and Shareholders

LO 22-3 Explain the role of a director, an officer, and a shareholder in a corporation.

Before discussing the duties and rights of each position within the corporation, it is important to understand the specific roles that directors, officers, and shareholders play. The rights and duties of each group of individuals depend on these roles.

Directors' Roles

Directors play a vital role within every corporation. When a corporation faces important decisions, the board of directors meets to decide what course of action the corporation will take.

Elections Typically, shareholders use a majority vote to elect directors. The only exception occurs during incorporation. Because there are no shareholders in the beginning, either the incorporators appoint board members or the corporate articles name the board members. This first board then serves until the first shareholder meeting, at which the shareholders elect a new board. The corporate articles or bylaws specify the number of corporate directors.

In the past, the minimum number of directors required was three, but today many states allow fewer. In fact, if a corporation has fewer than 50 shareholders, Section 7.32 of RMBCA allows companies to eliminate the board of directors altogether.

Interestingly, almost anyone can become a director. The legal requirements for director qualification are incredibly lax. In most states, directors are not even required to own stock in the corporation. In some cases, however, statutory law and corporate bylaws require not only ownership but also a minimum age.

Directors typically serve for one year, but most state statutes allow for longer terms if the terms of the various directors are staggered. These terms can be terminated and the directors can be removed from their positions *for cause*—for failing to perform a required duty. Directors who are removed for cause, however, can ask the courts to review the legality of the removal.

Meetings and Voting A minimum number of directors, or a quorum, must be present at each meeting of the board of directors for decisions made at the meeting to be valid. Quorum requirements are different in each state, but most states leave the decision up to the corporation itself. Because a quorum is required at each meeting, directors are notified whenever special meetings are called. Directors vote in person, and each director has one vote. Although ordinary decisions require a majority vote, more important decisions sometimes require a two-thirds majority vote.

Directors as Managers Although directors meet to vote on major decisions about the corporation, they are also responsible for many day-to-day managerial activities of the company. The directors appoint, supervise, and remove corporate officers as they see fit, and they declare and pay corporate dividends to shareholders. They are also responsible for making financial decisions and authorizing corporate policy decisions. Some directors are also officers or employees of the corporation; these directors are known as *inside directors*. Directors who are not officers or employees of the corporation are called *outside directors*. Outside directors are further divided into two groups: *affiliated directors* and *unaffiliated directors*. Affiliated directors have business contacts with the corporation; unaffiliated directors do not.

Because the day-to-day tasks of a corporation can be overwhelming for a small board of directors that has larger issues to address, directors often appoint an executive committee to handle day-to-day responsibilities.

Officers' Roles

Officers are executive managers whom the board of directors hires to run the organization. Although directors are in charge of major policy decisions, officers run the day-to-day business of the corporation. Officers act as agents of the corporation; thus, the rules of agency apply to their work.

Qualifications required of officers are set forth in the corporate articles and bylaws of each corporation, but, in most cases, an individual may serve as both a director and an officer. Many corporations find it beneficial to include an officer on their board of directors so that the board can stay in touch with the day-to-day operations of the company.

Shareholders' Roles

Shareholders own the firm. As soon as an individual purchases the stock of a particular corporation, she becomes an owner of the corporation. Although she is not legally recognized as an owner

of corporate property, she has an *equitable,* or ownership, interest in the company. Shareholders are not directly responsible for the daily management of the corporation, but they elect the directors who are responsible for that management.

Power of Shareholders The articles of incorporation established within each corporation and general incorporation law in each state grant shareholders certain powers within the institution. Because shareholders must approve major corporate decisions, they are, in a sense, empowered to make major decisions for the corporation.

Their most influential power, however, is the power to elect and remove members of the board of directors. The board of directors is responsible for making crucial policy decisions for the corporation, and the shareholders have the power to decide who the directors will be.

Shareholders also have the power to propose ideas for the corporation. If a shareholder feels that he has a worthwhile idea for the company, he can include his proposal in the *proxy materials* sent out to the shareholders before their annual meeting. The Securities and Exchange Commission (SEC) has established that any shareholder who owns more than $1,000 worth of stock in the corporation can submit proposals to be included in proxy materials.

Meetings Typically, shareholders meet once a year, but, in emergencies, they can meet more often. The board of directors, shareholders who own at least 10 percent of the corporation's outstanding shares, and others authorized in the articles of incorporation may call a special shareholder meeting. Before each meeting, each shareholder receives notice of the time and place of the meeting. If the meeting is a special meeting, the purpose of the meeting is also included in the notice.

Like directors' meetings, shareholder meetings require a quorum. Generally, a quorum of shareholders exists when shareholders holding more than 50 percent of the outstanding shares are present. Once a quorum is present, a majority vote of the shares represented at the meeting is required to pass resolutions. Occasionally, however, articles of incorporation include supermajority provisions, which state that more than a majority is needed to pass major corporate proposals, such as corporate merger or dissolution.

Voting Like directors, each shareholder is entitled to one vote per share in most instances. Corporations practice unique voting processes, however, that alter the influence of each shareholder's votes. These voting processes are especially important for minority shareholders within a corporation. One voting process required in most states, called *cumulative voting,* ensures that minority shareholders have a voice in electing the board of directors.

The cumulative-voting process divides shareholders into majority and minority shareholders, and each group has a certain number of votes to cast in the election. The number of votes is determined by multiplying the number of shares the group owns by the number of open director positions. If a company is electing eight directors and the minority shareholders own 2,000 shares, the minority shareholders get 16,000 votes to cast in the election. If the majority shareholders in the same corporation own 8,000 shares, they get 64,000 votes.

Although it may seem that the minority shareholders have little influence in the election, the cumulative-voting process permits them to vote at least one director onto the board because they can cast all of their votes for one candidate. If the majority shareholders wish to elect all eight directors from their nominees, each nominee must receive more than 16,000 votes to beat the 16,000 votes of the minority nominee. However, because the majority shareholders have only 64,000 votes to cast in the election, they cannot cast more than 16,000 votes for eight candidates ($16,000 \times 8 = 128,000$). Thus, if the minority shareholders cast all of their 16,000 votes for one candidate, they guarantee that candidate's election.

Cumulative voting is more egalitarian than simple majority voting because it ensures that every voice within the corporation is heard. Cumulative voting, however, occurs only when the corporation's articles of incorporation provide for it.

Exhibit 22-4 summarizes the respective roles of directors, officers, and shareholders.

Exhibit 22-4

Roles of Directors, Officers, and Shareholders

Shareholders
Role: the Owners
The shareholders own the stock of the company.
Goal: to increase the value of the company's stock

Directors
Role: the Decision Makers
The directors share in the power of deciding the course of action for the corporation.
Directors also appoint and supervise officers and declare and pay corporate dividends.
Goal: to ensure that the institution survives and that they keep their jobs

Officers
Role: the Managers
The officers manage the day-to-day activities of running the business,
including acting as agents for the corporation.
Goal: to ensure that the institution survives and that they keep their jobs

Duties of Directors, Officers, and Shareholders

LO 22-4 List the duties of a director, an officer, and a shareholder in a corporation.

Because all individuals within a corporation depend on one another, the law requires directors, officers, and shareholders to perform certain duties within the business. In other words, individuals within the corporation have legal responsibilities to the corporation. These duties to the corporation are called *fiduciary duties*. Because individuals play different roles within the corporation, the fiduciary duties of individuals depend on whether they are directors, officers, or shareholders.

Duties of Directors and Officers

Because the owners of the corporation, the shareholders, have little input in the day-to-day operations of the corporation, they trust the directors and officers to run the company to the best of their ability. Thus, directors and officers have duties to the shareholders and to the corporation. The two primary fiduciary duties of directors and officers are the duty of care and the duty of loyalty.

Duty of Care Directors and officers have a fiduciary *duty of care,* meaning that they must exercise *due care* when making decisions for the corporation. The phrase *due care* is ambiguous, and various courts have interpreted it differently over time. In general, however, acting with due care requires one to exercise the care that an ordinary prudent person would exercise in the management of her own assets.

Because directors and officers have a duty to act in the best interest of the company, they must supervise, to a reasonable extent, employees who work for the corporation. They also have a duty to attend director and corporate business meetings. Most important, however, directors and officers have a fiduciary duty to make informed and reasonable business decisions.

E-COMMERCE and the Law

When a B2B Company Cooks the Books

When a business student thinks about companies that cook the books, it is not surprising that companies such as Enron, Tyco, and Adelphia come to mind. Issues related to the duty of care of officers and directors are expressed in questions such as, "Why did the officers and directors of the company fail to realize accountants were cooking company books? Were any officers or directors involved in the fraud?"

Students may have to revise their images of corrupt companies by adding business-to-business (B2B) e-commerce companies to their Enron-Tyco-Adelphia list.

PurchasePro is one example of an e-commerce company that allegedly engaged in "overstating revenues, engaging in aggressive accounting practices and mismanaging corporate assets." Basically, the company manipulated its financial records to make it look far more successful than it really was. PurchasePro, which provided a business-to-business software application that gave companies access to an online marketplace, went bankrupt in September 2002.

Federal prosecutors charged company officers and directors with conspiracy, securities fraud, and obstruction of justice. Two senior officers, Jeffrey R. Anderson and Scott H. Miller, pleaded guilty to federal crimes in 2003. Their behavior was similar to that of officers of other, more well-known companies that have cooked the books—they had secret side deals with purchasers that gave the appearance of sales, but the sales did not exist; they misrepresented the company's financial health, so investors could not make informed decisions; and, when news of alleged fraud surfaced, the officers used their energy to shred incriminating documents.

For example, recall the dramatic fall of the Texas energy company, Enron. At Enron's peak, its shares were worth $90.75, but after the company declared bankruptcy on December 2, 2001, they plummeted to $0.67 by January 2002. Directors and officers of Enron failed in their duty of care with regard to their shareholders by not acting in the best interest of the company. The directors and officers continued to urge employees to invest in the employee stock-sharing options, even though it appears that the directors and officers knew the stock was drastically overpriced. Furthermore, the directors and officers failed in their duty of care regarding oversight. They either did not pay enough attention to see the collapse of the company's stock coming or purposely kept the information secret. Either way, the directors and officers at Enron breached their fiduciary duty of care and therefore are liable to their shareholders, many of whom were employees of Enron.

Directors and officers are expected to stay abreast of all important corporate matters. In other words, they must obtain information about business transactions, review contracts, read reports, and attend presentations. Because some directors are too busy to stay informed on every subject, RMBCA allows directors to make decisions on the basis of information gathered by other employees.

The decisions that corporate directors and officers make must not only be informed, however, but also be reasonable. If a director or officer is taken to court for breaching his duty of care by making an unreasonable decision, the court typically inquires whether the decision had any rational business purpose. In other words, the courts ask whether there was good reason to think that the decision *could* have helped the company.

Part of the duty of care a director or officer owes a corporation, as was previously stated, is to act in the corporation's best interest. Therefore, when the corporation is doing something the director or officer does not think is in its best interest, it is up to her to voice her dissent.

Duty of Loyalty Because directors and officers have great decision-making freedom, they have the power to make business decisions that will benefit themselves while harming the company. Thus, to protect shareholders, directors and officers have a fiduciary *duty of loyalty*. In other words, they have a fiduciary duty to put the corporation's interest above their own when making business decisions.

When directors or officers violate their duty of loyalty, they are **self-dealing**. There are two types of self-dealing in which a director or officer can engage. The first form of self-dealing, called *business self-dealing,* occurs when a director or officer makes decisions that benefit other companies with which he has a relationship. The second form of self-dealing, called *personal self-dealing,*

self-dealing
Any instance in which directors or officers make decisions that violate their corporate duty of loyalty.

occurs when a director or officer makes business decisions that benefit her personally. When Enron directors and officers advocated employees buying more Enron stock so that the directors and officers would make more money, they engaged in self-dealing in breach of their fiduciary duty of loyalty.

CASE Nugget

How Should Directors Handle Personal Interests That Conflict with Corporate Interests?

Patrick v. Allen

United States District Court, S.D. New York

355 F. Supp. 2d 704 (2005)

RPO, a privately traded corporation, rented land to a private golf course of which several of RPO's directors were members. The directors charged the golf course enough rent to cover only the property taxes on the land. Patrick, a shareholder of RPO, brought a suit against the directors of RPO, alleging that RPO's directors breached their fiduciary duty of loyalty to the corporation by failing to maximize the value of the corporation for shareholders. The directors argued that they were exempt from liability under the business judgment rule. (The business judgment rule is covered in the section entitled "Liabilities of Directors, Officers, and Shareholders," which follows shortly.)

The U.S. District Court for the Southern District of New York ruled against the RPO's directors, holding:

> The business judgment rule will not protect a decision that was the product of fraud, self-dealing, or bad faith. Directors may benefit from the rule only if they possess a disinterested independence and do not stand in a dual relation which prevents a nonprejudicial exercise of judgment. It is black-letter, settled law that when a corporate director or officer has an interest in a decision, the business judgment rule does not apply.... A director is considered interested in a transaction if the director stands to receive a direct financial benefit from the transaction which is different from the benefit to shareholders generally.... The duty of loyalty requires a director to subordinate his own personal interests to the interest of the corporation.

In many situations, when a director or officer is self-dealing, he forces the corporation into unfair business deals. Directors and officers can also breach their fiduciary duty of loyalty, however, by *preventing* corporate opportunity. This breach usually happens when directors or officers own other companies that compete with their corporation without the consent of the board of directors or the shareholders. If a director or officer uses corporate assets to start another business, goes into the same line of business, or uses her position in the company to develop a new business that the company might have pursued, she is preventing corporate opportunity and can be held liable for violating her fiduciary duty of loyalty.

A director or officer convicted of breaching his duty of loyalty is required to cede to the corporation all the profits he earned as a result of the breach. The goal of this rule is to create incentives that discourage breaches of the duty of loyalty by taking from the director or officer all the profits that he made.

The fiduciary duties of care and loyalty are rooted in ethics. Without these legal duties, directors and officers could pursue their own interests at the expense of others. Through these fiduciary duties, however, the law requires directors and officers to consider the interests of others. Think back to Chapter 2 and the different ethical guidelines used to make ethical decisions. Which ethical guideline is the legal system using when it delegates fiduciary duties?

Duty to Disclose Conflict of Interest Because there are many times when individual directors and officers may personally benefit from decisions made by the board of directors, the directors and officers have a fiduciary duty to disclose conflicts of interest fully that arise in corporate transactions. Moreover, if the board of directors addresses an issue that might personally benefit a particular director, the director is required not only to disclose the self-interest but also to abstain from voting on that issue. Note that a decision can be made that will personally benefit one director or officer as long as (1) there is full disclosure of the interest and (2) the disinterested board members and/or disinterested shareholders approve the decision.

Duties of Shareholders

Although shareholders typically have few legal duties, in rare instances, majority shareholders have fiduciary duties to the corporation and to minority shareholders. In some corporations, the majority shareholder owns such a significant portion of the corporation's stock that she essentially controls the firm. When the majority shareholder sells her shares, control of the company shifts to another individual. Thus, the majority shareholder in this situation has a fiduciary duty to act with care and loyalty when selling her shares.

As noted earlier, more than half of all U.S. publicly traded corporations are incorporated in Delaware. Thus, when Delaware courts rule on the duties of majority shareholders to minority shareholders, for example, the courts's rulings have a far-reaching impact.

In Case 22-2, a minority shareholder sued the majority shareholder of a corporation for violating its fiduciary duties.

CASE 22-2

MCCANN v. MCCANN
Supreme Court of Idaho
275 P. 3D 824 (2012)

FACTS: *McCann Ranch & Livestock Company Inc. was a closely held corporation created by William McCann, Sr. In the 1970s, Ron and Bill McCann, the creator's sons, were each given 36.7 percent of the shares of the corporation. The remaining shares were held by the creator, William, until those shares were transferred to a trust to benefit his wife, Gertrude. The trustee, Gary Meisner, was given the power and discretion to vote and sell Gertrude's shares under certain circumstances. Upon the death of the corporation's creator, William, Bill became the President and CEO of the corporation. Then, upon the death of Gertrude, Bill received her remaining shares.*

In June 2008, Ron, a minority shareholder, filed an initial complaint, alleging that the corporation breached its fiduciary duty owed to him as a shareholder by engaging in a squeeze-out and that such injury is grounds for a direct action. Ron's claim mentioned his treatment by the corporation as well as financial transactions of the corporation. The district court granted summary judgment in favor of the respondent, Bill. Ron appealed.

ISSUE: May a minority shareholder bring suit against the majority shareholder for breach of fiduciary duty?

REASONING: Squeeze-outs are actions taken by controlling shareholders to deprive a minority shareholder of his or her interest in a business or a fair return on an investment. Courts have analyzed the legality of squeeze-outs by referring to the conduct of the majority shareholders and how such conduct affects the minority shareholders. In other words, the law recognizes that directors owe a fiduciary duty to minority shareholders.

In this case, the court reasoned that the conduct of the corporation could not be excused under the business judgement rule. Specifically, the corporation had not paid dividends despite sufficient cash flow, not provided board membership to Ron, authorized phony transactions to Gertrude to avoid any benefit to Ron, and had made management decisions that benefited Bill and Gertrude at the expense of Ron. The court inferred from these observations, along with the fact that Ron was specifically hurt from the payments that the corporation made to Gertrude, that a material question existed with regard to the corporation engaging in a squeeze-out.

DECISION AND REMEDY: The court reversed in favor of Ron and remanded to address the questions of whether the alleged squeeze-out constituted a violation of fiduciary duties.

SIGNIFICANCE: This case highlights the legal obligations of majority stakeholders to minority stakeholders because of the overwhelming power majority stakeholders have over the operation of the corporation.

(continued)

> **CRITICAL THINKING**
>
> What reasons does the court give for its conclusion? Are you persuaded by those reasons?
>
> **ETHICAL DECISION MAKING**
>
> Clearly, the court emphasizes particular values in its ruling. What are these values? Which shareholders in the business community are affected by focusing on these values when assessing a corporation's fiduciary duties?

Liabilities of Directors, Officers, and Shareholders

LO 22-5 Summarize the liabilities of a director, an officer, and a shareholder in a corporation.

Because almost all individuals within a corporation have legal fiduciary duties to the corporation, almost all individuals within the firm can be held liable for harming the business by violating these duties. There are, however, certain instances in which directors, officers, and shareholders cannot be held liable for harming the business.

Liability of Directors and Officers

Directors and officers are held liable for many of the same actions because they have nearly identical fiduciary duties to the corporation. A shareholder can sue them if the shareholder feels that they have caused harm to the business by violating their fiduciary duties. Directors and officers can be held liable for the torts and crimes of their employees.

Liability for Torts and Crimes Although the corporations themselves are liable for the torts and crimes of their directors and officers, directors and officers can also be held personally responsible for their own torts and crimes. They can even be held personally responsible for the torts and crimes of other employees within the organization when they have failed to supervise the employees' behavior adequately.

According to the *responsible person doctrine,* a court may find a corporate officer criminally liable regardless of the extent to which the officer took part in the criminal activity. Even if the officer knew nothing about the criminal activity, the officer can still be held criminally liable if the court determines that a responsible person would have known about and could have prevented the illegal activity.

Corporate directors and officers may also be held liable for wrongful personal transactions involving company stock. Directors and officers who use inside information to trade the corporation's stock for a profit can be held liable for breaching their fiduciary duty to the shareholders from whom they purchase or to whom they sell the stock.

Business Judgment Rule Although directors and officers are expected to make decisions that are in the best interest of the corporation, they are not expected to make perfect decisions all the time. Many directors and officers make decisions that inadvertently harm the corporation. Although shareholders may want to hold their directors and officers liable for these decisions, the *business judgment rule* does not allow them to do so.

Under this rule, directors and officers are not liable for decisions that harm the corporation if they were acting in good faith at the time of the decision. In other words, if there was reason to believe that the decision was a good decision at the time, the directors and officers are not liable for the resulting harm.

Although the business judgment rule is not a statute, it is common law recognized by almost every court in the country. The rule is practical because it grants directors the freedom to work without constant fear of personal liability. The business judgment rule also encourages individuals to serve as directors by removing the threat of personal liability for inadvertent mistakes.

The Business Judgment Rule

Auerbach v. Bennett

N.Y. Court of Appeals

393 N.E.2d 994 (1979)

An internal audit of the GTE Corporation suggested that the corporation's management had paid more than $11 million in bribes and kickbacks both in the United States and abroad over a four-year period. Auerbach, a GTE shareholder, immediately initiated a shareholder derivative action (discussed below) against GTE's directors.

The Court of Appeals of New York, however, held that the business judgment rule exempted the GTE directors from liability for their poor business decisions. The court stated:

> [The business judgment doctrine] bars judicial inquiry into actions of corporation directors taken in good faith and in the exercise of honest judgment in the lawful and legitimate furtherance of corporate purposes. Questions of policy of management, expediency of contracts or action, adequacy of consideration, lawful appropriation of corporate funds to advance corporate interests, are left solely to their honest and unselfish decision, for their powers therein are without limitation and free from restraint, and the exercise of them for the common and general interests of the corporation may not be questioned, although the results show that what they did was unwise or inexpedient.

The ongoing importance of the business judgment rule is demonstrated by its recent application in Case 22-3.

KDW RESTRUCTURING & LIQUIDATION v. GREENFIELD
United States District Court, S.D. New York
874 F.Supp.2d 213 (2012)

FACTS: *KDW Restructuring & Liquidation Services LLC is the trustee of a litigation trust held on the behalf of debtor Jennifer Convertibles Inc. In 2009, Jennifer had losses from a transaction between Jennifer and Jara Enterprise Inc. As a result of the loss, KDW brought an action against the members of Jennifer's 2009 Board of Directors: Harley Greenfield, Edward G. Bohn, Kevin L. Coyle, Rami Abada, and Mark J. Berman. All defendants moved to dismiss the breach of fiduciary duty claim based on the business judgement rule.*

Abada is a director, but voted against all but the final agreement. He also consistently expressed reservations about transactions with Jara.

ISSUE: What are the standards whereby a director would be judged to have violated his duties as a corporate director?

REASONING: The court noted that Albada's behavior was consistent with his responsibilities as a director. They

(continued)

reasoned that the plaintiffs had not adequately pleaded specific facts about Abada to demonstrate that he was interested, acting in bad faith, lacking a rational purpose, or grossly negligent. Abada was not an interested party—he derived no unfair personal benefit from the transaction. His prior employment at Jara does not plausibly demonstrate that he was reaping a material personal benefit, particularly considering that he thought the deal ill-advised. Neither a bad faith nor duty of loyalty claim against this defendant is plausible.

Greenfield, another director, alleged that because of the collective nature of the complaint, the allegations against him were not sufficient to support a claim. However, there were sufficient specific facts in the complaint to support claims for breach of the duty of loyalty and breach of the duty to act in good faith against Greenfield. Greenfield claimed an overly broad interpretation of the business judgment rule that shields directors whenever a majority of the board is disinterested. This formulation does not take into account the possibility that certain actions can only be explained by bad faith.

Greenfield was a founder of Jara. Greenfield's sister, Jane Love, was the controlling shareholder of Jara during the 2009 transactions. He voted to continue shipping to Jara despite Jara's delinquency and corporate counsel's advice that he refrain from voting on Jara-related issues. Drawing reasonable inferences in the plaintiff's favor, these facts suggest that Greenfield may have been an interested party, on both sides of the transaction, reaping a personal benefit. As such, plaintiffs have adequately pled that Greenfield is not protected by the business judgment rule, and the breach of the duty of loyalty claim survives.

DECISION AND REMEDY: Abada's motion to dismiss is granted because he is protected by the business judgment rule. Greenfield's motion to dismiss is denied because the basis for the duty to act in good faith claim survives.

SIGNIFICANCE OF THE CASE: Corporate directors need to be fully aware of the standards they must meet if they wish to use the business judgment rule to shield them from liability for potentially violating one or more of the duties owed by every corporate director.

CRITICAL THINKING

Are there words or phrases in the court's argument that are ambiguous? Why are these ambiguous words important?

ETHICAL DECISION MAKING

Suppose you were on the board of directors in this case. The universalization test guides your ethical decisions. Would you have made a different decision?

Liability of Shareholders

The main liability that shareholders face is liability for the debts of the corporation. Because shareholders are the owners of the corporation, they are liable to the extent of their investment when the company loses money. There are rare instances, however, in which shareholders are personally liable.

For instance, in some cases, individuals sign stock subscription agreements before incorporation. Once an individual signs a stock subscription agreement, she is contractually obligated to purchase shares in the corporation. For **par-value shares,** or shares that have a fixed face value noted on the stock certificate, the shareholder must pay the corporation at least the par value of the stock. For **no-par shares,** or shares without a par value, the shareholder must pay the corporation the fair market value of the shares. The shareholder is personally liable for breach of contract if she does not buy shares.

Alternatively, a shareholder can be held personally responsible if he receives **watered stock.** Watered stock is stock issued to individuals below its fair market value. When watered stock is issued to a shareholder, the shareholder is individually liable for paying the difference between the price he paid for the shares and the stated corporate value of the shares.

par-value shares
Stock shares that have a fixed face value noted on the stock certificate.

no-par shares
Stock shares that do not have a par value.

watered stock
Stock that is issued to individuals below its fair market value.

Rights of Directors, Officers, and Shareholders

LO 22-6 Explain the rights of a director, an officer, and a shareholder in a corporation.

Along with their specific roles and duties, directors, officers, and shareholders have specific rights within the corporation. Because shareholders are in a position of limited decision-making power, they have rights that allow them to participate within the corporation. Directors and officers also have specific rights that allow them to perform their duties within the corporation to the best of their abilities.

Directors' Rights

The unique responsibilities of corporate directors call for unique rights. Directors have four major rights within a corporation: rights of compensation, participation, inspection, and indemnification.

All corporate directors have a *right to compensation* for their work, and different corporations grant compensation in different ways. Most directors hold other managerial positions within their companies and receive their compensation through those positions. Another common avenue for compensating directors is paying them nominal sums as honorariums for their contributions to the company.

Because directors are required to make informed business decisions, they have the *rights of participation and inspection*. In other words, they have the right to get involved in and understand every aspect of the business. A corporate director has the right to be notified of all meetings and has access to all books and records.

Finally, because directors are in positions of great legal vulnerability, they have the *right to indemnification*. That is, they have the right to reimbursement for any legal fees incurred in lawsuits against them.

Officers' Rights

Because corporate officers are technically employees of the corporation, their rights are defined by employment contracts drawn up by the board of directors or the incorporators. They are in a contractual relationship with the corporation, and if they are removed from their positions in violation of the terms of the contract, the corporation may be liable for breach of contract.

Shareholders' Rights

Although shareholders' most powerful right is the right to vote at shareholders' meetings, they also possess many other rights.

stock certificate
A document that serves as a stockholder's proof of ownership in a corporation.

Stock Certificates Some corporations issue **stock certificate** to shareholders as proof of ownership in the corporation. Each certificate includes the corporation's name and the number of shares represented by the certificate. An example of a stock certificate is shown in Exhibit 22-5. A shareholder's ownership in the corporation, however, does not depend on his possession of the physical stock certificate. For example, if the certificate is destroyed in a fire, the shareholder's ownership in the corporation is not destroyed.

Preemptive Rights Under common law, shareholders have *preemptive rights*. Preemptive rights give preference to shareholders to purchase shares of a new issue of stock. Each shareholder receives preference in proportion to the percentage of stock she already owns.

For example, suppose that a shareholder owns 1,500 shares in a corporation with 5,000 outstanding shares. She owns 30 percent of the corporation's outstanding stock. Now suppose that the corporation decides to issue an additional 10,000 shares. If the corporation does not grant preemptive rights, her relative control of the corporation will fall because she now owns only 1,500 of the corporation's 15,000 outstanding shares, or 10 percent of its stock. In contrast, if the corporation does reserve preemptive rights, and the shareholder elects to purchase 3,000 shares

GLOBAL Context

Criminal Liability in France

The French Penal Code does not adhere to a detailed or large list of corporate offenses but, instead, adopts what is called the specialty principal, which requires corporate criminal liability to be applicable only when an "express mention in the law or in a French regulation exists."

The French system of corporate criminal liability sharply contrasts with that of the United States in that it has often been described as scientific. This is because the French system requires heavy statistics outlining the frequency of a corporate officer's involvement in a crime before any criminal liability may be enforced. This approach differs from the approach here in the United States because our responsible-person doctrine may require a corporate officer to be held criminally liable regardless of the extent of participation in a crime.

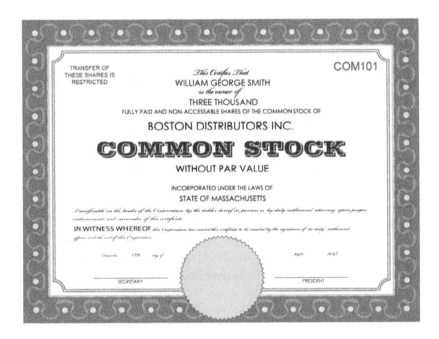

Exhibit 22-5

Example of a Stock Certificate

of the newly issued stock, then she owns 4,500 shares of the corporation's 15,000 outstanding shares (30 percent). Thus, with preemptive rights, she can maintain her proportionate control of the corporation.

In most states, a corporation's bylaws can negate preemptive rights, so the corporation determines whether to grant preemptive rights. Preemptive rights are especially important for individuals who own stock in closely held corporations due to the relatively small number of issued shares. If a closely held corporation issues additional shares, an individual shareholder may lose proportional control over the firm if he does not buy newly issued shares. However, if preemptive rights exist within the corporation, all shareholders receive **stock warrant**, which they can redeem for a certain number of shares at a specified price within a given time period.

Dividends If directors fail to declare and distribute dividends, shareholders have the right to take legal action to force the directors to declare the dividends. In many cases, however, directors have good reason to hold dividends for a limited amount of time to finance major undertakings such as research or expansion. Thus, shareholders must show that the directors are acting unreasonably and abusing their discretion in withholding the dividend.

Inspection Rights All shareholders have the right to inspection under both statutory and common law. A shareholder can, moreover, appoint an agent to conduct the inspection on her behalf. To prevent abuse, however, this right has many limitations. Shareholders can inspect records and

stock warrant
A type of security issued by a corporation (usually together with a bond or preferred stock) that gives the holder the right to purchase a certain amount of common stock at a stated price.

books only if they ask in advance and have a *proper purpose.* Many states have other statutory limitations to this right. For example, some states allow only shareholders with a minimum number of shares to inspect. Other states require the shareholder to own stock for a minimum amount of time before inspection. Additionally, corporations can deny shareholders' right to inspect confidential corporate information, such as trade secrets.

Share Transfer The law generally permits property owners to transfer their property to another person, and, in most cases, stock is considered transferable property. In closely held corporations, however, the transfer of stock is usually restricted so that shareholders can choose the corporation's other shareholders.

One method of restricting stock transferability is called the **right of first refusal**. If a corporation establishes this right in its bylaws, the corporation or its shareholders have the right to purchase any shares of stock offered for resale by a shareholder within a specified period of time.

right of first refusal
A method of restricting stock transferability whereby a corporation or its shareholders have the right to purchase any shares of stock offered for resale by a shareholder within a specified time frame.

Corporate Dissolution Shareholders have the right to petition the court to dissolve their corporation if they feel the company cannot continue to operate profitably. According to Section 14.30 of RMBCA, if a corporation engages in any of the following behaviors, shareholders have a legal right to initiate dissolution.

1. Directors are deadlocked in managerial decisions and harming the corporation.
2. Directors are acting in illegal, oppressive, or fraudulent ways.
3. Assets are being wasted or used improperly.
4. Shareholders are deadlocked and cannot elect directors.

Once dissolution has taken place and the corporation has settled its debts with its creditors, shareholders have a right to receive the remaining assets of the company in proportion to the number of shares they own.

Shareholder's Derivative Suit One of the shareholder's most important rights is the right to a derivative suit. If corporate directors fail to sue when the corporation has been harmed by an individual, another corporation, or a director, individual shareholders can file a **shareholder's derivative suit** on behalf of the corporation. In most jurisdictions, to be able to file a shareholder's derivative suit, a shareholder must have held stock at the time of the alleged wrongdoing. Before filing the suit, the shareholder must file a complaint with the board of directors. If nothing is done in response to the complaint, the shareholder can proceed with the suit. Enron shareholders brought shareholder's derivative suits against various directors and officers of Enron.

shareholder's derivative suit
A lawsuit filed by a shareholder on behalf of the corporation.

Derivative suits are important when shareholders believe that directors of the corporation are harming the corporation. It seems highly unlikely that directors will sue themselves for damages they cause in such cases. Thus, the shareholder's derivative suit is important for holding the directors accountable for their behavior. Because the shareholder files this suit on behalf of the corporation, all damages recovered are given to the corporation, not the individual shareholder.

Shareholder's Direct Suit In addition to their right to bring suit on behalf of the corporation, shareholders can also bring a direct suit against the corporation. In a **shareholder's direct suit**, the shareholder alleges that she has suffered damages caused by the corporation. For example, shareholders can bring a direct suit if the board of directors is improperly withholding dividends or if a shareholder is wrongly denied his right to inspect corporate records. If a court awards damages as a result of a shareholder's direct suit, the damages go to the shareholder personally.

shareholder's direct suit
A lawsuit filed by a shareholder against the corporation.

Exhibit 22-6 summarizes the respective rights of directors, officers, and shareholders.

Purchase of Stock

Corporations can extend their operations by purchasing another corporation's stock. An acquiring corporation, or *aggressor,* can buy any or all of another corporation's voting shares. Through such a stock purchase, the purchasing corporation gains control of the selling corporation in a corporate **takeover**.

takeover
The purchase of one corporation's stock by another corporation as a means of gaining control of the selling corporation.

POSITION	RIGHT
Directors	Right to compensation
	Right to participation
	Right to inspection
	Right to indemnification
Officers	Rights determined in employment contract
Shareholders	Right to have stock certificates
	Preemptive rights
	Right to dividends
	Right to transfer shares
	Inspection rights
	Right to corporate dissolution
	Right to file a derivative suit
	Right to file a direct suit

Exhibit 22-6

Rights of Directors, Officers, and Shareholders

The Nature of Takeovers

During the 1980s, not only did the number of corporate takeovers increase, but so too did the number of hostile takeovers. **Hostile takeovers** are takeovers to which the management of the target corporation objects. When a hostile takeover succeeds, the target corporation's management frequently compares the transition to a full-scale invasion characterized by layoffs and dramatic changes in company policy.

hostile takeover
A takeover to which the management of the target corporation objects.

Types of Takeovers

To initiate a stock purchase, the aggressor must appeal directly to the shareholders of the corporation it hopes to buy, known as the *target corporation*. The aggressor can offer several types of deals to the target shareholders. It can make a **tender offer,** in which it offers target shareholders a price above the current market value of the stock. Aggressors, however, often require them to receive a certain number of shares within a certain time frame.

Alternatively, the aggressor may make an exchange tender offer. In an **exchange tender offer,** the aggressor offers to exchange target shareholders' current stock for stock in the aggressor's corporation. The aggressor may also make a **cash tender offer** to the target shareholders, in which it pays them cash for their stock.

After acquiring a substantial number of the target corporation's shares, the aggressor initiates a proxy fight by fighting for control over target shareholders' proxies. Before an aggressor can gain control of the target corporation through proxies, it needs a key piece of information: a list of target shareholders. Although resistant target corporations often want to conceal this information, federal securities law requires target corporations to assist aggressors in some ways. Thus, to avoid lengthy and expensive lawsuits, target corporations often provide a list of shareholders voluntarily.

Those seeking to acquire corporations have developed tactics to overcome the law's rules encouraging cooperation from target corporations. Because contacting each individual target shareholder is expensive, aggressors often try to win the favor of a few institutional investors that own a large block of shares. If an aggressor can obtain the proxies of these investors, it can win control of the target corporation.

tender offer
A type of takeover in which the aggressor corporation offers the target shareholders a price above their stock's current market value.

exchange tender offer
A type of takeover in which the aggressor corporation offers to exchange the target shareholders' current stock for stock in the aggressor's corporation.

cash tender offer
A type of takeover in which the aggressor corporation offers to pay the target shareholders cash for their stock.

Response to Takeovers

Once an aggressor has presented its offer to the target corporation's shareholders, the target corporation's board of directors must inform shareholders of all facts pertinent to shareholders' votes. After reviewing these material facts, the directors vote to accept or reject the offer and advise shareholders accordingly.

E-COMMERCE and the Law

Hostile Takeovers Online

In the world of business, physical corporations are not the only ones who fear hostile takeovers. Companies such as Yahoo!, the Internet web browser and email hub, have recently fallen on hard times. Advertising companies have many more outlets now in which to fund advertising, which means that advertising business is being spread thin over the online realm. This market slowdown presents a problem for Yahoo because 90 percent of Yahoo's profits come from selling advertising space. Because Yahoo's stock prices have dropped dramatically, the company is worried about hostile takeover. To help prevent this from happening, Yahoo has created a poison pill. This measure means that if another company, such as Google, owns a majority of Yahoo's stock, other stockholders holding shares of Yahoo may purchase more Yahoo stock at a significantly lower price. This bargain drastically increases the price the (would be) takeover company would have to pay to complete the takeover. Yahoo may then feel a little more secure about its continued independence.

Sources: http://www.forbes.com/2001/03/02/0302ecommerce.html; http://business.yourdictionary.com/poison-pill.

If the directors conclude that a takeover is not in the company's best interest, the company may employ many methods of resistance. One common method is a **self-tender offer,** in which the target corporation offers to buy its shareholders' stock. If the shareholders accept the offer, the target corporation maintains control of the business.

Alternatively, target corporations may defend themselves using leveraged buyouts. A **leveraged buyout (LBO)** occurs when a group within a corporation (usually management) buys all the outstanding corporate stock held by the public. Thus, the group gains control over corporate operations by going private, or becoming a privately held corporation.

LBOs are usually high-risk endeavors, however, because the target corporation must borrow money to purchase the outstanding stock. It may have to borrow money from an investment bank or issue corporate bonds.

In Case 22-4, the U.S. Court of Appeals for the Sixth Circuit considered the legality of a golden parachute, a clause in an executive's contract specifying that he or she will receive large benefits if the executive's employment is terminated. Management of the target firm in this case made an agreement with its top managers to win their cooperation in resisting a takeover.

self-tender offer
A takeover-resistance strategy in which the target corporation offers to buy its shareholders' stock.

leveraged buyout (LBO)
A takeover-resistance strategy in which a group within the target corporation buys all the corporate stock held by the public, thereby turning the company into a privately held corporation.

CAMPBELL, KESSER, AND WILLIAMS v. POTHAS CORPORATION
U.S. Court of Appeals for the Sixth Circuit
238 F.3d 792 (2001)

FACTS: *PCS, a Saskatchewan fertilizer corporation, approached Arcadian, a Tennessee fertilizer corporation, about a possible merger. They decided to merge, and Arcadian and PCS negotiated the terms of the merger and the severance agreements. The Arcadian board approved employment agreements for nine senior executives that included so-called golden parachutes. The "golden parachute" portion of the severance package provided a formula to compensate senior executives in case of a change in corporate control accompanied by a material change in the executive's position at the new company. In such a circumstance, the executive could leave the company and receive an aggregate payment in one lump sum. Three of the executives elected to take the severance package but were not compensated accordingly. Campbell (the former president and CEO), Kesser (the former vice president and general counsel), and Williams (the former vice president and CFO) sued PCS for breach of contract because PCS refused to make those severance payments to the executives. In response, PCS argued that the golden parachute agreements were not enforceable because they violated public policy.*

ISSUE: Are the golden parachutes offered to the three executives enforceable, or do they violate public policy?

REASONING: PCS argued that the golden parachutes violated public policy and therefore that the assumption agreement promising them is void. However, PCS offered

(continued)

golden parachutes to its own top managers. Hypocrisy aside, PCS cited no circuit case law supporting this proposition. Although the court has frowned on golden parachutes in past dicta, the court has never held that such severance packages were per se unlawful. Furthermore, PCS did not provide much reason to equate this type of executive compensation with contracts prohibited by public policy, such as ones to perform illegal acts. Although PCS cited a congressional committee report saying that golden parachutes should be discouraged, Congress decided to tax golden parachutes and not prohibit them. Therefore, PCS did not cite compelling evidence that the contracts were prohibited by public policy.

DECISION AND REMEDY: The court ruled that the golden parachutes did not violate public policy, and it subsequently ordered enforcement of the severance agreements. PCS already received the benefit of the bargain it struck concerning golden parachutes in having an orderly change in corporate control, and it cannot refuse payment in return.

SIGNIFICANCE OF THE CASE: This case illustrates the trouble firms can get into when they engage in mergers and severance packages. PCS made an agreement to secure the loyalty of its top management, but when the time came to fulfill its part of the bargain, the company tried to argue that enforcement of such a contract would be unfair and unenforceable.

CRITICAL THINKING

Is there any missing information you would ask for when considering the facts of this case? If you were to argue to reverse the trial court's decision, what reasons would you offer?

ETHICAL DECISION MAKING

Think about the WH process of ethical decision making. What is the purpose of the court's decision? In other words, which value is upheld? What value is in conflict with the reasoning of the court?

Response to Termination

The death of a corporation occurs in two phases: *dissolution,* the legal termination of the corporation, and *liquidation,* the process by which the board of directors converts the corporation's assets into cash and distributes them among the corporation's creditors and shareholders.

Dissolution

Dissolution may be voluntary or involuntary, depending on who initiates and compels the dissolution. *Voluntary dissolution* occurs when the directors or shareholders trigger the dissolution procedures. The directors can initiate the proposal and submit it to the shareholders for a vote, or the shareholders can begin dissolution procedures. *Involuntary dissolution* results when government or private parties successfully get a court order dissolving the corporation. Either way, for dissolution to be successful, shareholders must unanimously vote for the proposal.

Regardless of whether the directors or shareholders initiate dissolution, the corporation must follow specific procedures. First, the directors must file articles of dissolution with the secretary of state. These articles must include the company name, the date of dissolution, and the method of authorization of dissolution. Next, the directors must notify the shareholders. If shareholders or creditors have claims against the corporation, they must make them known within a stipulated time frame. Although the corporation establishes the time frame, the period must extend at least 120 days after the date of dissolution.

CASE OPENER WRAP-UP

The Formation of the Facebook Corporation

Zuckerberg's actions in 2004 were instrumental in creating what is now Facebook today. There are currently over 250 million users on the website from all over the world. In 2007, Microsoft bought a 1.6 percent share of the corporation for $240 million. Facebook gained another investor in November of that year, a billionaire from Hong Kong, Li Ka-shing, for $60 million. Estimates put the corporation's value around $4 billion to $5 billion. Ultimately, its corporate status allows Facebook to enjoy perpetual existence, to sue and be sued, to acquire property, to make contracts, to borrow and lend money, to make charitable donations, and to establish rules for managing the corporation. Moreover, Facebook's shareholders enjoy limited liability.

PRACTICAL TIPS for Business Managers

1. One of the biggest advantages to forming a corporation is the liability "veil" it provides. Due to a corporation's status as a "legal person," most legal disputes are aimed at the corporation and not at its shareholders. Remember, however, that this veil is not entirely impenetrable; shareholders can, on occasion, be held liable for corporate actions. In other words, do not assume that forming a corporation makes you legally bulletproof.

2. Remember that most states will not allow you to do business in their state unless you file a certificate of authority in that state. Failure to do so can result in fines and other legal trouble.

3. It may be wise to create a shareholder agreement that outlines the rights and responsibilities of shareholders. Such an agreement limits the power of shareholders and may prevent the increasingly popular trend of shareholder activism.

SUMMARY

Characteristics of Corporations	• Legal entity. • Rights as person and citizen. • Creature of the state. • Limited liability of shareholders. • Unrestricted transferability of corporate shares. • Perpetual existence. • Centralized management. • Corporate taxation. • Liability for corporate agents.
Corporate Powers	Corporations have both express and implied powers. An *ultra vires* act is a corporate action that is beyond the scope of the corporation's authority.
Classification of Corporations	• Public or private. • For profit or nonprofit. • Domestic, foreign, or alien. • Publicly held or closely held. • S corporations. • Professional corporations.
Formation of the Corporation	To create a corporation: • Promoters organize corporate formation. • Subscribers offer to purchase stock in the corporation during the formation process. • A state is selected for incorporation.

Legal Process of Incorporation	The legal requirements are: • Selecting the corporate name. • Drafting and filing the articles of incorporation. • Holding the first organizational meeting.
Potential Problems with Formation of the Corporation	Remedies for defective incorporation include: • *De jure* corporations. • *De facto* corporations. • Corporations by estoppel. • Piercing the corporate veil.
Corporate Financing	• Debt securities (bonds). • Equity securities (preferred stock and common stock).
Roles of Directors, Officers, and Shareholders	Directors: • A minimum number must attend board of directors meetings. • They are legally responsible for day-to-day managerial activities. Officers: • Run the day-to-day business of the organization. • Act as agent of the corporation. Shareholders: • Approve major corporate decisions. • Elect and remove members of the board of directors.
Duties of Directors, Officers, and Shareholders	Directors and officers: • Duty of care. • Duty of loyalty. • Duty to disclose conflicts of interest. Majority shareholder: • Duty of care and loyalty when selling shares.
Liabilities of Directors, Officers, and Shareholders	Directors and officers: • Personally liable for their own torts and crimes. • Liable for torts and crimes of others when there was inadequate supervision. Shareholders: • Liable for debts of the corporation to the extent of their investment.
Rights of Directors, Officers, and Shareholders	Directors: • Compensation. • Participation. • Inspection. • Indemnification. Officers's rights are spelled out in their employment contract. Shareholders: • Right to vote at the shareholders' meeting. • Preemption. • Inspection. • Share transfer. • Petition for dissolution. • Direct and derivative suits against the corporation.
Purchase of Stock	An acquiring corporation can take control of another corporation by purchasing a substantial amount of its voting stock.

(continued)

(concluded)

The Nature of Takeovers

A corporation can expand its size and operations by purchasing the stock of another firm.

A *hostile takeover* is a takeover to which the management of the target corporation objects.

Types of takeovers:
- Tender offers.
- Exchange offers.
- Cash tender offers.

Response to Termination

Response to takeovers: Directors declare whether they accept or reject the offer. If they object to the offer, they can engage in methods of resistance.

Dissolution is the legal death of a corporation.

Point/Counterpoint

Is Shareholder Activism Good for Business?

Shareholder activism is a process in which a major shareholder or shareholders use their voting rights in the company to force the management to either change their practices or be ousted by the shareholders. Shareholders may take this approach due to their dissatisfaction with the corporation's environmental or social policies, or simply because they feel the management is poorly running the business.

YES	NO
Over the last thirty years, shareholder activism has become an increasingly common practice as shareholders expect more from the corporations they invest in. While corporations are often fearful of shareholder activism gaining a foothold in the company, it can sometimes be just what the business needs. When the management board of a corporation gets complacent or ineffective, it can be very difficult to get them to budge and work hard enough to get the maximum value out of their stock. That is where shareholder activists come in. Shareholder activists can hold the management's feet to the fire and force them to shape up or face hostile takeover. Another benefit of shareholder activism is that the practice can quickly increase the value of a corporation's stock, at least in the short run. Shareholder activists tend to snatch up a large percentage of a corporation's stock in a short period of time to gain a significant foothold in the company. The mass buy up of stock will likely catch the attention of other investors, and the share price will rise. Not all shareholder activism focuses on maximizing share value, however. Some groups of shareholder activists seek to hold a corporation's management accountable for harmful environmental or social practices that go beyond what regulatory oversight can prevent. Shareholder activism forces the management to shape up or face being replaced and is therefore a great tool for making drastic change within a corporation.	Shareholder activism has become a popular tool for shareholders to assert their influence over a corporation, but it often is not the right tool for the job when it comes to benefiting a corporation in the long run. First of all, shareholder activism often involves investors buying up enough of a corporation to try to control the management of that corporation. But this method only works if the shareholders know how to run the business better than the management, which seems unlikely when you consider that it is the management that has the daily hands-on experience of actually running the company. Shareholders may think they know what it takes to improve the business, but shareholders are not always right, and they are arguably not in the best position to know what to do. A second problem with shareholder activism is that many people who engage in the practice are not "activists" at all, but are corporate raiders trying to take over a company through buying up a lot of stock. Often, these corporate raiders will purchase enough stock to have large-scale voting rights, compel a hostile takeover, sell off the company's assets to boost the stock price, and then sell off all of their shares to make a quick profit. These two reasons highlight a final reason why shareholder activism is bad for business: the practice has only the short-term interests of a business in mind, not the long-term interests. The goal of almost all shareholder activists is to raise the value of the corporation's stock to increase the value of their own assets. These shareholder activists have little concern, however, for the company's well-being after they sell off their stock, and the decisions they influence in order to make short-term gains may not be what is best for the business in the long run.

Questions & Problems

1. Name at least three characteristics that distinguish corporations from other forms of business organization.

2. How is an *ultra vires* act related to a corporation's powers?

3. Compare and contrast common stock and preferred stock.

4. Explain the primary duties of officers and directors.

5. What is the business judgment rule?

6. Carnival Corporation & PLC is a dual-listed corporation, i.e., a corporate structure that joins separate corporations in a common economic enterprise. The dual-listed corporation (DLC) was formed by Carnival Corporation, a corporation headquartered in Florida, and Carnival PLC, a British company headquartered in Southampton, England. Zolt Sabo, Ilija Janev, and Stefan Vidojkovic (The Seafarers) were workers aboard a cruise ship, working for Cunard Celtic Hotel Services, Ltd., a company that operated under the corporate umbrella of Carnival Corporation & PLC. The Seafarers sustained back injuries while working aboard the cruise ship and collected benefits in accordance with their employment contract. However, the Seafarers became unsatisfied with their compensation and believed that their contracts unfairly limited it.

 In July 2012, Sabo, Janev, and Vidojkovic filed a class action lawsuit against Carnival Corporation and Carnival PLC, alleging failure to provide maintenance and cure in accordance with general US maritime law and the Jones Act, a federal statute that provides legal remedies not guaranteed under general maritime law. The defendants filed a motion to dismiss, asserting that the Seafarer's claims should be dismissed because the court lacked jurisdiction over Carnival PLC, and Carnival Corporation was shielded from liability due to its corporate form. In the Seafarer's final response, the plaintiffs made clear that they were only suing Carnival Corporation & PLC, the dual-listed corporation, and further asserted that the DLC was a suable entity because it was in reality a corporation, the Carnival Corporation & PLC was subject to corporation by estoppel, among other claims.

 For the first claim, the Seafarers argued that Carnival Corporation & PLC resembled a corporation in many ways and claimed that "Carnival Corporation & PLC operate[s] as a single enterprise sufficient to establish personal jurisdiction over both entities as a single operation." Is this enough to qualify Carnival Corporation & PLC as a corporation? What would the DLC need to have done to be considered as a corporation and thus suable by the plaintiffs? What are the requirements for corporation by estoppel? Does corporation by estoppel apply in this case?

 Source: *Sabo v. Carnival Corp.*, 762 F. 3d 1330 (2014).

7. The Chomiak family opened a seasonal summer resort on the shores of Lake George in 1957. To run the resort, the Chomiaks formed a closely held corporation called Twin Bay Village, Inc. One hundred shares were issued which were split between Stephan and Eleonara Chomiak and their two sons, Vladimir and Leon. Decades later in 2004, the shares passed hands and Vladimir's son and daughter were the beneficial owners of a total of 48 shares in Twin Bay Village. Leo and his two daughters owned the remaining 52 shares. Ownership of shares were not the only things that changed in Twin Bay Village. By this point, Vladimir no longer helped manage the corporation, and only Leo and his two daughters were still involved. In 2009, the majority shareholders, Leo and his two daughters, attempted to force the minority shareholders to sell their shares back to the corporation. Met with refusal, the majority shareholders filed a suit to force dissolution of the corporation. The minority shareholders subsequently countersued, alleging the majority shareholders had breached their fiduciary duty to them by engaging in oppressive behavior and looting corporate assets. The trial court ruled in favor of Vladimir and his two daughters, the minority shareholders, for all claims alleged by them. While Leo appealed the decision, the appellate court affirmed the trial court. Could the minority shareholders apply Section 14.30 of the RMBCA to initiate a dissolution based on the claims it won? In other words, is Section 14.30 applicable to the claims of this case? Why or why not? What kind of evidence did the appellate court look at to determine that corporate looting took place?

 Source: *Matter of Twin Bay Village, Inc.*, 153 AD 3d 998 (2017).

8. Indiana Electronic Workers Pension Trust Fund (IBEW) is a retirement system that provides benefits to electrical workers in Indiana. IBEW is also a stockholder of Wal-Mart corporation. Wal-Mart de Mexico (WalMex) is a subsidiary of Wal-Mart, in which Wal-Mart owns a controlling interest. In April 2012, The New York Times released an article titled *Vast Mexico Bribery Case Hushed Up by Wal-Mart After Top-Level Struggle*, which described a scheme of illegal bribery payments to Mexican officials that occurred between 2002-2005, at the direction of the then-director of WalMex. The article showed that Wal-Mart officials knew of the illegal conduct no later than September 2005 and suggested that Wal-Mart's response was deficient. In June 2012, IBEW sent a letter to Wal-Mart

requesting inspection of documents related to the WalMex incident so that the IBEW board could investigate if Wal-Mart had breached fiduciary duty and if a derivative suit was necessary. In August, Wal-Mart complied with IBEW's request, but most of the documents were highly redacted with no explanation for the redaction. IBEW subsequently filed a complaint with the court of chancery, alleging that certain documents within the scope of the demand had not been produced. After a series of court hearings, the court entered a final order in October 2013. In its final order, the court required Wal-Mart to produce documents protected by attorney-client privilege, as well as other privileged documents, among other kinds of documents. The court also granted the IBEW's use of whistleblower documents, which IBEW had anonymously received and promptly reported in May 2012. Both parties appealed to the Supreme Court of Delaware; Wal-Mart asserted that the court had ordered Wal-Mart to produce documents that far exceeded the scope of the request and IBEW cross-appealed in regards to the whistleblower documents, as well as other assertions.

What is the Garner Doctrine as described by the Supreme Court of Delaware, and how is it related to this case? What arguments did Wal-Mart make to assert that IBEW had not met its burden to show it needed the privileged information that IBEW sought? What do you think of Wal-Mart's arguments? How did the Supreme Court of Delaware decide?

Source: *Wal-Mart Stores v. Indian Elec. Workers,* 95 A.3d 1264 (2014).

9. In June 2009, Country Bank failed, and the Federal Deposit Insurance Corporation (FDIC) subsequently became the receiver for Country Bank. As the receiver, the FDIC sought to recover losses of $42 million created by the 2009 bank failure by filing an action against Country Bank's former bank officers. The FDIC alleged claims of negligence and breach of fiduciary detail. The defendants answered by asserting that the FDIC's claims were barred by the business judgement rule. In response, the FDIC argued that the business judgement rule does not apply to bank officers, therefore does not grant the defendants any immunity. How did the court rule? What reasons did the court provide? Do you agree with the court's reasoning? Why or why not?

Source: *Federal Deposit Insurance Corporation v. Hawker,* 2012 U.S. Dist. LEXIS 79320.

10. Several shareholders of Fannie Mae filed a derivative suit on behalf of the company after Fannie Mae had announced one of the largest corporate earning restatements in history. The first derivative suit was dismissed, and James Kellmer, along with Arthur Middleton and L. Jay Agnes, all of whom were shareholders, filed new derivative suits that were consolidated by the district court. Fannie Mae moved to dismiss the cases presented by Kellmer and Middleton, but not Agnes. In 2008, Congress passed the Housing and Economic Recovery Act (HERA), which established a new federal agency, FHFA, that became Fannie Mae's conservator. Soon after its creation, the FHFA intervened in the derivative actions, arguing that HERA endowed it with sole authority to litigate claims belonging to Fannie Mae. The district court granted the FHFA's motion and followed the court decision by moving for voluntary dismissal without prejudice or a 180-day stay of litigation, arguing that the FHFA needed more time to evaluate if proceeding with the suit would further the conservatorship's statutory purposes. The district court granted FHFA's motion to dismiss without prejudice for Agnes and granted Fannie Mae's motion to dismiss Kellmer and Middleton with prejudice. Shareholders Kellmer and Agnes appealed the substitution order, and the FHFA appealed the district court's favorable ruling for Fannie Mae.

Why would the FHFA appeal the district court's dismissal of Kellmer and Middleton when the FHFA filed a motion to voluntarily dismiss the cases? This case is about who has the right to file a derivative suit. What was the core reason behind the appeals court's decision in regards to Kellmer's and Agnes's appeal?

Source: *Kellmer v. Raines,* 674 F.3d 848 (2012).

11. Moto Inc. operates gasoline convenience stores throughout the American Midwest. It does business as MotoMart. In June 2011, the collective owners of at least 20 percent of Moto, Virgil Kirchoff, Kirchoff LP, and the Virgil Kirchoff Revocable Trust, launched a suit against Moto seeking corporate dissolution based on minority shareholder oppression. The plaintiffs asserted that by arbitrarily valuing its stock and failing to inform minority shareholders of Casey's General Stores, Inc.'s interest in acquiring Moto. The trial court granted Moto summary judgment and the plaintiffs appealed. According to the appellate court, to fulfill a shareholder oppression claim, a claimant must show "(1) burdensome, harsh, and wrongful conduct; (2) a lack of probity and fair dealing in the affairs of the company to the prejudice of some of its members; or (3) a visible departure from the standards of fair dealing and a violation of fair play on which every shareholder is entitled to rely when

entrusting her money to a company." In its appeal, the plaintiffs-now-appellants argued the trial court should not have granted Moto summary judgment because Moto had a fiduciary duty to use reasonable care in evaluating its stocks and there was enough prima facie evidence showing Moto failed to exercise reasonable care. Suppose the appellate court accepted the appellants's argument that there was a genuine dispute of material fact about Moto's actions. Would Moto's lack of reasonable care fulfill the three requirements of a shareholder oppression claim quoted above? Why or why not? List examples of evidence appellants could show to meet the criterion.

Source: *Virgil Kirchoff Revoc. Trust v. Moto, Inc.*, 482 SW 3d 834 (2016).

Part 7 Business Organizations

CHAPTER 23

Securities Regulation

CASE OPENER

The Martha Stewart Case

On December 27, 2001, Martha Stewart's stockbroker, Peter Bacanovic, informed Stewart that two of his clients, Samuel Waksal, CEO of the biopharmaceutical company ImClone, and Waksal's daughter, had just sold all of their ImClone stock. Waksal secretly knew that the FDA was about to reject Erbitux, a key cancer drug ImClone had developed. Stewart did not know about the impending FDA rejection of Erbitux, and information about Waksal's sale of ImClone's stock was not available to the public. After receiving the information about Waksal's transaction, Stewart instructed her broker to sell all of her shares of ImClone stock. The following day, the FDA announced its rejection of Erbitux, and ImClone's stock price plummeted 16 percent. Stewart's timely trade allowed her to avoid a $45,673 loss.

Eighteen months later, the Securities and Exchange Commission (SEC) filed charges against Stewart and her broker for illegal insider trading and securities fraud.[1]

1. Did Martha Stewart violate federal securities law? Why or why not?
2. Did Martha Stewart's broker violate securities law? Why or why not?

LEARNING OBJECTIVES

After reading this chapter, you will be able to answer the following questions:

LO 23-1 Identify types of securities.
LO 23-2 Outline the requirements imposed by the Securities Act of 1933.
LO 23-3 Explain how the Securities Exchange Act of 1934 regulates security trading.
LO 23-4 Recall the legislative Acts that regulate investment companies.
LO 23-5 Identify state laws that regulate securities.

[1] www.sec.gov/news/press/2003-69.htm.

Companies frequently need to raise money to expand. One way they raise money is by issuing securities. For example, corporations issue corporate securities—stocks and bonds—to raise capital for corporate expansion.

However, a security is simply a piece of paper; it has no intrinsic value. Consequently, without securities regulations, corporations could easily commit fraud by issuing large numbers of securities and then refusing to repay them. Thus, the government heavily regulates securities issuance and trading.

Securities regulation is a relatively recent body of law. Before the Great Depression, government did not regulate securities, and fraudulent transactions occurred frequently. After the stock market crash in 1929, Congress passed several laws to regulate securities markets.

This chapter begins by defining *security*. It then examines federal securities regulations. Finally, the chapter briefly discusses state securities regulations.

security
A financial instrument designated as a note, stock, or bond or any other instrument named in the Securities Act of 1933.

What Is a Security?

LO 23-1 Identify types of securities.

Physically, a security is merely a piece of paper, as illustrated in Figure 22-3 of the previous chapter. But when investors buy securities, they are not buying a piece of paper. They are buying what the paper represents. The value of the security is based on what the paper represents.

Earlier, we loosely defined securities as stocks and bonds. To be more precise, securities include stocks, bonds, debentures, and warrants. Furthermore, certain items mentioned in securities acts, such as interests in oil and gas rights, are securities. In some cases, courts have even defined cosmetics, vacuum cleaners, and cemetery lots as securities. Finally, investment contracts, contracts in which individuals invest money with the expectation of making a profit, are securities. How can so many different things be securities?

The Securities Act of 1933 offers a complicated definition of *security*, and, as a result, courts have struggled when determining whether a particular instrument is a security. In an effort to provide a framework for analysis, the U.S. Supreme Court stated in the 1985 case *Landreth Timber Co. v. Landreth* that courts should presumptively treat as a security any financial instrument designated as a note, stock, bond, or other instrument named in the 1933 act.

If the instrument in question does not have the characteristics of an instrument specifically named in the 1933 act, the courts apply a three-part test. In the classic 1946 case *SEC v. W.J. Howey Co.,*[2] the U.S. Supreme Court defined a security as an (1) investment in a common enterprise with the (2) reasonable expectation of profit gained (3) primarily or substantially from others' efforts. Anything that meets these three criteria is subject to security law. Case 23-1 illustrates how courts apply the *Howey* test.

[2] 28 U.S. 293, 66 S. Ct. 1100 (1946).

SECURITIES AND EXCHANGE COMMISSION, v. ARCTURUS CORPORATION
United States District Court, N.d. Texas, Dallas Division
171 F.SUPP.3D 512 (2016)

FACTS: Arcturus Corporation and Aschere Energy, LLC are two companies owned by Leon Parvizian. The companies were Texas corporations based in Dallas that offered and sold interests in gas and oil drilling ventures. Parvizian and his companies sold the drilling interests, which were officially called "Joint Ventures," through limited liability companies that cold-called United States residents.

In 2013, the SEC filed an action against Parvizian, Arcturus, Aschere, and the cold-calling contractors. The SEC alleged that the Joint Ventures offered and sold by Arcturus and Aschere were actual securities and Parvizian was trying to avoid securities regulation by naming them Joint Ventures. The defendants filed a motion to dismiss which was rejected by the trial court. Both the SEC and the defendants subsequently

(continued)

moved for summary judgement against each other. The crucial first question the court had to decide was whether the Joint Ventures were in fact securities regulated under the Securities Act of 1934.

The Securities and Exchange Commission (SEC) filed an action against the defendant based on believed violations of securities laws. The plaintiff argued that the defendant lied about the life expectancy figures independent doctors were producing and that most of the life expectancy numbers were false. At an evidentiary hearing, the issue of whether a viatical settlement investment is an investment contract under securities laws arose.

ISSUE: Were Parvizian's investments truly "Joint Ventures," or were they disguised securities?

REASONING: To determine whether Parvizian's investment scheme involved securities, the court used the *Howey* test. The Howey test has three prongs: (1) an investment of money; (2) in a common enterprise; (3) on an expectation of profits to be derived solely from the efforts of individuals other than the investor. There was no dispute that the scheme involved individuals investing money in oil and gas drilling, a common enterprise. To prove the third prong, the court used the Williamson test, which essentially analyzes how much control over business operations the so-called "venturers" have and how much experience and knowledge they have of the industry in question. The court determined that the venturers had practically no control over the operation of the "venture" and that the venturers knew nothing about the oil and gas industry. In addition, Parvizian had complete control over the investment accounts and the venturers had no power to replace him. Having satisfied all three factors of both tests, the court ruled that the "Joint Ventures" were indeed securities and therefore subject to the Securities Acts.

DECISION AND REMEDY: Affirmed in favor of the SEC.

SIGNIFICANCE OF THE CASE: If it looks and functions like a security, it's probably a security. Using creative names for securities doesn't mean they aren't subject to SEC regulations. This case is a good example of how the *Howey* test can detect securities by any other name.

CRITICAL THINKING

Suppose someone makes an inference about a court case based on the three *Williamson* factors and states "since joint ventures are not investment contracts, none of the three *Williamson* factors must have been met." Is the statement logically sound? Why do you think the creators of the three *Williamson* factors decided that any one criteria could identify the contracts as investment contracts?

ETHICAL DECISION MAKING

Do you think Parvizian would have sent the angry email to Ullrey if he had applied the disclosure test before he sent it?

Securities Regulation

Two crucial acts regulate securities transactions. The Securities Act of 1933[3] regulates how companies issue corporate securities; the Securities Exchange Act of 1934[4] oversees the purchase and sale of securities. Both attempt to provide greater stability to securities transactions. First, the acts mandate that investors must have access to certain information when deciding whether to buy or sell securities. Second, the acts strive to curb fraudulent securities transactions.

The Securities and Exchange Commission

The 1934 act created the Securities and Exchange Commission (SEC), an independent agency whose function is to administer federal securities laws. The SEC is headed by five individuals appointed by the president. These five individuals serve fixed five-year terms.

[3] 15 U.S.C. §§ 77a–77aa.
[4] 15 U.S.C. §§ 78a–78mm.

GLOBAL Context

Sweden's Securities Market

©Digital Archive Japan/Alamy

The Swedes divide securities into bonds and shares. Individuals invest in either the bond market or the stock market. Companies issue bonds to boost funds from sources outside their shareholders. Bondholders have a right to a fixed rate of interest regardless of whether the company earns a profit. Bondholders do not, however, have a right to take part in company decision making.

Companies issue shares to increase their equity capital. Unlike bond interest payments, shareholders' returns vary with the company's profits. Because shareholders' interests are closely tied to the success of the company, shareholders have a voice within the company. Generally, this voice comes in the form of a vote at general meetings.

Regulation of the stock and bond markets in Sweden is unique because Sweden has no equivalent of the SEC. Instead, banks themselves oversee the issuance of shares and bonds.

The SEC has a number of responsibilities. First, it is responsible for the enforcement of securities laws. Thus, in response to an allegation of violation of securities law, the SEC investigates and, if necessary, initiates an enforcement action against the violator. Enforcement actions often include penalties or injunctive remedies. If the violation is severe enough to warrant criminal prosecution, however, the Fraud Section of the Criminal Division of the Department of Justice will prosecute the alleged violator.

Second, the SEC interprets the securities acts and adopts rules to achieve the purposes of the acts. Thus, the SEC passes securities transaction regulations that have the force of law.

Third, the SEC regulates the activities of securities brokers, dealers, and advisers. All securities dealers and brokers must register with the SEC.

Fourth, the SEC regulates the trade of securities on securities exchanges. As you read about securities regulations in this chapter, remember that the SEC is responsible for their administration and enforcement. Exhibit 23-1 traces how the SEC's regulatory powers have evolved.

Exhibit 23-1
The Expansion of SEC Powers in the 1990s

Securities Enforcement Remedies and Penny Stock Reform Act of 1990
Permits the SEC to:
- Issue a cease-and-desist order against a violator of any federal securities law
- Seek civil money penalties against any violators
- Create rules to require brokers and dealers to provide information concerning prices and risk associated with the penny-stock market

Market Reform Act of 1990
Allows the SEC to suspend securities trading if prices vary excessively in a short time period

Securities Acts Amendments of 1990
Permit the SEC to seek punishment of violators of foreign securities laws

National Securities Markets Improvement Act of 1996
Permits the SEC to exempt persons, securities, and transactions from securities regulations

Sarbanes-Oxley Act of 2002
- Increases corporate disclosure requirements
- Penalizes violators of securities laws more heavily
- Holds corporate executives responsible for errors in corporate reports filed with the SEC
- Requires earlier filing of financial and stock transaction reports
- Creates and establishes SEC oversight over the Public Company Accounting Oversight Board to regulate public accounting firms

Other countries have bodies that serve functions similar to those of the SEC. In Singapore, for example, the government controls securities markets through the Securities Industry Council. The council is an advisory board to the minister of finance, who, on the basis of the council's advice, issues guidelines for exchanges, mergers, and trading.

The Securities Act of 1933

LO 23-2 Outline the requirements imposed by the Securities Act of 1933.

Congress passed the Securities Act of 1933 in reaction to the mistrust of securities transactions before the Great Depression. Thus, a function of the act was to legitimize security transactions by requiring the registration of securities offered to the public. Through the registration process, investors have more information to make better-informed decisions about their securities purchases.

Section 5 of the 1933 act mandates that any corporation, partnership, association, or individual that offers the sale of securities to the public through the use of mails or through any facility of interstate commerce must register that security, unless it qualifies for an exemption (discussed later in this section). Any issuer of securities must file a written registration statement and a prospectus with the SEC.

Registration Statement

Although the SEC requires different types of companies to complete slightly different registration forms, the SEC requires certain elements from all companies. The registration statement generally contains (1) a description of the securities offered for sale, (2) an explanation of how proceeds from the sale of the securities will be used, (3) a description of the registrant's business and properties, (4) information about the management of the company, (5) a description of any pending lawsuits in which the registrant is involved, and (6) financial statements certified by an independent public accountant.

Caution: Although the SEC requires the registration of securities, the SEC does not approve these securities. In other words, the SEC does not make any judgment about the worth of securities; it simply enforces the requirement that issuers must provide certain information to potential buyers.

Prospectus

prospectus
A written document filed with the SEC that contains a description of a security and other financial information regarding the company offering the security; also distributed as an advertising tool to potential investors.

Along with filing a registration statement, issuers of securities must also file a prospectus with the SEC. A **prospectus** is a written document that contains most of the information that appears in the registration statement. The prospectus is different from the registration statement, however, because it is an advertising tool that issuers distribute to potential investors, who rely on the prospectus to help decide whether they should buy the securities.

The SEC requires, but cannot guarantee, the accuracy of facts stated in the registration statement and prospectus. If an issuer makes a false or misleading statement in a registration statement, the issuer can be subject to criminal or civil penalties. These penalties are discussed later in the chapter.

Periods of the Filing Process

The filing process consists of three periods: prefiling, waiting, and posteffective.

Prefiling Period The prefiling period begins when an issuer starts to think about issuing securities, and it ends when the issuer files the registration statement and prospectus with the SEC. Before filing these documents, the issuer cannot make any offers to sell securities. The issuer can, however, negotiate with *underwriters,* investment banking firms that purchase securities from the issuing corporation with the intent to sell them to brokerage houses, which then sell them to the public.

Because they cannot offer to sell securities during the prefiling period, issuers, officers, directors, and underwriters usually try to avoid generating publicity about possibly issuing securities

GLOBAL Context

©Digital Archive Japan/ Alamy

China's Securities Regulation

The purpose of securities regulation in China is similar to that of the SEC in the United States. The Securities Law of the People's Republic of China, for example, contains 12 chapters that regulate items such as stock issuance, stock transactions, prohibited transactions, stock exchanges, securities companies, securities registration, and much more. The Securities Law of China does not approve registration of securities; instead, securities are approved by the regulatory body under the State Council of China.

because the SEC may construe such publicity as an attempt to condition the market to generate interest in the securities. In fact, the SEC could consider speeches or press releases mentioning the potential issuance of securities to be an offer. The issuing firm is permitted, however, to publish a notice about the prospective offering, including the name of the issuer as well as a description of the potential securities, but the notice may not include the name of the underwriter.

Waiting Period Once an issuer files a registration statement and prospectus, the waiting period begins. During this period, the SEC reviews the information filed by the issuing firm. Issuers must wait 20 days after the filing date to sell their securities. During the waiting period, issuers may make oral offers to sell the securities and may distribute a red-herring prospectus—a prospectus with a notice written in red print at the top of the page warning investors that the registration has been filed with the SEC but is not yet approved. Moreover, the issuer may publish a tombstone advertisement, a brief ad with a format similar to that of a tombstone.

Posteffective Period The posteffective period begins when the SEC declares the registration statement effective, and it ends when the issuer sells all securities offered or withdraws them from sale. During this time, investors may buy and sell the securities, but purchasers must receive a final prospectus, a version of the prospectus the buyer must receive at the point of sale. If an issuer does not send the purchaser a final prospectus, the issuer has violated the 1933 Securities Act.

red-herring prospectus
A prospectus with a warning written in red print at the top of the page telling investors that the registration has been filed with the SEC but is not yet approved.

tombstone advertisement
A print advertisement in a format similar to that of a tombstone that announces a forthcoming sale of securities.

BUT WHAT IF? . . .

WHAT IF THE FACTS OF THE CASE OPENER WERE DIFFERENT?

Let's say in the Case Opener that Waksal had submitted his registration statement to the SEC. He went through the waiting period and eventually the SEC declared his registration statement to be effective. Waksal thinks that he has now completed all the periods of the filing process. Is he correct? Is he missing anything?

Special Registration Provisions

Amendments to the 1933 act stipulate more relaxed registration requirements for companies that have consistently satisfied applicable registration requirements. For example, *shelf registrations* permit certain qualified issuers to register securities that they will sell off the shelf on a delayed or continuous basis in the future. SEC Rule 415 requires corporations using shelf registrations to keep information in the original registration accurate and up to date.

Exemptions under the 1933 Act

Although these complex registration requirements are standard for most securities, they do not apply to all securities in all situations. Some securities are exempt because of the nature of the securities themselves—exempt *securities*—and others are exempt when exchanged in certain ways—exempt *transactions*.

Exempt Securities The 1933 act provides that certain securities are exempt from the registration procedures previously described. These securities are *unregistered, unrestricted securities,* and they include:

1. Securities issued by governments, including municipal, state, and federal governments.
2. Securities issued by nonprofit issuers, such as religious institutions or charitable organizations.
3. Securities issued by financial institutions supervised by banking associations.
4. Securities issued as a result of corporation reorganization in which one security is exchanged for another security.
5. Stock dividends and stock splits.
6. Insurance and annuity contracts issued by insurance companies.
7. Securities issued by federally regulated carriers (i.e., railways).
8. Short-term notes with a maturity date that does not exceed nine months.
9. Securities sold before July 27, 1933.
10. An issuer's offer of up to $5 million in securities in a 12-month period.

Although the last exemption permits the issuer to avoid the complex requirements of full registration, Regulation A nevertheless requires the issuer to file certain reports with the SEC. For example, the issuer must file a notice of issue and an offering circular and provide the offering circular to investors before sale of the securities. Even with these requirements, however, the registration process under Regulation A is much less burdensome than the full registration process. For example, Regulation A permits issuers to test the waters for interest in their securities before preparing the offering circular, selling any securities, or gaining the commitment of interested buyers.

Exempt Transactions If a security does not fall under one of the exempt categories previously described, an issuer can nevertheless avoid registering the security by making certain exempt transactions. Such securities are *unregistered restricted securities.* Issuers might want to avoid the registration process because it can be costly, complicated, and time-consuming. Exempt transactions offer an opportunity to save time and money. Consequently, many issuers sell securities through exempt transactions.

Although these transactions are exempt from registration, issuers must still provide investors with information about the securities, such as annual reports and financial statements.

Exempt transactions include limited offers, intrastate issues, and resales of securities.

Limited Offers. Certain securities transactions known as *limited offers* are exempt from the registration process because they either involve small amounts of money or are offered only to sophisticated investors. The investors who participate in limited offers generally do not need the protection afforded by the registration process.

The SEC's Regulation D enumerates three exemptions for limited offers (Rules 504, 505, and 506). Section 4(6) of the 1933 act contains an additional limited-offer exemption.[5]

Private Placement Exemption (Rule 506) Issuers who make private offerings of securities are exempt from the registration process. These issuers, however, cannot advertise their private offerings to the general public. This exemption, usually referred to as the *private placement exemption,* allows firms to issue an unlimited number of securities to an unlimited number of accredited investors. Consequently, firms can easily raise large amounts of capital. Firms may not, however, issue to more than 35 unaccredited investors.

The SEC defines an **accredited investor** as:[6]

accredited investor
A private investor who is allowed to accept private securities offerings under certain specific guidelines the SEC sets.

1. Any natural person who has a net worth of at least $1 million.
2. Any natural person whose annual income has been at least $200,000 for the two previous years and expects to make at least $200,000 in the current year.
3. Any corporation or partnership with total assets in excess of $5 million.
4. Insiders of the issuers, such as executive officers or directors.
5. Registered investment companies, colleges and universities, banks, and insurance companies.

[5] http://securities-law-blog.com/2010/01/28/section-46-registration-exemption-for-accredited-investors/.
[6] SEC Rule 501.

The SEC assumes that accredited investors are better able to evaluate the financial risk associated with buying securities. Consequently, government protection in the form of required registration is less imperative.

The SEC also requires firms selling to unaccredited investors under Rule 506 to believe that these investors have expertise regarding securities trading. The firms must believe that the investors, or their representatives, have reasonable ability to evaluate the financial risk associated with purchasing securities. If a firm privately offers any unaccredited investor the opportunity to purchase securities, all investors must receive the basic information contained in the registration statement (i.e., information about the issuing company and the securities). If no unaccredited investors are involved, however, the issuer does not have to disclose any information. Although firms do not have to register these exempt securities with the SEC, they must notify the SEC of any sales made under the exemption.

The private placement exemption provides one of the easiest ways for firms to raise capital. Instead of making a public offering, firms can simply make a private offering to investors capable of ascertaining the risk associated with buying securities. The firms cannot, however, advertise the securities to the general public.

Rule 505 Rule 505 is very similar to Rule 506, but it has two important differences. First, according to Rule 505, a firm's private offerings may not exceed $5 million in a 12-month period, whereas under Rule 506, a firm may issue an unlimited number of securities. Second, according to Rule 506, if a firm issues securities to unaccredited investors, it must believe that they have the knowledge to evaluate the risk associated with the security. In contrast, under Rule 505, firms need not believe that investors have the reasonable ability to evaluate the risk of purchasing securities.

Rule 504 *Noninvestment firms*—firms that do not engage primarily in the buying and selling of securities—that offer no more than $1 million in securities over a 12-month period are exempt from registration. An unlimited number of both accredited and nonaccredited investors may purchase these securities; however, the firms must notify the SEC of their securities sales. Issuers need not disclose any information to investors.

Section 4(6) If a firm offers securities only to accredited investors for an amount less than $5 million, the issuer is exempt from registration. An unlimited number of accredited investors may participate in the transactions, but no unaccredited investors may buy these securities. Firms may not advertise these securities to the public. Issuers do not have to disclose any information to investors, but they must notify the SEC of any sales under this exemption. Moreover, investors who want to resell these securities must register them with the SEC.

Intrastate Issues. Under Section 3 of the 1933 act, any security offered or sold to a permanent resident of the single state where the issuer of the security resides and does business is exempt. Thus, local businesses can rely on local investors to raise an unlimited amount of capital without registration.

Courts and the SEC have interpreted this exemption very narrowly. Issuers must do at least 80 percent of their business within the state, receive at least 80 percent of their profits within the state, have at least 80 percent of their assets within the state, plan to use at least 80 percent of the profits within the state, and have their main offices in the state.

During the period of sale and at least nine months after the period of sale, buyers of these securities under the intrastate exemption may not resell them to nonresidents. Issuers must take precautions against interstate resales.

Resales. The 1933 Securities Act creates an exemption for "transactions by any person other than an issuer, underwriter, or dealer." Because another section of the act exempts dealers and brokers, only issuers and underwriters are not exempt from registering resales of securities.

Consequently, the average investor does not have to register securities when he or she wants to sell. If the investor acquired them through sales under Rule 505, Rule 506, or Section 4(6) (e.g., intrastate issues and limited offers), they are **restricted securities**. If an investor wants to resell a restricted security, she must register the securities unless she follows Rule 144 or Rule 144a.

Rule 144 Under Rule 144, any person wanting to resell a restricted security is exempt from registration if certain criteria are met. First, the public must have access to adequate current

restricted securities
Securities that have limited transferability and are usually issued in a private placement.

information about the issuer. Second, the person selling the restricted security must have owned the security for at least two years. Third, the seller must sell the restricted securities in limited amounts in unsolicited broker transactions. Fourth, the seller must notify the SEC of the sale.

If a seller is not an affiliate of the issuer and has owned the securities for three years, he or she may sell the securities in unlimited amounts and is not subject to any of the criteria under Rule 144 for an exemption. An *affiliate* is a person who controls, is controlled by, or is in common control with the issuer. The 1933 act restricts sales by affiliates much more than sales by nonaffiliates. If an affiliate sells restricted or nonrestricted securities, he must meet all but one of the criteria to be exempt from registration. The affiliate does not need to hold a nonrestricted security for two years before reselling.

Violations and Liability

If the SEC uncovers a potential violation of the 1933 act, it can (1) take administrative action, (2) take injunctive action, or (3) recommend criminal prosecution. If the SEC takes administrative action, it conducts a formal investigation of the potential violation by calling relevant witnesses to testify or produce evidence. If, through this investigation, the SEC uncovers evidence of a violation, the SEC may order an administrative proceeding before an administrative law judge, who can impose sanctions and even revoke a security's registration.

Usually the SEC takes injunctive action when it believes that a defendant is likely to continue to violate the law. Thus, the SEC may seek an injunction to prevent issuers from advertising securities by mail. In the Case Opener about Martha Stewart, the SEC sought an injunction prohibiting Stewart and her broker from violating securities laws and limiting her activities as an officer of a public company.

Finally, if the SEC recommends criminal action, the Department of Justice prosecutes criminal charges against the violators. Criminal penalties include a fine up to $10,000, imprisonment for up to five years, or both.

How might an issuing company violate the 1933 Securities Act? First, a company violates the act if it intentionally misleads investors by omitting or falsifying information on a registration statement or prospectus. Second, if a company is negligent in discovering a fraudulent statement, it can be liable. Third, an issuing company violates the act if it sells securities before the effective date of the registration statement.

Issuers charged with a violation of the 1933 act can raise several defenses. A company charged with the first or second violation previously described may claim that the omitted or false statement was immaterial to the sale of the security. Similarly, if the issuer can prove that the plaintiff was aware of the omission or false statement when she bought the security, the defendant can avoid liability.

Any defendant except the issuer can assert the **due diligence defense**. This defense requires the defendant to demonstrate that she investigated the registration statement and had reasonable grounds to believe that the statement was accurate and had no omission of material facts.

If a defendant sold securities before the effective date of registration, however, she will almost certainly be liable because the 1933 act provides no defenses for this violation.

If an investor purchased securities and suffered damages as a result of an issuer's false or misleading statement, the investor is entitled to bring a civil suit to recover his losses. The burden of proof falls on the investor to demonstrate this incomplete or inaccurate disclosure of facts.

due diligence defense A defense in which the defendant argues that he or she applied the appropriate degree of attention, care, and research expected of a party in a given situation and had reasonable grounds to believe that certain facts and statements were accurate and had no omission of material facts.

The Securities Exchange Act of 1934

LO 23-3 Explain how the Securities Exchange Act of 1934 regulates security trading.

Whereas the 1933 Securities Act regulates the issuance of securities, the 1934 Securities Exchange Act regulates the subsequent trading (resale) of securities, chiefly through the required registration of securities exchanges, brokers, dealers, and national securities associations. Moreover, the act requires certain issuers to file periodic reports with the SEC. The 1934 act also permits the SEC to monitor securities markets for fraud and market manipulation.

In one of the most closely watched business cases in years, the Supreme Court, in a 5–3 decision, protected those who are only secondary players in securities fraud, as long as the public did not rely on any explicit statements or actions by the secondary players. The case appears to limit the liability of accountants, lawyers, and others who may know about questionably fraudulent corporate behavior but can establish that they are not directly involved in it. The Case Nugget explains this case.

Aiders and Abettors Get a Free Pass from the Supreme Court

Stoneridge Inv. Partners, LLC v. Scientific-Atlanta, Inc. United States Supreme Court

128 S. Ct. 761 (2008)

The facts in the case are fairly straightforward. At the request of Charter, which in mid-2000 was falling short of its cash-flow target, Scientific-Atlanta and Motorola agreed to increase their prices for the cable boxes they sold to Charter and to use the extra money to buy advertising on Charter's cable stations. The arrangement allowed Charter to treat the advertising purchases as current revenue but list the money spent on cable boxes as a capital expense. By recording the advertising purchases as millions in extra revenue, Charter fooled its auditor into approving a financial statement that showed it met projected revenue and operating cash-flow numbers.

In rejecting the shareholders' claim against Scientific-Atlanta, the Court majority said that shareholders could not sue bankers, lawyers, and other third parties who did not directly mislead investors but worked with corporations that did. The Court reasoned that the key test for liability is whether investors relied on an entity's actions or statements, and, in this case, investors had no knowledge of deceptive acts during the time in question. The Court noted that the primary punishment of such secondary offenders would come from criminal prosecutions and enforcement actions by the SEC.

Shareholder activists are concerned that this decision, in effect, offers less protection to shareholders in a variety of situations. Their concerns were strengthened when less than a week after the Stoneridge case, the high court turned down an appeal in a class action case in which several plaintiffs alleged that three investment banks, Credit Suisse, Merrill Lynch, and Barclays Bank, had aided and abetted Enron in the fraudulent practices that led to the giant corporation's filing of bankruptcy. The plaintiffs were seeking up to $40 million in damages. The court of appeals in the case held that the shareholders could not sue the banks for fraud because the banks did not have primary liability for the matter under federal law.

Section 10(B) and Rule 10B-5

One of the most important sections of the 1934 act is Section 10(b), which prohibits the use of manipulative and deceptive devices to bypass SEC rules. Within this section, subsection 5 prohibits fraud associated with the purchase or sale of all securities. Thus, even though securities may be exempt from registration, they are still subject to Rule 10b-5:

It shall be unlawful for any person, directly or indirectly, by use of any means or instrumentality of interstate commerce or of the mails, or of any facility of any national securities exchange,

a. to employ any device, scheme, or artifice to defraud,
b. to make any untrue statement of a material fact or to omit to state a material fact necessary to make the statements made, in light of circumstances under which they were made, not misleading, or
c. to engage in any act, practice, or course of business that operates or would operate as a fraud or deceit upon any person, in connection with the purchase or sale of any security.

Both Section 10(b) and Rule 10b-5 play an important role in preventing insider trading.

Insider Trading

When a company employee or executive uses material inside information to make a profit, she or he is engaging in **insider trading.** Insider trading is illegal because it gives the violator an important advantage over the general public and shareholders.

Section 10(b) and Rule 10b-5 define *insiders* as corporate officers, directors, employees, lawyers, consultants, accountants, majority shareholders, or any other individuals who receive private information regarding the trading of securities.

A company employee or executive who has inside information may be liable, depending on whether the information is material. If there is a material omission or misrepresentation during a securities transaction, the individual has violated Section 10(b) and Rule 10b-5. If the omission or misrepresentation is not material, however, the individual is not liable. Examples of material information include the following:

1. A change in the status of litigation against the company.
2. A change in dividends.
3. A contract for the sale of corporate assets or for the purchase of assets.
4. A new product, process, or discovery.
5. A significant change in the financial status of the company.

According to the SEC, an individual with material inside information should either refrain from using the information or disclose the information to the other parties involved in the transaction.[7] Case 23-2 is a classic example of how courts address insider trading.

insider trading
The illegal buying or selling of a corporation's stock or other securities by corporate insiders, such as officers and directors, in breach of a fiduciary duty or some other relationship of trust and confidence, while in possession of material, nonpublic information about the security.

[7] Matter of Cady, Roberts & Co., 40 SEC 907 (1961).

CASE 23-2

SECURITIES AND EXCHANGE COMMISSION v. TEXAS GULF SULPHUR CO.

U.S. Court of Appeals for the Second Circuit

401 F.2d 833 (1968)

FACTS: In June 1963, Texas Gulf Sulphur (TGS) acquired the option to buy land in Timmons, Ontario. Preliminary drilling in November 1963 suggested that the land held great amounts of copper and zinc. This information was supposed to be kept secret; not even officers and executives at TGS were supposed to know about the results of the drilling. TGS acquired the land and resumed drilling in March 1964.

From November 1963 to March 1964, certain directors, officers, and employees of TGS received tips and purchased TGS stock or options to buy shares at a fixed price. When drilling began in November 1963, these people had owned 1,135 shares of TGS stock and possessed no calls; thereafter they owned a total of 8,235 shares and possessed 12,300 calls.

In April 1964, when rumors of a major mineral find appeared in newspapers, TGS publicly claimed that the rumors had no factual basis; a few days later TGS confirmed that the strike was expected to yield many million tons of ore. Between the days that TGS claimed the rumors were false and later confirmed the rumors, two defendants, Clayton and Crawford, ordered a combined total of 500 shares of TGS stock. The SEC brought suit against TGS and 13 of its directors, officers, and employees for violation of Section 10(b) of the Exchange Act and SEC Rule 10(b)-5, seeking an injunction to prevent TGS from publishing misleading press releases and requesting rescission of TGS's purchases and stock options. The district court dismissed the charges against all defendants but Clayton and Crawford, and the SEC appealed.

ISSUE: Did the corporate employees possess inside information that should have precluded them from trading in the company's stock?

REASONING: The essence of the Rule is that anyone in possession of material inside information must either disclose it to the investing public, or, if he cannot or chooses not to do so, must abstain from trading in or recommending the securities concerned while such inside information remains undisclosed.

> The basic test of materiality ". . . is whether a *reasonable man* would attach importance . . . in determining his choice of action in the transaction in question. . . ." This, of course, encompasses any fact ". . . which in reasonable and objective contemplation *might* affect the value of the corporation's stock or securities. . . ." Such a fact is a material fact and must be effectively disclosed to the investing public prior to the commencement of insider trading

(continued)

in the corporation's securities . . . disclosing the earnings and distributions of a company but also those facts which affect the probable future of the company and those which may affect the desire of investors to buy, sell, or hold the company's securities. . . .

Here . . . knowledge of the possibility . . . of the existence of a mine of the vast magnitude indicated by the remarkably rich drill core located rather close to the surface . . . might well have affected the price of TGS stock and would certainly have been an important fact to a reasonable, if speculative, investor in deciding whether he should buy, sell, or hold.

DECISION AND REMEDY: The court found that the "transactions in TGS stock or calls by individuals apprised of the drilling results were made in violation of Rule 10b-5." The judgment was reversed in favor of the plaintiff.

SIGNIFICANCE OF THE CASE: The case is significant because it clarifies the type of information a person may possess that qualifies him as an inside trader for purposes of the law.

CRITICAL THINKING

How does the reasoning in this case highlight the link between business law and business ethics?

ETHICAL DECISION MAKING

What values are being emphasized by the prohibition on insider trading?

BUT WHAT IF? . . .

WHAT IF THE FACTS OF THE CASE OPENER WERE DIFFERENT?

Let's say in the Case Opener that Stewart had heard on the news about the rejection of Waksal's new drug. She immediately had her agent sell her stock as fast as she could and she beat a lot of other people to it. Is there anything wrong with Stewart's action in this case?

The Private Securities Litigation Reform Act of 1995

Although Rule 10b-5 was designed to encourage disclosure of accurate information, one of its unintended side effects was the deterrence of forecasts. Shareholders who purchased stock in corporations with high earning forecasts often brought suit against the corporations' directors if actual corporate earnings fell short of forecast earnings, alleging that the directors violated Rule 10b-5 by disclosing misleading financial information.

Congress attempted to remedy this problematic side effect by passing the *Private Securities Litigation Reform Act (PSLRA) of 1995*. Among other things, the act provides a "safe harbor" from liability for publicly held issuers who make financial forecasts as long as the forecasts are "accompanied by meaningful cautionary statements identifying important factors that could cause actual results to differ materially from those in the forward-looking statement."[8]

Innovative shareholders attempted to subvert the PSLRA by suing corporate directors in state courts. Congress responded to these efforts by passing the *Securities Litigation Uniform Standards Act of 1998*. This act strictly limits shareholders' ability to bring class action suits against nationally traded corporations.

Outsiders and Insider Trading

Not only may insiders be liable for omitting or misrepresenting material information, but certain outsiders may also be liable through two theories: *misappropriation theory* and the *tipper/tippee theory*. Consider whether Martha Stewart is liable under either theory.

[8] 15 U.S.C. § 77z-2 (2005).

misappropriation theory
A theory of insider trading that holds that if an individual wrongfully acquires (misappropriates) and uses inside information for trading for his or her personal gain, that person is liable for insider trading.

Misappropriation Theory In addition to being a theory of tort (see Chapter 7), misappropriation can establish liability for insider trading. **Misappropriation theory** holds that if an individual wrongfully acquires (misappropriates) and uses inside information for trading for her personal gain, she is liable for insider trading. Because she wrongfully acquires inside information, she is essentially stealing information to use for her benefit.

Because it expands the SEC's power, application of misappropriation theory to insider trading has not gone uncontested. The *United States v. Carpenter* Case Nugget describes the first case in which the SEC convinced the courts to apply misappropriation theory to Rule 10b-5. Over 10 years later, in *United States v. O'Hagan*, the U.S. Supreme Court conclusively established that misappropriation theory is applicable to Rule 10b-5.[9]

tipper/tippee theory
A theory of insider trading that holds that any individual (tippee) who acquires material inside information as a result of an insider's (tipper's) breach of duty has engaged in insider trading.

tippee
An individual who receives confidential information from an insider.

tipper
An insider who gives inside information to someone.

Tipper/Tippee Theory The **tipper/tippee theory** holds that any individual who acquires material inside information as a result of an insider's breach of duty has engaged in insider trading. This individual, one who has received a tip from an insider, is called a **tippee**. The insider who gives the inside tip is called the **tipper**.

Suppose that Jerry (the tipper) gives material inside information to Elaine (the tippee). Jerry is liable because he is passing on inside information. Furthermore, Elaine is liable if she makes trading decisions based on information that she should know is not public. Jerry is liable for any profits made by Elaine. Now suppose that Elaine passes the tip on to George. Elaine is now a tipper, and George is a tippee. Both Elaine and Jerry are liable for the profits made by George. George is liable for the profits of his transactions if he knew or should have known that the material information was not public.

In some business areas, such as public accounting, employees may be exposed to sensitive information when working for a client. As a preventive measure, businesses will often impose strict rules to ensure that their employees keep client information private. However, regardless of the regulations, some unethical employees may attempt to use classified client information to make money through insider trading.

For example, in 2008 two PricewaterhouseCoopers employees were caught using their access to client information to engage in insider trading. One of the employees, Patrick Borchard, worked in PwC's Transaction Advisory Group assisting corporate clients interested in mergers or acquisitions. As a member of the Transaction Advisory Group, Borchard often heard about corporate takeovers before they were announced publicly. Rather than keeping the information private, Borchard tipped off his co-worker, Gregory Raben, to confidential client plans so that Raben could trade on the information and make a profit. Additionally, Raben tipped off two of his own acquaintances to the information provided by Borchard. Eventually, PwC noticed what Raben and Borchard were doing and alerted the SEC. Both Borchard and Raben were charged by the SEC with insider trading and were required to not only pay fines but also pay back the trading profits earned with client information.[10]

Sentences for the tipper or tippee can be harsh, as Case 23-3 illustrates.

[9] 117 S. Ct. 2199 (1997).
[10] http://www.reuters.com/article/pricewaterhousecoopers-sec-idUSN1553446820080115.

CASE 23-3

UNITED STATES OF AMERICA v. MATTHEW KLUGER
United States Court Of Appeals For The Third Circuit
722 F.3d 549; 2013 U.S. App. LEXIS 13880 (2013)

FACTS: *Defendant Kluger and his codefendant Bauer were conspirators in a 17-year-long, three-man insider-trading scheme, along with Kenneth Robinson. Kluger entered a guilty plea to four counts of offenses related to insider trading and was sentenced to a total custodial term of 144 months, a term thought to be the longest insider-trading sentence ever imposed. The sentence included a $400 special assessment, a three-year term of supervised release to follow the custodial term, and occupational restrictions relating to Kluger's employment in the securities industry.*

(continued)

Bauer, who actually made the trades using the inside information, was sentenced to a total custodial term of 108 months. Bauer's sentence also included the same additional penalties as Kluger's.

Robinson, who had received the information from Kluger and transferred it to Bauer, pled guilty to a three-count information for securities fraud and was sentenced to concurrent 27-month custodial terms on all counts.

Kluger appealed his sentence.

ISSUE: Can a tipper, who benefits less than the tippee, receive a longer sentence than the tippee?

REASONING: The district court attributed all of the gains from Bauer's trading to Kluger for purposes of sentencing, even though Kluger's gains were far less than Bauer's share when Bauer was trading on Kluger's information.

The Appeals court followed commentary on the relevant statute to reject Kluger's appeal of his sentence. They supported the district court's sentencing on Kluger because Bauer is an individual to whom Kluger provided insider information.

The court rejected Kluger's argument that a sentence based on unforeseeable conduct by a co-conspirator serves no useful purpose. They concluded that by punishing the conspirator who is the source of the information for all gains made by his co-conspirators, they were reinforcing the deterrence message sent to would-be tippers by many courts. Moreover, they were sending a clear warning to individuals, such as Kluger, who, in an attempt to limit their responsibility and the extent of their potential sentencing exposure, allege that they had agreements with their co-conspirators to cap the illicit gains. Would-be tippers would be forewarned that they cannot be certain that they will restrict their responsibility by coming to limiting agreements with their co-conspirators prior to commission of their offenses and will come to realize the inherent risk in leaking inside information. The court pointed out that both the long sentence imposed on Kluger and their opinion would be likely to come to the attention of would-be insider traders who may be better educated and informed than persons engaging in other criminal activity, particularly inasmuch as insider-trading cases seem to be well publicized in the general media.

DECISION AND REMEDY: The court upheld the sentence handed down by the lower court.

SIGNIFICANCE OF THE CASE: The case is significant because it demonstrates that the tipper can still receive a significant sentence, regardless of the fact that the tipper profited very little from the information he passed on to the tippee.

CRITICAL THINKING

What do you think is the strongest argument Kluger could make to justify his claim that his sentence was too harsh? What evidence, had it been before the court, might have swayed the appellate court to rule in Kluger's favor?

ETHICAL DECISION MAKING

What values are furthered by a policy that allows a defendant to receive a lower sentence in exchange for cooperating in a criminal investigation? Which stakeholders have an interest in this policy?

Section 16(B)

Under Section 16(a) of the 1934 act, certain large stockholders, executive officers, and directors are considered **statutory insiders.** All statutory insiders must file a report detailing their ownership and trading of the corporation's securities.

To prevent statutory insiders from using inside information for their personal gain, Section 16(b) of the 1934 act requires statutory insiders to return all **short-swing profits,** or profits made from the sale of company stock within any six-month period by a statutory insider, to the company. Even if the insider did not use the inside information to make the transaction, all short-swing profits belong to the corporation. Section 16(b) imposes strict liability on statutory insiders who earn short-swing profits. In other words, violators cannot use lack of intent or lack of knowledge as a defense. Certain transactions, however, such as bankruptcy proceedings, are exempt from Section 16(b).

statutory insiders
Certain large stockholders, executive officers, and directors who are deemed insiders by the Securities Exchange Act of 1934.

short-swing profits
Profits made from the sale of company stock within any six-month period by a statutory insider.

Using Misappropriation Theory to Establish a 10b-5 Violation

United States v. Carpenter U.S. Court of Appeals for the Second Circuit

791 F.2d 1024 (1985)

In 1981, R. Foster Winans wrote the "Heard on the Street" column for *The Wall Street Journal*. He was aware of the *Journal*'s policy that all news material employees gained in the course of their employment was confidential. However, Winans and Carpenter, a news clerk at *The Wall Street Journal*, participated in a scheme with stockbrokers Felis and Brant. Winans provided the brokers with securities-related information that was about to be printed in the "Heard on the Street" column. Using this information, Felis and Brant would buy or sell the relevant securities. Carpenter was a messenger between Winans and the stockbrokers. The net profit on the use of the information for securities trading was $690,000. Felis's brokerage firm noticed a correlation between the "Heard on the Street" column and the trading in Felis's account. Although Felis denied the connection, he began making plans to cover up the trading. Soon, the SEC began an investigation. Later, Winans and Carpenter revealed the entire scheme to the SEC. The SEC sued under Section 10(b) of the Securities Exchange Act of 1934, arguing that the securities information had been misappropriated. The district court found Winans and Felis guilty of misappropriation under 10(b), and Carpenter was convicted for aiding and abetting the commission of securities fraud. They appealed.

The appellate court held that the newspaper employees and brokers, acting in concert, criminally violated the federal securities laws by misappropriating material, nonpublic information, in contravention of established policies of the newspaper, for their own profit in the purchase and sale of securities. This type of fraudulent act is the type of deceptive act that the securities laws are designed to protect against. It therefore falls under the broad prohibition of "any device, scheme, or artifice to defraud."

Proxy Solicitations

proxy solicitation
The process of obtaining authority to vote on behalf of shareholders.

proxy
A writing signed by a shareholder that authorizes the individual named in the writing to exercise the shareholder's votes (corresponding to his or her shares of stock) at a shareholders' meeting.

A **proxy** is a writing signed by a shareholder that authorizes the individual named in the writing to exercise the shareholder's votes (corresponding to his shares of stock) at a shareholders' meeting. Corporate managers often contact shareholders to request them to give a certain manager authority to vote on their behalf in an upcoming meeting. This process of obtaining authority to vote on behalf of shareholders is called **proxy solicitation**. Because the proxy solicitation process is potentially susceptible to fraud, the SEC regulates the proxy solicitation process.

Section 14(a) states that an issuer making proxy solicitations must also furnish a written proxy statement to shareholders. This statement must disclose to the shareholder all facts pertinent to the voting that will occur in the meeting. See Exhibit 23-2 for an excerpt from an example of a proxy statement that explains the purpose of the statement.

Violations Of The 1934 Act

The 1934 act authorizes both civil and criminal penalties. If an individual engages in insider trading, a violation of Section 10(b) or Rule 10b-5, she has committed a criminal offense punishable with a fine up to $1 million, a prison sentence up to 10 years, or both. If a defendant did not know

> **Exhibit 23-2**
>
> Proxy Statement Opening
>
> **Proxy Statement Excerpt**
>
> General Electric Company
> 3135 Easton Turnpike
> Fairfield, Connecticut 06828
>
> This proxy statement is furnished in connection with the solicitation of proxies by General Electric Company on behalf of the Board of Directors for the 2010 Annual Meeting of Shareowners. Distribution of this proxy statement and a proxy form to shareowners is scheduled to begin on or about March 9, 2010.
>
> You can ensure that your shares are voted at the meeting by submitting your instructions by telephone or by Internet, or by completing, signing, dating and returning the enclosed proxy form in the envelope provided. Submitting your instructions or proxy by any of these methods will not affect your right to attend the meeting and vote. A shareowner who gives a proxy may revoke it at any time before it is exercised by voting in person at the annual meeting, by delivering a subsequent proxy or by notifying the inspectors of election in writing of such revocation.
>
> Source: General Electric, "Notice of 2010 Annual Meeting and Proxy Statement," www.ge.com/pdf/investors/financial_reporting/proxy_statements/ge_proxy_2010.pdf.

of the rule she violated, she cannot be imprisoned. If a partnership or corporation engages in insider trading, it is subject to fines up to $2.5 million.

Both the SEC and private parties can bring civil actions against violators of the 1934 act. The SEC investigates alleged violations under the 1934 act. In the process of an investigation, the SEC can enter into consent orders with defendants or seek injunctions to stop certain actions by defendants.

Perhaps the most useful tool the SEC has for punishing those engaging in insider trading is the *Insider Trading Sanctions Act of 1984*, which permits the SEC to sue any individual who violates the 1934 act or who helps another person to engage in insider trading in violation of that act. If the SEC succeeds in demonstrating its claim, courts may assess a civil penalty up to triple the profits gained or losses avoided by the defendant.

Congress expanded the SEC's authority to punish violators of the 1934 act when it passed the *Insider Trading and Securities Fraud Enforcement Act of 1988*. This act subjects more individuals to civil liability for insider trading and grants the SEC power to award **bounty payments** (government rewards for acts beneficial to the public) to insider-trading whistle-blowers.

bounty payment
A government reward for an act that is beneficial to the public.

Private parties may also sue violators of the 1934 act. A private party may seek rescission of a contract to buy securities or may recover damages based on the violator's profits. If the court rules that an individual is liable for violations of the 1934 act, he can seek contribution from others, such as accountants or lawyers, who shared responsibility for the violation.

Regulation of Investment Companies

LO 23-4 Recall the legislative Acts that regulate investment companies.

In the past century, smaller investors have become important players in securities markets. Investment companies facilitate their involvement in securities markets by purchasing a large, diverse portfolio of securities and managing it on behalf of small investor-owners. However, because small investors often lack the resources and ability to supervise investment companies adequately, they can be subject to fraud and exploitation.

E-COMMERCE and the Law

Marketing Securities on the Internet

Perhaps one of the most important and exciting developments in securities regulations is the explosion of the securities market on the Internet. The web permits companies to sell securities in an online initial public offering (IPO), thereby avoiding the expensive filing requirements of a traditional IPO. For example, some websites allow customers to buy and sell securities online. This technology, however, presents a greater opportunity for fraud because anyone can create a website. Several cases of securities fraud have occurred simply because a company has offered so-called free stock over the Internet. The SEC has issued cease-and-desist orders against companies that illegally offered free stock on their websites because they had not registered their stocks with the SEC.

Another type of online fraud, called pumping and dumping, occurs when an owner of a particular stock tells other investors about the virtues of the stock, artificially increasing demand for the stock and pumping up its price, only to sell (dump) it for a quick profit. Unregulated message boards on the Internet permit pumping and dumping on a fast and efficient basis. In response to the increasing use of this method of stock price manipulation, the SEC has pursued harsher and more frequent prosecution of individuals who engage in pumping and dumping online.

The web also provides potential investors with access to enormous amounts of information about securities. For example, the SEC maintains the Electronic Data Gathering, Analysis, and Retrieval (EDGAR) database to help investors access information about IPOs and other documents filed with the SEC. Visit www.sec.gov/edgar.shtml to learn more about EDGAR.

SEC regulations apply to online advertising and securities transactions. When a company delivers a paperless prospectus, it is subject to the following rules:

1. The company must provide timely and adequate notice of the delivery of information.
2. The company must use an easily accessible communication system, such as the Internet.
3. The company must create evidence of the delivery of information.

To prevent exploitation of small investors, Congress passed the *Investment Company Act of 1940,* providing for SEC regulation of investment companies. Congress later expanded the SEC's power to regulate investment companies under the *Investment Company Act Amendments of 1970* and the *National Securities Markets Improvement Act of 1996.*

The Investment Company Act defines an *investment company* as any entity (1) that is engaged primarily in the business of investing, reinvesting, or trading in securities or (2) that is engaged in such business and in which more than 40 percent of the company's assets are investment securities. The act excludes a number of institutions, however, including banks, insurance companies, savings and loans, and finance companies.

The act requires all investment companies to file a notification of registration with the SEC. In addition, they must file annual reports with the SEC and hold all securities in the custody of a bank or member of the stock exchange. The act prohibits investment companies from purchasing securities on the margin (borrowing money to purchase securities), selling short (selling securities that the company does not yet own), or participating in joint trading accounts.

State Securities Laws

LO 23-5 Identify state laws that regulate securities.

blue-sky law
A law that regulates the offering and sale of purely intrastate securities.

Not only must issuers obey federal securities regulations, but they are also subject to state securities laws, often referred to as blue-sky laws. These laws regulate the offering and sale of purely intrastate securities. Hence, although certain securities are exempt from federal securities regulation, they may be subject to state securities laws.

Many state securities laws serve functions similar to those of federal securities laws. For example, many states have laws requiring registration of securities issued within the state and requiring disclosure of certain information. Moreover, state laws regulate securities brokers and dealers. Although the purposes of state and federal securities regulations overlap, the specific regulations differ. Consequently, to encourage greater coordination between federal and state securities laws, most states have adopted the Uniform Securities Act.

CASE OPENER WRAP-UP

The Martha Stewart Case

Although the SEC sought to charge Martha Stewart with illegal insider trading, a grand jury did not indict Stewart on that charge. The grand jury did, however, indict Stewart and her broker on nine criminal charges, including securities fraud, obstruction of justice, and conspiracy. Stewart pled not guilty to all charges.

The district court threw out the securities fraud charge on grounds that "no reasonable jury could find it to be accurate." A week later, however, a jury convicted her of the four remaining counts against her, all of which related to her statements to SEC investigators in their investigation of her sale of ImClone stock. The jury sentenced her to five months in a minimum-security prison. A federal appellate court upheld her sentence.[11]

PRACTICAL TIPS for Business Managers

1. Insider business information can sometimes be easy to come upon and presents the temptation of using it to engage in insider trading. However, being convicted of insider trading can cost huge amounts of money and ruin a career. It may be best to think twice before using insider information for illegal gains.
2. As this chapter demonstrates, registering securities to be offered to the public is a complex and multi-step process. It may be wise to seek counsel before attempting to register securities to ensure that you make no fatal mistakes in the process.
3. If you manage an investment company, the SEC provides a whole other set of regulatory hoops to jump through. Be sure to keep up on the rules and regulations so that you do not, for example, miss an annual report and find yourself in trouble.

SUMMARY

What Is a Security?	As defined by the *Howey* test, a security is an investment in a common enterprise with the reasonable expectation of profit gained predominantly from others' efforts.
Securities Regulation	The Securities and Exchange Commission (SEC) was created in 1934 to enforce securities laws, interpret provisions of securities acts, and regulate the trade of securities as well as the activities of securities brokers, dealers, and advisers.
The Securities Act of 1933	A *registration statement* is a document containing a description of the securities offered for sale, an explanation of how proceeds from the sale will be used, a description of the registrant's business and properties, information about the management of the company, a description of any pending lawsuits, and certified financial statements.

A *prospectus* is a written document that is similar to a registration statement and is used as a selling tool for potential investors.

Periods of the filing process:

1. Prefiling period
2. Waiting period
3. Posteffective period

Unregistered, unrestricted securities are exempt from the registration process. They include:

1. Securities issued by governments, including municipal, state, and federal governments.
2. Securities issued by nonprofit issuers, such as religious institutions or charitable organizations.
3. Securities issued by financial institutions supervised by banking associations.
4. Securities issued as a result of corporation reorganization in which one security is exchanged for another security.

(continued)

[11] https://www.thoughtco.com/martha-stewarts-insider-trading-case-1146196.

(concluded)	5. Stock dividends and stock splits. 6. Insurance and annuity contracts issued by insurance companies. 7. Securities issued by federally regulated carriers (i.e., railways). 8. Short-term notes with a maturity date that does not exceed nine months. 9. Securities sold before July 27, 1933. 10. An issuer's offer of up to $5 million in securities in a 12-month period. Certain kinds of transactions are exempt, including the following: 1. *Limited offers* involve small amounts of money or are offered only to sophisticated investors. There are four possible exemptions: a. *Private placement exemption:* Exempts private offerings of securities. b. *Rule 505:* States that private offerings may not exceed $5 million in a 12-month period, and firms do not have to believe that investors have a reasonable ability to evaluate risk. c. *Rule 504:* Exempts noninvestment firms that offer no more than $1 million in securities in a 12-month period. d. *Section 4(6):* Exempts securities offered only to accredited investors for an amount less than $5 million. 2. *Intrastate issues* exempt local investors in local businesses. 3. *Resales* exempt transactions by any person other than an issuer, underwriter, or dealer. *Restricted securities* are securities acquired under Rule 505, Rule 506, or Section 4(6) that must be registered for resale unless the investor follows Rule 144 or Rule 144(a). *Violations may result in:* 1. Administrative action 2. Injunctive action 3. Criminal prosecution
The Securities Exchange Act of 1934	*Section 10(b):* Regulation that prohibits the use of manipulative and deceptive devices to bypass SEC rules. *Insider trading:* Trading in which a company employee or executive uses material inside information to make a profit. *Misappropriation theory:* Theory that an individual who wrongly acquires and uses inside information for profit is liable for insider trading. *Tipper/tippee theory:* Theory that an individual who receives material inside information as a result of an insider's breach of duty is guilty of insider trading. *Statutory insiders:* Certain stockholders, executive officers, and directors who must file a report detailing their ownership and trading of the corporation's securities. *Short-swing profits:* Profits made from the sale of company stock within any six-month period by a statutory insider; according to Section 16(b), these profits must be returned to the company. *Proxy:* Document that authorizes an individual to vote a shareholder's share of stocks at a shareholders' meeting. *Proxy solicitation:* Process of obtaining the authority to vote on behalf of a shareholder. *Violations may result in:* 1. Criminal penalties. 2. Civil penalties. 3. Suits against those involved in insider trading under the Insider Trading Sanctions Act of 1984.
Regulation of Investment Companies	The *Investment Company Act of 1940* provides for SEC regulation of investment companies. The act requires all investment companies to file a notification of registration with the SEC, file annual reports with the SEC, and hold all securities in the custody of a bank or member of the stock exchange.
State Securities Laws	State securities laws serve functions similar to those of federal securities laws and include blue-sky laws. *Blue-sky laws* regulate the offering and sale of securities within the state only.

Point/Counterpoint

High-frequency trading is an investment practice utilized by large firms in recent years. This process involves using a computer to transact millions of security transactions in a few seconds, which allows a firm to beat competing investors to the market with ease. Sometimes these computer programs are also used to rapidly make multiple bids on a particular security owned by a high-frequency trader to drive the price of the security up by a few pennies per share, followed by the trader pulling the bids and selling multitudes of shares at slightly higher prices per share that add up fast.

Should High-Frequency Trading of Securities be Banned?	
YES	**NO**
High-frequency trading is a toxic practice that can shatter the stock market and benefit the few at the expense of the many. First of all, high-frequency trading creates an unfair playing field in the securities exchange market. Smaller firms simply can't compete with larger firms who have the computing power to beat traditional traders to the market by milliseconds and get the better deal every time. In fact, this creates adverse selection, where investors are less willing to invest and/or take risks because they feel that a computer will beat them to it every time. Lower investor confidence hurts the market as a whole. Second, high-frequency transactions can create enormous amounts of volatility in the securities market. In fact, high-frequency trading was identified as the cause of the Flash Crash of 2010. This event saw the Dow Jones Industrial Average drop almost 1,000 points and then recover about half an hour later, after a mutual fund rapidly sold $4.1 billion worth of shares to other HFT firms that then rapidly bought and sold these shares. High-frequency trading creates volatility and chaos that is unhealthy for the market. Finally, despite arguments to the contrary, high-frequency trading can actually lead to less liquidity in the market. As previously mentioned, HFTs can beat traditional traders to the market with effectively no competition or risk. This phenomenon hurts investor confidence and makes them less willing to take risks, which can mean that they're only willing to invest in securities when they present less risk; that is, the price is lower. This can cause prices to lower until they meet the standards of traditional traders, which leads to less value for securities and less liquidity overall. High-frequency trading may help a few large firms obtain large amounts of wealth, but it does so at the expense of smaller firms and the market as a whole.	There's absolutely no reason that high-frequency trading should be outlawed. High-frequency traders (HFT's) employ the same strategies as other firms, only much faster. Everything an HFT can do can be done by a standard firm. The speed at which the transactions are performed does not make the transactions themselves any different. It can be argued that speed can give an advantage over other firms in securities trading, and this assertion is absolutely true. Speed always has and probably always will be an advantage in trading. The speed limit has simply been raised; what makes the fastest traders today any different than the fastest traders yesterday? Banning HFT transactions would effectively put an arbitrary limit on how expediently a firm can process its transactions. Not only is nothing inherently wrong with the practices of HFT's, but they actually benefit the securities market in some ways. One such way is by increasing the liquidity of the market. Liquidity is the ability to sell a security without first having to discount its price. Because HFT transactions account for about half of United States trading volume, a large portion of the market is constantly having its prices driven up in tiny amounts due to the nature of HFT transactions. This means that prices tend to go up instead of down, and, due to the increased liquidity, the value of the securities held by most investors do not greatly diminish. In other words, the overall value of the market increases due to HFT trading. However, even with benefits that HFT's can bring to the market, and even with the apparent advantage they have over other firms, they actually don't make that much more money than other firms. HFTs normally drive the price of securities up by a few cents, and so even when they are trading large volumes of securities, the firms don't make large sums of money. What makes HFT firms profitable is that, unlike normal firms, they rarely ever *lose* money on an investment because they're always driving the prices of their securities up by tiny amounts. For example, over a period of about four years, the HFT firm Virtu Financial Inc. made a profit on 1,237 out of 1,238 days. At any rate, HFTs don't bring in enough of a profit over traditional firms to justify banning the high-frequency trading process.

Questions & Problems

1. What was the stimulus for the creation of securities regulation? State the purposes of the two main federal securities laws.
2. What is the function of the SEC?
3. Explain the process of registering securities.
4. Why are certain securities transactions exempt from the registration process?
5. Why is it illegal to use inside information in making securities transactions?
6. How does the misappropriation theory apply to insider trading?
7. Two affiliates of a hugely successful hedge fund, SAC Capital Advisors, were charged with insider trading by the SEC. CR Intrinsic Investors and Sigma Capital Management apparently relied on "expert networks" through which, allegedly, people with information funnel it to traders. The biggest charge involved a doctor who had passed information to a trader regarding problems with a new drug that targeted Alzheimer's disease. After the information was passed, around $1 billion worth of shares of two pharmaceutical companies related to the drug were sold before the drug information was released to the public. The drug was still going through scientific trials, and the results had not been finalized. The doctor was a part of that project. How do you think this case turned out?

 Source: *Securities and Exchange Commission v. Sigma Capital Management, LLC, et al.*, C.A. No. 13-civ-(2013).

8. Newman and Chiasson were among a group of financial analysts who received, directly and indirectly, information from corporate insiders. With information about Dell, they traded that company's stock, making over $4 million. They were found by a jury to have engaged in insider trading of information about Dell based on the following chain of events. Rob Ray of Dell's investor relations department tipped information regarding Dell's consolidated earnings numbers to Sandy Goyal, an analyst at Neuberger Berman. Goyal in turn gave the information to Diamondback analyst Jesse Tortora. Tortora in turn relayed the information to his manager Newman as well as to other analysts including Level Global analyst Spyridon "Sam" Adondakis. Adondakis then passed along the Dell information to Chiasson, making Newman and Chiasson three and four levels removed from the inside tipper, respectively. Newsman and Chiasson appealed their convictions, arguing that they could not be guilty of insider trading because there was no evidence that the corporate insiders provided information for a personal benefit. How do you think the court ruled on the appeal?

 Source: *U.S. v. Todd Newman, Anthony Chiasson, Jon Horvath, Danny Kuo, Hyung Lim, Michael Steinberg*, 773 F.3d 438; 2014 U.S. App. LEXIS 2.

9. Stacey Greenfield sued the Cadian entities under Section 16(b) of the Securities Act of 1934. The Cadian entities which were a number of investment funds and partnerships. Cadian Fund LP and Cadian Master Fund were two investment funds which shared a general partner, Cadian GP, LLC, which Eric Bannasch was the sole managing member. Cadian Capital Management, LP, also owned by Bannasch, served as the investment manager of the investment funds. In addition, all financial decisions for all Cadian entities pertinent to this case were made by Bannasch.

 Recall that Section 16(b) requires all statutory insiders to return short-swing profits. Statutory insiders also include "beneficial owners" of more than 10 percent of any equity security of a company. In this case, Greenfield sued the Cadian entities as a group for buying and selling Infoblox common stock shares while collectively owners of at least 10 percent of Infoblox. In filing this complaint, Greenfield also asserted that the Cadian entities are "inexorably intertwined" and could therefore be held liable under Section 16(b). The Cadian entities moved for summary judgement for failure to state a claim.

 Do you think groups of individuals or other entities such as companies and investment funds should be collectively classifiable as beneficial owners and therefore liable under Section 16(b)? What are the implications if they are? What are the implications if they are not? How do you think the court decided this question?

 Source: *Greenfield v. Cadian Capital Management*, LP, 213 F. Supp. 3d 509 (2016).

10. The Garrett Group, a real estate company, was in financial trouble after the recession. It informed its creditor, CVB Financial Corporation that it could not make loan payments. After CVB restructured the Garrett Group's loan, Garrett indicated it still could not fulfill its debts to CVB. Despite the Garrett Group's statements, CVB represented to the SEC that the Garrett Group were financially solvent and would be able to pay its loans with ease.

In 2010, the SEC filed a subpoena seeking information about CVB's loan underwriting process. A month after the subpoena, CVB wrote off $34 million in loans to the Garrett Group. CVB's stock also dropped 22 percent as a result of the subpoena and writing off the loan. A class action was subsequently filed against CVB alleging violations of Section 10(b) and Rule 10b-5 of the Securities and Exchange Act of 1934.

Consider the following statements made by CVB: "[t]he overall credit quality of the loan portfolio is sound;" "CVB's credit metrics are superior" to those of its peers; "strong credit culture and underwriting integrity remain paramount at CVB"; and "CVB's culture has limited its exposure to problem credits." Do you think any of the four statements on their faces, could be considered materially misleading under Rule 10b-5? Why or why not? What do you think is necessary for a statement to be considered materially misleading?

Source: *Lloyd v. CVB Financial Corp.,* 811 F. 3d 1200 (2016).

11. Plaintiffs allege that the pharmaceutical company Sanofi, along with its predecessor and three company executives, made materially false or misleading statements regarding its breakthrough drug, Lemtrada, designed to treat multiple sclerosis (MS). In part because of Lemtrada's unique treatment design, Genzyme used a single-blind study in its early clinical trials. In a single-blind study, either the researcher or the patient does not know which drug was administered. The FDA prefers double-blind studies, and repeatedly told Defendant the lack of double-blind studies was troubling, but the FDA continued to provide the company encouragement. Plaintiffs allege that while Lemtrada was undergoing Phase III clinical trials prior to FDA approval, Sanofi misled investors by failing to disclose that the FDA had repeatedly expressed concern with Sanofi's use of single-blind studies and had encouraged Sanofi to use double-blind studies in its clinical trials. Plaintiffs allege that these omissions misled investors and artificially inflated the value of Plaintiffs's contingent value rights (CVRs), specialized financial instruments whose value is tied to the achievement of certain "milestones." While defendants remained optimistic, the FDA did not approve the drug before the milestone was met, causing the value of the stock to fall. Plaintiffs's filed their action alleging violations of Sections 10(b), 18, and 20(a) of the Securities Exchange Act of 1934, 15 U.S.C. Section 78a et seq., and Sections 11 and 12 of the Securities Act of 1933, 15 U.S.C. Sections 77a et seq. Defendants filed a motion to dismiss, which the trial court granted, finding that the defendants had made no false or misleading statements but had merely expressed their optimistic beliefs as to how the FDA would act in the future. How do you believe the appellate court ruled in the case? Why?

Source: *Gen. Partner Glenn TONGUE, Deerhaven Capital Management, et al v. Sanofi Pharmaceuticals, Inc., Consolidated, et.al,* caselaw.findlaw.com/Us-2d-circuit/1727858 (2016).

12. Compagnie Générale des Eaux was a centuries-old French utilities conglomerate led by Jean Marie-Messier. In 1998, Compagnie Générale des Eaux changed its name to Vivendi, S.A., and developed a plan to reinvent as an entertainment media company. Vivendi shed its utilities and environment divisions and acquired various media businesses. A significant point in Vivendi's plan occurred in 2000 when Vivendi merged with Canal Plus, S.A., a French film and television production company, and Seagram Company Ltd., a Canadian entertainment company.

The three way merger resulted in one of the world's largest media companies, rivaling AOL-Time Warner. Vivendi, S.A., now simply Vivendi, continued to acquire companies, spending cash and increasing its debts. Vivendi's debts blossomed from 3 billion euros to 21 billion euros between 2000 and 2002. Throughout Vivendi's buying frenzy, it continuously conveyed to the public it was in sound financial health. Despite Vivendi's assurances, Moody's and the S&P downgraded Vivendi's debt rating in 2002, rating it at just above junk level. Vivendi's stock subsequently fell 26 percent.

As it turned out, Vivendi's financial officers had known it was facing a cash-crunch long before the negative ratings were publicized. Chief Financial Officer Hannezo had informed Messier of the impending problems, who continued to acquire companies. A class action lawsuit was filed against Vivendi, Messier, and Hannezo in 2002 under Rule 10(b). The case moved to trial in 2009 after years of discovery, and, in 2011, a jury found under Rule (10b) violations of fifty seven statements, Vivendi liable, but not Messier or Hannezo.

Vivendi appealed the jury verdict on grounds that it could not be held liable for certain statements because they were forward-looking statements under the PSLRA. Under what conditions is a defendant not liable under 15 U.S.C. § 78u-5(c), the provision concerning forward-looking statements?

One of the alleged misstatements identified by Vivendi was the following given on February 14, 2001: "Vivendi Universal enters its first full year of operations with strong growth prospects and a very strong balance sheet. This new company is off to a fast start and we are very confident that we will meet the very aggressive growth targets we have set for ourselves both at the revenues and EBITDA levels." Do you think the statement qualifies as a forward-looking statement? Why or why not? What did the court think of this statement?

Source: In re Vivendi, SA Securities Litigation, 838 F. 3d 223 (2016).]

Part 8 Government Regulation

CHAPTER 24

Employment and Discrimination Law

CASE OPENER

Brad Gets Fired from So Clean!

Brad has worked in the marketing department of So Clean! for the past five years. So Clean! is a company that produces household cleaners. Brad is an excellent employee and was recently promoted. Shortly after his promotion, Brad decided to reveal publicly that he is homosexual. His family and most of his co-workers have been very supportive.

Soon after his promotion and announcement that he is gay, Brad began having problems with his boss, Jennifer. She began asking Brad questions about his personal life. At first it was small things, such as asking Brad whether he was a smoker (he is, although only outside the workplace). Jennifer then began asking Brad very personal questions about his sexuality and told him she did not like weak men. The final straw for Brad occurred when Jennifer announced that she would cure his homosexuality and told him to come home with her that night after work or be fired. Jennifer was careful that there were never any witnesses around when she asked Brad personal questions or propositioned him. Brad refused Jennifer's advances and was fired. He filed an administrative complaint with the Equal Employment Opportunity Commission (EEOC) alleging wrongful termination, retaliation, sexual harassment, and sexual discrimination. Jennifer's response was that she fired Brad because of creative differences about how to run the marketing department and because he is a smoker. Jennifer denied Brad's accusations about sexual discrimination and harassment.

1. May an employer discriminate against an employee because he is gay?
2. May an employer fire an employee because the employee smokes outside the workplace?
3. May a man file a claim of sexual discrimination? Sexual harassment?
4. What defenses does an employer have to a charge of discrimination?

LEARNING OBJECTIVES

LO 24-1 Describe employment-at-will.
LO 24-2 Outline the federal laws governing employment discrimination.
LO 24-3 Describe the legal requirements for a charge of sexual discrimination or sexual harassment.
LO 24-4 Identify the categories of people protected from discrimination.
LO 24-5 Summarize the laws that protect employees's wages, benefits, health and safety standards, and rights.

LO 24-6 Identify components of employer privacy policies.

LO 24-7 Recognize the circumstances under which social media may be used in employment decisions.

LO 24-8 Recall the agency and legislation that regulates the hiring of foreign workers.

LO 24-9 List the three major pieces of labor legislation.

When May an Employee Be Fired?

LO 24-1 Describe employment-at-will.

During the eighteenth and nineteenth centuries in the United States, employees had no protection in the workplace. An employee who was injured could be fired. In fact, an employer could fire a worker for no reason at all. This concept came to be known as *at-will employment*.[1] At-will employment applied in all states with no exceptions until 1959.[2]

Today, any employee who is not employed under a contract for a set duration or under a collective bargaining agreement is considered an at-will employee.[3] This means that the employee may quit at any time for any reason or no reason at all, with no required notice to the employer.[4] Similarly, an employer may fire the employee at any time, with no notice, for almost any reason. For example, your employer could decide he doesn't like the color of your shirt and fire you on the spot! The exception to the at-will rule is that an employer may not fire an employee for an illegal reason. What is an illegal reason? Broadly, any termination based on a violation of a state statute, a state constitution, a federal law, the U.S. Constitution, or a public policy is illegal. Exceptions to at-will employment have also been found through breaches of implied contracts with employees that were based on employee handbooks.[5]

Federal Employment Discrimination Laws Governing Employers

LO 24-2 Outline the federal laws governing employment discrimination.

Employees are protected in the workplace by a number of both federal and state laws. Federal laws apply to everyone in the United States. Federal law may be described as a minimum level of protection for all workers. State laws may give employees more, but not less, protection than federal laws. Exhibit 24-1 is an overview of some of the most important federal employment discrimination laws.

Title VII of the Civil Rights Act (CRA) of 1964
Federal law (amended by the Civil Rights Act of 1991) that protects employees against discrimination based on race, color, religion, national origin, and sex; also prohibits harassment based on the same protected categories.

Civil Rights Act—Title VII

The Civil Rights Act is divided into sections, called *titles*. **Title VII of the Civil Rights Act (CRA) of 1964** deals with discrimination in employment. It prohibits employers from hiring, firing, or otherwise discriminating in terms and conditions of employment, and it prohibits segregating

[1] See *Toussaint v. Blue Cross & Blue Shield of Mich.,* 408 Mich. 579, 600, 292 N.W.2d 880, 885 (1980), for an extended discussion on the at-will rule.

[2] Gantt v. Sentry Insurance, 1 Cal. 4th 1083 (1992).

[3] Union employees are covered by collective bargaining agreements.

[4] Most employees do give an employer notice before leaving a job as a matter of professional courtesy. Such action, however, is not required under the law.

[5] The At-Will Presumption and Exceptions to the Rule, http://www.ncsl.org/research/labor-and-employment/at-will-employment-overview.aspx

Exhibit 24-1
Federal Discrimination Laws

FEDERAL LEGISLATION	PURPOSE
Civil Rights Act (CRA) of 1964—Title VII (as amended by the Civil Rights Act of 1991)	Protects employees against discrimination based on race, color, religion, national origin, and sex; also prohibits harassment based on the same protected categories
Pregnancy Discrimination Act (PDA) of 1987	Amends Title VII of the CRA, expanding the definition of sex discrimination to include discrimination based on pregnancy
Age Discrimination in Employment Act (ADEA) of 1967	Prohibits employers from refusing to hire, discharging, or discriminating in terms and conditions of employment on the basis of an employee or applicant being age 40 or older
Americans with Disabilities Act (ADA) of 1990 (as amended in 2008)	Prohibits discrimination against employees and job applicants with disabilities
Equal Pay Act of 1963	Prohibits an employer from paying workers of one gender less than the wages paid to employees of the opposite gender for work that requires equal skill, effort, and responsibility

employees in a manner that would affect their employment opportunities on the basis of their race, color, religion, sex, or national origin. There are two ways to prove discrimination under Title VII—disparate treatment and disparate impact.

Title VII of CRA applies to employers who have 15 or more employees for 20 consecutive weeks within one year and who are engaged in a business that affects commerce. The U.S. government, corporations owned by the government, agencies of the District of Columbia, Indian tribes, private clubs, unions, and employment agencies are also covered by Title VII of CRA.

Disparate-Treatment Discrimination Under Title VII

To sue for **disparate treatment** under Title VII, the plaintiff must be a member of a protected class as listed in CRA. In other words, the employee must have been discriminated against on the basis of race, color, national origin, religion, or sex (i.e., gender). If the employee has been refused work, fired, denied a promotion, or the like, based on membership in a protected class, this is a form of intentional discrimination and qualifies the employee to sue for disparate-treatment discrimination. Proving disparate-treatment discrimination in employment under Title VII is a three-step process:

1. Plaintiff (the employee) must demonstrate a prima facie case of discrimination.
2. Defendant (the employer) must articulate a legitimate, nondiscriminatory business reason for the action.
3. Plaintiff (the employee) must show that the reason given by the defendant (the employer) is a mere pretext.[6]

disparate treatment
A form of intentional discrimination in which an employee is hired, fired, denied a promotion, or the like, based on membership in a protected class. This is a form of intentional discrimination.

After all the evidence has been presented, the trier of fact (a jury in most cases) must decide whether discrimination has occurred.[7] The burden of proof in a civil case is preponderance of the evidence (i.e., more likely than not). If the jury finds in favor of the plaintiff-employee, damages must be assessed. Damages under Title VII of CRA include up to two years of back pay, compensatory damages, punitive damages (limited in some cases), attorney fees, court costs, court orders (including reinstatement), and remedial seniority. If the jury finds in favor of the defendant-employer, the plaintiff-employee receives nothing.

[6] *McDonnell Douglas v. Green*, 411 U.S. 792, 802 (1973).
[7] In a bench trial, the judge becomes the trier of fact because no jury is impaneled. A discrimination case could also be decided by a judge on a motion for summary judgment.

CASE Nugget

Must an Employer Reasonably Accommodate an Employee's Religious Practices?

Kumar v. Gate Gourmet, Inc. Supreme Court of Washington

2014 Wash. LEXIS 387 (May 22, 2014)

The plaintiffs in this action (the employees) work near SeaTac airport for the defendant, Gate Gourmet, preparing meals for service on trains and airplanes. Due to security concerns, the employees can neither bring food with them to work nor leave the premises to obtain food during their 30-minute lunch break. Instead, Gate Gourmet provides meals for employees to consume during their break. These meals ostensibly consist of one vegetarian and one meat-based main dish. The employees allege, however, that Gate Gourmet uses animal by-products in the vegetarian option. They also allege that they informed Gate Gourmet that their various religious beliefs prohibited them from eating the beef-pork meatballs the company served, that Gate Gourmet responded by temporarily switching to turkey meatballs, that the company later switched back to the beef-pork mixture without notifying the employees, and that it now refuses to alter the employee meals.

The employees brought a class action against the employer, challenging the employer's meal policy, which barred employees from bringing in their own food for lunch (for security reasons), leaving only employer-provided food for the employees to eat. The employees claimed that the meal policy constituted a failure to accommodate their religious practices reasonably, had a disparate impact on employees who adhered to certain religions, and constituted battery and negligent infliction of emotional distress. The trial court dismissed the case, holding that there was no requirement to "reasonably accommodate" the employees's religious practices under Washington state law. The decision was affirmed by the appellate court.

Upon review by the Washington Supreme Court, the lower court decision was reversed. The Washington Supreme Court stated that under state rules of statutory interpretation and persuasive federal antidiscrimination case law (Title VII), the Washington Law Against Discrimination (WLAD) implies a requirement to accommodate religious practices reasonably. Moreover, both the disparate-impact and religious-accommodation doctrines bar facially neutral employment policies that have disproportionate adverse effects on a protected class.

Disparate-Impact Discrimination under Title VII

disparate impact
A form of discrimination that arises when an employer's policy or practice appears to apply to everyone equally but its actual effect is that it disproportionately limits employment opportunities for a protected class.

Disparate-impact cases are sometimes called *unintentional-discrimination* cases. Although it is very difficult to prove disparate treatment, it is even more difficult to prove disparate impact. Disparate-impact cases arise when a plaintiff attempts to establish that although an employer's policy or practice appears to apply to everyone equally, its actual effect is that it disproportionately limits employment opportunities for a protected class.

The plaintiff proves a case based on disparate impact by first establishing statistically that the rule disproportionately restricts employment opportunities for a protected class. The burden of proof then shifts to the defendant, who can avoid liability by demonstrating that the practice or policy is a business necessity. The plaintiff, at this point, can still recover by proving either that the necessity was promulgated as a pretext for discrimination or that an

Exhibit 24-2

Disparate Treatment and Disparate Impact Burden Shifting

alternative employment practice had less discriminatory effect and the employer failed to adopt such practice.

The initial steps for proving a prima facie case of disparate impact were set forth in *Griggs v. Duke Power Co.*[8] In that case, the employer-defendant required all applicants to have a high school diploma and a successful score on a professionally recognized intelligence test for all jobs except laborer. The stated purpose of these criteria was to upgrade the quality of the workforce.

The plaintiff statistically demonstrated the discriminatory impact by showing that 34 percent of the white males in the state had high school diplomas, whereas only 12 percent of the black males did, and by introducing evidence from an EEOC study showing that 58 percent of the whites, compared to 6 percent of the blacks, had passed tests similar to the one given by the defendant. Because the defendant could not demonstrate any business-related justification for either employment policy, the plaintiff was successful. Requiring a high IQ or high school or college diploma may be necessary for some jobs, but not for all jobs at Duke Power.

Sexual Harassment under Title VII

Harassment first developed in the context of discrimination based on sex, and it evolved to become applicable to other protected classes. The definition of **sexual harassment** stated in the Equal Employment Opportunity Commission (EEOC) guidelines and accepted by the U.S. Supreme Court is "unwelcome sexual advances, requests for sexual favors, and other verbal or physical conduct of a sexual nature" that implicitly or explicitly makes submission a term or condition of employment; makes employment decisions related to the individual dependent on submission to or rejection of such conduct; or has the purpose or effect of creating an intimidating, hostile, or offensive work environment. In the Case Opener, did the actions of Jennifer create a sexually hostile environment for Brad?

Two distinct forms of sexual harassment are recognized. The first, and generally easiest to prove, is *quid pro quo*, which occurs when a supervisor makes a sexual demand on someone and this demand is reasonably perceived as a term or condition of employment. The second form of sexual harassment involves the creation of a hostile work environment. Case 24-1 demonstrates how the "totality of circumstances" standard is applied to determine whether an employer's conduct has created a hostile work environment.

sexual harassment
Unwelcome sexual advances, requests for sexual favors, and other verbal or physical conduct of a sexual nature that makes submission a condition of employment or a factor in employment decisions or that creates an intimidating, hostile, or offensive work environment. The two types are hostile environment and *quid pro quo*.

[8] *Griggs v. Duke Power*, 401 U.S. 424 (1971).

CASE 24-1

BRENNAN v. TOWNSEND & O'LEARY ENTERPRISES, INC.
Court of Appeals of California, Fourth Appellate District
199 CAL.APP.4TH 1336 (2011)

FACTS: Plaintiff Stephanie Crowley Brennan sued her employer for sexual harassment. At trial, plaintiff testified that an August 2004 email was the only incident of harassment based on her gender directed at her. Although the August 2004 email, which referred to plaintiff as a "big-titted, mindless one" (although it did not mention her by name), was rude, insulting, and unprofessional, the record shows it was an isolated event as there was no evidence that the sender (or any employee of the agency) made any other derogatory remarks about plaintiff, sex based, or otherwise, in any other context. The evidence showed the email was not intended to be shared publicly. Instead, it was indisputably intended to have been only sent to one person. Plaintiff saw the August 2004 email only after it was inadvertently forwarded it to another employee, who in turn forwarded it to plaintiff. Plaintiff was a vice-president of the agency at the time of the August 2004 email, and the author of the email was never her supervisor. A jury found that plaintiff was subjected to severe or pervasive harassment based on her gender, in violation of Gov. Code, § 12940, of the Fair Employment and Housing Act. The Superior Court of Orange County, California, granted defendant employer's motion for judgment notwithstanding the verdict, and the employee appealed.

ISSUE: Whether a single email that was inadvertently forwarded to plaintiff may constitute hostile environment sexual harassment under the "totality of the circumstances" test?

REASONING: Like federal law, California law prohibits sexual harassment in the workplace. With respect to sexual harassment in the workplace, the prohibited conduct ranges from expressly or impliedly conditioning employment benefits on submission to, or tolerance of, unwelcome sexual advances to the creation of a work environment that is hostile or abusive to employees because of their sex. Thus, similar to the federal law's Title VII, California's FEHA recognizes two theories of liability for sexual harassment claims, quid pro quo harassment, where a term of employment is conditioned upon submission to unwelcome sexual advances and hostile work environment, where the harassment is sufficiently pervasive so as to alter the conditions of employment and create an abusive work environment.

In construing California's FEHA, this court has held that the hostile work environment form of sexual harassment is actionable only when the harassing behavior is *pervasive* or *severe*. This limitation mirrors the federal courts's interpretation of Title VII. To prevail on a hostile work environment claim under California's FEHA, an employee must show that the harassing conduct was severe enough or sufficiently pervasive to alter the conditions of employment and create a work environment that qualifies as hostile or abusive to employees because of their sex. There is no recovery for harassment that is occasional, isolated, sporadic, or trivial. Courts that have construed federal and California employment discrimination laws have held that an employee seeking to prove sexual harassment based on no more than a few isolated incidents of harassing conduct must show that the conduct was severe in the extreme. A single harassing incident involving physical violence or the threat thereof' may qualify as being severe in the extreme. Under California's FEHA, as under the federal law's Title VII, the existence of a hostile work environment depends upon the *totality of the circumstances*.

In evaluating the totality of the circumstances to determine the existence of a hostile work environment, the following factors can be considered: (1) the nature of the unwelcome sexual acts or works (generally, physical touching is more offensive than unwelcome verbal abuse); (2) the frequency of the offensive encounters; (3) the total number of days over which all of the offensive conduct occurs; and (4) the context in which the sexually harassing conduct occurred. In determining what constitutes sufficiently pervasive harassment, the courts have held that acts of harassment cannot be occasional, isolated, sporadic, or trivial, rather the plaintiff must show a concerted pattern of harassment of a repeated, routine or a generalized nature.

We begin our analysis by observing the trial evidence did not support a finding plaintiff was ever subjected to severe sexual harassment. Neither the August 2004 email nor any evidence at trial showed plaintiff was ever assaulted, subjected to "unwelcome physical contact," threatened, propositioned, or subjected to explicit language directed at her or at anyone else in her presence. As plaintiff acknowledged in her own testimony, the August 2004 email was the only sexual harassment she claims she directly experienced up to that time. Evidence of the August 2004 email, even when coupled with evidence of the few other sporadic claimed incidents of sexual harassment she witnessed over the years, her conversations with Steve O'Leary about her personal life, and her discovery of other incidents of sexual harassment that did not occur in her presence or immediate work environment, is insufficient as a matter of law to establish the existence of pervasive harassment based on gender.

DECISION AND REMEDY: Judgment for defendant is affirmed.

SIGNIFICANCE: To be actionable as hostile environment sexual harassment, plaintiff must be able to show a pattern of offensive and/or abusive conduct that is severe and pervasive. An isolated incident, unless extremely severe, is insufficient to support a claim of hostile environment sexual harassment.

(continued)

CRITICAL THINKING

Identify the Court's reasons for its finding. Do you think these reasons were sufficient for the court to uphold the lower court's judgment notwithstanding the [jury's] verdict? Why or why not?

ETHICAL DECISION MAKING

Imagine that the judge is operating under a duty-based system of ethics. What duty is she advocating in terms of employer-employee relationships? Would this ruling serve well as a universal standard?

LO 24-3 Describe the legal requirements for a charge of sexual discrimination or sexual harassment.

Same-Sex Harassment—The Supreme Court Speaks Initially, same-sex harassment was not covered under CRA. By 1997, however, the courts were split on the issue. This issue was resolved in 1998 in *Oncale v. Sundowner Offshore Services, Inc.*[9] Oncale, a male, was forcibly subjected to sex-related, humiliating actions against him by other male employees. He was physically sexually assaulted and threatened with rape. He complained, to no avail, so he filed a lawsuit alleging discrimination based on sex. The district court granted a motion for summary judgment for the defendant employer, holding that a male had no cause of action under Title VII for harassment by male co-workers. The case was appealed to the U.S. Supreme Court, which concluded that sex discrimination consisting of same-sex sexual harassment is actionable under Title VII. The Court reversed the appellate court's order and remanded the case for further proceedings.

Harassment by Nonemployees under Title VII Employers may be held liable for harassment of their employees by nonemployees under very limited circumstances. If an employer knows that a customer repeatedly harasses an employee yet the employer does nothing to remedy the situation, the employer may be liable. For example, in *Lockhard v. Pizza Hut, Inc.*, the franchise was held liable for the harassment of a waitress by two male customers because no steps had been taken to prevent the harassment.[10]

Harassment of Other Protected Classes Under Title VII

Hostile-environment issues have also been used in cases of discrimination based on religion and race. For example, in a 1986 case, *Snell v. Suffolk County*, Hispanic and black corrections workers demonstrated that a hostile work environment existed by proving that they had been subjected to continuing verbal abuse and racial harassment by co-workers and that the county sheriff's department had done nothing to prevent the abuse.[11] The white employees had continually used racial epithets and posted racially offensive materials on bulletin boards, such as a picture of a black man with a noose around his neck, cartoons favorably portraying the Ku Klux Klan, and a "black officers' study guide," consisting of children's puzzles. White officers once dressed a Hispanic inmate in a straw hat, sheet, and sign that said "spic." Such activities were found by the court to constitute a hostile work environment.

Pregnancy Discrimination Act of 1987— An Amendment to Title VII

In 1987, Title VII was amended by the Pregnancy Discrimination Act (PDA). This law expanded the definition of discrimination based on gender to include discrimination based on pregnancy.

[9] 523 U.S. 75, 118 S. Ct. 998 (1998).
[10] 162 F.3d 1062 (10th Cir. 1998).
[11] 782 F.2d 1094 (1986).

"Discrimination on the basis of pregnancy, childbirth or related medical conditions constitutes unlawful sex discrimination under Title VII."[12] Under the act, temporary disability caused by pregnancy must be treated the same as any other temporary disability.

Defenses to Claims Under Title VII

As a business owner or manager, how would you respond if one of your employees filed a lawsuit under Title VII? Are there any legal exceptions for discriminating against a protected class? The answer, surprising to many business owners and managers, is yes. The three most important defenses available to defendants in Title VII cases are the bona fide occupational qualification, merit, and seniority system defenses. These defenses are raised by the defendant after the plaintiff has established a prima facie case of discrimination based on either disparate treatment or disparate impact. They would obviously not be applicable to a claim based on harassment.

The Bona Fide Occupational Qualification Defense The bona fide occupational qualification (BFOQ) defense allows an employer to discriminate in hiring on the basis of sex, religion, or national origin (but not race or color) when doing so is necessary for the performance of the job. Necessity must be based on actual qualifications, not stereotypes about one group's abilities. For example, being a male cannot be a BFOQ for a job simply because it is a dirty job, although there may be a valid requirement that an applicant be able to lift a certain amount of weight if such lifting is part of the job. Conversely, being a female may be a BFOQ for modeling female clothing. An employer would not be required or expected to hire a male for such a job. Employer arguments about inconvenience to the employer, such as having to provide two sets of restroom facilities, have not been persuasive in the courts. Nor have customer preferences to be served by a particular gender or nationality. The only exception to customer preference is sexual privacy (e.g., female restroom attendants in the women's restroom and male attendants in the men's room).[13]

The Merit Defense The merit defense is usually raised when hiring or promotion decisions are partially based on test scores. Professionally developed ability tests that are not designed, intended, or used to discriminate may be used. Although these tests may have an adverse impact on a class, as long as they are manifestly related to job performance, they do not violate the act. Since 1978, the "Uniform Guidelines on Employee Selection Procedures" (UGESP) resource has guided government agencies charged with enforcing civil rights, and they provide guidance to employers and other interested persons about when ability tests are valid and job-related. Under these guidelines, tests must be validated in accordance with standards established by the American Psychological Association.

Three types of validation are acceptable: (1) *criterion-related validity*, which is the statistical relationship between test scores and objective criteria of job performance; (2) *content validity*, which isolates some skill used on the job and directly tests that skill; and (3) *construct validity*, wherein a psychological trait needed to perform the job is measured. A test that required a word processor to use a computer would be content-valid. A test of patience for a teacher would be construct-valid.

The Seniority System Defense A bona fide seniority system is a legal defense under Title VII. Even though a seniority system, in which employees are given preferential treatment based on their length of service, may perpetuate past discrimination, such systems are considered bona fide and are thus not illegal if (1) the system applies equally to all persons; (2) the seniority units follow industry practices; (3) the seniority system did not have its genesis in discrimination; and (4) the system is maintained free of any illegal discriminatory purpose.

Remedies under Title VII

A plaintiff may seek both equitable and legal remedies for violations of Title VII. Courts have ordered parties to engage in diverse activities ranging from publicizing their commitment to

[12] EEOC, "Facts about Pregnancy Discrimination," www.eeoc.gov/facts/fs-preg.html.
[13] *In the Matter of the Accusation of the Department of Fair Employment and Housing v. San Luis Obispo Coastal Unified School District, Respondent; Marlene Anne Mendes, Complainant,* Case No. E95-96 L-0725-00s, 98-14 (October 7, 1998). See www.dfeh.ca.gov.

minority hiring to establishing special training programs for minorities. A successful plaintiff may recover back pay for up to two years from the time of the discriminatory act.

A plaintiff who was not hired for a job because of a Title VII violation may also receive remedial seniority dating back to the time when the plaintiff was discriminated against; compensatory damages, including those for pain and suffering; and, in some cases, punitive damages.

In cases based on discrimination other than race, however, punitive damages are capped at $300,000 for employers of more than 500 employees, $100,000 for firms with 101 to 200 employees, and $50,000 for firms with 100 or fewer employees. An employer will not be held vicariously liable for punitive damages as long as it made good-faith efforts to comply with federal law.

Attorney fees may be awarded to a successful plaintiff in Title VII cases. They are typically denied only when special circumstances would render the award unjust. If it is determined that the plaintiff's action was frivolous, unreasonable, or without foundation, the courts may award attorney fees to the prevailing defendant.

For more information on Title VII, visit the EEOC website at www.eeoc.gov.

Procedure for Filing a Claim under Title VII

Filing a claim under Title VII is much more complicated than simply filing a lawsuit. Failure to follow the proper procedures within the strict time framework may result in the plaintiff losing his or her right to file a lawsuit under Title VII.

The first step in initiating a Title VII action is for the aggrieved party to file a charge with the state Equal Opportunity Commission or, if no such agency exists, the federal EEOC. A *charge* is a sworn statement that states the name of the charging party, the name(s) of the defendant(s), and the nature of the discriminatory act. Title VII charges must be filed with the EEOC within 180 days of the alleged discriminatory act.

Within 10 days of receiving the charge, the EEOC must notify the alleged violator of the charge. Then the EEOC investigates the matter to determine whether there is reasonable cause to believe that a violation has occurred. If the EEOC does find such reasonable cause, it attempts to eliminate the discriminatory practice through conciliation, that is, by trying to negotiate a settlement between the two parties. If unsuccessful, the EEOC *may* file suit against the alleged discriminator in federal district court. Failure to file suit does not necessarily mean that the EEOC thinks the plaintiff does not have a valid claim; the EEOC may simply feel that the claim is not the type it wishes to use its limited resources on pursuing.

If the EEOC decides not to sue, it notifies the plaintiff of his or her right to file an action and issues the plaintiff a *right-to-sue letter,* which is not intended to be anything other than a statement that the plaintiff has followed the proper initial procedures and therefore may file a lawsuit. The plaintiff must have this letter to file a private action.

Discrimination Based on Sexual Orientation—Actionable?

There is currently no federal legislation that prohibits discrimination based on sexual orientation. What does exist are individual state laws that prohibit such discrimination. Twenty states and the District of Columbia prohibit discrimination based on both sexual orientation and gender identity: California, Colorado, Connecticut, Delaware, Hawaii, Illinois, Iowa, Maine, Maryland, Massachusetts, Minnesota, New Jersey, New York, Nevada, New Mexico, Oregon, Rhode Island, Utah, Vermont, and Washington. An additional two states prohibit discrimination based solely on sexual orientation. Those states are New Hampshire and Wisconsin. Another 10 states provide such protections, but only to public employees.

What do these laws mean to Brad, the employee in our Case Opener? It depends on where Brad lives. If Brad lives in Texas and is fired for being gay, he has absolutely no legal rights and cannot sue his employer. Conversely, if Brad lives in California, he may sue Jennifer and So Clean, his employer, for discrimination based on sexual orientation. In California and a handful of other

states, discriminating in terms of employment (i.e., failure to hire, firing, failure to promote) based on a person's sexual orientation is prohibited.

In Case 24-2 below, the Equal Employment Opportunity Commission (EEOC) had to determine whether the agency would include discrimination based on sexual orientation within the meaning of "sex" under Title VII, based on a complaint by a federal government employee.

CASE 24-2

DONALD BALDWIN, COMPLAINANT v. ANTHONY FOXX
Secretary of Dept. of Transportation (FAA), Agency
2015 EEOPUB LEXIS 1905 (July 26, 2015)

FACTS: *Donald Baldwin worked as a Supervisory Air Traffic Control Specialist at the Agency's Southern Region in Miami, Florida. When he learned that he was not selected for a permanent position as a Front Line Manager (FLM) at the Miami tower, he filed an Equal Employment Opportunity (EEO) complaint alleging discrimination based on gender (male) and sexual orientation discrimination. The Agency accepted the complaint for investigation.*

Baldwin alleged that he was not selected because he is gay. He alleged that his supervisor, who was involved in the selection process for the permanent position, made several negative comments about Baldwin's sexual orientation. For example, Baldwin stated that in May 2011, when he mentioned that he and his partner had attended Mardi Gras in New Orleans, the supervisor said, "We don't need to hear about that gay stuff." He also alleged that the supervisor told him on a number of occasions that he was "a distraction in the radar room" when his participation in conversations included mention of his male partner.

ISSUE: The agency examined whether the Title VII requirement that personnel actions affecting federal employees be made free from any discrimination based on "sex" includes discrimination based on sexual orientation.

REASONING: In its decision, the Agency did not address the merits of Baldwin's claim. Instead, the Agency dismissed the complaint on the grounds that it had not been raised in a timely fashion with an EEO Counselor, as required by EEOC regulations. Additionally, Baldwin was notified that pursuant to the "Secretary's Policy on Sexual Orientation" and the "Departmental Office of Civil Rights' March 7, 1998 Procedures for Complaints of Discrimination based on Sexual Orientation," the "sexual orientation portion of the claim is appealable to the Agency and the portion of the claim involving reprisal is appealable to the EEOC." Baldwin appealed the Agency's decision.

Baldwin's complaint makes it clear that he believes that he was denied a permanent position because of his sexual orientation. The Agency erred in its decision to process this claim only under its internal procedures concerning sexual orientation discrimination and not through the EEO complaint process.

Title VII requires that all personnel actions affecting federal employees or applicants for employment ". . . shall be made free from any discrimination based on sex." This provision is analogous to the section of Title VII governing employment discrimination in the private sector in which it is unlawful for a covered employer to fail or refuse to hire or to discharge any individual, or otherwise to discriminate with respect to his compensation, terms, conditions, or privileges of employment, because of such individual's sex.

Title VII's prohibition of sex discrimination means that employers may not rely upon sex-based considerations or take gender into account when making employment decisions This prohibition applies equally in claims brought by lesbian, gay, and bisexual individuals under Title VII.

When an employee raises a claim of sexual orientation discrimination as sex discrimination under Title VII, the question is not whether sexual orientation is explicitly listed in Title VII as a prohibited basis for employment actions. It is not. Rather, the question for purposes of Title VII coverage of a sexual orientation claim is the same as any other Title VII case involving allegations of sex discrimination–whether the agency has relied on sex-based considerations or taken gender into account when taking the challenged employment action.

DECISION AND REMEDY: Complainant Baldwin has stated a claim of sex discrimination. We conclude that sexual orientation is inherently a "sex-based consideration," and an allegation of discrimination based on sexual orientation is necessarily an allegation of sex discrimination under Title VII. Allegations of discrimination on the basis of sexual orientation state a claim of discrimination on the basis of sex within the meaning of Title VII. Furthermore, we conclude that Complainant's initial EEO Counselor contact was timely. We remand Complainant's claim of discrimination to the Agency for further processing for a determination on the merits.

(continued)

SIGNIFICANCE OF THE CASE: This was the first time that the EEOC stated that a claim for sexual orientation discrimination is, by definition, also a claim for sex discrimination.

Since that 2015 agency decision, several courts have taken up the same issue. A 2017 case in the federal Seventh Circuit Court of Appeals determined that discrimination based on sex, under Title VII, includes sexual orientation discrimination. *Hively v. Ivy Tech Community College of Indiana* reverses decades of precedent.[14] In that case, an adjunct community college professor alleged that she was denied a permanent position on the faculty because she was a lesbian. She sued under Title VII, arguing that discrimination based on "sex" includes sexual orientation discrimination. The trial court dismissed her case with prejudice. On appeal, the Seventh Circuit reinstated the case. The majority opinion contains three basic arguments why sex discrimination under Title VII should include sexual orientation discrimination:

1. Discrimination against a lesbian is necessarily discrimination against a woman.

2. Title VII case precedent protects individuals who do not adhere to gender norms (see *Price Waterhouse v. Hopkins,* 409 U.S. 228 (1989)).

3. Title VII case precedent protects persons from discrimination based on their association with those of other races (see *Loving v. Virginia,* 388 U.S. 1 (1967)), therefore Title VII should also protect persons from discrimination based on the sex of the associate.

Other federal circuit courts have come to the opposite conclusion. Ultimately, this matter will either be decided by Congress, or more likely, the U.S. Supreme Court.

CRITICAL THINKING

The courts are divided about whether sexual orientation is an element of "sex" for purposes of enforcing Title VII. When the Supreme Court or the legislature acts to resolve this disagreement, what reason would you expect to be most persuasive if their eventual conclusion is that "sex" does include "sexual orientation"?

ETHICAL DECISION MAKING

When the debate addressed in this case is eventually decided, value preferences will play a major role in shaping the conclusion. What stakeholders other than LGBTQ will be affected by the eventual conclusion?

Discrimination Based on Marriage

On June 26, 2015, in a historic 5-4 ruling, the Supreme Court of the United States found bans on marriage equality to be unconstitutional—and that the right to marriage is a fundamental right for everyone.

Prior to the Supreme Court's ruling in *Obergefell v. Hodges,* a same-sex couple could marry in a state where such unions were legal, but if they moved to a state that did not recognize same-sex marriage, they were legally treated as if their marriage did not exist.[15] This was because of section 2 of the Defense of Marriage Act (DOMA), a federal law passed in 1996 providing that no state "shall be required to give effect" to same-sex marriages granted by other states. As a result, some states chose to recognize same-sex marriage, and other states chose not to. The Supreme Court's 2015 ruling struck down section 2 of DOMA, effectively ending the unequal treatment of the marriages of same-sex couples. Same-sex couples are now able to exercise the right to marry in all states, and there is no longer any basis for any state to refuse to recognize lawful same-sex marriages performed in other states.

[14] 2017 U.S. App. LEXIS 5839 (April 4, 2017).

[15] 135 S.Ct. 2584 (2015).

Importantly, the majority opinion of the Supreme Court held as follows:

1. Under the Due Process Clause of the Fourteenth Amendment, no state shall deprive any person of life, liberty, or property, without due process of law. The fundamental liberties protected by this clause include most of the rights enumerated in the Bill of Rights. In addition, these liberties extend to certain personal choices central to individual dignity and autonomy, including intimate choices that define personal identity and beliefs.
2. In interpreting the Equal Protection Clause, the U.S. Supreme Court has recognized that new insights and societal understandings can reveal unjustified inequality within the United States's most fundamental institutions that once passed unnoticed and unchallenged.
3. Under the Due Process and Equal Protection Clauses of the Fourteenth Amendment, same-sex couples have a fundamental right to marry.

Age Discrimination in Employment Act of 1967

Age Discrimination in Employment Act (ADEA) of 1967
Federal law that prohibits employers from refusing to hire, discharging, or discriminating in terms and conditions of employment on the basis of an employee or applicant being age 40 or older.

LO 24-4 Identify the categories of people protected from discrimination.

The **Age Discrimination in Employment Act (ADEA) of 1967** was enacted to prohibit employers from refusing to hire, discharging, or discriminating in terms and conditions of employment against employees or applicants age 40 or older. The language describing the prohibited conduct is virtually the same as that of Title VII, except that age is the prohibited basis for discrimination. ADEA applies to employers having 20 or more employees.

In Case 24-3 below, the jury had to decide whether there was substantial evidence to prove that the defendants' conduct was motivated by age discrimination and not by a justified business reason.

CASE 24-3 BOBBY DEAN NICKEL v. STAPLES CONTRACT & COMMERCIAL INC.

FACTS: *Plaintiff had worked for Corporate Express, an office-supplies company catering to businesses, for seven years when Staples Inc. purchased Corporate Express in June 2008 and renamed the company Staples Contract and Commercial Inc. Plaintiff's job was to manage the physical plant at the Corporate Express fulfillment facility in La Mirada. His duties included general repair and maintenance, equipment maintenance, and security. Plaintiff maintained this supervisory position after Staple's Inc.'s takeover of Corporate Express.*

In 2007, Plaintiff began reporting to a new manager, Lionel Marrero. After the 2008 takeover, Marrero worked for Defendants. There was evidence at trial that Marrero was on a mission to cut costs at the La Mirada facility by replacing older, higher paid employees with part-time and temporary employees. His method was to increase the workload for older employees, forcing them into retirement, or "[write] them up" for errors or ethical violations, creating a basis for their termination.

In 2011, when Plaintiff was 64 years old, Defendants terminated Plaintiff's employment based on allegations that he stole a bell pepper from the La Mirada facility's cafeteria, which was run by a third-party vendor. Plaintiff explained that he entered the cafeteria after hours to evaluate whether the refrigerators were working properly. When he opened the fridge, a salad fell out onto the floor. Plaintiff threw out the salad but ate the bell pepper. He testified in his deposition and at trial that there was an accepted practice of taking food from the cafeteria after hours and paying for it later. When he was investigated for the theft, Plaintiff reported that he had previously taken three Monster drinks after hours and subsequently paid for them. A cafeteria worker likewise told company investigators that she noticed the drinks missing and that Plaintiff had paid for them. Plaintiff initially stated in a written statement that he forgot to pay for the 68-cent bell pepper but later testified at his deposition and at trial that he did pay for the pepper.

Following his termination, Plaintiff filed this lawsuit against Defendants, alleging age-based discrimination. Plaintiff tried

(continued)

his age discrimination claim to a jury, who found in his favor and awarded him approximately $3.2 million in compensatory damages and $22.8 million in punitive damages ($13,053,664 against Staples, and $9,790,248 against Staples Inc.). Defendants filed motions for new trial and for judgment notwithstanding verdict. Partially granting Staples Inc.'s motion for judgment, the court reduced punitive damages to $13 million by striking the punitive damages against Staples Inc. The court entered judgment jointly against Defendants in the amount of $16,317,080.

ISSUE: The court examined whether there was insufficient evidence of discrimination to support the jury's verdict given defendants's assertions of a legitimate, nondiscriminatory business reason for the firing (i.e., alleged theft and unethical behavior)?

REASONING: Defendants argue they were entitled to judgment notwithstanding the verdict because Plaintiff failed to present sufficient evidence that age was a substantial factor motivating his termination. FEHA prohibits age discrimination by employers against employees. A plaintiff alleging disparate treatment discrimination has the burden to establish unlawful discrimination by showing that he was a member of the protected age group, he competently performed his job, he was terminated or suffered an adverse employment action, and there was a discriminatory motive for the termination. As to the last prong, "[t]here must . . . be evidence of a causal relationship between the [discriminatory] animus and the adverse employment action." If the employer shows that its action was taken for a legitimate, nondiscriminatory reason, the plaintiff must show that the employer's reason for termination was pretextual. Specifically, Defendants contend that there is insufficient evidence of discriminatory animus and persuasive evidence that Defendants based the termination on a legitimate, nondiscriminatory reason: Plaintiff's unethical behavior and violation of the anti-theft policy. We reject Defendants' argument that there was a dearth of evidence Plaintiff was fired because of his age. Based on testimony from Daniel Velasquez, Plaintiff, and other employees, the jury could reasonably find that Marrero was motivated by anti-age animus and Marrero's role in the decision to terminate Plaintiff was a substantial factor causing the termination.

Marrero's comments about older employees, made in the context of an intention to eliminate them from the workforce, provided substantial evidence of age-based animus. Plaintiff elicited testimony from Velasquez, a former Staples manager who worked with Plaintiff prior to and at the time of his termination. Velasquez testified he had witnessed a "concerted attempt to get rid of the higher paid, older workers" following Defendants' takeover. In management meetings, Velasquez heard Marrero talk about the company's desire to push out older employees and heard Marrero say he wanted to find reasons to discharge them. Specifically, Marrero told the managers to "'[t]ake a closer look at the older people. They are starting to drag and are slowing down. If they are not top performers, write them up and get rid of them.'" Marrero also said, "'[w]e need young energetic people. Walk around the facility with the older workers and if they cannot keep up then get rid of them.'" Plaintiff also testified that Marrero made similar ageist comments at managers' meetings.

Former employees also testified that they and their coworkers were pushed out of their jobs as a result of this discriminatory animus. The testimony of witnesses Velasquez, Vance, Munoz, and Evans provided substantial evidence of Marrero's discriminatory comments and Defendants' practice of forcing out older workers. This testimony provided ample evidence supporting the court's finding there was substantial evidence of discrimination.

DECISION AND REMEDY: The judgment is affirmed on all grounds. Plaintiff Bobby Dean Nickel is awarded his costs on appeal.

SIGNIFICANCE: Despite defendants' protestations that plaintiff was fired for a legitimate, nondiscriminatory reason (theft and unethical behavior), the jury had the right, based on the substantial evidence presented at trial, to reject defendants' reasons and determine that the firing was based on age discrimination.

CRITICAL THINKING

Do you believe that plaintiff's alleged theft of a bell pepper should constitute sufficient grounds for termination absent any evidence of discrimination? Does your opinion change when presented with the additional evidence of age discrimination? Why or why not?

ETHICAL DECISION MAKING

What social values are protected in the court's decision? Do you agree with the court's decision to protect these values?

Americans with Disabilities Act

Americans with Disabilities Act (ADA)
Federal law that prohibits discrimination against employees and job applicants with disabilities.

The goal of the **Americans with Disabilities Act (ADA)** is preventing employers from discriminating against employees and applicants with disabilities. ADA attempts to attain this objective by requiring employers to make reasonable accommodations to the known physical or mental disabilities of an otherwise qualified person with a disability, unless the necessary accommodation would impose an undue burden on the employer's business.

> When the ADA was before Congress, some members predicted a flood of lawsuits that would bankrupt or at least overburden business. . . . Studies have shown, however, that businesses have adapted to the ADA much more easily—and inexpensively—than the doomsayers predicted. . . . [One] survey found that three-quarters of all changes cost less than $100. Moreover, the predicted flood of lawsuits proved to be imaginary. Almost 90 percent of the cases brought before the Equal Employment Opportunity Commission are thrown out. The American Bar Association recently conducted a survey and learned that, of the cases that actually go to court, 98 percent are decided in favor of the defendants, usually businesses.[16]

ADA is enforced by the EEOC in the same way that Title VII is enforced. To bring a successful claim under ADA, the plaintiff must show that he or she meets all of the following:

- Had a disability.
- Was otherwise qualified for the job.
- Suffered an adverse employment decision because of that disability.

Remedies for ADA violations are similar to those available under Title VII. A successful plaintiff may recover reinstatement, back pay, and injunctive relief. In cases of intentional discrimination, limited compensatory and punitive damages are also available. An employer who has repeatedly violated the act may be subject to fines of up to $100,000.

May an Employer Discriminate against a Smoker?

In the Case Opener, Jennifer discovered that Brad was a smoker. Later, she fired him. One of Jennifer's given reasons for terminating Brad's employment was that he was a smoker. May Jennifer and So Clean! legally fire an employee for smoking outside the workplace? The answer is, "It depends!"

A recent trend has been for employers to consider a potential employee's lifestyle when deciding whether to hire that person. Employers argue that smokers have higher health care costs and miss more work, lowering productivity.

As a result, some companies either won't hire smokers or are threatening to fire current employees who will not or are unable to quit smoking. In 2005, Michigan-based Weyco Inc. announced that it would terminate all workers who did not stop smoking.[17] Many states have passed laws preventing companies from engaging in such action.

If Brad works in Michigan, Jennifer and So Clean! may legally terminate him for smoking outside the workplace. Conversely, if Brad works in the District of Columbia or any of the 29 states with smoker's rights laws, he could not be legally terminated for smoking outside the workplace.[18] Finally, employers should be aware that giving breaks on health care plans to employees who are nonsmokers could be in violation of a smoker's rights law.

[16] Center for an Accessible Society, "Disability issues information for Journalists", http://www.accessiblesociety.org/topics/ada/.

[17] Weyco Announces Plan to Charge Employees if their Spouses Smoke, http://tobaccoanalysis.blogspot.com/2005/12/weyco-announces-plan-to-charge.html.

[18] Twenty-nine states plus the District of Columbia have laws in effect elevating smokers to a protected class. American Lung Association, http://www.lungusa2.org/slati/appendixf.php.

LAW	IMPACT
Wage and hour laws	Federal and state laws that impose minimum-wage and hour requirements for employees.
Family and Medical Leave Act (FMLA)	Federal act requiring certain employers to establish a policy that provides all eligible employees with up to 12 weeks of unpaid leave during any 12-month period for several family-related occurrences (e.g., birth of a child, care of a sick spouse).
Unemployment compensation	State system, created by the Federal Unemployment Tax Act (FUTA), that provides unemployment compensation to qualified employees who lose their jobs.
Consolidated Omnibus Budget Reconciliation Act (COBRA)	Federal law that ensures that when employees lose their jobs or have their hours reduced to a level at which they would not be eligible to receive medical, dental, or optical benefits from their employer, the employees will be able to continue receiving benefits under the employer's policy for up to 18 months by paying the premiums for the policy.
Employee Retirement Income Security Act (ERISA)	Federal law that sets minimum standards for most voluntarily established pension and health plans in private industry to provide protection for individuals in these plans. HIPAA and COBRA are amendments to ERISA.
Occupational Safety and Health Act (OSHA) of 1970	Federal law that established the Occupational Safety and Health Administration, the agency responsible for setting safety standards under the act, as well as enforcing the act through inspections and levying fines against violators.

Exhibit 24-3

Significant Laws Governing the Employment Relationship

Equal Pay Act of 1963

Although we generally think of Title VII as the primary means of protecting women from discrimination, an earlier law actually was designed specifically to protect women from wage discrimination: the **Equal Pay Act (EPA) of 1963**. When EPA was passed, the average wages of women were less than 60 percent of those of men. The primary purpose of the law was to eliminate situations in which women, working alongside men or replacing men, would be paid lower wages for doing substantially the same job. EPA prohibits any employer from discriminating within any "establishment . . . between employees on the basis of sex by paying wages to employees in such establishment at a rate less than the rate at which he pays wages to employees of the opposite sex . . . for equal work on jobs the performance of which requires equal skill, effort, and responsibility, and which are performed under similar working conditions, except where payment is made pursuant to (i) a seniority system; (ii) a merit system; (iii) a system which measures earnings by quantity or quality of production; or (iv) differential based on any factor other than sex."[19]

Equal Pay Act (EPA) of 1963

Federal law that prohibits an employer from paying workers of one gender less than the wages paid to employees of the opposite gender for work that requires equal skill, effort, and responsibility.

Additional Laws Governing the Employment Relationship

LO 24-5 Summarize the laws that protect employees' wages, benefits, health and safety standards, and rights.

Several major categories of state and federal laws cover employee rights with regard to wages, benefits, and health and safety standards. Exhibit 24-3 lists significant state and federal legislation protecting employees.

[19] 29 U.S.C. § 206(d)1.

Fair Labor Standards Act

Fair Labor Standards Act (FLSA)
Federal law requiring a minimum wage of a specified amount be paid to all employees in covered industries; also mandates that employees who work more than 40 hours in a week be paid no less than 1½ times their regular wage for all hours beyond 40 worked in a given week.

The **Fair Labor Standards Act (FLSA)** requires a minimum wage of a specified amount to be paid to all employees in covered industries. Congress raises the specified amount periodically to compensate for increases in the cost of living caused by inflation. The current federal minimum wage is $7.25.

On May 18, 2016, the Department of Labor (DOL) published new FLSA regulations requiring that employers pay overtime (time and a half on anything over 40 hours worked in a workweek) to full-time salaried executive, administrative and professional employees making $47,476 annually ($913/week) or less. On August 31, 2017, a federal court invalidated the new rule. The court held that when the DOL set the $23,660 minimum salary level in 2004, it was consistent with Congress' intent because the level was set as a floor to screen out "obviously nonexempt employees, making an analysis of duties in such cases unnecessary." Experts say the minimum salary level should be "somewhere around $33,000" to account for the rate of inflation since 2004.[20] As of late 2018, no final decision had been announced.

Some states are now taking up the issue of minimum salaries for white collar workers. Currently, eight states have minimum compensation thresholds for overtime exemptions that are higher, or will eventually be higher, than then federal minimum of $23,660 per year: Alaska, California, Colorado, Connecticut, Iowa, Maine, New York, and Oregon.[21]

Family and Medical Leave Act

Family and Medical Leave Act (FMLA)
Federal act requiring employers to provide all eligible employees with up to 12 weeks of leave during any 12-month period for several family-related occurrences (e.g., birth of a child, care of a sick spouse).

When the **Family and Medical Leave Act (FMLA)** went into effect in 1993, it was hailed by its supporters as a breakthrough in American law and feared by its opponents as an unwieldy encumbrance on business. FMLA covers all public employers and private employers with 50 or more employees. It guarantees all eligible employees (those who have worked at least 25 hours a week for each of 12 months prior to the leave) up to 12 weeks of *unpaid* leave during any 12-month period for any of the following family-related occurrences:

- The birth of a child.
- The adoption of a child.
- The placement of a foster child in the employee's care.
- The care of a seriously ill spouse, parent, or child.
- A serious health condition that renders the employee unable to perform any of the essential functions of his or her job.

Unemployment Compensation

unemployment compensation
A state system, created by the Federal Unemployment Tax Act (FUTA), that provides unemployment compensation to qualified employees who lose their jobs.

Although FMLA in some cases helps employees retain their jobs, what happens if they lose their jobs? The Federal Unemployment Tax Act (FUTA), passed in 1935, created a state system to provide **unemployment compensation** to qualified employees who lose their jobs.[22] Under this law, employers pay taxes to the states, which deposit the money into the federal government's Unemployment Insurance Fund. Each state has an account from which it can access the money in the fund in accordance with the rules the state establishes for eligibility. States have different minimum standards for qualifying for unemployment compensation, although almost all states require that the applicant did not voluntarily quit or get fired for cause.

Workers's Compensation Laws

Unlike many other laws affecting the employment relationship, workers's compensation legislation is purely state law. Our coverage of this topic must therefore be rather generalized. Prudent

[20] https://www.xperthr.com/news/usd47476-minimum-salary-struck-down-new-minimum-of-about-usd33000-expected/28730/.

[21] https://www.xperthr.com/blog/2018/02/07/minimum-salaries-for-white-collar-employees-will-states-pick-up-where-us-labor-department-left-off/.

[22] 26 U.S.C. §§ 3301–3310.

businesspeople will familiarize themselves with the workers's compensation statutes of the states within which their companies operate.

Workers' compensation laws ensure that covered workers who are injured on the job can receive financial compensation through an administrative procedure rather than having to sue their employer. For administrative convenience, most states exclude certain types of businesses and small firms from coverage. Some states also allow businesses that have sufficient resources to be self-insured rather than participate in the state program.

workers' compensation laws
State laws that provide for financial compensation to employees or their dependents when a covered employee is injured on the job.

Consolidated Omnibus Budget Reconciliation Act of 1985

The **Consolidated Omnibus Budget Reconciliation Act (COBRA)** ensures that employees who lose their jobs or have their hours reduced to a level at which they are no longer eligible to receive medical, dental, or optical benefits can pay to continue receiving benefits for themselves and their dependents under the employer's policy. The employee must pay the premiums for the policy, plus up to a 2 percent administration fee, to maintain the coverage for up to 18 months, or 29 months for a disabled worker. This is often quite expensive. An employee has 60 days after coverage would ordinarily terminate to decide whether to maintain the coverage. This obligation does *not* arise under either of two conditions:

1. The employee is fired for gross misconduct.
2. The employer decides to eliminate benefits for all current employees.

Consolidated Omnibus Budget Reconciliation Act (COBRA)
Federal law that ensures that when employees lose their jobs or have their hours reduced to a level at which they would not be eligible to receive medical, dental, or optical benefits from their employer, the employees will be able to continue receiving benefits under the employer's policy for up to 18 months by paying the premiums for the policy.

Employee Retirement Income Security Act of 1974

The **Employee Retirement Income Security Act (ERISA)** is "a federal law that sets minimum standards for most voluntarily established pension and health plans in private industry to provide protection for individuals in these plans."[23] Under ERISA, employers must provide participants with all the following:

1. Plan information (i.e., features and funding).
2. Assurances of the fiduciary responsibility of those in charge of managing and controlling the plan assets.
3. A grievance and appeals process for participants to get benefits from their plans.
4. The right to sue for benefits and breaches of fiduciary duty.[24]

Employers are required to provide to employees, free of charge, a *summary plan description (SPD),* which provides information on how the plan operates, the benefits under the plan, how to apply for such benefits, when such benefits vest, and when benefits may be paid out.

Employee Retirement Income Security Act (ERISA)
Federal law that sets minimum standards for most voluntarily established pension and health plans in private industry to provide protection for individuals in these plans.

Occupational Safety and Health Act of 1970

The federal government regulates workplace safety primarily through the **Occupational Safety and Health Act (OSHA) of 1970**, which requires every employer to "furnish to each of his employees . . . employment . . . free from recognized hazards that are likely to cause death or serious physical harm." The Occupational Safety and Health Administration (abbreviated OSHA, the same as the act) is the agency that is responsible for promulgating workplace safety standards, inspecting facilities to ensure compliance with the standards, and bringing enforcement actions against violators.

In 2015, Ashley Furniture was fined almost $2.3 million for workplace safety violations. OSHA issued citations for 38 violations after more than 1,000 recordable work-related injuries in the previous 3.5 years at Ashley Furniture's two Wisconsin locations. Citations included failure to implement procedures to prevent machines from unintentional start-up when operators changed blades and cleared jams and exposing workers to dangerous machine-operating parts.[25]

In addition, OSHA undertakes educational programs for employers and employees.

Occupational Safety and Health Act (OSHA) of 1970
Federal law that established the Occupational Safety and Health Administration, the agency responsible for setting safety standards under the act and enforcing the act through inspections and the levying of fines against violators.

[23] Department of Labor, Employee Retirement Income Security Act, www.dol.gov/dol/topic/health-plans/erisa.htm.
[24] Ibid.
[25] OSHA Newsletter, October 26, 2015. http://www.safetynewsalert.com/ashley-furnitures-osha-fines-now-total-almost-2-3-million-in-2015/.

Employee Privacy in the Workplace

LO 24-6 Identify components of employer privacy policies.

With the rapid expansion of technology, new issues related to workplace privacy have evolved. Technology has stimulated the growth of such issues in a number of ways. On the one hand, technology has given employers new ways to gather information about employees. On the other hand, new technologies have provided more temptations for employees to be off the job at work, thus creating a stimulus for more employer monitoring. For example, surveys have shown that 90 percent of employees with Internet access at work look at non-work-related websites at least once a day,[26] 90 percent receive non-work related email,[27] and 84 percent send non-work related email.[28] Given such statistics, it is understandable that employers would want to monitor what employees are doing when they are supposed to be on the job.

Electronic Monitoring and Communication

Business owners need to understand the extent to which they can monitor employees' telephone conversations, read their email, and listen to their voice mail. Questions related to such monitoring primarily involve the common law tort of invasion of privacy and two federal statutes: the Omnibus Crime Control and Safe Streets Act of 1968,[29] as amended by the Electronic Communications Privacy Act of 1986.[30]

Under the **Omnibus Crime Control and Safe Streets Act of 1968,** employers cannot listen to the private telephone conversations of employees or disclose the contents of these conversations. They may, however, ban personal calls and monitor calls for compliance as long as they discontinue listening to any conversation once they determine it is personal. Violators may be subject to fines of up to $10,000. Under the **Electronic Communications Privacy Act (ECPA) of 1986,** employees' privacy rights were extended to electronic forms of communication, including email and cellular phones. ECPA outlaws the intentional interception of electronic communications and the intentional disclosure or use of the information obtained through such interception.

In recognition of the fact that employers may sometimes need to monitor employee conversations for the purpose of improving employee performance or protecting employees from harassing calls, ECPA includes a business-extension exemption that allows employers to monitor employee telephone conversations in the ordinary course of their employment. This exception is subject to the constraint preventing the employer from continuing to listen to conversations after it becomes apparent that they are personal. Also, monitoring is allowed when employees give their prior consent.

In Case 24-4, the court had to determine whether a woman's Facebook post was protected under the ECPA.

> **Omnibus Crime Control and Safe Streets Act of 1968**
> Federal statute banning employers from listening to the private telephone conversations of employees or disclosing the contents of these conversations. Employers may ban personal calls and monitor calls for compliance as long as they discontinue listening to any conversation once they determine it is personal.

> **Electronic Communications Privacy Act (ECPA) of 1986**
> Federal law that extended employees' privacy rights to electronic forms of communication, including email and cellular phones; outlaws the intentional interception of electronic communications and the intentional disclosure or use of the information obtained through such interception.

[26] Dyland Loeb McClain, "I'll Be Right with You, Boss, as Soon as I Finish My Shopping," *New York Times*, January 10, 2001, p.G1.
[27] Ibid.
[28] Ibid.
[29] 18 U.S.C. § 2210 et seq.
[30] 18 U.S.C. §§ 2510–2521.

EHLING v. MONMOUTH-OCEAN HOSP. SERV. CORP.
961 F. Supp. 2d 659 (Delaware 2013)

FACTS: *Plaintiff Deborah Ehling was hired by Monmouth-Ocean Hospital Service Corp. (MONOC) in 2004 as a registered nurse and paramedic. Plaintiff's claims in this case arise out of an incident involving her Facebook account. Ehling sued for violation of her privacy rights under the Federal Stored Communications Act (SCA) because another employee who had access to her social networking site copied items that were posted on the site and provided those copies to the her supervisor.*

(continued)

From 2008-2009, Plaintiff maintained a Facebook account and had approximately 300 Facebook friends. She selected privacy settings for her account that limited access to her Facebook wall to only her Facebook friends and did not add any MONOC managers as Facebook friends. However, Plaintiff added many of her MONOC coworkers as friends, including a paramedic named Tim Ronco. Plaintiff posted on Ronco's Facebook wall, and Ronco had access to Plaintiff's Facebook wall. Unbeknownst to Plaintiff, Ronco was taking screenshots of Plaintiff's Facebook wall and printing them or emailing them to MONOC manager Andrew Caruso. Ronco and Caruso became friends while working together at a previous job, but Ronco never worked in Caruso's division at MONOC. The evidence reflects that Ronco independently came up with the idea to provide Plaintiff's Facebook posts to Caruso. Caruso never asked Ronco for any information about Plaintiff, and never requested that Ronco keep him apprised of Plaintiff's Facebook activity. In fact, Caruso was surprised that Ronco showed him Plaintiff's Facebook posts. Caruso never had the password to Ronco's Facebook account, Plaintiff's Facebook account, or any other employee's Facebook account. Once Caruso received copies of Plaintiff's Facebook posts, he passed them on to Quagliana, MONOC's Executive Director of Administration.

On June 8, 2009, Plaintiff posted the following statement to her Facebook wall:

An 88 yr old sociopath white supremacist opened fire in the Wash D.C. Holocaust Museum this morning and killed an innocent guard (leaving children). Other guards opened fire. The 88 yr old was shot. He survived. I blame the DC paramedics. I want to say 2 things to the DC medics. 1. WHAT WERE YOU THINKING? and 2. This was your opportunity to really make a difference! WTF!!!! And to the other guards. . . .go to target practice.

*After MONOC management was alerted to the post, Plaintiff was temporarily suspended **with** pay, and received a memo stating that MONOC management was concerned that Plaintiff's comment reflected a "deliberate disregard for patient safety." In response, Plaintiff filed a complaint with the National Labor Relations Board ("NLRB"). After reviewing the evidence, the NLRB found that MONOC did not violate the National Labor Relations Act. The NLRB also found that there was no privacy violation because the post was sent, unsolicited, to MONOC management.*

ISSUE: Whether the privacy protections of the Federal Stored Communications Act (SCA) applies to Facebook wall posts?

REASONING: WILLIAM J. MARTINI, U.S. DISTRICT JUDGE: Plaintiff argues that Defendants violated the SCA by improperly accessing her Facebook wall post about the museum shooting and that her Facebook wall posts are covered by the SCA because she selected privacy settings limiting access to her Facebook page to her Facebook friends. The issue before the Court is whether the SCA applies to Facebook wall posts. Very few courts have addressed this issue. For the reasons set forth below, the Court finds that Facebook wall posts fall within the purview of the SCA.

In 1986, Congress passed the Electronic Communications Privacy Act, which was intended to afford privacy protection to electronic communications. Title II of the Electronic Communications Privacy Act contains the SCA. The SCA covers: (1) electronic communications, (2) that were transmitted via an electronic communication service, (3) that are in electronic storage, and (4) that are not public. Facebook wall posts that are configured to be private meet all four criteria. The Court finds that, when users make their Facebook wall posts inaccessible to the general public, the wall posts are "configured to be private" for purposes of the SCA. Accordingly, the Court finds that non-public Facebook wall posts are covered by the SCA. Because Plaintiff in this case chose privacy settings that limited access to her Facebook wall to only her Facebook friends, the Court finds that Plaintiff's Facebook wall posts are covered by the SCA.

The Court next evaluates whether either of the SCA's statutory exceptions apply. The SCA "does not apply with respect to conduct authorized (1) by the person or entity providing a wire or electronic communications service; [or] (2) by a user of that service with respect to a communication of or intended for that user." First, access to Plaintiff's Facebook wall post was "authorized." The undisputed evidence establishes that Ronco voluntarily provided Plaintiff's Facebook posts to MONOC management without any coercion or pressure. Caruso testified at his deposition that Plaintiff's Facebook friend Ronco voluntarily took screenshots of Plaintiff's Facebook page and either emailed those screenshots to Caruso or printed them out for him. Plaintiff provided no evidence to support her theory that access to her Facebook page was unauthorized. Second, access to Plaintiff's Facebook wall post was authorized "by a user of that service." Plaintiff acknowledged that she added Ronco as a Facebook friend and posted on Ronco's Facebook wall. Third, Plaintiff's Facebook wall post was "intended for that user." Based on the privacy settings that Plaintiff selected for her Facebook page, Plaintiff's wall posts were visible to, and intended to be viewed by, Plaintiff's Facebook friends. On June 8, 2009, when Plaintiff posted the comment about the museum shooting, Ronco was one of Plaintiff's Facebook friends. Thus, the post was intended for Ronco.

DECISION AND REMEDY: In conclusion, access to Plaintiff's Facebook wall post was authorized by a Facebook user with respect to a communication intended for that user. Therefore, the authorized user exception applies and defendants are not liable under the SCA. Motion for summary judgment GRANTED.

SIGNIFICANCE OF THE CASE: A third party with legitimate access to your Facebook page falls under the "authorized user exception" to the SCA and could cost you your right to privacy.

(continued)

> **CRITICAL THINKING**
>
> Do you agree with the court's reasoning? Would you have thought that actions by a third party could destroy your right to privacy?
>
> **ETHICAL DECISION MAKING**
>
> How do you feel about the third party employee who gave copies of the plaintiff's Facebook postings to management? Was his behavior ethical? Why or why not?

The key question in cases involving employer monitoring and interception of employee communications via email, telephone, or voice mail is whether the employee had a *reasonable expectation of privacy* with respect to the communication in question. Employers are in the strongest position if they have a clear policy establishing that there is no reasonable expectation of privacy. Employment law experts advise that having a clear employee privacy policy that employees sign will minimize the likelihood of being sued by employees for invasion of privacy. These policies should be spelled out in the employee handbook and explained to employees. At a minimum, employer privacy policies should cover the following issues:

1. Employer monitoring of telephone conversations.
2. Employer surveillance policies.
3. Employee access to medical and personnel records.
4. Drug testing policies.
5. Lie detector policies.
6. Ownership of computers and all issues unique to the electronic workplace.

Use of Social Media in Employment Decisions

LO 24-7 Recognize the circumstances under which social media may be used in employment decisions.

The growth of social media has changed how the world communicates, and this includes the business world. Companies now use social media for recruiting, employee engagement, knowledge sharing, branding, and more. Business use of Facebook, Twitter, LinkedIn, YouTube, and numerous other social media sites has become commonplace. With that use, however, comes legal responsibilities.

Use of Social Media in Hiring Decisions

May an employer use social media in hiring decisions? The answer is yes—if it's done carefully. Probably the biggest concern is discrimination. When accessing a potential employee's social media (e.g., Facebook), the employer may gain access to information in protected categories, such as age, race, ethnicity, religious affiliations, and disabilities. Such information must be ignored when making hiring decisions. Some experts suggest that employers delay social media searches until after they meet candidates. Another suggestion is to have someone other than the employment decision-maker do the social media screening, and then create a firewall between this person and the decision-maker. Employers should keep in mind, however, that background screening

companies that use social media sites are subject to the provisions of the Fair Credit Reporting Act, under which the applicants must give permission for pre-employment investigations.[31]

A potentially little-known problem with using social media—particularly the most popular sites like Facebook and LinkedIn—is that the subscribers do not represent the U.S. or global job-applicant pools. Recent surveys show that social media sites have lower percentages of Latino and black users than are in the general population. The burden, then, is on the employer to widen their search for new applicants to include job boards, newspapers, magazines, and job fairs.[32]

Some employers are waiting until the potential employee comes in for an interview and then asking the candidate for their Facebook password. Many job applicants claim it is an invasion of privacy. But is it legal? The answer varies by state. Since 2012, some sort of protective legislation has been passed in 20 states. As of 2016, similar legislation is pending in 13 additional states.[33] The amount of protection varies by state.

Use of Social Media in Firing Decisions

What happens if an employer wants to use information obtained from social media to fire an employee? The answer depends on the manner in which the employee is using social media. As an example, the National Labor Relations Act (NLRA) protects the rights of employees to act together to address conditions at work, with or without a union. This protection extends to certain work-related conversations conducted on social media, such as Facebook and Twitter.[34] Employees are permitted to use social media to initiate group action and employers are prohibited from restraining those rights. However, personal grievances not aimed to induce group action would not be protected. Over the last several years, there have been a number of National Labor Relations Board (NLRB) lawsuits filed, with mixed outcomes.

In response to employer requests for guidance, the NLRB's Acting General Counsel released three memos in 2011 and 2012 detailing the results of investigations in dozens of social media cases.[35] Several discharges were found to be unlawful because they flowed from unlawful social media policies. In one case, however, a discharge was upheld despite an unlawful policy because the employee's posting was not work-related. The report underscored two main points regarding the NLRB and social media:

1. Employer policies should not be so sweeping that they prohibit the kinds of activity protected by federal labor law, such as the discussion of wages or working conditions among employees.

2. An employee's comments on social media are generally not protected if they are mere gripes, not made in relation to group activity among employees.

In September 2012, the NLRB issued a decision that the firing of a BMW salesman for photos and comments posted to his Facebook page did not violate federal labor law. At issue was whether the salesman was fired exclusively for posting photos of an embarrassing accident at an adjacent Land Rover dealership, which did not involve fellow employees, or for posting mocking comments and photos with co-workers about serving hot dogs at a luxury BMW car event. Both sets of photos were posted to Facebook on the same day; a week later, the salesman was fired. The Board agreed with the Administrative Law Judge that the salesman was fired solely for the photos he posted of a Land Rover incident, which was not concerted activity and so was not protected.[36]

On December 14, 2012, the NLRB found it was unlawful for a non-profit organization to fire five employees who participated in Facebook postings about a co-worker who intended to complain to management about their work performance.[37] In its analysis, the Board majority applied

[31] Fair Credit Reporting Act, 15 USC § 1681 et seq.

[32] *Use Social Media Smartly When Hiring*, Steve Bates, March 19, 2013, retrieved from Society for Human Resource Management (SHRM) at https://www.shrm.org/hrdisciplines/technology/articles/pages/be-smart-when-using-social-media-for-hiring.aspx.

[33] *National Conference of State Legislators* (*NCSL*), Access to Social Media Usernames and Passwords, retrieved at http://www.ncsl.org/research/telecommunications-and-information-technology/employer-access-to-social-media-passwords-2013.aspx. According to the NCSL, the states with some sort of protective legislative are: Arkansas, California, Colorado, Connecticut, Delaware, Louisiana, Maine, Maryland, Michigan, Montana, Nevada, New Jersey, New Mexico, Oregon, Rhode Island, Tennessee, Utah, Vermont, Washington, and Wisconsin.

[34] *The NLRB and Social Media*, National Labor Relations Board, retrieved at https://www.nlrb.gov/news-outreach/fact-sheets/nlrb-and-social-media.

[35] *Ibid.*

[36] Karl Knauz Motors, Inc., d/b/a Knauz BMW *and* Robert Becker. Case 13-CA-046452 (September 28, 2012).

[37] Hispanics United of Buffalo, Inc. and Carlos Ortiz. Case 03-CA-027872 (December 14, 2012).

settled Board law to social media and found that the Facebook conversation was concerted activity and was protected by the National Labor Relations Act.

Employers may use social media in making employment decisions, but care must be taken to not run afoul of state and federal employment laws and regulations.

Responsibilities of Employers in Hiring Foreign Workers

LO 24-8 Recall the agency and legislation that regulates the hiring of foreign workers.

Until 1986, there was virtually no risk for employers who hired undocumented immigrants. But, beginning that year, Congress passed the Immigration Reform and Control Act. This act is an amendment to the Immigration and Nationality Act (INA) and requires employers to verify the identity and eligibility of all individuals hired in the United States after November 6, 1986.[38]

The INA designates Employment Eligibility Verification Form I-9 (Form I-9) as the means of documenting this verification.[39] For assistance in filling out Form I-9, the United States Customs and Immigration Service (USCIS) publishes an Employer's Handbook on their website. This 69-page handbook also provides a thorough explanation of an employer's legal obligations.

The federal agency responsible for immigration worksite enforcement is Immigration Customs and Enforcement (ICE).[40] As of 2012, the agency developed a "comprehensive worksite enforcement strategy."[41] If it is discovered that an employer knowingly employed undocumented immigrant workers, that employer will be in serious legal trouble. But what if the employer didn't know that the employee's papers were bad? Employers are required to confirm "work authorization," which means that every new employee, at the time of hire, must fill out Form I-9. The employer is not expected to be an expert on the documents that establish work authorization but must make a good faith effort to ensure that their employees are legally permitted to work in the United States. An employer should accept documents that "reasonably appear on their face to be genuine and to relate to the person presenting them."[42] Additionally, employers are required to retain I-9 forms for a period of at least three years from the date of hire or one year after the employee is no longer employed, whichever is longer.

Employers who fail to comply with the laws regarding the hiring of foreign workers risk criminal and civil sanctions. Civil penalties begin with a minimum of $375 per unauthorized worker for the first offense and range up to $3,000 or more per unauthorized worker for employers who engage in a "pattern and practice" of hiring undocumented workers. These employers may also be imprisoned for up to six months.

Labor Laws and Unions

LO 24-9 List the three major pieces of labor legislation.

Many people would argue that the laws that improved workers' conditions most significantly were not any of those previously discussed but, rather, were the laws that gave employees the right to organize and bargain collectively over wages and terms and conditions of employment. During the Great Depression, workers were first given the right to organize, and since that time, unionization rates have varied significantly. During the post-World War II period, more than one-third of American workers were organized. In 2016, the percentage of workers belonging to a union in the United States was 10.7 percent.[43]

[38] 8 U.S.C. section 1324a – Unlawful Employment of Aliens.

[39] You can find Form I-9 on the U.S. Citizenship and Immigration Services (UCSIS) website: http://uscis.gov/forms.

[40] The Employer's Handbook may be viewed at https://www.uscis.gov/sites/default/files/files/form/m-274.pdf.

[41] Legal Pitfalls of Hiring Undocumented Immigrants, located at http://www.nolo.com/legal-encyclopedia/legal-pitfalls-hiring-undocumented-immigrants.html.

[42] Ibid.

[43] Bureau of Labor Statistics (January, 2016), https://www.bls.gov/news.release/pdf/union2.pdf.

Employment Eligibility Verification
Department of Homeland Security
U.S. Citizenship and Immigration Services

USCIS Form I-9
OMB No. 1615-0047
Expires 08/31/2019

▶ **START HERE:** Read instructions carefully before completing this form. The instructions must be available, either in paper or electronically, during completion of this form. Employers are liable for errors in the completion of this form.

ANTI-DISCRIMINATION NOTICE: It is illegal to discriminate against work-authorized individuals. Employers **CANNOT** specify which document(s) an employee may present to establish employment authorization and identity. The refusal to hire or continue to employ an individual because the documentation presented has a future expiration date may also constitute illegal discrimination.

Section 1. Employee Information and Attestation *(Employees must complete and sign Section 1 of Form I-9 no later than the **first day of employment**, but not before accepting a job offer.)*

Last Name (Family Name)	First Name (Given Name)	Middle Initial	Other Last Names Used (if any)

Address (Street Number and Name)	Apt. Number	City or Town	State	ZIP Code

Date of Birth (mm/dd/yyyy)	U.S. Social Security Number	Employee's E-mail Address	Employee's Telephone Number

I am aware that federal law provides for imprisonment and/or fines for false statements or use of false documents in connection with the completion of this form.

I attest, under penalty of perjury, that I am (check one of the following boxes):

☐ 1. A citizen of the United States

☐ 2. A noncitizen national of the United States *(See instructions)*

☐ 3. A lawful permanent resident (Alien Registration Number/USCIS Number): _____

☐ 4. An alien authorized to work until (expiration date, if applicable, mm/dd/yyyy): _____
Some aliens may write "N/A" in the expiration date field. *(See instructions)*

*Aliens authorized to work must provide only one of the following document numbers to complete Form I-9:
An Alien Registration Number/USCIS Number OR Form I-94 Admission Number OR Foreign Passport Number.*

1. Alien Registration Number/USCIS Number: _____
OR
2. Form I-94 Admission Number: _____
OR
3. Foreign Passport Number: _____
 Country of Issuance: _____

QR Code - Section 1
Do Not Write In This Space

Signature of Employee	Today's Date (mm/dd/yyyy)

Preparer and/or Translator Certification (check one):

☐ I did not use a preparer or translator. ☐ A preparer(s) and/or translator(s) assisted the employee in completing Section 1.
(Fields below must be completed and signed when preparers and/or translators assist an employee in completing Section 1.)

I attest, under penalty of perjury, that I have assisted in the completion of Section 1 of this form and that to the best of my knowledge the information is true and correct.

Signature of Preparer or Translator	Today's Date (mm/dd/yyyy)

Last Name (Family Name)	First Name (Given Name)

Address (Street Number and Name)	City or Town	State	ZIP Code

 Employer Completes Next Page

Form I-9 11/14/2016 N

588 Part 8 Government Regulation

Employment Eligibility Verification
Department of Homeland Security
U.S. Citizenship and Immigration Services

USCIS Form I-9
OMB No. 1615-0047
Expires 08/31/2019

Section 2. Employer or Authorized Representative Review and Verification

(Employers or their authorized representative must complete and sign Section 2 within 3 business days of the employee's first day of employment. You must physically examine one document from List A OR a combination of one document from List B and one document from List C as listed on the "Lists of Acceptable Documents.")

Employee Info from Section 1	Last Name *(Family Name)*	First Name *(Given Name)*	M.I.	Citizenship/Immigration Status

List A Identity and Employment Authorization	OR	List B Identity	AND	List C Employment Authorization
Document Title		Document Title		Document Title
Issuing Authority		Issuing Authority		Issuing Authority
Document Number		Document Number		Document Number
Expiration Date *(if any)(mm/dd/yyyy)*		Expiration Date *(if any)(mm/dd/yyyy)*		Expiration Date *(if any)(mm/dd/yyyy)*
Document Title		Additional Information		QR Code - Sections 2 & 3 Do Not Write In This Space
Issuing Authority				
Document Number				
Expiration Date *(if any)(mm/dd/yyyy)*				
Document Title				
Issuing Authority				
Document Number				
Expiration Date *(if any)(mm/dd/yyyy)*				

Certification: I attest, under penalty of perjury, that (1) I have examined the document(s) presented by the above-named employee, (2) the above-listed document(s) appear to be genuine and to relate to the employee named, and (3) to the best of my knowledge the employee is authorized to work in the United States.

The employee's first day of employment *(mm/dd/yyyy)* : _____ *(See instructions for exemptions)*

Signature of Employer or Authorized Representative	Today's Date *(mm/dd/yyyy)*	Title of Employer or Authorized Representative	
Last Name of Employer or Authorized Representative	First Name of Employer or Authorized Representative	Employer's Business or Organization Name	
Employer's Business or Organization Address (Street Number and Name)	City or Town	State	ZIP Code

Section 3. Reverification and Rehires *(To be completed and signed by employer or authorized representative.)*

A. New Name *(if applicable)*			**B.** Date of Rehire *(if applicable)*
Last Name *(Family Name)*	First Name *(Given Name)*	Middle Initial	Date *(mm/dd/yyyy)*

C. If the employee's previous grant of employment authorization has expired, provide the information for the document or receipt that establishes continuing employment authorization in the space provided below.

Document Title	Document Number	Expiration Date *(if any) (mm/dd/yyyy)*

I attest, under penalty of perjury, that to the best of my knowledge, this employee is authorized to work in the United States, and if the employee presented document(s), the document(s) I have examined appear to be genuine and to relate to the individual.

Signature of Employer or Authorized Representative	Today's Date *(mm/dd/yyyy)*	Name of Employer or Authorized Representative

Form I-9 11/14/2016 N

LISTS OF ACCEPTABLE DOCUMENTS
All documents must be UNEXPIRED

Employees may present one selection from List A
or a combination of one selection from List B and one selection from List C.

LIST A Documents that Establish Both Identity and Employment Authorization	LIST B Documents that Establish Identity	LIST C Documents that Establish Employment Authorization
1. U.S. Passport or U.S. Passport Card	1. Driver's license or ID card issued by a State or outlying possession of the United States provided it contains a photograph or information such as name, date of birth, gender, height, eye color, and address	1. A Social Security Account Number card, unless the card includes one of the following restrictions: (1) NOT VALID FOR EMPLOYMENT (2) VALID FOR WORK ONLY WITH INS AUTHORIZATION (3) VALID FOR WORK ONLY WITH DHS AUTHORIZATION
2. Permanent Resident Card or Alien Registration Receipt Card (Form I-551)		
3. Foreign passport that contains a temporary I-551 stamp or temporary I-551 printed notation on a machine-readable immigrant visa	2. ID card issued by federal, state or local government agencies or entities, provided it contains a photograph or information such as name, date of birth, gender, height, eye color, and address	2. Certification of Birth Abroad issued by the Department of State (Form FS-545)
4. Employment Authorization Document that contains a photograph (Form I-766)		
	3. School ID card with a photograph	3. Certification of Report of Birth issued by the Department of State (Form DS-1350)
5. For a nonimmigrant alien authorized to work for a specific employer because of his or her status: a. Foreign passport; and b. Form I-94 or Form I-94A that has the following: (1) The same name as the passport; and (2) An endorsement of the alien's nonimmigrant status as long as that period of endorsement has not yet expired and the proposed employment is not in conflict with any restrictions or limitations identified on the form.	4. Voter's registration card	4. Original or certified copy of birth certificate issued by a State, county, municipal authority, or territory of the United States bearing an official seal
	5. U.S. Military card or draft record	
	6. Military dependent's ID card	
	7. U.S. Coast Guard Merchant Mariner Card	
	8. Native American tribal document	5. Native American tribal document
	9. Driver's license issued by a Canadian government authority	6. U.S. Citizen ID Card (Form I-197)
	For persons under age 18 who are unable to present a document listed above:	7. Identification Card for Use of Resident Citizen in the United States (Form I-179)
6. Passport from the Federated States of Micronesia (FSM) or the Republic of the Marshall Islands (RMI) with Form I-94 or Form I-94A indicating nonimmigrant admission under the Compact of Free Association Between the United States and the FSM or RMI	10. School record or report card	8. Employment authorization document issued by the Department of Homeland Security
	11. Clinic, doctor, or hospital record	
	12. Day-care or nursery school record	

Examples of many of these documents appear in Part 8 of the Handbook for Employers (M-274).

Refer to the instructions for more information about acceptable receipts.

May an Employee Be Fired for Using Medical Marijuana?

Roe v. Teletech Customer Care Management (COLO.) LLC Washington Court of Appeal

257 P.3d 586 (Wash. 2011)

Jane Roe suffered from debilitating migraine headaches that caused chronic pain, nausea, blurred vision, and sensitivity to light. Roe took over-the-counter pain medication and prescription drugs for her headaches, but she claims conventional medications did not provide significant relief. In 2006, Roe received a prescription for lawful medical marijuana pursuant to Washington state's Medical Use of Marijuana Act (MUMA). A few months later, TeleTech offered Roe a position as a customer service representative, incumbent upon her passing a drug test. Roe informed TeleTech of her use of medical marijuana and accepted the position. Two weeks after taking the drug test, Roe was fired because of the positive result from the drug test.

Roe sued TeleTech, alleging wrongful termination in violation of the public policy. Roe argues MUMA proclaims a broad policy that the medical use of marijuana by patients with terminal or debilitating illnesses is a personal, individual decision. TeleTech argued MUMA has a narrow purpose—namely, to provide users and physicians with an affirmative defense under state drug laws, not to broadly entitle users to employment protections. The trial court granted motion for summary judgment in favor of defendant TeleTech. The appellate court affirmed the decision, holding that MUMA does not prohibit an employer from discharging an employee for medical marijuana use, nor does it provide a civil remedy against the employer. It further held that MUMA also does not proclaim a sufficient public policy to give rise to a tort action for wrongful termination for authorized use of medical marijuana.

Sources: Steven Musil, "Company Settles Facebook Firing Case," CNet.com, February 7, 2011, http://news.cnet.com/8301-1023_3-20030955-93.html; see also Erline Aguiluz, "AMR Settles Dawnmarie Souza's Wrongful Termination Lawsuit," PhiladelphiaEmploymentLawNews.com, February 11, 2011, http://philadelphiaemploymentlawnews.com/2011/02/amr-settles-dawnmarie-souzas-wrongful-termination-lawsuit.html.

Not all occupations are equally organized. The union membership rate for public-sector workers (34.4%) was substantially higher than the rate for private-sector workers (6.4%). Within the public sector, the union membership rate was highest for local government (40.3%), which includes employees in heavily unionized occupations such as teachers, police officers, and firefighters. In the private sector, industries with high unionization rates included utilities (21.56%), transportation and warehousing (18.4%), telecommunications (14.6%), and construction (13.9%).[44]

Labor-management relations in the United States today are governed by three major pieces of legislation:

1. Wagner Act of 1935
2. Taft-Hartley Act of 1947
3. Landrum-Griffin Act of 1959

The Taft-Hartley Act amended the Wagner Act, and they are jointly referred to as the *National Labor Relations Act (NLRA)*.

The Wagner Act of 1935

collective bargaining The process whereby workers organize collectively and bargain with employers regarding the workplace.

The first major piece of federal legislation adopted explicitly to encourage the formation of labor unions and provide for **collective bargaining** between employers and unions as a means of

[44] Bureau of Labor Statistics, https://www.bls.gov/news.release/union2.nr0.htm.

obtaining the peaceful settlement of labor disputes was the **Wagner Act**. Collective bargaining "consists of negotiations between an employer and a group of employees so as to determine the conditions of employment."[45]

Wagner Act
The first major piece of federal legislation adopted explicitly to encourage the formation of labor unions and provide for collective bargaining between employers and unions as a means of obtaining the peaceful settlement of labor disputes.

CASE Nugget

May an Employee Be Fired for Trashing Her Employer on Facebook?

Fair Labor Standards Board [on behalf of Dawnmarie Souza] v. American Medical Response of Connecticut Unpublished

Case settled by unpublished agreement between parties, February 2011

Dawnmarie Souza was employed by American Medical Response (AMR) of Connecticut. A customer complained about her work, and she was required to prepare a response. Her supervisor refused to permit a representative of the Teamsters, the union representing the company's workers, to help prepare her response. Souza then mocked her supervisor on Facebook, reportedly using several vulgarities to ridicule him. According to copies of Souza's Facebook posts obtained by CNET, she called her supervisor a "dick" in one and "a scumbag" in another. The Connecticut ambulance company fired Souza on December 1, 2009. The National Labor Relations Board (NLRB) sued AMR of Connecticut on October 27, 2010, claiming the employee, Dawnmarie Souza, was illegally fired and denied union representation.

The closely watched case touched on whether employers have the right to discipline employees for comments they make on social-networking sites. The NLRB complaint said that Souza "engaged in concerted activities with other employees by criticizing respondent's supervisor . . . on her Facebook page" on November 8, 2009. In response to the NLRB complaint, AMR claimed Souza's comments were not protected activity. However, the NLRB contended that AMR's termination of Souza's employment violated the National Labor Relations Act, which allows employees to discuss the terms and conditions of their employment with co-workers and others.

In February 2011, the case was settled. Under the terms of the settlement, AMR agreed to revise "overly broad rules" in the employee handbook regarding how employees can communicate on the Internet and with co-workers regarding their work conditions. According to the Fair Labor Standards Board (FLSB), the company also agreed not to discipline employees for requesting union representation. The allegations regarding Souza's discharge were resolved in a separate, private agreement between Souza and AMR. Financial terms were not revealed.

Sources: Steven Musil, "Company Settles Facebook Firing Case," CNet.com, February 7, 2011, http://news.cnet.com/8301-1023_3-20030955-93.html; see also Erline Aguiluz, "AMR Settles Dawnmarie Souza's Wrongful Termination Lawsuit," PhiladelphiaEmploymentLawNews.com, February 11, 2011, http://philadelphiaemploymentlawnews.com/2011/02/amr-settles-dawnmarie-souzas-wrongful-termination-lawsuit.html.

The Wagner Act also created an administrative agency, the *National Labor Relations Board (NLRB)*, to interpret and enforce the NLRA. Finally, the act provides for judicial review in designated federal courts of appeal.

The Taft-Hartley Act of 1947

The 12 years between passage of the Wagner Act and that of the Taft-Hartley Act saw a huge growth in unionization, which resulted in an increase in workers' power. Public perception of this trend led

[45] Legal Information Institute, www.law.cornell.edu/topics/collective_bargaining.html.

Taft-Hartley Act
Federal legislation designed to curtail some of the powers that unions had acquired under the Wagner Act; designates certain union actions as unfair. Also called *Labor-Management Relations Act.*

Landrum-Griffin Act
Federal legislation that primarily governs the internal operations of labor unions. It requires financial disclosures by unions, establishes penalties for financial abuses by union officials, and includes "Labor's Bill of Rights" to protect employees from their own unions.

National Labor Relations Board (NLRB)
An administrative agency created by the Wagner Act to interpret and enforce the National Labor Relations Act.

to the passage of the Taft-Hartley Act, also known as the Labor-Management Relations Act, which was designed to curtail some of the powers the unions had acquired under the Wagner Act.

The Landrum-Griffin Act of 1959

The Landrum-Griffin Act primarily governs the internal operations of labor unions. This act, which was a response to evidence of certain undesirable internal labor union practices, requires certain financial disclosures by unions and establishes civil and criminal penalties for financial abuses by union officials. "Labor's Bill of Rights," contained in the act, protects employees from their own unions.

The National Labor Relations Board

The National Labor Relations Board (NLRB) is the administrative agency that interprets and enforces the *National Labor Relations Act (NLRA)*. The NLRB's three primary functions are to:

1. Monitor the conduct of the employer and the union during an election to determine whether workers want to be represented by a union.
2. Prevent and remedy unfair labor practices by employers or unions.
3. Establish rules interpreting the NLRA.

The NLRB has jurisdiction over all employees *except* those who work in federal, state, and local government and those covered by the Railway Labor Act (employees in the transportation industry); independent contractors; agricultural workers; household domestics; persons employed by a spouse or parent; and supervisors, managerial employees, and confidential employees. In the preceding Case Nugget, the NLRB brought a lawsuit against a company after the company fired an employee for posting negative comments about a supervisor on Facebook.

The Collective Bargaining Process

Once the union has been certified, union and management must begin to bargain in good faith about wages, hours, and other terms and conditions of work. Note that the NLRB can order only the parties to bargain in good faith; it cannot order them to reach an agreement with respect to any contract term.

The NLRA requires both the employer and the bargaining-unit representative to bargain collectively in good faith with respect to wages, hours, and other terms and conditions of employment.

CASE OPENER WRAP-UP

Brad Gets Fired from "So Clean!"

At the beginning of this chapter, you were confronted with the situation between Brad and Jennifer. By now you should be able to answer all the questions presented to you.

In many states, an employee can legally be fired on the basis of sexual orientation. Discrimination in this area is based solely on state law. There is no federal protection against discrimination based on being gay. Similarly, firing an employee for smoking (including off the job) is also a state law issue. In Michigan, for example, such a firing would be legal. Many states are now passing laws preventing employers from firing those who smoke outside the workplace.

Brad is an at-will employee, but that does not mean that he can be fired for an illegal (i.e., discriminatory) reason. Laws protecting employees against sex discrimination and sexual harassment are just as applicable to men as they are to women. Anyone who is treated in a discriminatory way "based on sex" may sue under the appropriate state or federal antidiscrimination laws. Most, though not all, states have their own state laws against discrimination and harassment. States may give more protection than federal laws but not less protection. There are still a few states that have no state laws against employment discrimination.

PRACTICAL TIPS for Business Managers

1. Employers should have privacy-related policies or handbooks for employees that are reviewed for compliance with the NLRA and state law.
2. Employee handbooks and policies should have a "disclaimer" that is sufficient to assure the employee's status remains at-will.
3. All policies and handbooks should be reviewed annually and updated by an employment law attorney.

SUMMARY

When May an Employee Be Fired?	*At-will employment* means that any employee who is not employed under a contract or a collective bargaining agreement may quit at any time for any reason or no reason at all, with no required notice to the employer. Moreover, the employer may fire the employee at any time, with no notice, for almost any reason.
Federal Employment Discrimination Laws Governing Employers	Federal law may be described as a minimum level of protection for all workers. State laws may give employees more, but not less, protection than federal laws. Some of the most important federal employment discrimination laws are the Civil Rights Act (CRA) of 1964— Title VII (as amended by the Civil Rights Act of 1991); Pregnancy Discrimination Act (PDA) of 1987; Age Discrimination in Employment Act (ADEA) of 1967; Americans with Disabilities Act; and Equal Pay Act of 1963.
Civil Rights Act-Title VII	Title VII of CRA (1964, as amended by the Civil Rights Act of 1991) protects employees against discrimination based on race, color, religion, national origin, and sex. It also prohibits harassment based on the same protected categories. Defenses to a charge of discrimination under Title VII include *merit, seniority system,* and *bona fide occupational qualification (BFOQ).*

- *Disparate treatment:* If the employee has been hired, fired, denied a promotion, and so on, based on membership in a protected class under Title VII, this is a form of intentional discrimination and qualifies the employee to sue for disparate-treatment discrimination.
- *Disparate impact:* Disparate-impact cases arise when a plaintiff attempts to establish that although an employer's policy or practice appears to apply to everyone equally, its actual effect is that it disproportionately limits employment opportunities for a protected class.
- *Sexual harassment:* Sexual harassment includes unwelcome sexual advances, requests for sexual favors, and other verbal or physical conduct of a sexual nature that implicitly or explicitly makes submission a term or condition of employment; makes employment decisions related to the individual dependent on submission to or rejection of such conduct; or has the purpose or effect of creating an intimidating, hostile, or offensive work environment. Two recognized forms are *hostile environment* and *quid pro quo harassment.*

Pregnancy Discrimination Act of 1987: PDA amended Title VII of CRA, expanding the definition of sex discrimination to include discrimination based on pregnancy.

Equal Employment Opportunity Commission: The EEOC is the federal agency charged with overseeing Title VII of the Civil Rights Act. Administrative complaints must be filed with the EEOC (or an equivalent state agency) prior to filing a lawsuit for discrimination or harassment.

Discrimination Based on Sexual Orientation— Actionable?	No federal legislation currently prohibits discrimination based on sexual orientation. What does exist are individual state laws that prohibit such discrimination. State laws prohibiting discrimination based on sexual orientation exist in 21 states and the District of Columbia.
Discrimination Based on Marriage	In 2015 the U.S. Supreme Court held that marriage is a fundamental right for all and struck down state laws forbidding same-sex marriage.
Age Discrimination in Employment Act of 1967	ADEA prohibits employers from refusing to hire, discharging, or discriminating in terms and conditions of employment on the basis of an employee or applicant being age 40 or older.

(continued)

Americans with Disabilities Act	ADA prohibits discrimination against employees and job applicants with disabilities.
May an Employer Discriminate against a Smoker?	Some companies either won't hire smokers or are threatening to fire current employees who will not or are unable to quit smoking. Many states have passed laws preventing companies from engaging in such action.
Equal Pay Act of 1963	EPA prohibits an employer from paying workers of one gender less than the wages paid to employees of the opposite gender for work that requires equal skill, effort, and responsibility.
Additional Laws Governing the Employment Relationship	*Fair Labor Standards Act:* Employers must follow federal minimum-wage and hour laws. FLSA covers all employers engaged in interstate commerce or the production of goods for interstate commerce and requires a minimum wage of a specified amount to be paid to all employees in covered industries. Congress periodically raises the specified amount to compensate for increases in the cost of living caused by inflation. The current minimum wage is $7.25. *Family and Medical Leave Act:* The FMLA covers all public employers, as well as private employers with 50 or more employees. It guarantees all eligible employees (those who have worked at least 25 hours a week for each of 12 months prior to the leave) up to 12 weeks of unpaid leave during any 12-month period for any of the following family-related occurrences: the birth of a child; the adoption of a child; the placement of a foster child in the employee's care; the care of a seriously ill spouse, parent, or child; or a serious health condition that renders the employee unable to perform any of the essential functions of his or her job. *Unemployment compensation:* The Federal Unemployment Tax Act (FUTA) created a state system that provides unemployment compensation to qualified employees who lose their jobs. *Workers's compensation laws:* State laws provide for financial compensation to employees or their dependents when a covered employee is injured on the job. *Consolidated Omnibus Budget Reconciliation Act:* COBRA ensures that when employees lose their jobs or have their hours reduced to a level at which they would not be eligible to receive medical, dental, or optical benefits from their employer, the employees will be able to continue receiving benefits under the employer's policy for up to 18 months by paying the premiums for the policy. *Employee Retirement Income Security Act:* ERISA sets minimum standards for most voluntarily established pension and health plans in private industry to provide protection for individuals in these plans. *Occupational Safety and Health Act of 1970:* The Occupational Safety and Health Administration is responsible for setting safety standards under OSHA and enforcing the act through inspections and levying fines against violators.
Employee Privacy in the Workplace	Privacy issues are of increasing importance in the workplace. Privacy policies should cover matters such as employer surveillance policies, control of and access to medical and personnel records, drug testing, and email policies. *Omnibus Crime Control and Safe Streets Act of 1968:* Employers cannot listen to the private telephone conversations of employees or disclose the contents of these conversations. They may, however, ban personal calls and monitor calls for compliance as long as they discontinue listening to any conversation once they determine it is personal. Violators may be subject to fines of up to $10,000. *Electronic Communications Privacy Act of 1986:* Under ECPA, employees's privacy rights were extended to electronic forms of communication, including email and cellular phones. ECPA outlaws the intentional interception of electronic communications and the intentional disclosure or use of the information obtained through such interception.
Use of Social Media in Hiring Decisions	May an employer use social media in hiring decisions? The answer is yes—if it's done carefully. Probably the biggest concern is discrimination. A potentially little-known problem with using social media—particularly the most popular sites like Facebook and LinkedIn—is that the subscribers do not represent the U.S. or global job-applicant pools. Recent surveys show that social media sites have lower percentages of Latino and black users than are in the general population. The burden, then, is on the employer to widen their search for new applicants to include job boards, newspapers, magazines, and job fairs.

(continued)

(concluded)

Hiring Foreign Workers	In 1986, Congress passed the Immigration Reform and Control Act. This act is an amendment to the Immigration and Nationality Act (INA) and requires employers to verify the identity and eligibility of all individuals hired in the United States after November 6, 1986. If it is discovered that an employer knowingly employed undocumented immigrant workers, that employer will be in serious legal trouble. Employers who fail to comply with the laws regarding the hiring of foreign workers risk criminal and civil sanctions.
Labor Laws and Unions	*Wagner Act of 1935:* The Wagner Act was the first major piece of federal legislation adopted explicitly to encourage the formation of labor unions and provide for collective bargaining between employers and unions as a means of obtaining the peaceful settlement of labor disputes. *Collective bargaining:* Collective bargaining consists of negotiations between an employer and a group of employees to determine the conditions of employment. *National Labor Relations Board:* The Wagner Act created the NLRB, an administrative agency, to interpret and enforce the National Labor Relations Act (NLRA) and to provide for judicial review in designated federal courts of appeal. *Taft-Hartley Act of 1947:* Also known as the Labor-Management Relations Act, the Taft-Hartley Act was designed to curtail some of the powers the unions had acquired under the Wagner Act. Just as the Wagner Act designated certain employer actions as unfair, the Taft-Hartley Act designated certain union actions as unfair. *Landrum-Griffin Act of 1959:* The Landrum-Griffin Act primarily governs the internal operations of labor unions. It requires certain financial disclosures by unions and establishes civil and criminal penalties for financial abuses by union officials. "Labor's Bill of Rights," contained in the act, protects employees from their own unions.

Point/Counterpoint

Should Employers Be Permitted to Use Social Media in Hiring and/or Firing Decisions?	
YES	**NO**
An employer's reputation may be tarnished if its employees present a public image contrary to the company's values. Use of social media could lead to more information about a candidate's fit with the corporate culture, thereby assisting in better hires. An employee who displays his/her depth of knowledge and interest on his/her public profile might increase his/her chances of getting hired.	An employee's online presence is personal to that employee and should not be considered a reflection of the company's values. Use of social media often reveals personal information such as race, gender, religion, sexual orientation, and more, all which could lead to discrimination in the hiring and/or firing process. You can't assess someone's suitability for a role based on an online profile designed to be purely social.

Questions & Problems

1. Name five statutes that prohibit discrimination in employment.
2. Does Title VII apply to same-sex harassment? Explain your response.
3. What is the current amount of the federal minimum wage? When was the last time the federal minimum wage was increased?
4. What is required for an employee to be eligible for benefits under the Family and Medical Leave Act?
5. List the protected classes under the Civil Rights Act of 1964 (as amended in 1991).
6. Marykate Ellingsworth lives with her husband in Allentown, Pennsylvania. In 2012, she was hired by The Hartford (an insurance company) as a customer

service representative. After completing her initial training, she was placed on a work team supervised by Angela Ferrier. Ms. Ferrier allegedly harassed Ms. Ellingsworth in various ways over the span of approximately one year. Ferrier would tell Ellingsworth that she "dresses like a dyke." Ferrier would also make fun of Ellingsworth's clothing, call her "stupid," and tell her that she "sucks." In addition to ridiculing Ms. Ellingsworth directly, Ms. Ferrier would also tell her coworkers that Ellingsworth "dresses like a dyke" and has a "lesbian tattoo." Ferrier went so far as to tell Ellingsworth's coworkers that Ellingsworth was a lesbian. These remarks were made in private to Ms. Ellingsworth and also, at other times, in front of coworkers. Because of this persistent harassment, Ellingsworth's coworkers began to adopt the belief that Ellingsworth was gay, even though she is not. Eventually, it became "generally accepted" in the workplace that Ellingsworth was gay. Ellingsworth felt compelled to explain to her coworkers that she was not a lesbian. This situation began to exacerbate Ellingsworth's pre-existing depression and anxiety. Due to her anxiety and depression, she took a leave of absence beginning January 6, 2014. On March 24, 2014, The Hartford wrote her a letter stating that she could either return to work or be terminated. Believing that nothing would be done to resolve the harassment, Ms. Ellingsworth was unable to return to work. She claims she was constructively discharged. Two months later, on May 23, 2014, Ms. Ellingsworth filed an administrative complaint with the Pennsylvania Human Relations Committee ("PHRC"). This complaint was cross-filed with the EEOC on or after that date. Ms. Ellingsworth filed a complaint against her former employer, alleging sexual harassment, gender discrimination, and retaliation. Defendant filed a motion to dismiss, alleging Title VII does not permit a claim based on sexual orientation. How should the court rule? Why?

Source: *Marykate Ellingsworth, Plaintiff, v. Hartford Fire Insurance Company,* 247 F.Supp.3d 546 (2017).

7. In October 2005, Blatt was diagnosed with "Gender Dysphoria, also known as Gender Identity Disorder," which she argues substantially limits one or more of her major life activities, including, but not limited to, interacting with others, reproducing, and social and occupational functioning. Blatt alleges that shortly after she was hired by Cabela's in September 2006, Cabela's began to discriminate against her on the basis of her sex and her disability, in violation of Title VII of the Civil Rights Act and the ADA, and that Cabela's retaliated against her for opposing this discrimination, also in violation of these statutes. Blatt further alleges that in February 2007, Cabela's terminated her employment based on her sex and disability. The stated purpose of the ADA is to "provide a clear and comprehensive national mandate for the elimination of discrimination against individuals with disabilities." Congress opted to define the scope of the statute's coverage by means of a flexible and broad definition of "disability," namely, "a physical or mental impairment that substantially limits one or more major life activities of [an] individual." Standing in contrast to this broad definition of disability, there are a few exceptions to the ADA's coverage. The provision at issue in this case excludes from ADA coverage approximately one dozen conditions, including gender identity disorders. Cabela's contends that the statute's reference to gender identity disorders applies to Blatt's condition and that the provision therefore excludes her condition from the ADA's scope and that the court should grant defendant's motion to dismiss the claims. Should the court grant the defendant's motion to dismiss? Should gender identity disorder be considered a disability under the ADA?

Source: *Blatt v. Cabela's Retail, Inc.,* 2017 U.S. District LEXIS 75665 (Eastern District PA 2017).

8. The Pregnancy Discrimination Act added new language to the definitions subsection of Title VII of the Civil Rights Act of 1964. The first clause of the Pregnancy Discrimination Act specifies that Title VII's prohibition against sex discrimination applies to discrimination "because of or on the basis of pregnancy, childbirth, or [1341] related medical conditions." The act's second clause says that employers must treat "women affected by pregnancy . . . the same for all employment-related purposes . . . as other persons not so affected but similar in their ability or inability to work."

Petitioner Young was a part-time driver for respondent United Parcel Service (UPS). When she became pregnant, her doctor advised her that she should not lift more than 20 pounds. UPS, however, required drivers like Young to be able to lift up to 70 pounds. UPS told Young that she could not work while under a lifting restriction. Young subsequently filed this federal lawsuit, claiming that UPS acted unlawfully in refusing to accommodate her pregnancy-related lifting restriction. She brought only a disparate-treatment claim of discrimination. Young pointed to UPS policies that accommodated workers who were injured on the job, had disabilities covered by the Americans with Disabilities Act of 1990 (ADA), or had lost Department of Transportation (DOT) certifications. Pursuant to these policies, Young contended, UPS had accommodated several individuals whose disabilities created work restrictions similar to hers. She argued that these policies showed that UPS discriminated

against its pregnant employees because it had a light-duty-for-injury policy for numerous "other persons," but not for pregnant workers. UPS responded that, since Young did not fall within the on-the-job injury, ADA, or DOT categories, it had not discriminated against Young on the basis of pregnancy, but had treated her just as it treated all "other" relevant "persons." UPS brought a motion for summary judgment, which the trial court granted against Young. The decision was affirmed on appeal by the Fourth Circuit. The case was then appealed to the U.S. Supreme Court. How should the court decide? Should Young be given the opportunity to argue before a trial court that treating pregnancy differently from other disabilities is disparate treatment discrimination? Why or why not?

Source: *Young v. UPS,* 135 S. Ct. 1338 (2015).

9. Exxon maintained a corporate policy that prohibited its pilots from flying corporate aircraft after they reached the age of 60 and forced such pilots to retire involuntarily at age 60. This rule mirrored a rule the Federal Aviation Administration applied to pilots flying for commercial airlines. Based on the age 60 rule, Exxon forced at least three pilots to retire in 2006 and 2007. The EEOC brought suit on behalf of these pilots and others alleging age discrimination. Exxon argued that the requirement for pilots to be under age 60 is a *bona fide occupational qualification.* Do you agree? Why or why not?

Source: *EEOC v. Exxon Mobil Corp.,* 2012 U.S. Dist. LEXIS 183101.

10. The plaintiff, Aileen Rizo, is an employee of the public schools in Fresno County. After discovering that the County paid her less than her male counterparts for the same work, she brought this action under the Equal Pay Act, and the California Fair Employment and Housing Act. When the County moved for summary judgment, it conceded that it paid the plaintiff less than comparable male employees for the same work. However, it argued that this result was lawful because the pay differential was "based on any other factor other than sex," an affirmative defense to a claim under the Equal Pay Act. This other factor was prior salary. The court granted motion for summary judgment for the County. Plaintiff appeals from that decision. Is "prior salary" alone a sufficient defense to violation of the Equal Pay Act? Why or why not?

Source: *Rizo v. Yovino,* 854 F.3d 1161 (2017).

11. Lisa Harrison (Harrison) was hired by Defendant at Family House of Louisiana (Family House), a long-term residential treatment facility for chemically dependent women and their children, on November 23, 1999. She was hired as a Prevention/Intervention Specialist, and her job included overseeing a day care program for the children of mothers staying at Family House. At the time she was hired, Harrison weighed more than 400 pounds. On September 6, 2007, Harrison was terminated from her position at Family House. At the time of her termination, she weighed 527 pounds. On October 17, 2007, Harrison filed a charge of discrimination with the Equal Employment Opportunity Commission (EEOC), alleging that she had been terminated because Defendant regarded her as disabled due to her obesity. Is obesity a disability under the Americans with Disabilities Act?

Source: *EEOC v. Res. for Human Dev., Inc.,* 25 Am. Disabilities Cas. (BNA) 964 (2012).

12. Cynthia Anderson worked for the city of Columbus, Georgia, answering telephone calls from citizens. Anderson knew that calls from citizens were recorded, but she was unaware that, due to a glitch, the telephone system continued to record her statements through her headset after a call was terminated. Anderson made disparaging remarks about the city manager's office, and her employment was terminated. Anderson sued the city, alleging that it had no right to record her private conversations with co-workers, which were unrelated to city business. Who should prevail? Why?

Source: *Anderson v. City of Columbus, Georgia,* 2005 U.S. Dist. LEXIS 12612.

13. Meadows worked as an assistant manager at a Dollar General store. While she was ringing up an order for a customer, he became verbally abusive. After she finished ringing up the sale, the customer threw the bag, containing a can of motor oil, at Meadows, hitting her in the eye. Meadows suffered a detached retina. The trial court awarded Meadows permanent partial disability under workers' compensation for her injury. Dollar General appealed the decision, arguing that the injury did not occur in the course of employment. Do you believe that Meadows' injury occurred during the course of her employment? Why or why not? Was the trial court correct in granting her permanent partial disability, or did the appeals court overturn that decision?

Source: *Dollar General Corp. v. Meadows,* 63 P.3d 548 (2002 WL 31991909, Okla. Ct. Civ. App. 2002).

14. Several employees of Wal-Mart Stores Inc. (Wal-Mart) were involved in physical confrontations with shoplifting customers. Wal-Mart's policy requires employees to disengage and withdraw from potentially violent situations.

The first incident involved plaintiffs Derek Holt and Eric Hunter, who were employed at Wal-Mart's West Valley City, Utah, store. Holt and Hunter confronted a

shoplifter. When the shoplifter tried to run away, they grabbed her arms. During the ensuing struggle, the shoplifter pulled out a small pocketknife and shouted that she was going to stab Holt and Hunter if they did not let go. Holt and Hunter maintained their hold, however, and a customer helped pry the knife out of the shoplifter's hand. Wal-Mart terminated Holt's and Hunter's employment for violating the policy.

The second incident involved plaintiffs Shawn Ray, Lori Poulsen, and Gabriel Stewart, who were employed at Wal-Mart's Layton, Utah, store. Several employees at that store, including Ray and Poulsen, approached a customer who was attempting to steal a laptop by concealing it in his pants and escorted him to the store's asset protection office, where they were joined by Stewart. There is some discrepancy regarding what happened next. According to Wal-Mart, the customer placed the laptop on a desk and stated, "You have your laptop, I am now going to leave, and I have something I am not supposed to have." Poulsen saw the customer move a gun from his back to his coat pocket. A physical struggle ensued, resulting in the Wal-Mart employees pinning the customer against a wall and grabbing the gun. The employees's account of the incident differs somewhat. According to them, after the customer removed the laptop from his pants he said, "I have something I shouldn't have. Don't make me do this!" Poulsen noticed the customer had a gun and yelled "Gun! Hand!" The customer rushed towards the door but then turned and shoved Stewart against the wall and pressed the gun to his back. A skirmish resulted, and the Wal-Mart employees managed to remove the gun from the customer's hands and force him to the ground. Ultimately, Ray, Poulsen, and Stewart were all fired following the incident for violating the policy.

The employees sued Wal-Mart in federal district court for wrongful termination, arguing that terminating a person's employment for exercising self-defense in the workplace violates Utah public policy. Is the right of self-defense a public policy exception to "at will" employment? In other words, were the employees within their rights to defend themselves? Or should Wal-Mart be permitted to terminate the employees for violation of the policy?

Source: *Ray v. Wal-Mart Stores, Inc.,* 2015 UT 83 (2015).

15. Anthony Romeo was an openly gay student at Seton Hall, a Catholic university. He claimed that he elected to attend Seton Hall in part because of its published anti-discrimination policy. In 2003, he applied to the Seton Hall University Department of Student Affairs for provisional recognition of a gay and lesbian student organization. Dr. Laura Wankel, vice president of student affairs at Seton Hall, responded to Romeo's application in a letter, stating in pertinent part: "I am informing you that your application for provisional recognition has been denied. No organization based solely upon sexual orientation may receive formal University recognition." Wankel went on to say that "the Division of Student Affairs remains prepared to work with gay and lesbian students to meet their needs. I am committed to working collaboratively with you and other students in fostering a positive, safe and caring community." Wankel then made a modest proposal giving the group some rights. The proposal, however, was not satisfactory to the students, and Romeo filed his complaint alleging violations of New Jersey's law against discrimination. He also cited Seton Hall's nondiscrimination policy: "No person may be denied employment or related benefits or admission to the University or to any of its programs or activities, either academic or nonacademic, curricular or extracurricular, because of race, color, religion, age, national origin, gender, sexual orientation, handicap and disability, or veteran's status." How should the court rule? May a Catholic university prevent recognition of a gay and lesbian student organization? Does the university's nondiscrimination policy make a difference? Why or why not?

Source: *Romeo v. Seton Hall University,* 2005 N.J. Super. LEXIS 197.

Part 8 Government Regulation

CHAPTER 25

Consumer Law

CASE OPENER

Deceptive Advertising and the Ultimate Weight Loss Cure

Defendant Trudeau acted in several infomercials promoting his product, the *Weight Loss Cure* book. Trudeau's weight loss plan consisted of four phases. These phases included instructions such as calorie intake, food restrictions, regular colonics, and avoiding skin creams. Above all else, the defendant's weight loss plan instructed dieters always to take daily doses of coral calcium. The defendant also claimed that his weight loss method had been suppressed from the mainstream by food and restaurant companies and government agencies.

The Federal Trade Commission (FTC) sought to hold Trudeau in contempt for violating a 2004 Consent Order command that he must not misrepresent the content of his book in his infomercials. The court ruled in favor of the FTC, finding that nowhere in Trudeau's infomercials did he mention colonics, organ cleanses, eating only organic food, and calorie restrictions. The court found that Trudeau's statements on the infomercials misled consumers and violated the Consent Order.

1. What did the court look for to determine whether Trudeau misled consumers?
2. How was Trudeau sanctioned by the court?

LEARNING OBJECTIVES

After reading this chapter, you will be able to answer the following questions:

LO 25-1 Recall the function of the Federal Trade Commission.
LO 25-2 Explain how the Federal Trade Commission determines what constitutes deceptive advertising.
LO 25-3 Recall the purpose of the federal laws that regulate product labeling and packaging.
LO 25-4 Describe the different acts that provide credit protection.
LO 25-5 Describe the different laws that help ensure consumer health and safety.

Consumers buy products and services from sellers every day. In some instances, however, consumers do not have as much power in the transaction as the seller has. As we see in the Trudeau case, the book author had much more knowledge about the product he was selling than did the consumers. Because Congress recognized the opportunities for sellers to take advantage of buyers in this way, it created laws that regulate transactions between consumers and sellers. A **consumer law** is a statute or administrative rule serving to protect consumer interests.

> **consumer law**
> A statute or administrative rule serving to protect consumer interests.

Various state and federal consumer laws protect consumers from unfair trade practices of sellers as well as from unsafe products. Although the laws differ among the states, many of the state laws provide consumer protection exceeding that guaranteed by federal law. This chapter explores a range of consumer laws concerning deceptive advertising, product labeling, sales procedures, health and product safety, and consumer credit. First, however, it discusses a federal agency that is one of the most important creators and enforcers of consumer protection laws—the Federal Trade Commission.

The Federal Trade Commission

LO 25-1 Recall the function of the Federal Trade Commission.

Congress created the Federal Trade Commission (FTC) through the Federal Trade Commission Act of 1914.[1]

The FTC is an independent federal agency with five commissioners appointed by the president and confirmed by the Senate. Each commissioner serves a seven-year term. The president chooses one commissioner to serve as chair of the FTC.

How does the FTC meet its goal of protecting consumers? It does so through two methods: (1) consumer education and (2) legal action. First, the FTC creates campaigns to educate consumers about laws that protect them. Second, the FTC educates businesses to help them comply voluntarily with consumer laws. For example, the FTC creates **industry guides,** interpretations of consumer laws, to encourage businesses to stop unlawful behavior. When businesses follow the FTC guidelines, they can cut potentially steep costs associated with violating consumer laws.

> **industry guides**
> Interpretations of consumer laws created by the FTC to encourage businesses to stop unlawful behavior.

How the FTC Brings an Action

The FTC receives a variety of complaints about businesses from consumer groups and individuals. When a consumer files a complaint with the FTC, he or she triggers a chain of events that could lead to an FTC action against the violator. The FTC typically begins a nonpublic investigation of the company.

If, after its investigation, the FTC believes that a company violated the law, the FTC sends a complaint to the alleged violator. At that time, the FTC may settle the complaint through a consent order with the company. A **consent order** is a statement in which the company agrees to stop the disputed behavior but does not admit that it broke the law. If the company violates the consent order, it will usually be forced to pay a fine.

> **consent order**
> A statement in which a company agrees to stop disputed behavior but does not admit that it broke the law.

If the company refuses to enter into a consent agreement, the FTC may then decide to issue a formal administrative complaint. Issuance of this complaint leads to a hearing before an administrative law judge. If the judge decides that the company has violated the law, the FTC issues a **cease-and-desist order,** requiring the company to stop the illegal behavior. However, the company may appeal this decision to the five commissioners. If the commissioners uphold the ruling, the company may appeal to the U.S. court of appeals and, finally, to the Supreme Court. If the courts uphold the FTC's decision, the company must follow the cease-and-desist order. If the company violates the order, the FTC can seek an injunction against the company or fine the company up to $10,000 per violation.

> **cease-and-desist order**
> An FTC order requiring that a company stop its illegal behavior.

[1] 15 U.S.C. §§41–58.

> **BUT WHAT IF? . . .**
>
> WHAT IF THE FACTS OF THE CASE OPENER WERE DIFFERENT?
>
> Let's say in the Case Opener that the FTC investigated Trudeau and decided that he was misrepresenting his products and possibly harming people. They sent him a complaint letter. What are different options the two parties can take to resolve the issue?

Trade Regulation Rules

Bringing an action may be effective in protecting consumers from the activities of one company, but how can the FTC protect consumers if most companies within one industry are using the same unfair or deceptive practices? Bringing actions against all of these companies would be costly and time-consuming.

An alternative method of addressing these practices is through trade regulation rules. If the FTC finds that deception is pervasive in an industry, it can recommend an administrative rule that has the effect of law. For example, while the marketing industry may be a substantial and mostly legitimate field, it is an industry where the power of deception can be easily abused. Thus, the ways in which a company can market its products is heavily regulated by the FTC. For example, a growing trend within the marketing industry has been to pass off paid advertisements as regular news stories or social media posts. The FTC is concerned that people will believe these posts to be legitimate news stories or recommendations, and thus has set rules requiring that paid posts clearly distinguish themselves as such. In April 2017, the FTC sent out over 90 letters to celebrities, athletes and other influencers on Instagram reminding them to clearly disclose their relationships to brands when endorsing products through social media. However, the FTC's power does not simply stop at recommendation letters; the FTC can bring legal action against those who violate FTC rules.

Deceptive Advertising

LO 25-2 Explain how the Federal Trade Commission determines what constitutes deceptive advertising.

Section 5 of the Federal Trade Commission Act prohibits deceptive and unfair acts in commercial settings, including consumer purchases. This section of the chapter specifically analyzes *deceptive advertising,* the practice of advertising with claims that mislead or could mislead a reasonable consumer. *Puffing,* the use of generalities and clear exaggerations, is permissible. Puffing tends to take the form of opinions and unverifiable claims about a product. For example, a pizza restaurant may claim that it has "the best pizza in the world." It is impossible to prove that the pizza is really the best, but it is also impossible to prove that it is not. Either way, the claim is so beyond belief that a reasonable consumer would not accept it as true. On the other hand, deceptive advertising tends to involve false claims regarding verifiable facts. For example, if a car manufacturer claims that a car gets 40 miles to the gallon when it really only gets 30 miles to the gallon, the manufacturer would be engaging in deceptive advertising. A car's gas mileage is a testable and verifiable fact that is not simply a matter of opinion.

The FTC decides whether an advertisement is deceptive on a case-by-case basis. Deceptive claims have three elements: (1) a *material* misrepresentation, omission, or practice that is (2) *likely to mislead* (3) a *reasonable* consumer.[2] When an advertised claim appears to be authentic but in fact is not, the advertising is deceptive. Although the FTC has the power to determine whether an advertisement is deceptive or just mere puffing, courts often have to make that determination as well. Sometimes companies seek injunctive relief for false advertising of their competitors under the Lanham Act. The Lanham Act was intended, in part, to protect persons engaged in commerce against false advertising and unfair competition.

[2] FTC, "Policy Statement on Deception," 1983.

GLOBAL Context

©Digital Archive Japan/ Alamy

Deceptive Advertising in China

Although puffery is allowed in advertising in the United States, not all countries universally accept puffery as an acceptable form of advertising. For example, China has greater restrictions on advertising than the United States. Problematic advertising claims in China fall largely into two groups: (1) unsubstantiated, clearly deceptive claims that are also unacceptable in the United States and (2) claims such as puffery, comparisons, and statistics, which are okay in the United States but are prohibited or restricted by Chinese law. Specifically, in China there is a statutory ban on puffery due to its deceptive nature. The ban has the potential to affect commerce and trade of global brands that have a standardized approach in their international advertising and do not take into careful consideration the Chinese regulations. In such cases, Chinese regulators of advertisements rarely accept the puffery defense.

However, the supreme court of China does exonerate commercial claims that employ obvious exaggerations but are unlikely to mislead the public, slightly contradicting the statutory ban on puffery (China Supreme People's Court, 2007). Because the terms *misleading* and *deceptive* are subjective, sometimes judges in China determine on a case-by-case basis what constitutes obvious exaggerations that are unlikely to mislead, just as U.S. judges often face this question. To help judges determine whether illegal puffing has occurred, the China Supreme People's Court advises the lower courts to consider factors such as daily life experience, the general understanding of the relevant public, people being misled, and the situation of the target consumer.

Source: Zhihong Gao, "Controlling Deceptive Advertising in China: An Overview," *Journal of Public Policy & Marketing*, Fall 2008, pp. 165–177.

half-truth
Information that is true but is not complete.

ad substantiation
An FTC standard requiring advertisers to have a reasonable basis for the claims made in their ads.

If the advertisement is a **half-truth**—that is, the information presented is true but incomplete—the advertiser is deceiving the consumer. To combat deceptive advertisements and half-truths, the FTC requires **ad substantiation**, requiring advertisers to have a reasonable basis for the claims made in their advertisements. As Exhibit 25-1 points out, many have argued that "Made in the USA" labels are deceptive advertising.

An illustration of an FTC claim of deceptive advertising involved Bayer HealthCare Pharmaceuticals in 2007. The FDA required Bayer to run corrective commercials that adjust assertions made in Bayer's original Yaz commercials. Federal laws state that drug advertising can promote only federally approved uses of a drug. Although the agency approved Yaz as a drug for birth control with a side value of treating premenstrual dysphoric disorder, the Yaz commercials implied that Yaz was a drug for acne and general mood problems. Bayer agreed in 2009 to run a $20 million marketing campaign for the next six years that was to be federally screened before being submitted for public viewing. The advertising of Yaz is a major concern because it is the leading oral contraceptive in the country. Sales of Yaz in 2008 totaled approximately $616 million.

Exhibit 25-1

"Made in the USA" Labels and the FTC

Many companies boast that their products are "Made in the USA." But what exactly does it mean to be made in the USA? The FTC is charged with setting forth standards to avoid deception and false advertising about products that are supposedly made in the United States.

In the past, the FTC stated that "Made in the USA" should not be used "unless all, or virtually all, of the components and labor are of U.S. origin." Thus, if a company assembled a product in the United States but shipped in some components from out of the country, the FTC would argue that "Made in the USA" should not be used.

The FTC's definition of the phrase was stricter than others's definitions. For example, NAFTA defines "Made in the USA" as a product for which at least 55 percent of the labor and components are from Canada, Mexico, or the United States. Customs draws the line at 50 percent.

In 1997, the FTC decided to continue enforcing the "all or virtually all" standard. Those with strong union ties had argued against any weaker definition, such as the 75 percent the FTC had been debating.

Source: Federal Trade Commission, FTC Consumer Report, "Complying with the Made in the USA Standard," www.ftc.gov/bcp/conline/pubs/buspubs/madeusa.shtm.

The existence of deceptive advertisements is not enough alone to prove damages for recovery when individual civil suits are filed. For example, in *Oliveira v. Amoco Oil Co.,* a class action suit was brought against Amoco for deceptive advertisements.[3] Amoco had made many claims over a seven-year period about the superiority of its premium gasoline. However, according to the plaintiffs, no scientific evidence supported Amoco's claims. The plaintiffs argued that the advertisements created a higher demand for Amoco gas, creating artificially high gas prices that hurt consumers regardless of specific reliance on the advertisements. The court rejected this argument, ruling that a market theory of causation is not enough to establish damages in an individual case. Moreover, reliance on the advertisements is a crucial element in establishing liability.

Case 25-1 provides a discussion of the questions involved in deciding whether a marketing claim is materially deceptive.

[3] 201 Ill. 2d 134 (2002).

CASE 25-1

GOTLIN v. LEDERMAN
United States Court of Appeals for the Second Circuit
483 FED. APPX. 583 (2012)

FACTS: Gary D. Gotlin, an administrator of the estate of a decedent, and the deceased's surviving spouse, Giuseppe Bono, alleged the misrepresentation of a particular form of cancer treatment, Fractionated Stereotactic Radiosurgery (FSR). The plaintiffs further asserted that this deceptive marketing led the decedent to "unnecessarily undergo an ineffective and harmful form of radiation therapy." According to the plaintiffs, the marketing of the cancer treatment, which included brochures, videos, advertisements, seminars, and Internet sites, made unrealistic claims about the treatment's success rates. Specifically, the defendants made deceiving claims that the FSR treatment had success rates of greater than 90 percent in treating pancreatic cancer. The district court dismissed the plaintiffs's claims, and the plaintiffs subsequently appealed.

ISSUE: Did the defendants engage in deceptive advertising when marketing their cancer treatment?

REASONING: New York business law states that deceptive practices are "those likely to mislead a reasonable consumer acting reasonably under the circumstances." The court reviewed the expert testimony provided in the lower court to determine whether the way in which the FSR treatment had been advertised was fraudulent or misleading. While reviewing the testimony of two medical experts, Judge Katzmann concluded that the therapy in question was not as highly successful as it had been marketed to be. The following excerpt from Judge Katzmann's court opinion highlights the evidence the judge relied on in reaching his conclusion:

> This claim as to FSR therapy's 94% success rate in treating pancreatic cancer is materially identical to claims made in defendants' marketing brochures. Moreover, while the brochures at one point define "success" in a relatively circumscribed manner, including cases in which the cancer stopped growing or shrunk but did not disappear altogether, at other points the brochures suggest that FSR treatment will yield much broader successes than merely arresting the growth of cancer (describing "possibilities never dreamt before," "superb results," "great effectiveness," and "superior outcomes").

In addition, Drs. Harrison and Gliedman's expert report states several times that FSR therapy was unnecessary, either because it had no "curative potential" with regard to a particular patient's circumstances or because the patient in question "presented with incurable disease" generally. Accordingly, in the opinion of Drs. Harrison and Gliedman, those patients had been "subjected to widespread radiation therapy without any chance of benefit." By making such statements, Drs. Harrison and Gliedman impliedly impugn the accuracy of defendants' brochure's representations that FSR therapy had achieved "superb results" in instances in which "normal radiation has not been successful." Importantly, Drs. Harrison and Gliedman did not merely represent that FSR treatment had not proven effective for the particular patients in question, but that defendants marketed FSR treatment as having "a very high rate of success," for "so-called 'hopeless cases,'" to patients who, in fact, had incurable cancer.

DECISION AND REMEDY: The court found that a genuine issue of material fact existed about whether the marketing of the FSR treatment's success rates was materially deceptive to a reasonable consumer and whether plaintiffs suffered injury as a result of the alleged misleading advertising. Consequently, the court vacated the district court's judgment and remanded.

SIGNIFICANCE OF THE CASE: Deceptive advertisements can lead consumers to make decisions that negatively affect their health. The court should be particularly vigilant in considering whether scientific evidence supports the contentions of medical advertisements.

(continued)

CRITICAL THINKING

What are the primary facts of this case? Is there any missing information you would call for to enable you to evaluate the court's reasoning better? What evidence does the court use to support its decision? Are you persuaded by this evidence?

ETHICAL DECISION MAKING

Do you think the company's advertising for the cancer treatment in this case was ethical, given the facts of the case? Review the WH process of ethical decision making in Chapter 2. What value did the court highlight in its decision?

Future business managers need to understand what kinds of advertising claims are permissible in our legal environment. To simplify this understanding for consumers and business owners, the FTC has classified certain types of deceptive or unfair advertising practices and created specific rules defining and prohibiting violations of these rules. The next section examines these classifications.

Bait-and-Switch Advertising

bait-and-switch advertising
A deceptive practice in which a seller advertises a low-priced item, generally unavailable to the consumer, and then pushes the consumer to buy a more expensive item.

When sellers advertise a low price for an item generally unavailable to the consumer and then push the consumer to buy a more expensive item, they are engaging in **bait-and-switch advertising**. The low advertised price baits the consumer. Then the salesperson switches the consumer to a higher-priced item. In 1968, the FTC prohibited bait-and-switch advertising.

According to the FTC's "Guides against Bait Advertising," a seller can engage in bait-and-switch advertising in several ways. For instance, the seller might advertise a low price but have too little of the advertised good in stock, or the seller might discourage employees from selling the advertised item. These bait-and-switch advertising techniques violate FTC rules.

CASE Nugget

Are "No-Annual-Fee" Offers Bait-and-Switch Advertising?

Paula E. Rossman, Individually and for All Others Similarly Situated v. Fleet Bank (R.I.) National Association et al. U.S. Court of Appeals for the Third Circuit

280 F.3d 384 (2002)

In 1999, Paula Rossman received a credit card offer from Fleet Bank, advertising the Fleet Platinum MasterCard with a low annual percentage rate and no annual fee. A few months after Rossman's account was opened, Fleet imposed an annual fee. Fleet sent a letter to Rossman stating that a $35 annual fee would be charged to her account annually on the anniversary of her account's opening; however, a second letter notified Rossman that the annual fee would be charged to her account within months. Rossman sued Fleet, claiming that Fleet had violated the disclosure requirements of the Truth in Lending Act (TILA) and engaged in a bait-and-switch advertising scheme. The district court dismissed Rossman's TILA claim for failing to state a claim on which relief could be granted. Rossman appealed to the Third Circuit Court of Appeals.

The appeals court applied a two-part test to determine whether Fleet's statement that the card had no annual fee was lawful. "First, it must have disclosed all of the information

(continued)

required by the statute. And second, it must have been true—that is, an accurate representation of the legal obligations of the parties at that time—when the relevant solicitation was mailed." Rossman claimed that Fleet was required to disclose all fees that were currently imposed as well as any fees that may be imposed later. Rossman also claimed that Fleet's disclosure was misleading by suggesting that there would never be an annual fee. Rossman further claimed that Fleet engaged in a bait-and-switch scheme, using the no-annual-fee provision to lure consumers into a contract when Fleet had no intention of honoring that provision.

Fleet argued that it was not required to disclose any future fees that may be imposed, but only those that currently existed. Fleet also turned to a clause in the solicitation disclosure insert that stated it "reserve[d] the right to change the benefit features associated with your Card at any time." The court found that Fleet's disclosure, as placed and worded, did not adequately link itself to the no-annual-fee provision. Thus, if the annual fee was permitted to be changed within the first annual term of the agreement, the court found that the statement "no annual fee" was an inadequate disclosure.

Turning its attention to Rossman's claim that Fleet had engaged in a bait-and-switch advertising scheme, the court did not see Fleet's behavior as resembling the classic bait-and-switch design. Ordinarily, consumers are baited with certain terms, but the switch for less enticing products or terms is made before the consumer enters into an agreement or contract. Rather, the court found Fleet's behavior to be even more egregious because Fleet actually bound the consumer in a contract before making the switch to charging an annual fee. Thus, Fleet had violated the provisions of TILA. The appeals court reversed the district court's ruling and remanded for proceedings.

FTC Actions against Deceptive Advertising

If the FTC takes actions against a company and proves the advertising is deceptive, the FTC may issue a cease-and-desist order. To go a step beyond cease-and-desist orders, the FTC may also issue multiple-product orders. A **multiple-product order** is a form of cease-and-desist order the FTC issues that applies not only to the product that was the subject of the action but also to other products produced by the same firm. Alternatively, the FTC may require the company to engage in **corrective advertising** (or *counteradvertising*), running advertisements in which the company explicitly states that the formerly advertised claims were untrue.

The rationale for corrective advertising is that consumers who saw the previous ads will see the new ads; the new ads will then correct the deceptive advertising. In a commission decision regarding the deceptive advertising of Doan's Pills, Commissioner Sheila Anthony described the characteristics of a situation requiring corrective advertising:

> Requiring the dissemination of a truthful message to counteract beliefs created or reinforced by a respondent's deceptive message is an appropriate method of restoring the *status quo ante* and denying a respondent the ability to continue to profit from its deception. . . . Corrective advertising is an appropriate remedy if (1) the challenged ads have substantially created or reinforced a misbelief; and (2) the misbelief is likely to linger into the future.[4]

However, some companies—and even one of the commissioners of the FTC—argued that corrective advertising is a form of compelled speech and thus violates the First Amendment right to engage in free speech.

multiple-product order
A form of cease-and-desist order issued by the FTC that applies not only to a specified product but also to other products produced by the same firm.

corrective advertising
Advertising in which the company explicitly states that formerly advertised claims were untrue. Also called *counteradvertising*.

Telemarketing and Electronic Advertising

Consumer law, like all law, is dynamic, and technology is often a driving force of change in consumer law. Although technological developments provide consumers with a host of benefits, these developments are often prone to abuses not covered by existing consumer law. For example, telephones and fax machines facilitate fast and inexpensive communication around the world, but

[4] FTC, "Doan's Pills Must Run Corrective Advertising: FTC Ads Claiming Doan's Is Superior in Treating Back Pain Were Unsubstantiated" (press release), May 27, 1999.

E-COMMERCE and the Law

Deceptive Ads Could Lead to Computer Consequences

Deceptive advertising in the physical world usually causes consumers to waste money on a worthless product or one that does not function as advertised. On the Internet, deceptive advertising can be much more dangerous to the individual. A very popular form of this deceptive advertising is known as bait and click advertising. Bait and click advertising involves Internet sites attracting Internet users by having them click ads that appear to be from well-known brands. These sites can then potentially install spyware or viruses on the individual's computer. Spyware is a program that is installed on the user's computer without his or her knowledge. The program can then store the individual's passwords or data and send them to outside sources. This consequence of deceptive online advertising could financially ruin the Internet user if the spy-ware user can access bank passwords or Social Security numbers. Therefore, it's important for Internet users to realize that advertisers are attempting to deceive the users online as well as in everyday life, and individuals should be cautious when clicking advertisements on the web.

Source: Linda Rosencrance, "E-Commerce Leaders Move to Fight Deceptive Online Ads," ComputerWorld.com, August 5, 2008, http://www.computerworld.com/s/article/9111682/E_commerce_leaders_move_to_fight_deceptive_online_ads.

telemarketers can use these technologies to deceive consumers and invade their privacy. Hence, lawmakers have passed new laws to address these issues.

Two major acts regulate advertising by telephone and fax, the Telephone Consumer Protection Act (TCPA) of 1991 and the Telemarketing and Consumer Fraud and Abuse Prevention Act of 1994.[5] TCPA, which the Federal Communications Commission (FCC) enforces, forbids phone solicitation using an automatic telephone dialing system or a prerecorded voice. TCPA also makes it illegal to transmit advertisements by fax unless the recipient agrees to the fax transmission. TCPA allows consumers to obtain a private right to legal action. In other words, the act gives consumers the right to recover for their losses. If a telemarketer violates TCPA, the consumer can recover either monetary losses or $500 per violation. However, if the telemarketer willfully violated the act, the court can decide to triple the amount owed to the consumer.

Data Mining and Privacy

The availability of huge amounts of electronic data has resulted in an understandable desire on the part of governments and marketers to access and use this data for national security and enhanced sales, respectively. Citizens and consumers often rebel at the thought that their personal information and communications are being used by governments and corporations; they see such *data mining* as an invasion of their privacy.

The Constitution Project, a bipartisan think tank, has thoroughly analyzed the conflicting interests in its 2010 booklet, *Preserving Civil Liberties in the Information Age*.[6] The culmination of their study was a list of guidelines that they offered to data miners. Studying a few of their suggestions provides future business managers with an introduction to this area of legal tension.

Consumers are often in the dark about the extent to which their behavior is tracked and used. They have no idea that each purchase is charted and the information is then sold to marketers.

For example, in Vermont, the legislature became so aroused about the selling of prescription information to data mining companies that they passed a statute in 2007 prohibiting pharmacies from engaging in this practice. Drug manufacturers who bought the information from data mining firms would urge their salespeople to use the information to convince physicians to prescribe more of the manufacturers' costly brand-name drugs. The prescription information purchased from the data miners enabled the manufacturers to target particular physicians who were not prescribing their drugs or who were prescribing competing drugs. On June 23, 2011, the United States Supreme Court, in *Sorrell v. IMS Health Inc.*, determined that this statute violated the free speech clause of the First Amendment. The Court reasoned that the state may not burden the speech of others to tilt public debate in a particular direction.

As technology becomes more sophisticated, marketers will be increasingly tempted to take advantage of the rich data they will have available to channel their marketing efforts. Courts and legislatures will be negotiating this thorny issue for many years. Business leaders need to watch and learn as this conflict unfolds.

[5] 47 U.S.C. § 227.
[6] Available at http://www.constitutionproject.org/wp-content/uploads/2012/09/DataMiningPublication.pdf.

- Prior to acquisition, clearly articulate, in writing, the purpose(s) of data acquisition and the intended use(s) of that data.
- Create a comprehensive data mining Plan ("Plan") covering data sources, data acquisition, system design and capabilities, and intended uses.
- Perform an internal evaluation of the program's expected effectiveness, costs, benefits, compliance with existing law, and impact on civil liberties and constitutional values.
- Establish and enforce penalties for misuse and abuse of data mining programs by operators or others.

Data Integrity and Security:
- Incorporate technical and administrative measures to limit access to or availability of personal data, particularly sensitive or personally identifiable data.
- Evaluate and improve data security, integrity, accuracy, and timeliness on a regular basis, including the use of audit trails.
- Conduct training and evaluation for employees with access to personal data or data mining systems.

Data Appropriateness/Minimization:
- Ensure that acquisition only includes data relevant to the purpose of the program and minimize the extent to which databases are aggregated.
- Set limited retention periods and ensure complete destruction of expired data.

Exhibit 25-2

Principles for the Development of Data Mining Programs:

The Federal Communications Commission (FCC) has an important role to play in the emerging threats to consumer privacy. In his first 100 days, President Donald Trump signed a bill that repealed Internet privacy rules established by the FCC in 2016. These FCC guidelines had not gone into effect, but would have given consumers greater control over what Internet service providers (ISPs) can do with the personal data of their respective users. The repealed guidelines required ISPs to disclose to their users what information and data was being collected and to receive affirmative consent from consumers before selling their data.

Supporters of the repeal argued that the FCC regulations put ISPs at a disadvantage because the FCC regulations did not apply to other companies that collect consumer data, like Google or Facebook. Under the new legislation, ISPs may continue to collect data like geolocation, browsing history, and other important personal information. Data mining provides a significant source of revenue for ISPs when they sell data to online advertising agencies. These advertising agencies use the data that the ISPs collected to tailor advertisements to specific consumers, based on each consumer's internet browsing history.

Sometimes laws are ineffective in achieving their goals, however, so legislatures draft new laws to supplement existing law. Thus, even though the goals of consumer laws may not change, specific rules and provisions do. For example, despite the protection of TCPA, consumers still lost an estimated $40 billion in telemarketing fraud after the act went into effect.

To give consumers more protection against deceptive and abusive telemarketing practices, Congress enacted the Telemarketing and Consumer Fraud and Abuse Prevention Act of 1994.[7] Through this act, Congress asked the FTC to define "deceptive and abusive" telemarketing practices and required the FTC to create and enforce rules governing telemarketing that would prohibit such practices. Consequently, the FTC created the Telemarketing Sales Rule of 1995, which requires telemarketers to (1) identify the call as a sales call; (2) identify the product name and seller; (3) tell the total cost of goods being sold; (4) notify the listener or reader of whether the sale is nonrefundable; and (5) remove the consumer's name from the potential contact list if the consumer so requests.[8]

[7] 15 U.S.C. §§6101–6108.
[8] 16 C.F.R. § 310.

GLOBAL Context

©Digital Archive Japan/Alamy

Advertising Law in China

For many years, advertising in China has been regulated by the People's Republic of China Advertising Law. However, in 2009, China's State Administration for Industry and Commerce (SAIC), the regulatory authority for advertising and marketing, released a Draft Revised Law for advertising regulation. This Draft Revised Law would include heavier restrictions on permissible advertising in China.

For example, under the Draft Revised Law of China, false advertising is defined as:

1. Material inconsistencies between the representations made in the advertisement and actual fact.
2. Concealing key information or providing misleading information about the product being advertised.
3. False statements concerning the prizes or awards that may be obtained upon purchasing a product or service.
4. Fabricating the experience of other participants with the products.

Although the Draft Revised Law mirrors legislation in the United States that regulates deceptive advertising, one major difference between the two countries is that neither the Chinese Draft Revised Law nor the original People's Republic of China Advertising Law mention the *type of person* that a misleading or deceptive advertisement would affect. For example, as you learned earlier in this chapter, in the United States, advertisements cannot contain a material misrepresentation or omission that is likely to mislead a *reasonable consumer*. The Draft Revised Law does not contain any such specifications.

Moreover, the FTC established that certain telemarketing behaviors are abusive, such as using profane or obscene language toward a customer and calling a person who previously requested to be taken off the particular seller's calling list. In addition, if the telemarketer calls a residence before 8 a.m. or after 9 p.m., the telemarketer is engaging in abusive behavior. If a telemarketer violates the rule, the FTC or the state attorney general may bring an action against the telemarketer. If the action is successful, the telemarketer can face fines up to $10,000 per violation.

While reviewing and amending the Telemarketing Sales Rule in 2002, the FTC created a Do Not Call registry. The FTC states that the purpose of the Do Not Call registry is to give consumers a choice regarding telemarketing calls. The registry makes it illegal for telemarketers to call any number that has been registered for more than 31 days. The registration lasts for five years, and it can be completed online through the FTC's website. Both the FTC and the FCC are responsible for maintaining the list, with help from local state law enforcement officials.

Spam is a major problem for consumers because it comprises around 90 percent of all email messages. Over the latter portion of 2007 and the majority of 2008, the spam organization Herbalking was responsible for sending consumers billions of messages over the Internet. At one time, Herbalking was behind one-third of all Internet spam. To send such a substantial amount of email, Herbalking used software that infected computers, usually without the knowledge of the owners. In fact, estimates indicate that the Herbalking spam group used as many as 35,000 computers, a network capable of sending 10 billion email messages a day. The spam group actually pulled in $400,000 from Visa charges during one month, and the group had ties to five countries. Such facts make this spam operation perhaps the most extensive spam setup the FTC has ever come across.

The Can-Spam Act of 2003 states that spammers may not send email messages containing false information or provide consumers with no option concerning whether they receive messages in the future. In October 2008, the Federal Trade Commission successfully convinced a Chicago court to freeze the assets and shut down the extensive Herbalking spam network for violating the act.

Tobacco Advertising

The tobacco industry's advertising is regulated through two acts: the Public Health Cigarette Smoking Act[9] of 1970 and the Smokeless Tobacco Health Education Act[10] of 1986. The Public Health Cigarette Smoking Act prohibits radio and television cigarette advertisements, and the Smokeless Tobacco Act imposes the same restrictions for smokeless tobacco ads.

[9] 15 U.S.C. § 1331.
[10] 15 U.S.C. §§ 4401–4408.

Labeling and Packaging Laws

LO 25-3 Recall the purpose of the federal laws that regulate product labeling and packaging.

When consumers examine a product to decide whether to buy it, the label often influences the decision to purchase. For example, many of us have purchased food because the label said the food was low fat. Unfortunately, manufacturers can include or omit information on labels that misleads consumers.

Consequently, federal and state governments have passed laws that regulate product labeling. These laws generally require the manufacturer to provide accurate, understandable information on the label. Furthermore, if the product is potentially harmful, the manufacturer must make the consumer aware of the harm.

Several federal laws regulate product labeling. The Wool Products Labeling Act of 1939 requires accurate labeling of wool products.[11] Similarly, the Fur Products Labeling Act of 1951 requires the accurate labeling of fur products.[12] The Flammable Fabrics Act of 1953 makes it illegal to produce or distribute clothing "so highly flammable as to be dangerous when worn."[13] The Fair Packaging and Labeling Act of 1966[14] requires products to carry labels that identify the product and provide specific information about the contents, such as the quantity of the contents and the size of a serving, if the number of servings is stated. Moreover, under this act, food product labels must show the nutritional content of the product. Similarly, the Nutrition Labeling and Education Act of 1990 requires standard nutrition information (i.e., calories and fat) to be provided on food labels.[15] Furthermore, this act defines the words *fresh* and *low fat*. In 1994, the FTC issued a statement saying that it would apply these label restrictions to food advertising to prevent deceptive advertising. Thus, not only do sellers need to be concerned about the use of *high, low,* and *light* on labels, but they are also required to use these words in particular ways in advertisements.

While the FTC has specifically regulated some terms used on labels, others, like the popular "all-natural" or "100% natural", often ride the line between puffery and deceptive advertising. For example, the 2014 case *In re Hain Celestial Seasonings Prods. Consumer Litig.,* involved a class of consumers suing Hain Celestial Group for falsely labeling its teas as "100% Natural" even though they contained traces of pesticides. In response to the allegations, Hain claimed that the "100% Natural" label was mere puffery. The court disagreed with Hain and argued that "100% Natural" is a verifiable term that could be proven false with evidence of any amount of artificial chemicals. Therefore, the label was deceptive and not merely puffery. However, in a 2013 case against Hain, a judge dismissed deceptive advertising claims regarding "natural" body wash and lotion produced by Hain. According to the judge, a reasonable consumer would not believe that the products were "produced by nature" because these products are not natural by definition.

The goal of these laws is to provide consumers with specific information about product content. However, Congress has passed several laws that require information about the potential harms associated with a product. For example, the Federal Hazardous Substances Act of 1960 requires all items containing dangerous substances, such as corrosive or flammable chemicals, to carry warning labels. The Act has been amended to include other hazardous products such as fireworks or child toys that have electrical or thermal dangers.

Canada's cigarette labeling requirements are even more stringent than the U.S. requirements. In Canada, approximately 40 percent of cigarette packaging must be devoted to health warnings.[16] The Canadian Bureau of Tobacco Control, however, has proposed a new rule requiring manufacturers to dedicate 50 percent of cigarette packages to such warnings, which would include explicit pictures and images of mouth cancer or other severe diseases caused by cigarette use.[17]

[11] 15 U.S.C. §§ 1331–1341.
[12] 15 U.S.C. § 69.
[13] 15 U.S.C. § 1191.
[14] 15 U.S.C. §§ 1451 et seq.
[15] 21 U.S.C. § 343-1.
[16] Tobacco Products Control Regulations, SOR/89-21.
[17] Action on Smoking and Health, "Majority of Canadians Want Larger Warnings on Cigarette Packages" (press release), October 2000.

Sales

The FTC and other government agencies have the power to regulate sales. For example, the Federal Reserve Board of Governors has the power to govern credit provisions related to sales contracts through its Regulation Z.[18] The FTC has created rules that govern specific types of sales in which the consumer is in a more vulnerable position compared to a consumer who walks into a traditional retail setting. This section examines FTC regulation of three of these uncommonly vulnerable commercial settings: door-to-door, telephone, and mail-order sales.

Door-to-Door Sales

Imagine that you hear a knock at your door, open it, and discover a salesperson for an Internet provider. The person who knocked knows that you have just purchased a computer and are interested in learning about the Internet. The salesperson explains the price of various programs by which you can become familiar with the Internet. You listen but decide that you would like to get additional information from an alternative provider. However, the salesperson is extremely pushy; to get the salesperson out of your house, you decide to purchase one of his plans.

In most door-to-door sales, the consumer does not have a chance to compare products and services to find the best service for his or her money. In addition, many consumers find it difficult to escape a salesperson in their homes. It is much easier to walk out of a store. Because the consumer is in a particularly vulnerable position in a door-to-door sale, the FTC has created special rules for such sales.

The FTC created the Cooling-Off Rule, giving consumers three days to cancel purchases they make from salespeople who come to their homes. Moreover, the salesperson must notify the consumer, both verbally and in writing, that the sales transaction may be canceled. The FTC rule also requires the consumer to be notified in writing in the same language in which the oral negotiations were conducted, thereby preventing unscrupulous businesses from taking advantage of non-English speakers.

The following case provides an example of the kinds of pressures that the FTC is trying to offset: Consolidated Promotions offered consumers a free gift for setting up a meeting in their home to discuss Consolidated products. This in-home meeting was in fact a sales pitch for Consolidated's photography packages, which included film and discounted photo processing. These packages cost from $1,200 to $2,500. When Consolidated Promotions refused to cancel some of the consumers's contracts, the FTC approved a complaint and referred it to the Department of Justice. The FTC alleged that the company violated the Cooling-Off Rule by "(1) failing to honor valid cancellation notices; (2) misrepresenting consumers' rights to cancel their contracts; and (3) failing to inform each buyer orally of his or her right to cancel the order."[19] The FTC proposed that Consolidated Promotions enter into a consent decree whereby Consolidated would be required to send notice to all customers who bought a photography package after July 1, 1996, giving them an opportunity to cancel their contracts.

Telephone and Mail-Order Sales

Telephone and mail-order purchases trigger more complaints than do traditional retail or door-to-door sales. Suppose the office manager of a small accounting firm ordered five new chairs for the office through a catalog. The writing in the catalog indicated that the chairs would arrive within two weeks. The office manager called in the chair order, but six weeks later, he had heard nothing from the company. What rights does he have in this situation?

The FTC originally addressed problems with mail-order sales through the 1975 Mail-Order Rule.[20] The Mail or Telephone Order Merchandise Rule of 1993 amended the 1975 Mail-Order Rule to extend protections to consumers who purchase goods over phone lines, including through computers and fax machines.

The 1993 rule established three key guidelines. First, sellers must ship items within the time promised. If they do not specify a time, the seller is limited to 30 days from receipt of the order.

[18] 12 C.F.R. § 225.

[19] FTC, "FTC Settlement Protects Door-to-Door Sales Consumers" (press release), May 2, 2000.

[20] 16 C.F.R. §§ 435.1–435.2.

Second, if the seller cannot ship the item within the promised time, the seller must notify the customer in writing and offer an opportunity to cancel. Third, if a customer decides to cancel the order, the seller must refund the customer's money within a specified period of time.

Unsolicited Merchandise When a consumer goes to her mailbox only to discover that a company has sent her a book, must she pay for the book? Anyone who receives unsolicited merchandise may treat the item as a gift. The recipient may keep or dispose of the merchandise without any obligation to the sender. In accordance with the Postal Reorganization Act of 1970, any unsolicited merchandise sent by mail is free to be used by the recipient as he or she sees fit with no obligation by the recipient to the sender.[21]

BUT WHAT IF? . . .

WHAT IF THE FACTS OF THE CASE OPENER WERE DIFFERENT?

Let's say in the Case Opener that customers call Trudeau with questions about his weight loss plan and he tells them he will send them information. He ends up mailing them his weight loss book and charging them for it. What kind of sales is this called? Is it legal?

FTC Regulation of Sales in Specific Industries

In certain industries, sellers have extensive opportunities to take advantage of customers. Thus, the FTC—and in some cases, Congress—creates special rules for certain sales practices specific to certain industries.

Used-Car Sales Consumers who purchase a used car often have very little information about the car's history. For instance, they do not know whether the car has been in an accident or whether there are any serious problems that are not visible.

To protect used-car buyers, Congress passed the Odometer Act of 1973, which protects against odometer fraud in used-car sales. The FTC extended that act's protection through the 1984 Used Motor Vehicle Registration Rule.[22] Under this rule, a dealer must attach a buyer's guide label to any used car he or she is attempting to sell. The label must state that the car is being sold as is. This label is a warning to the customer that the seller is not guaranteeing anything at all about the performance of the car. Furthermore, the label must include a suggestion for the buyer to obtain an inspection for the used car before any decision to purchase.

Consumer protections against fraud in used-car sales vary widely from state to state. During the 1960s and 1970s, there was widespread pressure to reform our legislative system to protect consumers from fraud; many states responded more favorably to that pressure than did others. All states enacted the Uniform Commercial Code (UCC), but in each case, the UCC was enacted with significant variations. The difference between states was also heightened in that each state had unique nonuniform consumer protection statutes. Consequently, the consumer protection laws against used-car fraud (also known as lemon laws) vary from state to state. Some states provide minimum protection, whereas other states, such as Minnesota, presume that one unsuccessful effort to repair a used car demonstrates nonrepairability. Minnesota's laws also extend statutory protection to potential buyers of returned vehicles by banning resale of automobiles returned because of major safety defects.[23]

Funeral Home Services Consumers who must purchase goods and services for a funeral and burial are often vulnerable for several reasons. The consumer is usually preoccupied with his or her loss of a relative or friend and is unlikely to comparison shop. In addition, grieving consumers can be more readily persuaded to purchase unnecessary, expensive items for this last tribute to their loved ones. To prevent funeral homes from taking advantage of these customers, the FTC created

[21] 16 C.F.R. § 256.
[22] 16 C.F.R. §§ 455.1–455.5.
[23] David A. Rice, "Product Quality Laws and the Economics of Federalism," *Boston University Law Review,* 65 (1985), p.1.

the 1984 Funeral Rule and revised it in 1994. The rule requires those who operate funeral homes to provide customers with itemized price information about funeral goods and services. Furthermore, funeral homes may not misrepresent legal or cemetery requirements or require the customer to buy certain funeral goods and services as a condition for receiving other funeral goods and services.

Real Estate Sales Because real estate purchases are probably one of the largest purchases a consumer will make, Congress passed several acts requiring sellers to disclose certain information about the property. First, the Interstate Land Sales Full Disclosure Act, passed in 1968, requires disclosure of information to consumers so that they can make informed decisions about real estate purchases.[24] Under this act, anyone planning to sell or lease 100 or more lots of unimproved land through a common promotional plan must file an initial statement of record with the Department of Housing and Urban Development's (HUD's) Office of Interstate Land Sales Registration. Before the developer can offer land for sale, HUD must approve the initial statement.

Congress provided more protection for people purchasing homes in the Real Estate Settlement Procedures Act of 1974 and its 1976 amendments.[25] This act requires the disclosure of information regarding mortgage loans to the buyer. For example, the lender must give the buyer an estimate of the costs for finalizing the real estate purchase.

Online Sales With the ever-expanding reach of the Internet, there has been an increase in business-to-consumer (B2C) sales transactions. Anyone with an Internet connection can make purchases from his or her favorite stores, from Barnes & Noble to Macy's. Most existing consumer protection laws were developed to protect consumers in their interactions with businesses face-to-face. Hence, protecting consumers online requires new approaches. Although not a specific industry, the Internet facilitates such a huge volume of commerce that additional focused protective legislation is needed.

Despite the difficulty of prosecuting online fraud, the FTC has brought a number of enforcement actions against online businesses. The federal statutes already in existence prohibiting wire fraud apply to online transactions. In addition, several states have begun to amend statutes to protect online consumers explicitly.

Credit Protection

LO 25-4 Describe the different acts that provide credit protection.

The widespread use of credit to purchase goods and services means that consumer credit protection has become increasingly important. This section explores three key federal laws regulating the credit industry to protect consumers: the Truth in Lending Act, the Fair Credit Reporting Act, and the Fair Debt Collection Practices Act.

The Truth in Lending Act

One of the earliest, most significant statutes regulating credit is Title I of the Consumer Credit Protection Act (CCPA), referred to as the *Truth in Lending Act (TILA)*.[26] The purpose of the act is to require sellers to disclose the terms of the credit or loan to help consumers compare a variety of credit lines or loans. More important, consumers must be able to understand this disclosure of terms. TILA is administered, in part, by the Federal Reserve Board through the previously mentioned Regulation Z.

Suppose your business extends credit to customers. Are your credit lines through your business subject to TILA? First, TILA applies to consumer loans only. Second, TILA applies to those who lend money or arrange for credit through the ordinary course of business. Third, the credit or loan must be in the amount of $25,000 or less, unless the loan is secured by a mortgage on real estate. Fourth, the creditor must be making the loan to a natural person, not a legal entity. Fifth, the credit or loan must be subject to a finance charge or must have repayments of more than four installments.

[24] 15 U.S.C. §§ 1701–1720.
[25] 12 U.S.C. §§ 2501–2517.
[26] 15 U.S.C. §§ 1601–1693r.

All creditors subject to TILA must disclose the finance charge and the annual percentage rate of the credit or loan. This information must be disclosed in a meaningful way, which prevents creditors from burying the information in a large paragraph unrelated to the credit terms.

Consider the following example: In May 1999, a jury heard *Carlisle v. Whirlpool Financial National Bank*.[27] According to the case facts, Gulf Coast Electric's salespeople traveled door to door, selling satellite dishes through financing. The price of the satellite dish was $1,100. However, consumers could purchase this dish for approximately $200 in stores. The consumers argued that the salespeople hid the number of payments the buyers would have to pay, thereby violating TILA. The jury found in favor of the customers and awarded them $581 million.

Types of Loans under TILA TILA includes three categories of loans: open-end credit, closed-end credit, and credit card applications and solicitations. Each category has specific disclosure requirements. For example, an *open-end credit line* permits repeated transactions and assesses a finance charge on unpaid balances. A creditor of an open-end credit line is required to disclose information in periodic statements. In contrast, a *closed-end credit line* is one for a loan given for a specific amount of time. The creditor of a closed-end credit line must disclose the total amount financed and the number, amount, and due dates of payments. Finally, *credit card applications and solicitations* must include the annual percentage rate (APR), annual fees, and the grace period for paying without a finance charge.

Unauthorized Charges and Disputes TILA establishes certain consumer protection rules regarding unauthorized charges to credit cards. If your credit card is stolen and someone makes unauthorized purchases on your account, your liability for those charges cannot exceed $50 per card if prompt notification of the theft is made to the credit card company. If you notify the credit card company before unauthorized charges are made, you cannot be held liable for any of the charges. Similarly, if a credit card company sends you an unsolicited card in the mail and the card is stolen, you cannot be held liable for any of the charges.

TILA offers another protection to consumers who unknowingly purchase damaged goods using a credit card. If three requirements are met, consumers will not be obligated to pay for the good. First, the consumer must purchase the item near her home (i.e., the business is in the same state as the consumer's home or within 100 miles of the home). Second, the item must cost more than $50. Third, the consumer must make a good-faith effort to resolve the dispute, such as asking the store for a refund. If these requirements are met, the credit card company cannot bill the consumer for the damaged item.

Consumer Leasing Act In 1988, the Consumer Leasing Act (CLA) amended TILA to provide greater protection for people leasing automobiles and other goods.[28] CLA applies to those who lease goods as part of their regular business. For CLA to apply, the lease must be for a minimum of four months and the price must not exceed $25,000. Under CLA and its controlling regulation, Regulation M, anyone leasing goods must disclose up front, in writing, all the material terms and conditions of the lease.[29]

Equal Credit Opportunity Act In the 1970s, a woman old enough to have children would have had difficulty securing credit because creditors believed that married women with children would be less likely to pay their debts. In response to this discrimination, Congress passed the Equal Credit Opportunity Act (ECOA) as a 1974 amendment to TILA.[30] This amendment makes it illegal for creditors to deny credit to individuals on the basis of race, religion, national origin, color, sex, marital status, or age. When determining the creditworthiness of a credit applicant, the creditor cannot use information about the applicant's marital status, nor can the creditor require a spouse to co-sign the application. Finally, the act prohibits creditors from denying credit on the basis of whether the applicant receives public assistance benefits.

[27] No. 97-068, Cir. Ct., Hale Co., Ala.
[28] 15 U.S.C. §§ 1667–1667e.
[29] 12 C.F.R. Part 213.
[30] 15 U.S.C. §§ 1691–1691f.

The Fair Credit Reporting Act

If you own a credit card, you also have a credit report. If you apply for a new credit card or a loan, the creditor will check your credit history to make a judgment about your creditworthiness by examining a copy of your credit report. This report contains information about your financial transactions, such as payments on credit, debt collection, and other financial information the creditor needs to know about if entering a business transaction with you.

CASE Nugget

What Does the Law Require of a Loan Company?

Alvin Ricciardi, Appellant, v. Ameriquest Mortgage Company U.S. Court of Appeals for the Third Circuit

164 Fed. Appx. 221 (2006)

Alvin Ricciardi bought a home in May 2001, took out a home equity loan, and then applied for a loan to refinance the two mortgages through Ameriquest. Ricciardi closed the loan in September 2002, at which point he received the "Borrower's Acknowledgment of Final Loan Terms" acknowledgment. The acknowledgment stated the terms of the loan that Ricciardi had originally requested as well as the final terms of the loan. Ricciardi signed all the documents presented during the closing. Eight months later, Ricciardi filed suit against Ameriquest for violating the Equal Credit Opportunity Act, the Truth in Lending Act, and the Pennsylvania Unfair Trade Practices Act and Consumer Protection Law. The district court granted summary judgment regarding Ricciardi's ECOA claim, finding for Ameriquest. At a later nonjury trial, the court ruled in favor of Ameriquest on all counts of Ricciardi's claim as well as Ameriquest's counterclaim. Ricciardi appealed.

Ricciardi's appeal put forth three arguments. First, Ricciardi argued that the district court erred in granting summary judgment in favor of Ameriquest on his ECOA claim because Ameriquest failed to provide notice of its counteroffer before the closing of the loan. Second, Ricciardi argued that the district court erred in its adverse credibility finding. Third, Ricciardi argued that Ameriquest violated TILA by overcharging him for insurance and thereby gave him the right to rescind the loan. The appeals court rejected each of Ricciardi's arguments, affirming the ruling of the district court.

The appeals court addressed each of Ricciardi's arguments in turn. First, TILA requires a creditor to respond to an applicant within 30 days of receipt of a completed loan application. Ameriquest did respond to Ricciardi within 30 days. Contrary to Ricciardi's claim, Ameriquest was not required to respond in the form of a counteroffer. In addition, there is no requirement for a counteroffer to be received before the closing of the loan. Thus, Ricciardi's argument was without merit. Second, the appeals court found that the district court had not erred in its adverse credibility finding. Ricciardi contradicted himself numerous times on the record, in addition to committing fraud by misrepresenting his occupation and income on the loan application to Ameriquest. Third, the appeals court found that Ricciardi had not presented sufficient evidence to substantiate his claim that Ameriquest had overcharged him for insurance. In Pennsylvania, state-mandated insurance rates are published in the *Rate Manual*. The rates decrease if the property is being refinanced and was previously insured. Ricciardi failed to show that the property was previously insured; thus, the district court correctly found no evidence that Ameriquest had overcharged Ricciardi.

FCRA establishes other duties for credit reporting. For example, a credit bureau cannot report obsolete information. (General credit information is considered obsolete after seven years.) However, a bankruptcy will be reported for 10 years. Also, any time a credit applicant is denied credit, the

applicant has a right to the name and address of the credit bureau that reported the information. Finally, FCRA imposes a duty on the reporting agency to ensure that its information is as accurate as possible. Consequently, if a consumer reports an error in his or her credit report, the credit bureau is required to investigate the error and make the corrections in the report if necessary.

The Fair Debt Collection Practices Act

Suppose a consumer owes $3,000 on his credit card and has not been able to make monthly payments for the past six months. The credit card company will likely refer the case to a collection agency, which will notify the consumer in an attempt to get him to pay the debt. The collection agency then may start calling the consumer regularly to discuss the debt. Next, the agency might start contacting the consumer's acquaintances, telling them about the debt in an effort to pressure the consumer into paying the debt.

This type of debt-collecting behavior is prohibited by the Fair Debt Collection Practices Act (FDCPA).[31] This act applies only to debt collectors who regularly attempt to collect debts on behalf of others. The following collection behaviors are expressly prohibited by FDCPA:

1. Contacting a debtor at work if the debtor's employer objects.
2. Contacting a debtor who has notified the collection agency that he or she wants no contact with the agency.
3. Contacting the debtor before 8 a.m. or after 9 p.m.
4. Contacting third parties about the debt (exceptions: contacting the debtor's parents, spouse, or financial adviser).
5. Using obscene or threatening language when communicating with the debtor.
6. Misrepresenting the collection agency as a lawyer or a police officer.

Case 25-2 provides an analysis of an application of the "least sophisticated consumer" standard for interpreting the scope of consumer protection statutes, as it applies to the Fair Debt Collection Practices Act.

[31] 15 U.S.C. § 1692.

CRAWFORD v. LVNV FUNDING, LLC
United States Court of Appeals, Eleventh Circuit
758 F.3d 1254 (2014)

FACTS: Stanley Crawford owed $2,037.99 to the Heilig-Meyers furniture company. Heilig-Meyers charged off the debt in 1999, and, in September 2001, a company affiliated with the defendant in this case, LVNV Funding LLC, acquired the debt. On October 2004, the debt became unenforceable due to Alabama's statute of limitations and because the last transaction on the debt occurred three years earlier in October 2001.

On October 2, 2008, Crawford filed for Chapter 13 bankruptcy in Alabama. During the proceeding, LVNV filed a proof of claim to collect the Heilig-Meyers debt, despite the fact that the statute of limitations had expired four years prior. Crawford filed a counterclaim against LVNV, alleging that LVNV filed stale claims as a routine business practice and that attempting to claim the Heilig-Meyers debt violated the FDCPA. The bankruptcy judge dismissed Crawford's adversary claims, and the district court affirmed the decision. Crawford appealed once again to the U.S. Court of Appeals in May 2013.

ISSUE: Using the "least-sophisticated consumer" standard, does the Fair Debt Collection Practices Act disallow the attempt to collect a debt when the claim is time-barred?

REASONING: The initial step of the Court's reasoning was to make it clear that it was going to rely on statutory interpretation. Thus, the first thing it did was review relevant parts of the governing law, the FDCPA, to see whether the statute is consistent with Crawford's claim. The Court said it found "abundant evidence" of the kind of practices that Congress wanted to stop, namely "the use of abusive, deceptive, and unfair debt collection practices." 15 U.S.C. § 1692(a). To enforce the FDCPA's prohibitions, Congress equipped consumer debtors with a private right of action, rendering "debt collectors who violate the Act liable for actual damages, statutory damages up to $1,000, and reasonable attorney's fees and costs."

(continued)

Section 1692e of the FDCPA provides that "[a] debt collector may not use any false, deceptive, or misleading representation or means in connection with the collection of any debt." Section 1692f states that "[a] debt collector may not use unfair or unconscionable means to collect or attempt to collect any debt."

Given the ambiguity of "unfair" and "unconscionable," the Court adopted a "least-sophisticated consumer" standard to evaluate whether a debt collector's conduct is "deceptive," "misleading," "unconscionable," or "unfair" under the statute. The inquiry is not whether the particular plaintiff-consumer was deceived or misled; instead, the question is "whether the 'least sophisticated consumer' would have been deceived" by the debt collector's conduct. The reason behind LVNV's practice of filing time-barred proofs of claim in bankruptcy court is simple. Absent an objection from either the Chapter 13 debtor or the trustee, the time-barred claim is automatically allowed against the debtor pursuant to 11 U.S.C. § 502(a)-(b) and Bankruptcy Rule 3001(f). As a result, the debtor must then pay the debt from his future wages as part of the Chapter 13 repayment plan, notwithstanding that the debt is time-barred and unenforceable in court. Neither the bankruptcy trustee nor Crawford objected to the claim by LVNV to collect the debt during the bankruptcy proceeding; instead, the trustee actually paid monies from the Chapter 13 estate to LVNV (or its surrogates) for the time-barred debt. It wasn't until four years later, in May 2012, that debtor Crawford—with the assistance of counsel—objected to LVNV's claim as unenforceable. A debt collector's filing of a time-barred proof of claim creates the misleading impression to the debtor that the debt collector can legally enforce the debt. The "least sophisticated" Chapter 13 debtor may be unaware that a claim is time-barred and unenforceable and thus fail to object to such a claim. Given the Bankruptcy Code's automatic allowance provision, the otherwise unenforceable time-barred debt will be paid from the debtor's future wages as part of his Chapter 13 repayment plan.

DECISION AND REMEDY: The Court concluded that LVNV's filing of the proof of claim fell well within the ambit of a "representation" or "means" used in "connection with the collection of any debt." It was an effort "to obtain payment" of Crawford's debt "by legal proceeding." In fact, payments to LVNV were made from Crawford's wages as a result of LVNV's claim. And, it was Crawford—not the trustee—who ultimately objected to defendants's claim as time-barred. Our conclusion that §§ 1692e and 1692f apply to LVNV's proof of claim is consistent with the FDCPA's definition of a debt-collector as "any person who . . . regularly collects or attempts to collect, *directly or indirectly,* debts owed or due or asserted to be owed or due another." 15 U.S.C. § 1692a(6) (emphasis added).

SIGNIFICANCE OF THE CASE: The Court is interpreting the FDCPA in a manner that makes it clear that the consumer protection statute here is not to protect an average consumer, but rather the vulnerable consumer.

CRITICAL THINKING

What ambiguity is central to this case? Do you agree with the way the court interpreted the ambiguous word or phrase? Can you think of an alternative reasonable definition?

ETHICAL DECISION MAKING

Who are the primary stakeholders affected by the court's ruling? Does the appeals court have a stronger ethical obligation to elevate the interests of some stakeholders over others? Explain.

The Credit Card Fraud Act

Credit card fraud is a serious problem in the United States, costing consumers millions of dollars per year. Accordingly, Congress passed the Credit Card Fraud Act of 1984 to close existing loopholes in federal laws that allowed credit card fraud to be pervasive.[32] The Credit Card Fraud Act states that it is unlawful to (1) possess an unauthorized credit card, (2) counterfeit or alter a credit card, (3) use the account number of another's credit card to perpetuate fraud, and (4) use a credit card obtained from a third party with his or her consent, even if the third party conspires to report the card as stolen. The act also increases the penalty for committing credit card fraud.

[32] 18 U.S.C. § 1029(a)(1-4).

The Fair Credit Billing Act

Did your credit card company fail to extend your credit when it informed you that your credit would be extended? Were you ever charged for merchandise you did not purchase or receive? Were you ever charged twice for one purchase? If so, you have been the victim of a credit billing error. The Fair Credit Billing Act (FCBA) of 1986 was created to handle such billing errors and many others.[33]

FCBA, enforced by the FTC, creates procedures consumers are to follow in filing complaints when billing errors occur. FCBA also requires the creditor to explain to the consumer and FTC why the error occurred and promptly fix the error. When a complaint is filed, the creditor may not try to collect on the disputed amount or take any action against the consumer until the complaint is answered.

The Fair and Accurate Credit Transactions Act

The Fair and Accurate Credit Transactions Act (FACTA) of 2003 was passed in response to the growing number of identity-theft cases.[34] If someone thinks he is a victim of identity theft, he may contact the FTC, and an alert will be placed in his credit files. The credit files then serve as a national fraud alert system to enhance authorities's ability to catch those who are stealing identities.

Several other requirements created by the act protect consumers. First, major credit reporting agencies are required to provide consumers with a free copy of their credit reports every 12 months. Second, receipts from credit card purchases are to list an abbreviated version of the card number to protect consumer accounts. Third, financial institutions must work with the FTC to red-flag suspicious transactions that might be a sign of identity theft. Fourth, assistance will be provided to victims of identity theft to help them rebuild their credit. Fifth, victims of identity theft may report fraud directly to creditors to protect their credit ratings.

The Credit Card Holders' Bill of Rights Act

Under FACTA, the three major agencies required to provide credit reports are Experian, Equifax, and TransUnion. Some websites or other credit bureaus have claimed to provide free credit reports, such as freecreditreport.com, but there is only one authorized site for government-required free credit reports from the three agencies: AnnualCreditReport.com. A visitor of the site may receive one free report per year, but the three agencies make money in other ways, such as providing credit numbers or additional reports in a year.

The FTC fined freecreditreport.com more than once during the Bush administration. The dishonest website claimed to give consumers a free credit report and then charged consumers who signed up for a report. Advertisements that deceive consumers about free credit reports are subject to more than mere wrist slaps now that the Credit Card Holders' Bill of Rights Act was signed by President Obama on May 22, 2009. Because of this act, the FTC may produce new rules that make free credit report advertisers affirm that only AnnualCreditReport.com provides free credit reports to consumers.

The act, also known as the CARD Act, has four provisions that target unfair credit card practices. The first provision mandates the adjustment of several credit practices. First, creditors are required to notify consumers of changes to fees and interest rates before such changes take place. Furthermore, contractual agreements must be made with clients if fees and interest rates are to be changed at all. Second, the limits of fees and interest rates of all credit companies will be strictly regulated by the FTC to avoid unfairly high maximums. Third, penalty fees such as late fees and over-the-limit fees must have reasonable maximums.

The second provision of the CARD Act covers notification and information. First, creditors must notify consumers about payoff timing. Second, all billing statements must conspicuously display whether and when the interest rate will increase. Third, creditors must inform consumers, up front, about the dates on which payments are considered late and the interest rates associated with late payments. Fourth, creditors must post all conditions associated with each credit arrangement option on the Internet. The last part of the provision modifies deceptive advertising associated with free credit reports.

[33] 15 U.S.C. § 1601.
[34] Pub. L. No. 108-159, 117 Stat. 1952.

The CARD Act's third provision prohibits credit card companies from extending credit offers to anyone under 21. Consumers under 21 may acquire credit only with a co-signer and proof of sufficient income. The third provision also blocks creditors from using tangible items to persuade college-age consumers to apply for credit. This provision also requires creditors to submit an annual report to a federal review board. Specifically, the annual report must include three pieces of information: (1) all memorandums or agreements between creditors and institutions of higher education, (2) the total number of payments and payment amounts creditors make to institutions of higher education, and (3) the number of credit accounts opened under an agreement between a credit card company and an institution of higher education per year.

The fourth provision of the CARD Act contains three directives. First, consumers will be charged fees for dormant or inactive gift cards. Second, only one fee per month may be charged to a consumer with a gift card that is inactive for 12 months. Third, gift cards, prepaid cards, and gift certificates must inform customers of three conditions before purchase: the existence of the dormancy fee, the amount of the dormancy fee, and the frequency of the dormancy charge.

Consumer Health and Safety

LO 25-5 Describe the different laws that help ensure consumer health and safety.

The legislation and rules we have examined regulate the advertising, labeling, and sale of products. Now we turn to legislation regarding product safety. The purpose of such regulations is to ensure that companies such as Firestone and Ford produce safe products for consumers who do not have all the information. The two main federal statutes that address product safety are the Federal Food, Drug, and Cosmetic Act and the Consumer Product Safety Act.

The Federal Food, Drug, and Cosmetic Act

In 1906, Congress created the first federal legislation regulating food and drugs, the Pure Food and Drugs Act. Subsequently, Congress amended the Pure Food and Drugs Act when it created the Federal Food, Drug, and Cosmetic Act (FFDCA) in 1938 to protect consumers against misbranded or adulterated food, drugs, medical devices, or cosmetics.[35] The U.S. Food and Drug Administration (FDA), the agency responsible for administering FFDCA, creates standards to regulate food and drugs, thus protecting consumers. Specifically, the FDA must ensure that food, drugs, cosmetics, and medical devices meet specific safety standards.

CASE 25-3 examines the regulatory scope of the FFDCA.

[35] 21 U.S.C. §§ 301–393.

UNITED STATES v. KAPLAN,
United States Court of Appeals, Ninth Circuit 836 F. 3D 1199 (2016)

FACTS: Doctor Michael Kaplan was a urologist who owned and operated two urology clinics in Henderson and Las Vegas, Nevada. Both clinics regularly performed prostate needle biopsies. The procedure is used to collect prostate tissue sample to examine for disease. To complete the prostate needle biopsy, an ultrasound probe is inserted into a patient's rectum to locate the prostate, after which a hollow needle is used to penetrate the rectal wall and gather the needed tissue sample. To help stabilize the needle, an enclosure which houses it, called a needle guide, is used. Needle guides come in reusable stainless steel which must be sterilized or in single-use disposable plastic forms. Kaplan's clinics had used a stainless steel needle guide with a corresponding sterilizing machine until December 2010, when the machine broke.

(continued)

Because of time constraints, the sales representative Kaplan worked with, Timothy Brandt, arranged to send disposable plastic guides to the clinics.

By January 2011, both clinics were low on plastic needle guides and additional shipments were on backorder. Kaplan instructed his medical assistant supervisor at the Henderson clinic to tell the medical assistants to sterilize the plastic needle guides using the same protocol they had previously used for the stainless steel needle guide. Medical assistants noticed blood and pinkish water left in the plastic needle guides after the sterilization process. Assistants also noticed brown scratches that would not come clean and noted that they could tell a used plastic needle guide from a new one due to its discoloration. In March 2011, Kaplan's medical assistants reported him to the Nevada State Medical Board. Following an investigation, a grand jury indicted Kaplan in October 2013 on two counts. The relevant count charged Kaplan with conspiracy to commit adulteration in violation of the FDCA. In September 2014, a jury found Kaplan guilty of conspiring to commit adulteration and a final judgement for a 48 months prison sentence was entered in May 2015. Kaplan appealed, contending that he could not be criminally prosecuted under the FDCA.

ISSUE: Can Kaplan be criminally prosecuted under the FDCA?

REASONING: The FDCA is intended to protect consumers from dangerous products. It contains a provision that covers products "held for sale." Kaplan argued that the phrase "held for sale" is different than "held for use." Kaplan contended that because title and possession of the needle guides were not transferred to patients, there was no sale and thus the use of the guides during biopsies falls outside of the FDCA.

The court disagreed, however. It argued that "held for sale" has a broader meaning than the one Kaplan assigns to it and that it focuses on "the commercial nature of the transaction, actors, and products." The court argued that the single-use nature of the guides is meant to be "consumed" in treating a patient similar to a single-use drug. Thus, when a physician uses a single-use medical device on a patient and the device is then disposed of, the device is "held for sale" provided that there is a commercial relationship between the doctor and the patient. The court found that Kaplan knowingly endangered public health in his practices and that the FDCA does apply.

DECISION: Affirmed.

SIGNIFICANCE: This case shows that the FFDCA can apply even in cases where traditional "products" and "sales" may not seem to be involved.

CRITICAL THINKING

How does this case increase your appreciation for the role that ambiguity plays in the interpretation of laws?

ETHICAL DECISION MAKING

How does this case demonstrate that the freedoms of businesses are in conflict with the freedoms of consumers?

The Food and Drug Administration Modernization Act of 1997 amended FFDCA to help improve the regulation process, particularly for drugs and medical devices. For example, the 1997 act reauthorized a drug program that cut the time required for a drug review from 30 months to 15 months. The act also increased patient access to experimental drugs and accelerated the review of important new medications.

The Consumer Product Safety Act

In the Consumer Product Safety Act of 1972, Congress created and directed the Consumer Product Safety Commission (CPSC) to "protect the public against unreasonable risks of injuries and deaths associated with consumer products."[36]

The CPSC protects the public from injuries associated with consumer products in several ways. First, the CPSC issues and enforces mandatory standards regarding product safety. Similarly, the commission works with industries to develop voluntary product standards. If the CPSC cannot establish a standard that would adequately protect the public, it can ban consumer products from

[36] 15 U.S.C. § 2051.

the market. In addition, the CPSC can administer existing product safety legislation. Examples of such legislation include the Child Protection and Toy Safety Act of 1969[37] and the Federal Hazardous Substance Act of 1960.[38]

Second, the CPSC can arrange for a recall of products. Although the CPSC has the authority to issue product recalls on its own, usually the CPSC works with companies that are voluntarily issuing recalls for dangerous products. For example, in August 2006, both Dell and Apple issued voluntary recalls, with the help of the CPSC, for lithium ion batteries sold in their laptops. Both companies received several separate complaints about their batteries overheating; thus, the CPSC aided the companies in the battery recall.

Third, the commission conducts research regarding potentially hazardous products. The National Highway Traffic Safety Administration (NHTSA) is similar to the CPSC in that it, too, conducts investigations about the safety of potentially hazardous products. The NHTSA, however, focuses primarily on motor vehicles. The NHTSA was the agency that investigated the Ford/Firestone case to determine exactly what was causing the tire blowouts in Ford Explorers.

Fourth, the CPSC educates consumers about product safety. One important way the CPSC offers this education is through the National Injury Information Clearinghouse.

This section examined two of the most significant laws governing the safety of consumer products. Exhibit 25-3 lists six additional consumer product safety laws.

Exhibit 25-3

Other Consumer Protection Laws

STATUTE	PURPOSE
Pure Food and Drugs Act of 1906 (amended by the Federal Food, Drug, and Cosmetic Act)	Forbids the adulteration and mislabeling of food and drugs
Meat Inspection Act of 1906	Provides for meat inspection
Wholesome Meat Act of 1967	Amended the Meat Inspection Act to promulgate standards for animal slaughter for meat
Flammable Fabrics Act of 1953	Prohibits the sale of highly flammable clothing
Child Protection and Toy Safety Act of 1969	Requires childproof devices and toy labeling
Toy Safety Act of 1984	Permits the Consumer Product Safety Commission to recall toys and other articles children use
Drug-Price Competition and Patent-Term Restoration Act of 1984	Simplifies the Food and Drug Administration's process of approving generic drugs

CASE OPENER WRAP-UP

Deceptive Advertising and the Ultimate Weight-Loss Cure

To determine that Trudeau misled consumers, the court reasoned that Trudeau failed to mention a single aspect of his weight-loss protocol. Further, the court found that Trudeau painted a picture of a safe and easy diet, when in reality the diet had extreme restrictions. Trudeau's selective quotations misled consumers because they created a false image of the diet.

The FTC preferred that Trudeau reimburse all consumers who purchased his book via the infomercial—a remedy that would have resulted in over $46 million in reimbursements. Alternatively, the FTC asserted that, at the very least, Trudeau should disgorge his profits, estimated at around $12 million. In the end, the court required Trudeau to pay the FTC a little over $5.1 million.

[37] Amendments to 15 U.S.C. §§ 1251, 1252, and 1274.
[38] 15 U.S.C. §§ 1251–1277.

CHAPTER 25 Consumer Law 621

PRACTICAL TIPS for Business Managers

1. While it may be tempting to tout your products beyond their abilities, the Bayer example shows why this can be a costly mistake in terms of both time and money. Always advertise your products accurately. While puffing is allowed, make sure the claim is clearly a joke or an exaggeration or you may find yourself in trouble.

2. Product recalls can be very expensive and can tarnish the reputation of your business. Instead of rushing a product out the door, it is usually a better idea to take your time to fully test it for defects to avoid spending more money on a recall later.

3. As this chapter explains, there are many different types of making sales and each one generally has its own set of rules attached to it. Before trying to make a certain type of sale, be sure you know any regulations attached to it so that you do not violate the law in the process.

SUMMARY

The Federal Trade Commission

How the FTC brings an action:

1. FTC conducts an investigation.
2. FTC sends a complaint to the violator.
3. FTC and the violator settle the complaint through a consent agreement.
4. If the company refuses to enter the consent agreement, the FTC may issue a formal administrative complaint, which leads to an administrative hearing.
5. If the company has violated the law, the FTC issues a cease-and-desist order.

If the company violates the order, the FTC can seek an injunction against the company or fine the company up to $10,000 per violation.

Deceptive Advertising

Bait-and-switch advertising: Advertising a low price to bait the consumer into the store only so that the salesperson can switch the consumer to another, higher-priced item.

FTC actions against deceptive advertising:

- *Cease-and-desist actions:* Court orders requiring firms to stop their current advertising behavior.
- *Multiple-product orders:* Court orders requiring firms to stop current advertisements on numerous products, as opposed to one specified product.
- *Corrective advertising:* Advertising in which the company explicitly states that the formerly advertised claims were untrue.

Telemarketing and electronic advertising:

- *1991 Telephone Consumer Protection Act:* Telemarketers cannot use an automatic telephone dialing system or a prerecorded voice.
- *Telemarketing and Consumer Fraud and Abuse Prevention Act of 1994:* This act created certain requirements regarding when and how telemarketers may make calls.
- *Federal Do Not Call registry:* Telemarketers cannot call consumers who have voluntarily placed their phone numbers on the federal Do Not Call list.

Tobacco advertising: Cigarette and smokeless-tobacco advertising is restricted.

Labeling and Packaging Laws

Federal and state governments have passed laws requiring manufacturers to provide accurate, understandable information on labels. Furthermore, if a product is potentially harmful, the manufacturer must make the consumer aware of the harm.

Sales

Door-to-door sales: The Cooling-Off Rule gives consumers three days to cancel purchases they make from salespeople who come to their homes.

Telephone and mail-order sales: The *Mail or Telephone Order Merchandise Rule of 1993* extends protections to those who purchase over the phone or by fax.

Unsolicited merchandise: Anyone receiving such merchandise may treat it as a gift.

(continued)

(concluded)

	FTC regulation of specific industries: 1. Used-car sales 2. Funeral home services 3. Real estate sales 4. Online sales
Credit Protection	*The Truth in Lending Act* requires sellers to disclose the terms of the credit or loan to facilitate the consumer's comparison of a variety of credit lines or loans. *The Fair Credit Reporting Act* ensures accurate credit reporting. *The Fair Debt Collection Practices Act* regulates the actions of debt collectors that regularly attempt to collect debts on behalf of others. *The Credit Card Fraud Act* closes loopholes in federal laws to punish people further who commit credit card fraud. *The Fair Credit Billing Act* seeks to rectify problems and abuses associated with credit billing errors. *The Fair and Accurate Credit Transactions Act* takes affirmative actions to control and prosecute identity theft.
Consumer Health and Safety	*The Federal Food, Drug, and Cosmetic Act* protects consumers against misbranded or adulterated food, drugs, medical devices, and cosmetics. *The Consumer Product Safety Act* created the Consumer Product Safety Commission (CPSC) to "protect the public against unreasonable risks of injuries and deaths associated with consumer products."

Point/Counterpoint

Should Prescription Drugs be Allowed to be Advertised Directly to Consumers?

YES	NO
The question of whether DTCA should be legal has been a heated debate in many countries. DTCA can be a great tool for informing consumers and improving lives, and should remain legal in the United States for several reasons. One reason DTCA should be allowed is that it can inform consumers about diseases they may have and possible treatments for those diseases. Being that the ads in question are prescription drug ads, DTCA also encourages consumers to seek help from their doctors before using the drug. DTCA can educate consumers about illnesses they may or may not have realized they were suffering from and encourage them to seek treatment and improve their lives. Another benefit of DTCA is that, like any other form of advertisement, it gets the word out about a product (in this case, prescription drugs) which can lead to more sales for the pharmaceutical companies that produce them. More sales leads to more revenue, which can be used to research new and better life-improving medications. In effect, the advertising of today can save lives tomorrow.	The advertisement of prescription drugs represents a potential threat to consumers everywhere. In fact, direct-to-consumer drug advertisement (DTCA) has been deemed so unsafe that every country besides the United States and New Zealand has banned it outright. There are many reasons why DTCA should be banned in the United States as well. First, DTCA ads can misinform consumers and can lead to consumers misdiagnosing themselves. A 2013 study published in the *Journal of General Internal Medicine* found that 55 percent of claims made in DTCA were "potentially misleading," while 2 percent were "false." DTCA can mislead consumers to request drugs and seek treatment for diseases that the ads lead them to believe they have, which in turn can mean adverse medical results from taking unnecessary medications. Second, drugs are often advertised through DTCA before their long-term health effects are fully known. Contrary to the belief of many consumers, drugs can be advertised and sold before long-term safety trials have been completed. The drug Vioxx was marketed, requested by patients and prescribed before being taken off the market after being linked to over 4,500 deaths related to strokes and heart attacks induced by the drug.

Perhaps one of the most straightforward reasons as to why DTCA should remain legal is that it should be allowed as protected free speech. Ford and Chrysler can advertise their new cars, so why can't pharmaceutical companies advertise their new legal, FDA-approved, life-improving drugs? The First Amendment should protect DTCA as it does almost all other forms of advertisement.

Consumers aren't the only people being negatively affected by DTCA, either. A large portion of medical doctors have reported having patients request advertised drugs that were inappropriate for their treatment, and many of these doctors have felt pressured to prescribe the inappropriate drugs. And even if a doctor refuses to prescribe the drug, a 2013 study found that 50 percent of patients were "disappointed" in their doctors for refusing to prescribe an advertised drug and 25 percent of patients surveyed would try to convince the doctor to give them the drug or acquire the drug somewhere else. Not even wary doctors can prevent consumers from falsely medicating themselves as a result of DTCA.

Questions & Problems

1. What is the goal of the FTC, and how does it achieve its goal? What are some pieces of legislation that enable the FTC to achieve its goal?

2. What are the elements of a deceptive advertisement, and how does the FTC prove an ad is deceptive?

3. What are the main provisions of the Truth in Lending Act, and how does it aid consumers?

4. On June 27, 2012, Abigail Strubel opened a Victoria's Secret brand credit card to purchase an item for $19.99. The card was issued by Comenity Bank and the credit card agreement provided by Comenity disclosed certain consumer rights under the Truth in Lending Act. On June 27, 2013, Strubel filed a class action against Comenity Bank seeking statutory damages under the TILA for problems in the disclosures contained in the credit card agreement. Strubel's contended that Comenity failed to disclose four items: Failing to clearly disclose that "(1) cardholders wishing to stop payment on an automatic payment plan had to satisfy certain obligations; (2) the bank was statutorily obliged not only to acknowledge billing error claims within 30 days of receipt but also to advise of any corrections made during that time; (3) certain identified rights pertained only to disputed credit card purchases for which full payment had not yet been made, and did not apply to cash advances or checks that accessed credit card accounts; and (4) consumers dissatisfied with a credit card purchase had to contact Comenity in writing or electronically." The district court granted Comenity's motion for summary judgment and Strubel appealed. For each of the four claims alleged by Strubel, write who the appellate court rule in favor of and what its reasoning was.

 Source: *Strubel v. Comenity Bank,* 842 F. 3d 181 (2016).

5. Fresh, Inc., is a cosmetics and skin care manufacturer. One of its product lines is Sugar Lip Treatment (Sugar), a lip balm. Sugar is a lip balm treatment that comes in various shades of color and comes in an oversized dispenser tube. Each tube of Sugar retails at stores and online at $22.50 to $25.00.

 Angela Ebener, a California resident that purchased tubes of Sugar over the four years preceding the trial, sued Fresh for deceptive advertising. Ebener asserted that she was led to believe there was more product than the packaging indicated because only 75 percent of the product was reasonably accessible. The remaining 25 percent was blocked by the tube's screw mechanism. Furthermore, although the Sugar tube label stated there was 4.3 grams of product, a metallic weight at the base of the tube combined with packaging box itself made the whole package weigh approximately 29 grams. Ebener argued that the extra weight misled her to think there was more product than there actually was.

 The district court ruled against Ebener, citing that California's Safe Harbor doctrine and federal preemption under the FDCA were fatal to her claims. Ebener appealed and the Ninth Circuit Court took the case and affirmed the district court decision. What is the Safe Harbor Doctrine and federal preemption under the FDCA according to the appellate court and how did they interact with California's deceptive advertising laws to lead to the case's conclusion? Do you think it is ethical for Fresh to label Sugar as having 4.3 grams of lip balm when only three-fourths of it is accessible?

 Source: *Ebner v. Fresh, Inc.,* 838 F. 3d 958 (2016).

6. In June 2013, Paul M. Stelmachers took a taxi cab ride in Las Vegas, Nevada, and paid for the cab fare with his credit card. A VeriFone Systems Inc. product was used to receive payment in the taxi cab. Stelmachers

noticed that the computer generated receipt displayed more than the last five digits of his credit card. In November 2014, Stelmachers filed an instant class action suit against VeriFone, asserting that VeriFone had violated the Fair and Accurate Credit Transaction Act of 2003.

VeriFone responded by filing a motion to dismiss, pursuant to Federal Rule of Civil Procedure 12(b)(6). In its motion, VeriFone stated three arguments, (1) FACTA does not prohibit merchants from printing the first digit of a credit card number; (2) Plaintiff cannot plausibly allege that VeriFone willfully violated FACTA, and thus cannot seek statutory damages; and (3) Plaintiff sued the wrong party because FACTA does not apply to point-of-sale system providers like VeriFone. What do you think of VeriFone's argument? Which argument did the court deem most important and why? How did the court rule? Explain the court's reasoning.

Source: *Stelmachers v. VeriFone Systems, Inc.,* 2015 U.S. Dist. LEXIS 163857.

7. Hewlett-Packard, a U.S.-based company, is famous for its reliable laptop computers and printers. Wilson purchased an HP Pavilion Notebook computer in the summer of 2004 for approximately $1,500. In the fall of 2006, over two years after its purchase and shortly after the limited warranty expired, Wilson's laptop began to display low-power warnings and would run on battery power even when plugged into an AC adapter. Wilson alleges that the problem became worse until he "was unable to utilize the laptop at all." When Wilson contacted HP about his laptop in or about December 2006, HP informed him that his warranty had expired on August 15, 2006, and that he could return the laptop to HP and have his motherboard replaced for over $400, plus shipping and taxes. He opted to have his battery and AC adapter changed for $150, to no effect. Wilson then proceeded to sue, in early 2009, claiming that HP had not disclosed the defect to its consumers. The case was later dismissed after it was presented that the laptop stopped working outside of its warranty span. One day, Wilson turned his laptop on, and it began to smoke and emit fire. When Wilson sued again, he received the same results. Furthermore, the court stated that "complaint's allegations were insufficient to plausibly allege a defect in the HP laptop computers that creates an unreasonable safety risk." Do you believe that HP violated consumer safety law? Why or why not?

Source: *Wilson v. Hewlett-Packard Co.,* 10-16249, (9th Cir. 2012).

8. Comcast Corporation is a worldwide cable company that provides cable services to businesses and individual consumers. In Philadelphia, the company purchased some of its competitors, which increased the percentage of its local ownership to about 75 percent. In early 2003, six individual Comcast customers sued the company, claiming that they paid too much for their cable service as a result of its recent acquisition. What aspects of consumer law does this case involve?

Source: *Behrend v. Comcast Corp.,* 655 F.3d 182, 185 (3d Cir. 2011).

9. On May 5, 2006, Rex Gilbert executed an adjusted rate note with First National Arizona to refinance the existing lien on his home in Ocracoke, North Carolina. Thereafter, First National Arizona transferred its interests in the Gilberts's mortgage to First National Bank of Nevada, First National Nevada transferred its interests to Residential Funding LLC (RFL), and RFL sold its interests to Deutsche Bank Trust Company Americas, as the trustee for Residential Accredit Loans Inc. (RAL). GMAC Mortgage LLC is the subservicer.

The Gilberts defaulted on their loan in 2008, and, on March 12, 2009, David Simpson, acting as the substitute trustee for Deutsche, filed a foreclosure action against the Gilberts in the Hyde County Superior Court. On April 5, 2009, the Gilberts's counsel wrote a letter to GMAC alleging violations of TILA and providing notice that the Gilberts were rescinding the mortgage transaction. GMAC replied, informing the Gilberts that they had not found any basis that would give rise to an extended right of rescission.

On September 14, 2009, while their appeal was pending, the Gilberts filed suit in the Hyde County Superior Court, seeking to enjoin the mortgage foreclosure sales and to rescind their May 6, 2006 loan. The defendants removed the suit to the district court and the district court granted the defendants's motion to dismiss.

However, in light of new evidence, the Gilberts filed a motion for relief pursuant to Rule 60(b) with the district court. On June 15, 2011, the district court filed an order stating that should the Fourth Circuit return jurisdiction to this court, the court would grant the Rule 60(b) motion, dismiss the federal claims for the reasons stated in the July 7, 2010, order (dismissing all of the Gilberts's claims), and remand all state-law claims to Hyde County Superior Court. The Gilberts subsequently filed a motion with the Fourth Circuit U.S. appeals court to reverse and remand the case to the district court.

The Gilberts argued that the district court erred in dismissing their TILA claim on the basis that they had failed to exercise their extended right to rescind in a timely manner. The question at hand is if a borrower must file a lawsuit within three years of closing the loan or if the borrower must only assert his or

her right to rescind through a written notice. What arguments can you make for each side? How did the Fourth Circuit court decide on this matter, and what was its reasoning?

Source: *Gilbert v. Residential Funding LLC,* 678 F. 3d 271 (2012).

10. Penny Morris took metoclopramide from early 2006 to July 2008. However, ingesting the drug for more than twelve weeks had already been contraindicated on FDA-approved labels since 2004 and by "black box" labeling since 2009. Morris developed tardive dyskinesia and akathisia, two movement disorders, as a result of taking the medication and filed suit against generic drug manufacturers PLIVA, TEVA, and Actavis, asserting theories of defective construction and composition of the drug, defective design, breach of express warranty, and inadequate warning. The suit was subsequently stayed to await the Supreme Court's decision in *Mensing,* which held that while state-law "failure to warn" claims are allowed against brand name manufacturers, they were preempted by federal laws against generic manufacturers.

 The district court took into account the Supreme Court decision and dismissed the complaint. Morris moved the district court under Rule 59(e) to amend its earlier ruling based on four assertions: (1) Appellant PLIVA failed to comply with the 2004 FDA-approved label change; (2) the generic defendants failed to properly test their products and report that information; (3) breach of express warranty; and (4) Appellant TEVA may be held liable for a "failure to warn" because of its status as a reference listed drug (RLD) holder.

 What is the "duty of sameness" as defined by the court, and how is it relevant in this case? The court cites the FDCA to show that Morris' claim that "the generic defendants failed to test and inspect the product according to federal law" failed. "[A]ll such proceedings for the enforcement, or to restrain violations of [the FDCA] shall be by and in the name of the United States." Why would legislatures limit claims that manufacturers failed to properly test their drugs to the government? Who are the relevant stakeholders in this case, and how would such a limitation affect the stakeholders?

 Source: *Morris v. Pliva, Inc.,* 713 F. 3d 774 (2013).

11. Verisign, Inc. is a long established tech company that sells Internet domain names. Verisign also runs popular .com and .net top-level domains. In 2014, XYZ.COM, LLC entered the domain business and launched a new top-level domain ending in ".xyz." Verisign subsequently sued XYZ and its founder, Daniel Negari. At the heart of this case is not XYZ's upstart desire to compete with Verisign, but in the alleged false advertisement that XYZ engaged in to market itself. Some of the allegations of improper marketing included blog posts where Negari touted .xyz's high registration numbers and popularity. Another supposedly misleading marketing statement was when XYZ paraphrased an interview between NPR and Negari, stating that an NPR reporter described ".xyz—the Next .com." In addition to the accusations of exaggerated claims, Verisign also contended that XYZ had falsely disparaged the .com domain, against Verisign's interests. For example, Negari said of .com domains in the NPR interview that "[a]ll of the good real estate is taken."

 The district court granted summary judgment to XYZ, proclaiming that Verisign had failed to establish a violation of the Lanham Act on a number of grounds. The district court reasoned that statements like "all of the good real estate is taken" and "it's impossible to find the domain name that you want," were not false or misleading and amounted to puffery. Additionally, the district court stated that even if the statements were false or misleading, it did not violate the Lanham Act unless it deceived consumers.

 Do you think the quoted statements should be considered to be puffery? Why do you think false statements are not actionable unless it deceives consumers? How would a plaintiff alleging false advertising prove to the court that consumers have been deceived?

 Source: *Verisign, Inc. v. XYZ. COM LLC,* 848 F. 3d 292 (2017).

12. Delores Villegas was having problems paying her debts, so she filed for bankruptcy. Weinstein, a debt collector handling some of her debt, sent a letter to the lawyer representing Villegas in bankruptcy. The letter discussed the debt and advised the attorney that the debt might be nondischargeable, and the company would like to settle the matter if counsel did not believe that there was a defense to the claim. After discovering a letter was sent to a third party instead of her, Villegas sued Weinstein for violation of the Fair Debt Collection Practices Act. Do you think the debt collector violated the FDCPA by contacting her bankruptcy attorney instead of Villegas? Why or why not?

 Source: *Villegas v. Weinstein & Riley,* 723 F. Supp. 2d 755 (2010).

13. As part of its ongoing efforts to uncover overhyped health claims in food advertising, the Federal Trade Commission issued an administrative complaint charging the maker of POM Wonderful 100% Pomegranate Juice with making false and unsubstantiated claims that its products will prevent or treat heart disease. The ads in question appeared in national publications such as *Parade, Fitness, The New York*

Times, and *Prevention.* Furthermore, executives at POM made statements, in some instances through an expert endorser, that clinical studies, research, and/or trials prove that drinking 8 ounces of POM juice daily prevents or reduces the risk of heart disease and treats heart disease by decreasing arterial plaque and improving blood flow to the heart. However, the FTC alleges that clinical studies, research, and/or trials do not prove these things. Among other things, large studies that POM funded showed no significant difference between consumption of pomegranate juice and that of a control beverage in thickness of arteries and reduction of plaque. Furthermore, two smaller studies funded by POM Wonderful or its agents showed no significant difference between consumption of pomegranate juice and that of a control beverage on measures of cardiovascular function, and multiple studies funded by POM Wonderful or its agents did not show that POM products reduce blood pressure. The FTC complaint alleges that POM Wonderful's heart disease claims are false and unsubstantiated because many of the scientific studies conducted by POM Wonderful did not show heart disease benefit from use of its products. How do you think this FTC case will turn out? In other words, do you think POM's actions violate Section 5 of the Federal Trade Commission Act? Why or why not? If so, what possible penalties could POM face?

Source: FTC complaint available at www.ftc.gov/os/caselist/0823122/100927pomcmpt.pdf (2010).

The Constitution of the United States of America

Preamble

We the People of the United States, in Order to form a more perfect Union, establish Justice, insure domestic Tranquility, provide for the common defense, promote the general Welfare, and secure the Blessings of Liberty to ourselves and our Posterity, do ordain and establish this Constitution for the United States of America.

Article I

Section 1 All legislative Powers herein granted shall be vested in a Congress of the United States, which shall consist of a Senate and House of Representatives.

Section 2 The House of Representatives shall be composed of Members chosen every second Year by the People of the several States, and the Electors in each State shall have the Qualifications requisite for Electors of the most numerous Branch of the State Legislature.

No Person shall be a Representative who shall not have attained to the age of twenty five Years, and been seven Years a Citizen of the United States, and who shall not, when elected, be an Inhabitant of that State in which he shall be chosen.

Representatives and direct Taxes shall be apportioned among the several States which may be included within this Union, according to their respective Numbers, which shall be determined by adding to the whole Number of free Persons, including those bound to Service for a Term of Years, and excluding Indians not taxed, three fifths of all other Persons.[1] The actual Enumeration shall be made within three Years after the first Meeting of the Congress of the United States, and within every subsequent Term of ten Years, in such Manner as they shall by Law direct. The Number of Representatives shall not exceed one for every thirty Thousand, but each State shall have at Least one Representative, and until such enumeration shall be made, the State of New Hampshire shall be entitled to choose three, Massachusetts eight, Rhode-Island and Providence Plantations one, Connecticut five, New York six, New Jersey four, Pennsylvania eight, Delaware one, Maryland six, Virginia ten, North Carolina five, South Carolina five, and Georgia three.

When vacancies happen in the Representation from any State, the Executive Authority thereof shall issue Writs of Election to fill such Vacancies.

The House of Representatives shall chuse their Speaker and other Officers; and shall have the sole Power of Impeachment.

Section 3 The Senate of the United States shall be composed of two Senators from each State, chosen by the Legislature thereof, for six Years; and each Senator shall have one Vote.[2]

Immediately after they shall be assembled in Consequence of the first Election, they shall be divided as equally as may be into three Classes. The Seats of the Senators of the first Class shall be vacated at the Expiration of the second Year, of the second Class at the Expiration of the fourth Year, and of the third Class at the Expiration of the sixth Year, so that one third may be chosen every second Year; and if Vacancies happen by Resignation, or otherwise, during the Recess of the Legislature of any State, the Executive thereof may make temporary Appointments until the next Meeting of the Legislature, which shall then fill such Vacancies.[3]

No Person shall be a Senator who shall not have attained to the Age of thirty Years, and been nine Years a Citizen of the United States, and who shall not, when elected, be an Inhabitant of that State for which he shall be chosen.

The Vice President of the United States shall be President of the Senate, but shall have no Vote, unless they be equally divided.

The Senate shall chuse their other Officers, and also a President pro tempore, in the Absence of the Vice President, or when he shall exercise the Office of President of the United States.

The Senate shall have the sole Power to try all Impeachments. When sitting for that Purpose, they shall be on Oath or Affirmation. When the President of the United States is tried, the Chief Justice shall preside: And no Person shall

[1] Changed by the Fourteenth Amendment.
[2] Changed by the Seventeenth Amendment.
[3] Changed by the Seventeenth Amendment.

be convicted without the Concurrence of two thirds of the Members present.

Judgment in Cases of Impeachment shall not extend further than to removal from Office, and disqualification to hold and enjoy any Office of honor, Trust or Profit under the United States: but the Party convicted shall nevertheless be liable and subject to Indictment, Trial, Judgment and Punishment, according to Law.

Section 4 The Times, Places and Manner of holding Elections for Senators and Representatives, shall be prescribed in each State by the Legislature thereof; but the Congress may at any time by Law make or alter such Regulations, except as to the Places of chusing Senators.

The Congress shall assemble at least once in every Year, and such Meeting shall be on the first Monday in December, unless they shall by Law appoint a different Day.[4]

Section 5 Each House shall be the Judge of the Elections, Returns and Qualifications of its own Members, and a Majority of each shall constitute a Quorum to do Business; but a smaller Number may adjourn from day to day, and may be authorized to compel the Attendance of absent Members, in such Manner, and under such Penalties as each House may provide.

Each House may determine the Rules of its Proceedings, punish its Members for disorderly Behavior, and with the Concurrence of two thirds, expel a Member.

Each House shall keep a Journal of its Proceedings, and from time to time publish the same, excepting such Parts as may in their Judgment require Secrecy; and the Yeas and Nays of the Members of either House on any question shall, at the Desire of one fifth of those Present, be entered on the Journal.

Neither House, during the Session of Congress, shall, without the Consent of the other, adjourn for more than three days, nor to any other Place than that in which the two Houses shall be sitting.

Section 6 The Senators and Representatives shall receive a Compensation for their Services, to be ascertained by Law, and paid out of the Treasury of the United States. They shall in all Cases, except Treason, Felony and Breach of the Peace, be privileged from Arrest during their Attendance at the Session of their respective Houses, and in going to and returning from the same; and for any Speech or Debate in either House, they shall not be questioned in any other Place.

No Senator or Representative shall, during the Time for which he was elected, be appointed to any civil Office under the Authority of the United States, which shall have been created, or the Emoluments whereof shall have been encreased during such time; and no Person holding any Office under the United States, shall be a Member of either House during his Continuance in Office.

Section 7 All Bills for raising Revenue shall originate in the House of Representatives; but the Senate may propose or concur with Amendments as on other Bills.

Every Bill which shall have passed the House of Representatives and the Senate, shall, before it becomes a Law, be presented to the President of the United States; If he approves he shall sign it, but if not he shall return it, with his Objections to that House in which it shall have originated, who shall enter the Objections at large on their Journal, and proceed to reconsider it. If after such Reconsideration two thirds of that House shall agree to pass the Bill, it shall be sent, together with the Objections, to the other House, by which it shall likewise be reconsidered, and if approved by two thirds of that House, it shall become a Law. But in all such Cases the Votes of both Houses shall be determined by Yeas and Nays, and the Names of the Persons voting for and against the Bill shall be entered on the Journal of each House respectively. If any Bill shall not be returned by the President within ten Days (Sundays excepted) after it shall have been presented to him, the Same shall be a Law, in like Manner as if he had signed it, unless the Congress by their Adjournment prevent its Return, in which Case it shall not be a Law.

Every Order, Resolution, or Vote to which the Concurrence of the Senate and House of Representatives may be necessary (except on a question of Adjournment) shall be presented to the President of the United States; and before the Same shall take Effect, shall be approved by him, or being disapproved by him, shall be repassed by two thirds of the Senate and House of Representatives, according to the Rules and Limitations prescribed in the Case of a Bill.

Section 8 The Congress shall have Power To lay and collect Taxes, Duties, Imposts and Excises, to pay the Debts and provide for the common Defence and general Welfare of the United States; but all Duties, Imposts and Excises shall be uniform throughout the United States.

To borrow Money on the credit of the United States;

To regulate Commerce with foreign Nations, and among the several States, and with the Indian Tribes;

To establish an uniform Rule of Naturalization, and uniform Laws on the subject of Bankruptcies throughout the United States;

To coin Money, regulate the Value thereof, and of foreign Coin, and fix the Standard of Weights and Measures;

To provide for the Punishment of counterfeiting the Securities and current Coin of the United States;

To establish Post Offices and post Roads;

To promote the Progress of Science and useful Arts, by securing for limited Times to Authors and Inventors the exclusive Right to their respective Writings and Discoveries;

To constitute Tribunals inferior to the supreme Court;

[4] Changed by the Twentieth Amendment.

To define and punish Piracies and Felonies committed on the high Seas, and Offences against the Law of Nations;

To declare War, grant Letters of Marque and Reprisal, and make Rules concerning Captures on Land and Water;

To raise and support Armies, but no Appropriation of Money to that Use shall be for a longer Term than two Years;

To provide and maintain a Navy;

To make Rules for the Government and Regulation of the land and naval Forces;

To provide for calling forth the Militia to execute the Laws of the Union, suppress Insurrections and repel Invasions;

To provide for organizing, arming, and disciplining, the Militia, and for governing such Part of them as may be employed in the Service of the United States, reserving to the States respectively, the Appointment of the Officers, and the Authority of training the Militia according to the discipline prescribed by Congress;

To exercise exclusive Legislation in all Cases whatsoever, over such District (not exceeding ten Miles square) as may, by Cession of particular States, and the Acceptance of Congress, become the Seat of the Government of the United States, and to exercise like Authority over all Places purchased by the Consent of the Legislature of the State in which the Same shall be, for the Erection of Forts, Magazines, Arsenals, dock-Yards, and other needful Buildings;—And

To make all Laws which shall be necessary and proper for carrying into Execution the foregoing Powers, and all other Powers vested by this Constitution in the Government of the United States, or in any Department or Officer thereof.

Section 9 The Migration or Importation of such Persons as any of the States now existing shall think proper to admit, shall not be prohibited by the Congress prior to the Year one thousand eight hundred and eight, but a Tax or duty may be imposed on such Importation, not exceeding ten dollars for each Person.

The Privilege of the Writ of Habeas Corpus shall not be suspended, unless when in Cases of Rebellion or Invasion the public Safety may require it.

No Bill of Attainder or ex post facto Law shall be passed.

No Capitation, or other direct, Tax shall be laid, unless in Proportion to the Census of Enumeration herein before directed to be taken.[5]

No Tax or Duty shall be laid on Articles exported from any State.

No Preference shall be given by any Regulation of Commerce or Revenue to the Ports of one State over those of another: nor shall Vessels bound to, or from, one State, be obliged to enter, clear, or pay Duties in another.

No Money shall be drawn from the Treasury, but in Consequence of Appropriations made by Law; and a regular Statement and Account of the Receipts and Expenditures of all public Money shall be published from time to time.

No Title of Nobility shall be granted by the United States: And no Person holding any Office of Profit or Trust under them, shall, without the Consent of the Congress, accept of any present, Emolument, Office, or Title, of any kind whatever, from any King, Prince, or foreign State.

Section 10 No State shall enter into any Treaty, Alliance, or Confederation; grant Letters of Marque and Reprisal; coin Money; emit Bills of Credit; make any Thing but gold and silver coin a Tender in Payment of Debts; pass any Bill of Attainder, ex post facto Law, or Law impairing the Obligation of Contracts, or grant any Title of Nobility.

No State shall, without the Consent of the Congress, lay any Imposts or Duties on Imports or Exports, except what may be absolutely necessary for executing its inspection Laws: and the net Produce of all Duties and Imposts, laid by any State on Imports or Exports, shall be for the Use of the Treasury of the United States; and all such Laws shall be subject to the Revision and Controul of the Congress.

No State shall, without the consent of Congress, lay any Duty of Tonnage, keep Troops, or Ships of War in time of Peace, enter into any Agreement or Compact with another State, or with a foreign Power, or engage in War, unless actually invaded, or in such imminent Danger as will not admit of delay.

Article II

Section 1 The executive Power shall be vested in a President of the United States of America. He shall hold his Office during the Term of four Years, and, together with the Vice President, chosen for the same Term, be elected, as follows

Each state shall appoint, in such Manner as the Legislature thereof may direct, a Number of Electors, equal to the whole Number of Senators and Representatives to which the State may be entitled in Congress: but no Senator or Representative, or Person holding an Office of Trust or Profit under the United States, shall be appointed an Elector.

The Electors shall meet in their respective States, and vote by Ballot for two Persons, of whom one at least shall not be an inhabitant of the same State with themselves. And they shall make a List of all the Persons voted for, and of the Number of Votes for each; which List they shall sign and certify, and transmit sealed to the Seat of the Government of the United States, directed to the President of the Senate. The President of the Senate shall, in the Presence of the Senate and House of Representatives, open all the Certificates, and the Votes shall then be counted. The Person having the greatest Number of Votes shall be the President, if such Number be a Majority of the whole Number of

[5] Changed by the Sixteenth Amendment.

Electors appointed; and if there be more than one who have such Majority, and have an equal Number of Votes, then the House of Representatives shall immediately chuse by Ballot one of them for President; and if no Person have a Majority, then from the five highest on the List the said House shall in like Manner chuse the President. But in chusing the President, the Votes shall be taken by States, the Representation from each State having one Vote; A quorum for this purpose shall consist of a Member or Members from two thirds of the States, and a Majority of all the States shall be necessary to a Choice. In every Case, after the Choice of the President, the Person having the greatest Number of Votes of the Electors shall be the Vice President. But if there should remain two or more who have equal Votes, the Senate shall chuse from them by Ballot the Vice President.[6]

The Congress may determine the Time of chusing the Electors, and the Day on which they shall give their Votes; which Day shall be the same throughout the United States.

No Person except a natural born Citizen, or a Citizen of the United States, at the time of the Adoption of this Constitution, shall be eligible to the Office of President; neither shall any Person be eligible to that Office who shall not have attained to the Age of thirty five Years, and been fourteen Years a Resident within the United States.

In Case of the Removal of the President from Office, or of his Death, Resignation, or Inability to discharge the Powers and Duties of the said Office, the Same shall devolve on the Vice President, and the Congress may by Law provide for the Case of Removal, Death, Resignation or Inability, both of the President and Vice President, declaring what Officer shall then act as President, and such Officer shall act accordingly, until the Disability be removed, or a President shall be elected.[7]

The President shall, at stated Times, receive for his Services, a Compensation, which shall neither be encreased nor diminished during the Period for which he shall have been elected, and he shall not receive within that Period any other Emolument from the United States, or any of them.

Before he enter on the Execution of his Office, he shall take the following Oath or Affirmation:—"I do solemnly swear (or affirm) that I will faithfully execute the Office of President of the United States, and will to the best of my Ability, preserve, protect, and defend the Constitution of the United States."

Section 2 The President shall be Commander in Chief of the Army and Navy of the United States, and of the Militia of the several States, when called into the actual Service of the United States; he may require the Opinion, in writing, of the principal Officer in each of the executive Departments, upon any Subject relating to the Duties of their respective Offices, and he shall have Power to grant Reprieves and Pardons for Offences against the United States, except in Cases of Impeachment.

He shall have Power, by and with the Advice and Consent of the Senate, to make Treaties, provided two thirds of the Senators present concur; and he shall nominate, and by and with the Advice and Consent of the Senate, shall appoint Ambassadors, other public Ministers and Consuls, Judges of the supreme Court, and all other Officers of the United States, whose Appointments are not herein otherwise provided for, and which shall be established by Law; but the Congress may by Law vest the Appointment of such inferior Officers, as they think proper, in the President alone, in the Courts of Law, or in the Heads of Departments.

The President shall have Power to fill up all Vacancies that may happen during the Recess of the Senate, by granting Commissions which shall expire at the End of their next Session.

Section 3 He shall from time to time give to the Congress Information of the State of the Union, and recommend to their Consideration such Measures as he shall judge necessary and expedient; he may, on extraordinary Occasions, convene both Houses, or either of them, and in Case of Disagreement between them, with Respect to the Time of Adjournment, he may adjourn them to such Time as he shall think proper; he shall receive Ambassadors and other public Ministers; he shall take Care that the Laws be faithfully executed, and shall Commission all the Officers of the United States.

Section 4 The President, Vice President and all civil Officers of the United States, shall be removed from Office on Impeachment for, and Conviction of, Treason, Bribery, or other high Crimes and Misdemeanors.

Article III

Section 1 The judicial Power of the United States, shall be vested in one supreme Court, and in such inferior Courts as the Congress may from time to time ordain and establish. The Judges, both of the supreme and inferior Courts, shall hold their Offices during good Behaviour, and shall, at stated Times, receive for their Services, a Compensation, which shall not be diminished during their Continuance in Office.

Section 2 The judicial Power shall extend to all Cases, in Law and Equity, arising under this Constitution, the Laws of the United States, and Treaties made, or which shall be made, under their Authority;—to all Cases affecting Ambassadors, other public Ministers and Consuls;—to all Cases of admiralty and maritime Jurisdiction;—to Controversies to which the United States shall be a party;— to Controversies between two or more States;—between a State and Citizens of another State;[8]—between Citizens of different

[6] Changed by the Twelfth Amendment.
[7] Changed by the Twenty-fifth Amendment.
[8] Changed by the Eleventh Amendment.

States;—between Citizens of the same State claiming Lands under Grants of different States, and between a State, or the Citizens thereof, and foreign States, Citizens or Subjects.

In all Cases affecting Ambassadors, other public Ministers and Consuls, and those in which a State shall be Party, the supreme Court shall have original Jurisdiction. In all the other Cases before mentioned, the supreme Court shall have appellate Jurisdiction, both as to Law and Fact, with such Exceptions, and under such Regulations as the Congress shall make.

The Trial of all Crimes, except in Cases of Impeachment, shall be by Jury: and such Trial shall be held in the State where the said Crimes shall have been committed; but when not committed within any State, the Trial shall be at such Place or Places as the Congress may by Law have directed.

Section 3 Treason against the United States, shall consist only in levying War against them, or in adhering to their Enemies, giving them Aid and Comfort. No Person shall be convicted of Treason unless on the Testimony of two Witnesses to the same overt Act, or on Confession in open Court.

The Congress shall have Power to declare the Punishment of Treason, but no Attainder of Treason shall work Corruption of Blood, or Forfeiture except during the Life of the Person attainted.

Article IV

Section 1 Full Faith and Credit shall be given in each State to the public Acts, Records, and judicial Proceedings of every other State. And the Congress may by general Laws prescribe the Manner in which such Acts, Records and Proceedings shall be proved, and the Effect thereof.

Section 2 The Citizens of each State shall be entitled to all Privileges and Immunities of Citizens in the several States.

A Person charged in any State with Treason, Felony, or other Crime, who shall flee from Justice, and be found in another State, shall on Demand of the executive Authority of the State from which he fled, be delivered up, to be removed to the State having Jurisdiction of the Crime.

No Person held to Service or Labour in one State, under the Laws thereof, escaping into another, shall, in Consequence of any Law or Regulation therein, be discharged from such Service or Labour, but shall be delivered up on Claim of the Party to whom such Service or Labour may be due.[9]

Section 3 New States may be admitted by the Congress into this Union; but no new State shall be formed or erected within the Jurisdiction of any other State; nor any State be formed by the Junction of two or more States, or Parts of States, without the Consent of the Legislatures of the States concerned as well as of the Congress.

The Congress shall have Power to dispose of and make all needful Rules and Regulations respecting the Territory or other Property belonging to the United States; and nothing in this Constitution shall be so construed as to Prejudice any Claims of the United States, or of any particular State.

Section 4 The United States shall guarantee to every State in this Union a Republican Form of Government, and shall protect each of them against Invasion; and on Application of the Legislature, or of the Executive (when the Legislature cannot be convened) against domestic Violence.

Article V

The Congress, whenever two thirds of both Houses shall deem it necessary, shall propose Amendments to this Constitution, or, on the Application of the Legislatures of two thirds of the several States, shall call a Convention for proposing Amendments, which, in either Case, shall be valid to all Intents and Purposes, as Part of this Constitution, when ratified by the legislatures of three fourths of the several States, or by Conventions in three fourths thereof, as the one or the other Mode of Ratification may be proposed by the Congress; Provided that no Amendment which may be made prior to the Year One thousand eight hundred and eight shall in any Manner affect the first and fourth Clauses in the Ninth Section of the first Article; and that no State, without its Consent, shall be deprived of its equal Suffrage in the Senate.

Article VI

All Debts contracted and Engagements entered into, before the Adoption of this Constitution, shall be as valid against the United States under this Constitution, as under the Confederation.

The Constitution, and the Laws of the United States which shall be made in Pursuance thereof; and all Treaties made, or which shall be made, under the Authority of the United States, shall be the supreme Law of the Land; and the Judges in every State shall be bound thereby, any Thing in the Constitution or Laws of any State to the Contrary notwithstanding.

The Senators and Representatives before mentioned, and the Members of the several State Legislatures, and all executive and judicial Officers, both of the United States and of the several States, shall be bound by Oath or Affirmation, to support this Constitution; but no religious Test shall ever be required as a Qualification to any Office or public Trust under the United States.

Article VII

The Ratification of the Conventions of nine States, shall be sufficient for the Establishment of this Constitution between the States so ratifying the Same.

[9] Changed by the Thirteenth Amendment.

Done in Convention by the Unanimous Consent of the States present the Seventeenth Day of September in the Year of our Lord one thousand seven hundred and eighty seven and of the Independence of the United States of America the Twelfth. In witness whereof We have hereunto subscribed our Names.

Amendments

[The first 10 amendments are known as the "Bill of Rights."]

Amendment 1 (Ratified 1791) Congress shall make no law respecting an establishment of religion, or prohibiting the free exercise thereof; or abridging the freedom of speech, or of the press; or the right of the people peaceably to assemble, and to petition the Government for a redress of grievances.

Amendment 2 (Ratified 1791) A well regulated Militia, being necessary to the security of a free State, the right of the people to keep and bear Arms, shall not be infringed.

Amendment 3 (Ratified 1791) No Soldier shall, in time of peace be quartered in any house, without the consent of the Owner, nor in time of war, but in a manner to be prescribed by law.

Amendment 4 (Ratified 1791) The right of the people to be secure in their persons, houses, papers, and effects, against unreasonable searches and seizures, shall not be violated, and no Warrants shall issue, but upon probable cause, supported by Oath or affirmation, and particularly describing the place to be searched, and the persons or things to be seized.

Amendment 5 (Ratified 1791) No person shall be held to answer for a capital, or otherwise infamous crime, unless on a presentment or indictment of a Grand Jury, except in cases arising in the land or naval forces, or in the Militia, when in actual service in time of War or public danger; nor shall any person be subject for the same offence to be twice put in jeopardy of life or limb; nor shall be compelled in any criminal case to be a witness against himself, nor be deprived of life, liberty, or property, without due process of law; nor shall private property be taken for public use, without just compensation.

Amendment 6 (Ratified 1791) In all criminal prosecutions, the accused shall enjoy the right to a speedy and public trial, by an impartial jury of the State and district wherein the crime shall have been committed, which district shall have been previously ascertained by law, and to be informed of the nature and cause of the accusation; to be confronted with the witnesses against him; to have compulsory process for obtaining Witnesses in his favor, and to have assistance of counsel for his defence.

Amendment 7 (Ratified 1791) In Suits at common law, where the value in controversy shall exceed twenty dollars, the right of trial by jury shall be preserved, and no fact tried by a jury, shall be otherwise re-examined in any Court of the United States, than according to the rules of the common law.

Amendment 8 (Ratified 1791) Excessive bail shall not be required, nor excessive fines imposed, nor cruel and unusual punishments inflicted.

Amendment 9 (Ratified 1791) The enumeration in the Constitution, of certain rights, shall not be construed to deny or disparage others retained by the people.

Amendment 10 (Ratified 1791) The powers not delegated to the United States by the Constitution, nor prohibited by it to the States, are reserved to the States respectively, or to the people.

Amendment 11 (Ratified 1795) The Judicial power of the United States shall not be construed to extend to any suit in law or equity, commenced or prosecuted against one of the United States by Citizens of another State, or by Citizens or Subjects of any Foreign State.

Amendment 12 (Ratified 1804) The Electors shall meet in their respective states, and vote by ballot for President and Vice-President, one of whom, at least, shall not be an inhabitant of the same state with themselves; they shall name in their ballots the person voted for as President, and in distinct ballots the person voted for as Vice-President, and they shall make distinct lists of all persons voted for as President, and of all persons voted for as Vice-President, and of the number of votes for each, which lists they shall sign and certify, and transmit sealed to the seat of the government of the United States, directed to the President of the Senate;—The President of the Senate shall, in the presence of the Senate and House of Representatives, open all the certificates and the votes shall then be counted;—The person having the greatest number of votes for President, shall be the President, if such number be a majority of the whole number of Electors appointed; and if no person have such majority, then from the persons having the highest numbers not exceeding three on the list of those voted for as President, the House of Representatives shall choose immediately, by ballot, the President. But in choosing the President, the votes shall be taken by states, the representation from each state having one vote; a quorum for this purpose shall consist of a member or members from two-thirds of the states, and a majority of all the states shall be necessary to a choice. And if the House of Representatives shall not choose a President whenever the right of choice shall devolve upon them, before the fourth day of March next following, then the Vice-President shall act as president, as in the case of the death or other constitutional disability of the

President.[10]—The person having the greatest number of votes as Vice-President, shall be the Vice-President, if such number be a majority of the whole number of Electors appointed, and if no person have a majority, then from the two highest numbers on the list, the Senate shall choose the Vice-President; a quorum for the purpose shall consist of two-thirds of the whole number of Senators, and a majority of the whole number shall be necessary to a choice. But no person constitutionally ineligible to the office of President shall be eligible to that of Vice-President of the United States.

Amendment 13 (Ratified 1865)
Section 1 Neither slavery nor involuntary servitude, except as a punishment for crime whereof the party shall have been duly convicted, shall exist within the United States, or any place subject to their jurisdiction.

Section 2 Congress shall have power to enforce this article by appropriate legislation.

Amendment 14 (Ratified 1868)
Section 1 All persons born or naturalized in the United States, and subject to the jurisdiction thereof, are citizens of the United States and of the State wherein they reside. No State shall make or enforce any law which shall abridge the privileges or immunities of citizens of the United States; nor shall any State deprive any person of life, liberty, or property, without due process of law; nor deny to any person within its jurisdiction the equal protection of the laws.

Section 2 Representatives shall be apportioned among the several States according to their respective numbers, counting the whole number of persons in each State, excluding Indians not taxed. But when the right to vote at any election for the choice of electors for President and Vice President of the United States, Representatives in Congress, the Executive and Judicial officers of a State, or the members of the Legislature thereof, is denied to any of the male inhabitants of such State, being twenty-one years of age, and citizens of the United States, or in any way abridged except for participation in rebellion, or other crime, the basis of representation therein shall be reduced in the proportion which the number of such male citizens shall bear to the whole number of male citizens twenty-one years of age in such State.[11]

Section 3 No person shall be a Senator or Representative in Congress, or elector of President and Vice President, or hold any office, civil or military, under the United States, or under any State, who, having previously taken an oath, as a member of Congress, or as an officer of the United States, or as a member of any State legislature, or as an executive or judicial officer of any State, to support the Constitution of the United States, shall have engaged in insurrection or rebellion against the same, or given aid or comfort to the enemies thereof. But Congress may by a vote of two-thirds of each House, remove such disability.

Section 4 The validity of the public debt of the United States, authorized by law, including debts incurred for payment of pensions and bounties for services in suppressing insurrection or rebellion, shall not be questioned. But neither the United States nor any State shall assume or pay any debt or obligation incurred in aid of insurrection or rebellion against the United States, or any claim for the loss or emancipation of any slave; but all such debts, obligations and claims shall be held illegal and void.

Section 5 The Congress shall have power to enforce, by appropriate legislation, the provisions of this article.

Amendment 15 (Ratified 1870)
Section 1 The right of citizens of the United States to vote shall not be denied or abridged by the United States or by any State on account of race, color, or previous condition of servitude.

Section 2 The Congress shall have power to enforce this article by appropriate legislation.

Amendment 16 (Ratified 1913)
The Congress shall have power to lay and collect taxes on incomes, from whatever source derived, without apportionment among the several States, and without regard to any census or enumeration.

Amendment 17 (Ratified 1913)
The Senate of the United States shall be composed of two Senators from each State, elected by the people thereof, for six years; and each Senator shall have one vote. The electors in each State shall have the qualifications requisite for electors of the most numerous branch of the State legislatures.

When vacancies happen in the representation of any State in the Senate, the executive authority of such State shall issue writs of election to fill such vacancies: Provided, That the legislature of any State may empower the executive thereof to make temporary appointments until the people fill the vacancies by election as the legislature may direct.

This amendment shall not be so construed as to affect the election or term of any Senator chosen before it becomes valid as part of the Constitution.

Amendment 18 (Ratified 1919; Repealed 1933)
Section 1 After one year from the ratification of this article the manufacture, sale, or transportation of intoxicating liquors within, the importation thereof into, or the exportation thereof from the United States and all territory subject to the jurisdiction thereof for beverage purposes is hereby prohibited.

[10] Changed by the Twentieth Amendment.
[11] Changed by the Twenty-sixth Amendment.

Section 2 The Congress and the several States shall have concurrent power to enforce this article by appropriate legislation.

Section 3 This article shall be inoperative unless it shall have been ratified as an amendment to the Constitution by the legislatures of the several States, as provided in the Constitution, within seven years from the date of the submission hereof to the States by the Congress.[12]

Amendment 19 (Ratified 1920)

The right of citizens of the United States to vote shall not be denied or abridged by the United States or by any State on account of sex.

Congress shall have power to enforce this article by appropriate legislation.

Amendment 20 (Ratified 1933)

Section 1 The terms of the President and Vice President shall end at noon on the 20th day of January, and the terms of Senators and Representatives at noon on the 3d day of January, of the years in which such terms would have ended if this article had not been ratified; and the terms of their successors shall then begin.

Section 2 The Congress shall assemble at least once in every year, and such meeting shall begin at noon on the 3d day of January, unless they shall by law appoint a different day.

Section 3 If, at the time fixed for the beginning of the term of the President, the President elect shall have died, the Vice President elect shall become President. If a President shall not have been chosen before the time fixed for the beginning of his term, or if the President elect shall have failed to qualify, then the Vice President elect shall act as President until a President shall have qualified; and the Congress may by law provide for the case wherein neither a President elect nor a Vice President elect shall have qualified, declaring who shall then act as President, or the manner in which one who is to act shall be selected, and such person shall act accordingly until a President or Vice President shall have qualified.

Section 4 The Congress may by law provide for the case of the death of any of the persons from whom the House of Representatives may choose a President whenever the right of choice shall have devolved upon them, and for the case of the death of any of the persons from whom the Senate may choose a Vice President whenever the right of choice shall have devolved upon them.

Section 5 Sections 1 and 2 shall take effect on the 15th day of October following the ratification of this article.

Section 6 This article shall be inoperative unless it shall have been ratified as an amendment to the Constitution by the legislatures of three-fourths of the several States within seven years from the date of its submission.

Amendment 21 (Ratified 1933)

Section 1 The eighteenth article of amendment to the Constitution of the United States is hereby repealed.

Section 2 The transportation or importation into any State, Territory, or possession of the United States for delivery or use therein of intoxicating liquors, in violation of the laws thereof, is hereby prohibited.

Section 3 This article shall be inoperative unless it shall have been ratified as an amendment to the Constitution by conventions in the several States, as provided in the Constitution, within seven years from the date of the submission hereof to the States by the Congress.

Amendment 22 (Ratified 1951)

Section 1 No person shall be elected to the office of the President more than twice, and no person who has held the office of President, or acted as President, for more than two years of a term to which some other person was elected President shall be elected to the office of the President more than once. But this Article shall not apply to any person holding the office of President when this Article was proposed by the Congress, and shall not prevent any person who may be holding the office of President, or acting as President, during the term within which this Article becomes operative from holding the office of President or acting as President during the remainder of such term.

Section 2 This Article shall be inoperative unless it shall have been ratified as an amendment to the Constitution by the legislatures of three-fourths of the several States within seven years from the date of its submission to the States by the Congress.

Amendment 23 (Ratified 1961)

Section 1 The District constituting the seat of Government of the United States shall appoint in such manner as the Congress may direct:

A number of electors of President and Vice President equal to the whole number of Senators and Representatives in Congress to which the District would be entitled if it were a State, but in no event more than the least populous State; they shall be in addition to those appointed by the States, but they shall be considered, for the purposes of the election of President and Vice President, to be electors appointed by a State; and they shall meet in the District and perform such duties as provided by the twelfth article of amendment.

[12] Repealed by the Twenty-first Amendment.

Section 2 The Congress shall have power to enforce this article by appropriate legislation.

Amendment 24 (Ratified 1964)

Section 1 The right of citizens of the United States to vote in any primary or other election for President or Vice President, for electors for President or Vice President, or for Senator or Representative in Congress, shall not be denied or abridged by the United States or any State by reason of failure to pay any poll tax or other tax.

Section 2 The Congress shall have power to enforce this article by appropriate legislation.

Amendment 25 (Ratified 1967)

Section 1 In case of the removal of the President from office or of his death or resignation, the Vice President shall become President.

Section 2 Whenever there is a vacancy in the office of the Vice President, the President shall nominate a Vice President who shall take office upon confirmation by a majority vote of both Houses of Congress.

Section 3 Whenever the President transmits to the President pro tempore of the Senate and the Speaker of the House of Representatives his written declaration that he is unable to discharge the powers and duties of his office, and until he transmits to them a written declaration to the contrary, such powers and duties shall be discharged by the Vice President as Acting President.

Section 4 Whenever the Vice President and a majority of either the principal officers of the executive departments or of such other body as Congress may by law provide, transmit to the President pro tempore of the Senate and the Speaker of the House of Representatives their written declaration that the President is unable to discharge the powers and duties of his office, the Vice President shall immediately assume the powers and duties of the office as Acting President.

Thereafter, when the President transmits to the President pro tempore of the Senate and the Speaker of the House of Representatives his written declaration that no inability exists, he shall resume the powers and duties of his office unless the Vice President and a majority of either the principal officers of the executive department or of such other body as Congress may by law provide, transmit within four days to the President pro tempore of the Senate and the Speaker of the House of Representatives their written declaration that the President is unable to discharge the powers and duties of his office. Thereupon Congress shall decide the issue, assembling within forty-eight hours for that purpose if not in session. If the Congress, within twenty-one days after receipt of the latter written declaration, or, if Congress is not in session, within twenty-one days after Congress is required to assemble, determines by two-thirds vote of both Houses that the President is unable to discharge the powers and duties of his office, the Vice President shall continue to discharge the same as Acting President; otherwise, the President shall resume the powers and duties of his office.

Amendment 26 (Ratified 1971)

Section 1 The right of citizens of the United States, who are eighteen years of age or older, to vote shall not be denied or abridged by the United States or by any State on account of age.

Section 2 The Congress shall have power to enforce this article by appropriate legislation.

Amendment 27 (Ratified 1992)

No law, varying the compensation for the services of the Senators and Representatives, shall take effect, until an election of Representatives shall have intervened.

Sarbanes-Oxley Act of 2002

Note: Only some sections are represented in this appendix. The full act can be found as a link on the text's website at www.mhhe.com/kubasekess2e.

Section 1. Short Title

(a) Short Title. This Act may be cited as the "Sarbanes-Oxley Act of 2002."

Title I—Public Company Accounting Oversight Board

Sec. 101. Establishment; administrative provisions.
Sec. 102. Registration with the Board.
Sec. 103. Auditing, quality control, and independence standards and rules.
Sec. 104. Inspections of registered public accounting firms.
Sec. 105. Investigations and disciplinary proceedings.
Sec. 106. Foreign public accounting firms.
Sec. 107. Commission oversight of the Board.
Sec. 108. Accounting standards. Sec. 109. Funding.

Title II—Auditor Independence

Sec. 201. Services outside the scope of practice of auditors.
Sec. 202. Preapproval requirements.
Sec. 203. Audit partner rotation.
Sec. 204. Auditor reports to audit committees.
Sec. 205. Conforming amendments.
Sec. 206. Conflicts of interest.
Sec. 207. Study of mandatory rotation of registered public accounting firms.
Sec. 208. Commission authority.
Sec. 209. Considerations by appropriate State regulatory authorities.

Title III—Corporate Responsibility

Sec. 301. Public company audit committees.
Sec. 302. Corporate responsibility for financial reports.
Sec. 303. Improper influence on conduct of audits.
Sec. 304. Forfeiture of certain bonuses and profits.
Sec. 305. Officer and director bars and penalties.
Sec. 306. Insider trades during pension fund blackout periods.
Sec. 307. Rules of professional responsibility for attorneys.
Sec. 308. Fair funds for investors.

Title IV—Enhanced Financial Disclosures

Sec. 401. Disclosures in periodic reports.
Sec. 402. Enhanced conflict of interest provisions.
Sec. 403. Disclosures of transactions involving management and principal stockholders.
Sec. 404. Management assessment of internal controls.
Sec. 405. Exemption.
Sec. 406. Code of ethics for senior financial officers.
Sec. 407. Disclosure of audit committee financial expert.
Sec. 408. Enhanced review of periodic disclosures by issuers.
Sec. 409. Real time issuer disclosures.

Title V—Analyst Conflicts of Interest

Sec. 501. Treatment of securities analysts by registered securities associations and national securities exchanges.

Title VI—Commission Resources and Authority

Sec. 601. Authorization of appropriations.
Sec. 602. Appearance and practice before the Commission.
Sec. 603. Federal court authority to impose penny stock bars.
Sec. 604. Qualifications of associated persons of brokers and dealers.

Title VII—Studies and Reports

Sec. 701. GAO study and report regarding consolidation of public accounting firms.

Sec. 702. Commission study and report regarding credit rating agencies.

Sec. 703. Study and report on violators and violations.

Sec. 704. Study of enforcement actions.

Sec. 705. Study of investment banks.

Title VIII—Corporate and Criminal Fraud Accountability

Sec. 801. Short title.

Sec. 802. Criminal penalties for altering documents.

Sec. 803. Debts nondischargeable if incurred in violation of securities fraud laws.

Sec. 804. Statute of limitations for securities fraud.

Sec. 805. Review of Federal Sentencing Guidelines for obstruction of justice and extensive criminal fraud.

Sec. 806. Protection for employees of publicly traded companies who provide evidence of fraud.

Sec. 807. Criminal penalties for defrauding shareholders of publicly traded companies.

Title IX—White-Collar Crime Penalty Enhancements

Sec. 901. Short title.

Sec. 902. Attempts and conspiracies to commit criminal fraud offenses.

Sec. 903. Criminal penalties for mail and wire fraud.

Sec. 904. Criminal penalties for violations of the Employee Retirement Income Security Act of 1974.

Sec. 905. Amendment to sentencing guidelines relating to certain white-collar offenses.

Sec. 906. Corporate responsibility for financial reports.

Title X—Corporate Tax Returns

Sec. 1001. Sense of the Senate regarding the signing of corporate tax returns by chief executive officers.

Title XI—Corporate Fraud and Accountability

Sec. 1101. Short title.

Sec. 1102. Tampering with a record or otherwise impeding an official proceeding.

Sec. 1103. Temporary freeze authority for the Securities and Exchange Commission.

Sec. 1104. Amendment to the Federal Sentencing Guidelines.

Sec. 1105. Authority of the Commission to prohibit persons from serving as officers or directors.

Sec. 1106. Increased criminal penalties under Securities Exchange Act of 1934.

Sec. 1107. Retaliation against informants.

Sec. 302. Corporate Responsibility for Financial Reports.

(a) Regulations Required. The Commission shall, by rule, require, for each company filing periodic reports under section 13(a) or 15(d) of the Securities Exchange Act of 1934 (15 U.S.C. 78m, 78o(d)), that the principal executive officer or officers and the principal financial officer or officers, or persons performing similar functions, certify in each annual or quarterly report filed or submitted under either such section of such Act that—

(1) the signing officer has reviewed the report;

(2) based on the officer's knowledge, the report does not contain any untrue statement of a material fact or omit to state a material fact necessary in order to make the statements made, in light of the circumstances under which such statements were made, not misleading;

(3) based on such officer's knowledge, the financial statements, and other financial information included in the report, fairly present in all material respects the financial condition and results of operations of the issuer as of, and for, the periods presented in the report;

(4) the signing officers—

 (A) are responsible for establishing and maintaining internal controls;

 (B) have designed such internal controls to ensure that material information relating to the issuer and its consolidated subsidiaries is made known to such officers by others within those entities, particularly during the period in which the periodic reports are being prepared;

 (C) have evaluated the effectiveness of the issuer's internal controls as of a date within 90 days prior to the report; and

 (D) have presented in the report their conclusions about the effectiveness of their internal controls based on their evaluation as of that date;

(5) the signing officers have disclosed to the issuer's auditors and the audit committee of the board of directors (or persons fulfilling the equivalent function)—

 (A) all significant deficiencies in the design or operation of internal controls which could adversely affect the issuer's ability to record, process, summarize, and report financial data and have identified for the issuer's auditors any material weaknesses in internal controls; and

 (B) any fraud, whether or not material, that involves management or other employees who have a significant role in the issuer's internal controls; and

(6) the signing officers have indicated in the report whether or not there were significant changes in internal controls or in other factors that could significantly affect internal controls subsequent to the date of their evaluation, including any corrective actions with regard to significant deficiencies and material weaknesses.

(b) Foreign Reincorporations Have No Effect. Nothing in this section 302 shall be interpreted or applied in any way to allow any issuer to lessen the legal force of the statement required under this section 302, by an issuer having reincorporated or having engaged in any other transaction that resulted in the transfer of the corporate domicile or offices of the issuer from inside the United States to outside of the United States.

(c) Deadline. The rules required by subsection (a) shall be effective not later than 30 days after the date of enactment of this Act.

Sec. 306. Insider Trades during Pension Fund Blackout Periods.

(a) Prohibition of Insider Trading during Pension Fund Blackout Periods.

(1) IN GENERAL—Except to the extent otherwise provided by rule of the Commission pursuant to paragraph (3), it shall be unlawful for any director or executive officer of an issuer of any equity security (other than an exempted security), directly or indirectly, to purchase, sell, or otherwise acquire or transfer any equity security of the issuer (other than an exempted security) during any blackout period with respect to such equity security if such director or officer acquires such equity security in connection with his or her service or employment as a director or executive officer.

(2) REMEDY—

 (A) IN GENERAL—Any profit realized by a director or executive officer referred to in paragraph (1) from any purchase, sale, or other acquisition or transfer in violation of this subsection shall inure to and be recoverable by the issuer, irrespective of any intention on the part of such director or executive officer in entering into the transaction.

 (B) ACTIONS TO RECOVER PROFITS—An action to recover profits in accordance with this subsection may be instituted at law or in equity in any court of competent jurisdiction by the issuer, or by the owner of any security of the issuer in the name and in behalf of the issuer if the issuer fails or refuses to bring such action within 60 days after the date of request, or fails diligently to prosecute the action thereafter, except that no such suit shall be brought more than 2 years after the date on which such profit was realized.

(3) RULEMAKING AUTHORIZED—The Commission shall, in consultation with the Secretary of Labor, issue rules to clarify the application of this subsection and to prevent evasion thereof. Such rules shall provide for the application of the requirements of paragraph (1) with respect to entities treated as a single employer with respect to an issuer under section 414(b), (c), (m), or (o) of the Internal Revenue Code of 1986 to the extent necessary to clarify the application of such requirements and to prevent evasion thereof. Such rules may also provide for appropriate exceptions from the requirements of this subsection, including exceptions for purchases pursuant to an automatic dividend reinvestment program or purchases or sales made pursuant to an advance election.

(4) BLACKOUT PERIOD—For purposes of this subsection, the term "blackout period," with respect to the equity securities of any issuer—

 (A) means any period of more than 3 consecutive business days during which the ability of not fewer than 50 percent of the participants or beneficiaries under all individual account plans maintained by the issuer to purchase, sell, or otherwise acquire or transfer an interest in any equity of such issuer held in such an individual account plan is temporarily suspended by the issuer or by a fiduciary of the plan; and

 (B) does not include, under regulations which shall be prescribed by the Commission—

 (i) a regularly scheduled period in which the participants and beneficiaries may not purchase, sell, or otherwise acquire or transfer an interest in any equity of such issuer, if such period is—

 (I) incorporated into the individual account plan; and

 (II) timely disclosed to employees before becoming participants under the individual account plan or as a subsequent amendment to the plan; or

 (ii) any suspension described in subparagraph (A) that is imposed solely in connection with persons becoming participants or beneficiaries, or ceasing to be participants or beneficiaries, in an individual account plan by reason of a corporate merger, acquisition, divestiture, or similar transaction involving the plan or plan sponsor.

(5) INDIVIDUAL ACCOUNT PLAN—For purposes of this subsection, the term "individual account plan" has the meaning provided in section 3(34) of the Employee Retirement Income Security Act of 1974 (29 U.S.C. 1002(34), except that such term shall

not include a one-participant retirement plan (within the meaning of section 101(i)(8)(B) of such Act (29 U.S.C. 1021(i)(8)(B))).

(6) NOTICE TO DIRECTORS, EXECUTIVE OFFICERS, AND THE COMMISSION—In any case in which a director or executive officer is subject to the requirements of this subsection in connection with a blackout period (as defined in paragraph (4)) with respect to any equity securities, the issuer of such equity securities shall timely notify such director or officer and the Securities and Exchange Commission of such blackout period.

(b) Notice Requirements to Participants and Beneficiaries under ERISA.

(1) IN GENERAL—Section 101 of the Employee Retirement Income Security Act of 1974 (29 U.S.C. 1021) is amended by redesignating the second subsection (h) as subsection (j), and by inserting after the first subsection (h) the following new subsection:

(i) Notice of Blackout Periods to Participant or Beneficiary under Individual Account Plan.

"**(1)** DUTIES OF PLAN ADMINISTRATOR—In advance of the commencement of any blackout period with respect to an individual account plan, the plan administrator shall notify the plan participants and beneficiaries who are affected by such action in accordance with this subsection.

"**(2)** NOTICE REQUIREMENTS—

"**(A)** IN GENERAL—The notices described in paragraph (1) shall be written in a manner calculated to be understood by the average plan participant and shall include—

"(i) the reasons for the blackout period,

"(ii) an identification of the investments and other rights affected,

"(iii) the expected beginning date and length of the blackout period,

"(iv) in the case of investments affected, a statement that the participant or beneficiary should evaluate the appropriateness of their current investment decisions in light of their inability to direct or diversify assets credited to their accounts during the blackout period, and

"(v) such other matters as the Secretary may require by regulation.

"**(B)** NOTICE TO PARTICIPANTS AND BENEFICIARIES—Except as otherwise provided in this subsection, notices described in paragraph (1) shall be furnished to all participants and beneficiaries under the plan to whom the blackout period applies at least 30 days in advance of the blackout period.

"**(C)** EXCEPTION TO 30-DAY NOTICE REQUIREMENT—In any case in which—

"(i) a deferral of the blackout period would violate the requirements of subparagraph (A) or (B) of section 404(a)(1), and a fiduciary of the plan reasonably so determines in writing, or

"(ii) the inability to provide the 30-day advance notice is due to events that were unforeseeable or circumstances beyond the reasonable control of the plan administrator, and a fiduciary of the plan reasonably so determines in writing, subparagraph (B) shall not apply, and the notice shall be furnished to all participants and beneficiaries under the plan to whom the blackout period applies as soon as reasonably possible under the circumstances unless such a notice in advance of the termination of the blackout period is impracticable.

"**(D)** WRITTEN NOTICE—The notice required to be provided under this subsection shall be in writing, except that such notice may be in electronic or other form to the extent that such form is reasonably accessible to the recipient.

"**(E)** NOTICE TO ISSUERS OF EMPLOYER SECURITIES SUBJECT TO BLACKOUT PERIOD—In the case of any blackout period in connection with an individual account plan, the plan administrator shall provide timely notice of such blackout period to the issuer of any employer securities subject to such blackout period.

"**(3)** EXCEPTION FOR BLACKOUT PERIODS WITH LIMITED APPLICABILITY—In any case in which the blackout period applies only to 1 or more participants or beneficiaries in connection with a merger, acquisition, divestiture, or similar transaction involving the plan or plan sponsor and occurs solely in connection with becoming or ceasing to be a participant or beneficiary under the plan by reason of such merger, acquisition, divestiture, or transaction, the requirement of this subsection that the notice be provided to all participants and beneficiaries shall be treated as met if the notice required under paragraph (1) is provided to such participants or beneficiaries to whom the blackout period applies as soon as reasonably practicable.

"**(4)** CHANGES IN LENGTH OF BLACKOUT PERIOD—If, following the furnishing of the notice pursuant to this subsection, there is a change in the beginning date or length of the blackout period (specified in such notice pursuant to paragraph (2) (A)(iii)), the

administrator shall provide affected participants and beneficiaries notice of the change as soon as reasonably practicable. In relation to the extended blackout period, such notice shall meet the requirements of paragraph (2)(D) and shall specify any material change in the matters referred to in clauses (i) through (v) of paragraph (2)(A).

"(5) REGULATORY EXCEPTIONS—The Secretary may provide by regulation for additional exceptions to the requirements of this subsection which the Secretary determines are in the interests of participants and beneficiaries.

"(6) GUIDANCE AND MODEL NOTICES—The Secretary shall issue guidance and model notices which meet the requirements of this subsection.

"(7) BLACKOUT PERIOD—For purposes of this subsection—

"(A) IN GENERAL—The term 'blackout period' means, in connection with an individual account plan, any period for which any ability of participants or beneficiaries under the plan, which is otherwise available under the terms of such plan, to direct or diversify assets credited to their accounts, to obtain loans from the plan, or to obtain distributions from the plan is temporarily suspended, limited, or restricted, if such suspension, limitation, or restriction is for any period of more than 3 consecutive business days.

"(B) EXCLUSIONS—The term 'blackout period' does not include a suspension, limitation, or restriction—

"(i) which occurs by reason of the application of the securities laws (as defined in section 3(a)(47) of the Securities Exchange Act of 1934),

"(ii) which is a change to the plan which provides for a regularly scheduled suspension, limitation, or restriction which is disclosed to participants or beneficiaries through any summary of material modifications, any materials describing specific investment alternatives under the plan, or any changes thereto, or

"(iii) which applies only to 1 or more individuals, each of whom is the participant, an alternate payee (as defined in section 206(d)(3)(K)), or any other beneficiary pursuant to a qualified domestic relations order (as defined in section 206(d)(3)(B)(i)).

"(8) INDIVIDUAL ACCOUNT PLAN—

"(A) IN GENERAL—For purposes of this subsection, the term 'individual account plan' shall have the meaning provided such term in section 3(34), except that such term shall not include a one-participant retirement plan.

"(B) ONE-PARTICIPANT RETIREMENT PLAN— For purposes of subparagraph (A), the term 'one-participant retirement plan' means a retirement plan that—

"(i) on the first day of the plan year—
"(I) covered only the employer (and the employer's spouse) and the employer owned the entire business (whether or not incorporated), or
"(II) covered only one or more partners (and their spouses) in a business partnership (including partners in an S or C corporation (as defined in section 1361(a) of the Internal Revenue Code of 1986)),

"(ii) meets the minimum coverage requirements of section 410(b) of the Internal Revenue Code of 1986 (as in effect on the date of the enactment of this paragraph) without being combined with any other plan of the business that covers the employees of the business,

"(iii) does not provide benefits to anyone except the employer (and the employer's spouse) or the partners (and their spouses),

"(iv) does not cover a business that is a member of an affiliated service group, a controlled group of corporations, or a group of businesses under common control, and

"(v) does not cover a business that leases employees."

(2) ISSUANCE OF INITIAL GUIDANCE AND MODEL NOTICE—The Secretary of Labor shall issue initial guidance and a model notice pursuant to section 101(i)(6) of the Employee Retirement Income Security Act of 1974 (as added by this subsection) not later than January 1, 2003. Not later than 75 days after the date of the enactment of this Act, the Secretary shall promulgate interim final rules necessary to carry out the amendments made by this subsection.

(3) CIVIL PENALTIES FOR FAILURE TO PROVIDE NOTICE—Section 502 of such Act (29 U.S.C. 1132) is amended—

(A) in subsection (a)(6), by striking "(5), or (6)" and inserting "(5), (6), or (7)";
(B) by redesignating paragraph (7) of subsection (c) as paragraph (8); and
(C) by inserting after paragraph (6) of subsection (c) the following new paragraph:

"(7) The Secretary may assess a civil penalty against a plan administrator of up to $100 a day from the date of the plan administrator's failure or refusal to provide notice to participants and beneficiaries in accordance with section 101(i). For purposes of this paragraph, each violation with respect to any single participant or beneficiary shall be treated as a separate violation."

(3) PLAN AMENDMENTS—If any amendment made by this subsection requires an amendment to any plan, such plan amendment shall not be required to be made before the first plan year beginning on or after the effective date of this section, if—

(A) during the period after such amendment made by this subsection takes effect and before such first plan year, the plan is operated in good faith compliance with the requirements of such amendment made by this subsection, and

(B) such plan amendment applies retroactively to the period after such amendment made by this subsection takes effect and before such first plan year.

(c) EFFECTIVE DATE—The provisions of this section (including the amendments made thereby) shall take effect 180 days after the date of the enactment of this Act. Good faith compliance with the requirements of such provisions in advance of the issuance of applicable regulations thereunder shall be treated as compliance with such provisions.

Sec. 402. Enhanced Conflict of Interest Provisions.

(a) **Prohibition on Personal Loans to Executives.** Section 13 of the Securities Exchange Act of 1934 (15 U.S.C. 78m), as amended by this Act, is amended by adding at the end the following:

"(k) **Prohibition on Personal Loans to Executives.**

"(1) IN GENERAL—It shall be unlawful for any issuer (as defined in section 2 of the Sarbanes-Oxley Act of 2002), directly or indirectly, including through any subsidiary, to extend or maintain credit, to arrange for the extension of credit, or to renew an extension of credit, in the form of a personal loan to or for any director or executive officer (or equivalent thereof) of that issuer. An extension of credit maintained by the issuer on the date of enactment of this subsection shall not be subject to the provisions of this subsection, provided that there is no material modification to any term of any such extension of credit or any renewal of any such extension of credit on or after that date of enactment.

"(2) LIMITATION—Paragraph (1) does not preclude any home improvement and manufactured home loans (as that term is defined in section 5 of the Home Owners' Loan Act (12 U.S.C. 1464)), consumer credit (as defined in section 103 of the Truth in Lending Act (15 U.S.C. 1602)), or any extension of credit under an open end credit plan (as defined in section 103 of the Truth in Lending Act (15 U.S.C. 1602)), or a charge card (as defined in section 127(c)(4)(e) of the Truth in Lending Act (15 U.S.C. 1637(c)(4)(e)), or any extension of credit by a broker or dealer registered under section 15 of this title to an employee of that broker or dealer to buy, trade, or carry securities, that is permitted under rules or regulations of the Board of Governors of the Federal Reserve System pursuant to section 7 of this title (other than an extension of credit that would be used to purchase the stock of that issuer), that is—

"(A) made or provided in the ordinary course of the consumer credit business of such issuer;

"(B) of a type that is generally made available by such issuer to the public; and

"(C) made by such issuer on market terms, or terms that are no more favorable than those offered by the issuer to the general public for such extensions of credit.

"(3) Rule of Construction for Certain Loans—Paragraph (1) does not apply to any loan made or maintained by an insured depository institution (as defined in section 3 of the Federal Deposit Insurance Act (12 U.S.C. 1813)), if the loan is subject to the insider lending restrictions of section 22(h) of the Federal Reserve Act (12 U.S.C. 375b)."

Sec. 403. Disclosures of Transactions Involving Management and Principal Stockholders.

(a) **Amendment.** Section 16 of the Securities Exchange Act of 1934 (15 U.S.C. 78p) is amended by striking the heading of such section and subsection (a) and inserting the following:

"**Sec. 16. Directors, Officers, and Principal Stockholders.**

"(a) **Disclosures Required.**

"(1) DIRECTORS, OFFICERS, AND PRINCIPAL STOCKHOLDERS REQUIRED TO FILE—Every person who is directly or indirectly the beneficial owner of more than 10 percent of any class of any equity security (other than an exempted security) which is registered pursuant to section 12, or who is a director or an officer of the issuer of such security, shall file

the statements required by this subsection with the Commission (and, if such security is registered on a national securities exchange, also with the exchange).

"(2) TIME OF FILING—The statements required by this subsection shall be filed—

"(A) at the time of the registration of such security on a national securities exchange or by the effective date of a registration statement filed pursuant to section 12(g);

"(B) within 10 days after he or she becomes such beneficial owner, director, or officer;

"(C) if there has been a change in such ownership, or if such person shall have purchased or sold a security-based swap agreement (as defined in section 206(b) of the Gramm-Leach-Bliley Act (15 U.S.C. 78c note)) involving such equity security, before the end of the second business day following the day on which the subject transaction has been executed, or at such other time as the Commission shall establish, by rule, in any case in which the Commission determines that such 2-day period is not feasible.

"(3) CONTENTS OF STATEMENTS—A statement filed—

"(A) under subparagraph (A) or (B) of paragraph (2) shall contain a statement of the amount of all equity securities of such issuer of which the filing person is the beneficial owner; and

"(B) under subparagraph (C) of such paragraph shall indicate ownership by the filing person at the date of filing, any such changes in such ownership, and such purchases and sales of the security-based swap agreements as have occurred since the most recent such filing under such subparagraph.

"(4) ELECTRONIC FILING AND AVAILABILITY—Beginning not later than 1 year after the date of enactment of the Sarbanes-Oxley Act of 2002—

"(A) a statement filed under subparagraph (C) of paragraph (2) shall be filed electronically;

"(B) the Commission shall provide each such statement on a publicly accessible Internet site not later than the end of the business day following that filing; and

"(C) the issuer (if the issuer maintains a corporate website) shall provide that statement on that corporate website, not later than the end of the business day following that filing."

(b) Effective Date. The amendment made by this section shall be effective 30 days after the date of the enactment of this Act.

Sec. 404. Management Assessment of Internal Controls.

(a) Rules Required. The Commission shall prescribe rules requiring each annual report required by section 13(a) or 15(d) of the Securities Exchange Act of 1934 (15 U.S.C. 78m or 78o(d)) to contain an internal control report, which shall—

(1) state the responsibility of management for establishing and maintaining an adequate internal control structure and procedures for financial reporting; and
(2) contain an assessment, as of the end of the most recent fiscal year of the issuer, of the effectiveness of the internal control structure and procedures of the issuer for financial reporting.

(b) Internal Control Evaluation and Reporting. With respect to the internal control assessment required by subsection (a), each registered public accounting firm that prepares or issues the audit report for the issuer shall attest to, and report on, the assessment made by the management of the issuer. An attestation made under this subsection shall be made in accordance with standards for attestation engagements issued or adopted by the Board. Any such attestation shall not be the subject of a separate engagement.

Sec. 802. Criminal Penalties for Altering Documents.

(a) In General. Chapter 73 of title 18, United States Code, is amended by adding at the end the following:

"Sec. 1519. Destruction, alteration, or falsification of records in Federal investigations and bankruptcy

"Whoever knowingly alters, destroys, mutilates, conceals, covers up, falsifies, or makes a false entry in any record, document, or tangible object with the intent to impede, obstruct, or influence the investigation or proper administration of any matter within the jurisdiction of any department or agency of the United States or any case filed under title 11, or in relation to or contemplation of any such matter or case, shall be fined under this title, imprisoned not more than 20 years, or both.

"Sec. 1520. Destruction of corporate audit records

"(a) (1) Any accountant who conducts an audit of an issuer of securities to which section 10A(a) of the Securities Exchange Act of 1934 (15 U.S.C. 78j-1(a)) applies, shall maintain all audit or review workpapers for a period of 5 years from the end of the fiscal period in which the audit or review was concluded.

"(2) The Securities and Exchange Commission shall promulgate, within 180 days, after adequate notice and an opportunity for comment, such rules and regulations, as are reasonably necessary, relating to the retention of relevant records such as workpapers, documents that form the basis of an audit or review, memoranda, correspondence, communications, other documents, and records (including electronic records) which are created, sent, or received in connection with an audit or review and contain conclusions, opinions, analyses, or financial data relating to such an audit or review, which is conducted by any accountant who conducts an audit of an issuer of securities to which section 10A(a) of the Securities Exchange Act of 1934 (15 U.S.C. 78j-1(a)) applies. The Commission may, from time to time, amend or supplement the rules and regulations that it is required to promulgate under this section, after adequate notice and an opportunity for comment, in order to ensure that such rules and regulations adequately comport with the purposes of this section.

"(b) Whoever knowingly and willfully violates subsection (a)(1), or any rule or regulation promulgated by the Securities and Exchange Commission under subsection (a)(2), shall be fined under this title, imprisoned not more than 10 years, or both.

"(c) Nothing in this section shall be deemed to diminish or relieve any person of any other duty or obligation imposed by Federal or State law or regulation to maintain, or refrain from destroying, any document."

Sec. 804. Statute of Limitations for Securities Fraud.

(a) In General. Section 1658 of title 28, United States Code, is amended–

(1) by inserting "(a)" before "Except"; and
(2) by adding at the end the following:

"(b) Notwithstanding subsection (a), a private right of action that involves a claim of fraud, deceit, manipulation, or contrivance in contravention of a regulatory requirement concerning the securities laws, as defined in section 3(a)(47) of the Securities Exchange Act of 1934 (15 U.S.C. 78c(a)(47)), may be brought not later than the earlier of–

"(1) 2 years after the discovery of the facts constituting the violation; or
"(2) 5 years after such violation."

(b) Effective Date. The limitations period provided by section 1658(b) of title 28, United States Code, as added by this section, shall apply to all proceedings addressed by this section that are commenced on or after the date of enactment of this Act.

(c) No Creation of Actions. Nothing in this section shall create a new, private right of action.

Sec. 806. Protection for Employees of Publicly Traded Companies Who Provide Evidence of Fraud.

(a) In General. Chapter 73 of title 18, United States Code, is amended by inserting after section 1514 the following:

"Sec. 1514A. Civil action to protect against retaliation in fraud cases

"(a) Whistleblower Protection for Employees of Publicly Traded Companies. No company with a class of securities registered under section 12 of the Securities Exchange Act of 1934 (15 U.S.C. 78l), or that is required to file reports under section 15(d) of the Securities Exchange Act of 1934 (15 U.S.C. 78o(d)), or any officer, employee, contractor, subcontractor, or agent of such company, may discharge, demote, suspend, threaten, harass, or in any other manner discriminate against an employee in the terms and conditions of employment because of any lawful act done by the employee–

"(1) to provide information, cause information to be provided, or otherwise assist in an investigation regarding any conduct which the employee reasonably believes constitutes a violation of section 1341, 1343, 1344, or 1348, any rule or regulation of the Securities and Exchange Commission, or any provision of Federal law relating to fraud against shareholders, when the information or assistance is provided to or the investigation is conducted by–

"(A) a Federal regulatory or law enforcement agency;
"(B) any Member of Congress or any committee of Congress; or
"(C) a person with supervisory authority over the employee (or such other person working for the employer who has the authority to investigate, discover, or terminate misconduct); or

"(2) to file, cause to be filed, testify, participate in, or otherwise assist in a proceeding filed or about to be filed (with any knowledge of the employer) relating to an alleged violation of section 1341, 1343, 1344, or 1348, any rule or regulation of the Securities and Exchange Commission, or any provision of Federal law relating to fraud against shareholders.

"(b) Enforcement Action.

"(1) IN GENERAL–A person who alleges discharge or other discrimination by any person in violation of subsection (a) may seek relief under subsection (c), by–

"(A) filing a complaint with the Secretary of Labor; or
"(B) if the Secretary has not issued a final decision within 180 days of the filing of the complaint and there is no showing that such delay is due to the bad faith of the claimant, bringing an action at law or equity for de novo review in the appropriate district court of the United States, which shall have jurisdiction over such an action without regard to the amount in controversy.

"(2) PROCEDURE–

"(A) IN GENERAL–An action under paragraph (1)(A) shall be governed under the rules and procedures set forth in section 42121(b) of title 49, United States Code.
"(B) EXCEPTION–Notification made under section 42121(b)(1) of title 49, United States Code, shall be made to the person named in the complaint and to the employer.
"(C) BURDENS OF PROOF–An action brought under paragraph (1)(B) shall be governed by the legal burdens of proof set forth in section 42121(b) of title 49, United States Code.
"(D) STATUTE OF LIMITATIONS–An action under paragraph (1) shall be commenced not later than 90 days after the date on which the violation occurs.

"(c) Remedies.

"(1) IN GENERAL–An employee prevailing in any action under subsection (b)(1) shall be entitled to all relief necessary to make the employee whole.

"(2) COMPENSATORY DAMAGES–Relief for any action under paragraph (1) shall include–

"(A) reinstatement with the same seniority status that the employee would have had, but for the discrimination;
"(B) the amount of back pay, with interest; and
"(C) compensation for any special damages sustained as a result of the discrimination, including litigation costs, expert witness fees, and reasonable attorney fees.

"(d) Rights Retained by Employee. Nothing in this section shall be deemed to diminish the rights, privileges, or remedies of any employee under any Federal or State law, or under any collective bargaining agreement."

(b) Clerical Amendment. The table of sections at the beginning of chapter 73 of title 18, United States Code, is amended by inserting after the item relating to section 1514 the following new item:

"1514A. Civil action to protect against retaliation in fraud cases."

Sec. 807. Criminal Penalties for Defrauding Shareholders of Publicly Traded Companies.

(a) In General. Chapter 63 of title 18, United States Code, is amended by adding at the end the following:

"Sec. 1348. Securities fraud

"Whoever knowingly executes, or attempts to execute, a scheme or artifice–

"(1) to defraud any person in connection with any security of an issuer with a class of securities registered under section 12 of the Securities Exchange Act of 1934 (15 U.S.C. 78l) or that is required to file reports under section 15(d) of the Securities Exchange Act of 1934 (15 U.S.C. 78o(d)); or

"(2) to obtain, by means of false or fraudulent pretenses, representations, or promises, any money or property in connection with the purchase or sale of any security of an issuer with a class of securities registered under section 12 of the Securities Exchange Act of 1934 (15 U.S.C. 78l) or that is required to file reports under section 15(d) of the Securities Exchange Act of 1934 (15 U.S.C. 78o(d));

shall be fined under this title, or imprisoned not more than 25 years, or both."

Sec. 906. Corporate Responsibility for Financial Reports.

(a) In General. Chapter 63 of title 18, United States Code, is amended by inserting after section 1349, as created by this Act, the following:

"Sec. 1350. Failure of corporate officers to certify financial reports

"(a) Certification of Periodic Financial Reports. Each periodic report containing financial statements filed by an issuer with the Securities Exchange Commission pursuant to section 13(a) or 15(d) of the Securities Exchange Act of 1934 (15 U.S.C. 78m(a) or 78o(d)) shall be accompanied by a written statement by the chief executive officer and chief financial officer (or equivalent thereof) of the issuer.

"**(b) Content.** The statement required under subsection (a) shall certify that the periodic report containing the financial statements fully complies with the requirements of section 13(a) or 15(d) of the Securities Exchange Act pf 1934 (15 U.S.C. 78m or 78o(d)) and that information contained in the periodic report fairly presents, in all material respects, the financial condition and results of operations of the issuer.

"**(c) Criminal Penalties.** Whoever—

"(1) certifies any statement as set forth in subsections (a) and (b) of this section knowing that the periodic report accompanying the statement does not comport with all the requirements set forth in this section shall be fined not more than $1,000,000 or imprisoned not more than 10 years, or both; or

"(2) willfully certifies any statement as set forth in subsections (a) and (b) of this section knowing that the periodic report accompanying the statement does not comport with all the requirements set forth in this section shall be fined not more than $5,000,000, or imprisoned not more than 20 years, or both."

(b) Clerical Amendment. The table of sections at the beginning of chapter 63 of title 18, United States Code, is amended by adding at the end the following:

"**1350. Failure of corporate officers to certify financial reports.**"

Glossary

A

absolute privilege A special right, immunity, permission, or benefit given to certain individuals that allows them to make any statements about someone without being held liable for defamation for any false statement made, regardless of intent or knowledge of the falsity of the claim.

acceptance A key factor in the agreement element of a contract; consists of the agreement of one party, the offeree, to the terms of the offer in the contract made by the other party, the offeror.

acceptor A person (drawee) who accepts and signs a draft to agree to pay the draft when it is presented.

accommodation party A party who signs an instrument to provide credit for another party who has also signed the instrument.

accord and satisfaction An arrangement between contracting parties whereby one of the parties substitutes a different performance for his or her original duty under the contract.

accredited investor A private investor who is allowed to accept private securities offerings under certain specific guidelines the SEC sets.

actual cause The determination that the defendant's breach of duty resulted directly in the plaintiff's injury.

actual malice In defamation, either a person's knowledge that his or her statement or published material is false or the person's reckless disregard for whether it is false.

actual notice Notice of agency termination that is given by directly informing third parties, either orally or in writing.

actus reus Latin for "guilty act"; a wrongful behavior that is associated with the physical act of a declared crime.

ad substantiation An FTC standard requiring advertisers to have a reasonable basis for the claims made in their ads.

adhesion contract A contract created by a party to an agreement that is presented to the other party on a take-it-or-leave-it basis. Such contracts are legal but are sometimes rescinded on the grounds of unconscionability and the absence of one party's free will to enter a contract.

administrative agency Any government body created by the legislative branch (e.g., Congress, a state legislature, or a city council) to carry out specific duties.

administrative law The collection of rules and decisions made by administrative agencies to fill in particular details missing from constitutions and statutes.

administrative law judge (ALJ) A judge who presides over an administrative hearing; may attempt to get the parties to settle but has the power to issue a binding decision.

Administrative Procedures Act (APA) Federal legislation that places limitations on how agencies are run and contains very specific guidelines on rule making by agencies.

admission A statement made in court, under oath, or at some stage during a legal proceeding in which a party against whom charges have been brought admits that an oral contract existed, even though the contract was required to be in writing.

adverse possession An involuntary property transfer in which a person acquires ownership of property by treating a piece of real property as his or her own, without protest or permission from the owner.

affirm An appellate court decision that accepts a lower court's judgment in a case that has been appealed.

after-acquired property Property acquired by a debtor after the security arrangement is made.

Age Discrimination in Employment Act (ADEA) of 1967 Federal law that prohibits employers from refusing to hire, discharging, or discriminating in terms and conditions of employment on the basis of an employee or applicant being age 40 or older.

agency The fiduciary relationship that arises when one person consents to have another act on his behalf and subject to his control and the other consents to do so.

agency by ratification An agency relationship that arises when an individual misrepresents himself or herself as an agent for another party and that party accepts the unauthorized act.

agency coupled with an interest An agency relationship that is created for the benefit of the agent, not the principal.

agency relationship The association between one party and an agent who acts on behalf of that party.

agent A party who has the authority to act on behalf of and bind another party.

agreement One of the four elements necessary for a contract; consists of an offer made by one party, the offeror, and the acceptance of the offer by another party, the offeree.

alien corporation A business that is incorporated in a foreign country.

allonge Accompanying a negotiable instrument, a piece of paper that provides room for an endorsement if no room is available on the instrument itself.

alteration An unauthorized change to an instrument that modifies the obligation of a party to the instrument.

alternative dispute resolution (ADR) The resolution of legal problems through methods other than litigation.

Americans with Disabilities Act (ADA) Federal law that prohibits discrimination against employees and job applicants with disabilities.

answer The response of the defendant to the plaintiff's complaint.

apparent agency An agency relationship created by operation of law when one party, by his actions, causes a third party to believe

someone is his agent when that person actually has no authority. Also called *agency by estoppel*.

appellate court A higher court, usually consisting of more than one judge, that reviews the decision and results of a lower court (either a trial court or a lower-level appellate court) when a losing party files for an appeal. Appellate courts do not hold trials but may request additional oral and written arguments from each party; they issue written decisions, which collectively constitute case law or the common law. Also called *court of appellate jurisdiction*.

arbitration A type of alternative dispute resolution in which disputes are submitted for resolution to private nonofficial persons selected in a manner provided by law or the agreement of the parties.

arraignment The first appearance in court by the defendant, at which the defendant is advised of the pending charges, the right to counsel, and the right to trial by jury and he or she enters a plea to the charge.

arrest The action in which a police officer or a person acting under the law takes a person into custody.

articles of incorporation A document that contains basic information about a corporation and is filed with the state.

articles of partnership The written agreement that creates the partnership.

assault A civil wrong that occurs when one person intentionally and voluntarily places another in fear or apprehension of an immediate, offensive physical harm. Assault does not require actual contact.

assignee In a contract, the party who receives the rights of another party (an assignor) to collect what was contractually agreed on in the original contract.

assignment A contracting party's transfer of his or her rights to a contract to a third party.

assignor In a contract, the party who transfers his or her rights to the contract to a third party (an assignee), giving the assignee the right to collect what was contractually agreed on in the original contract.

assumption of the risk A defense whereby the defendant must prove that the plaintiff voluntarily assumed the risk the defendant caused.

attachment The point at which a creditor becomes the secured party that has a security interest in the collateral.

automatic stay After bankruptcy has been filed, a moratorium during which creditors cannot bring or continue action against the debtor or his or her property.

B

bail A thing of value, such as a money bail bond or any other form of property, that is given to the court to allow a person's temporary release from jail and to ensure his or her appearance in court.

bait-and-switch advertising A deceptive practice in which a seller advertises a low-priced item, generally unavailable to the consumer, and then pushes the consumer to buy a more expensive item.

Bankruptcy Abuse Prevention and Consumer Protection Act (BAPCPA) of 2005 A renovation in the system of bankruptcy enacted to address increased numbers of bankruptcy filings, significant losses associated with bankruptcy filings, loopholes and incentives that allowed for abuse, and the financial ability of debtors.

bankruptcy estate The assets that are collected from a debtor who files for bankruptcy.

battery A civil wrong that occurs when one person intentionally and voluntarily brings about a non-consented harmful or offensive contact with a person or something closely associated with him or her. Battery requires an actual contact.

bearer instrument An instrument payable to cash or to whoever is in possession of the instrument.

bench trial A trial before a judge, with the judgment decided by the judge rather than a jury; occurs when the defendant has waived his or her right to a jury trial.

beneficiary A person who can expect to benefit from a relationship.

bilateral contract A promise exchanged for a promise.

binding arbitration clause A contract provision mandating that all disputes arising under the contract must be settled by arbitration.

blank endorsement A payee's or last endorsee's signature on a negotiable instrument.

blue-sky law A law that regulates the offering and sale of purely intrastate securities.

booking After an individual is arrested, the procedure during which the name of the defendant and the alleged crime are recorded in the investigating agency's or police department's records.

bounty payment A government reward for an act that is beneficial to the public.

bribery A corrupt and illegal activity in which a person offers, gives, solicits, or receives money, services, or anything of value to gain an illicit advantage.

brief A written legal argument, which a party presents to a court, that explains why that party to the case should prevail.

burden of proof The duty of the plaintiff or prosecution to establish a claim or allegation by admissible evidence and to prove that the defendant committed all the essential elements of the crime to the jury's or court's satisfaction beyond any reasonable doubt, in order to convict the defendant.

business ethics The use of ethics and ethical principles to solve business dilemmas.

business law The enforceable rules of conduct that govern the actions of buyers and sellers in market exchanges.

business trust A business organization governed by a group of trustees, who operate the trust for beneficiaries.

buyer in the ordinary course of business A person who routinely buys goods in good faith from a person who routinely sells such goods.

bylaws Rules and regulations that govern a corporation's internal management.

C

capacity The legal ability to enter into a binding contract.

case law The collection of legal interpretations made by judges. They are considered to be law unless otherwise revoked by a statutory law. Also known as *common law*.

case or controversy A term used in the U.S. Constitution to describe the structure and requirements of conflicting claims of individuals that can be brought before a federal court for resolution. A case or controversy requires an actual dispute between parties over their legal rights that remains in conflict at the time the case is presented and that is a proper matter for judicial determination. Also referred to as *justifiable controversy*.

cash tender offer A type of takeover in which the aggressor corporation offers to pay the target shareholders cash for their stock.

cashier's check A check for which both drawer and drawee are the same bank.

cease-and-desist order An FTC order requiring that a company stop its illegal behavior.

certificate of deposit A document whereby a bank promises to pay a payee a certain amount of money at a future time.

certificate of incorporation A document certifying that a corporation is incorporated in the state and is authorized to conduct business.

certified check A check that is accepted by the bank from which the funds are drawn.

chain-style business operation A type of franchise in which the franchise operates under the franchisor's business name and is required to follow the franchisor's standards and methods of business operation.

chattel paper A writing that indicates both a monetary obligation and a security interest in specific goods.

check A special draft that orders a bank (the drawee) to pay a specified sum of money to the payee from the drawer's account.

civil law The body of laws that govern the rights and responsibilities either between persons or between persons and their government.

closely held corporation A corporation that does not sell stock to the general public.

collateral The property that is subject to a secured interest.

collective bargaining The process whereby workers organize collectively and bargain with employers regarding the workplace.

commerce clause Clause 3 of Article I, Section 8, of the U.S. Constitution, which authorizes Congress to "regulate Commerce with foreign Nations, and among the several States, and with the Indian Tribes."

commercial reasonableness Reasonable commercial standards of fair dealing, required of merchants in addition to honesty in fact.

commercial speech Speech made by businesses about commercial matters, such as the sale of goods and services. It is protected by the First Amendment.

common stock Corporate stock that does not convey any preferences to its holders.

common-carrier delivery contract A type of contract in which purchased goods are delivered to the buyer via an independent contractor, such as a trucking line.

compensatory damages Money awarded to a plaintiff as reimbursement for her or his losses; based on the amount of actual damage or harm to property, lost wages or profits, pain and suffering, medical expenses, disability, and so on.

complaint A formal written document that begins a civil lawsuit; contains the plaintiff's list of allegations against the defendant along with the damages the plaintiff seeks.

complete performance Contract performance that occurs when all aspects of the parties' duties under the contract are carried out perfectly.

computer crime Crime committed using a computer.

concealment The active hiding of the truth about a material fact.

concurrent conditions In a contract, terms under which each party's performance is conditioned on the performance of the other; occur only when the parties are required to perform for each other simultaneously.

condemnation The legal process by which a transfer of property is made against the protest of the property owner.

condition precedent In a contract, an event that must occur for a party's duty to arise.

condition subsequent In a contract, a future event that terminates the obligations of the parties when it occurs.

conditional contract A contract that becomes enforceable only on the happening or termination of a specified condition.

conditional estate An ownership interest in which the holder has the same interest as that in a fee simple absolute except that this interest is subject to a condition.

conditional privilege A special right, immunity, permission, or benefit given to certain individuals that allows them to make any statements about someone without being held liable for defamation for any false statement made without actual malice.

conditional sales contract A type of contract in which the sale itself is contingent on approval; can be either a sale-on-approval contract or a sale-or-return contract.

conforming goods Goods that conform to contract specifications.

consent order A statement in which a company agrees to stop disputed behavior but does not admit that it broke the law.

consequential damages In a contract, foreseeable damages that result from special facts and circumstances arising outside the contract itself. The damages must be within the contemplation of the parties at the time the breach occurs. Also called *special damages*.

consideration The bargained-for exchange; what each party gets in exchange for his or her promise under a contract.

Consolidated Omnibus Budget Reconciliation Act (COBRA) Federal law that ensures that when employees lose their jobs or have their hours reduced to a level at which they would not be eligible to receive medical, dental, or optical benefits from their employer, the employees will be able to continue receiving

benefits under the employer's policy for up to 18 months by paying the premiums for the policy.

constitutional law The general limits and powers of a government as interpreted from its written constitution.

constructive notice Notice of agency termination that is usually given by publishing an announcement in a newspaper.

constructive trust An equitable trust imposed on a person who wrongfully obtains or holds legal right to property he or she should not possess.

consumer good A good used or bought for use primarily for personal, family, or household purposes.

consumer law A statute or administrative rule serving to protect consumer interests.

contract A promise or set of promises for the breach of which the law gives a remedy or the performance of which the law in some way recognizes a duty.

contract clause The clause in the U.S. Constitution that prohibits the government from unreasonably interfering with an existing contract.

contract under seal A contract that has a seal certifying its legality. Such contracts require no consideration for them to be legal.

contractual capacity The legal ability to enter into a binding agreement.

contributory negligence A defense to negligence whereby the defendant can escape all liability by proving that the plaintiff failed to act in a way that would protect him or her from an unreasonable risk of harm and that the plaintiff's negligent behavior contributed in some way to the plaintiff's accident.

conversion Permanent interference with another's use and enjoyment of his or her personal property.

cooperative An organization formed by individuals to market new products. Individuals in a cooperative pool their resources to gain an advantage in the market.

copyright The protection of the expression of a creative work; that is, protection of the fixed form that expresses the ideas.

corporation A legal entity formed by issuing stock to investors, who are the owners of the corporation.

corporation by estoppel A defective corporation that has conducted business with a third party and therefore cannot deny its status as a corporation to escape liability.

corrective advertising Advertising in which the company explicitly states that formerly advertised claims were untrue. Also called *counteradvertising*.

cost-benefit analysis An economic school of jurisprudence in which all costs and benefits of a law are given monetary values. Laws with the highest ratios of benefits to costs are then preferable to those with lower ratios.

counterclaim A claim made by the defendant against the plaintiff that is filed along with the defendant's answer.

course of dealing Previous commercial transactions between the same parties.

course of performance The history of dealings between the parties in the particular contract at issue.

covenant not to compete An agreement not to compete against a party for a set period of time within a designated geographic area.

cover A buyer's right to substitute goods for those due under a sales or lease agreement when the seller provides nonconforming goods.

creditor An entity to which a debtor owes money.

creditor beneficiary A third party who benefits from a contract in which the promisor agrees to pay the promisee's debt.

creditors' meeting A meeting of all the creditors listed in the Chapter 7 required schedule for liquidation.

criminal law The body of laws that involve the rights and responsibilities an individual has with respect to the public as a whole.

critical thinking in the legal context Examining a legal opinion to search for potential problems in the reasoning.

critical-thinking skills The ability to understand the structure and worth of an argument by evaluating the facts, issue, reasons, and conclusion of that argument.

cross-licensing An illegal contractual arrangement in which two or more parties license each other to use their specified intellectual property *only* on the condition that neither licenses anyone else to use the property without the other's consent.

cure A breaching party's right to provide conforming goods when nonconforming goods were initially delivered; subject to a reasonable time test.

cyber terrorist A hacker whose intention is the exploitation of a target computer or network to create a serious impact, such as the crippling of a communications network or the sabotage of a business or organization, which may have an impact on millions of citizens if the terrorist's attack is successful.

D

de facto **corporation** Latin for "corporation in fact" a corporation that has not substantially met the requirements of the state incorporation statutes.

de jure **corporation** Latin for "lawful corporation" a corporation that has met the mandatory statutory provisions and thus received its certificate of incorporation.

debt securities Securities that represent loans to a corporation.

debtor A party that owes money to another party.

defamation A false statement or an action that harms the reputation or character of an individual, business, product, group, government, or nation.

default Failure to make payments on a loan.

default judgment Judgment for the plaintiff that occurs when the defendant fails to respond to the complaint.

defective corporation A corporation about which an error or omission was made during its incorporation process.

defendant The person or party against whom a civil or criminal lawsuit is filed in a court of law.

delegatee A third party who is not part of the original contract but to whom duties to perform are transferred by one of the contracting parties (a delegator).

delegation A contracting party's (a delegator's) transfer of his or her duty to perform to a third party who is not part of the original contract (a delegatee).

delegator A party in a contract who transfers his or her duties to perform to a third party who is not part of the original contract (a delegatee).

demand instrument A type of draft that allows the payee to demand payment at any time from a holder.

deposition A pretrial sworn and recorded testimony of a witness that is acquired out of court with no judge present.

digital cash Money stored electronically in place of physical currency.

directed verdict A ruling by the judge, after the plaintiff has put forward his or her case but before any evidence is put forward by the defendant, in favor of the defendant because the plaintiff has failed to present the minimum amount of evidence necessary to establish his or her claim.

discharge A written federal court order signed by a bankruptcy judge stating that the debtor is immune from creditor actions to collect debt; that is, a release from liability.

discharged Released from liability for a negotiable instrument.

disclosed principal A principal whose identity is known to a third party. The third party is aware that the agent is making an agreement on behalf of the principal.

discovery The pretrial phase in a lawsuit during which each party requests relevant documents and other evidence from the other side in an attempt to "discover" pertinent facts and avoid any surprises in the courtroom during the trial. Discovery tools include requests for admissions, interrogatories, depositions, requests for inspection, and document production requests.

dishonored Refused; specifically, a payment that has been refused despite a holder presenting an instrument in a timely and proper manner.

disparagement A business tort that occurs when a statement is intentionally used to defame a business product or service.

disparate impact A form of discrimination that arises when an employer's policy or practice appears to apply to everyone equally but its actual effect is that it disproportionately limits employment opportunities for a protected class.

disparate treatment A form of intentional discrimination in which an employee is hired, fired, denied a promotion, or the like, based on membership in a protected class. This is a form of intentional discrimination.

dissolution The first stage in the termination of a partnership.

distributorship A type of franchise in which the franchisor manufactures a product and licenses a dealer to sell the product in an exclusive territory.

dividend A distribution of corporate profits or income ordered by the directors and paid to the shareholders.

domestic corporation A corporation in the state in which it is incorporated.

donee beneficiary A third party who benefits from a contract in which a promisor agrees to give a gift to the third party.

dormant commerce clause A restriction on states' authority that is implied in the commerce clause of the U.S. Constitution: The power given to Congress to enact legislation that affects interstate commerce in effect prohibits a state from passing legislation that improperly burdens interstate commerce.

draft An instrument validating an order by a drawer to a drawee to pay a payee.

drawee The party that must obey an order. In the context of banking, the drawee is the bank that must pay the funds ordered by a customer's check.

drawer The party that writes an order or the person who writes a check.

due diligence defense A defense in which the defendant argues that he or she applied the appropriate degree of attention, care, and research expected of a party in a given situation and had reasonable grounds to believe that certain facts and statements were accurate and had no omission of material facts.

due process clause A clause in the Fifth Amendment of the U.S. Constitution providing that the government cannot deprive an individual of life, liberty, or property without a fair and just hearing.

durable power of attorney A document that specifies that an agent's authority is intended to continue beyond the principal's incapacitation.

duress Any unlawful act or threat exercised on a person whereby the person is forced to enter into an agreement or to perform some other act against his or her will.

duty of accounting An agent's obligation to keep an accurate account of the transactions of money and property made on behalf of the principal.

duty of cooperation A principal's obligation to assist an agent in the performance of the agent's duties and not to interfere with reasonable conduct of the agent.

duty of loyalty An agent's obligation to act in the interest of the principal.

duty of notification An agent's obligation to inform the principal of the agent's actions on the principal's behalf and of all relevant information.

duty of obedience An agent's obligation to follow the lawful instruction and direction of the principal.

duty of performance An agent's obligation to perform the duties specified in the agency agreement and to do so with reasonable skill and care.

duty of reimbursement and indemnification A principal's obligation to reimburse an agent for any expenses or losses incurred by the agent on the principal's behalf.

duty to compensate A principal's obligation to pay an agent for his or her services.

duty to provide safe working conditions A principal's obligation to provide proper equipment, suitable premises, and a safe working environment in accordance with federal and state statutes.

E

e-money Electronic money.

easement An irrevocable right to use some part of another's land for a specific purpose without taking anything from it.

Electronic Communications Privacy Act (ECPA) of 1986 Federal law that extended employees' privacy rights to electronic forms of communication, including email and cellular phones; outlaws the intentional interception of electronic communications and the intentional disclosure or use of the information obtained through such interception.

electronic fund transfer (EFT) The transfer of funds by an electronic terminal, telephone, or computer.

embezzlement A wrongful conversion of another's funds or property by one who is lawfully in possession of those funds or that property.

Employee Retirement Income Security Act (ERISA) Federal law that sets minimum standards for most voluntarily established pension and health plans in private industry to provide protection for individuals in these plans.

enabling legislation A statute that specifies the name, functions, and specific powers of an administrative agency and grants the agency broad powers for the purpose of serving the public interest, convenience, and necessity.

endorsee One who receives an endorsement.

endorser One who issues an endorsement.

English rule A rule that states that the first assignee to give notice of assignment to the obligor is the party with rights to the contract.

entrustment The transfer of goods to a merchant who ordinarily deals in that type of goods. If the merchant subsequently sells them to a good-faith third-party purchaser, the buyer acquires good title to the goods.

equal dignity rule A rule requiring that contracts that would normally fall under the statute of frauds and need a writing if negotiated by the principal must be in writing even if negotiated by an agent.

Equal Pay Act (EPA) of 1963 Federal law that prohibits an employer from paying workers of one gender less than the wages paid to employees of the opposite gender for work that requires equal skill, effort, and responsibility.

equal protection clause A clause in the Fourteenth Amendment of the U.S. Constitution that prevents states from denying "the equal protection of the laws" to any citizen. This clause implies that all citizens are created equal.

equity securities Securities that represent ownership in a corporation.

establishment clause One of two provisions in the First Amendment of the U.S. Constitution that protect citizens' freedom of religion. It prohibits (i) the establishment of a national religion by Congress and (ii) the preference of one religion over another or of religion over nonreligious philosophies in general.

ethical dilemma A question about how one should behave that requires one to reflect on the advantages and disadvantages of the optional choices for various stakeholders.

ethical guideline A simple tool that helps determine whether an action is moral.

ethics The study and practice of decisions about what is good or right.

exchange tender offer A type of takeover in which the aggressor corporation offers to exchange the target shareholders' current stock for stock in the aggressor's corporation.

exculpatory clause A statement in a contract that frees one party (usually the drafter of the agreement) from all liability arising out of performance of the contract, generally based on factors such as consumer ignorance or a great deal of unexplained fine print that serve to deprive the less powerful party of a meaningful choice.

executed A term applied to a contract in which all of the terms have been fully performed.

executive agency An agency that is typically located within the executive branch, under one of the cabinet-level departments. The agency head is appointed by the president with the advice and consent of the Senate.

executory A term applied to a contract in which not all of the terms have been fully performed.

express authority The authority of an agent that arises when the principal explicitly instructs the agent to do something. Also called *actual authority*.

express condition A condition specifically and explicitly stated in a contract and usually preceded by words such as *conditioned on, if, provided that,* or *when.*

express contract A contract in which all the terms are clearly set forth in either written or spoken words.

express warranty Any description of a good's physical nature or its use, in either general or specific circumstances, that becomes part of a contract.

expressed agency An agency created in a written or oral agreement. Also called *agency by agreement.*

extortion A criminal offense in which a person obtains money, property, and/or services from another by wrongfully threatening or inflicting harm on the other's person, property, or reputation. Also called *blackmail.*

F

Fair Labor Standards Act (FLSA) Federal law requiring a minimum wage of a specified amount be paid to all employees in covered industries; also mandates that employees who work more than 40 hours in a week be paid no less than 1½ times their regular wage for all hours beyond 40 worked in a given week.

fair-use doctrine The lawful use of a limited portion of another's work for purposes of criticism, comment, news reporting, teaching, scholarship, and research that does not reduce the commercial value of the protected property.

False Claims Act An act that allows employees to sue employers on behalf of the federal government for fraud against the government. The employee retains a share of the recovery as a reward for his or her efforts.

false pretense The illegal obtaining of another's property by making a materially false representation of an existing fact, with knowledge of the falsity of the representation and with the intent to defraud.

Family and Medical Leave Act (FMLA) Federal act requiring employers to provide all eligible employees with up to 12 weeks of leave during any 12-month period for several family-related occurrences (e.g., birth of a child, care of a sick spouse).

federal preemption A principle asserting the supremacy of federal legislation over state legislation when both pertain to the same matter.

Federal Register The government publication in which an agency publishes each proposed rule, along with an explanation of the legal authority for issuing the rule and a description of how the public can participate in the rule-making process, and later publishes the final rule.

federalism A system of government in which power is divided between a central authority and constituent political units.

fee simple absolute An ownership interest in which the holder has exclusive rights to ownership and possession of the land; the most comprehensive type of estate.

felony A serious crime, such as murder, rape, or robbery, that is punishable by imprisonment for more than one year or by death.

fictitious payee Someone having no right to payment. The UCC fictitious-payee rule governs the liability of companies and their endorsed checks: any check made out to a fictitious payee and endorsed must be honored and is not considered a forgery.

fiduciary A person who has a duty to act primarily for another person's benefit.

financing statement A document that lists the names and addresses of all the parties involved in the transaction, a description of the collateral, and the signature of the debtor.

firm offer An offer made in writing and giving assurances that it will be irrevocable for not longer than three months despite a lack of consideration for the irrevocability.

first appearance The initial appearance of an arrested individual before a judge who determines whether there was probable cause for the arrest. If the judge ascertains that probable cause did not exist, the individual is freed.

first-assignment-in-time rule A rule that states that the first party granted the assignment is the party correctly entitled to the contractual right.

food disparagement A tort that provides ranchers and farmers with a cause of action when someone spreads false information about the safety of a food product.

for-profit corporation A corporation whose objective is to make a profit.

foreign corporation A corporation in a state in which it conducts business but is not incorporated.

forgery A criminal falsification by imitating, impersonating, or altering a real document with the intent to deceive or defraud.

formal contract A contract that must have a special form or must be created in a specific manner.

formal rule making A type of rule making that is used when legislation requires a formal hearing process with a complete transcript; consists of publication of the proposed rule in the *Federal Register*, a public hearing, publication of formal findings, and publication of the final rule if adopted.

franchise A business arrangement between an owner of a trade name or trademark and a person who sells goods or services under the trade name or trademark.

franchise agreement A contract whereby a company (the franchisor) grants permission (a license) to another entity (the franchisee) to use the franchisor's name, trademark, or copyright in the operation of a business and associated sale of goods in return for payment.

franchisee The seller of goods or services under a trade name or trademark in a franchise.

franchisor The owner of the trade name or trademark in a franchise.

fraud An intentional deception that causes harm to another.

fraud in the factum A liability defense available to a party who signs a negotiable instrument without knowing that it is a negotiable instrument.

fraudulent misrepresentation The tort that occurs when a misrepresentation is made with the intent to facilitate personal gain and with the knowledge that it is false.

fraudulent misrepresentation A false representation of a material fact that is consciously false and is intended to mislead the other party.

fraudulent transfer A transfer of property made with intent to defraud creditors or made for an amount significantly lower than the property's fair market value within two years of filing for bankruptcy.

free-exercise clause A clause in the First Amendment of the U.S. Constitution that states that government (state and federal) cannot make a law "prohibiting the free exercise" of religion. This clause is interpreted to include absolute freedom to believe and freedom to act that may face state restriction.

Freedom of Information Act (FOIA) Federal legislation that mandates and facilitates public access to government information and records, including records about oneself. Sensitive information (e.g., on national security) is excluded.

full faith and credit clause The clause of the U.S. Constitution (Article IV, Section 1) mandating that each state must recognize, respect, and enforce the public records, legislative acts, and judicial decisions of the other states.

future interest A person's present right to property ownership and possession in the future.

G

gambling An agreement in which a party pays consideration (money placed during bets) for the chance, or opportunity, to obtain an amount of money or property.

general partnership A partnership in which the partners divide profits and management responsibility and share unlimited personal liability for the partnership's debts.

good faith Honesty in fact.

Good Samaritan statute A statute that exempts from liability a person, such as a physician passerby, who voluntarily renders aid to an injured person but negligently, but not unreasonably negligently, causes injury while rendering the aid.

good title Title acquired from someone who already owns the goods free and clear.

goods All physically existing things that are movable at the time of identification to the contract for sale.

goods-in-bailment contract A type of contract in which the purchased goods are in some kind of storage under the control of a third party, such as a warehouseman.

gratuitous agent An agent who acts without being paid for his or her services.

gross negligence An action committed with extreme reckless disregard for the property or life of another person.

H

hacker A person who illegally accesses or enters another person's or a company's computer system to obtain information or to steal money.

half-truth Information that is true but is not complete.

historical school A school of jurisprudence that uses traditions as the model for future laws and behavior. Also called *tradition* or *custom.*

holder A party in possession of a negotiable instrument.

holder in due course (HDC) An individual who acquires a negotiable instrument in good faith.

hostile takeover A takeover to which the management of the target corporation objects.

hybrid agency An agency that has characteristics of both executive and independent agencies.

hybrid rule making A type of rule making that combines features of formal and informal rule making; consists of publication in the *Federal Register,* a written-comment period, and an informal public hearing with restricted cross-examination.

I

identification with the vulnerable A school of jurisprudence that holds that society should be fair. Particular attention is therefore paid to the poor, the ill, and the elderly.

illusory promise A situation in which a party appears to commit to something but really has not committed to anything. It is not a promise and thus not consideration.

implied authority The authority of an agent that arises by inference from the words and actions of the principal.

implied condition A condition that is not specifically and explicitly stated but is inferred from the nature and language of the contract.

implied contract A contract that arises not from words of agreement but from the conduct of the parties.

implied warranty of fitness for a particular purpose An assurance, inferred in any UCC sale, whereby when a seller or lessor knows or has reason to know (1) why the buyer or lessee is purchasing or leasing the goods in question and (2) that the buyer or lessee is relying on him or her to make the selection, the buyer or lessee has an enforceable warranty if such assurance is false.

implied warranty of merchantability An assurance, inferred in every sale unless clearly disclaimed, that merchantable goods will conform to a reasonable performance expectation. The purchaser must have purchased or leased the good from a merchant.

implied warranty of trade usage An assurance, inferred in the context of certain UCC sales, depending on the circumstances, that can be created through a well-accepted course of dealing or trade usage.

imposter rule A rule that holds that if one obtains a negotiable instrument by impersonating another and endorses it with the impersonated party's signature, the loss falls on the drawer of the instrument.

in pari delicto In equal fault.

in personam **jurisdiction** The power of a court to require a party (usually the defendant) or a witness to come before the court. The court must have personal jurisdiction to enforce its judgments or orders against a party. *In personam* jurisdiction extends only to the state's borders in the state court system and across the court's geographic district in the federal system.

in rem **jurisdiction** The power of a court over the property or status of an out-of-state defendant when that property or status is within the court's jurisdiction area.

incidental beneficiary One who unintentionally gains a benefit from a contract between other parties.

incorporator An individual who applies for incorporation on behalf of a corporation.

independent agency An agency that is typically not located within a government department. It is governed by a board of commissioners, who are appointed by the president with the advice and consent of the Senate.

independent contractor A person who is hired to perform a task for another but who is not the other's employee.

indictment Finding by the grand jury that there is adequate evidence to charge the defendant and bring him or her to trial.

indivisible contract A contract that requires complete performance by both parties.

industry guides Interpretations of consumer laws created by the FTC to encourage businesses to stop unlawful behavior.

informal contract A contract that requires no formalities.

informal rule making A type of rule making in which an agency publishes a proposed rule in the *Federal Register,* considers public comments, and then publishes the final rule. Also called *notice-and-comment rule making.*

information A document, prepared by the prosecutor and presented to the magistrate, which demonstrates that there is enough evidence to charge the defendant and bring him or her to trial.

injunction A court order either forcing a party to do something or prohibiting a party from doing something.

innocent misrepresentation A false statement about a fact material to an agreement that the person who made the statement believed to be true.

insider trading The illegal buying or selling of a corporation's stock or other securities by corporate insiders, such as officers and directors, in breach of a fiduciary duty or some other relationship of trust and confidence, while in possession of material, nonpublic information about the security.

insolvent debtor A debtor who cannot pay debts in a timely fashion.

instrument A writing that serves as evidence of the right to payment of money.

intellectual property Intangible property that is the product of one's mind and not one's hands.

intended beneficiary A third party to a contract whom the contracting parties intended to benefit directly from their contact.

intent The intended purpose or goal of an action, especially in a contract.

intentional interference with contract The tort that occurs when someone intentionally takes an action that will cause a person to breach a contract that she or he has with another.

intentional misrepresentation A false representation of a material fact made by a party who either knows or believes that the factual claim is false or knows that there is no basis for the assertion. Scienter is clear.

intentional tort A civil wrong resulting from an intentional act committed on the person, property, or economic interest of another. Intentional torts include assault, battery, conversion, false imprisonment, intentional infliction of emotional distress, trespass to land, and trespass to chattels.

intermediate scrutiny A standard of review under which a law must be necessary to achieve a substantial, or important, government interest and it must be narrowly tailored to that interest.

interpretive rule A rule that does not create any new rights or duties but is merely a detailed statement of an agency's interpretation of an existing law, including the actions a party is to take to be in compliance with the law.

interrogatories A formal set of written questions that one party to a lawsuit asks the opposing party as part of the pretrial discovery process in order to clarify matters of evidence and help determine in advance what facts will be presented at any trial in the case. The questions must be answered in writing under oath or under penalty of perjury within a specified time. Also called *requests for further information*.

Interstate Commerce Commission (ICC) The first federal administrative agency; created to regulate the anticompetitive conduct of railroads.

J

joint stock company A partnership agreement in which company members hold transferable shares, whereas all the goods of the company are held in the names of the partners.

joint venture An association between two or more parties wherein the parties share profits and management responsibilities with respect to a specific project.

judicial review The power of a court to review legislative and executive actions, such as a law or an official act of a government employee or agent, to determine whether they are constitutional.

L

lack of genuine assent A defense to the agreement of a contract in which the offeree claims that the offeror secured the agreement through improper means, such as duress, fraud, undue influence, or misrepresentation.

Landrum-Griffin Act Federal legislation that primarily governs the internal operations of labor unions. It requires financial disclosures by unions, establishes penalties for financial abuses by union officials, and includes "Labor's Bill of Rights" to protect employees from their own unions.

last-clear-chance doctrine A doctrine used by the plaintiff when the defendant establishes contributory negligence. If the plaintiff can establish that the defendant had the last opportunity to avoid the accident, the plaintiff may still recover, despite being contributorily negligent.

law Rules of conduct in any organized society that are enforced by the governing authority of the community.

lease A transfer of the right to possession and use of goods for a term in return for consideration.

leasehold A possessory interest, but not an ownership interest, transferred by contract.

legal object A purpose that does not violate a statute or public policy.

legal realism A school of jurisprudence that holds that context must be considered as well as law. Context includes factors such as economic conditions and social conditions.

lessee A person who acquires the right to possession and use of goods under a lease.

lessor A person who transfers the right to possession and use of goods under a lease.

letter of credit A binding document that a buyer can request from a bank to guarantee that the payment for goods will be made to the seller.

leveraged buyout (LBO) A takeover-resistance strategy in which a group within the target corporation buys all the corporate stock held by the public, thereby turning the company into a privately held corporation.

liability without fault A legal term that imposes responsibility for damages regardless of the existence of negligence. Also called *strict liability*.

license A revocable right to use another's property temporarily.

life estate An ownership interest in which the holder has the right to possess the property until her or his death.

limited liability company (LLC) An unincorporated business that is taxed like a partnership, with the members paying personal income taxes, and has the limited liability of a corporation.

limited liability partnership (LLP) A partnership in which all the partners assume liability for any partner's professional malpractice to the extent of the partnership's assets.

limited partnership (LP) A partnership consisting of at least one general partner and at least one limited partner in which the general partners assume all liability for the partnership's debts, and the limited partners assume no liability beyond their originally invested capital.

liquidated damages Damages specified as a term of the contract, before a breach of contract occurs.

liquidated debt Debt for which there is no dispute between the parties about the fact that money is owed and the amount of money owed.

liquidation The process in which a debtor turns over all assets to a trustee.

long-arm statute A statute that enables a court to obtain jurisdiction against an out-of-state defendant as long as the defendant has sufficient minimum contacts within the state, such as committing a tort or doing business in the state.

M

mailbox rule The principle that holds that an acceptance is valid when it is placed in the mailbox, whereas a revocation is effective only when received by the offeree. In some jurisdictions, the mailbox rule has been expanded to faxes.

maker A person who has promised to pay a set sum to the holder of a promissory note or certificate of deposit.

manufacturing arrangement A type of franchise in which the franchisor provides the franchisee with a formula or necessary ingredient to manufacture a product.

material breach A substantial breach of a significant term or terms of a contract that excuses the nonbreaching party from further performance under the contract and gives the nonbreaching party the right to recover damages.

med-arb A type of dispute resolution process in which both parties agree to start out in mediation and, if unsuccessful, to move on to arbitration.

mediation A type of intensive negotiation in which disputing parties select a neutral party to help facilitate communication and suggest ways for the parties to solve their dispute.

mens rea Latin for "guilty mind"; the mental state accompanying a wrongful behavior.

merchant Under the UCC, a person who regularly sells goods as his or her business or trade or employs people to sell those goods; anyone who works for a person selling those goods; or anyone who holds himself or herself out as a "merchant" or a "dealer."

merger clause A clause in a written agreement within the statute of frauds that states that the written agreement accurately reflects the final, complete version of the agreement.

minitrial A type of conflict resolution in which lawyers for each side present their arguments to a neutral adviser, who then offers an opinion as to what the verdict would be if the case went to trial. This decision is not binding.

Miranda rights The rights that are read to an arrested individual by a law enforcement agent before the individual is questioned about the commission of a crime.

mirror-image rule The principle that holds that the terms of the acceptance must mirror the terms of the offer; if the terms of the acceptance do not mirror the terms of the offer, no contract is formed, and the attempted acceptance is a counteroffer.

misappropriation theory A theory of insider trading that holds that if an individual wrongfully acquires (misappropriates) and uses inside information for trading for his or her personal gain, that person is liable for insider trading.

misdemeanor A crime that is less serious than a felony and is punishable by a fine and/or imprisonment for less than one year.

misrepresentation An untruthful assertion by one of the parties about a material fact.

mistake An erroneous belief about the facts of a contract at the time the contract is concluded. When a mistake occurs, legal assent is absent.

mixed sale A contract that combines a good with a service or real estate.

mock trial A contrived or imitation trial, with a jury recruited by a jury selection firm, that attorneys sometimes use in preparing for an actual trial to test theories, experiment with arguments, and try to predict the outcome of the real trial.

model laws Laws created to account for the variability of laws among states. These laws serve to standardize the otherwise different interstate laws. Also called *uniform laws*.

modified comparative negligence A defense accepted in some states whereby the defendant is not liable for the percentage of harm that he or she can prove is due to the plaintiff's own negligence if the plaintiff's negligence is responsible for less than 50 percent of the harm; if the defendant establishes that the plaintiff's negligence caused more than 50 percent of the harm, the defendant has no liability.

modify An appellate court decision that grants an alternative remedy in a case; rendered when the court finds that the decision of the lower court was correct but the remedy was not.

motion An application by a party to a judge or a court in a civil case requesting an order in favor of the applicant.

motion for judgment on the pleadings In a civil case, a request made by either party, after pleadings have been entered, that asks a judge or a court to issue a judgment.

motion for summary judgment In a civil case, a request made by either party that asks a judge or a court to promptly and expeditiously dispose of a case without a trial. Any evidence or information that would be admissible at trial under the rules of evidence, such as affidavits, interrogatories, depositions, and admissions, may be considered on a motion for summary judgment. A court may hold oral arguments or decide the motion on the basis of the parties' briefs and supporting documentation alone.

motion to dismiss A request by the defendant that asks a judge or a court in a civil case to dismiss the case because even if all the allegations are true, the plaintiff is not entitled to any legal relief. Also called *demurrer*.

movability The quality of a negotiable instrument that ensures it is mobile and available.

multiple-product order A form of cease-and-desist order issued by the FTC that applies not only to a specified product but also to other products produced by the same firm.

mutual mistake The result of an error by both parties about a material fact, that is, one that is important in the context of a particular contract.

N

National Labor Relations Board (NLRB) An administrative agency created by the Wagner Act to interpret and enforce the National Labor Relations Act.

negligence Behavior that creates an unreasonable risk of harm to others.

negligence per se A doctrine that allows a judge or jury to infer duty and breach of duty from the fact that a defendant violated a criminal statute that was designed to prevent the type of harm that the plaintiff incurred.

negligent misrepresentation A false statement of material fact made by a party who thinks it is true but who would have known the truth about the fact had he or she used reasonable care to discover or reveal it.

negligent tort A civil wrong that occurs when the defendant acts in a way that subjects other people to an unreasonable risk of harm (i.e., the defendant is careless, to someone else's detriment). Negligence claims are usually used to achieve compensation for accidents and injuries.

negotiable instrument A written document signed by a person who makes an unconditional promise to pay a specific sum of money on demand or at a certain time to the holder of the instrument; an acceptable medium for exchanging value from one person to another.

negotiation A bargaining process in which disputing parties interact informally to attempt to resolve their dispute. Also, the transfer of the rights to a negotiable instrument from one party to another.

no-par shares Stock shares that do not have a par value.

nolo contendere A plea in which the defendant does not admit guilt but agrees not to contest the charges.

nominal damages Monetary damages awarded to a plaintiff in a very small amount, typically $1 to $5, to signify that the plaintiff has been wronged by the defendant even though the plaintiff suffered no compensable harm.

nondisclosure A failure to provide pertinent information about a projected contract.

nonprofit corporation A corporation that operates for educational, charitable, social, religious, civic, or humanitarian purposes rather than to earn a profit.

note A promise by the maker of the note to pay the payee of the note.

novation In a contract, the substitution of a third party for one of the original parties; the duties remain the same under the contract, but one original party is discharged and the third party takes that original party's place.

O

obligee A contractual party who agreed to receive something from the other party.

obligor A contractual party who agreed to do something for the other party.

Occupational Safety and Health Act (OSHA) of 1970 Federal law that established the Occupational Safety and Health Administration, the agency responsible for setting safety standards under the act and enforcing the act through inspections and the levying of fines against violators.

offer A key factor in the agreement element of a contract; consists of the terms and conditions set by one party, the offeror, and presented to another party, the offeree.

Omnibus Crime Control and Safe Streets Act of 1968 Federal statute banning employers from listening to the private telephone conversations of employees or disclosing the contents of these conversations. Employers may ban personal calls and monitor calls for compliance as long as they discontinue listening to any conversation once they determine it is personal.

option contract An agreement whereby the offeree gives the offeror a piece of consideration in exchange for the offeror's agreement to hold the offer open for the specified period of time.

order A binding decision rendered by a judge.

order instrument An instrument payable to a specific, named payee.

order of relief An order stating that bankruptcy proceedings can continue.

P

par-value shares Stock shares that have a fixed face value noted on the stock certificate.

parol evidence rule A common law rule that states that oral evidence of an agreement made prior to or contemporaneously with a written agreement is inadmissible when the parties intend the written agreement to be the complete and final version of their agreement.

partial performance An exception to the statute of frauds in which the performance of portions of an unwritten agreement by one or both parties can constitute proof that an oral contract exists between the parties.

partially disclosed principal A principal whose identity is not known by a third party, although the third party is aware that the agent is making an agreement on behalf of a principal. Also called *unidentified principal*.

partnership A voluntary association between two or more people who co-own a business for profit.

patent Protection that grants the holder the exclusive right to produce, sell, and use the object of the patent for 20 years; can be obtained for a product, process, invention, machine, or a plant produced by asexual reproduction.

payee The party that receives the benefit of an order (check, etc.).

peremptory challenge In a jury trial, the right of the plaintiff and the defendant in jury selection to reject, without stating a reason, a certain number of potential jurors who appear to have an unfavorable bias.

perfect tender rule Requirement that the seller deliver the goods in conformity with the contract, down to the last detail.

perfection The series of legal steps a secured party takes to protect its right in the collateral from the other creditors who want their debt returned through the same collateral.

personal defense A liability defense that is not applicable to holders in due course.

personal property Tangible, movable objects.

petit jury A group of 6 or 12 citizens who are summoned to and sworn by the court to hear evidence presented by both sides and render a verdict in a trial.

petty offense A minor crime that is punishable by a small fine and/or imprisonment for less than six months in a jail.

plaintiff The person or party who initiates a lawsuit (also known as an *action*) before a court by filing a complaint with the clerk of the court against the defendant(s). Also known as *claimant* or *complainant*.

plea bargain An agreement in which the prosecutor agrees to reduce charges, drop charges, or recommend a certain sentence if the defendant pleads guilty.

pledge The transfer of collateral to the secured party.

police power Power retained by each state to pass laws that protect the health, safety, and welfare of its citizens.

policy statement A general statement about the directions in which an agency intends to proceed with respect to its rule-making or enforcement activities; has no binding impact on anyone.

political speech Speech that is used to support political candidates or referenda. It is given a high level of protection by the First Amendment as compared to other types of speech.

power of attorney A specific form of express authority that grants an agent specific powers.

precedent A tool used by judges to make rulings on cases on the basis of key similarities to previous cases.

preferential payment A payment made by an insolvent debtor that gives preferential treatment to one creditor over another.

preferred stock Stock that conveys preferences to its holders with respect to assets and dividends.

prejudicial error An error of law that is so significant that it affects the outcome of the case.

prenuptial agreement An agreement two parties enter into before marriage that clearly states the ownership rights each party enjoys in the other party's property. To be enforceable, the agreement must be in writing.

presentment The act of making a demand for the drawee to pay.

presentment warranty A warranty covering the parties accepting an instrument for payment; created to ensure that the accepting or paying party is paying the proper party.

pretrial conference An informal meeting of the judge with the attorneys representing the parties before the actual trial begins.

primarily liable Liable for paying the amount designated on an instrument when it is presented for payment.

principal The party that an agent's authority can bind or act on behalf of.

private corporation A corporation that is created by private persons and does not have government duties.

private law Law that involves suits between private individuals or groups.

private nuisance A nuisance that affects only a single person or a very limited number of individuals.

private trial An ADR method in which a referee is selected and paid by the disputing parties to offer a legally binding judgment in a dispute.

privileges and immunities clause Clause in the U.S. Constitution requiring a state to grant citizens of other states the same legal benefits that it grants its own citizens.

probable cause Any essential element and/or standard by which a lawful officer may make a valid arrest, conduct a personal or property search, or obtain a warrant.

procedural due process The requirement that a government must use fair procedures before depriving a person of his or her life, liberty, or property.

procedural unconscionability Unconscionability that derives from the process of making a contract.

proceeds Something that is exchanged for a debtor's sold collateral.

profit The right to go onto someone's land and take part of the land or a product of it away from the land.

promisee In a third-party beneficiary contract, the party to the contract who owes something to the promisor in exchange for the promise made to the third party.

promisor In a third-party beneficiary contract, the party to the contract who made the promise that benefits the third party.

promissory estoppel The legal enforcement of an otherwise unenforceable contract due to a party's detrimental reliance on the contract.

promoter A person who begins the corporate creation and organization process.

prospectus A written document filed with the SEC that contains a description of a security and other financial information regarding the company offering the security; also distributed as an advertising tool to potential investors.

proximate cause The extent to which, as a matter of policy, a defendant may be held liable for the consequences of his or her actions. In the majority of states, proximate cause requires the plaintiff and the plaintiff's damages to have been foreseeable at the time of the accident. In the minority of states, proximate cause exists if the defendant's actions led to the plaintiff's harm.

proxy A writing signed by a shareholder that authorizes the individual named in the writing to exercise the shareholder's votes (corresponding to his or her shares of stock) at a shareholders' meeting.

proxy solicitation The process of obtaining authority to vote on behalf of shareholders.

public corporation A corporation that is created by government to help administer law.

public figure privilege A special right, immunity, or permission given to people that allows them to make any statements about public figures, typically politicians and entertainers, without being held liable for defamation for any false statement made without malice.

public law Law that involves suits between private individuals or groups and their government.

publicly held corporation A corporation whose stock is available to the public.

punitive damages Compensation awarded to a plaintiff that goes beyond reimbursement for actual losses and is imposed to punish the defendant and deter such conduct in the future.

purchase-money security interest (PMSI) A security interest formed when a debtor uses borrowed money from the secured party to buy the collateral.

pure comparative negligence A defense accepted in some states whereby the defendant is not liable for the percentage of harm that he or she can prove is due to the plaintiff's own negligence.

Q

quasi-contract A court-imposed contractual obligation to prevent unjust enrichment.

question of fact A question about an event or characteristic in a case.

question of law An issue concerning the interpretation or application of a law.

R

Racketeer Influenced and Corrupt Organizations (RICO) Act A U.S. law that provides extended penalties for criminal acts performed as part of an ongoing criminal organization.

ratify To approve an unauthorized agent's signature on an instrument.

rational-basis test The lowest standard of review; requires a law to be designed to protect a legitimate state interest and be rationally related to that interest.

reaffirmation agreement An agreement in which the debtor agrees to pay a debt even though it could have been discharged in bankruptcy.

real defense A liability defense that applies universally to all parties.

real property Land and everything permanently attached to it.

reasonable person standard A measurement of the way members of society expect an individual to act in a given situation.

recognizance An obligation in which a party acknowledges in court that he or she will perform some specified act and/or pay a price on failure to do so.

red-herring prospectus A prospectus with a warning written in red print at the top of the page telling investors that the registration has been filed with the SEC but is not yet approved.

reg-neg A type of rule making in which representatives of concerned interest groups and of the involved government agency participate in mediated bargaining sessions to reach an agreement, which is forwarded to the agency.

relative permanence The quality of a negotiable instrument that ensures its longevity.

remand An appellate court decision that returns a case to the trial court for a new trial or for limited hearing on a specified subject matter; rendered when the court decides that an error was committed that may have affected the outcome of a case.

reply A response by the plaintiff to the defendant's counterclaim.

request to produce documents In a lawsuit, the right of a party to examine and copy papers of the opposing party that are relevant to the case. A legal request may be made, and the categories of the documents must be stated to allow the other party to know what documents he or she must produce.

res ipsa loquitur A doctrine that allows the judge or jury to infer that, more likely than not, the defendant's negligence was the cause of the plaintiff's harm, even though there is no direct evidence of the defendant's lack of due care.

rescind To cancel a contract.

rescission The termination of a contract.

respondeat superior Latin for "let the superior speak"; the principle by which liability for harm caused by an agent/employee is held by the principal/employer.

restatements of the law Summaries of common law rules in a particular area of the law. Restatements do not carry the weight of law but can be used to guide interpretations of particular cases.

restitution The return of any property given up under a contract.

restricted securities Securities that have limited transferability and are usually issued in a private placement.

retained earnings Profits that a corporation keeps.

reverse An appellate court decision that overturns the judgment of a lower court, concluding that the lower court was incorrect and its verdict cannot be allowed to stand.

right of first refusal A method of restricting stock transferability whereby a corporation or its shareholders have the right to purchase any shares of stock offered for resale by a shareholder within a specified time frame.

ripeness The readiness of a case for a decision to be made. The goal is to prevent premature litigation for a dispute that is insufficiently developed. A claim is not ripe for litigation if it rests on contingent future events that may not occur as anticipated or may not occur at all.

S

S corporation A corporation that enjoys the tax status of a partnership.

Sabbath laws Laws prohibiting the performance of certain activities on Sundays.

sale The passing of title from a seller to a buyer for a price.

sale-on-approval contract A contract in which the seller allows the buyer to take possession of the goods before deciding whether to complete the contract by making the purchase.

sale-or-return contract A contract in which the buyer and seller agree that the buyer may return the goods at a later time.

Sarbanes-Oxley Act An act that criminalizes specific nonaudit services when provided by a registered accounting firm to an audit client; also increases the punishment for a number of white-collar offenses. Also known as the *Public Company Accounting Reform and Investor Protection Act of 2002*.

scienter Intent or knowledge.

search warrant A court order that authorizes law enforcement agents to search for or seize items specifically described in the warrant.

secondarily liable Liable for paying the amount designated on an instrument if the primarily liable party defaults.

secured interest An interest in personal property or fixtures that secures payment or performance to a creditor.

secured party The party that holds the interest in a secured property.

secured transaction A transaction in which the payment of a debt is guaranteed by personal property the debtor owns.

security A financial instrument designated as a note, stock, or bond or any other instrument named in the Securities Act of 1933.

security agreement An agreement in which the debtor gives the secured interest to the secured party.

self-dealing Any instance in which directors or officers make decisions that violate their corporate duty of loyalty.

self-tender offer A takeover-resistance strategy in which the target corporation offers to buy its shareholders' stock.

service of process The procedure by which a court delivers a copy of the statement of claim or other legal documents, such as a summons, complaint, or subpoena, to the defendant.

severable contract A contract that contains multiple parts that can each be performed separately.

sexual harassment Unwelcome sexual advances, requests for sexual favors, and other verbal or physical conduct of a sexual nature that makes submission a condition of employment or a factor in employment decisions or that creates an intimidating, hostile, or offensive work environment. The two types are hostile environment and *quid pro quo.*

shadow jury An unofficial jury, provided by a jury selection firm, that sits in during the actual trial and deliberates at the end of each day to evaluate for the attorneys how each side is doing.

shareholder's derivative suit A lawsuit filed by a shareholder on behalf of the corporation.

shareholder's direct suit A lawsuit filed by a shareholder against the corporation.

shareholders Investors in a corporation, who own the corporation.

shelter principle The principle that when an item is transferred, the transferee acquires all the rights the transferor had to the item.

short-swing profits Profits made from the sale of company stock within any six-month period by a statutory insider.

signature liability Liability attributed because of a party's signature on an instrument.

simple delivery contract Type of contract in which purchased goods are transferred to the buyer from the seller either at the time of the sale or at some time later by the seller's delivery.

slander of quality A business tort that occurs when false spoken statements criticize a business product or service and result in a loss of sales.

slander of title A business tort that occurs when false published statements are related to the ownership of the business property.

social responsibility of business The expectations that a community places on the actions of firms inside that community's borders.

sole proprietor The single person at the head of a sole proprietorship.

sole proprietorship A business in which one person (sole proprietor) is in control of the management and profits.

special endorsement An endorser's signature accompanied by the name of the endorsee.

specific performance An order of the court requiring a breaching party to fulfill the terms of the contract.

stakeholders The groups of people affected by a firm's decisions.

stale check A check that is not presented to a bank within six months of its date.

standing The legal right of a party or an individual to bring a lawsuit by demonstrating to the court sufficient connection to and harm from the law or action challenged (i.e., the plaintiff has to demonstrate that he or she is harmed or will be harmed). Otherwise, the court will dismiss the case, ruling that the plaintiff "lacks standing" to bring the suit.

stare decisis "Standing by the decision" a principle stating that rulings made in higher courts are binding precedent for lower courts.

statute of frauds State-level legislation that addresses the enforceability of contracts that fail to meet the requirements set forth in the statute; serves to protect promisors from poorly considered oral contracts by requiring certain contracts to be in writing.

statutory insiders Certain large stockholders, executive officers, and directors who are deemed insiders by the Securities Exchange Act of 1934.

statutory law The assortment of rules and regulations put forth by legislatures.

stock certificate A document that serves as a stockholder's proof of ownership in a corporation.

stock warrant A type of security issued by a corporation (usually together with a bond or preferred stock) that gives the holder the right to purchase a certain amount of common stock at a stated price.

stop-payment order An order by a drawer that instructs the drawee bank not to pay an issued check.

stored-value card A plastic card that contains data regarding the value of the card, thereby allowing EFTs to be made.

strict liability Liability in which responsibility for damages is imposed regardless of the existence of negligence. Also called *liability without fault.*

strict scrutiny The most exacting standard of review used by the courts in determining the constitutionality of a statute; requires a compelling government interest and the least restrictive means of attaining that objective.

strict-liability offense Offense for which no *mens rea* is required.

strict-liability tort A civil wrong that occurs when the defendant takes an action that is inherently dangerous and cannot ever be undertaken safely, no matter what precautions the defendant takes. In such situations, a defendant is liable for the plaintiff's damages without any requirement for the plaintiff to prove that the defendant was negligent.

subject-matter jurisdiction The power of a court over the type of case presented to it.

subpoena *duces tecum* An order to appear and bring specified documents.

subpoena An order to appear at a particular time and place and provide testimony.

subscriber An investor who agrees to purchase stock in a new corporation.

subscription agreement An agreement between promoters (persons raising capital for a new corporation) and subscribers (investors), in which the subscribers agree to purchase stock in the new corporation.

substantial evidence Refers to evidence that a reasonable mind could accept as adequate to support a conclusion. It is defined as "more than a scintilla but less than preponderance" and consists of "such relevant evidence as a reasonable person would accept as adequate to support a conclusion."

substantial impairment A concept, used to modify the perfect tender rule, whereby a buyer can revoke acceptance of goods and, with installment contracts, a buyer/lessee can reject an installment of a particular item only if the defects substantially impair the value of the goods.

substantial performance Contract performance that occurs when nearly all the terms of the agreement have been met, there has been an honest effort to complete all the terms, and there has been no willful departure from the terms of the agreement.

substantive due process The requirement for laws depriving an individual of life, liberty, or property to be fair and not arbitrary.

substantive unconscionability Unconscionability that derives from one-sided, unjust, or overly harsh substance in a contract.

summary jury trial An abbreviated trial that leads to a nonbinding jury verdict.

summons A legal document issued by a court and addressed to a defendant that notifies him or her of the lawsuit and how and when to respond to the complaint. A summons may be used in both civil and criminal proceedings.

supremacy clause Article VI, Paragraph 2, of the U.S. Constitution, which states that the Constitution and all laws and treaties of the United States constitute the supreme law of the land. Thus, any state or local law that directly conflicts with the U.S. Constitution or federal laws or treaties is void.

syndicate An investment group that comes together for the explicit purpose of financing a specific large project.

T

Taft-Hartley Act Federal legislation designed to curtail some of the powers that unions had acquired under the Wagner Act; designates certain union actions as unfair. Also called *Labor-Management Relations Act*.

takeover The purchase of one corporation's stock by another corporation as a means of gaining control of the selling corporation.

takings clause A clause in the Fifth Amendment of the U.S. Constitution requiring that when government uses its power to take private property for public use, it must pay the owner just compensation, or fair market value, for the property.

tender of delivery Requirement that the seller or lessor have and hold conforming goods at the disposal of the buyer or lessee and give the buyer or lessee reasonable notification to enable him or her to take delivery.

tender offer A type of takeover in which the aggressor corporation offers the target shareholders a price above their stock's current market value.

tender An offer by a party in a contract to perform, along with being ready, willing, and able to perform, a duty outlined in that contract.

termination statement An amendment to a financing statement, which states that the debtor has no obligation to the secured party.

third-party beneficiary contract A contract created when two parties enter into a contract agreement with the intended purpose of benefiting a third party.

time instrument A type of draft that allows the payee to collect payment only at a specific time in the future.

tippee An individual who receives confidential information from an insider.

tipper/tippee theory A theory of insider trading that holds that any individual (tippee) who acquires material inside information as a result of an insider's (tipper's) breach of duty has engaged in insider trading.

tipper An insider who gives inside information to someone.

Title VII of the Civil Rights Act (CRA) of 1964 Federal law (amended by the Civil Rights Act of 1991) that protects employees against discrimination based on race, color, religion, national origin, and sex; also prohibits harassment based on the same protected categories.

tombstone advertisement A print advertisement in a format similar to that of a tombstone that announces a forthcoming sale of securities.

tort A violation of another person's rights or a civil wrongdoing that does not arise out of a contract or statute; primary types are intentional, negligent, and strict-liability torts.

tortfeasor A person who commits an intentional or through-negligence tort that causes a harm or loss for which a civil remedy may be sought.

trade dress The overall appearance and image of a product.

trade libel A business tort that occurs when false printed statements criticize a business product or service and result in a loss of sales.

trade secret A process, product, method of operation, or compilation of information that gives a businessperson an advantage over his or her competitors.

trademark A distinctive mark, word, design, picture, or arrangement that is used by a producer in conjunction with a product and tends to cause consumers to identify the product with the producer.

transfer warranty A warranty regarding a negotiable instrument and its transfer; created by a party who transfers the instrument.

traveler's check An order that is payable on demand, is drawn on or through a bank, is designated by the phrase *traveler's check*, and requires a counter-signature by the person whose signature appears on the check.

trespass to personal property Temporary interference with another's use or enjoyment of his or her personal property.

trespass to realty A tort that occurs when someone goes on another's property without permission or places something on another's property without permission.

trial court A court in which most civil or criminal cases start when they first enter the legal system. The parties present evidence and call witnesses to testify. Trial courts are referred to as *courts of common pleas* or *county courts* in state court systems and as *district courts* in the federal system. Also called *court of original jurisdiction* and *court of first instance*.

trustee A person who operates a trust for beneficiaries in a business trust. In bankruptcy proceedings, an individual who takes over administration of a debtor's estate.

tying arrangement An illegal agreement in which the sale of one product is tied to the sale of another.

U

unconscionable A term applied to a contract in which one party has so much more bargaining power than the other party that the powerful party dictates the terms of the agreement and eliminates the other party's free will.

undisclosed principal A principal whose existence is not known by a third party. That is, the third party does not know an agent is acting on behalf of a principal.

undue influence The situation in which one person has taken advantage of his or her dominant position in a relationship to persuade the other person unduly to the extent that the dominant person's persuasive efforts have interfered with the other person's ability to make his or her own decision.

unemployment compensation A state system, created by the Federal Unemployment Tax Act (FUTA), that provides unemployment compensation to qualified employees who lose their jobs.

unenforceable A term applied to a contract that, because of a law, cannot be enforced by the courts.

unfair competition The act of competing with another not to make a profit but for the sole purpose of driving that other out of business.

unforeseen circumstances Events that a reasonable person would not be expected to anticipate.

Uniform Commercial Code (UCC) A statutory source of contract law in the United States applicable to transactions involving the sale of goods. The UCC was created in 1952 and adopted by all 50 states, the District of Columbia, and the Virgin Islands; it may be modified by each state to reflect the wishes of the state legislature.

unilateral contract A promise exchanged for an act.

unilateral mistake The result of an error by one party about a material fact, that is, one that is important in the context of a particular contract.

United Nations Convention on Contracts for the International Sale of Goods (CISG) The legal structure for international sales, including business-to-business sales contracts.

universalization test The ethical guideline that urges us to consider, before we act, what the world would be like if everyone acted in this way.

unliquidated debt Debt for which the parties either dispute the fact that any money is owed or agree that some money is owed but dispute the amount.

usage of trade Any practice that members of an industry expect to be part of their dealings.

usury The lending of money at an exorbitant or unlawful rate of interest.

V

valid A term applied to a contract that includes all four elements of a contract: agreement (offer and acceptance), consideration, contractual capacity, and legal object.

values Positive abstractions that capture our sense of what is good and desirable.

venue (1) The place where a hearing takes place. Its geographic location is determined by each state's statutes and based on where the parties live or where the event occurred or the alleged wrong was committed. (2) A legal doctrine relating to the selection of a court with subject-matter and personal jurisdiction that is the most appropriate geographic location for the resolution of the dispute.

vest To mature, as in the maturation of rights that allow a party to act legally on the rights.

vicarious liability The liability or responsibility imposed on a person, a party, or an organization for damages caused by another; most commonly used in relation to employment, with the employer held vicariously liable for damages caused by its employees.

virus A computer program that rearranges, damages, destroys, or replaces computer data.

void A term applied to a contract that is not valid because its object is illegal or it has some defect that is so serious that it is not a contract.

void title Not true title; for example, the title held by someone who knowingly or unknowingly purchased stolen goods.

voidable A term applied to a contract that one or both parties have the ability to either withdraw from or enforce.

voidable title Title that occurs in situations in which the contract between the original parties would be void, but the goods have already been sold to a third party.

voir dire The process of questioning potential jurors to ensure that the jury will be made up of unbiased individuals.

W

Wagner Act The first major piece of federal legislation adopted explicitly to encourage the formation of labor unions and provide for collective bargaining between employers and unions as a means of obtaining the peaceful settlement of labor disputes.

warranty An assurance, either express or implied, by one party that the other party can rely on its representations of fact; in sales, a binding promise regarding a product in the event that the product does not meet the manufacturer's or seller's promises.

warranty liability Liability attributed when the transfer of an instrument breaches a warranty associated with an instrument.

warranty of title An assurance, inferred in every UCC sales transaction, that the seller has good and valid title to the goods and has the right to transfer title free and clear of any liens, judgments, or infringements of intellectual property rights of which the buyer does not have knowledge.

watered stock Stock that is issued to individuals below its fair market value.

WH approach (to ethical decision making) A set of ethical guidelines that urges us to consider whom an action affects, the purpose of the action, and how we view its morality.

white-collar crime A variety of nonviolent illegal acts against society that occur most frequently in the business context.

winding up The second stage in the termination of a partnership.

workers' compensation laws State laws that provide for financial compensation to employees or their dependents when a covered employee is injured on the job.

writ of certiorari A Supreme Court order, issued after the Court decides to hear an appeal, mandating that the lower court send to the Supreme Court the record of the appealed case.

writing A type of documentation that shows contractual intent and satisfies the statute of frauds requirement.

Name Index

A

Abado, Rami, 528-529
Abdulwahab, Adley, 383
Adondakis, Spyridon "Sam," 562
Agnes, L. Jay, 540
Albers, Emery, 415
Alito, Samuel, 42, 99
Allen, Tracy, 506
Allen, Woody, 150
Anderson, Cynthia, 597
Anderson, Jeffrey R., 524
Andrews, Erin, 150
Anthony, Casey, 401
Anthony, Sheila, 605
Arpaio, Joseph, 62
Artis, Theresa, 225
Aurigemma, Ralph M., 274

B

Bacanovic, Peter, 642
Baker, Jeremy, 261
Baker, Kim, 325
Baldwin, Donald, 574-575
Bannasch, Eric, 562
Barnett, Michael, 150
Barrie, Kelly, 507
Barrie, Matthew, 507
Baum, Corey, 404
Bayh, Birch, 271
Bentrim, Brent, 424
Berman, Mark J., 528
Beverly, Martina, 63
Bhayani, Dipak N., 486, 502, 504-505
Blakemore, Elaine, 149
Bogle, Rick, 221-222
Bohatch, Colette, 489
Bohn, Edward G., 528
Bolden, Shirley, 427-429, 432-433
Bongiorno, Annette, 121
Bongiovi, Joseph, Jr., 141-142, 144, 148-149, 163, 166
Bono, Giuseppe G., 603
Bonventure, Daniel, 121
Borchard, Patrick, 554
Borges, Joe, 451
Borges, Maria, 451
Bradshaw, Barbara, 170
Brandt, Timothy, 619
Breyer, Stephen G., 42, 182, 188
Brown, Walter, 63
Brumitt, Richard, 300-301
Brzonkala, Christy, 89-90
Budd, Linda, 371
Burton, Ken, 335
Bush, George W., 80, 178

C

Campbell, Richard, 400-4701
Caraway, Hugh L, 420-421
Chalfant, Robert, 30
Chambers, David, 223, 230
Chew, David, 385
Chomiak, Eleonora, 539
Chomiak, Leon, 539
Chomiak, Vladimir, 539
Christie, David, 498
Christmas, Rashad Jemal, 425
Clark, Dacorious, 140
Clark, Fred, Jr. "Dave," 123
Connaughton, Kyle, 169
Conwell, Dennis, 349
Cooper, T. Christian, 400-401
Coors, Cheryl, 270-271
Cortez, Carlos, 452-453
Cortez, Suzanne, 452-453
Coyle, Kevin L., 528
Crawford, James, 89
Crawford, Stanley, 615-616
Creek, Sara, 304
Crupi, Joann, 121

D

Darrell, Patrick, 317
Davis, Barbaralette, 483
Davis, Kim, 103
Dolezal, Todd Russell, 29
Don, Lawrence, 424
Dreifus, Claudia, 219-220
Drew, Lori, 264
Duffy, Tim, 300-301
Dusenbery, Larry Dean, 110-111
Dykstra, Lenny, 121

E

Ebbers, Bernard, 24
Ebener, Angela, 623
Ehling, Deborah, 582-583
Ellingsworth, Marykate, 595-596
Ells, Steven, 269
Eubanks, Carolene, 170

F

Ferrier, Angela, 596
Filipe, Luis, 484-485
Fitzgerald, Jr., Patrick, 76
Fitzgerald, Patrick, Jr., 76
Flaska, Kenneth, 426
Fontanetta, Adrian Philip, 158
Foxx, Anthony, 574
Francis, Joe, 237

G

Gallagher, Kathleen, 484
Gandee, Patty, 252
Garcia, Aurelio, 506
Garcia, Hugo V., 506
Garcia, Jason, 377
Garrison, Jay, 194
Garrison, Pamela, 194
Garrison, Penny, 458
Garrison, Thomas, 194
Gates, Bill, 18
Gerstenecker, Adam, 214
Gerstenecker, Janice, 214
Gerstenecker, Julie, 214
Gilbert, Rex, 624-625
Ginsburg, Ruth Bader, 42, 105-106
Glen, Alexander, 498
Glynn, Brian, 507
Gorsuch, Neil, 42
Gotlin, Gary D., 603
Goyal, Sandy, 562
Greenfield, Harley, 528-529
Greenfield, Stacey, 562
Griffin, Madeleine, 400
Groisman, Elias, 215
Groome, Odalis, 401
Groome, Sam, 401
Gyabaah, Adwoa, 215

H

Haber, Michael, 299
Hamer, Louisa, 217-218
Hammond, Keith, 507-508
Harrison, Lisa, 597
Hartman, Virginia, 149
Hirayama, Kazutoshi, 29
Holleyman, Robert, 330
Holt, Derek, 597-598
Holyfield, Evander, 272
Horvath, Jon, 562
Hosch, Edward, 300
Houser, Charles, 232
Hughes, Chris, 509
Huizenga, Amy, 401
Hunter, Eric, 597-598
Hurst, Natalie, 165

J

Jackson, Shantel, 169
Jacobus, Cheryl, 146-147
Jahn, Robert, 264
James Lewis, 31
Janev, Ilija, 539
Janousek, Christopher, 400
Jobs, Steve, 171
Johnson, Jeh, 62
Johnson, Marshall, 300
Jones, Julie, 141
Jones, Vernon T., 383
Jones, William, 253
Jordan, Michael, 195, 254-255, 260-261, 267, 418

K

Kagan, Elena, 42
Kaplan, Michael, 618-619
Kara, Maher, 118
Karimi, Mike, 506
Kavanaugh, Brett, 42
Kellmer, James, 540
Kelly, Jack, 270
Kelo, Susette, 177-178
Keri, Gabe, 149
King, Jim, 241
Kinney, Kenneth, 484
Kirchoff, Virgil, 540-541
Knafel, Karla, 254-255, 260-261, 267
Kozlowski, Dennis, 29
Kuykendall, David Chad, 436

L

Ladd, Ann, 168-169
Lake, Larry, 121
Lamb, Kip, 507
Laurent, Yves Saint, 182-183
Lawrence, Larry, 476
Leahy, Patrick, 17
Lefkowitz, Morris, 206-207
Lehner, James C., 506-507
Leonard, John, 196
Leonard, Ralph, 466
Lewandowsky, Corey, 146-147
Lewis, James, 31
Light, Randolph, J., 424
Lin, Jeremy, 184
Lissiak, Victor, 424
Lodge, Mary, 424-425
Logan, James, 400
Lopez, Adrian, 235

I-1

Louboutin, Christian, 182–183
Love, Jane, 529
Lyles, Bryan, 169–170

M

Madoff, Bernard, 18, 121, 258
Margeson, Jennifer, 225–226
Margeson, John, 225–226
Marie-Messier, Jean, 563
Marrero, Lionel, 576
Marshall, John, 86
Martin, Jonathan, 169
Martin, Rafael, 119
Matsuo, Ryuichi, 29
Mayfield, Christine, 404
Mayweather, Jr., Floyd, 169
McAdams, Tony, 75
McBride, John, 109
McCann, Gertrude, 526
McCann, Ron, 526
McCann, William, 526
McCollum, Andrew, 509
McDonald, John, 489
McDonald, Robert E., 112
McKinney, Michael, 503
McLemore, Sydney, 165
McMorris, Jerry, 466
Meier, Megan, 264
Meisner, Gary, 526
Merlo, Larry, 20
Meyer, Alan, 498
Middleton, Arthur, 540
Milam, David, 109–110
Miller, Scott H., 524
Milliken, Michael, 115
Mills, James, 402, 404–405
Minkow, Barry, 122
Mohammed, Raheek, 252
Moore, Mark, 349
Morris, Penny, 625
Morrison, Antonio J., 89
Moskovitz, Dustin, 509
Murray, Alan, 444
Murray, Catherine, 444

N

Napolitano, Janet, 62
Needham, Ella, 458
Negari, Daniel, 625
Nickel, Bobby Dean, 576–577
Nixon, Richard M., 79

O

Obama, Barack, 62, 78, 90
O'Connor, Sandra Day, 178
O'Guin, Alex, 160
O'Guin, Shaun, 160
O'Hara, Jerome, 121
Olesnyckyj, Myron F., 118
Olson, Aaron, 151
Oprea, Don, 300–301
Orbison, Roy, 188
Ortega, Selwin, 507–508
Orwell, George, 179
O'Shaugnessy, Jon, 270

P

Pacheco, Ali, 474
Packingham, Lester Gerard, 110
Paine, Louis, 489
Palmer, Joseph, 451
Parvizian, Leon, 543–544
Passante, Anthony, Jr., 216
Pearson, Monty J., 266
Perez, George, 121

Perez, Joana, 231
Peterson, Robert, 329–330
Peterson, William, 329
Pfeifer, Shabsi, 215
Phillips, Elgene, 227
Phillips, Gloria, 227
Phillips, James, J., 76
Piaget, Jean, 251
Plotkin, Leah, 138
Polleys, Nancy, 453
Pompa, Jill, 485
Pompa, William, 485
Poulsen, Lori, 598
Powers, Richard, 489
Pratt, John, 498–499

R

Raben, Gregory, 554
Ray, Rob, 562
Ray, Shawn, 598
Reddick, Timothy, 214
Rederford, Janelle, 322
Reed, Clyde, 109
Reed, Sheila, 194
Reeves, Aubrey, 372
Ricciardi, Alvin, 614
Rice, Mill, 434
Richmond, Kristin, 26
Riedl, Bryan, 241–242
Riedl, Roseann, 241–242
Riley, David Leon, 100–101
Riley, Larry, 246
Rizo, Aileen, 597
Roberts, John G., 42, 93, 101
Robinson, Kenneth, 554–555
Roe, Jane, 590
Rojas, Roxanne, 102
Romeo, Anthony, 598
Roopchand, Rosemarie, 252
Roosevelt, Franklin, 8
Rossman, Paula E., 604–605
Ruzicka, Jill, 219–220
Ryder, William, 401

S

Sabo, Zolt, 539
Samalia, Adrian, 193–194
Sandall, John, 329–330
Sanders, Carlous, 436
Savage, Michael, 483
Saverin, Eduardo, 509
Scalia, Anthony, 100
Scammel, Toby, 139–140
Scott, Mary W., 257
Segovia, Roberto, 400–401
Seilberling, Frank, 515
Shack, William E., Jr., 374, 382, 397
Shapiro, Morton, 76
Shkrell, Martin, 16
Sidway, Franklin, 217–218
Simmons, Dorothy, 223, 230–231
Simmons, Louise, 223, 230–231
Simmons, Vincent, 223, 230–231
Simpson, David, 624
Smith, Sammie, 140
Sotomayor, Sonia, 6, 42
Souza, Dawnmarie, 590–591
Soza, Andres, 453
Spitler, Timothy, 451
Stelmachers, Paul M, 623–624
Stenlund, Paula, 484
Stern, Donald, 28–29
Stern, Gillas, 364
Stern, Scott, 364
Stern, Staci, 364
Stewart, Gabriel, 598

Stewart, Martha, 138, 542, 550, 553–554, 559
Stone, Dale, 508
Story, William E., 217–218
Strickland, Jerry, 220–221
Strickland, Wanda, 220–221
Strubel, Abigail, 623
Stuart, Dan, 334
Stumpf, John, 19
Sullivan, Scott, 24
Sullivan, Walter, 141–142, 144, 148–149, 163, 166
Swift, Taylor, 184a
Swinehart, Suzanne, 149–150
Swoboda, Caroline, 158–159

T

Tam, Simon Shiao, 183
Tanke, Thomas, 119
Tebow, Tim, 184
Thomas, Clarence, 42
Thomas, Marilyn, 350
Thomas, Nancy, 350
Thrasher, Gail, 194
Thrasher, James Hall, 194
Thrasher, William S., 194
Tortora, Jesse, 562
Treadway, Lonnie, 507
Triffin, Robert, 411, 425
Trudeau, Kevin, 599–601, 611, 620
Trump, Donald J., 146–147, 607
Tucker, Dennis Dwayne, 138
Turner, James, 169
Tyson, Mike, 272, 289, 291, 296

U

Uecker, Bob, 168–169

V

Van Der Werf, Martin, 133
Vargas, Jose, 484–485
Velasquez, Daniel, 577
Vicidomine, Bernadette, 371
Vidojkovic, Stefan, 539
Villegas, Delores, 625
Vogel, Alice, 424
Vokes, Audrey, 270

W

Waksai, Samuel, 542
Wallingford, Jerry, 165
Wang, Yan Ju, 73–74
Wankel, Laura, 598
Weaver, Edward, 139
Webster, Pricilla, 359
Weidler, Charlotte, 338
Weimert, David, 138–139
Weinman, Bernie, 483
Weinstein, Harvey, 18
Wells, Theodore, 169
Wentworth, J. G., 300
Westheim, Paul, 338
Westheim-Frenk, Marianna, 338
Whittington, Don, 180
Wiggins, Matthew, Jr., 400
Wilhelm, Kenneth, 215
Williams, Jamie, 140
Winans, R. Foster, 556
Woodman, Trent, 247–248
Woskob, Helen, 491
Woskob, Leah Beth, 491
Woskob, Victor, 491
Wurie, Brima, 100–101

Z

Zucker, Kiren, 75n
Zuckerberg, Mark, 509, 512, 536

Subject Index

A

AAA. *See* American Arbitration Association (AAA)
Aaron Olson, Appellant, v. CenturyLink (2013), 151
Abandoned property, 180-181, 192
Abandonment doctrine, 193-194
Abbott Laboratories, 63
ABCW. *See* Anchor BanCorp Wisconsin (ABCW)
Abercrombie & Fitch, 461-462
Absolute privilege, 147
　applicability of, 149-150
　defined, 148
　See also Conditional privilege
ACA. *See* Affordable Care Act (ACA)
Acceptance
　elements of offer, 209-212
　of goods, 357
　mirror-image rule, 210-211
　offer, 197, 204, 212
　of property transfers, 176, 322
　revocation of, 367-369
　in sales and lease contracts, 333, 346
Acceptor, 412, 414
Accommodation party, 414, 422
Accord and satisfaction, 227-230, 309
Accounting
　demand an, 469, 481
　duty of, 468, 481
Accounts, 432
Accredited investor, 548
A & C Environmental Inc., 418
Ackerman v. Sobol Family Partnership, LLP (2010), 460
Acquiring corporation (aggressor), 533-544
Acquitted, 130
Acts of parties, discharge by, 477-478
Actual authority. *See* Express authority (actual authority)
Actual cause (cause in fact), 157, 167
Actual fraudulent transfer, 442
Actual malice, 141, 149
Actual notice of termination, 476-477
Actual possession, 177
Actus reus (guilty act), 113, 136
AC Wharton, 483
ADA. *See* Americans With Disabilities Act (ADA)
ADEA. *See* Age Discrimination in Employment (ADEA) of 1967
Adelphia, 524
Adequacy of consideration, 220-222, 229
Adhering party. *See* Presented party (adhering party)
Adhesion unconscionability, 245
Adjudicated a habitual drunkard, 250
Adjudicated insane, 250
Adjudication, power of, 65-68
Administrative action, 550, 560
Administrative agency, 82
　defined, 65, 81
　powers of, 65-68
　purpose of, 65-68
　types of, 68-69
　as unofficial fourth branch of U.S. government, 65
Administrative Dispute Resolution Act, 54
Administrative law
　defined, 7, 65, 81
　federal agencies, 7
Administrative law judge (ALJ), 67-68, 81
Administrative Procedures Act (APA), 69, 82, 84
Administrator's Interpretation, 83

Admiralty case, 35
Admission, 279, 297
ADR. *See* Alternative dispute resolution (ADR)
Adrian Lopez v. Kmart Corporation (2015), 235
Ad substantiation, 602
Adverse possession, 176-177
Advertising, 96
　direct-to-consumer drug ads, 622-623
　laws business ethics, in Sweden, 23
　of offer, 206-207, 212
　in Saudi Arabia, 204
　tombstone, 547
　See also Deceptive advertising; Federal Trade Commission (FTC)
Advisory opinion, 60
Affidavit, 48
Affiliate, 550
Affiliated director, 521
Affirm, 52
Affirmative defense, 45, 130, 148
Affirmative mark, 380, 398
Affordable Care Act (ACA)
　constitutionality of, 85, 105
　for-profit corporation and contraception, 98-99
　individual mandate provision, 85, 90
　as a tax, 93
Africa, special agency types in, 465
After-acquired property, 432, 449
Age Discrimination in Employment (ADEA) of 1967, 567, 576-577, 593
Agency
　by agreement. *See* Expressed agency (agency by agreement)
　defined, 456, 480
　by estoppel, 481
　by implied authority, 457-458, 470, 481
　and liability to third parties, 455-485
　limitations on power of, 77
　　informational, 74, 78, 82
　　judicial, 74-76, 82
　　political, 74-75, 82
　　statutory, 74-75, 82
　by ratification, 457, 461, 481
Agency coupled with an interest, 478
Agency law, 16, 456, 480
Agency relationship
　avoidance, 468
　constructive trust, 468
　creation of, 456
　　agency by ratification, 461
　　agent/employer law development, 461
　　apparent agency (agency by estoppel), 457, 459-460
　　by expressed agency, 457-458
　　by implied authority, 457-458, 481
　　by ratification, 457
　defined, 456, 481
　demand an accounting, 469
　employer-employee, 461-462, 481
　employer-independent contractor, 461-464, 481
　indemnification, 465, 468-469, 481
　principal-agent, 461-462, 481
　specific performance, 469
　termination of
　　by acts of parties, 477-478
　　actual notice, 476-477
　　agency coupled with an interest, 478
　　constructive notice, 477
　　fulfillment of purpose, 477
　　lapse of time, 477
　　mutual agreement by parties, 477
　　occurrence of specific event, 477

　　by operation of law, 478-479
　　renunciation by agent, 478
　　revocation of authority, 478
　tort and contract remedies, 469, 481
　tort liability and, 473-476, 482
　Agency Relationship Law, Chapter IV of, 477
Agent
　authority of, 469-470
　authorized acts, 471-472
　contractual liability of, 471-472
　defined, 456, 480
　duties to principal, 464-468
　principal's duties to, 465-466, 468
　renunciation by, 478
　unauthorized acts, 472
Agent liability, 471-472
Agent misrepresentation, 475, 482
Agent-principal relationship, 415
Aggressor, 533-534
Agreement, 197
　anticompetitive, 243-244
　to commit crime or tort, 240
　disagreement over, 267
　handwritten, 227
　offer and acceptance, 203-204, 212
　See also Contract(s)
AIG. *See* American International Group (AIG)
Air Jordan, 195
Airspace rights, 173-174
Air Wisconsin Airlines Corp. v. Hoeper (2014), 5-6
Alaska Pacific Trading Co. v. Eagon Forest Products Inc. (1997), 355
Albertson v. State (2013), 425
Alfonso Candela, Appellant, v. Port Motors Inc., Respondent, et al. (1994), 336
Alien corporation, 514
ALJ. *See* Administrative law judge (ALJ)
Allen v. Dackman (2009), 506
Allonge, 382
Alpacas of America, LLC, v. Sam Groome et al. (2014), 401
Alteration, 393-395, 398, 417-420
Alternative Dispute Resolution Act (1998), 54, 313
Alternative dispute resolution (ADR)
　arbitration, 55-57, 60
　defined, 53
　early neutral case evaluation, 58
　versus litigation, 53-54
　med-arb, 57
　mediation, 54-55, 60
　minitrial, 58
　negotiation, 54, 60
　private trial, 58-59
　summary jury trial, 58
Alvin Ricciardi, Appellant, v. Ameriquest Mortgage Company (2006), 614
AMAIA. *See* Medical Association Insurance Agency (AMAIA)
Amazon, 17, 179, 461
Ambiguity, 13
Ambiguous terminology, in contracts, 284, 297-298
Amegy Bank National Association, 435
Amendments
　Article V of U.S. Constitution, A-5
　Bill of Rights, A-6
　11th through 27th, A-6-A-9
　See also entries for specific amendments
America Invents Act, 190
American Arbitration Association (AAA), 55, 252
American Cancer Society, 123
American Express Travel Services, Inc., 404

I-3

American Intellectual Property Law Association, 54
American International Group (AIG), 218
American Law Institute, 5, 198
American Medical Response (AMR), 591
American Society for Industrial Security, 124
Americans With Disabilities Act (ADA), 322, 567, 578, 593, 596-597
Ameriquest, 614
Amplify Federal Credit Union v. Garcia (2017), 377
AMR. *See* American Medical Response (AMR)
Analogy, 13
Anchor BanCorp Wisconsin (ABCW), 138-139
Anchor Glass Container Corporation, 302, 320
Anderson v. City of Columbus, Georgia (2005), 597
Animal Farm, 179
AnnualCreditReport.com, 617
Annuity contract, 548, 560
Answer, defendant's, 45, 47
Anthony v. State of Florida (2012), 401
Antiassignment clause, 288
Anticipatory repudiation, 308
Anticompetitive agreement, 243, 252
AOL-Time Warner, 563
Apollo Jets, 252
Apparent agency (agency by estoppel), 457, 459-460, 481
Apparent authority, 470
Appeals, 60, 132, 136
Appellate court, 60
 overrule trial courts' decisions on questions of fact, 32
 See also Courts of appellate jurisdiction (appellate courts)
Appellate jurisdiction, original *versus,* 32, A-6
Appellate procedure, 51-53, 60
Apple, 620
Apple App Store, 2, 250
Apple Computer, 171, 233
Apple iPhone, 186, 189-191
Appropriation for commercial gain, 150
Arbitration
 arbitrator qualifications, 55
 arbitrator's award (decision), 56
 binding arbitration clause, 56-57
 defined, 55, 58
 electronic agreement, 235
 hearing for, 55
 versus litigation, 56
Arbitration agreement, 458
Arbitration heading, 55
Arbitrator's award, 56
Arbitrator's Code of Ethics, 55
Arbors Rehabilitation Center, 274
Arcturus Corporation, 543-544
Arlington Market, LLC., 201
Arms, right to, 95, A-6
Arraignment, 130, 132, 136
Arrest, 128, 132, 136
Arrest warrant, 128
Arthur Anderson, 133
Arthur Murray, 270
Articles of dissolution, 536
Articles of incorporation, 511-512, 516-517, 537
Articles of partnership, 490
Articles of the U.S. Constitutions. *See under* U.S. Constitution
Asbestos exposure, 63
Aschere Energy, LLC, 543
Ashcroft v. Iqbal, 261
Ashley Furniture, 581
Assault, 143-144, 167
Assembly, rights of, 94-98, 108-109, A-6
Asset and obligation transfer, 41
Assignee, 286-289, 297-298
Assignment(s), 287
 of contracts, 291
 defined, 286
 duty of, 290, 297

mirroring, 290
notice of, 289
of rights in Russia, 187
rights that cannot be assigned, 288-289, 297
Assignor, 286-289, 297
Assumption of the risk defense, 160-161, 167
Atlantic Marine Construction Company, Inc. v. United States District Court for the Western District of Texas et al., 38-39
ATM card, 396
ATM location, software to capture images at, 394, 398
Attachment, 429
At-will employment, 566, 593
Aubrey's R.V. Center, Inc. v. Tandy Corporation (1987), 372
Auction, 207
Auction without reserve, 207
Auction with reserve, 207
Auerbach v. Bennett (1979), 528
Aurigemma v. New Castle Care (2006), 274-275
Australian constitution, 94
Austria, 80
Authority
 of agent, 469-470
 revocation of, 478
Authorization of dissolution, 536
Authorized act, 471-472, 481
Autoliv, 227
Automated signature, 380, 398
Automatic perfection, 431-432, 449
Automatic stay (moratorium), 440
Automobile, perfection of security interests in, 431-432, 449
Automobile Dealers' Franchise Act of 1965, 501
Avoidance, 353, 468, 481
Award, arbitrator's, 56
Azteca, 119

B

Bad Frog Brewery v. New York Liquor Authority, 97
Badilla v. Wal-Mart Stores East Inc. (2015), 371-372
Bail and bail levels
 defined, 129
 excessive, 95, 128, A-6 (*See also* Eighth Amendment)
Bait-and-click advertising, 606
Bait-and-switch advertising, 604-605, 621
Baker v. Price (2015), 284-285
Bank
 charges to customer's account
 forgeries and alterations, 393-395, 398
 postdated check, 393, 398
 stale check, 393, 398
 stop-payment order, 393, 398
 wrongful dishonor, 392-393, 398
 check collection process, 390-391
 customer and, rights and responsibilities, 395
 deposits and collections, 5, 327
Bank fraud, 123
Banking, 385-386, 398-399. *See also* Check
Bank of America, 425
Bank reference, 122
Bankruptcy
 Chapter 7, 426, 438-445, 450, 452-454
 Chapter 9, 438
 Chapter 11, 438, 450, 452
 Chapter 12, 438, 450
 Chapter 13, 438, 450
 Chapter 15, 438
 and contract law, 302
 exclusive federal jurisdiction, 35
 and operation of law, 309
 proceedings, 438-439, 450
 as real defense to liability, 418-420
 and termination of agency relationship, 478, 482
 Title 11 of United States Code, 438
 types of relief available, 438-455

Bankruptcy Abuse Prevention and Consumer Protection Act of 2005 (BAPCPA), 438-440, 446
Bankruptcy Act, 438
Bankruptcy Code, 323
Bankruptcy estate, 439
Bankruptcy fraud, 120-121
BAPCPA. *See* Bankruptcy Abuse Prevention and Consumer Protection Act of 2005 (BAPCPA)
Barclays Bank, 551
Bargaining, 54. *See also* Negotiation
Barkley v. Olympic Mortgage Company (2014), 424-425
Barnes & Nobel, 612
Barth v. City of Peabody (2017), 110, 193
Batson v. Kentucky (1986), 48-50
Battery, 143-144, 167
Battle of the forms, 333, 346
Bayer Cropscience, LLC. v. Stearns Bank Nat'l Association (2016), 434-435
Bayer HealthCare Pharmaceuticals, 602
BB&T. *See* Branch Banking and Trust (BB&T)
Bean Timberland Inc., 452
Bearer instrument, 381-382
Behrend v. Comcast Corp. (2011), 624
Belarus, constitution of, 94-95
BellyUp.com, 115
Bench trial, 32, 130, 567
Beneficiary, in business trust, 497
Beneficiary contract, third-party, 291-298
Bentrim v. Wells Fargo Bank, NA, SC (2017), 424
Best Buy, 434
Better Business Bureau, 122
Beverly v. Abbott Laboratories (2016), 63
Beyond reasonable doubt, 130
BFOQ. *See* Bona fide Occupational Qualification Defense (BFOQ)
Big O Tire Dealers, 511
Bilateral contract
 defined, 199
 and lack of consideration, 218-219
 versus unilateral contract, 209
Bill of Rights
 and corporations, 510
 defined, 94
 and Fourteenth Amendment, 576
 and labor, 595
 summary of, 94-95
Bills of Exchange, 384
Binding arbitration clause, 56-57, 252
Birch Bayh, 271
Blackmail, 116. *See also* Extortion
Black Star Farms, LLC, et al. v. Jerry Oliver, Arizona Department of Liquor and Control et al. (2010), 91
Blank endorsement, 382
Blatt v. Cabela's Retail, Inc. (2017), 596
Blockchain technology, 213
Blue law, 143
Blue Ship Tea Room, Inc., 359
Blue-sky law, 558, 560
Bluffs LLC, The, 498
BMW of North America v. Ira Gore, Jr. (1995), 164n, 165
Board of directors, election of, 494, 512, 517, 537
Boats, perfection of security interests in, 431-432, 449
Bobby Dean Nickel v. Staples Contract Commercial Inc. (2014), 576-577
Bob Lyons and Sue Lyons v. Christine Collins (2001), 46
Bona fide occupational qualification defense (BFOQ), 572, 593, 597
Bondholder, 545
Booking, 129, 132, 136
Boskovich Farms, 153
Bounty payment, 557

Bouton v. Byers (2014), 231
Branch Banking and Trust (BB&T), 214-2158
Braswell v. United States (1998), 104
Breach, defined, 307. *See also entries for specific breaches*
Breach of contract
 buyer in, 345, 348
 discharge by material breach, 307-308, 313, 321
 and holder-in-due course (HDC) defense, 409
 as personal defense to liability, 419-420
 remedies for, 313-320
 risk of loss during, 344-345, 348
 seller in, 344-345, 348
Breach of duty, 155-156, 167
Brennan v. Townsend & O'Leary Enterprises, Inc. (2017), 570
Briarwood Farms, Inc. v. Toll Bros., Inc. (2011), 323
Bribery, 136
 charges for, at Tyson Foods, 25
 in China, 123-124
 commercial, 116
 defined, 115-116
 demonstrating, 115
 of foreign officials, 116
 of public official (statutory offense), 115
 as white-collar crime, 114, 136
Brief, 52
Brima Wurie, 100-101
Broadcasting Cos. v. Aereo, Inc. (2014), 188
Brock v. Local 375, Plumbers Int'l Union of America, 67
Brown, Cindy S., 63
Brown v. Board of Education (1954), 6, 105
Brown v. Lockheed Martin Corp. (2016), 63
Brzonkala, Christy, 89-90
BSH Home Appliances, 366
Bulk transfer, 327
Bullying, 169
Bundle of sticks (property), 172
Burden of persuasion, 130
Burden of production of evidence, 130
Burden of proof, 130
Burden shifting, disparate treatment and disparate impact, 569, 593-594
Burdick v. Superior Court (2015), 34
Bureau of Alcohol, Tobacco and Firearms, 128
Burger King, 500
Burton Construction, Inc., 335
Burwell v. Hobby Lobby, 98-99
Business ethics
 bad-for-business ethical failures, 19
 business law and, 16, 18-19
 defined, 15
 examples of unethical, 18
 extending beyond maximization of profits, 28
 primary values and, 21
 scope of, 16
 versus social responsibility, 15
 and Swedish advertising law, 23
 values as starting point for, 21-23
 WH framework, 23-24, 26-27
 WorldCom accounting scandal, 24
Business judgment rule, 527-528
Business law
 and business ethics, 16, 18-19
 defined, 3
 importance of understanding, 3
 sources of
 administrative law, 7
 cases, 5-7
 constitutions, 4
 executive orders, 8
 schools of legal interpretation, 8-9
 statutes, 5
 treaties, 7
Business organization
 business trust, 497-498
 cooperative, 496-497, 505

corporation, 494, 506
forms of, 486-508
forms of, on Internet, 487
franchise, 499-505
general partnership, 492-*493*, 497, 504
joint stock company, 497, 505
joint venture, 498
limited liability company (LLC), 494-496, 505
limited partnership, 483-*493*, 497, 504
partnership, 488-493, 505
sole proprietorship, 487-488, 504
syndicate, 499, 505
Business relationship. *See* Agency relationship
Business self-dealing, 525-526
Business-to-business e-commerce, 612
Business-to-business e-commerce company, 524
Business trust, 497-498
But-for causation, 157
Buyer
 of chattel paper and instruments, 434, 449
 of consumer goods, 434, 449
 in ordinary course of business, 434, 449
 secured party *versus*, 434, 449
Bylaw, 517, 521

C

Cabela's Retail, Inc., 5869
Cabinet-level agency, 68, 81
Cadian Capital Management, LP, 562
California Code of Regulations, 29
California Fair Employment and Housing Act, 597
California Family Code, 235
California Institute of Technology (Cal Tech), 78-79
California Labor Commission, 57
California's Safe Drinking Water and Toxic Enforcement Act (1986), 22
California Talent Agencies Act, 57
Cal Tech. *See* California Institute of Technology (Cal Tech)
Campbell, Kesser, and Williams v. Pothas Corporation (2001), 534-535
Canada, 80
 cigarette labeling requirements in, 609
 punitive damages in, 164
Canadian Bureau of Tobacco Control, 609
Canarsie Kiddie Shop, 185
Cancellation, discharge of liability on instruments through, 421, 423
Can-Spam Act of 2003, 608
CAPA. *See* Northwestern University and College Athletes Players Association (CAPA)
Capacity
 defined, 234, 250
 as element of contract, 197, 250
 of intoxicated persons, 239, 250
 of mentally incapacitated persons, 238, 250
 of minors, 234-237, 250-251
 See also Offer
CAPTCHA. *See* Completely Automated Public Turing Test to Tell Computers and Humans Apart (CAPTCHA)
Care, duty of
 directors, officers, and shareholders, 523-524, 537
 Enron failure in, 524
 in partnership, 489-490
Carlisle v. Whirlpool Financial National Bank (1999), 613
Carnival Corporation, 538
Case law, 5, 198. *See also* Common law; *and entries for specific cases*
Case-or-controversy (justifiable controversy), 43-44, 60. *See also* Ripeness threshold requirement
Cashier's check, 376, 387, 398
Cash tender, 533, 538
Cash tender offer, 533

Casino gambling, 242-243
Caterpillar, 33, 36-37
 questionable jurisdiction over, 31, 59, A-6
 under state jurisdiction, 35
Causa mortis gift, 180, 194
Causation, 155-157, 167
Cay Cubs, 123
CCIPS. *See* Computer Crime and Intellectual Property Section (CCIPS)
Cease-and-desist order, 600, 621
Cedar Creek Associates, 119
Celtic Bank, 34-305
Center for Science in the Public Interest, 26
Central Hudson Gas & Electric Corp. v. Public Service Commission of New York, 96-97
Central Hudson test, 107, 184
Centralized management, 512-513, 521, 536
CenturyLink, 151
CERS, 119
Certificate of authority, in limited liability company (LLC), 495
Certificate of deposit, 376
Certificate of incorporation, 516
Certificate of Probable Cause, 49
Certified, service by, 33
Certified check, 376, 388-389, 398
Certiorari, 49
CFAA. *See* Computer Fraud and Abuse Act (CFAA)
C&F (cost and freight), 343
CFPB. *See* Consumer Financial Protection Bureau (CFPB)
CFR. *See* Code of Federal Regulations (CFR)
CFTC. *See* Commodity Futures Trading Commission (CFTC)
Chad's Auto Recovery, 436
Chain-style business operation, 500
Chalfant v. Titan Distribution, Inc. (2007), 30
Chancery, in England's court structure, 42
Chaplinsky v. New Hampshire (1942), 97
Chapman University, 133
Chapter 7 bankruptcy, 426, 452
 automatic stay, 440
 creditors' claims, 442
 creditor's meeting, 440-441
 discharge, 442-444
 exempt property, 441
 fraudulent transfers, 442
 liquidation, 439
 order of relief, 440
 petitions in, 439-440, 453-454
 preferential payments, 441-442
 reaffirmation of debt, 445
 relief, 438
 trustee, 441
Chapter 9 bankruptcy, 438
Chapter 11 bankruptcy, 438, 445-446, 452, 466
Chapter 12 bankruptcy, 438, 447-448
Chapter 13 bankruptcy, 438, 444, 446-447, 615
Chapter 15 bankruptcy, 438
Charge, defined, 573
"Charges the jury," 51
Charitable donation, 513
Charles Schwab & Co. Inc. v. Douglas Castro (2013), 195
Chase Manhattan, 409
Chattel paper, 432, 434, 449
Check
 altered, 394-395, 398
 availability schedule for deposited, 391
 banking and, 385
 bearing forged endorsement, 394, 398
 bearing forged signature, 398, w*393*-394
 cashier's, 387, 389, 398
 cashier's *versus* certified, 388-389
 certified, 388-389
 collection between different banks, 390-391
 collection process, 390-391

Check—Cont.
 components of, 386-387
 defined, 376
 dishonored, 402
 drawer of, 387
 electronic presentment, 391
 example, 386
 key terms for, 386
 lost, stole, or destroyed, 389
 money order, 387-389, 398
 as negotiable instrument, 385, 397
 payee, 387
 postdated, 393
 requirements of negotiability in, 378-379
 stale, 393
 substitute, 391
 traveler's, 387-388, 398
 types of, 376
Check Clearing for the 21st Century Act (Check 21 or Check truncation Act), 391
Check collection
 between different banks, 390-391
 within same bank, 390
Check fraud, 401
Check kiting, 117
Checks and balances, 86-87
Check Truncation Act. *See* Check Clearing for the 21st Century Act (Check 21 or Check truncation Act)
Cheryl Jacobus v. Donald J. Trump, Corey Lewandowsky, and Donald J. Trump for President, Inc. (2017), 146-147
Chevron, 17, 20
Chicago Prime Packers, Inc. v. Northam Food Trading Co. (2005), 344-345
Child Protection and Toy Safety Act of 1969, 620
Child Support Enforcement Administration (CSEA), 425
Chimel v. California (1969), 101
China
 capacity of minor in, 236
 constitution of, 99
 corporate structure in, 518
 deceptive advertising in, 602
 embezzlement and bribery in, 123-124
 fraudulent crimes in, 260
 free speech in, 99
 government of, and cybersecurity, 9-10
 importance of contracts in, 224
 Securities Law in, 547
 white-collar crime in, 123-124
Chinese Criminal Code, 123
Chipotle, 169
CHIPS. *See* Clearing House Interbank Payments Systems (CHIPS)
Chiquita Brands International Inc., 22
Choice-of-law clause, 38-39
Chrietzberg Electric, 281
Christy Brzonkala v. Antonio J. Morrison et al. (2000), 89-90
Chrysler, 227, 623
CIF (cost, insurance, and freight), 343, 349
Cigarette labeling, 609
Cigna Insurance Company, 402, 404, 406
Cintas Corp., 232
Circuit courts of appeal, 43, 60
CISG. *See* Convention on Contracts for International Sales of Goods (CISG)
CISG. *See* United Nations Convention on Contracts for the International Sale of Goods (CISG)
Citibank, 112, 135
Citicapital Commercial Corporation, 300
Citicorp Leasing Corporation Inc., 300
Citigroup, 118
Citizenship, privileges of, Amendment 14, A-7
Citizenship, suspect classifications based on, 105
Citizens National Bank, 415
Citizens United v. FEC, 511-512

City of Arlington v. FCC (2013), 83-84
City of Memphis Public Works Division, 483
City of Peabody's Zoning Board of Appeals, 110
City of Pontiac Stadium Building Authority, 312-313
Civil cases, heard in England's county courts, 42
Civil law, defined, 4
Civil litigation, stages of, 52
 appellate procedure, 51-53, 60
 posttrial motions, 51, 53, 60
 pretrial stage, 44-48, 53, 60
 trial, 48-51, 53, 60
Civil penalty, 546, 560, 586
Civil Rights Act of 1964. *See* Title VII (Civil Rights Act of 1964)
CLA. *See* Consumer Leasing Act (CLA)
Claims or defenses, 409
Clark v. Zaid, Inc. (1971), 350
Clean Air Act, 64, 68, 75, 152
Clean Up, 122
Clean Water Act, 127
Clearing House Interbank Payments Systems (CHIPS), 395
Closed-end credit line, 613
Closely held corporation (close, family, or privately held), 514, 536
Closing argument, 51, 55, 60
Coca Cola, 182
Code of Federal Regulations (CFR), 8
Coercion, 263, 267-268, 334
Cohen v. McCutchin, 281
Colette Bohatch v. Butler & Binion, 489
Collateral, 428
 debtor rights in, 429, 449-450
 disposition of, 437, 450
 perfection of movable, 431-432
 retention of, 437, 450
 taking possession after default, 435-437, 450
Collateral promise, 276-277
Collective bargaining, 76
 and at-will employment, 566, 592-593
 defined, 446, 590-591
 process of, 592
 and union employees, 566, 595
 Wagner Act and, 590-591, 595
Collegiate body, 78
Colonial Pac. Leasing Corp. v. N&N Partners, LLC (2013), 451
Colonial Pacific Leasing Corporation, 300
Color, as trademark, 182-183
Colorado Supreme Court, 6
Colorado Wage Claim Act, 466
Columbia University, 33
Com (business website domain), 186
Comcast Corporation, 624
Comenity Bank, 623
Commerce clause, 106
 defined, 88, 107
 as restriction on state authority, 90-91
 as source of authority for federal government, 88-90
 Supreme Court determines constitutionality of statutes Congress passes, 88-90
Commercial bribery, 116
Commercial fund transfer, 395
Commercial impracticability, 310-311, 357, 369
Commercially reasonable sale, 427-429, 451
Commercial paper
 contracts as, 375
 problems with, 375, 398
Commercial reasonableness, 353
Commercial speech, 96-97
Commissions, 68, 81. *See also* Independent agency
Commodity Futures Trading Commission (CFTC), 7, 69
Common-carrier delivery contract, 340-343, 347
Common law, 5, 198
 and contracts, 203, 212
 "good," defined, 329

and judicial review, 86
and preemptive rights, 530, 533
requirements of writing sufficient to satisfy statute of frauds under, 282
trademark infringement, 194-195, 369
Uniform Commercial Code (UCC) or, 328
See also Restatements of the Law
Common stock (common shares), 520
Commonwealth, negotiable instruments in, 384
Commonwealth of Nations (formerly British Commonwealth), 384
Communication
 electronic monitoring and, 582-583, 594
 to offeree, 207, 212
Communications Decency Act (1996), 148
Community service, as punishment for white-collar crime, 115
Company Doe v. Inez Tenebaum et al., 16
Comparative law, 9-10
Comparative negligence defense, 160-161, 167
Compensation
 duty of, in agency-principal relationship, 465-466, 468, 481
 for members of Congress, A-9
 as objective for tort law, 142
 See also Unemployment compensation
Compensatory damages
 defined, 158
 as legal remedy, 314-315, 321
 punitive damages in addition to, 365
 as tort damage, 162-163, 167
Complaint
 defined, 33, 45
 example of typical, 46
 overview, including all steps in civil lawsuit, 53
Completely Automated Public Turing Test to Tell Computers and Humans Apart (CAPTCHA), 412
Complete performance, 305, 321, 323
Compromise, 55. *See also* Mediation
Computer crime, 114, 124-125, 136. *See also* Electronic fraud
Computer Crime and Intellectual Property Section (CCIPS), 124
Computer Fraud and Abuse Act (CFAA), 123
Computer game, 124
Concealment, 262
Conclusion, 11-12
Concurrent authority, 106
Concurrent condition, 303-304
Concurrent jurisdiction, 35-37, 60
Concurring opinion, 52
Condemnation, 177
Conditional contract, 303-305
Conditional estate, 174, 192
Conditional privilege, 147-148
Conditional sales contract, 341, 344
Condition precedent, 284, 303-304, 321, 324
Conditions *versus* covenants, 304-305. *See also* Conditional contract
Condition subsequent, 303, 321
Conflict of interest, 526
Conforming goods, 353, 369
Congressional taxes, 107
Connaughton v. CHIPOTLE INC. (2017), 269
Connected state, of e-government, 80
Conscious negligence, 157
Consent, as a defense to battery, 144
Consent order, 67, 600
Consequential damages
 case of, 316
 defined, 315
 suing to recover, 364
 UCC *versus* CISG, 331
Consideration
 defined, 217, 229
 examples of, 217
 illusory promise, 222
 lack of, 218-222, 409

lack of, as personal defense to liability, 419–420, 423
of offer, 197, 212
and partial payment of debt, 226–228
past, 222–223, 229
preexisting duty, 224, 226
required by sales and lease contracts, 333
rules of, 218–226
in transfer warranty, 416
types of, based on contract type, 219
under UCC, 333, 346
Consolidated Omnibus Budget Reconciliation Act (COBRA), 579, 581, 594
Constitution, U.S. *See* U.S. Constitution
Constitutional law, 4
Constitutional restrictions, on government, 93–94
Constitution Project, 606
Constructive fraud, 442
Constructive notice of termination, 477
Constructive trust, 468, 481
Construct validity, 572
Consumer contracts, in Japan, 263
Consumer Contracts Law, 263
Consumer Financial Protection Bureau (CFPB), 19, 84
Consumer good, 431, 434
Consumer health and safety
 Child Protection and Toy Safety Act of 1969, 620
 Consumer Product Safety Act of 1972, 619, 622
 Drug-Price Competition and Patent-term Restoration Act of 1984, 620
 Federal Food, Drug, and Cosmetic Act (FFDCA), 618–619
 Federal Hazardous Substance Act of 1960, 620
 Flammable Fabrics Act of 1953, 620
 Food and Drug Administration Modernization Act of 1997, 619
 Meat Inspection Act of 1906, 620
 National Highway Traffic Safety Administration (NHTSA), 620
 Pure Food and Drugs Act of 1906, 620
 Toy Safety Act of 1984, 620
 U.S. Food and Drug Administration (FDA), 618
 Wholesome Meat Act of 1967, 620
Consumer law
 credit protection, 612–618
 deceptive advertising, 599, 601–608
 defined, 600
 Federal Trade Commission (FTC), 600–612, 621–622
 health and safety, 618–620
 labeling and packaging laws, 609
 sales, 610–612
Consumer lease, 346
Consumer Leasing Act (CLA), 613
Consumer Packaging, Inc. (CPI), 302, 320
Consumer Product Safety Act of 1972, 619, 622
Consumer Product Safety Commission (CPSC), 7, 69, 619–620, 622
Consumer protection, 16
Consumer Protection Act, 88, 252
Content validity, 572
Context, and distinguishing fact from fiction, 146–148
Contingent value rights (CVRs), 563
Contraception, and right to privacy, 98–99, 104–105
Contract(s)
 acceptance, 204
 alteration of, 309
 ambiguous terms in, 284, 297–298
 assignment of, 291, 298
 assignments and delegations, 286, 289–291, 297
 bilateral *versus* unilateral, 199–200, 209
 breach of, or warranty, 419–420
 cancellation of, 255, 363
 classification of, 199–203, 212
 as commercial paper, 375, 398
 computer, 367
 conditional, 303–305
 condition precedent in, 284, 324
 consumer, in Japan, 263
 defenses to enforcement of, 197–198
 defined, 197, 212
 direct beneficiary, 292–293
 disaffirmance by minors, 234–237
 discharging
 by conditions, 303–305
 by performance, 305–313
 elements of, 197
 employment, 244
 executed *versus* executory, 203
 express beneficiary of, 292–293
 express power of corporation, 513
 express *versus* implied, 201, 212
 formal *versus* informal, 203, 212
 formation of, 204
 importance of, in other countries, 224
 incomplete, 284, 286, 297
 installment, 356
 insurance, 303
 intentional interference with, 153–154
 legal capacity, 209, 212
 liability of principal and agent, 471–472
 manifestation of intent to be bound to, 209–210
 mixed goods and services, 330
 modified, 283, 297
 nonfinalized, partially written and partially oral, 284, 297
 offer, 204–207
 option, 208
 orally agreed-on terms, 283–284, 297
 output, 332
 parents' liability for their children's, 238
 primary beneficiary of, 292–293
 quasi- (implied-in-law contract), 202
 real estate, 303
 requirements, 332
 requirements under UCC for sales, 281
 rights that cannot be assigned, 288–289, 297
 sales and lease, as formed under UCC, 332–336
 smart, 213
 substituted, 309
 third-party, 286–298
 third-party rights to, 286–291, 297
 types of sales, 341–344
 with typographical errors, 285, 297
 unconscionability effect on enforceability of, 353
 unenforceable, 202
 valid, 202, 212
 void, 202, 212
 voidable, 202–203
 written, 273–286, 297
Contract clause, 94, 107
Contract for a necessary, 237. *See also* Minor
Contract law
 agreement, 203–215
 capacity, 234–239
 consideration, 216–232
 contracts introduction, 196–203, 212–215
 contracts in writing, 273–284, 297
 discharge, 303–313, 321
 in European Union, 199
 in franchise relationship, 505
 legality, 240–247
 reality of assent, 254–271
 remedies, 313–321, 469
 sources of, 198, 212
 third-party contracts, 286–298
 See also Common law
Contract performance, 273–275, 278
Contractual capacity, 197, 212
Contractual liability, 471–472, 481
Contract under seal, 203
Contributory negligence, 160–161, 167
Convention, 7. *See also* Treaty
Convention on Contracts for International Sales of Goods (CISG), 198
 and battle of the forms, 333
 "good," defined, 329
 scope of, 346
 significance of, 346
 United States adopt, as substitute for UCC, 238
Conversion, 152–153, 167
Cooling-Off Rule, 610, 621
Cooperation, duty of, 466, 468, 481
Cooperative, 496–497
Cooper v. Campbell (2016), 400–401
Coors Executive Resources LTD, 270–271
Copeland Corporation, Plaintiff-Appellant, v. Choice Fabricators Incorporated Defendant-Appellee (2009), 350
Copyright, 35
 changes to law as technology evolves, 188
 criteria for, 187
 defined, 186, 192
 fair-use doctrine, 187–188, 192
 infringement case, 187, 189
 symbols indicating, 189
Copyright Act, 187–188
Copyrighted work, 181. *See also* Intellectual property
Corporate bylaw, 517
Corporate criminal liability, 136
 executive held liable for murder or manslaughter, 127
 strict-liability offenses, 125
 United States v. Park (1975), 126
Corporate dissolution, 532
Corporate election, 520
Corporate financing, 519–520
Corporate name, selecting, 516
Corporate officer. *See* Officer
Corporate shares, free transferability of, 511, 532, 536
Corporate transparency, 137
Corporation, 497
 advantages and disadvantages of, 494
 centralized (day-to-day) management of, 512, 536
 characteristics of, 510–512, 536
 classification of
 alien corporation, 514, 536
 closely held, 514
 domestic, 514, 536
 foreign, 514, 536
 for-profit, 514
 nonprofit, 514
 private, 513–514
 professional, 515
 public, 513–514
 publicly held, 514
 S corporation, 514–515
 creature of the state, 511–512, 536
 de facto corporation, 517–518
 defective, 517
 defined, 494, 505
 director roles, 520–521, 523
 directors' duties, 523–524
 by estoppel, 518–519, 536
 formation of
 Facebook, 509, 535
 organizing and promoting, 515–516, 536
 potential problems with, 517–519, 536
 preincorporation agreement, 515, 536
 selecting state for, 516, 536
 and Fourth Amendment, 107
 and gene patenting, 192–193
 infant, 515–518
 legal entity, 510, 512
 as legal entity under U.S. law, 510, 512
 as legal persons in courts, 510, 512
 liabilities in, 512
 limited liability, 511–512, 536

Corporation—*Cont.*
　limited liability company (LLC) *versus,* 505
　officers' duties, 523-525
　officers' roles, 520-521, 523
　perpetual existence of, 511-513, 536
　powers of, 513
　and "principal place of business," 36-39
　rights as person and citizen, 510, 512, 536
　shareholders' roles, 520-523
　share transfers, 511-512, 532, 536
　subject to *in personam* jurisdiction, 33
　taxation, 512, 536
　See also Incorporation
Corrective advertising (counteradvertising), 605, 621
Corso Properties, LLC and WC Investment Property, Inc. v. Branch Banking and Trust Company (2009), 214-215
Costa Rica, 22
Cost-benefit analysis, 9
Counterclaim, 45, 47, 53
Counterfeit Access Device and Computer Fraud and Abuse Act (1984), 124
Counterfeit money, 117
Countersignature, 376, 387-388, 398
Country Bank, 540
County court. *See* Courts of common pleas (county court)
Course of dealing, 356
Course of performance, 356
Court of appeal, in England, 42
Court of Appeals of Michigan (2010), 506
Court of equity, 313
Court of last resort, 41
Court of law, 313
Court order, as means of acquiring property, 181
Court reporter (stenographer), 48
Courts of appeal, 43. *See also* Courts of last resort; Intermediate courts of appeal
Courts of appellate jurisdiction (appellate courts), 32
Courts of common pleas (county court), 32, 41. *See also* State trial court
Courts of common pleas (trial court), 60
Courts of last resort, 40-41, 43. *See also* U.S. Supreme Court
Courts of original jurisdiction, 32
Court system, structure of
　in England, 42
　federal court system, 40-41
　state court systems, 40-41
　state system, 40-41, 43
　U.S. Supreme Court Justices, 42
Cousins Subs Systems, Inc. v. Michael R. McKinney (199), 503
Covenants, 7
　conditions *versus,* 304-305
　See also Treaty
Covenants not to compete (restrictive covenants), 243-244
Cover, 363
CPI. *See* Consumer Packaging, Inc. (CPI)
CPSC. *See* Consumer Product Safety Commission (CPSC)
"Cram down," 447
Cramming, 117, 120
Crawford, James, 89
Crawford, v. LVNV Funding, LLC (2014), 615-616
Creation, as means of acquiring property, 181
Creative Ventures, LLC v. Jim Ward & Associates (2011), 410
Creativity, 181. *See also* Intellectual property
Creature of the state, 511-512, 536
Credit card application and solicitations, 613
Credit card fraud, 116
Credit Card Fraud Act, 616, 622
Credit Card Holders' Bill of Rights Act, 617-618
Creditor, 438
　claims in Chapter 7 bankruptcy, 442

and fraud, 120
meeting in bankruptcy proceedings, 440-441, 447
Creditor beneficiary, 293-295, 298
Creditors' claims, to collateral, priority of, 433
Credit protection
　Credit Card Fraud Act, 616, 622
　Credit Card Holders' Bill of Rights Act, 617-618
　Fair and Accurate Credit Transactions Act (FACTA), 617, 622
　Fair Credit Billing Act (FCBA), 617
　Fair Credit Reporting Act, 61, 614-615, 622
　Fair Debt Collection Practices Act (FDCPA), 612, 615-616
　Truth in Lending Act, 612-613, 622
Credit Suisse, 551
Crime
　affirmative defenses to, 131
　based on violent media, 167-168
　classification of, 114
　elements of, 113-114, 136
　knowingly, purposefully, recklessly, 113-114
　See also entries for specific crimes
Criminal defendant's rights, 128
Criminal law, 4
Criminal penalty, 550, 556, 560
Criminal procedure
　overview of steps in, 132, 136
　posttrial procedure, 130-131
　pretrial procedure, 128-130
　trial procedure, 130
Criminal prosecution, 545, 550-551, 560
Criterion-related validity, 572
Critical thinking
　guidelines, 13
　in legal context, 12
　skills for, 11-13
　structure for, 12
Cross-examine, 50, 55, 71
Cross-licensing, 189, 192
Crown court, in England, 42
Cruel and unusual punishment, 95, A-6. *See also* Eighth Amendment
CryoUSA Import and Sales, LLC, 501
CSEA. *See* Child Support Enforcement Administration (CSEA)
Cubby v. CompuServ (1996), 148
Cumulative voting, 520, 522
Cunard Celtic Hotel Services, Ltd., 539
Cure, right to, 356, 369
Customer, bank rights and responsibilities to, 395
Customer data
　confidential, 121, 199-200
　financial privacy, 397
　and retailer privacy policy, 125
　security breach litigation, 199-200
　See also Fraud; Privacy, right to; Technology
CVB Financial Corporation, 562-563
CVRs. *See* Contingent value rights (CVRs)
CVS, 20-21
Cyberbullying, 264
Cyber terrorist, 124. *See also* Hacker

D

D. Light Design, Inc., et al., v. Boxin Solar Co. (2015), 62
DACA. *See* Deferred Action for Childhood Arrivals (DACA)
DaimlerChrysler, 501
Damages
　compensatory, 158, 162-163, 167, 314-315, 321
　consequential, 315-316, 318
　expectation, 314-315
　incidental, 314
　liquidated, 315-318, 321, 365
　mitigation of, 317-318
　of negligence action, 155, 157-158, 167

nominal, 315, 321
punitive, 315, 321
special, 315
stipulated-damage clause, 315-318
under Title VII of CRA, 567
See also entries for specific damages
Data. *See* Customer data
Data mining
　principles for development of programs for, 607
　and privacy, 606-608
Dawda, Mann, Mulcahy & Sadler, P.L.C. v. Bank of America, N.A. (2014), 426
DBCP. *See* Dibromochloropropane (DBCP)
Deangelis v. Deangelis (2007), 231-232
Death
　of corporation, 535, 538
　and dissolution of partnership, 491
　and termination of agency relationship, 478, 482
Death or incapacity
　of offeror, 209, 212, 230-231
　of party whose personal services are necessary, 310
Debt
　partial payment of, 226-228
　promises to pay, 273, 276, 278
Debtor
　defined, 428, 449-450
　and fraud, 120
　rights of, in collateral, 429
Debt securities (bonds), 519, 537
Deceptive advertising, 603
　ad substantiation, 602
　bait-and-switch, 604-605, 621
　and computer consequences, 606
　corrective advertising (counteradvertising), 605, 621
　data mining, 606-608
　defined, 601
　elements of, 601
　FTC actions against, 606
　half-truth, 602
　"Made in the USA" labels, 602
　multiple-product order, 605, 621
　privacy, 606-608
　puffing, 601
　telemarketing and electronic advertising, 605-606, 621
　tobacco advertising, 608, 621
　and weight loss cure, 599, 620
Decision, legality of, 16. *See also* Business ethics; Ethical dilemma
Deed, 174
Deere & Company, 451
Deere Credit, Inc. v. Spitler (2014), 451
Deerfield Project, 420
De facto corporation, 517-518, 537
Defalcation (fraud), 117
Defamation, 6, 145, 167, 169
　actual malice, 149
　defined, 144
　false statement made over information network, 148
　plastic surgeon case, 141-142, 144, 148-149, 163, 166
　public figure privilege, 149
　role of context to distinguish fact from fiction, 146-147
　truth and privilege, 147-148
　as unprotected speech, 97
Defamation per se, 145-146
Defamatory statement, 34
Default, on loan
　defined, 430
　disposition of collateral, 437, 450
　proceeding to judgment, 435, 437, 450
　retention of collateral, 437, 450
　taking possession of collateral, 435-437, 450
Default judgment, 45

Defective corporation, 517, 537
Defendant, 33, 130
Defendant's answer, 45, 47
Defense
 affirmative, 45, 130-131, 148
 comparative negligence defense, 160-*161*, 167
 consent as, to battery, 144
 contract, 420
 and holder-in-due course (HDC), 409
 to liability, 417-420, 423
 to negligence, 160-162
 of nonissuance, conditional insurance, or issuance for special purpose, 419-420
 of others, 144
 personal, 417, 419-420
 of property, 144
 real (universal), 417-420
 UCC Article 3, 419-420
Defense of Marriage Act (DOMA), 103, 575
Deferred Action for Childhood Arrivals (DACA), 62
Deficiency judgment, 448
Definite and certain terms, of offer, 207, 210-212
De jure corporation ("lawful corporation"), 517, 537
Delaware Court of Common Pleas, 448448
Delaware law, 516
Delegatee, 289, 297
Delegation, 290-291
 and assignment, 286-291, 297
 defined, 289
Delegator, 289, 297
Deliberate, 51
Deliver ex-ship (delivery from the carrying vessel), 343
Delivery
 of bearer instrument, 382
 of property transfers, 176
 stopping, 363
 withholding, 363, 419-420
Dell, 62, 562
Demand instrument, 376, 398, 408
Dempsey v. Rosenthal (1983), 372
Demurrer. *See* Motion
Denmark, 80
Dennis Conwell and Frank Splittorff, D/B/A Piece of America, Appellant, v. Gary Loon Outdoor Marketing Group, Inc., Appellee (2009), 348
Dennis Conwell et al. v. Gray Look Outdoor Marketing Group, Inc. (2009), 349
Department of Commerce, 79
Department of Energy, 69
Department of Health and Human Services, 68
Department of Housing and Urban Development (HUD), 612
Department of Justice
 and fraud, 120
 Fraud Section of Criminal Division, 545
Department of Labor (DOL), 461, 580
Department of Labor (DOL), Wage and Hour Division, 83
Department of Motor Vehicles (DMV), 73-74
Department of Revenue of Kentucky v. George W. Davis (2008), 108
Department of Transportation (DOT), 65n, 68, 596-597
Deposit
 accepting, 390-391, 398
 direct, 399
Deposition, 48, 53
Deposits and collections, bank, 5, 327
Derivative suit, shareholder's, 532, 537
Derogatory speech, as unprotected speech, 98
Destination contract, 342-343, 347
Destroyed goods, 356
Destruction of subject matter, 310
Determinate sentencing, 130
Detroit Radiant Products Company v. BSH Home Appliances Corporation (2007), 366

Deutsche Bank National Trust Co. v. Logan (2017), 400
Deutsche Bank Trust Company Americas, 624
Dibromochloropropane (DBCP), 22
Digital cash, 396, 399
Digital content, 179, 379
Digital Document Technologies, LLC (DDT), 349
Digital marketplace, 251
Direct beneficiary, of contract, 292-293, 297
Direct damages, 364
Direct deposit, 399
Directed verdict, 50
Direct examination, 50
Director
 duty of, 523-525
 as manager, 521
 meeting attendance, 521, 537
 personal interested conflicting with corporate interests, 525, 536
 rights of, 530, 533, 537
 roles of, 520-521, 523, 537
 voting to elect, 521, 537
 See also Corporation
Direct suit, shareholder's, 532-533, 537
Direct-to-consumer drug advertisement (DTCA), 622
Dirks v. SEC, 118
Discharge
 in Chapter 7 bankruptcy, 442-444
 in Chapter 13 bankruptcy, 447
 of contracts
 by conditions, 303-305, 313, 321
 by insolvency (bankruptcy), 302
 by material breach, 307-308, 313, 321
 by mutual agreement, 309-310, 313, 321
 by operation of law, 309-313, 321
 by performance, 305-313, 321
 of debts under repayment proceedings, 447
 defined, 419
 of liability on instruments
 by cancellation or renunciation, 421, 423
 by impairment of collateral, 421, 423
 by impairment of recourse, 421, 423
 by reacquisition, 421, 423
 through payment and tender of payment, 421, 423
Disclosed principal, 471
Discovery, 47-48, 60
Discrimination. *See* Employment discrimination
Dishonored instrument, 402, 409, 413, 422
Disloyalty of agent, 479, 482
Disney, 139-140
Disparagement, 153, 167. *See also* Defamation
Disparate-impact discrimination under Title VII, 568-569, 593
Disparate-treatment discrimination under Title VII, 567-569, 593
Dissenting opinion, 52
Dissociation, 496
Dissolution
 of LLC, 496
 or corporation, 532, 536-537
 of partnership, 490-492
 wrongful, 498
 See also Termination
Distributorship, 500
District court, 60. *See also* Trial court
Diversity-of-citizenship, 35-36
Dividends, 512, 531-533
Divisible contract, 249
DLC. *See* Dual-listed corporation (DLC)
DMV. *See* Department of Motor Vehicles (DMV)
Doan's Pills, 605
Documents of title, 327, 432
Dodd-Frank Wall Street Reform and Consumer Protection Act, 84
Doe v. Uber Technologies, Inc. (2016), 463
Doe v. United States SEC (2011), 66-67
Dofasco Steel Company, 27

DOL. *See* Department of Labor (DOL)
Dole Food Company (DOLE), 22-23
Dollar General Corp. v. Meadows (2002), 597
DOMA. *See* Defense of Marriage Act (DOMA)
Domain name, 186
Domestic corporation, 514, 536
Donald Baldwin, Complainant v. Anthony Foxx, Secretary of Dept. of Transportation (FAA), Agency (2015), 574-575
Donald J. Trump for President, Inc., 146-147
Donative intent, 180
Donee beneficiary, 293-295, 298
Do Not Call registry, 608, 621
Donovan v. RRL Corporation (2001), 353
Door-to-door sales, 610, 621
Dormancy fee, 618
DOT. *See* Department of Transportation (DOT)
Double jeopardy, 95, 102, 128, A-6. *See also* Fifth Amendment
Dovetail Builders, 498
Dow Chemical Company, 22
Dow Jones Sustainability Group Index, 27
Draft, 376, 386, 398
Dragon, 29-30
Drawee, 425
 altered checks, 394, 397
 cashier's check, 387, 397
 components of check, 386
 defined, 386-387, 412
 draft negotiable instrument, 376, 397
 forged endorsements, 394, 397
 money order, 389
 and money order, 387
 negotiable instruments, 379
 presentment, 413, 417
 presentment warranty, 417
 property payable rule, 392-394, 398
 and requirements of negotiability, 378, 397
 requirements of negotiability in check, 379
 secondary liability of, 412-413, 422
 signature of, 378-380, 386, 393, 397-398
 stop-payment order, 393, 397
 transfer warranty, 416
 traveler's check, 388
 of traveler's check, 388
 unauthorized signatures and endorsement, 414
 and words of negotiability, 381
Drawer, 425
 as accommodation party, 414
 altered checks, 394, 397
 cashier's check, 387, 397
 and certified checks, 388-389
 in check collection process, 390, 397
 components of check, 386
 defined, 386-387, 412
 and delivery of instrument, 382
 discharge by impairment of recourse, 421
 draft negotiable instrument, 376
 forged endorsements, 394, 397
 imposter rule, 415-416
 and money order, 387
 presentment warranty, 417
 property payable rule, 392-394, 398
 and requirements of negotiability, 378, 397
 secondary liability of, 412-413, 422
 signature of, 378-380, 386, 393, 397-398
 stop-payment order, 393, 397
 transfer warranty, 416
 of traveler's check, 387
 unauthorized signatures and endorsement, 414
 and words of negotiability, 381
 wrongful dishonor, 392
Drug advertising, 602
Drug-Price Competition and Patent-term Restoration Act of 1984, 620
Drug-sniffing dogs, 100
Drug test, 590, 594
DTCA. *See* Direct-to-consumer drug advertisement (DTCA)

DTG, 311
DTSG, Ltd., 300-301
Dual-listed corporation (DLC), 539
Due diligence defense, 550
Due process clause, 102, 105, 107
Due process rights, 95, 128, A-6. *See also* Fifth Amendment
Duly authorized agent, signature of, 380, 398
Dun & Bradstreet, 122
Dunkin' Donuts Corporation, 486, 493, 496, 502, 500, 504
Dunleavy v. Paris Ceramics (2002), 372
Durable power of attorney, 457-458, 481. *See also* Power of attorney
Duray Development, 506
Duress, 203, 270
 as affirmative defense to crime, 131
 defense to enforcement of contract, 198
 defined, 265, 267-268, 365
 economic, 265
 elements of, 265
 and holder-in-due course (HDC) defense, 409
 importance of genuine assent, 255
 in India, 265
 lack of genuine assent, 198
 as obstacle to genuine assent, 255
 as real defense to liability, 417
 real (universal) defenses to liability, 420
 rescission and restitution, 319
Durham Police Department, 110
Dusenbery v. United States (2002), 110-111
Duties, 155, 167
 of accounting, 468, 481
 of agent to principal, 464-468
 assignment of, 290, 297
 of care
 of directors, officers, and shareholders, 523-524, 537
 Enron failure in, 524
 in partnership, 489-490
 to compensate, 465-466, 468
 condition precedent, 303-304, 324
 of cooperation, 466, 468, 481
 delegation of, 289-291, 297
 of directors, 523-525
 to disclose conflict of interest, 526
 of loyalty
 of officers, directors, and shareholders, 524-525
 in partnerships, 467-468, 481
 of notification, 467-468, 481
 of obedience, 467-468, 481
 in partnership, 489
 of officers, 523-525
 of partners to one another, 489-490
 to perform
 to mitigate one's damages, 317
 of performance, 467-468, 481
 of principal to agent, 465
 of reimbursement and indemnification, 465, 468, 481
 of shareholders, 526, 537
 that cannot be delegated, 290-291, 297
 of working conditions, 466, 468, 481

E

Eagon Forest Products Inc., 355
Early neutral case evaluation, 58
Easement, 175-176, 192
Ebay, 124
Ebbers, Bernard, 24
Ebner v. Fresh, Inc. (2016), 623
E-books, 179
ECMC. *See* Educational Credit Management Corporation (ECMC)
ECOA. *See* Equal Credit Opportunity Act (ECOA)

E-commerce, and law
 business-to-business company, 524
 check-clearing process, 381
 computer contracts, 367
 computer use and ethics, 125
 and copyright law changes, 188
 deceptive ads, 606
 disclosure of definite and certain terms, 211
 domain names and trade names, 186
 electronic signature, 412
 forms of business organizations on Internet, 487
 fraudulent misrepresentation in social media, 264
 hostile takeovers online, 534
 marketing securities online, 558
 negligence on Internet, 156
 negotiable instruments in electronic form, 379
 observing Sabbath days online, 243
 relationship between government and citizens, 80
 sales in Middle East, 334
 sliding-scale standard for Internet transactions, 34
 software to capture images at ATM locations, 394
 technological records and ethics, 25
 Uniform Commercial Code (UCC) and Internet, 330
Economic duress, 265-266
Economic interests, intentional torts against, 153-154, 167
ECPA. *See* Electronic Communications Privacy Act (ECPA) of 1986
Ecuador
 Chevron in, 17
 Dole Food Company ad, 22
EDGAR. *See* Electronic Data Gathering Analysis and Retrieval (EDGAR)
Educational Credit Management Corp. v. Polleys (2004), 453
Educational Credit Management Corporation (ECMC), 444
Edu (educational website domain), 186
EEOC. *See* Equal Employment Opportunity Commission (EEOC)
EEOC v. Exxon Mobile Corp. (2012), 597
EEOC v. Res. for Human Dev., Inc. (2012), 597
Efficiency, 21
EFT. *See* Electronic fund transfer (EFT)
EFTA. *See* Electronic Fund Transfer Act (EFTA)
E-government, 80
Ehling v. Monmouth-Ocean Hosp. Serv. Corp. (2013), 582-583
Eighteenth Amendment, repeal of, A-8
Eighth Amendment, 95, 128, A-6
Election
 corporate, 104, 520, 537
 of President and Vice President, 12th Amendment, A-6-A-7
 of Senators by popular vote, 17th Amendment, A-7
Electrique Corporation, 281
Electronic advertising, 605-606, 621
Electronic check presentment, 391
Electronic Communications Privacy Act (ECPA) of 1986, 66-67, 582-583, 594
Electronic Data Gathering Analysis and Retrieval (EDGAR), 558
Electronic fraud, 124
Electronic Fund Transfer Act (EFTA), 395
Electronic fund transfer (EFT), 395, 399
Electronic monitoring and communication, 582-583, 594
Electronic negotiable instrument, 379
Eleventh Circuit of Appeals, 85
Elias Brothers Restaurants Inc., 312
Elliott Group, Inc. The, 201-202

Email, to communicate to offeror, 211, 214
Emancipation, 234
Embezzlement, 119
 in China, 123-124
 defined, 123
 as white-collar crime, 123-124
Emerging stage, of e-government, 80
Eminent domain authority, 114, 177-178
E-money
 defined, 396
 issuer's financial records, 396-397
 and online banking, 395-397, 399
E-money payment information, 396
Employee, liability of, 512
Employee Retirement Income Security Act (ERISA), 581, 594
Employer-contractor relationship, 462
Employer-employee relationship, 461-462, 481
Employer-independent contractor relationship, 461-464, 481
Employment-at-will, 566, 593
Employment contract, 244
Employment decisions, use of social media in, 584-586
Employment discrimination
 Age Discrimination in Employment (ADEA) of 1967, 576-577, 593
 Americans With Disabilities Act (ADA), 578, 594
 based on marriage, 575-576, 593
 based on sexual orientation, 573-575, 592-593
 Civil Rights Act of 1964, 566-573
 claims of, 55. *See also* Equal Employment Opportunity Commission (EEOC)
 defenses to claims under Title VII, 572, 593
 disparate-impact under Title VII (unintentional-discrimination), 568-569, 593
 disparate-treatment under Title VII, 567-569, 593
 Equal Pay Act of 1963, 579
 federal laws against, 567, 593
 harassment by nonemployees, under Title VII, 571
 and pre-employment investigations, 584-585
 Pregnancy Discrimination Act (PDA) of 1987, 571-572, 593
 procedure for filing claim under Title VII, 573, 593
 remedies under Title VII, 572-573
 sexual harassment under Title VII, 569-571, 592-593
 against smokers, 578, 594
 See also Fourteenth Amendment
Employment Eligibility Verification Department of Homeland Security, 587-589
Employment Eligibility Verification Form I-9 (Form I-9), 586-589
Enabling legislation, 65, 81
Encore Glass, Inc., 302, 320
Encumbrance, 340, 347
Endorsee, 382
Endorsement, 381
 checks bearing forged, 394
 defined, 382
 kinds of, 382
 noncriminal endorsement problems, 383-385
 See also Signature liability
Endorser
 as accommodation party, 414
 defined, 382
 and secondary liability, 412-413
Enforceability, unconscionability and, of contract, 353
England
 court structure in, 42
 negotiable instruments in, 411
 negotiable-instrument transactions in, 384

English Act for the Prevention of Frauds and Perjuries (1677), 273, 297
English law, 384
English rule, 289
Enhanced state, of e-government, 80
Enron, 75, 133, 137-138, 524-525
Enron scandal, 116
Entrapment, as affirmative defense to crime, 131
Entrustment, 338-340, 347
Environmental Management Agreement, 27
Environmental pollution, 17
Environmental Protection Agency (EPA), 7, 9, 64, 66, 68-69, 73, 82
Environment Canada, 27
Environs Inc., 324
EpiPen, 19
Equal Credit Opportunity Act (ECOA), 613-614
Equal Dignities Rule, 278
Equal dignity rule, 470
Equal Employment Opportunity Commission (EEOC), 7, 55, 69, 461, 597
and discrimination based on sexual orientation, 574, 593
and sexual harassment, 565, 569, 593
Equal Pay Act (EPA) of 1963, 567, 579, 593-594, 597
Equal Protection Clause, 105, 107, 576. *See also* Fourteenth Amendment
Equifax, 617
Equipment, 43.*2*
Equitable remedies, 314-320
distinction between legal and, 313
injunction, 319, 321
recovery based on quasi-contract, 320
reformation, 319
rescission and restitution, 318-319, 321, 369
specific performance, 319, 321
Equity securities (stock), 519-520, 537
Erbitux, 542
ERISA. *See* Employee Retirement Income Security Act (ERISA)
Escola v. Coca Cola, 158
Establishment clause, 98, 107
Estate. *See* Real property (realty)
Estoppel
agency by, 481
apparent authority and, 470
corporation by, 518-519, 536
partnership by, 490
Ethical dilemma, 15
Ethical guideline, 19, 24, 525
Ethics
in decision-making, 24, 26-27
Golden Rule, 26
public disclosure test, 26-27
universalization test, 27
See also Business ethics
Eubanks v. Northwest Herald Newspapers, 170
Europe, first-to-file rule, for patents, 190
European Commission, 199
European Union, 477
contract law in, 199
trade laws and restrictions, 9
Evaluation of the reasoning, 12
Evan Rothberg v. Walt Disney Pictures (1999), 264
Eves Express, Inc., 484-485
Evict, 175
Evidence, presentation of, 50
Examination of witnesses, 50, A-6
Excessive bail levels. *See* Bail and bail levels
Excessive fines. *See* Fines, excessive
Excessive pressure, 264
Exchange Act, 552-553
Exchanges of letters, 7. *See also* Treaty
Exchange tender offer, 533, 537
Exclusive federal jurisdiction, 60
admiralty cases, 35
bankruptcy cases, 35

claims against the United States, 35
claims involving federal statutes specifying, 35
federal criminal prosecutions, 35
Exculpatory clause, 245-248
Executed contract, 203, 212
Execution, of property transfers, 176
Executive agency, 68-69, 81
Executive branch, 81, 86, 106
Executive Branch (Article II of U.S. Constitution), A-3-A-4
Executive order, 8, 86-87
Executive power, of administrative agency, 65
Executory contract, 203, 212, 278, 323
Exemplary damage. *See* Punitive damages (exemplary damage)
Exempted rule making, 72-74
advising, researching, issuing permits, and managing property, 73
defined, 72, 82
problems associated with rule-making, 72-73
regulated negotiation, 72
Exempt property, 441
Exempt securities, 548, 560
Exide Technologies, 322-323
Expectation damages, 314-315
Expedited Funds Availability Act (1987), 391
Experian, 617
Express assumption of the risk, 161
Express authority (actual authority), 470, 481
Express beneficiary, of contract, 292-293
Express condition, 304
Express contract, 201, 212
Expressed agency (agency by agreement), 457-458, 481
Express powers, 513, 536
Express ratification, 237
Express warranty, 358, 361, 369
Extent of ownership, 173-174
Extortion, 116, 123, 136
Exxon Mobile Corp., 597

F

FAA. *See* Federal Arbitration Act (FAA)
FAA. *See* Federal Aviation Agency (FAA)
Facebook
and employment decisions, 584-586, 591-592, 594
and privacy, 583
Facebook accounts, service through, 33
Facebook Corporation, 61, 110, 512
formation of, 509, 536
and "principal place of business," 39
Fact, 12
as absolute defense, 148
role of context to distinguish fiction from, 146-147
FACTA. *See* Fair and Accurate Credit Transactions Act (FACTA)
Fair and Accurate Credit Transactions Act (FACTA), 617, 622, 624
Fair Credit Billing Act (FCBA), 617, 622
Fair Credit Reporting Act, 61, 585, 614-615
Fair Debt Collection Practices Act (FDCPA), 612, 615-616, 622, 625
Fair Employment and Housing Act, 570
Fair Housing Act (1968), 70-71
Fair Labor Standards Act (FLSA), 83, 461, 591, 594
Fair Labor Standards Act (FLSA) v. American Medical Response of Connecticut Unpublished (2011), 591
Fair Packaging and Labeling Act of 1966, 609
Fair-use doctrine, 187, 192
False assertion of fact, 262
False Claims Act, 131-133, 136
False designation of origin, 194-195
False entries (fraud), 117

False filings, 121
False identity, 264
False light, 150
False pretense (fraud), 117-118
False statement, 121, 123
False token (fraud), 117
Family and Medical Leave Act (FMLA), 8-9, 579, 594
Family cases, in England's court structure, 42
Family-farmer plan, 438, 447-448
Family-fisherman plan, 438, 447-448
Family Fitness Center, 161
Family House of Louisiana (Family House), 597
Fannie Mae, 424, 540
Fantastic Sams Salons Corp. v. PSTEVO, LLC and Jeremy Baker (2016), 261
Farm product, 432
Farris v. Country of Camden (1999), 56n
FAS (free alongside), 343
FBI. *See* Federal Bureau of Investigation (FBI)
FCBA. *See* Fair Credit Billing Act (FCBA)
FCC. *See* Federal Communications Commission (FCC)
FDA. *See* Food and Drug Administration (FDA)
FDIC. *See* Federal Deposit Insurance Corporation (FDIC)
FEC v. Beaumont, 511
Federal administrative agencies, list, 7, 82
Federal Arbitration Act (FAA), 56-57
Federal Aviation Administration, 597
Federal Aviation Agency (FAA), 68, 81
Federal bankruptcy exemptions, 441
Federal Bureau of Investigation (FBI), 25, 120, 128
Federal circuit courts of appeal, 43, 60
Federal Communications Commission (FCC), 7, 68-69, 83
and consumer privacy, 607
and Internet privacy rules, 607
and telemarketing, 606
Federal court system
circuits of, 41
and concurrent federal jurisdiction, 37
court of last resort, 41
and directed verdict, 50
exclusive federal jurisdiction, 35
intermediate courts of appeal, 40
and posttrial motions, 51
and state court systems, 35, 41
in structure of court system, 40-41
subject-matter jurisdiction, 35, 37
trial courts, 32
See also U.S. Supreme Court
Federal criminal prosecution, 35
Federal Deposit Insurance Corporation (FDIC), 7, 69, 513-614
Federal Deposit Insurance Corporation v. Hawker (2012), 540
Federal district court, 32, 54
courts of original jurisdiction, 32
procedure for filing claim under Title VII, 573
and questionable jurisdiction, 31, 37, 59
and same-sex couples' right to marry, 103
sliding scale standard, 34
and trademark infringement, 185
use of neutral case evaluation in, 58
Federal Do Not Call list, 608, 621
Federal Energy Regulation Commission (FERC), 69
Federal Express, 120
and independent contractors, 455-456, 462, 469, 480
questionable stock investment, 112, 135
Federal Food, Drug, and Cosmetic Act (FFDCA), 126, 618-619
Federal government, 86-87
Federal Hazardous Substance Act of 1960, 620
Federal Hazardous Substances Act of 1960, 609

Federal interest computer, 124
Federalism, 85-86, 106
Federal law
　against employment discrimination, 567
　and franchises, 501
　versus state law, 3-4
　See also entries for specific acts
Federal Mediation and Conciliation Services (FMCS), 55
Federal Mine Safety and Health Act, 88, 101
Federal preemption, 87, 106
Federal-question case, 35
Federal Register, 69-72, 82
Federal Reserve Board
　e-money payment information, 396
　enforcing EFTA, 395-396
　and Regulation E, 395-396
　wire transfer network, 395-396
Federal Reserve Board of Governors, 610
Federal Rule of Civil Procedure, 44, 261, 624
Federal securities law, 533
Federal statute
　application of laws in states, 326
　concurrent federal jurisdiction, 35
　consumer health and safety, 618-620
　contracts that violate, 240-243
　electronic monitoring and communication, 582
　and environmental protection, 7
　exclusive federal jurisdiction, 35
　and federal-question case, 35
　of frauds, 273
　fraudulent misrepresentation in social media, 264
　and government spending powers, 93
　online sales, 612
　rights that cannot be assigned, 389
　See also entries for individual acts
Federal Stored Communications Act (SCA), 582-583
Federal supremacy, 87, 90, 106
Federal Trade Commission (FTC)
　and abusive telemarketing behavior, 608
　cease-and-desist order, 600, 621
　computer use and ethics, 125
　consent order, 600, 621
　and consumer law, 599-612, 621-622
　deceptive advertising, 601-608, 620
　Do Not Call registry, 608, 621
　and franchise law, 501
　function of, 599-601
　as independent agency, 68, 73
　industry guides, 600
　labeling and packaging laws, 609
　"Made in the USA" labels and, 602
　as major federal administrative agency, 7, 68-69
　sales, 610-612
　trade regulation rules, 601, 621
Federal trademark dilution, 194-195
Federal treaty, and federal-question case, 35
Federal trial court, 40
Federal Unemployment Tax Act (FUTA), 580, 594
Fedex. See Federal Express
Fedwire. See Wire transfer network (Fedwire)
Fee simple absolute, 174, 192
Felony, 114, 129, 136
Felony stalking, 169
FERC. *See* Federal Energy Regulation Commission (FERC)
FFDCA. *See* Federal Food, Drug, and Cosmetic Act (FFDCA)
Fiction, role of context to distinguish fact from, 146-147
Fictitious-Payee Rule, 414, 416, 423
Fiduciary, 456, 466, 480
Fiduciary duty
　breach of, 498
　ethical guidelines, 525
　in partnerships, 489

Fife v. Facebook, 235
Fifth Amendment
　corporation rights under, 510, 512
　defined, 107
　double jeopardy, 95, 102, 128, A-6
　due process, 102-104, 128, A-6
　privilege against self-incrimination, 95, 104, 128, A-6
　prohibition against uncompensated takings, 95, 104, 107, 177-178, A-6
　substantive due process, 95, 103
　takings clause, 95, 107, 110, 177-178, 193, A-6
　text of, A-6
Fighting words, as unprotected speech, 97
Filing, perfection by, 430, 449
Filing process, periods of, 546-547, 559
Film Recovery System, 127
Finance lease, 346
Financial Advisory Fund, 122
Financial institution, technology to capture images at ATMs, 394
Financial Services Modernization Act, 397
Financing, corporate, 519-520
Financing statement, 430
Fines, excessive, 95, 128, A-6. *See also* Eighth Amendment
Firearm possessions, 89
"Fire!," yelling in crowded theater, 96
Firing decisions, social media in, 585-586, 595
Firm offer, 333
First Amendment, 67, 184
　corporation rights under, 510-511
　exercise religion, 110
　protection of technological records, 25
　Sabbath laws, 243
　summary of, 94-99, 107
　text of, A-6
　Twitter and, 268-269
First appearance, 129, 132, 136
First-assignment-in-time rule, 289
First Bank, 300-301
First Bank v. Brumitt (2017), 301
First National Arizona, 624
First National Bank of Boston v. Bellotti (1978), 96
First National Bank of Nevada, 624
First Pentecostal Church of Beaumont, 507
First-to-file rule, 190
First Union National Bank, 424
First United Pentecostal v. Parker (2017), 507
Fixture, 432
Flammable Fabrics Act of 1953, 609, 620
Flash Crash of 2010, 561
Fleet Bank, 604-605
Float time, 117, 381
Florida v. Harris, 100
Florida v. Jardines, 100
FLSA. *See* Fair Labor Standards Act (FLSA)
FMCS. *See* Federal Mediation and Conciliation Services (FMCS)
FMI, Inc., 484-485
FMLA. *See* Family and Medical Leave Act (FMLA)
FOB (free on board), 343
Food and Drug Administration (FDA), 68, 73, 81, 126, 618-620, 622
Food and Drug Administration Modernization Act of 1997, 619
Food disparagement, 153
Force, justifiable use of, 131
Ford, 623
Ford/Firestone case, 620
Fordyce Bank & Trust Co. v. Bean Timberland, Inc. (2007), 452
Foreclosure cast, 347-349
Foreign company, limited liability company (LLC) referred to as, 495
Foreign corporation, 514, 536
Foreign Corrupt Policies Act, 25, 116

Foreign workers, hiring, 586, 595
Forestal Guarani S.A., Appellant, v. Daros International, Inc. (2010), 349
Forest Practices Act (1974), 84
Forged endorsement, 394, 398
Forged signature, 393-395, 398
Forgery
　checks bearing forged signatures, 393-394, 398
　defined, 117, 119
　problems with endorsement, 383
　as real defense to liability, 395, 418-420
　and signature liability, 414-416
Formal contract, 203, 212
Formal record, 71
Formal rule making, 69-70, 82
Form I9. *See* Employment Eligibility Verification Form I-9 (Form I-9)
For-profit corporation, 514, 536
Forum non conveniens, 38
Forum-selection clause, 38-39
Fourteenth Amendment, 15, 49, 94, 105, 107, 128
　corporation rights under, 510, 512
　Due Process Clause of, 575
Fourth Amendment, 95, 99-102, 104, 107, 109, A-6
　corporation protection under, 510
　privacy, 108
　protection of corporations and places of business, 101
　search warrant, 99, 128
　warrantless entry, 110, 128
FOX television network, 147
Foxwood Estates Homeowners' Association, Inc. v. Foxwood Estates, LLC (2015), 322
Fractionated Stereotactic Radiosurgery (FSR), 603
France, 80, 531
Franchise
　advantages and disadvantages to start, 500
　agreement, 502
　creation of, 502
　defined, 499, 505
　termination of, 502-503
　top 10 global franchises, 500
　types of, 500
Franchise agreement, 486, 504
Franchisee, 499-500
Franchise law, 501
Franchisor, 499-500
Fraud, 18, 131, 334, 383
　bank, 123
　bankruptcy, 120-121
　check, 401
　check kiting, 117
　computer crime, 124-125
　concealment, 117
　credit card, 116
　defalcation, 117
　electronic, 124
　false filings, 121
　false pretense, 117, 119
　false statements, 121, 123
　false token, 117
　forgery, 117, 119
　health care, 117, 120
　and holder-in-due course (HDC) defense, 409
　insider trading, 116-118
　insurance, 116
　investor adviser, 121
　mail, 117, 119-121, 135
　money laundering, 121
　online, 558
　perjury, 121
　Ponzi scheme, 117
　Pretexting, 117
　prevention tips, 122
　as real defense to liability, 417
　Salman v. United States, 118
　securities, 116-117, 121

stock-option backdating, 118
telemarketing, 117
three elements of, 116-117
wire, 121, 135
Fraud Discovery Institute, 122
Fraud in the factum (fraud in execution and fraud in essence), 417-420
Fraud in the inducement, 419-420
Fraudulent concealment, 117
Fraudulent inducement, 268
Fraudulent misrepresentation, 153-154, 167, 255, 261, 267, 270-271
defined, 260
false assertion of fact, 262
intent to deceive, 263
justifiable reliance on false assertion, 263
in social media, 264
Fraudulent transfers, 442
Freecreditreport.com, 617
Freedom, 21, 268-269
Freedom Enterprises, 252
Freedom of Information Act (FOIA), 74, 78, 82
Freedom of religion, 98-99. *See also* First Amendment
Freedom of speech, copyright law and, 181
Freedom of speech and assembly, 94-98, 108-109
in China, 99
unprotected speech, 98-99
See also First Amendment
Free Enter. Fund v. Pub. Co. Accounting Oversight Bd. (2010), 84
Free-exercise clause, 98, 107. *See also* Freedom of speech and assembly
Freidberg & Parker, LLP, 349
French Penal Code, 531
Frenk v. Solomon (2018), 338-340
Fresh, Inc., 623
Fresh Del Monte Produce Inc., 22
Friedman v. Southern California Permanente Medical Group (2002), 11
Friendly Cash Checking Corporation, 425
Friends of the Earth v. Laidlaw Environmental Services (2000), 43-44
Frustration of purpose, 312-313
FSR. *See* Fractionated Stereotactic Radiosurgery (FSR)
FTC. *See* Federal Trade Commission (FTC)
Fuel Clothing Company, Inc. vs. Nike, Inc., 194-195
Fulfillment of purpose of termination, 477
Full faith and credit clause, 94, 107
Full-time employees, *versus* independent contractors, 482-483
Full warranty, 362, 369
Fully integrated agreement, 282
Fumazone, 22
Funeral home services, sales of, 611-612, 622
Funeral Rule of 1984, 612
Fungible goods (confusion), 181
Fur Products Labeling Act of 1951, 609
FUTA. *See* Federal Unemployment Tax Act (FUTA)
Future interest, 175

G

Galleon Group, 138
Gambling, 242-243
Gantt v. Sentry Insurance (1992), 566
Garage Solutions v. Monty J. Pearson (2016), 266
Garcia's Inc., 506
Garner Doctrine, 540
Garrett Group, 562-563
Garrison v. Garrison (2017), 194
Gate Gourmet, 568
Gateway 2000, 367
GATT. *See* General Agreement on Tariffs and Trade (GATT)
Ge Capital Commercial Inc., 300

Gen. Partner Glenn TONGUE, Deerhaven Capital Management, et al v. Sanofi Pharmaceuticals, Inc. Consolidated, et. al (2016), 563
Gender Dysphoria (Gender Identity Disorder), 596
Gender-motivated crime, 89
Gene patenting, 192-193
General Agreement on Tariffs and Trade (GATT), 9
General intangible, 432
General jurisdiction, state trial courts of, 41, A-6
General partnership, 492-493, 497, 504
General Services Administration (GSA), 7, 69
Genuine assent
defects in, in Philippines, 266
importance of, 255, 267
lack of, 197-198
obstacles to
duress, 255, 265-268
misrepresentation, 255, 259-263, 267
mistake, 255-259, 267
undue influence, 255, 263-264, 267
Germany
corporation law, 518
eminent domain in, 179
negligence in, 157
Gerstenecker v. Gerestenecker (2017), 214
GG v. Valve Corporation (2017), 252
Gharar, 204
Gideon Group Holdings, LLC, 325
Gift, property as, 179-180, 192, 194
Gift card, 618, 621
"Gig economy," 461
Gilbert v. Residential Funding LLC, 624-625
Girl Scouts, 123
Girls Gone Wild (GGW), 237
Global Crossing, 133
Global law. *See* International law
GMAC Mortgage LLC, 624
Golden parachute, 534-535
Golden Rule, 26, 67, 185, 383
Goldman Sachs, 30
Good faith, 337, 352-353
affect of, on perfect tender, 354, 369
as holder-in-due course (HDC) status requirement, 406-407
Good-faith purchaser, 338
Goods, 428
acceptance of, 357
defined, 327-329, 346
title *versus* other interests in, 340-341
Good Samaritan statute, 162
Goods-in-bailment contract, 341, 343-344, 347
Good title, 369
acquiring, 336-341, 347
defined, 336, 347
entrustment, 338-340
title *versus* other interests in good, 340-341
Goodyear Tire & Rubber Company, 511, 515
Google, 61, 66, 534
1-800-Got Junk? LLC v. Superior Court (Millennium Asset Recovery, Inc.) (2010), 507
Gotlin v. Lederman (2012), 603
Government in Sunshine Act, 74, 78, 82, 84
Gov (government website domain), 186
Grail Semiconductor, Inc., 28-29
Grail Semiconductor, Inc. v. Mitsubishi Electric & Electronics USA, Inc. (2014), 29
Gramm-Leach-Bliley Act, 397
Grand Duke of Luxembourg, 459
Grand jury
indicted by, 110
right to, A-6
subpoena for, 128
Gratuitous agent, 458
Great Clips, 500
Greenfield v. Cadian Capital Management, LP (2016), 562

Griggs v. Duke Power Co., 569
Griswold v. Connecticut (1965), 104-105
Gross negligence, 157
GSA. *See* General Services Administration (GSA)
GTE Corporation, 528
Guatemala, 22
Gulf Coast Electric, 613
Gun-Free School Zone Act, 88-89
Gyabaah v. Rivlab Transp. Corp. (2013), 215

H

Häagen-Dazs, 498
Habeas court, 49
Hacker, 124
Hadley v. Baxendale Court of Exchequer (1854), 316
Hain Celestial group, 609
Half-truth, 602
Hall Street Associates v. Mattel (2008), 56n
Hamer v. Sidway (1891), 217-218
Hamilton v. State Farm Fire & Casualty Insurance Company (2012), 307-308
Hancock v. Variyam, 145-146
Harassment by nonemployees, under Title VII, 571, 593. *See also* Employment discrimination; Sexual harassment
Harbour Town Hole 19, 186
Hard Assets LLC, 506
Hardee's, 500
Harlow Primate Psychology Laboratory, 221-222
Hartford, The, 595-596
Hartman and Swinehart v. Keri, 149-150
Hate speech, as unprotected speech, 98
Hawala, 465
Hawaladars, 465
Hayner Hoyt Corporation ethics case, 16
Haynes v. DaimlerChrysler Corp., 227
Haywood v. Baseline Construction Company, 155
HDC. *See* Holder-in-due course (HDC)
Health and safety, 618-620
Health care. *See* Affordable Care Act (ACA)
Health care fraud, 117, 120
Health insurance. *See* Affordable Care Act (ACA)
Hearsay, 50
Heat emissions, 99-100
HERA. *See* Housing and Economic Recovery Act (HERA)
Herbalking spam group, 608
Heritage Constructors, Inc. v. Chrietzberg Electric, Inc. and Richard Marc Chrietzberg (2015), 280-281
Hertz Corporation v. Friend, 36-37
Hewlett-Packard, 121, 624
Hexagon Holdings, Inc. v. Carlisle Syntec Inc. (2019), 300
HFT. *See* High-frequency traders (HFT)
High court, in England, 42
High Court of Chancery, 313
High-frequency traders (HFT), 561
Hill v. Gateway 2000 (1997), 367
Hilton, 112
Hiring decisions
of foreign workers, 586, 595
use of social media in, 584-585, 595
Historical school, 8
Hitachi, 29
Hively v. Ivy Tech Community College of Indiana (2017), 575
Holder
defined, 382, 404
and secondary liability, 413
Holder-in-due course (HDC), 402
claims or defects taken without notice for, 407-410, 422
claims or defenses, 409
complete and authentic negotiable instrument requirement for, 404-405
defined, 403

Holder-in-due course (HDC)—*Cont.*
 demand instrument, 408
 dishonored check and, 422
 failed attempt to become, 405
 good faith requirement for, 406-407
 overdue instruments and, 408
 personal defenses and, 417
 reason for, 403
 requirements for status, 403-410, 422
 shelter principle and, 410-411, 422
 time instrument, 408-409
 United States *versus* England, 411
 value requirement for, 405-406, 423
Home Building & Loan Association v. Blaisdell, 94
Honda, 5
Honda Motor Company v. Oberg, 163-164
Honduras, 22
Hosch v. Colonial Pacific Leasing Corp. (2012), 300
Hostile environment, 569-571, 593
Hostile takeover, 533-534, 538
House arrest, 115
House of Lords, in England's court structure, 42
House Oversight Committee, 19
Housing and Economic Recovery Act (HERA), 540
Howey test, 543-544, 559
HUD. *See* Department of Housing and Urban Development (HUD)
Hugh L. Caraway v. Land Design Studio (2001), 420-421
Human Rights Watch, 22
Hung jury, 130
Hybrid agency, 69, 82
Hybrid rule making, 68, 71, 82

I

IBEW. *See* Indiana Electronic Workers Pension Trust Fund (IBEW)
ICC. *See* Interstate Commerce Commission (ICC)
ICE. *See* Immigration Customer and Enforcement (ICE)
Ice Cream Partners, USA, 498
IDCA. *See* International Development Cooperative Agency (IDCA)
Identification with the vulnerable, 8
IFP. *See* In forma pauperis (IFP)
Iglesia Cristiana La Case Del Senor, Inc., Etc. v. L.M. (2001), 474
Illegal agreements, effect of, 248-250
Illegality
 and holder-in-due course (HDC) defense, 409, 417
 as personal defense to liability, 419-420
 as real defense, 419-420
Illusory promise, 222, 229-230
ImClone, 542, 559
Immigration, 83-84
Immigration and Nationality Act (INA), 586, 595
Immigration and Naturalization Service (INS), 83-84, 128
Immigration Customer and Enforcement (ICE), 586
Immigration Reform and Control Act, 586, 595
Impairment of collateral, discharge of liability on instruments by, 421, 423
Impairment of recourse, discharge of liability on instruments by, 421, 423
Impartial jury, 128, A-6
Implied assumption of the risk, 161
Implied authority, 457-458, 470, 481
Implied condition, 34
Implied contract, 201-202, 212
Implied-in-fact contract, 201
Implied-in-law contract, 202. *See also* Quasi-contract (implied-in-law contract)
Implied powers, 513, 536
Implied ratification, 237-238
Implied warranty
 defined, 358, 369
 diminishing fundamental free market and right to contract, 370-371
 disclaiming, 361
 fitness for particular purpose, 360, 369
 of merchantability, 359, 361, 368-369, 371
 of quality, 359
 of trade usage, 360
Impossibility, and termination of agency relationship, 479, 482
Imposter rule, 415-416, 423
INA. *See* Immigration and Nationality Act (INA)
Incapacity (incompetence)
 defined, 234
 discharge of contracts due to, 234-238
 general rules of, 239
 and holder-in-due course (HDC) defense, 409
 intoxication, 239, 250
 mental incapacitation, 238, 250
 of offeror, 209, 212
 as personal defense to liability, 419-420
 as real defense to liability, 417
 See also Capacity
Incarceration, for white-collar crime, 115
Inches-A-Weigh, 225
Incidental beneficiaries, 295-296
Incidental damages, 314-315, 364
Income taxes collection by Congress, 16th Amendment, A-7
Incompetence. *See* Incapacity (incompetence)
Incomplete contract, 284, 286, 297
Incorporated cooperative, 497
Incorporation
 articles of, 511-512, 516-517, 536
 certificate of, 516
 Delaware law for, 516
 incorporators, 516
 legal process of, 516-517, 537
 selecting state for, 516, 536
Incorporator, 517-518
Indemnification
 agency relationship, 465, 468-469, 481
 duties of reimbursement and, 465, 468-469, 481
 right of, 475, 533
Independent agency, 68-69, 81
Independent contractor
 defined, 462
 versus employee, 482
 Federal Express and, 455-456, 462, 469, 480
 principal's liability and, 475-476
 relationship with employer, 463-464
Indeterminate, 130
India, 380, 388
Indiana Electronic Workers Pension Trust Fund (IBEW), 539
Indiana Golf and Sports Turf, LLC, 451
Indiana Pacers, 254, 265
Indianapolis Motor Speedway Hall of Fame Museum, 180
Indian Contract Act, 65
Indian University-Purdue University Fort Wayne (IPFW), 149
 Indictment, grand jury, 95, 129-130, 132, 136. *See also* Fifth Amendment
Indispensable paper, 428
"Individual mandate," 84
Individual repayment plan (Chapter 13 bankruptcy proceedings), 446-447
Indivisible contract, 249
IndyMac Bank, F.S.B., 400
Infancy
 as affirmative defense to crime, 131
 as real defense to liability, 417, 419-420
 as type of incapacity, 239, 251
Infant corporation, 515-518
Informal contract (simple contract), 203, 212
Informal negotiation, 44-45
Informal rule making, 69-70, 82
In forma pauperis (IFP), 151
Informational limitations, on agency powers, 74-78, 82
Information (for misdemeanor), 129, 132
Initial public offering (IPO), 558
Injunction, 319
Injunctive action, 550, 560
Innocent misrepresentation, 259, 267, 334
In pari delicto, 248
In personam jurisdiction ("jurisdiction over the person"), 32-34, 45, 60
In re Borges (2010), 452
In re Cortez (2005), 453
In re Datamark, Inc., Realtor (2009), 231
In re Egebjerg (2009), 454
In re Exide Techs (2010), 323
In re Hain Celestial Seasonings Prods. Consumer Litig. (2014), 609
In re JW (2017), 271
In re Leah Bet Woskob, Debtor; Alex Woskob; Helen Woskob; the Estate of Victor Woskob v. Beth Woskob (2002), 491
In re Marilyn Thomas (1995), 350
In rem jurisdiction ("jurisdiction over the thing"), 35
In re Murray U.S. Bankruptcy Court (2016), 444-445
In re Palmer (2007), 451
In re Simon (2017), 452
In re Simpson (2012), 426
In re Soza (2008), 453
In re Vivendi, SA Securities Litigation (2016), 563-564
In re Vogeler (2008), 452
INS. *See* Immigration and Naturalization Service (INS)
Insanity, 239, 478, 482
Insanity, as affirmative defense to crime, 131
Insider director, 521
Insider trading
 case for legalizing, 137
 defined, 116, 552
 how courts address, 552-553
 Martha Stewart case, 138, 542, 550, 553-*554,* 559
 and Michael Milliken, 115
 outsiders and, 553-554
 Salman v. United States (2016), 118
 tipping, 117
Insider Trading and Securities Fraud Enforcement Act of 1988, 557
Insider Trading Sanctions Act of 1984, 557, 560
Insolvency. *See* Bankruptcy
Insolvent debtor, 438
Inspection of goods, 357, 369
Inspection rights, 531-532, 537
Installment contract, 356
Instrument, defined, 431. *See also entries for specific instruments*
Insubstantial performance, 306
Insufficient funds, 409
Insurable interest, 340-343, 347
Insurance contract, 303, 548, 560
Insurance fraud, 116
Insurance policy, 307-308
Intangible property, 179, 432
Intangibles, 48
Integrated agreement, 282, 297
Intellectual property
 America Invents Act, 190
 copyright infringement, 189
 copyrights, 186-188
 defined, 172, 181, 192
 nature of, 172-173
 patents, 188-190
 Samsung-Apple dispute, 171-172, 190-191
 trademarks, 181-186
 trade secrets, 190-191

Intended beneficiaries
 creditor beneficiary, 293-294, 298
 creditor *versus* donee beneficiaries, 294-295
 defined, 292
 donee beneficiary, 293-294, 298
 vesting of rights, 294, 298
Intent, of offer, 212
 advertisement, 206
 auction, 207
 defined, 205
 manifestation of, to be bound to contract, 209-210
 preliminary negotiation, 206
Intentional discrimination, 568-569
Intentional interference with contract, 153-154, 167
Intentional misrepresentation, 260
Intentional tort, 167
 defined, 143
 against economic interests, 153-154
 against persons, 143-151
 against property, 152-153
 and respondent superior, 475
Intent to deceive, 263
Interactive, of e-government, 80
Interest, in real property (realty), 174-176
Interim trustee, in Chapter 7 bankruptcy proceeding, 440-441
Intermediate courts of appeal, 40
Intermediate scrutiny, 105, 107
Internacional Realty, 420-421
Internal Revenue Service (IRS), 78, 85, 93
International agreements, 7, 346. *See also* Treaty
International Brotherhood of Teamsters, Local Union 25, 455
International Development Cooperative Agency (IDCA), 7, 69
International law
 assignment of rights in Russia, 287
 breach of contract in Latvia, 313
 capacity of minor in China, 236
 consumer contracts in Japan, 263
 contracts in Japan, 210
 corporate structure in China, 518
 criminal liability in France, 531
 deceptive advertising in China, 602
 defects in assent in Philippines, 266
 duress in India, 266
 eminent domain in Germany, 179
 England and Statute of frauds, 277
 formation of power of attorney in Luxembourg, 459
 importance of contracts, 224
 importance of understanding, 9-10
 liability of Japanese online service providers, 189
 limited liability partnership in Japan, 493
 money orders in India, 388
 national law *versus*, 3-4
 negligence in Germany, 157
 negotiable instruments in Commonwealth, 384
 negotiable instruments in England, 411
 property interests in Vietnam, 175
 punitive damages in Canada and Japan, 164
 religious laws, 354
 special agency types in Middle East and Africa, 465
 special negotiable instruments in India, 380
 Sweden's securities market, 545
 trial in Japan, 51
International Olympic Committee, 115. *See also* Bribery
Internet
 defamatory statement posted on, 149 (*See also* Conditional privilege)
 e-government, 80
 forms of business organizations on, 487
 hostile takeovers online, 534
 marketing securities of, 558
 minimum contacts via, 34
 negligence on, 156
 and risks to national security, 108
 Uniform Commercial Code (UCC) and, 330
 See also E-commerce, and law
Internet Network Information Center, 186
Internet service provider (ISP), 607
Internment camp, 8
Interpretive rule, 66, 72, 82
Interrogatories, 47-48, 53
Interstate Commerce Commission (ICC), 7, 65, 65n
Interstate Land Sales Full Disclosure Act, 612
Inter vivos gift, 180, 194
Intoxicated person, 239, 250
Intoxication
 as affirmative defense to crime, 131
 as type of incapacity, 239, 250
In transit, 363
Intrastate issue, 549, 560
Intrusion on an individual's affairs/seclusion, 150
Invasion of privacy torts, 150
Inventory, 432
Investigation, power of, 65-66
Investment company, 557-558, 560
Investment Company Act Amendments of 1970, 558
Investment Company Act of 1940, 558, 560
Investment Directions, Inc. (IDI), 138-139
Investment security, 327
Investor adviser fraud, 121-123
Involuntary dissolution, 536
Involuntary liquidation petition, 439-440
Involuntary manslaughter, 127
Involuntary transfer, of personal property, 180-181, 192
IP. *See* Intellectual property
IPO. *See* Initial public offering (IPO)
IRA, 122
Iron Gate Self Storage, 246
IRS. *See* Internal Revenue Service (IRS)
Isbell v. Friedman (2012), 300
ISP. *See* Internet service provider (ISP)

J

Jack Kelly LLC v. Zegelstein (2016), 270
Jackson Hosp. Corp. v. United Clinics of Ky., LLC (2018), 324
Jackson v. Mayweather (2017), 169
Jacobson Products Co., 182
James and Janet Baker; Paul G. Bamburg and Robert Roth v. Goldman Sachs & Co. (2014), 30
James v. Meow Media, 156
Jammin Java Corp., 66
Japan
 consumer contracts in, 263
 contracts in, 210
 first-to-file rule, for patents, 190
 liability of online service providers in, 189
 limited liability partnership in, 493
 liquidated damages in, 318
 punitive damages in, 164
Japanese legal system, 51
Japanese Social Policy Council, 263
Jara Enterprise Inc., 528-529
J-Crew Management, 38
Jersey Tractor Trailer Training (JITT) Inc., 406-407
Jetmar Properties LLC, 507-508
Jimenez v. Cintas Corp. (2015), 232
J.K.S.P Restaurant v. County of Nassau (1987), 349
Joint stock company, 497
Joint venture, 498. *See also* Syndicate
Joint venture agreement, 498
Joint Ventures, 543-544
Jones Act, 539
Jones v. Waffle house, Inc. (2017), 253
Jones v. Wells Fargo Bank (2012), 383
Jordan v. Nationstar Mortgage, LLC, 16

Joseph M. Arpaio v. Barack Obama (2015), 62
Jose Vargas, Plaintiff and Appellant, v. FMI, Inc., et al., Defendants and Respondents (2015), 484-485
Jose Vargas Plaintiff and Appellant, v. FMI, Inc., et al, Defendants and Respondents (2015), 485
Journal of General Internal Medicine, 622
J.P. Morgan, 112
JPL. *See* NASA's Jet Propulsion Laboratory (JPL)
JRM Hauling & Recycling Servs v. Newark Grp., Inc. (2017), 373
Judaism, 243. *See also* Religion
Judge, 32
Judgment, 45, 51, 53, 60
Judgment non obstante verdicto. See Motion
Judicial branch, 86-87, 106
Judicial Branch (Article III of U.S. Constitution), A-4-A-5
Judicial limits
 on agency powers, 75-76, 82
 11th Amendment, A-6
Judicial opinion, 5, 60
Judicial power, of administrative agency, 65
Judicial review, 86
Junction City Partners, 498
Jurisdiction
 of limited liability companies (LLCs), 495
 original *versus* appellate, 32
 over persons and property, 32-34
 questionable, over Caterpillar, 59
 right of removal, 31
 subject-matter, 35-37
 trial courts of general and limited, 41
Jurisprudence, schools of, 8
Jury
 impartial, 128
 instruction to, 51, 60
 selection of, 48-50, 60
Jury trial
 jury decides questions of fact, 32
Just compensation, 104
Justice, 21
Justifiable ignorance of facts, 248
Justifiable use of force, as affirmative defense to crime, 131

K

Karam Automotive Inc., 299
Karaoke, 189
KCS Financial, 138
KDW Restructuring & Liquidation v. Greenfield (2012), 528
Kellmer v. Raines (2012), 540
Kelo v. City of New London (2005), 177-178
Kenneth C. Kinney v. Kathleen M. Gallagher (2015), 484
Kentucky Registry of Election Finance, 510-511
Kenya Communications Act (2008), 379
Kera LLC, 247-248
Kergosien et al. v. Ocean Energy, Inc. (2004), 56n
Kindle, 173
King Home Services, 241
King v. Riedl (2010), 241-242
Kmart, 235, 494
Knowingly, committing crime, 113-114
Kraft Foods, 26
Krispy Kreme, 6
Kroger's, 26
Kumar v. Gate Gourmet, Inc. (2014), 568

L

Labeling and packaging laws, 609, 621
LaBella v. Charlies Thomas, Inc. and Mercedes-Benz of North America, 371
Labor Bill of Rights, 576, 595
Laborer's Pension Fund v. A & C Envtl., Inc. (2002), 418-419

Labor laws and unions, 586-592
 Employment Eligibility Verification Department of Homeland Security, 587-589
 history, 586, 595
 Landrum-Griffin Act of 1959, 590, 592, 595
 Taft-Hartley Act of 1947, 590-592, 595
 Wagner Act, 590-591, 595
Labor-Management Relations Act, 592. *See also* Taft-Hartley Act of 1947
"Labor's Bill of Rights," 592
Lack of genuine assent, 197-198
Lack of proper form, 197
Ladd v. Uecker (2010), 168-169
Lamb Law Firm, 57
Land, contracts related to interest in, 276, 278, 297. *See also* Real property (realty)
Land contracts, interest in, 273, 276-278, 297
Land Design, 420
Landlord (property owner), 175
Landreth Timber Co. v. Landreth (1985), 543
Landrum-Griffin Act of 1959, 590, 592, 595
Langeslag v. KYMN Inc, 151
Lanham Act, 182-183, 192, 601, 625
Lapse of time, 477
Larry Pettit et al. v. Hampton and Beech, Inc. et al. (2007), 306
Larry S. Lawrence v. Bainbridge Apartments et al. (1997), 472
Larson v. Burton Constr., Inc.(2018), 335
Last-clear-chance doctrine, 160-161
Latin American countries, 22
Latvia, breach of contract in, 313
Law
 and agency relationship termination, 479, 482
 classification of, 3-4
 purposes of, 3-4
Law Reform Act of 1954, 277
Lawrence v. Fox (1859), 283
Lawson v. FMR LLC (2012), 135
Lawyer's Weekly USA, 142
LBO. *See* Leveraged buyout (LBO)
Leading question, 50
Lease, 175
 consumer and finance, 346
 as defined by Uniform Commercial Code (UCC), 220-221, 327, 330-331, 346
Lease contract, 332-336, 346
Leasehold estate, 174-175, 192
"Least sophisticated consumer" standard, 615
Lefkowitz v. Great Minneapolis Surplus Store, Inc. (1957), 206-207
Legal capacity, 209, 212, 250, 419-420
Legal cause, 157
Legal entity, 510, 512, 536
Legal interpretation, schools of, 8-9
Legal issue, 11
Legality, 240-247
Legality of contracts
 agreements that violate public policy, 243-247, 250
 agreements to commit crime or tort, 240
 effects of illegality, 248-250
 exculpatory clauses and, 245-246
 federal and state statute violations of, 240-243, 250
 gambling and, 242-243
 justifiable ignorance of facts, 248
 licensing violations and, 240-241
 in restraint of trade, 243-247
 Sabbath laws and, 243
 severable contract, 249-250
 unconscionable provisions and, 244-245
 usury and, 242
Legal object, 197, 212
Legal purpose, 197. *See also* Offer
Legal realism, 8-9
Legal reasoning, 11-12

Legal remedies (monetary damages)
 for breach of contract, 313
 compensatory damages, 314-315, 321
 consequential damages, 315-316, 364
 distinction between equitable remedies, 313
 liquidated damages, 315-316, 321, 364
 mitigation of damages, 317-318
 nominal damages, 315, 321
 punitive damages, 315, 321, 364
 special damages, 315
Legends Partnership, 491
Leghner Family Business Trust, 506-507
Legislative branch, 86-87, 106
Legislative Branch (Article 1 of U.S. Constitution), A-1-A-3
Legislative power, of administrative agency, 65
Legislative rule, 66
Leitchfield Development Corp. v. Clark (1988), 372
Lemon law, 611
Lemon v. Kurtzman, 98
Lernout & Hauspie Speech Products (N.V.), 30
Lessee, 330-331
Lessee (tenant), 175
Lessor, 330-331
Lessor (property owner), 175
Letter of credit, 203
Letters of credit, 327
Letters of intent, 299
Leveraged buyout (LBO), 534
Levitt, Arthur, 75
Liability
 avoiding, for negotiable instruments, 417-423
 contractual, of principal and agent, 471-472
 defenses to, 417-419, 423
 of directors and officers, 527-529
 discharge of, on instruments, 419-423
 personal defenses to, 419-420, 423
 principal and independent contractor, 475-476, 481
 of shareholders, 529, 536
 signature, 411, 423
 for torts and crimes, 527
 warranty, 411, 416-417, 423
Liability waiver, 247
Liability without fault, 114
Libel, 144-145
Libel-proof, 148
License, 175-176, 192
 naked, 194-195
 of patent holder, 189, 192
Licensing statutes, 240-241
Life estate, 174-175
Liggett Restaurant Group, Inc. v. City of Pontiac Michigan (2003), 312-313
Likely to mislead, in deceptive advertising, 601
Lilo Ben Zicron, 325
Limited jurisdiction, 41, A-6
Limited liability company (LLC), 494-497, 512
 versus corporation, 505
 defined, 494, 505
 dissolution of, 496
 formation and management of, 495-496
 rapid acceptance of, 494-495
Limited Liability Partnership, in Japan, 493
Limited liability partnership (LLP), 492-493
Limited offer, 548, 560
Limited partnership, 483-493, 497, 504
Limited warranty, 352, 362, 369
Linda Budd, Appellant v. Maureen Quinlan et al, Respondents (2008), 371
LinkedIn, 584, 594
Liquidated damages, 315-318, 321, 364-365
Liquidated debt, 226-227, 229
Liquidation
 in bankruptcy proceeding, 439, 447-448
 of corporation, 536
 defined, 439
Liquidity, 561

Liquors, prohibition of manufacture and sale of intoxicating, Eighteenth Amendment, A-7-A-8
Lissiak v. SW LOAN OO, LP (2016), 424
Litigation, *versus* alternative dispute resolution, 53-54
Live Cryo, LLC v. Cryousa Import and Sales, LLC (2017), 501
LLC. *See* Limited liability company (LLC)
Lloyd v. CVB Financial Corp. (2016), 562-563
LLP. *See* Limited liability partnership (LLP)
L.M., 474
Lockhard v. Pizza Hut, Inc. (1998), 571
Lockheed Martin Corp., 63
Locus sigilli (L.S.), 203
Long-arm statute, 33
Loss, 340, 347
Lost property, 180, 192
Love Bug virus, 124
Loving v. Virginia (1967), 575
Loyalty, duty of
 of directors, officers, and shareholders, 524-525, 537
 in partnerships, 467-*468,* 481
L.S. (*locus sigilli* or the place for the seal), 203
Lucas v. South Carolina Coastal Commission, 104
Lucy v. Zehmer (1954), 205-206, 212
Luxembourg, formation of power of attorney in, 459
84 Lumber Co., L.P. v. Houser (2011), 232
Lumber Company L.P. (84 Lumber), 232
LVNV Funding LLC, 615-616
Lyles v. Medtronic Sofamor Danek (2017), 170

M

Mac Haik Dodge Chrysler Jeep, 377
Macy's, 612
"Made in the USA" labels, 602
Magistrates court, in England, 42
Magnani v. Cuggino (2017), 324
Magnetic ink character recognition (MICR), 391
Magnuson-Moss Act, 362, 368-369
Mailbox rule, 211
Mail fraud, 114, 117, 119-121, 135
Mail Fraud Act (1990), 119
Mail-Order Rule of 1975, 610
Mail-order sales, 610-611, 621
Mail or Telephone Order Merchandise Rule of 1993, 610
Main-purpose rule, 276
Majority opinion, 52
Major League Baseball Players Association (MLBPA), 216, 456
Maker
 as accommodation party, 414
 and secondary liability, 412
 signature of, 378, 380, 398
MAK LLC, 506
Malpractice, 156, 158-159, 493. *See also* Professional corporation
Manager, director as, 521
Manager-managed LLC, 496
Manheim Auto Auction, 427
Manslaughter, 127
Manufacturing arrangement, 500
Marbury v. Madison (1803), 86
Margeson v. Artis (2009), 225-226
Marijuana, medical, 590
Market Reform Act of 1990, 545
Market transparency, 137
Marriage
 consideration of, 273, 275, 278, 297
 discrimination based on, 575-576, 593
 and employment discrimination, 575-576
 of same sex couples, 103
Marriott, 112
Martha Rochelle v. Hartley Mead (2015), 269-270

Marvel, 139-140
Marykate Ellingsworth, Plaintiff, v. Hartford Fire Insurance Company (2017), 595-596
Mary W. Scott (Respondent-Appellant) v. Mid-Century Homes, Inc. (1987), 257
Massachusetts v. EPA (1999), 64, 75
Masson v. New Yorker Magazine, Inc. (1991), 6
Matal, Interim Director, United States Patent and Trademark Office v. Tam (2017), 183-184
Material alteration, 417-420
Material breach, 307-308, 323
Material information, examples of, 552
Material misrepresentation, 601
Material terms, 207
Matter of Twin Bay Village, Inc. (2017), 539
Maxim Solutions Group Inc., 385
Maya Development, LLC, 215
Mazda Motor Corporation v. Hurst (2017), 165
MBM Mowing Systems, LLC, 138
McCann Ranch & Livestock Company Inc., 526
McCann v. McCann (2012), 526
McDonald's, 6, 16, 182, 494, 500
McDonnell Douglas v. Green (1973), 567
McLellan v. Charly (2008), 221-222
MCPA. *See* Michigan Consumer Protection Act (MCPA)
McPherson v. First Presbyterian Church of Woodward et al. (2015), 170
MDT. *See* Montana Department of Transportation (MDT)
Means test, 440
Meat Inspection Act of 1906, 620
Med-arb (dispute resolution process), 57-58
MediaEdge, 425
Mediation, 54-55, 58
Mediation agreement, 54-55
Mediation brief, 54
Mediator, 72, 82
Medical Association Insurance Agency (AMAIA), 215
Medical malpractice, 158-159
Medical marijuana, 590
Medical User of Marijuana Act (MUMA), 590
Medinar Petroleum Corporation, 252
Medtronic Sofamor Danek, 169-170
Member-managed LLC, 496
Member (owner) in limited liability company (LLC), 495
Memorandum of understanding (MOU), 84
Mensing ("failure to warn") claim, 625
Mens rea (guilty mind), 113, 125, 136
Mentally incapacitated person, 238, 250, 419-420
Merchant, 330-331, 346
Merchantability, 368
 breach of, 372
 implied warranty of, 359, 361, 369
 waiver of, 371
Merchantable, 359
Merger clause, 283
Merit defense, 572, 593
Merrill Lynch, 551
MetLife, 300
Meyer v. Christie (2011), 498-499
MFSAs. *See* Motor Fuel Supply Agreements (MFSAs)
MGM Desert Inn Inc., 374, 382, 397
MHRA. *See* Minnesota Human Rights Act (MHRA)
Michael Savage v. City of Memphis (2015), 483
Michigan Consumer Protection Act (MCPA), 248
MICR. *See* Magnetic ink character recognition (MICR)
Microsoft, 5, 18, 536
Middle East
 e-commerce sales in, 334
 special types of agency in, 465
Migro Seed Company Inc., 329-330
Milam v. Commonwealth of Kentucky, 109-110

Militia, necessary for security, 95. *See also* Second Amendment
Millennium Asset Recovery Inc., 507
Miller v. California (1973), 97
Mind & Motion Utah Investments, LLC, Appellee v. Celtic Bank Corporation (2015), 304-305
Mineral rights, 173-174
Minimum salary, 580
Minimum wage, 580, 594
Minitrial, 58
Minnesota Human Rights Act (MHRA), 151
Minnesota's Mortgage Moratorium Act, 94
Minor
 capacity of, 234-237, 250-251
 defined, 234
 disaffirmance of contract by, 234-237
 liability of, for necessaries, 237
 parents' liability for children's contracts, necessaries, and torts, 238
 ratification, 237-238
Miranda rights, 128-129
Mirror-image rule, 333
Mirroring assignments, 290
Misappropriation, 29, 117, 153, 167
Misappropriation theory, 553-554, 556
Misdemeanor, 114, 129, 136
Mislaid property, 180, 192
Misrepresentation
 defined, 259, 267
 fraudulent, 260-263, 267
 innocent, 259, 267
 negligent, 260, 267
 as obstacle to genuine assent, 255, 267
 See also Fraudulent misrepresentation
Misrepresentation liability, 475
Mississippi Valley Silica Co., 4
Misspelled name, 384
Mistake, 334
 defined, 256
 mutual, 256-258, 267, 335
 as obstacle to genuine assent, 255-259
 unilateral, 256-257
Mistake of fact, as affirmative defense to crime, 131
Mitchell County v. Zimmerman (2012), 110
Mitchell v. Lath (1928), 299
Mitigation of damages, 317-318
Mitsubishi, 29
Mixed goods and services contract, 330, 346
Mixed sale, 330
MK Stucco, 485
MLBPA. *See* Major League Baseball Players Association (MLBPA)
Mock trial, 50
Model (uniform) law, 5
Modified comparative negligence, 161, 167
Modify, 52
Money, corporation's power to borrow and lend, 513
Money damages. *See* Legal remedies (monetary damages)
Money laundering, 121
Money order, 387-389, 398
Money Store, The, 424
Money Store Service Corporation, 424
Monmouth-Ocean Hospital Service Corp. (MONOC), 582-583
MONOC. *See* Monmouth-Ocean Hospital Service Corp. (MONOC)
Montana Department of Transportation (MDT), 349
Moore, Tiara, 71
Moore v. Goran, LLC (2017), 349
Moral coercion, 265
Morrison, Antonio, 89
Morris v. Plica, Inc. (2013), 625
Motion
 defined, 45

to dismiss (demurrer), 45
for judgment as a matter of law, 51
for a judgment as a matter of law, 50
for a judgment in accordance with the verdict *(judgment non obstante verdicto)*, 51
in judgment on the pleadings, 46
for a new trial, 51
overview, including all steps in civil lawsuit, 53
for summary judgment, 48, 53
Moto Inc., 540-541
Motor Fuel Supply Agreements (MFSAs), 364
Motorola, 551
Movability, 379, 398
Movable collateral, perfection of, 431-432, 449
M&T Bank, 432-433, 448
Mullen Company, The, 400-401
Multiple-product order, 605, 621
MUMA. *See* Medical User of Marijuana Act (MUMA)
Murray-Calloway Country Public Hospital Corporation, Plaintiff v. Samuel E. Sprehe M.D. Defendant/Third-Party Plaintiff V. Jon O'Shaughnessy (2012), 270-271
Mutual agreement, 308-309, 321
Mutual agreement by parties, 477
Mutual mistake, 256-258, 267, 335
Mutual rescission, 309
Myers, David, 24
Mylan, 19
MySpace, 264

N

NAFTA. *See* National American Free Trade Agreement (NAFTA)
Naked licensing, 194-195
Napster, 188
NASA. *See* National Aeronautics and Space Administration (NASA)
NASA's Jet Propulsion Laboratory (JPL), 78-79
NASA v. Nelson, 78-79
National Aeronautics and Space Administration (NASA), 7, 69, 78-79
National American Free Trade Agreement (NAFTA), 9
National Basketball Association, 254
National Collegiate Athletic Association (NCAA), 76
National Compressor Corp. v. Carrow and McGee (1969), 341
National Conference of Commissioners on Uniform State Laws (NCC/NCCUSL), 5, 198, 330, 384, 494
National Conference of Commissions on Uniform State Laws
National Federation of Independent Business, 85
National Fraud Information Center, 120
National Health Services, 125
National Highway Traffic Safety Administration (NHTSA), 620
National Information Infrastructure Protection Act (1996), 124
National Institutes of Health, 22
National Labor Relations Act (NLRA), 55, 76, 583, 585-586, 590-592, 595
National Labor Relations Board (NLRB), 69, 76-78, 455, 461, 583, 591-592
National law, *versus* international law, 3-4
National origin, suspect classifications based on, 15
National Science Foundation (NSF), 7, 69, 186
National Securities Markets Improvement Act of 1996, 545
National security, Internet and, 1108
National Transportation Safety Board (NTSB), 7
Nationstar mortgage, 16
NationsWay, 466
Nazi Holocaust, 338

NCAA. *See* National Collegiate Athletic Association (NCAA)
NCC. *See* National Conference of Commissioners on Uniform State Laws (NCC)
NCCUSL. *See* National Conference of Commissioners on Uniform State Laws (NCCUSL)
Necessaries, 237-238
Necessity, as affirmative defense to crime, 131
Nedlloyd v. Sanderson Farms, Inc. (2006), 350
Negligence, 143, 415, 423
 breach of duty, 155-156, 167
 causation (actual and proximate), 155-157, 167
 conscious and unconscious, 157
 contributory, 160-161
 damages, 155, 157-158, 167
 defenses to, 160-162, 167
 defined, 154
 duty, 155-156, 167
 elements of, 155-158, 167
 in Germany, 157
 Good Samaritan statutes, 162
 gross, 157
 plaintiff's doctrines, 158-160
 special defenses to, 162
 superseding cause, 162
Negligence per se
 case illustration, 160, 167
 defined, 159
 elements of, 160
Negligent, crime, 114
Negligent misrepresentation, 60
Negotiability, 376
 discharge of liability on, 419-422
 requirements for, 378-381, 398
 significance of, 398
 words of, 381-382, 398
Negotiable, potential complexity of, 376-377
Negotiable document of title, 343, 347
Negotiable instrument, 203, 327, 374
 avoiding liability for, 417-422
 contracts as commercial paper, 375
 defined, 375
 discharge of liability on instruments, 419-423
 in England, 411
 holder-in-due-course (HDC), 403-410
 liability defenses, 417-419, 423
 need for, 375, 398
 nondelivery of, 419-420
 oral agreement and, 397
 overview of law of, 376-382, 398
 problems with commercial paper, 375
 and promissory notes, 401
 types of, 376, 398
 unauthorized, nonfraudulent completion of, 419-420
Negotiable Instrument Act (1881), 380
Negotiation, 54
 defined, 382
 delivery, 382, 398
 endorsement, 382
 informal, 44-45, 58
 noncriminal endorsement problems, 383-385
Nerium International, 34
Nerve-center approach, 36-37
Nestlé, 498
Netherlands, 80
Network Solutions, Inc. (NSI), 186
Nevada State Bank, 381
Newark Group, Inc., The, 373
New Castle Care, LLC, 274-275
New Community Church, 109
Newnan Crossing Partnership, 400-401
New Wave Technologies, Inc. v. Legacy Bank of Texas (2008), 385
New York Stock Exchange, 84
New York v. Quarles (1984), 129
New Zealand, 622

NHTSA. *See* National Highway Traffic Safety Administration (NHTSA)
Nicaragua, 22
Nike, 182, 194
NIL. *See* Uniform Negotiable Instruments Law (NIL)
Ninth Amendment, 95, 104-105, 107, A-6
NLRA. *See* National Labor Relations Act (NLRA)
NLRB v. Jones & Laughlin Steel Corp. (1937), 88
NLRB v. Town & Country Electric (1995), 76
N&N Partners LLC, 451
Nolo contendere, 130
Nominal damages, 162-163, 167, 315, 321
Nonbinding jury verdict, 58
Noncompete agreement, 252, 266, 303, 323
Nonconforming goods, 367-369
Noncriminal endorsement problems, 383-385
Nondischargeable debt, under bankruptcy code, 442
Nondisclosure, 262
Nonemployees, harassment by, 571
Noninvestment firm, 549
Nonnegotiable document of title, 343, 347
Nonpossessory estate, 175-176
Nonprofit corporation, 514, 536
No-par shares, 529
NorthAM Food Trading Co., 344-345
Northwest Ecosystem Alliance v. Washington Forest Practices Board (2003), 84
Northwestern University and College Athletes Players Association (CAPA), 76
Norway, 80
Note, defined, 376, 398
Not guilty, 130
Notice
 actual, 476-477
 of assignment, 289
 of claim or defect, 407-410
 constructive, 477
 of dishonor, 409, 413-414, 422
 of franchise termination, 504
 of termination, 477
Notice-and-comment requirement, 83
Notice-and-comment rule making, 69-71
Notification, duty of, 467-468, 481
Notorious possession, 177
Novation, 309
NRC. *See* Nuclear Regulatory Commission (NRC)
NSF. *See* National Science Foundation (NSF)
NSI. *See* Network Solutions, Inc. (NSI)
NTSB. *See* National Transportation Safety Board (NTSB)
Nuclear Regulatory Commission (NRC), 7, 69
Nutrisoya v. Sunrich (2011), 323

O

Obedience, duty of, 467-468, 481
Obergefell v. Hodges (2015), 103, 575
Objective impossibility, 309-310
Objective satisfaction standard, 307
Obligation. *See* Performance obligation
Obligation transfer, 41
Obligee, 286-287, 290, 297
Obligor, 286-287, 289-290, 297-298
Obscenity, as unprotected speech, 97
Occupational Safety and Health Act (OSHA) of 1970, 101, 466, 579, 581, 594
Occupational Safety and Health Administration (OSHA), 7, 69, 73, 594
Occurrence of specific event, agency relationship termination, 478
Odometer Act of 1973, 611
OECD. *See* Organization for Economic Cooperation and Development (OECD)
Offer
 advertising, 206-207, 212
 defined, 197
 definite and certain, 207

 destruction or subsequent illegality of subject matter, 209, 212
 effect of acceptance after rejection, 211
 elements of, 205-207, 212
 elements of acceptance, 209-212
 lapse of time or failure of another condition specified in, 209, 212
 mailbox rule, 211
 rejection of, 208-209, 212
 in sales and lease contracts, 332-333, 346
 termination of, 208-209, 212
Offer and acceptance, 204
Offeree/offeror
 acceptance, 204, 209-210, 212, 333
 agreement, 203-204, 212
 bilateral *versus* unilateral contracts, 199-200
 communication to, 207, 211-212, 214
 death or incapacity of offeror, 209, 212
 defined, 197
 intent, 205-206
 lack of genuine assent, 198
 rejection of offer by, 208, 212
 revocation and rejection, 208-209, 212
 termination of offer, 208-209
Office of Personnel Management (OPM), 7, 69
Officer
 liability of, 512, 537
 rights of, 530, 533, 537
 roles of, 520-521, 523, 537
 See also Corporation
O'Guin v. Bingham County (2005), 160
Ohr, Peter Sung, 77
Old Mexican Foods, Inc. v. Hanson Staple Company (2009), 361
Old Order Groffdale Conference Mennonite Church, 110
Oliveira v. Amoco Oil Co. (2003), 603
Olympic Mortgage Company, 425
Omnibus Crime Control and Safe Streets Act of 1968, 582
Oncale v. Sundowner Offshore Services, Inc. (1998), 571
Oncology Tech, 299-300
Oncology Tech, LLC, Plaintiff v. Elekta, Inc. (2013), 299-300
One-tier board system, 518
Online banking, 395-397, 399
Online fraud, 558
Online sales, 612, 622
Online service provider (OSP), 189
Ontario's Ministry of the Environment, 27
Open Communities Alliance v. Carson (2017), 70-71
Open-end credit line, 613
Opening statement, 50, 60
Open possession, 177
Open terms, 333
Operating agreement, of limited liability company (LLC), 195
Operation of law, 321
 alteration of contract, 309, 313
 bankruptcy, 309, 313
 commercial impracticability, 310-311, 313
 frustration of purpose, 312-313
 impossibility of performance, 309-310, 313
 termination by, 478-479, 482
 tolling of statute of limitations, 309, 313
OPM. *See* Office of Personnel Management (OPM)
Oppenheimer v. Holt (2015), 195
Option contract, 208
Oracle, 117
Oral agreement, 374, 397
Oral contract
 case regarding breach of, 274-275, 300
 and parol evidence rule, 282
 terms, 284-285, 297
 versus written, 273 (*See also* Statute of frauds)
Oral joint venture agreement, 498

Oral testimony, 71
Order, 67
Order instrument, 382, 398
Order of relief, 440
Order to pay, 378, 380
Ordinary mail, 33
Organization. *See* Business organization
Organization for Economic Cooperation and Development (OECD), 116
Organized Crime Control Act (1970), 132
Origin contract, 343, 347
OSHA. *See* Occupational Safety and Health Act (OSHA) of 1970
OSP. *See* Online service provider (OSP)
Output contract, 332
Outrage, tort of, 150
Outside director, 521
Overdue instrument, 408
Ownership, 340, 347
Ownership, extent of, 173-174
Ownership interest, 175

P

PAA. *See* Pesticide Application Agreement (PAA)
Packaging. *See* Labeling and packaging laws
Packingham v. North Carolina (2017), 110
Palsgraf v. Long Island Railroad Company, 157
Pardanashin ladies, 265
Parents
 entering into contracts on minor's behalf, 247-248
 liability for their children's contracts, necessaries, and torts, 238
Parliament, 277
Parol evidence rule
 defined, 282, 297
 exceptions to, 283-286, 297
 fully integrated agreement, 282
 merger clause, 283
 and Uniform Commercial Code (UCC), 331
Partially disclosed principal, 471
Partial performance, 279, 297
Parties agreement, 356, 369
Partnership
 advantages and disadvantages of, 489
 articles of, 490
 Colette Bohatch v. Butler & Binion, 489
 defined, 488
 duties of partners to one another, 489-490
 by estoppel, 490
 formation of, 490
 general, 492-493
 limited, 483-493
 limited liability partnership (LLP), 492-493
 reasons for rightful dissolution of, 491
 termination of, 490-492
Par-value shares, 529
Passante, 228
Patagonia, 18
Patent
 application process, 190
 corporations and gene parenting, 192-193
 criteria for, 188-189, 192
 cross-licensing, 189, 192
 defined, 188-189, 192
 and exclusive federal jurisdiction, 35
 tying arrangement, 189
Patent infringement, 171, 190-191
Patent Office, 189-190
Patent search, 189
Paternity suit, 254-255
Patient Protection and Affordable Care Act, 84
Patrick v. Allen (2005), 525
Patty J. Gandee v. LDL Freedom Enterprises (2013), 252-253
Paul, Weiss, Rifkind, Wharton & Garrison, LLP, 169

Paula E. Rossman, Individuality and for All Others Similarly Situatd v. Fleet Bank (R.I.) National Association et al. (2002), 604-605
Paul Bishop et al. v. Wells Fargo & Co. et al., 16
Payable at time certain, 381, 398
Payee, 386, 398
Pay increases limits for Congress, 27th Amendment, A-9
Payment, 375, 398
Payment, default in, 400
"Pay to bearer," 381
"Pay to cash," 381
"Pay to the order of cash [or bearer]," 381
PDA. *See* Pregnancy Discrimination Act (PDA) of 1987
Peaceable assembly, 94. *See also* First Amendment
Peerless, 257
Pendergrass v. Ingram, 230
Pennsylvania Human Relations Committee (PHRC), 596
Penny Garrison et al. v. The Superior Court of Los Angeles et al. (2005), 458
People v. Diaz, 101
People v. Dolezal, 29
Pepsi, 196, 200, 206
PepsiCo, 200
Peremptory challenge, 48-49
Perez v. Mortgage Bankers Association (2015), 83
Perfected security interest
 automatic perfection, 431-432, 449
 by filing, 430, 432, 449
 of movable collateral, 431, 449
 by possession, 430-432, 449
 of security interests in automobiles and boats, 431-432, 449
Perfection, 430-432
Perfect tender rule, 353-357, 369
Performance
 accord and satisfaction, 309
 anticipatory repudiation, 308
 complete, 305
 discharge by, 305-313
 duty of, 467-468, 481
 impossibility of, 309-310
 impracticability of, 311
 insubstantial, 306
 job, and wage equality, 579, 594
 by material breach, 307-308
 by mutual agreement, 308-309
 mutual rescission, 309
 novation and, 309
 obtain specific, 366
 by operation of law, 309-313
 perfect tender rule, 353-355, 369
 as subject to satisfaction of contracting party, 307
 substantial, 305-306
 substituted contract, 309
 tender, 305
Performance obligation
 basic, 352-354, 369
 of buyers/lessees, 352-353, 357
 commercial reasonableness, 352
 conforming goods, 353, 369
 exceptions of, 357
 good faith, 352-353, 369
 perfect tender rule, 353-355, 369
 perfect tender rule exceptions
 commercial impracticability, 357, 369
 course of dealing, 356
 course of performance, 356
 destroyed goods, 356
 norms in industry and past dealings between parties, 356, 369
 outlined in parties' agreement, 356, 369
 seller's and lessor's right to cure, 356, 369
 substantial impairment, 356
 usage of trade, 356

 of sellers/lessors, 352-353
 tender of delivery, 353
Perjury, 121
Perpetual existence, 511-513, 536
Perrin Excavating, 506
Personal benefit, and tipping, 117
Personal defense, 417, 419-420, 423
Personal property
 court order, 181
 defined, 172, 192
 fungible goods (confusion), 181
 involuntary transfer of, 180-181, 192
 types of, 179, 192
 voluntary transfer of, 179-180, 192
Personal self-dealing, 525-526
Persons, intentional torts against, 143-151, 167
Persons and property, jurisdiction over, 32-35
Pervasive-regulation exception, 101
Pesticide Application Agreement (PAA), 322
Peterson v. North American Plant Breeders (1984), 329-330
Petition
 in Chapter 7 bankruptcy, 439-440, 453-454
 in Chapter 13 bankruptcy, 447
 for dissolution, 532, 537
 See also First Amendment
Petit jury, 130
Petroecuador, 17
Petty offense, 114, 136
Pfizer, Inc., 29
Pharmaceutical company advertising, 623
PHH Corp. v. Consumer Fln. Prot. Bureau (2018), 84
Philippines, defects in Assent in, 266
PHRC. *See* Pennsylvania Human Relations Committee (PHRC)
Physical coercion, 265, 267
Pierce the corporate veil, 519, 537
Plaintiff, 33-34
Plaintiff's doctrines
 defenses to, 160-162, 167
 negligence per se, 160-161, 167
 res ipsa loquitur, 158-160, 167
Plea bargain, 130
Pleading, 45-46, 53
Pledge, 431
PMC Corporation v. Houston Wire & Cable Company (2002), 332
PMSI. *See* Purchase-money security interest (PMSI)
Policy powers, 107
Policy statement, 72
Political limitations, on agency powers, 75, 82
Political speech, 96
POM Wonderful 100% Pomegranate Juice, 625-626
Ponzi scheme, 117, 121-123, 258
Port Motors Inc., 337
Portugues-Santana v. Rekomdiv International, Inc. (2013), 271
Possession, perfection, 430-432, 449
Possessory estates, in land, 174-175
Postal Reorganization Act of 1970, 611
Postdated check, 393, 398
Postdating check, 376
Posteffective period, 546-547, 559
Posttrial motion, 51, 53, 60, 136
Posttrial procedure, 130-131
Power of attorney, 457, 459, 481
Preamble, of U.S. Constitution, A-1
Preamble, to U.S. Constitution, A-1
Precedent, 5
Pre-employment investigations, 585
Preemptive rights, 530, 533, 537
Preexisting duty, 224-226, 229
Preferential payment, 441-442
Preferred stock (preferred shares), 519
Prefiling period, 546-547, 559

Pregnancy Discrimination Act (PDA) of 1987, 567, 571–572, 593, 596
Preincorporation agreement, 515
Prejudicial error, 52, 130
Preliminary negotiation, 206
Prenuptial agreement, 275
Prescription drugs, 622–623
Presentation of evidence, 50
Presented party (adhering party), 245
Presentment, 412–413, 422
Presentment warranty, 416–417, 423
Preserving Civil Liberties in the Information Age, 606
President, U.S., 8
Presidential succession, 25th Amendment, A-9
Presidential term limits, 23rd Amendment, A-8–A-9
Preston v. Ferrer (2008), 57
Pretexting, 117, 121
Pretrial conference, 48, 53, 60
Pretrial motion, 45–46
Pretrial procedure, 128–130, 136
Pretrial stage
 defendant's response, 45, 60
 discovery, 47–48
 informal negotiation, 44–45
 overview, including all stages, 53
 pleadings, 45–46
 pretrial motions, 45–46
 service of process, 45
Pre-UCC law, 340
Price Auto Sales, Inc. v. Sanders (2012), 436
PriceWaterhouseCoopers, 554
Price Waterhouse v. Hopkins (1989), 575
Pricilla D. Webster c. Blue Ship Tea Room, Inc. (1964), 359–360
Prima facie, 67
Primarily liable, 412, 422
Primary beneficiary, of contract, 292
Primary obligation, 276
Principal
 agent's duties to, 464–468, 481
 classification of, 471, 481
 contractual liability of, 471–472
 defined, 456, 480
 duties of, to agent, 465
 liability of, and independent contractor, 475–476, 481–482
 tortious conduct of, 473–475, 482
Principal-agent relationship, 461–462, 481
Principal liability, 471–472, 482
"Principal place of business," 36–37
Principles of European Contract Law, 198
Principles of Federal Prosecution, 129
Prior dealings exception to parol evidence rule, 285–286
Priority claim, in Chapter 7 bankruptcy proceeding, 442–443
Priority dispute
 buyers in ordinary course of business, 434
 defined, 449
 rights of secured party, 434
 secured party *versus* buyer, 434
 secured *versus* secured creditor, 433
 secured *versus* unsecured creditors, 433
Privacy
 data mining and, 606–608
 and deceptive advertising, 606–608
 reasonable expectation of, 584
 right to, 107–108
 contraception and Affordable Care Act (ACA), 98–99, 104–105
 technology and, 101
 and telemarketing, 606
 workplace, 582–584, 594
Privacy Act (1974), 74, 78–80, 82
Privacy policy, and customer data, 125
Privacy protection, 396–397
Privacy tort, 150–151
Private corporation, 513–514, 536

Private law, 3–4
Private nuisance, 152, 167
Private placement exemption (Rule 506), 548–549, 560
Private property taken without compensation, 95, 104, 107, 177. *See also* Fifth Amendment
Private Securities Litigation Reform Act of 1995, 553, 1995
Private trial, 58–59
Privilege, as affirmative defense, 148
Privileges and immunities clause, 93–94, 107
Privity of contract, 286
Probable cause, 99, 128
Probate court, 41
Procedural due process, 102
Procedural rule, 66
Procedural unconscionability, 245
Proceeds, 428, 432, 449–450
Process, service of, 33
Professional corporation, 515, 536
Professional ethics, 16
*Professional Sales, Inc. v. The Estate Of Joseph S. Brehaut, C. Christopher Moore, Jr. Executor And C. Christopher Moore, Jr. Individually And Joseph S. Brehaut And C. Christopher Moore, Jr. Partnership T/D/B/A/ Moll's Garag*e (2015), 299
Professional malpractice, 493
Profit, 175–176
Prohibition against uncompensated takings, 95, 104, 107, 177–178, A-6
Promisee, 292, 298
Promisor, 292, 298
Promissory estoppel, 219–220, 229, 279–281, 297
Promissory note, 384, 401
Promoter, 515, 518, 536
Proper form, lack of, 197–198
Properly payable rule, 392–394, 398
Proper purpose, 532
Property
 corporation's power to acquire, 513
 intentional torts against, 152–153, 167
Property interests, in Vietnam, 175
Property transfer, 176–179
Pro se complaint, 110, 193
Prosecution, 130
Prosecutor, 128
Prospectus, 546–547, 559
Protective sweep, 102
Protect My Check, Inc. v. Craig C. Dilger, John Steffen (2016), 510–511
Protocol, 7. *See also* Treaty
Proximate cause, 157, 167
Proxy, 556, 560
Proxy solicitation, 556, 560
PSTEVO, LLC, 261
Public Company Accounting Oversight Board, 84, 545
Public corporation, 513–514, 536
Public disclosure of private facts, 150
Public disclosure test, 26–27
Public figure privilege, 149
Public Health Cigarette Smoking Act of 1970, 608
Public hearing, 71
Public Housing Authorities, 71
Public law, 3–4
Publicly held corporation, 514, 536
Public nuisance, 152
Public official, 115–116
Public use, 177–178. *See also* Takings clause
Publishing Control Act, China's, 7
Puffing, 601–602. *See also* Deceptive advertising
"Pump and dump" scheme, 66
Pumping and dumping, 558
Punishment
 cruel and unusual, 95, A-6 (*See also* Eighth Amendment)
 for white-collar crime, 115

Punitive damages (exemplary damage)
 in addition to compensatory damages, 365
 in Canada and Japan, 164
 in cases based on discrimination other than race, 573, 578
 defined, 158, 315
 determining, 162–167
 as legal remedy, 321
 recovering, 364–365, 395, 440
Purchase-money security interest (PMSI), 429, 433–434, 449
PurchasePro, 524
Pure comparative negligence, 161, 167
Pure Food and Drugs Act of 1906, 620
Purposeful discrimination, 49
Purposefully, committing crime, 113–114

Q

Qiaodan Sports, 195
Qualification
 of corporate officers, 521
 in limited liability company (LLC), 495
Qualified endorsement, 382
Qualitex Co. v. Jacobson Products Co. (1995), 182–183
Quasi-contract (implied-in-law contract), 202, 320
Quasi-judicial, 149
Queen's Bench, in England, 42
Question of fact, 32
Question of law, 32
Quid pro quo harassment, 569–570, 593
Quill Corporation v. North Dakota (1992), 92

R

Race, suspect classifications based on, 105
Racketeer Influenced and Corrupt Organizations (RICO) Act, 131–132, 136
Racketeering, 132
Radio Shack, 372
RAI Entities, 112
Railway Labor Act, 592
Rakestraw v. Rodriguez (1972), 415
RAL. *See* Residential Accredit Loans Inc. (RAL)
Ralph T. Leonard et al. v Jerry D. McMorris et al. (2003), 466
"Ransomware" attack, 125
Rate Manual, 614
Ratification, 237–238
Ratification (Article VII of U.S. Constitution), A-5–A-6
Ratify, 414–415
Rational-basis test, 104, 107
Raul Morales-Izuierdo v. Alberto R. Gonazales, Attorney General (2007), 83–84
Ray v. Wal-Mart Stores, Inc. (2015), 598
Reacquisition, 421, 423
Reaffirmation agreement, 445
Real defense, 417–420
Real estate contract, 303
Real estate purchase contract (REPC), 305
Real estate sales, 612, 622
Real Estate Settlement Procedures Act of 1974, 612
Real Housewives of New Jersey, 120
Reality of assent, 254–271
Real property (realty)
 breach of contract, 321
 deed, 174
 defined, 172, 192
 extent of ownership, 173–174
 extent of rights, 174–175
 interests in, 174–176
 Jordan v. Nationstar Mortgage, LLC, 16
 property transfer, 176–179
Reasonable consumer, 601
Reasonable expectation of privacy, 584
Reasonable person standard, 156

Reasons and conclusions, 12-13
Recklessly, committing crime, 113-114
Reclaiming goods, 363
Recognizance, 203
Recording, of property transfers, 176
Recourse, discharge of liability on instruments by, 421, 423
Re-cross, 50
Reddick v. FDIC (2016), 214
Rederford v. US Airways, Inc. (2009), 322
Red-herring prospectus, 547
Redirect examination, 50
Red Lion Hotels Franchising Inc. v. MAK, LLC (2011), 506
Red Lodge Tied Projects, 349
Redress of grievances, petition government for, 94-95. *See also* First Amendment
Reed v. Gilbert, Arizona (2015), 108-109
Reformation, 319
Regents of the University of California v. Bakke, 6
Regional Bell Operating Companies, 24
Registration statement, 546, 559
Registration trademark symbol, 189
Reg-neg, 72, 82
Regulation E (enforcing EFTA), 395-396
Regulation Z, 610
Regulatory compliance, 396
Regulatory takings, 104. *See also* Takings clause
Reimbursement contract, 266, 465, 468, 481
Relative permanence, 379, 398
Religion
 freedom of, 95, A-6
 observing Sabbath days online, 243
 See also First Amendment
Religious Freedom Restoration Act of 1993 (RFRA), 98
Religious laws, 354, A-6
Remand, 52
RE/MAX, 500
Remedies
 accept nonconforming goods, 367, 369
 for ADA violations, 578
 in agency relationships
 avoidance, 468, 481
 constructive trust, 468, 481
 demand an accounting, 469, 481
 indemnification, 468-469, 481
 specific performance, 469, 481
 tort and contract, 469, 481
 available to buyers/lessees under UCC, 363-367, 369
 available to sellers/lessors and buyers/lessees, 362-363, 369
 breach of contract for, 313-320, 369
 alternative dispute resolution (ADR), 313
 equitable, 313-318, 321
 legal remedies (money damages), 313-318, 321
 cancellation, 255, 362, 368-369
 dispose of goods, 363, 369
 modifications or limitations to, under UCC, 367, 369
 obtain cover, 363, 369
 obtain specific performance, 366, 369
 reclaim goods, 363, 369
 reject nonconforming goods, 367, 369
 resell goods, 363, 369
 revoke acceptance of nonconforming goods, 367-369
 seek damages, 367, 369
 stop delivery, 363, 369
 suit to benefit bargain, 363, 369
 suit to recover damages, 364-365, 369
 under Title VII, 572-573
 ultra vires contracts, 513
 withhold delivery, 363, 369
Renasas Technology, 29
"Rent-a-judge," 59. *See also* Private trial
Rental Car Finance Corp., 311

Renunciation, 421, 423, 478
Reorganization proceeding (Chapter 11 bankruptcy), 445-446, 452
Repayment plan proceedings, 447
REPC. *See* Real estate purchase contract (REPC)
Reply, 45
Republic of Korea, 80
Request to admit certain facts, 47. *See also* Interrogatories
Request to produce documents, 47. *See also* Interrogatories
Requirements contract, 332
Resale, 549, 560
Rescinded, 255
Rescission, 318-319, 321, 369
Residential Accredit Loans Inc. (RAL), 624
Residential Funding LLC (RFL), 624
Residential service, 33
Res ipsa loquitur, 158-160, 167, 170
Respondent superior, 463, 473, 475, 482, 512
Responsible person doctrine, 527
Restatement of Agency
 definition of agency, 456
 and employer-independent contractor relationship, 462
 independent contractor or employer criteria, 464
Restatement of Contracts, 239
Restatement (Second) of Contracts
 and commercial impracticability, 310, 369
 distinction between intended and incidental beneficiaries, 295
 unconscionable contracts in, 245
Restatement (Second) of the Law Contract, 198, 203
Restatements of the Contract, 210
Restatements of the Law Contract, 7, 198
Restitution, 318-319, 321
Restraint of trade, contracts in, 243-244
Restricted securities, 549-550, 560
Restrictive covenants, 243-244
Restrictive endorsement, 382
Retained earnings, 512
Return promise, 209
Reverse, 52
Revised Model Business Corporation Act (RMBCA), 510, 513, 516, 539
Revocation, of authority, 478
Revolution Foods, 26
RFRA. *See* Religious Freedom Restoration Act of 1993 (RFRA)
Rice, Warden, et al. v. Collins (2006), 50n
RICO. *See* Racketeer Influenced and Corrupt Organizations (RICO) Act
Right of first refusal, 532
Right(s)
 to compensation, 530, 533
 of directors, 530, 533
 of indemnification, 475, 530
 of inspection, 530-533
 of officers, 530, 533
 of participation, 530, 533
 of removal, 31, 37, 59
 of shareholders, 530-533
 See also Bill of Rights; Constitution, U.S.
Rights and responsibilities
 customer and bank, 395
 See also Bill of Rights; Constitution, U.S.
Right to Financial Privacy Act, 396-397
Right-to-sue letter, 573
Riley and v. Stone Gate (2017), 246-247
Riley v. California (2009), 100-101
Ripeness threshold requirement, 43-44, 60
Risk of loss, 342-344, 347
Rival cause, 13
Rivlab Transportation Company, 215
Rizo v. Yovino (2017), 597
RMBCA. *See* Revised Model Business Corporation Act (RMBCA)

Robert J. Triffin v. Maryland Child Support Enforcement Administration, and Rashad Jemal Christmas (2014), *425*
Roe v. Teletech Customer Care Management (COLO.) LLC (2011), 590
Romeo v. Seton Hall University (2005), 598
Roopchand v. Mohammed (2017), 253
Royalty, 189
Rule 504, 549, 560
Rule 505, 549, 560
Rule making
 Administration Procedures Act (APA) guideline on, by agencies, 82
 exempted, 72-74
 hybrid, 68, 71
 informal or notice-and-comment, 69
 problems associated with, 72-73
Rule-making petition, 64, 66
Rule-making power, 65
Rule-making procedures, 71
Rule of four, 52
Rules of law, 12
Russia, 224, 287
Ruzicka v. Conde Nast Publishing (1993), 219-220

S

S. Brooke Purll, Inc., T/A Purll Construction v. Patrick Darrell (2004), 317
Sabbath laws, 243
Sabo v. Carnival Corp. (2014), 539
SAC Capital Advisors, 562
Safe harbor, 553
Salens v. Tubbs (2008), 194
Sale-on-approval contract, 344, 347
Sale-or-return contract, 344, 347
Sales
 defined, 327-328, 346
 door-to-door, 610, 621
 FTC regulation of, 610-612, 622
 funeral home services, 611-612, 622
 of goods more than $500, contracts for, 273, 278, 297
 online, 612, 622
 real estate, 612, 622
 of securities, 560
 telephone and mail-order sales, 610-611, 621
 used cars, 611
Sales contract
 common-carrier delivery contract, 340-341, 347
 conditional sales contract, 341, 344, 347
 goods-in-bailment contract, 341, 343-344, 347
 simple delivery contract, 341-342, 347
 under UCC, 332-336, 341-344, 346-347
Sales tax, 92
Salman v. United States (2016), 118
Salt Lake City Olympics Bid Committee, 115. *See also* Bribery
Same-sex harassment, 571, 593
Same-sex marriage, 103
Sammie Smith v. State of Mississippi (2014), 140
SAMS Hotel Group LLC, Inc. v. Environs (2013), 324
Samsung, 189
Samsung-Apple dispute, 171-172, 191
Sarasota Citizens for Responsible Government, 84
Sarasota Citizens for Responsible Gov't v. City of Sarasota, 84
Sarbanes-Oxley Act of 2002, 84, 131, 135-136, 545
 partial text of, B-1-B*10*
Satisfaction, 309
Saudi Arabia, 204
SBA. *See* Small Business Administration (SBA)
SBR. *See* Supervisory board report (SBR)
SCA. *See* Federal Stored Communications Act (SCA)
Schaeffer v. United Bank & Trust Co. (1976), 419
Schauer v. Mandarin Gems of California, Inc. (2005), 371

Scheerer v. Fisher (2010), 483-484
School segregation, 105
Scienter, 259
Scientific-Atlanta, Inc., 551
S corporation, 494-495, 536
Scrutiny. *See* Intermediate scrutiny
SCUTPA. *See* South Carolina Unfair Trade Practices Act (SCUTPA)
Seagram Company Ltd., 563
Search and seizure, unreasonable, 95, 99, 105, 107, 110, 128, A-6. *See also* Fourth Amendment
Search warrant, 110
SEC. *See* Securities and Exchange Commission (SEC)
Second Amendment, 95, A-6
Secondarily liable, 412-414, 422
Secondary obligation, 276
Secondary promise, 276-277
Secured creditor
 versus secured creditor, 433, 449
 versus unsecured creditor, 433, 449
Secured interest
 creation of, 429
 debtor rights in collateral, 429
 defined, 428, 449
 purchase-money, 429
 value, 429
 written agreement (security agreement), 429
Secured party (secured creditor), 428, 449
Secured party *versus* buyer, 434, 449
Secured perfected creditor
 versus secured perfected creditor, 433, 449
 versus secured unperfected creditor, 433, 449
Secured transaction
 defined, 428
 definitions associated with, 428, 449
 and Uniform Commercial Code (UCC), 327
Secured unperfected creditor, 433
Securities Act of 1933, 563
 defining, 543
 exemptions under, 547-550
 intrastate issues, 549
 issuing corporate securities, 544
 liability, 550
 limited offers, 548
 periods of filing process, 546-547, 559
 private placement exemption (Rule 506), 548-549
 prospectus, 546, 559
 registration statement, 546
 resales, 549
 Rule 144, 549
 special registration provision, 547
 violations, 550
Securities Act of 1934, 562-563
Securities Act of 1990, 545
Securities and Exchange Commission (SEC), 326
 defined, 559
 expansion of powers, 545
 as federal administration agency, 7, 68-69
 headed by, 544
 Martha Stewart case, 138, 542, 550, 553-554, 559
 and political limitations, 75
 responsibilities of, 545-546
Securities and Exchange Commission v. Arcturus Corporation (2016), 543-544
Securities and Exchange Commission v. Gulf Sulfur Co. (1968), 552-553
Securities and Exchange Commission v. Sigma Capital Management, LLC, et al. (2013), 562
Securities Enforcement Remedies and Penny Stock Reform Act of 1990, 545
Securities Exchange, 16
Securities Exchange Act of 1934, 118
 defined, 544, 550-551, 560
 insider trading, 552-554, 560
 misappropriation theory, 554, 556, 560

 outsiders and insider trading, 553-554
 and Private Securities Litigation Reform Act of 1995, 553
 proxy solicitations, 556-557, 560
 Section 16(B), 555
 Section 10(B) and Rule 10B(5), 551-554
 tipper/tippee theory, 554, 560
 violations of, 556-557, 560
Securities exchange market, 561
Securities fraud, 116-117, 121
Securities Industry Council, 546
Securities Law, People's Republic of China, 547
Securities Litigation Uniform Standards Act of 1998, 553
Securities regulation, 542, 544-546
Security, 21, 543, 559
Security agreement, 428, 449
Security breach, customer data, 199-200
Security interest
 after-acquired property, 432
 perfection by filing, 430
 proceeds, 432
 scope of, 431-432, 449
 termination of, 432, 449
SEC v. Scammel, 139-140
SEC v. W.J. Howey Co. (1946), 543
Seized possessions, 111
Self-dealing, 525-526
Self-defense, 144
Self-incrimination, protection against, 95, 102, 107, 128, A-6. *See also* Fifth Amendment
Self-tender offer, 534
Sell, 245
Sellers
 and contract remedies, 362, 369
 exceptions to basic obligation, 357, 369
 exceptions to perfect tender rule, 356, 369
 obligations of, 356, 369
 problems with acceptance, 357, 369
 remedies available to, under UCC, 362
 resell or dispose of goods, 363
 right to cure, 356, 369
 warranties of title, 358, 369
 warranty disclaimers and waivers, 362, 369
 withhold delivery of goods when buyer or lessee in breach, 362, 369
Seniority System Defense, 572, 593
Sentencing, 130, 136
Sentient Jet v. Mackenzie (2013), 252
Separation of powers, 84, 106, 108
Service by certified, 33
Service by email, 33
Service by social media, 60-61
Service of process, 33, 45, 60-61
Service through Facebook accounts, 33
Seton Hall University Department of Student Affairs, 598
7-Eleven Inc., 500
Seventh Amendment, A-6
Severable contract, 249-250
Several States (Article IV of U.S. Constitution), A-5
Severance package, 534-535
Severn Peanut Co., Inc. et al v. Industrial Fumigant Co. & Rollins, Inc. (2015), 322
Sex offender, registered, 110
Sexual harassment, 149
 defined, 593
 same-sex harassment, 571
 under Title VII of CRA, 569-571
Sexual orientation, 573-575, 592-593
SFT. *See* South Florida Transport (SFT)
Shabsi Pfeifer v. Elias A. Groisman, et al. (2014), 215
Shadow jury, 50
Shah, 380
Shahjog hundis, 380
Shareholder
 defined, 494

 derivative suit, 532-533
 direct suit, 532-533
 duties of, 526, 537
 liability of, 529, 537
 meeting attendance, 522, 537
 power of, 522
 rights of, 530-533, 537
 roles, 520-523, 537
 voting to elect, 522
 See also Corporation
Shareholder activism, 538
Share transfer, 511-512, 532, 537
Sharia law, 204, 354
Shelby's, Inc. v. Sierra Bravo, Inc. (2002), 277
Shelf registration, 547
Shell Chemical Company, 22
Shelter principle, 402, 410-411, 422
Shipment contract, 342, 347
Shipping terms, 343
Short-swing profit, 555, 560
Sigma Capital Management, 562
Signature
 automated, 380
 checks bearing forged, 393-395
 and Completely Automated Public Turing Test to Tell Computers and Humans Apart (CAPTCHA), 412
 of duly authorized agent, 380
 electronic, 412
 of maker or drawer, 378, 380, 398
Signature liability
 accommodation parties, 414, 423
 defined, 411, 422-423
 primary liability of makers and acceptors, 412, 422
 secondary liability of drawers and endorsers, 412-414
 unauthorized signatures and endorsements, 414-416
Signers, of U.S. Constitution, A-6
Silence, as form of offer acceptance, 209-210
Simkin v. Blank (2012), 258-259
Simmons v. Simmons (2007), 223, 230-231
Simple delivery contract, 341-342, 347
Sixth Amendment, 95, 128, A-6
Slamming, 117, 120
Slander, 144-145
Slander of quality, 153
Slander of title, 153
Slanderous per se, 141
Slander per se, 145, 169
Slavery abolished, 13th Amendment, A-7
"Sliding scale" standard, 34
Small business, 16
Small Business Administration (SBA), 7, 69
Small claim court, 41
Smart contract, 213
Smart Way Janitorial, 476
Smokeless Tobacco Health Education Act of 1986, 608
Smoking, employment discrimination based on, 578, 594
SNB. *See* Stillwater National Bank (SNB)
Snell v. Suffolk County (1986), 571
"Snitching," 169
Social media
 fraudulent misrepresentation in, 264
 service of process by, 60-61
 use of, in employment decisions, 584-586, 594-595
Social responsibility of business, 15, 28
So Clean!, 565, 578, 592
Soldiers in private residencies, 95. *See also* Third Amendment
Sole proprietor, 487-488, 497
Sole proprietorship, 487-488, 504
Sonic Drive-In, 500
Sony, 5, 188

Sorrell v. IMS Health Inc. (2011), 606
South Carolina Manufactured Housing Board, 257
South Carolina Unfair Trade Practices Act (SCUTPA), 194-195
South Dakota v. Dole, 93
South Dakota v. Wayfair, 92
South Florida Transport (SFT), 311
Southway Systems, Inc., 300-301
Southwest Bank v. Information Support Concepts, Inc., 383
Spam, 608
SPD. *See* Summary plan description (SPD)
Special damages, 315
Special endorsement, 382, 398
Special registration provisions, 547
Specific performance (specific enforcement), 319, 321, 469
Speech, freedom of, 94-95, 107. *See also* First Amendment
Speedy trial with impartial jury, 95, 128, A-6. *See also* Sixth Amendment
Speiser v. Randall, 184
Spending powers, 93
Sport Clips, 500
Spyware, 606
Squeeze-out, 526
SSA. *See* Structured settlement agreement (SSA)
St. John's Holdings, LLC v. Two Electronics LLC (2016), 299
Stakeholder, 23
Stale check, 393, 398
Standards, of business conduct, 15
Standing (or standing to sue) threshold requirement, 43-44, 60
Staples Contract and Commercial Inc., 576-577
Starbucks, 18
Stare decisis ("standing by the decision"), 6, 8
Star Machine, Inc. v. Ford Motor Co. (1998), 350
State authority, commerce clause as restriction on, 90-91
State Council of China, 547
State court system
 court of last resort, 43
 intermediate courts of appeal, 43, 60
 of limited jurisdiction, 41
 state trial court, 41
 and structure of U.S. legal system, 40
State Farm, 300
State government, 86
State governor, 8
State jurisdiction, 35, 60
State law
 discrimination based on sexual orientation, 573-575, 592
 and employment discrimination, 593
 federal law *versus*, 3-4
 and franchises, 501
 as source of contract law, 203, 212
 statute of frauds as, 278
 and workers' compensation, 594
 See also Tort law
State liquor Control Board, 91
State Of North Carolina v. Denni Dwayne Tucker (2013), 138
State sales tax
 and on-line sellers without physical presence in state, 92
State securities laws, 558, 560
State supreme court, 60
State trial court, 41
State v. Dumstrey (2016), 110
State v. Samalia (2017), 193-194
State whistle-blower protection laws, 134-135
Statute, 5. *See also entries for specific statutes*
Statute of frauds, 299
 assignments covered by, 287
 contracts falling within, 273-278

and consideration of marriage, 273, 297, 375, 378
 on contracts preventing possible performance within one year, 273-275, 278, 281
 on interest in land contracts, 273, 276-278, 297
 on promises to pay debt, 273, 276, 278
 defined, 273, 297
 England and, 277
 exceptions to, 279-280, 297
 merger clause, 283
 prenuptial agreement, 275
 purposes of, 273
 requirements of writing sufficient to satisfy, under Common law, 282
 requirements under, 334
 for sales of goods more than $500, 273, 278, 297, 331, 346
 sufficiency of writing, 180-282
Statute of limitations, 309, 369, 372
Statutory insider, 555, 560
Statutory law, 5
Statutory limitations, on agency powers, 75, 82
Stearns Bank National Association, 435
Stelmachers v. VeriFone Systems, Inc., 623-624
Stenlund v. Marriott International, Inc. (2016), 484
Stenographer. *See* Court reporter (stenographer)
Sterile Technologies Inc., 372
Stern Oil Co. v. Brown (2018), 364
Stewart Organization, Inc. v. Ricoh Corp., 38
Stillwater National Bank (SNB), 424
Stipulated-damage clause, 315-318
Stock certificate, 530-531, 533
Stock dividend, 548, 560
Stock purchase, 532, 537
Stock split, 548, 560
Stock warrant, 531
Stolichnaya vodka, 29
Stoneridge Inv. Partners, LLC v. Scientific-Atlanta, Inc. (2008), 551
Stone v. Jetmar Properties, LLC (2007), 507-508
Stop-payment order, 393, 398, 402
Storage Act, 246
Store-value card, 396
Straight bankruptcy, 439. *See also* Liquidation
Strict liability, 114, 162
Strict liability offenses, 125
Strict liability tort, 143, 162, 167
Strict scrutiny, 105, 107
Strubel v. Comenity Bank (2016), 623
Structured settlement agreement (SSA), 300
Subjective impossibility, 309-310
Subject-matter jurisdiction, 35-37, 60
Subleasing, 175
Subpoena, 66, 128
Subpoena *duces tecum,* 66
Subscriber, 515, 518
Subscription agreement, 515
Subsequent illegality, 310
Substantial evidence, 75
Substantial impairment, 356
Substantial performance, 305-306, 321
Substantive due process, 103, 107
Substantive unconscionability, 245
Substitute check, 391
Substituted contract, 309
Substitute for cash. *See* Negotiable instrument
Subsurface rights, 173-174
Suburban Tractor Company, 451
Suing to get benefit of bargain, 363
Suing to recover damages, 364-365
Sum certain in money, 378, 380, 398
Summary Judgment, 133
Summary jury trial, 58
Summary plan description (SPD), 581
Summons, 33
Sun Microsystems, 117

Sunrich, 323
Sunrish, 323
Sun's Market, 408, 411, 422
Superior court, 43. *See also* Intermediate courts of appeal
Superseding cause, 162
Supervisory board report (SBR), 518
Supremacy clause, 87, 90, 106
Suretyship promise, 276-278
Suspect classification, 105
Sweden, 80
 business ethics and advertising law, 23
 securities market in, 545
Swoboda v. Fontanetta (2015), 158-159
Syndicate, 499

T

Taco Bell, 153, 500
Taft-Hartley Act of 1947, 590-592
Takeover, 554
 defined, 532-533
 nature of, 533-535
 protection against, 516
 response to, 533-534, 538
 types of, 533
Takings clause
 of Fifth Amendment, 95, 104, 107, 177-178, 193, A-6
 in Germany, 179
Tandy Corporation, 372
Tangible goods (confusion), as means of acquiring property, 181
Tangible property, 179
TAP. *See* Telephone Assistance Plan (TAP)
Target, 26
Taxation and taxes
 on business trusts, 498
 collection of, 93
 corporate, 512, 536
 on corporate dividends, 494
 dormant commerce clause, 107
 evasion of, 121
 of federal government, 93
 on profits, 494
 state sales, 92
 See also Internal Revenue Service (IRS) *and entries for specific taxes*
Taxing and spending powers, of federal government, 92
TCPA. *See* Telephone Consumer Protection Act (TCPA) of 1991
Technology
 blockchain, 213
 to capture images at ATM locations, 394
 changes in, and unlawful search, 99-100
 electronic fund transfer, 395-396, 399
 electronic signature, 412
 e-money and online banking, 395-397, 399
 infancy doctrine in digital marketplace, 251
 intellectual property law (IP), 171
 online banking, 395-397, 399
 and privacy rights, 101
 records and ethics, 25
 and sentences for nonviolent offenders, 115
 unlawful search, 108
 See also Digital content; Intellectual property
Telemarketing, 605-606, 621
Telemarketing and Consumer Fraud and Abuse Prevention Act of 1994, 606-607, 621
Telemarketing fraud, 117
Telemarketing Sales Rule of 1995, 607
Telemarketing Sales Rule of 2002, 608
Telephone Assistance Plan (TAP), 151
Telephone Consumer Protection Act (TCPA) of 1991, 606-607, 621
Telephone Records and Privacy Protection Act (2006), 121

Telephone sales, 610-611, 621. *See also* Telemarketing
TeleTech, 590
Temp. Alternatives, Inc. v. Jamrowski (2014), 230
Tenant (lessee), 175
Tender, 305, 331
Tender of delivery, 342, 353, 369
Tender offer, 533, 538
Tender of payment, 421, 423
Tenth Amendment, 95, A-6
Termination
 of agency relationships
 by acts of parties, 477-478, 482
 actual notice, 476-477
 agency coupled with an interest, 478
 constructive notice, 477
 fulfillment of purpose, 477
 lapse of time, 477
 mutual agreement by parties, 477
 occurrence of specific event, 477
 by operation of law, 478-479, 482
 renunciation by agent, 478
 revocation of authority, 478
 of corporation, 535, 538
 of franchise, 502-504
 of partnership, 490-492
Termination statement, 432, 449
Terms of office for congress and president, 20th Amendment, A-8
Terra Venture, Inc. v. JDN Real-Estate Overland Park, L.P., 498
Texaco Petroleum. *See* Chevron
TexAmericas Center-East, 280-281
Texana Rice, Inc., 434
Texas Business and Commerce Code Ann., 385
Texas Business Forms, 487
Texas Gulf Sulfur (TGS), 552-553
Texas Insurance Code, 453
TGS. *See* Texas Gulf Sulfur (TGS)
The Goodyear Tire & Rubber Co. v. Hager et al., 16
The Inn Between, Inc. v. Romanco Metropolitan, Inc. (1997), 373
Thelma Agnes Smith v. David Phillip Riley (2002), 220-221
Thermal-imaging instrument, 100
The Slants, 183
Third Amendment, 95, 104, 107, A-6
Third parties
 assignment of rights to, 286-291
 assignment of contract, 291
 incidental beneficiaries, 295-296
 third-party beneficiary contracts, 291-296
 defined, 286
 satisfaction of, 307
Third-party beneficiary contract
 assignment of contract, 291-292
 defined, 291
 expert witnesses and, 300
 incidental beneficiaries, 295-296
 intended beneficiaries, 292-295
Third-party beneficiary of warranties, 360
Third-party beneficiary rights, 272, 297
Third-party contract, 286-298
Thompson Maple Products v. Citizens National Bank (1967), 415
Thrasher v. Thrasher (2014), 194
Threshold requirements, 43-44
Thrifty Rent-a-Car System v. South Florida Transport (2005), 311
TILA. *See* Truth in Lending Act (TILA)
Time instrument, 398, 408-409
 defined, 376
Timothy Tyrone Foster, Petitioner v. Bruce Chatman, Warden (2016), 49
Tipper/Tippee, 118, 553-555
"Tipping," 117
TISA. *See* Truth-in-Savings Act (TISA)
Titan Distribution Inc., 30

Title
 documents of, 327
 good title, 336-341
 versus other interests in goods, 340-341
 as ownership of property, 179-180
 void, 336
 voidable, 336
Title 11, U.S.C.-Bankruptcy, 438
Title transfer, 342, 347
Title VII (Civil Rights Act of 1964), 566-573, 593, 596. *See also* Employment discrimination
Tobacco advertising, 608, 621
Toll Brothers, 323
Tombstone advertisement, 547
Tomlinson v. Board of Education (2010), 460
Tor Project, 25
Tort(s)
 in agency relationships, 469, 473-475, 481
 classification of, 143, 167
 damages available in cases, 162-166
 defined, 142
 intentional, 143-154, 167, 475
 negligence, 143, 154-162, 167
 parents' liability for their children's, 238
 strict liability, 143, 162, 167
Tortfeasor, 163
Tort law, 141
 defined, 142, 167
 primary objective of, 142
Tort liability, 473-476, 482
Tort of outrage, 150
"Totality of circumstances" standard, 569-570
"To the order of," 381-382
Tour 18, Ltd., 186
Toussaint v. Blue Cross & Blue Shield of Michigan (1980), 566n
Toy Safety Act of 1984, 620
Toysmart.com, 125
Toys "R" Us, Inc. v. Canarsie Kiddie Shop, Inc., 185
Trade dress, 186, 191
Trade-dress infringement, 186
Trade fixture, 328
Trade libel, 153
Trademark
 celebrities seeking protection, 184
 color as, 182-183
 defined, 181, 192
 dilution of, federal, 194-195
 and domain and trade names, 186
 exclusive federal jurisdiction over cases of, 35
 infringement of, 182-185, 194-195
 interstate *versus* intrastate, 183
 offensive, 184
 symbols indicating, 189
Trademark Trial and Appeal Board, 183
Trade secret, 190-191, 195, 532
Trade usage, 285-286
Tradition (custom), 8. *See also* Historical school
Traffic ticket, 110
Transaction, high-frequency, 561
Transactional, of e-government, 80
Transference of property rights, 10, 192. *See also* Personal property
Transfer warranty, 416-417, 423
Transportation Department. *See* Department of Transportation
Transportation Security Administration (TSA), 6
Trapani Const. v. Elliott Group Inc., 201-202
Trapani Construction Company, 201
Traveler's check, 376, 387-388, 398
Travel freely from state to state, right to, 104
Treaty, 7, 35
Treibacher Industrie, A.G., Plaintiff-Appellee, v. Allegheny Technologies, Inc., a Pennsylvania Corporation et al., Defendants, TDY Industries, Inc., Defendant-Appellant (2006), 350
Trent Woodman v. Kera LLC (2010), 247-248
Trespassing, as an intentional tort, 152, 167

Trespass to personal property, 152, 167
Trespass to realty (trespass to real property), 152, 167
Trial, 53
 closing arguments, 51, 60
 examination of witnesses, 50, A-6
 by impartial jury of peers, 128, A-6
 in Japan, 51
 judgment, 60
 jury deliberations, 60, 136
 jury instructions, 51, 60
 jury selection, 48-50, 60
 jury verdict, 60, 136
 opening statements, 50, 60
 presentation of evidence, 50
 right to speedy and public, 95, 128, A-6
 as step in criminal procedure, 132
Trial court, 31, 40, 60
 determine questions of fact, 32
 of general jurisdiction, 41
 state, 41
 U.S. Supreme Court as, 41
 See also Courts of original jurisdiction
Trial procedure, 130, 136
Tri-M, 109
Tri-M Group v. Sharp (2011), 109
Troy Boiler Works, Inc. v. Sterile Technologies, Inc. (2003), 372
Trustee
 in business trust, 497
 in Chapter 7 bankruptcy proceeding, 440-441
Trust fund, misappropriation of, 117
Truth, as absolute defense, 148
Truth in Lending Act (TILA), 623
 case of bait-and-switch advertising, 604-605
 Consumer leasing Act (CLA), 613
 defined, 622
 Equal Credit Opportunity Act (ECOA), 613-614
 types of loans under, 613
 unauthorized charges and disputes, 613
Truth-in-Savings Act (TISA), 392
TSA. *See* Transportation Security Administration (TSA)
Turner v. Theodore V. Wells, Jr., Paul, Weiss, Rifkind, Wharton & Garrison, LLP (2018), 169
24 Art Holdings, LLC v. Berry-Hill Galleries, Inc. (2014), 371
Twin Bay Village, Inc., 539
Twitter
 and employment decisions, 584-585
 and freedom of speech, 268-269
 and use of service, 61
Two-tier board system, 518
Ty, Inc. v. GMA Accessories, Inc. (1998), 187
Tyco International, 29, 524
Tyco Intl, Ltd. v. Kozlowski (2005), 29
Tying arrangement, 189, 192
Typographical errors, in contracts, 285, 297
Tyson International, 25

U

UAAA. *See* Uniform Athlete Agents Act (UAAA)
UAE. *See* United Arab Emirates (UAE)
Uber, 461
Uber Technologies, Inc., 463
UCC. *See* Uniform Commercial Code (UCC)
UGESP. *See* Uniform Guidelines on Employee Selection Procedures (UGESP)
ULLCA. *See* Uniform Limited Liability Company Act (ULLCA)
Ultra vires ("beyond powers") act, 513, 536
Unaffiliated director, 521
Unauthorized act, 472, 481
Unauthorized agent
 article 3 (personal defenses), 419-420
 and signature liability, 414-416

Unauthorized endorsement, 414-416
Unauthorized signature, 414-416, 423
Unauthorized transfer, 395
Unconditional promise, 378, 380, 398
Unconscionability, 353
Unconscionable
 contract or contract provision, 334
 defined, 245
Unconscionable bargain, 265
Unconscionable contract provision, 56, 244-245, 347
Unconscious negligence, 157
Under seal. *See* contract under seal
Underwriter, 546-547
Undisclosed principal, 471
Undocumented immigrant worker, 586, 595
Undue influence, 334
 defined, 263, 267
 and moral coercion, 265, 267
 as obstacle to genuine assent, 255
Undue susceptibility, 264
Unemployment compensation, 579, 594
Unemployment Insurance Fund, 580
Unenforceable contract, 202, 240-247, 250, 255. *See also* Legality of contracts
Unfair competition, 154, 167, 194-195
Unfair Trade Practice Act and Consumer Protection Law, 614
Unforeseen circumstances, 226
Uniform Athlete Agents Act (UAAA), 456
Uniform Commercial Code (UCC)
 and after-acquired property, 432
 Article 2
 goods, 327-329, 331, 346
 merchants, 330, 346
 mixed goods and services contracts, 330-331, 346
 sale, 327-328, 346
 Article 2(A) of, 327, 330-331, 346
 Article 4 (bank account), 386, 378385
 Article 4(A) (commercial fund transfers), 395
 Article 3 (checks as negotiable instruments), 375, 378, 384, 386, 419-420
 bank cutoff hours, 391
 and buyer in ordinance course of business, 434
 and checks, 385-387, 397
 and commercial impracticability, 311
 common law or, 328
 and consequential damage, 331
 consumer good, 431
 versus contracts for international sale of goods, 331
 contracts for sale of goods governed by, 314, 319
 and contracts for sale of goods totaling more than $500, 273, 297, 331
 defined, 198, 326
 definition of bank under, 387
 definition of sale according to, 345
 exceptions to statute of frauds, 280, 297
 formation of sales and lease contracts
 battle of the forms, 333, 346
 consideration, 333, 346
 formation in general, 332, 346
 offer and acceptance, 332-333, 346
 requirements under statute of frauds, 334, 347
 unconscionability, 334, 347
 and franchise relationships, 505
 good faith, defined, 407
 implied warranties, 362
 imposter's rule, 415-416
 and Internet, 330
 and letters of credit, 203
 modifications or limitations to remedies otherwise provided by, 367
 negotiable instrument, 377
 and negotiable instruments, 471
 on negotiable instruments. *See* Negotiable instrument
 notice of fact, 408
 obligation of buyers/lessees and sellers/lessors, 352-353
 and open terms, 333
 outline of, 327
 parol evidence rule, 331
 personal defenses to liability, 419-420
 priority dispute, 434
 proceeds under, 432
 remedies available to buyers/lessees under, 363-367
 requirements for negotiability, 378
 requirements under, for sales contracts, 281, 297
 resell or dispose of goods, 363
 and revocation of offer by offeror, 208, 331
 scope of, 326, 346
 and secured transactions, 428
 significance of, 326-327, 346
 as source of contract law, 198, 212
 unconscionable contracts in, 245, 347
 United States adopt CISG as substitute for, 348
 and used-car fraud, 611
Uniform Guidelines on Employee Selection Procedures (UGESP), 572
Uniform Limited Liability Company Act (ULLCA), 494
Uniform Negotiable Instruments Law (NIL), 384
Uniform Partnership Act (UPA)
 dissolution defined under, 490, 492
 partnership defined under, 488
Uniform Securities Act, 558
Unilateral contract, 199-200, 202, 209, 212, 218-219
Unilateral mistake, 256-257, 267
Unincorporated cooperative, 497
Unintentional-discrimination, 568, 593. *See also* Disparate-impact discrimination under Title VII
Union employees, and collective bargaining, 566, 595
Union membership rate, 590
Unions. *See* Labor laws and unions
United Arab Emirates (UAE), 334
United Bank, 419
United Homes LLC, 424-425
United Kingdom
 and e-government, 80
 protecting whistleblowers in, 134
United Leasing Corp. v. Lehner Family Business Trust (2010), 506-507
United Methodist Church, 22
United Nations, 80
United Nations Convention on Contracts for the International Sale of Goods (CISG), 331
 scope and significance of, 331
 Uniform Commercial Code (UCC) *versus*, 331
United Parcel Service (UPS), 120, 135, 596-597
United States, 80
United States Code, Title 11 (Bankruptcy Code), 438
United States Customs and Immigration Service (USCIS), 586
United States of America v. Thomas Tanke (2014), 119-120
United States Patent and Trademark Office (USPTO), 190
United States Trustee Program (USTP), 120
United States v. Andrews (2015), 139
United States v. Carpenter, 554
United States v. Carpenter (1985), 556
United States v. Iverson (1998), 127n
United States v. Kaplan (2016), 618-619
United States v. Kuczora (2018), 138
United States v. Lopez (1995), 88-90
United States v. Lori Drew (2009), 264
United States v. O'Hagan, 554
United States v. Park (1975), 126
United States v. Weaver (2017), 139
United States v. Weimert (2016), 138-139
United States v. Windsor (2013), 103
United States v. Wurie (2014), 100-101
Universal City Studios Credit Union v. CUMIS Insurance Society, Inc. (2012), 401
Universalization test, 27
University of Phoenix, 133
Unlawful search. *See* Search and seizure, unreasonable
Unliquidated debt, 226-227, 229
Unpaid leave, 580
Unprotected speech, 97-98. *See also entries for specific types of speech*
Unqualified endorsement, 382
Unregistered, unrestricted securities, 548, 559
Unregistered restricted security, 548, 559
Unregistered security, 548, 559
Unrestricted security, 548, 559
Unsecured creditor, 433, 450
Unsolicited merchandise, 611, 621
UPA. *See* Uniform Partnership Act (UPA)
Upper Deck Company, 216, 228
UPS. *See* United Parcel Service (UPS)
UPS Store, The, 500
U.S. Bank N.A. v. Phillips (2006), 401
U.S. Constitution
 Amendments, 94-106, 426, A-5-A-9
 Articles, A-1-A-6
 Article I (Legislative Branch), of U.S. Constitution, A-1-A-3
 Article II (Executive Branch), of U.S. Constitution, A-3-A-4
 Article III (Judicial Branch), of U.S. Constitution, A-4-A-5
 Article IV (Several States), of U.S. Constitution, A-5
 Article V (Amendments of U.S. Constitution), A-5
 Article VI (Legal Status of Constitution), A-5
 Article VII (Ratification), of U.S. Constitution, A-5-A-6
 commerce clause, 88-90
 contract clause, 94
 criminal defendant's rights, 128
 federalism, 86, 106
 and federal-question cases, 35
 full faith and credit clause, 94
 judicial review, 86
 Preamble, A-1
 privileges and immunities clause, 93-94
 separation of powers, 84, 106-108
 signers of, A-6
 supremacy clause and federal preemption, 87
 text of, A-1-A-9
 See also Bill of Rights; *and entries for specific amendments*
U.S. Department of Housing and Urban Development (HUD), 71
U.S. Department of Justice, 124
U.S. Food and Drug Administration (FDA), 618
U.S. Patent and Trademark Office, 183-184
U.S. Supreme Court, 40, 60
 appeal to, 52
 justices of, 42
U.S. Treasury Department, 401
U.S. v. Todd Newman, Anthony Chiasson, Jon Horvath, Danny Keu, Jyung Lim, Michael Steinberg (2014), 562
Usage of trade, 285-286, 297, 356, 360
US Airways, Inc., 322
USA v. Shkreli et al., 16
USCIS. *See* United States Customs and Immigration Service (USCIS)
Used-car sales, 611, 622
"Use of right," 10
User Motor Vehicle Registration Rule of 1984, 611

US Life Insurance Company v. Wilson (2011), 215
USPTO. *See* United States Patent and Trademark Office (USPTO)
USTP. *See* United States Trustee Program (USTP)
Usury, 242

V

VA. *See* Veterans Administration (VA)
Valid contract, 202, 212
Value
 of Adidas, 20
 of American Express, 20
 defined, 19
 secured interest and, 429, 449
 as starting point for business ethics, 21–23
Veggie libel, 153
Vendstar, 139
Venue, 37–38, 60
Verdict, 50–51, 60
VeriFone Systems Inc., 623–624
Verisign, Inc. v. XYZ. COM LLC (2017), 625
Vest, 294, 298
Veterans Administration (VA), 69
 place and duration of filing, 430
 sexual harassment until Title VII, 569–571
Vicarious liability, 127, 473
Victoria's Secret, 271
Video game, 156
Vietnam, 10, 175
Village of Arlington Heights, 201–202
Villegas v. Weinstein & Riley (2010), 625
Violence Against Women Act, 89
Violent media, 167–168
Vioxx, 622
Virgil Kirchoff Revoc. Trust v. Moto, Inc. (2016), 541
Virginia Polytechnic Institute (Virginia Tech), 89
Virtu Financial Inc., 561
Virus, 124–125
Vivendi, S.A., 563
Vogel v. Wells Fargo Bank NA, 424
Voidable contract, 202–203, 212, 255, 267, 285, 297
Voidable title, 336–338, 347
Void contract, 202, 212, 285, 297
Void title, 336, 347
Voir dire, 48, 50
Vokes v. Arthur Murray (1968), 270
Voluntary dissolution, 536
Voluntary liquidation petition, 439–440
Voluntary transfer, of personal property, 179–180, 192
Voting and voting rights
 limited rights of preferred stock owners, 520
 not denied due to any poll tax, 24th Amendment, A-9
 not denied due to race, 15th Amendment, A-7
 not denied due to sex, 19th Amendment, A-8
 of president and vice president, A-6–A-7
 of senators by popular vote, Amendment 17, A-7
 set at 18 years or older, 26th Amendment, A-9
 shareholder, at shareholder meetings, 537
 substantive due process, 103

W

Wachovia Bank v. Foster Bancshares (2006), 425
Wachovia Bank (Wells Fargo), 16, 424–425
Wachovia Corporation, 424
Waffle House, 253
Wage and hour laws, 36, 579, 594
Wage discrimination, 579, 594
Wagner Act of 1935, 590–591, 595
Waiting period, 546–547, 559
Waiver
 of implied and express warranties, 362, 369
 of warranties, 371
Walgreen, 304

Wall Systems, Inc. v. Pompa (2017), 485
Walmart, 7, 494
Wal-Mart, 371–372, 539–540
Wal-Mart de Mexico (WalMex), 539–540
Wal-Mart Stores Inc., 597
Wal-Mart Stores v. Indian Elec. Workers (2014), 539–540
Walt Disney Pictures, 264
Walter Fernandez v. California (2014), 102
"Wanna Cry," 125
War, 479
Warranties
 defined, 358
 disclaimers and waivers, 361–362
 express, 358
 implied, 359–361
 of merchantability, 358–361, 369
 third party rights, 360–361
 of title, 358
Warrantless search, 101, 110. *See also* Search and seizure, unreasonable
Warranty, waiver of, 371
Warranty liability, 411
 defined, 416, 423
 presentment warranty, 417, 423
Washington Department of Ecology, 84
Washington Department of Natural Resources (agencies), 84
Washington Forest Practices Board, 84
Washington Law Against Discrimination (WLAD), 568
Washington Redskins, 184
Watered stock, 529
Water rights, 173–174
Watson Coatings, Inc. v. American Express Travel Services, Inc. (2006), 404
Wawel Savings Bank v. Jersey Tractor Trailer Training, Inc. (2009), 406–407
Wells Fargo bank, 409
Wells Fargo Bank, 19, 383
Wells Fargo & Company, 424
Western Nassau Orthopedic Associates, P.C., 158–159
Weyco Inc., 578
WH approach, 19, 21, 30
Whayne Supply Company, 31
WH framework, 23–24, 26–27
Whistleblower, 25, 540
Whistle-blower
 and False Claim Act, 133
 in good faith, 489
 protecting, in United Kingdom, 134
 and Sarbanes-Oxley Act, 134, B-7
 state protection laws, 134–135
White-collar crime
 blue-collar worker committing, 138
 bribery, 115–116
 in China, 123–124
 computer games, 124–125
 consequences of, 114–115
 defined, 114
 embezzlement, 123–124
 extortion, 116
 fraud, 116–123
 tools for fighting
 False Claims Act, 132–133, 136
 Sarbanes-Oxley Act, 133–136, B-1
 state whistle-blower protection laws, 134–135
Whole Foods, 26
Wholesome Meat Act of 1967, 620
Why Not, LLC, 507
Wickard v. Filburn, 106
Wiggins v. Janousek (2017), 400
Wikileaks, 25
Williamson test, 544
Wilson v. Hewlett-Packard Co. (2012), 624
Winding up, of partnership, 490–492
Wire fraud, 121, 135

Wiretapping, 108, 136
Wire transfer, 327
Wire transfer network (Fedwire), 395–396
Wisconsin National Primate Research Center, 221–222
Withhold delivery of goods, 362
Without warranty, 417
Witnesses, examination of, 50, 128, A-6
WLAD. *See* Washington Law Against Discrimination (WLAD)
Wolfe v. Gooding Co., 62
Woodcock v. Chemical Bank (1998), 453
Wool Products Labeling Act of 1939, 609
"Work authorization," 586
Workers' compensation, 402, 594
Workers' compensation laws, 580–581
Working conditions, duty to provide safe, 466, 468, 481
Workplace privacy, 582–584
WorldCom, 24, 133
World Savings Bank, 424
World Trade Organization (WTO), 9
World-Wide Volkswagen Corp. v. Woodson, 62
Writ of certiorari, 52
Written agreement (security agreement), 429, 449
Written contract, 299
 alteration of contract, 309
 assumption of the risk, 161
 contacts for sale of goods totaling more than $500, 278
 and creation of agency relationship, 456
 integrated, 282, 297
 merger clause, 283
 versus oral contract, 273
 parol evidence rule, 282–283, 297
 and promissory estoppel, 279
 reformation, 319
 versus speech-based communication, 269
 statute of frauds requiring, 273, 331
 sufficiency of writing, 280–282, 297
 See also Parol evidence rule; Statute of frauds
Written document, 379, 398
Written negotiable instrument, 380
Wrongful dishonor, 392–393, 398
WTO. *See* World Trade Organization (WTO)

X

XYZ.Com, LLC, 625

Y

Yahoo, 219
Yahoo!, 534
Yale Factors LLC, 407
Yan Ju Wang v. George Valverde (2008), 73–74
Y & N Furniture Inc. v. Nwabuoku (2001), 350, 354–355
Yomanty v. Mincola, 167
Young v. UPS (2015), 596–597
YouTube, 584
Yumilicious Franchise, LLC v. Barrie (2016), 507

Z

Zappos.com Inc., 199–200
Zinner v. Advanced Parking Servs. (2015), 170
Zippo Mfg. Co. v. Zippo Dot Com, Inc. (1997), 34
Zoning ordinance, 110
Zurich American Insurance Co., Subscribing to Policy Number IM 8338602, Certificate Number D279FL, Plaintiff-Counter-Defendant-Appellee, v. Felipe Grimberg Fine Art, Fine Arts Services, Inc., Doing Business as Felipe Grimberg Fine Art, Felipe Grimberg, John Does, A-Z, Defendants-Counter-Claimants-Appellants (2009), 372–373
ZZZZ Best Carpet Cleaning Company, 122